WATER-ROCK INTERACTION

VOLUME 2

PROCEEDINGS OF THE 7TH INTERNATIONAL SYMPOSIUM ON WATER-ROCK INTERACTION – WRI-7 / PARK CITY / UTAH / USA / 13-18 JULY 1992

Water-Rock Interaction

Edited by

YOUSIF K. KHARAKA & ANN S. MAEST

U.S. Geological Survey, Menlo Park, California, USA

VOLUME 2

MODERATE AND HIGH TEMPERATURE ENVIRONMENTS

A.A. BALKEMA / ROTTERDAM / BROOKFIELD / 1992

WATER-ROCK INTERACTION

INTERNATIONAL

The texts of the various papers in this volume were set individually by typists under the supervision of each of the authors concerned.

Published by
A.A. Balkema, P.O. Box 1675, 3000 BR Rotterdam, Netherlands
A.A. Balkema Publishers, Old Post Road, Brookfield, VT 05036, USA

For the complete set of two volumes, ISBN 90 5410 075 3
For Volume 1, ISBN 90 5410 076 1
For Volume 2, ISBN 90 5410 077 X

Printed in the Netherlands

Water-Rock Interaction, Kharaka & Maest (eds) © 1992 Balkema, Rotterdam. ISBN 90 5410 075 3

Scheme of the work

VOLUME 1
LOW TEMPERATURE ENVIRONMENTS

Keynote lectures
Mineral-fluid interface geochemistry
Adsorption and cation exchange
Kinetics of geochemical reactions
Recent advances in techniques and methodology
Geochemical modeling
Solution chemistry and thermodynamics
Solute transport
General applications
Organic geochemistry
Processes involving organics and microbes
Organic-inorganic interactions
Natural hazards and environmental contamination
Global processes
Oxidation-reduction reactions in surface and ground waters
Weathering processes and surface water environments
Saline lakes and evaporite deposits
Unsaturated zone environments
Groundwater environments

VOLUME 2
MODERATE AND HIGH TEMPERATURE ENVIRONMENTS

Keynote lectures
Stable and radioactive isotopes
Water and solutes
Exchange reactions
Geochemical modeling
Solution chemistry and thermodynamics
Solute transport
General applications
Sedimentary basins
Fluid evolution
Mineral diagenesis
Mineral transformations and interactions
Geothermal systems
Fluid chemistry
Mineral interactions
Geothermometry
Metamorphic environments
Hydrothermal ore deposits
Seawater-rock interactions

Water-Rock Interaction, Kharaka & Maest (eds) © 1992 Balkema, Rotterdam. ISBN 90 5410 075 3

Table of contents

Keynote lectures

Stable isotope exchange in fluid-mineral systems: Old problems, new solutions 861
J.R.O'Neil

Saline formation waters in sedimentary basins: Connate or diagenetic? 865
Lynton S.Land

Early diagenesis in sediments: A global view? 869
Max L.Coleman

The composition of gases in geothermal and volcanic systems as a function of tectonic setting 873
W.F.Giggenbach

Ore deposits in the Oquirrh and Wasatch Mountains, Utah: Examples of large-scale water-rock interaction 879
M.T.Einaudi

Results of water rock interactions in mid-ocean ridges 885
J.M.Edmond

Stable and radioactive isotopes

Water and solutes

Boron isotopic composition of Icelandic hydrothermal systems 893
J.K.Aggarwal, M.R.Palmer & K.V.Ragnarsdottir

Isotopic composition of Sr, Nd and Li in thermal waters from the Norris-Mammoth corridor, Yellowstone National Park and surrounding region 897
T.D.Bullen & Y.K.Kharaka

Combined noble gas and quadrupole mass spectrometer analysis of volatiles trapped in fluid inclusions 903
P.Burnard, L.Ayliffe, F.Stuart, G.Turner & C.Curtis

The signature of water-rock interactions in the formation waters of sedimentary basins: Some new evidence 907
Sam Chaudhuri, S.Furlan & N.Clauer

Stable isotopes in volcanic fluids and rocks at Vulcano (Sicily, Italy) 911
G.Cortecci, G.Ferrara, A.Maiorani & B.Turi

Unconventional stable isotopes: Iron 915
P.R.Dixon, D.R.Janecky, R.E.Perrin, D.J.Rokop, P.L.Unkefer, W.D.Spall & R.Maeck

Mantle derived helium in fluid inclusions of a 2.5 Ga old granulite, Nilgiri Hills, Southern India 919
T.J.Dunai, J.R.L.Touret & I.M.Villa

Noble gas geochemistry of mineralization in the Pennines 923
Jacqueline L.Ford, Finlay M.Stuart & Grenville Turner

Geochemical modelling of Sr isotopic signatures in the interaction between granitic rocks and natural solutions 927
B.Fritz, L.Richard & R.H.McNutt

Chlorine-36 in groundwaters in the Lac du Bonnet granite, southeastern Manitoba, Canada 931
M.Gascoyne, D.C.Kamineni & J.Fabryka-Martin

An oxygen isotope study of silicates at Larderello, Italy 935
P.Iacumin, E.Petrucci, G.Gianelli & M.Puxeddu

Oxygen isotope studies of hydrothermally altered basalts from Surtsey, Iceland 939
H.R.Karlsson, S.J.Jakobsson & J.G.Moore

Chlorine stable isotope distribution of Michigan Basin and Canadian Shield formation waters 943
Ronald S.Kaufmann, S.K.Frape, Robert McNutt & Christopher Eastoe

Anomalous ^{3}He contents of CH_4-rich gases in sedimentary basins 947
B.Mack Kennedy, J.Poths & H.Hiyagon

Changes of stable istopes and chemistry of fluids in the low-temperature geothermal field at Bakki-Thóroddstadir, Ölfus, SW-Iceland 951
Hrefna Kristmannsdóttir & Árný Erla Sveinbjörnsdóttir

Stable isotope studies of diagenesis in berthierine-bearing oil sands, Clearwater Formation, Alberta, Canada 955
F.J.Longstaffe, A.Ayalon & M.A.Racki

Mantle helium signal in the compressed feature, Outer Carpathians, Poland 959
P.M.Leśniak, H.Sakai, J.Ishibashi & H.Wakita

Chemical composition and stable isotopes of thermal waters, Norris-Mammoth corridor, Yellowstone National Park, USA 963
R.H.Mariner, Y.K.Kharaka, Gil Ambats & L.D.White

Hydrogen isotopic ratio of hornblende from Quaternary volcanic rocks in Japan 967
Osamu Matsubaya & Isoji Miyagi

Lead and strontium isotopic data on brines from the Michigan basin, Ontario and Michigan 971
C.M.McKenna, R.H.McNutt & S.K.Frape

The sulphur isotopic composition of mafic-intermediate volcanic rocks, Taupo Volcanic Zone, New Zealand 975
B.W.Robinson & I.J.Graham

The McLaughlin hot-spring gold-mercury deposit and its relationship to hydrothermal systems in the Coast Ranges of northern California, USA 979
R.L.Sherlock & E.C.Jowett

^{10}Be and ^{9}Be in the Salton Sea (USA) and other geothermal systems 983
Nathalie J.Valette-Silver, Fouad Tera & Milan Pavich

Exchange reactions

Isotope-geochemical effects of water-rock interactions in the Atesina-Cima d'Asta volcano-plutonic complex, Italy 989
S.Barth, F.Oberli, M.Meier, P.Blattner, G.M.Bargossi & G.Di Battistini

Possible evidence for fluid-rock oxygen isotope disequilibrium in hydrothermal systems 995
David R.Cole

The effects of dissolved salts on oxygen and hydrogen isotope fractionation between water vapor and brines at 50-100°C 999
J.Horita, D.J.Wesolowski & D.R.Cole

The effect of pressure on equilibrium isotope fractionation 1003
N.N.Kharlashina & V.B.Polyakov

Boron isotopic fractionation between coexisting vapor and liquid in natural geothermal systems 1007
William P.Leeman, Robert D.Vocke & Mike A.McKibben

Geochemical modeling

Solution chemistry and thermodynamics

Experimental study of aluminum hydrolysis and complexation with acetate and sodium ions in hydrothermal solutions 1013
S.Castet, J.L.Dandurand, J.Schott & R.Gout

Solubility of Pt, Pd sulfides and Au in aqueous bisulfide solutions to 350°C 1017
Pujing Pan & Scott A.Wood

Calculation of the thermodynamic properties of mononuclear aqueous Al species in the system Al_2O_3-H_2O-NaCl at temperatures to 1000°C and pressures to 5 kb 1021
Vitalii A.Pokrovskii & Harold C.Helgeson

Excess molar volumes for CO_2-CH_4-N_2 mixtures 1025
J.C.Seitz, J.G.Blencoe, D.B.Joyce & R.J.Bodnar

Calculation of the solubility of minerals in H_2O-CO_2 solutions at mid and lower crustal depths 1029
John V.Walther

Thermodynamic modeling of Be-mineralization: Preliminary calculations 1033
Maria Woitsekhovskaya

Solute transport

The natural evolution of geothermal systems in light of infiltration fronts and a look back at fluid-to-rock ratios 1039
Peter Blattner & Keith R.Lassey

Lattice gas automata for flow and transport in geochemical systems 1043
D.R.Janecky, S.Chen, S.Dawson, K.C.Eggert & B.J.Travis

CIRF: A general reaction-transport code: Mineralization fronts due to the infiltration of reactive fluids 1047
J.L.Potdevin, W.Chen, A.Park, Y.Chen & P.Ortoleva

Dissolved salt removal from Illinois Basin waters 1051
V.Ranganathan

Mechano-chemistry of compartmented basins 1055
E.L.Sonnenthal, J.M.Maxwell, C.Qin & P.J.Ortoleva

General applications

A quantitative first principles approach to supergene enrichment of a porphyry copper protore 1061
Giuseppe G.Biino & Peter C.Lichtner

Thermodynamic modelling of geochemical processes: The CSIRO-Monash THERMOCHEMISTRY System 1065
M.S.Bloom, M.W.Wadsley & Y.Gu

Thermodynamic model of interaction between sulphides and oxidizing solutions with special attention to intermediate conditions 1071
G.R.Kolonin & O.L.Gas'kova

Aluminium – catalyst, crystal deformer, and dye: Self-organizational crystallization of quartz in agates and amethyst geodes, Paraná flood basalts, Brazil 1075
Enrique Merino, Yifeng Wang & Etienne Deloule

Integration of intensive variable diagrams and fluid phase equilibrium with SOLMINEQ.88 pc/shell 1079
E.H.Perkins

Collapse of acid waters into boiling hydrothermal system and the origin of late stage pyrite and related kaolinite 1083
Mark H.Reed & Geoff Plumlee

Microcomputer-based modeling of speciation and water-mineral-gas reactions using programs SOLVEQ and CHILLER 1087
N.F.Spycher & M.H.Reed

Reaction path modelling of diagenesis in the Eocene Wilcox Group of the Texas Gulf Coast 1091
R.N.Tempel & W.J.Harrison

Analysis of heat source for the Xiangshan fossil hydrothermal system, China: A modeling approach 1095
Zhou Wenbin

Sedimentary basins

Fluid evolution

Regional hydrogeology and fluid geochemistry of the Mannville Group, Western Canada Sedimentary Basin: Synthesis and reinterpretation 1101
H.J.Abercrombie & E.G.Fullmer

Preservation of ancient basinal fluids in halite fluid inclusions 1105
D.Chipley & T.K.Kyser

Water composition variation within one formation 1109
Max L.Coleman

The effects of combined changes in pore fluid chemistry and stress state on reservoir permeability: Design and recirculating flow rig 1113
S.C.Elphick, G.B.Shimmield, B.T.Ngwenya, I.G.Main, B.G.D.Smart & B.R.Crawford

Chemical composition and origin of saline formation waters from the Konrad Mine, Germany 1117
H.Klinge, P.Vogel & K.Schelkes

Geochemical modeling of salt dome cap rock-brine reactions 1121
James A.Saunders

Contrasts between steam stimulation and in-situ combustion process: An analysis of co-produced fluid compositional data 1125
Maurice Shevalier, Ian Hutcheon, Cynthia Nahnybida & Hugh Abercrombie

Spatial variations in North Sea formation water compositions: Preliminary results 1129
E.A.Warren & P.C.Smalley

Mixing of groundwater with isotopically distinct signatures: A multi-component mixing model for the Dundee Formation of southwestern Ontario, Canada 1133
T.R.Weaver, S.K.Frape & J.A.Cherry

Halogen compositions of fluid inclusions as tracers of crustal fluid behaviour 1137
B.W.D.Yardley, D.A.Banks & I.A.Munz

Hydrogeochemical patterns of major stages of the sedimentary process 1141
V.P.Zverev

Mineral diagenesis

North Sea clastic diagenesis and formation water constraints 1147
P.Aagaard, P.Kr.Egeberg & J.S.Jahren

Cementation of North Sea sands: Constraints from fluid compositions 1153
Andrew C.Aplin, Edward A.Warren & Shona M.Grant

Silica mass transport during shale diagenesis: Are shales a source or sink for silica? 1157
J.Bloch & I.E.Hutcheon

Overpressure release, cross-formational porewater flow and diagenesis 1161
Patrice de Caritat & Julian C.Baker

Three zones for burial metamorphism 1165
D.D.Eberl

Secondary K-feldspar at the Precambrian-Paleozoic boundary, Ontario, Canada 1169
D.A.Harper, F.J.Longstaffe & M.A.Wadleigh

Illite/smectite diagenesis in overpressured and normally pressured environments, Beaufort-Mackenzie Basin, Arctic Canada 1173
J.Ko & R.Hesse

Isotopic, mineralogic and chemical evidence for multiple episodes of fluid movement during prograde and retrograde diagenesis in a Proterozoic basin 1177
T.G.Kotzer & T.K.Kyser

The mechanism of porosity reduction, styolite development and quartz cementation in North Sea sandstones 1183
E.H.Oelkers, P.A.Bjorkum, & W.M.Murphy

Mathematical simulation of water-mineral interaction and fluid flow with respect to the diagenetic evolution of sandstones 1187
R.Ondrak

Experimental studies of mudstone diagenesis 1193
Elspeth I.C.Rae & David A.C.Manning

Small scale trace element zonations in carbonates: An ion microprobe study of diagenesis in the Upper Devonian Nisku Formation, Western Canada Sedimentary Basin 1197
Lee R.Riciputi, David R.Cole, Hans G.Machel, Warner H.Christie & Thomas M.Rosseel

Coupled predictive models of diagenesis in sand-shale successions during progressive burial 1205
R.C.Surdam & D.B.MacGowan

A model for diagenesis in the Upper Wilcox reservoir sandstones at Fandango Field, south Texas, USA 1209
Glenn A.Wilson, Charles E.Harvie & David T.Lawrence

Evolution of $CaCl_2$ brine in Silurian aged formations of the Michigan basin, USA: The role of mineralogic reactions and evaporite diagenesis 1213
T.P.Wilson & D.T.Long

Mineral transformations and interactions

Constraints for an evaporitic sedimentation modeling: Microanalysis of frozen fluid inclusions in halite 1219
A.Canals, B.Carpentier, A.Y.Huc & B.Durand

Computer modelling of steam flood experiments 1223
Jon S.Dudley & Craig H.Moore

A stable isotope study of clay minerals from the Sancerre-Couy deep drill hole, Paris Basin, France 1227
A.M.Fouillac & D.Beaufort

Experimental study of reactions between selected clays and micas and aqueous fluid analogues of well-bore engineering fluids 1231
C.R.Hughes & P.I.Hill

Hydrodynamic and chemical controls of Paleozoic phosphorite and barite in the western Cordillera of North America 1235
P.W.Jewell

Access of pore waters to carbonate precipitation sites during concretion growth 1239
M.M.Jordan, A.C.Aplin, C.D.Curtis & M.L.Coleman

Wettability of petroleum reservoir rock minerals 1243
H.Lemke & K.Schwochau

Replacement of albite by K-feldspar and vice versa 1247
Radomir Petrovich & Philip M.Orville

Wettability of petroleum reservoir rock minerals 1243
H.Lemke & K.Schwochau

Replacement of albite by K-feldspar and vice versa 1247
Radomir Petrovich & Philip M.Orville

Illite age analysis, a new tool for basin thermal history analysis 1251
D.R.Pevear

Geothermal systems

Fluid chemistry

The occurrence of acidic fluids in the Leirbotnar field, Krafla, Iceland 1257
Halldór Ármannsson & Gestur Gíslason

Hydrochemistry of Olkaria and Eburru geothermal fields, Kenyan Rift Valley 1261
M.K.Arusei

Geochemistry of waters from three geothermal boreholes in the upper Rhein graben, Switzerland 1267
L.Calmbach

Thermal springs in the Precambrian crystallines of the western continental margin of India: Field and experimental results 1271
D.Chandrasekharam, A.Ramanathan & R.L.Selvakumar

Gas geochemistry for the Beppu hydrothermal system, Kyushu, Japan 1275
F.D'Amore, G.Gianelli, A.Caprai, Y.Yusa, K.Kitaoka & K.Takemura

Thermal waters and mineralization in the Geysers-Clear Lake area, California, USA 1279
Julie M.Donnelly-Nolan

The Salton Sea Hydrothermal System, California, USA: A review 1283
W.A.Elders, M.A.McKibben & A.E.Williams

The geochemistry of hot spring waters at Norris Geyser Basin, Yellowstone National Park, USA 1289
Robert O.Fournier, J.Michael Thompson & Roderick A.Hutchinson

Exploration progress of high enthalpy geothermal prospects in Indonesia 1293
Sugiarto Ganda, Djoko Sunaryo & Tanda Tampubolon

Hydrothermal systems of Hengduan Mountains, China 1297
Guo Guoying

Composition of gases from the Norris-Mammoth corridor, Yellowstone National Park, USA: Evidence for a magmatic source near Mammoth Hot Springs 1303
Y.K.Kharaka, R.H.Mariner, W.C.Evans & B.M.Kennedy

The origin of thermal waters in crystalline rocks of the North China fault block 1309
Li Guiru, Wang Jihua, Zhang Peiren & Jiang Fengliang

Genesis of mineral waters of the Rhenish Massif, central Europe 1313
Franz May, Stephan Hoernes & Horst J.Neugebauer

Environmental isotope study of saline geothermal systems in western Anatolia, Canakkale, Turkey 1317
Stefan Mützenberg, Werner Balderer & Werner Rauert

Trace element geochemistry of the system 'rock – thermal water – suspended matter – deposits' in a granitic environment 1321
Elka N.Pentcheva, Els Veldeman, Luc Van't dack & Renaat Gijbels

Thermal springs in granitic rocks from the Grimsel Pass (Swiss Alps): The late stage of a hydrothermal system related to Alpine Orogeny 1327
H.-R.Pfeifer, A.Sanchez & C.Degueldre

Thermal waters in México 1333
R.M.Prol-Ledesma

REE distribution in calcites from the Lisbon volcanic complex, Portugal 1337
M.I.Prudêncio, J.M.P.Cabral & M.A.Sequeira Braga

Petrology and fluid chemistry of magmatic-hydrothermal systems in the Philippines 1341
A.G.Reyes & W.F.Giggenbach

The question of recharge to the geysers and hot springs of Yellowstone National Park, Wyoming, USA 1345
Robert O.Rye & Alfred H.Truesdell

Secondary effects of reinjection on the Larderello, Italy geothermal system evaluated by the study of the isotopic and chemical composition of the gas mixtures 1349
G.Scandiffio, F.Baccarin, M.Valenti, C.Panichi, P.Noto & L.Bellucci

Inactive gas ratios of geothermal gases from the Sengan field, Northeast Japan, and their implications in origins of hydrothermal systems 1353
H.Shigeno, M.Takahashi & T.Noda

A conceptual model of the hydrothermal system in Long Valley caldera, California, USA 1357
Michael L.Sorey & Christopher D.Farrar

Temporal changes in mud volcanos, Salton Sea Geothermal Area 1363
A.A.Sturz, R.L.Kamps & P.J.Earley

Carbon stable isotope geochemistry of a geothermal system: Los Azufres, Mexico 1367
Mahendra P.Verma, Enrique Portugal & David Nieva

Geothermal systems in continental area of China 1371
Wang Ji-yang, Chen Moxing, Xiong Liang-ping & Pang Zhong-he

Determination of cation exchange parameters for a hot dry rock geothermal energy reservoir 1375
Luke Wilkinson, John H.Tellam, John R.Ashworth & Hugh G.Richards

The rare-alkali metal characteristics of Tibetan geothermal systems and the potential for exploitation and utilization 1379
Zheng Yaxin

Mineral interactions

Calcite deposition in geothermal wells caused by infiltrating seawater 1389
Stefán Arnórsson

An autoclave reinjection study of a vitreous rhyolite tuff 1393
Cheng Yan, A.C.Lee, V.J.Grassi, D.E.Grandstaff & Gene C.Ulmer

Experimental approach to the self-sealing of rock-water systems 1397
Masahiro Chigira & Masakazu Watanabe

Travertines deposited by CO_2-rich springs in the geothermal province of northern Latium, Italy 1401
G.Chiodini, A.R.Zanzari & L.Fanfani

Alteration mineralogy of the Wayang Windu geothermal field, West Java, Indonesia 1405
Sugiarto Ganda & Djoko Hantono

Augustine volcano fumarole wall rock alteration: Mineralogy, zoning and numerical models of its formation process 1411
Aberra Getahun, Mark H.Reed & Robert B.Symonds

Water-rock hydrothermal experiments: Influence of rock type on solution pH and oxygen fugacity 1415
V.J.Grassi, G.C.Ulmer & D.E.Grandstaff

The travertine totem forest of Yellowstone National Park, USA: Geologic controls and geochemistry 1419
R.A.Hutchinson & J.Michael Thompson

A look at silica phases in evolving hydrothermal systems 1423
T.E.C.Keith

Yellowstone travertines: U-series ages and isotope ratios (C, O, Sr, U) 1427
N.C.Sturchio, M.T.Murrell, K.L.Pierce & M.L.Sorey

Magnesium silicate scales in geothermal utilization 1431
Guðrún Sverrisdóttir, Hrefna Kristmannsdóttir & Magnús Ólafsson

Geothermometry

Correlation of the Na/K ratio in geothermal well waters with the thermodynamic properties of low albite and potash feldspar 1437
John A.Apps & Gee Minn Chang

Hydrocarbon gas ratio geothermometry in the East African Rift System 1441
W.G.Darling & J.C.Talbot

Past and present subsurface geothermal structure revealed by fluid inclusion thermometry of borehole SN-1, Sengan geothermal area, Japan 1445
B.De Vivo, F.Raia, M.Sasada & T.Sawaki

Silica geothermometers for use in predicting temperatures in Alberta Oil Sand Reservoirs during steam-assisted thermal recovery 1451
W.D.Gunter, Brian Wiwchar, Larry Holloway & E.H.Perkins

Calibration and reequilibration kinetics of cation geothermometers: An experimental study 1455
Houssein Ibrahim, Pierpaolo Zuddas, Gil Michard & Daniel Grimaud

Chemical reactions in upflow zones of geothermal systems affecting subsurface temperature estimates by chemical geothermometry 1459
G.Th.Magnúsdóttir, S.Arnórsson & Audur Andrésdóttir

Theoretical calibration of chemical geothermometers and its application to the granitic geothermal areas of SE China 1463
Pang Zhong-he

Alteration and temperature distribution within and at the margin of the volcanic zone on the Reykjanes peninsula, SW-Iceland 1467
Jens Tómasson & Hjalti Franzson

Metamorphic environments

Formation of zoned epidote in hydrothermal systems 1473
John G.Arnason & Dennis K.Bird

Dependence of hydrothermal alteration on rock texture and microporosity in mafic crystalline rocks 1477
Mark E.Brandriss, Dennis K.Bird & James R.O'Neil

A revised model for the formation of dequartzified and alkalinized granites (episyenites) 1481
C.Casquet, F.Tornos, J.M.Caballero & C.Galindo

Origin of the Maqarin, Jordan hyperalkaline groundwaters: Isotopic and geochemical evidence for in situ combustion, calcination and recarbonation of bituminous marls 1485
Ian D.Clark, Hani N.Khoury, Elias Salameh, Peter Fritz, Harald K.Seidlitz & Tony E.Milodowski

Hydrothermal alteration in two Palaeoproterozoic volcanic associations in northern Finland 1491
P.Eilu

Chemical variations accompanying detachment faulting in the Whipple Mountains, California, USA 1495
Harold Ekstrom & Jean Morrison

Evidence for the controlling effect of the high-pressure metamorphic PTt path on the mass transfer of major elements 1499
B.Goffé & O.Vidal

A different approach to fluid-rock-interactions applied to microtextures in metacarbonates 1503
S.Heuss-Aßbichler & C.Schneider

Kinetic aspects of formation of Japanese bentonite deposits 1507
Daizo Ishiyama, Masakazu Ito & Osamu Matsubaya

Composition and temperature effects on mass transfer during serpentinization of ultramafic rocks 1511
C.F.Jove

High temperature metamorphism and low temperature retrograde alteration of spontaneously combusted marls: The Maqarin cement analogue, Jordan 1515
Hani N.Khoury & Tony E.Milodowski

Fluid regime of glaucophane metamorphism 1519
G.G.Likhoidov & L.P.Plyusnina

Thermochemical determination of water activities relevant to eclogitic rocks 1523
H.-J.Massonne

Chemical and oxygen isotopic variations in upper-plate rocks of the Whipple Mountains detachment system, California, USA 1527
D.Mayo, J.Morrison & J.L.Anderson

Enhancement of porosity in gabbros by dike intrusion 1533
Richard J.Nevle & Dennis K.Bird

Constraining the temperature, pressure, and fluid compositional evolution of an accretionary complex by an integrated study of cement stratigraphy and tectonics 1537
David L.Orange, Duncan Geddes, J.Casey Moore, Randy Cox & Michael Underwood

Structural controls of fluid flow during regional metamorphism: Application of reaction front advection theory to greenschist facies metabasites in the Scottish Dalradian 1543
A.D.L.Skelton, C.M.Graham & M.J.Bickle

Fluid-rock interaction in skarns from La Cerdanya, Central Pyrenees, Spain: Evidence from carbon and oxygen isotopes 1547
A.Soler, J.Delgado, E.Cardellach & C.Ayora

Fluid interaction and deformation, the Brevard Shear Zone, Rosman, North Carolina, USA 1551
Jane R.Stockey & Eileen L.McLellan

Deposition from skarn-forming solution: The genesis of some vein-like skarns (with an example from the Longjiaoshan Cu-W deposit; southeastern Hubei, P.R. China) 1555
Yanxin Wang, Zhaoli Shen & Shaozhuo Xu

Fluid flow during thrusting: A Sr isotopic study 1559
D.M.Wayne & A.M.McCaig

Hydrothermal ore deposits

Hydrothermal replacement of carbonates by sulfides 1565
H.L.Barnes & W.W.Gould

Reaction path models for hydrothermal Au-PGE mineralization at Coronation Hill and similar deposits of the South Alligator Mineral Field, Australia 1569
M.S.Bloom, D.J.Gilbert, C.H.Gammons & A.R.Wilde

Computer models of formation of hydrothermal ore bodies at the expense of metals of the country rocks 1575
H.V.Borisov & Vict. L.Barsukov

Numerical models of mineral deposition for the Acupan vein system, Philippines 1579
David R.Cooke

Mineral paragenesis and regional zonation of granite-related Sn-As-Cu-Pb-Zn deposits: A chemical model for the Mole Granite district (Australia) based on PIXE fluid inclusion analyses 1583
Christoph A.Heinrich & Christopher G.Ryan

Water-rock interaction in the forming process of Dongchuan copper deposit, China 1589
Renmin Hua, Huichu Ruan, Pei Ni & Dennis P.Cox

Subsurface hydrothermal genesis of ores and minerals in recent hydrothermal system of Uzon caldera in Kamchatka, Russia 1593
Gennady A.Karpov

Relations between terrestrial heat flow and uranium deposits in Jiangxi Province, China 1597
Li Xueli, Zhou Wenbin, Zhang Weimin & Sun Zhanxue

Stable isotopic geochemistry of ore-forming solutions and genesis of a gold deposit in Maopai, Jiangxi Province, China 1601
Liu Xiaodong, Xu Haijiang & Lai Shaocong

Au sorption by arsenic compounds and organic matter 1605
I.Ya.Nekrasov

Hydrothermal systems related to epithermal gold deposits in Mesozoic volcanic areas in eastern China 1609
Qijang Ren, Chongze Zhang, Rongyong Yang, Xiaojun Xie & Zhaowen Xu

Metal transport and deposition in hydrothermal systems buffered by mineral assemblages 1613
Gilpin R.Robinson, Jr

Fluid-rock interaction at the Kaneuchi tungsten deposit in Japan 1617
Yasuhiro Shibue, Masami Noto & Kohei Kazahaya

Isotope geochemistry of oxygen and hydrogen for ore-forming hydrothermal solution in Xiangshan uranium orefield, China 1621
Sun Zhanxue, Li Xueli & Shi Weijun

Fluid-rock interaction and cassiterite deposition 1625
T.M.Sushchevskaya & M.V.Borisov

Si, Al, and W behavior in HF aqueous solutions in relation to the greisen problem 1629
George P.Zaraisky

Evidence of ancient hydrothermal activity and the relation to gold mineralization in the Rehai geothermal system, Yunnan, China 1633
Zhu Meixiang

Seawater-rock interactions

Oceanic metamorphism in the Kuyul ophiolitic terrane of north Kamchatka, Russia 1639
O.V.Chudaev & V.B.Kurnosov

Spilitic alteration as the source of the world-class Proterozoic manganese deposit 1643
D.H.Cornell & S.S.Schütte

Alkali basalt/seawater hydrothermal alteration 1647
I.V.Kholodkevich, J.L.Bischoff & R.J.Rosenbauer

Hydrothermal water-basalts interaction in the ocean 1651
Victor B.Kurnosov

Oceanic borehole fluid chemistry from Site 504B: Seawater/rock reaction within the borehole environment 1655
Andrew J.Magenheim, Greg Bayhurst, Joris M.Gieskes & Jeff Alt

Formation of authigenic smectite and zeolite and associated major element behavior during early diagenesis of volcanic ash in the Nankai Trough, Japan, ODP leg 131 1659
H.Masuda, H.Tanaka, T.Gamo, J.R.O'Neil, D.R.Peacor & W.-T.Jiang

Geochemistry of subduction-induced cold seepage at the eastern Nankai Trough, Japan 1663
H.Sakai, T.Gamo, Y.Ogawa & J.Boulegue

Peridotite-seawater interaction at Zabargad Island, Red Sea 1669
P.F.Sciuto

The isotopic composition of helium and sulphur in hydrothermal fluids trapped in ancient and modern sea floor deposits 1671
F.Stuart, G.Turner & J.Maynard

Microbial green marine clays in the hydrothermally altered deep sea sediments 1675
Kazue Tazaki, Kōichi Tazaki & W.S.Fyfe

Short-term variability, phase separation and water-rock reaction in hydrothermal fluids from 9-10°N, East Pacific Rise 1679
K.L.Von Damm

Author index 1681

Keynote lectures

Water-Rock Interaction, Kharaka & Maest (eds) © 1992 Balkema, Rotterdam. ISBN 90 5410 075 3

Stable isotope exchange in fluid-mineral systems: Old problems, new solutions

J.R.O'Neil
Department of Geological Sciences, University of Michigan, Mich., USA

ABSTRACT: Our knowledge and understanding of equilibrium and kinetic fractionation factors among the important rock-forming minerals and C-O-H-S fluids is incomplete but sufficient to conduct meaningful stable isotope investigations of the processes operative in hydrothermal, metamorphic and igneous systems. The same cannot be said for low-temperature systems except for processes involving only a few important minerals, most notably carbonates. For other important groups of minerals including the hydrous phyllosilicates and phosphates, we must rely on fractionation factors estimated from measurements of natural samples or from questionable extrapolations of the results of experiments conducted at high temperatures. New methods of semi-empirical calculations and novel experimental techniques, including bacterial syntheses, are now being used to obtain low-temperature fractionation factors of geological interest.

Of major concern in stable isotope studies is the question of retention of original isotopic ratios of minerals during diagenesis or exposure to surficial conditions through time. By careful selection and characterization of samples and recent developments in analytical techniques, we are coming to a much clearer understanding of this problem. Interactions between rocks, fluids and organic matter may explain seemingly enigmatic isotopic variations in a variety of geologic environments.

1. INTRODUCTION

Stable isotope measurements of rocks, minerals and fluids have played an important role in understanding key aspects of water/rock interactions that occur in both natural and laboratory systems. Essential to the interpretation of natural data is knowledge of the equilibrium and kinetic fractionation factors among the major mineral groups and C-O-H-S fluids. These fractionation factors are obtained by: (1) laboratory calibration experiments, (2) semi-empirical calculations using spectroscopic data and the methods of statistical mechanics, and (3) isotopic measurements of natural materials that are forming slowly under well-constrained environmental conditions.

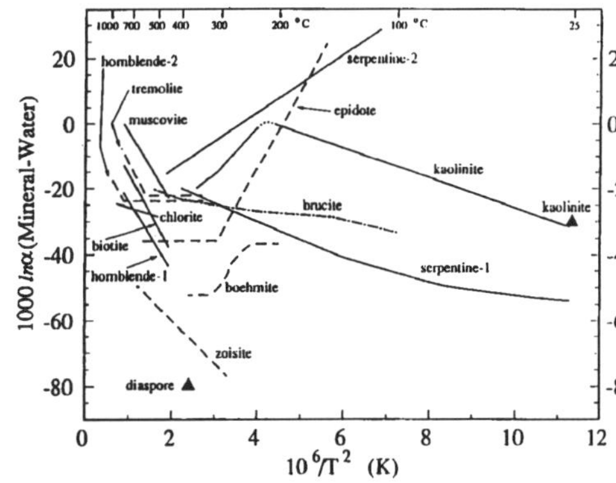

Figure 1: Variation in hydrogen isotope fractionations with temperature between various hydrous minerals and water.

The first method is the most reliable at the present time because, if laboratory exchange rates are rapid enough, it is possible to demonstrate that equilibrium has been atttained by approaching the equilibrium distribution of stable isotopes between the two substances from opposite directions. Unfortunately, for most systems, the problem of slow rates of isotopic exchange limits the feasibility of such experiments to relatively high temperatures. The second method is the method of choice but is plagued with uncertainties. With our present understanding (admittedly incomplete) of the theory of isotopic fractionation, it should be possible to calculate the fractionation factors of interest at any temperature and obviate the problems of laboratory experimentation. While progress has been made in this area (*e.g.* Clayton and Kieffer, 1991), it remains a formidable task to deal with bonding characteristics in materials as complex as aluminosilicate minerals. The third method has provided us with the only estimates we have of the fractionation factors between important mineral systems and water at low temperatures. Recognizing the serious drawbacks and uncertainties of using this method, it is often the only means available to obtain information on the sign and magnitude of equilbrium fractionations among substances of geologic interest.

Equal in importance to having a consistent set of fractionation factors to interpret natural data is an understanding of the factors that control the preservation of original stable isotope ratios during low-temperature geological processes like weathering, diagenesis and low rank metamorphism (O'Neil, 1987). Hydrogen isotope compositions of minerals can often provide the definitive evidence for the origin of environmental fluids because there is little chance of such fluids undergoing a "hydrogen isotope shift" prior to the formation of the mineral. Of particular recent concern, however, is the integrity of hydrogen isotope compositions of fine-grained hydrous minerals through time. Anomalous hydrogen isotope compositions of natural fluids and minerals have been explained by facile hydrogen isotope exchange between aqueous fluids and fine-grained hydrous minerals, by rock/water interactions of various types and by the participation of organic matter in the system. Many of these questions can be resolved by laboratory experiments.

2. FRACTIONATION FACTORS AT LOW TEMPERATURES

There are several unresolved ambiguities in published fractionation factors at low temperatures and reliable data are sorely lacking for many important mineral systems. For example, there are major discrepancies in the published determinations of D/H fractionation factors between hydrous minerals and water. Some features of the fractionation curves (Figure 1) are contradictory and others are difficult to understand in light of present theory. One glaring problem is the difference between the fractionation curve for serpentine-water determined by laboratory experiment (Sakai and Tsutsumi, 1978) and by regularities in natural data (Wenner and Taylor, 1973). In addition there are several unusual minima, maxima, inflections and crossovers in the curves shown in Figure 1 that are all theoretically possible but unlikely in the opinion of most workers in the field. The apparent lack of significant temperature dependence of the isotopic fractionation between certain minerals (*e.g.* epidote, clinozoisite, boehmite) and water over the geologically important temperature range of around 250° to 450°C has important implications and merits further investigation and corroboration. The serpentine problem and other irregularities have remained unresolved for over 15 years. Exchange experiments with water as the exchange medium will probably never provide the answer because, depending on the mineral, the exchange rates between minerals and water are prohibitively slow at temperatures lower than several hundreds of degrees.

2.1 H_2 as an exchange medium

During the past year we searched for a non-aqueous exchange medium that can be used for hydrogen isotope exchange experiments and discovered that exchange rates between dry molecular H_2 and hydrous minerals are remarkably fast at temperatures of 200-300°C and probably lower (Vennemann and O'Neil, unpub.). As examples of the enormous amounts of exchange observed in systems with relatively high mineral/H_2 ratios, δD of H_2 in contact with muscovite (δD = – 46) at 300°C changed from – 23 to – 260 in only 16 hours and, in another experiment with biotite (δD = – 84), δD of H_2 changed from + 2 to – 390 in 66 hours at the same temperature. For biotite companion runs the extent of exchange (to equilibrium) was 87% or higher in just three days! For the longest kaolinite run of 155 hours at 240°C, the extent of exchange was estimated at more than 50%. Rates of hydrogen isotope exchange in systems with water as the exchange medium are orders of magnitude lower. For comparison, O'Neil and Kharaka (1976) observed only 29% hydrogen isotope exchange between kaolinite and water at 200°C after 5,640 hours. Clearly the exchange rates between molecular hydrogen and minerals are extraordinarily high and this system is inherently cleaner and more simple in many respects. Because of these rapid exchange rates, it should be possible to obtain, in reasonable times, isotopic partition function ratios of the common hydrous minerals down to low temperatures and thereby establish a self-consistent set of fractionation factors for application to a variety of geologic problems.

2.2 Enzymatic activity

While it is possible to synthesize minerals like clays and certain other minerals from gels and other precursor substances at relatively low temperatures, attainment of isotopic equilibrium cannot be demonstrated using such methods (*e.g.* O'Neil, 1986). From oxygen isotope analyses of bones and teeth of modern fish and environmental waters, Kolodny *et al.* (1983) suggested that oxygen isotope equilbrium between bone apatite and body fluids was promoted by the enzymatic activity present in living organisms. It may be that such biologically catalyzed reactions can be exploited in the laboratory to obtain critical fractionaton factors at low temperatures. Lucas and Prévot (1981) have prepared well crystallized apatite from RNA and a $CaCO_3$ substrate by the action of bacteria. We are carrying out experiments based on this procedure at the present time in order to determine an oxygen isotope fractionation curve for the apatite-water system at low temperatures. We are also attempting to effect oxygen isotope exchange between dissolved phosphate and water with the aid of phosphatase enzymes. Such enzymatically catalyzed reactions can be applied to other systems as well.

2.3 Empirical bond-strength calculations

Beginning with Taylor and Epstein (1961) several researchers have made crude estimates of the oxygen isotope properties of complictated minerals by taking into account the number and bond strengths of the different bond types present in the minerals. The addition of more fundamental characteristics of the minerals, like vibrational frequencies, geometrical arrangements, and anion potentials have put such empirical calculations on a much firmer theoretical basis (*e.g.* Shutze, 1980). Owing to the pressing need for additional reliable fractionation factors, particularly at low temperatures, there is considerable interest in recent times in refining such calculations.

Savin and Lee (1988) developed an empirical bond-strength model that is very useful in predicting the isotopic behavior of phyllosilicate models. To accomplish this, the oxygen isotope fractionations between various bond types like Si-O-Si, Si-O-Al, Al-OH, M-O-M, *etc.*, were estimated from published experimental data on the oxygen isotope fractionations between water and the various end-member substances like quartz, magnetite, gibbsite and the like. Then, using a crystal model of the mineral and a minimum number of sensible assumptions (*e.g.* bonding between T-O-T layers is ignored), they calculated mineral-water fractionation expressions from the abundances of the different bond types in the mineral of interest. As data on the end-member substances improve, we will be in a better position to evaluate these types of approximations. At present they are the best thing we have.

3. RETENTION OF D/H RATIOS BY FINE-GRAINED MINERALS

In recent years there has been a heightened awareness of the power of hydrogen isotope measurements of minerals and fluids in resolving a variety of important petrologic processes. Overshadowing this development is a concern on the part of many workers in the field that original D/H ratios of minerals, particularly fine-grained minerals, may not be retained when exposed to surficial processes. For example, Bird and Chivas (1988) and Kyser and Kerrich (1991) present convincing evidence for fairly rapid hydrogen isotope exchange between meteoric fluids and fine-grained clay and serpentine minerals . On the other hand, there are numerous cases of hydrous minerals that are seemingly very retentive of their original isotopic compositions. A striking example of this is provided by the recent work of Arehart and O'Neil (1992) who measured variations of over 60 per mil in δD values of extremely fine-grained samples of supergene alunite of different Tertiary ages from nearby locations in Nevada.

The experimental work of O'Neil and Kharaka (1976) on the relative rates of D and ^{18}O exchange between clay minerals and water is often cited in support of the concept of facile hydrogen isotope exchange that can take place between solid minerals and water. Hydrogen isotope exchange in such systems was shown to take place by a mechanism (proton exchange?) different from the mechanism (recrystallization?) by which sluggish oxygen isotope exchange occurs in the same system. However, the use of the self-diffusion rates for hydrogen and oxygen isotopes in hydrous minerals determined from the results of laboratory experiments may not be applicable to natural systems except as a limiting case. The reason for this is that, in laboratory systems, unusually fine-grained materials are chosen for experimentation to augment exchange rates and, more importantly, the "effective" water/rock ratios are <u>much</u> higher than normally attained in natural systems. Certainly effective water/rock ratios and duration of exposure to surficial fluids in nature are going to play a major role in understanding the severity of this problem of post-formational exchange in nature. Again such parameters can be varied in laboratory systems and evaluated.

4. REFERENCES

Arehart, G.B. and O'Neil, J.R. (1992) Absolute age of global climate changes: isotopic analysis of alunite from continental settings. Nature, in press.

Bird, M.I. and Chivas, A.R. (1988) Stable isotope evidence for low temperature kaolinitic weathering and post-formational hydrogen isotope exchange in Permian kaolinites. Chem. Geol. (Isotope Geosci. Sect.), **72**, 249-265.

Clayton, R.N. and Kieffer, S.W. (1991) Oxygen isotopic thermometer calibrations In *Stable Isotope Geochemistry: A Tribute to Samuel Epstein*. The Geochemical Society, Special Publication No.3 (Eds: H.P. Taylor, Jr., J.R. O'Neil and I.R. Kaplan) p. 3-10.

Kolodny, Y., Luz, B. and Navon, O. (1983) Oxygen isotope variations in phosphate of biogenic apatites, I. Fish bone apatite - rechecking the rules of the game. Earth Planet. Sci. Lett. **64**, 398-404.

Kyser, T.K. and Kerrich, R. (1991) Retrograde exchange of hydrogen isotopes between hydrous minerals and water at low temperatures In *Stable Isotope Geochemistry: A Tribute to Samuel Epstein*. The Geochemical Society, Special Publication No.3 (Eds: H.P. Taylor, Jr., J.R. O'Neil and I.R. Kaplan) p. 409-422.

Lucas, J. and Prévot, L. (1981) Synthese d'apatite a partir de matiere organique phosphoree (ARN) et de calcite par voie bacterienne. C.R. Acad. Sci. Paris, **292**, 1203-1208.

O'Neil, J.R. (1986) Theoretical and experimental aspects of isotopic fractionation. <u>In</u> Stable Isotopes in High Temperature Geological Processes (eds. J.W. Valley, H.P. Taylor, Jr., and J.R. O'Neil) Rev. Mineral., **16**, 1-40

O'Neil, J.R. (1987) Preservation of H, C, and O isotopic ratios in the low temperature environment. <u>In</u>, Stable Isotope Geochemistry of Low Temperature Fluids, (ed., T.K. Kyser, MAC Short Course Handbook **13**, 85-128.

O'Neil, J.R. and Kharaka, Y.K. (1976) Hydrogen and oxygen isotope exchange between clay minerals and water. Geochim. Cosmochim. Acta, **40**, 241-246.

Sakai, H. and Tsutsumi, M. (1978) D/H fractionation factors between serpentine and water at 100 to 500°C and 2000 bars water pressure and the D/H ratio of natural serpentines. Earth Planet. Sci. Lett., **40**, 231-242.

Savin, S.M. and Lee, M. (1988) Isotopic studies of phyllosilicates. <u>In</u>, Hydrous Phyllosilicates (ed. S.W. Bailey), **19**, 189-223

Schutze, V.H. (1980) Der isotopenindex—eine inkrementenmethode zur naherugsweisen berechnung von isotopenaustauschgleichgewichten zwischen kristallinen substanzen. Chem. Erde, **39**, 321-334.

Taylor, Jr., H.P. and Epstein (1961) Relationships between $^{18}O/^{16}O$ ratios incoexisting minerals of igneous and metamorphic rocks, Part 2. Application to petrologic problems. Geol. Soc. Am. Bull., **73**, 675-694.

Wenner, D.B. and Taylor, Jr., H.P. (1973) Oxygen and hydrogen isotope studies of the serpentinization of ultramafic rocks in oceanic environments and continental ophiolites. Am. J. Sci., **273**, 207-239.

Water-Rock Interaction, Kharaka & Maest (eds) © 1992 Balkema, Rotterdam. ISBN 90 5410 075 3

Saline formation waters in sedimentary basins: Connate or diagenetic?

Lynton S. Land
Department of Geological Sciences, University of Texas, Austin, Tex., USA

ABSTRACT: High calcium concentrations in saline brines from reservoirs in Cenozoic clastic sediments from the Gulf of Mexico sedimentary basin can be explained by extensive dissolution of detrital plagioclase. $^{87}Sr/^{86}Sr$ ratios are controlled primarily by silicate dissolution, and are typically outside the range of coeval marine skeletal carbonate. Bromide concentrations can be explained by incongruent dissolution of halite, especially during deformation. Dissolution/recrystallization of bittern salts, including the transformation of carnallite to sylvite, are also processes which locally increase the Br/Cl ratio of the water. The two current reigning hypotheses for the origin of saline formation waters in sedimentary basins, connate as opposed to diagenetic, are unlikely to be resolved without quantitative consideration of the kinds and magnitudes of reactions recorded in the rocks.

1 INTRODUCTION

Saline formation waters in sedimentary basins typically contain concentrations of bromide and divalent ions, especially calcium, which are higher than can be reached by the congruent dissolution of halite. Two arguments have convinced many workers that such brines are modified seawater from which halite once precipitated: 1) $MeCl_2$ concentrations can resemble evaporated seawater, and 2) observed Br/Cl ratios are commonly similar to those characteristic of seawater during halite precipitation.

2 $MeCl_2$ CONTENT

The $MeCl_2$ concentration of a saline formation water (Ca + Mg + Sr - SO_4 - HCO_3 in equivalents/liter) "...provides a basis for determining if a brine has originated from the evaporation of seawater followed by simple diagenetic interactions with sedimentary rocks,..." (Carpenter, 1978, p. 73). This statement is certainly true for processes such as $CaCO_3$ and $CaSO_4$ precipitation, dolomitization, and sulfate reduction. But it is not true for processes which "exchange" monovalent for divalent ions. Albitization is such a process (equation 1), and one

$$CaAl_2Si_2O_8 + 4H_4SiO_4 + 2Na^+ \rightarrow 2NaAlSi_3O_8 + Ca^{++} + 8H_2O \quad (1)$$

which is characteristic of burial diagenesis. Sandstones and shales commonly contain detrital plagioclase at the time of deposition. Plagioclase is metastable with respect to albite, and is dissolved during burial. No Ca-silicate phase characterizes low-rank metamorphic rocks, which are dominated by the assemblage albite, quartz, calcite, illite, and chlorite. The large volume of calcium originally present in detrital plagioclase, if not retained in the rocks as calcite (or ankerite, laumontite, etc.), must have been lost, potentially contributing to the $MeCl_2$ content of the aqueous solutions.

Strontium isotopes aid in "tagging" the behavior of calcium, and nearly universally deviate from coeval seawater. In many cases values are quite radiogenic, demonstrating the destruction of crustal silicates (Stueber et al., 1987; Moldovanyi and Walter, 1992). In cases where the detrital plagioclase is derived from first cycle volcanics, very unradiogenic compositions are observed (Figure 1).

$^{87}Sr/^{86}Sr$ ratios clearly substantiate the importance of silicate reactions in forming saline brines, even at relatively low temperatures, where values inconsistent with seawater have been considered "anomalous" (Faure and Jones, 1969).

In areas such as south Texas, where Ca-rich saline brines are found in clastic reservoirs, the process of plagioclase dissolution (Milliken et al., 1989) is quantitatively sufficient to account for the calcium content and strontium isotopic composition of the water. In south Texas, many Ca-rich formation waters have Br/Cl ratios below 0.001. These samples, indistinguishable ionically from more Br-rich samples, have evolved to their present composition by albitization following emplacement of NaCl-type brines in the reservoirs. The $MeCl_2$ content of low-bromide waters in this area bears no relationship to an evaporated seawater signature, and some low-bromide samples have $MeCl_2$ contents similar to seawater evaporated to bittern salt stage. The $MeCl_2$ content of saline formation waters is not indicative of origin if interaction with silicates has occurred, especially if $^{87}Sr/^{86}Sr$ ratios deviate from possible seawater values.

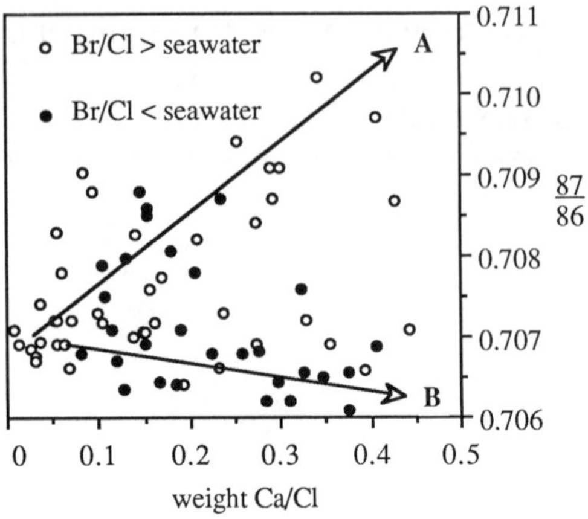

Fig. 1 $^{87}Sr/^{86}Sr$ ratio versus Ca/Cl (weight) ratio for formation waters from Oligocene clastic rocks, south Texas (Land and Macpherson, 1992). Increasing calcium content is due either to interaction with radiogenic crustal plagioclase deep in the basin (trend A), or with unradiogenic first cycle volcanics in the clastic reservoirs (trend B). Strontium concentrations in very Ca-rich brines can exceed 2 g/L.

3 BROMIDE CONTENT

In areas of the Gulf Coast sedimentary basin dominated by diapiric salt, the (weight) Br/Cl of formation water (near 0.001, Land and Macpherson, 1989) is never as low as the salt itself (near 0.0001, Land et al., 1988), even when salinities exceed 100 g Cl/L and more than 95% of the solutes are NaCl. Simple congruent dissolution of the salt does not occur, and the data are best explained by incongruent dissolution during recrystallization caused by diapirism. Many samples of formation water from the Gulf of Mexico, and from other basins, have much higher Br/Cl ratios, however. Although high Br/Cl ratios have been conventionally explained by the evaporation of seawater, several additional sources of Br other then the incongruent dissolution of halite can be invoked.

Release of bromide from kerogen during maturation, and from fluid inclusions, especially those in minerals from mafic rocks, are probably processes of insufficient magnitude to modify the Br/Cl ratios of very saline water. A potentially large reservoir of bromide exists in bittern salts, deposited during the late-stage evaporation of seawater. Subsequent dissolution and/or incongruent dissolution of bittern salts, including the release of bromide as carnallite is transformed to sylvite (Kuhn, 1968; Wardlaw, 1968) are sources for bromide which have not commonly been considered. In the case of the Mississippi salt basin, a small basin partly isolated from the open Gulf of Mexico in Jurassic time, bittern salt dissolution/recrystallization is a plausible explanation for high bromide (and lithium and rubidium) concentrations characteristic of formation waters from this area (Figure 2). Br/Cl ratios are probably the least rock-dominated components commonly measured, yet similar (equilibrium?) values apparently can be reached from both oversaturation (evaporation of seawater) and undersaturation (incongruent dissolution of marine evaporites).

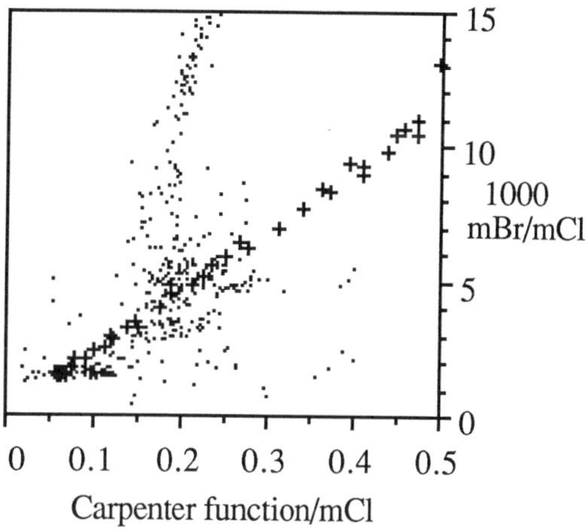

Fig. 2 Molar Br/Cl ratio versus divalent cation/chloride ratio for formation waters from the Gulf of Mexico sedimentary basin, the Illinois basin, and the Western Canada sedimentary basin. Some formation waters fall on the seawater trajectory (+'s), but most do not. Most very bromide-rich samples are from the Mississippi salt basin of the Gulf of Mexico, and may be due to bittern salt dissolution/recrystallization.

4 CONCLUSIONS

The calcium concentration and Br/Cl ratio of many saline formation waters can be consistent with modified "connate" brines (Figure 2). But brines formed solely in the subsurface by evaporite dissolution and subsequent water/rock interaction (diagenetic brines) can yield identical ionic compositions. Isotopic compositions (B, H, O, and Sr), prove the necessity for massive water/rock interaction. Diagenetic brines are additionally favored by the large volumes of halite which must be recycled through the Earth's crust (ca. 2 x 10^7 km^3 halite/225 Ma - Garrels and Mackenzie, 1981). Large scale salt recycling via saline brine discharge also expains the high chloride load of the world's rivers (Feth, 1981). Diagenetic brines formed by halite dissolution and water/rock interaction with feldspars can also provide large volumes of Pb- and Zn-bearing water, which is apparently necessary to account for economic deposits of Mississippi Valley-type ores.

It is unlikely that more analyses of saline formation waters will unequivocally distinguish between connate and diagenetic brines. If a reasonable case can be made that the Br/Cl ratio of a water sample can be evaporite-buffered during burial, irrespective of the source of water, then aqueous components which are more rock-dominated cannot possibly retain a connate signature during burial.

In order to resolve the two competing hypotheses, it will be necessary to quantitatively examine the rocks which have been in contact with the saline formation waters. If formation waters originate as connate seawater-derived brines, they must be stored subsequent to halite precipitation in units outside the evaporites them selves, since salt has such low porosity. Sufficiently large volumes of dolomite and authigenic $CaSO_4$ must then be formed to account for the observed volume of formation waters.

In the case of diagenetic brines, salt and/or bitterns must be dissolved/recrystallized. Unfortunately the modified evaporites which would result have a low preservation potential. Volumes of detrital plagioclase which have been dissolved can be difficult to quantify, but are unlikely to be limiting. If bittern salts are dissolved in the subsurface then authigenic K-silicates and late dolomite should occur in the rocks.

In the case of either hypothesis, the degree of water/rock interaction, as evidenced in the water by increased Ca, Li, and B concentrations, $^{87}Sr/^{86}Sr$ ratios, $^{11}B/^{10}B$ ratios, etc., must be balanced against the composition and abundances of diagenetic minerals in the rocks. Water/rock interactions must be integrated over the flow path of the water through time, and the scale of the processes involved (e.g. Oliver, 1990) are likely to require comprehensive geographic and cross-disciplinary studies.

REFERENCES

Carpenter, A.B. 1978. Origin and chemical evolution of brines in sedimentary basins. *Oklahoma Geol. Survey* 79: 60-77.

Faure, G.A. and Jones, L.M. 1969. Anomalous strontium in the Red Sea brines. In E.T. Degens and D.A. Ross (eds.), *Hot brines and recent heavy metal deposits in the Red Sea.* New York: Springer-Verlag: 243-250.

Feth, J.H. 1981. Chloride in natural continental water - a review. *U.S. Geol. Survey Water Supply Paper* 2176: 1-30.

Garrels, R. M. and Mackenzie, F.T. 1981. *Evolution of Sedimentary Rocks*. New York: Norton.

Kuhn, R. 1968. Geochemistry of the German potash deposits. *Geo. Soc. America Spec.l Paper* 88: 427-504.

Land, L.S., Kupecz, J.A. and Mack, L.E. 1988. Louann salt geochemistry (Gulf of Mexico sedimentary basin, U.S.A.): A preliminary synthesis. *Chem. Geol.* 74: 25-35.

Land, L.S. and Macpherson, G.L. 1989. Geochemistry of formation water, Plio-Pleistocene reservoirs, offshore La. *Trans. Gulf Coast Geol. Soc.* 39: 421-430.

Land, L.S. and Macpherson, G.L. 1992. The origin of saline formation waters, Cenozoic section, Gulf of Mexico sedimentary basin. *Bull. Amer. Assoc. Petrol. Geol.*:

Milliken, K.L., McBride, E.F. and Land, L.S. 1989. Numerical assessment of dissolution versus replacement in the subsurface destruction of detrital feldspars, Oligocene Frio Formation, S. Tx. *Jour. Sed. Petrol.* 59: 740-757.

Moldovanyi, E.P. and Walter, L.W. 1992. Regional trends in water chemistry, Smackover Formation, Southwest Arkansas: Geochemical and physical controls. *Bull. Amer. Assoc. Petrol. Geol.* 76:

Oliver, J.E. 1990. COCORP and fluids in the crust. In Studies in Geophysics, *The role of fluids in crustal processes.* Washington: National Academy Press: 128-139.

Stueber, A.M., Pushkar, P. and Heatherington, E.A. 1987. A strontium isotopic study of formation waters from the Illinois basin, U.S.A. *Applied Geochem.* 2: 477-494.

Wardlaw, N.A. 1968. Carnallite-sylvite relationships in the Middle Devonian Prairie evaporite Formation, Saskatchewan. *Bull. Geol. Soc. America 79*: 1273-1294.

Water-Rock Interaction, Kharaka & Maest (eds) © 1992 Balkema, Rotterdam. ISBN 90 5410 075 3

Early diagenesis in sediments: A global view?

Max L.Coleman
BP Research Centre, Sunbury-on-Thames & Postgraduate Research Institute for Sedimentology, Reading University, UK

ABSTRACT: The best way to take a global view of anything is to stand back from it. Seen from space, the processes of early diagenesis make an insignificant contribution to the earth's crust. However, they may control basin-scale processes which operate later in the burial history of the rocks; in particular, fluid-flow, which in turn controls movement of material and heat. Most significant reactions will occur where there is the greatest chemical potential to be released. Organic-rich mudrocks host the C-Fe-S system's microbially mediated redox reactions, which control the subsequent ability of the rock to deliver petroleum or material for mineral cements in other rocks. Emergent and near-emergent surfaces suffer extensive diagenesis, carbonates may be karstified, vastly increasing permeability, while both carbonates and clastics may show intense cementation at hardground surfaces. Sequence boundaries contain characteristic diagenetic effects and, more importantly, these can be predicted using the concepts of sequence stratigraphy.

1 SEDIMENTARY BASIN SETTING

When asked to put together this overview of early diagenesis, the initial reaction was to define what the problem was and then to stand back from the problem to get a better perspective on it. Firstly, since the subject is early, as opposed to late, diagenesis most of the relevant processes will be chemical rather than physical (assuming that initial compaction of mudrocks does not cause any major mineralogical change). Secondly, when viewed in the context of the earth's crust, the processes of early diagenesis seem relatively insignificant. However, the chain of geological events, which start with early diagenesis, are dependent on the earliest links in that chain. All the effects of water-rock interaction depend on the ability of water to move through the crust. The initial diagenetic processes affect the subsequent development of porosity and permeability and thereby control many later events, especially movement of fluids; petroleum, metalliferous and other basinal brines. During later diagenesis it is the basinal brines which may dissolve or precipitate minerals, which in turn modifies porosity and permeability. It would not be practical to set out a comprehensive view of early diagenesis, but by means of an idiosyncratic selection of examples it should be possible to demonstrate some of the common truths in an apparently disparate range of environments.

The unifying concept which links this look at early diagenesis is that a sedimentary basin must be viewed as an holistic system. The controls on sedimentation and environment of deposition (e.g. climate cycles) operate simultaneously on different facies, where they have different effects. Although these controls produce a variety of seemingly disconnected results, each may have a profound effect on the subsequent history of the basin. In other words, the influences on a sediment, shortly after its birth can affect its later life and even influence the whole community (heard that one before?).

The causal link is that the supposedly separate parts of a basin are connected. Most people accept that petroleum, which originates in the more deeply buried, distal mudrock facies of the basin fill, may migrate through permeable formations to eventually occupy the more porous sediments of the proximal coarse sandstones (e.g. Mackenzie et al., 1986). Equally, it has been shown that in some cases intergranular cements in sandstones do not originate from local redistribution of material but also are sourced from outside the formation (Gluyas and Coleman, 1992). In this case it is aqueous fluids which are migrating.

Some of the examples which are presented below show that early diagenetic processes affect the subsequent diagenetic history of the rock and have a significant effect in two ways. Firstly, they determine availability of material for migration to other parts of the basin. Secondly, precipitation of diagenetic mineral cements depends on the ability of host sediments to allow through passage of material, and to have sites suitable for its precipitation.

The driving force behind these reactions is the chemical disequilibrium of the assemblage that makes up the sedimentary input at deposition, or the introduction of components soon after burial which are not in equilibrium with the sedimentary system. This leads to confrontational chemical reactions. The approaches to equilibrium produce the diverse effects of early diagenesis. For example, the heterogeneous mixture in organic-rich mudrocks consists of sedimentary particulate matter resulting from erosion and transport under the oxidizing conditions of an aerobic atmosphere and oxygenated waters. At deposition, organic matter, much of which is locally produced or re-worked, provides a reducing agent in disequilibrium with the oxidized species in the sediment or waters. In a like way, exposure of marine sandstones or carbonates to sub-aerial exposure allows ingress of meteoric water unsaturated with respect to the rock mineralogy and leads to dissolution and precipitation of new phases. This produces the characteristic effects of unconformity leaching and cementation.

2 MUDROCKS

The processes which occur early during the burial history of mudrocks are important because they may control the subsequent diagenesis. Mudrocks make up the largest part of sedimentary basin fills and Evans (1989) showed that during deep burial mudrocks lost silica. This process has been studied in greater detail by Jordan et al. (1992) who showed that the sedimentary system was probably an open one. The exact relationship of early to late diagenesis has not been determined in detail but, any early approach to equilibrium must reduce the scope for its later occurrence, even given that the change in environment during burial will alter the equilibrium conditions to be reached. One aspect of this is the role of carbonate as a pH buffer. The processes which control carbonate dissolution (Canfield and Raiswell, 1991)and early diagenetic replacement of carbonate fossils by pyrite have been described (Coleman and Raiswell, 1992). Early exhaustion of the carbonate to buffer other reactions, clearly precludes its availability during later diagenesis.

The successive reactions whereby organic matter in mudrocks is oxidized by inorganic oxidants via microbially mediated reactions was summarized by Froelich et al. (1979). All the reactions give a common product, bicarbonate which may form a diagenetic cement. Prominent among these reactions are those which involve microbial sulphate reduction and production of pyrite. Considerable information about the depositional and early diagenetic environments, especially degree of oxygenation, can be deduced from the C-Fe-S system (Berner and Raiswell, 1983; Raiswell et al., 1988). The other resultant, inorganic products which also form early diagenetic cements can occur as concretionary nodules in mudrocks. Controls on the localization of growth of nodules within beds and, more importantly, between beds have been proposed (Coleman and Raiswell, 1992). The stratigraphic control relates to rate of sediment accumulation and appropriate conditions for growth of cyano-bacterial mats at the sediment-water interface.

Mossmann et al. (1990) showed that availability of reduced sulphur during early burial of mudrocks allowed sulphurization of kerogen (organic matter) only after most pyrite had formed. This is an early diagenetic control on organic matter composition and quality which in turn will affect the composition of petroleum sourced from it and the by-products which may react with other rocks.

3 COARSE CLASTICS

Preservation of large amounts of organic matter in these sediments is generally less likely than in mudrocks because they tend to be more permeable and therefore better oxygenated. Thus, aerobic oxidation processes effectively remove most of the organic-matter. However, local micro-environments may allow anaerobic processes to develop. In particular, burrows which have high concentrations of organic matter may act as nuclei for diagenetic mineral growth. Other nuclei for carbonate precipitation may be the mineralogically more stable carbonate fossils. More carbonate is provided by dissolution of less stable aragonitic or high Mg calcite. However, in this predominantly kinetically controlled process, preferential dissolution is controlled to a large extent by surface area rather than by thermodynamic stability. Thus fine-grained lime mud and fossils with high specific surface area may provide most early diagenetic carbonate (e.g. Walter and Morse, 1984). In addition to the redistributed carbonate, microbial reactions can contribute more material from oxidation of organic matter, in the same way as they do in mudrock systems. Carbonate-rich horizons, which may consist of discrete spheroidal nodules or laterally extensive doggers, can be shown by their carbon isotope compositions to have a significant input from degradation of organic-matter. Their early diagenetic origin is confirmed by the fact that they may enclose exceptionally-preserved fauna or flora, in some cases the putative cause of the concretionary growth.

Instability of silicate phases in the presence of an undersaturated pore-fluid can lead to dramatic changes. Emery et al. (1990) showed that palaeo-emergence of a felspathic sandstone led to dramatic changes in the uppermost part, most likely to have been flushed by meteoric water. There was dissolution of potassic feldspar, precipitation of kaolinite and a significant increase of porosity and permeability. Similarly, Suchecki et al. (1988) showed that early quartz cementation occurred in a meteoric water environment and was specific to a formation deposited during a transgressive highstand. It was thus liable to be emergent.

In contrast to these cementation reactions, the presence of very minor quantities of an early diagenetic mineral actually may prevent later growth of more extensive intergranular cements. Chlorite often forms a minor, early, grain-rimming cement in aeolian sandstones. The presence of chlorite preserves anomalously high porosity and permeability by apparently inhibiting later growth of pore-filling quartz cement (e.g. Dixon et al., 1989). For predictive purposes these early cements may be related to palaeo-water table.

4 CARBONATES

All the processes described above, which operate to dissolve or precipitate carbonate in sandstones, attempt to play the same role in limestones. However, their effects are not as easy to discern given the predominance of carbonate minerals in these rocks anyway. Frequently it can be seen that vigorous bioturbation prevents development of carbonate supersaturation and net dissolution results. Degradation of organic-matter may not cause precipitation of additional carbonate, as has been documented by Walter and Burton (1990).

The most dramatic effects on carbonate rocks occur when they are emergent or when there is a significant decrease in accumulation rate for those sediments which continue to be submerged. Carbonate hardgrounds are frequently adorned with phosphatic and iron-mineral cements. These surfaces may be considerable barriers to subsurface flow. Actual emergence of a carbonate as a beach rock leads to dramatic textural changes producing considerable porosity, however, pore-filling cements may be precipitated in it. Prolonged emergence, even if not always an early diagenetic event, leads to karstification and extremes of dissolution and chemical brecciation. These very porous and permeable formations have long been known as hosts of metalliferous ores (Siebenthal, 1915) but can also serve as petroleum reservoirs (Brennand and van Veen, 1975). The significance of both these occurrences is that they relate to basin-scale changes to sedimentation or erosion. Thus, they can be related to sequence stratigraphical analysis of the basin.

5 CONCLUSIONS

Taking a preliminary, "global" view of early diagenesis, it seems that nothing very much is happening. A closer examination of the topic suggests that some of the early diagenetic processes predispose the rock to act in a specific way during its later burial history. This may affect the quality of the petroleum sourced from it or its subsequent development of porosity and permeability.

Since most significant reactions in buried rocks involve movement of fluids, their behaviour during burial is modified, if not controlled, by early diagenetic processes.

The effect of early modification of porosity and permeability does not have a simple, linear relationship with later reactions because transfer of fluids, transport of mass, transfer of heat, precipitation and dissolution are all inter-coupled in a complex fashion.

Most of the early diagenetic processes relate

to larger-scale depositional cycles and may be able to be related by sequence stratigraphic analysis.

6 ACKNOWLEDGMENTS

I am grateful to my colleagues at BP Research Centre for their helpful discussions and particularly to Andrew Hogg and Norman Oxtoby.

7 REFERENCES

Berner, R A & R W Raiswell, 1983. Burial of organic carbon and pyrite sulfur over Phanerozoic time: a new theory. *Geochim. Cosmochim. Acta* 47: 855-862.

Blatt, H & V M Brown, 1974. Prophylactic separation of heavy minerals. *J. Sedim. Petrol.* 44: 260-261.

Brennand, T P & F R van Veen, 1975. The Auk Oilfield. In A W Woodland (ed.) *Petroleum and the Continental Shelf of North West Europe, Vol. 1, Geology*: 275-281. Appl. Sci. Publ.

Canfield, D E & R Raiswell, 1991. Carbonate precipitation and dissolution, its relevance to fossil preservation. In P A Allison & D E G Briggs (eds.) *Taphonomy: releasing the data locked in the fossil record*: 411-453. New York, Plenum Press.

Coleman and Raiswell, 1992. Source of carbonate and origin of zonation in pyritiferous carbonate concretions. Am. J. Sci. (in press).

Coleman, M L, 1985. Geochemistry of diagenetic non-silicate minerals: kinetic considerations. *Phil. Trans. Roy. Soc. Lond.* A315: 39-56.

Dixon, S A , D M Summers & R C Surdam, 1989. Diagenesis and preservation of porosity in Norphlet Formation (Upper Jurassic), Southern Alabama. *Am. Assoc. Petrol. Geol. Bull.* 73: 707-728.

Emery, D, K J Myers & R Young, 1990. Ancient subaerial exposure and freshwater leaching in sandstones. *Geology* 18: 1178-1181.

Evans, I J, 1989. Geochemical fluxes during shale diagenesis, an example from the Ordovician of Morocco. In Miles, D L (ed.) *Water-Rock Interaction WRI-6*: 219-222. Rotterdam: Balkema.

Froelich, P N, G P Klinkhammer, M L Bender, N A Luedke, G R Heath, D Cullen, P Dauphin, D Hammond, B Hartman & V Maynard, 1979. Early oxidation of organic matter in pelagic sediments of the eastern equatorial Atlantic: suboxic diagenesis. *Geochim. Cosmochim. Acta* 43: 1075-1090.

Gluyas, J G & M L Coleman, 1992. Material flux and porosity changes during sediment diagenesis. *Nature* 356: 5.3.92

Jordan, M M, A C Aplin, C D Curtis & M L Coleman, 1992. Access of porewaters to carbonate precipitation sites during concretion growth. *Water-Rock Interaction WRI-7*: this volume.

Mackenzie, A S, I Price, D Leythaeuser, P Müller, M Radke & R G Schaefer, 1986. The expulsion of petroleum from Kimmeridge Clay source-rocks in the area of the Brae Oilfield, UK continental shelf. In J Brooks & K W Glennie (eds.) *Petroleum Geology of North West Europe*:.865-877. London, Graham and Trotman.

Mossmann J R, A C Aplin, C D Curtis & ML Coleman, 1990. Sulfur geochemistry at sites 680 and 686 on the Peru Margin. *Proc. ODP, Sci. Results* 112: 455-464.

Raiswell, R, F Buckley, R A Berner & T F Anderson 1988. Degree of pyritization of iron as a paleoenvironmental indicator of bottom water oxygenation. *J Sedimentol. Petrol.* 58; 812-819.

Siebenthal, C E, 1915. Origin of the zinc and lead deposist of the Joplin region, Missouri, Kansas and Oklahoma. *US Geological Survey Bulletin* 606, 283pp.

Walter, L M, & E A Burton, 1990. Dissolution of Recent platform carbonate sediments in marine pore fluids. *Am. J. Sci.* 290: 601-643.

Walter, L M, & J W Morse, 1984. Reactive surface area of skeletal carbonates during dissolution: Effect of grain size. *J. Sediment. Petrol.* 54: 1081-1090.

Water-Rock Interaction, Kharaka & Maest (eds) © 1992 Balkema, Rotterdam. ISBN 90 5410 075 3

The composition of gases in geothermal and volcanic systems as a function of tectonic setting

W.F.Giggenbach
DSIR Chemistry, Petone, New Zealand

ABSTRACT: On the basis of relative He, Ar, N_2 and CO_2 contents of gas samples from a range of geothermal and volcanic systems associated with basaltic and andesitic magmatism, four major source components can be distinguished: a meteoric component contributing effectively all the Ar; a CO_2- and He-rich mantle component, predominating in "basaltic" gases; and two quite similar CO_2- and N_2-rich, He-poor components derived from the thermal decomposition of subducted and crustal sediments respectively. In the subducted, "andesitic" endmember component He, Ar, N_2 and CO_2 are proposed to be present in the proportions 0.015, 0.005, 25 and 975; in the thermogenic crustal component 0.001, 0.002, 5 and 995; and in the mantle component 0.05, 0.017, 0.5 and 999. The mantle component is to varying degrees contaminated with the thermogenic crustal component.

1 INTRODUCTION

In spite of a considerable body of information on the chemical composition of geothermal and volcanic gases, correlations with their geological or tectonic settings are only vaguely defined. Most attempts in this direction are based on the isotopic compositions of one or more selected species such as C, N, Ar and He. Because of the generally quite different origins of the various constituents, isotopic proof of the origin of one component can generally not be used as evidence for the origin of another. For instance, a high $^3He/^4He$-ratio can in no way be used as evidence for a mantle origin of associated CH_4 (Giggenbach et al. 1992).

Of the chemical constituents analysed on geothermal and volcanic gas samples, some are chemically reactive and therefore provide the basis for a number of geoindicators. Others are chemically inert, such as the noble gases, or chemically only little reactive, such as CO_2 and N_2, and therefore may initially be considered useful tracers to their origins. As in the case of the isotopes, chemical constituents are also likely to have a number of sources, each possibly characterised by a specific constellation of species. The main aim of this study is the delineation of compositional patterns typical of various tectonic environments.

2 CHEMICAL COMPOSITIONS OF VOLCANIC AND GEOTHERMAL GASES

Table 1 reports partial analyses of gas samples from a wide range of geothermal and volcanic systems associated both with basaltic and andesitic magmatism. The investigation is largely based on relative He, Ar, N_2 and CO_2 contents. Ne and O_2 contents are also given to allow the origins of the other two atmospheric gases, Ar and N_2, to be assessed. Samples containing O_2 were corrected by subtracting corresponding proportions from the Ne, Ar and N_2 results. The concentrations in Table 1, in mmol/mol, do not necessarily add up to 1000 due to the presence of other major constituents, such as SO_2, H_2S, HCl and H_2. For most of the gases listed in Table 1, isotopic analyses of He are available; the $^3He/^4He$-ratios for all these samples are >4 R_A, suggesting a magmatic origin of, at least, the He. Sampling temperatures for pools and fumaroles are those measured; in the case of well discharges measured or estimated deep temperatures are given.

Table 1. He, Ne, Ar, O_2, N_2 and CO_2 contents (in mmol/mol dry gas) of gases from geothermal and volcanic systems associated with basaltic and andesitic magmatism.

Symb.	Location	t(°C)	He	Ne	Ar	O_2	N_2	CO_2
	"Basaltic gases"							
KL	Kilauea, Hawaii	120	.0110	-	0.019	<.01	1.93	702
KR	Krafla, Iceland	260	.0210	-	1.180	1.20	83.6	875
LS	Laacher See, Germany	23	.0360	.00014	0.130	1.20	12.1	985
LN	Lake Nyos, Cameroon	23	.0102	.00080	0.520	<.01	25.0	965
OL	Olkaria, Kenya	250	.0070	.00070	0.350	<.06	20.0	835
RE	Cilaos, Reunion Island	39	.0180	.00014	0.080	0.09	4.1	996
AZ	Azufrales, Galapagos I.	280	.0270	.00005	0.020	0.01	1.5	971
	New Zealand, TVZ-W							
NP	Ngapuna, Rotorua	98	.0202	.00017	0.120	<.04	11.0	736
TK	Tikitere	81	.0188	-	0.310	<.01	26.7	794
MK	Mokai, well	250	.0156	.00028	0.220	<.03	14.0	883
WK	Wairakei, vent	104	.0220	.00025	0.240	<.04	13.2	975
	New Zealand, TVZ-E							
RK	Rotokawa, well	260	.0023	.00010	0.052	<.02	16.8	905
OH	Ohaaki, well	260	.0025	.00006	0.032	<.01	10.8	958
KT	Ketetahi, vent	119	.0040	.00006	0.020	<.01	9.5	826
NR	Ngauruhoe, vent	350	.0067	.00005	0.032	<.01	23.7	675
WI	White Island, vent	830	.0018	.00002	0.007	<.04	3.1	480
	"Andesitic gases"							
RI	Raoul Island, Kermadecs	98	.0060	.00010	0.102	<.03	14.0	984
CG	Cagua, Philippines	300	.0074	.00012	0.079	<.04	17.2	915
MH	Mahagnao, Philippines	300	.0024	.00004	0.018	<.01	8.4	970
AP	Alto Peak, Philippines	300	.0017	.00005	0.017	<.02	3.0	938
PP	Papandayan, Indonesia	400	.0060	.000012	0.010	<.05	13.5	716
DA	Darajat, Indonesia	104	.0033	.00006	0.041	<.05	7.1	931
US	Mt Usu, Japan	650	.0026	.00006	0.050	<.09	15.5	572
OS	Osorezan, Japan	97	.0029	.000013	0.012	<.03	16.2	780
TA	Tamagawa, Japan	94	.0112	.00003	0.026	<.02	18.7	910
ST	Satsuma Iwojima, Japan	880	.0023	.00007	0.022	<.20	2.9	193
ZN	Zunil, Guatemala	280	.0020	.00014	0.091	<.06	12.6	961
TC	Tecuamburro, Guatemala	95	.0029	.00018	0.120	<.01	35.0	850
NE	Nereidas, Colombia	85	.0007	.00004	0.020	<.01	5.9	949
ER	El Ruiz, Colombia	82	.0037	.00006	0.015	<.03	5.8	520
PC	Pichincha, Ecuador	86	.0028	.00001	0.009	<.04	21.5	928
CF	Campi Flegrei, Italy	156	.0090	<.00002	0.005	<.01	3.7	989
VU	Vulcano, Italy	580	.0099	.000005	0.008	<.01	7.8	849
AIR	Air	20	.0052	.0182	9.34	210	781	0.33
ASW	Air saturated water	20	.0024	.0098	16.4	336	633	14.9
	Inferred endmember components							
AS	Andesitic, subducted		.015	-	0.005	-	25.0	975
AT	Andesitic, thermogenic		.001	-	0.002	-	5.0	995
BC	Basaltic, contaminated		.018	-	0.006	-	4.0	996
BM	Basaltic, mantle		.050	-	0.017	-	0.5	999

3 RELATIVE N_2, He AND Ar CONTENTS

Diagrams involving relative N_2, He and Ar contents are now in use since 1978 (Giggenbach, unpublished results). In earlier versions, based largely on analytical data for gases associated with andesitic magmatism along convergent plate boundaries, three major endmember components, meteoric, "magmatic" and "crustal", were assumed to describe the variabilty in the relative contents of the three species adequately (Giggenbach et al. 1983). As analytical data became available for gases from systems associated with basaltic magmatism, their data points were found to overlap with the position of the high He "crustal" gases. A reliable distinction between high He "crustal" and "basaltic" gases is possible on the basis of the isotopic composition of He, the crustal gases having R/R_A-values <0.1 R_A, the basaltic gases >4 R_A. In Table 1 only gases known, or likely, to have $^3He/^4He$-ratios >4 R_A are given. Relative N_2, He and Ar contents are plotted in Fig. 1.

Gases associated with andesitic and basaltic magmatism clearly occupy distinct positions. The latter include samples from the western, "rift" part of the Taupo Volcanic Zone (TVZ-W). There active spreading is thought to lead to crustal thinning (Cole and Lewis 1981) or even emplacement of new lithosphere (Stern 1985) allowing basaltic magmatism to dominate fluid compositions (Giggenbach 1989). Gases from geothermal and volcanic systems associated with the "andesitic arc" along the eastern margin of the TVZ (TVZ-E), occupy distinctly different positions well within the field delineated by the other "andesitic" gases.

The spread in compositions for both basaltic and andesitic gases corresponds to mixing of respective endmember components, with the "meteoric" component, essentially air saturated groundwater. The N_2/Ar-ratio of the meteoric component (ME) is assumed to be 50, thus allowing for the presence of some "excess" air. The distribution of data points suggests that the magmatic components have to be represented by ranges of compositions rather

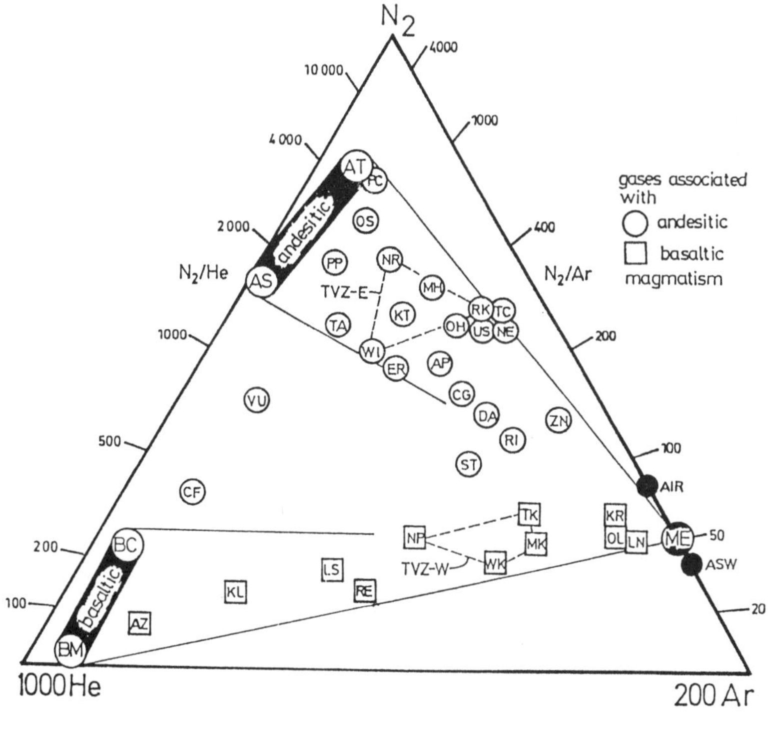

Fig. 1 Relative N_2, He and Ar contents

than single points. Both magmatic endmember components are low in Ar. The He/Ar-ratio of the basaltic component can be assumed to correspond to the $^4He/^{40}Ar$ production ratio of the mantle of about 3 (Ozima and Podosek 1983). N_2/He-ratios for the basaltic gases range from about 10 to 220.

The isotopic composition of He in andesitic systems indicates that it is also largely of mantle origin and, in the absence of obvious other sources, so is Ar. Taking into account the relative positions of the two endmember basaltic gases, the somewhat N_2-enriched one is likely to reflect some contamination by the higher N_2 gases, as represented by the andesitic ones. The least N_2-enriched component, then represents the least contaminated fluid, probably that released from the mantle itself. This assignment is supported by the fact that sample AZ is a gas from a high temperature fumarole (300°C) on the Galapagos hot spot. With a $^3He/^4He$-ratio of 17.5 R_A it is, of the gases listed in Table 1, probably the one most representative of mantle volatiles.

The compositions of the endmember "andesitic" gases then are likely to correspond to N_2/He-ratios of 1700 and 5000. Again, the higher He endmember gas (AS) is likely to be more closely related to deeper, possibly mantle vapors. As suggested for CO_2 (Marty and Giggenbach 1990), the most likely source of the excess N_2 are marine sediments carried to the zones of arc magma generation on top of the subducting slab. The He/Ar-ratio of this component is again assumed to be that of the mantle of 3. The comparatively He-poor endmember (AT) is likely to represent vapors produced away from the mantle, most probably through decomposition of crustal sedimentary material in contact with, or assimilated by the rising andesitic magmas. The N_2/He-ratio of the most N_2-enriched component approaches the high values suggested by Jenden et al. (1988) for "metasedimentary" gases produced within the California Great Valley. Component AT, therefore, is tentatively assumed to be a thermogenic, crustally derived gas. Its He/Ar-ratio is taken to correspond to that observed

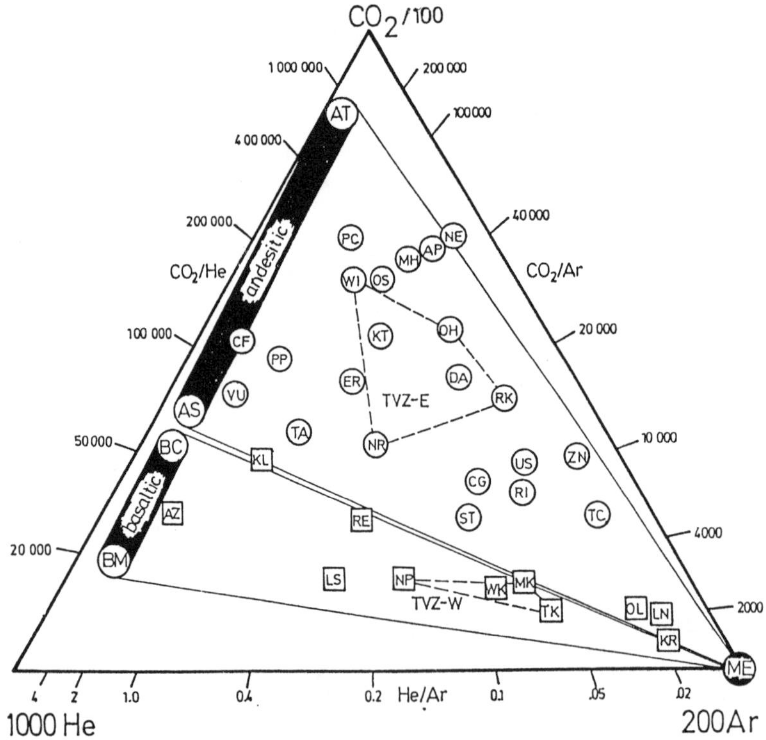

Fig. 2 Relative CO_2, He and Ar contents

for the New Zealand crust of about 0.5 (Giggenbach et al. 1992).

Most data points lie within the two quite well separated areas outlined by the two sets of magmatic endmembers and the meteoric component. A few fall inbetween. They are the two Italian samples from Vulcano Island (VU) and the Campi Flegrei(CF). Their positions are likely to reflect specific tectonic conditions prevailing along this plate margin. The identification and assignment of endmember components to specific sources on the basis of Fig. 1 alone is, of course, highly speculative. An independent check on possible origins may be made by taking into account the behaviour of another major constituent, CO_2.

4 RELATIVE CO_2, He AND Ar CONTENTS

A CO_2, He, Ar diagram similar to Fig. 2 was used by Marty and Giggenbach (1990) to discuss the possible origin of excess CO_2 in gas discharges from the White Island volcano. The most likely source was again thought to be subducted sediments. Following the above reasoning, the He-rich limiting andesitic component (AS) is again likely to originate from deeper levels, affected by the presence of mantle volatiles. The comparatively He-poor component (AT) then represents volatiles generated in a He depleted environment, e. g. by thermal decomposition of crustal sediments subject to heating from rising andesitic magmas.

The composition of the high He basaltic component corresponds to that derived for MORB gases (Marty and Giggenbach 1990). At a $CO_2/^3He$-ratio of $2{*}10^9$ and a $^3He/^4He$-ratio of around 10^5, the $CO_2/^4He$- or CO_2/He-ratio becomes 20 000. The He/Ar-ratio is again assumed to be that of the mantle $^4He/^{40}Ar$ production ratio of 3. The apparent upper limit for "contaminated" basaltic gases (BC) corresponds to a CO_2/He-ratio of 55 000. Again, the "rift-type" gases from the TVZ (TVZ-W) are well within the basaltic field, the "arc-type" gases (TVZ-E) in the andesitic field.

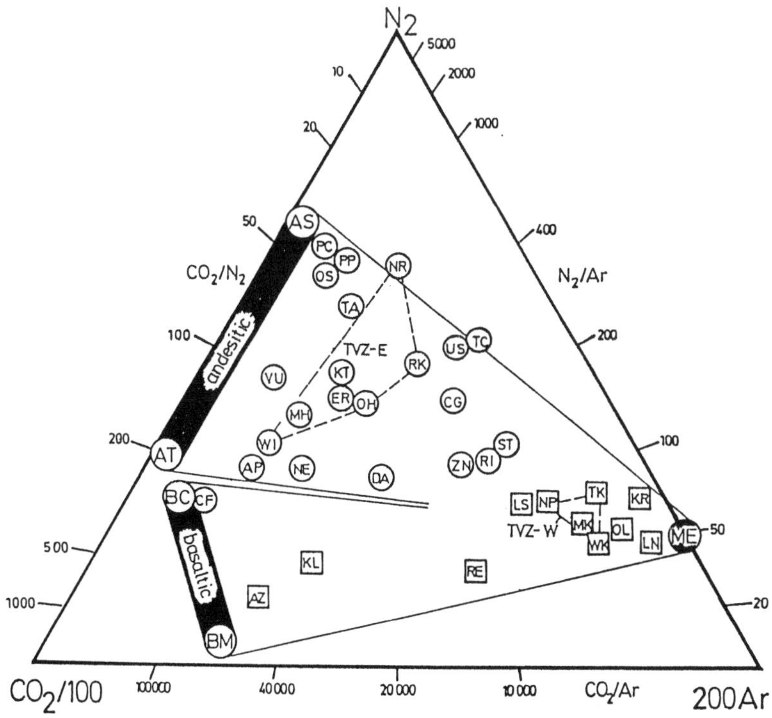

Fig. 3 Relative N_2, CO_2 and Ar contents

DISCUSSION AND CONCLUSIONS

The above evaluation treats both CO_2 and N_2 as chemically inert tracers, disregarding any possible conversion of CO_2 to carbonate or CH_4, of N_2 to NH_3, or the effects of differences in their solubilities. This approach is probably justified in the case of high temperature systems, at lower temperatures, the above processes may cause considerable deviations.

A check on the internal consistency of the approach, and the validity of the sets of endmember components derived above, is carried out by use of Fig. 3. There the data points are projected onto the N_2-CO_2-Ar-face of the N_2-He-Ar-CO_2-tetrahedron. It should be noted that the relative positions of AS and AT with respect to the basaltic fields are now inverted and so are the positions of many of the andesitic data points. The assignment of some of the Ar-rich basaltic vapors to the basaltic field is not as clear-cut anymore.

The trend in BM and BC towards AT, and not AS, clearly suggests that the gas phase "contaminating" the basaltic gases is more likely the thermogenic crustal (AT), rather than the subducted (AS) component, as expected. CO_2/He-ratios increase only by a factor of five on going from mantle (BM) to subducted gases (AS), but by a factor of 200 for N_2/He. It is the latter which renders Fig. 1 especially useful in the classification of geothermal and volcanic gases. The large increase in relative N_2-contents is due to very low N_2-contents of the mantle in the first place. The use of CO_2/N_2-ratios for classification purposes is somewhat impaired by the fact that, on addition of crustal gases, the quite widely separated CO_2/N_2-ratios of the mantle (2000) and subducted (40) gases converge towards a common intermediate, crustal value of around 200.

The above, albeit quite tentative identification of "andesitic" and "basaltic" endmember components can be expected to be of use in delineating the origins of both present-day and "fossil" gases, such as those extracted, with increasing success, from fluid inclusions. Because of likely interference from lower temperature processes, the above findings should only be applied to gases with close volcanic-magmatic associations and $^3He/^4He$-ratios well above those of air.

REFERENCES

Cole, J. W. and Lewis K. B. 1981. Evolution of the Taupo-Hikurangi subduction system. Tectonophys. 72: 1-21

Giggenbach, W. F. 1989. The chemical and isotopic position of the Ohaaki Field within the Taupo Volcanic Zone. Proc. NZ Geotherm. Workshop, Univ. of Auckland, New Zealand 11: 81-88

Giggenbach, W. F., Gonfiantini, R., Jangi, B. L. and Truesdell, A. H. 1983. Isotopic and chemical composition of Parbati Valley geothermal discharges, NW Himalaya, India. Geothermics 12: 199-222

Giggenbach, W. F., Sano Y. and Wakita H. 1992. Isotopic composition of He, and CO_2 and CH_4 contents in gases produced along the New Zealand part of a convergent plate boundary. Geochim. Cosmochim. Acta (submitted)

Jenden, P. D., Kaplan, I. R., Poreda, R. J. and Craig, H. 1988. Origin of N_2-rich natural gases in the California Great Valley: Evidence from He, C and N isotope ratios. Geochim. Cosmochim. Acta 52: 852-861

Marty, B. and Giggenbach, W. F. 1990. Major and rare gases at White Island volcano, New Zealand: Origin and flux of volatiles. Geophys. Res. Let. 17: 247-250.

Ozima, M. and Podosek, F. A. 1983. Noble gas geochemistry. Cambridge University Press 376

Stern, T. 1985. A back-arc basin formed within continental lithosphere: The Central Volcanic Region of New Zealand. Tectonophys. 112: 385 -409

Water-Rock Interaction, Kharaka & Maest (eds) © 1992 Balkema, Rotterdam. ISBN 90 5410 075 3

Ore deposits in the Oquirrh and Wasatch Mountains, Utah: Examples of large-scale water-rock interaction

M.T.Einaudi
Stanford University, Calif., USA

ABSTRACT: Mid-Tertiary ores in the Oquirrh Mountains (e.g., Bingham) and Wasatch Mountains (e.g., Park City) are products of magmatic-geothermal systems affecting up to 200 km^2 of uppermost crust. Early, ultra-saline hydrothermal fluids were generated at 700 to 600°C during final crystallization of plutons at paleodepths of 5 to 3 km. On cooling and reacting with the plutons, these magmatic-hydrothermal fluids exchanged K for Na and Ca, and precipitated quartz, Cu-Fe-sulfides, and MoS2 in fractures at temperatures greater than 400°C. On reacting with limestone, these early fluids formed anhydrous Cu-Fe-skarn. Pb-Zn-Ag ores beyond skarn in limestone may have formed during this early time frame as fluids mixed with formation water and/or meteoric water and cooled to below 300°C. Late acid-sulfate hydrothermal fluids of meteoric origin were concentrated within 1 or 2 km of the paleosurface and caused locally intense hydrolytic alteration of skarn to silica-pyrite and of plutons to sericitic and advanced argillic assemblages, accompanied by an increase in gold/base-metal ratios of ores.

1 INTRODUCTION

The two largest producers of base- and precious-metals in Utah are the Bingham and Park City mining districts.

The Bingham Cu-Au-Mo-(Pb-Zn-Ag) district (Fig. 1) ranks first in copper and second in gold production in North America. Through 1991, the district has produced (in metric tons) 10.9 million of Cu, 2.9 million of Pb + Zn, 490,000 of Mo, 10,000 of Ag, and 603 of Au from 1,500 million metric tons of porphyry, skarn, and carbonate-replacement ores. Present ore reserves are similar in grade and tonnage to past production.

The inactive Park City Ag-Pb-Zn-(Au-Cu) district is the largest metal producer in the Wasatch Mountains (Fig. 1). Through 1982, Park City produced (in metric tons) 1.9 million of Pb + Zn, 58,500 of Cu, 9,575 of Ag, and 55 of Au from 15 million metric tons of ore.

The ore deposits at Bingham and Park City differ in detail and commonly are classified as distinct deposit "types" (Bingham a porphyry-copper deposit, Park City a Ag-Pb-Zn carbonate-replacement deposit). However, the two districts share many of the same geological and geochemical features and serve as excellent examples of variations on the theme of porphyry-related hydrothermal systems.

2 CENTRAL WASATCH MINERAL BELT

Numerous mid-Tertiary granodiorite plutons and related mineral deposits form an E-W-trending belt 45 km long through the central Wasatch Mountains, coincident with the Uinta Arch (Fig. 1). Although K-Ar ages of igneous rocks in the western portion of the belt span 41 - 32 Ma, emplacement of mineralized plutons and hydrothermal ore formation most likely occurred toward the end of this period at 36 - 33 Ma (Bromfield et al., 1977). The Little Cottonwood stock and its molybdenum prospect at White Pine Canyon represent the youngest magmatic-hydrothermal event at about 25 Ma (Crittendon et al., 1973).

The central portion of the Wasatch Mountains was tilted eastward 20° during

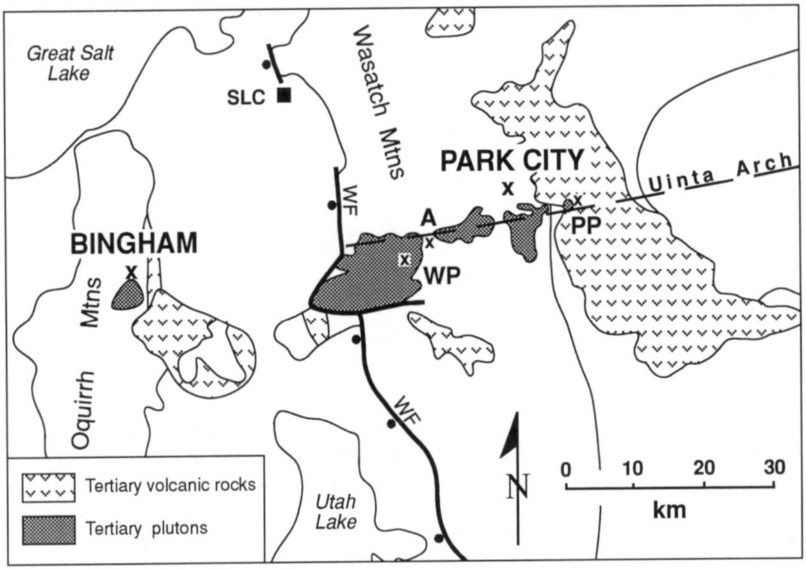

Fig. 1 Geologic sketch map of Oquirrh and Wasatch Mountains, illustrating allignment of igneous rocks and ore deposits along the westward extension of the Uinta arch. X = ore deposits mentioned in text. SLC = Salt Lake City, WF = Wasatch fault (heavy line, ball on down-thrown side), WP = White Pine Canyon Mo deposit, A = Alta, PP = Park Premier stock.

Basin and Range extension and down-to-the-west normal faulting (e.g., Wasatch fault, Fig. 1) during late Tertiary time (Lawton et al., 1980; Parry and Bruhn, 1987; John, 1989a). The result is that intrusions and their metasomatic aureoles presently are exposed at successively decreasing paleodepths from 11 km on the west to 0.5 km on the east. Variations in hydrothermal features of the mineral belt are best understood in terms of these varying depths of exposure.

At paleodepths of ~10-7 km on the west end of the belt (White Pine prospect, Fig. 1) low-salinity, CO2-rich hydrothermal fluids formed stockwork Mo-quartz-muscovite veins in the Little Cottonwood quartz monzonite stock (John, 1989a).

At paleodepths of ~5-6 km in the central portion of the belt near Alta (Fig. 1), low- to intermediate-salinity hydrothermal fluids of magmatic origin at 600-400°C interacted with medium-grained granodiorite, resulting in weak potassic alteration and sparse Cu-quartz veins (John, 1991). Small Fe-Cu skarns (Bowman and Cook, 1981) and Ag-Pb-Zn replacement deposits formed on the margins of stocks in Paleozoic limestone.

At shallower depths of ~2-3 km in the area of Park City (Fig. 1), high-salinity hydrothermal fluids were derived from aplitic groundmass porphyry stocks at a boiling horizon located at 5 km paleodepths (John, 1989a). The most productive mineral deposits in the Wasatch Mountains are located in the vicinity of porphyry stocks containing high-salinity fluid inclusions. In the Park City district, ores are zoned over a distance of 6 km from Cu-Au ores (Ag:Au = 16, Pb:Cu = 8) in potassically and argillically altered Mayflower stock, to peripheral Ag-Pb-Zn ores (Ag:Au = 300 - 1000, Pb:Cu = 40) in limestone at the Silver King and other mines (Nash, 1973; Bromfield, 1989).

At shallowest paleodepths of <1 km on the east end of the mineral belt, the small Park Premier granodiorite porphyry stock (Fig. 1) was emplaced into coeval volcanic rocks. Early high-salinity brines at 600°C introduced low-grade Cu-Au and altered the stock to potassic and sodic (plagioclase-actinolite) mineral assemblages. Incursion of dilute, 300°C, acid-sulfate (meteoric ?) water resulted in widespread sericitic and advanced argillic (pyrophyllite-alunite) alteration accompanied by locally anomalous Au values but no base metals (John, 1989b).

3 BINGHAM MINING DISTRICT

As is the case in the Wasatch Mountains, Basin and Range extension resulted in eastward tilting of the Oquirrh Mountains approximately 25° (Atkinson and Einaudi, 1978). The pattern of epizonal plutons on the west and coeval volcanic rocks on the east in both mountain ranges (Fig. 1) is the expression of this common tilt. The present surface in the Utah Copper pit at Bingham is estimated to lie at paleodepths of 2-3 km (Moore and Nash, 1974; Atkinson and Einaudi, 1978), equivalent to paleodepths exposed at Park City.

Plutonic and volcanic rocks of the Oquirrh Mountains were emplaced at 41 - 33 Ma (Warnaars et al., 1978). Ores in the Bingham district are genetically related to a 38 Ma old quartz monzonite porphyry stock emplaced into comagmatic equigranular monzonite and Pennsylvanian limestone and quartzite (Einaudi et al., 1978). The district is zoned over a horizontal distance of 3.5 km from central Cu-Mo-(Au) in quartz monzonite, to Cu-(Au) in skarn, to peripheral Pb-Zn-Ag vein and replacement bodies in limestone (Hunt, 1924).

3.1 Early potassic alteration and skarn

The central disseminated Cu-Mo-(Au) ore in porphyry (Ag:Au = 5-10; Pb:Cu = <0.01) mined in the Utah Copper pit originated during final crystallization of porphyry. Intense biotitization (±K-feldspar) and abundant quartz-sulfide veinlets with bornite-chalcopyrite or molybdenite characterize the zone of major influx of magmatic-hydrothermal fluids (Sheppard et al., 1971) containing 30-60 wt % NaCl(equiv) at temperatures of 400 to 600°C (Roedder, 1971; Moore and Nash, 1974; Bowman et al., 1987). Limestone beds near porphyry were replaced by anhydrous, andradite-salite skarn accompanied by magnetite, chalcopyrite, and pyrite. Skarn contains Cu-(Au) ores characterized by Ag:Au = 15-25 and Pb:Cu = 0.05. In both porphyry and anhydrous skarn, highest grades of Au correlate with highest grades of Cu.

Monzonite and quartzite responded in similar ways to interaction with these early hydrothermal fluids (Atkinson and Einaudi, 1978; Lanier et al., 1978): hydrothermal biotite gives way outward to hydrothermal actinolite at 500 to 800 m from porphyry; quartz veinlet and sulfide abundance diminish; pyrite:chalcopyrite ratios increase beyond 1:1; and copper grades drop below 0.4 wt % in monzonite and below 0.2 wt % in quartzite. In the actinolite zone, which extends up to 1500 m from porphyry, actinolite replaces hornblende in monzonite and diopside in quartzite.

Biotite and actinolite zones in quartzite and andradite-diopside skarn in limestone are superimposed on earlier diopside-wollastonite assemblages formed during contact metamorphism (?). These metasomatic effects also mark the first introduction of Si, Fe, Cu, Mo, and S into the contact aureole and are thought to have resulted from outward flow of magmatic-hydrothermal fluids (Atkinson and Einaudi, 1978). In monzonite, the transition from biotite to actinolite has been interpreted on the basis of O and H isotope data to be the result of progressive mixing of magmatic-hydrothermal fluids with formation waters or with meteoric waters that underwent isotopic exchange with igneous rocks at low water:rock ratios below the present level of exposure (Bowman et al., 1987). Lateral chemical changes in the fluid were toward increasing activity ratios of (alkali and alkaline-earth metal cations):(hydrogen ion) and (Fe):(Cu) as temperature declined from 600 to 300°C at constant oxidation and sulfidation state.

3.2 Zoning from skarn to limestone

Calcic skarn displays a classic zonal pattern. Andradite>salite near the stock gives way through salite>andradite outward to wollastonite at the marble front. Opaque minerals are zoned from magnetite-chalcopyrite to chalcopyrite± pyrite. Wollastonite zone contains minor disseminated chalcopyrite, bornite, and sphalerite. Retrograde alteration of skarn (see below) is concentrated near the stock at higher elevations and is accompanied by increased pyrite content.

Pb-Zn-Ag replacement and vein deposits in limestone extend outward 1.5 km from the outer limit of skarn. In the western contact zone, Pb-Zn-Ag veins locally cut skarn, indicating that, at least in part, Pb-Zn-Ag ore was late relative Cu-skarn ore.

The overall metal zoning pattern, from (1) skarn, through (2) fissures superimposed on outer skarn, to (3) replacement bodies in limestone, is: (1) Cu-(Au); (2) Cu-Pb-(Au-Ag), with Ag:Au = 10-30, Pb:Cu = 1.0, and Zn-Pb-Cu-(Ag) with Ag:Au = 50, Pb:Cu = 3.0; and (3) Pb-Zn-Ag with Ag:Au = 100, Pb:Cu = 15.

3.3 Late hydrolytic alteration

Late sericite-pyrite veins in porphyry are linked to influx of low salinity, isotopically unevolved meteoric water at temperatures near 350-250°C (Sheppard et al. 1971; Bowman et al. 1987). Pb-Zn-Ag replacement and vein deposits in limestone (Rubright and Hart, 1968) and in argillized monzonite on the fringe of the district (Hunt, 1957) may belong in part to this time frame. Skarn was retrograded to chlorite-clay-pyrite or converted to silica-pyrite rock along faults. Hydrogen ion metasomatism dominated this late stage, accompanied by oxidation (magnetite to hematite), sulfidation (mafic silicate minerals to pyrite), apparent leaching of base-metals deposited during earlier events, and local deposition of As-, Sb-, and Hg-bearing minerals. Ag:Au (0.5) and Cu:Au declined to their lowest values in silica-pyrite rock in skarn, which locally contains 10-20 ppm Au.

4 CONCLUSIONS

Geological features shared by both mineral districts include: (1) a genetic relation to 38 to 33 Ma-old porphyry-textured, granitoid stocks; (2) central Cu-(Au) ores in stocks, accompanied by hydrothermal biotite ± K-feldspar locally overprinted by weak hydrolytic alteration; (3) peripheral Pb-Zn-Ag ores in bleached and recrystallized Paleozoic limestone; and (4) local development of late alteration features indicative of acid-sulfate waters, accompanied by anomalous gold. In both mountains ranges, data from fluid inclusions and light stable isotopes indicate that stocks were sources for saline magmatic-hydrothermal fluids that carried metals and volatile components of magmatic origin, and that meteoric waters dominated the upper 2-3 km of the hydrothermal systems late in their evolutionary history.

Unusually extensive vertical exposure through the Wasatch Mountains, in conjunction with deep mining and drilling at Bingham, yield data bearing on the importance of level of emplacement of magmas in generating contrasting porphyry-related ores. Differences in metal associations between porphyry Mo (e.g., White Pine), porphyry Cu-Mo (e.g., Bingham), and porphyry Cu-(Au) (e.g., Park Premier) commonly have been ascribed to differences in magma composition and/or source regions (e.g., Kesler, 1973). In the present case, these differences may relate more importantly to variations in depth of emplacement of cogenetic magmas and the effects of such variations on volatile phase evolution and efficiency of extraction of Mo and Cu from silicate melts (cf., Candela, 1989).

Metal zoning is the result of the complex interplay of chemical and temperature changes in hydrothermal solutions through space and time. Chemical changes in the ore fluid brought on by water-rock interaction clearly are important for ore deposition in limestone and skarn environments, but the importance of such changes in causing ore deposition in other environments remains to be documented. The dominant zoning in early stages from central Cu at ~500°C to peripheral Pb-Zn at 300°C is consistent with transport of base metals as chloride complexes at concentrations of 100's to 1,000's ppm metal (Hemley and Hunt, 1992). The positive correlation between copper and gold in district centers is consistent with transport of gold as a chloride complex at high temperatures.

Late meteoric hydrothermal fluids locally modified earlier metal concentrations and may have introduced new components from non-magmatic sources. The chief chemical contrasts between these late fluids and earlier fluids are lower salinities, higher oxidation-sulfidation states, and lower alkali/hydrogen ion activity ratios. The change at some localities from early ores dominated by base-metals to late ores depleted in base-metals and enriched in gold (e.g., Park Premier, and silica-pyrite bodies in skarn at Bingham) may reflect a shift from chloride- to bisulfide-complexing for gold at temperatures below 350°C. Such localities serve as examples of the evolutionary link between porphyry-related ores and epithermal precious metal deposits.

REFERENCES

Atkinson, W.W., Jr. & M.T. Einaudi 1978. Skarn formation and mineralization in the contact aureole at Carr Fork, Bingham, Utah. Econ. Geol. 73: 1326-1365.

Bowman, J.R. & S.J. Cook 1981. Physical-chemical conditions of contact skarn formation at Alta, Utah (abst.). Geol. Soc. of America Abstracts with Programs 13: 414.

Bowman, J.R., W.T. Parry, W.P. Kropp & S.A. Kruer 1987. Chemical and isotopic evolution of hydrothermal solutions at Bingham, Utah. Econ. Geol. 82: 395-428.

Bromfield, C.S. 1989. Gold deposits in the Park City mining district, Utah. U.S. Geol. Survey Bull. 1857-C: C14-C26.

Bromfield, C.S., A.J. Erickson Jr., M.A. Haddadin & H.H. Mehnert 1977. Potassium argon ages of intrusion, extrusion, and associated ore deposits, Park City mining district, Utah. Econ. Geol. 72: 837-848.

Candela, P.A. 1989. Magmatic ore-forming fluids: thermodynamic and mass transfer calculations of metal concentrations. Reviews in Econ. Geol. 4: 203-221.

Crittenden, M.D., Jr., J.S. Stuckless, R.N. Kistler & T.W. Stern 1973. Radiometric dating of intrusive rocks in the Cottonwood area, Utah. U.S. Geol. Survey Jour. Research 1: 173-178.

Einaudi, M.T., W.J. Moore, & J.C. Wilson (eds.) 1978. An issue devoted to the Bingham mining district. Econ. Geol. 73: 1213-1365.

Hemley, J.J. & J.P. Hunt 1992. Hydrothermal ore-forming processes in the light of studies in rock-buffered systems: II. Some general geologic applications. Econ. Geol. 87: 23-43.

Hunt, J.P. 1957. Rock alteration, mica, and clay minerals in certain areas in the United States and Lark mines, Bingham, Utah. PhD thesis University of California, Berkeley, USA.

Hunt, R.N. 1924. The ores in limestone at Bingham, Utah. American Institute of Mining Metall. Engineers Trans. 70: 856-883.

John, D.A. 1989a. Geologic setting, depths of emplacement, and regional distribution of fluid inclusions in intrusions of the central Wasatch Mountains, Utah. Econ. Geol. 84: 386-409.

John, D.A. 1989b. Evolution of hydrothermal fluids in the Park Premier stock, central Wasatch Mountains, Utah. Econ. Geol. 84: 879-902.

John, D.A. 1991. Evolution of hydrothermal fluids in the Alta stock, central Wasatch Mountains, Utah. U.S. Geol. Survey Bull. 1977.

Kesler, S.E. 1973. Copper, gold, and molybdenum abundances in porphyry copper deposits. Econ. Geol. 73: 106-112.

Lanier, G., W.J. Raab, R.B. Folsom, & S. Cone 1978. Alteration of equigranular monzonite, Bingham mining district, Utah. Econ. Geol. 73: 1270-1286.

Lawton, T.F., D.A. John, & W.J. Moore 1980. Levels of emplacement of mid-Tertiary plutons in the central Wasatch Mountains, Utah (abs.). Geol. Soc. America Abstracts with Programs 12: 278.

Moore, W.J. & J.T. Nash 1974. Alteration and fluid inclusion studies of the porphyry copper ore body at Bingham, Utah. Econ. Geol. 69: 631-645.

Nash, J.T. 1973. Geochemical studies in the Park City district, I. Ore fluids in the Mayflower mine. Econ. Geol. 68: 34-51.

Parry, W.T. & R.L. Bruhn 1987. Fluid inclusion evidence for minimum 11 km vertical offset on the Wasatch fault, Utah. Geology 15: 67-70.

Roedder, E. 1971. Fluid inclusion studies on the porphyry-type ore deposits at Bingham, Utah, Butte, Montana, and Climax, Colorado. Econ. Geol. 66, 98-120.

Rubright, R.D. & O.J. Hart 1968. Non-porphyry ores in the Bingham district, Utah. *In* Ridge, J.D. (ed.) Ore deposits of the United States, 1933-1967 (Graton-Sales Vol.), New York, American Institute of Mining Metall. Engineers: 1: 886-907.

Sheppard, S.M.F., R.L. Nielsen, & H.P. Taylor Jr. 1971. Hydrogen and oxygen isotope ratios in minerals from porphyry copper deposits. Econ. Geol. 66: 515-542.

Warnaars, F.W., W.H. Smith, R.E. Bary, G. Lanier & M. Shafiqullah 1978. Geochronology of igneous intrusions and porphyry copper mineralization at Bingham, Utah. Econ. Geol. 73: 1242-1249.

Water-Rock Interaction, Kharaka & Maest (eds) © 1992 Balkema, Rotterdam. ISBN 90 5410 075 3

Results of water rock interactions in mid-ocean ridges

J.M. Edmond
MIT, Cambridge, Mass., USA

ABSTRACT: Hydrothermal activity and associated water-rock interactions have now been observed at mid-ocean ridges over the entire range of spreading rates and the complete spectrum of tectonic environments, open ocean, back arc and sediment buried. Several hydrothermally active seamounts have also been investigated. There are two outstanding problems: estimates of the total oceanic hydrothermal flux based on geochemical and geophysical considerations differ by at least a factor of five; none of the proposed mechanisms appears adequate to explain the very large range in salinity (chloride) observed in axial vent systems.

INTRODUCTION

Hydrothermal activity in the deep sea was proposed early in the history of oceanography to explain the origin of abyssal metalliferous sediments. This proposal was supported by the subsequent finding of high temperature alteration products in oceanic basalts, by the discovery of the He-3 excess in the deep water column and, of course, by the hypothesis of sea floor spreading itself. The technology necessary to explore for and sample hot springs in the deep sea was developed in the early seventies. Vent fields were discovered on the Galapagos Spreading Center in the late seventies in what was the first concerted exploration effort (Corliss et al, 1979). It is now known that hydrothermal activity is intrinsic to all submarine volcanism and, in particular, that it is as much a characteristic of the sea floor spreading process as pillow basalts and transform faults.

CHEMISTRY OF SUBMARINE HOT SPRINGS

Submarine hydrothermal fluids from open ocean ridges are entirely composed of chemically altered seawater; there is no evidence of a magmatic component. There is some evidence that subduction-related systems do contain water derived from the down-going slab. The temperatures of the venting fluids usually fall in a narrow range around 350 degrees C. Relative to seawater the vent fluids are completely depleted in Mg and sulphate, strongly but variably enriched in the alkalies, Li, K, Rb and Cs, the alkaline earths, Be, Ca, Ba and Ra and many of the transition elements and the REE. The solutions are acid, pH ~ 3.5, and enriched in H_2S and in mantle-derived CO_2, CH_4 and He-3 (Von Damm et al, 1985). Thermodynamic calculations suggest that the mineral-forming elements are at equilibrium with the greenschist facies assemblage (Bowers et al, 1988). This conclusion is supported by the observation that the composition of the fluids in a given vent field is stable over a period of years (Campbell et al, 1988). Chloride is the only significant anion in the vent fluids and hence its concentration determines the transport properties of the solutions by both electroneutrality and complex formation. However the chloride concentrations vary over a very wide range from more than double to less than one third the seawater value. Where observations exist the chloride concentration is also very stable over time.

On intermediate and fast spreading ridges the chemical and isotopic data can be interpreted in terms of a simple two-component reaction, seawater-tholeiite. On slow spreading systems there is good evidence from the heavy alkalies and the boron isotopes for the incorporation of altered basalts in the reaction system (Palmer and Edmond, 1989). Presumably the episodes of volcanism are sufficiently rare that low temperature alteration extends to the "zero-age" zone of recent intrusion. Conductive heating reactivates the altered country rock around the new volcanic center.

The few examples reported to date from back arc spreading centers show great compositional variability with some quite exotic features. The vents in the Marianas back-arc appear to have a sedimentary signal in Pb and B, presumably derived from the subducting slab. In the Okinawa Trough the fluids are so enriched in CO_2 that clathrates form in the vent fields (Sakai et al, 1990). It is much too early to be able to make generalisations about these fascinating systems.

Where the ridge axis is buried by sediment the volcanism is expressed solely as intrusions into the depositional column. This serves as a reactive and insulating blanket. The 100°C isotherm commonly lies at a depth of a few centimeters below the sediment surface. The sediments themselves are incorporated in the hydrothermal reaction process. That the fluids have interacted with the basaltic heat source is evidenced by their elevated He-3 concentrations. However their general chemistry is usually dominated by the sedimentary reactions and hence reflects the composition of this overburden. Thus in Guaymas Basin, in the central part of the Gulf of California, the fluids are alkaline and hence strongly depleted in the sulphide forming elements. They contain the complete range of hydrocarbon fractions often as separate phases. These are produced by the thermocatalytic breakdown of planktonic carbon enriched in the sediments (~4% by weight) under this area of very high biological productivity. The fluids are also enormoushy enriched in ammonia (>15 mM) of the same origin (Von Damm et al, 1985). The high Be-10 in the sediments allows estimates to be made of its mobility under hydrothermal conditions. It is very low (~0.5% mobilised), consistent with the results from the stable isotope, Be-9 (Bourles et al, 1991). In the Escanaba Trough, a turbidite filled segment of the southern Gorda Ridge, only waning stage (T ~ 110° C) activity has been found. This was associated with sediment hosted layers of massive sulphide of exotic mineralogy e.g. elemental Bi, indicating that the ore grade of such deposits depends much more on the composition of the host sediment than on that of the tholeiitic heat source. Recent (September, 1991) ODP drilling of Middle Valley, a sediment buried segment of the northern Endeavour Ridge, revealed very extensive sulphide mineralisation in the depositional column. ALVIN dives recovered vent fluids at 270° C; chemical data are not yet available.

Three hydrothermally active seamounts have been investigated, Loihi at the end of the Hawaiian chain, Red off-axis on the EPR at 21°North and Axial on the Juan de Fuca Ridge. The first two gave fluid temperatures <30° C from small chimneys located in heavily weathered pillowed terrain mantled with bright orange deposits of iron oxyhydroxide. This is apparently produced bacterially. The material is so flocculent and so reactive with the overlying seawater (~2% phosphate, very high REE) that it was impossible to obtain uncontaminated samples from the very narrow and fragile orifices. The reaction chemistry in these peculiar systems remains to be determined. Phase separated fluids were observed on Axial Seamount with very low salinity fluids venting from anhydrite chimneys (Butterfield et al, 1990).

Recently (April 1991) a huge hydrothermal system, extending over 2 kilometers along strike, was discovered at 9° north on the EPR associated with very young (days) volcanics. Phase separated fluids with salinities <1% of seawater at temperatures of 403° C issued from small chimneys in fractures in the axial collapse pits. These persisted in pristine form for less than a year; a recent re-visit to the site (March 1992) found well developed chimney growth and temperatures of ~380° C (K.L. Von Damm, pers. comm.). The relationship between phase separation and the general variability of chloride in vent fluids at lower temperatures will be discussed below.

PROBLEMS

The most fundamental problem is to understand the geology, physics and chemistry of the reaction zone at depth. A variety of processes are not understood including the mechanism of heat transfer to the fluid, the four dimensional structure of the reaction zone and the reaction pathways followed by the fluids. The chemistry of the vent fluids is the resultant of these processes. Despite detailed information on a wide range of elements and their isotopes it has not been possible to deconvolute the chemical signature to give mechanisms. Since thermodynamic equilibrium destroys the memory of prior disequilibrium trajectories the information may not be there. This problem will probably yield only to in situ measurements and sampling in deep drill holes on new crust.

Two crucial chemical problems are the magnitude of the axial hydrothermal flux and the mechanism responsible for the chloride variations. A third problem, of especial interest to ore geochemists, is the mechanism of ore deposition in the ophiolite-type massive sulphides presumably formed in an axial hydrothermal environment.

The first estimate of the global axial hydrothermal flux came from a consideration of the oceanic systematics of He-3 (Jenkins et al, 1978). Helium is lost to space by thermal escape from the high atmosphere. It is also injected into the atmosphere by the solar wind; this is especially important for the light isotope. Thus the global mass balance is highly uncertain. In the oceans there is a substantial concentration anomaly for He-3 relative to atmospheric saturation. The oceanic ventilation time, as calibrated by the C-14 decay in dissolved CO_2, is about 1,000 years. Hence the flux of He-3 into the deep ocean necessary to maintain the anomaly against loss to the atmosphere can be calculated; the value is ~1,000 moles/yr. The maximum anomaly in the ocean is centered at the regional depth of the ridge axes. If it is therefore assumed that all the He-3 is injected by axial hydrothermal systems then the flux of associated heat, water and dissolved species can be estimated from their measured ratios to dissolved He-3. This assumes no preferential He-3 loss by direct outgassing early in an eruptive cycle. The derived flux for water is equivalent to the passage of a volume of seawater equivalent to the entire ocean through the 350° C isotherm every 8 million years. The fluxes of many elements, positive and negative, are comparable to the fluvial inputs. From this perspective the ridge axes act as a geochemical "flywheel" stabilising seawater composition against variations in the continental inputs caused by climatic and tectonic variations. The mass balance for the strontium isotopes in seawater gives a compatible result (Palmer and Edmond, 1989). The continental input has a ratio of about 0.7119 whereas the vents contribute predominantly basaltic Sr at 0.7035. The seawater composition, at 0.7091, is very sensitive to variations in the hydrothermal flux. Geophysical estimates based on models of the thermal budget of the processes of crustal formation give flux estimates between five and ten times less than those based on the isotopic balances (Morton and Sleep, 1985). From this viewpoint axial hydrothermal activity is an interesting process but of no real significance to the overall geochemical cycle of the elements. This disagreement has existed since the first estimates based on He-3 were made using the early Galapagos data; it appears to be irreconcilable. It is symptomatic of a profound ignorance somewhere.

As the only important anion, chloride can be thought of as the "master variable" governing the chemistry of axial hot springs. Over the time interval since the discovery of the chloride variability in the original Galapagos hot springs (Edmond et al, 1979) a number of constraints have been developed on the causal mechanism. The process shows extreme temporal stability, i.e. to within ~1%, on time-scales of at least five years at a given site (Campbell et al, 1988). The anomalies occur at the intermediate temperatures, ~350° C, encountered at most vents which themselves are quite stable in time. Large positive and negative anomalies can be observed on a particular ridge segment but not in any obvious spatial relationship to one another. The ratio Cl:Br is identical to the seawater value at all chloride concentrations (Campbell and Edmond, 1989). This latter observation appears to rule out involvement of solid phases containing Cl since these should discriminate against Br. The temporal stability of both chloride and temperature makes it difficult to conceive of any mixing process involving the two components of a higher temperature (~405° C), phase separated fluid and waters of normal seawater salinity and intermediate temperature that would generate the anomalies. The problem remains open and intractable.

The hydrothermal fluids vent as a buoyancy driven flow at velocities of around 1.5 meters/sec. These clear, homogeneous fluids then entrain ambient seawater to form a turbulent plume. This entrainment induces the rapid precipitation of extremely fine-grained sulphide and oxide phases predominantly of Fe. The plume rises to its level of neutral buoyancy, a few hundred meters above the vents, and spreads laterally. On intermediate and fast spreading ridges this height is usually well above any topographic obstructions and the plume is dispersed, often over very long distances, in the ambient mid-depth circulation of the oceans (Edmond et al, 1982). Metalliferous, Fe-Mn-oxide sediments accumulate on the seafloor as the "shadow" of the plumes. On slow spreading ridges the plumes are confined within the rift valley walls leading to local sedimentation on the lower walls and floor of the rift. The proportion of the total flux retained in the spectacular chimneys and their talus mounds as sulphide and oxidised precipitates is extremely small. How is material to be retained and accumulated in this very high energy environment with enough efficiency to form the large ore bodies observed in ophiolites? The only feasable solution appears to be that, upon reaching a sufficient size, the mounds develop their own internal hydrology suffering either enough conductive heat loss to induce internal precipitation or experiencing inward leaks of seawater to accomplish the same thing. Small scale conductive heat loss has been observed at curious onion dome structures on the flanks of the main chimneys at the MARK site on the MAR (Campbell et al, 1988). The internal solutions are at about 325° C and are depleted in Fe. The domes are predominantly sphalerite. Similar domes are found at the TAG site on the MAR, a very large deposit close to the boundary fault of the rift. Here the temperatures are 300° C, the fluids have Mg concentrations indicating sewater entrainment and are Fe-depleted with pH < 3 indicative of sulphide precipitation. Large deposits require long periods of growth. It is possible that the uniformly small deposits found on the EPR are limited in their potential for growth by the relatively short repeat times of destructive eruptions. At the faster spreading ridges small "shows" may be present throughout the upper extrusive section while at slower rates much larger deposits may have time to form. Of course this raises questions as to the heat source for sustained, localised hydrothermal activity on slow spreading ridges. The ore deposit that is finally preserved must be very different in composition and other properties from the material originally accumulated from direct precipitation. The initial deposition process can be highly disequilibrium involving mixtures of sulphates, sulphides and oxides. Oxidative weathering by seawater of the surficial chimney material and talus produces iron oxyhydroxides that scavenge additional material from the seawater. As the talus is buried the isotherms rise through it driving the assembly of phases to equilibrium. Many of the minor constituents will be refluxed to the deposit surface to be reprecipitated. Thus the high grades of minor elements commonly observed in the surficial samples from oceanic deposits and in the gossans in ophiolites may be formed in this manner.

FUTURE DIRECTIONS

Future research contains several components. Time-series observations should be maintained, and initiated, to determine how the systems evolve. Exploration should continue, particularily on the slow-spreading ridges where large deposits are probably more likely. Sustained effort will be required to establish the range of hydrothermal phenomena occurring in back-arc and other subduction related environments. Drilling of large basalt hosted deposits and of sediment covered axes should be extended once a high temperature down-hole measurement and sampling capability has been developed. Much more experimental work needs to be done on the solution thermodynamics and phase relationships in the relatively unexplored region between 250° and 450° C and 200 to 700 atmospheres. Efforts towards close cooperation between geochemists and the geologists, geophysicists, biologists and physical oceanographers interested in oceanic hydrothermal phenomena need to be maintained. Collaborative work with land-based Earth Scientists needs to be extended.

REFERENCES

Bourles, D.L., G.M. Raisbeck, E.T. Brown, F. Yiou and J.M. Edmond 1991. Beryllium isotope systematics of submarine hydrothermal systems. Earth Planet. Aci. Lett. 105:534-542.

Bowers, T.S., A.C. Campbell, C.I. Measures, A.J. Spivack, M. Khadem and J.M. Edmond 1988. Chemical controls on the composition of vent fluids at 11-13° N and 21° N, East Pacific Rise. J. Geophys. Res. 93:4522-4536.

Butterfield, D.S., G.J. Massoth, R.E. McDuff, J.E. Lupton and M.D. Lilley 1990. The geochemistry of hydrothermal fluids from ASHES vent field, Axial Seamount, Juan de Fuca Ridge: subseafloor boiling and subsequent fluid-rock interaction. J. Geophys. Res. 95:12,895-12,921.

Campbell, A.C., T.S. Bowers, C.I. Measures, K.K. Falkner, M. Khadem and J.M. Edmond 1988. A time-series of vent fluid compositions from 21° N, East

Pacific Rise (1979, 1981, 1985) and Guaymas Basin (1982, 1985). J. Geophys. Res. 93:4537-4550.
Campbell, A.C., M.R. Palmer, G.P. Klinkhammer, T.S. Bowers, J.M. Edmond, J. Lawrence, J.F. Casey, G. Thompson, S. Humphris, P. Rona and J.A. Carson 1988. Chemistry of hot springs on the Mid-Atlantic Ridge. Nature 335:514-519.
Campbell, A.C and J.M. Edmond 1989. Halide systematics of submarine hydrothermal systems. Nature 342:168-170.
Corliss, J.B., J. Dymond, L. Gordon, J.M. Edmond, R.P. Von Herzen, R.D. Ballard, K. Green, D. Williams, A. Bainbridge, K. Crane and Tj. H. Van Andel 1979. Submarine thermal springs on the Galapagos Rift. Science 203:1073-1083.
Edmond, J.M., C.Measures, R.E. McDuff, L.H. Chan, R. Collier, B. Grant, L. Gordon and J.B. Corliss 1979. Ridgecrest hydrothermal activity and the balances of the major and minor elements in the ocean: the Galapagos data. Earth Planet. Sci. Lett. 46:1-18.
Edmond, J.M., K.L. Von Damm, R.E. McDuff and C.I. Measures 1982. The chemistry of hot springs on the East Pacific Rise and their effluent dispersal. Nature 296: 187-191.
Jenkins, W.J., J.M. Edmond and J.B. Corliss 1978. Excess He-3 and He-4 in Galapagos submarine hydrothermal waters. Nature 272:156-158.
Morton, J.L and N.H. Sleep 1985. A mid-ocean ridge thermal model: constraints on the volume of axial hydrothermal heat flux. J. Geophys. Res. 90:11,345-11,353.
Palmer, M.R. and J.M. Edmond 1989. The Sr isotope budget of the modern oceans. Earth Planet. Sci. Lett. 92:11-26.
Palmer, M.R. and J.M. Edmond 1989. Cesium and rubidium in hydrothermal fluids: evidence for recycling of alkali elements. Earth Planet. Sci. Lett. 95: 8-14.
Sakai, H., T. Gamo, E.-S. Kim, M. Tsutsumi, T. Tanaka, J. Ishibashi, H. Wakita, M. Yamano ans T. Oomori 1990. Venting of carbon dioxide-rich fluid and hydrate formation in the mid-Okinawa Trough backarc basin. Science 248: 1,093-1,096.
Von Damm, K.L., J.M. Edmond, C.I. Measures and B. Grant 1985. Chemistry of submarine hydrothermal solutions at Guaymas Basin, Gulf of California. Geochim. Cosmochim. Acta 49: 2221-2237.

Stable and radioactive isotopes

Water and solutes

Water-Rock Interaction, Kharaka & Maest (eds) © 1992 Balkema, Rotterdam. ISBN 90 5410 075 3

Boron isotopic composition of Icelandic hydrothermal systems

J.K.Aggarwal, M.R.Palmer & K.V.Ragnarsdottir
University of Bristol, UK

ABSTRACT: Icelandic hydrothermal systems were analysed for their boron isotopic composition. Apart from the seawater dominated Svartsengi field, the $\delta^{11}B$ for the analysed hydrothermal systems all lie in the range of -6.7 to +3.1‰. This reflects the isotopic composition of the rocks in contact with the fluids.

INTRODUCTION

The large percentage mass difference between the two stable isotopes of boron means that boron isotope ratios are sensitive to geological processes. Boron is a very mobile element and is strongly partitioned into the aqueous phase during water-rock interaction at high temperature. Studies to date have indicated that no fractionation of the isotopes occurs during the leaching of boron from basalts at high temperatures.

Many experiments have shown that clay minerals significantly alter the boron isotopic composition of the fluids in contact with them. At low temperature (<150°C), ^{10}B is preferentially absorbed as the hydrated tetrahedral complex onto the surface of clays. Examination of boron isotopes of fluids from sediment covered, sub-marine hydrothermal fluids from Guaymas Basin shows $\delta^{11}B$ values in the range of 16.5 to 23.2 ‰ (Spivack et al., 1987). Hydrothermal systems from mid-ocean ridges show higher $\delta^{11}B$ values (+30 to +36.8‰). The lighter $\delta^{11}B$ in the fluids from sediment rich systems is due to the incorporation of isotopically lighter boron into secondary minerals forming in the sediments . Boron isotope composition have been examined from Yellowstone, a rhyolite hosted, subaerial hydrothermal system. The geothermal fluids emerge with a $\delta^{11}B$ of -9.3 to +4.4‰ which is similar to that of the unaltered rhyolite and local sediments (Palmer and Sturchio, 1990). A complete evaluation of the cause of the boron isotopic composition variations of these fluids has not been completely determined and it is hoped that the examination of hydrothermal fluids from Iceland may reveal more detail on the constraints of boron isotope fractionation in geothermal systems.

The hydrothermal systems from Iceland can be considered to be intermediate between those of the mid-ocean ridges and those of the continental Yellowstone geothermal field. Many of the high temperature geothermal fields in Iceland are associated with acidic centres, yet are surrounded by tholeitic basalts. Previous studies suggest that the boron isotope composition of the fluids may distinguish the source rocks within the geothermal system. Some of the Icelandic geothermal systems are further complicated by the influence of seawater. This work will examine the factors that control the boron isotope composition in several Icelandic geothermal systems.

METHODS

Hydrothermal fluids were collected from a number of different localities across Iceland (Fig1). On the Snaefellsnes peninsular, two carbonate springs at Olkelda and Lysoholl were sampled. Hydrothermal fluids from the high temperature wells at Svartsengi, Nesjavellir, Krafla and Namafjall were collected. A boiling spring from the Landmannulaugar region was also sampled. Collected samples were acidified with sub-boiling hydrochloric acid. Boron concentrations were determined by a modified curcumin technique. Determination of boron isotopes was carried out by thermal ionization mass spectrometry of $Cs_2B_4O_7$. Results are reported relative to N.B.S. 951 and calculated as;

$$\delta^{11}B = \left(\left[\frac{(^{11}B/^{10}B)_{sample}}{(^{11}B/^{10}B)_{standard}}\right] - 1\right) \times 1000 - 0.19$$

RESULTS AND DISCUSSION

Boron concentrations and $\delta^{11}B$ values are shown in Table 1. Apart from the seawater dominated hydrothermal system at Svartsengi with its $\delta^{11}B$ of +30 ‰, the geothermal systems sampled, all lie in the region of -6.7‰ to +3.1‰. Cl/B ratios of the basalts all lie within the range of 100 to 150 (Arnorsson et al., 1989). Cl/B of most of the geothermal fluids collected show a lower ratio.

Figure 1. Locations of geothermal wells sampled.

This indicates that boron is being preferentially incorporated into the aqueous phase during leaching of the basalt. Studies on Yellowstone (Palmer and Sturchio 1990), indicate that the $\delta^{11}B$ of the fluids are similar to those of the rock that was in contact with the hydrothermal fluid. It is therefore suggested that the rocks in contact with the hydrothermal fluids have a $\delta^{11}B$ value in a similar range to those of the geothermal fluids.

The exact $\delta^{11}B$ of the rocks can not be deduced from the fluid composition as there are a number of processes which have been identified that modify the boron isotopic composition. These processes include the addition of primary boron during the degassing of magmas. No work cited to date has examined the $\delta^{11}B$ of magmatic gasses. During phase separation, ^{11}B is preferentially partitioned into the gas phase. As a result of this partitioning, the gas phase will have a higher $\delta^{11}B$ value. Studies so far have found that this increase in $\delta^{11}B$ is no greater than 0.5 ‰, but more extensive boiling may result in more significant isotope fractionation (Spivack et al, 1990).

Table 1. Boron composition of Icelandic geothermal waters.

	Boron concentration µM/kg	$\delta^{11}B$ ‰
Lysoholl	35	-6.7
Olkelda	90	+3.1
Landmannulaugar	291	-1.4
Nesjavellir 6	345	-6.2
Nesjavellir 16	205	-1.9
Krafla 12	222	+0.3
Krafla 15	528	-1.8
Namafjall 4	73	-1.6
Namafjall 11	349	-0.2
Svartsengi 7	447	+29.6
Svartsengi 8	846	+30.7
Seawater	445	+39.6

Previous work has shown the importance of clay minerals in the isotope fractionation of boron. This adsorption process is pH dependent (Palmer et al, 1987). At low pH (<8) isotope fractionation is more extensive and as the pH increases the degree of boron isotope fractionation decreases. Clay minerals occur in abundance in the upper parts of Icelandic geothermal systems, (Kristmannsdottir 1982). Furthermore, the pH of the issuing fluids from these geothermal systems are all below 8. This suggests that the boron isotope composition of the fluids may have been altered from their primary values, and the fluids seen at the surface are more positive than the fluids at depth. Absorption of boron onto other minerals has not been conclusively examined. It is possible that in the calcite supersaturated wells at Lysoholl, ^{11}B is absorbed into the calcite. This explains the negative $\delta^{11}B$ of this well. Calcite has been identified in drill chips from other geothermal wells in Iceland. The occurrence of calcite in these wells may further reduce the $\delta^{11}B$ of the issuing fluids. The majority of the hydrothermal systems in Iceland are dominated by meteoric water. In contrast, Svartsengi contains both meteroic water and seawater. Examination of chloride concentrations indicated that the geothermal waters contain about 60% seawater with a $\delta^{11}B$ of +39.6‰ and concentration of 420μM of boron.. This influence of seawater is also seen in the high $\delta^{11}B$ of the two Svartsengi wells at +30‰. The other geothermal systems in Iceland are meteroic water dominated. The boron isotopic composition of these meteroic waters is quite variable, but the low boron concentration of meteroic waters means that the boron isotope systemmatics of these systems are dominated by boron derived from the rock.

CONCLUSIONS

The boron isotope compositions of meteroic dominated Icelandic geothermal systems lie in the range of -6.7 ‰ to +3.1‰. This is in the same range as continental and oceanic crust, (Palmer and Slack 1989). A number of constraints have been proposed. These include the incorporation of boron onto clay minerals and carbonates such as calcite, as well as phase separation. Once these processes have been fully understood it may then be possible to identify the rocks that interact with the fluids in the geothermal systems.

REFERENCES

Arnorsson S., Gislason S.R., Gestdottir K. 1989. Chlorine and Boron in natural waters in Iceland.*Water-Rock Interaction.* D.L.Miles (ed). 37-40. Rotterdam: Balkema.

Kristmannsdottir H. 1982. Alteration in the IRDP drill hole compared with other drill holes in Iceland. *Journal of Geophysical Research.* 87: 6525-6531.

Palmer M.R, Sturchio N.C. 1990. The boron isotope systematics of the Yellowstone National Park (Wyoming) hydrothermal system: A reconnaissance. *Geochemica et Cosmochimica Acta.* 54: 2811-2815.

Palmer M.R., Slack J.F. 1989. Boron isotopic composition of tourmaline from massive tourmaline deposites and tourmalinites.Boron isotopic composition of Icelandic hydrothermal systems. *Contributions to Mineralogy and Petrology* 103: 434-451.

Palmer M.R., Spivack A.J., Edmond J.M. 1987.Temperature and pH controls over isotopic fractionation during adsorption of boron on marine clay. *Geochimica et Cosmochimica Acta* 51: 2319-2323.

Spivack A.J., Palmer M.R., Edmond J.M. 1 1987. The sedimentary cycle of the boron isotopes. *Geochimica et Cosmochimica Acta* 51: 1939-1949.

Spivack A.J., Berndt M.E., Seyfried W.E. 1990 Boron isotope fractionation during supercritical phase separation. *Geochimica et Cosmochimica Acta* 54: 2337-2339.

Water-Rock Interaction, Kharaka & Maest (eds) © 1992 Balkema, Rotterdam. ISBN 90 5410 075 3

Isotopic composition of Sr, Nd and Li in thermal waters from the Norris-Mammoth corridor, Yellowstone National Park and surrounding region

T.D. Bullen & Y.K. Kharaka
US Geological Survey, Menlo Park, Calif., USA

ABSTRACT: Thermal waters from the Norris-Mammoth corridor (NMC) and surrounding region comprise a diverse array of water types. The isotopic compositions of Sr, Nd and Li in 0.1μm-filtered samples of those waters provide a basis for distinguishing among the various lithologic sources of solute cations in this geologically complex region. Sr, Nd and Li display contrasting geochemical behavior, and thus their isotopic compositions each provide unique information concerning thermal water evolution. In the three-component system $^{87}Sr/^{86}Sr$-ε_{Nd}-δ^7Li, the waters fall into groupings that have geographic significance. Differences in $^{87}Sr/^{86}Sr$ can be attributed to local variations in the dominant rock types within the shallow fault systems that act as conduits to the thermal springs. Differences in ε_{Nd} suggest high-temperature inheritance of the rare earths primarily from Paleozoic to Cenozoic lithologies within NMC, and from Precambrian to early Paleozoic lithologies outside NMC. Differences in δ^7Li reflect a dominant contribution of relatively light Li from hydrothermally-altered, layer silicate-rich igneous rocks in the higher-temperature systems, and relatively heavy Li from marine sediments in the cooler waters.

1 INTRODUCTION

Thermal waters from the Norris-Mammoth corridor (NMC) and surrounding region north of the 0.6 Ma. Yellowstone caldera rim comprise a diverse array of water types, and numerous studies have detailed their chemical compositions (for example, White and others, 1988; Fournier, 1989; Kharaka and others, 1991). These studies reveal that in a general sense, with increasing distance from the caldera rim, temperatures of the hydrothermal systems decrease, and thermal waters progress from alkali chloride- to alkali earth-carbonate/sulfate compositions. Clearly the thermal waters have interacted with a variety of rock types. Because Sr, Nd and Li exhibit contrasting geochemical behaviors in thermal systems, each can be a powerful isotopic tracer of the sources of dissolved lithogenic cations. In order to better understand the lithologic controls on water chemistries, we have analyzed the isotopic composition of Sr, Nd and Li in both 0.1-μm filtered thermal water samples and rocks from the region. Much of the data has been reported previously (Kharaka and others, 1991); additional unpublished analyses are included for this report.

2 ISOTOPIC COMPOSITION OF Sr, Nd AND Li IN THE THERMAL WATERS

The isotopic data for the thermal waters are plotted in Figure 1; sample locations are shown on Figure 2. Four distinct groups that have geographic significance are identified on the isotope plots: samples from NMC, samples from east of NMC, samples from the Corwin Springs Known Geothermal Resources Area (CSKGRA) along the northern margin of NMC, and a sample from relatively dilute Chico Hot Springs, located 30 km north of NMC.

Two samples that provide important exceptions to the orderly geographic distribution of isotopic compositions are those from the cool Bear Creek

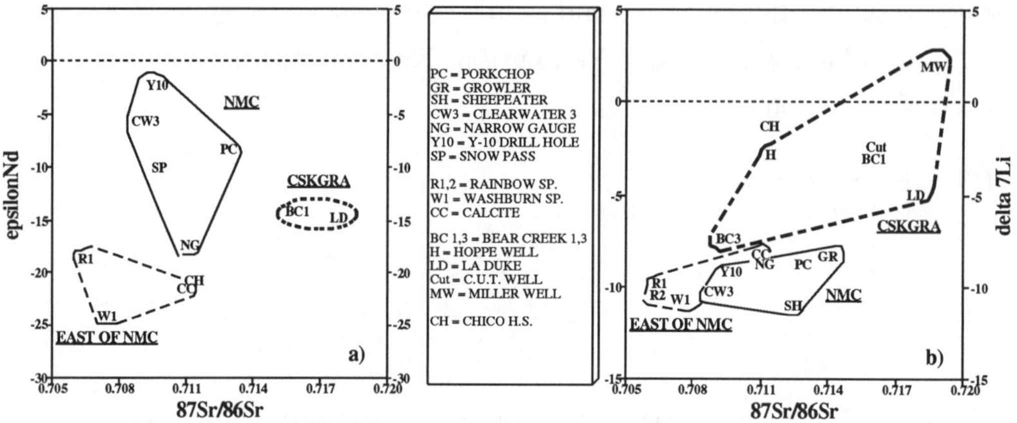

Figure 1. Isotopic composition of Sr, Nd and Li in Yellowstone region thermal waters. ε_{Nd} (epsilonNd) is the fractional deviation of $^{143}Nd/^{144}Nd$ from 0.51236, the inferred value of "bulk earth", times 10^4. δ^7Li (delta ^7Li) is the per mil deviation of $^7Li/^6Li$ from 12.1803, our long-term average measurement of the L-SVEC Li standard. Precision on the Sr- and Nd-isotopic measurements is better than .005% at the 95% confidence level; Li-isotopic measurements have been duplicated within 1‰.

spring (BC3) and Hoppe well (H) sites, both at the eastern end of CSKGRA. Although only $^{87}Sr/^{86}Sr$ and δ^7Li are available for these samples, the values are significantly different from the other Corwin Springs waters. In terms of Sr and Li isotopes (Fig. 1b), these two samples are intermediate between the Chico and Mammoth Hot Springs samples.

3 LITHOGENIC SOURCES OF Sr, Nd AND Li IN THE THERMAL WATERS

Sr, Nd and Li are particularly suitable elements with which to study thermal water evolution due to their contrasting geochemical behavior. For example, Sr is a reactive solute that exchanges with cations sorbed on mineral surface sites (Grim, 1968), and thus approaches equilibrium in terms of $^{87}Sr/^{86}Sr$ with Sr on those sites. Because Sr tends to be concentrated in Ca-bearing phases, and due to the low solubility of Ca- and Sr-sulfates and carbonates at high temperatures, Sr concentrations are low in high-temperature reservoir fluids (T>200°C). Moreover, because the solubility of these Sr-rich minerals increases with decreasing temperature, Sr isotopic signatures attained at high temperatures can be easily overprinted as temperatures decrease in the upflow conduits if those minerals are present. In contrast, Nd is partitioned into the fluid phase only under high-temperature (>250°C) geothermal conditions (Piepgras and Wasserburg, 1985). At the neutral pH values of most Yellowstone region thermal waters, Nd and the other rare earths readily sorb on mineral surface sites. Limited exchange of Nd from those sites can occur, but the isotopic signature of the dissolved Nd will be dominated by the lithogenic source at the depth of highest-temperature interaction. In contrast to both Sr and Nd, Li behaves as one of the most non-reactive species in thermal waters. Li is partitioned into the fluid phase, with increasing efficiency at high temperatures. Li/Cl is essentially constant at .01 in thermal waters from within NMC, but varies from 0.001 to .025 in samples from outside NMC. This indicates that Li can be fractionated from other "non-reactive" species, and/or that Li/Cl is highly variable in the possible lithogenic sources.

Considering these factors in light of the regional geology, we suggest that Sr in the thermal waters is derived from the shallow lithologies that line the upflow conduits (see Hildreth and others (1991) for a regional geologic map). In the Norris Geyser Basin, the wall rocks of the high-temperature reservoirs and upflow conduits are rhyolites (White and others, 1988; Fournier, 1989), and thus Sr in the thermal fluids is likely derived from the reaction of feldspar in the rhyolites with carbonic acid. At the travertine-precipitating Mammoth Hot Springs, the upflow conduits traverse Paleozoic and Mesozoic limestones and other sedimentary units (Pierce and others, 1991), allowing the thermal waters to equilibrate with these Ca- and Sr-rich rocks. To the east of NMC, compositionally-diverse Cenozoic and Quaternary igneous rocks dominate the shallow lithologies, although at Calcite Springs carbonate dissolution is obviously important to thermal water evolution. Along CSKGRA, upflow zones are intimately associated with the Gardiner reverse fault that juxtaposes Precambrian rocks of the Beartooth Range with Paleozoic and Mesozoic sediments correlative with those beneath Mammoth Hot Springs (Pierce and others, 1991). North of CSKGRA, the Tertiary Absaroka volcanics blanket much of the older sedimentary units and Precambrian basement rocks.

The deeper sources of dissolved Nd are more difficult to constrain. North of the 0.6 Ma caldera, Precambrian rocks are typically encountered at depths of less than a few kilometers (Pierce and others, 1991). Thus, it is probable that the thermal waters encounter these lithologies during much of their high-temperature evolution, and the Nd inherited from these rocks should have a Precambrian isotopic signature. That signature will be overprinted if substantial higher-temperature interaction with younger overlying rocks occurs, such as with the Quaternary rhyolites at Norris Geyser Basin, or if considerable mineral dissolution occurs, such as at Mammoth Hot Springs and CSKGRA where limestone-dolomite and gypsum dissolution, respectively, play important roles (Kharaka and others, 1991). Note, however, that Calcite Springs (CC) has a significant carbonate component, but Precambrian Nd.

The most striking feature of the Li isotope data is that samples from and to the east of NMC have similar δ^7Li, whereas samples from north of NMC diverge toward heavier compositions. The most important sources and sinks of Li pertinent to Yellowstone region thermal systems are probably the clays and other secondary minerals in marine sediments and hydrothermally-altered rocks. In a general sense, these two rock types will have relatively heavy and light Li-isotopic compositions, respectively, compared to primary igneous values (Chan and Edmond, 1988).

4 Sr, Nd AND Li ISOTOPES SUPPORT ASSESSMENT OF CATION SOURCES

The isotopic compositions of Sr, Nd and Li in the thermal waters support the above assessment of probable lithogenic sources for these cations. This conclusion is based on a comparison of the isotopic data to an extensive data set on the Sr- and Nd-isotopic compositions of volcanic, granitic, sedimentary and metamorphic rocks of all ages from the Yellowstone region (Doe and others, 1982; Wooden and Mueller, 1988; Hildreth and others, 1991). We are currently generating a data set for Li isotopes on those rock types, and incorporate some of the preliminary analyses in the following discussion.

On a regional scale, ^{87}Sr/^{86}Sr in rocks varies from 0.706 to >0.750. In general, the least radiogenic rocks are the compositionally diverse Absaroka volcanics, whereas the most radiogenic rocks are the Precambrian granitoids and metasediments. The Quaternary rhyolite tuffs and Paleozoic-Mesozoic sediments have intermediate values. Clearly, however, there is considerable overlap in values within this grouping. For the thermal waters, ^{87}Sr/^{86}Sr is least radiogenic in samples from east of NMC (except for Calcite Springs), most radiogenic in samples from CSKGRA, and intermediate in samples from NMC. The samples from Chico Hot Springs, Bear Creek #3 and Hoppe Well have ^{87}Sr/^{86}Sr like that at Mammoth Hot Springs (Narrow Gauge and Y-10), implying a similar Paleozoic-Mesozoic sedimentary source.

In terms of ε_{Nd}, Precambrian rocks have the most negative values, typically <-19, whereas Cenozoic-Quaternary volcanics and Paleozoic limestones (and presumably gypsums) have values ranging from 0 to -10. As suggested above, thermal water samples from east of NMC have the Precambrian signature. Samples from within NMC display a range of ε_{Nd} suggestive of mixing between Precambrian and the younger volcanic-sedimentary sources, although the latter clearly dominate. Samples from CSKGRA have intermediate values, consistent with the

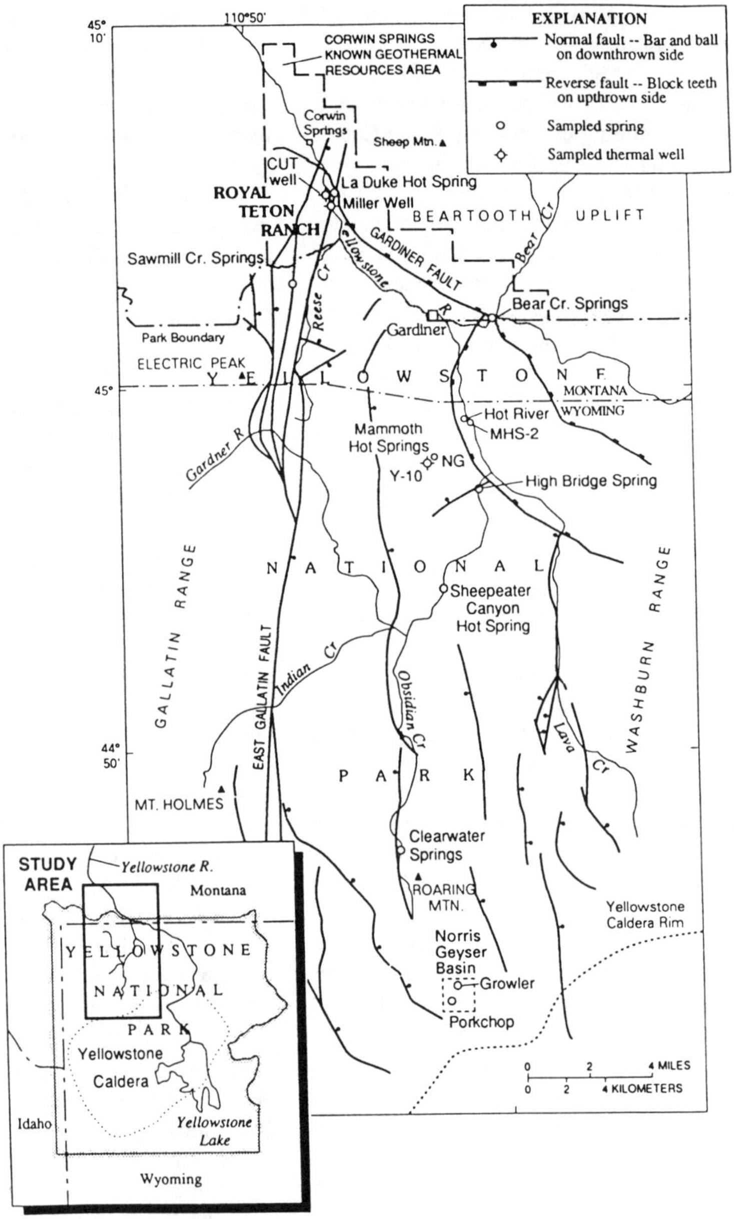

Figure 2. Map of principal geologic structures and locations of thermal water samples (adapted from Kharaka and others, 1991).

juxtaposition of Precambrian and younger lithologies along the Gardiner reverse fault.

In an initial effort to determine the isotopic composition of Li in rocks from this region, we have analyzed a fresh Quaternary rhyolite tuff and a Paleozoic limestone. δ^7Li in both samples is about -3.5‰, significantly heavier than any of the thermal waters from and to the east of NMC (Figure 1b). Chan and Edmond (1988) report a Li-fractionation factor of 1.019 between fluid and solid phases for the high-temperature thermal system at 21°N, East Pacific Rise. Fractionation of 19‰ requires that the source phase(s) in and to the east of NMC have δ^7Li≈-28‰ in the case of equilibrium fractionation, or -9‰ in the case of total dissolution. In either case, the fresh tuff and limestone are unlikely sources of the Li. Alternatively, hydrothermally-altered rock and/or rocks rich in non-marine layer-silicate minerals might provide the reservoir of light Li necessary to explain the compositions of these waters. The upper lithologies in and to the east of NMC are dominated by Cenozoic-Quaternary igneous rocks, a large proportion of which may have undergone considerable hydrothermal alteration. Finally, the trend toward more positive δ^7Li in thermal waters from and to the north of CSKGRA is consistent with both lesser geothermal activity and a greater contribution of Li from marine sediments in these areas.

REFERENCES

Chan, L.H. and J.M. Edmond, 1988. Variation of lithium isotope composition in the marine environment: a preliminary report. Geochim. Cosmochim. Acta 52: 1711-1717.

Doe, B.R., W.P. Leeman, R.L. Christiansen and C.E. Hedge, 1982. Lead and strontium isotopes and related trace elements as genetic tracers in the upper Cenozoic rhyolite-basalt association of the Yellowstone Plateau volcanic field. J. Geophys. Res. 87: 4785-4806.

Fournier, R.O., 1989. Geochemistry and dynamics of the Yellowstone National Park hydrothermal system, Wyoming. Ann. Rev. Earth Planet. Sci. 17: 13-53.

Grim, R.E., 1968. Clay Mineralogy, 2nd ed. New York: McGraw-Hill.

Hildreth, W., A.N. Halliday and R.L. Christiansen, 1991. Isotopic and chemical evidence concerning the genesis and contamination of basaltic and rhyolitic magma beneath the Yellowstone Plateau volcanic field. J. Petrology 32: 63-138.

Kharaka, Y.K., R.H. Mariner, T.D. Bullen, B.M. Kennedy and N.C. Sturchio, 1991. Geochemical investigations of hydraulic connections between the Corwin Springs Known Geothermal Resources Area and adjacent parts of Yellowstone National Park. In M.L. Sorey (ed), U.S.G.S. Water Resources Investigations Report 91-4052: F1-F38.

Piepgras, D.J. and G.J. Wasserburg, 1985. Strontium and neodymium isotopes in hot springs on the East Pacific Rise and Guaymas Basin. Earth Planet. Sci. Lett. 72: 341-356.

Pierce, K.L., K.D. Adams and N.C. Sturchio, 1991. Geologic setting of the Corwin Springs Known Geothermal Resources Area-Mammoth Hot Springs area in and adjacent to Yellowstone National Park. In M.L. Sorey (ed), op. cit.: C1-C37.

White, D.E., R. Hutchinson and T.E.C. Keith, 1988. The geology and remarkable thermal activity of Norris Geyser Basin, Yellowstone National Park, Wyoming. U.S.G.S. Prof. Paper 1456, 84p.

Wooden, J.L. and P.A. Mueller, 1988. Pb, Sr and Nd isotopic compositions of a suite of Late Archean igneous rocks, eastern Beartooth Mountains: implications for crust-mantle evolution: Earth Planet. Sci. Lett. 87: 59-73.

Water-Rock Interaction, Kharaka & Maest (eds) © 1992 Balkema, Rotterdam. ISBN 90 5410 075 3

Combined noble gas and quadrupole mass spectrometer analysis of volatiles trapped in fluid inclusions

P. Burnard, L. Ayliffe, F. Stuart, G. Turner & C. Curtis
Department of Geology, Manchester University, UK

ABSTRACT: Noble gases abundances in surface waters depend primarily on salinity and temperature at the time of equilibration with the atmosphere; subsequent interactions produce characteristic isotopic modifications. Previous techniques for the analysis of palaeofluids have relied on step-heating or crushing of fluid inclusions. Four improvements have been added to the VG5400-based system in Manchester: 1) pulsed laser extraction which releases gas from targetted fluid inclusion(s); 2) a quadrupole residual gas analyser to determine major gas compositions; 3) a capacitance manometer to determine absolute water contents and 4) the addition of Ne, Ar, Kr, and Xe spikes for elemental analysis by isotope dilution. Two preliminary analyses are presented here: 1) a vein quartz from Panasqueira Sn - W deposit, Portugal displays low elemental abundances, slight $^{84}Kr/^{36}Ar$ enrichment, and high $^{40}Ar/^{36}Ar$; 2) late diagenetic baryte from the North Sea Oil Province with slight $^{40}Ar/^{36}Ar$ but considerable $^{84}Kr/^{36}Ar$ enrichment.

1 INTRODUCTION

Fluid inclusions have long been studied as the only direct samples of ancient fluids. Noble gases (released by crushing or step heating) have been shown to provide information on the source and history of inclusion-trapped fluids (e.g. Turner and Bannon 1992). Recent advances in laser microprobe techniques allow analysis of particular inclusions or inclusion populations (e.g. Bohlke and Irwin 1992; Gibson and Carr 1989). A combined laser extraction system, capacitance manometer, quadrupole mass spectrometer and noble gas mass spectrometer at Manchester, England, allows analysis of noble gases and volatile species from single fluid inclusions or populations of inclusions. The equipment, experimental procedure and some preliminary results are described below.

The absolute and relative concentrations of noble gases in meteoric fluids is dependant on the temperature (T), pressure (P) and salinity (X) of the fluid at incorporation; assuming negligible diffusion loss, noble gas relative and absolute abundances in meteoric groundwaters can be related to different fluid sources and to T, P and X at entrapment (Ozima and Podosek, 1983). Subsequent fluid/rock interaction may lead to incorporation of radiogenic ^{4}He and ^{40}Ar (from U, Th and K decay or from juvenile magmatic fluids), indicating the net pre-entrapment interaction. In non-aqueous systems, such as hydrocarbon or hydrocarbon-water regimes, it is anticipated that partitioning between phases will occur. The hydrocarbon phase is enriched in heavier noble gases (Kharaka and Specht, 1988).

Analysis of volatile species such as water, methane (and other volatile hydrocarbons), carbon dioxide and sulphur species augments the information provided by noble gases on fluid source and history; deep-sourced fluids would have high CO_2:CH_4 ratios, with the lowest CO_2:CH_4 coinciding with hydrocarbon generation. Sulphur species (SO_2 or H_2S) could provide direct evidence of oxygen fugacities at the time of formation.

2. EQUIPMENT

In vacuo gas extraction procedures consist of either crushing rock samples or laser ablation of rock wafers. Crushing is performed in modified Nupro valves; the sample is placed in a stainless steel mortar, screwing a cylindrical stainless steel pestle

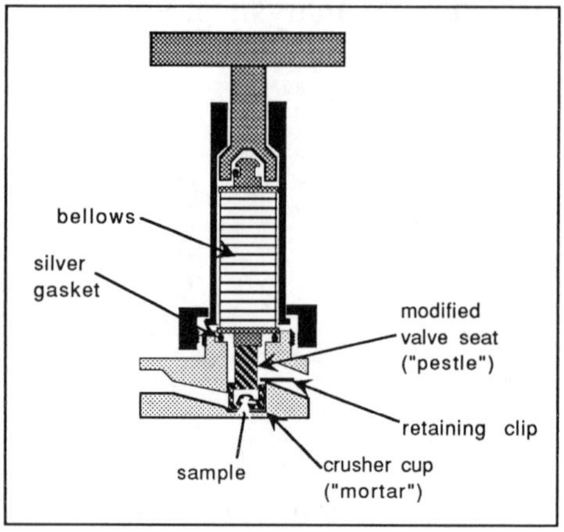

Figure 1. Schematic Diagram of crusher

down (see figure 1) in order to crush the sample. A 1060 nm, 5J Nd glass pulsed laser is used for lasering, the cleaned sample (a polished rock wafer) placed in an evacuated chamber with a U.H.V. viewport.

Total pressure of the released gases is determined on a MKS Baratron capacitance manometer. The minimum detectable pressure change is 1×10^{-6} torr, although in practice pressure changes are in the order 0.001 - 0.1 torr. Major gas analysis is performed on a Hiden HAL 200 residual gas analyser which uses a 125mm mass filter (c. 0.1 amu mass resolution) and a Faraday collector with a detection limit of $\approx 1 \times 10^{-11}$ torr.

After gettering, the noble gases are analysed on a VG5400 mass spectrometer, a 27 cm radius extended geometry, 60° sector instrument with a modified Nier-type electron impact source operating at 500 μA filament emission. Three collectors are available, two Faraday cups (off-axis and on axis) and a Daly collector. ^{40}Ar sensitivity under these conditions is 1.2×10^{-11} cm^3 mV^{-1} STP (10^{11} Ω resistor) for the axial Faraday cup. The Daly collector (used in current mode) response increases with increasing mass; at mass 40, the Daly signal (under present operating conditions) is 1247 ±21 times that of the Faraday.

3 PROCEDURE

The volume of the total volatiles released from the fluid inclusion(s) is determined as a pressure change on the capacitance manometer in a known volume (which is kept above 100° C.), and is used to determine absolute concentrations of noble gases in the inclusions. Approx. 0.5% of the gas is then separated, introduced into the quadrupole mass spectrometer and analysed for major gas composition (i.e. H_2O, CO_2, CH_4, N_2 etc.). The remaining gas is purified by a TiZr getter at 250°C which removes all reactive gases. Ar, Kr and Xe are separated from He and Ne by absorption onto an activated charcoal finger at liquid nitrogen temperature; a second charcoal finger (also at -196°C) is left open to the spectrometer during Ne analysis to minimize $^{40}Ar^{++}$ contributions at m/z = 20. The remaining $^{40}Ar^{++}$ contribution at m/z = 20 is corrected by monitoring $^{40}Ar^{+}$ during Ne analysis, and using a pre-determined $^{40}Ar^{++}/^{40}Ar^{+}$ factor of 0.158 ±0.0002. Ne, Ar, Kr and Xe abundances and Ar isotopic composition are determined on the VG5400 mass spectrometer. A "spiking" system, encompassing a reference gas enriched in particular isotopes will enable precise noble gas abundance determinations by isotope dilution in the near future.

Noble gas isotopes and abundances are currently calibrated against a standard aliquot of air, which is analysed with satisfactory reproducibility ($^{40}Ar/^{36}Ar$ = 300.5 ±3.7). Calibration against an artificial gas enriched in He (with respect to air) is planned.

At present, only semi-quantitative volatile species abundances are permitted due to the lack of a reference gas for the quadrupole mass spectrometer; the system is due to be upgraded shortly. We are confident, however, that ratios of volatile species presented here are correct as mass discriminations for the gases reported have been determined; abundances are semi-quantitative at best.

4 RESULTS

A study of noble gas and major volatile species in crustal fluids is underway. Preliminary results are available from two samples.

1.Panasqueira. The Panasqueira Sn-W deposit of northern Portugal is an approx. 300 m.y. vein deposit associated with a Hercynian granite, and hosted in Cambrian schists and metapelites (Snee et al, 1988). Fluid inclusion studies suggest 5 - 10 eq. wt% NaCl salinities and temperatures between 250 - 350°C (Kelly and Rye, 1979) for the ore-forming fluids.

A sample of clear vein quartz was both crushed and laser decrepitated. The results of six successive crushes are given in table 1 and figure 2; the concentration of ^{36}Ar,

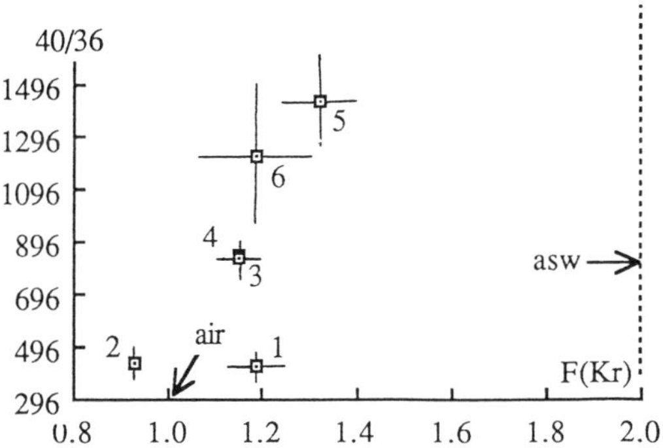

Figure 2. $^{40}Ar/^{36}Ar$ vs F(Kr) for Panasqueira quartz; crush nos indicated

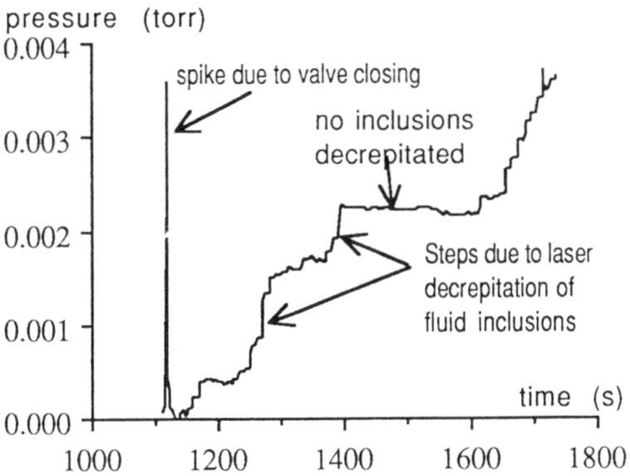

Figure 3. Baratron response to lasering of inclusions (Panasqueira vein quartz).

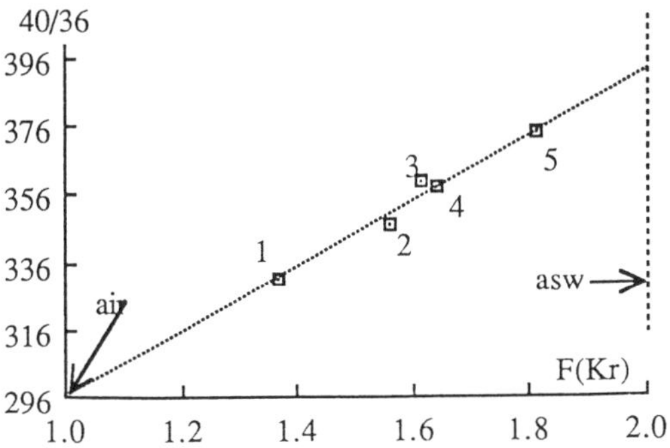

Figure 4. $^{40}Ar/^{36}Ar$ vs F(Kr) for North Sea baryte; crush nos indicated.

$^{40}Ar/^{36}Ar$, F(Kr) and F(Xe) all increase with successive crushing (where F(Kr) = $(^{84}Kr/^{36}Ar)_{sample}/(^{84}Kr/^{36}Ar)_{air}$ etc. after Smith and Kennedy, 1983).

Gas released by laser decrepitation had very low $H_2O:CO_2$ ratios (> 1:10) with CH_4, N_2 and SO_2 below detection. A $^{40}Ar/^{36}Ar$ ratio of 1141 ±6 was obtained from the noble gas fraction, although insufficient gas was released to satisfactorily analyse the remaining noble gases. Incremental pressure increases between 2×10^{-4} and 6.6×10^{-4} torr were observed during lasering (see figure 3). Assuming ideal gas behaviour this suggests individual fluid inclusions between 0.9×10^{-7} and 3×10^{-7} cm^3 were being burst; if the fluid inclusions were perfectly spherical, this would correspond to inclusions 40 - 80μm in diameter, in reasonable agreement with the observed inclusion sizes (10 - 100μm).

2. North Sea Baryte. Late diagenetic baryte (early - to - mid tertiary in age) from a fracture infill in the Piper Sandstone Formation (Upper Jurassic) of the Scott Oil Field, (Witch Ground Graben, Northern North Sea) was crushed. Analysis of the volatile species showed $H_2O:CO_2$ ratios of ≈5 and $CO_2:CH_4$ ratios of ≈1. No N_2 or sulphur species were detectable. Noble gas data is presented in table 2 and figure 4. With the exception of the first crush, $^{40}Ar/^{36}Ar$, F(Ne), F(Kr) and F(Xe) increase linearly with progressive crushing.

Table 1. Panasqueira Vein Quartz

Crush No	1	2	3	4	5	6
^{36}Ar conc*	0.12	0.14	0.175	0.22	0.13	0.21
$^{40}Ar/^{36}Ar$	431	436	834	852	1440	1230
F(Kr)	1.19	0.93	1.15	1.15	1.32	1.19
F(Xe)	1.37	nd	1.07	1.28	1.72	2.19

Table 2. North Sea Baryte

Crush No	1	2	3	4	5
^{36}Ar conc*	nd	1.63	1.66	1.27	1.05
$^{40}Ar/^{36}Ar$	330.6	346.8	359.3	358.3	374.1
F(Kr)	1.37	1.56	1.61	1.64	1.81
F(Xe)	2.14	2.51	2.59	2.86	3.21

$*x10^{-6}cm^3cm^{-3}STP$

5 DISCUSSION

1. Panasqueira. Progressive crushes show a reduction in the air component (i.e. early crushes have low $^{40}Ar/^{36}Ar$, F(Kr) and F(Xe)), suggesting that the final fluid composition is enriched in Kr and Xe relative to Ar (F(84) ≈1.5 and F(129) ≈2.5) and with $^{40}Ar/^{36}Ar$ >1450. The absolute abundance of the noble gases is also very low; the average ^{36}Ar concentration = 1.6×10^{-7} cm^3g^{-1}STP, in comparison to asw = 1×10^{-6} cm^3g^{-1} STP (Ozima and Podosek, 1983).

It would appear that the analysed gas is a mixture of at least 3 noble gas sources: a) a fluid with similar relative, but highly depleted, absolute, concentrations of the noble gases compared to air saturated water (asw); b) a component enriched in ^{40}Ar relative to the other isotopes of Ar and to the other noble gases and c) contamination by present day atmosphere. The Ar isotopic composition of component b) is consistent with a fluid which resided in the granitic pluton. The origin of component a) is more difficult to ascertain, although the noble gas depletion suggests phase separation has occurred. The high CO_2:H_2O ratios obtained may be a result of cold spots within the extraction line condensing H_2O; however, if the CO_2:H_2O ratio measured is correct, this would suggest that the majority of fluid inclusions opened while crushing are metamorphic. This can only be resolved with further time spent on lasering these samples.

2 North Sea Baryte. The noble gas concentrations and relative fractionation suggest that modified meteoric water is the principle fluid involved. Assuming that a fluid with an asw fractionation pattern and contemporary atmosphere contaminant are the only two components involved, it is possible to obtain the $^{40}Ar/^{36}Ar$ ratio of the fluid by extrapolating the data until $F(Kr)_{sample} = F(Kr)_{asw}$ (figure 4). Although the $^{40}Ar/^{36}Ar$ ratio is distinct from that of air ($\approx$390), this can be accounted for by accumulation of radiogenic ^{40}Ar by interaction of the fluid during burial with K - bearing minerals.

6 FURTHER STUDY

The ability to laser decrepitate and analyse the products for a variety of volatile species is ideally suited to the study of ancient fluids. A study of diagenetic and low and high temperature hydrothermal fluids is underway, concentrating on the following areas:

i. diagenetic: the Scott Field, North Sea and the Penrith Sandstones, Cumbria, U.K.;

ii. low temperature hydrothermal: Silvermines Pb - Zn mineralisation, Ireland;

iii. high temperature hydrothermal: Panasqueira Sn - W deposit, N. Portugal;

iv. He and Ar concentrations in fluids trapped in products of mid-ocean ridge hydrothermal circulation: Troodos Ophiolite complex, Cyprus

The fluid compositional data will permit more precise modelling of noble gas solubilities, enhancing our understanding of both the fluid source and history and the behaviour of noble gases in subsurface environments.

REFERENCES

Bohlke J.K. and Irwin J.J. 1992. Laser microprobe analyses of Cl, Br, I and K in fluid inclusions: Implications for sources of salinity in some ancient hydrothermal fluids. Geochim. Cosmochim. Acta 56: 203-206.

Gibson E.K. and Carr R.H. 1989. Laser microprobe-quadrupole mass spectrometer system for the analysis of gases and volatiles from geologic materials. U.S.G.S. Bull 1890: 35-50

Kharaka Y.K. and Specht 1988. The solubility of noble gases in crude oil at 25 - 100°C. App. Geochem. 3: 137-144.

Kelly W.C. and Rye R.O. 1979. Geologic fluid inclusion, and stable isotope studies of the tin-tungsten deposits of Panasqueira, Portugal. Econ. Geol. 74:.1721-1822

Ozima M. and Podosek F. 1983. Noble Gas Geochemistry. Cambridge University Press.

Smith S.P. and Kennedy B.M. 1983. The solubility of noble gases in water and NaCl brine. Geochim. Cosmochim. Acta 47: 503-515.

Snee L.W., Sutter J.F., and Kelly W.C. 1988. Thermochronology of Economic Mineral Deposits: Dating the stages of mineralisation at Panasqueira, Portugal, by high-precision $^{40}Ar/^{39}Ar$ age spectrum techniques on muscovite. Econ. Geol. 83: 335- 354.

Turner G. and Bannon M.P. 1992. Argon isotope geochemistry of inclusion fluids from granite associated mineral veins in southwest and northeast England.Geochim. Cosmochim. Acta 56: 204 - 226

Water-Rock Interaction, Kharaka & Maest (eds) © 1992 Balkema, Rotterdam. ISBN 90 5410 075 3

The signature of water-rock interactions in the formation waters of sedimentary basins: Some new evidence

Sam Chaudhuri
Kansas State University, Manhattan, Kans., USA

S.Furlan & N.Clauer
Centre de Géochimie de la Surface, Strasbourg, France

ABSTRACT: Sr-isotope and K-Rb data signify important roles of dissolution of alkali feldspar-mica and illitization of smectite in the chemical evolution of formation waters in sedimentary basins. Albitization may not be a major factor in the enrichment of Ca and Sr in many formation waters.

1 INTRODUCTION

Formation waters in sedimentary basins are commonly of Na-Ca-Cl to Na-Cl types. By comparison with sea waters at the same Cl concentrations, formation waters are very often enriched in Ca and Sr, but depleted in Mg. Enrichment of Ca and Sr has been very commonly explained along two different lines of argument, one of dolomitization (Collins, 1975; Carpenter, 1978) and the other of albitization of plagioclase (Land and Prezbindowski, 1981; Land, 1987). Formation waters from the Mahakam deltaic deposits of Miocene age in Indonesia provide a new perspective on the relative concentrations of these elements. These waters have a limited continental history with evolution from varied mixing between sea and meteoric waters, and they are depleted not only in Mg but also in Ca and Sr (Fig. 1).

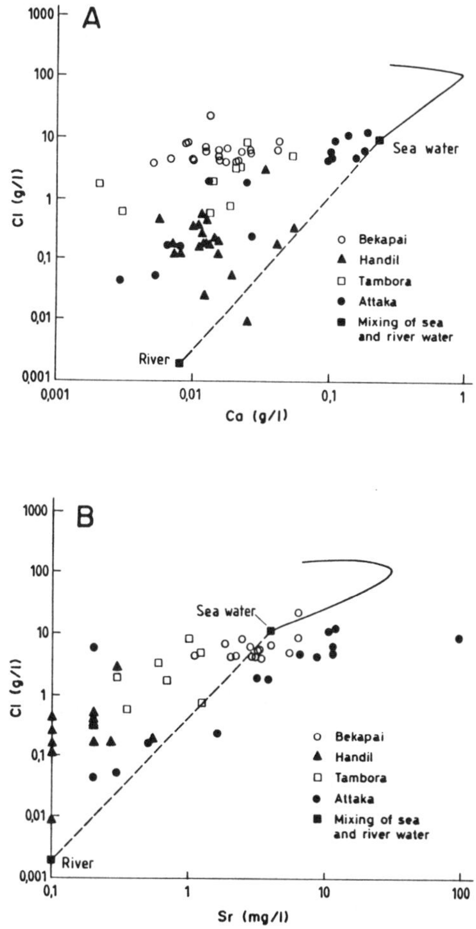

Fig. 1. (A) Ca and (B) Sr contents relative to Cl contents of Mahakam delta formation waters from different fields, as compared to that of river and sea waters (the curved line is the sea water evaporation curve of Zherebtsova and Volkova, 1966).

As Sr in the waters may be reasonably assumed to have come from the same source as Ca, and since Sr isotopes have proved as a useful natural tracer (Chaudhuri, 1978), many studies have analyzed Sr isotopic compositions of formation waters from sedimentary basins to determine the source of Sr and seek evidence for potential mineral-water reactions explaining the evolution of Ca in these waters (Sunwall and Pushkar, 1979; Starinsky et al., 1983; Stueber et al., 1984; Chaudhuri et al., 1987; Burtner, 1987; McNutt et al., 1987; Fisher and Kreitler, 1987; Lowry et al., 1988; Notsu, 1988; Land et al., 1988; Land and Macpherson, 1989; Egeberg and Aagaard, 1989).

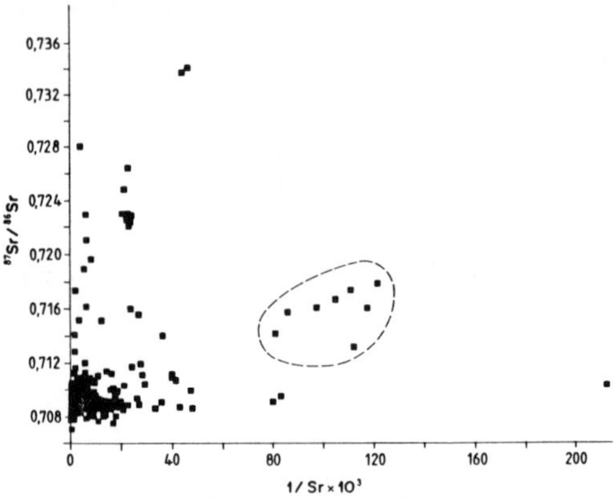

Fig. 2. Sr isotope ratios vs. 1/Sr values for saline formation waters from widely different locations and stratigraphic units across different continents. The encircled area includes data for samples with very low salinities.

2 DISCUSSION

We have reconstructed the Sr-isotope data published in the literature for all formation waters whose salinities are at least close to or higher than that of sea water (Fig. 2). Three features should be noted from this figure. Nearly sixty per cent of all formation waters have $^{87}Sr/^{86}Sr$ ratios higher than the Phanerozoic sea waters, which range from 0.7068 to 0.7092. In some cases, as in Japan, formation waters have Sr isotopic values less than Phanerozoic sea water, the more radiogenic Sr being found in those waters which are generally low in Sr contents. A very rough linear trend between the Sr-isotope and 1/Sr values is still apparent for formation waters whose Sr contents in all likelihood have been affected in various degrees by precipitation-dissolution-recrystallization of minerals and dilution. This array of data suggest that one of the sources of Sr in many formation waters is mainly the alteration of alkali feldspars and micas providing radiogenic Sr to the waters. Another important source is the alteration of carbonate and sulfate minerals of marine origin. The collection of data is for formation waters from widely different geographic, stratigraphic, and lithologic distributions. Since plagioclase minerals in crystalline rocks often have Sr isotopic values lower than sea water (Faure, 1986), and not too often any higher than sea water, we discount the importance of albitization as being a major source of Sr, and hence of Ca, in the waters.

Another potential source of Sr and Ca is from reaction with the clay minerals. However, this source cannot be confirmed by the Sr isotopic data because ion-exchange removes only the exchangeable Sr of the clays, which has isotopic values that are commonly very similar to or little more radiogenic than that of marine Sr. The exchangeable Sr certainly does not yield isotopic values as high as some of the highest values (about 0.725-0.735) which have been found for some formation waters. The importance of the role of clay mineral formation and/or recrystallization becomes evident from the K and Rb data of formation waters. The waters are often depleted in K, but enriched in Rb compared to sea waters. The data on formation waters from the Mahakam deltaic deposits illustrate these chemical characteristics (Fig. 3). This relative difference most likely relates to illitization in sedimentary basins, a process which should also increase the Ca and Sr contents in the waters.

Field data are needed to substantiate this hypothesis. Preliminary results of a mineralogical and K-Ar isotopic study on clay and feldspar fractions of the Mahakam delta deposits indicate that: (1) no major mineral-water interactions are detectable in the upper 1000 m of the sediment section, (2) much of the illitization process occurs at depths between 1000 and 1500 m, and (3) it proceeds slowly downward. As alkali-feldspar alteration seems to start only below 2500 m, this means that the illitization process in the zone between 1000 and 1500 m needed a supply of external K, possibly transported by upward migrating formation waters. Consequently, the contents of K and Rb of the formation waters seem to be controlled by illitization and feldspar dissolution, followed by illitization. The process of illitization in deep sedimentary basins most likely causes preferential removal of K over Rb, which might account for the apparent enrichment of Rb for many formation waters.

3 CONCLUSION

The evidence from the Sr-isotope data and K-Rb data suggests that dissolution of alkali feldspars, illitization, dissolution of halite, and recrystallization-dissolution of marine carbonate-sulfate minerals, which includes dolomitization, are important processes in the chemical evolution of formation waters. In many studies that have been made so far, Sr isotope data have not been fully integrated with major chemical and stable isotope data of formation waters, as for instance it has not been easy to distinguish between chemical signatures obtained before and after entrapment in the reservoir rocks. This information is essential to construct a fully developed model which can explain the chemical evolution of formation waters in sedimentary basins.

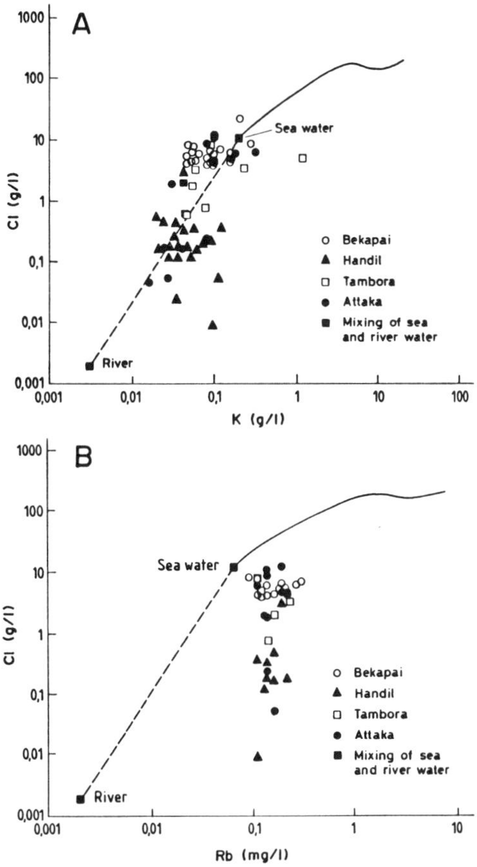

Fig. 3. (A) K and (B) Rb contents relative to Cl contents of Mahakam delta formation waters from different fields, as compared to that of sea waters (the curved line is the sea water evaporation curve of Zherebtsova and Volkova, 1966).

REFERENCES

Carpenter A. B. (1978) Origin and chemical evolution of brines in sedimentary basins. Oklahoma Geol. Surv. Circ. 79, 60-77.

Chaudhuri S. (1978) Strontium isoto pic composition of several oil-field brines from Kansas and Colorado. Geochim. Cosmochim. Acta 42, 329-332.

Chaudhuri S., Broedel V. and Clauer N. (1987) Strontium isotopic evolution of oil-field waters from carbonate reservoir rocks in Bindley field, central Kansas, U.S.A. Geochim. Cosmochim. Acta 51, 45-54.

Collins A. G. (1975) Geochemistry of oilfield waters. Elsevier, New York, 496p.

Egeberg P. K. and Aagaard P. (1989)

Origin and evolution of formation waters from oil fields on the Norwegian shelf. Appl. Geochem. 4, 131-142.

Faure G. (1986) Principles of Isotope Geology. Second Edition, J. Wiley & Sons, Chichester, 589 p.

Fisher R. S. and Kreitler C. W. (1987) Geochemistry and hydrodynamics of deep-basin brines, Palo Duro Basin Texas, U.S.A. Appl. Geochem. 2, 459-476.

Land L. S. and Prezbindowski D. R. (1981) The origin and evolution of saline formation water, Lower Cretaceous carbonates, south-central Texas, U.S.A. Jour. Hydrol. 54, 51-74.

Land L. S. (1987) The major ion chemistry of saline brines in sedimentary basins. In Physics and Chemistry of Porous Media II, J. R. Banavar, J. Koplik, and K. W. Winkler (eds.), Am. Inst. of Phys. Conf. Proc. 154, 160-179.

Land L. S., Macpherson G. L. and Mack L. E. (1988) The geochemistry of saline formation waters, Mioce-ne, offshore Louisiana. Trans. Gulf Coast Ass. Geol. Soc. 38, 503-511.

Land L. S. and Macpherson G. L. (1989) Geochemistry of formation water, Plio-Pleistocene reservoirs, off-shore Louisiana. Trans. Gulf Coast Ass. Geol. Soc. 39, 421-430.

Lowry R. M., Faure G., Mullet D. I. and Jones L. M. (1988) Interpretation of chemical and isotopic compositions of brines based on mixing and dilution, Clinton sandstones, eastern Ohio, U.S.A. Appl. Geochem. 3, 177-184.

McNutt R. H., Frape S. K. and Dollar P. (1987) A strontium, oxygen and hydrogen isotopic composition of brines, Michigan and Appalachian basins, Ontario and Michigan. Appl. Geochem. 2, 495- 505.

Notsu K., Wakita H. and Nakamura Y. (1988) Strontium isotopic composition of oil-field and gas-field waters, Japan. Appl. Geochem. 3, 173-176.

Starinsky A., Bielski M., Lazar B. Steinitz G. and Raab M. (1983) Strontium isotope evidence on the history of oilfield brines, Mediterranean coastal plain, Israel. Geochim. Cosmochim. Acta, 47, 687-695.

Stueber A. M., Pushkar P. and Hetherington E. A. (1984) A strontium isotopic study of Smackover brines and associated solids, southern Arkansas. Geochim. Cosmochim. Acta 48, 1637-1649.

Stueber A. M., Pushkar P. and Hetherington E. A. (1987) A strontium isotopic study of formation waters from the Illinois basin. Appl. Geochem. 2, 477-494.

Sunwall M. T. and Pushkar P. (1979) The isotopic composition of strontium in brines from petroleum fields of southeastern Ohio. Chem. Geol. 24, 189-197.

Zherebtsova I.K. and Volkova N.N. (1966) Experimental study of behavior of trace elements in the process of natural solar evaporation of Black Sea water and Sasyk-Sivash brine. Geochem. Intl. 3, 656-670.

Water-Rock Interaction, Kharaka & Maest (eds) © 1992 Balkema, Rotterdam. ISBN 90 5410 075 3

Stable isotopes in volcanic fluids and rocks at Vulcano (Sicily, Italy)

G. Cortecci & G. Ferrara
Dipartimento di Scienze della Terra, Pisa, Italy

A. Maiorani & B. Turi
Dipartimento di Scienze della Terra, Roma, Italy

ABSTRACT: Based on a considerable number of ∂34S and ∂13C data on gas samples from 1978 to 1991, and on sulfur bearing minerals from well cuttings and vents, sulfur in crater fumaroles at Vulcano should basically derive from magma degassing and water/rock interaction, this latter source being noteworthy during crisis periods; carbon in the gases should derive from magma degassing and decomposition of carbonate rocks beneath the volcano. Sulfur and carbon in beach fumaroles may be provided by crater-type fluids. An additional source of sulfur from reduction of seawater sulfate in hot aquifers can be inferred at the beach, especially during crisis events.

1 INTRODUCTION

Vulcano is the southernmost island of the volcanic Aeolian archipelago in the Tyrrhenian Sea north of Sicily, under which active subduction is postulated. The northern part of the island is occupied by a volcanic cone, with a summit crater named Gran Cratere or La Fossa. After the last eruption in 1888-1890, the volcanic activity was restricted to a fumarolic stage. Maximum temperature of crater fumaroles increased from 300°C in 1978 to over 600°C at present; beach fumaroles have steady temperature of about 100°C. After the strong tectonic earthquake on April 1978, a number of seismic events were recorded from 1979 to 1988. Basically, the island consists of lava flows and pyroclasts belonging to the shoshonitic magma series (Keller, 1980). A drill-hole in the La Fossa caldera penetrated a monzogabbroic intrusion from 1160 m down to 2050 m depth, where measured temperature exceeded 420°C (Faraone et al.,1986). No exploitable fluids or important aquifers were found during drilling. Seismic data support the existence of a magma body at 2-3 Km beneath the crater (Ferrucci et al., 1991).

Carbon dioxide, sulfur dioxide and/or hydrogen sulfide are major components of the dry gas fraction of fumaroles. Apart from magma degassing, additional important sources for SO2 and H2S at the crater may be leaching of the substantial amounts of hydrothermal pyrite and anhydrite found in well cuttings and recycling of elemental sulfur. Seawater sulfate and the few ppm of primary sulfur detected in a number of lava specimens should be minor sources only, when compared with the sulfur contents of the fumaroles. Carbon dioxide in the crater fumaroles was interpreted as mainly deriving from decomposition of marine carbonate rocks beneath the volcano (e.g. Mazor et al., 1988; Cannata et al., 1988); however, the contribution of magmatic CO2 may be important. At the beach, fumaroles are generally interpreted as being fed by a crater-type fluid, which is largely modified in its original composition while passing through shallow aquifers (e.g. Mazor et al., 1988 and references therein; Cioni et al., 1988).

A total of more than one hundred gas samples collected at the crater and at the beach between 1983 and 1991 have been analyzed for ∂34S(ΣS = SO2 + H2S, SO2, H2S) and ∂13C(CO2), and the data compared with the few and sporadic ones available from other workers for 1978-1987 samples (Allard, 1986 and personal communication; Cannata et al. 1988). These data, along with those on elemental sulfur at the fumarole outlets and on hydrothermal pyrite and anhydrite from well cuttings, are discussed below in terms of genetic implications for elements and possible volcanological meaning.

2 RESULTS AND DISCUSSION

Results for crater gases mostly deal with fumaroles F5 and FA. At the beach, they refer to fumaroles bubbling through seawater, sand and mud pool.

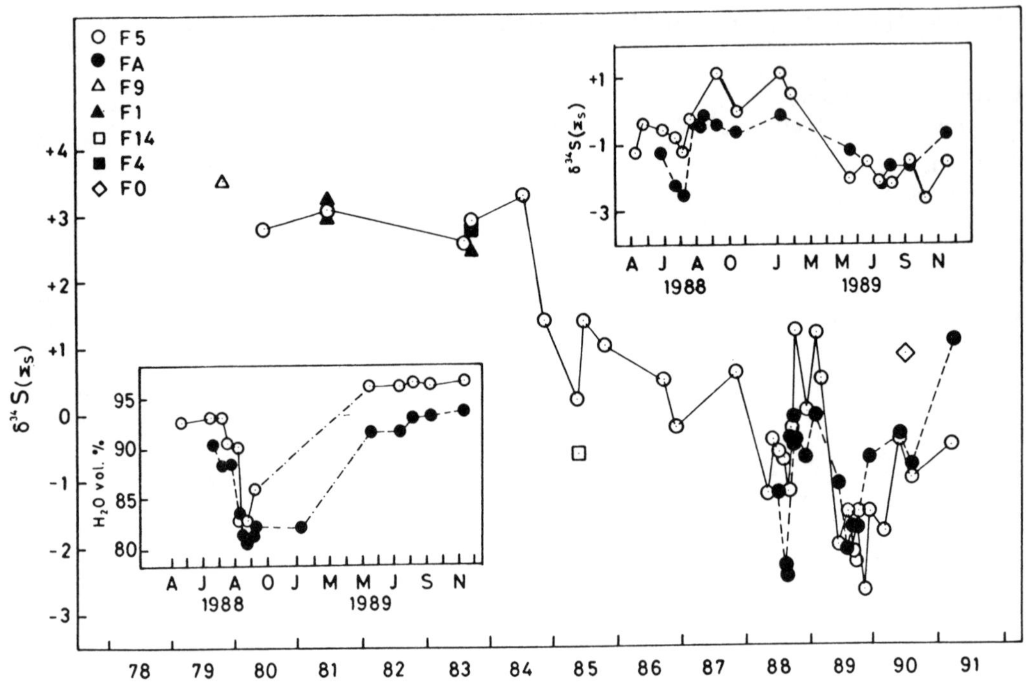

Fig.1 Chronogram of total sulfur ∂34S(∑S) values from 1979 to 1991 for a number of fumaroles (F) at the crater. In the insets, time variations of ∂34S and H2O content in the 1988-1990 fluids are compared.

2.1 ∂34S and ∂13C in crater fumaroles

The chronogram of total sulfur ∂34S(∑S) values is depicted in Fig. 1. Summing up, the ∂34S values kept quite constant close to +3‰ from 1979 to 1984, then decreased to about 0‰ during 1984-1985, and afterwards fluctuated between -3‰ and +1‰.

Values of +3‰ during 1979-1984 are consistent with those expected for volcanic gases and rocks from arc zones (Taylor, 1986 and references therein), which tend to be somewhat richer in 34S than primary mantle material (∂34S ≈ 0‰) as a consequence of possible assimilation of isotopically heavy sulfur from subducted marine sediments. Direct magma degassing beneath Vulcano may thus be considered the principal source of sulfur in fumaroles. In addition, the uniformity and constancy of values support steady degassing conditions at depth, followed by preservation of ∂34S in the released fluids during ascent. Seismic activity did not change the origin of sulfur until mid-1984. Finally, sulfur isotope differences in analyzed SO2-H2S pairs provide equilibrium thermometric estimates (Richet et al., 1977) of 300-400°C, a few tens of degrees higher than the outlet temperatures of corresponding fumaroles.

From the above considerations, the remarkable decrease of ∂34S(∑S) in 1984-1985 and the persisting low isotopic signature during the subsequent five-six years can be interpreted as due to contributions of isotopically light sulfur to the fluids. The low ∂34S on November 1984 preceeded by about six months the seismic crisis of May-August 1985.

The ∂34S values on fumaroles F5 and FA from 1988 to 1990 are especially noteworthy. They display the largest ranges of variation, anti-correlate well with the water content of fluids, and vary with time in a nearly synchronous manner. Temperature in fumarole F5 remained steady, close to 330°C whereas in fumarole FA it progressively increased from about 400 to 550°C. When plotted on a ∂34S vs XH2S (= mH2S/m∑S) diagram (Fig.2), data points for 1985-1990 samples match with a mixing model involving rising magmatic fluids rich in SO2 and 34S and shallower hydrothermal fluids rich in H2S and 32S. These latter fluids may derive from penetration of seawater into the volcano as a consequence of opening of fractures due to seismic shocks,

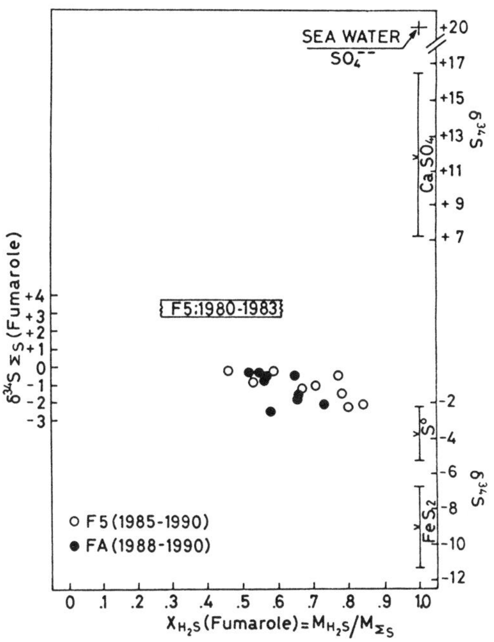

Fig.2 ∂34S - XH2S plot relating the 1985-1990 data points of crater fumaroles F5 and FA with pre-1985 ones for F5 and those expected for volatile sulfur (XH2S ≈ 1) from redox processes on shown compounds. The mineral ∂34S ranges refer to native sulfur (S°) at the crater and to hydrothermal pyrite and anhydrite from drill-cuttings (Vulcano Porto 1 well) at depths of 580 to 945 m.

accompanied by water/rock interaction processes and generation of gaseous sulfur species (H2S » SO2) via redox reactions upon sulfur-bearing compounds. Based on Fig. 2, the fraction of this sulfur from leaching of pyrite and elemental sulfur should largely exceed that from chemical reduction of anhydrite and seawater sulfate, and its ∂34S values may have been as low as -12‰ or even lower depending on the degree of isotopic equilibration with aqueous sulfate. Interestingly, the sharp and simultaneous increase of total sulfur ∂34S (and the water content decrease) observed in July-September 1988 (see insets of Fig. 1), in concomitance with a peak in the local seismicity and fumarolic activity, can be interpreted as due to a remarkable increase of the deep magmatic component in the fluids. Finally, the high temperature of fumarole FA excludes disproportionation of SO2 by interaction with a liquid water phase and removal of formed sulfate as a possible process for the observed 34S-depletions in the gases. In this connection, isotopic temperatures of 500-600°C are calculated from SO2-H2S pairs during 1988-1990.

All ∂13C(CO2) data discussed below refer to fumarole F5. Most of the sporadic values from 1978 to 1987 are between -0.5‰ and +0.5‰, only a few being as low as -2‰. On the other hand, temporally close fluctuations between -3‰ and +0.5‰ are displayed by the great number of 1988-1990 values. These values and features agree with a mixed origin of carbon in the fluids from two major sources, i.e.magmatic CO2 with generally accepted ∂13C of -5 to -8‰ and CO2 with ∂13C > 0‰ as from thermal decomposition of marine carbonate rocks beneath the volcano. Seawater has to be considered a negligible source of carbon. Note that the time fluctuations of ∂13C during 1988-1990 superimpose quite well on those observed for ∂34S.

2.2 ∂34S and ∂13C in beach fumaroles

Hydrogen sulfide is the exclusive sulfur species in the gases. Its ∂34S values cover with gaps the 1978-1990 interval and vary between -5‰ and +5‰. During 1978-1984, values are lower than those of total sulfur in crater fumaroles, as one can expect from a common origin of the fluids followed by complete disproportionation of the isotopically heavier SO2 in shallow aquifers beneath the beach. On the other hand, 1985-1990 values are even considerably higher than at the crater, suggesting chemical reduction of seawater sulfate (∂34S = +20‰) as additional source of sulfur.

Most ∂13C(CO2) values are significantly lower than those at the crater, and do not show any obvious correlation with the latter. Following previous interpretations (e.g. Cannata et al., 1988), the 13C-depletion in beach CO2 may account for partial removal of crater-type CO2 as 13C-enriched carbonate phases in aquifers.

3 CONCLUSIONS

Sulfur in crater fumaroles should basically derive from magma degassing and redox reactions on sulfur-bearing compounds during water/rock interaction processes. The role of this latter source likely depends on open/closed conditions of the volcano relative to seawater, and then on the opening of fractures in concomitance with seismicity and fumarolic activity crises. Carbon in the fumaroles likely derives from magma degassing and decomposition of carbonate rocks within a presumable basement at depth.

At the beach, crater-type fluids can be sources of sulfur and carbon in the fumarolic gases. Additional sulfur may derive from seawater sulfate reduction in hot aquifers, especially during crisis periods.

Acknowledgements

The research was financially supported by MURST, CNR and GNV of Italy. It greatly benefited from the collaboration of Armando Maiorani, co-author in this paper and suddenly dead on July 1991. Many thanks are due to peoples who provided us with a number of gas and mineral samples, as well as with unpublished chemical and isotopic data.

REFERENCES

Allard, P. 1986. Isotopic composition and origin of water, carbon and sulfur in volcanic gases: rift zones, continental margins and island arcs. State thesis, Paris University, 240pp.

Cannata S., Hauser, S., Parello, F. & M. Valenza 1988. Caratterizzazione isotopica della CO2 presente nelle manifestazioni gassose dell'Isola di Vulcano. Rend. Soc. It. Mineral. Petrol. 43: 153-161.

Cioni, R., Corazza, E., Magro, G., Guidi, M. & L. Marino 1988. Reactive and inert gases in low temperature fumaroles (Aeolian Islands, Italy). Rend. Soc. It. Mineral. Petrol. 43: 1003-1011.

Faraone, D., Silvano, A. & G. Verdiani 1986. The monzogabbroic intrusion in the island of Vulcano, Aeolian Archipelago, Italy. Bull. Volcanol. 48: 431-439.

Ferrucci, F., Gaudiosi, G., Milano, G., Nercessian, G., Vilardo, G. & G. Luongo 1991. Seismological exploration of Vulcano (Aeolian Islands, southern Tyrrhenian Sea): case history. Acta Vulcanol. 1: 143-152.

Keller, J. 1980. The island of Vulcano. Rend. Soc. It. Mineral. Petrol. 36: 369-414.

Mazor, E., Cioni, R., Corazza, E., Fratta, M., Magro, G., Matsuo, S.,Hirabayashi, J., Shinohara, H., Martini, M., Piccardi, G. & P. Cellini Legittimo 1988. Evolution of fumarolic gases-boundary conditions set of measured parameters: case study at Vulcano, Italy. Bull. Volcanol. 50: 71-85.

Richet, P., Bottinga, Y. & M. Javoy 1977. A review of hydrogen, carbon, nitrogen, oxygen, sulfur, and chlorine stable isotope fractionation among gaseous molecules. Ann. Rev. Earth Planet. Sci. 5: 65-110.

Taylor, B.E. 1986. Magmatic volatiles: isotopic variation of C, H and S. In J. W. Valley, H.P. Taylor & O'Neil J.R. (eds.), Stable Isotopes in High Temperature Geological Processes: 185-225. Min. Soc. Am.: Review in Mineralogy.

Water-Rock Interaction, Kharaka & Maest (eds) © 1992 Balkema, Rotterdam. ISBN 90 5410 075 3

Unconventional stable isotopes: Iron

P.R. Dixon, D.R. Janecky, R.E. Perrin, D.J. Rokop, P.L. Unkefer & W.D. Spall
Los Alamos National Laboratory, N.Mex., USA

R. Maeck
Central Bureau of Nuclear Measurements, Geel, Belgium

ABSTRACT: Iron isotopes offer an excellent opportunity for detailed tracing of natural processes, because of the range of stable isotopes (54, 56, 57, 58) and the importance of iron redox transitions in biochemical and inorganic processes. We have significantly improved the mass spectrometric technique for iron isotope analysis and calibrated with an absolute iron isotope abundance standard. 100-1000 nanogram filament loads of iron are analyzed using the low temperature (1200°C) silica gel/boric acid ionization technique with better than 0.1% precision and accuracy on the major isotopes. A survey of natural isotopic compositions is in progress to provide the first evaluation of magnitude and range of iron isotopic fractionation in Earth crustal systems. Initial samples indicate that iron isotope fractionation does occur in terrestrial processes.

1 INTRODUCTION

Stable isotope fractionation is a powerful indicator of a wide range of processes, including cosmochemical processes in extraterrestrial samples, and igneous, metamorphic, geothermal and sedimentary processes in terrestrial samples. Isotopes of oxygen, carbon, sulfur and hydrogen have seen extensive use in such studies (Hoef, 1973; Valley and others, 1986). In addition to the isotopes used in such well established studies, there are other major elements with several stable isotopes which may also be significantly fractionated by biogeochemical processes (Figure 1). In particular, terrestrial fractionation of naturally occurring isotopes of iron has great potential for unraveling biogeochemical processes in the Earth's crust.

Several factors have led us to develop the techniques necessary to quantify iron isotope fractionation. Iron is ubiquitous in terrestrial systems (fourth most abundant element in the Earth's crust), is an important biochemical nutrient and its transformations between redox states are used as an energy source or storage mechanism by many organisms. In addition, some bacteria form iron minerals (*e.g.* magnetite and pyrite) in a wide variety of environments (Farina and others, 1990; Mann and others, 1990; Fassbinder and others, 1990). Biologic isotope selectivity has been extensively demonstrated for carbon, oxygen, hydrogen, sulfur, nitrogen and selenium (*e.g.* Hoefs, 1973) and applied to numerous geologic problems. Iron isotopic fractionation in biogeochemical systems, in contrast, remains unexplored. Demonstration of iron isotopic fractionation by biologic systems would provide an important tool with which to test these models and compare iron minerals produced by modern bacteria with fossil mineral structures that may be either biologically produced or biologic templates. It has also been recently proposed that Antarctic ocean productivity is limited by iron nutrient availability and that CO_2 removal by the oceans could be strongly enhanced by iron seeding (Martin, 1991). This suggestion has generated significant controversy, and iron isotopic signatures of natural fractionation and/or tracer experiments may also be an essential part of critically evaluating the important processes involved.

While inorganic fractionation of terrestrial iron has not been observed, biologic fractionation is a much stronger possibility based on comparisons with other elements. Elements within a wide mass range exhibit diverse isotope fractionation behavior. Isotope fractionation of heavier elements (e.g. selenium - Rashid and others, 1978) are observed and lighter elements with substantial isotopic mass differences (*e.g.* calcium) are not observed to fractionate (Hoefs, 1973; Valley and others, 1986). An important factor that has been identified with measurable isotope fractionation is the structural environment of the atoms. For example, sulfur

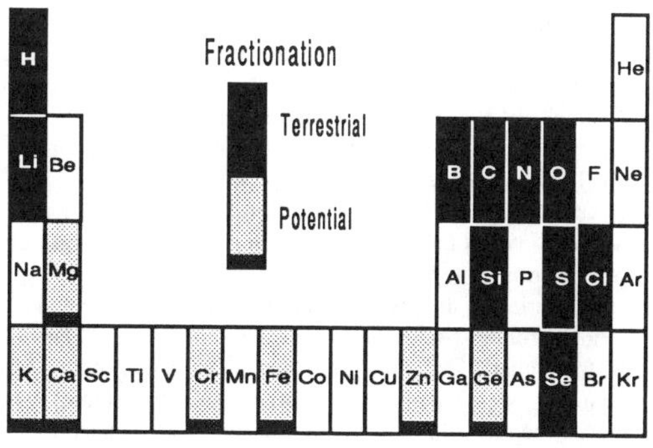

Fig.1 Light elements with multiple stable isotopes which are known or could potentially fractionate in terrestrial systems.

isotopes in processes that involve sulfide are strongly fractionated, while those processes involving only sulfate molecules do not result in significant sulfur isotope fractionation. Inorganic iron behaves similarly to sulfate, magnesium, and inorganic silicon (all chemical species that are strongly tied to oxygen) and is apparently not significantly isotopically fractionated. However, in biologic or organic systems, sulfur, nitrogen, selenium, and silicon are all isotopically fractionated (Hoefs, 1973; Valley and others, 1986; Rashid and others, 1978; Douthitt, 1982). More importantly, isotope fractionation effects are larger in elements with several redox states. Strong chelation of iron by biologic molecules and utilization of transformations in iron redox state for energy in biochemical processes indicates a significant potential for isotope fractionation in natural systems. Our efforts have, therefore, been directed towards enhancing the analytical techniques for iron isotopic analysis and beginning an analytical survey to assess the existence and occurrence of iron isotopic fractionation in terrestrial/biologic processes.

2 APPROACH AND TECHNIQUES

Thermal ionization mass spectrometry and ultra clean chemical separation techniques have been the primary focus of our efforts. Previous work using these techniques involved relatively large filament loads (10 to 50μg) and high temperature ionization (1350° to 1450°C) to facilitate analysis with faraday cup detectors (*e.g.* Maeck, 1992; Volkening and Papanatassiou, 1989). Isobaric corrections for chromium and nickel have typically been large (0.1 to 0.5%) or double spike techniques at 54 and 58 have been utilized. In this section we describe improved mass spectrometric analysis utilizing ion pulse counting detection and accompanying separations techniques for unspiked samples that allow quantification of all four stable iron isotopes, reduced filament loads, reduced isobaric interferences and stable ionization at reduced temperatures.

2.1 *Mass spectrometry*

Our analyses utilize a low temperature (1200°C) silica-gel/boric ionization technique on a NBS 12 inch 90° mass spectrometer. Sample solutions (in 1.5M HCl) are loaded on high purity platinum ribbon filaments with a silica-gel/boric acid mixture (4:1). Filament loads range from 100ηg to 1μg of iron.

Initial survey experiments on standard NBS filament assemblies (stainless steel posts with glass insulators) proved totally unacceptable for iron isotope analysis by pulse counting (Fig. 2). Very high backgrounds across the mass range of interest (51 to 61) were observed. This background was determined to be scattering of alkali ions generated by volatilization of the glass insulators. A secondary, but also serious problem was ionization of the stainless steel filament posts. Isobaric ^{54}Cr and ^{58}Ni, as well as Fe (see ^{56}Fe peak in Figure 2a) were characteristic. Dramatic improvements were found when the source filament assembly was changed to platinum posts in a high density Coors ceramic insulator. Alkali scattering backgrounds were reduced 1 to 2 orders of magnitude (to 10 counts or less) and stainless steel ionization was eliminated. Isotopic corrections to ^{54}Fe and ^{58}Fe for isobaric ^{54}Cr and ^{58}Ni have been reduced to less than 0.05%. In addition, source stability and ion focusing were significantly improved through utilization of a dimpled Delmore filament design. These radical improvements made to the ion spectra allowed development of a high precision (~0.1%) measurement technique for microgram and submicrogram quantities of iron.

Calibrating the machine fractionation followed, to allow absolute isotopic abundance measurement of all four stable iron isotopes. Collaboration between Los Alamos National Laboratory (LANL) and the Central Bureau of Nuclear Measurements (CBNM), Belgium provided high purity iron isotope samples of ^{54}Fe (99.8%), ^{56}Fe (99.9%), and ^{57}Fe (99.5%), as well as several mixtures of the ^{54}Fe and ^{56}Fe isotope samples (Maeck, 1992). Machine fractionation per unit mass for our analysis approach was determined to be 0.99341 ± 0.0015. In addition, this machine fractionation has been determined to be constant over a relatively wide range of filament temperatures and for multiple hour runs. In contrast, Maeck (1992) determined a machine fractionation of 1.00941 ± 0.00060, for larger sample sizes (10 to 20μg) and higher filament temperatures (1350° to 1400°C) using a Finnigan MAT-262 and Faraday cup detection. An

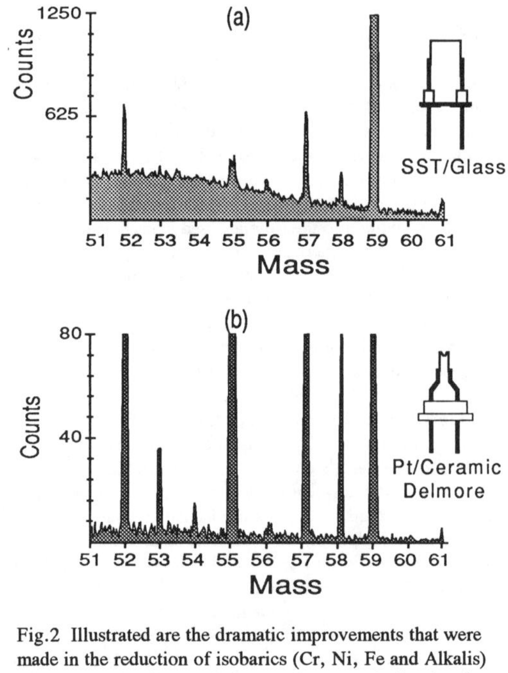

Fig.2 Illustrated are the dramatic improvements that were made in the reduction of isobarics (Cr, Ni, Fe and Alkalis) with conversion from stainless steel posts and glass insulators (**a**) to platinum posts in a Coors high density ceramic (**b**, note change in scale). Both examples are for blank filaments with enhancer at a measurement temperature of 1200°C. Peaks at masses 52, 53 and 54 are Cr and at mass 55 is Mn. Both Cr and Mn are impurities in the Pt filament ribbon. At masses 59, 61, 62 and 63 are CaF peaks, while 56.15, 57.15 and 58.15 are organic peaks from contaminants in the enhancer which are resolvable from the iron isotopic spectra.

interlaboratory comparison of absolute isotope abundance using an iron wire standard prepared at CBNM (Maeck, 1992) indicates both accuracy of the analytical procedures and machine fractionation determinations ($^{54}Fe/^{56}Fe$ LANL / $^{54}Fe/^{56}Fe$ CBNM = 1.00024 ± 0.00031).

2.2 *Sample processing*

Samples and standards for iron isotope analysis are processed in ultra clean laboratories. Iron is quantitatively separated using anion exchange resins. Inorganic (oxide and/or silicate) samples are digested in Teflon bombs in a CEM microwave using either concentrated HCl or a concentrated HCl:HF mixture. Organic (>1%) samples are digested in a Teflon bomb in a CEM microwave by sequential concentrated HNO3 and final H_2O_2 decompositions. Following digestion both inorganic and organic samples are handled similarly. The digested samples are dried down and redissolved in 6M HCl. Iron concentrations are determined spectrophotometrically. Then, 200 to 300μg of iron in the HCl solution are loaded onto an anion exchange column (Biorad AG-1X). The column is rinsed several times and then iron is eluted with 0.5M HCl. The eluted iron sample is dried down and again dissolved in 6M HCl and rerun through a second anion column as described above. The eluted iron from the second column (95% recovery) is dried down and dissolved at a concentration of 5μg/μl in 1.5 M HCl for loading onto filaments for analysis.

2.3 *Filament loading*

Loading of sample solutions onto filaments is accomplished using a microsample loader developed at Los Alamos. Using this tool, enhancer (silica-gel/boric acid) or iron solution in reproducable 100 nanoliter drops are consistently loaded onto a specific area of a filament. This allows variability in the quantity of iron loaded on filaments to be greatly reduced, resulting in more consistent analytical results. Furthermore, with the small sample size, the iron solutions can be easily loaded in a tiny spot on the filament, to give a near point source configuration for thermal ionization. An appropriate "sandwich" of enhancer and iron sample is produced by loading enhancer, then iron solution, and then more enhancer. Upon filament heating, this load produces a homogeneous glass bead. These improved loading techniques provide significant consistency between filaments and operators, and has resulted in substantially improved external reproducibility.

3 INITIAL RESULTS FOR TERRESTRIAL SAMPLES

Several terrestrial samples have been processed to date for iron isotope analysis. These samples include an iron wire, block iron, banded iron formation (BIF) from northern Michigan, cultured E. Coli bacteria, and a hydrothermal bacterial mat from Loihi seamount. The natural samples were selected to span the range of ancient sedimentary iron oxides, iron sequestered by biochemical processes, and iron oxide sequestered and precipitated in association with bacteria.

The commercial high purity (99.999%) iron wire (Johnson Matthey, processed at LANL) and an aliquot of iron metal

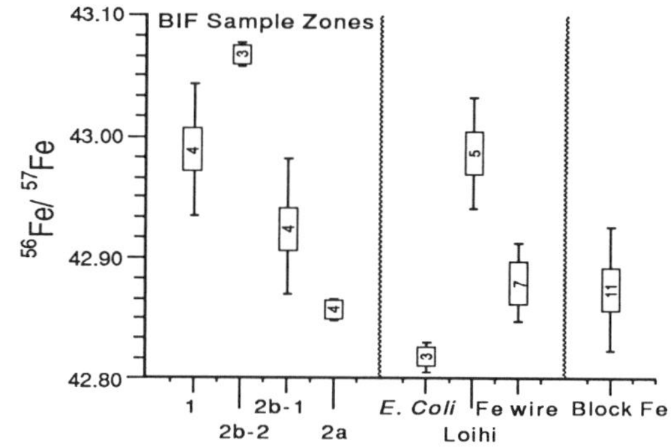

Fig.3 Iron isotopic ratio results for 56 mass to 57 mass. External 1s precision represented by vertical bar for each sample, for the indicated number of filaments analyzed.

processed by Maeck (1992) have been used as primary test and internal laboratory standards in this work. Within the precision of our analyses, these samples are indistinguishable (Figures 3, 4 and 5). Repeated analyses as our procedures were developed, verified the consistency of our mass spectrometric approach, and provided part of the interlaboratory comparison discussed above. In addition, these samples have been used to evaluate our sample processing procedures and demonstrated significant improvements in separation of isobarics as the laboratory procedures evolved.

Four consecutive zones in the BIF sample have been analyzed to date for their iron isotopic compositions (Dixon and others, in prep). Iron concentrations in each of these subsamples range from 70% to 10%. Each of these zones contains at least one distinct compositional band, but may contain several bands of varying composition. Both $^{56}Fe/^{57}Fe$ and $^{54}Fe/^{57}Fe$ ratios exhibit isotopic fractionation larger than the precision of the analyses and distinct from the processed metallic iron samples (Figures 3 and 4). The precision of our $^{58}Fe/^{57}Fe$ ratio analyses does not allow identification of fractionation (Figure 5). Covariation of $^{56}Fe/^{57}Fe$ and $^{54}Fe/^{57}Fe$ ratios in this sample suite is consistent with a mass dependent fractionation component.

The E. Coli bacteria sample represents one third of the intended analysis results of a culturing experiment. Samples still to be analyzed are the starting E. Coli growth media which was spiked with iron sulfate, as well as the media separated from the bacteria at the end of the experiment. The sequestering of iron by the bacteria does not indicate fractionation (Figures 3, 4 and 5). The experiment did demonstrate the capability of our chemical separations techniques for purifying trace amounts of iron contained in large quantities of organic material.

The sample from Loihi seamount (Hawaii) was collected from a bacterial mat associated with thermal springs (Sedwick and others, in press). The iron isotopic composition of this sample is measurably fractionated in $^{56}Fe/^{57}Fe$ ratio from the iron metal samples (Figure 3). However, the $^{54}Fe/^{57}Fe$ and $^{58}Fe/^{57}Fe$ ratios are indistinguishable from the iron metal samples (Figure 4 and 5). Such fractionation is significantly different from that observed in the BIF samples.

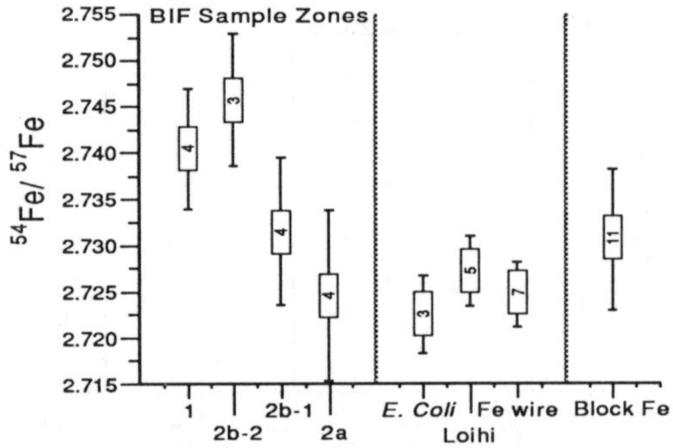

Fig.4 Iron isotopic ratio results for 54 mass to 57 mass. External 1s precision represented by vertical bar for each sample, for the indicated number of filaments analyzed.

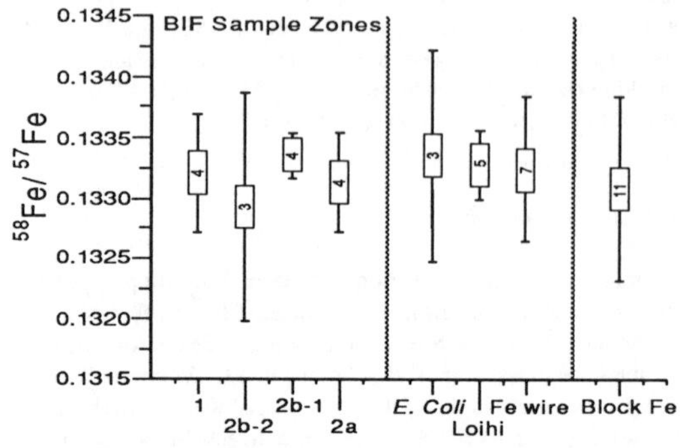

Figure 5: Iron isotopic ratio results for 58 mass to 57 mass. External 1s precision represented by vertical bar for each sample, for the indicated number of filaments analyzed.

4 CONCLUSIONS

We have successfully developed a set of procedures to analyze iron isotope compositions of terrestrial/biologic samples. In particular, this approach effectively eliminates Ni and Cr isobarics, interference from up mass tailing of alkali peaks, organic contaminants, and the necessity of internal isotopic spikes and normalization. These factors combined allow analysis of the major iron isotopes to a precision of ~0.1%. This effort has allowed the first observation of iron isotopic fractionation in terrestrial/biologic materials, both in ancient Banded Iron Formation and in a modern hydrothermal bacterial mat sample from Loihi volcano, Hawaii.

ACKNOWLEDGMENTS

These efforts have been funded by a LANL Laboratory Directed Research and Development grant and a LANL Director's funded Post Doc (PRD). P. DeBievre (CBNM) supported and facilitated the LANL/CBNM collaboration. G. McMurtry (Hawaii) provided the Loihi hydrothermal bacterial mat materials. We are also significantly benefiting from interactions with a growing number of collaborators and interested scientists inside and outside LANL.

REFERENCES

Douthitt, C.B. (1982) The geochemistry of the stable isotopes of silicon. *Geochimica Cosmochimica Acta* **46**, 1449-1458.

Farina, M., Esquivel, D.M., and Barros, H. (1990) Magnetic iron-sulfur crystals from magnetotactic microorganisms. *Nature*, **343**, 256-258.

Fassbinder, J.W., Stanjek, H. and Vali, H. (1990) Occurrence of magnetic bacteria in soil. *Nature*, **343**, 261-263.

Hoefs, J. (1973) "Stable Isotope Geochemistry," Springer-Verlag, New York, 140p.

Meack, R. (1992) The Absolute Aboundance of Iron Isotope. Unpublished Ph.D. Thesis, Central Bureau of Nuclear Measurements, Geel, Belgium.

Mann, S., Sparks, N., Frankel, R.B., Bazylinski, D.A., and Jannasch, H.W. (1990) Biomineralization of ferrimagnetic greigite (Fe_3S_4) and iron pyrite (FeS_2) in magnetotactic bacterium. *Nature*, **343**, 258-261.

Martin, J.H. (1991) Iron, Liebig's law and the greenhouse. *Oceanography*, **4**, 52-55.

Rashid, K., Krouse, H.R., and McCready, R.G.L. (1978) Selenium isotope fractionation during bacterial reduction. *US Geol. Survey Open-File Rept.* **78-701**, 347-348.

Sedwick, P.N., McMurtry, G.M. and MacDougall, J.D. (in press) Chemistry of Pele's vent warm springs, Loihi seamount, Hawaii. *Geochimica Cosmochimica Acta.*

Valley, J.D., Taylor, Jr., H.P., and O'Neil, J.R., Editors (1986) "Stable Isotopes in High Temperature Geological Processes," Reviews in Mineralogy Volume 16, Mineralogical Society of America, 570p.

Volkening, J. and Papanatassiou, D.A. (1989) Iron isotope anomalies. *The Astrophysical Journal*, **347**, L43-L46.

Water-Rock Interaction, Kharaka & Maest (eds) © 1992 Balkema, Rotterdam. ISBN 90 5410 075 3

Mantle derived helium in fluid inclusions of a 2.5 Ga old granulite, Nilgiri Hills, Southern India

T.J. Dunai
ETH-Zürich, Switzerland

J.R.L.Touret
Vrije Universiteit, Amsterdam, Netherlands

I.M.Villa
Istituto di Geocronologia CNR, Pisa, Italy (Presently: Universität Bern, Switzerland)

ABSTRACT: Excess ^3He was found in a 2.5 Ga old enderbite sample from Doddabetta, Nilgiri Hills, Southern India. The high $^3He/^4He$ ratios (up to $2.34R_a$) measured in the CO_2 inclusions, indicate a large portion of mantle derived He in our sample, as there is no evidence for a significant contribution of atmospheric, cosmogenic or nucleogenic ^3He. We conclude that the CO_2 has a common origin with the mantle He trapped in the fluid inclusions.

1. INTRODUCTION

Numerous hypotheses were proposed to explain the typical low a_{H2O} mineral assemblage of granulites (review by Newton et al. 1986). One of them suggests that pervasive influx of CO_2 into the lower crust is responsible for the dehydration of the granulites (Newton et al., 1980). A variation of this model, in which deep CO_2 is liberated by magma crystallization (Frost and Frost, 1987), has been proposed for South Indian granulites, which contain abundant high density CO_2 inlusions (Srikantappa et al., in press). The origin of this CO_2 remains unclear. Several explanations have been proposed (see Santosh et al. 1991) among which is a mantle origin of the CO_2 (Hoefs & Touret, 1975). The aim of the present study is to use helium as a tracer to test a mantle origin of the CO_2 in one of the South Indian granulites.

2. SAMPLE DESCRIPTION

We studied an enderbite sample from the Doddabetta quarry, Nilgiri Hills, which is known for its abundant high density CO_2 fluid inclusions. A detailed petrographic description and fluid inclusion data of this sample can be found in Touret and Hansteen (1988). They found that the CO_2 inclusions were trapped at or near the p-T conditions of the granulite formation and its isobaric cooling path.

The main target of our investigation were the garnets, as they contain numerous primary CO_2 inclusions and also retain He very well. Besides garnet, we studied a whole rock sample as well as plagioclase and pyroxene separates. We did not analyse quartz despite of its abundant secondary fluid inclusions, because this mineral does not retain He (Cerling, 1990). The same applies to plagioclase; nevertheless we measured a sample for screening purposes. The high helium retentivity of pyroxene for He qualified it for our measurements despite its lack of visible fluid inclusions. Due to the large grain size of the minerals (0.5-7mm) and the small size of our whole rock sample, the latter does not represent the bulk compositon of the enderbite.

3. ANALYTICAL PROCEDURE

We extracted He, Ne and Ar either by heating the samples to 1800ºC for 15min or by crushing in a hydraulic press. The extracted gases were cleaned by various getters. Ar was separated from He and Ne on two charcoal traps cooled with liquid nitrogen and Ne was subsequently frozen on a sintered metal filter (0.5µm grain size) at 14 K. Ne and Ar were measured in a Nier type mass spectrometer with 120 mm radius. The He spectrometer (r=270mm) is equipped with an Y-shaped flight tube with an electron multiplier used in ion counting mode for ^{3}He and a faraday cup for ^{4}He. The mass resolution of 750 (calculated for a valley of 50%) allows complete separation of $^3He^+$ and HD^+. Both mass spectrometers are equipped with a Baur-Signer ion source. The ^{3}He blank is about $1*10^{-14}$ccSTP. Here we report only the He data.

For complete extraction of the gases we heated samples to 1800ºC for 15 min. To discriminate noble gases residing in fluid inclusions from those in the lattice, we crushed several samples in an hydraulic press. Crushing of samples, to release ^{3}He trapped in fluid inclusions, is used by various groups measuring cosmogenic ^{3}He for surface exposure age dating (Marti & Craig, 1987; Cerling, 1990).

Our hydraulic press opens fluid inclusions quite effectively in coarse-grained or whole rock samples. However, mineral slabs such as the garnet sample removed from a doubly polished thin section used here may avoid complete destruction by distributing shear stress. Therefore crushing liberated only a small fraction of the He in fluid inclusions and for further calculations we consider only the isotopic composition of the He measured in the crushing experiment.

4. RESULTS AND INTERPRETATION

The results of our measurements expressed as $^3He/^4He$ ratios, absolute as well as relative to air (R/R_a) versus the amount of ^{4}He extracted are presented in Fig.1.

The most surprising feature of the data is that all minerals of this 2.5 Ga old rock (Peucat et al., 1989) have excess ^{3}He, which results in R/R_a values well above typical crustal values (0.01-0.02, Mamyrin & Tolstikhin, 1984). This requires the presence of a ^{3}He-rich component besides crustal He. Possible candidates are: i) cosmogenic $^3He_{cos}$, ii) nucleogenic $^3He_{nuc}$ (in minerals with high Li content), iii) atmospheric He_{atm} (R/R_a=1), iv) mantle He_{mantle} (MORB R/R_a=6-10, plume type $R/R_a \approx 30$).

$^3He_{cos}$ contributes significantly only if the sample was exposed to cosmic rays for more than 1000 years. The $^3He_{cos}$ production rate varies with altitude, latitude and overburden of the sample, but it is quite independent of the mineral chemistry (Cerling, 1989). Therefore the concentration of $^3He_{cos}$ in different minerals in a given sample should be the same. Thus, if all minerals in a given sample have the same initial (e.g. crustal) R/R_a value, this ratio will increase much faster in minerals with little inherited He than in those with much He. Hence, if $^3He_{cos}$ is responsible for the $^3He/^4He$ ratios obtained in our heating experiments, pyroxene should have higher R/R_a values than garnet. As the contrary is true we can exclude the presence of a significant amount of $^3He_{cos}$ in our sample.

Next we consider the results of the crushing experiments to distinguish between the three remaining He components. All the whole rock data and also the garnet point lie clearly above the atmospheric ratio (stippled line in Fig.1). This excludes atmospheric He as cause for the R/R_a ratios above the crustal values.

$^3He_{nuc}$ would be predominantely produced in the crystal lattice as it probably contains more Li than the CO_2 inclusions and would thus not be released by crushing. In heating experiments the garnet ratios are much lower than the ratios obtained in the crushing experiment. We therefore exclude unusually large amounts of $^3He_{nuc}$ in the garnet. This also applies to the whole rock sample.

The above considerations lead to the conclusion, that the observed $^3He/^4He$ ratios result from mixing of He_{mantle} (R/R_a=6-30) with He_{crust} (R/R_a=0.01-0.02). There are at least two scenarios describing this mixing:

1. The closing temperature for He in garnet is reached at time t_C, less than 100Ma after the entrapment (t_t). After t_C the He in the fluid inclusions retains its isotopic composition, whereas the He in the lattice evolves towards a crustal $^3He/^4He$ ratio. Therefore the He in the fluid inclusions we measure today represents the He at t_t (≈2.5Ga). In this case, mixing of He_{crust} and He_{mantle} must have occurred prior to entrapment, while He_{mantle} was ascending. Depending on the R/R_a value assumed for the tapped mantle reservoir (R/R_a=6-30), we obtain mixing ratios for He_{mantle}:He_{crust} between 1:2.5 and 1:12, respectively.

2. The closing temperature is only reached 0.1-2.5Ga after fluid inclusion entrapment (t_t). During that period the He in the fluid inclusions, as well as that in the garnet's lattice and presumably even in the whole rock has the same isotopic composition. Only after t_C the He in the fluid becomes isolated and stops evolving, whereas the $^3He/^4He$ ratio in the lattice decreases further. The later t_C is reached, the less mixing prior to t_t is needed to explain the measured $^3He/^4He$ ratios. If t_t-t_C reaches a certain value,

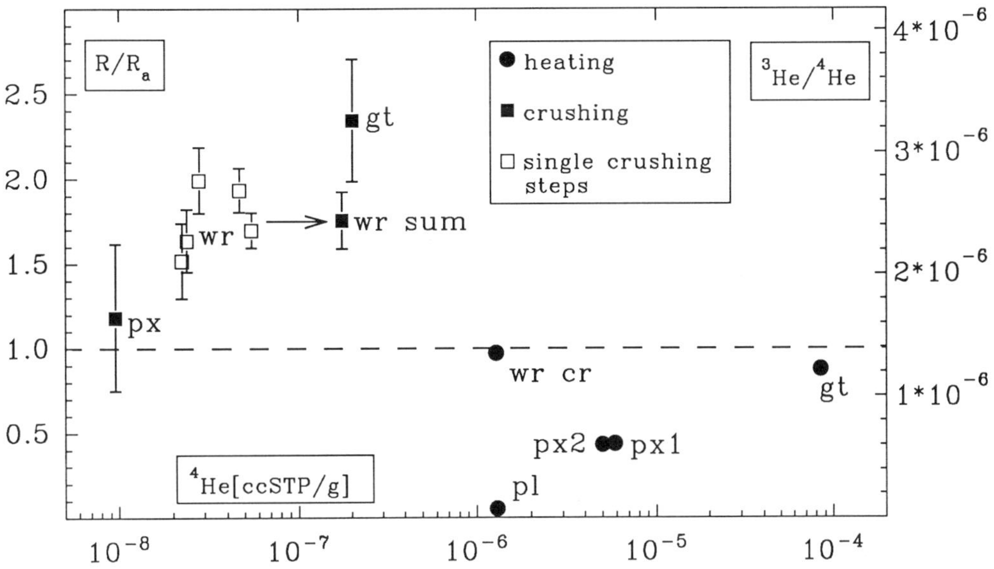

Fig. 1. Results of the heating and crushing experiments expressed as $^3He/^4He$ ratios, absolute and relative to atmospheric ratio R_a, versus the amout of extracted ^4He. The stippled line indicates the atmospheric $^3He/^4He$ ratio. Error bars show 1σ values. Explanations are given in the text. Abrevations used in the figure: wr=whole rock, wr cr=whole rock residual after crushing, wr sum=sum of wr crushing steps, px1 and px2=two pyroxene fractions, gt=garnet and pl=plagioclase

which depends on the contents of U, Th, Sm and Li, pure He_{mantle} must be trapped at t_t to explain todays He isotopic composition.

Scenario 2 is favoured by a Pan-African Ar-Ar age (uplift event?) determined in a plagioclase of our sample (t_C(Ar)<676Ma, saddle minimum). If we assume that this age dates a major cooling of the rock below T_C(He) for garnet, then probably almost pure He_{mantle} was trapped 2.5Ga ago.

But how did the He migrate from the mantle to the lower crust? To answer this question, we have to estimate the He concentrations in the fluid inclusions of the garnets. Using the R/R_a ratio of the fluid inclusions (2.34±.36) and the bulk R/R_a (0.876±.007) we calculate that 37±5% of the bulk He ($8.41*10^{-5}$ccSTP/g) resides in the fluid inclusions. Assuming an average CO_2 content in the garnet of 0.5wt% (estimated from Touret & Hansteen, 1988), the He concentration in the fluid inclusions today is roughly 1 ppm. As a trace element He can hardly have migrated on its own because active crack propagation by low He partial pressure is unlikely und diffusion is too slow. We expect therefore that He was transported by a carrier phase. It is most likely that the CO_2 in the fluid inclusions was this carrier, as there is no trace of an alternative carrier. Therefore at least a part of the CO_2 in our sample has a common mantle origin with He_{mantle} and thus mantle derived CO_2 was present during the peak of granulite metamorphism in this portion of the Nilgiri Hills.

5. CONCLUSIONS

1. The Doddabetta granulite contains excess ^3He. The high $^3He/^4He$ ratios of our sample cannot be attributed to cosmogenic, spallogenic or atmospheric He. Therefore mantle derived He must be the responsible ^3He rich component.

2. Mantle signatures of He may be preserved over long periods (>2.5Ga) of time, if large amounts of mantle He are trapped in fluid inclusions and if the host mineral retains He well.

3. The $^3He/^4He$ ratios in the fluid inclusions today are the results of mixtures between He_{crust} and He_{mantle}. Depending on the time of closure of the host mineral for He, mixing occurred either

prior to trapping, or after trapping in the host mineral. Presumably closing for He occurred late in the granulite evolution. If so, probably pure mantle helium was trapped during granulite metamorphism.

4. The mantle helium required a carrier phase to migrate from the mantle to the crust, as it is a trace element. In the absence of any alternative carrier, it is assumed that the carrier was CO_2 now occurring in the fluid inclusions. Therefore a part or even all of the CO_2 in the fluid inclusions has a common origin with He_{mantle}.

REFERENCES

Cerling, T.E. 1990. Dating of geomorphologic surfaces using cosmogenic ^{3}He. *Quatern. Res.* 33: 148-156.

Frost, B.R. & Frost, C.D. 1987. CO_2, melts and granulite metamorphism, *Nature* 327: 503-506.

Hoefs, J. & Touret. J.R.L. 1975. Fluid inclusion and carbon isotope study from Bamble granulites (S. Norway). *Contrib. Mineral. Petrol.* 52: 165-174.

Mamyrin, B.A. & Tolstikhin, I.N. 1984. *Helium isotopes in nature.* Amsterdam: Elsevier

Marti, K. & Craig. H. 1987. Cosmic ray produced neon and helium in the summit lavas of Maui. *Nature* 325: 335-337

Newton, R.C., Smith, J.V. & Windley, B.F. 1980. Carbonic metamorphism, granulites and crustal growth. *Nature* 288: 45-50.

Newton, R.C. 1986. Fluids of granulite metamorphism. In: J.M. Walther & B.J. Wood (eds.) *Fluid rock interaction during metamorphism. Advances in Physical Geochemistry 5*: 36-59. Berlin: Springer.

Peucat, J.J., Vidal, P., Bernard-Griffith, J. & Condie K.C. 1989. Sr, Nd and Pb isotopic systematics in the Archean low- to high-grade transition zone of Southern India: Syn-accretion vs. post-accretion granulites. *J. Geol.* 97: 537-550

Santosh, M., Jackson, D.H., Harris, N.B.W. & Mattey, D.P. 1991. Carbonic fluid inclusions in South Indian granulites: evidence for entrapment during charnockite formation. *Contrib. Mineral. Petrol.* 108: 318-330.

Srikantappa, C., Raith, M. & Touret J.R.L. (in press). Synmetamorphic high density fluid inclusions in the lower crust: evidence from the Nilgiri granulites, Southern India. *Contrib. Mineral. Petrol.*

Touret, J.R.L. & Hansteen T.H. 1988. Geothermobarometry and fluid inclusions in a rock from the Doddabetta charnockite complex, Southwest India. *Rend. Soc. Ital. Mineral. Petrol.* 43: 65-82.

Water-Rock Interaction, Kharaka & Maest (eds) © 1992 Balkema, Rotterdam. ISBN 90 5410 075 3

Noble gas geochemistry of mineralization in the Pennines

Jacqueline L. Ford, Finlay M. Stuart & Grenville Turner
Department of Geology, University of Manchester, UK

ABSTRACT : The measurement of noble gas abundances can be used to identify sources and interactions of groundwaters which have subsequently become trapped in fluid inclusions. Quartz, fluorite and baryte samples were collected from Mississippi Valley-type mineralization in the North and South Pennines, to determine the possible influence the Weardale granite had on the North Pennine mineralizing fluids which ascended through this unit. He, Ar, Kr and Xe isotopes measured show a distinct similarity between the orefluids of the two orefields. He concentrations in the fluids can be accounted for by simple crustal production. Excess ^{40}Ar and Xe can be accounted for in both fields due to the permeation of the orefluids through shales in the sedimentary source areas.

1 INTRODUCTION

The chemical inertness of the noble gases He, Ne, Ar, Kr and Xe makes them ideal physical tracers of geological fluid processes. Measuring the elemental and isotopic abundances can potentially be used to identify sources and interactions of hydrothermal fluids which have subsequently become trapped in fluid inclusions. The relative abundances of dissolved atmospheric gases in groundwater depends on the temperature and salinity of recharge. This may be futher modified by physical processes such as gas, oil and water partitioning, and by the introduction of radiogenic gases from the crust and ^{3}He from the mantle.

Research in progress is being used to assess the use of noble gases in the characterization of ancient hydrothermal fluids in ore deposits. Little systematic work has been carried out in this area, thus deposits chosen need to contain well characterized fluids. The fluids responsible for the Pennine mineralization have been characterized previously by fluid inclusion and isotope studies. This paper compares and contrasts the noble gas signatures of mineralization in the North and South Pennines. Results have been used to look at the possible influence of the Weardale granite which underlies the North Pennine orefield.

2 GEOLOGICAL BACKGROUND

The Pennine Orefields are situated in northern England as shown in figure 1. The North Pennines, comprising the Alston and Askrigg Orefields are areas of mineralization overlying tectonically stable, fault -bounded basement blocks (Dunham,1990; Ixer & Vaughan, 1992). Stratigraphy comprises a

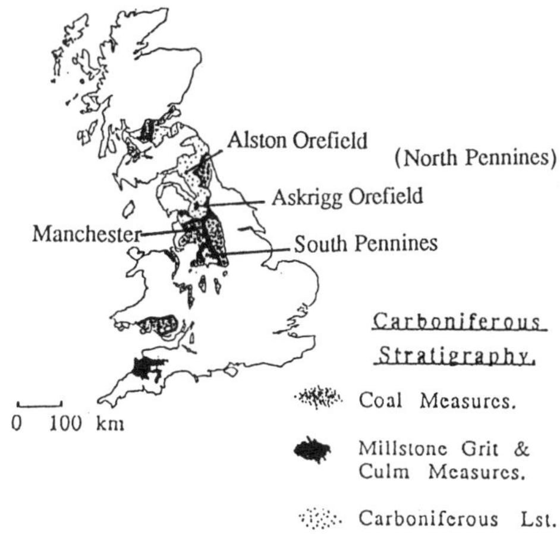

Figure 1. Position of the Pennine orefields.

basement of Lower Palaeozoic slates, volcanics and intrusives, which in turn have been intruded by the Weardale and Wensleydale granites, lying beneath the Alston and Askrigg fields respectively. The granites are Caledonian in age (410 ± 10 Ma), unconformably overlain by the Lower Carboniferous limestones, sandstones and shales, that host the ore deposits. A series of dyke and sheet-like structures collectively referred to as the Whin Sill was intruded into this sequence during late Carboniferous to early Permian times; the mineralization postdates these intrusions. Fluorite-baryte-Pb-Zn mineralization within the Lower Carboniferous carbonates occurs predominantly as narrow fissure veins. Fluorite veins are restricted to Dinantian and Namurian units,

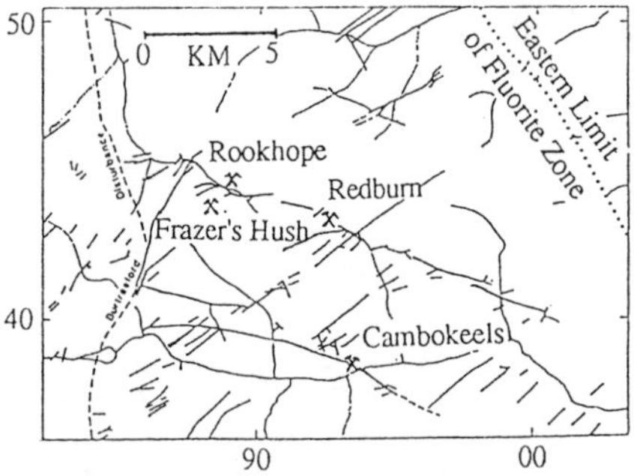

Figure 2. North Pennine sample locations (map after Rankin & Graham, 1988).

baryte mineralization is more ubiquitous. The Alston Block displays a distinct mineralogical zonation, which centres on cupolas of the Weardale granite, in Rookhope and Tynehead. A central fluorite zone passes abruptly outwards into a barium minerals zone (baryte and witherite). Sulphide mineralogy reflects this zonation also, as a chalcopyrite-pyrite-pyrrhotite assemblage passes out into a galena-sphalerite assemblage. Figure 2 shows sample locations within the Alston orefield.

South Pennine mineralization occurs in equivalent Carboniferous units further south in Derbyshire. The orefield occupies an upland area known as the Derbyshire Dome (Rogers, 1977). The major veins or 'rakes' occur in E-W trending faults. The mineralization comprises baryte-calcite-galena-sphalerite-fluorite, and shows a poor mineral zonation. Fluorite is restricted to the eastern edge of the Orefield, with an increase in baryte and calcite to the west. The location of the sample area in the South Pennine Orefield is shown in figure 3.

The North Pennine Orefields are recognised as being Mississippi Valley-Type Pb-Zn deposits (Dunham, 1983), similar to the Illinois and Kentucky deposits. The South Pennines are a more distal manifestation of this type of mineralization, being lower in temperature. In the North Pennines the granites acted as foci for the emanative fluids, no such units are present at depth in the South Pennines.

3 FLUID HISTORY

Isotopic data suggest their to be two orefluids responsible for the North Pennine mineralization: a shallow (>1km) connate brine derived from Lower Carboniferous sea water; and a deeper connate groundwater derived from deep sediments, which is richer in chloride than the shallower fluids (Solomon et al., 1971). Salinities of the fluids have been determined by fluid inclusion analyses which give, 7-27 wt.% NaCl equiv. for the North Pennines (Sawkins, 1966), and a comparable 18-25 wt. % NaCl equiv. for the South Pennines (Rogers, 1977). Homogenization temperatures of fluorite in the South Pennines are between 103-165°C. The North Pennine data shows the presence of hot-spots within

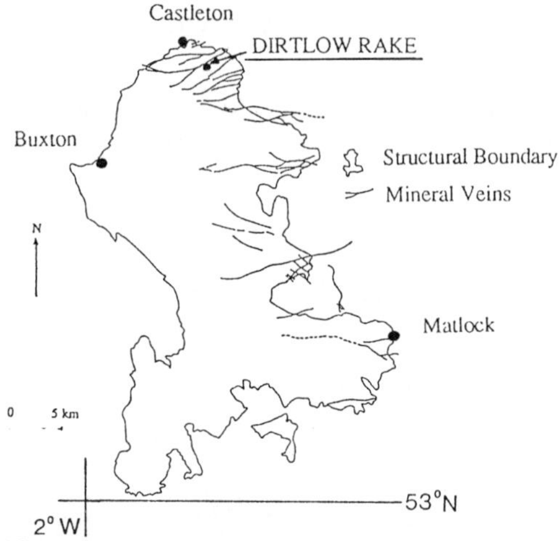

Figure 3. South Pennine sample location.

the fluorite zone (eg. the Great Sulphur Vein at 175-216°C.), which are thought to represent feeder zones through which the hydrothermal fluids rose. Temperatures decrease outwards from these centres, to as low as 50°C in the baryte zone.

4 TECHNIQUE AND PROCEDURE

Noble gas elemental and isotopic abundances in fluid inclusions were determined using a MAP 215 mass spectrometer (Mass Analyser Products, Northwich, Cheshire, U.K). Samples of quartz, fluorite and baryte were crushed with a stainless steel pestle and mortar and sieved. The 0.5 to 1mm size fraction was hand picked to eliminate contamination from other mineral species and to remove shard-like grains which do not crush readily. Approximately 0.5g of each sample was used. Gas is extracted from samples using crushing apparatus, which are constructed from modified Nupro vacuum taps. The valve seats are bored out to a diameter of 1.6cm and stainless steel mortars manufactured to fit this space. Samples were crushed by the closing of the valves which lowers a steel pestle into the mortar.

Active gases released during the crushing process were removed by reaction for twenty minutes with a hot (250°C) Zr-Al alloy getter. The Ar, Kr and Xe fraction is separated from the He and Ne by trapping on activated charcoal at -196°C. A second getter, maintained at room temperature, removed hydrogen from the He and Ne fraction prior to its admission into the spectrometer. A second charcoal finger close to the spectrometer inlet, is held at -196°C to minimise the pressure of argon in the spectrometer during the analyses of the He and Ne.

The Ar, Kr and Xe fraction is released by heating the first charcoal finger to 100°C and analysed after the He and Ne fraction. Data produced was corrected for background blank levels and mass discrimination using the blank and calibration measurements made before and after each crush.

5 RESULTS

Fluid inclusions in all the samples are dominantly two-phase liquid + vapour aqueous inclusions ranging in size from 5 to 50 µm. Primary inclusions are noted in all the mineral samples and are volumetrically more important than the secondary and pseudosecondary inclusions. Secondary inclusions are more prevalent in baryte than in the quartz and fluorite samples. As crushing averages the total fluid content, and is unable to distinguish different generations of inclusions, samples were carefully chosen so as to best represent the primary mineralizing fluids. Thermometric data from Cambokeels shows there to be no significant difference in salinity or homogenization temperature between the primary and secondary fluid inclusions in the quartz and fluorite (Bannon, 1989).

5.1 *Helium Isotope Ratios.*

$^3He/^4He$ ratios provide a useful signature to distinguish the relative importance of volatiles in flux from mantle and crustal processes. The ratio is reported as R/Ra where R is the $^3He/^4He$ ratio of the sample, and Ra is the ratio measured in air. Determinations of R/Ra at oceanic ridges indicate $R/Ra = 8 \pm 1$ is typical of the upper mantle. R/Ra for crustal (radiogenic) helium is determined by Li content and is typically in the range of 0.01 to 0.05. The fluorite samples released from 10 to 100 times more ^4He than the quartz samples, a consequence of larger and more abundant fluid inclusions in the fluorite. The $^3He/^4He$ ratios for all but one of the North Pennines samples cluster tightly around a value of 0.08. The $^3He/^4He$ ratios for the South Pennine fluorites are centred on 0.06 overlapping the range observed in the North (figure 4). The values for both the orefields are at the high end of crustally derived radiogenic helium, either as a result of high Li content in the source rocks (cf. Rankin and Graham, 1988), or the presence of a minor mantle component. The measured ^4He contents are higher than those observed in recent recharge waters, being more comparable to pore waters from deep within granite plutons and sedimentary basins. The correlation between the North and South Pennine data, and the absence of a granitic body in the south, suggests that He is probably derived from deep sedimentary units. Both ^3He and ^4He are produced in such aquifer rocks; ^4He by the decay of U and Th, ^3He from neutron interactions with ^6Li.

5.2 *Argon Isotope Ratios.*

An increase in the $^{40}Ar/^{36}Ar$ ratio above that of the atmospheric value of 295.5 arises from the presence of radiogenic ^{40}Ar released from the crust and is a function of the amount of time the fluid has spent in the system, and the age and composition of the rocks with which it has interacted. The North Pennine data ranges from 320 to 520, with the South Pennine data

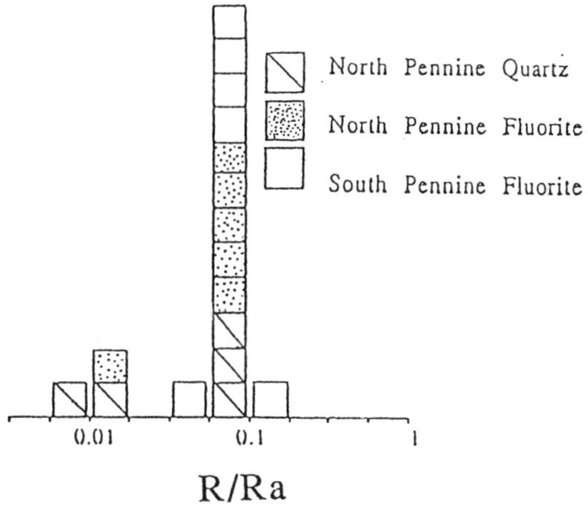

Figure 4. R/Ra ratios for North and South Pennines.

points slightly lower at 300 to 320. These values are typical of the range found in modern deep basinal brines. All of the samples were found to contain excess ^{40}Ar ($^{40}Ar^*$). Of the potential sources for this $^{40}Ar^*$, local Carboniferous shales are the most likely. Pore fluids expelled from such shales can accumulate enough radiogenic ^{40}Ar to account for the $^{40}Ar^*$. Its release into the fluid may result in part from the conversion of detrital K feldspars within the shales to illite. The granite underlying the North Pennine veins has been pervasively altered, at which time $^{40}Ar^*$ may have been released into the fluids during the destruction of K-feldspar and biotite. Much of this alteration predates the mineralization and it is not possible to say whether this process contributes significantly to the $^{40}Ar^*$ observed.

5.3 *$^{40}Ar^*/^4He$ Ratios.*

The ratio of ^4He and excess $^{40}Ar^*$ is used to provide information concerning the (U/Th)/K ratio of the source region. The ratios for the North Pennines show a strong separation between the quartz and fluorite values. It is not easy to account for this difference in terms of differences in the sources of the fluids since we know from thermometric data that the fluids trapped in the two minerals are very similar. ^4He has a higher diffusion coefficient in quartz than in fluorite, thus loss of He from the quartz is the most likely cause. Figure 5 shows the correlation between the radiogenic $^{40}Ar^*$ and ^4He in the fluorite samples from the two Orefields. The $^{40}Ar^*/^4He$ ratio is an order of magnitude lower than the production ratio and suggests that it is controlled by preferential loss of ^4He from the source rocks. This in turn suggests control by differential diffusive loss of He and Ar at relatively low temperatures. Release by chemical dissolution or igneous heating would give rise to a ratio closer to the average crustal production ratio. Comparison of the ratios from the inclusion fluids derived from the North and South

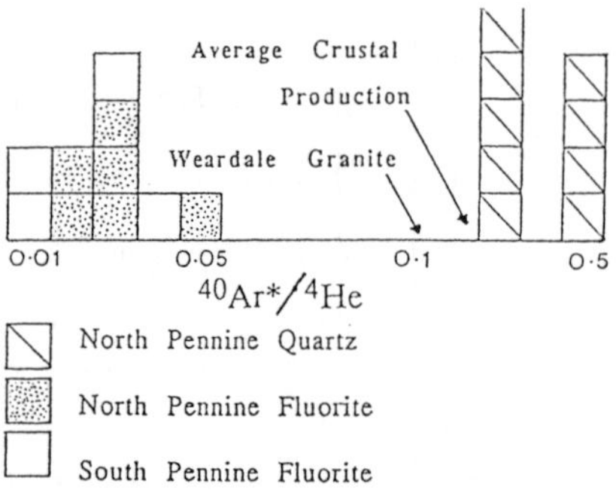

Figure 5. $^{40}Ar^{*}/^{4}He$ ratios.

Pennine ores shows a similarity which argues against any obvious involvement of the Weardale granite.

5.4 *Kr /Ar and Xe /Ar Ratio.*

Most of the budget of the noble gases in natural groundwaters arises from dissolution from the atmosphere. The solubility of the heavier noble gases increases with mass such that Ar<Kr<Xe, and Kr/Ar and Xe/Ar ratios can be used to test for an origin in air -saturated water. The results for fluorite and baryte from the South Pennines are plotted in figure 6 as F-values, in which the Kr/Ar and Xe/Ar ratios normalised to the atmospheric ratios. The baryte samples lie on a simple mixing line between air and air -saturated water. The fluorite samples have a higher proportion of air and in two cases excess xenon. The (unfractionated) air component may arise from a recent (subaerial) trapping and is probably unrelated to the inclusion fluids. The high noble gas content of shales makes them a possible source of excess Xe in deep sedimentary brines.

6 CONCLUSIONS

The fluids responsible for the mineralization in the North and South Pennine orefluids essentially have noble gas isotopic compositions which are characteristic of long-lived deep crustal fluids. The sources of the gases can be attributed to normal crustal production of radiogenic gases, with some possible additional involvement of noble gas-rich Lower Carboniferous shales in adjacent sedimentary basins. Average R/Ra values for the North and South Pennines of 0.079 ± 0.007 and 0.062 ± 0.006 respectively, are indistinguishable and typical of crustal production ratios with no detectable mantle component. Radiogenic $^{40}Ar^{*}/^{4}He$ ratios are an order of magnetude less than the typical production ratios, suggesting control by diffusive loss from the parent minerals at relatively low temperatures. Kr/Ar and Xe/Ar ratios are intermediate between air and air -saturated water. The air component possibly reflects recent trapping of (modern) atmosphere with no geological significance. Kr/Ar and Xe/Ar ratios

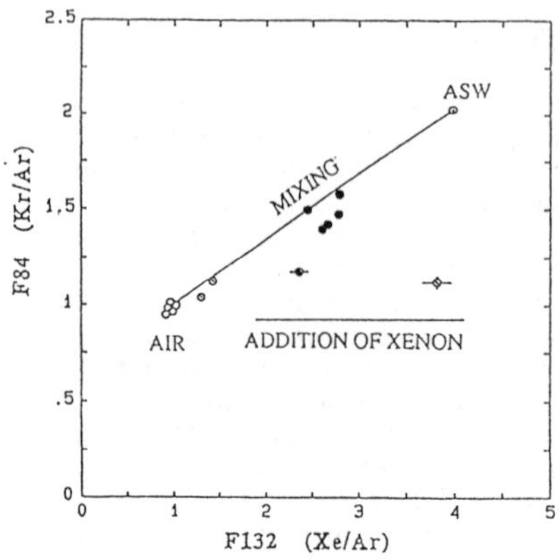

Figure 6. Kr/Ar v Xe/Ar ratios: ● baryte; ⊖ fluorite; ⊙ ASW at 10°C; o air calibrations (South Pennines).

in baryte are closest to air saturated water but contain little or no radiogenic component. The baryte data is probably dominated by late stage secondary fluid inclusions. As yet it is not possible to totally rule out the Weardale granite as a contributor of gases in the North. Overall there is no obvious difference between the North and the South Pennine data which could be ascribed to the involvement in the North Pennines of the Weardale granite, other than a pathway for the fluids.

REFERENCES.

Bannon, M.P. 1989. *Argon isotopes of fluid inclusions in quartz and fluorite from areas of mineralization.* Unpublished Phd Thesis University of Sheffield.

Dunham, K.C. 1990. *Geology of the North Pennine Orefields. Vol.1 Tyne to Stainmore.* Mem. Geol. Surv. 2nd Ed.

Dunham, K.C. 1983. Ore genesis in the English Pennines: A fluoritic subtype. *International conf. on M.V.T Pb-Zn deposits,* 86-112, University Missouri Press.

Ixer, R.A. & D.J.Vaughan. 1992. Pb-Zn-fluorite-baryte deposits in: *Mineralization in the British Isles* Chapman & Hall, London.

Rankin, A.H. & M.J.Graham. 1988. Na, K and Li contents of mineralizing fluids in the Northern Pennine orefield, England, and their genetic significance. *Trans I.M.M. 97, B99-B107.*

Rogers, P.J. 1977. Fluid inclusion studies in fluorite from the Derbyshire orefield. *Trans I.M.M. 86, B128-B132.*

Sawkins, F.J. 1966. Ore genesis in the North Pennine orefield in the light of fluid inclusion studies. *Econ. Geol. 61, 385-401.*

Solomon, M. et al. 1971. Sulphur and oxygen isotope studies in the North Pennines in relation to ore genesis. *Trans. I.M.M. 80, B259-B275.*

Water-Rock Interaction, Kharaka & Maest (eds) © 1992 Balkema, Rotterdam. ISBN 90 5410 075 3

Geochemical modelling of Sr isotopic signatures in the interaction between granitic rocks and natural solutions

B. Fritz & L. Richard
Centre de Géochimie de la Surface, CNRS, Université Louis Pasteur, Strasbourg, France

R. H. McNutt
Department of Geology, University of McMaster, Hamilton, Ont., Canada

ABSTRACT : The geochemical code Kindis has been extended in order to predict isotopic signatures of solutions and secondary minerals during interactions between natural solutions and rock-forming minerals each containing a specific strontium amount and 87/86 Sr signature. The simulations show very interesting results during the granite-dilute solution interaction, with an initial rapid increase of the 87Sr/86Sr ratio (0.710 to about 0.740 in several days) and then a decrease on longer term (months to years) down to 0.710 and even less . These values fit with data obtained on a natural weathering system. The effect of an increase of temperature from 25 to 100°C is also described. In these simulations, the necessity to combine thermodynamic and kinetic approaches for geochemical modelling is clearly underlined.

1 INTRODUCTION

The geochemical modelling approach has been extensively developped during the last twenty years for calculating the evolution of rock-water systems. This approach is of great interest for the geochemists who try to explain how rock-water systems behave as chemical reactors because of the frequent thermodynamic unstability related to the fluids which circulate in rocks in the water cycle on the surface of the Earth. The way these unstabilities are solved by natural systems which always tend to recover equilibrium conditions depends first on thermodynamic conditions for the reactant and product minerals and on kinetic rate laws. In this general approach, models used to refer mostly to major chemical compounds without considering isotopes. In order to explain or predict isotopic signatures, we have adapted existing codes by combining thermodynamics and kinetics of mineral-solution interactions (Madé 1990; Madé and Fritz this issue). This paper will only describe the results concerning the approach of strontium isotopes in the weathering of silicate minerals.

2 THE MODELLING APPROACH

The Model used for the simulation is the geochemical code Kindis (Madé et al. 1990; Madé 1991; Madé and Fritz this issue), which allows different reaction rates of dissolution for silicate minerals in contact with aqueous solutions to be calculated.

In this code we introduced the possibility to describe the behaviour of strontium major isotopes (Sr 88,87,86) as the 87Sr/86Sr ratio seems to be a good indicator of water-rock interaction in very different geochemical environments and particularly under weathering conditions (Clauer 1981, Mc Nutt et al. 1984).

The way to introduce these isotopes in the model is the following : 1. Each host mineral considered in the reacting rock contains a specific amount of total Sr with specific 87 and 86 Sr signatures. During the simulated dissolution of the rock, the composition of each reactant mineral is fixed for all elements including isotopes minerals. A specific kinetic rate law is calculated for each rock-forming mineral.

2. Secondary minerals predicted by the simulation may contain strontium. This can be taken into account in the model using either given Sr content and 87Sr/86Sr ratio in secondary phases predefined in the data bank or a solid solution model for a X-Sr substitution (Ca-Sr in calcite or clays as an example). To obtain these possibilities in the modelling tools, the standard software package KINDIS has been significantly transformed (Fritz et al. 1992, in prep.).

3 SIMULATION OF GRANITE-WATER INTERACTION UNDER WEATHERING CONDITIONS. RESULTS AND DISCUSSION.

3.1. The reacting rock

The composition of the granite considered in the simulated weathering process is given in table 1 in terms of surface of the different minerals in contact with a given amount of solution (1kg H2O) and reaction rates for dissolution. Each mineral contains a specific amount of Sr and a given 87Sr/86Sr signature.

Table 1. The reacting rock : dissolution rates, interaction surfaces, Sr content and isotopic composition of the minerals.

mineral	k_{dm}	surface	Sr	$^{87}Sr/^{86}Sr$
quartz	$4.998X10^{-7}$	9817.1	-	-
microcline	$5.872X10^{-5}$	63484.4	80	0.8764
albite	$9.973X10^{-5}$	74573.6	69	0.7057
anorthite	$3.154X10^{-4}$	18740.7	69	0.7057
biotite	$8.073X10^{-6}$	33384.2	3	2.3495

k_{dm} in mole.$m^{-1}.y^{-1}$, surface in cm^2, and Sr in ppm.

3.2 The reacting solution

The weathering solution is supposed to be a rain water type diluted solution (pH = 5.5 or 4.5 for an acid rain water) in equilibrium with the atmospheric CO2 (log pCO2 = -3.5). The total dissolved salts of these initial solutions are very low (TDS = 12 mg/l).

Near surface weathering processes are considered here for a temperature of 25°C.

3.3 Results of the simulations

The complete set of chemical results will not be detailed here. It allows to describe the predicted stability sequence of reactant minerals with successive saturations of quartz and K-spar, long term unstability of plagioclases (albite, anorthite end-members) and micas. The sequence of formation of secondary minerals is also described with successively, goethite, kaolinite, a complex 2:1 clay phase, and finally calcite.

The evolution of the solution chemistry is also followed in the modelling, showing a general increase of pH, alkalinity and major cation concentrations, if we exclude those which are early precipitated (Al, Fe).

3.4 The evolution of strontium in solution: Thermodynamic or kinetic modelling ?

We will focus in this paper on the results concerning strontium. For all simulations of interaction between diluted initial solutions and the granite, the evolution trend for the strontium was the same : the solution, initially poor in strontium (less than 5 ppb) with a marine type signature (87Sr/86Sr = 0.710) is enriched in strontium by the dissolution of micas, feldspars and plagioclases.

The 87Sr/86Sr isotopic ratio increases initially as long as the solution is able to dissolve the potassium feldspar. When the solution becomes saturated with respect to this mineral phase the same isotopic ratio decreases strongly (Fig. 1 to 3). This result has been obtained using either the model Dissol (thermodynamic model precursor of Kindis) without kinetic effects on the dissolution process, or the model Kindis with kinetic laws for the minerals dissolution. The two curves are represented on figure 1. If the trends of evolution are comparable, we will see later that the isotopic ratios calculated on a kinetic base appeared systematically more realistic if compared to field data (see also Probst et al., this issue) and we will now consider only the results of kinetic modelling which give in addition an estimation of time dependance of the corresponding processes.

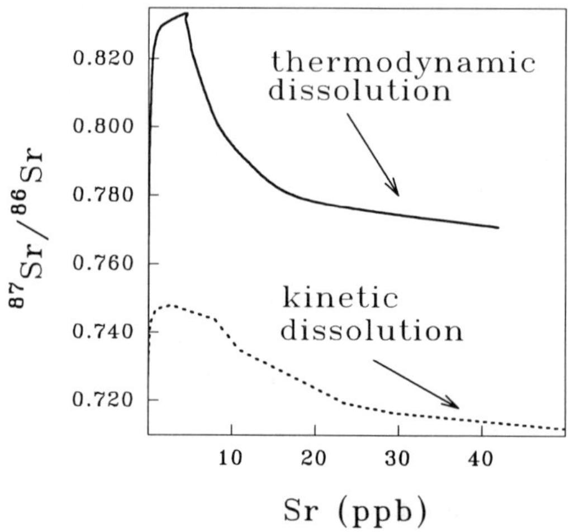

Fig. 1 Evolution of the 87Sr/86Sr isotopic ratio versus total Sr in solution in both thermodynamic and kinetic modellings.

3.5 Time dependance of the dissolution

The evolution of aqueous strontium is detailed in Figure 2 with comparison to a natural water collected in a granitic area (Probst et al., this issue). This point is representative of about ten samples in the same area (springs, small streams in granitic watersheds). The simulated curve (solid line) shows the initial increase of isotopic ratio from the initial value in rain water (0.710) up to a value of about 0.741. This part of the curve corresponds to the dissolution of all minerals from the granite with a major contribution of K-spar (87Sr/86Sr = 0.87) to the radiogenic strontium signature of the solution. The time required to reach the solution saturation point with respect to K-spar is about a week.

After this initial stage of the reaction the dissolution continues without any contribution of K-spar and the main contributor to the evolution of Sr in the solution is now the plagioclase phase (0.707) which lowers progressively the 87Sr/86Sr ratio in the solution down to values as low as in the initial sea water type rain water (0.710) and even less. In this part of the calculated curve the time dependance of the process is estimated in months of reaction. The solution becomes saturated successively with respect to kaolinite and a 2:1 clay phase.

The data obtained in the field for rain water and one typical water from a stream draining a granitic watershed are plotted on Figure 2. The river water Sr content and signature result from the mixing between a weathering solution and the local rain water (dotted line). Several solutions collected in the north-east of France on granitic drainage basins (Probst et al., 1990) confirm this type of Sr content (Probst et al., this issue).

4 SIMULATION OF GRANITE-WATER INTERACTION TEMPERATURE. RESULTS AND DISCUSSION

4.1 The reacting rock

The calculations were repeated at 60 and 100°C using the same initial granite composition. The kinetic constants kdm are deduced from data at 25°C using Arrhenius Law (Madé 1990 and Madé et al.,this issue).

4.2 The reacting solution

The same diluted initial solution is used for these calculations (pH = 5.5) with the same pCO2 value (log pCO2 = -3.5).

4.3 Results of simulations at 60 and 100°C.

Here again we will focuse only on the results concerning the strontium as tracer of the alteration process. The calculations show a chemical trend for strontium similar to that obtained at 25°C with an interesting evolution tendancy as the temperature increases up to 100°C, as shown on figure 3.

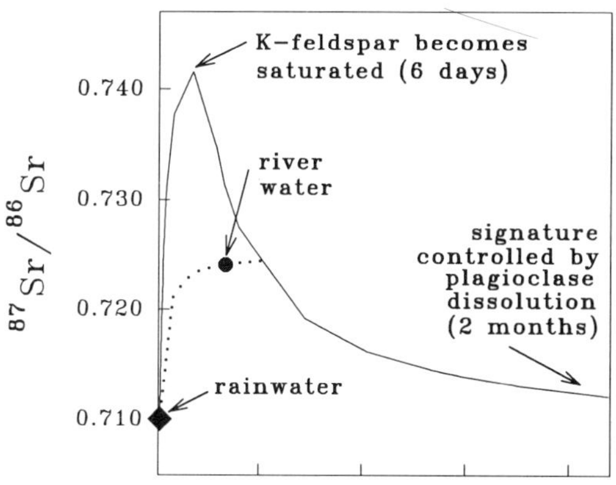

Figure 2. Evolution of the strontium characteristics of the solution during the simulated dissolution of granite.

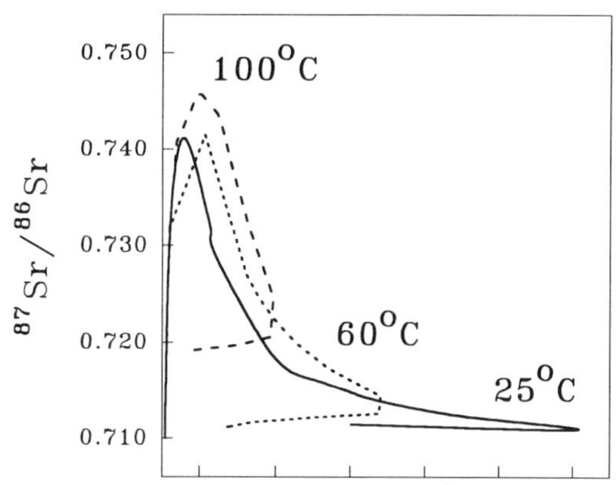

Figure 3. Strontium characteristics of the solution during the granite simulated dissolution at 60 and 100°C.

The rapid increase of 87Sr/86Sr in the initial alteration stage is confirmed at higher temperature and even enhanced, as is predicted a decrease of the same ratio in the following stages after saturation with respect to K-spar in the reacting solution.

However, the decrease of the 87Sr/86Sr ratio seems to be less important with increasing temperature and one additional saturation state had been reached by the solutions at the end of these simulations. It concerns the calcite which is here considered as a binary solid solution with a Ca-Sr substitution. The precipitation of calcite containing Sr as a minor element is responsible of the decrease of the total amount of aqueous Sr (Fig. 3) while the 87Sr/86Sr seems to be buffered at a fixed value slightly increasing with temperature from about 0.710 at 25°C and 0.720 at 100°C.

This result from the simulation is particularly interesting if it can be validated by experimental data or field data from hydrothermal systems. It could be a kind of geothermometer by using the Sr signature of calcite formed in weathering or alteration processes in granitic rocks. The Sr measurements on secondary phases could be used for estimating the tenperature of precipitation of these phases provided the strontium isotopic signature and content are known for the feldspars and micas of the host granitic rock. Additional research is being done to assess this preliminary result. The origin of this effect on strontium has to be found in the evolution with temperature of the kinetic constants kdm for feldspars and micas respectively. The large differences present at 25°C between plagioclases K-spars and micas tend to decrease with increasing temperature.

5 CONCLUSIONS

The water rock interaction modelling code Kindis has been adapted and is able to take into account Sr content and isotopic signatures for both mineral phases and aqueous solutions. The application to granite-dilute solutions interactions shows that the modelling has to be done on thermodynamic and kinetic bases to be efficient. It is then able to predict the behavior of strontium in solutions from springs and small rivers in granitic weathering environment. A predicted initial stage of rapid increase of 87Sr/86Sr ratio has to be validated by experimental or field approach. The temperature effect between 25 and 100°C could be used as a possible geothermometer from the Sr signature of secondary phases in rock-solution alteration.

REFERENCES

Clauer, N., 1981. Strontium and argon isotopes in naturally weathered biotites, muscovites and feldspars. Chem. Geol.,31: 325-334.

Madé, B., Clément, A. & Fritz, B. 1990. Modélisation thermodynamique et cinétique de l'altération : le modèle KINDIS. C.R. Acad. Sci. Paris, 310 : 31-36.

Madé, B. 1991. Modélisation thermodynamique et cinétique des réactions géochimiques dans les interactions eau-roche. Thèse, Univ. L. Pasteur,Strasbourg,308p.

Madé, B. & Fritz, B. 1992. Theoretical approach and modelling of the dissolution and precipitation of minerals under kinetic controls. WRI-7 Symposium. This issue.

Mc Nutt, R.H., Frape, S.K., & Fritz, P. 1984. Strontium isotopic composition of some brines from the Precambrian Shield of Canada. Isotope Geosci. 2: 494-505.

Probst, A., Dambrine, E., Viville, D. & Fritz, B. 1990. Influence of acid atmospheric inputs on surface water chemistry and mineral fluxes in a declining spruce stand within a small granitic catchment (Vosges Massif, France). J. Hydrol., 116: 101-124.

Probst, A., Fritz, B. & Stille, P. 1992. Consequence of acid deposition on natural weathering processes : field studies and modelling. WRI-7 symposium. This issue.

Water-Rock Interaction, Kharaka & Maest (eds) © 1992 Balkema, Rotterdam. ISBN 90 5410 075 3

Chlorine-36 in groundwaters in the Lac du Bonnet granite, southeastern Manitoba, Canada

M.Gascoyne & D.C.Kamineni
AECL Research, Whiteshell Laboratories, Pinawa, Man., Canada

J.Fabryka-Martin
Isotope Geochemistry Group, Los Alamos National Laboratory, N.Mex., USA

ABSTRACT: The values of $^{36}Cl/Cl$ ratios in brackish and saline groundwaters in a granite on the Canadian Shield have been found to be relatively constant (40-60 x 10^{-15}) irrespective of the Cl concentration. Calculations of in situ production rates of ^{36}Cl give comparable values. These results and rock leaching evidence indicate that most of the Cl in the groundwater is derived from the soluble salts in the rock matrix.

INTRODUCTION

The hydrochemistry of groundwaters in the Lac du Bonnet granite batholith, southeastern Manitoba, is being determined as part of the geoscience studies done by AECL Research for the Canadian Nuclear Fuel Waste Management Program. The Program has been described by Dormuth and Nuttall (1987) and details of the geoscience research have been given by Whitaker (1987).

Groundwaters from depths of up to 1 km in permeable fractures and fault zones in the granite have been analysed for major and minor ion composition and isotopic content to determine the source of dissolved salts, groundwater residence time and principal rock-water interactions. Some results have already been given by Gascoyne et al. (1987, 1989a,b). This paper describes the results of analyses of the ^{36}Cl content of selected groundwaters from the batholith and compares the results with expected levels of ^{36}Cl produced naturally by nuclear reactions in the granite.

METHODS

Groundwaters were selected for analysis for ^{36}Cl so that a wide range of salinities (mainly caused by the presence of Na, Ca and Cl) and low ^{3}H levels (indicating minimal drillwater contamination) were represented. Some of these groundwaters were especially of interest because of their location in the groundwater flow system (i.e., in recharge and discharge areas).

Groundwaters were analysed for their ionic composition by standard analytical techniques by AECL Research, Pinawa, and their ^{3}H content was determined under contract by the University of Waterloo Isotope Laboratory using ^{3}H enrichment by electrolysis. The atomic abundance ratio, ^{36}Cl/total Cl, was determined under contract by accelerator mass spectrometry at the University of Rochester, New York.

RESULTS

Table 1 shows the $^{36}Cl/Cl$ ratios and Cl and ^{3}H concentrations of groundwaters from 19 permeable zones in boreholes in the Lac du Bonnet granite. Tritium concentrations are low or below detection limits in most groundwaters. Contamination by residual drillwater or flow within the borehole before the permeable zones were isolated are the probable explanations for ^{3}H levels above 1-2 TU.

It can be seen that the Cl concentration generally increases with the depth of sampling, except in areas of groundwater recharge and discharge where dilute groundwater penetrates to depths of up to 360 m (sample M8-3- 7) and brackish water occurs at shallow depths (B34-2-4). The relationship between the $^{36}Cl/Cl$ ratio and Cl concentration is shown in Figure 1.

Table 1. Values of ^{36}Cl/Cl ratio, ^{3}H and Cl concentration in groundwaters in the Lac du Bonnet granite.

Sample #	Depth (m)	Cl (mg/L)	^{36}Cl/Cl (x10^{-15})	^{3}H (TU)
M14-1-4	50	17	107 ± 11	1.3
M8-3-7	360	18	203 ± 19	7.4
URL8-7-7	230	28	85 ± 9	<0.8
M5B-IN-9	120	30	120 ± 11	<8*
M2A-3-12	310	37	88 ± 9	<0.8
M1A-3-7	265	207	41 ± 4	<0.8
HC29-10	250	235	41 ± 8	<0.8
HC7	250	430	60 ± 9	ND
HC17-12	250	454	49 ± 5	<0.8
B34-2-4	40	510	45 ± 8	1.1
M13-2-5	250	1240	42 ± 5	<0.8
URL12-13-13	605	3070	59 ± 6	14.2
WN1-8-17	380	3430	57 ± 8	2.6
M10-3-8	410	3710	34 ± 7	<0.8
M14-4-5	370	6010	49 ± 5	0.9
M7-4-9	390	8700	41 ± 7	1.8
WD3-895-10	895	11500	43 ± 6	5.2
WN11-17-14	1000	19000	57 ± 7	3.1
WB1-7-7	1000	29500	44 ± 8	2.1

ND = Not Determined
* = Determined without enrichment by electrolysis

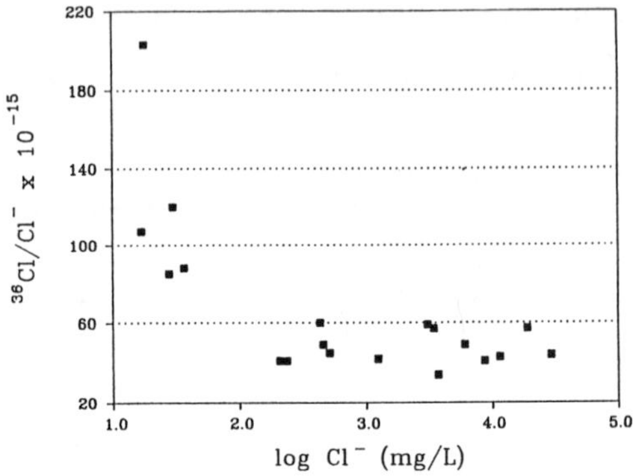

Figure 1. Variation of ^{36}Cl/Cl ratio with Cl concentration in groundwaters from the Lac du Bonnet granite.

DISCUSSION

Dilute groundwaters with the concentration of Cl < 50 mg/L have ^{36}Cl/Cl ratios ranging between 85 and 200 (x 10^{-15}). In these samples, ^{36}Cl is probably derived largely from cosmogenic sources. Analyses of ^{14}C and ^{2}H/^{18}O contents (Gascoyne, unpublished results) indicate that these groundwaters are <10^4 a old, and this is supported by the relatively high ^{36}Cl/Cl ratios.

Groundwaters with higher Cl concentrations (between 200 and 30 000 mg/L) have lower ^{36}Cl/Cl ratios (40-60 x 10^{-15}). No correlation is seen between the ^{36}Cl/Cl ratio and the groundwater Cl content (Fig. 1). In these samples, the surface-derived ^{36}Cl component either has decayed to very low levels or, more probably, has been masked by ^{36}Cl produced in situ by (η,γ) reactions with the ^{35}Cl present in soluble salts in the rock matrix. Previous studies of the content of soluble salts in various phases of the Lac du Bonnet granite (Gascoyne et al. 1989b) have shown that Cl concentrations are about 50 mg/kg in unaltered grey granite, with lesser amounts in pink, altered phases. These salts are believed to exist in micropores and as fluid inclusions and grain boundary deposits, which are accessible to permeating groundwater.

Chlorine-36 is produced in rocks by neutron-capture by ^{35}Cl and, to a minor extent, by ^{39}K. To calculate the in situ ^{36}Cl production rate, the neutron production rate and initial neutron energy spectrum were first estimated using the SOURCES code (Wilson et al. 1988). Neutrons in the subsurface are produced during spontaneous fission of ^{238}U and from (α,n) reactions on light elements in the granite matrix (mainly O, Na, Al and Si), where the α particles derive from the α-decay of ^{238}U, ^{235}U and ^{232}Th. The neutron flux, and consequently the production rate of ^{36}Cl, was then calculated using the Monte Carlo neutron transport code, MCNP (Briesmeister 1986), for the average elemental composition of the granite. Elemental concentrations were obtained from Kamineni et al. (1986), Stone et al. (1989) and Kamineni (unpublished results). A range of ^{36}Cl/Cl values was generated by sensitivity analyses with the MCNP code in which key elements controlling the ^{36}Cl production rate (H, Sm + Gd, U, Th) were varied within reasonable limits. The results show that ^{36}Cl/Cl ratios in the range of 40-100 x 10^{-15}, with an average of 65 x 10^{-15}, are predicted for soluble Cl in the unaltered granite matrix under conditions of secular equilibrium with the in situ neutron flux. Because the half-life of ^{36}Cl is 301 000 a, 90% of secular equilibrium is attained in about 1 Ma.

The largest source of uncertainty in these predictions is the variation of U and Th concentrations observed in rock from the fracture zones and the matrix (one standard deviation of the concentration distribution of these elements is ±35%). Nevertheless, it can be seen that the $^{36}Cl/Cl$ ratios in most groundwaters containing >50 mg/L Cl lie within the calculated range of values. This suggests that ^{36}Cl, and hence the dissolved Cl concentration in the groundwater, is derived largely from micropores containing saline water in the rock matrix and not from external sources.

Other elemental and isotopic evidence (e.g., Na/Ca and Br/Cl ratios, and ^{34}S concentrations) has previously indicated the intrusion of basin brines into the granite in this area (Gascoyne 1989a). Because the $^{36}Cl/Cl$ ratios of all the saline groundwaters are fairly constant, any recharge to the granite fracture system by brines derived from the lower U and Th environments of the Paleozoic rocks to the west must have occurred over ~1 Ma ago for secular equilibrium with the granite to be established.

CONCLUSIONS

$^{36}Cl/Cl$ ratios cannot be used to date groundwater in the Lac du Bonnet granite because of the high levels of in situ production of ^{36}Cl. In the present study, ^{36}Cl is found to be most useful to verify hydrogeological evidence for groundwater recharge locations and to determine the depth and extent of penetration of recent recharge into fracture systems in the granite. The results described here add support to the evidence indicating that soluble salts in rock are the principal source of groundwater salinity.

ACKNOWLEDGEMENTS

P. Sharma, P. Kubik and their colleagues at the University of Rochester, NY, provided the $^{36}Cl/Cl$ abundance ratios under contract to AECL Research. R. Watson (AECL Research) prepared and purified AgCl precipitates for ^{36}Cl analysis. The Canadian Nuclear Fuel Waste Management Program is funded jointly by AECL and Ontario Hydro under the auspices of the CANDU Owners Group. The reviews of this manuscript by J.D. Ross and J.J. Cramer are greatly appreciated.

REFERENCES

Briesmeister, J. 1986. MCNP - A general Monte Carlo code for neutron and photon transport, Version 3A, Los Alamos National Laboratories Report, LA-7396-M, Rev. 2.

Dormuth, K.W. & K. Nuttall 1987. The Canadian Nuclear Fuel Waste Management Program. Rad. Waste Man. and the Nucl. Fuel Cycle 8(2-3): 93-104.

Gascoyne, M., C.C. Davison, J.D. Ross & R. Pearson 1987. Saline groundwaters and brines in the Canadian Shield. *Saline Water and Gases in Crystalline Rocks.* Geol. Assoc. of Can. Spec. Paper No. 33: 53-68.

Gascoyne, M., A. Purdy, P. Fritz, J.D. Ross, S.K. Frape, R.J. Drimmie & R.N. Betcher 1989a. Evidence for penetration of sedimentary basin brines into an Archean granite of the Canadian Shield. *Proc. 6th Int. Symp. Water-Rock Interaction:* 243-245. Rotterdam: Balkema.

Gascoyne, M. J.D. Ross, R.L. Watson & D.C. Kamineni 1989b. Soluble salts in a Canadian Shield granite as contributors to groundwater salinity. *Proc. 6th Int. Symp. Water-Rock Interaction:* 247-249. Rotterdam: Balkema.

Kamineni, D.C., C.F. Chung, J.J.B. Dugal & R.B. Ejeckam 1986. Distribution of U and Th in core samples from the Underground Research Laboratory lease area, southeastern Manitoba, Canada. *Chem. Geol.* 54: 97-111.

Stone, D., D.C. Kamineni, A. Brown & R. Everitt 1989. A comparison of fracture styles in two granite bodies of the Superior Province. *Can. J. Earth Sci.* 26: 387-403.

Whitaker, S.H. 1987. Geoscience research for the Canadian Nuclear Fuel Waste Management Program. *Rad. Waste Man. and the Nucl. Fuel Cycle* 8(2-3): 145-196.

Wilson, W.B., M. Bozoian & R.T. Perry. 1988. Calculated α-induced thick target neutron yields and spectra, with comparison to measured data. *Nucl. Data for Sci. and Tech.* (1988 MITO): 1193-1197.

Water-Rock Interaction, Kharaka & Maest (eds) © 1992 Balkema, Rotterdam. ISBN 90 5410 075 3

An oxygen isotope study of silicates at Larderello, Italy

P. Iacumin
Istituto di Mineralogia e Petrografia Università di Trieste, Italy

E. Petrucci
Dipartimento di Scienze della Terra, Sezione di Geochimica, Università di Roma, Italy

G. Gianelli & M. Puxeddu
Istituto Internazionale per le Ricerche Geotermiche, CNR, Pisa, Italy

ABSTRACT: Stable isotope analyses have been performed on quartz, biotite, muscovite and one amphibole from basement samples cored at Larderello. $\delta^{18}O$ values obtained for minerals, suggest equilibration with a magmatic-metamorphic fluid and its progressive mixing with shallow meteoric waters.

1. INTRODUCTION

In the last fifteen years petrologic, geochronologic and fluid inclusions studies revealed that during and soon after the climax of the thermal event (3-4 Ma b.p.) that gave origin to the field, hydrothermal-magmatic fluids (Cathelineau et al. 1989) interacted with the metamorphic basement producing a 600-700 meters thick aureole characterized by biotite-tourmaline rich assemblages (Cavarretta & Puxeddu, 1990). A geopressurized HT fluid still exists in some tectonic traps of the basement as indicated by two blow-outs of well San Pompeo 2 and by extrapolated temperatures up to 450°C in the San Pompeo area at 3 Km depth (Cavarretta & Puxeddu, 1990).

The magmatic fluids were characterized, during the early stage of the hydrothermal metamorphism, by very high temperatures (T_t > 560°C), lithostatic pressures, very high salinities (up to 60% wt NaCl eq) and compositions typically magmatic as indicated by the abundance of KCl, NaCl, LiCl (Cathelineau et al 1989). Further evidences of the presence of a magmatic intrusion at 3-8 Km depth are given by the occurrence of blue corundum, andalusite, sanidine and muscovite (T=620°C) at 3 Km depth in well San Pompeo 2 (SP2) (Puxeddu, 1984) and by teleseismic P wave delays up to 1 sec suggesting the presence of an anomalous body rooted in the mantle (Foley et al 1990).

Therefore a study of stable isotopes on minerals and rocks would be useful in order to complete information about the deep Larderello system below the main shallow (500-1500 m b.g.l.) reservoir now exploited. Previous isotope data on well-head samples of fluids, tapped from the shallow reservoir (Panichi et al 1974), when plotted in a $\delta^{18}O$ diagram meet the M.W.L. in correspondence of $\delta^{18}O$=-7 (per mil versus SMOW) which is the isotopic composition of local meteoric water. Therefore fluids from the main reservoir have always been considered of meteoric origin.

2. ISOTOPE DATA

Oxygen isotope analyses have been performed on mineral samples separated from basement rocks cored in deep wells (3-4 Km) drilled in the southwestern part of the field. All the values are reported using the standard permil notation relative to SMOW: the $\delta^{18}O$ obtained for quartz, biotite, muscovite and amphibole are listed in table 1.

Quartz samples (Petrucci, 1989 and this paper) show a clustering of 9 values in the range + 13.4 + 14.7 with only two values as low as + 9.4 and + 12.0. These values are 1 to 3 permil higher than those reported by Hoernes and Friedrichsen (1978) for low to medium grade matapelites. The narrow range observed indicates that all the samples from different wells equilibrated with a very homogeneous deep fuid.

Quartz, being the most retentive mineral analysed, supplies valuable informa-

Table 1. $\delta^{18}O$ (per mil SMOW) and T (°C) values for metamorphic minerals and inferred deep fluid composition at Larderello.

Sample (depth)	min.	$\delta^{18}O$	min.pair	T_m	T_i	T_t	$\delta^{18}O$(fluid) BJ	RH	
Sasso22 (2502)	qtz	+14.7	qtz-mu	300	390		+9.1	+9.5 (qtz-mu)	▲
	mu	+10.3	qtz-bio		252		+9.1	+10.1 (mu-w)	▲
	bio	+ 1.9	mu-bio		148				
Sasso22 (2636)	qtz	+12.0	qtz-am	310	423	355	+7.2	+7.5 (qtz-w)	▲
	am	+ 5.8						+6.9 (am-w)	▲
							+5.3	+6.5 (am-w)	o
Sasso22 (3529)	qtz	+14.5		355		492	+11.2	+11.4 (qtz-w)	o
Sasso22 (3800)	qtz	+12.2	qtz-bio	370	298	534	+9.6	+9.7 (qtz-w)	o
	bio	+ 1.5							
Sasso22 (4028)	qtz	+14.6	qtz-bio	380	643	570	+13.4	+13.4 (qtz-w)	▲
	bio	+10.8						+13.1 (bio-w)	▲
							+12.5	+12.6 (qtz-w)	o
							+13.3	+13.1 (bio-w)	o
SP2 (2962)	mu	+ 8.2		450		560	+8.6	+9.3 (mu-w)	o
VC11	bio	+ 4.0		320		450	+6.3	+6.2 (bio-w)	o

Abbreviations: qtz=quartz, mu=muscovite, bio=biotite, am=amphibole, w=water. T_m = in hole measured and/or interpolated temperatures; T_i = temperatures calculated using isotope geothermometers; T_t = trapping temperatures. ▲ $\delta^{18}O$ values of fluids calculated using T_i; o $\delta^{18}O$ values of fluids calculated using T_t. Geothermometric equations used: BJ = Bottinga & Javoy (1973); RH = Richter & Hoernes (1988).

tion about the original $\delta^{18}O$ of the deep fluid. Biotite shows two very low $\delta^{18}O$ values of +1.5 and +1.9 an intermediate value of +4.0 and a very high value of +10.8 in the deepest sample (4 Km depth) from well Sasso 22. These data likely corresponds to the original isotopic composition of biotite system and interaction with later ^{18}O-depleted fluids. Muscovite yields two values of +10.3 and +8.2, comparable to those observed in metapelitic rocks (Hoernes and Friedrichsen, 1978).

The temperatures estimated by means of fluid inclusions (FI) data range from 255° to 570°C, while those calculated using isotope geothermometers (IG) fall in the range 148° - 643°C. Among them, three values - 148°C, 252°C and 298°C - are lower than present day in hole measured temperatures (300 - 370 °C): such an unrealistic situation can be explaind by late reequilibration processes which likely took place between minerals and a low $\delta^{18}O$ fluid. Temperatures calculated either by FI and IG were then introduced in the geothermometric equations (Bottinga & Javoy, 1973; Richter & Hoernes, 1988) in order to calculate the $\delta^{18}O$ of the fluid in equilibrium with the minerals. The $\delta^{18}O$ range is +5.3—+13.4 which is far from the initial $\delta^{18}O$ of + 2 to - 7 obtained by Panichi et al (1974). These positive values could indicate mixing between meteoric waters and magmatic fluids, or equilibration of early meteoric waters with the metapelitic basement under HT and low W/R ratios.

The likely presence of a magmatic source supplying fluids to the overlying country rocks is suggested by the very high temperature of 643°C calculated for the quartz biotite pair of sample 4028 from well Sasso 22. This value is in perfect agreement with T=620°C indicated by the appearance of corundum and with T_t>560°C found for hydrothermal magmatic fluids trapped as fluid inclusions at the bottom of the well San Pompeo 2. The Li-rich composition of some fluid inclusions is also evidence of an early

magmatic fluid.
During the slow, monotonic cooling phase of the Larderello history, a gradual dilution of deep fluids by external probably meteoric-waters strongly lowered $\delta^{18}O$ values of the original fluid, leading sometimes to the unrealistic low equilibrium temperatures measured for mineral pairs including biotite of Sasso 22 (2502 and 3800 m b.g.l.). Probably the strong decrease of fracture permeability, produced by self-sealing in the deepest part of well Sasso 22 prevents the inflow of shallow meteoric waters and then dilution of deep metamorphic magmatic fluids whose imprinting is still evident in $\delta^{18}O$ values of the sample Sasso 22 (4028 m b.g.l.).

REFERENCES

Bottinga, Y & Javoy, M. 1973. Comments on oxygen isotope geothermometry. Earth Planet. Sc. Lett. 20, 250-265.

Cathelineau, M., Marignac, C., Valori, A., Dubessy, J., Gianelli, G., & Puxeddu, M. 1989. Pressure-temperature-fluid composition changes from magmatic to present day stages in the Larderello geothermal field (Italy): Internat. Symposium Water-Rock Interaction, 6th, Malvern, UK, August 3-13, 1989, Proc., 137-140.

Cavarretta, G. & Puxeddu, M. 1990. Schorl-dravite ferridravite tourmalines deposited by hydrothermal magmatic fluids during early evolution of the Larderello geothermal field, Italy. Econ. Geol., 85, 1236-1251.

Foley, J.E., Toksöz, M.N., & Batini, F. 1990. Three-dimensional inversion of teleseismic travel times for velocity structure in the Larderello geothermal field, Italy. Geoth. Res. Coun. Trans., 14, 1413-1419.

Hoernes, S. & Friedrichsen H. 1978. Oxygen and hydrogen isotope study of the polymetamorphic area of the northern Ötztal-Stubai Alps (Tyrol) Contrib. Mineral. Petrol 67, 305-315.

Panichi, C., Celati, R., Noto, P., Squarci, P., Taffi, L., & Tongiorgi, E. 1974. Oxygen and hydrogen isotope studies of the Larderello (Italy) geothermal system. Isotope techniques in groundwater hydrology, vol II IAEA Vienna, 4-28.

Petrucci, E. 1989. Geochimica degli isotopi stabili, petrografia e considerazioni strutturali sulle rocce del basamento paleozoico toscano: dati dai pozzi Monteverdi 2 e San Giovanni 1 bis. Tesi di Dottorato, Università di Roma "La Sapienza", unpublished.

Puxeddu, M. 1984. Structure and late Cenozoic evolution of the upper litosphere in southwest Tuscany (Italy). Tectonophysics 101, 357-382.

Richter, R. & Hoernes, S. 1988. The application of the increment method in comparison with experimentally derived and calculated O-isotope fractionations. Chem. Erde 48, 1-18.

Sheppard, S.M.F. 1986. Characterization and isotopic variations in natural waters. In Stable Isotopes in high temperature geological processes, J.W. Valley, H.P. Taylor Jr & J.R. O'Neil, Eds, Reviews in Mineralogy 16, 165-181.

Water-Rock Interaction, Kharaka & Maest (eds) © 1992 Balkema, Rotterdam. ISBN 90 5410 075 3

Oxygen isotope studies of hydrothermally altered basalts from Surtsey, Iceland

H. R. Karlsson
Texas Tech University, Lubbock, Tex., USA

S. J. Jakobsson
Icelandic Museum of Natural History, Reykjavik, Iceland

J. G. Moore
US Geological Survey, Menlo Park, Calif., USA

ABSTRACT: Oxygen isotope ratios of hydrothermally altered rocks from a 181 m core through Surtsey island record the cumulate effects of 12 years of low-temperature alteration of basaltic tephra. $\delta^{18}O$ values (6 to 12‰) appear to be governed by formation of palagonite and its subsequent isotopic exchange with seawater. This is reflected in two trends in the $\delta^{18}O$-H_2O^+ relationship of the samples. Trend A (<100°C), shows a positive correlation between $\delta^{18}O$ and H_2O^+ similar to that of altered deep-sea basalts. This trend is dominated by the formation of palagonite from sideromelane. Trend B (>100°C) shows little change in H_2O^+ and a decrease in $\delta^{18}O$. This downward trend is consistent with oxygen isotopic exchange between palagonite and progessively hotter fluid ($\delta^{18}O$ = 0 to -2‰). Assuming that this exchange was controlled by the smectite-water fractionation curve, maximum calculated fluid temperatures range from 150°C to 183°C.

1 INTRODUCTION

Oxygen isotopes are a useful tool for studying water-rock interaction. Studies of rocks dredged or drilled from the sea floor reveal that a positive correlation exists between $\delta^{18}O$ and water content of altered basalts. This correlation was explained by Muehlenbachs and Clayton (1972) in terms of a mixing relationship between fresh anhydrous basalt as one end-member and hydrous minerals, formed at low temperatures, as the other end-member. Subsequent work shows that the two component mixing model is a good first approximation and that the $\delta^{18}O$- water relationship is often more complex because other effects, such as variation in secondary minerals, changes in water/rock ratio, duration of alteration and temperature interfer (Böhlke 1984, Muehelenbachs 1987).

The island of Surtsey, south of Iceland, provides us with a natural laboratory and an unparalleled opportunity to study the processes involved in basalt-seawater interaction. Three variables that are either poorly constrained or undetermined in the case of submarine basalts, are available for Surtsey, namely time, temperature and fluid composition. In addition, extensive mineralogical and chemical data are available (Jakobsson and Moore 1986).

Surtsey formed during a basaltic submarine eruption between 1963 and 1967. In the summer of 1979 a 181m cored hole was drilled through the island to study subsurface stratigraphy and hydrothermal alteration. The core extends 123 m below sea level about to the pre-Surtsey sea floor and records 12 years of alteration of basaltic tephra (Jakobsson and Moore 1980, 1982).

In order to enhance our understanding of water-rock interaction in Surtsey, we obtained $\delta^{18}O$ and water contents released above 110°C (H_2O^+) from 17 whole rocks sampled throughout the length of the core. Except for isotopic values for analcime (Karlsson and Clayton, 1990), $\delta^{18}O$ values have not been reported previously for Surtsey rocks or minerals.

2 ALTERATION OBSERVED IN DRILL CORE

Mineralogical and chemical effects resulting from the water-rock interaction have been extensively documented both for surface samples (Jakobsson 1972, 1978) and for the drill core (Jakobsson and Moore 1986). The primary material cored was vesiculated glassy basaltic tuff and alteration products thereof. At depths between 72-85 m the core exhibits an irregular dike complex. The intrusion of these dikes marks the onset of a hydrothermal system 12± 1 years prior to drilling. Jakobsson and Moore (1986) have estimated an effective average temperature profile during the 12 years of hydrothermal activity in the bore hole region. Estimated temperatures range from 60°C at the top of the hole, peak around 150°C in the midsection (100-120 m), and then taper off in an irregular fashion towards the bottom of the hole.

As a result of the hydrothermal alteration, the fresh glass of the tephra (sideromelane) altered partly or completely to palagonite; rinds of smectite developed on olivine phenocrysts and authigenic minerals formed. Ten different alteration minerals have been identified amongst which smectite, analcime, phillipsite and tobermorite are the most abundant (Jakobsson and Moore 1986).

3 RESULTS

The oxygen isotopic data and water contents for our 17 whole rock samples (circles) are shown in Fig. 1. Also included is an analcime sample collected from a vein occurring at 90.1 m depth, where the temperature was 145 ± 5°C (Karlsson and Clayton 1990). Isotopic results are reported in the conventional δ(‰) notation using SMOW as a standard (Craig 1957).

Whole-rock $\delta^{18}O$ values for Surtsey samples range from 6 to 12‰. Overall, as in the case of deep-sea basalts, whole-rock $\delta^{18}O$ values for Surtsey increase with increasing water content. However, at Surtsey we are able to resolve two trends: There is a general increase in $\delta^{18}O$ up to ~5 wt% H_2O^+, where a maximum of ~12‰ is reached (trend A). These rocks experienced temperatures up to 100°C. Above 5 wt% H_2O^+ there is a marked drop in $\delta^{18}O$ values and they scatter between 7.5 and 9‰, whereas the water content increases only slightly (trend B). These rocks were exposed to temperatures between 100 and 150°C.

4 DISCUSSION

Trend A is consistent with a fresh basalt-hydrous mineral mixing model analogous to that observed in altered submarine basalts. Although the analcime (8wt%, 12‰) appears to give a satisfactory mixing line with the fresh basalt, it must be ruled out because it formed at 145°C, whereas the samples comprising trend A were altered below 100°C. A more plausible explanation, consistent with mineralogy of the core and the temperature profile in the drill hole, is that trend A is primarily due to formation of palagonite. In Surtsey, the earliest signs of alteration were the transformation of fresh basaltic glass (sideromelane), which constituted ~90 vol% of the original rock, into palagonite (Jakobsson 1972, 1978; Jakobsson and Moore 1986). Palagonite is first detected in samples whose temperature exceeded ~40°C. The rate of palagonitization increases rapidly with temperature, doubling every 12°C, such that in 12 years virtually all the sideromelane has been converted to palagonite where the temperature exceeded approximately 100°C (Jakobsson and Moore 1986). Because the main constituent of the palagonite is smectite, the low-temperature trend can be interpreted as an array of mixing lines between fresh basalt and smectite formed between 40 and 100°C (Fig. 1). The two smectite $\delta^{18}O$ mixing lines were calculated using the smectite-water fractionation curve of Yeh and Savin (1977), assuming pure seawater ($\delta^{18}O$ = 0‰). As expected from the higher temperatures, the Surtsey trend A is below that of deep-sea basalts, which are generally thought to have been altered at temperatures below 40°C.

For whole-rocks samples that comprise trend B (>5 wt% H_2O^+), the $\delta^{18}O$-H_2O^+ relationship cannot be explained in terms of the fresh basalt-hydrous mineral mixing model. Sideromelane in these rocks has been completely replaced by palagonite, such that little further smectite is added. None of the abundant alteration minerals which appear at these temperatures, analcime, phillipsite or tobermorite, can account for the sudden decrease in $\delta^{18}O$ experienced by these rocks. The reversal in whole-rock $\delta^{18}O$ values at ~5 wt% H_2O^+ requires addition of a component that has both a lower water content and a lower $\delta^{18}O$ value than can be achieved by clays or zeolites.

A plausible explanation is that trend B resulted from oxygen isotope exchange between palagonite and progressively hotter fluid. Since the palagonite consists mainly of smectite, it is reasonable to assume a bulk fractionation factor for palagonite-water similar to that of smectite-water. In order to calculate a temperature, the fluid isotopic composition must be estimated. We assume that the fluid was pure or nearly pure seawater for the several reasons. First, hydrogen isotopic analysis yielded δD values as low as -4.2‰ suggesting nearly pure seawater with little mixing with precipitation (Jakobsson 1978). Second, the high permeability allowed infiltration of seawater through the tuff pile resulting in a high water/rock ratio. Hence the fluid must have been nearly pure seawater and was little effected by water-rock interaction. If equilibrium is assumed, and a fluid with $\delta^{18}O$ = 0‰ (seawater) is also assumed, then the smectite-water fractionation curve (Yeh and Savin, 1977) yields a maximum equilibrium temperature of 183°C for a rock $\delta^{18}O \approx 8$‰. This temperature is 33°C higher than the highest estimated temperature in the drill hole at 150°C (Jakobsson and Moore, 1986). Although the maximum temperature attainable in the drill core region according to the boiling-point curve of seawater may have been as high as 180°C at depths greater than 150 m, none of these samples came from such depths. Assuming that the smectite-water fractionation curve holds, these results indicate that temperatures may have been somewhat higher in the drill hole than at present and/or that the fluid was depleted in $\delta^{18}O$ due to slight mixing with local meteoric waters.

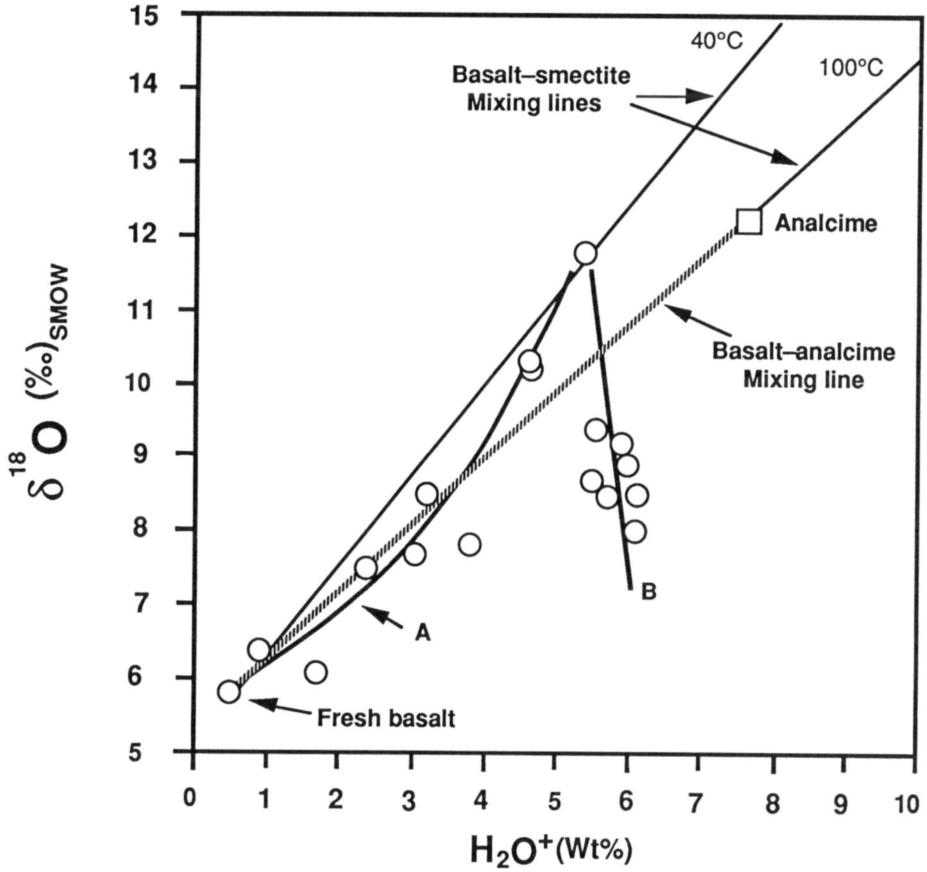

Fig. 1 $\delta^{18}O$ versus water content in Surtsey drill core samples

According to the smectite-water fractionation curve, a rock with $\delta^{18}O$ = 8‰ is in equilibrium with a fluid of -2‰ at 150°C.

5 CONCLUSIONS

The Surtsey core records extensive hydrothermal alteration and associated oxygen isotopic exchange of basaltic tephra with seawater that occurred over a period of 12 years. The rates of hydrothermal alteration and isotopic exchange at Surtsey exceed those encountered in deep-sea basalts because of higher alteration temperatures and predomanince of glass in the Sustey samples compared with holocrystalline deep-sea basalts. Three major observations are apparent in the data form the Surtsey core. First, although a hydrous secondary mineral may fall on the mixing line for altered basalts, that in itself does not prove conclusively that the mixing line is controlled by that particular mineral. Here, the analcime from Surtsey is a case in point. Paths in $\delta^{18}O$-H_2O^+ plots shoulds be interpreted with caution, especially when detailed mineralogical and petrographic information is lacking.

Second, palagonitization, dominated by the formation of smectite, can be the sole process governing $\delta^{18}O$ enrichment in basalts below 100°C. Third, once palagonitization is complete, the trend is reversed and basalts can become depleted in $\delta^{18}O$. The time required to reach this stage of reversal depends on the rate of palagonitization, which in turn is highly temperature-dependent. At Surtsey oxygen isotope exchange between the palagonite and seawater resulted in lower whole-rock $\delta^{18}O$ values. The lower limit on the whole-rock values is set by the composition of seawater. Thus, the Surtsey basalts cannot obtain negative $\delta^{18}O$ whole rock values. In order to obtain negative values, meteoric waters are required.

REFERENCES

Craig, H. 1957. Isotopic standards for carbon and oxygen and correction factors mass-spectrometric analysis of carbon dioxide. *Geochimica et Cosmochimica Acta* 12: 133-149.

Böhlke, J. K., J. C. Alt & K. Muhelenbachs 1984. Oxygen isotope - water relations in altered deep-sea basalts: low-temperature mineralogical controls. *Canadian Journal of Earth Science* 21: 67-77.

Jakobsson, S. P. 1972. On the consolidation and palagonitization of the tephra of Surtsey volcanic island, Iceland. *Surtsey Progress Report* VI: 121-128.

Jakobson, S. P. 1978. Environmental factors controlling the palagonitization of the Surtsey tephra, Iceland. *Bulletin of the Geological Society of Denmark* 27: 91-105.

Jakobsson, S. P. & J. G. Moore 1980. Unique hole show how volcano grew. *Geotimes* 25: 14-16.

Jakobsson, S. P. & J. G. Moore 1982. The Surtsey Research Drilling Project of 1979. *Surtsey Progress Report* IX: 76-93.

Jakobsson, S. P. & J. G. Moore 1986. Hydrothermal alteration minerals and alteration rates at Surstey volcano, Iceland. *Bulletin of the Geological Society of America* 97: 648-659.

Karlsson, H. R. & R. N. Clayton,1990. Oxygen and hydrogen isotope geochemistry of zeolites. *Geochimica et Cosmochimica Acta* 54: 1369-1386.

Muehlenbachs, K. 1987. Oxygen isotope exchange during weathering and low-temperature alteration. *Short Course in Stable Isotope Geochemistry* 13. Toronto: Mineralogical Association of Canada.

Muehelenbachs, K. & R. N. Clayton 1972. Oxygen isotope studies of fresh and weathered submarine basalts. *Canadian Journal of Earth Science* 9: 172-184.

Yeh H. W. & Savin S. M. 1977. The mechanism of burial metamorphism of argillaceous sediments. *Bulletin of the Geological Society of America* 88: 1321-1330.

Water-Rock Interaction, Kharaka & Maest (eds) © 1992 Balkema, Rotterdam. ISBN 90 5410 075 3

Chlorine stable isotope distribution of Michigan Basin and Canadian Shield formation waters

Ronald S. Kaufmann
Northern Illinois University, DeKalb, Ill., USA

S. K. Frape
University of Waterloo, Ont., Canada

Robert McNutt
McMaster University, Hamilton, Ont., Canada

Christopher Eastoe
University of Arizona, Tucson, Ariz., USA

ABSTRACT: The chlorine isotope ratio, mass 37/35, was measured on 14 formation waters from the Michigan Basin margin in southern Ontario, on eight formation waters from deep within the Basin, and on 11 Canadian Shield formation waters from the Sudbury, Thompson and Yellowknife mining districts. These samples were used to find evidence of chemical relationships between the Basin and the Shield. Composition of the Basin $\delta^{37}Cl$ varied from +0.01 to -1.2 (per mil difference from SMOC, precision of 0.16) and decreased with stratigraphic age to the Ordovician formation deep in the Basin, then increased to -0.4 in the Cambrian formation. Nine of the Shield waters were between +0.3 and +0.77, the other two were 0.0 and -0.45. Canadian Shield waters, the Silurian F-salt and Deep Michigan Basin waters appear as data field end members when $\delta^{37}Cl$ is graphed against chloride concentration, Br/Cl and $^{87}Sr/^{86}Sr$. The F-salt end member suggests the effect of salt dissolution in Cambrian, Ordovician and Silurian samples on the Basin margin. Ordovician waters derived from carbonates deep within the Michigan Basin appear to be independent of salt dissolution. The samples define correlations between the Deep Basin and Shield samples when $\delta^{37}Cl$ is plotted against K/Cl and Ca/Cl, suggesting possible mixing between the Shield and the Basin. Because they occur as data field end members on all of these plots, it is possible that Canadian Shield waters contributed to the geochemistry of waters at the Basin margin.

1 DISCUSSION AND NEW OBSERVATIONS

The possibility of a relationship between formation waters of the Michigan Basin and the Canadian Shield has been raised by McNutt et al. (1987). In terms of concentration, the Michigan Basin generally has higher K concentrations and lower Ca/Na ratios. In addition, on a hydrogen-oxygen isotope graph, the Shield waters plot above GMWL (Frape et al. 1984) and the Michigan Basin waters plot below (Clayton et al. 1966; Breen et al. 1985; McNutt et al. 1987). However, similar concentrations and trends in other chemical parameters suggest Shield-Basin mixing.

Mixing relationships among most chemical parameters are difficult to interpret with certainty in the Michigan Basin because of water-rock interactions. There is evidence that dolomitization and feldspar decomposition have affected many of the parameters (Wadleigh et al, in preparation). The problem of finding evidence of Basin-Shield mixing is further complicated by salt dissolution (Lowry et al. 1988) and the intrusion into the basin of Na-Ca-Cl water from the south which mixed with in-situ Ca-Na-Cl water (Frape et al. 1989).

For this study we looked for a chemical relationship between the Basin and the Shield through consideration of one parameter which may be minimally influenced by most water-rock interactions in low temperature sedimentary environments, the chlorine stable isotope ratio (37/35). A representative group of samples were taken from the Canadian Shield and deep within the Michigan Basin. These were compared with samples from southern Ontario in the northeastern part of the Basin where mixing is likely to occur. The results do suggest that the Shield may have influenced the chemistry of the formation waters at the Basin margin.

1.1 Chlorine stable isotope geochemistry

The range of chlorine stable isotope ratios of natural materials is small compared to most other commonly used isotope ratios, only 2-3 per mil. Chlorine has few oxidation states or chemical forms in the geologic environment, occurring almost exclusively as dissolved and precipitated salts, a fact that may account for the small isotope ratio range.

In a sedimentary basin such as the Michigan Basin, the primary fractionation process at low temperatures for chlorine isotopes is molecular diffusion (Desaulniers et al. 1986). In addition, the isotope signature of a formation water may be modified by salt dissolution. Thus to identify mixing in the Michigan basin, diffusion and salt dissolution must first be taken into account.

1.2 Sample collection and analysis

The 14 samples selected from the Michigan Basin margin in southern Ontario include brines from the Silurian F-salt, the Devonian Dundee formation, Silurian reefs of the Geulph formation, Ordovician carbonates of the Trenton formation, and clastic rocks of Cambrian age. Three Mississippian Berea formation waters and five of the Ordovician Trenton formation are from Michigan, deep within the Basin. The 11 Canadian Shield samples are from mining districts in Sudbury, Thompson and Yellowknife (Kaufmann et al. 1987).

Samples in the Basin margin were selected to minimize the effect of diffusion. Using diffusion-fractionation theory of Senftle and Bracken (1955) and models of Kaufmann et al. (1988), the estimated fractionation of chlorine originating from a saturated NaCl brine (210 g/l) would be undetectable (0.2 per mil fractionation or less) when the diffusing chloride concentration is above 150 g/l. At 100 g/l the maximum isotope difference between the diffusing chloride and the reservoir is about 0.5 per mil. Of the 14 Basin margin samples, only four have chloride concentrations less than 148 g/l.

Complete major ion chemistry has been determined for the samples as well as oxygen, hydrogen and strontium stable isotopes and are described elsewhere (McNutt et al. 1987; Frape et al. 1989). The chlorine isotope measurements were performed at the University of Arizona. Sample preparation followed the silver chloride-methyl chloride conversion method of Taylor and Grimsrud (1969) as modified by Kaufmann (1984). Isotope ratio measurements were performed on a modified VG Micromass 602C mass spectrometer with a two standard deviation precision of 0.16 per mil (based on repeated measurements of a SMOC standard). All results are expressed as per mil difference from SMOC.

1.3 Results

Chlorine isotope results are given in Figure 1. Deep Michigan Basin samples from the Ordovician formation are isotopically lightest, with values between -1.2 and -0.7. Basin margin samples from the Cambrian and Ordovician formations are next heaviest ranging from -0.5 to -0.3. Cambrian sample C-4 is isotopically heavier than the others at +0.1. Silurian samples cluster in a range from -0.5 to -0.2. Last are the Dundee and Berea samples at values of 0.0 and +0.1. Of the nine Canadian Shield samples, only two are lower than the range +0.77 to +0.3; these have values of 0.0 and -0.45.

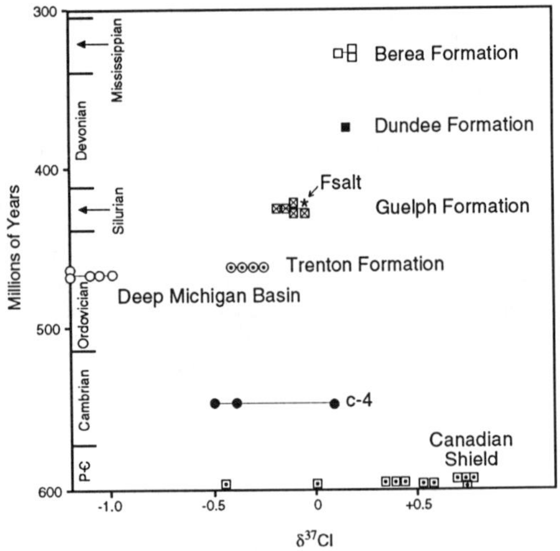

Figure 1. Chlorine isotope compositions for formation waters in the Michigan Basin and the Canadian Shield

1.4 Discussion

The chlorine isotope compositions of the Canadian Shield and Deep Michigan Basin samples bracket the compositions of samples from the Basin margin. Canadian Shield samples are generally heavier than the Michigan basin, and the Deep Michigan basin Ordovician samples are significantly lighter than the samples from the Basin margin. The total range of variation is almost 2.0 per mil, about ten times the precision.

Figures 2, 3 and 4 show plots of $\delta^{37}Cl$ and chloride concentration, $^{87}Sr/^{86}Sr$, and Br/Cl. In all of these plots, basin margin samples plot within a data field bounded by Deep Michigan Basin waters, Canadian Shield waters and the F-salt on three sides. The

F-salt and Deep Basin waters appear to have some influence on the chemistry of the Basin margin, possibly through mixing. The meaning of the Canadian Shield as end member is difficult to determine from these plots because the Mississippian samples, which fall between the salt and the Shield are not hydrologically connected to the Shield. However, the C-4 and Dundee samples may indicate mixing.

Figures 5 and 6 show plots of $\delta^{37}Cl$ with K/Cl and Ca/Cl. On both figures almost all samples form a linear pattern between the Shield and the Deep Michigan Basin. The F-salt falls along this trend as well in Figure 5. Figures 5 and 6 indicate a possible mixing relationship between Basin margin and Shield.

1.5 Summary and conclusion

Chlorine stable isotope analyses of formation water samples from deep within the Michigan Basin and Canadian Shield were compared to samples from the Basin margin nearest the Shield. The total range of variation is almost 2.0 per mil, about ten times the precision. Almost all of the Canadian Shield samples were significantly heavier than Basin samples, and the Deep Basin waters were isotopically lightest of all samples.

Strong correlations on plots of $\delta^{37}Cl$

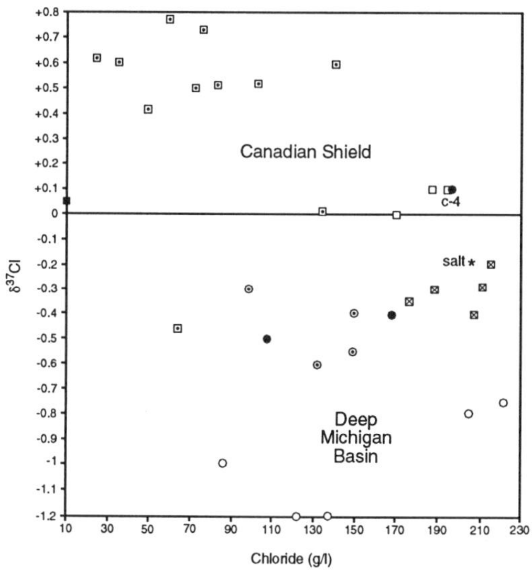

Figure 2. Chloride concentration plotted with chlorine isotope composition for Michigan Basin and Canadian Shield waters

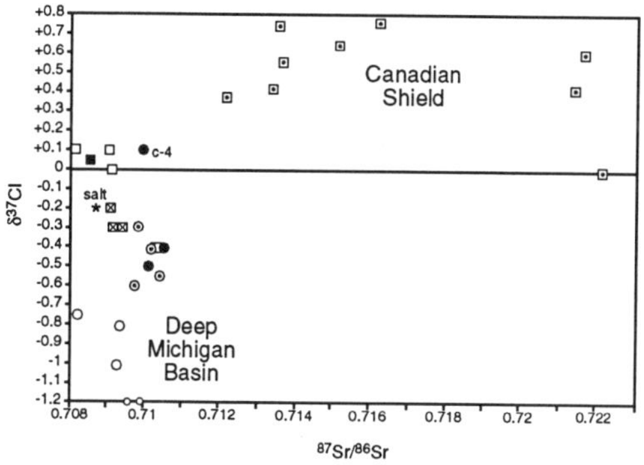

Figure 3. Strontium isotope ratio plotted with chlorine isotope composition for Michigan Basin and Canadian Shield waters

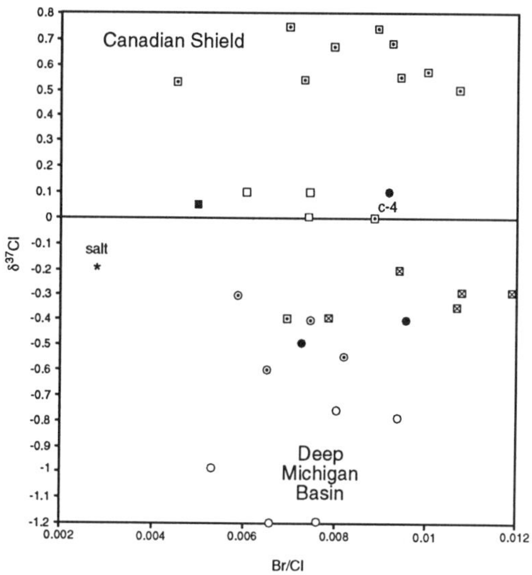

Figure 4. Bromide/chloride ratio plotted with chlorine isotope composition for Michigan Basin and Canadian Shield waters

versus K/Cl and Ca/Cl suggest that the Shield water is a data field end member and may be mixing with water at the Basin margin. Plots of $\delta^{37}Cl$ versus chloride concentration, $^{87}Sr/^{86}Sr$ and Br/Cl also have the Shield as a data field end member.

On plots with Michigan Basin data, Canadian Shield waters often occur as mixing end members. Although the meaning of this relationship is not yet clear, the hypothesis that Canadian Shield-type waters have influenced either the origin or the evolu-

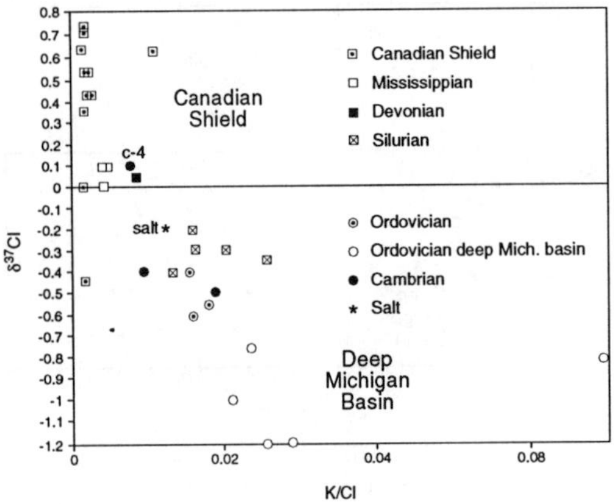

Figure 5. Potassium/chloride ratio plotted with chlorine isotope composition for Michigan Basin and Canadian Shield waters

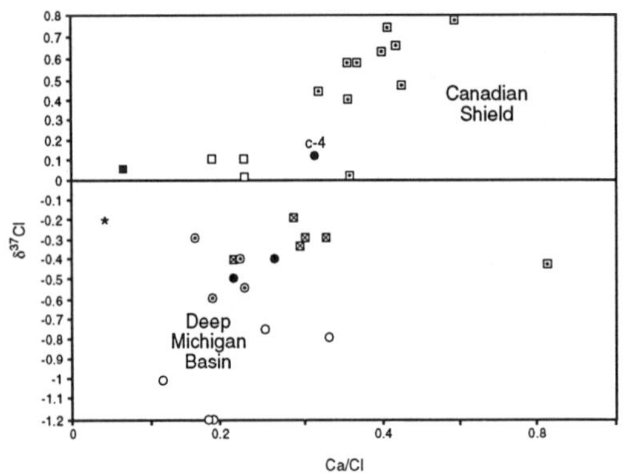

Figure 6. Calcium/chloride ratio plotted with chlorine isotope composition for Michigan Basin and Canadian Shield waters

tion of waters at the margin of the Michigan Basin should be considered.

REFERENCES

Breen, K.J., G.A. Clifford, R.W. Masters, & R.C. Sedam 1985. Chemical and isotopic characteristics of brines from three oil and gas producing sandstones in eastern Ohio, with applications to the geochemical tracing of brine sources. U.S. Geological Survey Water Resources Investigations Report: 84-4313.

Clayton, R.N., I. Friedman, D.L. Graf, T.K. Mayeda, W.F. Meents, & N.F. Shimp 1966. The origin of saline formation waters, 1. Isotopic composition. Journal of Geophysical Research. 71: 3869-3882.

Desaulniers, D.E., R.S. Kaufmann, J.A. Cherry, & H.W. Bentley. 1986. 37Cl-35Cl variations in a diffusion-controlled groundwater system. Geochimica et Cosmochimica Acta 31: 1757-1764.

Frape, S.K., P.S. Dollar, & B.S. Sherwood Lollar 1989. Mixing of saline basinal fluids in Southern Ontario: Implications of rock-water interaction, hydrocarbon emplacement and Canadian Shield brines. Proceedings of the 6th International Symposium on Water-rock Interaction, D.L. Miles (ed.), pg. 223-226.

Frape, S.K., P. Fritz, & R.H. McNutt 1984. The role of water-rock interaction in the chemical evolution of groundwaters from the Canadian Shield. Geochimica et Cosmochimica Acta 48: 1617-1627.

Kaufmann, R.S., A. Long, & D.J. Campbell 1988. Chlorine isotope distribution in formation waters, Texas and Louisiana. American Association of Petroleum Geologists Bulletin. 72: 839-844.

Kaufmann, R.S., S.K. Frape, P. Fritz, & H.W. Bentley 1987. Chlorine stable isotope composition of Canadian Shield brines. In: Saline waters and gases in crystalline rocks. P. Fritz & S.K. Frape (eds.), Geological Association of Canada Special Paper. 33: 89-93.

Kaufmann, R.S. 1984. Chlorine in groundwater: Stable isotope distribution. Ph.D. dissertation, University of Arizona, Tucson, Arizona: 132.

Lowry, R.M., G. Faure, D.I. Muller, & L.M. Jones 1988. Interpretation of chemical and isotopic compositions of brines based on mixing and dilution, "Clinton" sandstones, eastern Ohio, U.S.A. Applied Geochemistry. 3: 177-184.

McNutt, R.H., S.K. Frape, & P. Dollar 1987. A strontium-, oxygen- and hydrogen-isotopic composition of brines, Michigan and Appalachian basins, Ontario and Michigan. Applied Geochemistry. 2: 495-505.

Senftle, F.E. & J.T. Bracken 1955. Theoretical effect of diffusion on isotopic-abundance ratios in rocks and associated fluids. Geochimica et Cosmochimica Acta. 7: 61-75.

Taylor, J.W. & E.P. Grimsrud 1969. Chlorine isotopic ratios by negative ion mass spectrometry. Analytical Chemistry. 41: 805-810.

Water-Rock Interaction, Kharaka & Maest (eds) © 1992 Balkema, Rotterdam. ISBN 90 5410 075 3

Anomalous ^3He contents of CH_4-rich gases in sedimentary basins

B. Mack Kennedy
Department of Physics, University of California, Berkeley, Calif., USA

J. Poths
Los Alamos National Laboratory, N.Mex., USA

H. Hiyagon
Geophysics Institute, University of Tokyo, Japan

ABSTRACT: In typical crustal rocks, reflecting the predominance of helium from radiodecay of U and Th, $^3He/^4He$ ratios of ~10^{-7} or lower are expected. $^3He/^4He$ ratios of many CH_4-rich reservoirs are entirely consistent with this expectation. However, just as many reservoirs, often from the same field, contain helium with $^3He/^4He$ ratios as much as 10X the expected value. The utility of ^3He as a tracer for mantle volatiles input via magmatic activity is well documented. However, many of the anomalous reservoirs are in sedimentary basins which have not experienced significant tectono-magmatic activity during the last ~200 ma.

1. INTRODUCTION

Collectively, elemental and isotopic compositions of noble gases can be expected to contain information regarding source and migration of volatiles accumulated in crustal traps. Terrestrial noble gases are of two genetic types: those indigenous to the earth, inherited from meteoritic material through accretion, and those produced since earth formation by natural processes such as radiodecay or nuclear interactions. In earth models, reduced to simplest form, the two genetic types are contained in three primary reservoirs. The first is the mantle which, believed to be partially or non-degassed, is enriched in the indigenous or primordial component (eg. ^3He). The second is the crust which tends to be enriched in the radiogenic and nucleogenic component reflecting high parent/daughter ratios generated during crust formation. The third reservoir is the atmo-hydrosphere produced through the natural degassing of the other two. Each of these reservoirs can be recognized by their distinct isotopic compositions. Respective contributions to a natural gas trap, therefore, can be readily identified through the occurrence of specific isotopic signatures. The utility of mantle derived ^3He as a tracer for the injection and transport of magmatic volatiles in the crustal regime is well documented.

Early studies of noble gases associated with CH_4-rich natural gases found large excesses of total helium, radiogenic ^{40}Ar, nucleogenic ^{21}Ne, and in some cases fissiogenic $^{131-136}Xe$; consistent with the dominance of a crustal source (eg. Zartman et al., 1961). In more recent studies, helium isotopic compositions ranging from 2-10 X 10^{-8} confirm the dominance of the radiogenic component. Several CH_4-rich fields, however, contain helium with isotopic compositions significantly greater than expected for "normal" radiogenic production. In some cases, the ratios are as high 10^{-6} (eg. Oxburgh et al., 1986, Poreda et al., 1986). In this context, "normal" radiogenic helium is defined as that expected for production in average continental crust. ^3He is not produced directly by radiodecay, but is a nucleogenic component produced predominantly by the thermal neutron induced reaction $^6Li(n,\alpha)^3H$. Except very near the surface of the earth (<300 gm/cm^2) where cosmic ray interactions are important, the primary neutron sources are (α,n) reactions induced in light elements by a very small fraction of the alpha particles as they come to rest as ^4He atoms. The radiogenic $^3He/^4He$ ratio, therefore, will be independent of the U and Th concentrations but will vary linearly with Li concentration within the range of the thermalized neutrons. The calculated "normal" radiogenic helium isotopic composition (2×10^{-8}) corresponds to an average crustal Li of 60 ppm. Assuming that He (and by inference other radiogenic and nucleogenic components) in gases associated with large natural gas traps is averaged over many crustal generation sites and should, therefore, reflect production in average continental crust, $^3He/^4He$ ratios in excess of $\sim 10^{-7}$ have been interpreted as evidence for the injection of magmatic or mantle-derived helium. If true, then the occurrence of excess ^3He is related to the thermal history of the natural gas bearing sedimentary basins; the magnitude of the excess would depend on the enhanced heat flux associated with the input of mantle volatiles and time since last tectono-magmatic activity (Polyak and Tolstikhin, 1985 and Oxburgh et al., 1986).

2. ALBERTA IN A NUT SHELL

The analyses of noble gases in >20 CH_4-rich natural gases from mostly Devonian reservoirs throughout Alberta in the Western Canada Sedimentary Basin revealed two distinct noble gas components mixed in varying proportion (Hiyagon and Kennedy, 1992). One component has a "normal" radiogenic and nucleogenic composition and is derived from the Precambrian basement. This follows from (1) a uniform composition for reservoirs scattered over large lateral distances (>200 km) implying a spatially extensive and homogeneous source; (2) a nucleogenic-radiogenic composition as expected for production in average crust and identical to that in brines produced from the Precambrian basement (Bottomley et al, 1984); (3) an increase in the proportion of this component with proximity to the Precambrian basement; and (4) an exponential decrease in He concentration with increasing distance from the basement.

The second component appears to be a ubiquitous background feature of the sedimentary basin and compositionally is characterized by lower total helium and anomalously high $^3He/^4He$ ratios. As the basement component infiltrates the basin it dilutes the anomalous background component. Mixing of these two components is clearly demonstrated in Figure 1 ($^{21*}Ne$ is neon in excess of atmospheric abundance as defined by the measured $^{21}Ne/^{20}Ne$ ratio). Although only those reservoirs most enriched in the anomalous component are shown, the solid line was generated by a weighted least squares method using the entire data set. As there have not been any recent (<200 ma) volcanic events in Alberta of significant magnitude capable of providing the large abundance of "excess" ^3He estimated for the basin, the source must be exotic.

3. POTENTIAL SOURCES FOR EXCESS ^{3}HE

What ever the source and/or process, it must be a common and widespread phenomenon. We have found identical mixtures of "normal"

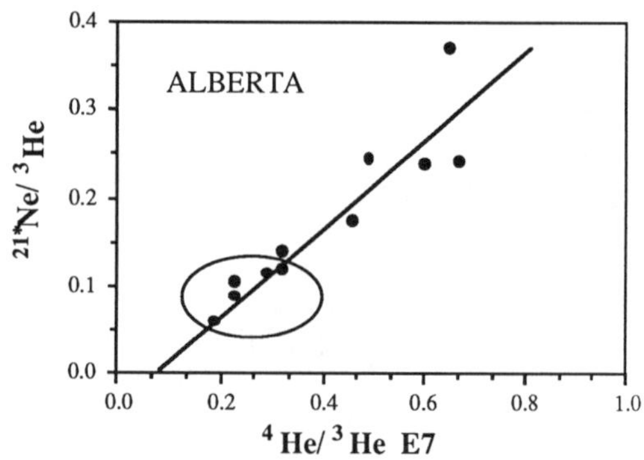

Figure 1: Covariant excess $^{21*}Ne$ and helium in Alberta natural gas wells.

radiogenic and anomalous ^3He-rich fluids in several other tectonically stable and volcanically quiet sedimentary basins, as demonstrated in Figure 2 for the Pecos Slope in southeast New Mexico (immediately north of the Delaware Basin) and the Panhandle-Hugoton Field of Texas, Oklahoma, and Kansas. Several possibilities exist: surface cosmogenic ^3He production, cosmic dust, nucleogenic production in a Li-enriched environment, or the mantle.

3.1 Cosmogenic production

Operating on the assumption that excess ^{21}Ne and He fractionation is not significant during extraction from their respective sources and subsequent migration, $^{21*}Ne/^{3}He$ ratios lower than 0.3, as observed in all three localities, rule out a cosmogenic source.

3.2 Cosmic dust

Cosmic dust particles (IDP's), being saturated in solar wind gases, can have He concentrations as high as 0.1% with $^{3}He/^{4}He$ ratios of 2 X 10^{-4}. The dust flux is such that this constitutes a significant source of ^{3}He at the surface of the earth. Identifiable IDP's have been extracted magnetically from recent deep ocean sediments (Ozima et al., 1984 and Amari and Ozima, 1985). However, to maintain the observed high $^{3}He/^{4}He$ ratios in the 200-300 ma old reservoir-source-cap rocks would require an IDP/(U+Th) ratio in the bulk sediments much higher than anticipated given the cosmic dust flux and high sedimentation rates in the shelf environment.

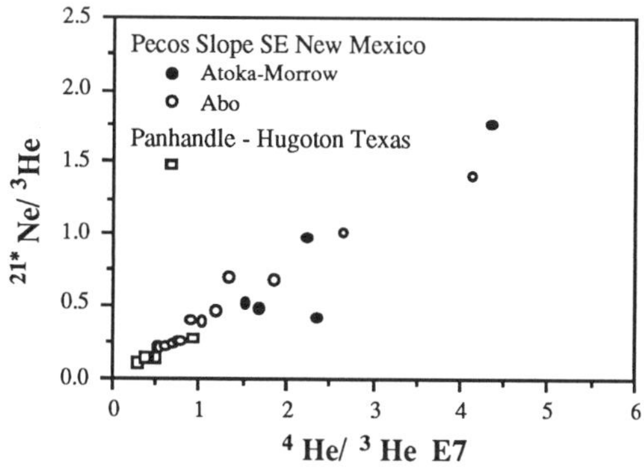

Figure 2: Covariant excess $^{21*}Ne$ and helium in natural gases from the Pecos Slope, New Mexico and the Panhandle-Hugoton Field of Texas and Oklahoma.

3.3 Production in Li-rich environments

$^{3}He/^{4}He$ ratios of 10-70 X 10^{-8} produced in sedimentary minerals such as carbonates (limestones and dolomites), anhydrites, and various clays require Li concentrations in excess of 1000 ppm or 10-100 times higher than typically encountered. However gas reservoirs are commonly capped by anhydrites and evaporites and Tardy et al (1972) found that Mg-rich clays formed under evaporitic conditions can be extremely enriched in Li, with concentrations as high as 6000 ppm depending on MgO content. If similar Li-enriched phases are associated with the gas reservoirs of this study and they preferentially release their ^{3}He-rich radiogenic component, then they may provide an adequate source for the apparent excess ^{3}He. In Figure 1, the five samples within the ellipse are all from southeast Alberta, where the carbonate reservoirs have been completely dolomitized and evaporite/carbonate ratios are high ranging from ~1-3 (Prather and McCourt, 1968). The reservoirs with the highest $^{3}He/^{4}He$ ratios are dolomitized biostromal reefs completely encased by evaporites. Anhydrite nodules extracted from the Hollenburg Formation of the Hugoton Field have $^{3}He/^{4}He$ ratios of ~60 X 10^{-8}. Li concentrations in the bulk nodules, however, are only ~120 ppm and too low to account for the high ^{3}He. Obviously, a more thorough understanding of Li, U, Th, and He siting in these and similar rocks is needed.

3.4 A mantle source

As argued convincingly by Oxburgh et al. (1986), young sedimentary basins formed by extensional forces, accompanied by lithospheric thinning and athenospheric upwelling, all contain significant quantities of mantle-derived excess ^{3}He. In this study, however, none of the basins associated with the gas reservoirs were formed by extension. The Alberta Basin, in fact, is a classic example of a continental terrace wedge, with crustal subsidence driven by an imposed load from sedimentation and accumulation of thrust sheets (the Laramide orogeny). In this and similar environs, crustal thinning and mantle melting are not expected. High $^{3}He/^{4}He$ ratios may reflect either hereto unknown zones in the local crust of high permeability to mantle volatiles or some other tectono-magmatic processes. For example, during orogenies, such as the Laramide or Wichita, plate collisions generate large compression forces, extensive thrust faulting, and presumably injection of mantle-derived magmas deep in the crust. Combined these could provide the energy, the source, and pathway for transport of copious volumes of fluids enriched in mantle volatiles deep into the bordering sedimentary basins (eg. Fyfe and Kerrich, 1985). A remarkable correspondence in time and space between remagnetization events in foreland carbonates, induced by fluid-rock interaction, and nearby orogenic events provides further evidence for the injection of thermal fluids deep into craton interiors (Van der Voo, 1989).

REFERENCES

Amari, S. & M. Ozima 1985. Search for the origin of exotic helium in deep-sea sediments. Nature 317: 520-522.

Bottomley, D. J., J. D. Ross, and W. B. Clarke 1984. Helium and neon isotope geochemistry of some ground waters from the Canadian Precambrian Shield. Geochim. Cosmochim. Acta 48: 1973-1985.

Fyfe, W. S. & R. Kerrich 1985. Fluids and thrusting. Chem. Geol. 49:353-362.

Hiyagon, H. & B. M. Kennedy 1992. Noble gases in CH_4-rich gas fields, Alberta, Canada. Geochim. Cosmochim. Acta, in press.

Oxburgh, E. R., R. K. O'Nions, and R. I. Hill 1986. Helium isotopes in sedimentary basins. Nature 324: 632-635.

Ozima, M., M. Takayanagi, S. Zashu, and S. Amari 1984. High $^3He/^4He$ ratio in ocean sediments. Nature 311: 448-450.

Polyak, B. G. & I. N. Tolstikhin 1985. Isotopic composition of the earth's helium and the problem of the motive forces of tectogenesis. Chem. Geol. (Isotope Geosciences Section) 52: 9-33.

Poreda, R. P. D. Jenden, I. R. Kaplan, and H. Craig 1986. Mantle helium in Sacramento Basin natural gas wells. Geochim. Cosmochim. Acta 50: 2847-2853.

Prather, R. W. & G. B. McCourt 1968. Geology of gas accumulations in Paleozoic rocks of Alberta plains. In B. W. Beebe & B. F. Curtis (eds.), Natural Gases of North America. Amer. Assoc. Petrol. Geol. Mem. 9: 1238-1284.

Tardy, Y., G. Krempp, & N. Trauth 1972. Le lithium dans les mineraux argileux des sediments et des sols. Geochim. Cosmochim. Acta 36: 397-412.

Van der Voo, R. 1989. Paleomagnetism of North America: The craton, its margins and the Appalachian belt. Mem. Geol. Soc. Am. 172: 447-470.

Zartman, R. E., G. J. Wasserburg, & J. H. Reynolds 1961. Helium, argon, and carbon in some natural gases. J. Geophys. Res. 66: 277-306

Water-Rock Interaction, Kharaka & Maest (eds) © 1992 Balkema, Rotterdam. ISBN 90 5410 075 3

Changes of stable istopes and chemistry of fluids in the low-temperature geothermal field at Bakki-Thóroddstadir, Ölfus, SW-Iceland

Hrefna Kristmannsdóttir
Orkustofnun, Reykjavík, Iceland

Árný Erla Sveinbjörnsdóttir
Science Institute, University of Iceland, Reykjavík, Iceland

ABSTRACT: Stable isotope results in the Bakki-Thóroddsstadir field suggest that the thermal fluids originate from the Langjökull ice cap some 70 km away. The original δD is still reflected in the fluid whereas an oxygen shift of approximately 1‰ has occurred.

The field has been exploited for 15 years and no changes in temperature or chemistry were observed until 1988 when a new well was put into production. The chemical composition and stable isotope ratios of the water from the well soon showed signs of dilution with cold freshwater.

An inflow of cooler water was subsequently localized by logging and downhole sampling. After successful repair of the well, the temperature and chemical composition of the water has gradually been restored, whereas previous stable isotopes ratios have not been approached.

1 INTRODUCTION

The Bakki - Thóroddsstadir geothermal field in Ölfus, SW Iceland, is located along a NE elongated strip in the direction of the main fault swarms in the region (Fig. 1). The field is about 15 km south of the borders of the Hengill central volcano, which comprises three major high-temperature geothermal areas. The Ölfus region is on the flank of the western active volcanic zone with high heat flow and many low-temperature geothermal fields (Sæmundsson, 1979). Natural hot springs occurred both NE and SW of the drilled area at Bakki (Fig. 1). The ones NE of the field had temperatures of up to 80°C, but they dried up soon after well BA-01 was drilled.

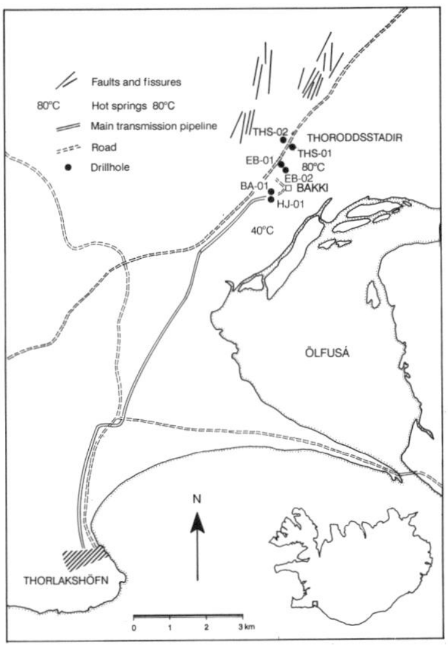

Figure 1. The location of the geothermal field at Bakki-Thóroddsstaðir.

The bedrock in the field is composed of basalt lavas, basaltic hyaloclastites, and clastic sediments of young Quaternary age. The sea has repeatedly submerged the area during the build-up of the rock formations, forming sediments and raising the salt content of the reservoir rocks and groundwater in the region (Kristmannsdóttir et al. 1990). The permeability of the rock formations is generally high, and the field is fractured with NE trending faults and fissures, resulting in high transmissivity and productive geothermal wells.

The water from the geothermal field is utilized for a heating system, *hitaveita*, in the town of Thorlákshöfn with about 1200 inhabitants, and also for nearby fish farms.

Four geothermal wells have been drilled into the Bakki - Thóroddsstadir geothermal field (Fig. 1). The wells are free flowing and the temperature at the wellhead is about 116-120°C. The water is flashed in a wellhead separator to atmospheric pressure. The maximum downhole temperatures are 135°C.

The pressure and temperature at the wellhead and the discharge have been monitored regularly in wells BA-01 and HJ-01. The geothermal water was analyzed just after drilling, and during the last five years the chemical composition of the geothermal water has been monitored regularly. No changes of temperature or pressure at the wellhead have been observed. Samples for chemical analysis were taken from THS-01 in 1977, 1978, and 1991; and from EB-01 in 1989 and 1991.

2 WATER CHEMISTRY AND STABLE ISOTOPES

The water is slightly saline. At the outset, fluids from wells BA-01 and HJ-01 contained approximately 650 to 700 mg/l of chloride (Table 1), whereas in wells EB-01 and THS-01 the salinity is lower, 400-470 mg/l. The fluids are in equilibrium with chalcedony and calcite at high pH, as is typical for low-temperature fields in Iceland. They contain 0.1-0.3 mg/l of H_2S.

Table 1. Chemical composition of water from several wells in the Ölfus area. Concentrations in (mg/l).

	BA-01	HJ-01	HJ-01	EB-01	THS-01	Bakki cold
Date	850115	850115	910124	910731	910731	910731
Number	0006	0007	0003	0165	0166	9082
°C	116,0	108,0	99,5	~100,0	113	5,5
pH/°C	8,7/22	8,7/22	9,1/24	9,0/25	8,6/25	8,0/25
SiO_2	132,6	118,3	108,7	118,4	118,4	15,6
Na	386	402	260	302	255	8,3
K	19,2	17,3	11,1	14,2	10,5	0,7
Ca	74,4	86,8	35,1	49,9	29,2	5,3
Mg	0,02	0,02	0,001	0,013	0,026	1,95
CO_2	7,7	8,7	7,8	6,2	17,7	19,7
SO_4	121,2	127,3	75,3	120,0	82,2	2,4
H_2S	0,29	0,29	0,12	0,19	0,11	<0,03
Cl	634	682	396	473	390	10
F	0,50	0,47	0,56	0,55	0,69	0,05
TDS	1470	1546	860	1138	938	48
B	0,27	0,26	-	0,28	0,31	-
Br	-	-	1,34	1,63	1,38	0,03
Al	0,08	0,06	-	-	-	-
$\delta^{18}O$ ‰	-9,35	-9,39	-8,67	-9,14	-9,12	-8,25
δD ‰			-64,6	-71,4	-70,7	-56,1

Figure 2 shows the isotopic composition of the original geothermal water at Bakki-Thóroddsstadir ($\delta^{18}O$ = -9.3‰ and δD = -71‰) and of the local precipitation ($\delta^{18}O$ = -8.3‰ and δD = -56‰). Also shown is the meteoric line (Craig, 1961) and, for comparison, the isotopic composition of water from the Seltjarnarnes geothermal field, SW-Iceland, which has almost the same base temperature as Bakki. The low δD values for the thermal water at Bakki-Thóroddsstadir suggest a distant origin of the water; most likely it originates from the Langjökull ice cap, some 70 km north of the area. The water at the Bakki-Thóroddsstadir field shows a distinct oxygen shift of about 1‰, as demonstrated in Fig. 2. Oxygen shifts have not previously been reported for waters from low-temperature geothermal fields in Iceland. The reason for the oxygen shift in the Bakki-Thóroddsstaðir field is not known. One possibility is

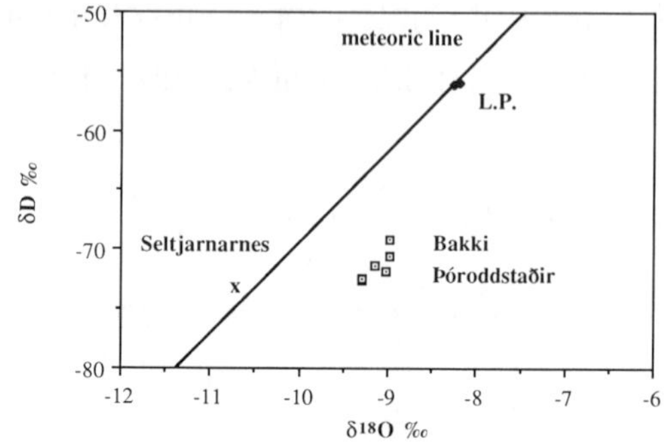

Figure 2. The relationship between δD and $\delta^{18}O$ for samples at Bakki-Thóroddsstaðir. Also shown is the meteoric line (Craig, 1961), local annual mean precipitation (L.P.), and, for comparison, the δD against $\delta^{18}O$ for geothermal water from the Seltjarnarnes field, SW-Iceland.

that the underground rocks are very young and contain much fresh and reactive glass that might allow an effective oxygen isotope exchange even at those low temperatures. The salinity of the water may also increase its reactivity and therefore enhance isotopic exchange. Thirdly, the recharge to the Ölfus low-temperature fields could partly be a runoff from the high-temperature fields within the Hengill volcanic area, which is situated only 15 km to the north of the Bakki-Thóroddsstaðir field. As shown in Fig. 3 the water from the Kolviðarhóll high-temperature geothermal field (in the Hengill area) has stable isotopic ratios almost identical to those of the Bakki-Thóroddsstaðir low-temperature geothermal field.

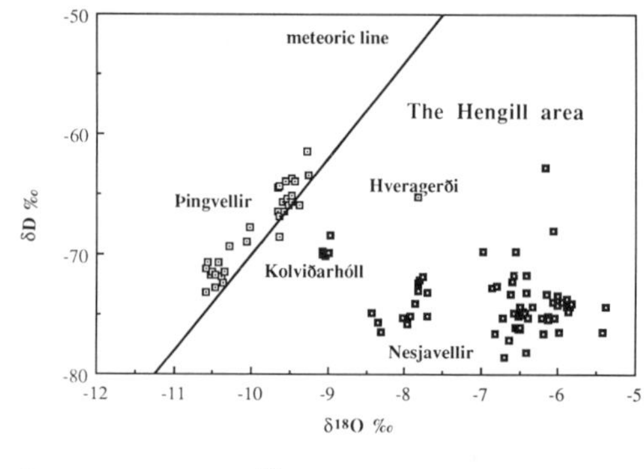

Figure 3. δD against $\delta^{18}O$ for geothermal water at three high-temperature fields in the Hengill volcanic area: Kolviðarhóll, Hveragerði, and Nesjavellir (Sveinbjörnsdóttir, 1989). Also shown is cold groundwater in the Thingvellir national park just north of Hengill.

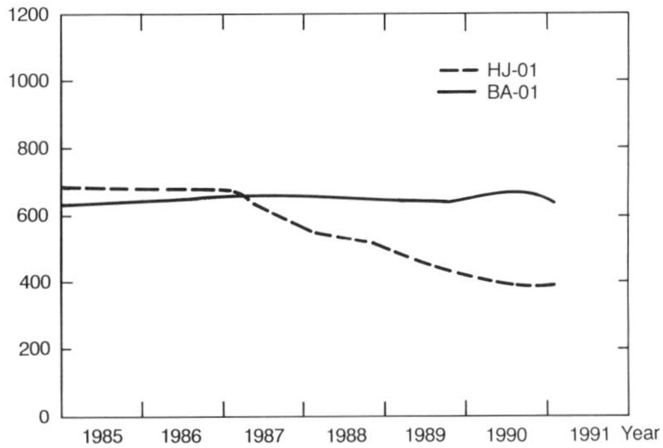

Figure 4. Changes in chloride concentration in the geothermal water with time.

The chemistry of geothermal water from wells THS-01 and BA-01 has not changed during the 15 years of production, whereas just a few months after the connection of well HJ-01 remarkable changes in the chemistry of its fluids were noticed (Fig. 4 and Table 1). As the main change appeared to be the dilution of the geothermal water with fresh water, it was suspected that an inflow of cold groundwater through damaged casing was the cause. Therefore the well was logged for temperature and pressure in September 1988, but only minor temperature changes were seen, and no sign of damage to the casing was observed.

In addition to the changes in the chemistry of the water, some sudden and drastic changes in the stable isotope ratio of the water were observed (Fig. 5 and 6), which also indicated an inflow of local groundwater.

As shown in Figs. 5 and 6 the samples from well HJ-01 at Bakki fall on a dilution line between the values for the local precipitation and cold spring water on one hand, and the geothermal water from Thóroddsstaðir and undiluted Bakki water on the other.

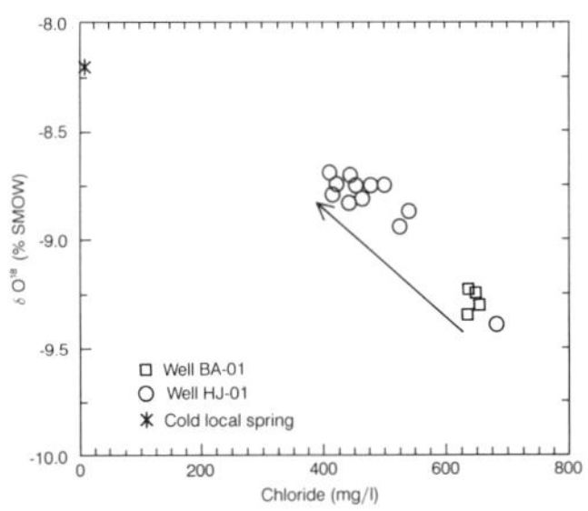

Figure 5. Oxygen isotope and chloride composition in wells BA-01, HJ-01 and a cold spring.

The $\delta^{18}O$ values of well HJ-01 changed very quickly to a fixed value after the well was connected, but the salinity of the water has been changing more gradually. No significant changes in stable isotopes were recorded in the water from well BA-01 or the other wells in the field, however. The chemical changes in well HJ-01 continued without any apparent changes of wellhead temperature or pressure. As seen in Table 1 and Fig. 7, the drop in chalcedony geothermometer temperature had reached about 5°C at the beginning of 1990. In February of that year new logs revealed cooling at the bottom of the well (Fig. 8).

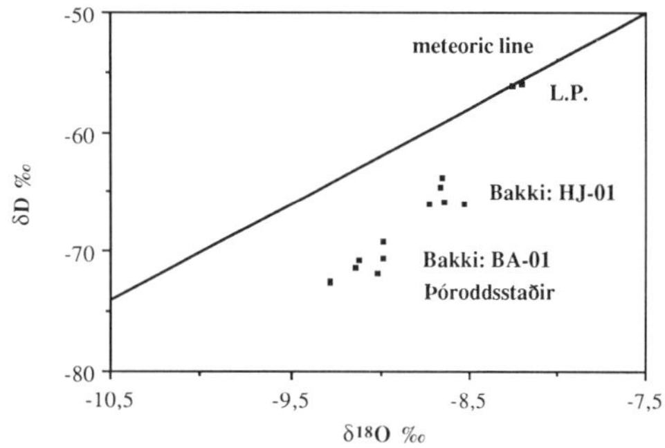

Figure 6. The relationship between δD and $\delta^{18}O$ for samples from Bakki-Thóroddsstaðir wells.

At the same time, five downhole samples were taken from different aquifers in the well, verifying the inflow of cold local groundwater at the bottom (Figs. 9 and 10).

Analyses of water sampled at the wellhead at different flow rates also reflected the variable chemical composition of the aquifers (Kristmannsdóttir et al. 1990). Taking the inflow at the bottom as one end-member and water from well BA-01 (or original water from well HJ-01) as the other, we find that the well water was mixed approximately in the ratio 70:30. The temperatures of the different aquifers indicate, however, that approximately 20% of the discharge was from the bottom aquifer, suggesting

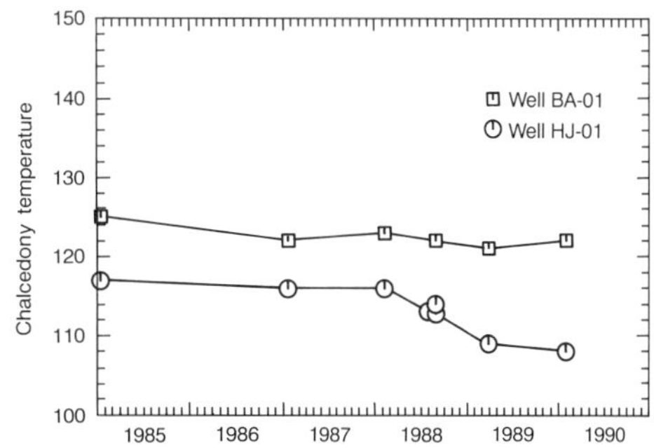

Figure 7. The calculated chalcedony temperature vs. time in wells BA-01 and HJ-01.

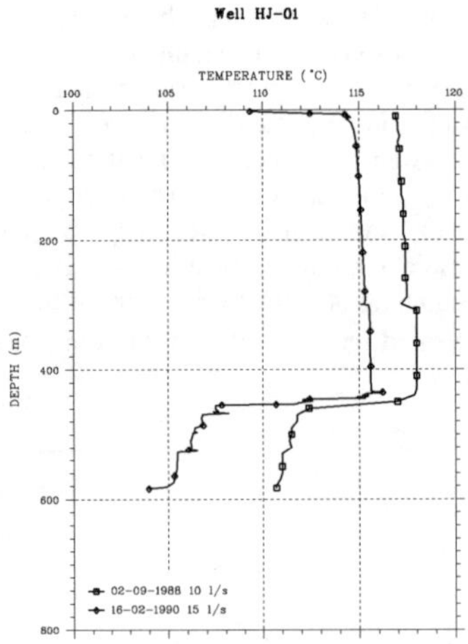

Figure 8. Temperature logs from well HJ-01.

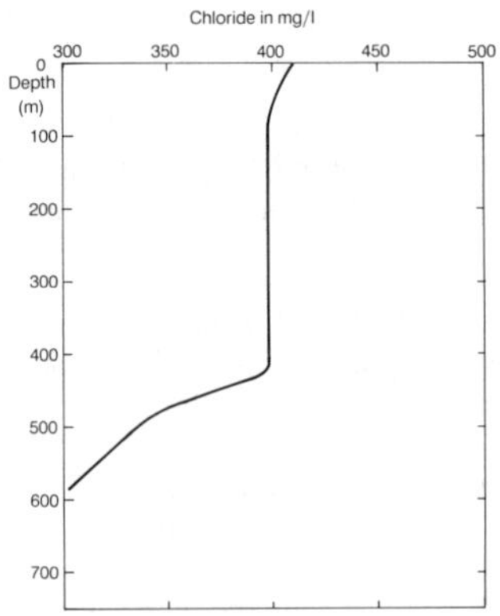

Figure 9. The concentration of chloride in downhole samples.

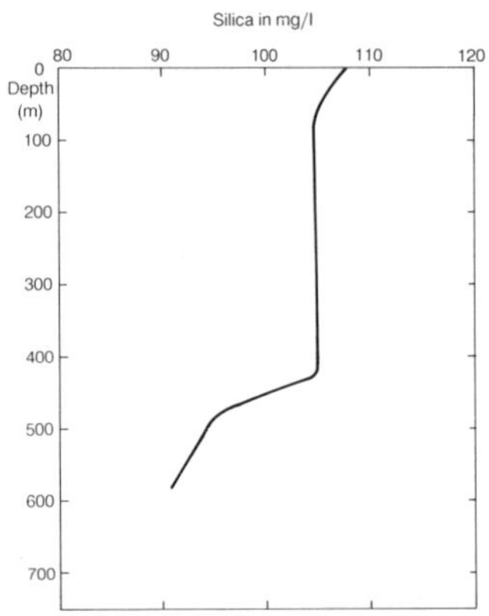

Figure 10. The silica concentration in downhole samples from well HJ-01.

that the upper aquifers were also affected by the cold inflow.

In late 1990 the bottom aquifer in well HJ-01 was sealed off by cementing, and the cold inflow now appears to have been stopped entirely. No recent temperature logs are available, but the chemistry is slowly changing back to its former condition as can be inferred from the change in slope of Cl versus time in Fig. 5. Even so, the stable isotope ratios in well HJ-01 show no change. A slight change in $\delta^{18}O$, was recently observed in wells BA-01 and EB-01. This may presage increased recharge of local groundwater.

3 CONCLUSION

During 15 years of exploitation the first utilized wells in the Bakki-Thóroddsstadir geothermal field have not exhibited noticeable changes. Inflow of colder water to a more recent well, however, disturbed the chemical composition and stable isotope ratios of its fluids in a short time after the well was put into production.

The inflow into the system affected only one of the two very closely spaced wells, which were, at the outset, fed by water of almost identical chemical composition and of the same origin. The cooling starts from the bottom aquifer, which is unique in the history of the exploitation of Icelandic geothermal fields, and is explained by an inflow through a fracture or fracture system directly connected to the reservoir feeding the bottom aquifer. The other aquifers were also somewhat affected.

After the bottom aquifer was sealed off, the water has slowly been regaining its original chemical composition, but so far its former isotopic value has not been approached. Small changes in the stable isotope ratios in other wells may indicate that these are a starting to recharge from local groundwater.

ACKNOWLEDGEMENTS

Our colleagues Magnús Ólafsson, Hilmar Sigvaldason, and the manager of Hitaveita Thorlákshafnar, Guðlaugur Sveinsson, are thanked for support and cooperation during this project.

REFERENCES

Craig, H. 1961. Isotope variations in meteoric waters. *Science* 133: 10703-10703.

Kristmannsdóttir, H., M. Ólafsson, H. Sigvaldason & H. Tulinius 1990. Exploitation induced influx of cold local groundwater into the geothermal system at Bakki, Ölfus SW-Iceland. *Geothermal Res. Counc. Transactions,* 14: 715-719.

Sveinbjörnsdóttir, Á.E. 1989. Stable isotope measurements on the thermal water at the Nesjavellir and Mosfellssveit geothermal fields, Iceland. In: *Miles D.L. (ed.), Water-Rock Interactions:* 665-668. A.A. Balkema Rotterdam.

Sæmundsson, K. 1979. Outline of the geology of Iceland. *Jökull,* 29: 7-28.

Water-Rock Interaction, Kharaka & Maest (eds) © 1992 Balkema, Rotterdam. ISBN 90 5410 075 3

Stable isotope studies of diagenesis in berthierine-bearing oil sands, Clearwater Formation, Alberta, Canada

F.J. Longstaffe & M.A. Racki
Department of Geology, University of Western Ontario, London, Ont., Canada

A. Ayalon
Department of Geology, University of Western Ontario, London, Ont., Canada (Presently: Geological Survey of Israel, Jerusalem, Israel)

ABSTRACT: Diagenetic minerals in thick oil sands from the deltaic Lower Cretaceous Clearwater Formation, Cold Lake, Alberta include early diagenetic smectitic clays and berthierine, and several generations of calcite. Oxygen-isotope data suggest that early diagenesis occurred in brackish to fresh water. Calcite crystallization was initially controlled by carbon dioxide of inorganic origin, but later was driven by microbial fermentation as temperatures rose. The association between berthierine and ^{13}C-rich calcite may indicate that microbial activity played a role in berthierine genesis. Berthierine survived because of the relatively low maximum temperatures attained during burial.

1 INTRODUCTION

This paper describes stable isotope results for diagenetic minerals from oil sands of the Lower Cretaceous Clearwater Formation in the Cold Lake area, east-central Alberta. This unit contains an estimated 69 billion barrels of heavy oil or bitumen in place, and is part of Canada's enormous Mannville Group oil sand reserves. We have focused on the water-organic-mineral interactions and temperatures that controlled diagenesis in these sediments. The abundance of diagenetic berthierine is of special interest, as are ^{13}C-rich calcite cements, earlier noted by Dimitrakopoulos and Muehlenbachs (1987).

2 GEOLOGICAL BACKGROUND

The Clearwater Formation was deposited in a nearshore, deltaic complex, which prograded in a north-northeasterly direction, forming a series of overlapping lobes (Harrison et al., 1981; Wightman et al., 1989). It varies in thickness from 35 to 65 m, thicker intervals resulting from stacking of successive deltaic lobes. Some of the highest quality oil sands occur in distributary channel and distributary mouth-bar facies of the complex in the south-central portions of the study area. Further seaward, the sands thin and shales become more prominent.

The sands consist mainly of bitumen-saturated or water-saturated, fine or medium-grained, (feldspathic) litharenites. They are dominated by rock fragments, mostly volcanic, chert and shale; plagioclase and quartz grains are less abundant. Thin calcite ± siderite-cemented intervals are ubiquitous.

Present burial depths are ≈400 to 500 m. Maximum overburden removed during post-Laramide erosion is unlikely to have exceeded ≈1000 m (Hacquebard, 1977; Beaumont et al., 1985). Calculations based on reasonable geothermal gradients for Eocene maximum burial suggest that temperatures never exceeded 80°C, and may have been lower. Nevertheless, because of their reactive mineralogy, the sands experienced extensive diagenesis.

3 MINERALOGY AND DIAGENESIS

In bitumen-saturated sands, early diagenetic minerals include pelloidal glauconite, Fe-bearing, smectitic clays (smectite-chlorite, smectite-illite), berthierine (Fe-rich 1:1 layer silicate similar in chemical composition to chamosite), pyrite, siderite and calcite. Early clay crystallization was partly preceded by concretionary calcite cementation, which replaced framework grains and filled pores. Glauconite/illite is commonly abundant in, and adjacent to, those intervals which were calcite cemented. Many sands also show a progressive increase in smectitic or illitic clay content and a decrease in berthierine as the calcite-cemented intervals are approached. Early framboidal pyrite formed following this stage of calcite cementation.

Fe-poor, grain-coating calcite with high $\delta^{13}C$ values formed somewhat later in some bitumen-saturated sands and calcite-cemented sandstones from the south-central portions of the study area, the same general area where berthierine is also abundant. Burial diagenesis was also accompanied by continued dissolution of feldspars and volcanic rock fragments, and crystallization of euhedral pyrite, K-feldspar and silica overgrowths, siderite, zeolites, and in the northern portion of the study area, kaolinite. The quartz-K feldspar-zeolite assemblage represents diagenesis at or near maximum burial temperatures (Abercrombie, 1989).

Precipitation of diagenetic, Fe-rich calcite dominated later stages of diagenesis. In the fringes of the deltaic complex, pore-filling kaolinite also continued to crystallize. Kaolinite is much rarer within the thick, bitumen-saturated sands from the central portions of the deltaic complex, occurring locally only within calcite and/or siderite-rich zones of low bitumen-saturation, or in sands situated close to the bitumen-water contact. Diagenesis was then greatly attenuated by emplacement of hydrocarbons, which were subsequently altered.

Water-saturated sands show the same general diagenetic pattern, except that grain-coating clays like berthierine and Fe-rich smectitic clays are rare, and early diagenetic grain-coating calcite with high $\delta^{13}C$ values is almost never encountered. Broadly similar paragenetic sequences have been reported in localized studies of the Clearwater Formation (e.g., Abercrombie 1989; Hutcheon et al., 1989).

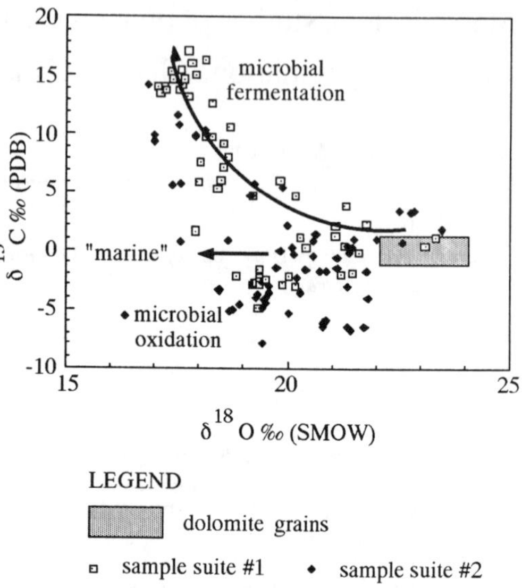

Fig. 1 Carbon- and oxygen-isotope compositions of diagenetic calcite. Suite #1 samples were collected only from oil-sands pilot sites; suite #2 represents the entire deltaic complex.

4 STABLE ISOTOPE GEOCHEMISTRY

Carbon- and/or oxygen-isotope results have been obtained for several diagenetic phases, including calcite (Figure 1), smectitic clays ($\delta^{18}O$ = +18.0 to +18.8 ‰), berthierine (+6.6 to +12.5 ‰), K-feldspar (+15.9 to +17.0 ‰) and quartz (+21.0 to +23.2 ‰). The O-isotope data have been interpreted using the quartz-water (Clayton et al., 1972), alkali feldspar-water (O'Neil and Taylor, 1967) and calcite- water (O'Neil et al., 1969) geothermometers, after Friedman and O'Neil (1977), and the smectite-water equation of Savin and Lee (1988), adjusted for interstratification. A berthierine-water, O-isotope geothermometer was calculated using chemical compositions for Clearwater berthierine, after Brindley (1982), and the method of Savin and Lee (1988). The resulting equation (T in °K) is:

$$1000\ln\alpha = 4.375(10^3)T^{-1} + 2.602(10^6)T^{-2} - 0.422(10^9)T^{-3} + 0.038(10^{12})T^{-4} - 12.95. \quad (1)$$

4.1 Isotopic evolution of diagenetic calcite

Diagenetic calcite shows a large range in $\delta^{13}C$ (-7.9 to +17.1 ‰) and $\delta^{18}O$ (+16.3 to +23.1 ‰) values (Figure 1). Samples with $\delta^{13}C$ values ≈ 0±2 ‰ are most typical of early calcite in bitumen-poor, water-saturated and calcite-cemented sands from the fringes of the deltaic complex. However, they also also occur in south-central portions, mostly in upper parts of the bitumen-bearing interval. Dominantly inorganic sources of carbon (i.e., marine pore-waters, dolomite grains that are locally abundant) are indicated by these C-isotope values. The cements also have $\delta^{18}O$ values of +20 to +23 ‰ (Figure 1), lower than expected for normal seawater during Cretaceous time. Porewaters during very early diagenesis appear to have been very brackish.

The second generation of diagenetic calcite is characterized by dramatically increasing $\delta^{13}C$ and gradually decreasing $\delta^{18}O$ values. It occurs in bitumen-free and bitumen-bearing sandstones, and bitumen-saturated sands, with $\delta^{13}C$ values generally increasing in that order. Normally, the ^{13}C-rich calcite is most abundant in lower portions of the bitumen-saturated interval, interbedded with bitumen-saturated sands that contain berthierine. The high, positive $\delta^{13}C$ values suggest generation of CO_2 by microbial fermentation of organic matter. The lowering of $\delta^{18}O$ values as $\delta^{13}C$ values rose probably resulted from increasing temperature as burial and fermentation progressed.

A third, more poorly defined generation of generally Fe-rich calcite cement shows a broad tendency towards low $\delta^{13}C$ and (in some cases) $\delta^{18}O$ values. Its isotopic composition reflects crystallization from formation waters that contained a sizeable fraction of meteoric water, and carried a mixture of carbonate ions, including some derived from low ^{13}C sources (perhaps oxidation of hydrocarbons by aerobic microbial action during water-washing).

Dimitrakopoulos and Muehlenbachs (1987) suggested that the ^{13}C-rich carbonate cements were derived from Clearwater hydrocarbons through a two stage process, degradation of hydrocarbons by aerobic microbes, followed by microbial fermentation of the resulting biomass and its by-products. In our samples, ^{13}C-rich calcite commonly predates local hydrocarbon emplacement on a pore by pore basis. But hydrocarbon emplacement in the Clearwater Formation may have occurred over a long period of time. Alteration of hydrocarbons trapped earlier could promote crystallization of isotopically appropriate calcite cements in parts of the system that were not yet bitumen saturated.

4.2 Oxygen-isotope evolution of porewaters

The paragenesis and O-isotope compositions of diagenetic minerals have been combined in Figure 2 to produce a model for porewater evolution. The curves (thin lines) are solutions to the mineral-water equations for the range of $\delta^{18}O$ values obtained for each diagenetic silicate mineral. The idealized porewater evolution pathway (heavy arrows) has been drawn to (i) intersect the mineral-water curves in a fashion compatible with the paragenetic sequence,

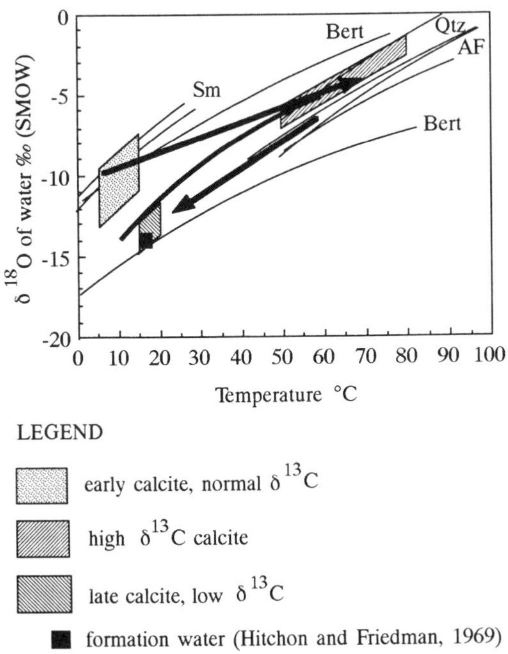

Fig. 2 Idealized porewater evolution pathway. The shaded areas indicate proposed conditions for crystallization of calcite. Sm = early smectitic clays; Bert = berthierine; Qtz = quartz; AF = K-feldspar.

(ii) not exceed burial temperatures of 70-80°C, and (iii) end at the current composition of Mannville Group formation water.

To begin crystallization of the early diagenetic smectitic clays at ≈10±5°C, porewater compositions of about -10 ‰ (i.e., brackish to freshwaters) are required. Such values overlap those needed to produce normal ^{13}C (≈ 0 ‰) diagenetic calcite formed at about the same time. But even lower $\delta^{18}O$ values (still fresher water) are needed to account for at least some O-isotope values obtained for berthierine.

Crystallization of quartz and alkali feldspar at or near maximum temperatures required ^{18}O enrichment of the porewater during burial. Continued formation of berthierine during burial is consistent with this porewater pathway, as are the O-isotope compositions of the high-^{13}C calcite (shown for 50 to 80°C) associated with berthierine-bearing sands.

Substantial cooling and lowering of porewater $\delta^{18}O$ values are required to achieve the current porewater composition. Both features are consistent with post-Laramide erosion, and infiltration of meteoric water. The O-isotope results for the latest diagenetic calcite are consistent with crystallization from such water at current subsurface temperatures.

5 DISCUSSION AND CONCLUSIONS

The deltaic setting of the Clearwater sands accounts in a general way for the dominance of brackish to meteoric porewaters throughout diagenesis. For example, the early diagenetic calcite concretions with $\delta^{13}C$ values ≈ 0 ‰ and $\delta^{18}O$ values that are the highest for calcite in the study area (but still low relative to normal marine carbonate), are most abundant in the fringes of the deltaic complex. These compositions can be explained by low temperature mixing between seawater and freshwater during very early diagenesis. The presence of glauconite provides further evidence for the transitional nature of this environment.

The early diagenetic evolution of the berthierine-bearing bitumen reservoirs in the thickest portion of the deltaic complex to the south is also probably related to sedimentation. Reducing conditions would be quickly achieved as earlier sands were rapidly buried by later deposits, trapping dominantly meteoric porewaters. Degradation of organic matter would then proceed primarily by microbial fermentation, leading to production of the ^{13}C-rich CO_2 that typifies grain-coating calcite in this locality. The lower calcite $\delta^{18}O$ values probably reflect at least in part higher temperatures resulting from burial and the fermentation process itself.

Where the low-Fe, high $\delta^{13}C$ grain-coating calcite cements did not crystallize, smectitic clays or Fe-rich, grain-coating berthierine developed. The O-isotope data suggest that berthierine (rather than smectitic clays) preferentially developed in facies where freshwater was most abundant.

Berthierine is not common in the diagenetic record. Its primary crystallization appears to require a supply of Fe, and a reducing environment, or at least one in which the supply of oxygen has been greatly restricted (Curtis and Spears, 1968). Formation of Fe-rich precursors to berthierine can be favoured in certain parts of the deltaic environment, such as distributary channels and stream-mouth bars. River waters can have high Fe-contents. Fe-oxides and hydroxides deposited by such waters under oxidizing conditions become unstable under reducing conditions, providing a supply of Fe. In the Clearwater Formation, dissolution of Fe-rich volcanic rock fragments would have provided an adequate supply of iron. Growth of berthierine as a grain-coating may simply reflect recrystallization of "Fe-oozes" which had been accreted about sand grains by current movement in a distributary-channel type environment. However, the association between the ^{13}C-rich calcite cements and berthierine may be more than coincidence. Microbes that contributed to formation of ^{13}C-rich calcite elsewhere in this system may also have concentrated metals such as Fe on their membranes. The resulting, metal-rich microbial slimes that coated grains may have led to preferential nucleation of berthierine (e.g., Ferris et al., 1987).

Curtis et al. (1985), among others, have noted that Fe-rich chlorite replaces berthierine and smectite-chlorite as temperatures rise. Jahren and Aagaard (1989) described pore-filling berthierine from sandstones that were undergoing transformation to chamosite at temperatures as low as 70°C. Perhaps temperatures needed to transform berthierine to Fe-chlorite in the Clearwater Formation were never achieved.

ACKNOWLEDGMENTS

We thank AOSTRA and NSERC for funding.

REFERENCES

Abercrombie, H. J. 1989. Water-rock interaction during diagenesis and thermal recovery, Cold Lake, Alberta. Ph.D. thesis, University of Calgary, Canada.

Beaumont, C., R. Boutilier, A.S. MacKenzie & J. Rullkotter 1985. Isomerization and aromatization of hydrocarbons and the paleothermometry and burial history of the Alberta foreland basin. Amer. Assoc. Petrol. Geol. Bull. 69: 546-566.

Brindley, G. W. 1982. Chemical compositions of berthierines - a review. Clays Clay Min. 30: 153-155.

Clayton, R.N., J.R. O'Neil & T.K. Mayeda 1972. Oxygen isotope exchange between quartz and water. J. Geophys. Res. 77: 3057-3067.

Curtis, C.D. & D.A. Spears 1968. The formation of sedimentary iron minerals. Econ. Geol. 24: 257-270.

Curtis, C.D., C.R. Hughes, J.A.Whiteman & C.K. Whittle 1985. Compositional variations within some sedimentary chlorites and some comments on their origin. Mineral. Mag. 49: 375-386.

Dimitrakopoulos, R.G. & K. Muehlenbachs 1987. Biodegradation of petroleum as a source of ^{13}C-enriched carbon dioxide in the formation of carbonate cement. Chem. Geol. (Isotope Geosci. Sect.) 65: 283-291.

Ferris, F.G., W.S. Fyfe & T.J. Beveridge 1987. Bacteria as nucleation sites for authigenic minerals in a metal-contaminated lake sediment. Chem. Geol. 63: 225-232.

Friedman, I. & J. R. O'Neil 1977. Compilation of stable isotope fractionation factors of geochemical interest. In M. Fleischer (ed.), Data of Geochemistry, 6th edition. U.S.G.S. Prof. Paper 440-KK.

Hacquebard, P. A. 1977. Rank of coal as an index of organic metamorphism for oil and gas in Alberta. In G. Deroo, T. Powell, B. Tissot & R. McCrossan (eds.), The origin and migration of petroleum in the Western Canadian sedimentary basin, Alberta. Geol. Surv. Can. Bull. 262: 11-22.

Harrison, D. B., R. P. Glaister & H.W. Nelson 1981. Reservoir description of the Clearwater oil sand, Cold Lake, Alberta Canada. In R. F. Meyer & C. T. Steele (eds.), The Future of Heavy Crude Oils and Tar Sands. p. 264-279. McGraw Hill, New York.

Hitchon, B. & I. Friedman 1969. Geochemistry and origin of formation waters in the western Canada sedimentary basin - I. Stable isotopes of hydrogen and oxygen. Geochim. Cosmochim. Acta 33: 1321-1349.

Hutcheon, I., H. Abercrombie, P. Putnam, R. Gardner & H.R. Krouse 1989. Diagenesis and sedimentology of the Clearwater Formation at Tucker Lake. Bull. Can. Petrol. Geol. 37: 83-97.

Jahren, J.S. & P. Aagaard 1989. Compositional variations in diagenetic chlorites and illites, and relationships with formation-water chemistry. Clay Min. 24: 157-170.

O'Neil, J. R. & H.P. Taylor, Jr. 1967. The oxygen isotope and cation-exchange chemistry of feldspars. Amer. Mineral. 52: 1414-1437.

O'Neil, J.R., R.N. Clayton & T.K. Mayeda 1969. Oxygen isotope fractionation in divalent metal carbonates. J. Chem. Physics 51: 5547-5558.

Savin, S. M. & M-C. Lee 1988. Isotopic studies of phyllosilicates. In S.W. Bailey (ed.), Hydrous Phyllosilicates (exclusive of micas). Reviews in Mineralogy 19: 189-223.

Wightman, D., B. Rottenfusser, J. Kramers & R. Harrison, R. 1989. Geology of the Alberta oil sands deposits. In L. G. Hepler & C. Hsi (eds.), AOSTRA Technical Handbook on Oil Sands, Bitumens and Heavy Oils. AOSTRA Technical Publication Series 6: 1-9.

Water-Rock Interaction, Kharaka & Maest (eds) © 1992 Balkema, Rotterdam. ISBN 90 5410 075 3

Mantle helium signal in the compressed feature, Outer Carpathians, Poland

P.M. Leśniak
Institute of Geological Sciences, Polish Academy of Science, Zwirki i Wigury, Warsaw, Poland

H. Sakai
Department of Earth Science, Yamagata University, Japan

J. Ishibashi & H. Wakita
Laboratory for Earthquake Chemistry, University of Tokyo, Japan

ABSTRACT: Mantle-derived helium (^3He) is thought to escape through the continental lithosphere in areas undergoing extension. We present evidence that preferential leakage of ^3He is also occurring in the Outer Carpathians, a compressed subduction-collisional feature of accretionary wedge and overriding Pannonian microplate. Mantle helium is accompanying cold CO_2 emanations and CO_2-rich carbonated waters. A maximum contribution of mantle helium was calculated to be 11%, the other components being deep crustal and atmospheric helium.

1 EXPERIMENTAL AND FIELD PROCEDURE

In September 1989, in the area of the collisional-subduction zone of the West Carpathians (Lefeld and Jankowski 1985; Oszczypko and Ślączka 1985; Fig. 1), 9 samples of water and one of gas were taken from springs and boreholes situated in the flysch rocks. In Szczawnica, boreholes Jan and B-4 are located at a close contact with the fractured andesite dykes. Other samples also included CO_2-CH_4 (Rymanów, Tytus) and methane (Rabka, R-18) dominated waters. Two sampling techniques were used for helium isotope analyses. Bubbling gases in waters were transferred to evacuated lead-glass bottles by means of two manifold stopcocks connected with a syringe. In some samples gases were first reacted with 10N KOH to remove CO_2, and residual gases were transferred to the bottles. The procedures were repeated until a few ml STP of residual gases were accumulated. In the Laboratory of the Earthquake Chemistry, University of Tokyo 0.5 cc STP of the gas sample extracted from water was admitted into the metallic vacuum line. Samples were purified using a Ti-Zr getter and two charcoal traps at liquid nitrogen temperature before introduction to a mass spectrometer.

The $^4He/^{20}Ne$ ratios were measured by the quadrupole mass spectrometer (QMG112, Balzers). Helium was completely separated from neon using a cryogenic charcoal trap held at 40K and transferred into a high precision mass spectrometer (VG5400). $^3He/^4He$ ratios were measured and normalized against atmospheric helium (R/R_a, Fig. 1). Other technical details are given in Sano and Wakita (1988).

2 RESULTS AND DISCUSSION

$^3He/^4He$ ratios in all waters are lower than the atmospheric ratio $^3He/^4He=1.4x10^{-6}$ but considerably higher than the average crustal helium ratio $1.5x10^{-8}$ (Sano et al. 1982), in-situ production ratio $1.6x10^{-8}$ (Torgersen 1989) or crustal degassing helium ratio $6.6x10^{-8}$ (Torgersen and Clarke 1987). The helium to neon ratios, ranging from 2.2 to 497.5, show that samples were not substantially contaminated with air and the ratios reflect the extent of the local contact with air-saturated waters (ASW). The lower limit of helium $^3He/^4He$ ratio in the Carpathian carbonated waters is $3.5x10^{-7}$ and that for $^3He/^{20}Ne$ is $1x10^{-7}$ (Fig. 2). The $^4He/^{20}Ne$ ratio in all samples (except Tytus in Rymanów, Fig. 1) is higher than in the atmospheric samples ($1.4x10^{-6}$). The samples of the Carpathians do not fit the mixing curve of radiogenic helium source and helium of the air, suggesting the necessity of inclusion of mantle helium. The nearly linear relation between the two quantities indicates mixing of mantle helium with a crustal component prior to uniform dilution with ASW (Fig. 2). Taking into account 3 simultaneous equations on mass balance for helium and neon on the condition that $\Sigma X = 1$, where X (i=1,2,3) is a partial contribution to the the total helium from 3

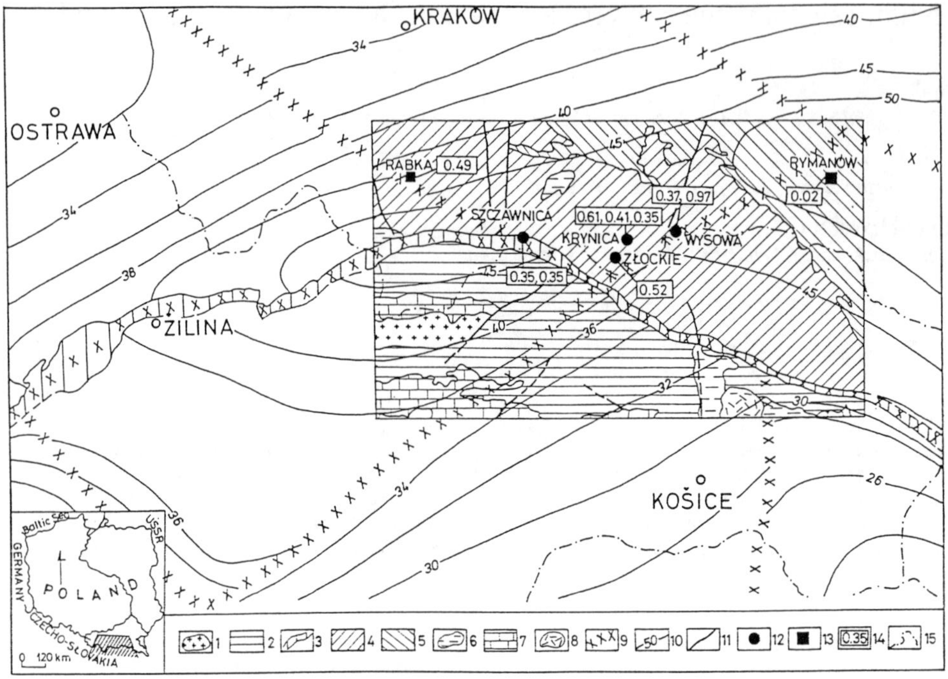

Figure 1 - Sketch of the Outer-West Carpathians, Poland.

1-Crystalline complex, 2-Inner Carpathian Paleogene, 3-Pieniny Klippen Belt, 4-Magura Nappe, 5-Silesian Nappe, 6-Neogene depression, 7-Triassic carbonates, 8-Neogene volcanics, 9-Moho associated discontinuity (Fusan, 1984), 10-Isopach map of depth to Moho (Fusan, 1984), 11-Major faults, 12-CO_2 dominated waters, 13-CH_4 dominated waters, 14-R/R_a ration, 15-State boundary.

sources (Sano and Wakita 1985), where $^3He/^4He$ and $^4He/^{20}Ne$ of mantle, ASW at 10°C and crustal radiogenic helium were taken as $11.2x10^{-6}$ and $>10^8$, $1.4x10^{-6}$ and 0.318 (10°C), $1.5x10^{-8}$, and 10^8, respectively (Sano et al. 1982; Sano and Wakita 1985), it was found that the maximum fraction of mantle helium in the present samples amounts to 11% in Aleksandra water (Wysowa) that has a R/R_a ratio of 0.97.

The relation between R/R_a ratio and the chemistry of the waters at depth in the wells is complex. Główne of Krynica had relatively low total dissolved carbonate content (TDC=108±25m mol/kg H_2O) but displayed R/R_a ratio nearly two times higher than the waters from deep boreholes of Krynica K-9 and Zuber-3 (TDC=115±26 and 390±90 m mol/kg H_2, respectively). Of special significance is the result obtained for Zuber-3 in Krynica because this water is characterized by the highest ratio of $^4He/^{20}Ne$=497.5 and a high concentration of helium reaching 2700 times the saturation level at 10°C. It probably reflects the highest accumulation of helium $1.2x10^{-4}$ cc/gH_2O (Janiak 1980), which is consistent with the highest alkalinity and sodium concentration in the Carpathian waters. This result places an upper limit on the helium concentration in the waters of interest. Such high concentration of the crustal helium might significantly lower the R/R_a ratio which was 0.35. Samples Główne and K-9 of Krynica contained one third of the total dissolved carbonate of the sample Zuber-3 but had nearly two times higher the R/R_a ratio of sample Zuber-3. The sample taken from Złockie represents dry cold emanation with R/R_a ratio of 0.52. CH_4-dominated water R-18 in Rabka (TDC=27±7 m mol/kg H_2O) had similar R/R_a isotope ratio as CO_2 saturated waters. It could suggest that both CO_2 and CH_4 are related to the crustal source.

To account for the mantle helium signature in the Outer Carpathians, let's suppose that a helium flux is maintained since the dyking episodes started some 20 Ma. The usual process which decreases the viscosity

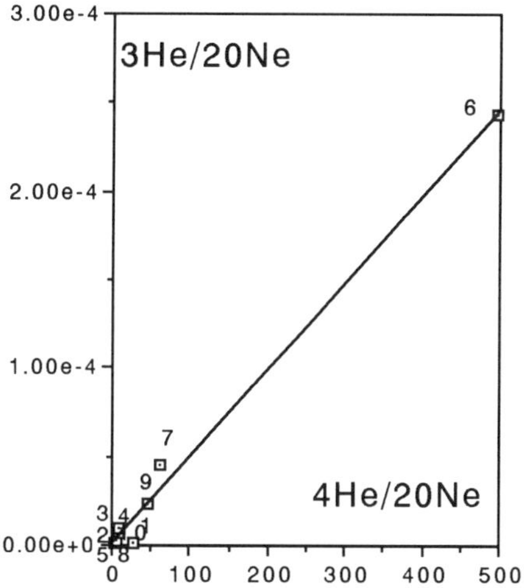

Figure 2 - The relation between helium and neon isotopes in the Outer Carpathians.

of melt allowing the extraction of mantle gases may be the deep penetration of substantial amounts of fluids into the lower crust (O'Nions and McKenzie 1988). Disappearance of the flysch sea during a subduction-collision event (Tomek et al. 1987) could supply such a mass of solution. Overthrusting of the hot deeper crust over the cold sinking slab can induce important convection directed downward (Etheridge et al. 1983). The subduction ended some 17-14 Ma ago in the West Carpathians (Tomek et al. 1987; Oszczypko and Ślączka 1985) though dyking was still active at 13 Ma (Birkenmajer 1986). Connection between the mantle helium occurrence and the subduction zone has its weak point in view that mantle helium is not limited to the proximity of the presumed subduction zone, nor to the localized occurrence of the andesite volcanics, but is pervasively distributed within the Magura Nappe.

3 CONCLUSIONS

- There is an apparent mantle helium signal at the compressional-collisional zone of the Carpathians.
- Similar R/R_a ratios of helium in CO_2 and CH_4 dominated gases suggest that both gases are of similar, crustal origin.
- Mantle helium contribution is estimated as 11%, and is pervasively distributed within the localized occurrence of the andesite volcanics.

ACKNOWLEDGEMENTS

We thank Prof. Lefeld for the critical reading of the previous manuscript and acknowledge the financial support of the Project 4.2.91 of the Institute of Science to the senior author.

REFERENCES

Birkenmajer, K. 1986. Stages of the structural evolution of the Pieniny Klippen Belt, Carpathians. Stud. Geol. Polonica: 7-32.

Etheridge, M.A., W.J. Wall & R.K. Vernon 1983. The role of fluid phase during regional metamorphism and deformation. Journ. of Metamorphic Geology 1: 205-226.

Fusan, O. 1984. Outline of the West Carpathian deep structure. Mineralia Slovakia 17: 195-204.

Janiak, E. 1980. The characteristic of gaseous composition of Zuber type water in Krynica with particular consideration of noble gases (in Polish). Proc. "Contemporary problems of regional hydrogeology:" 303-313.

Lefeld, J. & J. Jankowski 1985. Model of the deep structure of the Polish Inner Carpathians. Publs. Inst. Geophys. Pol. Acad. Science 175: 71-100.

O'Nions, R.K. & D. McKenzie 1988. Melting and continent generation. Earth. Planet. Sci. Letters 90: 448-456.

Oszczypko, N. & A. Ślączka 1985. Ar attempt to palinspatic reconstruction of Neogene basins in the Carpathian foredeep. Ann. Soc. Geol. Poland 55: 55-75.

Sano, Y. & H. Wakita 1985. Geographical distribution of $^3He/^4He$ ratios in Japan: Implications for arc tectonics and incipient magmatism. Jour. of Geophys. Research 90: 8729-8741.

Sano, Y. & H. Wakita 1988. Precise measurement of helium isotopes in terrestrial gases. Bull. Chem. Soc. Japan 61: 1153-115.

Sano, Y., T. Tominaga, Y. Nakamura & H. Wakita 1982. $^3He/^4He$ ratios of methane rich natural gas in Japan. Geochem. Journal 16: 337-345.

Torgersen, T. & W.B. Clarke 1987. Helium accumulation in groundwater. III. Limits on helium transfer across the mantle crust boundary beneath Australia and the magnitude of mantle degassing. Earth Planet. Sci. Letters 84: 345-355.

Tomek, C., L. Dvorakova, I. Ibrmajer, R. Jiricek & T. Korab 1987. Crustal profiles of active continental collisional belt: Czechoslovak deep seismic reflection profiling in the West Carpathians.

Geophys. J. R. Astr. Society 89: 383-388.
Torgersen, T. 1989. Terrestrial helium degassing and the atmospheric helium budget: Implication with the respect to the degassing processes of the continental crust. Chem. Geology 79: 1-17.

Water-Rock Interaction, Kharaka & Maest (eds) © 1992 Balkema, Rotterdam. ISBN 90 5410 075 3

Chemical composition and stable isotopes of thermal waters, Norris-Mammoth corridor, Yellowstone National Park, USA

R.H. Mariner, Y.K. Kharaka, Gil Ambats & L.D. White
US Geological Survey, Menlo Park, Calif., USA

ABSTRACT: The dominately Na-Cl waters of the Norris area are apparently due to water-rock reaction at high temperatures (up to 300°C) in volcanic rock. Northward in the corridor, aquifer temperatures decrease to as low as 70°C and progressively more sedimentary components (calcium, magnesium, sulfate, and bicarbonate) are overprinted on the initial Na-Cl character. Mixing of low TDS groundwater with high temperature thermal water may reduce chloride from 600 mg/L at Norris to as little as 43 mg/L at the northern end of the corridor. From Norris northward through Mammoth to Bear Creek, all thermal waters have essentially the same Br/Cl value. The much higher Br/Cl value of La Duke Hot Spring water indicates that it is not related to the other waters by dilution. Thermal waters within the corridor have δD and $\delta^{18}O$ values attributable to regional meteoric water which has been modified by isotopic exchange with rocks at high temperatures, boiling, and dilution.

1 INTRODUCTION

The geologic, geophysical, volcanic, glacial, and geothermal characteristics of Yellowstone National Park have recently been reviewed by Fournier (1989). Norris Geyser Basin has been described and discussed in Fournier (1989) and White et al. (1988). Basic isotopic characteristics of Yellowstone thermal waters were discussed by Truesdell et al. (1977).

The Norris-Mammoth corridor (Figure 1) is a complex subsidence structure about 40 km long and 8 km or less wide, extending northward from the Yellowstone Caldera to the base of the Beartooth Uplift (Pierce, 1991). The corridor contains most of the young rhyolitic domes and basaltic vents that occur outside of the 0.6 Ma Yellowstone Caldera. Several N-S faults or lineaments bound or occur within the corridor; major reverse faults terminate the corridor to the north and northeast. Hot springs within the corridor typically discharge at fault intersections or where faults intersect topographic lows (White et al., 1988; Fournier, 1989). Within the corridor, Cenozoic volcanic and volcaniclastic rocks overlie Mesozoic and Paleozoic marine sediments (Pierce, 1991).

The source of heat for the geothermal system at Yellowstone National Park is apparently magmatic although the water is at least 99.6 to 99.8% meteoric (Fournier, 1989). Recharge to the major geothermal system within the Park has been attributed to precipitation in the mountain ranges along the north and northwest sides of the Park (Truesdell et al. 1977; Fournier, 1989). These meteoric waters are heated to temperatures of as much as 360 °C either within the Caldera (Fournier, 1989) or beneath Roaring Mountain within the Norris-Mammoth corridor (White et al., 1988). These deep fluids are cooled adiabatically, conductively, or by mixing and then react with the various volcanic and sedimentary rocks at different temperatures to produce the range in chemical and stable isotope compositions seen in the hot spring waters.

2 CHEMICAL AND ISOTOPIC DATA

The chemical and isotopic composition of selected hot springs in the Norris-Mammoth corridor are given in Table 1. Chemically the thermal waters range from slightly saline (2000

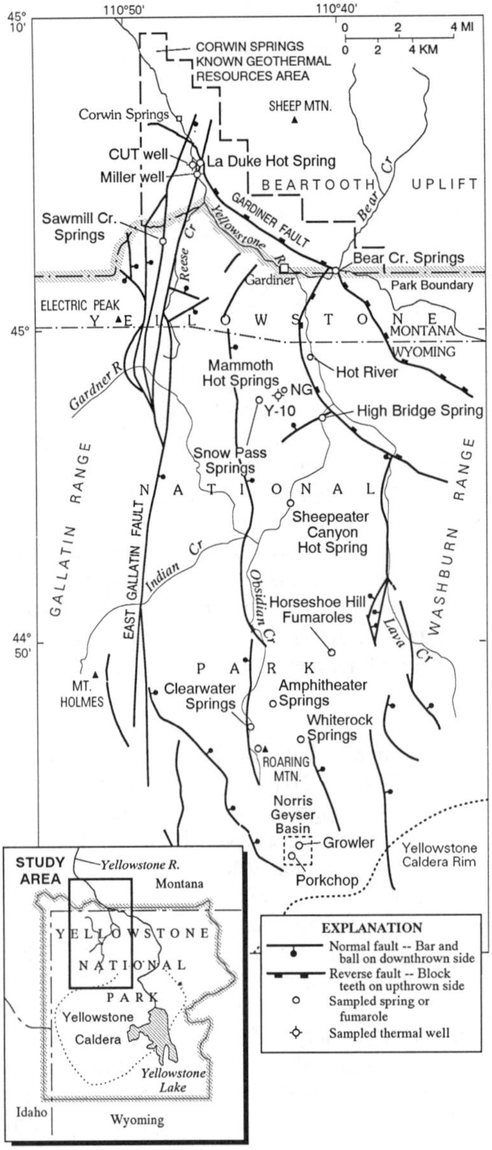

Figure 1. Location map of sampled hot springs and wells in the Norris Mammoth corridor, Yellowstone National Park.

mg/L TDS) Na-Cl waters to slightly saline Ca-Na-SO_4 waters. Chloride concentrations decrease from about 700 mg/L at Norris to 43 mg/L at Bear Creek and La Duke hot springs at the north end of the corridor. As chloride decreases, sulfate increases to about 1250 mg/L and bicarbonate increases to about 1500 mg/L (Table 1). Isotopic data (Table 1) show a narrow range in deuterium (-140 ‰ to -149 ‰) but a wide range in $\delta^{18}O$ values (-19.5 ‰ to -15.0 ‰). Large oxygen shifts are associated with spring waters in the southern end of the corridor; little or no oxygen shift is observed in the thermal waters at the northern end of the corridor.

3 CHEMICAL AND ISOTOPE RELATIONS

The wide range in rock types (volcanic to sedimentary), differences in aquifer-temperatures, and various amounts of dilution with local meteoric water may account for the differences observed in the major element concentrations and water stable isotope data.

3.1 *Reactions which might convert Norris water to Mammoth water*

Changes in dissolved constituent concentrations from Norris northward can be attributed to dilution which reduces the chloride concentrations and dissolution of dolomite and gypsum or anhydrite which increases the dissolved calcium, sulfate, and bicarbonate concentrations. Some cation exchange or mineral dissolution-precipitation is also required to adjust sodium, potassium, and calcium concentrations. Mammoth water (Narrow Gauge) could be generated from Norris water (Growler) by the following reaction: (Growler water x 0.297 + 0.703 meteoric water) + 2.757 millimoles $CaMg(CO_3)_2$ + 5.868 millimoles $CaSO_4$ + 1.610 millimoles $NaAlSi_3O_8$ + 1 millimole $CaAl_2Si_2O_8$ + 0.685 millimoles $KAlSi_3O_8$ + 3.295 CO_2 = Narrow Gauge water + clays. Apparently little or no sulfate reduction is occurring as $\delta^{34}S_{sulfate}$ is +21 ‰ (Kharaka et al., 1991), very near the +22 ‰ value of anhydrite from the Madison (Busby et al., 1991). A purely sedimentary origin for non-Norris dissolved constituents could be achieved by similar reactions and cation exchange.

Busby et al. (1991) in studying the Madison aquifer in Montana, Wyoming, and South Dakota demonstrated that sulfate reduction and pyrite precipitation was common and that appreciable CO_2 was added from an organic source. In the Norris-Mammoth corridor it is likely that the CO_2 which is consumed by various reactions is from the deep geothermal fluid.

Table 1. Chemical composition of selected springs and wells in the Norris-Mammoth Corridor
[Concentrations are in milligrams per liter; isotopes are in permil relative to SMOW]

Sample Site	pH	T(°C)	Li	Na	K	Mg	Ca	Cl	Br	HCO_3	SO_4	SiO_2	B	δD	$\delta^{18}O$
Norris Geyser Basin															
Growler	6.8	93	6.7	394	103	0.01	2.4	695	2.1	32	41	602	9.9	-140	-13.5
Porkchop Geyser	8.5	93	6.5	388	91	.06	3.5	687	2.1	62	23	741	9.9	-142	-15.7
Between Norris and Mammoth															
Clearwater 3C	6.7	93	4.1	278	64	.12	14	467	1.4	33	65	304	7.6	-142	-15.7
Sheepeater	6.5	73	1.1	94	36	13	141	97	.32	570	84	84	2.4	-149	-18.8
Mammoth Area															
High Bridge	6.3	34	.74	458	35	44	156	140	.49	1500	218	18	5.2	-150	-18.6
Narrow Gauge	6.5	72	1.6	130	56	70	384	164	.51	897	573	57	3.8	-147	-18.5
Y-10 (well)	6.2	73	1.7	156	64	76	492	174	.54	1040	808	95	4.2	-149	-18.4
Bear Creek Area															
Bear Creek 1	5.8	32	.42	113	49	91	507	43	.13	1240	869	35	1.0	-146	-19.4
La Duke Area															
CUT[1] (well)	6.7	53	.21	210	26	73	334	38	.19	250	1230	76	.5	-145	-19.4
La Duke	6.5	68	.28	240	24	64	330	45	.21	295	1250	56	.5	-144	-19.3

[1] Church Universal and Triumphant well at La Duke Hot Springs

3.2 *Geothermometry*

Geothermometry, which gives the subsurface temperature of last equilibrium between water and minerals, shows that the hottest waters occur at Norris (250-300 °C) at the south end of the corridor and the coolest (70-80 °C) at Bear Creek and La Duke hot springs.

3.3 *Possible effects of dilution on the thermal waters*

If dilution is the principal factor controlling conservative constituent concentrations in the thermal waters of the Norris-Mammoth corridor then conservative constituent ratios must be the same in all thermal waters of the corridor. Cold groundwaters are so low in chloride and bromide that diluting a thermal water with them produces a mixed water with the same conservative constituent ratio as the initial thermal water. The two most conservative chemical constituents are chloride and bromide, and Br/Cl values are virtually identical in the thermal waters of the corridor from Norris northward through Mammoth to Bear Creek (Figure 2). The markedly different Br/Cl values at La Duke Hot Springs shows that it is not related to the Norris-Mammoth waters by simple mixing. However, Br/Cl values vary even within the Madison Limestone. Br/Cl values calculated from the data of Busby et al. (1991) for the Madison in eastern Montana, South Dakota, and Wyoming range from 2.9 x 10^{-3} to 26 x 10^{-3} (w/w). The higher values are associated with presumably hypersaline parts of the formation in South Dakota. Values near 3 x 10^{-3} are more common in Montana. The source of the chloride (and bromide) in the thermal waters of the Norris-Mammoth corridor is uncertain. Fournier and Pitt (1985) in discussing the Yellowstone Park magmatic-hydrothermal system favored a saline brine located above a crystallizing magma

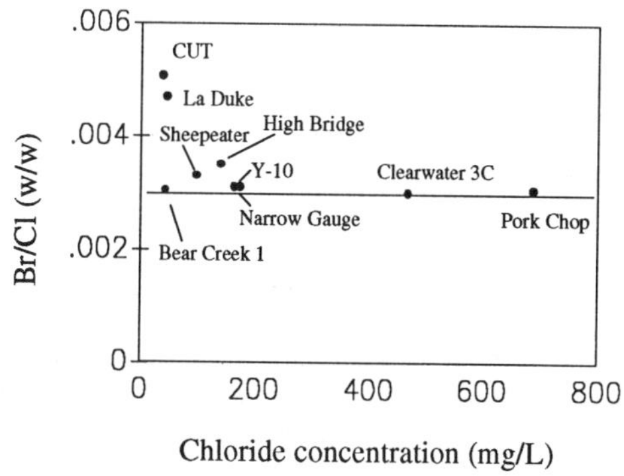

Figure 2. Br/Cl as a function of chloride concentration, note consistancy of values excluding samples from the La Duke area.

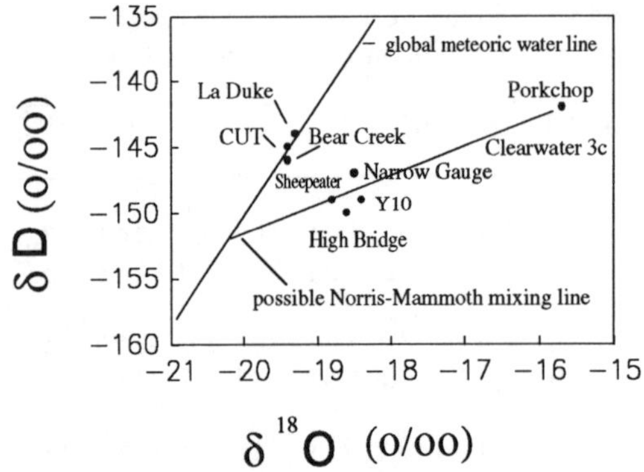

Figure 3. δD-δ¹⁸O plot, note large oxygen shift for Norris waters, little or no oxygen shift for La Duke and Bear Creek hot springs at the north end of the Norris-Mammoth corridor.

and beneath the shallow very dilute geothermal systems which feeds the hot springs of the park.

Dilution, if it is occurring, must also show in the δD-$\delta^{18}O$, δD-Cl, and $\delta^{18}O$-Cl plots (Truesdell et al., 1977). The δD-$\delta^{18}O$ plot (Figure 3) shows that Norris water could be diluted with a fresh water of about -152 ‰ δD and -20.3 ‰ $\delta^{18}O$ to produce all of the sampled thermal waters between Norris and Mammoth. δD-Cl and $\delta^{18}O$-Cl plots (not shown) show similar isotopic compositions for the diluting water.

4 CONCLUSIONS

Thermal waters of the Norris-Mammoth corridor change from Na-Cl in character at Norris to Ca-Na-HCO_3-SO_4 at the northern end of the corridor because of progressively cooler aquifer temperatures, interaction with sedimentary rather than volcanic rocks, and possibly dilution with a low total dissolved solids meteoric water. Waters at the south end of the corridor (Norris) have chemical and isotopic characteristics typical of a high temperature water in continental volcanic rock (large oxygen shifts, appreciable H_2S, high dissolved silica concentrations but low bicarbonate and sulfate. Thermal springs at the northern end of the corridor are cooler and are enriched in the sedimentary components (calcium, magnesium, sulfate, and bicarbonate), they show little oxygen shift, and are low in dissolved silica. The similarity in Br/Cl values and δD, $\delta^{18}O$, Cl relations permit, but do not prove, dilution of thermal water flowing northward from Norris to Mammoth, and Bear Creek.

REFERENCES

Busby, J.F., Plummer, L.N., Lee, R.W., and Hanshaw, B.B., 1991, Geochemical evolution of water in the Madison aquifer in parts of Montana, South Dakota, and Wyoming, *U.S. Geol. Surv. Prof. Pap. 1273-F*, F1 - F89.

Fournier, R.O., 1989, Geochemistry and dynamics of the Yellowstone National Park hydrothermal system, *Ann. Rev. Earth Planet. Sci., v.* 17, 13-53.

Fournier, R.O., and Pitt, A.M., 1985, The Yellowstone magmatic-hydrothermal system, *Trans, Geotherm. Counc. Int. Symp. Geotherm. Energy*, ed. C. Stone, Int. Vol. 319-327.

Kharaka, Y.K., Mariner, R.H., Bullen, T.D., Kennedy, B.M., and Sturchio, N.C., 1991, Geochemical investigations of hydraulic connections between the Corwin Springs Known Geothermal Resources Area and adjacent parts of Yellowstone National Park, in M.L. Sorey (ed.), Effects of Potential Geothermal Development in the Corwin Springs Known Geothermal Resources Area, Montana, on the thermal features of Yellowstone National Park, *U.S. Geol. Surv. Water-Res. Inv. Rep. 91-4052*, F-1 F-38.

Pierce, K.L., Adams, K.D., and Sturchio, N.C., 1991, Geologic setting of the Corwin Springs Known Geothermal Resources Area-Mammoth Hot Springs Area in and adjacent to Yellowstone National Park, in M.L. Sorey (ed.), Effects of Potential Geothermal Development in the Corwin Springs Known Geothermal Resources Area, Montana, on the thermal features of Yellowstone National Park, *U.S. Geol. Surv. Water-Res. Inv. Rep. 91-4052*, C-1 - C-37.

Truesdell, A.H., Nathenson, M., and Rye, R.O., 1977, The effects of subsurface boiling and dilution on the isotopic compositions of Yellowstone thermal waters, *Jour. Geophys. Res., v.* 82, 3694-3704.

White, D.E., Hutchinson, R.A., and Keith, T.E.C., 1988, The geology and remarkable thermal activity of Norris Geyser Basin, Yellowstone National Park, Wyoming, *U.S. Geol. Surv. Prof. Pap. 1456*, 84 p.

Water-Rock Interaction, Kharaka & Maest (eds) © 1992 Balkema, Rotterdam. ISBN 90 5410 075 3

Hydrogen isotopic ratio of hornblende from Quaternary volcanic rocks in Japan

Osamu Matsubaya & Isoji Miyagi
Research Institute of Natural Resources, Mining College, Akita University, Japan

ABSTRACT: Hornblendes from 16 volcanoes in Japan were analyzed for hydrogen isotopic ratios. Amphiboles from pumice flow and pumice fall, which must have been quickly cooled, have rather uniform D/H ratio, while those from lava and pyroclastic flow show a large variation. Because such large variations are considered caused by isotopic fractionation through dehydration from magma during its cooling and also through oxidation to basaltic hornblende, pumice seems to keep an original isotopic values. Most D/H ratios of amphiboles from pumice have δD values that range from -40 to -60 $^\circ/_{\circ\circ}$ SMOW. This range is similar to the range of volcanic steam in Japan rather than the range of mantle water, and suggests that some water derived from other than mantle may involve the magma genesis of island arc volcanoes.

1 INTRODUCTION

Most of Quaternary volcanoes in Japan occur in extrusive rocks of mainly andesitic and dacitic composition. Steam from these volcanoes have rather uniform hydrogen isotopic ratio in the range of -15~-40 $^\circ/_{\circ\circ}$ in δD r.t. SMOW, although they have been more or less diluted with meteoric water. This range of δD values is obviously higher than the range of magmatic water (-70~-100 $^\circ/_{\circ\circ}$) estimated for basaltic magma of mantle origin (Taylor et al. 1984). If δD values of the volcanic steam reflect the isotopic composition of magmatic water, it can be estimated that water involved in magma genesis of island arc volcano has higher δD than the mantle water. In order to obtain information about the magmatic water of island arc volcanoes, hornblende in extrusive rocks may be useful. In this study, therefore, D/H ratios of hornblendes of various extrusive rocks from Quaternary volcanoes were measured. The localities of these volcanoes are shown in Fig. 1.

2 RESULTS AND DISCUSSION

In Fig. 2, the results of measured δD values are summarized. These δD values vary in a wide range of -20~-110 $^\circ/_{\circ\circ}$, but a much smaller variation (30~60 $^\circ/_{\circ\circ}$) was found in each volcano. Myoko Volcano has the heaviest value (-20~-50 $^\circ/_{\circ\circ}$), while Chokai Volcano has the lighest (-80~-110 $^\circ/_{\circ\circ}$). Hornblendes of higher D are mostly reddish basaltic hornblende, or oxidized hornblende.

Figure 1 - Localities of volcanoes studied.

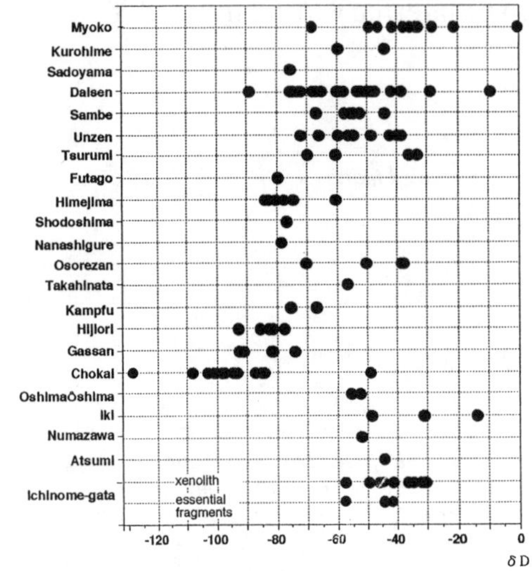

Figure 2 - δD values of hornblende and also biotite from the Quaternary volcanoes in Japan.

2.1 Hydrogen isotopic change through the formation of basaltic hornblende

Basaltic hornblende is considered to be derive from greenish one through oxidation. Barnes (1930) experimentally observed the color change from green to red through heating of hornblende in air. To understand the observed trend, heating of green hornblende was experimentally carried out both in vacuum and in air. The color of hornblende changed to reddish after both heatings. In dehydration of hornblende through the heating in vacuum, because considerable amounts of hydrogen (H_2) was released, the color change is considered the result of oxidation of Fe^{2+} to Fe^{3+} by the reduction of OH^- to H_2.

The δD value of residual water in the hornblende increased drastically with progress of dehydration during the heating in vacuum, and finally reached to +200 ‰. This large isotopic fractionation is caused by the hydrogen generation, because 230~240 ‰ fractionation was observed between the hydrogen and the water simultaneously removed from the hornblende.

In the case of heating in air at 700°C, isotopic fractionation is similar to that of heating in vacuum. However, above 800°C, dehydration and isotopic fractionation are quite different from those in vacuum. The amount of residual water and its δD values were kept nearly constant during temperature increase up to 1000°C. In this process, the δD values discontinuously decreased from a higher value raised in dehydration at 700°C to a lower value of -100~-140 ‰. Because the δD values of -100~-140 ‰ is the range for air moisture, this suggests that the exchange may have taken place between hornblende and air moisture.

On the basis of results discussed above, basaltic hornblende is considered to have been formed through dehydration of greenish hornblende, when H_2O pressure of the magma decreased after ascending to near surface or after extrusion. In this process, the δD values of hornblende is increased.

2.2 Original δD value of the magmatic water.

As shown in Fig. 2, the observed δD values of hornblende are widely scattered; it seems that some of these hornblendes are subjected to secondary modification of their δD values. Therefore, it is necessary to determine the original δD value of the magmatic water.

Fig. 3 shows δD values of hornblendes from Daisen Volcano. Although the values range from -10 to -90 ‰, the higher two values are for basaltic hornblende, and can be excluded from the discussion of the magmatic water. In this figure, hornblendes and biotite from pumice flows and tephra have fairly similar δD values, while those of pyroclastic flows have δD ranging widely from values similar to that of pumices to much lower values. Because these pyroclastic rocks may be cooled more slowly than the pumices, it is expected that dehydration progresses further in the cooling of pyroclastic rocks than those of pumices. In the dehydration of magma rocks, large drops in δD values of residual water were actually observed (Taylor et al. 1984; Newman et al. 1988). Therefore the pyroclastic rocks of lower δD values can be eliminated from the primary one. If these two exclusions are taken into consideration, δD of the pumices have rather a restricted range of -40~-60 ‰, and are consistent with the original value of magmatic water in Daisen Volcano.

If the same exclusion is applied to other volcanoes, most volcanoes in Fig. 2 have rather common original values similar to the value of Daisen Volcano. This range is similar to the range of volcanic steam in Japan rather than the range of mantle water. However, some volcanoes, for instance, Chokai Volcano have δD values as light as the mantle water. Although this difference cannot be explained at the present, the higher δD values of primary

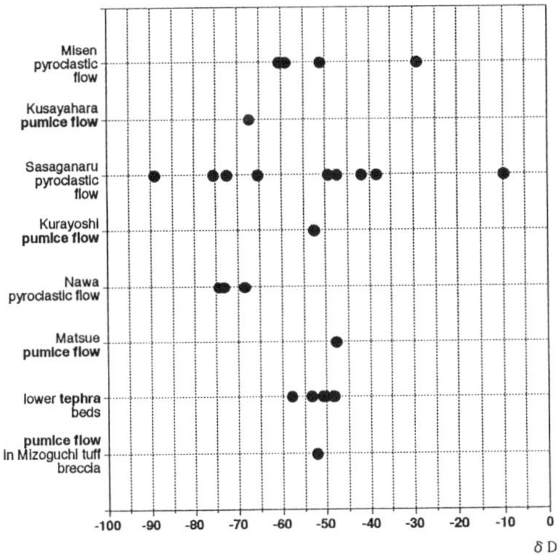

Figure 3 - δD of hornblende and biotite from Daisen Volcano.

magmatic water suggests that some water originated from other than mantle may be contributed to the magma genesis of island arc volcanoes.

2.3 Possible sources of magmatic water of island arc volcanoes

As regards to the magma genesis in island arc setting, several processes have been considered. In these processes, possible sources of the magmatic water may be hydrous minerals in subducting oceanic crust or in lower crust. The former is favored by the higher δD values, because hydrous minerals with -15~-40 ‰ have been commonly found in hydrothermally altered oceanic crust (Stakes and O'Neil 1982; Kawahata et al. 1987).

If the lower crust is composed of granitic rocks or metamorphic rocks, this possibility is not likely, because δD values of hydrous minerals in these rocks range from -50 to -140 ‰ (e.g., Kuroda et al. 1986). However, if amphibolite similar to the xenoliths found in extrusive rock of Ichinimegata Volcano, are involved in magma genesis (Takahashi 1986) (Fig. 2), and because amphiboles have heavier δD values, they can contribute to the heavier δD values of magmatic water in island arc volcanoes.

REFERENCES

Barnes, V.E. 1930. Changes in hornblende at about 800°C. J. Mineral. Soc. Amer. 15: 393-417.

Kawahata, H., M.Kusakabe & Y.Kikuchi 1987. Strontium, oxygen, and hydrogen isotope geochemistry of hydrothermally altered and weathered rocks in DSDP Hole 504B, Costa Rica Rift. Earth Planet. Sci. Letter 85: 343-355.

Kuroda, Y., T.Yamada, H.Kobayashi, Y.Ohtomo, M.Yagi & S.Matsuo 1986. Hydrogen isotope study of the granitic rocks of the Ryoke Belt, central Japan. Chem. Geol. (Isotope Geosci. Sec.) 58: 283-302.

Newman, S., S.Epstein & E.Stolper 1988. Water, carbon dioxide, and hydrogen isotopes in glasses from ca 1340 A.D. eruption of the Mono Craters, California: constraints on degassing phenomena and initial volatile content. J. Volvanol. Geotherm. Res. 35: 75-96.

Stakes, D.S. & J.R.O'Neil 1982. Mineralogy and stable isotope geochemistry of hydrothermally altered oceanic rocks. Earth Planet. Sci. Letter 57: 285-304.

Taylor, B.E., J.C.Eichelberger & H.R.Westrich 1984, Hydrogen isotopic evidence of rhyolitic magma degassing during shallow intrusion and eruption. Nature 306: 541-545.

Water-Rock Interaction, Kharaka & Maest (eds) © 1992 Balkema, Rotterdam. ISBN 90 5410 075 3

Lead and strontium isotopic data on brines from the Michigan basin, Ontario and Michigan

C.M.McKenna & R.H.McNutt
Department of Geology, McMaster University, Hamilton, Canada

S.K.Frape
Department of Earth Sciences, University of Waterloo, Canada

ABSTRACT: The present-day, Pb and Sr isotopic data from brines and host rocks in producing and dry fields in Ontario and Michigan, indicate that water/rock interaction occurred in the basin and/or in the Precambrian basement, followed by fluid migration and fluid mixing. Diagenetic and late-stage mineral phases retain isotopic signatures reflecting water/rock reaction with brines no longer in the formation.

1 INTRODUCTION.

The study area is located in southwestern Ontario and Michigan (Figure 1). The entire sequence is 5 km thick and ranges in age from Lower Cambrian to Upper Jurassic. The Michigan basin is a relatively "bowl shaped" intercratonic sag basin which is thickest in the centre and thins to the sides in all directions. It is virtually undisturbed Phanerozoic sedimentary succession that rests unconformably on Precambrian basement rocks.

Sanford et al. (1985) postulated that tectonic activity throughout Palaeozoic time in southwestern Ontario has resulted in extensive fracture networks. The tectonically related fault block readjustments provide the structural control for the origin and development of southwestern Ontario's petroleum traps. In addition, it is probable that basement structural features and major fractures exist in the Precambrian basement and influence fluid movement in the eastern part of the study area (Easton, pers. comm.).

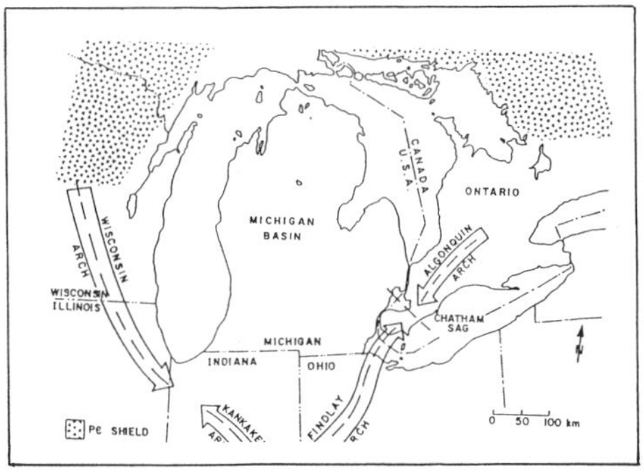

Fig. 1: Location map of Michigan showing the tectonic features of the basement. The Appalachian Basin is located to the east of the Algonquin Arch (modified from Ellis, 1969).

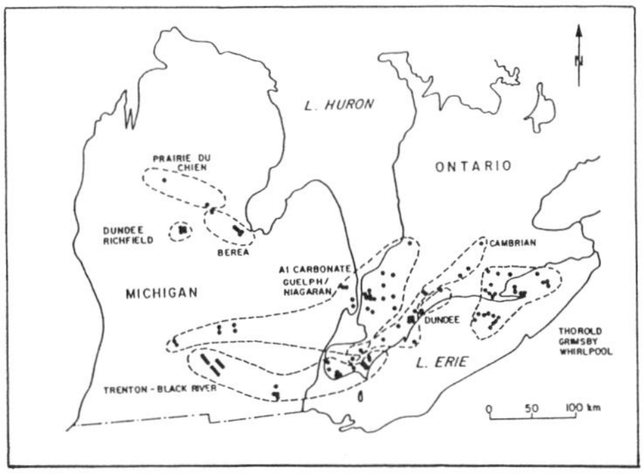

Fig. 2: Sample location map of water sampling sites (modified from Dollar, 1988).

2 ANALYTICAL PROCEDURES.

The brines were sampled by Dollar (1988), from producing oil and gas wells and dry holes and cover an area from the centre of the Michigan basin to the its eastern margin and eastward into the Appalachian basin (Figure 2). They were collected in pre-cleaned polyethylene bottles, and prior to Pb and Sr isotopic analysis, 125 ml aliquots were filtered through a .45 μm

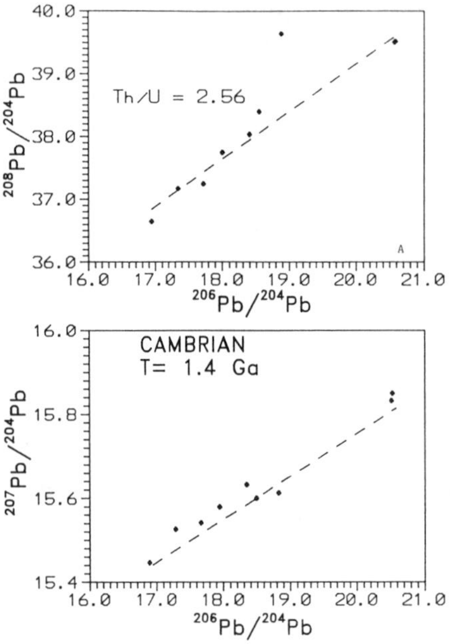

Fig. 3: Plot of $^{208}Pb/^{204}Pb$ vs $^{206}Pb/^{204}Pb$ (top) and $^{207}Pb/^{204}Pb$ vs $^{206}Pb/204Pb$ (bottom) for oilfield brines from Cambrian aged host rocks. A best-fit line through the data is dashed in.

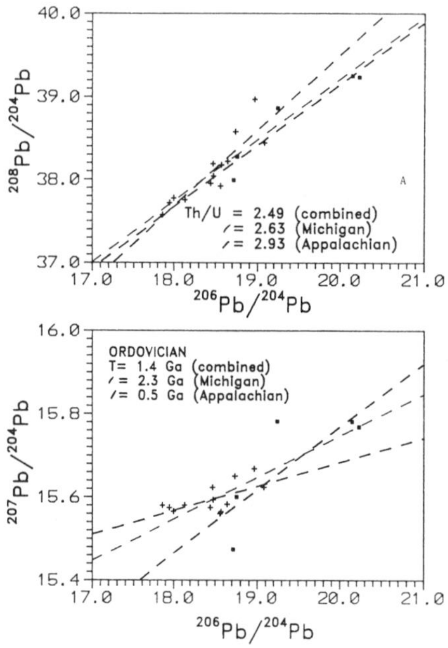

Fig. 4: Graph of $^{208}Pb/^{204}Pb$ vs $^{206}Pb/^{204}Pb$ (top) and $^{207}Pb/^{204}Pb$ vs $^{206}Pb/^{204}Pb$ (bottom) for oilfield brines from Ordovician aged host rocks. Best-fit lines through the data are dashed in.
X = Ontario samples; ■ = Michigan samples

filter system and acidified.

Calcite, dolomite, pyrite and anhydrite were removed from core slabs using a dental drill to get approximately 100 mg powdered samples.

The Pb was eluted with 1M HBr using anion exchange columns, and loaded onto a single Re filament bead into a mixture of 0.3M H_3PO_4 and silica gel. An operating temperature of 1300°C was controlled by a pyrometer. The total procedural blank was 0.5 to 1.2 ng. The Sr was extracted with 2.5N HCl using a standard cation exchange column procedure.

Repeated runs of NBS SRM 981 gave mean values and standard errors (σ) as follows: $^{208}Pb/^{204}Pb$ = 36.528 ± 0.0855; $^{207}Pb/^{204}Pb$ = 15.452 ± 0.0289; $^{206}Pb/^{204}Pb$ = 16.887 ± 0.0259. Repeated runs of the NBS 987 Standard gave a mean value of .71022 ± 0.000048 (2σ).

3 RESULTS AND DISCUSSION.

For some of the brines, there is a strong collinearity in $^{207}Pb/^{204}Pb$ vs $^{206}Pb/^{204}Pb$ and $^{208}Pb/^{204}Pb$ vs $^{206}Pb/^{204}Pb$ diagrams when plotted by stratigraphic unit (e.g., Figure 3 for the Cambrian). For others, it is not so obvious as exemplified by the Ordovician data (Figure 4). Overall, the data give a calculated model age circa 1.2 Ga (Figure 5). In detail, the brines from specific stratigraphic units show both older model ages (2.0–2.3 Ga for the Ordovician carbonate-hosted in Michigan; Silurian, sandstone-hosted in Ontario; Silurian, carbonate-hosted in Michigan, and the Mississippian) and younger (0.52–0.76 Ga for the Ordovician, carbonate-hosted in Ontario, and Silurian, carbonate-hosted in Ontario). The Th/U ratios of the brines range from 2.45 to 3.35. If these model ages have any meaning, they probably reflect the age of the source for the Pb.

There is a general east-west trend with the more radiogenic Pb in the west in the Michigan Basin (Figure 6). This is best exemplified by the Ordovician carbonate-hosted brines where they have distinctly more radiogenic Pb isotopic ratios from the centre of the Michigan Basin than those from the margin of the Appalachian Basin. In contrast, the more radiogenic Sr is located in the eastern portion of the study area. This probably reflects the influence of the basement there, because the sedimentary cover is much thinner.

Comparing the data to the evolution curves defined by the plumbotectonics model of Zartman and Doe (1981) (see Figure 5), the data imply a crustal signature for the

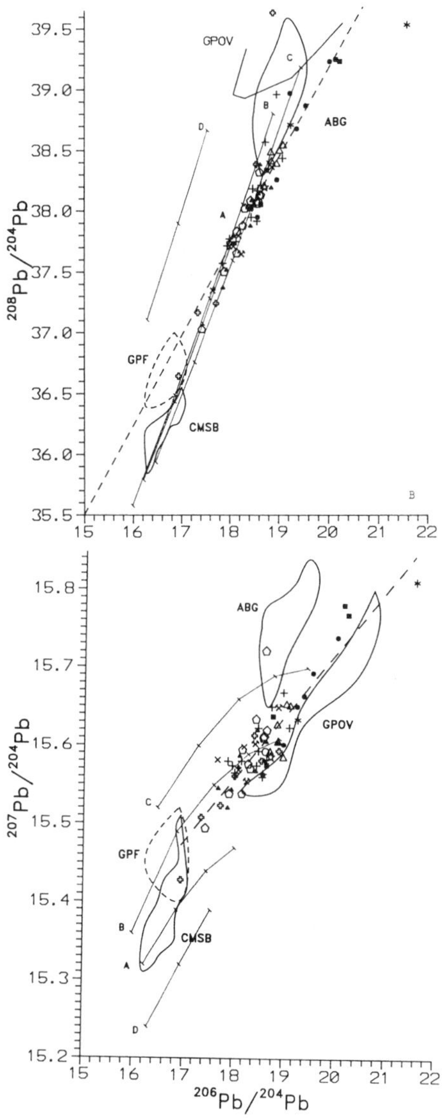

Fig. 5: Summary plot of $^{208}Pb/^{204}Pb$ vs $^{206}Pb/^{204}Pb$ and $^{207}Pb/^{204}Pb$ vs $^{206}Pb/^{204}Pb$ for oilfield brines from Paleozoic units of the Michigan and Appalachian Basins. For comparison, the figure shows the fields for galenas and feldspars from the Central Metasedimentary Belt, CMSB (Fletcher and Farquhar, 1982); feldspars from the Grenville Province, the western Adirondacks, GPF (Zartman, 1969); galenas from Phanerozoic vein deposits, GPOV (Fletcher and Farquhar, 1982); and galenas from the Appalachian Basin, ABG (Heyl et al., 1966). Letters refer to lead evolution lines described by Zartman and Doe (1981): A. mantle; B. orogene; C. upper crust; and D. lower crust.

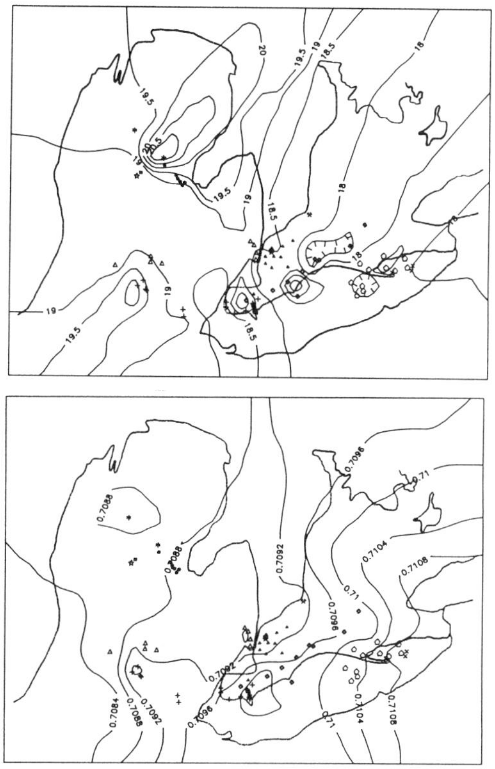

Fig. 6: Contour of $^{206}Pb/^{204}Pb$ isotopic ratios for brines in the study area (top). Contour of $^{87}Sr/^{86}Sr$ isotopic ratios (bottom).

Pb. In addition, the bulk of the data points lie between the areas defined by feldspars and galenas from Grenville rocks (CMSB and GPF) and those of Phanerozoic age (GPOV and ABG). Therefore, the trends may represent mixing between a minimum of two sources. The present study area is underlain by the Grenville and Southern Provinces with rocks ranging from 2.7 to 1.0 Ga in age. Here is an obvious source for some of the Pb, either leached directly from the basement by invading fluids now residing in the Phanerozoic section, or by water/rock reaction of aggregates of mechanically reworked Grenville rock (ie.

HOST ROCK	SYMBOL
Mississippian Berea	•
Devonian Dundee	○
Devonian Richfield	✫
Silurian Salina	⁑
Silurian Guelph	▴
Silurian Niagaran	△
Silurian Thorold-Grimsby	⬠
Silurian Whirlpool	×
Ordovician Trenton-Michigan	■
Ordovician Trenton-Ontario	+
Ordovician Praire Du Chien	*
Cambrian	◇

Paleozoic sedimentary rocks). Galenas from the Paleozoic rocks are a possible source for the more radiogenic Pb along with other sulphides, gypsum and halite (Haynes and Mostaghel, 1982). Note that a single galena from the Michigan Basin was more radiogenic in $^{206}Pb/^{204}Pb$ and less radiogenic in $^{208}Pb/^{204}Pb$ than other galenas from the Niagara region of the Appalachian Basin consistent with the brine data.

Another obvious possible source for the Pb in the brines is the host carbonate. Lead and Sr isotopic values in solid phases were compared to the brine values using two wells in the Ordovician, carbonate-hosted Hillman field in southwestern Ontario. Late stage mineral phases anhydrite, saddle dolomite and calcite have similar present-day $^{87}Sr/^{86}Sr$ ratios to the matrix dolomite (.7083-.7086 vs .7086-.7089) and distinctly less radiogenic that the present day brine (.7103). The $^{206}Pb/^{204}Pb$ ratio of the anhydrite is 18.65 compared to the dolomite value of 19.62,and to the brine value of 18.93. In another well, the late stage minerals anhydrite (18.65), calcite (18.22) and pyrite (19.22) have more radiogenic Pb than the brine (18.00). While most of the brine data lie close to the more radiogenic Paleozoic sulphides and carbonates, it appears that the rock and secondary mineral phases are not in Pb or Sr isotopic equilibrium with the present day brine, and they reflect earlier phases of water/rock interaction, perhaps even back to the time of diagenesis.

4 CONCLUSIONS.

The present-day Pb and Sr isotopic composition of brines in the Michigan basin reflect water/rock interaction, that could be explained by in-situ reaction in the sedimentary rocks and/or reaction with the Precambrian basement. Mixing of fluids during migration across stratigraphic boundaries cannot be ruled out. The host rocks and their late-stage minerals have Pb and Sr "fossil" isotopic signatures compared to the brines now in the rock formations, reflecting reaction with older fluid systems in the basin.

5 REFERENCES.

Dollar, P.S. 1988. Geochemistry of formation waters, southwestern Ontario, Canada and southern Michigan, U.S.A.-Implications for origin and evolution. *Unpublished M.Sc. Thesis, Dept. of Earth Sciences, University of Waterloo, Waterloo, Canada.*

Ellis, G.D. 1969. Architecture of the Michigan Basin. In *Studies of the Precambrian Michigan Basin, Michigan Basin Geol. Soc. Guidebook* 60-80.

Fletcher, I.R.& R.M.Farquhar 1982. The protocontinental nature and regional variability of the Central Metasedimentary Belt of the Grenville Province: lead isotope evidence. *Canadian Journal of Earth Science* 19: 239-253.

Haynes, S.J. & M.A. Mostaghel 1982. Present-day precipitation of lead and zinc from groundwater. *Mineralium Deposita* 17: 213-228.

Heyl A.V., M.H.Delevaux, R.E. Zartman & M.R. Brock 1966.Isotopic study of galenas from the Upper Mississippi Valley, the Illinois-Kentucky, and some Appalachian Valley mineral districts. *Economic Geology* 61: 933-961.

McNutt, R.H. S.K. Frape & P.S. Dollar 1987. A strontium, oxygen and hydrogen isotopic composition of brines, Michigan and Appalachian Basins, Ontario and Michigan. *Applied Geochemistry* 2: 495-505.

Sanford, B.V., F.J. Thomson & G.H. McFall 1985. Platetectonics - A possible controlling mechanism in the development of hydrocarbon traps in southwestern Ontario. *Bull. of Can. Pet. Geol.* 33: 52-71.

Zartman, R.E. 1969. Lead isotopes in igneous rocks of the Grenville province as a possible clue to the presence of older crust. *Geological Association of Canada Spec. Paper* 5: 193-205.

Zartman, R.E. & B.R. Doe 1981. Plumbotectonics - The model. *Tectonophysics* 75: 135-162.

Water-Rock Interaction, Kharaka & Maest (eds) © 1992 Balkema, Rotterdam. ISBN 90 5410 075 3

The sulphur isotopic composition of mafic-intermediate volcanic rocks, Taupo Volcanic Zone, New Zealand

B.W. Robinson & I.J. Graham
Nuclear Sciences Group, DSIR Physical Sciences, Lower Hutt, New Zealand

ABSTRACT: Forty-four samples of young calc-alkaline volcanic rocks from the Taupo Volcanic Zone, New Zealand, have been analysed for S and $\delta^{34}S$. Total S contents range from <10 to 250 mg/kg with most samples being <40 mg/kg and isotope compositions range from 0‰ (basalts) to +12.5‰ (andesites). These values are similar to other volcanic arc systems such as Japan and the Marianas, and contrast with much higher S contents and typically lower $\delta^{34}S$ values for oceanic basalts.

Straightforward relations between $\delta^{34}S$, ΣS and other geochemical parameters have not been found. It appears that sulphur loss and $\delta^{34}S$ changes during magma degassing are superimposed on a primary variation involving mixing of primordial mantle sulphur (0‰) with crustal sulphur of >+12‰ either from lower crustal rocks or subducted arc sediments.

1 INTRODUCTION

As part of an overall study of the sulphur cycle in the New Zealand lithosphere a programme of analysing Late Cenozoic volcanic rocks from the Taupo Volcanic Zone (TVZ) for total sulphur and its isotopic composition has been initiated. We report here the first phase of the study, focusing mainly on andesite-dacite lavas from the active volcanoes of Ruapehu, Ngauruhoe and White Island (Fig. 1), plus a smaller number of basalts from scattered localities within the TVZ.

The TVZ is a southward extension of the Pleistocene Tonga-Kermadec oceanic volcanic arc into the continental crustal environment of New Zealand (Fig. 1: inset) and results from the oblique subduction of the Pacific plate beneath the Indian plate. The TVZ consists of an andesite-dacite volcanic front along the eastern margin from White Island to Ruapehu, and an extensional region of mainly rhyolitic volcanism with minor volumes of high alumina basalt to the west. The majority of TVZ mafic-intermediate lavas are calc-alkaline and on the basis of high alkali and light REE contents and elevated radiogenic and stable isotope ratios are thought to result from crustal contamination of mantle-derived basaltic magmas (Graham and Hackett, 1987; Graham et al., 1992).

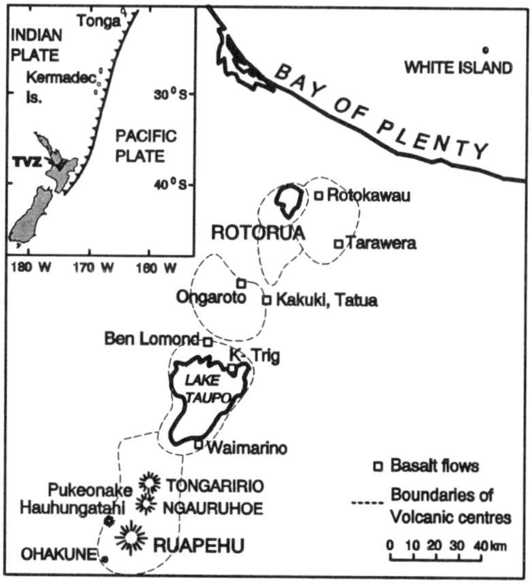

Fig. 1: The TVZ showing the location of the main andesite volcanoes and basalt flows.

2 ANALYTICAL METHODS AND RESULTS

Ground rock samples were initially subjected to the treatment of weak Kiba reagent under vacuum (Ueda and Sakai, 1983) to separate sulphate from sulphide sulphur. Since many of the samples have less than 20 mg/kg total sulphur, problems in reproducibility and processing large amounts of sample prompted extraction of all the sulphur with strong Kiba. The evolved H_2S was precipitated as Ag_2S and reacted with Cu_2O for isotopic analyses with an overall error of ±0.2% (1σ). The ΣS (mg/kg) values were determined gravimetrically on the weight of Ag_2S obtained with a reproducibility of ±2% (1σ) and an insignificant blank of 1 mg/kg.

Table 1 gives a brief description of the samples analysed as well as their ΣS and $\delta^{34}S$ values and Fig. 1 shows the sample localities. Total sulphur contents range from <10 to 250 mg/kg with most samples being <40 mg/kg (Fig. 2). Isotope compositions range from 0 to +12.5‰ CDT with an average of 4.3±3.0‰ (n = 44). TVZ basalts have typically low $\delta^{34}S$ values, the most primitive rocks (e.g., BH4) being close to 0‰. However, Waimarino basalt has an unusually high $\delta^{34}S$ value of +6.4‰. In discussing the data for the andesites and dacites we have adopted the nomenclature of Graham and Hackett (1987) for Ruapehu rocks: type 1 andesites and their accumulative variants (types 2, 3) make up more than 85% of TVZ mafic-intermediate volcanics and have wide ranging $\delta^{34}S$ values (+0.1 to 12.5‰). Although three samples from the 1954 eruption of Ngauruhoe show excellent agreement, this contrasts with three similar Ruapehu lavas which have widely differing $\delta^{34}S$ (Table 1). Type 5 andesites make up about 8% of total lavas and have similar $\delta^{34}S$ and S contents to many type 1 andesites. The volumetrically rare type 6 andesites on the other hand have nearly identical low values. Recent eruptives from White Island have variable but typically high S contents and relatively high $\delta^{34}S$ values.

Table 1: Sulphur isotope compositions

Sample	Locality	ΣS (ppm)	$\delta^{13}S_{CDT}$ (‰)
BASALTS			
BH4	Kakuki	60	+0.3
22996	K-Trig	6	+1.8
PK3	K-Trig	11	+1.5
22993	Tatua	43	+0.5
22998	Ongaroto	6	+2.7
22994	Ben Lomond	9	+1.9
PK4	Rotokawau	34	+3.5
L10	Ruapehu	7	+1.7
WB	Waimarino	10	+6.4
11965	Red Crater	78	+4.2
TYPES 1 & 3 ANDESITES & DACITES			
L20	Ruapehu	15	+7.2
L21	"	13	+4.9
L22	"	15	+0.1
X14	"	14	+3.5
L6	"	17	+4.0
29247)	Ngauruhoe (1954)		+3.8
29248)	"	66	+3.9
29244)	"	49	+3.6
A8	Ruapehu	19	+9.8
A36	"	23	+2.7
A13	"	16	+6.0
L1	"	37	+4.8
B3	"	8	+1.8
N4	"	4	+1.8
B2	"	8	+4.9
W1	"	28	+6.9
N2	"	33	+12.5
G4	"	18	+2.5
N6	"	14	+8.6
E5	"	18	+1.6
11242	White Island	38	+12.1
N8	Ruapehu	25	+6.0
E4	"	109	+8.8
N112A	"	52	+2.6
TYPE 5 ANDESITES			
H5	Hauhungatahi	6	+2.8
F4	Ohakune	9	+3.5
H6	Hauhungatahi	12	+5.6
TYPE 6 ANDESITES			
H7	Pukeonake	8	+1.3
M17	Ruapehu	7	+1.5
TYPE 9 ANDESITES			
26592	White Is (1977)	36	+5.4
26589	"	72	+9.2
-	"	26	+7.4
-	" (1987)	70	+7.5
-	" (1988)	250	+6.3
METASEDIMENTARY BASEMENT			
Mean (n=70)		565	+4.6

3 DISCUSSION

Mafic-intermediate lavas from volcanic arc systems are characterised by variable but typically higher $\delta^{34}S$ values and much lower S contents than those of oceanic basalts (Fig. 2). Although few comparative studies have been undertaken globally, the results from the Japan arc (Ueda and Sakai, 1984) and Mariana arc (Woodhead et al., 1987) are similar to those from the TVZ, and show a range in $\delta^{34}S$ from +4 to +20‰ and S contents typically less than 50 mg/kg. Several explanations are possible for the higher isotopic ratios of arc lavas including source inhomogeneity and/or contamination, crustal contamination, near-surface hydrothermal alteration, sulphide precipitation and isotope fractionation due to degassing.

Mantle xenoliths have $\delta^{34}S$ values similar to the narrow range in meteorites and the close agreement between oceanic basalts (i.e., MORB and OIB) and continental tholeiites suggests that the mantle is homogeneous in $\delta^{34}S$ at around 0‰ (Sakai et al., 1984). Subarc mantle, although being otherwise similar to the MORB-source, typically has an additional subduction component which may involve pelagic sediments and/or seawater (e.g. Ellam and Hawkesworth, 1988). Woodhead et al. (1987) attributed ^{34}S enrichment in Mariana arc parental magmas to recycling of seawater sulphate (~20‰), suggesting that proportions of between 10 and 100% could explain the data. Likewise, Ueda and Sakai (1984) attributed the sulphur isotope compositions of Japan arc lavas to mantle inhomogeneities. Although such mechanisms could explain the range in $\delta^{34}S$ observed in TVZ lavas it is interesting to note that the most primitive basaltic rocks (e.g., BH4; Table 1) have the lowest $\delta^{34}S$ values. These rocks are considered to best reflect the isotopic composition of the mantle source region beneath the TVZ (Graham et al., 1992) since they are the least affected by subsequent crustal contamination and crystal fractionation processes. Although some of the high alumina basalts show a correlation between $\delta^{34}S$ and $^{87}Sr/^{86}Sr$, most of the andesites show only a very weak correlation (Table 1 and Fig. 3). Sulphide precipitation at depth is a mechanism which could increase the $\delta^{34}S$ values of the residual magma. Sulphur saturation from mixing of mantle and high-sulphur wedge material could lead to sulphide precipitation but we have no evidence for this, e.g. from xenoliths. Similarly hydrothermal alteration has not affected most of the rocks particularly those with low S content.

Good correlations between stable and radiogenic isotopes and tight geochemical trends for TVZ andesitic lavas suggest that they are probably the result of crustal contamination of mantle derived primary basalts (e.g. Graham and Hackett, 1987). However, there is no clear correlation between $\delta^{34}S$ and the degree of contamination as deduced from combined assimilation and fractional crystallisation models (Graham and Hackett, 1987). Probable surface equivalents of the metasedimentary basement rocks, the presumed assimilant for TVZ andesites (Graham et al., 1992) yield a range of $\delta^{34}S$ values from -13.5 to +42.5‰, but with an average of +4.6‰, make crustal contamination an unlikely explanation of the high $\delta^{34}S$ values.

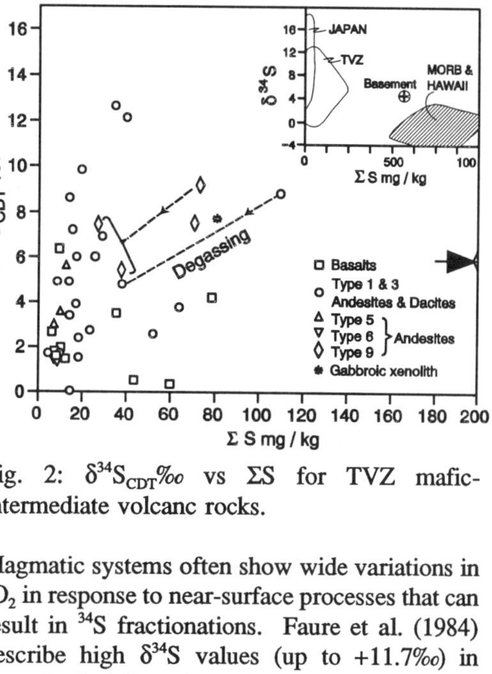

Fig. 2: $\delta^{34}S_{CDT}$‰ vs ΣS for TVZ mafic-intermediate volcanc rocks.

Magmatic systems often show wide variations in fO_2 in response to near-surface processes that can result in ^{34}S fractionations. Faure et al. (1984) describe high $\delta^{34}S$ values (up to +11.7‰) in Jurassic tholeiites from Antarctica and explain this as SO_2 degassing under very high fO_2 conditions. This, however, is probably a special case and it is more likely that SO_2 degassing will result in ^{34}S loss (Sakai et al., 1982). SO_2 is the principal sulphur component in low pressure gases from mafic-intermediate magmas whereas H_2S is more important at higher pressure and

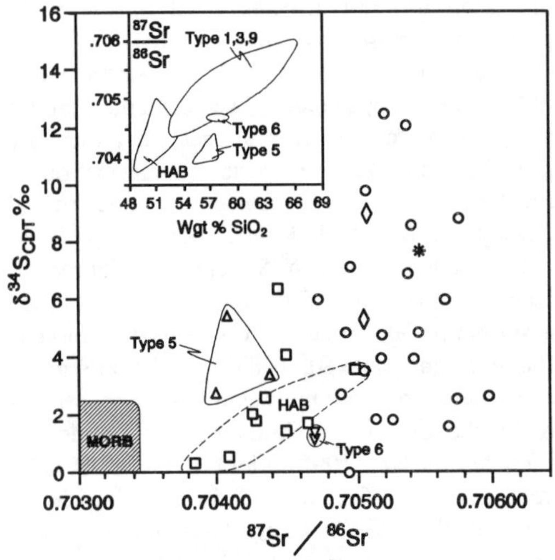

Fig. 3: $\delta^{34}S$ vs $^{87}Sr/^{86}Sr$ for TVZ mafic-intermediate volcanic rocks. Inset shows correlation between $^{87}Sr/^{86}Sr$ and SiO_2.

lower temperatures. Some TVZ andesites show apparent SO_2 degassing trends: the 1977 eruptives from White Island show a broad decreasing trend from bomb to block to vesicular lava (Fig 2), as do Ruapehu dacites E4 and E5 (Table 1). However, three similar (and probably coeval) Ruapehu andesites L20-L22 show a wide range in $\delta^{34}S$ but no significant difference in S content. SO_2 degassing can account for some of the unexpectedly low $\delta^{34}S$ values and the low S contents. The magmas may have had higher $\delta^{34}S$ values than the data would suggest but degassing is not the cause of such values.

$\delta^{34}S$ values for total sulphur from Ngauruhoe eruptions in 1989 were +12.3‰, some 8.5‰ higher than the 1954 lava. In contrast, White Island lavas show typically high $\delta^{34}S$ values for 1977-1988 eruptives (+5.4 to +9.2‰) but much lower values of +4.7‰ for SO_2 and +0.3‰ for H_2S gases over the same period of time (Robinson and Giggenbach, unpublished data). Some geothermal fields of the TVZ (e.g., Wairakei, Ohaaki) have relatively homogeneous H_2S gas (+4.3±1‰ and +5.5±0.7‰ respectively), while others (e.g., Rotorua) range more widely (+3.9‰ to +8.7‰). However, the average geothermal H_2S reflects average silicic magma values of +4.8±1.2‰ (n = 83).

4 CONCLUSIONS

Straightforward relations between $\delta^{34}S$, ΣS and other geochemical parameters have not been found. It appears that sulphur loss and $\delta^{34}S$ changes during degassing are superimposed on a primary variation involving mixing of primordial mantle sulphur (0‰) with crustal sulphur of >+12‰ either from lower crustal rocks (supported by other geochemical evidence) or from subducted arc sediments. Although the latter source is favoured to explain the high $\delta^{34}S$ values, the basalts only exhibit a maximum $\delta^{34}S$ of +6.4‰

REFERENCES

Ellam, R.M. and Hawkesworth, C.J. (1988) Elemental and isotopic variations in subduction related basalts: evidence for a three component model. Contrib. Min. Petrol. 98: 72-80.

Faure, G., Hoefs, J. and Mensing, T.M. (1984) Effect of oxygen fugacity on sulphur isotope compositions and magnetite concentrations in the Kirkpatrick basalt, Mount Falla, Queen Alexandra Range, Antarctica. Isot. Geosci. 2: 301-311.

Graham, I.J. and Hackett, W.R. (1987) Petrology of calc-alkaline lavas from Ruapehu volcano and related vents, Taupo Volcanic Zone, New Zealand. J. Petrol. 28(3): 531-567.

Graham, I.J., Gulson, B.L., Hedenquist, J.W. and Mizon, K. (1992) Petrogenesis of Cenozoic volcanic rocks from the Taupo Volcanic Zone, New Zealand, in the light of new Pb isotope data. Geochim. Cosmochim. Acta (in press).

Sakai, H., Casadevall, T.J. and Moore, J.G. (1982) Chemistry and isotope ratios of sulphur in basalts and volcanic gases at Kilauea Volcano, Hawaii. Geochim. Cosmochim. Acta 46: 729-738.

Sakai, H., Marais, D.J.D., Ueda, A. and Moore, J.G. (1984) Concentrations and isotope ratios of carbon nitrogen and sulphur in ocean-floor basalts. Geochim. Cosmochim. Acta 48: 2433-2441.

Ueda, A. and Sakai, H. (1983) Simultaneous determination of the concentration and isotope ratio of sulphide- and sulphate-sulphur and carbonate carbon in geological samples. Geochem. J. 17: 185-196.

Ueda, A. and Sakai, H. (1984) Sulphur isotope study of Quaternary volcanic rocks from the Japanese Islands Arc. Geochim. Cosmochim. Acta 48: 1837-1848.

Woodhead, J.D., Harmon, R.S. and Fraser, D.G. (1987) O, S, Sr and Pb isotope variations in volcanic rocks from the Northern Mariana Islands: implications for crustal recycling in intra-oceanic arcs. Earth Planet. Sci. Lett. 83: 39-52.

Water-Rock Interaction, Kharaka & Maest (eds) © 1992 Balkema, Rotterdam. ISBN 90 5410 075 3

The McLaughlin hot-spring gold-mercury deposit and its relationship to hydrothermal systems in the Coast Ranges of northern California, USA

R. L. Sherlock & E. C. Jowett
Department of Earth Sciences, University of Waterloo, Ont., Canada

ABSTRACT: Active hot-springs in the Wilbur Springs District appear to be modern analogues of the McLaughlin gold-mercury mine, based on isotopic and geochemical similarities. The active Coast Range hydrothermal systems show a well defined mixing trend from hot-spring composition ($\delta^{18}O$ 8 to 10 ‰ , δD -15 to -30 ‰) to meteoric water composition ($\delta^{18}O$ -9.5 ‰ , δD -61 ‰), with mineral-springs intermediate on the trend. The Geysers Geothermal System shows a similar isotopic trend as the rest of the Coast Range hydrothermal systems, suggesting a similar origin. The origin of the hot-spring end-member fluid is consistent with connate fluid derived from Franciscan sediments.

INTRODUCTION

The Coast Ranges of northern California have long been recognized as an ideal location to study water rock interaction (Barnes et al., 1973a,b; White et al., 1973). Active auriferous hot-springs, mineral-springs, and vapor-dominated systems are all isotopically and chemically related. The recent discovery of the McLaughlin gold-mercury mine has enabled this fossil hot-spring system to be examined at various depths and compared to active hydrothermal systems.

McLAUGHLIN MINE

The McLaughlin Mine (17.5 M tonnes @ 5.21 g/T Au; Lehrman, 1986) and historic mercury producer is a premier example of a hot-spring gold deposit, with a well preserved subaerial siliceous sinter, and Hg, Au, Ag and base-metal mineralization telescoped to less than 800 m. Base-metals occur in the deepest veining (~800 m below the sinter), Ag dominates from < 800 m to ~200 m, Au is dominant from ~200 m to the sinter, and Hg is dominant within the sinter. The deepest veining has fluid inclusion homogenization temperatures of ~250°C, which is below the hydrostatic boiling curve for water at that depth. Evidence for boiling is first suggested at ~200 m, where coeval fluid inclusions homogenize to liquid or vapor at the hydrostatic boiling temperature. Above 200 m, fluid inclusion homogenization temperatures are generally equal to or greater than the hydrostatic boiling temperatures. Locally, fluid inclusion homogenization temperatures exceed the lithostatic boiling temperatures, suggesting fluid pressures in excess of the lithostatic load, and in the case of hydrofracturing exceeding the tensile strength of the host rock. The coincident onset of boiling and gold mineralization suggests that the physical-chemical changes as a result of boiling are responsible for gold precipitation. Fluid salinities of 0 to 4 wt.% NaCl eq. are calculated from final melting temperatures. These salinities are similar to those of the active hot-springs in the Wilbur Springs District (Peters, 1990).

Preliminary oxygen isotope data (Fig. 1) on vein quartz show increasingly heavy values toward the surface, from ~18 ‰ at 800 m, to ~31 ‰ at the sinter. Using fluid inclusion homogenization temperatures the $\delta^{18}O$ value for the hydrothermal fluid is calculated to be constant at ~8 to 10‰ . Deuterium analysis of montmorillonite (n=4) shows that the δD ratio of the altering fluid is -15 to -30 ‰ .

ACTIVE HYDROTHERMAL SYSTEMS

Since the discovery of gold mineralization within these hot-spring systems their origin

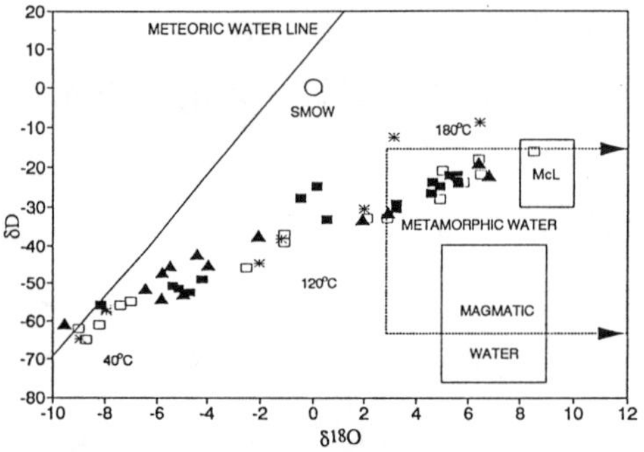

Fig. 1 $\delta^{18}O$ vs δD for Coast Range hydrothermal systems and the McLaughlin Mine composition. Empty squares Peters 1990, Astrix Barnes et al. 1973a, Filled squares White et al. 1973, Filled triangles this study

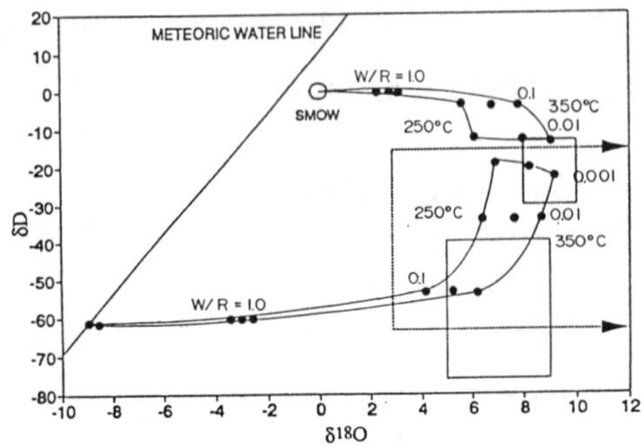

Fig. 2 Calculated water rock ratios to evolve hot-spring fluids from meteoric water and ocean water.

acquires new importance. Not only are these fluids capable of forming relatively small hot-springs such as the Wilbur Springs District, they also occur on a much grander scale, capable of forming large precious metal deposits. The Coast Range hydrothermal systems show a well defined mixing trend (Fig. 1) from isotopically heavy hot-spring fluids to meteoric water compositions. This trend is defined by linear relationships on the $\delta^{18}O$ - δD plot as well as on Cl^--Br^-, δD-Cl^-, Cl^--Na^+, and $\delta^{18}O$-Cl^- plots (Barnes et al., 1973; Peters, 1990). Mineral-springs plot intermediate between hot-springs and meteoric water. Reservoir temperatures were calculated using the Mg-Li geothermometer of Kharaka and Mariner (1989). Hot-springs have resevoir temperatures of ~180°C, and the mineral-springs range from ~120°C for the isotopically heavy end-member to ~40°C for springs similar to meteoric water (Fig. 1).

There are two ways of forming the hot-spring end member: (1) the original fluid may have originated as meteoric water and evolved to an isotopically heavy fluid, (2) the fluid may have originated as an isotopically heavy fluid such as magmatic, metamorphic, or connate fluids. Water rock ratios needed to evolve such an isotopically heavy fluid are calculated using the method of Ohmoto and Rye (1974). Boundary conditions are well defined with meteoric water composition from fresh water springs, the isotopic composition of the rocks from Margaritz and Taylor's (1976) average values for Franciscan greywacke, the final water composition from hot-springs, and the fractionation factors of albite-H_2O for oxygen (O'Neil and Taylor, 1976) and chlorite-H_2O for deuterium (Taylor, 1974). Temperatures of 250°C to 350°C are used consistent with the highest fluid inclusion temperatures. Waters of hot-spring composition can be produced from meteoric water, but water rock ratios are 0.001 (Fig. 2), an unreasonably low value compared to 0.2 to 2 characteristic of meteoric water dominated epithermal deposits (Field and Fifarek, 1985). Water of hot-spring composition can be produced from SMOW, using the same boundary conditions, if water rock ratios are 0.01 (Fig. 2). This ratio is also very low, unusual in a hydrothermal system. However if the water originated as trapped pore fluid in Franciscan sediments and evolved to an isotopically heavy fluid during burial and low grade metamorphism, then the calculated water rock ratio of 0.01 corresponds to a porosity of ~2.5% which is the water rock ratio present within the original sediments.

There is no geologic evidence for the fluids being magmatic in origin. Numerous hot-springs are active in the Wilbur Springs District with no geologic or geophysical evidence of recent igneous activity. Hot-spring fluid isotopic composition plot outside the field for magmatic fluids, and is only on the margin of the metamorphic fluid field (Fig. 1). It is difficult to discriminate between evolved sea water and metamorphic fluids strictly on the basis of their isotopic composition. However the isotopic composition of the hot-springs are very similar to California oil field brines (White et al., 1973). The ubiquitous presence of hydrocarbons within the Coast Range hydrothermal systems, and the high sulfur and methane content of the hot-springs suggest a connate fluid origin.

The Geysers is a vapor-dominated hydrothermal system within the Franciscan Complex, and is heated by a large siliceous magma chamber related to the Clear Lake Volcanic Field. The geologic setting of the Geysers is very similar to the rest of the

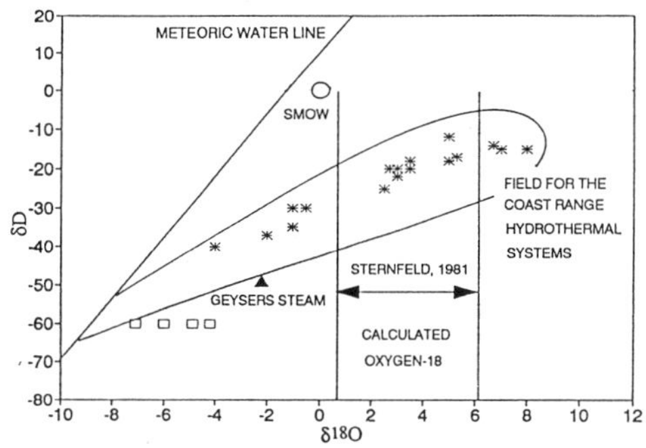

Fig. 3 $\delta^{18}O$ vs δD for springs from the Geysers Geothermal Field. Astrix Craig 1963, Squares Truesdell and Hulston 1980. Also shown is the field for the Coast Range hydrothermal systems and calculated $\delta^{18}O$ for hydrothermal fluid from Sternfeld 1980.

Coast Range hydrothermal systems, with the main difference being the size of the geothermal field and the close proximity of the Geysers to a siliceous magma chamber.

Isotopic data from hot-springs in the Geysers (Craig, 1963; Truesdell and Hulston, 1980) plot on the same trend as the other springs from the Coast Range hydrothermal systems (Fig. 3), suggesting a similar origin. Sternfeld (1980) calculated the oxygen isotopic composition of the hydrothermal fluids that precipitated vein quartz and carbonates. The bulk of Sternfeld's oxygen isotope data plots between 2 and 6 ‰ . Lighter values of 0.8 ‰ were seen in one sample. Geologically and geochemically the Geysers is very similar to the rest of the Coast Range hydrothermal systems and although speculative it is possible that the Geysers have a significant component of connate water.

CONCLUSIONS

Preliminary data from the McLaughlin Mine indicates that the ore-forming fluid had an isotopic composition of $\delta^{18}O$ 8 to 10‰ and δD -15 to -30‰ (Fig. 1), identical to active hot-springs in the Wilbur Springs District. The fluid originated at depths of ~250 to 300°C and migrated upwards, where it intersected the hydrostatic boiling curve, precipitating gold over the upper 200 m.

The similarity in isotopic composition and geologic setting of the Coast Range hydrothermal systems, the McLaughlin Mine, and the Geysers suggest similar origins. The McLaughlin deposit and smaller active auriferous hot-springs are formed by fluids which define the isotopically heavy end-member of the mixing trend, whereas mineral springs are formed from fluids with a greater meteoric water component. The isotopically heavy fluids are interpreted to be evolved sea water originally trapped in Franciscan sediments. The Geysers Geothermal System defines a similar mixing trend as the Coast Range hydrothermal systems, suggesting a component of the isotopically heavy end-member and a large component of meteoric water, either by natural or artificial recharge. These isotopically heavy fluids have the potential to form Au-Hg deposits, mineral-springs, or geothermal systems depending on local geology, and degree of dilution by meteoric water.

REFERENCES

Barnes, I., Hinkle, M.E., Rapp, J.B., Heropoulos, C., and Vaughn, W.W., 1973a. Chemical Composition of Naturally Ocurring Fluids in Relation to Mercury Deposits in Part of North Central California. U.S.G.S. Bul. 1382-A, 19pp.

Barnes, I., O'Neil, J.R., Rapp, J.B., and White, D.E., 1973b. Silica Carbonate Alteration of Serpentinite: Wall Rock Alteration in Mercury Deposits of the California Coast Ranges. Econ. Geol., vol. 68, pp 388-398

Craig, H., 1963. The Isotopic Geochemistry of Water and Carbon in Geothermal Areas. in Nuclear Geology on Geothermal Areas, edited by E. Tongiorgi, Pisa. Congiglio Nazionale della Ricerche, Laboratorio de geologia nuclear. pp 17-55

Field, C.W., and Fifarek, R.H., 1985. Light Stable Isotope Systematics in the Epithermal Environment. in Reviews in Economic Geology vol. 2 edited by B.R. Berger and P.M. Bethke p. 99-128

Kharaka, Y.K., and Mariner, R.H., 1989. Chemical Geothermometers and their Application to Formation Waters from Sedimentary Basins. in Thermal History of Sedimentary Basins Methods and Case Histories edited by N.D. Naeser and T.H. McCulloh, Springer-Verlag pp 99-119

Lehrman, N.J., 1986. The McLaughlin Mine, Napa and Yolo Counties, California. in Nevada Bureau of Mines and Geology Report 41 edited by J.V. Tingley and H.F. Bonham, Jr. pp 85-89

Magaritz, M., and Taylor, H.P. Jr., 1976. Oxygen Hydrogen and Carbon Isotope Studies of the Franciscan Formation, Coast Ranges, California. Geochim. et Cosmochim. Acta, vol. 40, pp 215-234

Ohmoto, H., and Rye, R.O., 1974. Hydrogen and Oxygen Isotopic Compositions of Fluid Inclusions in the Kuroko Deposits, Japan. Econ. Geol. vol. 69, pp 947-953

O'Neil, J.R., and Taylor, H.P.Jr., 1967. The Oxygen Isotope and Cation Exchange Chemistry of Feldspars. American Mineralogist, vol. 52, pp 1414-1437

Peters, E.K., 1990. The Aqueous Geochemistry of the Clear Lake Area, Northern California, and its Implications for Hot-Spring Gold Deposits. Unpub. Ph.D. thesis Harvard University, 237 pp

Sternfeld, J.N., 1980. The Hydrothermal and Stable Isotope Geochemistry of Two Wells in the Geysers Geothermal Field, Sonoma County, California. Unpub. M.Sc. thesis University of California, Riverside, 202 p

Taylor, H.P.Jr., 1974. The Application of Oxygen and Hydrogen Isotope Studies to Problems of Hydrothermal Alteration and Ore Deposition. Econ. Geol. vol. 69 pp 843-883

Truesdell, A.H., and Hulston, J.R., 1980. Isototopic Evidence on Enviroments of Geothermal Systems, Chapter 5 in Hand-Book of Environmental Isotope Geochemistry vol. 1, edited by P. Fritz and J.C. Fonts, Elsevier pp 170-226

White, D., Barnes, I., and O'Neil, J.R., 1973. Thermal and Mineral Waters of Nonmeteoric Origin, California Coast Ranges, Geol. Soc. America Bul. vol. 84 pp 547-560

Water-Rock Interaction, Kharaka & Maest (eds) © 1992 Balkema, Rotterdam. ISBN 90 5410 075 3

^{10}Be and ^{9}Be in the Salton Sea (USA) and other geothermal systems

Nathalie J. Valette-Silver
University of Maryland, College Park, Md., USA

Fouad Tera
Carnegie Institution of Washington, D.C., USA

Milan Pavich
USGS, Reston, Va., USA

ABSTRACT: ^{10}Be and ^{9}Be were analyzed in rocks and fluids from the Salton Sea geothermal system. The ^{10}Be inventory reveals that the sediments hosting the geothermal field are younger than a few million years. ^{10}Be/^{9}Be in rocks decreases near the 1837 and 2956 m aquifers. A reverse trend exists in the fluids. ^{10}Be and ^{10}Be K_{d}s decrease from the surface to the bottom of the well. Occupying adsorption sites, ^{10}Be is more easily removed than ^{9}Be from the metamorphic rocks. Comparison of Salton Sea with other marine and continental natural systems, shows that Be solubility is strongly influenced by pH and salinity.

INTRODUCTION

The Salton Sea geothermal system (Southern California, USA) is located in the Salton Trough (Fig. 1), rift valley filled with Colorado River sediments that represents the on-land extension of the East Pacific Rise (Elders and Biehler, 1975). In 1986, the SSSDP California State 2-14 well , was drilled in order to investigate the roots of the geothermal field. Cores were collected down to 3250 m and two flow tests at 1850 m and 3250 m were performed. We studied fluids and rocks from this deep hole, surface emissions and a shallower well, Fee #5, for their ^{10}Be and ^{9}Be concentrations.

^{10}Be is a radioactive isotope ($T_{1/2}=1.5\times10^{6}$ years) formed in the upper atmosphere by cosmic ray spallation of N and O nuclei (Arnold, 1956). After formation, the isotope is delivered to the earth's surface by rain, aerosols and snow with a flux of about 0.04 atoms/cm^{2}s (Monaghan et al., 1986). For ^{10}Be, the chemical separation used is based on the isotopic dilution technique (Valette-Silver et al., 1986, 1987). ^{10}Be measurements were performed by accelerator mass spectrometry. ^{9}Be was analyzed after chemical pre-concentration using the same technique as for ^{10}Be.

Several kinds of in situ or flowline fluid samples were collected: filtered (on 0.45 μm

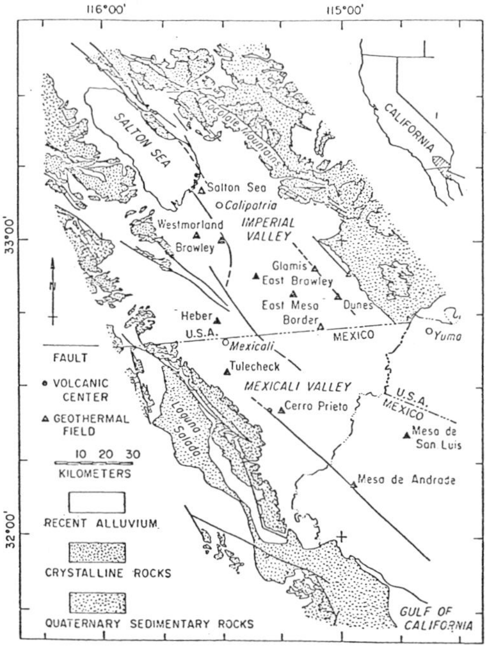

Figure 1. Location of the Salton Sea Geothermal system after Elders and Cohen (1983)

pore diameter nucleopore membrane) and unfiltered samples, straight and diluted (1:9 ratio) with distilled water, and diluted (1:9) with a 2N HCl acid solution. All Fee #5 flowline samples were filtered, diluted with distilled water (1:9) and HNO_3 acidified. Samples collected from the surface emissions (mud pots and cool pools) were only filtered and acidified.

RESULTS

In the SSSDP rock samples, ^{10}Be (Fig. 2) display a large peak in concentration (259×10^6 at/g) around 1000m depth, followed by a decrease (58×10^6 at/g) at 2000 m (depth of the first flow test). An apparent oscillation of maxima and minima around an "average" value of $100 \pm 20 \times 10^6$ at/g extends across the whole length of the core. The distribution of ^9Be in the core shows (Fig, 3) a very jagged profile on a scale far outside experimental errors (high values close to 2 ppm and lows > 1ppm). A distinct minimum (0.5 ppm) exists at a depth of 2500 m. In the rocks, ^{10}Be and ^9Be are not well correlated but the correlation is best at low concentration. $^{10}Be/^9Be$ ratios vary In the rocks, from 38 to 189×10^{-11}. In general, the average value of this ratio decreases from the upper to the lower aquifer.

During both flow tests of well SSSDP, sampling problems were encountered. They were related to the high temperature and high salinity of the fluids collected, and also to the lack of reinjection capabilities . However, clean

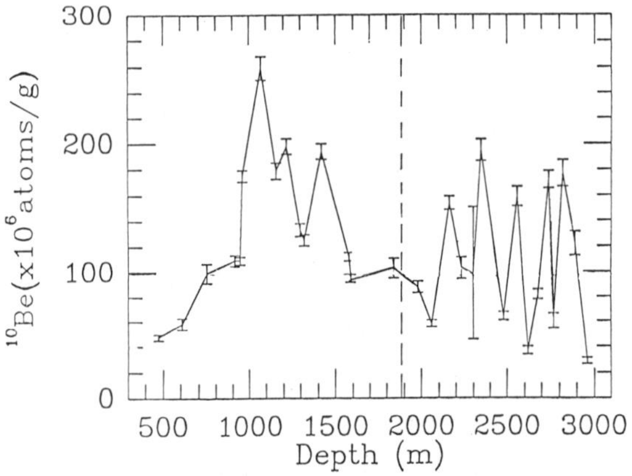

Figure 2. Vertical profile of the ^{10}Be concentrations and error bars in the core samples from the SSSDP well. The horizontal line represents the depth of the 1st flow test.

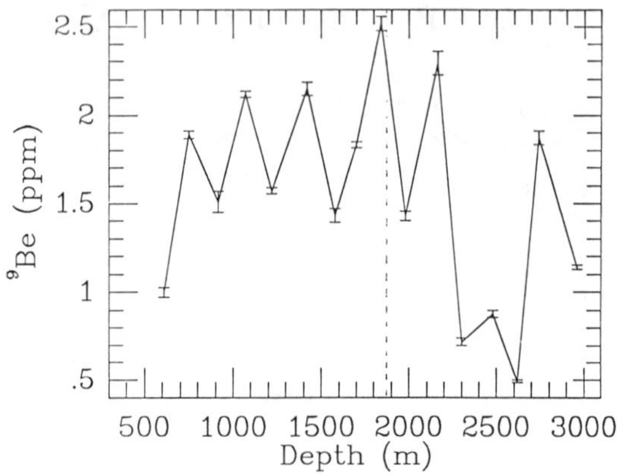

Figure 3. Vertical profile of the ^9Be in the rocks (same legend as Fig.2).

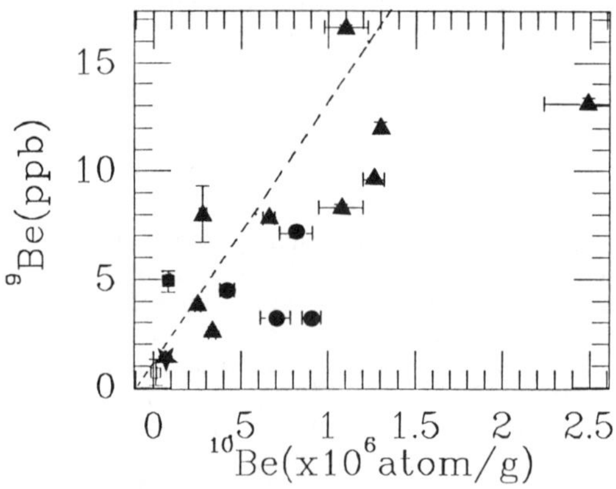

Figure 4. Relation between ^{10}Be and ^9Be in the fluids collected in the Salton Sea Geothermal System. Dashed line represents theoretical mixing of suspended sediment and pure water.

- □ : mud pot;
- ■ : in situ sample;
- ▲ : samples from the 1st flow test
- ● : samples from the 2nd flow test;
- ★ : a sample collected in the well Fee #5.

samples could be collected later (Williams,1987).

^{10}Be and ^9Be concentrations in the SSSDP fluids are shown in Fig. 4. ^{10}Be concentrations vary widely 0.341 to 2.490 x 10^6 at/g.and ^9Be concentrations from 0.7 to 16.6 ppb with an average of all the values of 4.1 ppb. ^{10}Be and ^9Be concentrations do not correlate well. Their ratio increases from 144 to 310 $x10^{-11}$ when going from the upper to the lower aquifer. Average ^{10}Be and ^{9}Be distribution coefficients (K_{ds}) generally decrease from top to bottom.

The average ^{10}Be concentration is 1.4 $x10^6$ at/g n the samples collected at 600m depth in well Fee#5, and 0.009 $x10^6$ at/g in the surface emissions (mudpots and active pools). For ^{9}Be, average concentrations of 3.8 ppb were measured in well Fee#5 and of 1.1 ppb in the surface emissions.

DISCUSSION

From in situ rain deposition the steady state ^{10}Be inventory is N_o=1.3 $x10^{13}$ at/cm^2. Compared to this inventory, the ^{10}Be inventory measured in the core reaches about 1 $x10^{15}$ at/cm^2. This is 2 orders of magnitude higher showing that these sediments are young (less than few half lives of ^{10}Be).

The differences in concentrations and patterns of ^{10}Be and ^{9}Be probably reflect the fact that ^{10}Be is transported as an adsorbed cation on fine-grained particles and ^{9}Be is probably contained in the lattices of detrital minerals from igneous or metamorphic rocks.

^{10}Be concentrations decrease near the depth of the two aquifers. This suggests the leaching of ^{10}Be from the host rocks by the hydrothermal fluid at the depth of both aquifers.

As shown in Fig. 4, ^{9}Be and ^{10}Be are poorly correlated. The mixing line shown on the figure, represents a model of simple mixing through suspension of sediment containing average concentrations of ^{10}Be and ^{9}Be (^{10}Be=121 at/g and ^{9}Be = 1.55 ppm) in pure water. Several of the observed fluid compositions plot near this theoretical mixing line, most of the others plot in the field that corresponds to enrichment with ^{10}Be. This suggests, that fine, non-filterable particles (<0.45 μm) suspended in the fluid as well as leaching contribute to the measured ^{10}Be concentrations. These values represent upper limits for the isotope concentrations.

Compared to other natural waters (Fig.5) ^{10}Be concentrations in the Salton Sea geothermal waters are high (Valette-Silver et al., 1988 b). Surface waters (Potomac and Choptank rivers, and aquifers) have very low

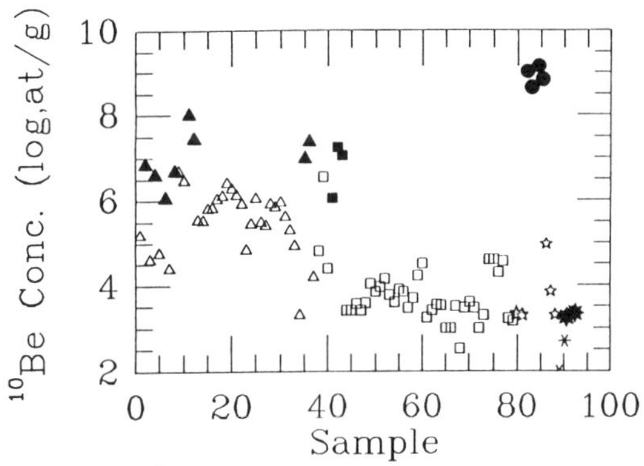

Figure 5. ^{10}Be concentrations in the Salton Sea fluids compared to other natural waters.
-Salton Sea geothermal system fluids (△), and associated rocks (▲).
-Fluids (□) from submarine hot springs and terrestrial geothermal systems (21 and 13°N EPR, Juan de Fuca,Valles Caldera, Long Valley Caldera, Pozzuolli,Cerro Prieto) and associated sediments (■).
-Waters(☆) from rivers and aquifers (Potomac Choptank rivers,and Spalding) , and associated sediments (●) .
-The average ocean and the average equatorial rain (✸) and Blanks (✻).

^{10}Be concentrations (2 $x10^3$ at/g). The same range of values is observed for continental geothermal waters. Hydrothermal fluids collected on the East Pacific Rise (EPR) and on the Juan de Fuca Ridge (Valette-Silver et al., 1987) exhibit also low ^{10}Be values ranging from 10^3 to 10^4 at/g. Concentrations in the Salton Sea fluids are 2 orders of magnitude higher (average 0.5 $x10^6$ at/g). In the rivers, the difference between ^{10}Be concentrations in the fluid and in the associated sediments reaches 7 orders of magnitude. In the Salton Sea, a difference of only 1 to 2 orders of magnitude is observed. This is explained by the fact that ^{10}Be is highly soluble and easily desorbed from sediments in acidic solutions (pH<5.5) with high salinity and high organic matter contents (Valette-Silver et al., 1986, 1987, 1992).

The Salton Sea fluids are also very rich in ^{9}Be when compared to other natural systems. Average natural waters exhibit ^{9}Be concentrations that are 2 to 6 orders of magnitude lower than the average value (4 ppb)

measured in the SSSDP fluids.

^{10}Be and ^{9}Be K_{ds} decrease from the surface to the bottom of the SSSDP well and ^{10}Be/^{9}Be in the rocks decrease abruptly near the first aquifer. This is further evidence that leaching of beryllium from solid phases by hydrothermal waters is occurring in the Salton Sea System but that ^{10}Be is more actively removed. The difference in behavior of the two isotopes is probably due to the difference in the sites that the two isotopes occupy in the solid phases.

CONCLUSION

In the Salton Sea Geothermal System:
1) ^{10}Be inventory suggests that the sediments are young (a few million years).
2) Both ^{10}Be and ^{9}Be are highly enriched in the fluids relative to other surface waters. ^{10}Be-^{9}Be correlation suggests enrichment by suspended ultrafine material and/or through leaching of the isotopes from the solid phase.
3) Distribution coefficient systematics indicate, unambiguously, strong preferential leaching of ^{10}Be relative to ^{9}Be. This observation is consistent with the fact that the exogenic ^{10}Be occupies primarily adsorption and exchange sites, and ^{9}Be resides in the mineral lattice sites.

REFERENCES

Arnold, M. H., ^{10}Be produced by Cosmic Rays, *Science*, V. 1 IV, 584-585, 1956.

Elders, W.A., and S. Biehler, Gulf of California Rift System and its implication for the tectonics of western North America, *Geology*, 3, 85-87, 1975.

Monaghan, M.C., S. Krishnaswani, and K.K. Turekian, The global production rate of ^{10}Be, *Earth Planet. Sci. Lett.*, 76, 279-288, 1986.

Valette-Silver, J.N., F. Tera, J. Klein, and R. Middleton, ^{10}Be in geothermal systems, *Geothermal Resources Council Transactions,* 10,161-166, 1986.

Valette-Silver, J.N., F. Tera, J. Klein, and R. Middleton, ^{10}Be in hydrothermal vent deposits from the East Pacific Ridges: Role of sediments in the hydrothermal process, *J of Geophys. Res.*, 92, 11367-11372, 1987a

Valette-Silver, J.N., F. Tera, M. Pavich, J. Klein, and R. Middleton, ^{10}Be-^{9}Be in the Salton Sea geothermal system (U.S.A.), *EOS* , Trans. Am. Geophys. Un., 69, 527, 1988a

Valette-Silver, J.N., F. Tera, M. Pavich, J. Klein, and R. Middleton, ^{10}Be in hydrothermal systems and natural waters, *Proceedings of the 1st Goldschmidt Conference*, 79, 1988b.

Valette-Silver, J.N., Behavior of Al, Cu, Cr and Be during the percolation of acid coal leachate through soils. *Report to State of Maryland PPRP,* 1992,in press.

Williams, A.E., Chemical and isotopic variations and the distribution of brines across the Salton Sea geothermal field, California, *EOS , Trans. Am. Geophys. Un.*, 68, 438, 1987.

Stable and radioactive isotopes

Exchange reactions

Water-Rock Interaction, Kharaka & Maest (eds) © 1992 Balkema, Rotterdam. ISBN 90 5410 075 3

Isotope-geochemical effects of water-rock interactions in the Atesina-Cima d'Asta volcano-plutonic complex, Italy

S. Barth, F. Oberli & M. Meier
Institut für Krist. und Petrogr., ETH Zürich, Switzerland

P. Blattner
Institute of Geosphere Research, DSIR Lower Hutt, New Zealand

G. M. Bargossi
Dipartimento di Scienze Mineral., Università Bologna, Italy

G. Di Battistini
Istituto di Petrografia, Università di Parma, Italy

ABSTRACT: Basaltic andesites to rhyolites and quartz norites to granites from the Lower Permian Atesina-Cima d'Asta volcano-plutonic complex (N Italy) show variable degrees of secondary overprint. Fluid infiltration led to enrichment in K_2O, Rb, depletion in Na_2O, CaO, Sr, and mobility of Ba in the Atesina volcanics. Geochemically least modified volcanics yield a "primary" Lower Permian Rb-Sr total-rock errorchron age of 267 ± 35 Ma (95% c.l., ext.), whereas strongly altered volcanics yield a "rejuvenated" age of 255 ± 4 Ma. $^{18}O/^{16}O$ isotope shift of the Atesina volcanics and Cima d'Asta intrusives is a result of medium- to low-temperature water-rock interaction. REE patterns and Sm-Nd isotope signatures as well as some of the $\delta^{18}O$ values (quartz) were not affected by hydrothermal overprint.

1 INTRODUCTION

The Lower Permian calc-alkaline Atesina-Cima d'Asta volcano-plutonic complex (Southern Alps, N Italy) represents a Late Hercynian cogenetic magmatic province. Stratigraphically, the ~2000 m thick Atesina volcanic sequence ("Bozener Quarzporphyr") is divided into a Lower Group composed of (basaltic) andesites to rhyolites, and an Upper Group dominated by rhyodacites to rhyolites. The Cima d'Asta intrusives range in composition from quartz norites to granites.

The Atesina volcanics and Cima d'Asta intrusives show variable degrees of secondary overprint by fluid infiltration. Primary magmatic characteristics and alteration effects are discussed in the light of new geochemical and Rb-Sr, Sm-Nd, and oxygen isotope data.

2 MAJOR/TRACE ELEMENT MOBILITY

Basaltic andesitic to rhyolitic lavas and rhyodacitic to rhyolitic ignimbrites from the Atesina volcanic complex, and quartz norites to granites from the Cima d'Asta intrusion show well defined correlation of most major/trace elements with degree of magma differentiation. However, some highly mobile major and trace elements of the Atesina volcanics, e.g., K_2O, Na_2O, CaO, Rb, Sr, Ba display large scatter compared to the linear trends of the Cima d'Asta intrusives. Systematic enrichment in K_2O, Rb, and depletion in Na_2O, CaO, and Sr within rocks of similar lithology reflect increasing degree of alteration, e.g., in the andesitic group (A6 to A4), in the dacitic-rhyodacitic group (D20 to RD7), and in the rhyolitic group (R12 to R11-R15-R16-R17 to R3-R21; Table 1). Ba variations follow no systematic enrichment/ depletion trend. Very low Ba concentrations of some rhyolitic ignimbrites (R11, R15; Table 1) are probably a result of both primary depletion caused by thermogravitational diffusion in the high-silica roof-zone of a magma chamber (Hildreth 1979), and of mobility during alteration.

Total-rock alteration is accompanied by break-down of feldspar. Degree of albitization of plagioclase increases with progressive alteration; the Ab component shows an increase in the

Table 1. XRF total-rock major (wt.%) and trace (ppm) element concentrations of the Atesina volcanics. Different lithologies are separated by solid vertical lines. Dashed lines separate less altered samples (left) from more altered samples (right).

	Atesina volcanic complex												
	BA1	BA2	A6	A4	D20	RD7	R12	R11	R15	R16	R17	R3	R21
SiO_2	56.46	55.09	58.27	58.12	64.83	64.76	70.97	74.65	74.35	71.77	71.27	70.89	73.77
TiO_2	0.69	0.69	0.79	0.76	0.54	0.55	0.42	0.06	0.06	0.20	0.21	0.24	0.24
Al_2O_3	15.45	15.48	16.60	16.10	15.81	15.79	14.43	12.97	12.98	14.23	14.02	14.45	13.77
Fe_2O_3	7.50	7.55	7.11	6.46	4.63	4.82	3.40	0.98	0.96	1.60	1.79	2.08	2.53
MnO	0.13	0.14	0.13	0.09	0.05	0.06	0.05	0.02	0.02	0.02	0.03	0.05	0.04
MgO	7.55	8.20	5.07	4.16	2.99	1.85	1.68	0.35	0.32	0.47	0.61	0.89	0.61
CaO	8.55	8.98	6.91	5.66	1.91	0.58	1.89	0.94	0.91	1.35	1.21	1.30	0.45
Na_2O	1.60	1.29	2.51	2.13	2.95	2.76	4.39	3.77	4.11	3.91	4.26	2.23	0.68
K_2O	0.47	0.27	0.73	2.78	3.02	7.44	1.48	3.22	3.26	3.59	3.30	5.64	6.09
P_2O_5	0.14	0.14	0.16	0.16	0.15	0.13	0.12	0.02	0.03	0.06	0.07	0.08	0.05
L.O.I.	1.45	2.17	1.72	3.58	3.12	1.28	1.17	2.99	3.00	2.78	3.23	2.14	1.78
Ba	345	290	358	418	649	609	335	55	56	394	360	317	470
Cr	663	682	211	146	28	15	15	5	3	4	5	9	11
Nb	13	10	14	16	18	15	15	20	18	15	16	16	15
Ni	40	42	19	24	9	8	5	3	4	7	5	5	5
Rb	10	0	59	105	109	272	75	228	218	288	261	261	254
Sr	474	821	393	327	125	60	187	101	48	129	142	52	34
V	163	177	162	136	68	64	50	5	3	12	14	22	22
Y	27	26	29	26	39	37	25	57	56	37	38	36	37
Zr	148	150	169	150	185	166	151	84	86	153	161	151	139
Th	9	7	10	7	14	15	16	26	26	23	26	21	16
Co	26	26	20	17	8	9	5	4	2	8	6	4	6
Sc	30	32	25	23	19	16	13	7	7	6	6	7	7

dacitic-rhyodacitic group from D20 (52-58 mole%) to RD7 (98), and in the rhyolitic group from R12 (48-59) to R11-R15-R16-R17 (63-78) to R3-R21 (99).

3 OPEN-SYSTEM BEHAVIOR OF Rb-Sr

Geochemically least modified andesitic (A6), dacitic (D20) and rhyolitic (R12) samples define a "primary" Rb-Sr total-rock errorchron of 267 ± 35 Ma (Fig. 1) with a $(^{87}Sr/^{86}Sr)_i$ ratio of 0.7105 ± 5 (95% c.l., ext.; MSWD=3.3). The scatter of the data may be due to minor secondary shifts and/or to somewhat different initial Sr compositions and possibly slightly different emplacement ages. The Lower Permian "age" defined by these geochemically least modified samples is compatible with literature data (see discussion) and suggests that the primary magmatic characteristics of these samples are relatively well preserved.

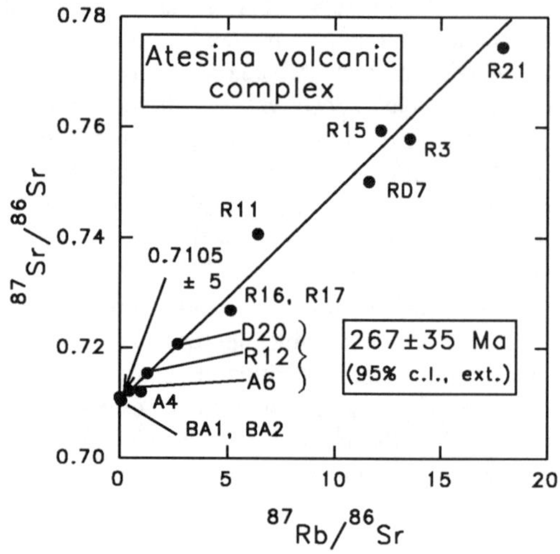

Fig. 1 $^{87}Sr/^{86}Sr$ vs $^{87}Rb/^{86}Sr$ diagram for the Atesina volcanics.

Strongly altered andesitic (A4), rhyodacitic (RD7) and rhyolitic (R16, R17, R3, R21) samples displaying large variations in highly mobile major/trace elements plot below this "primary" errorchron. These samples define a "rejuvenated" Rb-Sr total-rock errorchron of 255 ± 4 Ma with a $(^{87}Sr/^{86}Sr)_i$ ratio of 0.7084 ± 2 (MSWD=6.8). Rhyolitic samples (R11, R15) plotting above the "primary" errorchron may reflect both enrichment in $(^{87}Sr/^{86}Sr)_i$ due to isotopic gradients in the high-silica roof-zone of a magma chamber (Hildreth 1981), and secondary mobility of Rb and Sr.

4 Nd-Sr ISOTOPE SYSTEMATICS

Initial ε_{Nd} and $^{87}Sr/^{86}Sr$ isotope signatures of the Atesina volcanics and Cima d'Asta intrusives at 270 Ma are shown in Fig. 2. The Cima d'Asta intrusives are characterized by crustal initial ε_{Nd} and $^{87}Sr/^{86}Sr$ values ranging from -2.7 to -6.5 and from 0.7057 to 0.7092, respectively. The initial ε_{Nd} values of the Atesina volcanics range from -3.6 to -6.5, overlapping with the values of the intrusives. Atesina volcanics with "closed" Rb-Sr system cluster at initial ε_{Nd} values of -5.5 to -6.2 and $^{87}Sr/^{86}Sr$ values of 0.7103 to 0.7105, whereas some of the Atesina volcanics characterized by "open-system" behavior of Rb-Sr show considerable scatter with respect to the Cima d'Asta data array.

Although the basaltic andesites have high contents of MgO, Cr and Ni (see Table 1) indicative for mantle derivation, they show highly crustal initial ε_{Nd} values of -6.2 to -6.4 and $^{87}Sr/^{86}Sr$ values of 0.7100 to 0.7109. The partial decoupling of the Nd isotope signatures of the Atesina volcanics from a typical differentiation trend, however, is not the result of REE mobility. The REE patterns of the Atesina volcanics overlap with those of the Cima d'Asta intrusives and are characterized by regular primary magmatic LREE enrichment, HREE depletion and negative Eu anomalies (Barth et al. 1990); Ce anomalies are absent. The Cima d'Asta intrusives also display partially inverse correlations between isotope signatures and magma differentiation. Nd-Sr isotope data of the Atesina volcanics and Cima d'Asta intrusives therefore do not support a model restricted to single-stage AFC-type magma evolution, but favor a MASH-type (Hildreth and Moorbath 1988) process (Barth et al. 1992).

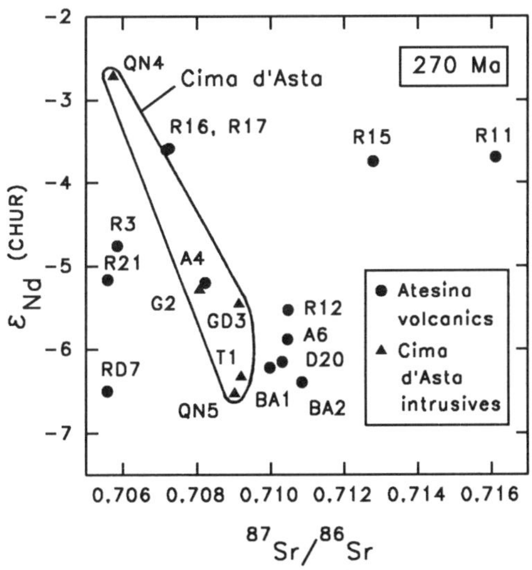

Fig. 2 Initial ε_{Nd} vs $^{87}Sr/^{86}Sr$ isotope ratios of the Atesina volcanics and Cima d'Asta intrusives at 270 Ma.

5 $^{18}O/^{16}O$ ISOTOPE SHIFT

Oxygen isotope data of the Atesina volcanics and Cima d'Asta intrusives are shown in Fig. 3. Total-rock $\delta^{18}O$ values of the Atesina volcanics ranging from 10.0-13.6‰ for basaltic andesitic and andesitic lavas, and from 11.9-16.6‰ for dacitic, rhyodacitic and rhyolitic lavas and ignimbrites indicate secondary shift to higher $^{18}O/^{16}O$ values. In addition, quartz and sanidine from a vitrophyric rhyolitic ignimbrite (R11) yield $\delta^{18}O$ values of 9.6‰ and 9.2‰ ($\Delta^{18}O_{(Qz-Sa)}$=0.5‰), respectively; perlitic glass matrix from the same sample furnishes a very high value of 17.3‰. Quartz from a rhyolitic lava (R21) yields 10.5‰, whereas the devitrified matrix is also characterized by a very high value of 17.8‰.

The total-rock $\delta^{18}O$ values reflect increasing degree of alteration within the dacitic-rhyodacitic

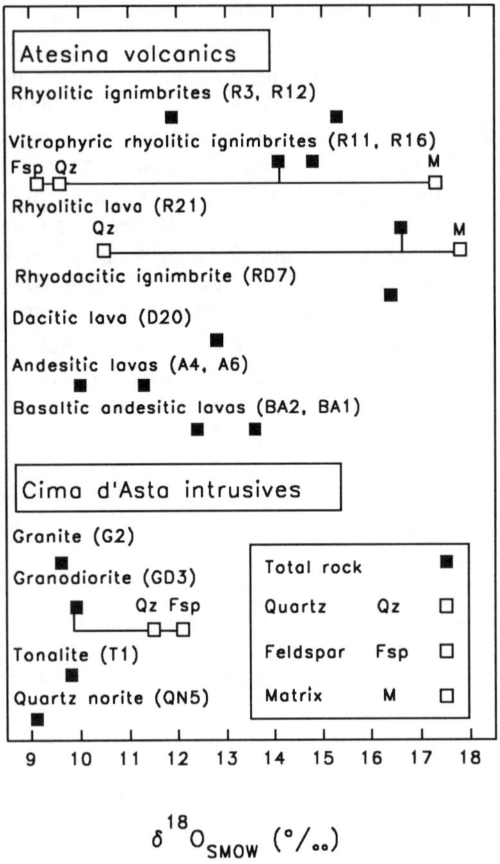

Fig. 3 $\delta^{18}O$ values of the Atesina volcanics and Cima d'Asta intrusives.

group from D20 to RD7. However, within the andesitic and rhyolitic groups the total-rock $\delta^{18}O$ values do not strictly follow the alteration pattern displayed by mobile major/trace element and Rb-Sr isotope data. Local temperature differences of secondary processes may have predominantly controlled the strongly temperature-dependent $^{18}O/^{16}O$ isotope shifts.

Total-rock $\delta^{18}O$ values of the Cima d'Asta intrusives display a narrow range from 9.1‰ for quartz norite to 9.9‰ for granodiorite. Quartz and feldspar from granodiorite yield $\delta^{18}O$ values of 11.5‰ and 12.1‰, respectively, and thus indicate isotopic disequilibrium by a negative $\Delta^{18}O_{(Qz\text{-}Fs)}$ value of -0.6‰.

Quartz is considered to be relatively inert during the exchange process. Indeed, $\delta^{18}O$ of quartz from vitrophyric rhyolitic ignimbrite, rhyolitic lava and granodiorite correlates inversely with initial ε_{Nd} values, reflecting significant crustal contribution during magma evolution.

Elevated $\delta^{18}O$ total-rock values, very high $\delta^{18}O$ values of perlitic and devitrified matrix, and isotopic disequilibrium between quartz and feldspar are a typical result of water-rock interaction at low temperature. Assuming a plausible $\delta^{18}O$ value near -5‰ for meteoric water involved in the exchange process and complete isotopic equilibration between feldspar and water, extrapolation based on typical feldspar-H_2O isotope fractionation parameters (O'Neil and Taylor 1967) results in temperature estimates of ~130° C for vitrophyric rhyolitic ignimbrite (R11) and of ~100° C for granodiorite (GD3).

6 DISCUSSION AND CONCLUSIONS

6.1 *Cima d'Asta intrusion*

The Cima d'Asta intrusives are far less affected by major/trace element mobility as compared to the Atesina volcanics. Lower Permian Rb-Sr total-rock isochron ages of 274 ± 2 Ma and 275 ± 9 Ma (1σ, recalculated with λ (^{87}Rb) $=1.42\times10^{-11}$ a^{-1}) agree with biotite model ages and are interpreted as emplacement ages (Borsi et al. 1974).

However, clear isotopic disequilibrium between quartz and feldspar from granodiorite indicates medium- to low-temperature water-rock exchange for at least this sample.

6.2 *Atesina volcanic complex*

The Atesina volcanics display variable degrees of secondary overprint by fluid infiltration. This is documented by major and trace element mobility, e.g., enrichment in K_2O, Rb, depletion in Na_2O, CaO, Sr, mobility of Ba (Table 1), open-system behavior of Rb-Sr (Figs. 1, 2), and shift in $^{18}O/^{16}O$ (Fig. 3).

Elevated $\delta^{18}O$ total-rock values (10.0-16.6‰),

very high $\delta^{18}O$ values of perlitic glass and devitrified matrix (17.3 and 17.8‰), and isotopic disequilibrium between quartz and sanidine (Δ=0.5‰) support water-rock exchange at low temperature. We note that two early values of Taylor (1968) conform to this relationship.

Some of the $\delta^{18}O$ values (quartz) as well as the REE patterns and Sm-Nd isotope signatures of the Atesina volcanics were not affected by medium- to low-temperature hydrothermal overprint and are thus valuable indicators for primary magma evolution.

Geochemically least modified volcanics yield a "primary" Rb-Sr total-rock errorchron age of 267 ± 35 Ma (95% c.l., ext.). This age is compatible with published Rb-Sr biotite ages of 267-276 Ma for rhyodacitic and rhyolitic ignimbrites, which indicate a Lower Permian age for the volcanic activity (D'Amico et al. 1980, D'Amico and Del Moro 1988). Geochemically altered volcanics displaying large variations in highly mobile major/trace elements, in contrast, yield a "rejuvenated" Rb-Sr total-rock errorchron age of 255 ± 4 Ma. Similar and younger Rb-Sr total-rock and feldspar errorchron ages of 226-252 Ma were obtained by D'Amico et al. (1980) and D'Amico and Del Moro (1988) for rhyodacitic and rhyolitic ignimbrites. The isotope-geochemical data suggest that both Rb enrichment and Sr depletion related to break-down of feldspar account for the opening of the Rb-Sr system.

Hydrothermal activity associated with the Middle/Upper Triassic alkaline magmatism in the South Alpine domain may be responsible for a large part of the alteration effects. On the other hand, biotite of the Atesina volcanics displays closed-system behavior of Rb-Sr, with the exception of some rhyodacitic ignimbrites at the immediate contact with Middle Triassic intrusives (D'Amico et al. 1980, D'Amico and Del Moro 1988). This suggests a temperature of < 300° C (blocking temperature of biotite) for hydrothermal overprint.

Fission-track data of volcanic glasses yield additional evidence for a complex thermal history. Corrected fission track ages range from 268-276 Ma to 231-235 Ma (Storzer 1970) and indicate the age of (Lower Permian) formation and of major (Triassic) annealing, respectively. According to Storzer (1970), thermally unaffected fission track ages ranging from 149 Ma (~130° C) in the south of the Atesina volcanic complex to 21 Ma (~190° C) in the north determine the end of the last thermal events.

Geochemical and isotope studies carried out on Pb-Zn deposits in the Eastern/Southern Alps give further information on the Triassic hydrothermal event. Lead isotope data of Pb-Zn deposits hosted in Permian/Triassic sediments indicate that a substantial amount of ore lead derived from basement rocks and to some extent from overlying clastic sediments, which were subject to large-scale extensive leaching by low-temperature solutions (Köppel and Schroll 1988). The Ga/Ge ratios in sphalerites suggest equilibrium leaching of silicates at > 180° C (Möller 1985). Based on sedimentological and stable isotope data a temperature of < 100° C is estimated for ore deposition (Kappel and Schroll 1982).

We propose that the Eastern/South Alpine Triassic hydrothermal activity is part of a superregional hydrothermal cycle that was active in a variety of locations, for example in Spain (210-230 Ma, Halliday and Mitchell 1984), Germany (220-225 Ma, Mertz et al. 1990), and England (210-220 Ma, Halliday and Mitchell 1976). This cycle is associated with the break-up of the Pangaean megacontinent and is characterized by initial rifting of the continental crust during extensional tectonic regimes, uprise of mantle-derived magmas, increase in the geothermal gradient, and hydrothermal circulation.

REFERENCES

Barth, S., Bargossi, G.M., Di Battistini, G. and Mohr, B. 1990. Petrogenesis and age of the Late Hercynian volcano-plutonic complex of Trentino-Alto Adige (Southern Alps, N Italy). *Intern. Volcanol. Congress, IAVCEI*, Mainz.

Barth, S., Oberli, F., Meier, M., Blattner, P., Bargossi, G.M. and Di Battistini, G. 1992. Nd-Sr isotope evidence for mantle-crust interaction: a study of the Late Hercynian calc-alkaline Atesina-Cima d'Asta volcano-plutonic complex, N Italy. *Intern. Geol. Congress, IGC,* Kyoto (submitted).

Borsi, S., D'Amico, C. and Del Moro, A. 1974. Studio radiometrico delle rocce intrusive del massiccio di Cima d'Asta (Trentino). *Mem. Soc. Geol. It.* 13/1: 145-159.

D'Amico, C., Del Moro, A., Freddo, A. and Pardini, G. 1980. Studio radiometrico delle ignimbriti riolitiche atesine, gruppo superiore. *Rend. Soc. Ital. Miner. Petrol.* 36: 703-716.

D'Amico, C. and Del Moro, A. 1988. Permian and Triassic Rb-Sr dating in the Permian rhyodacitic ignimbrites of Trentino (Southern Alps). *Rend. Soc. Ital. Miner. Petrol.* 43: 171-180.

Halliday, A.N. and Mitchell, J.G. 1976. Structural, K-Ar and ^{40}Ar-^{39}Ar age studies of adularia K-feldspars from the Lizard complex, England. *Earth Planet. Sci. Lett.* 29: 227-237.

Halliday, A.N. and Mitchell, J.G. 1984. K-Ar ages of clay-size concentrates from the mineralisation of the Pedroches Batholith, Spain, and evidence for Mesozoic hydrothermal activity associated with the break up of Pangaea. *Earth Planet. Sci. Lett.* 68: 229-239.

Hildreth, W. 1979. The Bishop Tuff: Evidence for the origin of compositional zonation in silicic magma chambers. *Geol. Soc. Am. Spec. Pap.* 180: 43-75.

Hildreth, W. 1981. Gradients in silicic magma chambers: Implications for lithospheric magmatism. *J. Geophys. Res.* 86: 10153-10192.

Hildreth, W. and Moorbath, St. 1988. Crustal contributions to arc magmatism in the Andes of Central Chile. *Contrib. Mineral. Petrol.* 98: 455-489.

Kappel, E. and Schroll, E. 1982. Ablauf und Bildungstemperatur der Blei-Zink-Vererzung von Bleiberg-Kreuth / Kärnten. *Carinthia II* 172: 49-62.

Köppel, V. and Schroll, E. 1988. Pb-isotope evidence for the origin of lead in strata-bound Pb-Zn deposits in Triassic carbonates of the Eastern and Southern Alps. *Mineral. Deposita* 23: 96-103.

Mertz, D.F., Lippolt, H.J. and Müller, G. 1990. Isotopic dating of adularia-bearing epigenetic mineralizations. I. Saar-Nahe region / Southwest Germany. *Eur. J. Mineral.* 2: 89-102.

Möller, P. 1985. Development and application of the Ga/Ge-geothermometer for sphalerite from sediment-hosted deposits. *Monograph Series on Mineral Deposits* 25: 15-30.

O'Neil, J.R. and Taylor, H.P. Jr. 1967. The oxygen isotope and cation exchange chemistry of feldspars. *Am. Mineralogist* 52: 1414-1437.

Storzer, D. 1970. Fission track dating of volcanic glasses and the thermal history of rocks. *Earth Planet. Sci. Lett.* 8: 55-60.

Taylor, H.P. Jr. 1968. The oxygen isotope geochemistry of igneous rocks. *Contrib. Mineral. Petrol.* 19: 1-71.

Water-Rock Interaction, Kharaka & Maest (eds) © 1992 Balkema, Rotterdam. ISBN 90 5410 075 3

Possible evidence for fluid-rock oxygen isotope disequilibrium in hydrothermal systems

David R.Cole
Oak Ridge National Laboratory, Tenn., USA

ABSTRACT: There is ample evidence from geothermal systems that isotope temperatures estimated from the oxygen isotope fractionation between alteration phases and coexisting aquifer fluids agree closely with measured bore-hole temperatures. Similar, but limited evidence is found in epithermal vein deposits where isotope temperatures agree well with fluid inclusion homogenization temperatures. Conversely, many hydrothermal systems exhibit varying degrees of fluid-rock oxygen isotope equilibration (measured $\alpha_{r-w} \neq$ equilibrium α_{r-w}). There appears to be a crude relationship between increasing degree of equilibration and increasing temperature and salinity. The observed variations in the degree of exchange may have resulted from local, self-sealing of the fracture network prior to equilibration. The ability for fractures to remain open or to propogate allowing continued fluid flow may be the deciding factor in the attainment of isotopic equilibration.

1. INTRODUCTION

During fluid flow, the most important process by which the oxygen isotopic composition of the hydrothermal fluid may be altered prior to mixing and boiling at or near the surface, is through oxygen isotope exchange with silicates and carbonates at depth (Craig, 1963). The magnitude of the oxygen isotopic shifts in waters and rocks is controlled by the temperature of the interaction, the relative proportions of rock and fluid, the initial difference in the $\delta^{18}O$ values between the rock and fluid, and whether the system is open or closed to flow. The final rock-fluid isotope fractionation (α_{r-w}) is dependent only upon temperature for systems that have attained equilibrium. For non-equilibrium rock-fluid interaction, isotopic exchange is influenced not only by temperature, but also fluid composition, mineralogy, fluid/rock ratio, surface area (permeability) and time. The purpose of this communication is three fold. First, we will assess the degree of isotopic equilibration between alteration phases (veins, pore-filling, matrix) and co-existing fluids. If we can demonstrate that a fluid of known isotopic composition has equilibrated with a given alteration assemblage, then we can assess the reaction progress or degree of exchange between bulk rock and fluid, knowing that the two are related temporally. Upon demonstration that a given system exhibits fluid-rock isotope disequilibrium, we can begin to examine the relationships between the degree of exchange and variables such as temperature, solution composition and permeability. This approach differs from that described by Blattner (1985), who used an equilibrium model to assess the magnitude of incremental isotope shifts in fluids and rocks from selected geothermal systems.

2. ISOTOPIC DATA

Isotopic data on fluids and solids were selected from both active and fossil hydrothermal systems where a single fluid-type dominated (e.g., meteoric water, seawater) and the host rocks were igneous derived - i.e., volcanic, plutonic, or their weathered equivalents. For the fossil systems, we assumed that vein fluids approximated wall-rock fluids, thereby allowing use of fluid inclusion information (temperatures, compositions). Wherever possible, the effects of boiling and mixing on isotope budgets have been accounted for so that reservoir isotope values for the fluids were used in the model calculations. The isotope compositions of unreacted fluids and solids are known or can be inferred quite accurately for all of the systems studied. Isotope compositions of bore-hole fluids are used for the geothermal systems, whereas calculated fluid isotope compositions (in equilibrium with vein assemblages) are used for the epithermal ore deposits. With the exception of the Reykjanes, Iceland geothermal system, fluids from the remaining 11 geothermal areas and 7 ore systems exhibit characteristic ^{18}O shifts, with little or no D/H shift.

3. MINERAL-FLUID ISOTOPE FRACTIONATION

In order to directly compare fluid isotope shifts with those measured in the bulk rock, we need to demonstrate that some or preferably all of the alteration phases have equilibrated with a common fluid of known origin and composition. This will give us a strong indication that the fluid and altered rock are related temporally. Figure 1 shows isotopic temperatures plotted against measured temperatures for a variety of geothermal systems. An agreement of ±30°C (dashed lines) is considered evidence for equilibration given the errors in both the fractionation factors and measured down-hole temperatures. Data on Fig. 1a-c indicate that the agreement between isotopic and measured temperatures is good for all phases examined from most of the geothermal systems suggesting that the bore-hole fluids and the observed alteration are temporally related. This is particularly true for Wairakei, Cerro Prieto, Reykjanes, Svartsengi, and Krafla. Notable exceptions include the lower chlorite and calcite-fluid temperatures estimated for the Roosevelt Hot Springs system and lower quartz-fluid isotope temperatures for Broadlands and Salton Sea.

4. ROCK-FLUID ISOTOPE FRACTIONATION

The oxygen isotope fractionations (α_{r-w}) between the altered rocks and coexisting fluids have been calculated for the geothermal systems and epithermal vein deposits, and are plotted against measured temperatures on Figs. 2a and 2b, respectively. Also shown for reference are the equilibrium isotope fractionation curves for the systems: (1) granite/ rhyolite-, (2) granodiorite/dacite-, and (3) andesite/basalt-fluid. These curves are based on equilibrium alteration mineral assemblages predicted from rock-water

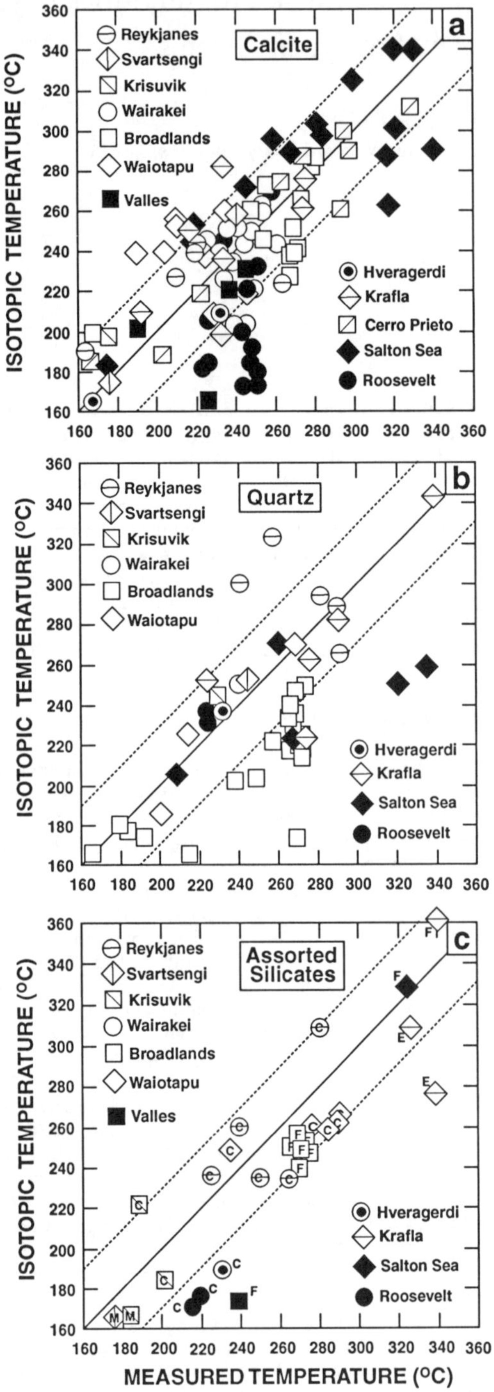

Fig. 1 Mineral-fluid isotope temperatures versus bore-hole temperatures. (a) calcite, (b) quartz, (c) assorted silicates.

interaction simulated through mass transfer computations (Bowers and Taylor, 1985; Cole et al, 1987, 1992). With the exception of the Salton Sea, all of the geothermal systems exhibit some degree of deviation from equilibrium isotope fractionation. For example, equilibrium fractionation between altered granite/rhyolite and water is about 6 per mil at 250°C, but rhyolitic systems such as Wairakei and Valles yield maximum fractionations of 10 and 14 per mil, respectively. For any given system, typical deviations from equilibrium rock-fluid fractionation range from a few per mil to over 8 per mil.

The observed rock-fluid isotopic fractionations in the vein-type ore deposits, detailed in Fig. 2b, also show apparent disequilibrium tendencies. For example, fractionation at Goldfield ranges from 8.3 to 14.4 per mil, whereas the expected equilibrium value at 250°C is 5.1 per mil. Most of these deposits exhibit a significant range in rock-fluid fractionation, similar in magnitude to ranges observed in the geothermal systems. We might conclude from this analysis that within a given hydrothermal system, isotope equilibrium may occur in some locales (or depths) but not others. In some deposits, such as Sunnyside, the degree of equilibration varies considerably over distances of only a few meters.

5. DISCUSSION

The extent of fractional approach to oxygen isotope equilibrium in these hydrothermal systems can be expressed in the familiar form of F, as derived by Northrop and Clayton (1966):

$$F = (\alpha^i - \alpha^f)/(\alpha^i - \alpha^{eq}) \qquad (1)$$

where the α's are the isotopic fractionation factors between rock and fluid at time t (α^f), at time=0 (α^i) and t= (α^{eq}). We have used the following equilibrium isotope fractionation equations to estimate the α's for granite/rhyolite (Cole et al, 1992) and andesite/basalt (Cole et al, 1987), respectively,

$$1000\ln\alpha_{gran-f} = 2.68(10^6/T^2) - 3.57 \qquad (2)$$

$$1000\ln\alpha_{bas-f} = 2.75(10^6/T^2) - 7.25 \qquad (3)$$

Curves have also been calculated for compositions intermediate to these two end-members.

Values for the degree of equilibration (F) fall between about 0.45 and 1.0 for the hydrothermal systems under consideration, with each system exhibiting its own distinct range. For example, F values from Hveragerdi, Iceland vary from 0.45 to about

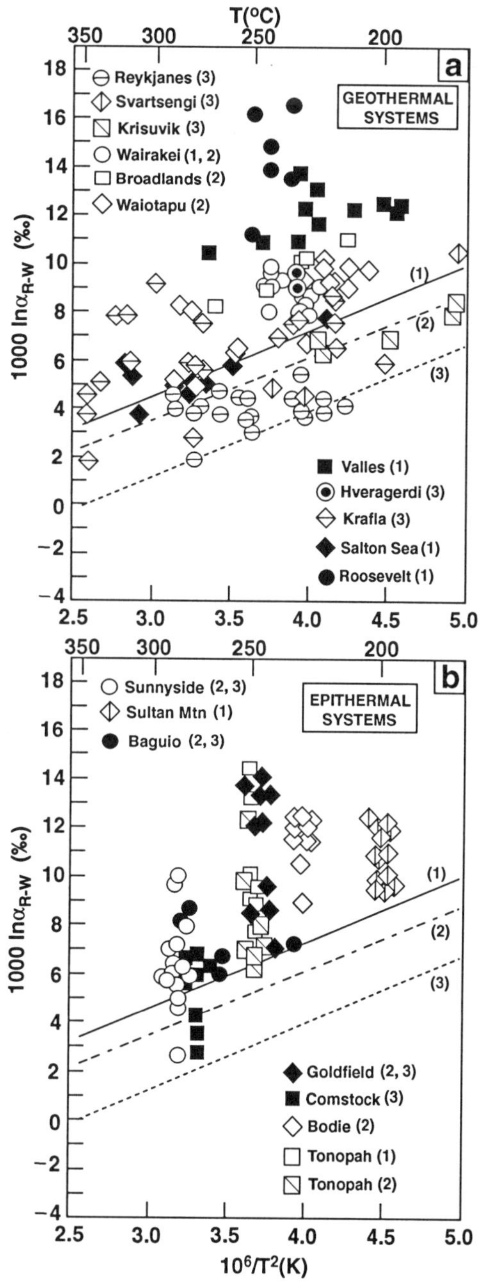

Fig. 2 Isotope fractionation between rocks and fluids in (a) geothermal or (b) epithermal systems.

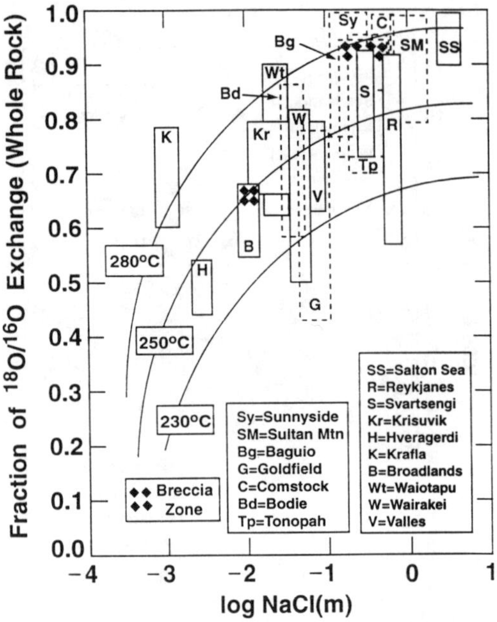

Fig. 3 Degree of isotopic equilibrium (F) versus salinity.

0.55, whereas Reykjanes F values range from 0.65 to slightly over 0.9. The epithermal vein deposits exhibit a broad range in F values with most systems very near, if not at equilibrium. Closer inspection of the data indicates that F increases with increasing temperature and salinity. These relationships are shown in Fig. 3. Note that the higher temperature (>250°C) - higher salinity (>0.5m) systems have F values commonly in excess of 0.75 (e.g., Salton Sea, Reykjanes, Sunnyside, Comstock, Sultan Mtn., Baguio), as compared to more dilute systems such as Hveragerdi, Broadlands, and Krisuvik. Higher temperature, dilute systems such as Krafla, exhibit more exchange than their lower temperature, similarly diluted counterparts, such as Hveragerdi. This tendency of increasing degree of equilibration with increasing salinity is analogous to results obtained from experimental rock (mineral)-fluid systems (Cole et al, 1987; 1992).

The fact that silicate reactions and accompanying isotopic exchange are fast (Cole et al, 1987; 1992) leads us to suggest that observed, small-scale isotopic disequilibrium has resulted from local self-sealing of the fracture network prior to final isotopic (or chemical) equilibration. The ability for fractures to remain open or to continue to propogate may be a deciding factor in whether isotopic equilibration is achieved. Interestingly, there is some preliminary indication that the most fractured (or brecciated) zones exhibiting the greatest ^{18}O shifts also yield greater F values. This is observed for Tonopah, Baguio and Broadlands (Fig. 3). Once a local domain seals, it may remain closed to flow for the duration of hydrothermal activity, or at the least, until the thermal regime has declined in magnitude to a point where reaction rates are too slow to allow for significant exchange (or re-equilibration). The self-sealing cap rocks observed on the margins and tops of certain geothermal systems are manifestations of the this phenomenon.

ACKNOWLEDGEMENT

Research sponsored by the Geosciences Research Program of BES/DOE under contract DE-AC05-84OR21400 with Martin Marietta Energy System, Inc.

REFERENCES

Blattner, P. 1985. Isotope shift data and the natural evolution of geothermal system. Chem. Geol. 49: 197-203.

Bowers, T.S. and Taylor, H.P. Jr. 1985. An integrated chemical and stable-isotope model of the origin of mid-ocean ridge hot spring systems. J. Geophys. Res. 90: 12, 583-12,606.

Cole, D.R., Mottl, M.J. and Ohmoto, H. 1987. Isotopic exchange in mineral-fluid systems. II. Oxygen and hydrogen isotopic investigation of the experimental basalt-seawater systems. Geochim. Cosmochim. Acta 51: 1523-1538.

Cole, D.R., Ohmoto, H. and Jacobs, G.K. 1992. Isotopic exchange in mineral-fluid systems. III. Rates and mechanisms of oxygen isotope exchange in the system granite-H_2O±NaCl±KCl at hydrothermal conditions. Geochim. Cosmochim. Acta 56: 445-446.

Craig, H. 1963. The isotopic geochemistry of water and carbon in geothermal areas. In Nuclear Geology in Geothermal Areas (E. Tongiorgi, ed.), Consiglio Nazionale delle Richerche, Pisa: 17-53.

Northrop, D.A. and Clayton, R.N. 1966. Oxygen-isotope fractionation in systems containing dolomite J. Geol. 74: 174-196

Water-Rock Interaction, Kharaka & Maest (eds) © 1992 Balkema, Rotterdam. ISBN 90 5410 075 3

The effects of dissolved salts on oxygen and hydrogen isotope fractionation between water vapor and brines at 50-100°C

J. Horita, D.J. Wesolowski & D.R. Cole
Chemistry Division, Oak Ridge National Laboratory, Tenn., USA

ABSTRACT: A series of vapor-liquid water equilibration experiments were carried out in order to determine the effects of dissolved salts (NaCl, KCl, $MgCl_2$, $CaCl_2$, $MgSO_4$, Na_2SO_4, and their mixtures) on oxygen and hydrogen isotope activity ratios in brines at 50-100°C. NaCl, KCl, and Na_2SO_4 have little or no effect on oxygen isotope activity ratios. However, oxygen isotope activity ratios in $CaCl_2$, $MgCl_2$, and $MgSO_4$ solutions are significantly lower than their concentration ratios. In all of the salt solutions studied, hydrogen isotope activity ratios are appreciably higher than their concentration ratios. The isotope effects are all linear with the molalities of the solutions. Mixed salt solutions, including those mimicking natural brines, exhibit the isotope effects additive of those of individual salts. The isotope salt effects observed in this study have profound implications to isotope geochemistry.

1 INTRODUCTION

Saline natural water with wide ranges of chemical compositions and temperatures are encountered in groundwaters, saline lakes, sedimentary basins, crystalline rocks, and geothermal systems. It has long been recognized that oxygen and hydrogen isotope activity ratios in salt solutions differ from their concentration ratios (Taube, 1954; Sofer and Gat, 1972, 1975; Stewart and Friedman, 1975), but this effect has not been well understood at elevated temperatures (Truesdell, 1974). Knowledge of the isotope salt effects is essential to better understanding of brine-dominated geochemical processes such as evaporation, boiling, mineral precipitation, and rock alteration. In this paper we report precise data on the difference between isotope activity and concentration ratios in the system Na-K-Ca-Mg-Cl-SO_4-H_2O between 50 and 100°C.

2 EXPERIMENTAL

The isotope salt effects in an aqueous salt solution can be defined as,

$$\Gamma \equiv \frac{R_{activity}}{R_{concentration}} = \frac{aHDO/aH_2O}{mHDO/mH_2O} \text{ or } \frac{aH_2{}^{18}O/aH_2{}^{16}O}{mH_2{}^{18}O/mH_2{}^{16}O} = \frac{\gamma HDO}{\gamma H_2O} \text{ or } \frac{\gamma H_2{}^{18}O}{\gamma H_2{}^{16}O}$$

where a, m, and γ denote activity, concentration, and activity coefficient of isotopic water molecules, respectively. The Γ can also be defined by isotope activity and concentration fractionation factors between water vapor and a salt solution as,

$$\Gamma = \frac{\alpha_{vap\text{-}soln(conc.)}}{\alpha_{vap\text{-}soln(activity)}} = \frac{R_{vap}/R_{soln(conc.)}}{R_{vap}/R_{soln(activity)}}$$

When pure water ($R_{activity}=R_{conc.}$) and salt solution are separately equilibrated with water vapor,

$$\Gamma = \frac{\alpha_{vap\text{-}salt\ soln(conc.)}}{\alpha_{vap\text{-}pure\ water}} = \frac{R_{vap}/R_{salt\ soln(conc.)}}{R_{vap}/R_{pure\ water}}$$

When $R_{salt\ soln(conc.)}=R_{pure\ water}$, this

equation can be reduced to,

$$\Gamma = \frac{R_{vap\text{-}salt\ soln}}{R_{vap\text{-}pure\ water}}$$

Then,

$$10^3 \ln\Gamma \simeq \delta_{vap\text{-}salt\ soln} - \delta_{vap\text{-}pure\ water}$$

We report values of $10^3\ln\Gamma$ in this paper (Fig. 1-3 and Table 1). A series of the vapor-liquid water equilibration experiments were carried out at 50, 70, 90, and 100°C. Our results of oxygen and hydrogen isotope fractionation factors between vapor and pure water are in excellent agreement with Majoube (1971).

3 RESULTS

3.1 Single salt solutions

The δD values of water vapor in equilibrium with NaCl solutions are consistently higher than those of pure water of the same isotopic composition (positive $10^3\ln\Gamma$ values). In other words, hydrogen isotope activity ratios in NaCl solutions are higher than the concentration ratios. The effects are linear with the molality of NaCl solutions, and decrease with increasing temperature. Our results are in quantitative agreement with Kakiuchi et al. (1988), who employed an H_2-water equilibration method. Our and his results confirm that values of $10^3\ln\Gamma(D)$ for NaCl solutions monotonously decrease from 2.7/molal at 10°C to 1.2/molal at 100°C. Our results are also consistent with those by Stewart and Friedman (1975) at 20°C, but not with those of Sofer and Gat (1975) at 20°C, which show large scatter in the data. Dissolved NaCl has no effect on oxygen isotope activity ratios between 50 and 100°C within ±0.2 in $10^3\ln\Gamma$. Taube (1954) and O'Neil and Truesdell (1991) also reported no effect in NaCl solutions at room temperature. However, Truesdell (1974) found very large effects of NaCl on oxygen isotope activity ratios at 102°C.

The effects of KCl on oxygen and hydrogen isotope activity ratios are very similar to those of NaCl, except for slight positive values of $10^3\ln\Gamma(^{18}O)$ in KCl solutions at 50°C, which are consistent with results of Sofer and Gat (1972), Götz and Heinzinger (1973), Truesdell (1974), and O'Neil and Truesdell (1991) at 25°C.

The effects in $MgCl_2$ and $CaCl_2$ solutions are very large for both oxygen and hydrogen isotope ratios, but are still linear with their molalities. Hydrogen isotope activity ratios in these two salt solutions are higher than their concentration ratios, and oxygen isotope activity ratios are lower than concentration ratios. Our results for oxygen isotopes in the both salt solutions are consistent with those of several investigators near room temperatures. However, those of hydrogen isotope ratios at 25°C by Sofer and Gat (1975) and Stewart and Friedman (1975) contradict each other.

The oxygen isotope effects in $MgSO_4$ solutions are very similar to those in $MgCl_2$ solutions, while the hydrogen isotope effects are much smaller than in $MgCl_2$ solutions.

The oxygen and hydrogen isotope effect in Na_2SO_4 solutions are similar to those observed in NaCl solutions.

3.2 Mixed salt solutions

In order to examine mixing rules of the isotope salt effects, several mixed salt solutions, including ones which mimic natural brines, were studied at 100°C. Both oxygen and hydrogen isotope effects in all of the mixed salt solutions studied are additive with respect to the isotope effects of individual salt solutions (Table 1). This simple mixing rule in mixed salt solutions was also observed at room temperatures by Sofer and Gat (1972, 1975).

4 DISCUSSION

The results in this paper and those from the literature clearly demonstrate that the oxygen and hydrogen isotope salt effects in many salt solutions of geochemical interests are too large to be ignored, at least up to 100°C. The simple rules still hold; linearity with molality and additive property of the effects of single salts. However, there are some indications that the sign of the isotope salt effect may change and its magnitude may increase with

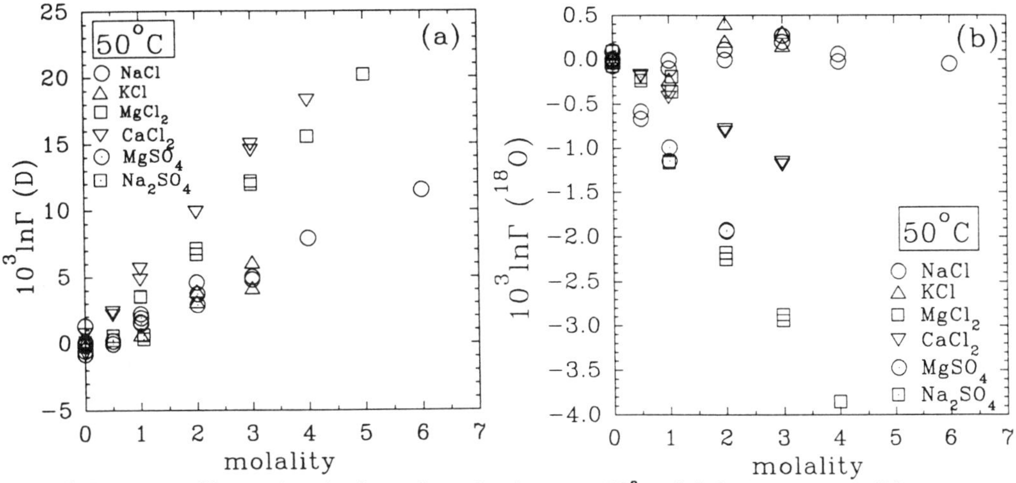

Fig. 1 Isotope effects in single salt solutions at 50°C; (a) hydrogen and (b) oxygen

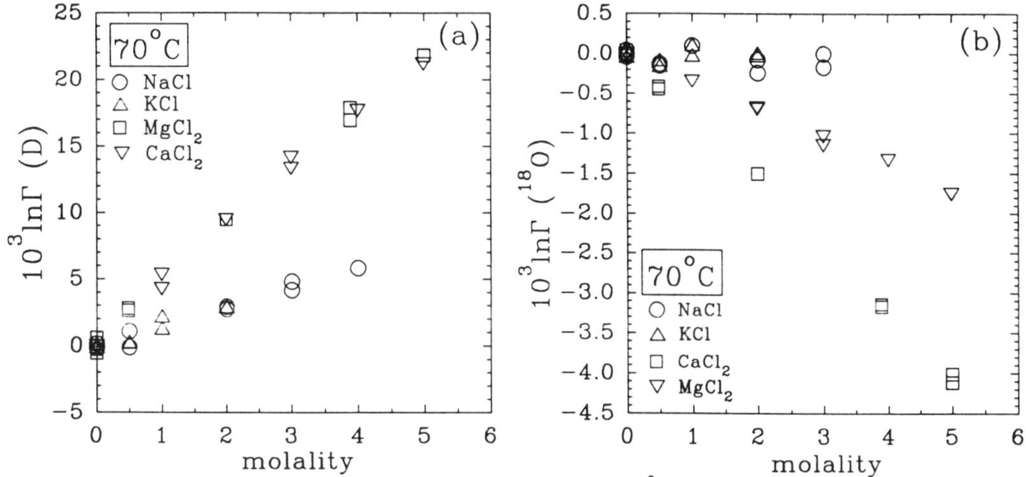

Fig. 2 Isotope effects in single salt solutions at 70°C; (a) hydrogen and (b) oxygen

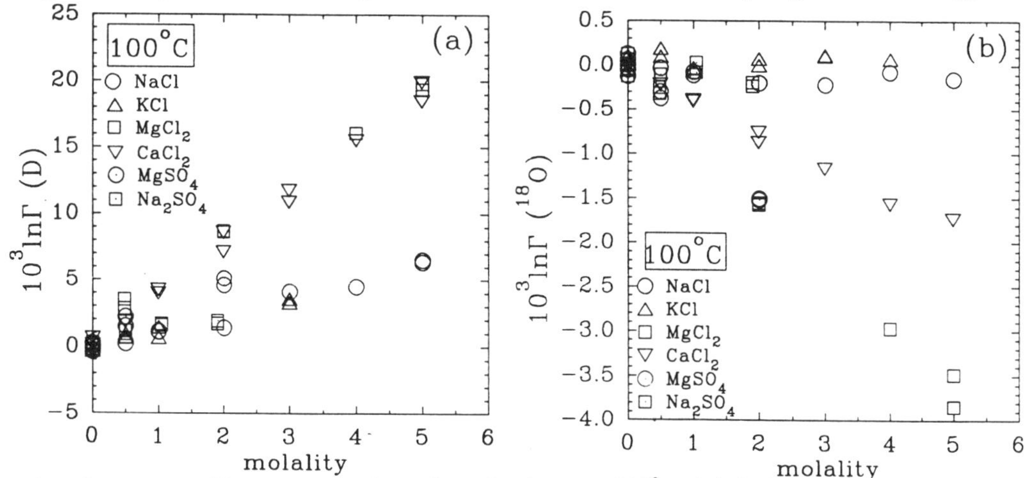

Fig. 3 Isotope effects in single salt solutions at 100°C; (a) hydrogen and (b) oxygen

increasing temperature.

Examples of implications of the isotope salt effects to isotope geochemistry were discussed by Sofer and Gat (1972, 1975) and Truesdell (1974). In principle, isotope activity ratios of brines should be used for discussion of any isotopic exchange reactions between brines and other phases (minerals, vapor, and gases), while their isotope concentration ratios are more useful in delineating geochemical processes in which oxygen and hydrogen isotopes are conserved, such as dissolution and precipitation of salts and mixing of different brine bodies.

Table 1. Results of the isotope salt effects in mixed salt solutions at 100°C

salt	molality	$10^3 \ln\Gamma$ (D)	$10^3 \ln\Gamma$ (^{18}O)
NaCl	1.00	+2.2	-0.01
KCl	1.00		
$MgCl_2$	1.00	+8.9	-1.21
$CaCl_2$	1.00		
$MgCl_2$	1.98	+10.6	-1.63
NaCl	2.01		
$MgSO_4$	1.01	+2.9	-0.89
Na_2SO_4	1.00		
NaCl	2.00	+3.1	-0.11
$NaSO_4$	1.00		
NaCl	2.01	+9.1	-1.60
$MgCl_2$	0.99		
$MgSO_4$	1.01		
Salton Sea brine			
NaCl	3.00	+9.2	-0.45
KCl	0.50		
$CaCl_2$	0.99		

ACKNOWLEDGEMENT

Funding for the studies of single electrolytes is provided by the U. S. Department of Energy (DOE), Office of Basic Energy Sciences, Division of Engineering and Geosciences. Measurement of salt mixtures and natural brine compositions is funded by DOE's Office of Conservation and Renewable Energy, Division of Geothermal Technology Development. These projects are conducted under contract number DE-AC05-84OR21400 with Martin Marietta Energy Systems, Inc.

REFERENCES

Götz, D. & Heinzinger, K. 1973. Sauersltoffisotopieeffeckte und hydratstruktur von alkalihalogenid-lösungen in H_2O and D_2O. Z. Naturforsch. 28a: 137-141.

Kakiuchi, M., Okutsu, K. & Niki, T. 1988. Fractionation of hydrogen isotopes between vapor and liquid in aqueous salt solutions. Abst with program, Geochem. Soc. of Japan.

Majoube, M. 1971. Fractionment en oxygene 18 et en deuterium entre l'eau et sa vapeur. J. Chim. Phys. 68: 1423 1436.

O'Neil, J.R. & Truesdell, A.H. 1991. Oxygen isotope fractionation studies of solute-water interactions. In Epstein's special volume. Geochem. Society

Sofer, Z. & Gat, J.R. 1972. Activities and concentrations of oxygen-18 in concentrated aqueous salt solutions: analytical and geophysical implications. Earth PLanet. Sci. Lett. 15: 232-238.

Sofer, Z. & Gat, J.R. 1975. The isotope composition of evaporating brines: effect of the isotope activity ratios in saline solutions. Earth PLanet. Sci.Lett. 58: 523-528.

Stewart, M.K. & Friedman, I. 1975. Deuterium fractionation between aqueous salt solutions and water vapor. J. Geophys. Res. 80: 3812-3818.

Taube, H. 1954. Use of oxygen isotope effects in the study of hydration of ions. J. Chem. Phys. 58: 523-528.

Truesdell, A.H. 1974. Oxygen isotope activities and concentrations in aqueous salt solutions at elevated temperatures: consequences for isotope geochemistry. Earth Planet. Sci. Lett. 23: 387-396.

Water-Rock Interaction, Kharaka & Maest (eds) © 1992 Balkema, Rotterdam. ISBN 90 5410 075 3

The effect of pressure on equilibrium isotope fractionation

N.N. Kharlashina & V.B. Polyakov
Vernadsky Institute, Moscow, Russia

ABSTRACT: The effect of pressure on equilibrium isotope fractionation is examined theoretically. A quasi-harmonic approximation technique is developed for examination of the pressure effect on equilibrium isotope fractionation in solids. The application of the technique to some silicates (quartz, albite, muscovite, enstatite, pyrope, forsterite), rutile and calcite (^{16}O-^{18}O fractionation), graphite-calcite-diamond (^{12}C-^{13}C fractionation) and brucite-water (H-D fractionation) shows that a pressure of ten kbars may produce a measurable effect on the equilibrium isotope fractionation and even change the sign of isotopic shift. Taking into account the pressure effect may be important for correct interpretation of high-pressure experimental isotopic data and for correct calculation of equilibrium isotopic constants.

1 INTRODUCTION

The determination of the pressure effect on the equilibrium isotope fractionation is important not only in physics and chemistry of isotopes but also for correct interpretation of isotope distributions in natural samples and experimental calibrations.

We consider the direct effect of pressure on equilibrium isotope fractionation constants of minerals. The effect of pressure on structure, chemical and phase composition, as well as kinetic isotope effects, are not examined in this study.

The first quantitative estimations of the pressure effect were conflicting. Joy and Libby (1961) concluded that the effect may be determined; however, the results of theoretical and experimental studies of Clayton concluded that the pressure effect was negligible. The effect of pressure on equilibrium isotopic fractionation was suggested to explain experimental and natural sample data by Garlick (1971) and Clayton (1988).

The quasi-harmonic approximation method for calculation of the pressure effect has been developed by Polyakov (1989, 1991) and Kharlashina (1992) for minerals. We follow it in this study.

2 THE EFFECT OF PRESSURE ON THE β-FACTOR IN SOLIDS

The equilibrium constant for an isotope exchange reaction between two substances A and B may be written in terms of β-factors (reduced isotopic partition function ratio) of these substances:

$$\alpha = \frac{\beta_A}{\beta_B} . \tag{1}$$

To examine the pressure effect on isotopic fractionation it is essential to estimate the pressure dependence of β-factors on the substance.

In harmonic approximation the motion of atoms in an N-atomic crystal is presented as 3N-6 harmonic vibrations. In this case the β-factor may be expressed by the G. Urey formula:

$$\beta = \prod_{i=1}^{3N-6} \frac{u_i^* \, sh(0.5u_i)}{u_i \, sh(0.5u_i^*)} . \tag{2}$$

Here $u_i = \frac{h\nu_i}{kT}$ is the dimensionless frequency of the ith mode, ν_i refers to the ith harmonic normal mode frequency, T is absolute temperature, h and k have their usual meanings, and the asterisk * denotes isotopically heavy substance. The harmonic approximation does not describe the frequency dependence on pressure or volume. To take into account the effect of pressure the quasi-harmonic approximation was used (Leibfrid 1961), in which the vibrational frequency ν_i is assumed to shift with volume according to:

$$\gamma_i = -\left(\frac{\partial \ln \nu_i}{\partial \ln V}\right) . \quad (3)$$

The quantity γ_i is the mode Gruneisen parameter of the lattice vibration at frequency ν_i. Comparing the frequency derivation on pressure and temperature one can obtain:

$$\left(\frac{\partial u_i}{\partial p}\right)_T = -\frac{\gamma_i T}{B_T}\left(\frac{\partial u_i}{\partial T}\right)_V , \quad (4a)$$

where $B_T = -\left(\frac{\partial \ln V}{\partial p}\right)^{-1}$ is the isothermal bulkmodulus. A similar expression holds for the isotopically heavy form:

$$\left(\frac{\partial u_i^*}{\partial p}\right)_T = -\frac{\gamma_i^* T}{B_T}\left(\frac{\partial u_i^*}{\partial T}\right)_V . \quad (4b)$$

Differentiating (2) on pressure and taking into account (4a) and (4b) one can obtain:

$$\left(\frac{\partial \beta}{\partial p}\right) = -\frac{\gamma T}{B_T}\left(\frac{\partial \beta}{\partial T}\right)_V , \quad (5)$$

where

$$\gamma = \frac{\sum_i [\gamma_i u_i \mathrm{cth}(0.5\, u_i) - \gamma_i^* u_i^* \mathrm{cth}(0.5\, u_i^*)]}{\sum_i [u_i \mathrm{cth}(0.5\, u_i) - u_i^* \mathrm{cth}(0.5\, u_i^*)]} \quad (6)$$

Equation (5) with definition (6) is exact in approximation and may be useful for calculating the pressure effect on the β-factor of minerals.

The dependence β(T) is usually known from experiment or calculation. The value of the parameter γ was discussed in Polyakov (1989). It was proved that in some physical models (e.g., Debye and Einstein models), γi is approximated to be a constant. For solids consisting of similar atoms the parameter γ is equal to the thermal Gruneisen parameter:

$$\gamma_{th} = \frac{\sum_i \gamma_i c_{vi}}{C_v} = \frac{\alpha V}{C_v B_T} , \quad (7)$$

where α is the thermal expansion, V is the molar volume, $c_{\nu i}$ is heat capacity referring to the ith mode, c_V is the molar heat capacity. To calculate the pressure dependence of the β-factor for simple solids (e.g., graphite, diamond), γ_{th} and the dependence of β(T) are needed to apply Equation (5).

The general case of a solid consisting of different atoms is more difficult because it is necessary to calculate the parameter γ. The model approach was used to solve this problem, and it was assumed that $\gamma_i = \gamma_o^*$. The model suggested by Kieffer (1982) for vibrational spectrum of minerals was applied. It was completed by the experimental parameters γ_i for high-frequency modes, which are the most important in these calculations. Raman and IR-spectroscopic and thermodynamic data were used. The parameters of the model are given in Kharlashina (1992).

3 CALCULATION OF THE EFFECT OF PRESSURE ON β-FACTORS OF MINERALS

The dependence β(T,P) and γ(T,P) were calculated at temperatures of 0-1000°C and pressures of 0-50 kbar for a set of minerals: quartz, calcite, albite, muscovite, enstatite, pyrope, forsterite and rutile, for ^{16}O-^{18}O substitution.

Calculated γ values for minerals at different temperatures are shown in Fig. 1. One can see that for studied solids γ has the value 0.1-1.5. The value of γ depends slightly on temperature and is essentially independent of pressure (the change of the value is less than 0.05 for 50 kbar). Values of γ and γ_{th} vary for different minerals.

Pressure-induced corrections at P=10 kbar for the set of minerals are given in Table 1. The change in lnβ is strongly temperature dependent, and the correction for the equilibrium constant α measured in experiments is more significant at lower temperatures.

The pressure-induced correction may be rather great in the systems of minerals with different crystal structure, for example, rutile-silicate or calcite-silicate. For the silicate pairs with similar crystal structure the effects are small.

Pressure effects may be important for correct interpretation of isotopic data. One can demonstrate this using the system quartz-calcite. The system was studied experimentally by Clayton (1989) at temperatures of 600, 700, 800 and 1000°C and a pressure of 15 kbar. The equilibrium isotope constants from these experiments were used for extrapolation to low temperatures;

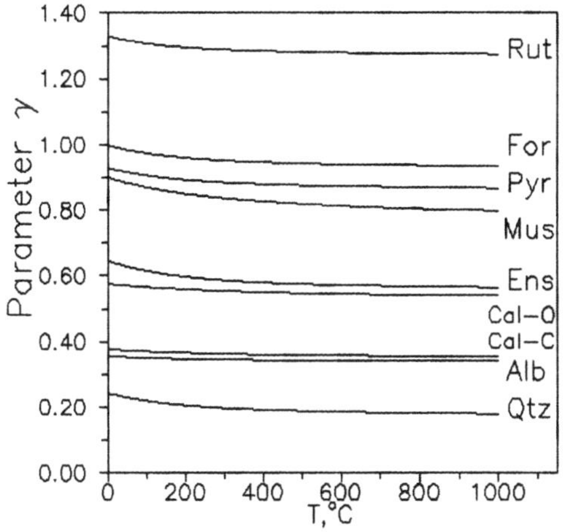

Figure 1 - The dependence γ(T) of minerals: Qtz-quartz, Alb-albite, Cal-O-calcite, oxygen fractionation, Ens-enstatite, Mus-muscovite, Pyr-pyrope, For-forsterite, Rut-rutile, Cal-C-calcite, carbon fractionation.

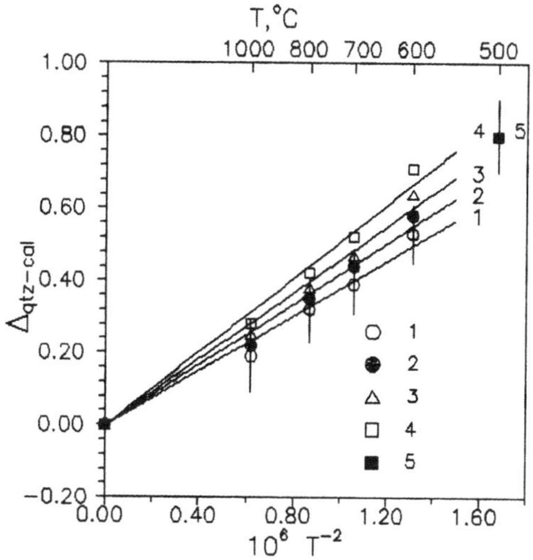

Figure 2 - The interpretation of experimental data on equilibrium oxygen isotope fractionation in quartz-calcite system. The isotopic shift Δ on temperature: 1 - experimental data of Clayton (1989), anhydrous exchange, p=15 kbar; 2-4 - pressure corrected values of Δ, different variations of the model (Kharlashina, 1992), 5 - hydrothermal experiment data.

the authors consider the pressure effects to be negligible. The results are shown in Fig. 2. Taking into account the effect of pressure, one can obtain the corrections for isotope exchange constants. The difference is significant, especially at the extrapolation to low temperatures. For example, the correction at T=600°C corresponds to changing the temperature 100 degrees. This is rather large for isotopic geothermometry. One can note that for this system the corrected line coincides better with the result of hydrothermal experiments.

The other system of minerals with different crystal structure is calcite-graphite-diamond for carbon fractionation (Polyakov 1989). Graphite and diamond consist of carbon atoms only, so $\gamma=\gamma_{th}$ for these substances. The model spectrum was used for calcite. One can note that γ values for calcite for ^{12}C-^{13}C and ^{16}O-^{18}O are not equal. It was found that the β-factor for graphite increases faster with increasing pressure than it does for calcite or diamond. This creates an interesting situation: if at ambient pressure the isotopic shift Δ of the system diamond-graphite is more than zero at any

Table 1. The correction value to $10^3\ln\beta$ at P=10 kbar and different temperatures.

Mineral	$10^3[\ln\beta(P=10kbar)-\ln\beta(P=1bar)]$ at T°C						
	0	100	200	300	500	700	900
Quartz 1	1.12	0.66	0.43	0.30	0.17	0.11	0.08
Quartz 2	1.41	0.83	0.54	0.38	0.21	0.14	0.09
Albite	1.16	0.73	0.50	0.36	0.21	0.14	0.10
Muscovite	1.68	1.01	0.67	0.47	0.27	0.17	0.12
Enstatite	1.11	0.66	0.43	0.30	0.17	0.11	0.08
Pyrope	0.93	0.56	0.37	0.26	0.15	0.09	0.07
Forsterite	1.29	0.77	0.51	0.36	0.20	0.13	0.09
Rutile	0.74	0.44	0.29	0.20	0.12	0.07	0.05
Calcite	1.28	0.82	0.56	0.41	0.24	0.16	0.11

temperature, then at elevated pressure Δ becomes less than zero. For example, at T=1000°-1200°C and P=40-50 kbar (temperatures and pressures of interest in geochemical and experimental applications), the calculated isotopic shift is about Δ=-0.5‰ instead of the harmonic value Δ=+0.5‰.

Pressure corrections at P=10 kbar for calcite, graphite and diamond are a few per mil, which allow an experimental check of the effect on the calcite - graphite system.

The calculation of hydrogen fractionation is less precise because the isotopic model spectra of minerals have not been developed. Fortunately, the pressure correction values are practically independent of details of the spectrum but are dependent on γ_i. Using the silicates isotope model spectra as an analogy, the model spectra for brucite was developed. The corrections to lnβ at P=1 kbar are about 2.5‰ at T=0°C and 0.4‰ at T=1000°C.

Water corrections were calculated using a thermodynamic approach (Clayton 1975). It is interesting that for water, the molar volume of H_2O is greater than that of D_2O, so the β-factor of water decreases with increasing pressure. The correction to $10^3 \ln\beta$ induced by P=1 kbar has a minimum value, ~ -0.6‰ at T=110°C, increases to -2 to -3‰ at lower and higher temperatures and is not determined at the critical point. Thus, in the brucite-water system the pressure correction for δD is about 3-4‰ $kbar^{-1}$.

4 CONCLUSIONS

The results of theoretical study of the pressure effect on equilibrium isotope fractionation show that a pressure of tens kbars may produce a measurable correction to equilibrium isotopic constants and change the sign of the isotopic shift. Thus, it is necessary to estimate the pressure effect on high-pressure isotopic calibrations. The most significant effect is in the case of H - D substitution in the mineral - water system, where the β-factor of minerals increases with pressure, while the β-factor for water decreases.

REFERENCES

Clayton, R.N., J.R.Goldsmith, K.J.Karel, T.K.Mayeda & R.C.Newton 1975. Limits on the effect of pressure on isotopic fractionation. Geochimica et Cosmochimica Acta 39: 1197-1201.

Clayton, R.N. 1981. Isotopic thermometry. In: Thermodynamics of Minerals and Melts, Adv. Phys. Geochem., Vol. 1, edited by R.C.Newton, A.Navrotsky & B.J.Wood, New York, Springer-Verlag: 85-109.

Clayton, R.N., T.K. Mayeda, J.R.Goldsmith, H.Chiba & T.Chacko 1988. Oxygen isotope fractionation factors among rock-forming minerals at high temperatures. Chemical Geology 70: 183.

Garlick, G.D., I.D. MacGregor & P.E.Vogel 1971. Oxygen isotope ratios in eclogites from Kimberlites. Science 172: 1025-1027.

Joy, H.W. & W.F. Libby 1960. Size effect among isotopic molecules. Journal of Chemical Physics 33: 1276.

Kharlashina, N.N. & V.B.Polyakov 1992. The effect of pressure on the equilibrium isotopic fractionation: silicates, calcite, rutile. Geokhimiya 2: 189-200.

Kieffer, S.W. 1982. Thermodynamics and lattice vibrations of minerals: Application to phase equilibria, isotopic fractionation, ad high-pressure thermodynamic properties. Reviews of Geophysics and Space Physics 20/4: 827-849.

Leibfrid, G. & W.Ludwig 1961. Theory of anharmonic effects in crystals. Solid State Physics 12: New York.

Polyakov, V.B. & N.N. Kharlashina 1989. The effect of pressure on the equilibrium isotopic fractionation in solids. Proc. 5th Working Meeting Isotopes in Nature, Leipzig.

Polyakov, V.B. & N.N. Kharlashina 1991. The effect of pressure on the equilibrium isotopic fractionation in solids: Geokhimiya 11: 1605-1612.

Water-Rock Interaction, Kharaka & Maest (eds) © 1992 Balkema, Rotterdam. ISBN 90 5410 075 3

Boron isotopic fractionation between coexisting vapor and liquid in natural geothermal systems

William P. Leeman
Keith-Wiess Geological Laboratories, Rice University, Houston, Tex., USA

Robert D. Vocke
Inorganic Analytical Chemistry Division, National Institute of Standards and Technology, Gaithersburg, Md., USA

Mike A. McKibben
Center for Geothermal Resources Research, IGPP, University of California, Riverside, Calif., USA

ABSTRACT: Boron isotopic fractionation between coexisting aqueous vapor and liquid has been evaluated via analyses of natural geothermal fluids. ^{11}B is selectively enriched in the vapor phase in all wells sampled. With increasing temperature, the magnitude of isotopic fractionation decreases, whereas the concentration of B in steam relative to brine increases. The effects of steam loss on the isotopic composition of residual brines may be assessed using these preliminary results.

1 INTRODUCTION

Boron isotopic compositions of thermal waters have been reported for various localities (e.g., Mushashi et al. 1988; Palmer and Sturchio 1990; Palmer 1991), but the effects of B loss in 'flashed' steam have not been determined. Published analyses of coexisting liquid and steam samples suggest that under certain conditions B may be significantly concentrated in the steam phase (Smith et al. 1987). If significant B isotopic fractionation occurs between these phases, steam losses may result in isotopic shifts in the liquids that have been studied. To evaluate the importance of this effect, B isotopic compositions and concentrations were measured in hydrothermal brines and steam condensates collected from commercial and scientific wells from the Salton Sea, Coso, and Geysers geothermal fields, all located in California.

2 THE GEOTHERMAL SYSTEMS STUDIED

The Salton Sea Geothermal Field (SSGF) occurs in the Imperial Valley in a region of active continental rifting and volcanism (Elders and Sass, 1988). The hydrothermal reservoir includes deltaic, lacustrine, and evaporite sediments derived largely from the Colorado River drainage. SSGS fluids are stratified into a deeper hypersaline brine (up to 365°C, > 20 wt. % total dissolved solids) and a shallower lower salinity brine (<260°C, <10% TDS) (Thompson and Fournier, 1988; Williams and McKibben 1989). The deeper brines are associated with alteration and low pressure (<3.2 km depth) greenschist to amphibolite metamorphism of the host rocks (Muffler and White 1969), as well as active base metal ore formation (Skinner et al. 1967; McKibben et al. 1987, 1988).

The Coso Geothermal Field (CGF) is situated in the Coso Range east of Owens Valley. The thermal area is associated with Pliocene-Holocene rhyolite domes and basalt flows extruded upon a basement of intrusive and metamorphic rocks (Duffield et al. 1980). Fournier et al. (1980) present representative brine analyses; maximum fluid temperatures measured at <2 km depths range between 200-340°C. Sulfur isotope, gas and fluid chemistry data suggest that CGS fluids likely represent mixing between at least three dissimilar fluids (Williams and McKibben 1990a, b); the system is partly liquid-dominated, but boiling of ascending fluid has produced shallow steam caps.

The Geysers Geothermal Field (GGF) is situated between the Plio-Pleistocene Sonoma and Clear Lake volcanic fields, which are related to local extension along the strike-slip transform plate margin (Thompson 1989). The basement comprises graywackes and ultramafic rocks of the Franciscan Formation. Shallow felsic intrusions apparently are heat sources for the hydrothermal activity. Tourmaline-rich aureoles have been cored in proximity to these intrusions. The thermal system is vapor-dominated, with most wells producing dry steam from a high temperature (>200°C) reservoir (Truesdell et al., 1989).

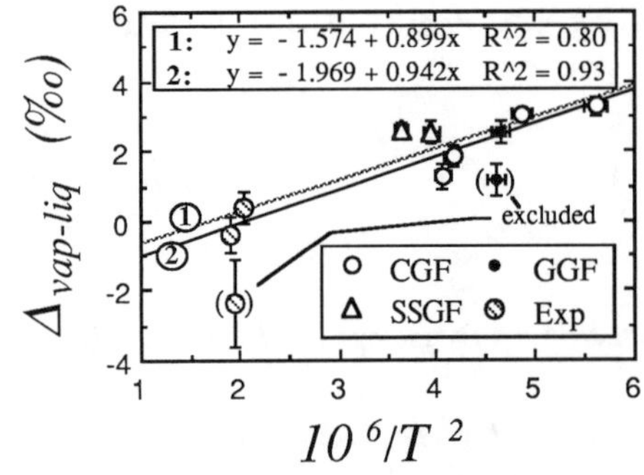

Figure 1. Temperature dependence of $\Delta_{vap\text{-}liq}$; Δ is defined as $[\delta^{11}B_{steam} - \delta^{11}B_{liquid}]$, where $\delta^{11}B = [(^{11}B/^{10}B)_{sample}/(^{11}B/^{10}B)_{standard} - 1]\bullet 1000$. Data points represent replicate analyses of coexisting steam - brine pairs from this study; experimental data from Spivack et al. (1990) are shown for comparison. Regression lines and equations are shown both (1) including and (2) excluding the SSGF pairs (see discussion).

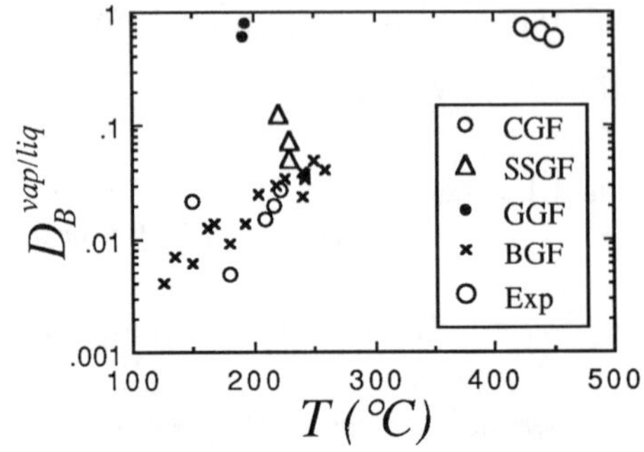

Figure 2. Apparent partitioning of B contents (by mass) between vapor and liquid phases shown as a function of separation temperature. Data from this study except for Broadlands geothermal field (BGF; Glover 1988, pers. comm.). With exception of two GGF samples, relative concentration of B in the vapor phase decreases sharply with temperature.

3 BORON ISOTOPIC ANALYSIS

Samples were obtained at well-head temperatures using a Webre mini-separator. Filtered fluids were analyzed for B content via ICP emission spectroscopy using the method of standard additions to minimize matrix effects. B was separated for isotopic analysis via an ion exchange method and analyzed mass spectrometrically as dicesium borate (Leeman et al. 1991). Average in-run precision ($2\sigma_m$) for 123 analyses was 0.12±0.07‰. Replicate analyses of either the same sample solution or samples reprocessed through the full separation procedure indicate an average external precision ($2\sigma_m$) of 0.17 ±0.10‰. Analyses of standards (NBS-951 boric acid) processed through full chemistry with each batch of samples indicate long-term external precision of better than 0.2‰ (2σ). Data are presented in conventional delta notation relative to the NBS-951 standard ($^{11}B/^{10}B$ certified as 4.04362±0.00021; Catanzaro et al. 1970).

4 VAPOR-LIQUID FRACTIONATION

Brine-steam pairs from all but one well studied yielded consistent B isotopic fractionation, with the steam phase being enriched in $\delta^{11}B$. This consistency is remarkable, considering the broad ranges in fluid chemistry (TDS from 25% to <<1%) and temperatures of phase separation (260-170°C) encountered.

Enrichment of ^{11}B in the vapor cannot be explained simply in terms of speciation of boron between the relatively ^{11}B-rich trigonal $B(OH)_3$ complex and the tetrahedral $B(OH)_4^-$ complex because, under slightly acidic to neutral pH conditions in the natural fluids, the trigonal species is predominant in both phases (cf. Kakihana et al. 1977; Spivack and Edmond 1987; Leeman unpub.).

In contrast to our work, Spivack et al. (1990) reported negligible fractionation between vapor and brine produced in experiments at 425-450°C. Differences between these studies exceed analytical uncertainty, and may reflect influences of bulk composition, temperature, or pH of the respective fluids, or possibly kinetic or other disequilibrium effects accompanying sampling. The consistent fractionation observed in most of our sample pairs seems at odds with such essentially random effects.

5 TEMPERATURE DEPENDENCE OF ISOTOPIC FRACTIONATION

The variation in $\Delta_{vap\text{-}liq}$ (= $\delta^{11}B_{vap} - \delta^{11}B_{liq}$) with temperature is shown in Fig. 1. There is a good correlation between $\Delta_{vap\text{-}liq}$ and $10^6/T^2$. Excluding one sample with large analytical uncertainty from Spivack et al. (1990), and one of our GGF samples which was contaminated, we obtain a correlation coefficient of 0.80. All other pairs except those from the Salton Sea field lie within error of the regression line. Regression excluding the SSGF pairs yields a correlation coefficient of 0.93. These observations suggest that temperature variations largely control the isotopic fractionation, but that chemical effects may also be significant in extremely concentrated brines. There is no apparent systematic relation between $\Delta_{vap-liq}$ and B content in the fluids. The data base is too limited at present to evaluate in detail possible variations with other constituents.

Our results indicate that isotopic effects are not negligible during phase separation in hydrothermal systems. The magnitude of such effects depends on the extent of steam loss (i.e., 'flashing') that occurs upon decompression of the liquid as well as on the partitioning of boron between coexisting fluids. Available data suggest that the vapor phase can have significant boron contents (cf. Smith et al. 1987; Glover 1988, personal comm.), but mass partition coefficients for boron ($D_{steam/liquid}$) are quite variable (0.02 to ≥ 0.66, Fig. 2). There is also a crude correlation between $D_{steam/liquid}$ and fluid temperature, but the effects of fluid composition are unknown and may be significant. In summary, isotopic shifts are more significant at lower T, but the proportion of total B in the vapor decreases with T. Hence, these effects partly compensate.

6 APPLICATION TO PHASE SEPARATION

Analyses of flashed brines from the State 2-14 well (SSGF) exhibit small decreases in $\delta^{11}B$ that are consistent with steam loss (cf. Palmer, 1991). The flash fraction (f) can be estimated [eqn (1)] from the observed B enrichment in going from a well-head sample (260°C, $\delta^{11}B$ = -1.9‰, 380 ppm B) to a Weir box sample (90°C, $\delta^{11}B$ = -3.3‰, 490 ppm B).

$$[B_f]/[B_i] = (1-f)^{\{D-1\}} \quad \text{or} \quad = 1/[(D-1)\cdot f + 1] \quad (1)$$

Rayleigh eqn. Mass balance

Assuming $D_{steam/liquid} \approx 0.1$ (representative of several SSGF steam-liquid pairs; Fig. 2), this degree of B enrichment corresponds to ~25% steam loss from the well-head sample which is in reasonable agreement with independent estimates (ca. 30%) based on enthalpy change. Assuming a Rayleigh fractionation model for this process [eqn (2)], and using the empirically determined fractionation factor for this field [$\Delta_{vap-liq} \sim 2.5 = 1000\cdot(\alpha_{vap-liq} - 1)$], the calculated isotope shift (ca. -0.72‰) is just within error of the actual difference (-1.4‰) in isotopic composition for these samples.

$$(\delta^{11}B_f + 1000)/(\delta^{11}B_i + 1000) = (1-f)^{(\alpha-1)} \quad (2)$$

This example illustrates that small but potentially significant isotopic fractionation may occur during phase separation at relatively low temperatures. It appears that this may be negligible at temperatures much above 300-400°C. However, protracted steam loss in open thermal systems may lead to a significant cumulative isotope shift.

7 CHARACTERIZATION OF HYDROTHERMAL RESERVOIRS

This study illustrates the utility of B isotopic tracer studies in distinguishing lithologies in fluid reservoirs within the crust. Despite small isotope shifts within individual reservoirs, the relatively large differences seen between these reservoirs are consistent with the known geologic framework in each locality. At Coso, SSGF, and Yellowstone Park (Palmer and Sturchio 1990), B isotopic compositions remain relatively constant for fluids exhibiting large temperature (and compositional) ranges. $\delta^{11}B$ in the fluids appears to be dominated by B extracted from the local reservoir rocks in each case and steam losses do not significantly modify the water compositions.

Sediments in the SSGF and basement rocks in the Coso area are dominantly of continental affinity - very light B in these sources is consistent with little or no marine component (Spivack et al. 1987; Swihart et al. 1988). Predominantly oceanic basement rocks contribute heavy B to the Geysers thermal fluids, although $\delta^{11}B$ is considerably lower than seawater values (ca. 39-40‰, Spivack and Edmond 1987). Many thermal springs in Japan have high $\delta^{11}B$ approaching oceanic values, but in detail, compositions of those fluids also exhibit regional control by the local sediments and basement rocks, with high values occurring in the Izu-Bonin arc (Musashi et al. 1988).

Boron isotopic studies hold considerable promise for unravelling processes of hydrologic transport, particularly where fluids are derived from two or more isotopically distinct reservoirs.

8 ACKNOWLEDGEMENTS

This project was supported by National Science Foundation grant EAR-90-15619 awarded to WPL. We thank E.S. Beary, P.J. Paulsen, and J.D. Fassett for support and suggestions in carrying out portions of this study.

REFERENCES

Catanzaro, E.J., C.E. Champion, E.L. Garner, O. Marienko, K.M. Sappenfield & W.R.Shields 1970. Boric acid isotopic and assay standard reference materials. *U.S. Nat. Bur. Stand. Spec. Pub. 260-17.*

Duffield, W.A., C.R. Bacon & G.B. Dalrymple 1980. Late Cenozoic volcanism, geochronology, and structure of the Coso Range, Inyo County, California. *J. Geophys. Res.* 85: 2381-2404.

Elders, W.A. & J.H. Sass 1988. The Salton Sea Drilling Project. *J. Geophys. Res.* 93: 12953-12968.

Fournier, R.O., J.M. Thompson & C.F. Austin 1980. Interpretation of chemical analyses of waters collected from two geothermal wells at Coso, California. *J. Geophys. Res.* 85: 2405-2410.

Glover, R.B. 1988. Boron distribution between liquid and vapour in geothermal fluids. *Proc. 10th New Zealand Geotherm. Workshop*: 223-227.

Kakihana, H., M. Kotaka, S. Shohei, M. Nomura & M. Okamoto 1977. Fundamental studies oon the ion-exchange separation of boron isotopes. *Bull. Chem. Soc. Jap.* 50: 158-163.

Leeman, W.P., R.D. Vocke, E.S. Beary & P.J. Paulsen 1991. Precise boron isotopic analysis of aqueous samples: ion exchange extraction and mass spectrometry. *Geochim. Cosmochim. Acta* 55: 3901-3907.

McKibben, M.A., A.E. Williams, W.A. Elders & C.S. Eldridge 1987. Saline brines and metallogenesis in a modern sediment-filled rift: the Salton Sea geothermal system, California, U.S.A. *Appl. Geochem.* 2: 563-578.

McKibben M.A., J.P. Andes, Jr. & A.E. Williams 1988. Active ore formation at a brine interface in metamorphosed deltaic lacustrine sediments: the Salton Sea Geothermal System, California. *Econ. Geol.* 83: 511-523.

Muffler, L.J.P. & D.E. White 1969. Active metamorphism of upper Cenozoic sediments in the Salton Sea Geothermal Field and the Salton Trough, southeastern California. *Geol. Soc. Amer. Bull.* 80: 157-182.

Mushashi, M., M. Nomura, M. Okamoto, T. Ossaka, T. Oi & H. Kakihana 1988. Regional variation in the boron isotopic composition of hot spring waters from central Japan. *Geochem. J.* 22: 205-214.

Palmer, M.R. & N.C. Sturchio 1990. The boron isotope systematics of the Yellowstone National Park (Wyoming) hydrothermal system: a reconnaissance. *Geochim. Cosmochim. Acta* 54: 2811-2815.

Palmer, M.R. 1991. Boron isotope systematics of hydrothermal fluids and tourmalines: a synthesis. *Chem. Geol. (Isotope Geosci. Sec.)* 94: 11-121.

Skinner, B.J., D.E. White, H.J. Rose & R.E. Mays 1967. Sulfides associated with the Salton Sea geothermal brine. *Econ. Geol.* 62: 316-330.

Smith, C.L., W.H. Ficklin & J.M Thompson. 1987. Concentrations of arsenic, antimony, and boron in steam and steam condensate at the Geysers, California. *J. Volcanol. Geotherm. Res.* 32: 329-341.

Spivack, A.J. & J.M. Edmond 1987. Boron isotope exchange between seawater and the oceanic crust. *Geochim. Cosmochim. Acta* 51: 1033-1043.

Spivack, A.J., M.R. Palmer & J.M. Edmond 1987. The sedimentary cycle of the boron isotopes. *Geochim. Cosmochim. Acta* 51: 1939-1949.

Spivack, A.J., M.E. Berndt & W.E. Seyfried, Jr. 1990. Boron isotope fractionation during supercritical phase separation. *Geochim. Cosmochim. Acta* 54: 2337-2339.

Swihart, G.H., P.B. Moore & E.L. Callis 1988. Boron isotopic composition of marine and nonmarine evaporite borates. *Geochim. Cosmochim. Acta* 50: 1297-1301.

Thompson, J.M. & R.O. Fournier 1988. Chemistry and geothermometry of brine produced from the Salton Sea Scientific Drill Hole, Imperial Valley, California. *J. Geophys. Res.* 93: 13165-13173.

Thompson, R.C. 1989. Structural stratigraphy and intrusive rocks at the Geysers geothermal field. *Trans. Geotherm. Resour. Council* 13: 1-5.

Truesdell, A.H., J.R. Haizlip, H. Armansson, & F. D'Amore 1989. Origin and transport of chloride in superheated geothermal steam. *Geothermics* 18: 295-304.

Williams, A.E. & M.A. McKibben 1989. A brine interface in the Salton Sea Geothermal System, California: fluid geochemical and isotopic characteristics. *Geochim. Cosmochim. Acta* 53: 1905-1920.

Williams, A.E. & M.A. McKibben 1990a. Isotopic and chemical characteristics of reservoir fluids from the Coso geothermal field; China Lake Naval Weapons Center., California. In J.L. Moore & M.E. Erskine (eds.) *Amer. Assoc. Petrol. Geol. field trip guidebook*, AAPG EMD #1:85-302.

Geochemical modeling

Solution chemistry and thermodynamics

Water-Rock Interaction, Kharaka & Maest (eds) © 1992 Balkema, Rotterdam. ISBN 90 5410 075 3

Experimental study of aluminum hydrolysis and complexation with acetate and sodium ions in hydrothermal solutions

S.Castet, J.L.Dandurand, J.Schott & R.Gout
Laboratoire de Géochimie, CNRS, Université Paul-Sabatier, Toulouse, France

ABSTRACT: Boehmite solubility was measured in non-complexing solutions (pH=2 to 9) from 90 to 350°C. A least-squares regression of the data was used to calculate values for the boehmite dissociation constants (K^*s_0 to K^*s_4) and of the Al hydrolysis constants (K^*_1 to K^*_4 and β^*_4) between 90 and 350°C. Solubilities measured in acidic and neutral solutions are lower than those reported by previous authors. This discrepancy is believed to be due to the formation of Al-acetate and Al-fluorine complexes in previous studies. A study of Al-acetate complexation was undertaken at 170 and 200°C. It allowed us to characterize the stoichiometry of 1:1 Al:AcO complexes and to calculate the stability constants of $Al(OH)_2AcO$, $Al(OH)AcO^+$ and $AlAcO^{2+}$. Experiments carried out at 300°C in alkaline solutions with NaCl concentrations up to 1 M demonstrate the formation of a complex between Na^+ and $Al(OH)_4^-$.

1 INTRODUCTION

Despite extensive studies of aqueous aluminum complexes, the speciation of this element is not yet well known. Literature shows many inconsistencies and an obvious lack of data at high temperature, especially in acidic and neutral media. Moreover, recent works show that the aluminum can easly form complexes with numerous organic and inorganic ligands, which were not taken into account in previous works. The purpose of this study was first to determine the speciation of aluminum in non complexing hydrothermal solutions. The results are then used to interpret new experiments on Al-acetate and Al-sodium complexation.

2 EXPERIMENTAL PROCEDURE

Solubility experiments were carried out at saturated vapor pressure between 90 and 350°C. Synthetic boehmite was obtained via a hydrothermal treatment (300°C in deionized water) of pure synthetic gibbsite. Solubility experiments at 90°C were conducted in polypropylene flasks immersed in a water-thermostated bath; the solution was stirred with a magnetic stirrer. At higher temperatures stainless steel reactors were used for solubility measurements in basic solutions.

In acidic and neutral solutions the experiments were carried out in pure titanum autoclaves which allowed the injection of the mineral and solution sampling at the temperature of the run. Solutions were mixed by rocking the ovens.

We have checked the reversibility of equilibria at 170°C in acidic solutions from under- and supersaturated experiments. Precipitation and dissolution experiments both lead to the same aluminum concentration, but precipitation experiments take 10 times longer to attain equilibrium. In view of this, the stability of the aluminum concentration obtained by dissolution was the equilibrium criterion in most of our experiments. During each run, the solution was sampled regularly in order to measure the evolution of pH and aluminum concentration in the solution. The aluminum content was analysed by flame and flameless atomic absorption, colorimetry and ionic chromatography.

3 RESULTS

3.1 Boehmite solubility in non-complexing media

In order to cover a wide range of pH (2 to 9), three types of solutions were used: HCl/NaCl,

NH_4OH/NH_4Cl and NaOH/NaCl at constant ionic strength (I=0.025 at 90°C and 0.01 at other temperatures). The high temperature pH's were calculated by solving, with an iterative method, a system of non-linear equations comprising equations for dissociation equilibria for all the components of the solution, mass balance and charge balance equations. Activity coefficients were calculated from the Debye-Huckel equation with an å parameter of 4.10^{-8} cm for all the charged species in solution. Figure 1 shows the results obtained at 300°C. Similar results have been obtained at 90, 170, 200, 250 and 350°C.

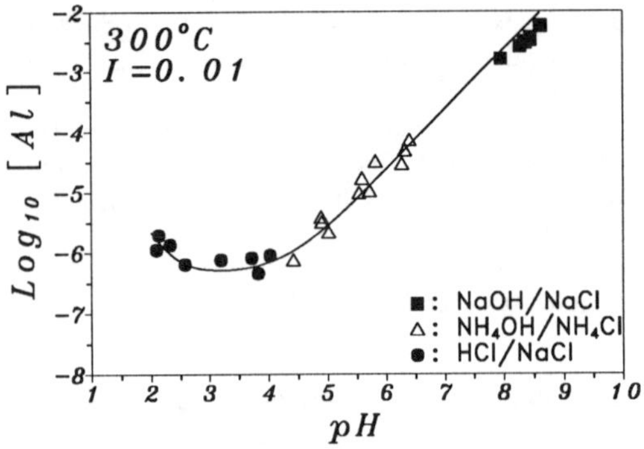

Fig 1. Boehmite solubility at 300°C. Experimental results and calculated curve.

These experimental data have been fitted using a least-squares method (see curve in figure 1) in order to calculate the dissociation constants of boehmite at each temperature. Hydrolysis constants of aluminum were then deduced from dissociation constants. The results are listed in table 1.

The data show that in basic solutions solubility increases with increasing temperature, whereas it decreases in acidic solutions. Also, the minimum in solubility shifts to more acidic pH as temperature increases. This evolution is illustrated in figure 2 where the percentage of aqueous species is reported as a function of pH at 170 and 350°C.

An important observation is that the neutral species is largely predominant (more than 70%) at a pH more acidic than the neutrality. At neutral pH, the charged species $Al(OH)_4^-$ is dominant.

Comparison of our results with other published data shows good agreement in basic media while the solubilities measured in acidic solutions are lower than those reported by previous authors (figure 3).

Table 1. Calculated boehmite dissociation constants and Al hydrolysis constants.

T°C	pK^*_{s0}	pK^*_{s1}	pK^*_{s2}	pK^*_{s3}	pK^*_{s4}
90	-	-1.77	2.60	7.15	13.17
170	-0.81	0.95	3.78	6.76	11.75
200	0.028	1.41	4.00	6.65	11.39
250	-	2.10	4.26	6.56	10.99
300	-	2.30	4.50	6.35	10.74
350	-	2.20	4.37	6.10	10.77
T°C	pK^*_1	pK^*_2	pK^*_3	pK^*_4	$p\beta^*_4$
90	-	4.37	4.55	6.02	-
170	1.765	2.82	2.98	4.99	12.56
200	1.379	2.59	2.65	4.74	11.36
250	-	2.16	2.30	4.43	-
300	-	2.20	1.85	4.39	-
350	-	2.17	1.75	4.67	-

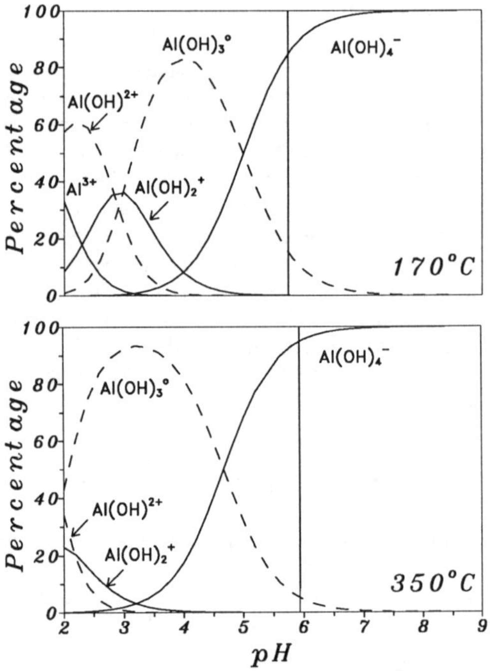

Fig 2. Distribution of aluminum species vs pH at 170 and 350°C. Vertical lines show the neutral pH's.

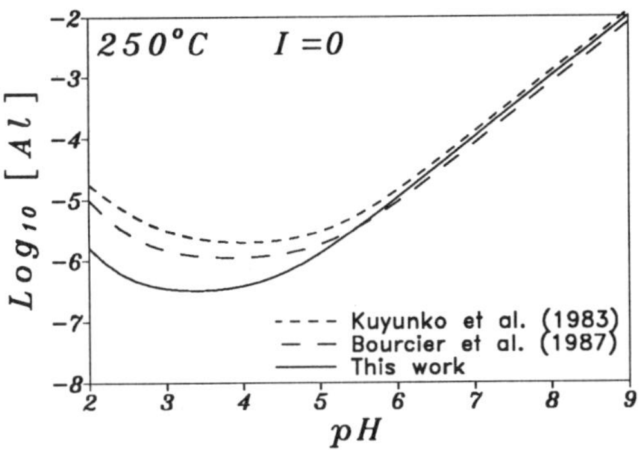

Fig 3. Comparison of the results from this study with published data at 250°C and I=0.

This discrepancy can be explained by the use of apparatus involving PTFE and/or CH_3COOH/CH_3COONa buffer solutions in the previous works. Indeed, measuring the solubility of boehmite at 200°C in presence of teflon, we found a simultaneous increase in the concentration of fluorine and aluminum in solution as a function of time. The aluminum content was more than 50 times higher than that obtained in the absence of fluorine. This result agrees with studies of Michard et al. (1979), Couturier et al. (1986) and Sanjuan and Michard (1987) which showed the formation of Al:F complexes.

3.2 Aluminum-Acetate complexation

We have studied the influence of acetate ion on the solubility of the boehmite. Experiments were performed at 170°C with AcO^- concentrations ranging from 0 to 10^{-2} mol/l, in solutions of constant pH (3.4) and ionic stength (0.01). The linear relation observed in figure 4, between the aluminum and the acetate concentrations in solution indicates the formation of 1:1 Al:AcO complexes.

Other solubility experiments in CH_3COOH/CH_3COONa medium (CH_3COO^- =0.01 M) were undertaken at 170 and 200°C and various pH. The association constants of complexes according to the reaction:

$$Al(OH)_n^{(3-n)} + AcO^- = Al(OH)_nAcO^{(2-n)}$$

where n=0, 1 or 2, are derived from the difference between the solubility of boehmite in the presence of acetate ion and its solubility in HCl/NaCl media. The obtained values (table 2) show the high stability of the complexes; this stability increases with the charge of the aqueous aluminum species and the temperature. Unfortunately, there are no

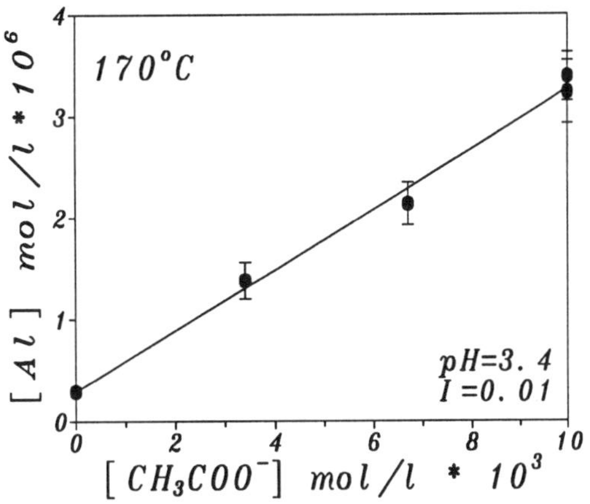

Fig 4. Boehmite solubility vs acetate concentration.

Table 2. Values of the stability constants K_n of three Al:AcO complexes at 170 and 200°C.

T°C	K_0	K_1	K_2
170	240000	3500	1300
200	-	(11700)	4500

(): estimated value

previously published data which could allow the comparison of our values of K_1 and K_2 constants. Our value of K_0 is coherent with the value proposed by Fein (1991) at 80°C and by Martell and Smith (1977) at 25°C.

3.3 Aluminum-Sodium complexation

The purpose of the last part of this study was to test for the formation of complexes between the aqueous aluminum and sodium ion at hydrothermal conditions. The formation of such complexes has been proposed at high temperature and pressure (750°C, 2kbars). In such experimental conditions the quantitative interpretation of the results is very difficult.

Experiments consisted of boehmite solubility measurements at 300°C in NaOH/NaCl solutions with NaCl concentrations up to 1 mol/l. Calculation of the speciation of these solutions were as previously described but the activity coefficients were calculated with the equation proposed by Helgeson et al. (1981). The pH's of the solutions were chosen so that the only species present was

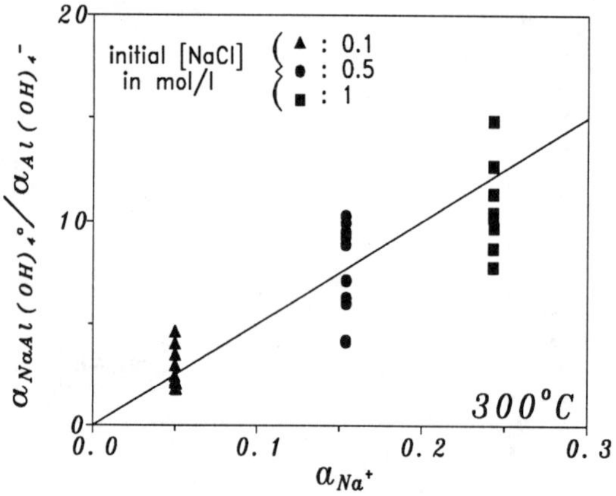

Fig 5. Variation of the ratio $a_{NaAl(OH)_4^\circ}/a_{Al(OH)_4^-}$ vs Na^+ activity.

$Al(OH)_4^-$. The difference between the solubility measured under these conditions and the solubility calculated using the K^*s_4 constant previously determined, gives the amount of complexed Al according to the reaction:

$Al(OH)_4^- + Na^+ = NaAl(OH)_4^\circ$.

The linear increase of the ratio $a_{NaAl(OH)_4^\circ}/a_{Al(OH)_4^-}$ as a function of Na^+ activity (figure 5) is strong evidence for a 1:1 stoichiometry of the complex.

The slope of this line gives the value of the stability constant of the $NaAl(OH)_4^\circ$ complex: K=50 at 300°C.

REFERENCES

BOURCIER, W.L., KNAUSS K.G. & JACKSON K.J. 1987. Aluminium hydrolysis constants to 250°C determined from boehmite solubility measurements. *Geological Society of America Abstracts of Programs* 19: 596.

COUTURIER, Y. 1986. Contribution à l'étude des complexes mixtes de l'aluminium (III) avec les ions fluorure et hydroxyde. *Bull. Soc. Chimique de France* 2: 171-177.

FEIN, J.B. 1991. Experimental study of Al- Ca- and Mg- acetate complexing at 80°C. *Geochim. Cosmochim. Acta* 55: 955-964.

HELGESON, H.C., KIRKHAM D.H. & FLOWERS G.C. 1981. Theoretical prediction of the thermodynamic behaviour of aqueous electrolytes at high pressures and temperatures; IV. Calculation of the activity coefficients, osmotic coefficients and apparent molal and standard and relative partial molal properties to 600°C and 5 kb. *Am. J. Sci.* 281: 1249-1516.

KUYUNKO, N.S., MALININ S.D. & KHODAKOVSKI I.L. 1983. An experimental study of aluminum ion hydrolysis at 150, 200 and 250°C. *Geochim. Int.* 20: 76-86.

MARTELL, A.E. & SMITH R.M. 1977. *Critical stability constants, III; other organic ligands.* Plenum Press, N.Y.

MICHARD, G., OUZOUNIAN G., FOUILLAC C. & SARAZIN G. 1979. Contrôle des concentrations d'aluminium dissous dans les eaux des sources thermales. *Geochim. Cosmochim. Acta* 43: 147-155.

SANJUAN, B. & MICHARD G. 1987. Aluminum hydroxide solubility in aqueous solution containing fluoride ions at 50°C. *Geochim. Cosmochim. Acta* 51: 1823-1831.

Water-Rock Interaction, Kharaka & Maest (eds) © 1992 Balkema, Rotterdam. ISBN 90 5410 075 3

Solubility of Pt, Pd sulfides and Au in aqueous bisulfide solutions to 350°C

Pujing Pan
Department of Geological Sciences, McGill University, Montreal, Canada

Scott A.Wood
Department of Geology & Geological Engineering, University of Idaho, Idaho, USA

ABSTRACT: The solubility of Pt, Pd sulfides and Au metal was measured from 200°C to 350°C at saturated water vapor pressure in a gold-lined titanium autoclave. Oxidation state and pH were buffered by the coexistence of $H_2S/HS^-/SO_4^{2-}$. The buffered pH was in the range 5.91-9.43 at elevated temperature and ΣS was 0.3 - 2.2 m. Under these conditions the solubilities measured were 4-800 μg/L Pt, 1-600 μg/L Pd and 2-300 mg/L Au, with Au always 2 to 3 orders of magnitude more soluble than Pt or Pd. The species $Au(HS)_2^-$, $Pt(HS)_2^0$ and $Pd(HS)_2^0$ were found to adequately represent the observed solubility behavior. The solubility of Pt, Pd sulfides and Au was also measured from 25° to 90°C under H_2S-saturated conditions at 1 atm. Solubilities measured ranged from 11 μg/L Au, 3.7 μg/L Pt and 0.5 μg/L Pd at 25°C to 75 μg/L Au, 19 μg/L Pt and 9 μg/L Pd at 90°C.

1 INTRODUCTION

Recently, Mountain and Wood (1988) and Wood et al. (1989, 1992) have suggested, on the basis of theoretical calculations, that Pt and Pd could be transported in a variety of geological environments, similar to Au, as bisulfide complexes. However, these authors pointed out that their calculations had a rather large degree of uncertainty, and that experimental testing of this hypothesis was required. We report here the results of experimental studies designed to provide stoichiometric and thermodynamic data on Pt and Pd bisulfide complexes in hydrothermal solutions from 200° to 350°C at saturated water-vapor pressure. We also report results on the solubility of these metals in H_2S-saturated solutions from 25° to 90°C. We have included Au in our study to make a direct comparison of Au solubility with that of Pt and Pd and to allow cross-checking of our results with the measurements of Seward (1973), Renders and Seward (1989) and Shenberger and Barnes (1989) on Au bisulfide complexation.

2 EXPERIMENTAL

The solid phases employed in this study were PtS, PdS and Au metal. The synthesis of PtS was accomplished by dissolving Pt in aqua regia, adjusting the pH to near 2 with NaOH and then bubbling H_2S through the solution. The black precipitate was then vacuum sealed into a pyrex tube and annealed at 450°C for 70 hrs. Palladium sulfide was synthesized by mixing Pd powder with resublimed sulfur powder, vacuum sealing this mixture into a pyrex tube and heating at 450°C for a week.

The identities of all solids before and after all runs were confirmed by powder x-ray diffraction, optical microscopy and SEM observation. Semi-quantitative energy dispersive analysis demonstrated that no significant alloying or solid solution occurred among the minerals present.

2.1 *High temperature experiments*

A hydrothermal system designed by Autoclave Engineers was employed for the study from 200° to 350°C. This system included a gold-lined, titanium autoclave with a volume of 500 mL. The autoclave was equipped with a magnetic stirrer to effect mixing during a run. A Ti frit was welded onto the solution-wetted end of a Ti sampling tube in order to minimize the carryover of solid particles during sampling. A thermocouple was inserted into a Ti thermowell in direct contact with the solution. Temperature was controlled and measured to within $\pm 3°C$.

Solutions were extracted from the autoclave at temperature and pressure using a technique described by Bourcier and Barnes (1987). All samples were immediately introduced into a 0.5 m NaOH solution to minimize loss of H_2S. Samples for total sulfide were stabilized by forming a zinc sulfide precipitate. Samples taken for metal and total sulfur analysis were oxidized by adding 5-10mL H_2O_2 solution and leaving them overnight. The samples for metal analysis were then acidified by addition of 5 mL concentrated HCl. Analyses of metals were performed after Te coprecipitation (Fryer and Kerrich, 1978) by Zeeman-corrected graphite furnace (Pt, Pd) and flame (Au) atomic absorption spectrophotometry. Total sulfide was determined by iodometric titration and total sulfur by a gravimetric method involving precipitation of the oxidized sulfur as $BaSO_4$.

Attainment of equilibrium was demonstrated by the agreement of results obtained by approach from both supersaturated (higher temperature) and undersaturated (lower temperature) conditions, and by the invariance of measured metal concentrations with increasing time. Preliminary tests showed that equilibrium was attained after 200 hours at 200°C (Wood et al., 1992). It was assumed that equilibrium was attained in a similar time period at all higher temperatures.

2.2 *Low temperature experiments*

Experiments between 25° and 90°C were conducted in a three-necked Pyrex flask of 2 liter capacity that was heated with an electric mantle heater. Temperature was controlled to $\pm 0.1°C$. Gold metal, PtS and PdS were reacted with pure deionized water through which H_2S was continually bubbled at 1 atm. Solution pH was monitored daily at temperature. Fluid samples were withdrawn through a syringe-mounted 0.45µm Millipore filter. Aliquots of this sample were analysed for metals, total sulfur and total sulfide as described above. Equilibrium was approached only from supersaturation, i.e., high temperature. That equilibrium was attained was demonstrated by the lack of variation of pH and metal concentrations with time after 8-10 days (Figure 1).

3 RESULTS

At temperatures from 200° to 350°C under our experimental conditions the measured solubilities ranged between 4 and 800 $\mu g/L$ Pt, 1 and 600 $\mu g/L$ Pd and 2 and 300 mg/L Au. Our gold solubilities agree well with those measured by Seward (1973) and Shenberger and Barnes (1989) under similar conditions. A few reconaissance experiments conducted at temperatures between 300° and 500°C showed solubilities to decrease dramatically above 300°C, suggesting that bisulfide transport of these metals is unlikely at these higher temperatures.

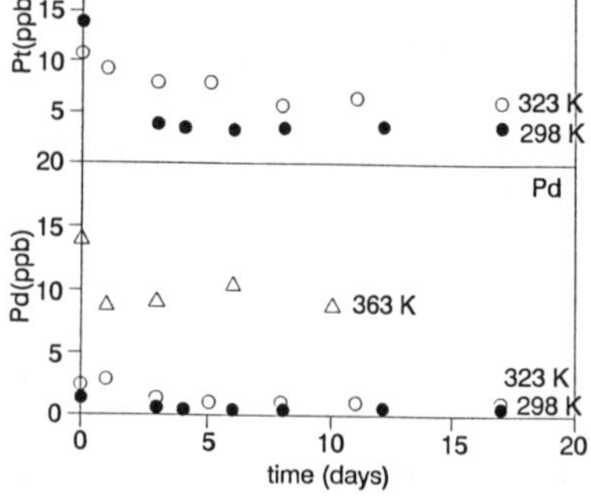

Figure 1: Measured solubilities of Au, Pt and Pd as a function of time for the low temperature experiments.

The high temperature pH and redox state were calculated from the partial pressure of H_2S and the measured concentrations of total sulfur and total sulfide. The pH at temperature ranged from 5.9 to 9.4 and ΣS = 0.3 to 2.2 molal. Reconnaissance measurements of the solubility of Pt and Pd in NaOH solutions showed that the contribution of hydroxide complexes to the measured Pt and Pd solubilities is negligible at these pH values.

Graphical analysis of the data indicates that the results are best explained by the following predominant species: $Au(HS)_2^-$, $Pt(HS)_2^0$ and $Pd(HS)_2^0$ (Figs. 2 and 3). Log K values for the formation of these species according to:

$$Au + H_2S(aq) + HS^- = Au(HS)_2^- + 1/2H_2(g) \quad (1)$$

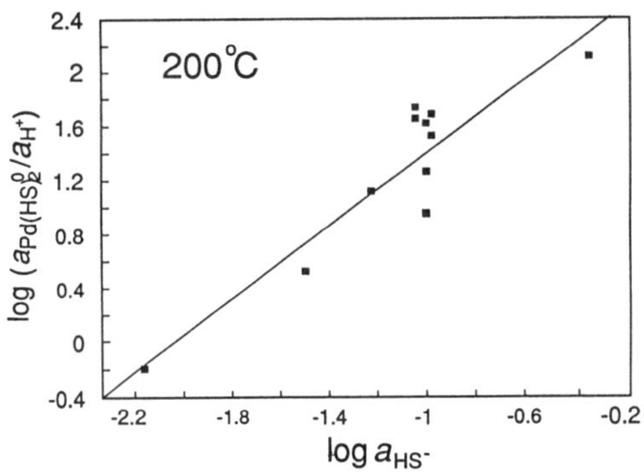

Figure 2: Summary plot of Pd solubility data at 200°C showing the fit to a model assuming $Pd(HS)_2(aq)$ is the predominant species.

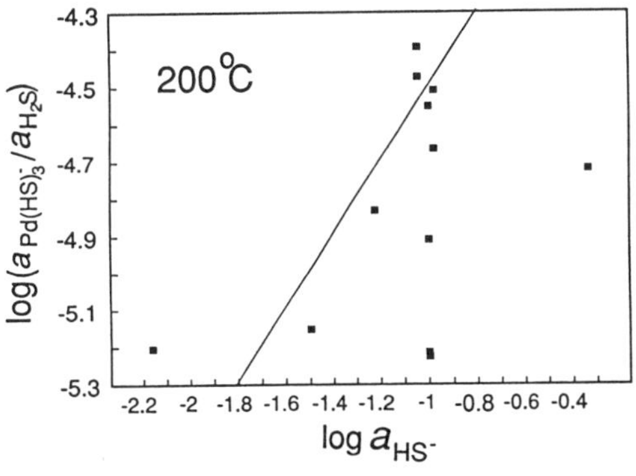

Figure 3: Summary plot of Pd solubility data at 200°C showing the fit to a model assuming $Pd(HS)_3^-$ is the predominant species.

$$PtS + HS^- + H^+ = Pt(HS)_2^0 \quad (2)$$

$$PdS + HS^- + H^+ = Pd(HS)_2^0 \quad (3)$$

were determined from the measured solubilities (Table 1).

Table 1. Equilibrium constants for dissolution of Au, Pt and Pd as bisulfide complexes.

T(°C)	log K_1	log K_2	log K_3
200	-2.63	2.35	2.51
250	-2.53	2.49	2.62
300	-2.44	2.60	2.77
350	-2.40	2.85	2.99

The lower temperature experiments yielded metal solubilities increasing steadily with temperature. Measured pH was 4.05 at 25°C, 3.52 at 50°C and 3.30 at 90°C. The measured solubilities were (in μg/L): for Au, 11, 38 and 75; for Pt, 3.7, 6.2, and 18.7; and for Pd, 0.5, 1.1, and 8.6; at 25°, 50° and 90°C, respectively. There are presently insufficient data with which to determine the speciation at lower temperature.

4 DISCUSSION

Using the equilibrium constants calculated from our high-temperature experimental data, the solubility of Pt and Pd in natural hydrothermal solutions may be calculated. At 300°C, ΣS = 10^{-3} molal, and pH = 5.4, conditions near those of the Salton Sea Brines (McKibben et al., 1990), we calculate Pt and Pd solubilities as bisulfide complexes of 0.5 and 0.4 ppb respectively, assuming saturation with respect to PtS and PdS. These are very close to the maximum values measured by McKibben et al. (1990) in the Salton Sea. On the other hand, theoretical and experimental evidence suggest that the solubility of Pt and Pd as chloride complexes is many orders of magnitude lower than the measured concentrations under these conditions (Wood et al., 1992). Therefore, we conclude that Pt and Pd are present in these brines predominantly as bisulfide complexes. Other geological processes where Pt and Pd transport may occur as bisulfide complexes include: seafloor hydrothermal activity, serpentinization, and the formation of a wide variety of hydrothermal Pt and Pd deposits.

Our low temperature solubility data suggest that Pt and Pd are also quite mobile as bisulfide complexes in H_2S-saturated solutions. This may also permit transport of these metals in anoxic environments.

REFERENCES

Bourcier, W.L. & Barnes, H.L. 1987. Rocking autoclaves for hydrothermal experiments. I. Fixed-volume systems. In G.C. Ulmer & H.L. Barnes (eds.), *Hydrothermal Experimental Techniques*: 189-215. New York: Wiley-Interscience.

Fryer, B.J. & Kerrich, R. 1978. Determination of precious metals at ppb levels in rocks by a combined wet chemical and flameless atomic absorption method. Atomic Absorption Newsletter 17: 4-6.

McKibben, M.A., Williams, A.E., and Hall, G.E.M. 1990. Solubility and transport of platinum-group elements and Au in saline hydrothermal fluids: Constraints from geothermal brine data. Econ. Geol. 85: 1926-1934.

Mountain, B.W. & Wood, S.A. 1988. Chemical controls on the solubility, transport, and deposition of platinum and palladium in hydrothermal solutions: A thermodynamic approach. Econ. Geol. 83: 492-510.

Renders, P.J. & Seward, T.M. 1989. The stability of hydrosulphido- and sulphido-complexes of Au(I) and Ag(I) at 25°C. Geochim. Cosmochim. Acta 53: 245-254.

Seward, T.M. 1973. Thiocomplexes of gold and the transport of gold in hydrothermal ore solutions. Geochim. Cosmochim. Acta 37: 379-399.

Shenberger, D.M. & Barnes, H.L. 1989. Solubility of gold in aqueous sulfide solutions from 150°C to 350°C. Geochim. Cosmochim. Acta 53: 1269-1278.

Wood, S.A., Mountain, B.W. & Fenlon, B.J. 1989. Thermodynamic constraints on the solubility, transport, and deposition of platinum and palladium in hydrothermal solutions: reassessment of hydroxide, bisulfide and ammonia complexing. Econ. Geol. 84: 2020-2028.

Wood, S.A., Mountain, B.W. and Pan, P. 1992. The aqueous geochemistry of platinum, palladium and gold: Recent experimental constraints and a re-evaluation of theoretical predictions. Can. Min. 30: (in press).

Water-Rock Interaction, Kharaka & Maest (eds) © 1992 Balkema, Rotterdam. ISBN 90 5410 075 3

Calculation of the thermodynamic properties of mononuclear aqueous Al species in the system Al_2O_3-H_2O-NaCl at temperatures to 1000°C and pressures to 5 kb

Vitalii A. Pokrovskii & Harold C. Helgeson
Department of Geology & Geophysics, University of California, Berkeley, Calif., USA

ABSTRACT: Thermodynamic properties and equations-of-state coefficients generated from experimental solubility data reported in the literature for gibbsite, boehmite, diaspore, and corundum (Pokrovskii & Helgeson 1989, 1992) were used together with the revised HKF equation of state for aqueous species (Tanger & Helgeson 1988; Shock & Helgeson 1988; Shock et al. 1989) to calculate the dissociation constants for $NaAl(OH)_4^0$ and equilibrium constants for the overall hydrolysis of Al^{3+} to yield $Al(OH)^{2+}$, $Al(OH)_2^+$, $Al(OH)_3^0$, and $Al(OH)_4^-$ at temperatures to 1000°C and pressures to 5 kb.

INTRODUCTION

Reliable values of the thermodynamic properties of aqueous Al species are required to calculate hydrothermal solubilities of aluminosilicates and the chemical mass transfer of aluminum during water-rock interaction. Recent advances in estimating and predicting the standard and relative partial molal properties of aqueous species and electrolytes with the aid of the revised HKF equation of state (Tanger & Helgeson 1988; Shock & Helgeson 1988; Shock et al. 1989) facilitate evaluation and regression of mineral solubilities at both high and low pressures and temperatures to retrieve values of the thermodynamic properties of minerals and aqueous species. Recent evaluation of experimental solubilities reported in the literature for gibbsite, boehmite, diaspore, and corundum resulted in a revised set of standard thermodynamic properties and the HKF equations-of-state coefficients for mononuclear aqueous Al species in the system Al_2O_3–H_2O–NaCl (Pokrovskii & Helgeson 1989, 1992). The purpose of the present communication is to summarize the equilibrium constants for the dissociation of $NaAl(OH)_4^0$ and the overall hydrolysis of Al^{3+} to yield $Al(OH)^{2+}$, $Al(OH)_2^+$, $Al(OH)_3^0$, and $Al(OH)_4^-$ at temperatures to 1000°C and pressures to 5 kb.

CALCULATION OF EQUILIBRIUM CONSTANTS

The hydrolysis of the aluminum ion to yield mononuclear hydroxide complexes can be expressed as (Baes & Mesmer 1976)

$$Al^{3+} + nH_2O = Al(OH)_n^{3-n} + nH^+, \qquad (1)$$

where n ranges from 1 to 4. The equilibrium constant for the nth hydrolysis reaction can be written as

$$K_{1n} = \frac{a_{Al(OH)_n^{3-n}} a_{H^+}^n}{a_{Al^{3+}} a_{H_2O}^n}, \qquad (2)$$

where a_i refers to the activity of the ith species. The dissociation of $NaAl(OH)_4^0$ is given by

$$NaAl(OH)_4^0 = Na^+ + Al(OH)_4^-. \qquad (3)$$

Equilibrium constants for reactions (1) and (3) for temperatures to 1000°C and pressures to 5 kb are given in Table 1. The values shown in this table were computed by combining the standard thermodynamic properties and the HKF equations-of-state coefficients for Al^{3+}, $Al(OH)^{2+}$, $Al(OH)_2^+$, $Al(OH)_3^0$, $Al(OH)_4^-$, and $NaAl(OH)_4^0$ (Pokrovskii &

Helgeson 1989, 1992) with those for Na^+ and H_2O given by Tanger & Helgeson (1988) and Johnson & Norton (1991), respectively, and the relation

$$\log K = -\frac{\Delta \overline{G}_n^\circ}{2.303RT}, \quad (4)$$

where R stands for the gas constant (1.9872 cal $mol^{-1}K^{-1}$) and

$$\Delta \overline{G}_n^\circ = \sum_i \hat{n}_{i,n} \Delta \overline{G}_i^\circ, \quad (5)$$

where $\Delta \overline{G}_i^\circ$ refers to the apparent standard partial molal Gibbs free energy of the ith species and $\hat{n}_{i,n}$ denotes the reaction coefficient for the ith species in the nth reaction, which is positive for products and negative for reactants.

Equilibrium constants given in Table 1 can be used together with apparent standard partial molal Gibbs free energies of formation of the aluminum ion listed in Table 2 and those for H_2O

TABLE 1: Equilibrium constants for reactions (1) and (3).

T°C	log K11						
	PRESSURE, KB						
	SAT	0.5	1	2	3	4	5
25	-4.99	-5.11	-5.21	-5.37	-5.49	-5.59	-5.67
50	-4.29	-4.42	-4.52	-4.69	-4.83	-4.94	-5.04
100	-3.08	-3.21	-3.31	-3.48	-3.62	-3.74	-3.84
150	-2.05	-2.18	-2.29	-2.46	-2.60	-2.71	-2.81
200	-1.15	-1.29	-1.40	-1.58	-1.72	-1.83	-1.93
250	-0.34	-0.49	-0.62	-0.81	-0.95	-1.07	-1.16
300	0.40	0.23	0.08	-0.12	-0.27	-0.39	-0.48
350	0.87	0.79	0.69	0.49	0.34	0.22	0.13
400		1.07	1.21	1.06	0.91	0.79	0.69
450			1.61	1.57	1.43	1.31	1.21
500			1.79	2.03	1.91	1.79	1.68
550			1.74	2.42	2.34	2.21	2.11
600				2.78	2.74	2.62	2.50
700				3.42	3.49	3.36	3.23
800				4.06	4.24	4.07	3.89
900					5.00	4.75	4.51
1000					5.78	5.40	5.10

TABLE 1 (cont.)

T°C	log K12						
	PRESSURE, KB						
	SAT	0.5	1	2	3	4	5
25	-10.13	-10.29	-10.42	-10.61	-10.74	-10.84	-10.91
50	-8.78	-8.95	-9.08	-9.29	-9.45	-9.57	-9.66
100	-6.54	-6.72	-6.86	-7.08	-7.24	-7.37	-7.48
150	-4.73	-4.92	-5.07	-5.30	-5.47	-5.60	-5.71
200	-3.20	-3.41	-3.58	-3.83	-4.01	-4.15	-4.26
250	-1.86	-2.10	-2.31	-2.59	-2.79	-2.94	-3.05
300	-0.67	-0.96	-1.21	-1.52	-1.74	-1.90	-2.01
350	0.03	-0.12	-0.27	-0.59	-0.81	-0.98	-1.10
400		0.19	0.49	0.25	0.02	-0.15	-0.27
450			1.02	0.99	0.78	0.61	0.46
500			1.15	1.64	1.46	1.27	1.11
550			0.85	2.17	2.05	1.86	1.70
600				2.64	2.59	2.40	2.23
700				3.45	3.62	3.40	3.18
800				4.29	4.65	4.34	4.04
900					5.74	5.27	4.85
1000					6.86	6.17	5.60

TABLE 1 (cont.)

T°C	log K13						
	PRESSURE, KB						
	SAT	0.5	1	2	3	4	5
25	-16.13	-16.30	-16.44	-16.66	-16.81	-16.92	-17.01
50	-14.19	-14.37	-14.52	-14.76	-14.93	-15.07	-15.18
100	-10.99	-11.19	-11.35	-11.59	-11.78	-11.93	-12.06
150	-8.43	-8.65	-8.82	-9.08	-9.28	-9.44	-9.57
200	-6.29	-6.54	-6.74	-7.03	-7.25	-7.41	-7.54
250	-4.45	-4.73	-4.97	-5.32	-5.55	-5.73	-5.87
300	-2.85	-3.17	-3.47	-3.85	-4.11	-4.30	-4.44
350	-2.00	-2.06	-2.20	-2.57	-2.85	-3.05	-3.30
400		-1.73	-1.20	-1.45	-1.72	-1.93	-2.09
450			-0.53	-0.46	-0.70	-0.91	-1.10
500			-0.43	0.39	0.20	-0.03	-0.23
550			-0.94	1.08	0.97	0.74	0.54
600				1.66	1.68	1.46	1.24
700				2.66	3.00	2.76	2.49
800				3.67	4.32	3.98	3.60
900					5.70	5.16	4.64
1000					7.13	6.30	5.60

TABLE 1 (cont.)

T°C	log K_{14}						
	PRESSURE, KB						
	SAT	0.5	1	2	3	4	5
25	-22.87	-23.01	-23.11	-23.25	-23.32	-23.36	-23.37
50	-20.43	-20.58	-20.70	-20.87	-20.99	-21.07	-21.12
100	-16.57	-16.73	-16.85	-17.04	-17.18	-17.29	-17.37
150	-13.66	-13.82	-13.94	-14.14	-14.28	-14.38	-14.47
200	-11.39	-11.54	-11.67	-11.87	-12.01	-12.12	-12.20
250	-9.59	-9.73	-9.86	-10.06	-10.20	-10.30	-10.39
300	-8.22	-8.28	-8.41	-8.58	-8.72	-8.83	-8.91
350	-7.77	-7.35	-7.27	-7.38	-7.50	-7.60	-7.68
400		-7.27	-6.45	-6.39	-6.47	-6.56	-6.63
450			-6.04	-5.58	-5.60	-5.67	-5.75
500			-6.18	-4.94	-4.87	-4.93	-4.99
550			-6.97	-4.47	-4.27	-4.29	-4.35
600				-4.13	-3.76	-3.74	-3.79
700				-3.70	-2.89	-2.81	-2.86
800				-3.37	-2.13	-2.03	-2.09
900					-1.37	-1.32	-1.44
1000					-0.61	-0.68	-0.89

TABLE 1 (cont.)

T°C	log $K_{NaAl(OH)_4^o}$						
	PRESSURE, KB						
	SAT	0.5	1	2	3	4	5
25	0.75	0.84	0.92	1.04	1.13	1.19	1.24
50	0.66	0.74	0.81	0.91	0.98	1.02	1.06
100	0.40	0.48	0.55	0.65	0.71	0.76	0.79
150	0.06	0.17	0.26	0.39	0.47	0.53	0.57
200	-0.36	-0.19	-0.05	0.12	0.24	0.32	0.37
250	-0.88	-0.60	-0.40	-0.15	0.01	0.11	0.18
300	-1.60	-1.10	-0.78	-0.43	-0.22	-0.09	0.00
350	-2.74	-1.69	-1.21	-0.73	-0.47	-0.31	-0.20
400		-2.50	-1.72	-1.07	-0.75	-0.54	-0.42
450			-2.34	-1.46	-1.05	-0.81	-0.63
500			-3.02	-1.87	-1.36	-1.04	-0.83
550			-3.84	-2.27	-1.64	-1.27	-1.01
600				-2.70	-1.93	-1.49	-1.20
700				-3.67	-2.59	-1.96	-1.56
800				-4.78	-3.34	-2.49	-1.95
900					-4.18	-3.06	-2.35
1000					-5.04	-3.64	-2.76

TABLE 2: Apparent standard partial molal Gibbs free energies of formation ($\Delta\overline{G}^o$) of Al^{3+} calculated using thermodynamic data and parameters reported by Pokrovskii & Helgeson (1989, 1992).

T°C	$\Delta\overline{G}^o$, kcal mol^{-1}						
	PRESSURE, KB						
	SAT	0.5	1	2	3	4	5
25	-116.54	-117.03	-117.46	-118.19	-118.79	-119.32	-119.78
50	-114.49	-115.01	-115.47	-116.25	-116.92	-117.50	-118.02
100	-110.22	-110.81	-111.33	-112.21	-112.96	-113.62	-114.22
150	-105.71	-106.41	-107.01	-108.02	-108.86	-109.60	-110.26
200	-100.91	-101.77	-102.50	-103.69	-104.65	-105.48	-106.20
250	-95.80	-96.85	-97.79	-99.23	-100.35	-101.27	-102.07
300	-90.36	-91.70	-92.93	-94.63	-95.93	-96.98	-97.86
350	-85.90	-86.97	-88.00	-89.90	-91.38	-92.56	-93.53
400		-83.75	-83.18	-85.04	-86.66	-87.97	-89.04
450			-78.84	-80.04	-81.77	-83.20	-84.46
500			-75.98	-75.05	-76.81	-78.47	-79.88
550			-75.03	-70.26	-71.92	-73.71	-75.27
600				-65.47	-66.92	-68.87	-70.60
700				-55.69	-56.32	-58.77	-61.02
800				-44.71	-44.46	-47.86	-50.94
900					-30.98	-36.00	-40.32
1000					-15.96	-23.32	-29.22

generated from equations and parameters reported by Johnson & Norton (1991) to calculate values of $\Delta\overline{G}^{\circ}$ for $Al(OH)^{2+}$, $Al(OH)_2^+$, $Al(OH)_3^\circ$, $Al(OH)_4^-$, and $NaAl(OH)_4^0$ over a wide range of pressures and temperatures of geologic interest.

ACKNOWLEDGEMENTS. This research was supported by the Department of Energy (DOE Grant DE-FG03-85ER-13419), and in part by the Petroleum Research Fund, administered by the American Chemical Society (ACS-PRF Grant 23026-AC2).

REFERENCES

Baes C.F., Jr. & R.E. Mesmer 1976. *The hydrolysis of cations.* New York: J. Wiley & Sons, xxi, 489 pp.

Johnson J.W & D. Norton 1991. Critical phenomena in hydrothermal systems: State, thermodynamic, electrostatic, and transport properties of H_2O in the critical region. *Amer. J. Sci.* 291: 541-648.

Pokrovskii V.A. & H.C. Helgeson 1989. New estimates of thermodynamic properties of aqueous aluminum species in the system Al_2O_3–H_2O–NaCl to 5 kb and 1000°C, using the modified Helgeson-Kirkham-Flowers equation of state. *Geol. Soc. Amer. Annu. Meet. Abstr. Progr.* 21: A156.

Pokrovskii V.A. & H.C. Helgeson 1992. Thermodynamic properties of aqueous species and the solubilities of minerals at high pressures and temperatures: The system Al_2O_3–H_2O–NaCl. *Amer. J. Sci.*: in preparation.

Shock E.L. & H.C. Helgeson 1988. Calculation of the thermodynamic and transport properties of aqueous species at high pressures and temperatures: Correlation algorithms for ionic species and equation of state predictions to 5 kb and 1000°C. *Geochim. Cosmochim. Acta* 52: 2009-2036.

Shock E.L., H.C. Helgeson & D.A. Sverjensky 1989. Calculation of the thermodynamic and transport properties of aqueous species at high pressures and temperatures: Standard partial molal properties of inorganic neutral species. *Geochim. Cosmochim. Acta* 53: 2157-2183.

Tanger J.C. & H.C. Helgeson 1988. Calculation of the thermodynamic and transport properties of aqueous species at high pressures and temperatures: Revised equation of state for the standard partial properties of ions and electrolytes. *Amer. J. Sci.* 288: 19-98.

Water-Rock Interaction, Kharaka & Maest (eds) © 1992 Balkema, Rotterdam. ISBN 90 5410 075 3

Excess molar volumes for CO_2-CH_4-N_2 mixtures

J.C. Seitz
Oak Ridge National Laboratory, Oak Ridge, Tenn., USA and Department of Geological Services, Virginia Polytechnic Institute, Blacksburg, Va., USA

J.G. Blencoe & D.B. Joyce
Oak Ridge National Laboratory, Oak Ridge, Tenn., USA

R.J. Bodnar
Department of Geological Sciences, Virginia Polytechnic Institute, Blacksburg, Va., USA

Abstract: Vibrating-tube densimetry experiments are being performed to determine the excess molar volumes of single-phase CO_2-CH_4-N_2 gas mixtures at pressures as high as 3500 bars and temperatures up to 500 °C. In our initial experiments, we determined the P-V-T properties of: (1) CO_2-CH_4, CO_2-N_2, CH_4-N_2, and CO_2-CH_4-N_2 mixtures at 1000 bars, 50 °C; and (2) CO_2-CH_4 mixtures from 100 to 1000 bars at 100 °C. Excess molar volumes in the binary subsystems are very accurately represented by two-parameter Margules equations. Experimentally determined excess molar volumes are in fair to poor agreement with predictions from published equations of state. Geometric projection techniques based on binary system data yield calculated excess molar volumes for CO_2-CH_4-N_2 mixtures that are in good agreement with our experimental data.

1 INTRODUCTION

H_2O, CO_2, CH_4, and N_2 are key volatile constituents of the earth; therefore, thermochemical data on these gas species and their mixtures find application in geochemical modeling of crustal fluids in a wide variety of geologic environments. With the ultimate goal of developing new thermodynamic equations of state for gas-rich crustal fluids, we are currently collecting P-V-T data on binary and ternary CO_2-CH_4-N_2 gas mixtures at 100-1000 bars and 50-400 °C. In this article we describe our initial results at 50 and 100 °C.

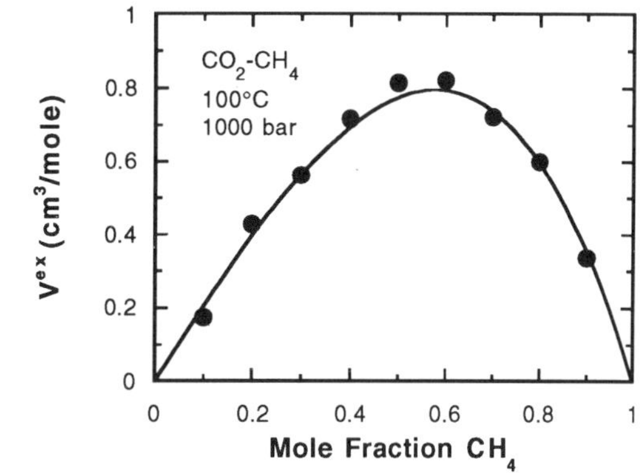

Figure 1. V^{ex} for CO_2-CH_4 mixtures at 1000 bars, 100 °C. Plot symbols and the V^{ex} curve illustrate, respectively, our experimental data and two-parameter Margules model.

2 EXPERIMENTAL METHODS

P-V-T data are being collected using a vibrating-tube densimeter designed to measure the P-V-T properties of fluids at P-T conditions as high as 3500 bars and 500 °C. An isobaric, isothermal flow-through method was employed in all of our experiments. The fundamental equation for calculating densities of gases by vibrating-tube densimetry is

$$\rho - \rho_o = K(\tau^2 - \tau_0^2), \quad (1)$$

where ρ is the density of an *unknown* gas; ρ_o is the density of a *known* (standard) gas; K is a parameter that depends on the physical properties of the vibrating tube; τ is the period of vibration for the *unknown* gas; and τ_o is the period of vibration for the *known* (standard) gas. K is evaluated using three standard gases (Ar, He, and N_2) and has been found to be a constant at a given pressure and temperature (i.e., the vibrating tube behaves ideally). Therefore, equation (1) represents a *linear* relationship between ρ and τ^2 where the density of an unknown gas is measured relative to standard gases.

Short term instrumental drift, although rarely observed, was checked for by repeated measurements on Ar during each run. Precisions achieved in experimentation were: P ± 0.1 bar, and T ± 0.01 °C. Conservative estimates of accuracy are: P ± 2.0 bar, T ± 0.2 °C. The precisions of τ^2, ρ, and V^{ex} derive from counting statistics and vary as a function of the density and composition of the gas mixture. The nominal uncertainty in V^{ex} is approximately ± 0.1 cm³/mole; however, the disposition of our data suggests that the actual uncertainty in V^{ex} may be considerably less than that.

3 THEORY AND DATA

Excess molar volumes were calculated using measured and ideal molar volumes for the mixtures, the latter being derived from the molar volumes of the end member gases measured during the same experiment. V^{ex} is calculated using the relation

$$V^{ex} = V - V^{id} = V - \sum_{i=1}^{n} X_i i . \quad (2)$$

V^{ex} data for the binary systems were fit to an equation of the form

$$\frac{V^{ex}}{X_1 X_2} = a + bX_2 , \quad (3)$$

which yields the W_{V_1} and W_{V_2} parameters of the two-parameter Margules formulation according to the identities $W_{V_2} = a$ and $W_{V_1} = b + W_{V_2}$. Finally, V^{ex} for any binary composition can be calculated from the relation

$$V^{ex} = W_{V_1} X_1 X_2^2 + W_{V_2} X_1^2 X_2 . \quad (4)$$

Our results for binary systems are summarized graphically in Figs. 1-4. Figure 1 shows that V^{ex} for CO_2-CH_4 mixtures is continuously positive and only slightly asymmetric at 100 °C and 1000 bars. Figure 2 illustrates the pressure dependence of $W_{V_{CO2}}$ and $W_{V_{CH4}}$ at 1-1000 bars, 100 °C. Values for these W_V's can be substituted into equation (4) to calculate V^{ex} for CO_2-CH_4 mixtures at 100 °C. Figures 3 and 4 show V^{ex} for CH_4-N_2 and CO_2-N_2 mixtures, respectively, at 50 °C and 1000 bars. V^{ex} for the CH_4-N_2 system is continuously positive, essentially symmetric, and smaller than the V^{ex} for the CO_2-CH_4 system (cf. Fig. 1). The CO_2-N_2 system, in contrast, exhibits a small V^{ex} that is mostly negative and strongly asymmetric. At 1000 bars, V^{ex} for the CO_2-CH_4 mixtures is unchanged between 50 °C (not illustrated) and 100 °C (Fig. 1). Despite the disparate

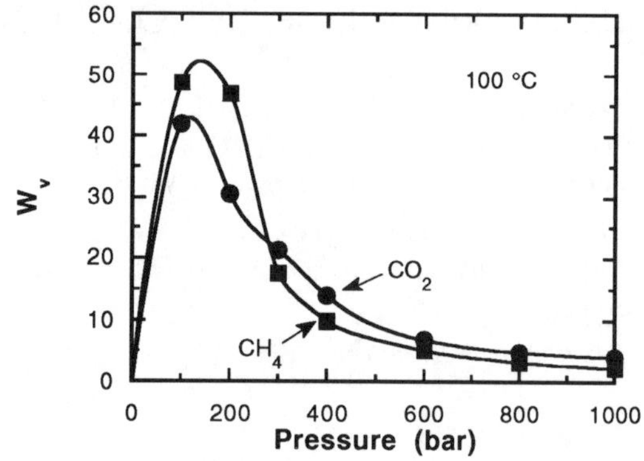

Figure 2. Margules parameters (W_V's) for CO_2 (circles) and CH_4 (squares) from 1 to 1000 bars at 100 °C. Plot symbols represent experimental data.

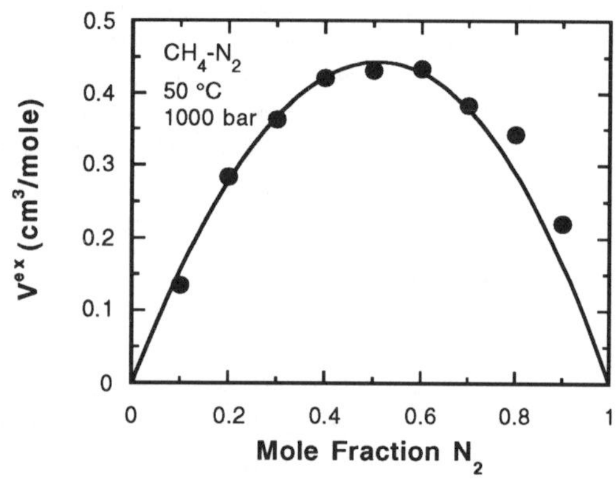

Figure 3. V^{ex} for CH_4-N_2 mixtures at 1000 bars, 50 °C. Plot symbols and the V^{ex} curve illustrate, respectively, our experimental data and two-parameter Margules model.

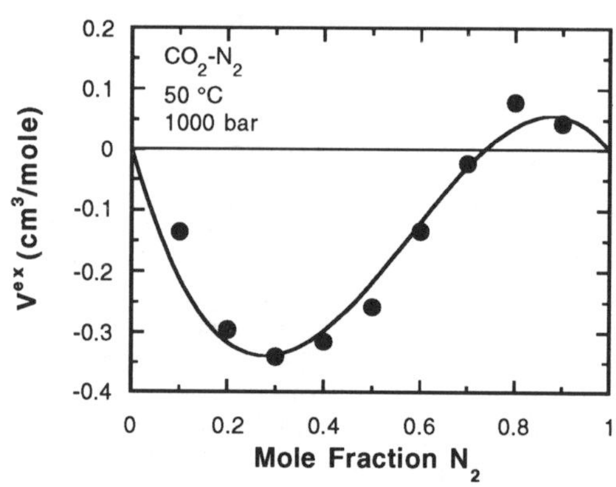

Figure 4. V^{ex} for CO_2-N_2 mixtures at 1000 bars, 50 °C. Plot symbols and the V^{ex} curve illustrate, respectively, our experimental data and two-parameter Margules model.

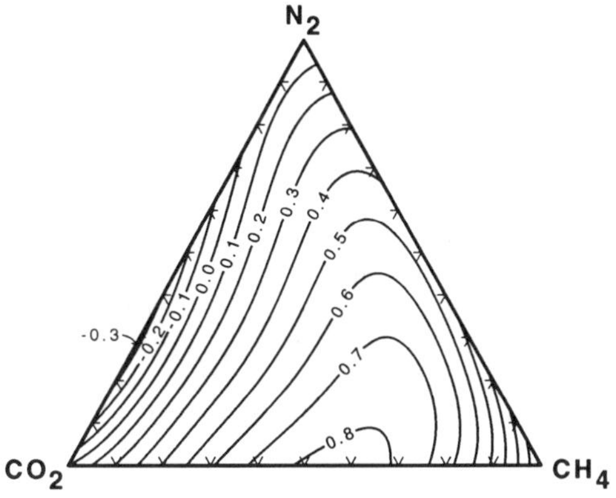

Figure 5. Experimentally determined V^{ex} for CO_2-CH_4-N_2 mixtures at 1000 bars, 50 °C. Contours indicate mixtures of constant excess volume (cm^3/mole).

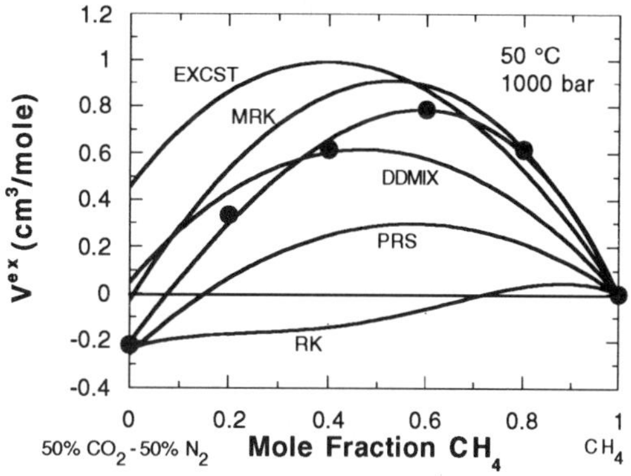

Figure 6. Calculated and experimentally determined excess volumes for CO_2-CH_4-N_2 mixtures at 1000 bars, 50 °C (see text for explanation). Filled circles represent experimental data along the $X_{CO_2} = X_{N_2}$ pseudobinary join.

volumetric properties of the three binary systems, their V^{ex} relations can be represented very accurately by two-parameter Margules equations.

We have also collected P-V-T data on CO_2-CH_4-N_2 mixtures at 50 °C and 1000 bars. A contoured V^{ex} surface has been generated from these data (Fig. 5), and it has a complex topology with a region of negative V^{ex} near the CO_2-N_2 binary.

4 EQUATIONS OF STATE

We have compared our experimental data with excess volumes calculated from various equations of state, and representative results are illustrated in Fig. 6. There is generally poor agreement between excess volumes calculated from the Redlich-Kwong (RK) equation of state (Redlich and Kwong, 1949) and our experimentally determined values, and molar volumes calculated from this model differ from our data by as much as 4%. The modified Redlich-Kwong (MRK) equation of state developed by Holloway (1977) yields accurate calculated values of V^{ex} for gas mixtures with compositions illustrated in Fig. 6, but calculated excess volumes for other CO_2-CH_4-N_2 mixtures at 1000 bars, 50 °C are not nearly as accurate. Also, molar volumes calculated from the Holloway MRK equation differ from our experimentally determined values by as much as 6%.

We have also evaluated equations of state in two software packages that are commercially available from the National Institute of Standards and Technology (NIST): Supertrapp (Ely and Huber, 1990) and DDMIX (Ely, 1988). With Supertrapp,

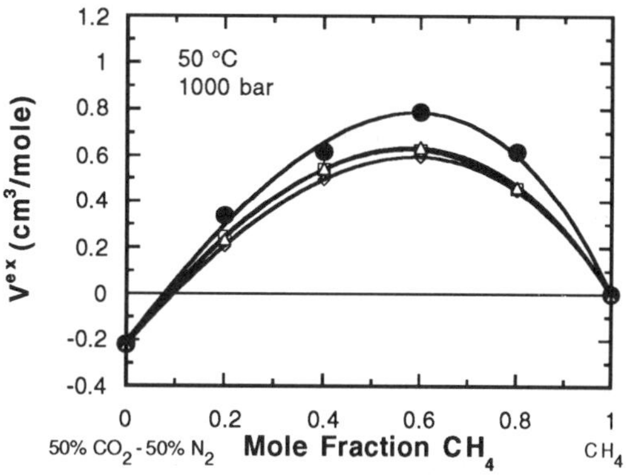

Figure 7. Experimentally determined (filled circles, this study) and calculated (different binary projection methods; diamonds - method of Hillert, triangles - methods of Colinet and Muggianu, squares - method of Kohler) excess volumes for CO_2-CH_4-N_2 mixtures at 1000 bars, 50 °C along the $X_{CO_2} = X_{N_2}$ pseudobinary join.

the volumetric properties of mixtures can be calculated using either a Peng-Robinson (PRS) model or the NIST extended corresponding states (EXCST) model. The PRS model yields calculated values for V^{ex} that are in poor to fair agreement with our experimental data; however, the molar volumes derived from this model differ from our experimentally determined values by as much as 10%. Likewise, the EXCST model yields calculated values for V^{ex} that are in poor to fair agreement with our experimentally determined data, but calculated molar volumes differ from our experimentally

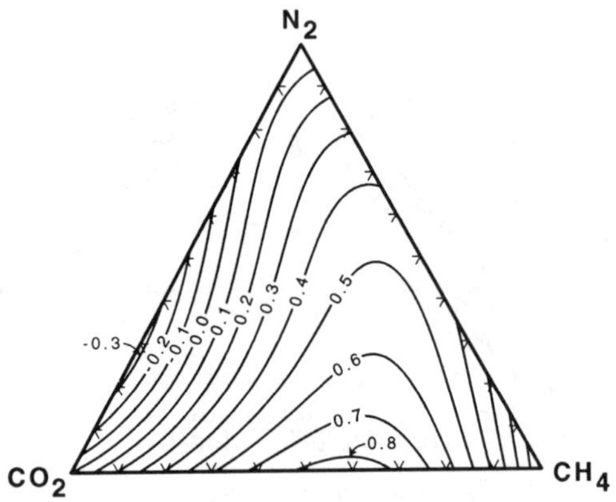

Figure 8. Excess volumes for CO_2-CH_4-N_2 mixtures at 1000 bars, 50 °C, calculated using the method of Muggianu (1975). Contours indicate mixtures of constant excess volume (cm^3/mole).

determined values by less than 3%. Finally, DDMIX yields calculated values for V^{ex} that are in fair to good agreement with our experimental data, whereas molar volumes calculated from this model differ from our experimentally determined values by less than 1%.

5 GEOMETRIC PROJECTION METHODS

We have evaluated various empirical geometric methods for predicting the volumetric properties of a ternary system using the excess molar volumes of the binary subsystems (Kohler, 1960; Colinet, 1967 as referenced by Hillert, 1980; Hillert, 1980; Muggianu et al., 1975). Each of these methods estimates the ternary properties by projection from the binary subsystems, but they employ different projection vectors. The reader is referred to Hillert (1980, and references therein) for a review of these techniques. Geometric projection equations have the distinct advantage of reducing to a two-parameter equation along each binary sideline.

Figure 7 shows that the various projection methods yield similar results for our data. The methods of Colinet (1967) and Muggianu et al. (1975) yield essentially identical results. Figure 8 depicts the contoured V^{ex} surface of ternary CO_2-CH_4-N_2 mixtures calculated using the method of Muggianu et al. (1975). Comparison of these results with experimental data (cf. Figs. 5 and 8) indicates that this methodology is useful in predicting excess properties for ternary fluid mixtures. In fact, V^{ex} predicted by the Muggianu method can be more accurate than predictions from published equations of state (cf. Figs. 6 and 7).

ACKNOWLEDGEMENTS

Research sponsored by the Division of Engineering and Geosciences, Office of Basic Energy Sciences, U.S. Department of Energy, under contract DE-AC05-84OR21400 with Martin Marietta Energy Systems, Inc. Partial funding has been provided by NSF grant EAR-8857778-03PYI to RJB.

REFERENCES

Colinet, C. 1967. D.E.S. University of Grenoble, France.

Ely, J. F. 1988. NIST mixture property program (DDMIX). *NIST Standard Reference Database* 14.

Ely, J. F. & M. L. Huber. 1990. NIST thermophysical properties of hydrocarbon mixtures database (SUPERTRAPP). *NIST Standard Reference Database* 4.

Hillert, M. 1980. Empirical methods of predicting and representing thermophysical properties of ternary solution phases. *CALPHAD* 4: 1-12.

Holloway, J. R. 1977. Fugacity and activity of molecular species in supercritical fluids. In *Thermodynamics in Geology*. Dordrecht: Reidel. 161-181.

Kohler, F. 1960. Zur Berechnung der thermodynamischen Daten eines ternären Systems aus den zugehörigen binären Systemen. *Monat. Chem.* 91: 738-740.

Muggianu, Y.-M., M. Gambino & J.-P Bros. 1975. Enthalpies de formation des alliages liquides bismuth-étain-gallium a 732 K. Choix d'une représetation analytique des grandeurs d'excès intégrals et partielles de mélange. *J. Chim. Physique* 72: 83-88.

Water-Rock Interaction, Kharaka & Maest (eds) © 1992 Balkema, Rotterdam. ISBN 90 5410 075 3

Calculation of the solubility of minerals in H_2O-CO_2 solutions at mid and lower crustal depths

John V.Walther
Department of Geological Sciences, Northwestern University, Evanston, Ill., USA

ABSTRACT: In order to understand the process of metamorphism, we need to be able to calculate the solubility of minerals as a function of pressure, temperature and fluid composition. In CO_2 rich environments this is difficult because of the lack of experiments measuring thermodynamic properties in these solutions. It is proposed to use the Born equation to calculate the change in energetics in going from an infinitely dilute H_2O solution to that composed of a mixture of all the uncharged species in solution. These include associated electrolytes as well as the molecular gases such as CO_2 and CH_4. Introduction of molecular gases will lower the dielectric constant, ε, of the solution while the uncharged ion pairs such as NaCl and KCl with their large dipole moment will increase ε (Walther, 1991).

The limited data on KCl association at elevated temperatures in solvent mixtures indicate that a Gilkerson-Fuoss type model (Brady and Walther, 1990) can be used to predict log K of association as ε decreases. Such an analysis gives Born coefficients for charged as well as uncharged species. The long range electrostatic interactions of charged species in the solution can be calculated with a Debye-Hückel type model from the mixed molecular gas and associated electrolyte solvent density and dielectric constant.

Using the approach outlined here one can begin to understand mineral solubilities in CO_2 rich fluids at mid and lower crustal depth.

INTRODUCTION

The most common approach to understanding mineral solubilities and solution species in metamorphic fluids has been to obtain the molal Gibbs free energy of each species in a hypothetical solution where each molecule of the species has the energetic properties as though it is in a pure H_2O solution at the temperature and pressure of interest by making use of pure H_2O equilibrium constants for ion association. Association of ions to form species of lower charge becomes particularly important above the H_2O critical temperature where the dielectric properties of H_2O are smaller and molecular gas species with small dielectric constants such as CO_2 can occur in abundance. For charged species a Debye-Hückel type formulation with an extended term in ionic strength is then used to calculate the changes in Gibbs free energy with changing solution composition from this pure H_2O solution. Using this approach, one also needs to calculate the Gibbs free energy change of uncharged species as the solution departs from a pure H_2O solution with addition of uncharged ion pairs or molecular gas species.

In fluids with large concentrations of CO_2 using this procedure to obtain the molal Gibbs free energy in solution is unreliable, because CO_2 doesn't contribute to the ionic strength, yet changes in its concentration dramatically affects the properties of charged as well as uncharged species in solution.

IONIC ASSOCIATION IN PURE H_2O

As discussed in Brady and Walther (1990) a Fuoss (1958) type equation, based on simple electrostatics of ion pair formation in a dielectric continuum, can be used to calculate the logarithm of the equilibrium constant for association, log K_a, of ion pairs in pure H_2O. A term, K_a', suggested by Gilkerson (1970) needs to be included to account for non-electrostatic solvent-solute interactions:

$$\log K_a = -2.60 + 3\log d + 72{,}800 Z^+ Z^- /(d\varepsilon T) + \log K_a' \quad (1)$$

whrere d is the center to center distance in Å between ions in the complex, Z stands for the ionic charge of the superscripted ion, T is temperature in K and ε represents the dielectric constant of the fluid. Walther and Schott (1988) and Brady and Walther (1990) have demonstrated that d and K_a' can be regarded as independent of temperature at constant pressure in supercritical solutions so that log K_a at constant pressure is a linear function of $1/(\varepsilon T)$.

ε can be calculated with the Kirkwood (1939) equation:

$$(2\varepsilon+1)(\varepsilon-1)/9\varepsilon = 4\pi N^o/3M(\alpha+\mu^2 g/3\kappa T) \qquad (2)$$

where M, α, μ, and **g** are the molecular weight, molecular polarizability (1.444 x 10^{-24} cm^3 for H_2O), molecular dipole moment (1.84 x 10^{-18} statcoulomb cm^{-1}), and the Kirkwood correlation factor for the species, respectively. N^o stands for Avogadro's number and κ represents Boltzman's constant. **g** for H_2O can be calculated from the equation given by Pitzer (1983):

$$\mathbf{g} = 1 + 2.68\rho + 6.69\rho^5 ((565/T)^{0.3} - 1) \qquad (3)$$

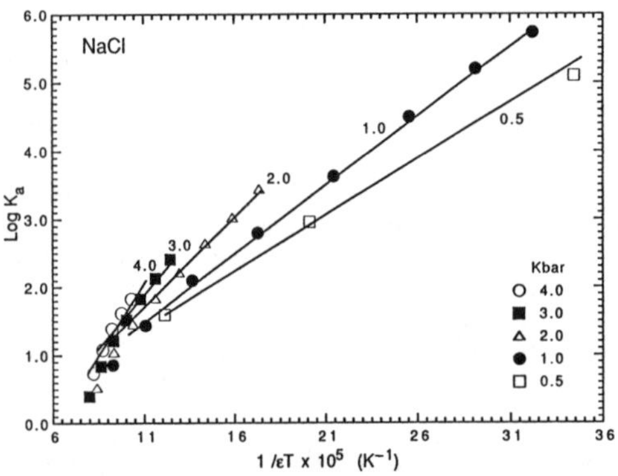

Fig. 1. Logarithm of the association constant of NaCl in H_2O as a function of the inverse of the dielectric constant of H_2O times temperature in K at constant pressure. Symbols are the association constants from Quist and Marshall (1968).

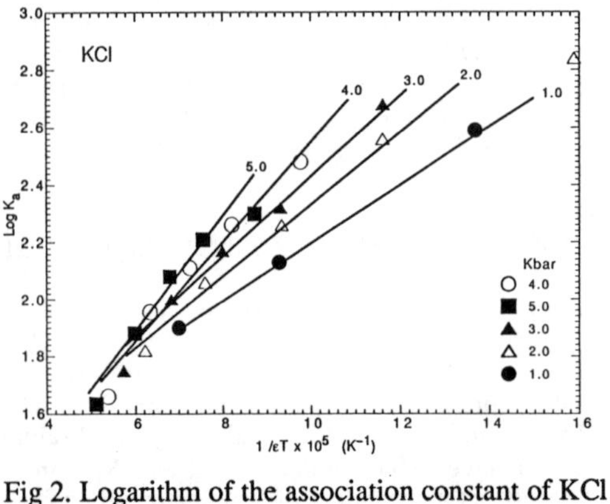

Fig 2. Logarithm of the association constant of KCl in H_2O times temperature in K at constant pressure. Symbols are the association constants from Ritzert and Franck (1968).

where ρ is the density of H_2O in gm cm^{-3}. Note in fig. (1) and (2) that log K_a of NaCl and KCl at constant pressure are good linear functions of $1/(\varepsilon T)$.

ION ASSOCIATION IN H_2O-AR MIXTURES

We can evaluate Hartmann and Franck's (1969) measurements of log K_a for KCl in H_2O-Ar mixtures by calculating ε of the mixture, ε_{mix} from the equation of Looyenga (1965):

$$\varepsilon_{mix} = ((\varepsilon_{Ar}^{1/3} - \varepsilon_{H_2O}^{1/3})V_f^{Ar} + \varepsilon_{H2O}^{1/3})^3 \qquad (4)$$

where ε_{Ar}, ε_{H_2O}, and V_f^{Ar} are the dielectric constant of Ar and H_2O and the volume fraction of Ar in the solution, respectively .

Shown in fig. (3) are the measured values of log K_a for KCl in H_2O-Ar mixtures as a function of $1/(\varepsilon_{mix}T)$. While there is significant scatter of log K_a, the values tend to cluster around the pure H_2O values. The problem in determining the relationship between log K_a and $1/(\varepsilon_{mix}T)$ with addition of Ar is assessing the uncertainty of both calculating ε_{mix} and determining K_a in H_2O-Ar mixtures. As a first approximation it is probably best to assume that values of log K_a for KCl at similar values of $1/(\varepsilon_{mix}T)$ and pressure are the same. This implies that electrostatic interactions prevail and short range solute-solvent interactions are similar in H_2O-Ar solutions when $1/(\varepsilon_{mix}T)$ and pressure are similar. If this is the case, we can determine the association constant of KCl in any fluid mixture with H_2O for which ε_{mix} can be determined.

Note that K_a, by definition of the standard state,

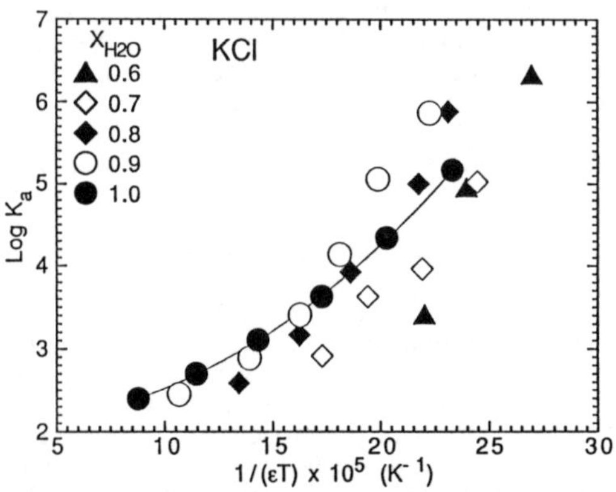

Fig. 3. Logarithm of the association constant of KCl at 440 °C as a function of the inverse of the dielectic constant of H_2O-Ar mixtures times temperature in K. Symbols are the association constants of Hartmann and Franck (1969) at the indicated mole fraction of H_2O in the solution,

is a function of CO_2 concentration at constant temperature and pressure. The standard state for aqueous species is then defined as a mole of particles at infinite dilution in 1000 gms of a mixed solvent of uncharged species at the temperature and pressure of interest. The difference of Gibbs free energy per 1000 gms of solvent between this standard state for ions and that of the more common standard state of infinite dilution in pure H_2O, ΔG_{Born}, can be evaluated with the Born (1920) equation:

$$\Delta G_{Born} = \omega(1/\varepsilon_{mix} - 1/\varepsilon_{H2O}) \quad (5)$$

where ω is the Born coefficient which is independent of concentration but can vary as a function of temperature and pressure. One can compare the values of ω calculated here with those reported by Fein and Walther (1989) and Walther and Schott (1988).

ION ASSOCIATION IN H_2O-CO_2 MIXTURES

One can calculate the ε_{mix} of H_2O-CO_2 mixtures using the procedure outlined above for H_2O-Ar mixtures. These calculations indicate that at 2 kbar and constant temperature between 400 and 600 °C K_a for KCl increases about 2 orders of magnitude while that of NaCl increases by over 4 orders of magnitude as the CO_2 mole fraction increases from 0.0 to 0.35. Such changes in log K_a will have dramatic affects on the solubility of minerals in CO_2 rich environments.

CONCLUDING REMARKS

Using the approach outlined here one can compute the thermodynamics of solute species association in aqueous fluids with large concentrations of CO_2 or other molecular gases. In such solutions, because of the increase in K_a as ε_{mix} decreases, the concentration of charged species is lower than it would be in pure H_2O at the same pressure and temperature, This association will increase the concentration of ion pairs like KCl and NaCl. Because of the large dipole moments, ion pairs will have the opposite effect on ε_{mix} as does CO_2 or CH_4 and increases ε_{mix} somewhat. If this effect is large then it must be accounted for in calculating the distribution of species. These changes in ε_{mix} should be taken into account before the effects of charged species are computed. Therefore, the use of a Debye-Hückel type model to account for the energetics of charged species in solution should use the ε_{mix} and ρ_{mix} of all the uncharged species in solution.

To calculate mineral solubilities in these fluids requires that the chemical potential of k components at the pressure and temperature of interest be solved from a set of k heterogeneous equations (e.g. Van Zeggeren and Storey, 1970). With increasing temperature as depth or proximity to an igneous intrusion increases, carbonate minerals will dissolve to a greater extent in metamorphic fluids and the concentration of carbon bearing species such as CO_2 will increase. It is only by using approaches such as the one outlined here that the thermodynamic properties of CO_2 rich solutions in the earth's crust are going to be understood and solubilites of minerals reliably calculated.

REFERENCES

Born, M., (1920) Volumen und Hydrationswarme der Ionen. Zeitschrift Physikica, 1, 45-48.

Brady, P. V. and Walther, J. V. (1990) Algorithms for predicting ion association in supcritical H_2O fluids. Geochim. Cosmochim. Acta, 54, 1555-1561.

Fein, J. B. and Walther, J. V. (1989) Portlandite solubilities in supercritical Ar-H_2O mixtures: Implications for quantifying solvent effects. Amer. J. Sci., 289, 975-993.

Fuoss, R. M. (1958) Ionic association. III. The equilibrium between ion pairs and free ions. J. Amer. Chem. Soc., 80, 5059-5061

Gilkerson, W. R. (1970) The importance of the effect of the solvent dielectric constant on ion-pair formation in water at high temperatures and pressures. J. Phy. Chem., 74, 746-750.

Hartmann, D. and Franck, E. U. (1969) Elektrische Leitfähigkeit wässriger Lösungen bei hohen Temperaturen und Drucken. III Kaliumchlorid in überkritischen Wasser-Argon-Mischungen. Berichte Bunseng. phy. Chem., 73, 514-521.

Kirkwood, J. (1939) The dielectric polarizability of polar liquids. J. Chem. Phy., 7, 911-919.

Looyenga, H. (1965) Dielectric constants of heterogeneous mixtures. Physica, 31, 401-406.

Pitzer, K. S. (1983) Dielectric constant of water at very high temperature and pressure. Nat. Acad. Sci. Proc., 80, 4575-4576.

Quist, A. S. and Marshall, W. L. (1968) Electrical conductances of aqueous sodium chloride solutions from 0 to 800 °C and at pressures to 4000 bars. J. Phy. Chem., 72, 684-703.

Ritzert, G. and Franck, E. U. (1968) Elektrische Leitfähigkeit wässriger Lösungen bei hohen Temperaturen und Drucken. I KCl, $BaCl_2$, $Ba(OH)_2$, und $MgSO_4$ bis 750 °C und 6 kbar. Berichte Bun. Phy. Chem., 72, 798-808.

Van Zeggeren, F. and Storey, S. H. (1970) The Computation of Chemical Equilibrium: Cambridge Mass., Cambridge University Press, 425p.

Walther, J. V. (1991) Determining the thermodynamic properties of solutes in crustal fluids. Amer. J. Sci., 291, 453-472,

Walther, J. V. and Schott, J. (1988) The dielectric constant approach to speciation and ion pairing at high temperature and pressure. Nature, 332, 635-638.

Water-Rock Interaction, Kharaka & Maest (eds) © 1992 Balkema, Rotterdam. ISBN 90 5410 075 3

Thermodynamic modeling of Be-mineralization: Preliminary calculations

Maria Woitsekhovskaya
Institute of Mineralogy and Petrography, Novosibirsk, Russia

ABSTRACT: The importance of investigating the mechanism of the dissolution, transport and deposition of the beryllium minerals such as beryl, danalite, helvite in hydrothermal systems is self-evident. Multicomponent heterogeneous chemical equilibrium computations were used to simulate Be-transport and precipitation of Helvite-group minerals in hydrothermal systems. The Be-containing hydrothermal system was investigated as an 11-component system (H, O, Na, Cl, F, Be, Mn, S, Si, Al, Fe) between 300° to 500°C at P = 1000 bar. Calculations were accomplished with the program GIBBS, which is based on Gibbs energy minimization. To model the solubility of beryllium-bearing phases, the method of reaction progress was used. The results of these calculations show that the application of modeling to characterize mineral zonation, transport and precipitation of metal provides a useful means for examining the T-pH-aO_2-aS_2 paths of an individual deposit. Because beryllium minerals most commonly occur in metasomatic rocks, perhaps their most useful petrologic attribute is as indicators of the activities of sulfur and oxygen of a hydrothermal system.

1 INTRODUCTION

The general formula of helvite is (Fe, Zn, Mn)$Be_6(SiO_4)_6S_2$ (Hassan et al. 1985). A solid solution is formed between three end members: helvite (Mn), danalite (Fe) and genthelvite (Zn).

Recent studies have shown that Be deposits containing Helvite-group minerals can be considered a world Be resource, similar to beryl or bertrandite-phenakite deposits. Helvite mineralization may take place over a wide range of temperatures, pressures and fluid compositions, comprising diverse environments that extend from magmatic to lower temperature hydrothermal mineralization. Most of the helvite mineralization appears to be present in hydrothermal paragenetic association as Bilal and Fontielles (1991) have suggested:

a) Feldspathization zones: danalite appears in zones of progressive albitization of the Sucuri granite (subprovince of Rio-Parana, State of Goias, Brazil). It is associated with biotite, sulfides and sometimes fluorite;

b) Different types of hydrothermal veins, pegmatite and greisen where helvite minerals are associated with albite, phenakite, sulfide, and quartz.

Deposition of helvite minerals occurs in response to changes in the chemical and physical environment through which the hydrothermal fluid are migrating. Helvite formation is mainly controlled by the sulphur activity of the fluid.

The importance of the theoretical approach to the study of the genesis of helvite-containing deposits and experimental systems is self-evident. The rapid accumulation and increased sophistication of field and laboratory observations combined with theoretical modeling based on experimental investigation of hydrothermal processes increases the understanding of ore deposit genesis.

The objective of this study was to verify the possibilities of using theoretical techniques to characterize the solubility of Helvite-group minerals in hydrothermal solutions by comparing the obtained results with experimental data and the observed mineralogical zonation in Be deposits.

2 CALCULATIONS AND RESULTS

Multicomponent heterogeneous chemical equilibrium computations were used to simulate Helvite-group mineral solubility, Be transport and precipitation.

The 11-component theoretical system (H-O-

Na-Cl-F-Be-Mn-S-Si-Al-Fe) was investigated between 300° to 500°C at 1 Kbar. The system is described by 37 solid phases (minerals) and 64 solutes.

Calculations were accomplished using the GIBBS program of Shvarov (1976). This program is designed to model various chemical interactions in heterogeneous systems and is based on minimizing Gibbs energy. Available, experimentally-derived thermodynamic data were used in conjunction with theoretical extrapolation and prediction methods. The method of reaction progress was applied to calculate the speciation of Be and the solubility of helvite and danalite minerals.

High temperature data exist for a relatively large number of aqueous beryllium species, including hydroxy complexes (Renders and Anderson 1987), fluoride complexes (Mesmer and Baes 1969), and mixed complexes (Soboleva et al. 1984). These were employed by Wood (1991) to predict the speciation of Be and the solubility of bertrandite-phenakite minerals in hydrothermal solutions. Wood underlined the importance of Cl- complexes with increasing temperature compared with the simple ion and the fluoride and hydroxide complexes. According to Burt (1980), Fursenko (1989) and Bilal (1991) helvite is the stable phase at temperatures > 300°C and danalite above 450°C.

The next step was to estimate the solubility product of helvite and danalite under given hydrothermal conditions. Finally, the speciation of BE and the solubility of Be-containing minerals were calculated in hydrothermal solutions with a total chloride concentration of 1-2 molal and total fluoride concentrations of 1 molal, pH 2.3-7.8. Theoretical chemical compositions of helvite and danalite were used for calculations as well as the natural composition of danalite.

The first observation that can be made from speciation and solubility calculations is that mixed hydroxo-fluoride complexes and hydroxo-complexes are generally stable in aqueous solutions at near-neutral pH and at temperatures between 300° to 500°C and play an important role in hydrothermal beryllium transport. Furthermore, the solubility of Be as the chloride or fluoride complexes at a given pH decreases with temperature, owing to increased association of fluoride to form $HF_{(aq)}$ or $HCl_{(aq)}$. In the presence of excess chloride, $BeCl^{+}$ should become more stable, but the concentrations of the chloride complex is still unimportant over most geologically reasonable pH ranges.

Calculated data have shown that in a reducing hydrothermal solution the stability of danalite could be summarized by stating that at low H_2S activities, danalite oxidizes to a mixture of magnetite, phenakite, troilite and quartz as shown theoretically by the following equation:

$$2Fe_4Be_3(SiO_4)_3S + O_2 =$$

$$6\ FeS + 3Be_2SiO_4 + 3SiO_4 + Fe_2O_3 + 0.25S_2$$

Calculated data have shown that danalite has an extremely narrow stability field centered near the pyrrotite-magnetite field boundary. Outside this field, phenakite is stable with magnetite and quartz. The unique feature of danalite, as compared with Zn and Mn end members of the Helvite group, is its sensitivity to oxidation. The cooling of hydrothermal solutions is also important for danalite deposition because its field of stability lies above 450°C.

On the contrary, helvite forms in a large domain of oxygen and sulphur fugacity. The calculated data of speciation of Be and solubility of danalite-helvite minerals are presented in Table 1. These data are compatible with those that have been obtained experimentally by Fursenko (1989) and with observed mineralogical assemblages suggested by Bilal and Fonteilles (1991).

CONCLUSIONS

The fields of stability of helvite and danalite minerals occur in various $loga_{S2}$-a_{O2}-pH-T spaces. Helvite precipitates from near-neutral solutions at low activities of total sulfur and elevated activities of hydrogen at T>300°C and most commonly occurs under given hydrothermal conditions. Furthermore, the domain of danalite stability is very limited by sulphur and oxygen activities. Danalite is stable only above 450°C. Finally, it appears that the compositional zonation of Helvite group minerals could control physicochemical parameters of the hydrothermal depositional environment.

The predicted behavior of Be complexes was briefly discussed above. The mixed hydroxo-fluoride complexes can contribute up to 100 ppm Be at 450°C, if total fluoride is sufficiently high (1.0 molal). In most geologic environments, externally imposed chemical potentials govern the stability of beryllium minerals. Thus, they may be good indicators of the sulfur and oxygen activities and provide a suitable basis for the prediction of equilibria in more complex systems. Silica and alumina activities, for example, provide useful

Table 1. Predicted physicochemical parameters of Be speciation and precipitation of Helvite group minerals (HF 1.0, NaOH 1.0, HCl 1.0 mole/kg H_2O.

450°C 1000 bar					
products of danalite dissolution					
	mg	mg	Q+mg	Q+mg	Q+mg+phen+trl+dan
pH	3.7	3.7	4.1	4.2	4.6
f/r	10^{-3}	10^{-2}	10^{-1}	$10^{-0.5}$	10^{0}
$lgaO_2$	-24.0	-26.0	-28.0	-28.0	-29.0
$lgaS_2$	-3.0	-2.0	-2.0	-1.0	-1.0
lgaBeOHF	-3.0	-2.0	-1.0	-1.0	-1.0
$lgaBe(OH)_2F$	-3.0	-2.0	-2.0	-2.0	-2.0
$lgaBe(OH)_3^-$	-4.0	-3.0	-3.0	-3.0	-3.0

Table 1. (continued, HF 1.0, NaOH 2.0, HCl 1.0 mole/kg H_2O.

450°C 1000 bar					
products of helvite dissolution					
	tph	tph+helv	tph+helv	Q+tph+hel	Q+tph+helv+phen
pH	8.3	8.3	8.3	8.2	6.6
f/r	10^{-3}	10^{-2}	10^{-1}	$10^{-0.5}$	10^{0}
$lgaO_2$	-22.0	-26.0	-24.0	-24.0	-21.0
$lgaS_2$	-13.0	-4.0	-6.0	-7.0	-7.0
lgaBeOHF	-5.0	-2.0	-2.0	-1.0	-1.0
$lgaBe(OH)_2F$	-5.0	-3.0	-2.0	-2.0	-3.0
$lgBe(OH)_3$	-6.0	-4.0	-3.0	-3.0	-4.0

Q:quartz, mg:magnetite, trl:troilite, danl:danalite, phen:phenakite, tph:tephroite, hel:helvite.

constraints for Be mineral assemblages and require further investigations.

REFERENCES

Burt, D.M. 1980. The solubility of danalite $Fe_4Be_3(SiO_4)_3S_2$. Am. Mineral. 65: 355-360.

Bilal, E. & M. Fonteilles 1991. Chemical properties of Helvite group minerals in different types of occurrences. Source, Transport and Deposition Metals. Proceedings of the 25 years SGA Anniversary Meeting. Rotterdam:Balkema.

Fursenko, D.M. 1989. Uslovija sinteza mineralov gruppy gel'vina. Izd. Nauka, Sibirskoe Otdelenie, Novosibirsk: 76 p.

Hassan, I. & H.D. Grundy 1985. The crystal structures of helvite-group $(Mn,Fe,Zn)_8Be_6(SiO_4)_6S_2$. Amer. Mineral. 70: 186-192.

Mesmer, R.E. & C.F. Baes, Jr. 1969. Fluoride complexes of beryllium (II) in aqueous media. Inorg. Chem. 8: 618-626.

Renders, P.J. & G.M. Anderson 1987. Solubility of kaolinite and beryl to 573 K. Appl. Geochem. 2: 193-203.

Shvarov, Y.V. 1976. Algorithm for calculations equilibria compositions in a multicomponent heterogeneous system. Dokl. Akad. Nauk SSSR. 229.5: 1224-1225.

Soboleva, G.I., I.A. Tugarinov, N.S. Golitsina & I.L. Khodakovskiy 1984. The behavior of beryllium in fluorine-bearing hydrothermal solutions at 150-250°C. Geochem. Int.: 20-30.

Geochemical modeling

Solute transport

Water-Rock Interaction, Kharaka & Maest (eds) © 1992 Balkema, Rotterdam. ISBN 90 5410 075 3

The natural evolution of geothermal systems in light of infiltration fronts and a look back at fluid-to-rock ratios

Peter Blattner
Geosphere Research, DSIR, Lower Hutt, New Zealand

Keith R. Lassey
Atmosphere Research, DSIR, Lower Hutt, New Zealand

ABSTRACT: Persistent infiltration eventually brings porous rock into isotopic and chemical equilibrium with the fluid. Dropping $\delta^{18}O$ values clock up fluid throughput, and although the dropping away may mimic a box model, it is the effect of a passing isotope geochemical front. For all components the basis for calculated cumulative fluid throughputs is front progress. For rapid equilibration (high Damköhler numbers) a large advance portion of invading fluid may be buffered by rock sufficiently to be completely untraceable by geochemical means. For several New Zealand systems $\delta^{18}O$ values and geothermal gas chemistry appear to evolve in parallel.

1 BOX (0-D) MODELS

Active, high throughput geothermal systems are among the most inspiring examples of fluid-rock interaction. While there clearly has to be a mass balance between the oxygen isotopic alteration of the recharge water and that of the rock, little quantitative use has been made of that relationship. Clayton, Muffler and White (1968) used a closed (zero dimensional, 0-D) box model on the Salton Sea system and Taylor (e.g. 1977) did the same basic calculation for a totally open box. The hydrogen contents of rock have too often been considered to be negligible, whereas in fact δD is simply one of the earliest signals of infiltrating fluids to imprint itself. Exactly as with $\delta^{18}O$, the fluid δD will depend on the provenance and pre-history of the fluid. Box models always require the contents of the box to be thoroughly mixed and have been discussed in detail in Blattner (1985) and Blattner and Lassey (1990). For the ideal, completely open box, the cumulative volumetric fluid-to-rock ratio equals

$$\Phi_V = 1.3 \ln [(\sigma_w - \sigma_R)/\sigma_w] \qquad (1)$$

where σ denotes an isotope shift from an initial composition δ_i to a final value δ_f (so that $\sigma = \delta_f - \delta_i$), w refers to the fluid at a given time and R to the overall rock reservoir. The sum σ_w-σ_R equals the total isotopic shifting potential P. In such a model, the incremental isotope shift of the discharge water at time t is a precise indicator of the integrated or cumulative opposing shift suffered by the reservoir rock up to time t.

2 FLOW (1-D) MODELS AND INFILTRATION FRONTS

Unlike in manufacturing and cooking, fluids in the lithosphere will generally follow 1-D flow lines and interact with a long succession of rock cross sections. Our general view of fluid rock interaction should therefore be based on equations of motion or transport and initially we learn most by assuming instantaneous equilibration throughout. Looking at the infiltration of homogeneous permeable rock by a fluid with a constant externally controlled isotopic composition, such an equation may look as follows

$$\frac{\partial}{\partial t} \delta_f + \Psi q \frac{\partial}{\partial z} \delta_f = 0 \qquad (2)$$

where z is infiltration distance (z=0 at time t=0), q the interstitial fluid velocity (Darcy velocity divided by porosity) and Ψ is the ^{18}O-porosity or approximately the oxygen porosity, i.e., the volumetric fraction of total oxygen that resides in the fluid. Now consider what happens. In the case of a completely inert rock, or if a tracer is rejected by the solid (e.g. He) the tracer front will move exactly with the velocity q of the physical fluid. On the other hand, if the fluid, at z=0, starts to share the tracer with the rock, the tracer front will lag behind. For a chemical tracer it is easy to see just how slowly that front will move, namely, with the fluid velocity q times the proportion of tracer that is hosted by the fluid, or (with φ = porosity, ρ = density, C = concentration, f = fluid, s = solid)

$$F_d = \frac{\varphi \rho_f C_f}{\varphi \rho_f C_f + (1-\varphi)\, \rho_s C_s} \tag{3}$$

For ^{18}O, some algebra turns this into

$$F_d = \Psi \approx \frac{1}{1 + \frac{1-\varphi}{\varphi} \frac{\rho_s}{\rho_f} K_d\, \alpha} \tag{4}$$

which gives also the position of the ^{18}O infiltration front obtained by solving eq (2) (Blattner and Lassey 1989). K_d is the carrier (oxygen) distribution coefficient. F_d or Ψ is also the reciprocal retardation factor of chromatography.

For a given input fluid and aquifer there will therefore arise a sequence of chemical and isotopic fronts lagging behind the actual infiltration front at unit distance. For each species such a front subdivides the flow path into a fluid dominated and a rock dominated section (Blattner and Lassey 1989). First we could expect negative (wash out) fronts for inert gases or soluble salts, and these would be followed by

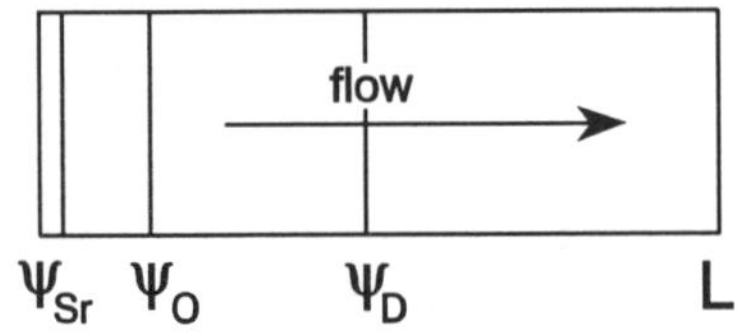

Fig. 1 Sequence of fronts according to effective porosities Ψ, starting from aquifer entrance on left. Porosity 20%, L = fluid front at unit length, approx. Ψ for respective stable isotopes.

D/H, $^{18}O/^{16}O$, and $^{87}Sr/^{86}Sr$ fronts among others (Fig. 1). The passage of a front at a certain distance determines fluid-to-rock ratios for any given section of aquifer. With reference to the aquifer length already traversed by the compositional front, the fluid to rock ratio is $1/\Psi$, wrt half that length $2\ \Psi^{-1}$, and for 1/n of front progress

$$W/R = \Phi = n\ \Psi^{-1}. \tag{5}$$

3 REAL FRONT SHAPES

If, during infiltration, equilibrium is not established rapidly, and in case of dispersion in the fluid, fronts will become diffuse or 'flattened' (Fig. 2).

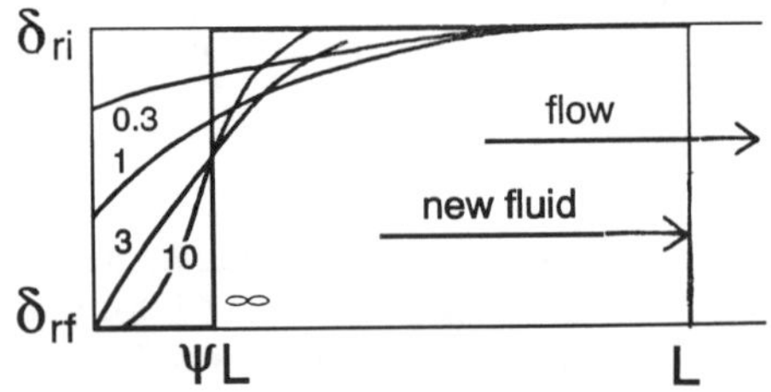

Fig. 2 Isothermal isotopic front profiles as a function of Damköhler number $N_D = \kappa L/q$.

Profile shapes, but not front positions, depend on the Damköhler and Péclet numbers, N_D and P_e, which are the non-dimensionalised ratios between a kinetic exchange rate constant κ, or a diffusivity, and actual fluid velocities (Lassey and Blattner 1988, 1989; Bickle and McKenzie 1987; Blattner and Lassey 1989).

Even a fragment of actual compositional profile may, in a given aquifer frame, determine the Damköhler number for a given infiltration and that in turn would determine the total, reacted and unreacted fluid throughput. In this way Blattner and Absar (1988) in detailed modelling of the Ohaaki-Broadlands geothermal field, estimated a $\delta^{18}O$ front position and N_D = 4, which put the fluid infiltration front about 12 times 'ahead' of the surface discharge. The fluid throughput for the system behind (below) the discharge area was thus 12 times the pore space, and the distance-dependent fluid-to-rock ratio, $12\varphi/(1-\varphi)$. Cf Fig. 4.

4 FLAT FRONTS AND CHANNELLING - THE BOX MODEL FAKED

Channelled flow effectively removes part of the rock from circulation, thereby increasing the porosity of the remainder. The message of the fluid is carried further into the aquifer, but leaves a weaker signal in the average rock. The effect is that a front becomes still more weakened than by kinetics and dispersion alone, and the average rock behind the front will reach equilibrium usually much later than for a homogeneous permeable solid.

For small N_D and P_e, and especially when channelling seems to further weaken it, the δ-profile of a given section of aquifer may resemble a horizontal level, dropping away like that of a box model. Now, however, the flatness is not due to cumbersome mixing in a box, nor to the small size of an elemental volume, but is an artefact of exchange kinetics, dispersion and channelling. (Only infinitely small values of N_D and P_e, especially the latter, would simulate complete mixing in the whole aquifer-"box".)

The errors afflicting 0-D box modelling are such that one could accidentally arrive at almost any throughput value, including a true one. However, the intercomparison of geologically similar systems with short aquifers can be valid, in that hydrodynamic and exchange kinetic shortcomings may cancel out and a realistic order of evolution may emerge.

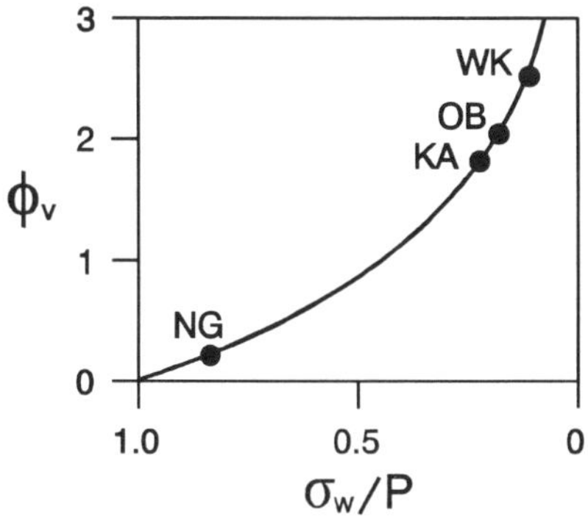

Fig. 3 Cumulative H_2O throughput of New Zealand geothermal systems according to open box model, eq 1 (Blattner 1985).

5 NGAWHA, KAWERAU, OHAAKI-BROADLANDS AND WAIRAKEI (NG, KA, OB, WK)

Based on the 0-D model of eq 1, Fig. 3 shows throughputs of several greywacke-hosted geothermal systems of New Zealand, based on eq 1. In the case of Ngawha (NG), σ_r is so small that there is no actual evidence for a meteoric water input "as yet", whereas in Wairakei (WK) most evidence for original magmatic and/or pore

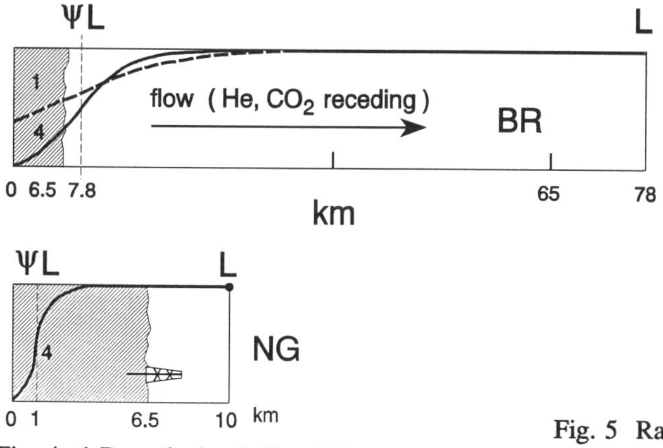

Fig. 4 1-D analysis of Ohaaki-Broadlands with Ngawha for comparison. Ψ is for $\delta^{18}O$. Fronts of degassing may lie well in advance of $\delta^{18}O$ fronts. Damköhler numbers marked.

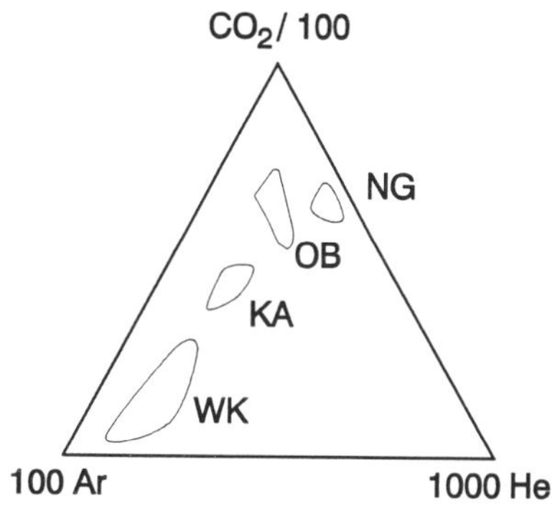

Fig. 5 Ratios of Ar, CO_2 and He in geothermal gas discharges from the systems in Fig. 3 according to Giggenbach (1989). The evolution goes from CO_2 rich subduction-related to Ar rich meteoric-dominated gases.

waters has by now been washed out. The box model thus yields a sequence of integrated fluid throughputs that corresponds to a very simple natural evolution for these systems.

For Ohaaki-Broadlands (OB) more detailed work is available in Blattner and Absar (1988) and Fig. 4 shows the more realistic 1-D interpretations of this field and of Ngawha (adapted from 0-D data). It is easy to see that a new batch of magma at depth may 'rejuvenate' a system with regard to magmatic components.

On the basis of gas contents of geothermal discharges Giggenbach (1986, 1989)also finds a compositional sequence for New Zealand geothermal systems including the above four. It is of interest here that in his Ar-CO_2-He and Ar-CO_2-N_2 plots the sequence is NG, OB, KA, WK (Fig. 5) and according to Giggenbach stretches from close to a "magmatic subduction gas" for Ngawha, towards the Ar rich corner, dominated by meteoric water. Here, too, initial contributions and residues of magmatic fluid seem to have been washed out by increasing meteoric throughput. KA and OB are intermediate in Fig. 3 and their relative positions in terms of $\delta^{18}O$ are not well defined. On the whole, for the four systems available for comparison, evolution of $\delta^{18}O$ of water on one hand and of gas chemistry on the other complement each other remarkably well: the system output becomes increasingly surface dominated.

6 CRUSTAL BUFFERING

The case of near surface geothermal systems is relatively simple; a lot of surface water does eventually get turned over and fronts, however flat, eventually pass through the discharge area. Beyond that, it has always been tempting to use "exotic" fluids to explain unusual rock compositions, however deep in the crust, by single stage WRI. It is well to remember that such fluids do not appear out of nowhere, and only rarely through pipe-like channels. By the time they reach the desired scene of action they will have been buffered by overwhelming amounts of rock, largely by the rules of 1-D advection and according to effective porosities Ψ.

REFERENCES

Bickle, M.J. and McKenzie, D. 1987. The transport of heat and matter by fluids during metamorphism. Contrib. Mineral. Petrol. 95: 384-392.

Blattner, P. 1985. Isotope shift data and the natural evolution of geothermal systems. Chem. Geol. 49: 187-203. (WRI-4 issue)

Blattner, P. & Absar, A. 1988. Mass balance and Damköhler numbers in flow systems - Fluid throughput at Ohaaki-Broadlands. Proc. 10th NZ Geothermal Workshop, Univ. of Auckland: 17-22.

Blattner, P. & Lassey, K.R. 1989. Stable isotope exchange fronts, Damköhler numbers and fluid to rock ratios. Chem. Geol. 78: 381-392.

Blattner, P. & Lassey, K.R. 1990. Transport of stable isotopes, kinetics, dispersive advection, etc. (Discussion) Contrib. Mineral. Petrol. 105: 486-490.

Clayton, R.N., Muffler, L.J.P. and White,, D.E. 1968. Oxygen isotope study of calcite and silicates of the River Ranch No. 1 well, Salton Sea geothermal field, California. Am. J. Sci. 266: 968-979.

Giggenbach, W.F. 1986. The use of gas chemistry in delineating the origin of fluids discharged over the Taupo Volcanic Zone. Proc. 5th Intern. Volcanol. Congress, Auckland, New Zealand, 47-50.

Giggenbach, W.F. 1989. The chemical and isotopic position of the Ohaaki field within the Taupo Volcanic Zone. Proc. 11th New Zealand Geothermal Workshop, Auckland, N.Z., 81-88.

Lassey, K.R. and Blattner, P. 1988. Kinetically controlled oxygen isotope exchange between fluid and rock in one-dimensional advective flow. Geochim. Cosmochim. Acta 52: 2169-2175.

Lassey, K.R. and Blattner, P. 1989. Unidimensional tranposrt of stable isotopes: the shapes of geochemical fronts. Proc. WRI-6 (Malvern) Balkema, Rotterdam, 421-424.

Taylor, H.P. Jr, 1977. Water/rock interactions and the origins of H_2O in granitic batholiths. J. Geol. Soc. London 133: 509-558.

Water-Rock Interaction, Kharaka & Maest (eds) © 1992 Balkema, Rotterdam. ISBN 90 5410 075 3

Lattice gas automata for flow and transport in geochemical systems

D.R.Janecky, S.Chen, S.Dawson, K.C.Eggert & B.J.Travis
Los Alamos National Laboratory, Los Alamos, N.Mex., USA

ABSTRACT: Lattice gas automata models are described, which couple solute transport with chemical reactions at mineral surfaces within pore networks. Diffusion in a box calculations are illustrated, which compare directly with Fickian diffusion. Chemical reactions at solid surfaces, including precipitation/dissolution, sorption, and catalytic reaction, can be examined with the model because hydrodynamic transport, solute diffusion and mineral surface processes are all treated explicitly. The simplicity and flexibility of the approach provides the ability to study the interrelationship between fluid flow and chemical reactions in porous materials, at a level of complexity that has not previously been computationally possible.

1 INTRODUCTION

Chemical reaction processes are often spatially distributed in both geologic and engineered systems. Intrinsically coupled flow and chemical reaction processes are also a common factor. In geologic systems, a detailed description and simulation capability for mineral-fluid interaction as a function of time is critical to our understanding of a wide variety of geochemical processes, including weathering, diagenesis, and hydrothermal alteration. There are also a variety of practical applications which demand sophisticated modeling capabilities. In petroleum reservoirs, applications of interest include coupling between oil migration, secondary recovery processes, and the evolution and/or manipulation of porosity. Similarly, reactions coupled to ground water flow and contaminant migration are important in environmental systems.

As chemical reaction models have become increasingly comprehensive to include such complexities, it has been recognized that new modeling approaches are required to better understand many pressing issues. Interrelated factors such as rock texture, mineral distribution, multiphase flow, and pore or fracture network geometry all affect the simultaneously interrelated processes of mineral dissolution, deposition, and mass transport. In addition, reaction kinetics can significantly affect the integrated results of such processes (*e.g.* Wells and Ghiorso, 1991). For bulk reactions or large scales of tens to thousands of meters, models of geochemical processes have provided significant insights using averaged or phenomenological descriptions of the permeability, fluid flow fields, mineral distributions, and fluid composition (Bassett and Melchior, 1990; Lichtner, 1988; Ague and Brimhall, 1989; and references therein). There are also models for reactions on the molecular scale (*e.g.* Lasaga and Gibbs, 1987). Between these two scales, however, there is a paucity of general and flexible models of fluid-rock interaction. This intermediate scale is particularly important to a quantitative understanding of both geochemical and flow processes in natural porous media because it is precisely the scale at which most detailed analytical, experimental, and descriptive methods provide information.

Rothman (1988) and Chen and others (1991) examined applications of lattice gas automata (LGA) methods to geological problems by formulating models to simulate flow through complex porous media. Wells and others (1991) examined coupling to the LGA models both surface reactions and fluid transport through flow and diffusion for a single solute. These studies demonstrated the potential of this class of models for examining complex pore network processes explicitly. In this paper, we report further developments and refinements of the techniques.

2 LATTICE-BASED HYDRODYNAMIC MODELS

The core of these models is streaming and collision of "particles" on a lattice. The lattice can be developed in a variety of structures to represent two or three dimensional flow systems (see examples in Diemer and others, 1990). A hexagonal symmetric lattice has been shown to provide a numerical solution to the two-dimensional Navier-Stokes equations (Frisch and others, 1986, 1987; Wolfram, 1986)), and thus is capable of modeling real systems Hydrodynamic operation of the model is divided into two processes: streaming of "particles" from one node to the next during each time step, and calculation of collision results at the new nodes. Constraining the "particles" to single node streaming results in an isothermal model.

The calculations of Wells and others (1991) used the minimal collision 2-D FHP-I model of Frisch and others (1986) in which each lattice node can contain up to six solvent "particles". Exclusion rules allow only one "particle" traveling in any one direction to occupy a node. This model was subsequently upgraded to include a full set of collisions and the possibility of a rest "particle" at each node, which results in more physically reasonable viscosity (Chen and others, 1991). The binary basis of these approaches (a "particle" either exists or is absent at each of the six or seven possible states of a node) results in very fast, parallel computations. However, the models intrinsically are noisy, due to short term fluctuations of density on individual nodes caused by the limited number of accessible states.

To eliminate most of the noise of the LGA systems, lattice Boltzmann equation (LBE) approaches have recently been developed (Chen and others, 1992). In LBE models, real numbers are used instead of bits to represent "particle" distributions. Thus, the LBE method is a finite-difference approach to solving a set of kinetic equations in discrete space and discrete time.

3 CHEMICAL REACTION SUB-MODELS

The simplest models for chemical processes in a porous media involve diffusion in the solvent occupying pores, and reactions at solid surfaces. Our initial interest is in systems involving solutes in an aqueous solvent. Thus, the distinction here between "particles" in the LGA or LBE sense and the chemical perspective of units of solution containing both solvent and solutes. Homogeneous solution transport is commonly assumed for geochemical calculations (*e.g.* Helgeson and others, 1984) and will be used here. However, such single fluid phase systems are only part of the range of interest with respect to chemical interactions to which these methods are being examined, including oil or gas saturated systems or even two-phase systems (Rothman and Keller, 1988; Somers and Rem, 1991; Chen and others, 1991).

In a geologic pore network, chemical reactions at surfaces may take several forms. Surface reactions may involve significant mass transfer via dissolution and/or precipitation, which modifies the pore network structure and thus the hydrodynamic environment. When the components of interest are at minor or trace concentrations, sorption and desorption reactions become important and the pore network may be either fixed or variable. Catalytic reactions of solutes at solid surfaces could also be modeled using this approach.

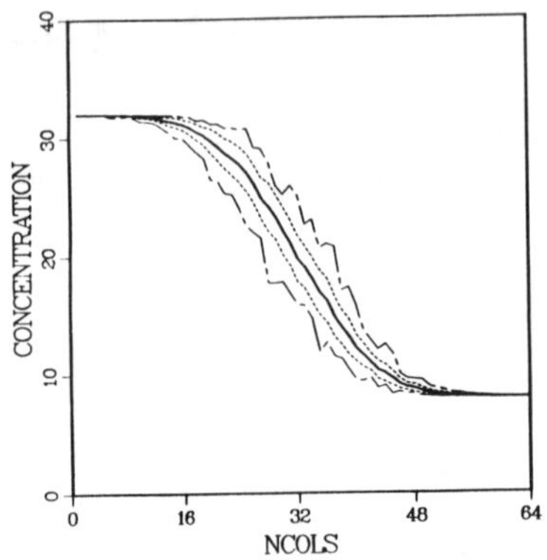

Figure 1. Diffusion profiles calculated by LGA methods at 60 steps for 64^2 box with initial concentration step of 32 to 8 units. Shown are the average profile (solid line), one standard deviation from average for each row (dotted lines), and extrema for each row (dashed lines).

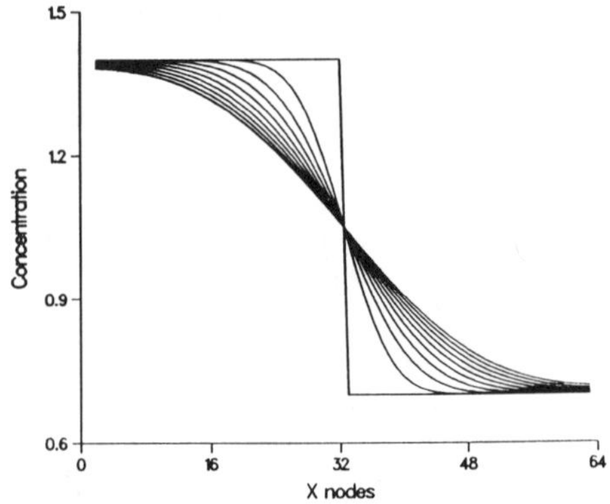

Figure 2. Diffusion profiles calculated using the LBE method for a 64^2 box and initial concentration step of 1.4 to 0.7 units. Shown are concentration profiles for initial conditions and nine subsequent conditions (50 step intervals).

3.1 *Diffusion*

Solutes are transported down a concentration gradient by diffusion. To simulate diffusion in an LGA model, a concentration of a solute is assigned to each open space node (Wells and others, 1991). If the mass of solute or changes in the mass of solute is small relative to the total mass of solution, only minor perturbations of momentum for the lattice "particles" are involved which can be ignored. This approach to solution composition tracking is similar to the energy tracking approach of Sero-Guillaume and Bernardin (1989). In LGA models, each "particle" is assigned a concentration equal to that of the node at which it is residing. During each time step, "particles" travel to new nodes and the concentration of those nodes is assigned the concentration of the arriving "particle". If more than one "particle" arrives at a node during a time step, the resulting concentration at that node becomes the mean concentration of all "particles". During the next time step, all "particles" carry the mean concentration to their next destination. This method of simulating diffusion contrasts with models for inter-diffusion of different types or species of "particles", primarily by considering solvent flow separately from solute diffusion and transport. This distinction between the solvent flow in the form of "particles" and solutes as variable terms attached to the "particles" results in considerable flexibility for the models, including multicomponent solute extensions. This approach was examined by performing a set of calculations of diffusion in a closed box (Figure 1). The initial concentration field consists of a step function in which the concentration equals 32 arbitrary units in the left half of the box, and 8 units in the right half. Mean concentration profiles at different time steps for the LGA calculation are in good agreement with Fickian diffusion profiles calculated using a finite difference technique. Each point on the observed profiles constitute the average value of each column in the box. This calculation also illustrates the level of noise inherent in LGA calculations (as discussed above). Even the average is not completely smooth for a 64x64 node box. This rate of diffusion may be used to delineate the space-time relationships of models that are being applied to real physical problems. For instance, Wells and others (1991) calculated a space to time ratio of 0.002 nodal spacing in meters per second, using a diffusion coefficient of 10^{-9} m^2sec^{-1}.

In comparison to the LGA results, the LBE diffusion calculation produces smooth profiles (Figure 2), with no variations along the diffusion front. Extension of these models to multicomponent-multiphase systems with independent diffusion behavior is relatively straight forward. In fact, the present LBE model includes potential for up to 200 solute components and binary solute reactions (Dawson and Chen, unpublished).

3.2 *Surface reactions*

Surface reactions involving dissolution and precipitation are simulated by allowing wall nodes to serve as sources or sinks for mass of a dissolved component (Wells and others, 1991). Whenever a "particle" collides with a wall, mass may be exchanged, thus increasing or decreasing the local concentration in solution depending upon the saturation state of the fluid. Any kinetic reaction model can be integrated into the lattice models and for this work, a representation of a transition state rate law was applied (Helgeson and others, 1984). In addition, pressure dependent affects on surface reactions as a function of spatial distribution in the pore network are included in the method.

After some number of mass transfer events in an undersaturated fluid, a wall node becomes an open space node, simulating dissolution. Similarly, after some number of supersaturated mass transfer events, an open space node becomes a wall node, simulating mineral precipitation. The number of events required to create or destroy a wall node depends on the relative concentrations of solute species per unit volume in the solvent and solid phases.

Many mineral dissolution reactions are spatially heterogeneous. For example, crystallographic defects are sites of excess strain energy which constitute regions especially favorable to dissolution (Lasaga and Blum, 1986). Such heterogeneous surface reactions can be simulated with a lattice model by defining zones of higher solubility (Figure 3). Over the course of this simulation the entire wall recedes slightly, but the central region dissolves more readily, resulting in an etch pit. The lattice models can effectively model transport-controlled dissolution in which a significant rate-limiting step is the diffusion of the dissolved component away from a mineral surface. This is possible because the model explicitly treats both the fluid and the mineral surface. In this example, the local solution concentrations within an etch pit are much higher than the concentration in the bulk fluid. With increasing diffusion control in the etch pit, the local concentration approaches saturation. Thus, the local probability of mass transfer decreases and continued dissolution relies on the ability of the solute to diffuse out of the etch pit.

4 COUPLED FLOW AND CHEMICAL REACTIONS

An example of a simple chemically reactive porous flow model is shown in Figure 4. The concentration field that

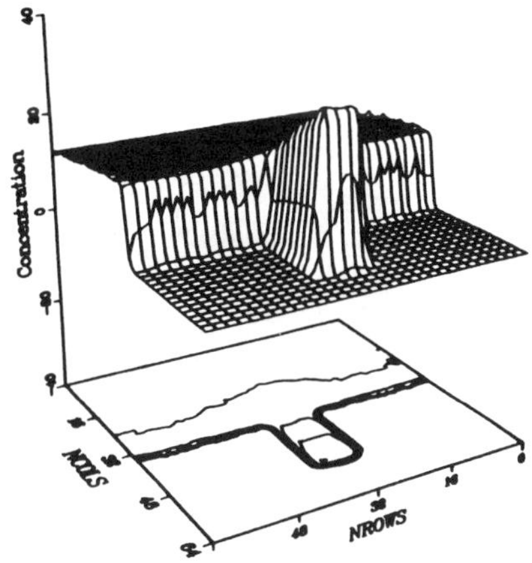

Figure 3. Static dissolution LGA experiment for a hypothetical single component solid into a solution. An 8 node wide line defect in the center of the solid phase dissolves more rapidly to create an etch pit (after Wells and others, 1991).

results from the reaction of a wall mineral with an initially undersaturated fluid is clearly inhomogeneous. The concentration within the principle flow channel is relatively low, whereas the concentration in dead-end pores and around obstructions is relatively high. This type of simulation has important implications for increasing our understanding of the relationship between porosity and permeability in dynamic systems in which the porous structure of a rock is altered by mineral dissolution and growth. The affect of reaction and transport heterogeneities on the overall evolution of fluid composition and rock porosity and permeability is not well understood at this time, but is certainly a function of fluid velocity, rock composition, and the geometry of the porous network. The lattice approaches have great potential to aid in investigations of these phenomena.

5 SUMMARY

We have presented a general approach to lattice models (LGA and LBE) which couple solute transport with chemical reactions at mineral surfaces and in pore networks. The models can treat both surface-controlled and transport controlled mineral reactions because solute diffusion, fluid flow, and mineral surface detachment processes are all treated explicitly. This will be particularly useful for the analysis of diagenetic and alteration processes observed in thin section and hand specimen samples in which fluid flow and mineral reaction processes are heterogeneous in space and time. The simplicity and flexibility of this lattice approaches provides the ability to study the interrelationship between fluid flow and chemical reactions in porous materials at a level of complexity that has not been computationally possible in the past.

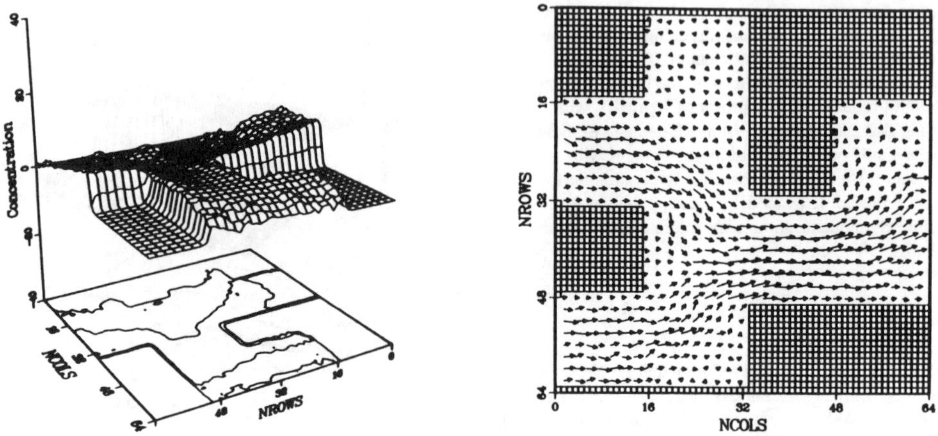

Figure 4. An example of kinetically controlled dissolution reactions in a pore network with flowing solution. Heterogeneous concentration field (a) in a simple structure (b) in which initial and input concentration is zero units with soluble walls (after Wells and others, 1991). In this figure, the vectors represent the local fluid velocity.

ACKNOWLEDGMENTS

This work has been supported by funding from the Institute of Geophysics and Planetary Physics (Los Alamos National Laboratory, University of California), the Geosciences Research Program in US DOE's Office of Basic Energy Science, and the Isotope and Nuclear Chemistry Division of Los Alamos National Laboratory. Funding for developing and evaluating hydrodynamics of fluids in porous media has been provided by the Fossil Energy Program of US DOE (to KCE and SC). Computations and development of these models relies intensively on the parallel processing resources of the Advanced Computing Laboratory of Los Alamos National Laboratory.

REFERENCES

Ague, J.J., and Brimhall, G.H. (1989) Geochemical modeling of steady state fluid flow and chemical reaction during supergene enrichment of porphyry copper deposits. *Econ. Geol.*, **84**, 506-528.

Bassett, R.L., and Melchior, D.C. (eds.) (1990) Chemical Modeling of Aqueous Systems II. *Amer. Chem. Soc., Washington, DC*, vol. **416**.

Chen, S., Diemer, K., Doolen, G.D., Eggert, K., Fu, C., Gutman, S., and Travis, B.J. (1991) Lattice gas automata for flow through porous media *Physica D* **47**, 72-84

Chen, S., Wang, Z., Shan, X., and Doolen, G.D. (in press, 1992) Lattice Boltzmann Computational Fluid Dynamics in Three-Dimensions. *J. Stat. Phys.*

Diemer, K., Hunt, K., Chen, S., Shimomura, T., and Doolen, G. (1990) Density and Velocity Dependence of Reynolds Numbers for Several Lattice Gas Models. in: *Lattice Gas Methods for Partial Differential Equations*, G.D. Doolen, (ed.) Addison-Wesley, 137-178.

Frisch, U., d'Humieres, D., Hasslacher, B., Lallemand, P., Pomeau, Y., and Rivet, J.-P. (1987) Lattice gas hydrodynamics in two and three dimensions. *Complex Systems*, **1**, 649-707

Frisch, U., Hasslacher, B., and Pomeau, Y. (1986) Lattice-gas automata for the Navier-Stokes equation. *Physical Review Letters*, **56**, 1505-1508

Helgeson, H.C., Murphy, W.M., and Aagaard, P. (1984) Thermodynamic and kinetic constraints on reaction rates among minerals and aqueous solutions; II, Rate constants, effective surface area, and the hydrolysis of feldspar. *Geochim. Cos. Acta*, **48**, 2405-2432.

Lasaga, A.C., and Blum, A.E. (1986) Surface chemistry, etch pits and mineral-water reactions. *Geochim. Cos. Acta*, **50**, 2263-2379.

Lasaga, A.C., and Gibbs, G.V. (1987) Applications of quantum mechanical potential surfaces to mineral physics calculations. *Physics and Chemistry of Minerals* **14** 107-117.

Lichtner, P.C. (1988) The quasi-stationary state approximation to coupled mass transport and fluid-rock interaction in a porous medium. *Geochim. Cos. Acta*, **52**, 143-165.

Rothman, D.H. (1988) Cellular-automaton fluids: a model for flow in porous media. *Geophysics*, **53**, 509-518

Rothman, D.H., and Keller, J.M. (1988) Immiscible cellular-automaton fluids. *J.Stat. Phys.*, **52**, 1119-27

Sero-Guillaume, O.E., and Bernardin, D. (1990) A lattice gases model for heat transfer and chemical reaction. *Euro. J. Mech., B/Fluids*, **9**, 177-196

Somers, J.A., and Rem, P.C. (1991) Analysis of surface tension in two-phase lattice gases. *Physica D*, **47**, 39-46.

Wells, J.T., and Ghiorso, M.S. (1991) Coupled fluid flow and reaction in mid-ocean ridge hydrothermal systems: The behavior of silica. *Geochim. Cos. Acta*, **55**, 2467-2481.

Wells, J.T., Janecky, D.R., and Travis, B.J. (1991) A lattice gas automata model for heterogeneous chemical reactions at mineral surfaces and in pore networks. *Physica D*, **47**, 115-123

Wolfram, S. (1986) Cellular automation fluids. I. Basic theory. *J. Stat. Phys.*, **45**, 471-526

Water-Rock Interaction, Kharaka & Maest (eds) © 1992 Balkema, Rotterdam. ISBN 90 5410 075 3

CIRF: A general reaction-transport code: Mineralization fronts due to the infiltration of reactive fluids

J. L. Potdevin
Department of Chemistry and Geological Sciences, Indiana University, Bloomington, Ind., USA and Institut Français du Pétrole and URA, Rueil-Malmaison cedex, France

W. Chen, A. Park, Y. Chen & P. Ortoleva
Departments of Chemistry and Geological Sciences, Indiana University, Bloomington, Ind., USA

ABSTRACT: The numerical reaction-transport code CIRF allows the simulation in two spatial dimensions of mass transfer, porosity and permeability changes and reaction front formation induced by infiltration of reactive fluids in rocks. The code draws from a large thermodynamic and kinetic database for minerals and aqueous species. The initial rock and the inlet fluid composition, as well as the parameters which control the fluid infiltration process are specified by the user. Numerous powerful computational techniques have been developed to increase the speed and the accuracy of CIRF. Most notable of these is the adaptive gridding technique which allows the capture of mm-thick reaction fronts in hundred-meter scale systems. The application of CIRF to dolomitization, genesis of seals in sandstones, acid stimulation of petroleum reservoirs, and genesis of gold deposits are discussed.

1 INTRODUCTION

Mass transfer of chemical elements and reaction fronts induced by the infiltration of reactive fluids in rocks are commonplace in weathering, diagenetic, metamorphic, and igneous environments or in association with petroleum engineering, in situ mining, heap leaching and waste repository evolution. They can lead to dramatic changes in the texture, the composition and the mechano-chemical properties of the initial rock. Here we outline our attempt to study some of these problems via a chemically general, reaction-transport model and simulation code and illustrate its applications to a range of fluid infiltration problems.

2 THE CIRF SIMULATOR

For a simulator to be a practical research tool it must be sufficiently general and flexible. Our code draws from a base of thermodynamic data for more than fourteen hundred minerals and aqueous species. This database contains equilibrium constants for both the mineral dissolution reactions and aqueous reactions. Aqueous reactions are assumed to be fast enough to allow equilibrium to be attained instantaneously everywhere in the fluid. Such equilibrium behavior is not always attained for dissolution of minerals. When experimental kinetics data are available, these reaction rate coefficients and activation energies have been added to the database. The code allows for the input of new reactions and reaction rates, or modified data. The thermodynamic model presently used for the aqueous chemistry is the Debye-Hückel theory. New minerals can nucleate via a nucleation threshold supersaturation law. By allowing for multiple reactions per mineral, the pH or other dependence of rate mechanisms may be accurately modeled. The composition, temperature, the flux direction (lateral, vertical, upward or downward) and influx rate of the fluid imposed is determined by the user; one may also specify with which minerals, if any, the inlet fluid is in equilibrium. The initial distribution of the grain size and number of grains per unit volume may be input by the user using a graphical interface. Permeability is linked to the grain size and to geometrical parameters which describe the mineral grain shape. Original and authigenic minerals have spherical, needle-like, fractal, or other grain morphologies. One or two

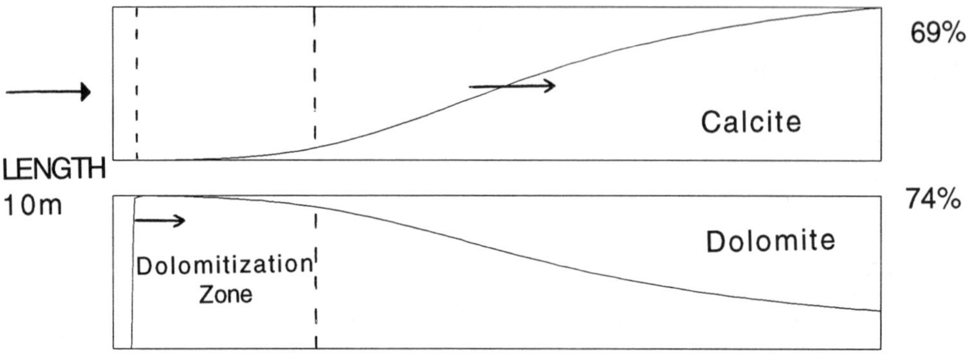

Fig.1 A CIRF simulation on dolomitization. The inlet fluid is undersaturated with respect to both calcite and dolomite. After one thousand years a dolomitic zone develops between the two dissolution fronts. The direction of fluid flow and the reaction fronts movements are shown by the arrows.

spatial dimensional systems may be simulated. The distribution of pressures or fluid velocity at the boundaries of the system may be specified. The spatial distribution of temperature may either be explicitly specified by the user, take on a constant specified geothermal gradient, or calculated by an energy balance equation; in the latter case, the program automatically draws from a large database of specific heats, partial specific heats, and thermal conductivities. A complex subsidence history during the infiltration event could also be introduced. Built-in graphical interfaces allow for easy interpretation of simulation results.

Developing a simulator which is so general requires the use of a number of powerful computational techniques to increase the speed of the program. Most notable of these is the adaptive gridding technique which allows the capture of very narrow reaction fronts or other small-scale features. For example, mm-thick fronts in hundred meter-scale systems can be captured by the program. Equations are solved using finite difference or finite element techniques depending on conditions. Furthermore, the nonlinearity of the problem requires a range of iteration techniques. As these iteration schemes typically converge only when the initial guess is very close to the actual answer, we employ a number of techniques that help the program find a good starting guess. We are continuously developing faster and more reliable algorithms. CIRF, written in ANSI77 FORTRAN, can be executed on VAX, CONVEX, STARDENT, and CRAY mainframe computers and also on powerful workstations, such as SunsparcII, IBM RS6000, and Stardent Titan (Ortoleva et al 1992). The time needed to perform a simulation depends on a number of parameters, including system size, the number of aqueous species and minerals, and the degree to which the imposed fluid is out of equilibrium with the initial rock.

3 CIRF APPLICATIONS

CIRF is used on several infiltration problems involving (1) dolomitization and reaction front fingering in carbonate rocks (Chen and Ortoleva 1990), (2) genesis of seals and reservoir rock in sandstones (Chen and Ortoleva 1992), (3) development of hydrothermal gold deposits, and (4) acid stimulation of sandstone reservoirs using HCl/HF mixtures.

Dolomitization is a typical infiltration problem. Dolomitic rocks form mainly by replacement of calcite when large volumes of magnesium rich water flow through existing carbonate sediments or rocks. Although widely studied, the dolomitization process is not yet completely understood. The infiltrative fluid composition, the role of kinetic control, the location and the time when the process occurs still are controversial topics (Hardie,1987; Braithwaite, 1991). CIRF is a powerful tool to test dolomitization models. A first attempt to study dolomitization has been to simulate a system of ten meter long and one meter wide. Its initial composition was fixed to be 80% calcite,

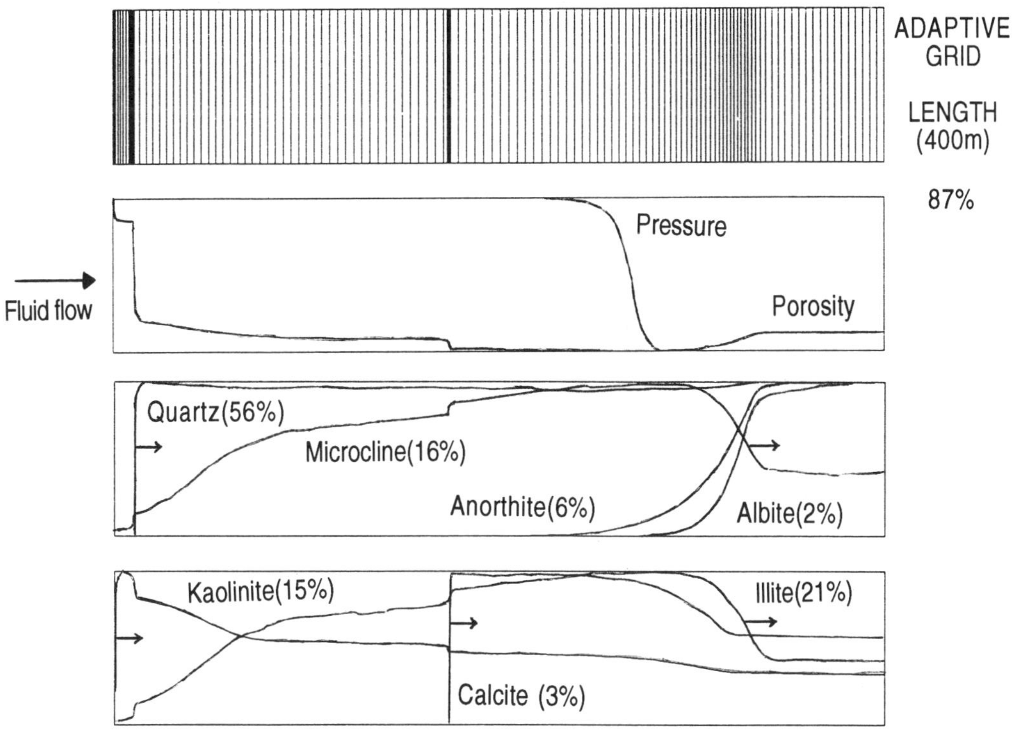

Fig.2 A CIRF simulation on the infiltration-driven diagenesis of an arkose. The initial rock composition was fixed to be 55% quartz, 11% microcline, 7% anorthite, 6% kaolinite, 2% calcite, 1% illite, 1% albite, and 17% porosity. Maximum amounts observed are given in parentheses.

10% dolomite and 10% porosity. A range of composition and velocity for the inlet fluid were investigated. For instance, fluids in equilibrium with calcite and dolomite near the surface have been chosen for the inlet fluid which flow through the system located at a deeper depth. As the solubility of carbonates decrease with temperature and thus with depth, such fluid can become highly supersaturated with respect to calcite and dolomite, as it descends. Rocks which suffered infiltration by such fluid are rapidly cemented by dolomite which prevents further infiltration by decreasing the porosity and the permeability. Such flow self-arresting process occurs in a thin zone at the inlet and could explain some of the narrow low permeability zones in sedimentary layers. On the other hand, if the disequilibrium between the fluid and the rock is produced by the temperature gradient along the fluid path alone, dolomite cementation is very slow. Even with high fluid velocities the process needs several ma to produce the same flow self-arresting phenomena. In both cases a higher reaction rate and supersaturation will allow the dolomite to precipitate earlier and to prevent precipitation of calcite. Under some conditions, calcite can dissolve, even if the inlet fluid is supersaturated with respect to this mineral. To permit dolomitization through the infiltration of magnesium-rich, dolomite undersaturated fluid, calcite must dissolve so that flow self-arresting does not occur and cause an early termination of dolomitization. If the velocity of the calcite dissolution front is faster than the velocity of the dolomite dissolution front a dolomitization zone can form (Fig. 1). Dolomite precipitation between the two fronts results from the calcite dissolution which releases calcium and carbonates in the fluid. The length of the dolomitization zone and thus the significant development of dolomite bodies is not only dependent on the fluid composition, but also on the fluid velocity, the initial rock composition, and the reaction rates (if the width

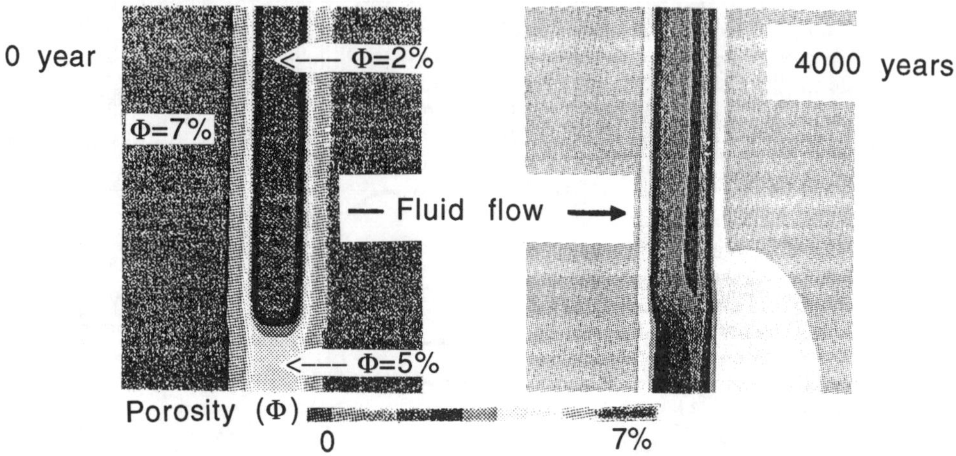

Fig.3 An initially breached seal is "healed" by calcite precipitation in 4000 years as seen by the contour map of porosity at the initial time and after 4000 years. The seal is 100m long and the fluid pressure applied across the domain is one hundred of bars.

of the dolomitization zone is comparable to the width of the reaction fronts).

The diagenesis of an arkose can lead to regions of enhanced permeability and "seals" (i.e., low permeability barriers). In Fig. 2 we see a simulation of the infiltration of fresh water into an arkose containing quartz, microcline, anorthite, kaolinite and minor amounts of calcite, albite, and illite. After 400,000 years there are three distinct reaction fronts. There is also a broad illite transition zone. A seal develops at the peak of illite precipitation; note that where this occurs the fluid pressure changes dramatically. Enhanced porosity/permeability rock (reservoir) is created near the inlet. Because CIRF presently does not account for mechanical collapse, we note the unreasonably high permeability in the more permeable zone.

Results will also be presented in the HCl/HF stimulation of an arkose near the well bore to simulate formation damage remediation by highly acid (pH=-3) fluids. CIRF can be used to design ideal composition and velocity of imposed remediation fluids which can minimize the authigenic mineral precipitation.

The ability of CIRF to analyze two dimensional systems is shown in Fig. 3, where a breached pressure seal deep in a sedimentary basin is shown to "self-heal" via a mechanism of calcite precipitation in the host sandstone. A feedback of pressure gradient self-enhancement due to pressure mediated reactions is the dominant mechanism represented in this last example.

ACKNOWLEDGEMENTS

Research supported in part by a contract from the Gas Research Institute and grants from the Basic Energy Sciences Program of the U.S. Department of Energy, Shell Development Company, AMOCO Production Company, Mobil Exploration and Producing Services, IBM and Sun Microsystems.

REFERENCES

Braithwaite, C.J.R. 1991. Dolomites, a review of origins, geometry and textures. *Trans. Roy. Soc. Edinburgh* 82: 99-112.

Hardie, L.A. 1987. Perspectives: dolomitization, a critical review of some current views. *J. Sedimentary Petrol.* 57: 166-183.

Chen, Y. & P. Ortoleva 1992. Self-enhancing diagenetic trapping through pressure sensitive reactions. Submitted to *AAPG Memoir.*

Chen, W. & P. Ortoleva 1990. Reaction front fingering in carbonate-cemented sandstones. *Earth-Sci. Rev.* 29: 183-198.

Ortoleva, P., W. Chen, Y. Chen, P. Heng, J. Kim, J. Mu, A. Park & C. Wang 1992. *CIRF, Models, numerical techniques and graphical analysis.* Geochem. Res. assoc., Bloomington.

Water-Rock Interaction, Kharaka & Maest (eds) © 1992 Balkema, Rotterdam. ISBN 90 5410 075 3

Dissolved salt removal from Illinois Basin waters

V. Ranganathan
Indiana University, Bloomington, Ind., USA

ABSTRACT: The TDS content of interstitial waters above the New Albany Shale in the Illinois Basin, mostly composed of dissolved sodium and chloride, increases at an average rate of 150,000 ppm/km. Calculations show that if the present day concentration gradient is used as an initial condition, and if no salt is transported upward across the New Albany Shale, vertical diffusion would remove 45% to 60% of the dissolved salt in pore waters above the New Albany Shale over 200 Ma. The present day concentration gradient can be maintained only if extremely high formation factors are assumed [$F=\phi^{-5}$ where ϕ is porosity], an unlikely case, or if salt dissolves in the rocks above the New Albany Shale at an average rate amounting to removal of a 12m-thick halite column over 200 Ma. Upward transport of dissolved salt across the New Albany Shale, a third possible mechanism for maintaining high salinities, was not modeled here.

1 INTRODUCTION

Despite the fact that no salt beds are known in the Illinois Basin, the TDS [total dissolved solids] content of interstitial waters increases with depth to as much as 170,000 ppm, most of which is due to dissolved sodium and chloride [data of Meents et al., 1952]. Three brine-forming mechanisms have been proposed to explain the high TDS of these interstitial waters:

1) membrane filtration [Bredehoeft et al., 1963],
2) burial of evaporated sea water in the Devonian to Ordovician section [Walter et al., 1991], and
3) halite dissolution in the Pennsylvanian to Mississippian section [Walter et al., 1991].

Regardless of which brine forming mechanism is invoked, a question of interest arises. How long can interstitial waters above the New Albany Shale [henceforth referred to as "shallow waters"], remain high in TDS in the absence of a brine-producing mechanism ? The youngest marine deposits known in the Illinois Basin are Pennsylvanian. If burial of evaporated seawater was the principal brine-forming mechanism, this should have occurred at least 200 Ma b.p. and the high TDS content of waters at the present should be explicable without invoking subsurface halite dissolution.

Although there is some scatter in the data, the TDS content of shallow waters increases with depth at an average rate of 150,000 ppm/km or 166 g/l per km [Figure 1]. Vertical diffusion of dissolved salt over the past 200 Ma should have lowered the TDS content of these waters from some initially higher values if no salt dissolved in the shallow part of the basin or if no dissolved salt was transported across the

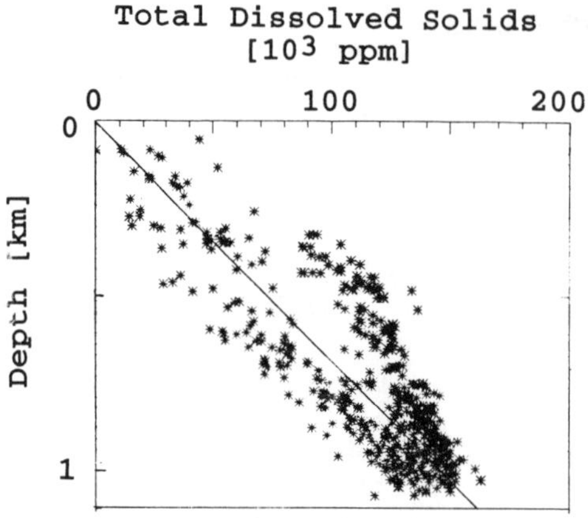

Fig. 1. TDS content of interstitial waters above the New Albany Shale [data of Meents et al., 1952]

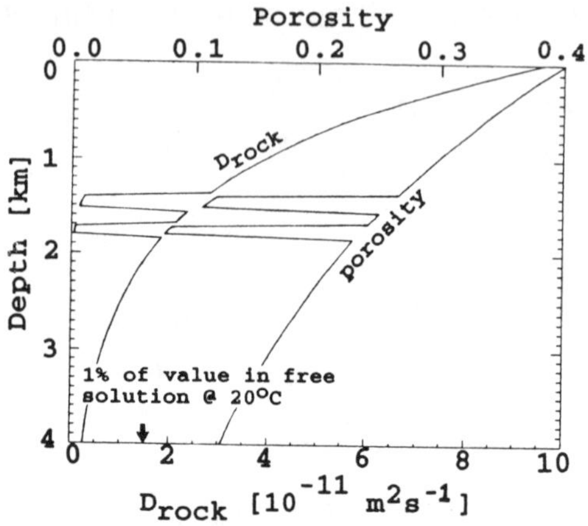

Fig. 2. Porosities and rock diffusion cofficients assumed when $F = ø^{-2.8}$.

New Albany Shale from deeper rocks. Bethke [1986] postulated that the uplift of the Pascola Arch at the southern end of the Illinois Basin in Permian/Mesozoic time resulted in regional groundwater flow northward. Because groundwater flow would flush out dissolved salts more rapidly than diffusion alone, diffusive fluxes are likely to provide conservatively low estimates of the rates of dissolved salt removal in the interstitial water column.

2 VERTICAL DIFFUSION OF DISSOLVED SODIUM CHLORIDE

The mass conservation equation for vertical diffusion and reaction in a sediment column was solved by a finite difference method to obtain transient profiles of dissolved sodium chloride concentration versus depth for a point in the Illinois Basin where the New Albany Shale is deepest [1.4 km]. A subsidence curve at this point shows that there has been little sediment accumulation since the Pennsylvanian [Ranganathan, 1991]. Some Permian and Mesozoic sediments may have been deposited and later been eroded, but the rates and durations of post-Pennsylvanian vertical movement are unknown. The numerical solution was checked by comparing it with the analytical solution of Carslaw and Jaeger [1959] for the heat conduction analog. Model assumptions follow:

1) The thickness of the rock column, 4 km, was assumed to be constant for the 200 Ma simulation period.
2) Two shale layers were modeled from 1.4 km to 1.55 km deep and from 1.7 km to 1.8 km deep, corresponding to the New Albany and Maquoketa Shales [Figure 2].
3) The non-shale section was represented by a single curve of porosity [ø] vs. depth [x] and the two shales were represented by a second such curve [Figure 2].
4) Formation factors for the non-shale section [$F=ø^{-2.8}$], were higher than for the two shales [$F=ø^{-4}$; Figure 2]. The F-ø relation used for the non-shale section yields high formation factors [Lerman, 1979, p. 91], probably characteristic of well cemented sediments, and should therefore yield low estimates of the rate of dissolved salt removal. Archie's Law [$F=ø^{-2}$], a relation commonly used for

sandstones and carbonates, yields much higher diffusive fluxes and rates of dissolved salt removal.

4) The concentration at the top was 0 g/l and no flux was allowed past the base.

5) A constant geothermal gradient of 20°C/km was assumed. Temperature and viscosity effects were incorporated via the Stokes-Einstein equation [Lerman, 1979].

3 RESULTS

Two different initial concentration profiles were used. In the first case, initial concentrations above the New Albany Shale were fixed at values shown by the trend of average salinities known at the present time [from Figure 1]. In the second case, initial concentrations were higher. In both cases, the effect of vertical diffusion over a 200 Ma time span was to reduce maximum concentrations of interstitial waters just above the New Albany Shale to between 40% and 55% of the initial value, 235 kg/m^3 or 235 g/l [Figure 3]. The initial concentration gradient was maintained without salt dissolution when extremely high formation factors, $F=\phi^{-5}$ were assumed [Figure 4]. However, the relation $F=\phi^{-5}$ represents the high end of values measured for clays [Lerman, 1979] and it is doubtful whether such formation factors apply to the whole section above the New Albany Shale. For $F=\phi^{-2.8}$, the initial concentration gradient was maintained when rocks 1.36 km to 1.40 km deep released sodium chloride into solution a t a rate of 10 $kg.m^{-3}.Ma^{-1}$, which amounted to the removal of a halite column 12 m thick over 200 Ma [Figure 4].

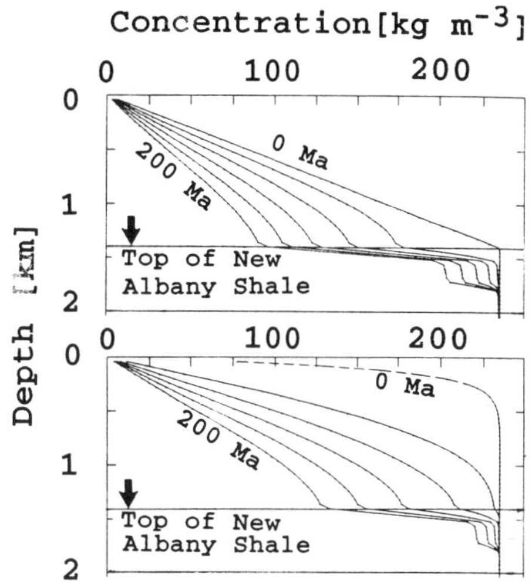

Fig. 3. Concentration profiles at 40 Ma increments. Concentrations below 2 km are almost time-invariant and are not shown.

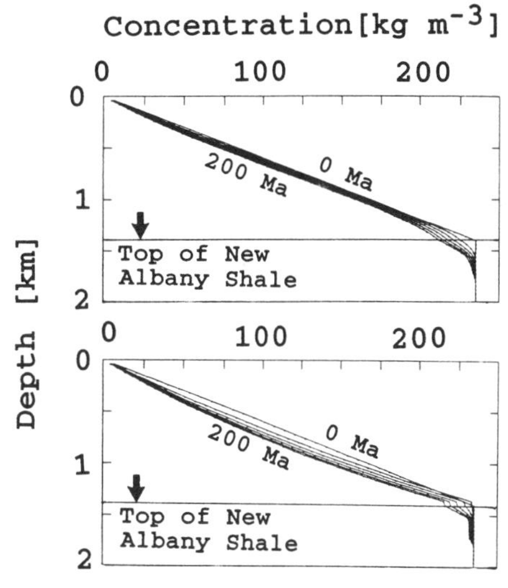

Fig. 4. Calculated concentration profiles changed little over 200 Ma when a high formation factor, $F=\phi^{-5}$ was used and no salt dissolved [Top]. Profiles were similar with $F=\phi^{-2.8}$, provided salt dissolved between 1.36 km and 1.4 km in depth [Bottom].

4 CONCLUSIONS

Over the past 200 Ma, vertical diffusion should have reduced the TDS content of interstitial waters above the New Albany Shale. The high concentration gradient at the present time can be explained by requiring: 1) paleosalinities to have been at least twice as high 200 Ma b.p. [possibly approaching a level for halite saturation], 2) transport of dissolved salt across the New Albany Shale from deeper rocks [a hypothesis not tested here], or 3) subsurface halite dissolution in the section above above the New Albany Shale. The fact that no halite beds are known in the Illinois Basin may reflect the difficulty of detecting thin halite beds while drilling and preserving halite while coring. Even small rates of groundwater flow would greatly accelerate the removal of dissolved salt.

ACKNOWLEDGEMENTS

Acknowledgement is made to the donors of the Petroleum Research Fund administered by the American Chemical Society for support of this work through grant ACS-PRF 21667-G8.

REFERENCES

Bethke, C.M. 1986. Hydrologic constraints on the genesis of upper Mississippi Valley mineral district from the Illinois Basin brines. *Economic Geology* 81: 233 - 249.

Bredehoeft, J.D., C.R. Blythe, W.A. White, and G.B. Maxey. 1963. Possible mechanisms for concentration of brines in subsurface formations. *American Association of Petroleum Geologists Bulletin* 47: 257 - 269.

Carslaw H.S. and J.C. Jaeger. 1959. Conduction of heat in solids. 2nd edition. Oxford University Press, Oxford. 510 pp.

Lerman, A. 1979. Geochemical processes: Water and sediment environments. R.E. Krieger Publishing Company, Malabar, Florida. 481 pp.

Meents W.F., A.H. Bell, O.W. Rees and W.G. Tilbury, 1952. Illinois oil-field brines: Their geologic occurrence and chemical composition. Illinois GeologicalSurvey, *Illinois Petroleum* No. 66.

Ranganathan, V. 1991. Salt diffusion interstitial waters and halite removal from sediments: examples from the Red Sea and Illinois Basin. *Geochimica et Cosmochimica Acta* 55: 1615 - 1625.

Walter, L.M., A.M. Stueber, and T.J. Huston. 1991. Br-Cl-Na systematics in Illinois Basin fluids: constraints on fluid origin and evolution. *Geology* 18: 315 - 318.

Water-Rock Interaction, Kharaka & Maest (eds) © 1992 Balkema, Rotterdam. ISBN 90 5410 075 3

Mechano-chemistry of compartmented basins

E.L.Sonnenthal, J.M.Maxwell, C.Qin & P.J.Ortoleva
Departments of Chemistry and Geological Sciences, Indiana University, Bloomington, Ind., USA

ABSTRACT: A reaction-transport-mechanical model for basin diagenesis and compartmentation is described. It involves the coupled equations of pressure solution-mediated compaction, fluid flow, fracture kinetics, heat transport, sedimentation and subsidence. A homogenization technique captures the effect of small-scale features such as cm-scale laminations on the permeability and pressure distribution in a km-scale basin. Numerical simulations in two-dimensions show the development of low permeability seals surrounding overpressured hydrofractured rock with high permeability. In simulations using the homogenization technique, the intensification of small textural nonuniformities resulted in low-permeability diagenetically-banded seals, sustaining a large pressure gradient.

1 INTRODUCTION

An important aspect of water-rock interactions deep in a sedimentary basin is the coupling of fluid flow and mechano-chemical processes (e.g. pressure solution and rock fracturing). Many sedimentary basins have compartments characterized by abnormal fluid pressure, a pervasive fracture network, and diagenetically banded seals. Some seals have alternating bands of enhanced cementation and compaction, cutting across stratigraphic contacts (Bradley, 1975; Powley, 1990; Tigert and Al-Shaieb, 1990).

To describe the dynamical evolution of a compartmentalized basin we developed a reaction-transport-mechanical (RTM) model incorporating fluid flow, pressure solution, fracture kinetics, heat transport, sedimentation and subsidence. In this model, the mechanisms of overpressuring are pressure solution-mediated compaction and thermal expansion of the fluid. Simulation of textural variations on the centimeter- to meter-scale and the influence of these small heterogeneities on compartmentation and other processes on the kilometer-scale are accomplished by a homogenization technique (Qin and Ortoleva, 1992).

2 THE RTM MODEL

The reaction-transport-mechanical model describes the fluid-rock system in the deeper part of a basin; i.e. below the shallow depths of early mechanical compaction (of clastic rocks), and of infiltration of meteoric or other surface waters. Although the latter processes may be important for the shallow hydrogeological system, the depth of compartmentation is commonly 3 km or more and has been suggested to be generally decoupled from the upper fluid flow regime (Powley, 1990).

In our RTM model the rock matrix is characterized by the grain size, shape, and number density. The matrix is also given an initial statistical distribution of fracture nucleii. Rates of grain growth/dissolution and fracture growth and healing are dependent on the pore fluid composition, fluid pressure, temperature, local average stress and grain-scale stresses. Because the fluid pressure is dependent on the rate of compaction and the permeability, the latter affected by the pressure-temperature dependent fracture kinetics, it is clear that the system is tightly coupled and the chemical and physical processes may not be adequately modeled independently.

2.1 *Macroscopic and microscopic stresses*

The vertical stress on a given rock volume is calculated using force balance and assuming the stress tensor to be triaxial. The horizontal stress is assumed to be a fraction of the vertical stress; values are from Gulf coast fracture gradient studies (Althaus, 1975).

Micro-stresses are those applied to grain faces, whether in contact with other grains or the pore fluid. The rock matrix consists of a single stress-supporting mineral (quartz) whose grains are truncated spheres in rectangular packing (Weyl, 1959). The normal stress to the z-contact (face supporting vertical stresses) is calculated from the grain geometry, fluid pressure and macroscopic stress (Sonnenthal and Ortoleva, 1992; see, however, Dewers and Ortoleva, 1990b; Rutter, 1976, 1983).

2.2 *Grain growth/dissolution rate laws*

Grain growth and dissolution rates are dictated by the kinetics at each of the sites on the grain surface (those supporting vertical or horizontal stresses, or in contact with the pore fluid). In this RTM model, we consider only the reaction quartz $\Leftrightarrow$ SiO_2 (aq). The grain height (L_z) changes over time owing to vertical rock flow (tectonic subsidence) and by pressure solution:

$$\partial L_z / \partial t = - \underline{u} \cdot \underline{grad}\, L_z + G_z \qquad (1)$$

where $\underline{u}$ is the rock flow velocity and G_z is the rate of dissolution of the grain in the vertical direction. The first-order kinetic law for the stress-supporting z-face is given by:

$$G_z = k_z(c - c_z^{eq}) \qquad (2)$$

where c_z^{eq} is the equilibrium concentration of SiO_2 in the water film between stress-supporting contacts (Weyl, 1959; Rutter, 1976, 1983; Dewers and Ortoleva, 1990a,b), c is the pore fluid concentration of SiO_2 (aqueous) and k_z is the rate constant.

The rate constant k_z is written in terms of the water-film diffusion constant and thickness (Sonnenthal and Ortoleva, 1992; Dewers and Ortoleva, 1990a). Modifications to these rate laws arise from coatings such as clay which tend to inhibit precipitation on faces in contact with pore fluid. In a locally closed system this results in a reduction in the compaction rate by the water-film diffusion pressure solution mechanism (Sonnenthal and Ortoleva, 1992).

2.3 *Fracture kinetics*

The kinetic model for hydrofracture growth and healing is that described in Sonnenthal and Ortoleva (1992) (see also Dewers and Ortoleva, 1992). Initially, the rock has a statistical distribution of fracture nucleii that grow (subcritically) in length and width if the fluid pressure exceeds the sum of the horizontal stress and the local rock tensile strength.

Once a fracture network develops it affects the rock permeability through the fracture size and the degree of connectivity. Using percolation theory the fracture permeability can be obtained by consideration of the mean size and spacing of fractures (Sonnenthal and Ortoleva, 1992; Gueguen et al., 1990).

2.4 *Fluid flow and overpressuring*

To solve for the fluid overpressure we combine an equation for mass conservation of water with Darcy's Law and a linearized equation of state for water (Sonnenthal and Ortoleva, 1992; Palciauskas and Domenico, 1989). Permeability is taken as the sum of matrix and fracture contributions; the former calculated using a modified Kozeny-Carman relation. These are expanded into the two-dimensional forms and solved on a rectangular grid which has been transformed from a basin-adapted grid.

2.5 *Heat transport*

The transport of heat by fluid may play an important role when the fluid flow velocities are sufficiently high to disturb the local geothermal gradient or when nearby rock thermal conductivities are markedly different. For low permeability rocks the fluid velocities may be on the order of centimeters per year and thus heat transport may be negligible; however, in the vicinity of permeable faults or in highly fractured or porous rocks, heat transport cannot be neglected. Therefore, we coupled the heat transport equations of Bear (1972) (modified for rock motion) to the equations for fluid pressure and rock deformation. We assume that the rock and fluid attain thermal equilibrium in a time that is short compared to the fluid flow rate. In the simulation shown in this paper we assumed a constant geothermal gradient because of the slow flows encountered.

3 HOMOGENIZATION THEORY

Sedimentary basins provide a very challenging problem in modeling and computer simulation because they are multiple length-scale systems. For example, we are interested in the kilometer or even basin scale fluid pressure yet the permeability varies on the meter or centimeter scale. That this is important is evidenced by banded pressure seals (Powley, 1990; Bradley, 1975; see Dewers and Ortoleva, 1992), described by Tigert and Al-Shaieb (1990) in a sandstone top seal in the Anadarko basin and also in a bottom shale seal in the same basin (Ortoleva and Al-Shaieb, 1992). This short-scale feature is of interest because it formed by diagenesis, apparently via mechanisms involving mechano-chemical feedback (Dewers and Ortoleva, 1990a,b; 1992a,b; Qin and Ortoleva, 1992; Ortoleva, 1992b). In this context, the pressure distribution is affected by the banded cements and the rate of development of the bands is affected (through the pressure solution dynamics) by the pressure. Another aspect of the multi-scale nature of basin dynamics is depositional layering, which is affected by diagenesis and, through its effect on permeability, influences the large-scale pattern of diagenesis.

The direct simulation of submeter-scale features in basin-scale systems is clearly prohibited by limitations of grid resolution. However, homogenization theory considers the two scales self-consistently but independently. As the resulting equation can be solved readily on a parallel computer, we believe that this approach is an important new technical advance for the modeling of basin diagenesis.

4 RESULTS OF SIMULATIONS

Simulation of a 200km-wide basin using the RTM model described in section 2 shows the development of an overpressured hydrofractured compartment sealed by enhanced compaction on its margins (Fig 1). The maximum sedimentation rate was 500m per my and the burial depth was 6.8km.

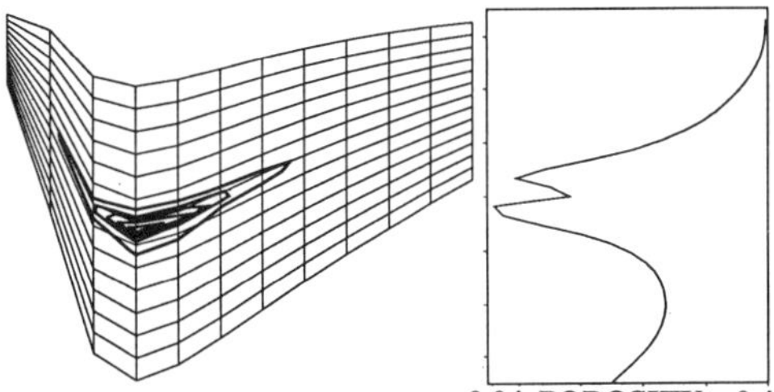

Fig. 1 Overpressure contours (max. = 524 bars) outlining a high pressure auto-isolated compartment. Vertical porosity profile (right) through center shows porosity lows (seals) encasing a high porosity fractured interior. Below is a higher porosity coarser-grained bed.

Because banded seals are an important aspect of compartments we use homogenization theory to simulate their effect on the larger-scale pressure distribution. The model is similar to that in section 2 except that a full stress calculation was carried out, SiO_2 (aq) was allowed to diffuse and flow, and free face reactions driven by strain energy were faster than pressure solution at grain-grain contacts. Simulations showed that initial grainsize variations (i.e. sedimentary laminations) were intensified by mechano-chemical feedback.

A developing overpressure was applied to the bottom of a two-dimensional domain. There the texture is amplified less because of stress reduction at grain contacts by higher fluid pressure. Shallower, mechano-chemical feedback is also slowed because overburden is less. Thus, a zone developed where the banding is most intense. The locally averaged ("homogenized") permeability is lowest in the banded region because of the low porosity bands and the strong nonlinear dependence of permeability on porosity. Note, in Fig. 2, the pressure gradient becomes focussed across this "banded seal".

The depth of this seal at 3.8 to 4 km agrees with that of the Anadarko basin top seal. Although the latter seal has a more complex texture (Tigert and Al-Shaieb, 1990), other mechano-chemical feedback mechanisms can generate complex banded seals (Qin and Ortoleva, 1992; Ortoleva and Al-Shaieb, 1992; Ortoleva, 1992a,b).

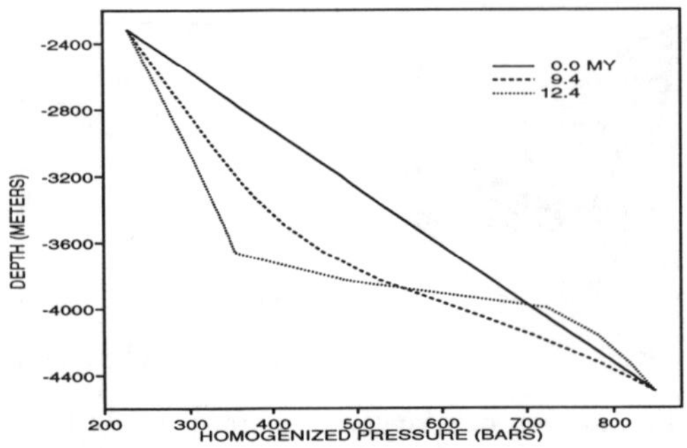

Fig. 2 Homogenized pressure at 0, 9, and 12 my. At 12 my and above 3600m, the pressure gradient approaches the hydrostatic gradient (not shown), and below the seal zone (3800-4000m) the pressure increased to greater than the imposed high pressure gradient.

5 CONCLUSIONS

Diagenesis and compartmentation in sedimentary basins involves the coupling of physical and chemical processes such as fluid flow, rock fracture, and pressure solution-mediated compaction. Simulations using the described reaction-transport-mechanical model show the development of fractured overpressured compartments at depths similar to those observed in many basins. The formation of small-scale banded seals in a km-scale domain and their effect on the large-scale fluid pressure was modeled using homogenization theory.

REFERENCES

Althaus, V.E. 1975. A new model for fracture gradient. *SPWLA Sixteenth Annual Logging Symposium.* 1-19.

Bear, J. 1972. *Dynamics of fluids in porous media.* New York: American Elsevier.

Bradley, J.S. 1975. Abnormal formation pressure. *AAPG Bull.* 59: 957-973.

Dewers, T. & P. Ortoleva 1990a. A coupled reaction/transport/mechanical model for intergranular pressure solution, stylolites, and differential compaction and cementation in clean sandstones. *Geoch. Cosmoch. Acta.* 54: 1609-1625.

Dewers, T. & P. Ortoleva 1990b, Geochemical self-organization III: A mean field, pressure solution model of spaced cleavage and metamorphic segregational layering. *Amer. J. Sci.* 290: 473-521.

Dewers, T. & P. Ortoleva 1992. Nonlinear dynamical aspects of deep basin hydrology: Fluid compartment formation and episodic fluid release. *Amer. J. Sci.*, in press.

Gueguen, Y., T. Reuschle, & M. Darot 1990. Single-crack behavior and crack statistics. In D.J. Barber & P.J. Meredith, eds., *Deformation Processses in Minerals, Ceramics and Rocks.* 48-71. Unwin-Hyman.

Ortoleva, P. 1992a. Development of diagenetic differentiated structure through reaction-transport feedback. In K.H. Wolf and G.V. Chilingarian, eds., *Diagenesis* 4, in press.

Ortoleva, P. 1992b. *Geochemical Self-Organization.* Oxford University Press, in press.

Ortoleva, P. & Z. Al-Shaieb 1992. Genesis and dynamics of basin compartments and seals. *Amer. J. Sci.*, accepted.

Palciauskas, V.V. & P.A. Domenico 1989. Fluid pressures in deforming porous rocks. *Water Resources Research.* 25: 203-213.

Powley, D.E. 1990. Pressures and hydrogeology in petroleum basins: *Earth Sci. Rev.* 29: 215-226.

Qin, C. & P. Ortoleva 1992. Banded diagenetic pressure seals: Types, mechanisms and homogenized basin dynamics. *AAPG Memoir*, submitted.

Rutter, E.H. 1976. The kinetics of rock deformation by pressure solution. *Phil. Trans. R. Soc. London.* A283: 203-219.

Rutter, E.H. 1983. Pressure solution in nature, theory and experiment. *J. Geol. Soc. London. 140*: 725-740.

Sonnenthal, E.L. & P.J. Ortoleva 1992. Numerical simulation of overpressured compartments in sedimentary basins. *AAPG Memoir*, submitted.

Tigert, V. & Z. Al-Shaieb 1990. Pressure seals: Their diagenetic banding patterns. In P. Ortoleva, B. Hallet, A. McBirney, I. Meshri, R. Reeder & P. Williams, eds., Self-Organization in Geological Systems, Univ. of Cal. at Santa Barbara. *Earth Sci. Rev.* 29: 227-240.

Weyl, P.K. 1959. Pressure solution and the force of crystallization. *J. Geophys. Res.* 64: 2001-2025.

Geochemical modeling

General applications

Water-Rock Interaction, Kharaka & Maest (eds) © 1992 Balkema, Rotterdam. ISBN 90 5410 075 3

A quantitative first principles approach to supergene enrichment of a porphyry copper protore

Giuseppe G. Biino & Peter C. Lichtner
Mineralogisch-Petrographisches Institut, Bern, Switzerland

ABSTRACT: The supergene weathering of a porphyry copper deposit is described quantitatively based on the multiple reaction path formulation of fluid transport and chemical reaction in a porous media. The evolution in time of the fluid chemistry and minerals within the supergene weathering column is completely determined by the governing equations once initial and boundary conditions specifying the composition of rain water and the modal composition of the protore are given. Numerical calculations are presented for both fully saturated and unsaturated conditions in the presence of a deep water table. The effects due to variation of water table level, the ratio of pyrite to chalcopyrite, and the presence of gangue minerals are investigated. The precipitation of supergene minerals produces a complex mineralogic zoning without the necessity of appealing to periodic variations in the water table or cyclic repetition of the supergene weathering process. An enrichment blanket is formed in the saturated zone below the water table which migrates downward with increasing time. A peak in copper grade occurs at the top of the blanket in agreement with field observations. The distance a packet of fluid travels to reach equilibrium with protore sulfide minerals is proportional to the amount of oxygen present in the inlet fluid , but the protore oxidation rate is independent of the initial oxygen fugacity. Oxidation of the protore takes place near the surface through reactions with dissolved oxygen. Deeper in the weathering profile the protore reacts with other redox sensitive species. The calculations show that to form an enriched blanket 50 meters thick for a Darcy flow velocity of one meter per year requires approximately one million years for fully saturated conditions. In the case of a deep water table (100m), a blanket zone three times as thick is produced during the same time span. Under the same conditions in an argillic altered protore the blanket front moves twice as fast compared to an unaltered granitic protore.

1 INTRODUCTION

This paper investigates quantitatively supergene enrichment of a porphyry copper protore based on a first principles approach of fluid transport and fluid-rock interaction. The evolution in time and space of solute species concentrations and mineral abundances due to fluid-rock interaction can be determined by combining thermodynamic data with kinetic rate laws and mass transport equations. The goal of this study is to define the timing and formation of the enrichment blanket produced during weathering of a porphyry copper protore. Previous quantitative studies of fluid-rock interaction have utilized a single reaction path model based on the work of Helgeson (1968). With this approach, however, it is not possible to treat time and space as independent variables. As a consequence it is not possible to determine changes in the abundances of minerals or porosity at a point fixed in space. This study quantitatively describes supergene enrichment as a metasomatic process using a multiple reaction path approach based on the computer code MPATH. This method enables a description of the reaction between fluid and minerals in both time and space. In contrast to the work of Ague and Brimhall (1989), the computational method used here employs an adaptive grid technique allowing the movement of mineral reaction fronts to be traced in time. Details of the theoretical model and the computational method are presented elsewhere (Lichtner, 1988; 1992). A previous study (Lichtner and Biino, 1992) focused on supergene enrichment in the Cu-Fe-S-O-H subsystem. This contribution investigates the effect of gangue minerals on the supergene enrichment process. The evolution in space and time in the system Na-K-Ca-C-Cu-Fe-Al-Si-S-O-H is considered.

2 THEORETICAL ASSUMPTIONS

A quantitative description of supergene enrichment determines the sequence of reaction zones, solute concentrations and mineral abundances as functions of time and distance. For given initial and boundary conditions the model completely predicts the future evolution of the system without any additional assumptions. Minerals can become undersaturated, saturated or supersaturated according to their affinity at different points along the flow path. The mass transport equations are solved taking into account the movement of chemical reaction fronts and mineral alteration zones. Thus the description of supergene enrichment forms a moving boundary problem (Stefan problem). The concentrations of solute species may vary by many orders of magnitude along the flow path as the pH and redox state change. Within the multiple reaction path formulation, the supergene enrichment process is susceptible to numerical solution for a one-dimensional porous medium in reasonable computational time for virtually any number of reacting species.

3 THE CHEMICAL SYSTEM

The fluid composition is fixed at the inlet of the weathering column by the composition of rain water with a pH of 4.5, corrected for transport through a soil zone at the top of the deposit.

Two different host rock compositions were considered. The first rock column consists of a granitic rock formed by the following primary minerals: 20% quartz (Qtz), 25% plagioclase, 20% K-feldspar (Kfs), 5% muscovite (Ms), 10% annite (Ann). A second set of calculations were performed considering a hydrothermally altered granitic rock formed by primary minerals: 30% muscovite, 20% quartz, 30% paragonite. The protore consists of 2% magnetite (Mag), pyrite (Py) and 2% chalcopyrite (Ccp) with two different pyrite to chalcopyrite ratios of 2:1 and 5:1. In addition to these primary minerals, the following secondary minerals precipitated from solution: albite (ab) kaolinite (Kao), gibbsite (Gbs), bornite (Bn), chalcocite (Cc), covellite (Cv), cuprite (Cpr), copper (Cu), brochantite (Brc), alunite (Alu), ferrihydrite (Fhr), chalcedony (Cld), and jarosite (Jrs).

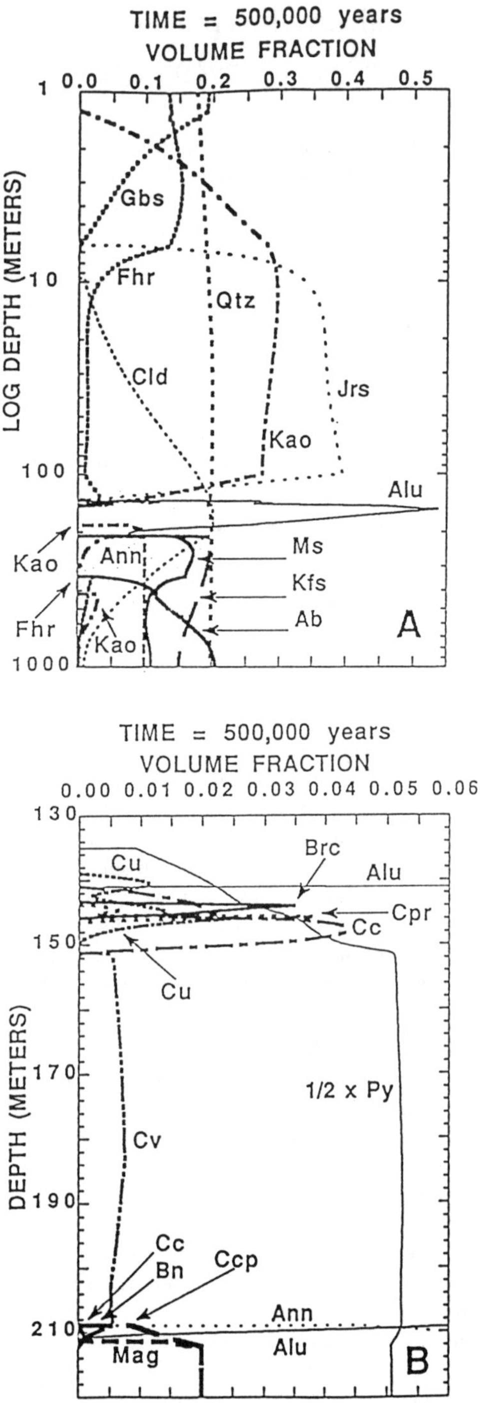

Fig. 1A,B Mineral volume fractions plotted as a function of depth at 500,000 years. Abbreviations explained in text.

4 PHYSICAL CONDITIONS

Rain water percolates through the weathering column with a Darcy flow velocity of one meter per year. The supergene weathering of the protore was considered with the water table close to the surface (oxidation of the protore occurring under fully saturated conditions) and with the water table at 100 meters depth. In the later case, the weathering profile is divided in two parts. The upper part of the column is formed in the partially saturated zone above the water table, and the lower part in the fully saturated zone below the water table. In the partially saturated part of the column, the fluid is assumed to be in equilibrium with the atmosphere.

5 RESULTS

The weathering profile for the granitic protore (py:ccp = 5:1) with the water table at a depth of 100 meters is shown in Figs. 1A and B, 2 and 3 after an elapsed time of 500,000 years. A complex sequence of alteration zones is formed with copper minerals enriched in a blanket zone lying below the water table (Fig. 1A, B). The thickness of the blanket zone is approximately 80 meters, with a maximum in copper grade occurring in the upper 10 meters of the blanket in agreement with field observations (Ransome, 1919; Alpers and Brimhall, 1989). The top of the blanket zone is formed by cuprite, native copper and brochantite consistent with field observations (Guilbert and Park, 1986). In the blanket, chalcocite forms two zones. In the upper zone chalcocite precipitates at the pyrite dissolution front, and probably produces the commonly described texture of chalcocite coating pyrite. The highest copper content in the weathering profile corresponds to this upper chalcocite zone. A lower, but smaller chalcocite (+covellite+bornite) zone is formed at the chalcopyrite dissolution front. The two zones are separated by a covellite zone.

The gangue minerals form a complex zoning pattern. Gibbsite precipitates near the surface of the weathering profile together with ferrihydrite. Ferrihydrite also forms a deeper zone (approximately at 340 meters) as annite is oxidized. The existence of two ferrihydrite zones predicted by the model is in agreement with field observations (Ransome 1919; Schwartz 1953). Kaolinite and jarosite are formed on the top of the supergene enrichment zone, but two additional deeper kaolinite zones are also predicted to occur. The alunite zone lies between the kaolinite and jarosite zones and the hydrolytic front formed by dissolution of K-feldspar, muscovite and annite. The replacement of alunite by jarosite is described by Alpers and Brimhall (1989). The presence of kaolinite throughout the weathering column is also confirmed by field observations (Alpers and Brimhall, 1989). The pH and pe show important changes along the weathering profile. The pe drops in the upper part of the blanket zone as pyrite is oxidized. The pH increases to 5.5 at approximately 140 meters (Fig. 2) corresponding to the top of the blanket zone and to the alunite precipitation front. Further downstream, the pH drops to 3 at the transition between the upper chalcocite zone and the covellite and pyrite zones resulting from pyrite oxidation. The pH increases at the annite, muscovite, K-feldspar hydrolytic front, and further downstream at the dissolution front of ferrihydrite and albite. Equilibrium between fluid and the host rock is achieved further downstream (at approximately 3.5 km, not shown) at a pH of 9.8. The drop in concentration

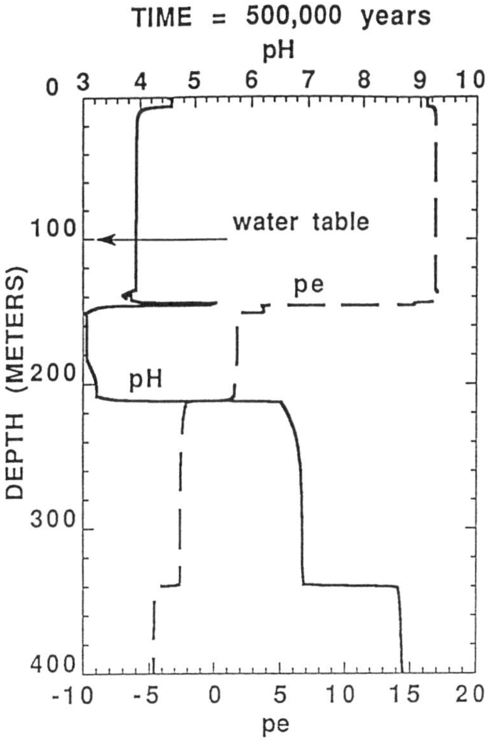

Fig. 2 The pH and pe plotted as function of depth at 500,000 years

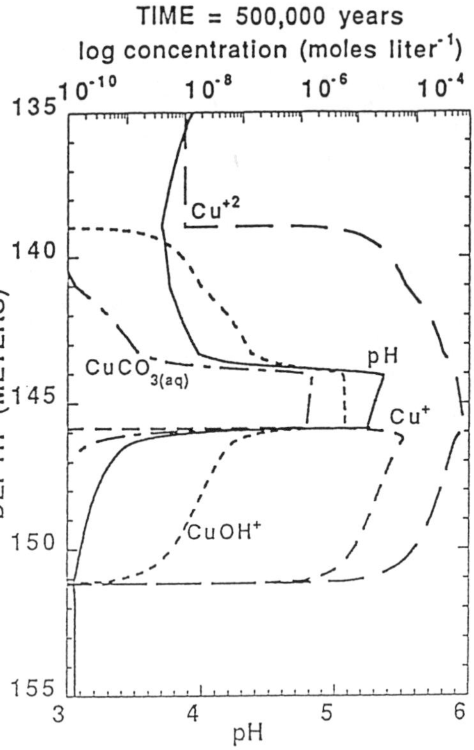

Fig. 3 Copper concentration plotted as function of depth at 500,000 years

of the Cu aqueous species in the upper blanket zones (Fig. 3) shows that the system is closed with respect to copper. In the partially saturated zone the main oxidizing aqueous species is oxygen. Below the saturated zone oxygen is completely consumed and other oxidizing species are responsible for oxidation of the protore.

Under fully saturated conditions a blanket zone of 20 meters thick is formed after 500,000 years. Also in this case the highest Cu grade corresponds to the top of the blanket zone but the maximum Cu grade is only half that obtained in the presence of the water table.

For the case of a hydrothermally altered protore under fully saturated conditions, the model predicts the formation of an enriched blanket zone approximately 50 meters thick after 500,000 years. The change in Cu grade in the weathering profile is comparable to the latter case.

The calculations with different pyrite to chalcopyrite ratios support the conclusion that the most favorable conditions for the formation of a blanket zone corresponds to a higher pyrite to chalcopyrite ratio (Anderson, 1982).

6 CONCLUSION

The calculations predict an evolution of the supergene enrichment process that is in excellent agreement with field observations. The results indicate that is possible to understand quantitatively the supergene enrichment process.

References

Ague, J. J. and Brimhall, G. H. 1989. Geochemical modelling of steady state fluid flow and chemical reaction during supergene enrichment of porphyry copper deposits. Econ. Geol. 84: 506–528.

Alpers, C. N. and Brimhall, G. H. 1989. Paleohydrologic evolution and geochemical dynamics of cumulative supergene metal enrichment at La Escondida, Atacama desert, northern Chile. Econ. Geol. 84: 229–254.

Anderson, J. A. 1982. Characteristics of leached capping and techniques of appraisal. In Advances in the geology of porphyry copper deposits, Southwestern North America (ed. S. R. Titley), Univ. Arizona press, Tucson: 275-295.

Guilbert, J.M. and Park, C.F. 1986. The geology of ore deposits. New York: W.H. Freeman and Company.

Helgeson, H. C. 1968. Evaluation of irreversible reactions in geochemical process involving minerals and aqueous solutions I. Thermodynamic relations. Geochim. Cosmochim. Acta 32: 853–877.

Lichtner, P. C. 1988. The quasi-stationary approximation to coupled mass transport and fluid-rock interaction in a porous medium. Geochim. Cosmochim. Acta 52: 143–165.

Lichtner, P. C. 1992. Time-space continuum description of fluid/rock interaction in permeable media. Water Res. Res. Submitted for publication.

Lichtner, P. C. and Biino, G. G. 1992. A first principles approach of a porphyry copper protore. I. Cu-Fe-S-H_2O subsystem. Geochim. Cosmochim. Acta in press.

Ransome, F. L. 1919. The copper deposits of Ray and Miami, Arizona. USGS Prof. Pap. 115.

Schwartz, G. M. 1953. The copper deposits of the San Manuel district, Arizona. USGS Prof. Pap. 256.

Water-Rock Interaction, Kharaka & Maest (eds) © 1992 Balkema, Rotterdam. ISBN 90 5410 075 3

Thermodynamic modelling of geochemical processes: The CSIRO-Monash THERMOCHEMISTRY System

M.S.Bloom & Y.Gu
Department of Earth Sciences, Monash University, Clayton, Vic., Australia

M.W.Wadsley
Department of Chemical Engineering, Monash University, Clayton, Vic., Australia

ABSTRACT: The CSIRO-Monash THERMOCHEMISTRY System is a package of applications for the Macintosh and IBM-PC platforms. The system couples a common thermodynamic database with a modular set of applications, each using a common directive language. Applications perform data management, calculation of thermodynamic properties for species and reactions, estimation of unknown thermodynamic properties, construction of predominance area and related diagrams and equilibrium and energy balances for complex processes. Macintosh applications may be launched individually, or they may be launched using MacThermo, a software tool which is configured for launching the CSIRO-Monash as well as other aqueous speciation and reaction path modelling applications. MacThermo also provides on-line help for the CSIRO-Monash Thermochemistry System and for other selected software packages. QuickDraw™ graphics on the Macintosh and VGA, EGA, CGA, Tektronics™ and HPGL™ graphics on the PC are supported for graphical output.

INTRODUCTION

The CSIRO-Monash Thermochemistry System consists of a thermodynamic data base for chemical substances, and an integrated package of compatible applications to carry out thermodynamic calculations using these data. It provides a readily accessible source of critically assessed, up-to-date and self-consistent thermochemical data, and the facility to use these data to predict the equilibrium products of chemical and metallurgical processes.

The system is divided into eight major applications (ESTIMA, EXERGY, CHEMIX, FILER, REACT, REFER, SYSTEM, VAPOUR) each of which accesses a common directive language and files of the thermochemical data base. The system is an improved and extended microcomputer implementation of the CSIRO-SGTE THERMODATA software package (Turnbull and Wadsley, 1984; 1986) for VAX and main-frame computers.

GENERAL FEATURES

The CSIRO-Monash Thermochemistry System has been designed to be as general as possible, to facilitate calculations with many types of chemical species or systems, a wide range of thermodynamic properties and a variety of equilibrium simulations. Modular construction based on a shared subroutine library and a generalized file structure ensure that extension to include new properties and algorithms may be achieved with minimal effort.

Control of the system is provided by a set of directive words, common to all programs, which constitutes a high-level language. Each directive activates a program module for a particular task, and on input may be followed by a set of pre-defined option and data entries in free format. An essential feature of the system is the ability to read chemical formulae in a variety of free formats, and to convert them to unique internal identifiers for solid, aqueous and gas species which may then be assigned to one or more phases. A variety of units is available for input and output, but a fixed set of units is used for internal computations. Other useful utilities include a comprehensive set of example applications available from file and a full set of on-line guides to the use of directives (on the Macintosh, this feature is available in the desk accessory MacThermo).

THERMOCHEMICAL DATA

Data necessary for thermodynamic calculations are stored in a data base which is subdivided into separate files on the basis of the type and source of the contained data. For calculations involving geochemical processes, standard molal thermodynamic properties at 25°C and 1 bar, equation of state parameters, heat capacity coeffients and phase transition data for minerals, aqueous species and gases in *SUPCRT92* (Johnson *et al.*, 1992 and references therein) are accessible. The internally-consistent thermodynamic data sets of Berman (1988) for aluminosilicate minerals and Mills (1974) for sulfides, selenides and tellurides are also available. Other sources which have been recast into CSIRO-Monash Thermochemistry formats to facilitate intra-program consistency include SOLTHERM (Spycher and Reed, 1989 and updates thereof), amendments to the data for aluminosilicates by Sverjenski *et al.* (1991), and for halogen partitioning between minerals and hydrothermal fluids (Zhu and Sverjenski, 1991).

Data sets may be retrieved from the data base or from users private files, and if necessary are converted at run-time to Gibbs free energies. All or part of the data for a specific execution may also be entered directly to the application, either as full data sets or in the form of compact Gibbs free energy-temperature relations. The responsibilty for ensuring consistent usage of HKF and non-HKF data for aqueous species, and more generally for avoiding introduction of internal inconsistencies when additional data external to the data base are provided, is devolved to the user.

APPLICATIONS

The CSIRO-Monash Thermochemistry System may be applied to a number of the specific computational requirements for modelling geochemical processes. Among these are aqueous speciation-solubility, phase distribution and reaction path modelling, and construction of diverse chemical potential diagrams. These calculations are possible using the modules CHEMIX, REACT and SYSTEM, for which overviews are presented below.

program CHEMIX

In order to model complex geochemical processes it is necessary to calculate the equilibrium composition of multicomponent systems having many elemental components, dependent species and solution phases which may exhibit either ideal or highly non-ideal behaviour. CHEMIX is a general-purpose equilibration and mass transfer application which performs these calculations, and also provides powerful methods for defining the chemical system, the constraints and the results required.

CHEMIX defines a set of chemical species and phases present in a system in terms of selected elements or species and their quantities using standard chemical nomenclature. Thermodynamic data for the set of component and dependent species are retrieved from the data base, or entered directly by the user. For non-ideal solution phases, activity coefficients are defined as functions of temperature and composition. A number of methods are provided for this purpose and include Redlich-Kwong and virial equations of state for gaseous species, extended Debye-Hückel and Pitzer formulations for aqueous species, and a variety of Margules and site-substitution formulations for solids.

CHEMIX calculates the number of moles of all species that minimize the total Gibbs free energy of a system of given bulk composition, temperature and pressure, using the method of Eriksson (1975, 1979). CHEMIX differs from EQ3NR (Wolery, 1983) and SOLVEQ (Reed and Spycher, 1984) in that it does not require numeric solution of a set of non-linear equations using Newton-Raphson iteration, but is similar to the single-step minimization method presented by De Capitani and Brown (1987). CHEMIX also calculates the enthalpy balance between initial and equilibrium states, the distributions of the elements among the equilibrium products, and the thermodynamic properties of the total system and of individual phases and species. There are options to fix, maximize or minimize chosen properties of the system by systematically varying others.

A powerful feature which makes CHEMIX particularly amenable to modelling dynamic geochemical processes is the ability to configure numerical experiments for repetitive or step-wise free energy minimizations. Single-step isothermal equilibration of an aqueous fluid with a given bulk composition (typically representing a rock, but in principal any combination of solid, liquid and gaseous components) can be performed at a specified temperature and pressure, or as a function temperature and pressure. A series of equilibrium state calculations for variable ratios of input fluid reacting with a protolith is equivalent thermodynamically to the "titration" models for heterogeneous fluid-rock equilibration (Heinrich *et al.*,1989) performed by EQ6 (Wolery, 1983) and CHILLER (Reed and Spycher, 1984).

Equilibration of a single, finite quantity of rock with a single, finite quantity of fluid does not describe water-rock interaction during advective mass transfer in a geologically realistic fashion. Numerical experiments can, however, be configured with CHEMIX to approximate fluid-rock interaction in a flow-through system. The equilibrium stage-wise reactor model allows up to ten segments linked by up to thirty matter streams (which provide for input, output and flow between segments). Each segment has independent temperature, pressure and volume, or may be described by a defined energy flow. Each stream contains defined phases of the system with independently-defined quantities and temperatures. Almost any kind of advective geochemical process having fixed or moving streams (co-current or counter-current), with bypass and(or) recycle options, can be defined. The interaction of segments and streams in the equilibrium stage-wise reactor model can be configured to simulate one-dimensional fluid advection, in which local equilibrium obtains between successive small batches of fluid and segments of a column of reactive rock along a fluid low line. Similar models have been applied to specific situations of hydrothermal transport and redistribution of metals (Bloom *et al.*, this volume; Heinrich, 1990).

CHEMIX can readily be configured to calculate the consequences of fluid-fluid mixing and conductive cooling, or combinations of these processes, using a stage-wise reactor model. CHEMIX may also be coerced to calculate fluid speciation and mineral precipitation or dissolution as a function of volatile loss from a fluid undergoing phase separation, either "boiling" of hydrothermal fluids (Reed and Spycher, 1984) or "pressure-induced fluid immiscibility" (Bowers, 1991). Further code development is necessary for computations of the latter types, and the advanced features of CHILLER (Spycher and Reed, 1989) for Rayleigh fractionation and EQ6 (as modified by Bowers, 1991) for the associated stable isotope signatures are not as yet implemented. The existing capabilities with Redlich-Kwong and virial expansions, however, will facilitate simple extension of H_2O-CO_2 to derive volumes and fugacities for other gases which conform to the corresponding states principle, using a compensated Redlich-Kwong formulation (Holland and Powell, 1991).

program REACT

REACT calculates a range of thermodynamic properties for chemical species and changes in properties for chemical reactions. It allows a reaction or a list of species to be defined using standard chemical terminology and if necessary for a reaction to be balanced. The revised HKF equation of state for aqueous solute species and data implemented in *SUPCRT92* (Johnson *et al.*, 1992) may be employed, but it is also possible to compute properties for any other species or reactions for which conventional free energies are available or can be calculated. Some twenty-two thermodynamic properties may be tabulated, plotted or regressed as a function of temperature in any chosen units. There is also the inverse procedure of tabulating the temperatures which correspond to selected values of thermodynamic properties.

For the defined reaction, the equilibrium yield and the composition of reaction products may be calculated by an iterative solution of the equilibrium constant expression. The mass and heat balances between initial and final states may be calculated, and if the reaction is considered as an electrochemical cell, the cell emf and other properties may also be calculated. Another option allows basic thermodynamic data for a species to be derived from a set of equilibrium data for a reaction by the "thirdlaw" method. The input data may be the Gibbs energy change, equilibrium constant or cell emf, and the output property the standard enthalpy of formation, entropy or heat capacity of the species or phase.

program SYSTEM

The equilibrium relations of chemical species within a defined chemical system may be compactly represented on a variety of chemical potential diagrams. The most common of these used in modelling geochemical processes are predominance, activity-activity and fugacity diagrams. SYSTEM automatically calculates, displays and provides hard-copy output for these and other chemical potential diagrams.

The properties available for use as the axes of a diagram may be chosen from among eight options (log fugacity of species *i*, log activity of species *i*, Eh, pH, pe, ΔG, temperature and inverse temperature). The log fugacity and log activity options may involve ratios of species on the axes, or may represent all species formed from one thermodynamic component. The system for a diagram is defined by entering the set of species to be involved in the axes, areas and overlay lines. Chemical reactions involving all valid combinations of these species are set up and balanced. The basic thermodynamic data for the species are retrieved from the data base or are entered directly, and the Gibbs energy changes of the reactions calculated, including the effects of fixed activities and molar volumes.

The predominance areas are calculated by a revised and extended algorithm of Froning, Shanley and Verinc (1976). Up to five principal (metal) components and five ligand components are allowed in the system. A stable area is defined as a closed, convex polygon surrounded by reaction or axis lines. SYSTEM first calculates stable ligand areas, and then stable species areas within each ligand field. Overlay lines represent reactions which are involved with the axes but not with the stable areas.

When more than one metal component is present in a system, the diagram is often complicated by overlapping areas which result from compounds containing more than one component. This may be avoided by defining the limiting upper and lower ratios of the components, thereby restricting the system to certain areas of stoichiometry. Two or more sub-diagrams may then be required to define the entire system.

The computed diagram, with the axes scaled, marked and labelled and the areas identified, is presented in a standard Macintosh GUI window and may be saved to a PICT file and(or) sent to a laser or other postscript printer.

Although the data base has extensive coverage, it is often the case that data are required for one or more species of a reaction or system, for which there has been no thermodynamic measurement. The separate program ESTIMA provides for this eventuality by facilitating estimates of basic thermodynamic data for organic gases, inorganic solids (including silicates), and aqueous ions and complexed species.

REFERENCES

Berman, R.G. 1988. Internally-consistent thermodynamic data for minerals in the system Na_2O K_2O CaO MgO FeO Fe_2O_3 Al_2O_3 SiO_2 TiO_2 H_2O CO_2. *Jour. Petrol* 29: 445-522.

Bloom, M.S., Gammons, C.H., Gilbert, D. & Wilde, A.R. 1992. Reaction path models for hydrothermal Au-PGE mineralization of the South Alligator Valley Mineral Field, Australia. *7th International Symposium on Water-Rock Interaction*, 9-23 July, 1992, Park City, Utah. (this volume)

Bowers, T.S. 1991. The deposition of gold and other metals: pressure-induced fluid immiscibility and associated stable isotope signatures. *Geochim. Cosmochim Acta* 55: 2417-2434.

De Capitani, C. & Brown, T.H. 1987. The computation of chemical equilibrium in complex systems containing non-ideal solutions.*Geochim. Cosmochim. Acta* 51: 2639-2652

Eriksson, G.A. 1975. Thermodynamic studies of high-temperature equilibria, XII. SOLGASMIX a computer program for calculation of equilibrium compositions in multiphase systems. *Chimica Scr.* 8: 100-103.

Eriksson, G.A. 1979. An algorithm for the computation of aqueous multicomponent, multi-phase equilibria. *Anal. Chim. Acta* 112: 375-383.

Froning, M.H., Shanley, M.E. & Verink, E.D, 1976. An improved method of calculation of potential-pH diagrams of metal ion-water systems by computer, *Corrosion Sci.* 16:371-377.

Heinrich, C.A. 1990. The chemistry of hydrothermal tin(-tungsten) ore deposition. *Econ. Geol.* 85: 457-481.

Heinrich, C.A., Henley, R.W. & Seward, T.M. 1989. Hydrothermal systems: Adelaide, *Austra-lian Mineral Foundation*, 74 p.

Holland, T. & Powell, R. 1991. A compensated Redlich-Kwong (CORK) equation for volumes and fugacities of CO_2 and H_2O in the range 1 bar to 50 kbar and 100-1600°C. *Contrib. Mineral. Petrol.* 109: 265-273.

Johnson, J.W., Oelkers, E.H. & Helgeson, H.C. 1992. *SUPCRT92:* A software package for calculating the standard molal thermodynamic properties of minerals, gases, aqueous species and reactions from 1 to 5000 bars and 0° to 1000°C. *Computers & Geosciences* (in press).

Mills, K.C. 1974. Thermodynamic data for inorganic sulphides, selenides and tellurides. Butterworths, London, 845 p.

Reed, M.H. 1982. Calculation of multicomponent chemical equilibria and reaction processes in systems involving minerals, gases and an aqueous phase. *Geochim. Cosmochim. Acta* 46: 513-528.

Reed, M.H. & Spycher, N.F. 1984. Calculation of pH and mineral equilibria in hydrothermal waters with applications to geothermometry and studies of boiling and dilution. *Geochim. Cosmochim. Acta* 48: 1479-1492.

Spycher, N.F. & Reed, M.H. 1989. Evolution of a Broadlands-type epithermal ore fluid along alternative P-T paths: Implications for the transport and deposition of base, precious and volatile metals. *Econ. Geol.* 84: 328-359.

Sverjenski, D.A., Hemley, J.J. & D'Angelo, W.M. 1991. Thermodynamic assessment of hydrothermal alkali feldspar-mica-aluminosilicate equilibria. *Geochim. Cosmochim.. Acta* 55: 989-1004.

Turnbull, A. G. & Wadsley, M.W. 1986. The CSIRO-SGTE THERMODATA System, version V: Melbourne, CSIRO Div. Mineral Chemistry Commun., v. 1-7, 413 p.

Turnbull, A. G. & Wadsley, M.W. 1984. Thermodynamic modelling of metallurgical process-es by the CSIRO-SGTE THERMODATA system. Melbourne, *Aus.I.M.M. Symposium on extractive metallurgy*, 79-85.

Wolery, T. 1983. EQ3NR, a computer program for geochemical aqueous speciation-solubility calculations, User's guide and documentation. Lawrence Livermore National Laboratory, UCRL-53414, 191p.

Zhu, C. & Sverjenski, D.A. 1991. Partitioning of F-Cl-OH between minerals and hydrothermal fluids.*Geochim. Cosmochim. Acta* 55: 1837-1858.

Water-Rock Interaction, Kharaka & Maest (eds) © 1992 Balkema, Rotterdam. ISBN 90 5410 075 3

Thermodynamic model of interaction between sulphides and oxidizing solutions with special attention to intermediate conditions

G.R. Kolonin & O.L. Gas'kova
Associated Institute of Geology, Geophysics and Mineralogy, Academy of Sciences, Novosibirsk, Russia

ABSTRACT: A thermodynamic model of oxidizing dissolution of sulphides is presented using Helgeson's ξ-reactor approach. It describes sulphide behavior both under natural conditions and in some artificial situations, e.g., at sulphide-bearing mine dumps. Three stages of progressive process are emphasized depending on the ratio of the dissolved constituents in solution as well as the red-ox conditions. They are: 1) the reduction stage, taking place at the film solution on the sulphide surface; 2) the intermediate stage caused as further reactions progress, characterized by non-oxidized components both in solution and at solid phases; and 3) the equilibrium stage corresponding to oxidized conditions similar to atmospheric conditions.

1 INTRODUCTION

There are an impressive number of studies devoted to geochemical and physical-chemical conditions of the formation of oxidized zones in sulphide deposits. In this paper processes are considered in terms of Helgeson's ξ-reactor (1979) approach, which makes it possible to estimate dynamics of development of reactions on the boundary between sulphide and the solution. Reaction progress of oxidizing dissolution, in non-dimensional time, can be joined with the space coordinate and be connected directly with reduction inside the mineral. The order of the intermediate conditions of the system and their features should not be changed sharply by changing the scale, e.g., transition from individual grains to ore pieces and even to the oxidizing zones of single ore bodies or ore deposits. The simultaneous occurrence of different states of local equilibrium may be presumed for any of these scales. A physical-chemical model of an oxidizing deposit can thus be generalized in time and space.

2 THE COMPUTATION TREATMENT AND INITIAL THERMODYNAMIC DATA

Simulation has been carried out with the use of the free energy minimizing code "GIBBS" (Shvarov 1976) at 25°C and 1 bar pressure. The main features of the separate numerical experiments are described in detail by M. Borisov and Y. Shvarov (1992). The quantity of sulphide reacted with 1.0 kg H_2O increased from 10^{-6} to 0.1 g-mol. In most cases, in the initial solid phase, pyrite, chalcopyrite, sphalerite and galena were present simultaneously. In some cases there were pyrite and chalcopyrite, or only pyrite or chalcopyrite. The handbook by Naumov et al. (1971) was used as a source of free energy constants for these minerals as well as for 30 other minerals, the appearance of which was the result of the reactions in solution. Thermodynamic properties of the aqueous species were taken from Tanger and Helgeson (1988). Thermodynamic properties of sulfate complexes were taken from Ryzhenko et al. (1985).

3 RESULTS OF THE FIRST STAGE OF DISSOLUTION (REDUCED CONDITIONS)

Figure 1 shows the main features at the beginning of sulphide grain dissolution, when the first drops of water touch their surface inside of the ore body where contact with atmospheric oxygen is absent. One can see formation of chalcocite and bornite as new phases under these conditions. Subzones of sulphide enrichment of many ore deposits are well known and exhibit high stability of these copper sulphides under similar natural conditions. The appearance of elemental sulfur during slow dissolution is the next important feature of the process, where sulphide-

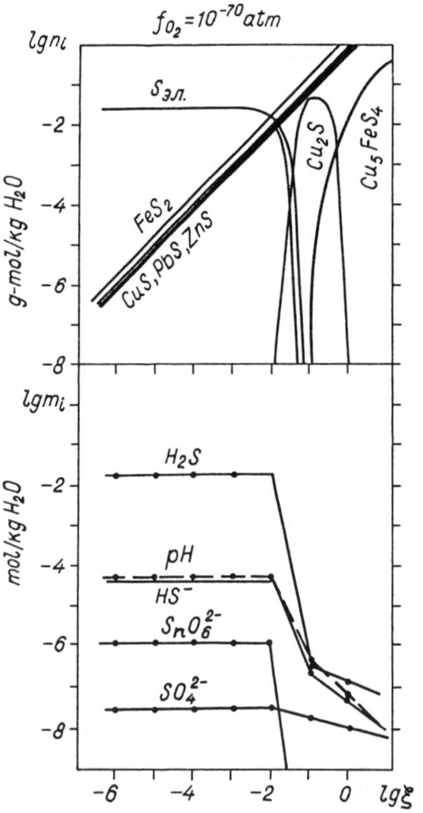

Figure 1 - Calculated composition of solid phase (above) and solution (at the bottom) under reduced conditions ($fO2=10^{-70}$) after interaction of FeS2 + CuFeS2 + ZnS + PbS assemblages with initial H2SO4 0.05-molality as function of solid phase/solution ratios (log ξ).

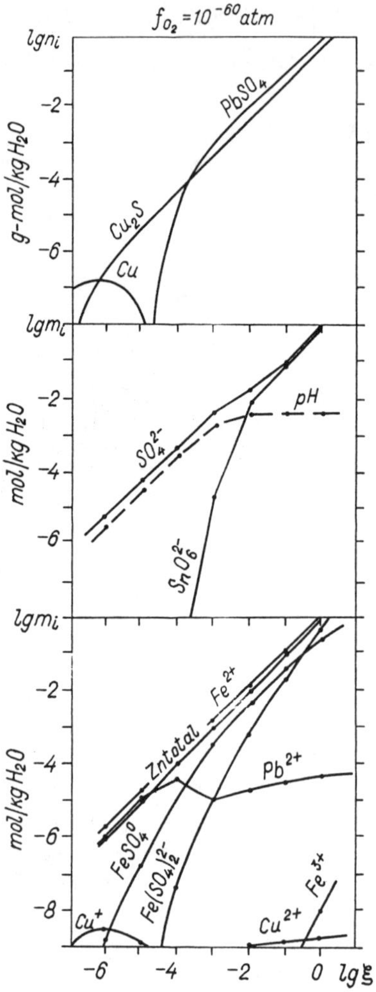

Figure 2 - Calculated composition of solid phase (above), possible concentration of sulfur complexes (middle) and the metal complexes (at the bottom) in solution under intermediate conditions ($fO2 = 10^{-60}$) after interaction of FeS_2 + ZnS + PbS assemblages with water as a function of log ξ.

dominating conditions are characterized by high values of solid sulphide:solution ratios. These phase charges are in good correspondence with a solution composition where H_2S predominates and polythionate as well as sulfate sulfur are present only in negligible amounts. Pyrite, sphalerite and galena remain stable minerals under these conditions and, as a result, Fe, Zn and Pb in the solutions do not reach measurable concentrations.

4 SECOND STAGE OF DISSOLUTION (INTERMEDIATE CONDITIONS)

When the system is open, extremely limited quantities of oxygen ($f_{O2} = 10^{-60}$ bar) are needed for full dissolution of appreciable amounts of pyrite and sphalerite, with Fe and Zn going into the solution phase, partly as ferro-sulfate complexes (Fig. 2). Copper remains in the solid phase as chalcocite. The predominant feature of this stage is the oxidation of sulfur from the initial sulphides into sulfate and the possible occurrence of significant amounts of more exotic intermediate sulfur forms (polythionate and thiosulfate). The transformation of galena to anglesite, and its very restricted solubility above pH2,

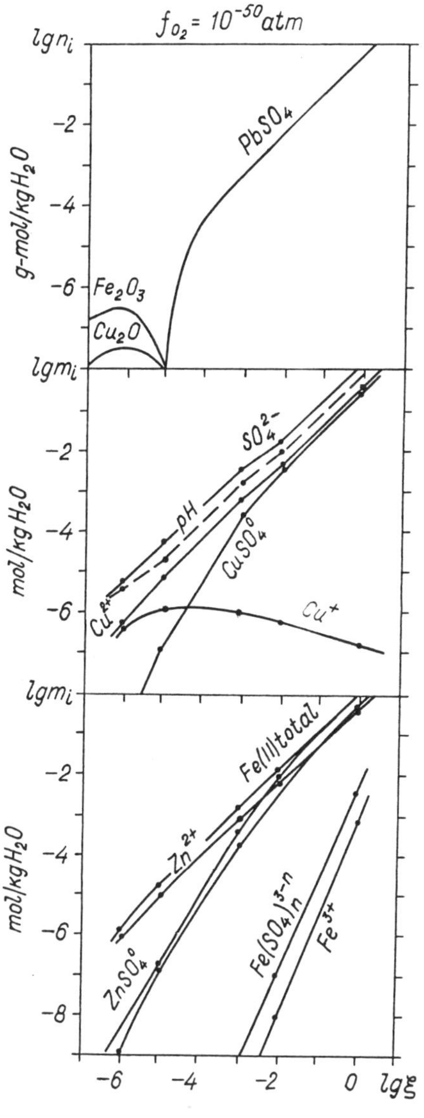

Figure 3 - Calculated compositions for oxygen fugacity=10^{-50} atm. Concentration of Cu complexes are shown in the middle plot.

explains the complete absence of Pb in solution up to an oxygen fugacity of 10^{-50} atm (Fig. 3). At the same time the full oxidizing dissolution of copper sulfides takes place under these conditions, but the description of the buffering action of surrounding rocks at this pH is beyond the scope of the model. Appearance of high Cu concentration in the form of the aqueous ion and sulfate complex as well as ferri-sulfate complexes are the result of reaction progress at intermediate oxygen conditions.

5 THE THIRD STAGE OF DISSOLUTION (OXIDIZING CONDITIONS)

An open system containing sulfides should come to equilibrium at an oxygen fugacity of about 10^{-30} bar. Unlike the solution under intermediate conditions, the initial composition of the sulfide assemblage has great importance in this stage. The dissolution of sphalerite and pyrrhotite, just like pyrite, favors high solution acidity and makes the preservation of newly formed lead and copper materials more problematical. In general, chalcopyrite ore composition is preferable, in order for native copper as well as tenorite and cuprite to appear. The iron sulphides should be transformed completely into goethite or hematite and jarosite. The latter is represented as hydronium jarosite (Fig. 4). The position of these phases is unusual in the diagram because goethite or other iron hydroxides are the final oxidizing products under natural paragenesis (e.g., Shadlun 1948). The high sulfate concentrations within the framework of the model may explain the formation of jarosite. Perhaps the interaction between sulphuric acid solutions and silicate or carbonate rocks can explain the more restricted occurrences of natural jarosite. Similar conditions should favor lower copper concentration due to precipitation of chrysocolla, malachite and azurite.

6 CONCLUSION

A simplified description of changing stages of the "sulphide-oxygen-water" system is presented in Fig. 5, constructed for pyrite. It corresponds to constant sulfate values of 10^{-3} M, e.g., to a rather high dissolution reaction rate, and reflects the possible states of local equilibria during imaginary movement from inner zones of the ore piece to its surface (from left to right). Three parts of the abscissa correspond to main stages of oxidizing dissolution: 1) conditions of pyrite stability in sulphide (reduced) solutions; 2) conditions of unstability of any iron solid phase when the unrestricted dissolution of ferro-forms takes place; and 3) conditions of goethite stability formed as a result of oxidizing of Fe(II) and precipitation of Fe(III), with a decrease in total iron concentration by about three orders of magnitude. Similar diagrams are especially interesting for

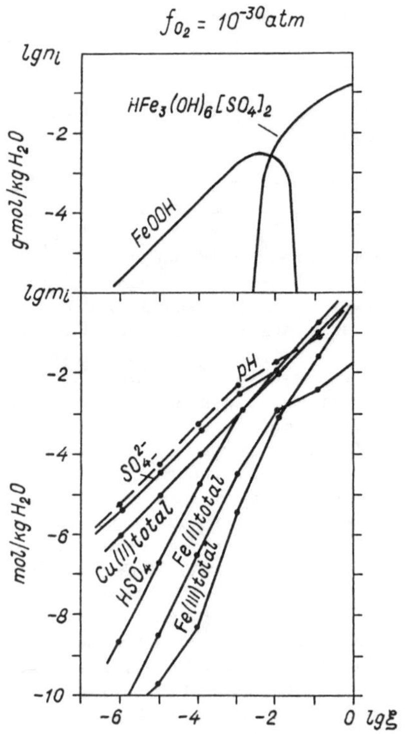

Figure 4 - Possible result of pyrite interaction with water under rather oxidizing conditions (f02 = 10^{-30} atm).

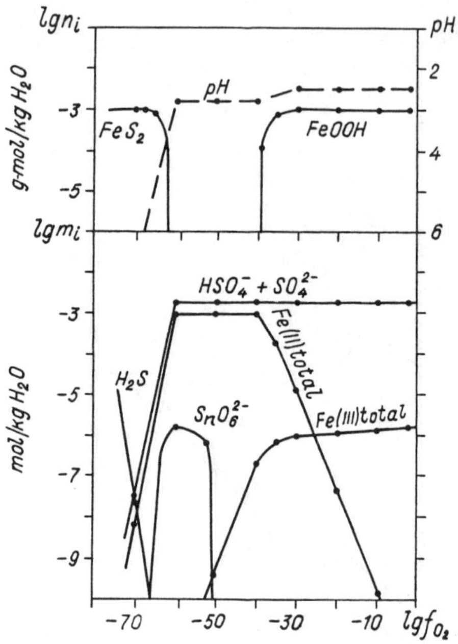

Figure 5 - Dependence of FeS2 - water system on oxygen fugacity at constant log ξ = -3.

sulphides containing changeable valency elements (arsenopyrite, chalcopyrite) because they show the step-by-step sequences of increasing oxidation state both in solid phases and in solution, which are often shifted relative to one another.

It is obvious that the results presented will remain a rather rough approximation of natural reality until interaction of the solution with sulphide-containing rocks is taken into account. Nevertheless, including additional phases in the initial system composition, such as calcite, does not change the overall features of oxidizing dissolution of sulphides.

REFERENCES

Borisov, M.V. 1992. Thermodynamics of geochemical processes. Moscow: MGU.

Helgeson, H.C. 1979. Mass exchange between minerals and hydrothermal solutions. In H. Barnes (ed.), Geochemistry of Hydrothermal ore deposits: 451-480. N.Y.-Chichester-Brisban-Toronto: J. Wiley and Sons.

Naumov, G.B., B.N. Ryzhenko & I.L. Khodakovsky 1971. Handbook of thermodynamic values. Moscow: Atomizdat. 239 p.

Ryzhenko, B.N., A.I. Shapkin & O.V. Bryzgalin 1985. Electrolytic dissociation of sulfates of metals in aqueous solution. Geochemistry 4: 534-539.

Shadlun, T.N. 1948. Mineralogy of oxidizing zone of Blyava polymetallic sulphide deposit, S. Ural. In Tr. of Inst. Geol. Sci. 96/11: 1-102.

Shvarov, Yu.V. 1976. Calculation of equilibrium composition in the multi-component heterogeneous system. Docl. AN SSSR 229: 1224-1227.

Tanger, J.C. & H.C. Helgeson 1988. Calculation of the thermodynamic and transport properties of aqueous species at high pressures and temperatures. Am. J. Sci. 288: 19-98.

Water-Rock Interaction, Kharaka & Maest (eds) © 1992 Balkema, Rotterdam. ISBN 90 5410 075 3

Aluminium: Catalyst, crystal deformer, and dye: Self-organizational crystallization of quartz in agates and amethyst geodes, Paraná flood basalts, Brazil

Enrique Merino & Yifeng Wang
Department of Geological Sciences, Indiana University, Bloomington, Ind., USA

Etienne Deloule
Centre de Recherches Pétrographiques et Géochimiques, Vandœuvre-lès-Nancy, France

ABSTRACT: Associated repetitive textures and composition of agate quartz require closed-system growth via Al catalysis at a morphologically unstable growth front. These instabilities operate only for very-silica-rich drops, not aqueous solutions. Trace element contents confirm this model. Quartz fibers that incorporate Al become twisted. The quartz becomes a new polymorph, one with no symmetry. Agates are all similar because textures/compositions are determined by the Al and Fe partitioned into the original silica drops, themselves probably formed by liquid immiscibility.

1 INTRODUCTION

Agates in flood basalts display striking repetitive quartz textures and composition: layers made of low-Al, fine, untwisted quartz fibers systematically alternate with layers made of high-Al, finer, *twisted* fibers. All fibers are length-fast. The fibrous layers are followed inward by a layer of coarse, length-slow quartz, and by a central void (Michel-Lévy and Munier-Chalmas, 1892; Frondel, 1978; Merino, 1984; and Wang & Merino, 1990). The repetitive textures and compositions observed must be produced by a self-organizational process -- one governed by its own dynamics, not by externally imposed causes (Nicolis and Prigogine, 1977). The genesis of agate textures and compositions brings out remarkable problems in crystal growth, quartz crystal-chemistry, the geochemistry of authigenic textures (Merino 1990), and the properties of igneous melts. Agates have been thought to form by alteration: silica-rich waters come from somewhere and precipitate quartz in basalt vesicles. We think this idea cannot explain the genesis of the textures and compositions of agate quartz. The crystallization model below plus field and ion-probe evidence suggest that agates must be part and parcel of the igneous phenomenon, and must form under closed-system conditions from silica that came from nowhere -- it was there all along since it separated by (we infer) liquid immiscibility from liquid basalt.

2 MODEL

All the textures described above are beautifully explained by a self-organizational reaction/transport model based on the autocatalytic reaction,

$$\text{Initial Medium} \overset{Al}{\Longrightarrow} \text{Quartz} + Al^{3+},$$

and on a morphologically unstable quartz crystallization front. The reaction makes concentrations of Al (and other traces) at the growth front oscillate (Fig. 1 and Wang

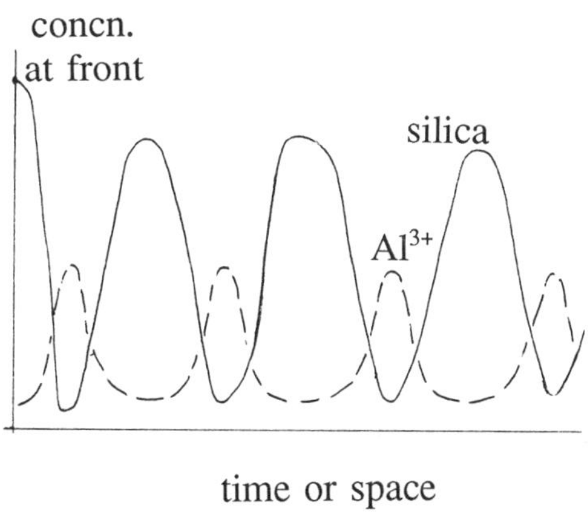

Fig. 1. Oscillation of concentrations of silica and Al^{3+} at front, produced by autocatalysis -- see reaction in text and Wang & Merino (1990, figs. 5,6).

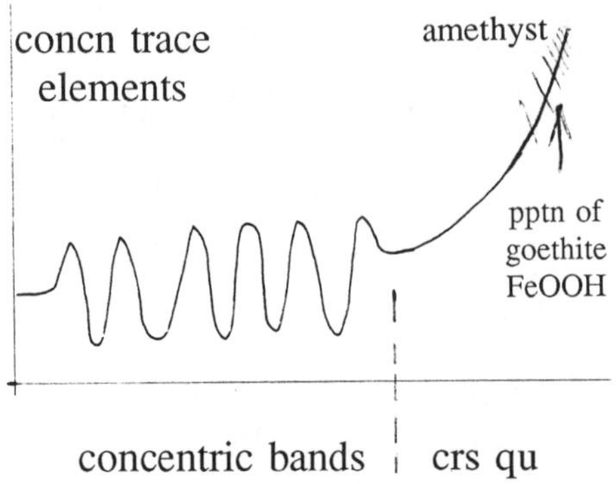

Fig. 2. Ion-probe profile of trace-element contents is evidence of closed-system growth. Towards end, increasing Fe^{3+} produces increasingly darker purple amethyst, and may produce "igneous" goethite within it.

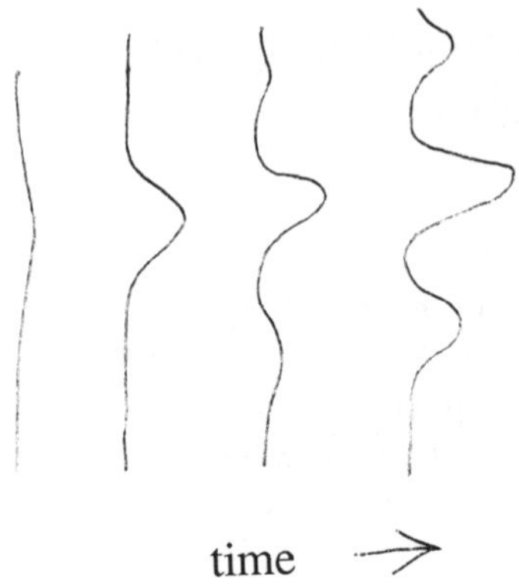

Fig. 3. Morphologically unstable crystallization front grows fingers. Fingers become fibrous texture.

& Merino, 1990). Thus, trace-element contents incorporated by the growing quartz also oscillate. This is confirmed by cathodo-luminescence and by electron- and ion-probe analyses for H, Na, K, Al, Fe, Ti, and Ni (Fig. 2). Predicted band spacings coincide with observed ones. Predicted formation time of each fibrous band is a few tens of hours.

3 FIBER GROWTH AND TWISTING: NEW MECHANISM

Quartz grows as fibers because the growth front is morphologically unstable (Fig. 3). Any bump in an unstable growth front lengthens into a finger and produces other bumps nearby, which also lengthen. Each finger in the front is a quartz fiber that grows by addition of conical sleeves (Fig. 4). As fibers grow, the sleeves pierce through decreasing, and oscillating, profiles of Al concentration. The bases of these sleeves thus take up more Al than their tips do, so the fiber ends up having more Al in its periphery than in its core. This radial gradient makes the fiber grow twisted to house excess Al ions (which are larger than Si ions). The proposed mode of fiber growth and twisting mechanism are new. Calculated twist periods coincide with observed twist periods (Merino, Wang, & Deloule, unpubl). A quantitative model of these ideas yields which conditions produce different authigenic textures (Wang & Merino, in progress; Merino 1990).

Twisted quartz can be viewed as a new polymorph having much less symmetry than quartz does, both because of Al^{3+} ordering and because of the twisting it produces.

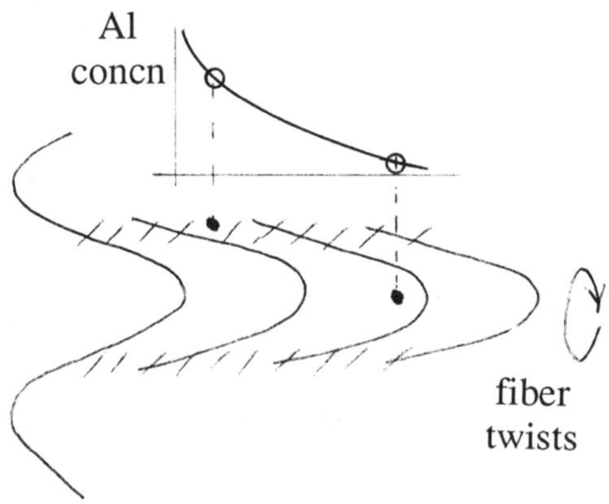

Fig. 4. Proposed mode of growth of a fiber by addition of conical sleeves that (because of the autocatalytic instability of Fig. 1) pick up more Al at their bases than at tips. The periphery of the fiber becomes richer in Al (ruled) than the fiber core. This radial gradient makes fibers twist to make room for excess peripheric Al. The twisted quartz has greatly lowered symmetry.

4 FIELD EVIDENCE

In a well-exposed basalt flow within the Paraná basalt, at Salto de Jacuí, Rio Grande do Sul, Brazil, numerous agates occur *in situ* in the bottom 3 m of the flow whereas all the vesicles, which are 12 m above the flow bottom, are empty. The agates did not grow in vesicles.

5 CRYSTALLIZATION SEQUENCE

The model sketched above predicts (Wang & Merino, 1990) that the initial material must have had huge silica concentration (> 1 g SiO_2/cm^3!) to give rise to the oscillatory textures and compositions observed. This points to growth not from aqueous solutions but from silica-very-rich lumps. The model, plus field and ion-probe evidence (below), suggests that the silica-rich, trace-element-bearing, hydrous lumps could have separated from the basalt only by liquid immiscibility. (Melting diagrams do show a narrow 2-liquid field with an end at 97% silica, but we do not know how the basalt could have reached it.) Regardless of how the lumps were emplaced, as the fibrous layers crystallize, silica in the residuum eventually falls below the 1-g-SiO_2/cm^3 threshold, while trace elements and water necessarily go up in concentration. Thus, growth naturally switches to coarse, non-banded, length-slow quartz. Optical orientation of the quartz fibers flips from length-fast to slow because silica concentration (and thus degree of polymerization) decreases -- Folk & Pittman (1971). The agate's central void reflects the difference between the SiO2 concentration of the initial melt and quartz density. The magmatic water in that central void eventually leaks or flashes out, altering surrounding basalt to layer silicates, which mistakenly have been taken as evidence of *open*-system alteration (Meunier et al, 1988).

6 AGATE QUARTZ CRYSTAL CHEMISTRY

Under closed-system growth, as quartz layers crystallize, trace elements and water must go up in concentration in the residuum. Thus, towards the end quartz must take up increasing Al^{3+} and Fe^{3+}, which make it increasingly darker smoky and purple toward the agate's center, as observed. The Fe^{3+} is produced, we think, by oxidation of original Fe^{2+} by H_2O via

$$Fe^{2+} + H_2O = Fe^{3+} + 0.5H_2 + OH^-,$$

driven by the very removal of SiO2 to make chalcedony layers. The OH^- produced helps grow zeolites, often observed in agates. Residual magmatic water and Fe^{3+} may even form goethite needles within the quartz.

Cathodo-luminescence, electron- and ion-probe analyses of agates (Fig. 2) all show the alternating trace-element content of agate layers predicted by the model, and the exponential increase in trace-elements one would expect to find toward the end of closed-system crystallization.

REFERENCES

Folk R L & Pittman J S, 1971. Length-slow chalcedony: A new testament for vanished evaporites. *J Sediment Petrol* 41: 1045-58.

Frondel C, 1978. Character of quartz fibers. *Amer Mineral* 63: 17-27.

Merino E, 1984 Survey of self-patterning textures. In *Chemical Instabilities* (eds. Nicolis G & Baras F), pp 305-328. Reidel.

Merino E, 1990. The geochemistry of quartz authigenic textures. *Chem Geol* 84:233-234.

Meunier A, Formoso M L L, Patrier P & Chies J O, 1988. Altération hydrothermale de roches volcaniques liée à la genèse des améthystes, bassin du Paraná, sud du Brésil. *Geochim Brasiliensis* 2: 127-142.

Nicolis G & Prigogine I, 1977. *Self-organization in Non-equilibrium Systems.* Wiley, New York.

Michel-Lévy A & Munier-Chalmas E, 1892. Mémoire sur diverses formes affectées par le réseau élémentaire du quartz. *Bull Soc Franç Minéral* 15:161-190.

Wang Y & Merino E, 1990. Self-organizational origin of agates: Banding, fiber twisting, composition, and dynamic crystallization model. *Geochim Cosmochim Acta* 54: 1627-1638.

Water-Rock Interaction, Kharaka & Maest (eds) © 1992 Balkema, Rotterdam. ISBN 90 5410 075 3

Integration of intensive variable diagrams and fluid phase equilibrium with SOLMINEQ.88 pc/shell

E. H. Perkins
Oil Sands and Hydrocarbon Recovery Department, Alberta Research Council, Edmonton, Alb., Canada

ABSTRACT

Assessment of the chemical interactions which have occurred between aqueous fluids and minerals can be undertaken by two different, but complimentary methods. One is to calculate the distribution of mass among the aqueous species and complexes and then to calculate the saturation state of all possible minerals. The other is to plot the stability fields of all possible minerals as a function of various intensive variables, specifically the activity of various components in the fluid and/or temperature and/or pressure, and then to plot the activity of the aqueous components for a given water on these diagrams.

A new release of SOLMINEQ.88 pc/shell allows both methods to be used in a quick and user friendly environment on a personal computer. The first method has been present in SOLMINEQ.88 since it's inception, along with a large number of various other options which allows the user to model geochemical processes. The use of intensive variable diagrams in SOLMINEQ.88 pc/shell is new, and is accomplished by precalculating a number of diagrams by using Ge0-Calc (the preferred method), by using any other program or by hand, and then storing them in a database integral to SOLMINEQ.88 pc/shell. At any point, these activity diagrams can be recalled and have plotted on them any number of water analyses which have been reduced or calculated by SOLMINEQ.88 pc/shell.

INTRODUCTION

SOLMINEQ.88 pc/shell (Perkins, et al, 1989) is a interactive program for a IBM personal computer (or clone) which was specifically developed to provide a user friendly interface to the input, running and output of the geochemical modelling program, SOLMINEQ.88 (Kharaka et al, 1988). Although there are a great number of possible applications, one of the most common uses of these programs is to examine the saturation states (SI) of minerals using a given water analyses. From the mineral SI's and from field observations, the water-rock reactions which controlled the fluid composition can be deduced. The advantage of this approach is the rapidity with which individual analyses can be entered into SOLMINEQ.88 pc/shell, the resulting SI's calculated and examined. The main disadvantage is the difficulties encountered when examining a large number of water analyses and looking for correlations, especially when the water compositions and/or source temperatures are significantly different.

Intensive variable diagrams have been used for many years to display the stability relationships between phases and/or dominate species (aqueous, gas or solid) as a function of the log activity of various components (or their ratios), temperature or pressure. These diagrams are more commonly named after their axis titles, for example: log activity diagrams where both axis represent the log activities of components; temperature versus log activity diagrams where one axis represent temperature and the other the log activity of a component; etc. The areas on these diagrams which are enclosed by the line represent the stability field of a phase or assemblage. The lines represent equilibrium reaction between the two different phases or assemblages. Ge0-Calc (Berman et al, 1989) allows the rapid, thermodynamic valid calculation of these types of diagrams in the user friendly environment of an IBM personal computer or clone. If the log activity of an aqueous species (which has the same composition and thermodynamic properties as the axis component) is calculated from a given water analysis, it can be plotted on the appropriate diagram. This identifies the assemblage or equilibrium which controls that component in the fluid. Large amounts of data can be easily plotted on these figures, trends and reaction controls are very easy to spot. The main disadvantage of these figures is the difficulty of calculating the activities of the component species and mechanically adding them to the diagrams.

A new release of SOLMINEQ.88 pc/shell removes

the disadvantages of both methods by integrating the functionality of both into a single user friendly program.

PROGRAM DETAILS

Intensive variable diagrams can take considerable computer time to calculate, but once calculated, they can be used to present many different sets of data. SOLMINEQ.88 pc/shell takes this into account by storing precalculated intensive variable diagrams in an internal data base. Thus the general procedure is to initially calculate a standard suite of diagrams that may be needed for a given project and to enter them into the SOLMINEQ.88 pc/shell activity diagram data base. Then once a series of water analyses have been input and run, the appropriate diagram can be displayed on the screen with the water analyses plotted on them.

To accommodate these extra functions, the menus and functions of SOLMINEQ.88 pc/shell have been expanded. The main command screen now includes a new command, PLOT ACTIVITIES. When this is selected, a new screen with six options appears. The first option allows you to select an intensive variable diagram and plot data from selected SOLMINEQ.88 pc/shell run products. This option will be discussed in more detail below. The next three maintain and modify the intensive variable data base. The first of these adds a new intensive variable diagram to the SOLMINEQ.88 pc/shell data base. The user is prompted for a file containing the diagram, which must be stored in a Ge0-Calc plot diagram format. This format is simple, and if necessary can be prepared with any editor. The axis labels are parsed to recover information about the diagram and the user is prompted to correct the parsing and to enter ancillary information before entering the diagram into the SOLMINEQ.88 pc/shell intensive variable data base. The second of these three options allows you to mark an intensive variable diagram to be removed (deleted) from the data base or to remove the deletion flag from a diagram marked for deletion. These diagrams can be recovered until the next selection, the Re-Index function is selected. The Re-Index function cleans up and compresses the data base by removing all of the intensive variable diagrams which have been flagged for deletion. After this option has been run, deleted diagrams can NOT be recovered. The next option allows the user to save an intensive variable diagram in a ASCII file using the Ge0-Calc format, thus allowing manual modification of the figure. The final option on this screen is to return to the main menu.

If the first option (plot a diagram) is chosen, a screen similar to figure 1 appears. This screen lists the number of diagrams in the database, the number which are current (not marked for deletion) and the number of the diagram currently shown. Information on the current diagram follows.

Using the cursor keys, you can page through the

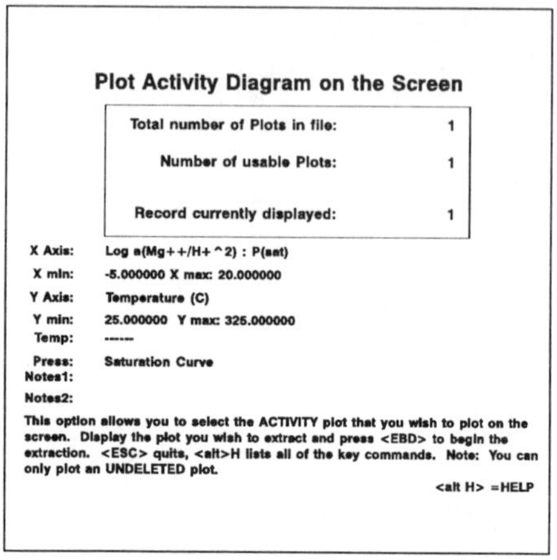

Figure 1: Screen display for the selection of which intensive variable diagram is to be plotted.

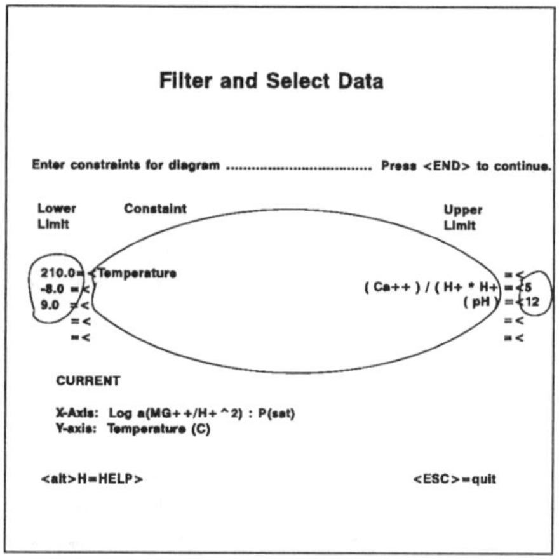

Figure 2: Screen which allows the user to set constraints on the data to be included in the diagram.

diagram data base until the diagram you wish to use appears. Select it by pressing the <END> key. The diagram currently displayed would plot the log activity of magnesium over hydrogen ion squared versus temperature. The next screen allows you to select the data files to be plotted. You can specify which symbols are used for the data from each of the files. The next screen normally seen is shown as figure 2. This screen allows the user to enter various constraints on what data will be extracted from the SOLMINEQ.88 pc/shell data files and plotted. For clarity, examples of three different types of constrains have been shown and they are highlighted on the figure. The first constraint indicates that only data sets with a temperature greater than or equal to 210 degrees Centigrade are to be plotted. The second indicates that the log activity ratio of calcium over hydrogen ion squared must be equal to or between -8 and 5 to be plotted, while the third constraint indicates that only samples with a pH equal to or between 9 and 12 will be plotted. As can be seen from this strange set of constraints, they can refer to components or species which are not present as axis variables. This allows a very flexible set of filters to be applied to any set of data.

Figure 3 is typical of the SOLMINEQ.88 pc/shell output when the Plot Activities option is chosen.

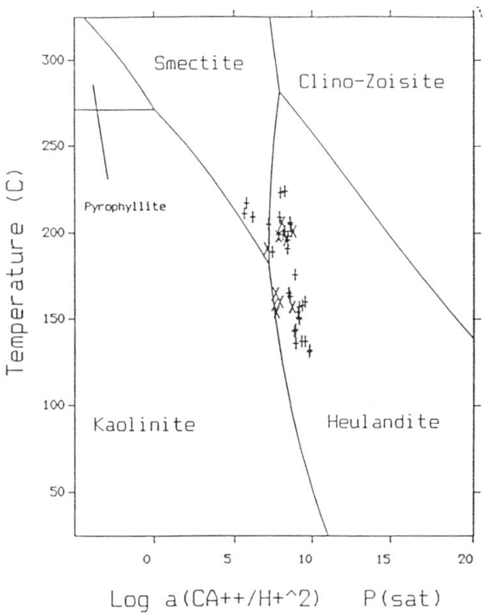

Figure 3: A typical temperature versus activity diagram with data as plotted using SOLMINEQ.88 pc/shell.

The figure shows the stability relationships in the calcium, silicon, oxygen and hydrogen system, balancing on aluminum and projecting from both quartz and water. The data is from produced water data collected from the MOCO Monarch Anticlinal Area Steam Drive Pilot in the South Midway-Sunset Field, Kern County, California (Perkins, et al, 1992).

REFERENCES

Berman, R.G., T.H. Brown, and E.H. Perkins. PTA-SYSTEM: A Ge0-Calc Software Package for the Calculation and Display of Activity-Temperature-Presssure Phase Diagrams. American Mineralogist, 74, p 485-487.

Kharaka, Y.K., W.D. Gunter, P.K. Aggarwal, E.H. Perkins, and J.D. DeBraal. 1988. Solmineq.88 - A Computer Program for Geochemical Modeling of Water - Rock Interactions. USGS Water Resources Investigations Report 88-4227. 420 pages.

Perkins, E.H., W.D. Gunter, M.L. Barrett, and J. Turner. 1992. Rock-Steam Reactions at a South Midway-Sunset Heavy Oil Reservoir. Published in the Conference Proceedings of the Symposium on California Fossil Fuel Resources, American Chemical Society, San Francisco Meeting, April 5-10, 1992.

Perkins, E.H., W.D. Gunter and W.D. Bird. Application of SOLMINEQ.88 and SOLMINEQ.88 pc/shell, in the Proceedings of the 6th International Symposium on Water-Rock Interactions (WRI-6), editor D.L. Miles, Aug. 3-8, 1989, pages 553-556.

Water-Rock Interaction, Kharaka & Maest (eds) © 1992 Balkema, Rotterdam. ISBN 90 5410 075 3

Collapse of acid waters into boiling hydrothermal system and the origin of late stage pyrite and related kaolinite

Mark H. Reed
University of Oregon, Eugene, Oreg., USA

Geoff Plumlee
US Geological Survey, Denver, Colo., USA

ABSTRACT: Many hydrothermal systems contain late-stage and peripheral mineral assemblages dominated by botryoidal pyrite and marcasite, with or without additional sulfosalts, realgar, stibnite, gold, cinnabar, barite, alunite and kaolinite. Numerical models of mixing of acidic, Fe and Al-rich shallow ground waters with ascending gases and liquids from the boiling hydrothermal system show that pyrite/marcasite alone form where H_2S-rich gases condense into Fe-rich ground water and that trace element-rich pyrite with accessory sulfosalts, barite, etc. form where ascending liquids mix with Al and Fe-rich groundwaters.

1 INTRODUCTION

Botryoidal pyrite and marcasite have been documented in the late stages of many epithermal systems, including Creede and Lake City, CO, Julcani and Orocopampa, Peru (Plumlee, 1989) and Bohemia, OR (Katsura, 1988; McChesney, 1987). In all deposits studied, some pyrite and marcasite growth zones are extremely enriched in certain trace elements, particularly As, Sb, Ag, Au, Hg, Mn and Te. These growth zones typically contain intergrown stibnite, cinnabar, realgar, Pb-As sulfide glass, gratonite, sphalerite, acanthite, barite and sulfosalts. Other growth zones lack such accessory sulfides, particularly those with stalactite and drape textures, which suggest deposition of pyrite/marcasite from fluids flowing down the walls of vapor-filled cavities (Plumlee, 1989). In addition to accessory sulfides, in some examples Al-minerals are admixed or interlayered with pyrite, e.g. kaolinite at Bohemia (Katsura, 1988) and alunite at Lake City (Plumlee and Rye, 1992). In some cases the late sulfides and kaolinite precipitated on corroded sulfides deposited at an earlier stage (Katsura, 1988; Plumlee and Rye, 1992).

Similar mineralization has also been identified on the low-temperature fringes of diverse hydrothermal deposits (Plumlee and Rye, 1992), and is thought to result from the mixing of low-temperature hydrothermal fluids or gases with cool, dilute groundwaters that overlie or surround hydrothermal systems. Plumlee and Rye (1992) have classified this fringe mixing environment as the "perithermal" environment. In epithermal systems, late stage perithermal mineralization forms as oxidized, acidic waters collapse into the central, main-stage ore zones; mineral precipitation may result from condensation of H_2S-rich steam into acidic groundwaters or from mixing of acidic groundwaters with low-T hydrothermal fluids.

2 NUMERICAL EXPERIMENTS

We have explored possible origins of the late-stage mineralization by numerically modeling (program CHILLER) several hypothetical mixing processes, of which we report three here. In each of the mixing schemes, we start with a descending acid-sulfate water generated by atmospheric oxidation of hydrothermal gas condensates above the boiling water table (Reed and Spycher, 1985). The acidic waters react with a dacitic wall rock during descent, producing alunite-kaolinite alteration and leaching iron and aluminum (Reed, 1992). Depending on their initial degree of oxidation and the extent of their reaction with wall rock, the descending waters may be strongly enriched in Fe or Al or both (up to 60 ppm Al, 120 ppm Fe).

We envision the acidic water trickling down a fracture. In some instances, the trickling water reacts with ascending gases or gas condensate on the walls of a gas-filled cavity; in other instances it reacts with the boiling liquid-gas mixture at and beneath the boiling water table. The three mixing calculations are:

i) Mixing at 99°C of an Fe-rich, Al-poor ground water (pH 4.65, 121 ppm Fe, .016 ppm Al; in equilibrium with quartz, pyrite, hematite, kaolinite, barite, cinnabar) with H_2S-rich gas condensate (pH 4.91) plus uncondensed gas. Results shown in Figure 1.

ii) Mixing of the same Fe-rich water with boiled, trace metal-rich liquid plus gas. Results shown in Figure 2.

iii) Mixing of an Al and Fe-rich acidic water (pH 2.74, 50 ppm Al, 48 ppm Fe; in equilibrium with kaolinite, barite, alunite, quartz, cinnabar) with trace metal-rich boiled fluids (as in case *ii*), pure silica phases except amorphous silica are not allowed to precipitate, thereby simulating the common condition of quartz supersaturation at low temperature. Results shown in Figure 3.

3 RESULTS

Case i. Mixing of the Fe-rich water with gas plus gas condensate produces nearly pure pyrite with traces (three orders of magnitude less) of cinnabar, sphalerite, galena and chlorite (fig. 1). The pyrite forms from reaction of gaseous H_2S with iron in the descending water. Pb and Zn were leached from dacite then precipitate in trace sphalerite and galena upon reacting with gaseous sulfur. In natural systems, there could be much more sphalerite and galena if the descending acidic waters leach early-formed vein sulfides (e.g. Creede sphalerite stalactites, Plumlee, 1989). Cinnabar forms as sulfur supplied by the gas reacts with Hg from both of the end member mixing fluids. Chlorite forms from constituents in the descending fluid when pH increases.

Case ii. Mixing of the descending Fe-rich water with ascending liquid and gas produces the same minerals as in case *i* (for the same reasons), plus substantial barite and trace electrum, stibnite, boulangerite, enargite, bournonite and microcline. Barite forms from Ba in the ascending liquid reacting with sulfate in the descending water. Precipitation of the Au, Ag, As and Sb minerals reflects the supply of these elements in the ascending liquid. Microcline forms when pH increases toward the end of the mixing calculation.

Case iii. Mixing of the Al and Fe-rich water with ascending silica-bearing liquid (plus gas) yields the same sulfide and sulfosalt assemblage as in case *ii*, but the concentrated Al in the descending water produces added alunite, pyrophyllite and microcline. Alunite forms under the acidic conditions of a large ratio of acidic water to boiled fluids (initial pH 2.74) by the addition of Al and sulfate in the descending fluid to K in the ascending liquid. (The descending fluid is depleted in K by precipitation of alteration alunite where the acidic water reacts with dacite above the fluid-fluid mixing zone; Reed, 1992.) Pyrophyllite replaces alunite when the pH increases to 3.1, then microcline replaces pyrophyllite when pH is 4.2 at a small ratio of acidic fluid to the boiled fluids. In the natural setting, one or another of these Al phases would be expected to form, depending on the actual mixing ratio at a given site in the shallow hot spring environment. Muscovite and illite fail to form in these numerical experiments.

4 DISCUSSION

Most of the observed mineral assemblages of natural systems are matched by the numerical model results, including the pyrite, sulfosalts, cinnabar, barite and alunite. We do not perfectly match the Pb-As sulfide glass, but Pb-Sb minerals, bournonite and boulangerite do form. It appears that we can account for the pyrite/marcasite precipitation by reaction of descending Fe-rich waters with sulfur from ascending gases, with or without an accompanying liquid. The difference between the results of cases *i* and *ii* show that natural pyrite/marcasite bands lacking barite, As, Sb and Au probably formed above the boiling water table where only the ascending gas phase could mix with descending waters. This conclusion is further supported by the common occurrence of stalactite and drape textures in the trace element-poor late pyrite (Plumlee, 1989). It is further apparent from cases *i* and *ii* that the trace element-enriched pyrite bands may form by acidification of ascending waters, as discussed by Spycher and Reed (1989).

The results of mixing of the Al-rich fluid with boiled water (case *iii*) indicate that alunite and

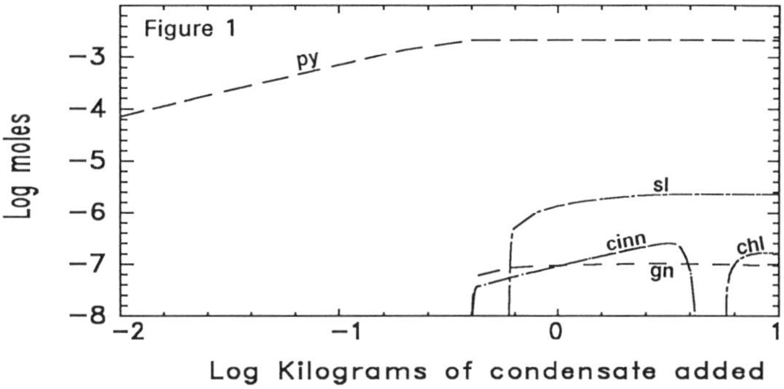

Figure 1. Minerals produced by mixing of Fe-rich acid sulfate water with the H_2S-rich gas phase plus condensate (case *i*). Abscissa shows the quantity of gas plus condensate added to one kilogram of the Fe-rich water.

Mineral abbreviations are as follows: alun, alunite; anh, anhydrite; bar, barite; boul, boulangerite ($Pb_5Sb_4S_{11}$); bour, bournonite ($PbCuSbS_3$); cinn, cinnabar (HgS); chl, chlorite; en, enargite; gn, galena (PbS); kaol, kaolinite; micr, microcline; py, pyrite; pyro, pyrophyllite; sl, sphalerite (ZnS); stib, stibnite.

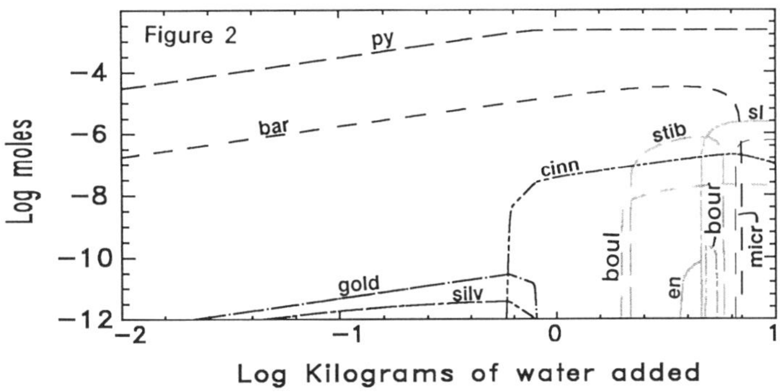

Figure 2. Mixing of Fe-rich acid sulfate water with trace metal-bearing, boiled liquid plus H_2S-rich gas (Case *ii*). Abscissa shows the quantity of liquid plus gas added to one kilogram of the Fe-rich water.

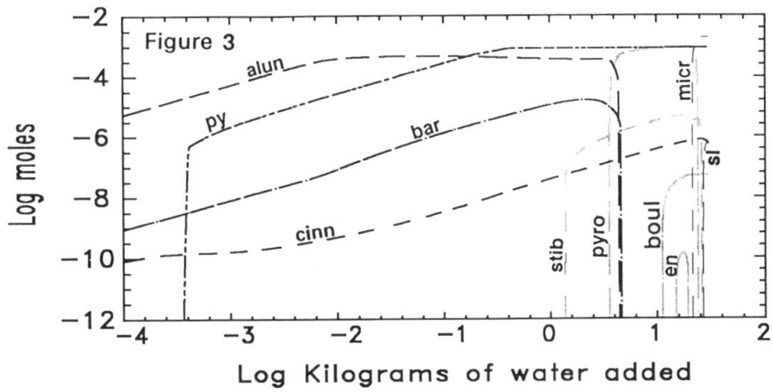

Figure 3. Mixing of Al- and Fe-rich acid sulfate water with trace metal-bearing liquid plus H_2S-rich gas (Case *iii*). Abscissa shows the quantity of liquid plus gas added to one kilogram of the Fe-rich water.

kaolinite admixed with late pyrite probably develops from a mixing of descending Al and Fe-enriched acidic waters with ascending silica-rich waters. In the case *iii* results, pyrophyllite precipitated instead of kaolinite, but this is a consequence of the elevated silica activity that prevails in the absence of quartz saturation. In separate model runs where quartz, chalcedony or cristobalite equilibrium is allowed, kaolinite forms in place of pyrophyllite.

The calculated precipitation in the quartz-supersaturated calculations of microcline (a proxy for adularia) with pyrophyllite but not muscovite (illite) resembles another observation, the association of opal, kaolinite and adularia in epithermal precious metal deposits such as Hishikari, Japan (Izawa, 1990). A similar association of adularia with kaolinite is found in some of the volcanic-hosted epithermal gold deposits of northern Nevada. These findings suggest the possibility that at least some of the adularia that is a nearly universal constituent of the banded opal vein ores of the precious metal hot spring environment originates in *descending* acidic waters rather than ascending waters and that some or all of the gold in this setting precipitates owing to acidification of gold-bearing ascending waters (Reed and Spycher, 1985; Spycher and Reed, 1989).

5 CONCLUSION

The numerical experiments show that perithermal mineralization formed in late-stage epithermal systems could likely have formed as a result of two processes: mixing of low-T hydrothermal fluids (liquid and gas) with cool, oxidized, acidic, steam-heated groundwaters or by condensation of H_2S-rich gases into such groundwaters. Mixing of boiled hydrothermal fluids with Fe-rich groundwaters results in precipitation of pyrite with abundant accessory minerals such as sphalerite, galena, barite and sulfosalts; enrichments of As, Sb, Ag, Au, Mn, etc. in the pyrite itself would also likely occur. Mixing of the boiled hydrothermal fluids with Al-rich groundwaters produces alunite, kaolinite and, at mixing ratios with a large proportion of boiled fluid, K-feldspar. In contrast, condensation of H_2S into groundwaters produces pyrite or marcasite with few trace elements. Thus, our modeling calculations reproduce quite well the range of mineral assemblages documented by Plumlee and Rye (1992) in the late-stage perithermal environment.

REFERENCES

Izawa, E., Urashima,Y., Ibaraki, K., Suzuki, R., Yokoyama, T., Kawasaki, K., Koga, A. and Taguchi, S., 1990, The Hishikari gold deposit: high-grade epithermal veins in Quaternary volcanics of southern Kyushu, Japan, Jour. Geochem. Explor., 36, 1-56.

Katsura, K., 1988, The geology and epithermal vein mineralization at the Champion Mine, Bohemia Mining District, Oregon. M.S. Thesis, University of Oregon, 254 p.

McChesney, S., 1987, Geology of the Sultana Vein, Bohemia Mining District, Oregon. M.S. Thesis, University of Oregon, 160 p.

Plumlee, G., 1989, Processes controlling epithermal mineral distribution in the Creede Mining District, Colorado. Ph.D. Thesis, Harvard University, 378 p.

Plumlee, G. and Rye, R., 1992, Mineralogic, isotopic and other characteristics of the fringes of diverse hydrothermal systems: the perithermal environment; abstract. Proceedings of the 1992 V.M. Goldschmidt Conference.

Reed, M.H., 1992, Computer modeling of chemical processes in geothermal systems: Examples of water-rock reaction, boiling and mixing. In *Applications of Geochemistry in Geothermal Reservoir Development*, United Nations Institute for Training and Research, United Nations Development Program.

Reed, M.H., and N. Spycher, 1985, Boiling, cooling and oxidation in epithermal systems: A numerical modeling approach. Ch. 11 of Berger, B.R. and Bethke, P.M. (eds.), Geology and Geochemistry of Epithermal Systems, Reviews in Economic Geology, Volume 2, p. 249-272.

Spycher, N. F. and Reed, M. H., 1989, Evolution of a Broadlands-type epithermal ore fluid along alternative P-T paths: Implications for the transport and deposition of base, precious, and volatile metals. Economic Geology, v.84, 328-359.

Water-Rock Interaction, Kharaka & Maest (eds) © 1992 Balkema, Rotterdam. ISBN 90 5410 075 3

Microcomputer-based modeling of speciation and water-mineral-gas reactions using programs SOLVEQ and CHILLER

N. F. Spycher
University of Oregon, Eugene, Oreg., USA (Currently: International Technology Corporation, Irvine, Calif., USA)
M. H. Reed
University of Oregon, Eugene, Oreg., USA

ABSTRACT: Chemical equilibrium computations are performed with a microcomputer using two geochemical programs originally designed for main-frame systems. Because the programs are compact, and memory-efficient, they run fast on most microcomputers. Example calculations are presented for simple aqueous speciation problems in homogeneous systems, and two complex water-solid-gas equilibrium problems. Typical applications include pH calculations, analytical data processing, geothermometry, boiling, condensation, evaporation, fluid-fluid mixing, water-gas-rock interactions, and other various reaction path simulations.

1 INTRODUCTION

Programs SOLVEQ and CHILLER compute multicomponent chemical equilibria in chemical systems that include an aqueous phase. Details of the programs have been given by Reed (1982), Reed and Spycher (1988), and Spycher and Reed (1988). The main features of the programs are summarized below.

For a given temperature and total composition of a homogeneous aqueous solution, SOLVEQ computes the distribution of dissolved species, saturation indices of solids, and fugacities of gases. Various options include setting the activity of hydrogen ion or another component to compute the total concentration of that component, or forcing mineral and gas equilibrium to compute the aqueous concentration of any component present in the specified mineral or gas. The temperature and pH of the solution can be changed interactively during calculations. SOLVEQ is useful to compute pH at high temperature from low temperature measurements, and to evaluate metal speciation and mineral saturation indices as a function of temperature and composition. SOLVEQ is also useful to process water chemistry data from field and laboratory measurements.

CHILLER is more complex than SOLVEQ. This program is designed to model the effects of variation of temperature, pressure, enthalpy, and composition of heterogenous chemical systems including an aqueous phase, solids, and gases, in both open and closed systems. CHILLER simulations include cooling, heating, and mixing of solutions, various reaction paths (i.e. water/gas/rock interactions), hydrothermal boiling, gas condensation, and aqueous phase evaporation. In gas-saturated systems, the program optionally solves coupled heat balance, mass balance, and mass action equations as a function of temperature, pressure or heat, and system composition, at temperatures ranging from 25 to 300 C and pressures from 0 to 500 bars. A virial equation treatment is implemented to allow for non-ideal mixtures of real gases (Spycher and Reed, 1988). The program also incorporates full redox equilibrium, and computational schemes for both ideal and non-ideal solid solutions.

The data base for both programs,

SOLTHERM, provides reaction stoichiometries, equilibrium constants, and other thermodynamic data needed for the computations. The current data base contains data for approximately 250 species and 200 minerals at temperatures ranging from 25 to 300 C along the water saturation pressure curve. The data base can be easily expanded by the user to include more compounds, or to change the temperature and pressure ranges. The data in SOLTHERM were compiled with special emphasis on metal species, and are internally consistent.

Both programs were originally developed for use on mainframe computers. In the last few years, they have been adapted for use on microcomputers. Because the calculation methods (Reed, 1982) and programming are efficient, the programs do not require extensive memory and run easily on most microcomputers (i.e. 80286, 80386, 80486, Macintosh). Examples of geochemical modeling applications using both programs are presented below.

2 HOMOGENOUS EQUILIBRIUM

Typically, cases not requiring heterogenous calculations are processed using SOLVEQ. Even though these cases could also be run with CHILLER, this program is reserved for more complex simulations.

2.1 *Metal aqueous speciation*

The speciation of zinc and arsenic species in two separate cases assuming sphalerite and orpiment equilibrium, respectively, are shown in Figure 1. For both cases, the systems are initially specified as a solution containing 0.01m *total* aqueous sulfide at a pH of 1 and a temperature of 100 C. The program is set to compute metal activities for forced equilibration with the given solid phase. For the first equilibration point at pH = 1, the program is set to adjust charge balance automatically with chloride, thus simulating the use of HCl to establish the initial low pH. After equilibration at unit pH, the pH is interactively changed to higher values, this time using sodium for charge balance. This essentially simulates the effect of adding NaOH to the solution, keeping the solution in equilibrium with the solid, and keeping the total concentration of sulfide and chloride constant (0.01m and 0.15m, respectively). Dominant complexes and resulting total solubility curves are shown for each case. In the case of arsenic, the concentration of free bisulfide decreases at higher pH, as the concentration of arsenic sulfide complexes increase but the total sulfide content of the solution is constant. Figure 1 is provided as an example only, and reflects the species and equilibrium constants used in the calculations (in this case Ruaya and Seward 1986; Bourcier and Barnes, 1987; and Spycher and Reed, 1989a and 1989b and references therein).

Speciation calculations are useful not only to process experimental data, but are also a good tool to interpret groundwater quality data (for example during environmental investigations). However, because such calculations assume aqueous phase equilibrium and because chemical reactions (notably redox) are generally more sluggish at low than at high temperatures, applications at low temperature or reactions occurring in a short time frame must be carefully evaluated.

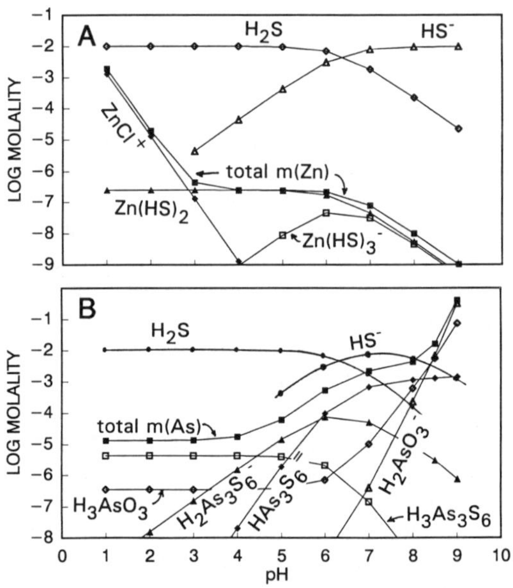

Fig.1 Speciation calculations: sphalerite (a) and orpiment (b) solubilities at 100 C, 0.01m total sulfide, 0.15m HCl, and 0-0.5m NaOH

2.2 *Mineral saturation index calculations*

Given a water composition and temperature, the saturation index (log Q/K) describes the tendency of a mineral to precipitate (log Q/K > 0) or remain dissolved (log Q/K < 0) in that solution. Figure 2 shows an example of saturation index calculations applied to geothermometry (Reed and Spycher, 1984). In this example, saturation indexes are computed over a 25-300 C temperature range for a geothermal water. The initial equilibration is calculated at the known pH and composition of the solution at low temperature. Temperature is then changed interactively and equilibration recomputed. Minerals shown on Figure 2 equilibrate near the same temperature (point at log Q/K = 0). These minerals are host rock minerals, and the temperature at which their curves intersect is similar to the measured temperature of the reservoir.

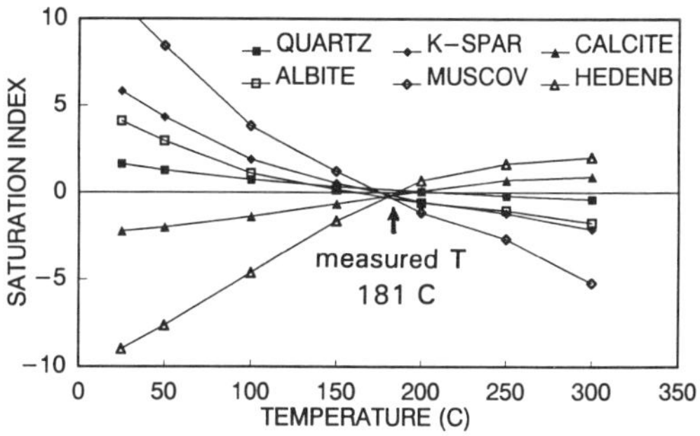

Fig.2 Homogenous equilibrium: saturation index calculations applied to geothermometry

3 HETEROGENOUS EQUILIBRIUM

Complex systems that include solid and gas phases in addition to an aqueous phase require use of CHILLER. The authors have used this program to model geochemical water-gas-rock interactions including wall rock alteration, boiling of geothermal waters and oil field brines, groundwater mixing with hot waters, acid mixing with metal bearing waters, evaporation of saline lake waters to produce evaporite minerals, weathering of basalt and granite under oxidizing conditions, reduction of metal bearing waters by reaction with carbon, and diagenesis of arkosic sand. Although the program was designed to simulate geochemical systems, it is suitable to model other aqueous-solid-gas systems. Examples of simulations follow.

3.1 *Hydrothermal boiling*

Figure 3 shows the precipitation sequence and amounts of sulfide minerals and electrum computed with CHILLER by simulating isoenthalpic boiling of a geothermal water starting at 280 C (one kilogram of initial solution). The simulation and water composition used are similar to those discussed by Spycher and Reed (1989a). For the present case, the program is set to decrease temperature at a given constant enthaply equal to the heat of the starting solution. For each temperature step, the pressure of the system, and resulting fraction of boiled gases, is computed from the given enthalpy and system composition by solving a heat balance equation together with chemical mass-balance and mass-action equations.

This simulation assumes fractionation of minerals (open system for solids), but gases are kept in contact with the solution. If desired, gas (Raleigh) fractionation can be optionally selected. Other schemes of enthalpy, temperature, and pressure variations can be specified.

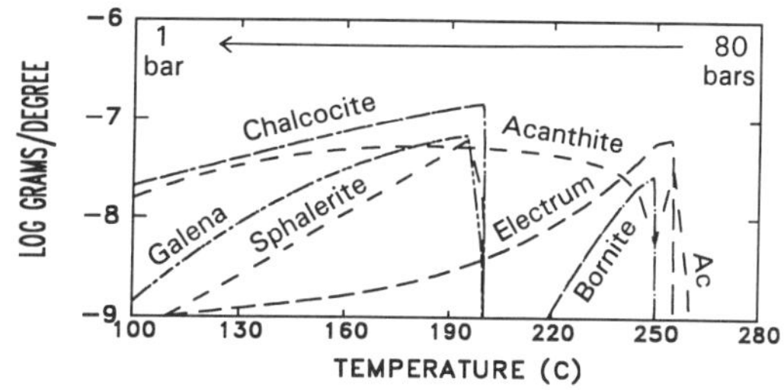

Fig.3 Isoenthalpic boiling simulation from 280 C: calculated mineral precipitation sequence

3.2 *Water-rock reaction*

As an example of water-rock interaction, we model the reaction of mine tailings with surface water, keeping the system in contact with atmosphere. The simulation is performed by titrating solids consisting primarily of silicates and pyrite (1/1 wt ratio) and small amounts of copper, zinc, lead, and arsenic sulfides into a water similar to surface water. To simulate contact with atmosphere, both oxygen and carbon dioxide fugacities are buffered at atmospheric values. Some of the reaction results are shown in Figure 4, as a function of the mass of tailings reacted per one kilogram of original solution. Until goethite starts to dissolve at low pH, iron remains small. In this simulation, using arbitrary reactants, Cu, Zn, and As accumulate in the aqueous phase, but Pb is limited by the precipitation of anglesite.

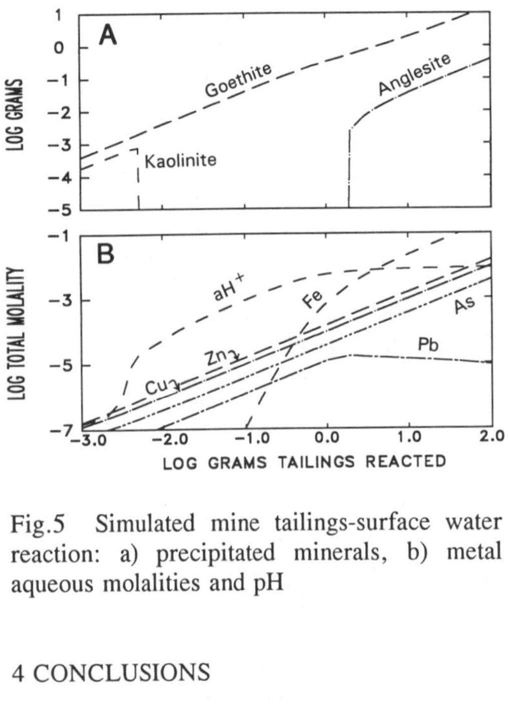

Fig.5 Simulated mine tailings-surface water reaction: a) precipitated minerals, b) metal aqueous molalities and pH

4 CONCLUSIONS

The geochemical simulations presented above are examples of typical microcomputer applications of programs SOLVEQ and CHILLER. These models are powerful tools to study many types of chemical reactions. However, as with any chemical model, the quality of the results is as good as the quality of the thermodynamic data used and as valid the assumptions made. Geochemical models should not be applied blindly, and their use should be carefully evaluated and compared with natural examples when possible. This is particularly true for simulations at low temperatures relevant to short periods of time.

REFERENCES

Bourcier, W.L. & H.L. Barnes 1987. Ore solution chemistry-VII: the stabilities of chloride and bisulfide complexes of zinc to 350 C. *Econ. Geol.*: 82, 1839-1863

Reed, M.H. 1982. Calculation of multicomponent chemical equilibria and reaction processes in systems involving minerals, gases, and an aqueous phase. *Geochim. Cosmochim. Acta*: 46, 513-528.

Reed, M.H. & N.S. Spycher 1988. Chemical modeling of boiling, condensation, fluid-fluid mixing and water-rock reactions using programs CHILLER and SOLVEQ. *Abstract, ACS Symposium: Chemical Modeling in Aqueous Systems II*, Los Angeles, California, September 25-30.

Reed, M.H. & N.S. Spycher 1984. Calculation of pH and mineral equilibria in hydrothermal waters with applications to geothermometry and studies of boiling and dilution. *Geochim. Cosmochim. Acta*: 48, 1479-1492.

Ruaya, J.R. & T.M. Seward 1986. The stability of chlorozinc II complexes in hydrothermal solutions up to 350 C. *Geochim. Cosmochim. Acta*: 50, 651-661.

Spycher, N.S. & M.H. Reed 1989a. Evolution of a Broadlands epithermal ore fluid along alternative P-T paths: implications for the transport and deposition of base, precious, and volatile metals. *Econ. Geol.*: 84, 328-359.

Spycher, N.S. & M.H. Reed 1989b. As(III) and Sb(III) sulfide complexes: an evaluation of stoichiometry and stability from existing experimental data. *Geochim. Cosmochim. Acta*: 53, 2185-2194.

Spycher, N.S. & M.H. Reed 1988. Fugacity coefficients of H_2, H_2O, CO_2, CH_4 and H_2O-CO_2-CH_4 mixtures: a virial equation treatment applicable to calculations of hydrothermal boiling. *Geochim. Cosmochim. Acta*: 52, 739-749.

Water-Rock Interaction, Kharaka & Maest (eds) © 1992 Balkema, Rotterdam. ISBN 90 5410 075 3

Reaction path modelling of diagenesis in the Eocene Wilcox Group of the Texas Gulf Coast

R.N.Tempel & W.J.Harrison
Colorado School of Mines, Golden, Colo., USA

ABSTRACT: Diagenesis of the Eocene Wilcox Group of the Texas Gulf Coast can be simulated by combining the separate outputs from chemical reaction path calculations and paleohydrologic reconstructions. The reaction path model, EQ3/6 (Wolery, 1979), calculates chemical equilibria during mineral-gas-fluid interactions, and paleohydrologic simulations by Harrison and Summa (1991) constrain temperature, evolution of pressure, and the total pore volumes of fluid. This integrated approach determines paragenesis at each stage of burial history, yielding volumes and chemical compositions of precipitated and dissolved minerals.

1. INTRODUCTION

Galloway (1984) defined three major hydrologic regimes in sedimentary basins - meteoric, compactional and thermobaric, and each regime is characterized by chemical composition, flow rate, and distribution within a basin. Galloway further identified the associated diagenetic reactions in each regime based on Gulf of Mexico basin petrographic data.

Diagenetic reactions that take place during burial of sedimentary rocks can be simulated by combining the results from geochemical reaction path calculations and paleohydrologic reconstructions by means of a conceptual model.

Calibration of this model can be accomplished by comparisons to petrographic observations described by previous workers of the Wilcox in the Gulf of Mexico basin (Fisher and Land, 1986; Land, et al., 1987; Loucks, et al., 1984).

2 PALEOHYDROLOGY

Hydrologic evolution of the Gulf of Mexico basin (Harrison and Summa, 1991) is characterized by three stages: the first stage from Jurassic to early Tertiary times was characterized by circulating meteoric waters and mixed at depth with saline waters expelled from compacting Jurassic salts. The second stage, during much of Tertiary time, involved development of geopressures and restricted meteoric water infiltration as the clastic sediment supply increased during deposition of prograding deltaic sequences. In the final stages of widespread geopressure, from late Miocene to the present time, nearly lithostatic fluid pressure gradients and minimal meteoric water infiltration have been active.

Three diagenetic pathways in the Wilcox, each with a characteristic burial history and paleohydrology, have been chosen for study. The pathways are onshore, nearshore, and offshore facies and range from shallow to deep burial, respectively. Onshore facies were flushed with meteoric fluid during early burial and later flushed with shale derived compactional fluids. Nearshore and offshore facies were reacted only with compactional fluids, and differ in that the more deeply buried offshore facies were reacted with fewer pore volumes at higher temperatures than the nearshore facies.

3 METHODS

Simulation of diagenesis in clastic formations is achieved through use of a conceptual model which combines the results of both geochemical reaction path calculations and paleohydrologic reconstructions of the Gulf of Mexico basin.

The reaction path model, EQ3/6 (Wolery, 1979), calculates chemical equilibria during mineral-gas-fluid interactions, and the results from paleohydrologic reconstructions of the Gulf of Mexico basin by Harrison and Summa (1991) constrain temperature, evolution of pressure, and fluid composition, which are used as input parameters for EQ3/6. Additionally, paleohydrologic reconstructions determine the total numbers of pore volumes of fluid flowing through a rock during a stage of burial.

The following procedure is used to combine the results of both calculations. First, the fixed fluid volume that EQ3/6 reacts with a rock is considered equal to one pore volume so that the resulting volumes of minerals precipitated and dissolved during one calculation equal the amount of diagenetic reaction between the rock and one pore volume of fluid. Secondly, diagenesis is simulated for a complete stage of burial by obtaining the number of pore volumes flowing through the rock during that stage from the paleohydrologic reconstruction and multiplying that value times the result of a single EQ3/6 calculation. A stage of burial is defined as any given time step during which the physical conditions of the system are isobaric and isothermal. This procedure is repeated for each stage until the entire diagenetic history is reconstructed.

In addition to the constraints defined by the paleohydrologic reconstructions, the following input parameters have been used in EQ3/6 calculations: (1) three sandstone compositions represent the Wilcox sandstones; lithic arkoses, feldspathic litharenites, and quartzose subarkoses (Fisher and Land, 1986), (2) partly evolved meteoric water is represented by a river water composition (Wolery, 1989), (3) compactional water composition is represented by NaCl water-type (Morton and Land,1987), (4) a starting sandstone is assumed to have a stable intergranular volume of 25%, and (5) partial pressures of CO_2 are determined from depth plots by Franks and Forrester (1984).

A number of assumptions and uncertainties underlie this approach including: (1) extent of rock reaction allowed, (2) the assumption of isothermal /isobaric conditions for any given time step, (3) each pore volume of water is "new", not evolved, (4) no precipitation kinetics have been included at this time, (5) pressure solution is not taken into account, (6) volumes of minerals precipitated and dissolved are independent of the effects of compaction, and (7) total pressure is used to constrain partial pressures of CO_2 and CH_4 gases.

4 RESULTS

Results of the diagenetic pathway calculations for the onshore facies are presented in Figures 1 - 4.

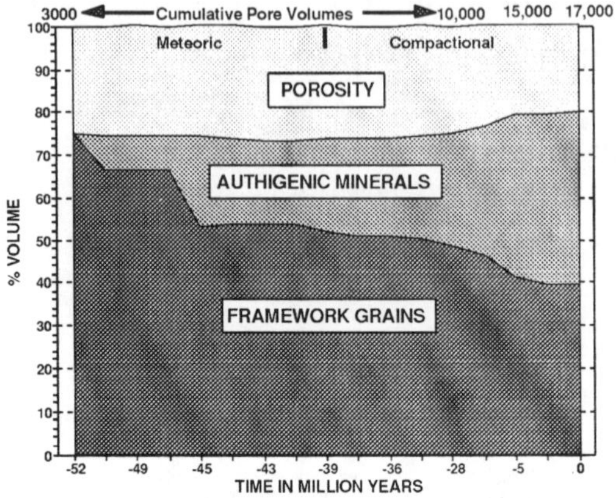

Fig. 1 Calculated diagenetic pathway for onshore Wilcox Group sandstone.

Early stages of diagenesis occurred while the sandstone was being invaded by meteoric water, and later stages occurred as compactional water flushed the sandstone.

Figure 1 shows the changes in composition of the model sandstone from time of deposition to the present day. Porosity decreases from 25% to 20% as a result of authigenic mineral formation. These authigenic phases replace framework grains, which initially comprised 75% of the rock, so that the framework grains and authigenic phases each comprise 40 % of the rock at the present time.

Figure 2 shows that plagioclase feldspar is the least stable component and is completely dissolved during meteoric invasion. Potassium feldspar persists over the duration of the pathway as does

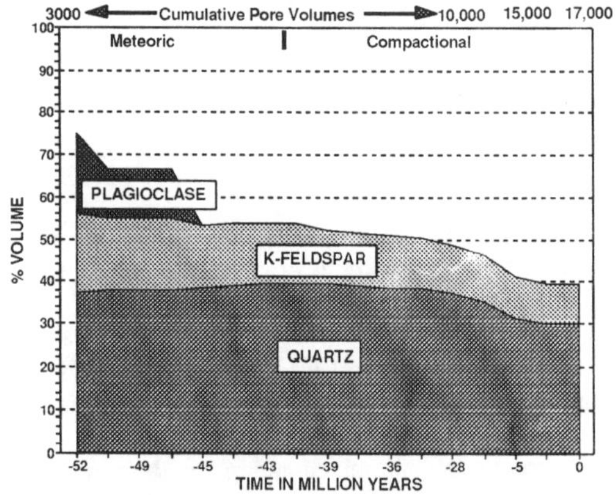

Fig. 2 Calculated changes in detrital mineralogy for onshore Wilcox sandstone.

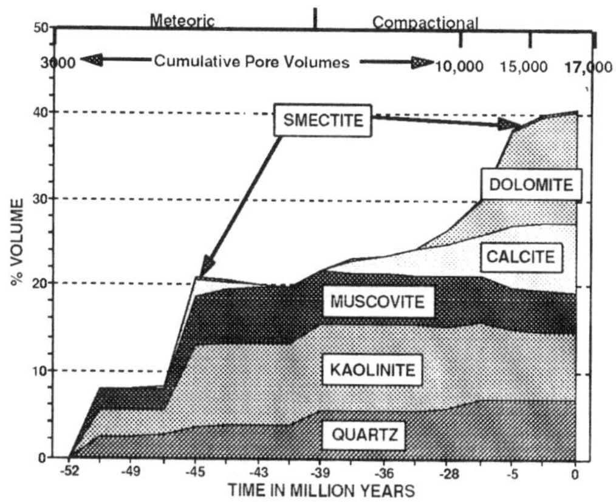

Fig. 3 Authigenic phases produced during diagenesis.

detrital quartz. Although the volume of quartz remains constant throughout the pathway, its percentage varies slightly over time as a function of changing porosity and authigenic mineral volumes.

Figure 3 shows more detail of the diagenetic pathway for the onshore Wilcox facies. Quartz cement, kaolinite and muscovite (a thermodynamic proxy for illite) form continuously throughout the burial history as a result of K-feldspar dissolution. Two episodes of carbonate cementation are observed. From 43-47 mybp, a few percent calcite cement is produced as a result of plagioclase dissolution in the presence of pCO_2. This calcite is later dissolved during the transition from meteoric to compactional systems. A second volumetrically significant carbonate cementation event occurs involving calcite initially(beginning at 39 mybp) and

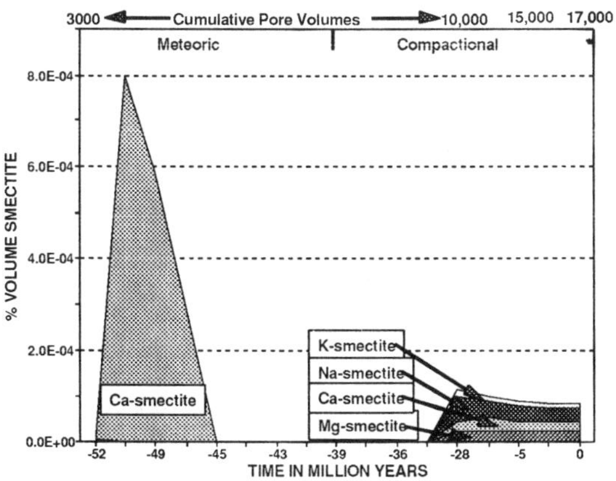

Fig. 4 Composition of diagenetic smectite.

then calcite and dolomite (beginning at 37 mybp). These carbonate cements form during later stages of burial after the replacement of the meteoric water by compactional waters is complete. The final authigenic assemblage in this onshore Wilcox sandstone is thus 13% dolomite, 7% calcite, 4% illite, 10% kaolinite and 6% quartz.

Figure 4 shows compositions and volumes of smectites which form as plagioclase dissolves. Smectites precipitated in the meteoric regime are dominantly a calcic composition whereas the compactional regime produces a Na-smectite.

Petrographic descriptions of paragenetic sequences in the Wilcox Group provide the calibration for our model diagenetic pathway. Our simplified pathway is in good agreement with observations of reactions, although we are limited by lack of petrographic data for shallow Wilcox Group sandstones since most of the classic diagenetic studies (Boles and Franks, 1978; Loucks et al., 1984; Fisher and Land, 1986) involve deeper rocks (generally 2->4 km burial depth) which do not have comparable burial histories. Maximum temperature of burial is only 80 ^{o}C in this pathway and depths greater than 1.5 km are not reached. Reactions such as albitization and smectite conversion to illite are not predicted in this pathway.

A maximum of 6% quartz cement is predicted, which is reasonable given the temperatures and depths of this pathway, the arkosic composition of the starting composition, and the fact that the quartz is derived from feldspar dissolution. Significantly higher amounts of quartz cement are not commonly reported at depths shallower than 1.5 km (McBride, 1989). Our predicted smectite compositions are comparable to those analysed by Hanor and McManus (1985) in shallow diagenesis in the Louisiana Cretaceous aquifers.

The dissolution of plagioclase and K-feldspars and the precipitation of calcite, illite and kaolinite as either replacement or as discrete pore-filling phases are commonly reported diagenetic consequences of meteoric water infiltration into feldspathic sandstones. Dolomite is rarely reported in the Gulf Coast Tertiary formations, however, ferroan calcite and ankerite cements are common in the early Tertiary rocks (Boles and Franks, 1978; Fisher and Land, 1986; Land et al., 1987). Dolomite is probably being precipitated in place of the ferroan carbonates in our model because iron was not included in the chemical system in order to simplify preliminary calculations and because the appropriate thermodynamic data for ankerite or ferroan calcite does not exist in the database. These limitations are the subject of ongoing studies.

Although the onshore facies pathway was chosen for detailed discussion in this paper, nearshore and offshore facies pathway calculations show different diagenetic events as a result of deeper burial and invasion by only shale derived compactional fluids. Authigenic phases observed from calculations of these deeper pathways include the precipitation of: (1) chlorite, (2) albite, (3) significant volumes of quartz, and (4) assemblages of clay minerals that differed from those of the onshore pathway.

5 DISCUSSION AND CONCLUSIONS

The combination of chemical calculations constrained by physical data from paleohydrologic reconstructions makes possible the quantification of the diagenetic products that may be produced along specific pressure and temperature paths during sediment burial. Although our calculations are incomplete, the results to date demonstrate that diagenesis is episodic in nature, major cementation events are related to significant changes in hydrologic evolution, and types and amounts of clay minerals precipitated are dependent on hydrologic regime.

Continued work with this method will use sensitivity analyses to accomplish the following tasks: (1) determine the relative importance of the chemical and physical parameters that are involved in diagenesis, (2) determine extent of reactivity of mineral surfaces, (3) constrain cement volumes, (4) quantify dissolution, (5) determine the changes in pore water chemistry with increasing mineral interaction, (6) determine the effects of organic acids on diagenesis, and (7) quantitatively evaluate the significance of pCO_2 to the diagenetic process.

Additional research will involve the calculation of thermodynamic properties of clay minerals in order to augment the existing database of EQ3/6.

6 ACKNOWLEDGEMENTS

This work has been supported by grants to the second author from Texaco Exploration Technology Division, Mobil Research and Development Company, and the National Science Foundation NSF-8916440.

REFERENCES

Boles J.R. and Franks S.G. (1978) Clay diagenesis in Wilcox sandstones of Southwest Texas: implications of smectite diagenesis on sandstone cementation. Journ. Sed. Petrol. 49, p. 55-70.

Fisher R.S. and Land L.S. (1986) Diagenetic history of Eocene Wilcox sandstones, south central Texas: Geochim. Cosmochim. Acta, 50, p. 551-563.

Galloway W.E. (1984) Hydrogeologic regimes of sandstone diagenesis. In: McDonald D.A. and Surdam R.C. (eds.) Clastic Diagenesis: American Assoc. Petrol Geol. Memoir 37, p. 3-13.

Hanor J.S. and McManus K.M. (1985) Sediment alteration and clay mineral diagenesis in a regional groundwater flow system, Mississippi Gulf Coastal Plain, Gulf Coast Assoc. Geol. Socs. Trans., 38, p. 495-502.

Harrison W.J. and Summa L.L. (1991) Paleohydrology of the Gulf of Mexico Basin. Amer. Journ. Sci., 291, 109-176.

Land L.S., Milliken K.L and McBride E.F. (1987) Diagenetic evolution of Cenozoic sandstones, Gulf of Mexico basin. Sed. Geol., 50, p. 195-225.

Loucks R.G., Dodge M.M. and Galloway W.E. (1984) Regional controls on diagenesis and reservoir quality in Lower Tertiary sandstones along the Texas Gulf Coast. In: In: McDonald D.A. and Surdam R.C. (eds.) Clastic Diagenesis: American Assoc. Petrol Geol. Memoir 37,p. 15-47.

McBride E.F. (1989) Quartz cement in sandstones: a review. Earth-Science Reviews 26, p. 69-112.

Morton R.A. and Land L.S. (1987) Regional variations in formation water chemistry, Frio Formation (Oligocene) Texas Gulf Coast. Amer. Assoc. Petrol. Geol. Bull., 71, p. 191-206.

Wolery T.J. (1979) Calculation of chemical equilibria between aqueous solutions and minerals: The EQ3/6 software package. LLNL Report UCRL-52658.

Wolery T.J. and Daveler S.A. (1989) EQ6: A computer program for reaction path modeling of aqueous geochemical systems: User's guide and documentation. LLNL Report.

Water-Rock Interaction, Kharaka & Maest (eds) © 1992 Balkema, Rotterdam. ISBN 90 5410 075 3

Analysis of heat source for the Xiangshan fossil hydrothermal system, China: A modeling approach

Zhou Wenbin
East China Geological Institute, Fuzhou, Jiangxi, People's Republic of China

ABSTRACT: Finite-element modeling was used to analyze the heat source for the Xiangshan fossil hydrothermal system in China. The modeling results have shown that heat from the volcanic lava could affect the geothermal field for only about 3 million years and could not produce enough heat for the ore-forming hydrothermal system that formed about 40 Ma later. The water in the ore-forming system was most likely heated while circulating at depths in the crust.

1 INTRODUCTION

An extensive fossil geothermal system with metallic mineralization was found in Xiangshan volcanic basin in southeastern China. Among the mysteries of the hydrothermal system and mineralization is the source of heat. This paper attempts to find a reasonable solution to this question by employing a finite-element model of the paleogeothermal field after the volcanic eruption.

2 PRINCIPLES OF THE MODELING

Heat transfer within the earth is a complicated thermophysical process that can be described with the following equation:

$$\frac{\partial}{\partial x}(k\frac{\partial T}{\partial x}) + \frac{\partial}{\partial y}(k\frac{\partial T}{\partial y}) + \frac{\partial}{\partial z}(k\frac{\partial T}{\partial z}) - C_\omega \cdot \rho_\omega[\frac{\partial}{\partial x}(V_X \cdot T) + \frac{\partial}{\partial y}(V_Y \cdot T) + \frac{\partial}{\partial z}(V_Z \cdot T)] + Q = C \cdot \rho\frac{\partial T}{\partial t} \quad (1)$$

Where, k, C, ρ and μ separately represent heat conductivity, capacity, density and specific heat for the rock; C_ω and ρ_ω are the specific heat and density of the geothermal fluids; V_X, V_Y, V_Z are the component velocities in the X, Y, and Z direction; Q is the heat source (or converge); T is the temperature; and t is time.

The equation shows that the temperature change with time in a certain section of the crust is a result of heat conduction, convection and additional heat sources or converges. In principle, Equation (1) has a solution that is dependent upon a given set of thermophysical parameters of the rocks and fluids and given boundary and initial conditions. As a result, temperatures can be obtained in this manner for any point in the study area at any time.

Under most circumstances, however, it is difficult to arrive at the analytical solution for Equation 1 because of complicated geological conditions, inhomogeneity in rocks, thermophysical parameters, and irregularity in boundaries. Consequently, numerical methods have been used to solve the equation, and finite-element analysis is among the mostly commonly used numerical methods.

Essentially, the finite-element modeling of a geothermal field divides a continuous thermal field into finite temperature points, then calculates the temperatures for the points of interest and applies them as an approximate representation of the actual thermal field. The steps are (1) dividing the continuous medium into finite polygonal elements with common temperature points, with the thermophysical parameters being constants within an element; (2) building up the relationship between temperature and thermophysical properties for each element, resulting in a set of equations including unknowns equivalent to the number of nodes; and (3) solving the equation set and obtaining temperature values for the nodes.

3 MODELING OF XIANGSHAN PALEOGEOTHERMAL FIELD

Xiangshan ore field consists of a group of hydrothermal ore deposits in a volcanic rock region in southeastern China. Before and during the mineralization processes, there were a series of tectonic-geothermal events, such as tectonic activities, volcanic eruptions, and small rock body and vein rock intrusions. Hydrothermal alterations can frequently be found in this area, which indicates that the mineralization was closely related to both the geothermal background at that time and the favorable tectonic-hydrogeological and geochemical conditions. Therefore, it is very important to research the likely geothermal condition during the metallization stage in order to explain the genesis of the ore field. Some geologists have given opinions on the heat source for the mineralization in the past based on non-quantitative methods. By using a finite-element model, the author attempts to present a quantitative analysis of paleogeothermal evolution after the volcanic eruption.

3.1 *Background Data*

There were two volcanic cycles in the Xiangshan area. The second volcanic cycle erupted about 160 million years ago and is commonly considered to be the most important geothermal event responsible for the formation of the ore deposits. The temperature of the lava was around 1200°C, according to melt inclusion thermometry. The main mineralization stages were between 119 and 99 million years ago. The age difference between the volcanic eruption and the mineralization was 40 to 50 years.

It is evident that the general structural framework had been formed in the Xiangshan area after the second volcanic cycle, that is the geological structure at that time is similar to the present day one to a great extent. Thus, it is reasonable to take the measured section as the basis for calculations for the modeling after proper modification.

3.2 *Model Set-up*

3.2.1 Section selection for the calculation

The main cross-section of the Xiangshan area, which cuts across the whole Xiangshan rock body and the major structures, is chosen as the calculation section. The lower limit was extended to the level of -6.3 kilometers in order to get an isothermic or isoheat-flow boundary, while the upper limit was lifted by 1 kilometer on one side and 0.5 kilometer in the middle of the section, in order to represent landform changes caused by weathering and erosion.

Since the original section is rather complex in petrology, some simplification has been made: the rocks are generally classified into two units, that is, the bed rocks (predominantly phyllite) and cover rocks (mainly porphyroclastic lava). The bed rocks are the basic type of rock in the region, whereas the cover rocks make up the anomalous rock area. The actual section used in the model is shown in Figure 1.

3.2.2 Element division

The calculated region is divided into more than 200 elements as in Figure 1, which has been proved rational by the calculation results.

3.2.3 Mathematical Model

The basic differential equation describing the heat transfer has been given above. As conduction was the major means of heat transfer in the Xiangshan area, and heat convention only existed in some limited parts with no regional significance, convection was not considered. The radiogenic heat from the rocks, the heat source item Q, was so low that it can be ignored, compared to the enormous heat generated by volcanic activities. Since the calculation region is a plane, the thermal field is two dimensional. Consequently, the basic differential equation used in the modeling is as follows:

$$\frac{\partial}{\partial x}\left(K\frac{\partial T}{\partial x}\right) + \frac{\partial}{\partial y}\left(K\frac{\partial T}{\partial y}\right) = C \cdot \rho\frac{\partial T}{\partial t} \qquad (2)$$

Where all symbols have the same meaning as those in Equation 1.

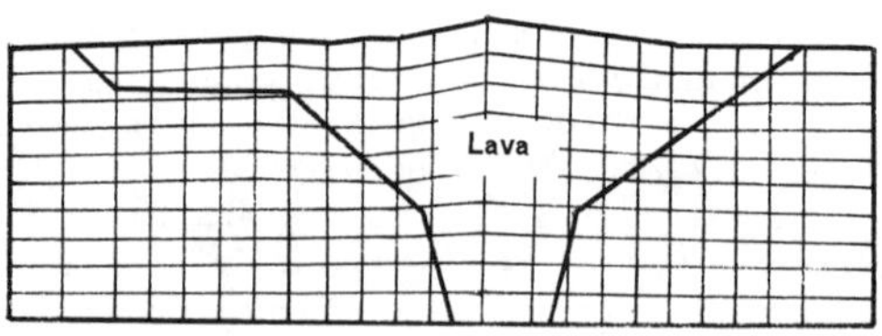

Figure 1 - The modeling cross section and element division.

3.2.4 Boundary Condition

The calculation region has four boundaries, the upper, the lower, the left and the right, which were treated as follows:

1. The left and right boundary were assumed to be heat-insulated, that is

$$q_x = -K\frac{\partial T}{\partial x} = 0 \qquad (3)$$

2. The upper boundary was treated as a constant-heat boundary, in which the temperatures were equal to yearly mean air temperatures for each point.
3. The lower boundary was an iso- heat-flow boundary, with an assumed value of 2.0 H.F.U. about 84 mW/m^2).

3.2.5 Initial Condition

1. The volcanic lava region was assumed to be an isothermal region with a temperature of 1200°C, based on the mineral inclusion data.
2. Within the normal region (bed rock region), the initial temperatures were estimated according to the surface temperatures and the geothermal gradient.

3.3 *Results*

Calculations have been made with the model which was set up as described above, and the results are presented in Figure 2.

4 ANALYSIS OF THE HEAT SOURCE FOR THE ORE-FORMING HYDROTHERMAL SYSTEM

The modeling results shown above revealed that the energy from the volcanic lava could affect the geothermal field for only about 3 million years. Because the main mineralization stage was about 40 Ma later than the volcanic activity, it was impossible for the volcanic rock bodies to offer heat energy to the ore-forming hydrothermal system. The water in the system was considered to be heated while circulating in the deep crust. As the study area was situated on the plate boundary circling the Pacific Ocean at that time, the geothermal gradient for the area was estimated to be up to 80°C/km, based on the active hydrothermal systems in Taiwan and Japan. The temperature of the hydrothermal solution has been measured to be some 200°C, so the required circulating depth would be no less than 3 kilometers, assuming a geothermal gradient of 80°C/km.

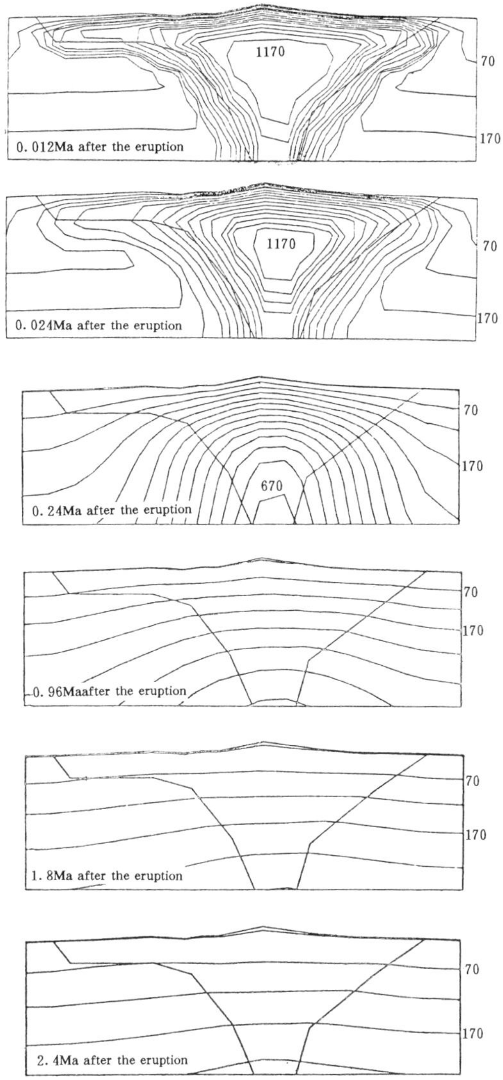

Figure 2 - The evolution of the geothermal field after the volcanic eruption.

5 CONCLUSIONS

Based on the modeling results shown in Fig. 2 and the above analysis, we can made the following conclusions:

1. The volcanic lava affected the geothermal situation in Xiangshan area for about 2.4 million years.

2. The mineralization was about 40 million years later than the volcanic activity, so the volcanic rock body could not heat the hydrothermal system where ore deposits were formed.

3. The water in the hydrothermal system could be heated when it circulated at depth in the crust. Because the hydrothermal solution was about 200°C during mineralization, the circulating depth would be at least 3 kilometers if the geothermal gradient was 80°C/km at that time.

REFERENCES

Lee, T. et al. 1980. Application of finite-element analysis to terrestrial heat flow. Department of Natural Resources Geological Survey Occasional Paper 29.

Zhang, J.M. 1986. Application of finite-element method to geothermal research. Publishing House of Science, China.

Sedimentary basins

Fluid evolution

Water-Rock Interaction, Kharaka & Maest (eds) © 1992 Balkema, Rotterdam. ISBN 90 5410 075 3

Regional hydrogeology and fluid geochemistry of the Mannville Group, Western Canada Sedimentary Basin: Synthesis and reinterpretation

H.J.Abercrombie & E.G.Fullmer
Geological Survey of Canada, Calgary, Alb., Canada

ABSTRACT: Evidence of significant flow and mass transport within the Lower Cretaceous Mannville Group of the Western Canada Sedimentary Basin is recorded in the giant oil sand deposits of Alberta and Saskatchewan. Regional mapping of present-day structural and potentiometric surfaces, and water, oil and gas compositions across the basin demonstrates that regional flow and geochemical patterns can not be explained by simple two dimensional models. These data, and constraints provided by the distribution and geochemistry of oils and waters in the basin, suggest that present and past flow in the Mannville Group involves two active flow systems. In the west a deep flow system transports fluids eastwards, up and out of the basin. In the east a shallow flow system moves low salinity waters of meteoric origin westwards and down into the basin. The two flow systems meet along an axial trough that is roughly coincident with the -800 m elevation contour on the base Mannville Group surface. Prior to erosional unroofing of the WCSB, and coincident with major hydrocarbon migration, the axial trough may have been located to the east, near the eastern limit of the oil sand deposits. Biodegradation of the oil sands occurred as the axial trough moved westwards in response to unroofing of the basin allowing incursion of low salinity, oxygenated waters.

1 INTRODUCTION

The Mannville Group and equivalent strata of the Western Canada Sedimentary Basin (WCSB) form part of a Jurassic through Tertiary foreland basin succession that formed in response to orogenic activity along the western margin of the North American Craton. These rocks contain signficant quantitites of conventional oil and gas, and giant oil sand deposits (Fig. 1). Numerous models have been advanced to explain the origin of the oil sand deposits and the evolution of flow systems within the WCSB (Jones, 1980; Garven and Freeze, 1984a, b; Hitchon, 1984; Masters, 1984; Garven, 1989). Present day flow has been modelled in 2-D by Hitchon (1969a, b; 1984) and in the pseudo 3-D by Hitchon and co-workers (1989a, b; 1991).

The purpose of the present study, which is presented as a preliminary report, is to integrate regional structure and hydrogeology with geochemical data for fluids contained within the Mannville Group. Large amounts of data gathered during hydrocarbon exploration are available and have been used to construct regional potentiometric and structural surface maps, and maps of water, gas and oil compositions for the Mannville Group in the Alberta portion of the WCSB. The maps are used to provide constraints on the nature and distribution of hydrogeological and geochemical processes active in the basin at the present and, when combined with geological and geochemical constraints, can be used to test models of flow in the basin. The results have been instructive and show that simple two dimensional modelling can not fully explain the present distribution and, by extension, past evolution of flow in the Mannville Group.

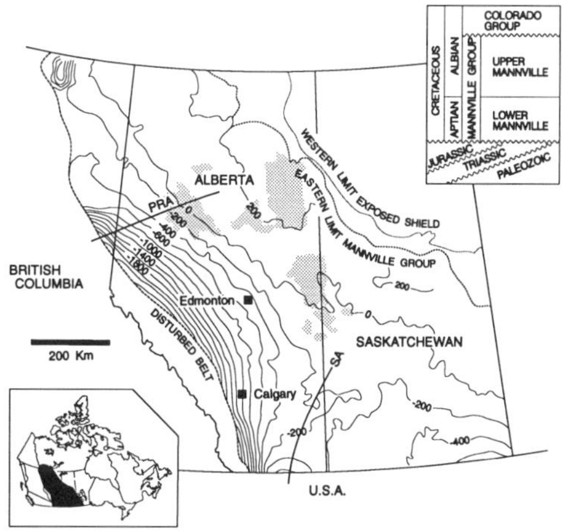

Figure 1. Elevation of the base of the Mannville Group, Alberta and Saskatchewan. The location of the Western Canada Sedimentary Basin is given in the lower inset and the regional stratigraphic framework is illustrated in cartoon form in the upper inset. Shaded areas are the locations of oil sand deposits within the Mannville Group. Axes of the Peace River Arch (PRA) and Sweetgrass Arch (SA) are shown.

2 METHODS

Over 200,000 exploration wells have been drilled within the WCSB and more than 30,000 oil and gas pools are defined in Alberta alone. Formation tables for each well drilled, and temperatures, pressures and average gas and oil compositions for each oil pool are available. More than 130,000 water analyses for Alberta have recently been made available in digital form and more than 30,000 individual gas tests are also available. Due to the advanced state of data curation in Alberta, results reported here are for Alberta only. Future work will integrate data from other provinces.

The sources of data used in constructing Figures 1 to 4 are given in Table 1. Based on the results of Hitchon et al., 1990, which showed evidence for separate Upper and Lower Mannville flow systems, individual and pool averaged data were sorted by formation into Lower and Upper Mannville Groups. In Alberta, the Lower Mannville was defined as base Mannville Group to top Ostracode Zone; the Upper Mannville was defined as top Ostracode Zone to base Colorado Group (refer to Jackson, 1984). Once sorted, the data were gridded and contoured using SURFACE II.

TABLE 1

FIGURE	MAP	DATA USED	SOURCE
Fig. 1	Base Mannville Group	Elevation	ERCB[1] well file
Fig. 2	Potentiometric surface	Initial pool pressure elevation	ERCB pool file
Fig. 3	TDS	TDS@110°C	ERCB water analysis file[2]
Fig. 4	Mole fraction CO_2	XCO_2	ERCB pool file

[1]Energy Resources Conservation Board, Alberta
[2]Digitized version from Rakhit Petroleum Consulting Ltd., Calgary, Alberta.

3 EVIDENCE OF FLOW

The giant oil sand deposits contained within the Mannville Group attest to massive flow and mass transport. Additional geological information provides constraints on the nature of this flow:

- Although no unequivocal source for Mannville oils has been identified (Brooks et al., 1988; Creaney et al., 1992), volumetric constraints have been used to rule out significant contributions from Paleozoic source rocks (Moshier and Waples (1985).
- No oils belonging to families found in or below Mannville Group reservoirs are known to occur within or above the overlying, dominantly shale Colorado Group (Fig. 1; Allan and Creaney, 1991);
- Chemically and isotopically, waters from strata above the Colorado Group have distinctly different compositions from waters in the Mannville Group and underlying Paleozoic strata (Connolly et al., 1990a, b).

Together these data strongly suggest that significant fluid flow and mass transport has occurred within the Mannville Group with little, if any, upward flow through the Colorado Group which apparently has acted as a regional seal for fluids flowing within the Mannville Group.

Additional constraints on the evolution of fluid flow in the Mannville Group can be inferred from studies of water-rock and water-oil-bacterial interactions. Diagenetic studies show that the Mannville Group records a protracted diagenetic history (Abercrombie, 1989) indicative of a long lasting flow system. New insights into the role of biofilms in the degradation of oil (Van Elsberg and Abercrombie, 1991), now present as heavy oil and bitumen, suggest that oil migrated in non-degraded form from the west where thermal maturation data show oil generation to have occurred (Creaney et al., 1992). Once trapped, the oil was subsequently degraded by aerobic and anaerobic microbial processes. Oxygen is a pre-requisite for microbiological activity and was likely supplied by incursion of meteoric waters into the Mannville reservoirs.

4 REGIONAL FLOW AND FLUID CHEMISTRY

Two flow systems appear to be active at present. In the east, hydraulic potentials for both the Upper and Lower Mannville Groups decrease westwards (Fig. 2) implying flow down and into the basin. As was previously observed by Hitchon et al. (1990), hydraulic potentials for the Upper Mannville in east-central Alberta are up to 100 m higher than those in the Lower Mannville, indicating a potential for downward flow in this area. In the west, high water potentials associated with deeply buried Mannville strata decrease towards the east, implying flow up and out of the deepest part of the basin. The two flow systems meet in an axial trough which consists of a series of closed lows that coincide, roughly, with the -800 m elevation contour on the base of the Mannville Group (Fig. 1).

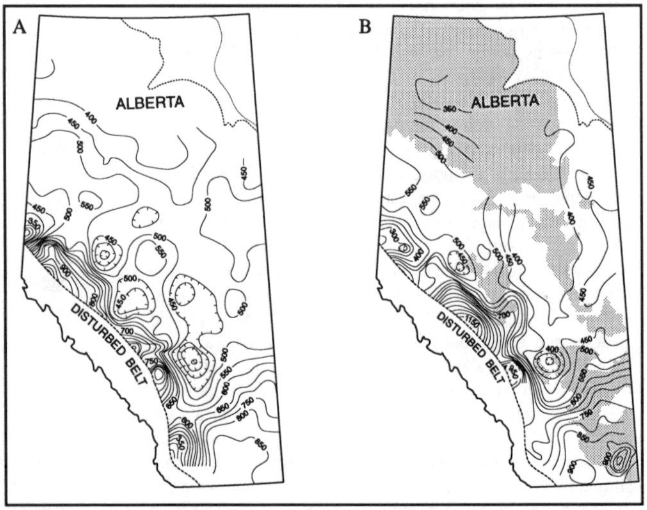

Figure 2. Regional fresh water potentiometric surface maps for the Mannville Group. A. Upper Mannville. B. Lower Mannville. Shaded regions in the Lower Mannville area areas where Lower Mannville strata are partly missing due to non-deposition on topographic highs in the pre-Mannville erosion surface.

Regional salinity patterns in both the Upper and Lower Mannville (Fig. 3) are consistent with the suggested flow patterns. In the north-central part of eastern Alberta the highest hydraulic potentials (Fig. 1) are associated with the salinities less than 10,000 mg/L TDS (total dissolved solids; Fig. 3). Towards the west salinity increases in both the Upper and Lower Mannville reaching >80,000 mg/L TDS in the Lower Mannville in the region of the axial trough. Farther to the west similar behaviour is observed - highest hydraulic potentials correspond to salinities <30,000 mg/L TDS and lowest potentials correspond to salinities >60,000 mg/L TDS. Low salinities are also observed in the extreme south and in the north-central part of western Alberta where they reflect recharge along the Sweetgrass and Peace River Arches (Fig. 1).

The CO_2 contents of Mannville Group gas pools are included (Fig. 4) to test whether or not regional variations in gas compositions reflect the regional trends in the potentiometric surface and water compositions. With the exception of southwesternmost Alberta, low salinity, recharge areas are associated with low CO_2 contents whereas CO_2 contents >2.5 mol% are observed with high salinity waters in closed hydraulic potential lows associated with the axial trough. No explanation for high CO_2 contents in gas pools in southwestern Alberta (Fig. 4) is known. Elevated CO_2 contents are observed in Paleozoic pools in this region and Cretaceous through early Tertiary igneous rocks are present. Any genetic link remains speculative and awaits results of work in progress by J. Cody, an M.Sc .researcher at The University of Calgary.

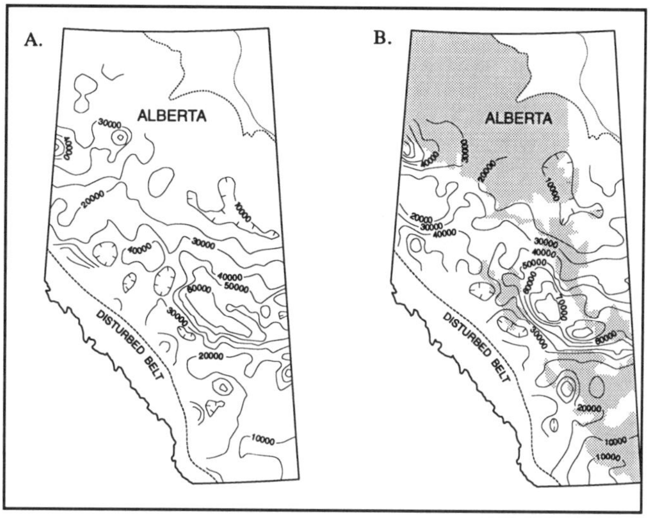

Figure 3. Preliminary regional salinity maps for the Mannville Group. A. Upper Mannville; B. Lower Mannville. See explanation for Figure 2.

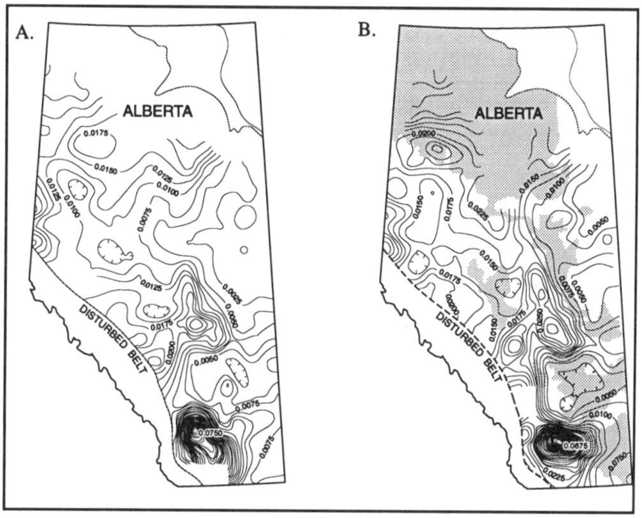

Figure 4. Mole fraction CO_2 in Mannville Group gas pools. A. Upper Mannville; B. Lower Mannvile. See explanation for Figure 2.

5 CONCLUSIONS AND IMPLICATIONS

The Mannville Group has experienced significant fluid flux associated with the migration, then biodegradation of the oil sands. Regional mapping of the present day potentiometric surface for the Mannville Group shows that two flow systems, an eastern, down-dip flow system and a western, up-dip flow system, meet along an interface, or axial trough, which is roughly coincident with the -800 m elevation contour on the base Mannville Group surface. The axial trough is associated with the lowest hydraulic potentials, highest salinities and, with the exception of southwest Alberta, highest CO_2 contents in gas pools. The coincidence of hydrogeology and fluid chemistry implies that regional flow and evolution of the potentiometric surface through time have acted to control mass transport and, in all probability, water-rock-organic interactions.

The recognition of two fundamentally different flow systems within the Mannville Group has significant implications for the evolution of the flow system and the origin of the oil sands. If, as has been suggested, the biodegraded oils arrived as light oils it is probable that the axial trough at the time of migration lay near the present eastern limit of oil sand deposits. Erosional unroofing of the WCSB would have resulted in westward movement of the axial trough which would allow downward and westward flowing, low salinity, oxygenated waters of meteoric origin access to the reservoired oils. Alternatively, it is conceivable that the oils sands formed at the interface between the two flow systems and biodegradation proceeded more or less simultaneously with arrival of oil.

These and other aspects of flow in the WCSB are currently under investigation. Future work will focus on evolution of the potentiometric surface through time and on the role of hydrogeology in controlling hydrocarbon migration and the nature and distribution of diagenetic water-rock-organic interactions.

6 REFERENCES

Abercrombie, H.J., 1989. Water-rock interactions during diagenesis and thermal recovery, Cold Lake, Alberta. Ph.D. Thesis, University of Calgary, 183 p.

Allan, J. and Creaney, S., 1991. Oil families in the Western Canada Basin. Bulletin of Canadian Petroleum Geology, **39**, 107-122.

Connolly, C.A., Walter, L.M., Baadsgaard, H. and Longstaffe, F.J., 1990a. Origin and evolution of formation waters, Alberta Basin, Western Canada Sedimentary Basin. I. Chemistry. Applied Geochemistry, **5**, 375-396.

_____, 1990b. Origin and evolution of formation waters, Alberta Basin, Western Canada Sedimentary Basin. II. Isotope systematics and water mixing. Applied Geochemistry, **5**, 397-414.

Creaney, S., Allan, J., Cole, K.S., Fowler, M.G., Brooks, P.W., Osadetz, K.G., Macqueen, R.W., Snowdon, L.R. and Riediger, C.L., 1992. Petroleum generation and migration in the Western Canada Sedimentary Basin. Geological Atlas of the Western Canada Sedimentary Basin, Alberta Research Council/Canadian Society of Petroleum Geology joint project, in press.

Garven, G. and Freeze, A., 1984a. Theoretical analysis of the role of groundwater flow in the genesis of stratabound ore deposits. 1. Mathematical and numerical model. American Journal of Science, **284**, 1085-1174.

_____, 1984b. Theoretical analysis of the role of groundwater flow in the genesis of stratabound ore deposits. 2. Quantitative results. American Journal of Science, **284**, 1085-1174.

Garven, G., 1989. A hydrogeologic model for the formation of the giant oil sand deposits of the Western Canada Sedimentary Basin. American Journal of Science, **289**, 105-166.

Hitchon, B., 1969a. Fluid flow in the Western Canada Sedimentary Basin. 1. Effect of topography. Water Resources Research, **5**, 186-195.

_____, 1969b. Fluid flow in the Western Canada Sedimentary Basin. 2. Effect of Geology. Water Resources Research, **5**, 460-469.

_____, 1984. Geothermal gradients, hydrodynamics, and hydrocarbon occurrences, Alberta, Canada. American Association of Petroleum Geologists, Bulletin, **68**, 713-743.

_____, Bachu, S., Sauveplane, C.M., Ing, A., Lytviak, A.T. and Underschultz, J.R., 1989a. Hydrogeological and geothermal regimes in the Phanerozoic succession, Cold Lake area, Alberta and Saskatchewan. Alberta Research Council, Bulletin 59.

_____, Sauveplane, C.M., Bachu, S., Koster, E.H. and Lytviak, A.T., 1989b. Hydrogeology of the Swan Hills area, Alberta: Evaluation for deep waste injection. Alberta Research Council, Bulletin 58.

_____, Bachu, S. and Underschultz, J.R., 1990. Regional subsurface hydrogeology, Peace River Arch area, Alberta and British Columbia. Bulletin of Canadian Petroleum Geology, **38A**, 196-217.

Jackson, P.C., 1984. Paleogeography of the Lower Cretaceous Mannville Group of Western Canada. *In* J.A. Masters (ed.) Elmworth - Case Study of a Deep Basin Gas Field, American Association of Petroleum Geologists, Memoir 38, 49-77.

Jones, R.W., 1980. Some mass balance and geological constraints on migration mechanisms. *In* W.H. Roberts III and R.J. Cordell (eds.) Problems of Migration, American Association of Petroleum Geologists, Studies in Geology, **10**, 47-68.

Masters, J.A., 1984. Lower Cretaceous oil and gas in western Canada. *In* J.A. Masters (ed.) Elmworth - Case Study of a Deep Basin Gas Field, American Association of Petroleum Geologists, Memoir 38, 1-34.

Moshier, S.O. and Waples, D.W., 1985. Quantitative Evaluation of Lower Cretaceous Mannville Group as Source Rock for Alberta's Oil Sands. American Association of Petroleum Geologists Bulletin, **69**, 161-172.

Van Elsberg, J.N. and Abercrombie, H.J., 1991. A microbiological model of heavy oil. Proceedings, Eighth Annual Heavy Oil and Oil Sands Technical Symposium, Canadian Heavy Oil Association/University of Calgary, March 14, 1991.

Water-Rock Interaction, Kharaka & Maest (eds) © 1992 Balkema, Rotterdam. ISBN 90 5410 075 3

Preservation of ancient basinal fluids in halite fluid inclusions

D.Chipley & T.K.Kyser
University of Saskatchewan, Saskatoon, Sask., Canada

ABSTRACT: Fluids are intimately involved in the deposition and diagenesis of minerals in sedimentary basins, but the chemical and isotopic compositions of these fluids are usually inferred indirectly from analyses of diagenetic minerals. However, fluid inclusions, trapped in halite during recrystallization of evaporites in basins may, in some cases, preserve the temperatures and isotopic compositions of the fluids. The Devonian Prairie Evaporite Formation in the Western Canadian Sedimentary Basin in Saskatchewan contains numerous generations of halite which record distinct fluid events. During Pennsylvanian time, the evaporite minerals were recrystallized by a warm (35°C) fluid with a δD value near -100 and $\delta^{18}O$ value near -9, similar to an evolved meteoric water from a near-coastal environment. A later recrystallization event occurred at about 80°C during the Late Cretaceous Period, involving a fluid with a δD value near -140 and a $\delta^{18}O$ value near -15, similar to an evolved meteoric water originating at high latitudes or altitudes. Such a fluid exists presently within overlying Cretaceous aquifers in the basin but at temperatures of only 15°C. The chemical compositions of the inclusion fluids result from dissolution of evaporite minerals, including carnallite ($KMgCl_3 \cdot 6H_2O$), so that a magnesium-rich brine produced from diagenesis of the evaporite could dolomitize limestone elsewhere in the basin.

1 INTRODUCTION AND GEOLOGIC SETTING

Reactions between fluids migrating in sedimentary basins and the various mineral phases encountered result in dissolution, recrystallization, and replacement by new mineral phases and may deposit economic concentrations of metallic minerals and affect petroleum migration (Bethke and Marshak, 1990). The extent of these fluid-mineral reactions are determined by the pressure, temperature, and chemical characteristics of the sediment and fluid, the latter of which is usually inferred indirectly by analysis of the minerals formed (e.g. Carpenter and Lohmann, 1989). Migration of fluids in sedimentary basins can be induced by compaction of water-saturated sediments shortly after deposition, or by thrusting, compression, and tectonic uplift in adjacent orogens (Garven, 1989; Bethke and Marshak, 1990).

The Western Canada Sedimentary Basin (Fig. 1) has been subjected to episodes of tectonic activity producing topographic uplift during Lower Devonian, Upper Mississippian, Lower Pennsylvanian to Jurassic time, and also during formation of the Cordillera from Middle Jurassic through Miocene time (Porter *et al.*, 1982). Expulsion of water by sediment compaction in the Late Pennsylvanian (295 Ma) has been credited with formation of the Pine Point Pb-Zn deposit (Cumming *et al.*, 1990). Later fluid migration due to Cretaceous uplift has been postulated as the mechanism for the formation of massive oil sands deposits in Alberta (Garven, 1989). The fluid migration events also have resulted in dissolution and recrystallization of evaporite minerals in the Prairie Evaporite Formation in Saskatchewan (Chipley and Kyser, 1991).

The Prairie Evaporite Formation is a Middle Devonian evaporite sequence about 200 m thick that is buried, beneath Devonian carbonates and Cretaceous sandstones and shales, to about 1000 m below the present surface in the Saskatoon area, (Fig. 1; Fuzesy, 1982). The sequence is composed of a lower portion of alternating halite and anhydrite beds overlain by about 70 m of halite containing four potash ore horizons deposited during the Middle Devonian, when the continent was near to the equator. The ore horizons consist of varying proportions of sylvite (KCl) and carnallite ($KMgCl_3 \cdot 6H_2O$)

Fig. 1. Map showing the locations of the PCS Cory, Rocanville, and Lanigan potash mines in Saskatchewan. The areal extent of potash horizons in the Prairie Evaporite Formation is ruled.

intermixed with halite (NaCl), interrupted intermittently by leach and collapse anomalies. Leach anomalies are areas of partial or total replacement of the ore minerals by halite whereas, in collapse anomalies, the evaporite horizon has been partially or completely removed by dissolution allowing collapse of overlying rocks into the voids (DeMille *et al.*, 1964). Vertical extent of collapse structures allows determination of the ages of dissolution and, therefore, timing of fluid migrations, (McTavish and Vigrass, 1987) which correlate broadly with tectonic events defined by Porter *et al.*, (1982).

Recrystallization of the evaporite minerals of the Prairie Evaporite Formation resulted in preservation of post-Middle Devonian fluids by entrapment as fluid inclusions in halite. Halites, therefore, contain not only a record of fluid migration events, but also a sample of the migrating fluid itself. The temperatures at which these fluids were trapped and the chemical and stable isotopic compositions of these trapped fluids are invaluable in determining both fluid source and reactions that may have occurred during fluid migration.

Ore samples from the potash horizons of the Prairie Evaporite Formation were obtained from Potash Corporation of Saskatchewan (PCS) mines (Fig. 1) to determine the number of fluids that have affected the evaporite, as well as the temperatures, and chemical and isotopic compositions of the fluids. The purpose of this study was to acquire a better understanding of the temperatures, chemical compositions, and sources of fluids involved in the diagenetic history of the Prairie Evaporite as well as the Western Canada Sedimentary Basin.

2 DISCUSSION OF RESULTS

Analyses were carried out as outlined in Chipley and Kyser (1991) and Koehler *et al.* (1991). Stable isotopic compositions are reported in δ-notation relative to Standard Mean Ocean Water (SMOW).

2.1 *Stable isotopes of hydrogen and oxygen in waters, basinal brines, and evaporite brines*

Hydrogen and oxygen isotopic compositions of precipitation, resulting from condensation of atmospheric water vapour, define the meteoric water line (Fig. 2). The atmospheric water vapour is enriched in H and ^{16}O relative to its source, the degree of enrichment enhanced by lower mean annual temperature, *i.e.* higher latitude (Craig, 1961; Friedman *et al.*, 1964). D and ^{18}O in precipitation are further depleted with increased distance from the ocean and higher latitude (Friedman *et al.*, 1964). The extreme evaporation of seawater required for evaporite deposition results in the isotopic composition of the brine following the trajectory shown in figure 2. Basinal fluids have isotopic compositions that lie on basin lines that intersect the meteoric water line at approximately the isotopic composition of local modern meteoric water and the fluids usually become progressively enriched in D and ^{18}O with depth (Clayton *et al.*, 1966; Hitchon and Friedman, 1969).

Modern meteoric water in Saskatchewan has δD values near -150 and $\delta^{18}O$ values near -20 (Wittrup and Kyser, 1990), distinct from deep basinal fluids in the Western Canada Sedimentary Basin, evaporated seawater and Devonian meteoric water precipitated near the equator, the location of the Prairie Evaporite Formation during deposition (Fig. 2). Ancient and modern meteoric waters and basinal fluids may have recrystallized the evaporites in the past, and are preserved in halite fluid inclusions. The distinct isotopic compositions of these waters allows them to be distinguished.

2.2 *Fluids in the Western Canada Sedimentary Basin*

Halite from the Prairie Evaporite Formation has two distinct textures formed by fluid inclusions, chevron-texture and randomly distributed fluid inclusions in clear halite. Where both types of inclusions are present in the same halite crystal, chevron texture always occurs at the center of the crystal surrounded by clear halite with randomly distributed fluid inclusions. Therefore, chevron fluid inclusions contain older fluids than randomly distributed fluid inclusions.

Fluids that leak into potash mines and recrystallize halite quickly reach chemical equilibrium by dissolution of evaporite minerals (Wittrup and Kyser, 1990; Chipley and Kyser, 1991) and are thus saturated with the evaporite phases in the environs of the halite when trapped as fluid inclusions. Chemical compositions of the fluids in inclusions, therefore, do not represent the composition of the source of the fluid but a fluid that has dissolved and recrystallized evaporites. Halite fluid inclusions from PCS Cory (Fig. 1), contain sylvite daughter minerals and those from PCS Lanigan and PCS Rocanville, contain carnallite daughter minerals, similar to the mineralogy of the respective ore horizons. The melting characteristics of the frozen fluids in halite fluid inclusions from all mines indicate that they contain Na-K-Ca-Mg-Cl brine, although fluids from the PCS Lanigan and Rocanville mines are richer in magnesium as indicated by saturation with carnallite. Magnesium-rich fluids have been proposed as agents in the dolomitization of limestones and some dolomitizing fluids may have had their origins in dissolution and recrystallization of evaporite minerals such as in the Prairie Evaporite Formation.

Minimum trapping temperatures of the fluids, determined by redissolving the sylvite and carnallite daughter minerals, range from 35°C to 110°C and are distributed bimodally (Chipley and Kyser, 1991). In chevron fluid inclusions, daughter mineral dissolution temperatures range from 35°C to 50°C and daughter minerals in randomly distributed fluid inclusions dissolve between 50°C and 110 °C. These temperatures are greater than the present temperature of about 27°C for the evaporites in the Saskatoon area, therefore the Prairie Evaporite Formation must have been buried deeper, or the geothermal gradient was greater, in the past.

The fluids from the chevron fluid inclusions have δD values near -92 and $\delta^{18}O$ values near -8, which do not resemble the stable isotopic compositions of evaporated seawater, residual brine from seawater evaporation, or water expelled by compaction of marine sediments (Fig. 2). An isotopically depleted water component, similar to meteoric waters from low latitudes, as well as reaction with sediments may have been involved to produce the isotopic compositions of the fluids in chevron inclusions. The chevron halites preserve a single fluid event that occurred shortly after deposition, most likely during Late Devonian or Early Mississippian time when the area was at low latitudes and being eroded.

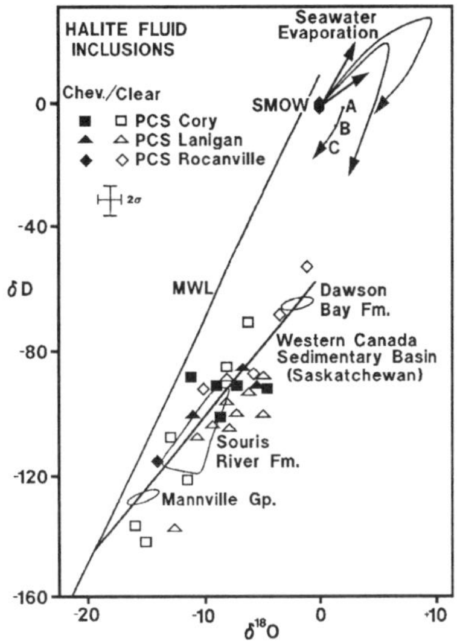

Fig. 2. δD and $\delta^{18}O$ values for meteoric water (MWL), seawater (SMOW), basinal brines from the Western Canada Sedimentary Basin in the Saskatoon area (Wittrup and Kyser, 1990), chevron (chev.) and clear inclusion fluids from halite from the Prairie Evaporite Formation, and the isotopic composition of seawater during evaporation of natural and experimental systems under various atmospheric conditions, including evaporated seawater from Bonaire Island crystallizers, A - pre-gypsum, B - pre-halite, and C - post-halite, pre-sylvite (Chipley and Kyser, 1991, and references therein).

Fluids extracted from the randomly distributed fluid inclusions belong to two distinct fluid events (Chipley and Kyser, 1991), a low temperature event near 50°C and a higher temperature event near 80°C. The lower temperature event fluid had δD near -70 and $\delta^{18}O$ near -6, similar in isotopic composition to the most evolved basinal fluids in the Western Canada Sedimentary Basin and fluids presently in the carbonates of the Dawson Bay Formation which overlies the Prairie Evaporite Formation (Fig. 2). The high temperature fluid has δD near -141 and $\delta^{18}O$ near -15, similar to relatively modern meteoric water and to fluids currently within the overlying Cretaceous aquifers. The similarity between the isotopic composition of Cretaceous formation fluids and the high temperature fluid in halite inclusions probably indicates that this recrystallization event occurred before erosion near the end of the Cretaceous. Access of the fluids to the Prairie Evaporite Formation from overlying aquifers may have been due to reactivation of deep structures during tectonic events that caused fracturing of the sedimentary rocks (Dunn, 1982).

3 SYNOPSIS

Fluid inclusions in halite from the Prairie Evaporite Formation contain a valuable record of post-Middle Devonian fluids that may have been involved in deposition of diagenetic minerals in the Western Canada Sedimentary Basin. Temperatures of the fluids were higher than at present and may reflect deeper burial of the evaporite when diagenesis occurred or a greater geothermal gradient in the past.

Stable isotopic compositions reveal that the fluids that affected the evaporite are complex mixtures of meteoric waters and evolved basinal brines, but seawater or evaporated seawater was not found in the halite fluid inclusions. Chemical compositions of the inclusion fluids represent fluids that dissolved and recrystallized evaporites and migrated farther. Some of these fluids are magnesium-rich and may have been responsible for dolomitization of limestones elsewhere in the basin.

Fluid migration events in the Western Canada Sedimentary Basin were probably induced by tectonic activity. Reactivation of deep structures during these tectonic events probably caused fracturing of the Prairie Evaporite Formation and provided access for the fluids to the evaporite minerals.

ACKNOWLEDGEMENTS

This project was funded by Saskatchewan Potash Producers Association and NSERC Operating Grants to T.K.K..

REFERENCES

Bethke, C.M. & S. Marshak 1990. Brine migrations across North America - the plate tectonics of ground water. *Annu. Rev. Earth Planet. Sci.* 18: 287-315.

Carpenter, S.J. & K.C. Lohmann 1989. $\delta^{18}O$ and $\delta^{13}C$ variations in Late Devonian marine cements from the Golden Spike and Nevis Reefs, Alberta, Canada *J. Sediment. Petrol.* 59: 792-814.

Chipley, D. & T.K. Kyser 1991. Large scale fluid movements in the Western Canada Sedimentary Basin as recorded by fluid inclusions in evaporites. In J.E. Christopher & F.M. Haidl (eds.), *Sixth International Williston Basin Symposium*, Saskatchewan Geological Society, Spec. Pub. 11: 265-269.

Clayton, R.N., I. Friedman, D.L. Graf, W.F. Mayeda & N.F. Shimp 1966. The origin of saline formation waters 1. Isotopic composition. *J. Geophys. Res.* 71: 3869-3882.

Craig, H. 1961. Isotopic variations in meteoric waters. *Science* 133: 1702-1703.

Cumming, G.L., J.R. Kyle & D.F. Sangster 1990. Pine Point: a case history of lead isotope homogeneity in a Mississippi Valley-Type district. *Econ. Geol.* 85: 133-144.

DeMille, G., J.R. Scholdice & H.W. Nelson 1964. Collapse structures related to the evaporites of the Prairie Evaporite Formation, Saskatchewan. *Geol. Soc. Am. Bull.* 75: 307-316.

Dunn, C.E. 1982. Geology of the Middle Devonian Dawson Bay Formation in the Saskatoon potash mining district, Saskatchewan. *Saskatchewan Energy and Mines*, Report 194: 117 p.

Friedman, I., A.C. Redfield, B. Schoen & J. Harris 1964. The variation in the deuterium content of natural waters in the hydrologic cycle. *Rev. Geophys.* 2: 177-224.

Fuzesy, A. 1982. Potash in Saskatchewan. *Saskatchewan Energy and Mines*, Report 181: 44 p.

Garven, G. 1989. A hydrologic model for the formation of the giant oil sands deposit of the Western Canada Sedimentary Basin. *Am. J. Sci.* 289: 105-166.

Hitchon, B. & I. Friedman 1969. Geochemistry and origin of formation waters in the western Canada sedimentary basin -I. Stable isotopes of hydrogen and oxygen. *Geochim. Cosmochim. Acta* 33: 1321-1349.

Koehler, G.D., D. Chipley & T.K. Kyser 1991. Measurement of the hydrogen and oxygen isotopic compositions of concentrated chloride brines and brines from fluid inclusions in halite. *Chem. Geol.* 94: 45-54.

McTavish, G.J. & L.W. Vigrass 1987. Salt dissolution and tectonics, south central Saskatchewan. In C.G. Carlson & J.E. Christopher (eds.), *Fifth International Williston Basin Symposium*: 157-168.

Porter, J.W., R.A. Price & R.G. McCrossan 1982. The Western Canada Sedimentary Basin. *Roy. Soc. London Phil. Trans.* Ser. A305: 169-192.

Wittrup, M.B. & T.K. Kyser 1990. The petrogenesis of brines in Devonian potash deposits of western Canada. *Chem. Geol.* 82: 103-128.

Water-Rock Interaction, Kharaka & Maest (eds) © 1992 Balkema, Rotterdam. ISBN 90 5410 075 3

Water composition variation within one formation

Max L.Coleman
Postgraduate Research Institute for Sedimentology, Reading University, UK & BP Research Centre, Sunbury-on-Thames, UK

ABSTRACT: Contrary to the assumption that composition varies little within a single, continuous formation, salinity varies by a factor of 2.5 for waters co-produced with oil from a large off-shore field. The usual cause for variation is that reservoir exploitation involves injection of water into the formation from the surface. Stable isotope and most chemical components vary linearly between the extreme compositions. Variation occurs within individual wells and between wells. These data can be explained only as mixtures of two end-member compositions. However, the stable isotope values preclude the possibility that injected sea-water is one of the two components. Analysis of data from a single well implies that there is vertical variation in composition. Possibly, the more dilute pore-waters in the oil-zone of the reservoir represent fossilised palaeo-porewater while the brine below the oil-water contact has become more saline in the time since emplacement of the oil.

1 ACCESS TO DEEP FORMATION WATERS WATERS

Formation water is co-produced with oil from many reservoirs and is viewed as a problem by companies whose main interest is production of hydrocarbons. However, it gives unique access to deep formation waters and its study can help to alleviate some of the problems which water production brings, in particular, the precipitation of chemical scale. This may occur, for example, when the oil-field production strategy requires injection of water from the surface to maintain pressure in the reservoir. If the injected water and the native formation water have incompatible compositions then precipitation may occur. For example, precipitation of barite ($BaSO_4$) results from reaction between formation waters rich in Ba with sea-water, which contains dissolved sulphate.

Initial exploration usually results in samples being taken for analysis from the water-zone (aquifer) beneath the oil-filled part of the reservoir. If water is co-produced with the oil later during the field's history then it is assumed to result from coning up of the aquifer caused by differential pressure reduction in the vicinity of a producing well. Only if there are likely to be problems from break-through of injected water are further routine analyses likely to be made.

2 VARIATION OF COMPOSITION WITHIN ONE FIELD

The instigation of this work was the observation that Cl contents from various routinely-taken produced-water samples showed considerable variation. Analyses were made of a restricted number of samples from various wells in the field which gave Cl contents from 48,600 mg/l to 31,500 mg/l. These data were initially believed to result from mixing of formation water with injected sea-water (Cl ≈ 20,000 mg/l). The aquifer had been sampled during initial exploration drilling and Cl had been measured at 60,000 mg/l.

Samples were selected initially to cover a range of Cl contents and a complete set of analyses was made of Na, K, Mg, Ca, Sr, Ba, Br, SO_4. However, SO_4 was seen to be lower than the contribution expected to be seen from a sea-water component. Stable isotope analyses (δD and $\delta^{18}O$) are capable of identifying sea-water, which has a very distinctive and uniform isotopic signature (Craig, 1960) and were performed on the same sample set.

All the measured components, with the exception of Ba and SO_4 showed linearly correlated results. Correlation coefficient values ranged from 0.69 for $\delta^{18}O$-K to 0.99 for Ca-Na.

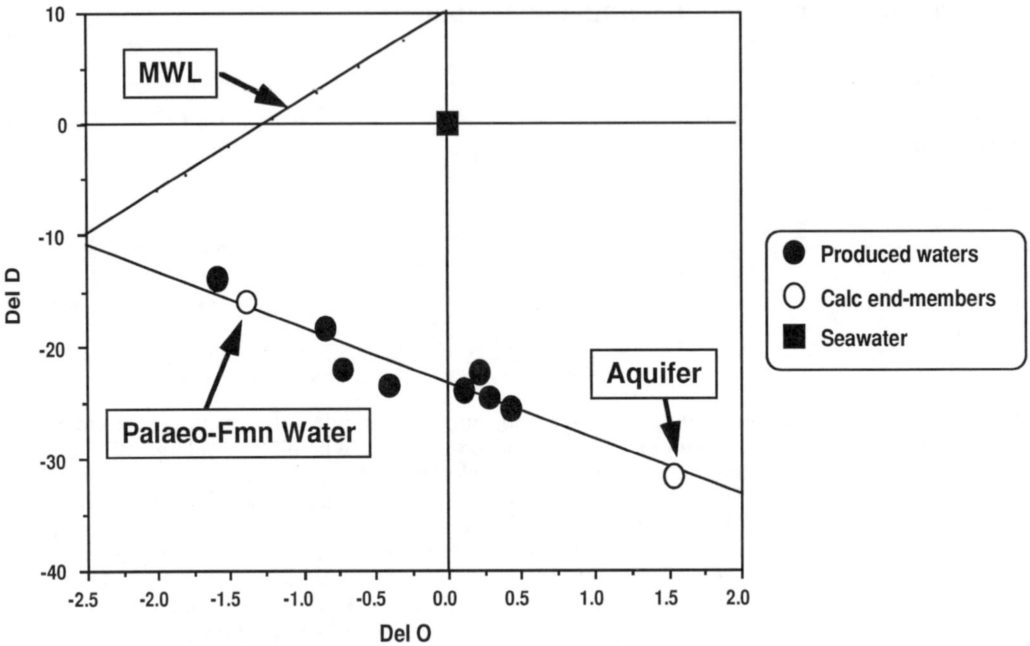

Figure 1. Isotopic characterisation of formation waters, showing that they consist of two-component mixtures and that sea-water is not one the components. The meteoric water line is shown for reference.

These results suggested that the produced waters were mixtures of two end-member components. However, it was easy to show that sea-water was not one of those components. A cross plot of δD and $\delta^{18}O$ shows a linear trend for the produced waters which is completely separated from the sea-water value (see Fig. 1).

3 VARIATION OF CHLORIDE CONTENT WITHIN ONE WELL IN THE FIELD

In a few cases further analyses were made of subsequent samplings from the same well. In one case the Cl value had changed from 40,600 mg/l to 26,400 mg/l. In all cases all the components were measured and fell perfectly on the same linear trends defined by the analyses from different wells. This suggested that the same process was controlling both between-well and within-well variation.

Although oil companies do not analyse produced water samples comprehensively in a routine fashion, it is common for a restricted set of analyses to be performed. Because of this there is an extensive record of chloride analyses for this field. It can be seen that for many individual wells there was a repeated pattern of Cl variation. As water was first produced in small amounts with the oil, the Cl content was relatively low. As water-cut (the percentage of water in the total fluid) increased, so did the Cl content.

Water cut tends to increase as a cone of underlying aquifer is drawn into the vicinity of the producing part of the well. This is the depth range over which the well is perforated, allowing access to the formation fluids. When the water cut had risen to unacceptably high levels in any well, the well would be taken out of production and worked over. This involved plugging the well just above the depth from which water was being produced and re-perforating it at a higher level, thus restoring production from within the oil-zone. The resulting lower water-cut would be accompanied by a reduction in Cl content. This relationship is shown in Figure 2, It can be seen that there is an excellent, but possibly fortuitous, correlation between the two parameters. However, it should be noted that there is no reason why water-cut and Cl content should define a linear correlation, as opposed to any other.

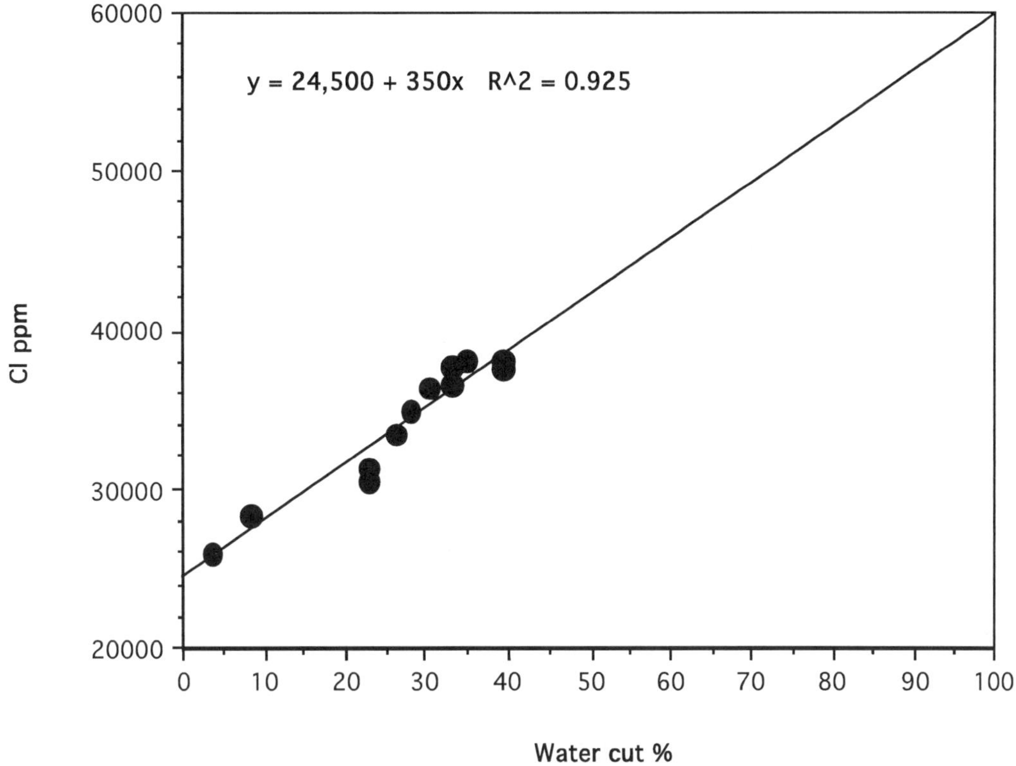

Figure 2. Relationship between water-cut and Cl content.

4 CHARACTERISATION OF END-MEMBER COMPOSITIONS

Because of the excellent correlation between Cl and other components in the fully analysed samples, it is reasonable to assume that similar relations would apply to the historical samples. This implies that the concept of the produced water being a mixture of two components is also related to variation of salinity with water-cut. Water-cut increases as a result of coning of the underlying aquifer. Thus, it seems likely that the more saline end member represents the aquifer.

The composition of the less saline end-member is expressed to the greatest extent when there is low water-cut after re-perforating a well higher up in the oil-zone. It is possible that the residual water in the oil column is contributing this composition to the mixtures. However the mixing two waters of different salinity should not produce a linear relationship. Nevertheless, it is interesting to note that extrapolation of the line to 100% water-cut would give a Cl content of 60,000 mg/l (Fig. 2). This value is that which had been recorded historically for aquifer water. Further and more recent confirmation of this composition was recorded by Coleman et al. (1990) who extracted water from a core taken from the transition zone between oil and aquifer. The Cl content was found to be 59,300 mg/l.

Although the less saline end-member cannot be defined perfectly, using the linear relationship it is probable that its maximum salinity is given by the zero water-cut intercept of the linear relationship. This would be equivalent to 24,500 mg/l Cl. However, the trend expressed by all the points, apart from the two lowest values, implies mixing with a very low salinity end-member. The end-member compositions are indicated on Figure 1, and were calculated from the linear relationship between water-cut and stable isotopic compositions. The values for the less saline component indicate that it probably had a meteoric water origin.

Vertical stratification of formation waters in an oil reservoir is likely to be the norm, if it can be detected. Formation waters may originate from a meteoric source, as shown by their stable isotope compositions, but subsequent water-rock interaction can produce a saline brine. Consequently, it is to be expected that if part of a formation water is separated during this process of evolution, then it will preserve a less evolved composition. Emplacement of oil into a porous and permeable formation, previously saturated with water, should trap samples of that water in the residual pore-space not filled by oil. This may be a considerable volume of water because even in clean sandstones oil may saturate only 70% of the available pore-space.

5 CONCLUSIONS

Oil production allows access to water which normally would not be available for sampling and analysis. In the case described here there is evidence of preservation of a palaeo-formation water of meteoric origin trapped by oil emplacement. The aquifer continued to evolve in composition to give its current, much more saline values.

There are practical implications from these results. The water sampled during exploration and development of an oil-field by necessity is taken from the underlying aquifer. This composition is used as the basis for all subsequent predictions. It seems likely that, at least in some cases, the water in the oil-zone may have a different composition. This may be either more or less saline than the aquifer, depending on the subsequent history of formation water evolution. The effect on prediction of precipitation resulting from reactions between incompatible waters is self-evident.

6 REFERENCES

Coleman, M L, M R O Jones & M A Cox, 1990. Analysis of formation water sampled from core. *North Sea Oil and Gas Reservoirs - II:* 165-171. London: Graham and Trotman.

Craig, H. 1961. Isotopic variations in meteoric waters. *Science* 133: 1702-1703.

Water-Rock Interaction, Kharaka & Maest (eds) © 1992 Balkema, Rotterdam. ISBN 90 5410 075 3

The effects of combined changes in pore fluid chemistry and stress state on reservoir permeability: Design and recirculating flow rig

S.C. Elphick, G.B. Shimmield, B.T. Ngwenya & I.G. Main
Department of Geology & Geophysics, University of Edinburgh, UK

B.G.D. Smart & B.R. Crawford
Department of Petroleum Engineering, Heriot Watt University, Riccarton Campus, Edinburgh, UK

ABSTRACT: Two separate rigs have been constructed for studying the role of pore fluid chemistry (Edinburgh University) and stress state (Heriot Watt University) on the permeability of a hydrocarbon reservoir at in situ pressures and temperatures and therefore simulate changes during production. The final design will integrate the two rigs so that the combined effects of pore fluid chemistry and stress-state can be quantified. We discuss the design and construction of the recirculating flow-rig at Edinburgh and present preliminary results. The rig has been designed on the basis of a well-mixed flow reactor but with the capability to sample and analyse microlitre fluid volumes at the pressures and temperatures of reaction (nominally 80°C and 150-350 bars) using high performance liquid chromatography. Initial trials on onshore analogues of North Sea reservoir sandstones show rapid increases in the solute content of the fluid over one week experiments. These increases are accompanied by a marked increase in the permeability of the core resulting from dissolution processes.

1 INTRODUCTION.

During the operational lifetime of a reservoir significant portions of the host rock will be subjected to sequential variation in pore fluid chemistry, temperature, pressure and stress state. These changes lead to alteration in permeability through the generation and sealing of microfractures, solution and precipitation reactions. The in-situ permeability is therefore a function of the complex interaction of chemical and physical processes. We simulate and examine the combined effects of changes in stress state and pore fluid chemistry on the permeability of typical reservoir rocks, using two separate rigs. A recirculating flow rig designed to monitor the evolution of pore fluid chemistry in a rock core maintained at in situ reservoir conditions is now operating at Edinburgh University. A parallel rig has been developed at Heriot Watt University where true-trixial stress and permeability measurements are coupled to real-time recording of acoustic emissions from stressed reservoir samples. In the final design, the two rigs will be integrated so that the combined effects of pore fluid chemistry and stress state can be quantified. In this paper, we focus on the design and construction of the recirculating flow rig and present preliminary results on analogues of North Sea reservoir rocks.

2 DESIGN PRINCIPLES

The design of an experimental rig is dictated by the theories of flow operation which have been reviewed in detail by Posey-Dowty et al (1986). Most of the rigs used to date (e.g. Rimstidt & Dove, 1986; Dove & Crerar, 1990) allow sampling only in the semi-static mode, requiring corrections for quenching, pressure drops and dilution. Our rig has been designed on the basis of a well-mixed flow reactor which must:

1. provide continuous recirculation of a fluid through a standard rock core at in situ pressures and temperatures without any reverse flow or pressure pulses;

2. be able to work with any fluid found in oilfield environments and to withstand major corrosion by aggressive brines and organic acids;

3. allow analysis, on line, of the fluid chemistry on microlitre samples without resorting to quench techniques to avoid corrections to the data;

4. allow introduction of tracers in order to effect monitoring of fluid pathways and reaction sites in the rock using an ion microprobe;

5. be fully automated in order to provide continuous uninterrupted sampling.

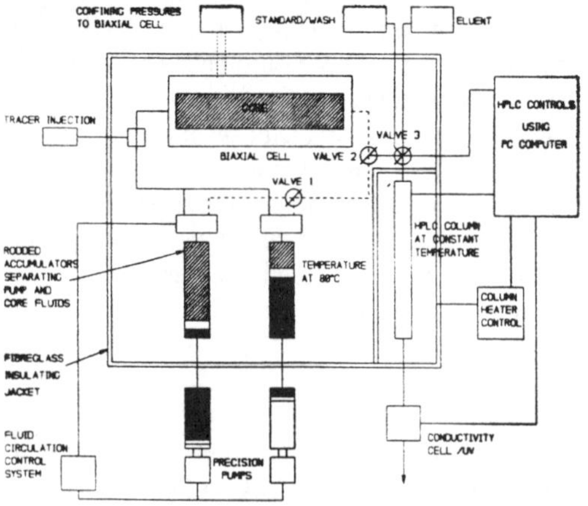

Fig. 1 Setup of recirculating rig.

Figure 1 is a schematic diagram of the final design of the rig. Fluid analysis is by high performance liquid chromatography (HPLC) using a Waters™ Action Analyser 600E pump with conductivity detection for alkalis, alkaline earths and silica, and UV detection for transition metals. Detection limits range from 2-200 ppb which means that any trace metals released from apparently pure 316 stainless by corrosive fluids can interfere with the analysis. Only Grade 1 titanium, PEEK (a high temperature thermoplastic) and PTFE can therefore be used.

Fluid circulation is driven by a pair of stepper-driven M2-1000 Hydratron pumps supplied by Scotia Oilfield Sevices Ltd. The pumping fluid is separated from the analytical fluid by isolation chambers machined in-house from high tensile stainless steel and fitted with PEEK liners and end plugs. One end of the chambers is connected to the pumps, the other end to the coreholder and its attendant valving. The top closure of the isolation chambers have been fitted with a 2 μm titanium frit to prevent migration of plastic particles dislodged from the chamber walls into the flow lines. The fluids in the chamber are separated by a PEEK slider.

A conventional coreholder with a choice of 316 stainless steel or Hastalloy platens was supplied by Edinburgh Petroleum Services (EPS). The original platens were replaced by Grade 1 titanium platens machined in our workshop.

Connections between the various components of the rig (isolators, coreholder, HPLC etc) is by 0.06" OD and 0.02" ID PEEK tubing rated to 5,000 psi. To control fines migration into the fine bore tubing, the outlet platen of the core holder has been re-designed to hold a filter stack. The original stainless end-piece supplied by the manufacturer consisted of a single bored stainless rod, with a stainless platen screwed to its inner end. The new design consists of a titanium platen with an extended rear threaded section, which screws into a chamber in a shortened backing rod. The filter stack is housed in this chamber, in the order (starting from the down flow side) 2 μm titanium frit, 0.45 μm hydrophilised polypropylene Millipore™ filter, a stack of twenty 5 μm nylon filters and another 2 μm titanium frit. Finally, a 316 stainless end shoe is screwed to the outer end of the titanium section to spread the end-load of the confining collar.

Analysis is automated by programming via a computer that manages the HPLC and all components attached to it, including detectors, a column heater controller, column heaters and air actuated rotary PEEK valves. The sample taken is typically 20 μl and can be reduced further, depending on the predicted level of analytes. As sample is diverted into the HPLC , the rig is automatically topped up by exactly the same volume from a wash line that is switched in and out of line through the valve sequences. Because of the small volume taken compared with total in the isolation chambers (typically 200 ml) no correction is applied to the data.

3 APPLICATION.

We report here preliminary results on the Clashach Sandstone, an analogue of North Sea reservoirs of Permo-Triassic age retrieved from open cast mining sites, chosen because of unlimited availability and favourable physical and chemical characteristics. It is a pale-fawn, well sorted quartz-feldspar sandstone with minor kaolinite, traces of dolomite and a relatively high permeability. The first test was run at room temperature, with core confined to 5,000 psi and fluid pressure maintained at 3,000 psi. Ultrapure water was used as reaction fluid flowing at 0.3 ml/min. The second run was carried out at 80°C under pressure conditions similar to the first. The reaction fluid, flowing at 0.1 ml/min, was 0.5mM lithium hydroxide of pH 10.3 chosen to test enhancement of silica dissolution. All analytical columns were maintained at 80°C. At the time of writing this contribution, the second test was still in progress so that only partial interpretations are presented.

Results from the first test (Fig. 2) show that there is rapid release of both Mg and Ca into solution in the first 150 hours of the reaction followed by a quasi-equilibrium solution composition. The absolute values are low and are probably limited by the trace quantities of dolomite presumed to be their source.

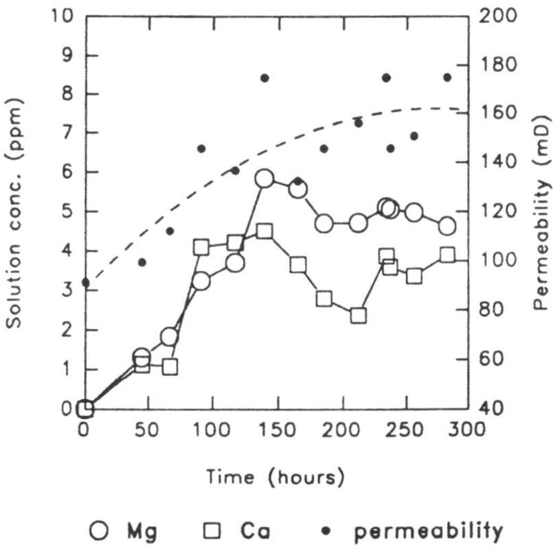

Fig. 2 Alkaline earth elements and permeability trends at room temperature.

The permeability of the core increases markedly from 91 to 176 mD during the course of the experiment. These values were calculated from the pressure differential across the core recorded during sampling from pressure transducers monitoring the individual pumps. The values varied depending on (a) which transducer was the forward stroke and (b) at what stage in the changeover cycle the readings were taken. We find that if Pump2 is on a forward stroke a higher value is recorded. Secondly, this value increases when the pumps approach their end-limits. Sampling near these periods therefore yields an apparently low permeability, explaining the irregularities in the permeability curve, also noted by other workers.

Post-reaction characterisation of the core shows that permeability increase may be correlated with a dissolution process. In

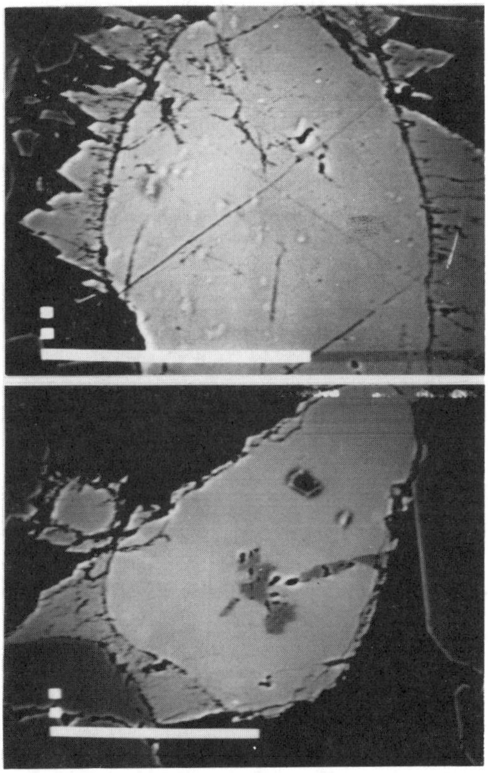

Fig.3 K-feldspar overgrowths (a) in unreacted sample and (b) removed by dissolution in reacted sample. Scale bars = 100 μm

particular, we note a reduced incidence of K-feldspar overgrowths on detrital orthoclase (Figure 3) where the overgrowths are exposed to large pore spaces where the fluid/rock contact is likely to have been maximised.

The concentration of silica in solution in the second run is compared with theoretical dissolved silica values (Fig 4) in pure water in equilibrium with quartz at 80°C, 200 bars (3,000 psi) calculated using SUPCRT92 (Johnson et al, 1991). Once again, rapid release of silica into solution is noticed over the first 180 hours. When a steady-state composition is reached, the Si values are maintained at slightly above the theoretical limits. We attribute this to two possible reasons. Firstly, SUPCRT92 only calculates values in equilibrium with quartz at neutral conditions, while silica is known to be more soluble at a pH>9 (Krauskopf, 1981) where quartz may be in equilibrium with solutes other than aqueous SiO_2. The high values may thus be due to enhanced solubility due to the artificially high pH. However, since the system is closed to the atmosphere, there is some likelihood that it will eventually be buffered towards a neutral pH with silica concentrations approaching theoretical values. Secondly, there is a possibility that the extra silica is generated from dissolution of other minerals and in particular, K-feldspar. As in TEST2, the

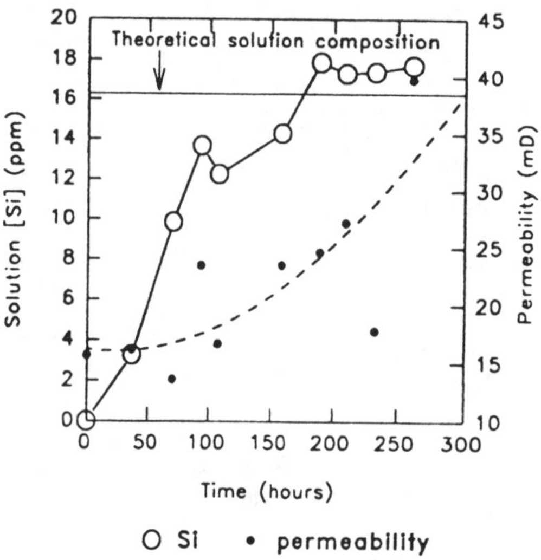

Fig. 4 Dissolved silica and permeability trends at 80°C.

solubility of silica is accompanied by a marked increase in the permeability of the core from 15 to 40 mD.

4 CONCLUSIONS

A recirculating flow rig for studying reactions in rock cores at in situ hydrocarbon reservoir conditions has been constructed with the capability to provide on line analysis of small volumes at pressures and temperatures of reaction. Preliminary results on reservoir analogues show rapid increases in the concentration of solutes in solution accompanied by a marked increase in the permeability of the core.

REFERENCES.

Dove, P.M. and Crerar, D.A. (1990). Kinetics of quartz dissolution in electrolytre solutions using a hydrothermal mixed flow reactor. Geochimica et Cosmochimica Acta 54: 955-969.

Johnson, J.W., Oelkers, E.H., and Helgeson, H.C. 1991. SUPCRT92: A software package for calculating the standard molal thermodynamic properties of minerals, gases, aqueous species, and reactions from 1 to 5000 bars and 0° to 1000°C. Computers and Geosciences.

Krauskopf, K.B. 1981. Introduction to Geochemistry. McGraw-Hill, London 617pp.

Rimstidt, J.D., and Dove, P.M. 1986. Mineral/solution reaction rates in a mixed flow reactor: Wollastonite hydrolysis. Geochimica et Cosmochimica Acta 50: 2509-2516.

Water-Rock Interaction, Kharaka & Maest (eds) © 1992 Balkema, Rotterdam. ISBN 90 5410 075 3

Chemical composition and origin of saline formation waters from the Konrad Mine, Germany

H. Klinge, P. Vogel & K. Schelkes
Federal Institute for Geosciences and Natural Resources, Hannover, Germany

ABSTRACT: Distribution of formation water salinities in the abandoned Konrad iron mine in the North German Lowlands indicates a horizontal salinity stratification and a linear increase in salinity with depth. The vertical linear increase of TDS and of the Ca/Cl and Br/Cl ratios indicates that the water was formed by mixture of a residual brine and a more dilute Na-Cl brine. Two-dimensional modeling of salinity-dependent regional groundwater movement reveals only minor groundwater movement in deep aquifers. The salinity gradient is in reasonable agreement with empirical data. This indicates that the saltwater system is controlled by diffusion. The assumed mixing of fluids is therefore predominantly a result of vertical diffusive mass transport and not of saline water convection.

1 INTRODUCTION

The abandoned Konrad iron ore mine at Salzgitter in Northern Germany has been investigated to serve as a repository for low- and medium-level radioactive waste.

As part of an extensive site characterization program chemical analyses and isotope studies of mine waters were carried out over a period of seven years.

2 HYDROGEOLOGIC FRAME

The investigation area is part of the North German Basin containing thick accumulations of Mesozoic marine sediments and of about 1000 m of Permian and Triassic evaporites deposited during periods of restricted marine circulation. The evaporite units have formed diapirs that have migrated towards the land surface.

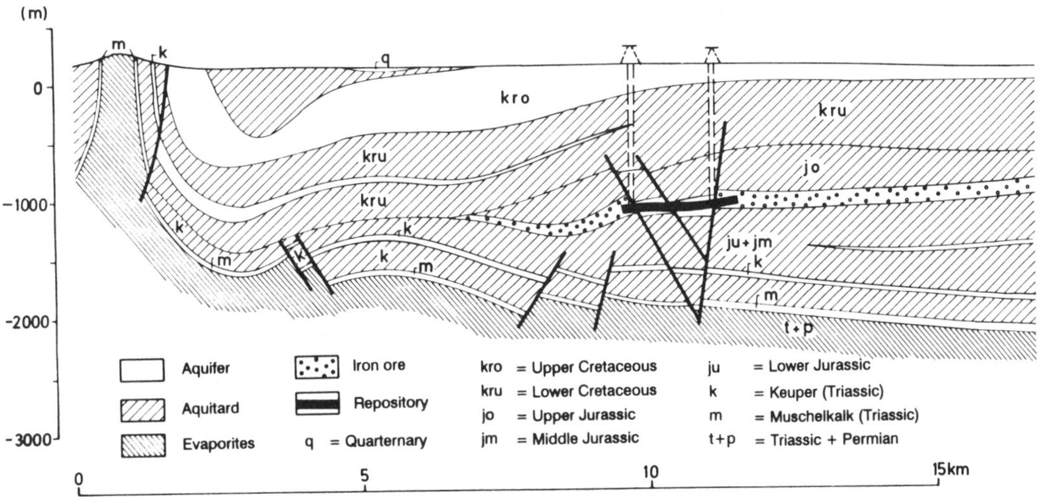

Figure 1: Hydrogeologic cross section of the Konrad investigation area

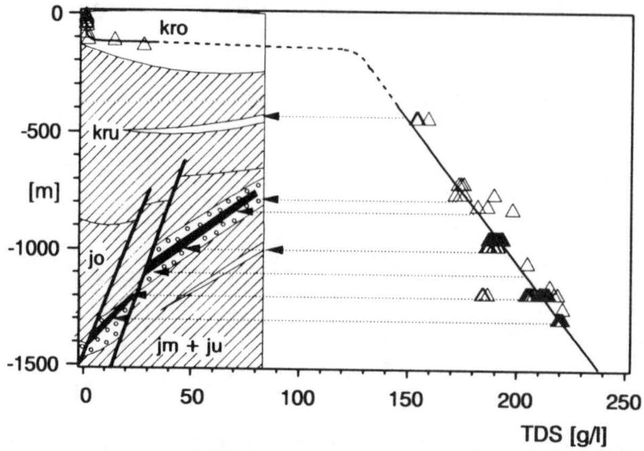

Figure 2: Salt concentration in water samples as a function of depth (For abbreviations and signatures see fig. 1).

The structural development is therefore predominantly a result of halokinetic processes. The Triassic and Jurassic strata in the investigation area have a total thickness of 2000 m. The sediment sequence is built up to the major part of low permeable marine marls and claystones and of permeable sandstones and carbonate sedimentary rocks. During the Upper Jurassic (Oxfordian) the sedimentary iron ores of the Konrad mine have been accumulated in rim synclines of salt diapirs. The Jurassic sediments are covered by low permeable Lower Cretaceous claystones up to 500 m thick.

Where salt diapirs have reached the land surface, they have sometimes displaced the overlying sedimentary strata and generate topographic heights where meteoric fluids can infiltrate into displaced Mesozoic strata (Fig. 1).

3 CHEMICAL AND ISOTOPE ANALYSES

The Jurassic iron ore of the Konrad mine is a medium to low permeable fissured rock with a generally low matrix permeability. Water discharge is therefore restricted to a limited number of permeable fractures. In total 93 samples have been analyzed for their chemical constituents and isotopes. This includes samples obtained from a Lower Cretaceous sandstone and a Middle Jurassic carbonate sandstone layer from one of the mine shafts (Fig. 2).

4 DISCUSSION AND RESULTS

At the mine an upper regime of low mineralized freshwater is present down to a depth of 150 m. Below this depth water is generally saline. There is a linear concentration increase from 160 g/l TDS at a depth of 450 m to 220 g/l TDS at depth of 1300 m. (Fig. 2). The distribution of the salt concentration indicates a horizontal salinity stratification (Fig. 2).

All brines are Na-Ca-Cl type waters. Except for SO_4^{--}, HCO_3^-, which are generally low, and Sr^{++}, and J^-, all constituents positively correlate with depth. There is a linear vertical increase of the Ca/Cl-, Mg/Cl-, and Br/Cl ratio and a corresponding decrease of the Na/Cl ratio (Fig. 3).

The stable isotope composition (H and O) shows two groups of samples with a distinctive different isotope signature. Samples from the upper freshwater regime fit with the global meteoric water line, while formation waters show typical stable isotopic composition of basinal brines plotting to the right of the meteoric water line.

The formation waters get progressively enriched with depth in the heavy oxygen isotope, whereas Deuterium erratically varies between -20 and -30 ‰ SMOW. This indicates an isotope exchange between water and carbonate rock minerals.

The trend of the regression line in Fig. 2 indicates that saturated brines can be expected at a depth of approximately 2300 m at the top of the first bedded evaporite layer. We therefore conclude, that the origin of the mine waters is related to these evaporites.

The linear correlation between TDS and chloride as well as the correlation between TDS and the ion/Cl ratios of Fig. 3 we interpret as a result of a two component mixture between a saturated brine and a more dilute water.

The vertical trend of these ion ratios define the saturated end member as a Ca-Cl, bromide-enriched residual brine.

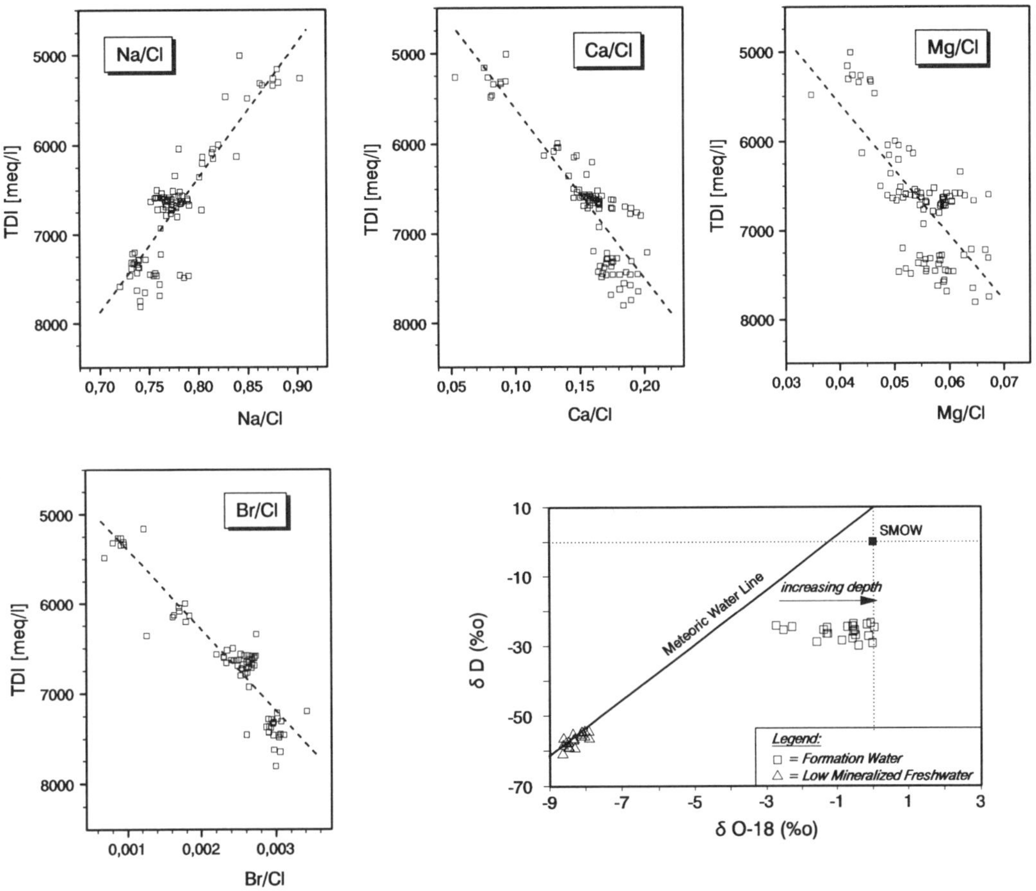

Figure 3: Chemical composition and oxygen and hydrogen isotopic composition of Konrad mine waters

This is supported by the Br-Cl plot in Fig. 4, in which we have compared the Br/Cl brine data with the evaporation line of modern sea water (Herrmann et al. 1973). The regression line of the mine water data intersects the evaporation line beyond the point of halite precipitation.

A brine at this stage of evaporation has a Na/Cl ratio of 0,55. This roughly corresponds with the trend of the Na/Cl-plot in Fig. 3.

The second component of the mixing line is an approximately half concentrated Na-Cl brine with a low Br/Cl ratio which typically is found in brines formed by dissolution of halite. The Br/Cl data are not conclusive to identify the dilute end member which may be of meteoric or of marine origin. The isotope signature defines the mine waters as formation waters which have undergone isotope exchange between water and rock minerals. They originally may have been old meteoric waters, which infiltrated during a warmer climatic period, or a mixture of meteoric water and sea water, and/or of a residual brine.

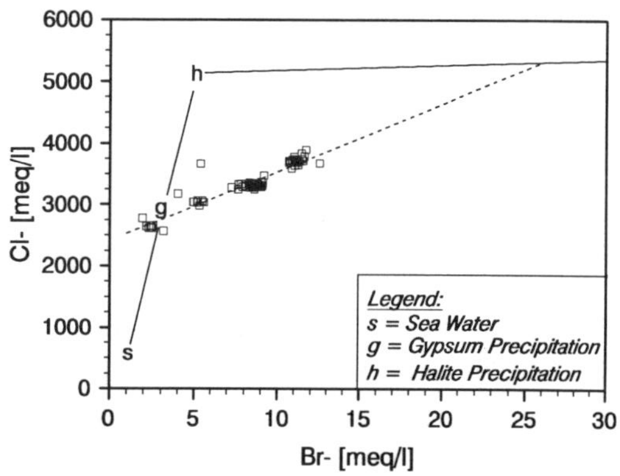

Figure 4: Bromide and chloride concentration in mine waters in comparison with the sea water evaporation line

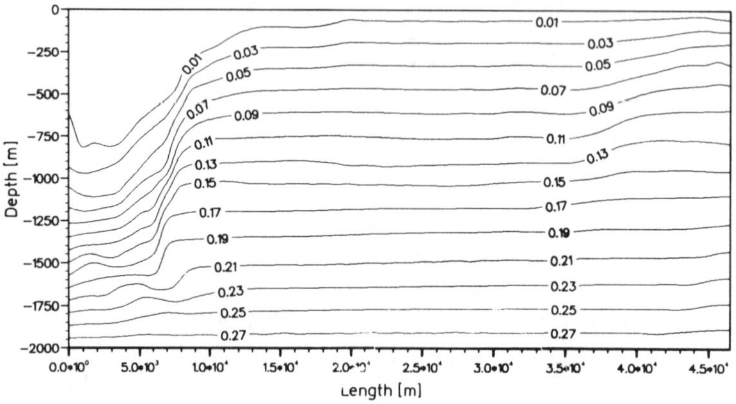

Figure 5: Isolines of solute mass fraction at steady state

5 HYDRAULIC MODELING

Numerical hydraulic models can serve as a useful tool for a better understanding concerning dynamic processes such as mixing of fluids of different density and salinity. Since the observed density stratification has an important impact on the basin hydraulics, conventional hydraulic models which consider freshwater conditions are of a limited value, only.

We have studied the behavior of a groundwater system of a variable water density using the SUTRA code. For the modelled 2D-section the cross section exhibited in Fig. 1 was taken into consideration. It was assumed that the water at the base of the model is in contact with salt formations.

A comparison of the results with corresponding freshwater systems reveals a major difference in the deep groundwater hydraulics. The freshwater system is dominated by groundwater convection. The saltwater system, apart from a fresh water lens below the recharge area, exhibits a linear increase in density and salinity with depth (Fig. 5). It is characterized by a reduction of the interstitial velocity by one to two orders of magnitude relative to the freshwater system. A linear vertical increase of density indicates that in the saltwater system, diffusion is the dominant mechanism of solute transport from the bedded salt to the land surface.

The calculated density stratification is in reasonable agreement with the empirical data.

6 CONCLUSIONS

The linear vertical increase of TDS and of relevant ion ratios in the Konrad mine indicates that the mine waters have been formed by a two component mixture of a saturated residual brine expelled from deep lying bedded evaporites and a more dilute Na-Cl brine.

From two-dimensional numerical hydraulic models, which take into account the variable density of fluids it can be concluded that this gradient is a result of a predominantly diffusive mass transport from deep lying bedded salt to the land surface.

A linear depth-dependant trend in the salinity and a chemical composition of brines similar to that of the Konrad brines has been frequently reported from sedimentary basins with deep lying evaporites.

We therefore conclude that the observed linear increase of porewater salinity and the mixing of fluids, which has been presumed by many authors to interpret the linear depth-dependant trend in brine chemistry is in general a result of predominantly diffusive processes and only to a minor extend a result of vertical saline water convection.

REFERENCES

Herrmann, A.G., Knake, D., Schneider, J. & Peters, H. 1973. Geochemistry of modern seawater and brines from salt pans: main components and bromine distribution. Contr. Mineral. and Petrol. **35**: 567-598.

Schelkes, K., Vogel, P., Klinge, H., Knoop, R.-M. 1991. Modelling of variable-density groundwater flow with respect to planned radioactive waste disposal sites in West Germany - Validation activities and first results - Proceedings Geoval-1990: 328-335. Stockholm.

Voss, C.I. 1984. SUTRA: A finite-element simulation model for saturated-unsaturated, fluid-density-dependent ground-water flow with energy transport or chemically-reactive single species solute transport. U.S. Geol. Survey, Water Resour. Invest. Rep. **84-4369**.

Water-Rock Interaction, Kharaka & Maest (eds) © 1992 Balkema, Rotterdam. ISBN 90 5410 075 3

Geochemical modeling of salt dome cap rock-brine reactions

James A. Saunders
Department of Geology, Auburn University, Ala., USA

ABSTRACT: Na-Ca-Cl oil field brines from the Mississippi Salt Dome Basin comprise some of the best documented and most metal-enriched natural waters known, typically containing up to 2000 mg/l Sr, >100 mg/l Pb+Zn along with low concentrations of reduced sulfur (<0.1 mg/l). Reaction-path modeling of reactions involving a typical Mississippi oil field brine at the anhydrite-calcite interface in two salt dome cap rocks generally reproduces the "exotic" mineral suites present. At Tatum Salt Dome, where sulfate reduction apparently ceased late in the development of the cap rock, the modeling predicts the observed abundance of celestite, strontianite, and barite. Alternatively, modeling of hypothesized processes at Hazlehurst Salt Dome, where sulfate reduction was a long-lived process, predicts the observed mineral assemblage of Fe-Pb-Zn sulfides with calcite and the absence of Cu and Ag sulfides.

1 INTRODUCTION

Oil field brines from the Mississippi Salt Dome Basin are probably the best-characterized metal-rich sedimentary formation waters known (Carpenter, et al., 1974; Kharaka et al., 1987; Saunders and Swann, 1990; Saunders and Rowan, 1990). Na-Ca-Cl oil field brines hosted by Jurassic and Cretaceous units in central Mississippi exceed 100°C in temperature, have total dissolved solids of 200,000 to >300,000 mg/l (resulting in densities of 1.1-1.25 g/ml), and typically contain 1000-2000 mg/l Sr. In addition, where dissolved sulfide concentrations are low (usually in Cretaceous siliciclastics) combined Pb+Zn values of several hundred mg/l are not uncommon. Quantitative modeling by Sverjensky (1984, 1987) has shown that these brines can evolve by water-rock reactions to become ore-forming solutions responsible for producing Mississippi Valley-type (MVT) and other sediment-hosted deposits located adjacent to major basins. In addition, the discovery of salt dome-hosted Zn-Pb-Ag-Ba mineralization at Hockley dome, Texas (Kyle and Price, 1986) indicates that brines can evolve to become ore-forming solutions within basins.

In this study, hypothesized water-rock reactions between a typical Mississippi Salt Dome Basin formation water and developing cap rocks at Tatum and Hazlehurst Salt Domes, Mississippi are modelled. In the process, the role that sulfate reduction plays in determining the ultimate cap rock mineralogy is evaluated.

2 GEOLOGIC SETTING

The Mississippi Salt Dome Basin is a rift basin that developed at the cratonic margin of the greater Gulf Coast basin, and contains up to 8000 m of Jurassic to Tertiary siliciclastic, carbonate, and evaporite units. The basin contains at least 60 piercement salt domes that are rooted in the 1500 m-thick Jurassic Louann Salt. Tatum Salt Dome lies at the southern edge of the Mississippi Salt Dome Basin (Figure 1) and the upper portion of the cap rock contains a 20-60 m thick horizontally-banded calcite zone. This banding is typical of most cap rock calcite zones, which form by the progressive dissolution of the underlying anhydrite (Posey et al., 1987). However, a 10 m-thick zone at the base of the calcite zone at Tatum dome contains massive amounts (>50% of the rock) of celestite ($SrSO_4$) and strontianite ($SrCO_3$), with minor amounts of barite.

Hazlehurst Salt Dome lies 35 km SSW of Jackson, (Figure 1) and recent sulfur exploration (early 1991) has provided a continuous core through the 50 m-thick

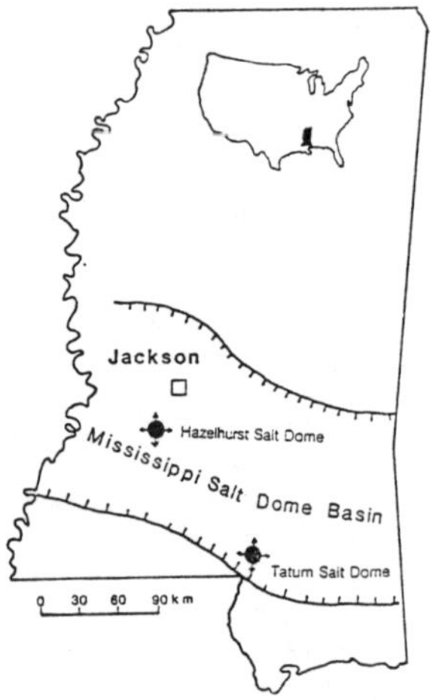

Figure 1. Map showing the location of Tatum and Hazlehurst Salt Domes.

calcite portion of the cap rock, which contains zones of massive pyrite and marcasite, and locally abundant sphalerite, galena, barite, celestite, and strontianite. These minerals occur in two distinct settings: 1) in sub-horizontal bands; and 2) steeply dipping banded veins up to 20 cm wide containing alternating layers of white and petroleum-bearing dark-brown calcite, with variable amounts of sulfide minerals. Geochemical analyses indicate that there are significant local enrichments of Zn and Pb, particularly near the base of the calcite zone. One 0.3 m interval contains >4% Zn+Pb, 2% Mn, and only trace amounts of copper (<100 ppm) and silver (<1 ppm). Molar Zn:Pb ratios average about 9:1, a value similar to present-day oil field brines in Mississippi. Sulfide textures and the presence of marcasite suggest a low temperature of formation, apparently low enough for sulfate-reducing bacteria to operate. Preliminary data from fluid inclusions in barite and celestite adjacent to sulfides indicate that they were precipitated at temperatures <60°C.

3 GEOCHEMICAL MODELING

Geologic relationships and mineralogic, trace-element geochemical, and isotopic data strongly indicate that metal-rich basin brines played an important role in the genesis of cap rock at Tatum and Hazlehurst salt domes. However, it appears that differences in the extent to which sulfate reduction operated in the domes caused major differences in minerals precipitated. To evaluate the relative importance of the presence or absence of sulfate reduction on cap rock mineralogy, the computer programs SOLVEQ and CHILLER were employed (Reed, 1982; Spycher and Reed, 1989). SOLVEQ is an aqueous speciation program, capable of calculating chemical equilibrium conditions over a wide range of temperature and pressure. CHILLER is an equilibrium reaction path model applicable to heterogeneous systems with aqueous phase true ionic strengths up to five. The program solves a series of mass balance and mass action equations by a Newton-Raphson numerical solution to the algebraic form of the governing equations. CHILLER computes the heterogeneous equilibrium conditions (aqueous speciation and amounts of minerals precipitated) as a reaction progress variable (e.g., composition, temperature, or pressure) is changed incrementally.

A brine from the Raleigh Oil Field, Mississippi (sample 84-MS-11 from Kharaka et al., 1987), was chosen as a starting model solution. This Na-Ca-Cl brine contains 316,000 mg/l TDS, with 1920 mg/l Sr, 60 mg/l Ba; 465 mg/l Fe; 243 mg/l Zn; 70 mg/l Pb; 212 mg/l Mn; and 0.021 mg/l Cu. In addition, an arbitrary value of 0.001 mg/l for Ag was chosen based on brine analyses from the Raleigh Oil Field by Saunders and Swann (1990). SOLVEQ was used to recalculate pH and alkalinity under estimated reservoir conditions of 120°C and 800 bars. CHILLER was then used to cool the model brine to reflect physicochemical changes associated with brine migration up the flanks of the salt domes and into the shallow cap rock setting. Cooling results in precipitation of muscovite, quartz, strontianite, and barite (Figure 2), the latter of which commonly forms annular scale deposits in oil wells in Mississippi because of cooling (e.g., Carpenter et al., 1974; Saunders and Rowan, 1990). After cooling, the model solution is undersaturated with respect to anhydrite and calcite, only slightly undersaturated with respect to celestite, and is saturated with respect to barite.

At Tatum dome, where sulfate reduction ceased late in the development of the calcite cap rock, incoming brine apparently dissolved anhydrite and calcite. This process can be modeled using CHILLER to titrate anhydrite and a minor amount of calcite (to account for the lower solubility of calcite) into the 50°C brine. Model results show that aqueous SO_4 and HCO_3 increase as anhydrite and calcite dissolve, celestite, strontianite, and barite are precipitated (Figure 3). These minerals are found in

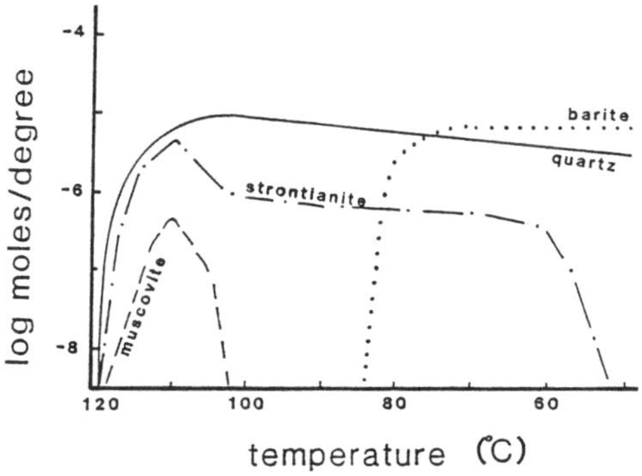

Figure 2. Plot showing the relative order and amounts of minerals precipitated as a function of cooling the model brine from 120° to 50°C.

this relative order of abundance in the lower 10 m of the "calcite" cap rock at Tatum dome, and textural evidence indicates that, at least locally, all have replaced calcite or fill open cavities (Saunders et al., 1988).

At Hazlehurst dome, bacterially-mediated sulfate reduction was an important and long-lived process, and can be modelled by using CHILLER to titrate the 50°C brine with the products of sulfate-reduction. The brine becomes more reducing as H_2S is added, and existing sulfate is reduced to sulfide in the process, resulting in barite and celestite undersaturation. Strontianite precipitates (Figure 4) due to increased HCO_3 from the oxidation of organic carbon attending sulfate reduction. In addition, the onset of sulfide mineral precipitation proceeds in the following order (Figure 4): pyrite, galena, sphalerite, alabandite. Calcite begins to precipitate near the end of sulfide deposition.

Textural evidence from the sulfide-rich portions of Hazlehurst dome generally suggests that pyrite, marcasite, galena, and sphalerite precipitated more or less contemporaneously, with some indications that locally Pb-Zn sulfides precipitated after earlier Fe-sulfides. In addition, minor amounts of strontianite are locally intergrown with sulfides, an association predicted by the model. In the model, very small amounts of chalcopyrite ($CuFeS_2$) and acanthite (Ag_2S) precipitate briefly, but subsequently dissolve at higher aqueous sulfide contents. No Cu or Ag sulfides have been observed at Hazlehurst. The presence of locally abundant barite and celestite, which commonly are present as vug fillings or intergrown with sulfides, is not

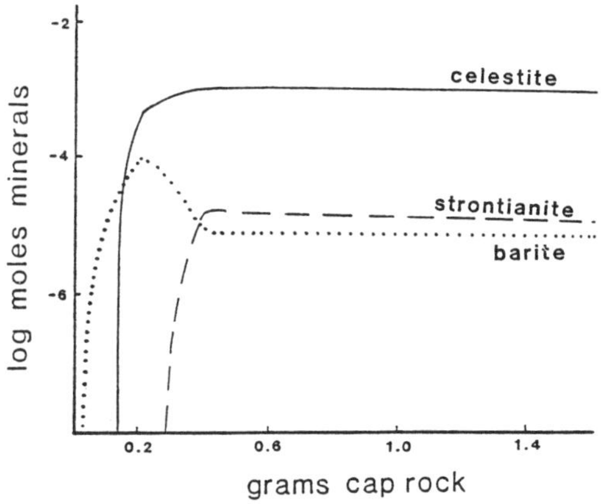

Figure 3. Plot showing the relative order and amounts of minerals precipitated as the cap rock minerals anhydrite and calcite are titrated into 1 kg of the model brine.

compatible with the simple sulfate reduction model. Precipitation of these minerals requires that sulfate reduction either stops or proceeds at a much slower rate than anhydrite dissolution. In either case, it appears that mineral precipitation may briefly follow the path modelled for Tatum dome, e.g., deposition

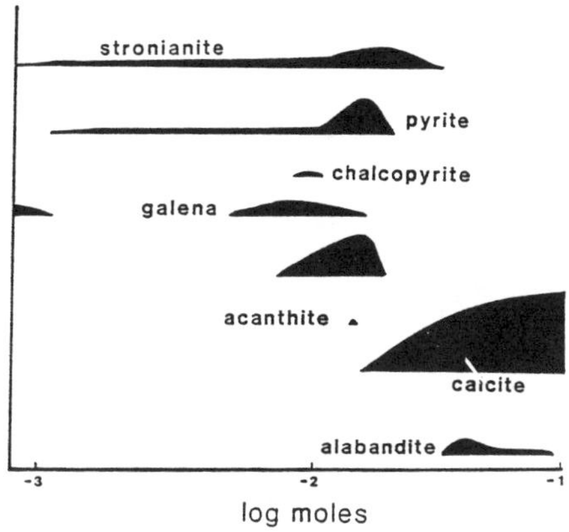

Figure 4. Paragenesis-style diagram illustrating relative order and abundances of mineral precipitation as the products of sulfate reduction are titrated into 1 kg of the model brine.

of celestite and barite. Given the abundant supply of sulfate in cap rock, it appears likely that the depletion of a finite supply of hydrocarbons may have temporarily shut down the sulfate reduction process at Hazlehurst dome, causing local sulfate mineral deposition. Moreover, a lack of abundant hydrocarbons late in the history of "calcite" cap rock growth at Tatum dome could explain the predominance of celestite and barite over sulfides near the base of the calcite cap rock.

4 CONCLUSIONS

Modeling of cap rock water-rock reactions using a typical oil field brine generally reproduces both the minerals present and their relative abundances in two salt dome cap rocks from Mississippi. Model results indicate that sulfate reduction leads to metal sulfide precipitation in the same relative abundances as the metals in the model brine. Results also show how the presence or absence of sulfate reduction can lead to substantially different cap rock mineralogy. In addition, this study provides additional support for the concept that formation waters play an integral role in cap rock formation in general, and in particular, in the genesis of the Sr-enriched cap rock in the Mississippi Salt Dome Basin.

REFERENCES

Carpenter, A.B., M.L. Trout, & E.E. Pickett 1974. Preliminary report on the origin and chemical evolution of lead- and zinc-rich oil field brines in central Mississippi. Econ. Geol. 69:1191-1206.

Kharaka, Y.K., A.S. Maest, W.W. Carothers, L.M. Law, P.J. Lamothe, & T.L. Fries 1987. Geochemistry of metal-rich brines from central Mississippi Salt Dome basin, U.S.A.: Appl. Geochem. 2:543-561.

Kyle, J.R., & P.E. Price 1986. Metallic sulphide mineralization in salt-dome caprocks, Gulf Coast, U.S.A.: Instit. Min. Metall. Trans. 95:B6-B16.

Posey, H.H., P.E. Price, & J.R. Kyle 1987. Mixed carbon sources for calcite cap rocks of Gulf Coast salt domes. In I. Lerche and J.J. O'Brien (eds.), Dynamical Geology of Salt Domes and Related Structures: 593-630. Austin: Academic Press.

Reed, M.H. 1982. Calculation of multicomponent chemical equilibria and reaction processes in systems involving minerals, gases, and an aqueous phase. Geochim. Cosmochim. Acta 46:513-528.

Saunders, J.A., J.D. Prikryl, & H.H. Posey 1988. Mineralogic and isotopic constraints on the origin of Sr-rich caprock, Tatum salt dome, Mississippi, U.S.A. Chem. Geol. 74:137-152.

Saunders, J.A., & C.T. Swann 1990, Trace metal content of Mississippi oil field brines. J. Geochem. Explor. 37:171-183.

Saunders, J.A., & E.L. Rowan 1990. Mineralogy and geochemistry of metallic well scale, Raleigh and Boykin Church oil fields, Mississippi, U.S.A. Inst. Min. Metall. Trans. 99:B54-B58.

Spycher, N.F., & M.H. Reed 1989. Evolution of a Broadlands-type epithermal ore fluid along alternative paths: implications for the transport and deposition of base, precious, and volatile metals. Econ. Geol. 84:328-359.

Sverjensky, D.A. 1984. Oil field brines as ore-forming solutions: Econ. Geol. 79:23-37.

Sverjensky, D.A. 1987. The role of migrating oil field brines in the formation of sediment-hosted Cu-rich deposits. Econ. Geol. 82:1130-1141.

Water-Rock Interaction, Kharaka & Maest (eds) © 1992 Balkema, Rotterdam. ISBN 90 5410 075 3

Contrasts between steam stimulation and in-situ combustion process: An analysis of co-produced fluid compositional data

Maurice Shevalier, Ian Hutcheon & Cynthia Nahnybida
Department of Geology and Geophysics, University of Calgary, Alb., Canada

Hugh Abercrombie
Geological Survey of Canada, Calgary, Alb., Canada

ABSTRACT: Water sources during steam stimulation and *in-situ* combustion can be identified by co-produced water compositions. Water compositions vary with time due to fundamental differences between the steam stimulation and *in-situ* combustion processes. Steam stimulation and *in-situ* combustion pilots in the Cold Lake area of Alberta highlight these differences.

During steam stimulation, hot condensed steam mixes with cool formation water, causing the salinity of the co-produced water to increase as temperature and silica concentration decrease. In contrast, complex processes occur during *in-situ* combustion. NaCl concentration increases as the combustion front approaches a well. As the hot combustion zone passes, salinity decreases due to steam condensation, and the SiO_2 concentration is high.

Thermodynamic modelling shows that water compositions fall along different mineral reaction boundaries. Waters from steam stimulation tend to fall close to reaction boundaries, while there is a much wider scatter for combustion waters. This results from the physical and chemical complexity of the combustion process.

1. INTRODUCTION

Analyses of the chemical composition of co-produced waters are rarely used to monitor the *in-situ* combustion process (Bojes and Wright 1986, Parker 1984) and more commonly to monitor steam assisted recovery (Gunter et al. 1988, Abercrombie and Hutcheon 1986, Hutcheon et al. 1988, Hutcheon and Abercrombie 1990). In this paper, steam injection and combustion are compared in terms of variations observed in the co-produced fluids as well as the different processes, both physical and chemical, that these variations reflect.

The Wolf Lake Phase "A" pilot project is an oxygen injection pilot near Cold Lake, Alberta,that started as a steam stimulation pilot. Steam injection began in 1979 with most wells experiencing four or five steam cycles until March 1983, and was followed by oxygen combustion. The majority of the water samples reported here were collected in 1987 and 1988. Wolf Lake is compared to the steam assisted recovery pilots studied at Tucker Lake pilot and Esso Leming pilot, both near Cold Lake. The data from Tucker Lake are first cycle wells and those from the Leming pilot are from wells up to the ninth cycle.

As a result of the steam injection in the Wolf Lake wells, the pore water sampled was a mixture of formation water and condensed steam. A third potential source is "bottom" water (Abercrombie 1991). Also complicating matters is the potential for invasion of water from adjacent wells, since pathways between the wells have been shown to exist (Hallam et al. 1985). As a consequence, the interpretation of co-produced water compositions from *in-situ* combustion is a great deal more complicated than interpretation of steam assisted recovery co-produced water composition.

2. CO-PRODUCED WATER CHEMISTRY

Since steam is fresh water, the early stages of steam stimulation production have low total concentrations of dissolved salts, which increases as production proceeds. Also, the temperatures are relatively high in the early stages, decreasing as production continues. Figure 1a shows a mixing diagram of Cl vs Na for all the steam injection wells. This trend can be explained in terms of mixing the formation water with the condensed injected steam. Closer examination of the trend shows that these data lie slightly above the simple mixing line between pure water (approximate composition of condensed steam) and the composition of Clearwater Formation water (Bachu et al. 1989), indicating that there may be invasion of "bottom" water from below the

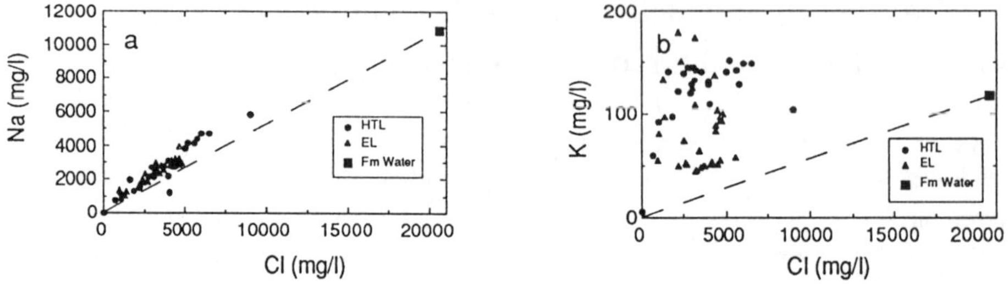

Figure 1. Mixing diagrams for the co-produced waters from the steam stimulation pilots. The dashed line shows the ideal mixing line between Clearwater Formation water and condensed steam.
(a) Cl versus Na.
(b) Cl versus K.

reservoir (Abercrombie 1991). Figure 1b shows shows that K is not acting as a conservative species for the steam injection wells, i.e. the data does not fall along the mixing line. This would indicate that the K concentration may be influenced by reactions with the rock matrix or some other process.

Figure 2a shows a mixing diagram for the co-produced waters from Wolf Lake Phase "A". Water compositions from wells on the communication paths are shown as open symbols and those not along a communication path are shown as closed symbols. The composition of the Clearwater Formation (Bachu et al. 1989) and condensed steam (Abercrombie 1991) are shown as large symbols, with the straight line between the two symbols representing a simple mixture between the two waters. Many of the points lie well away from the mixing line, with some waters enriched in Na relative to Cl, while others are depleted. There are a number of possible explanations for the observed relationship: 1) because of the steam injection process, which preceded *in-situ* combustion; 2) the co-produced waters may not be simple mixtures of steam and formation water, concentrated waters may have been co-produced by the evaporation of water at the combustion front; 3) water-rock interactions may have modified the composition.

Data is not available for the K concentration of condensed steam, but an estimate can be made using the sample with the lowest Cl value which is not far from the Na-Cl mixing line. The K concentration from this well was used to approximate the K-Cl mixing line between Clearwater Formation water and condensed steam. Figure 2b shows that the co-produced waters are all enriched in K relative to Cl, indicating extensive water-rock interactions. The water-rock interactions are more obvious, since the concentration of K is much less than the Na concentration and the amount of K released due to the water-rock interaction represents a larger portion of the K present before combustion.

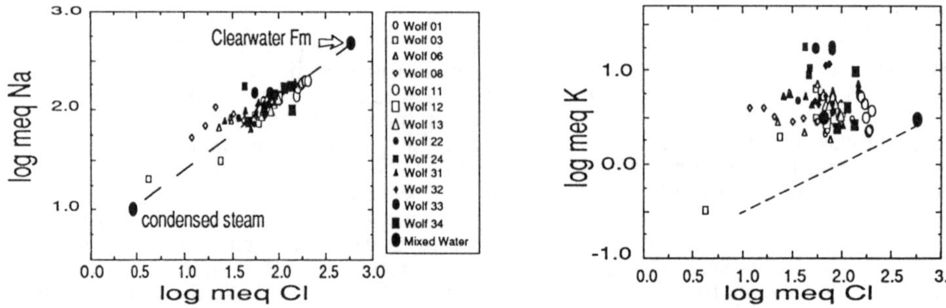

Figure 2. Mixing diagrams for the co-produced waters from Wolf Lake Phase "A". The dashed line shows the ideal mixing line between Clearwater Formation water and condensed steam.
(a) Cl versus Na.
(b) Cl versus K.

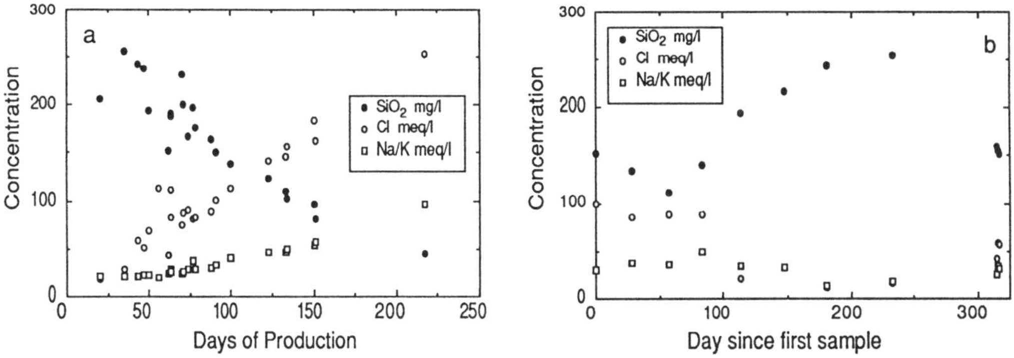

Figure 3. Trends in SiO_2, Na/K and Cl with time.
(a) The co-produced waters from steam stimulated wells from Tucker Lake.
(b) The co-produced waters from Wolf Lake well 08 in the combustion pilot.

In co-produced waters, low salinity is associated with high temperatures and high steam fractions. Also, the solubility of various SiO_2 minerals increases as temperature increase. Thus salinity, temperature and SiO_2 concentration correlate in steam co-produced waters. Since there are a number of exchange reactions with silicates, Na/K ratios tend to decrease with increasing temperature, SiO_2 content and decreasing salinity. Figure 3a shows the co-produced water compositions from three steam stimulated wells in Tucker Lake. The patterns of the fluids are very consistent, i.e. SiO_2 concentration decreases, reflecting a temperature decrease, salinity and Na/K ratio increase as production proceeds and the waters cool. A similar plot for the Wolf Lake wells do not show a similar trend, in fact no obvious trend is observed, reflecting the complexity of fluid flow.

During combustion, wells become hotter then cooler as the combustion front advances and at any given time the wells will all be experiencing different processes. Figure 3b shows the co-produced water compositions for a single well at Wolf Lake (Wolf 08), which shows a trend, but not the same as observed for steam stimulation. The increase in SiO_2 and decrease in salinity and the Na/K ratio probably reflects the increasing temperature and the approach of heated water associated with the combustion zone. There is a clear maximum in SiO_2 and a minimum in Na/K and Cl at approximately 240 days.

3. MINERAL REACTIONS

The relationship between the co-produced fluid compositions and mineral reactions can be studied by superimposing different chemical ratios in the co-produced water on thermodynamically calculated activity-temperature plots. In Figure 4, the log of the activity ratio of Na/K is plotted against temperature for both the steam stimulated and the combustion wells. In general, Na/K activity ratios follow a parallel trend with temperature. Figure 4a shows that the samples from early cycles at Husky Tucker Lake parallel the reaction boundary between albite and Na-smectite. The samples from Esso Leming, which are later cycle wells, are close to the kaolinite/Na-smectite reaction boundary. For the combustion wells (Figure 4b), the spread of data is much wider than for the steam stimulation wells, but the data appear to fall along the albite/Na-smectite reaction boundary.

Figure 5 shows the K/H activity ratio of the co-produced waters and the stability of various aluminosilicate minerals. This data follows the same trends that were observed for Na/H, i.e. there are two groups of points. Figure 5a shows that most of the data from the steam stimulated wells fall along the K-feldspar/illite reaction boundary. Some data points, from the Esso Leming pilot, also follow the metastable K-feldspar/smectite boundary and the illite/smectite boundary. The co-produced waters from the combustion wells are plotted in Figure 5b. Here, as for Na/H, there is a spread of data, but most points falls along the K-feldspar/illite reaction boundary.

The data from steam stimulated and combustion wells all tend to fall along calculated reaction boundaries. The agreement from the combustion wells is not as good as for the steam stimulated wells, probably indicating that the complex history of the wells, i.e. steam injection followed by combustion, has resulted in complex water composition histories.

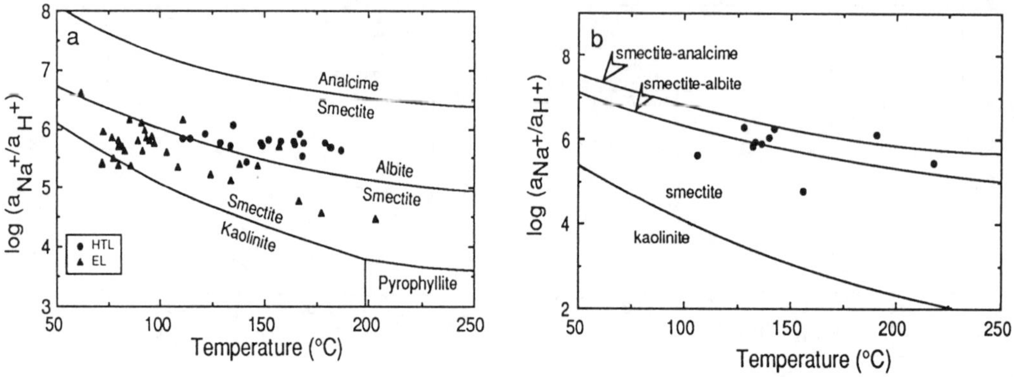

Figure 4. Thermodynamic stability of minerals compared to Na/H activity from the co-produced waters.
(a) for the steam stimulated wells.
(b) for the combustion wells.

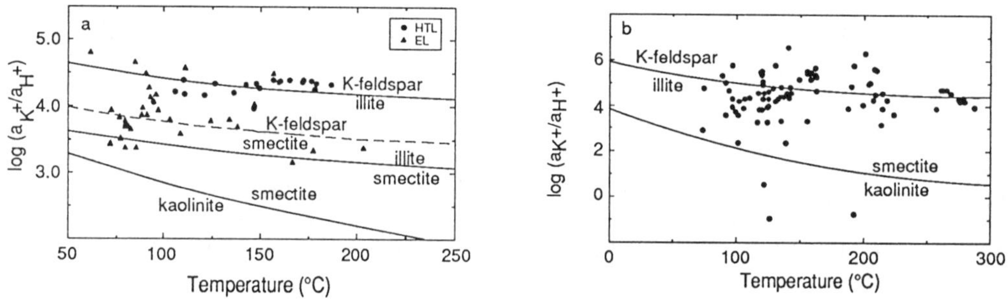

Figure 5. Thermodynamic stability of minerals compared to K/H activity from the co-produced waters.
(a) for the steam stimulated wells.
(b) for the combustion wells.

4. CONCLUSIONS

Examination of water compositions from *in-situ* combustion and steam stimulation processes show that the *in-situ* combustion is a more complex process. Combustion waters are more complex than the simple mixtures of formation water and condensed steam, in steam stimulation. There is significant enrichment and depletion of chemical species due to extensive water-rock interactions in combustion. The compositional data does generally fall along reaction boundaries,the scatter indicates that there are much more complex interactions occurring.

REFERENCES

Abercrombie, H., Applied Geochem., **6**, (1991)

Abercrombie, H. and Hutcheon, I., 5th Int. Symp. Water-Rock Interaction, p. 1-4. (1986)

Bachu, S, Perkins, E.H., Hitchon, B., Lytviak, A.T. and Undershultz, J.R., Alta. Rsch. Council, Geol. Surv. Bull., **60** (1989).

Bojes, J.M. and Wright, G.B., Ann. Tech. Mtg., Petrol. Soc. CIM, 86-37-61, p. 315-349, (1986)

Gunter, W.D., Perkins, E.H., Young, B., and Bird, G.W.,4th UNITAR/UNDP Conf. Heavy Crude Tar Sands, Paper 12, (1988)

Hallam, R.J., Hajdo, L.E. and Staples, S., Heavy Oils and Tar Sands, Producing, Upgrading and Economics Conference, Dec. 9-10, 1984. (1985)

Hutcheon, I., Abercrombie, H., Shevalier, M. and Nahnybida, C., 4th UNITAR/UNDP Conf. Heavy Crude and Tar Sands, paper 235. (1988)

Hutcheon, I. and Abercrombie, H.J., AAPG mem **49**, (Eds) Meshri and Ortoleva), p. 161-170. (1990).

Parker, R.J., AOSTRA Jour. Rsch., l, p. 37-42, (1984)

Water-Rock Interaction, Kharaka & Maest (eds) © 1992 Balkema, Rotterdam. ISBN 90 5410 075 3

Spatial variations in North Sea formation water compositions: Preliminary results

E.A.Warren & P.C.Smalley
BP Research, Sunbury-on-Thames, Middlesex, UK

ABSTRACT: Significant variations in the salinity, chemical and stable isotopic composition of formation waters have been observed in the North Sea basin. Variations occur at all scales, from intra-formational and inter-reservoir variations within fields to basin-wide variations in water chemistry within and between formations. Clearly, the present-day North Sea basin waters are highly heterogeneous. However, our understanding of the processes contributing to this is limited by the sparse coverage of the existing published data, consequently a new effort is being made to compile formation water data for all fields in the greatest detail possible. Interim results from this compilation are reported.

1. INTRODUCTION

Sampling formation waters is a routine operation during exploration, appraisal and production operations in the North Sea. However, unlike the comprehensive geological and petroleum data widely available in the open literature, very few of the water data have been published. This is in stark contrast to the wealth of water data available for other basins, e.g. Gulf Coast (Collins 1975; Land et al. 1988). This contribution reviews our present understanding of formation water compositions and variations in the North Sea using the present literature and describes the progress made so far in the compilation of a new atlas of North Sea formation water data currently in preparation (Warren et al. 1993).

There are three main sources of formation water data in the open literature: those for individual fields, the compilation of Norwegian shelf data by Egeberg & Aagaard (1989), and the North Sea Rw catalogue (SPWLA 1989). Inevitably, the quality and amount of information provided in these three sources are variable. However, these data do offer insight into the type and magnitude of variability in formation water chemistry in the North Sea. Chiefly, the data document four main variations: intra-formational variations within fields, variations between different reservoir horizons within fields, regional variations within formations and regional variations between different formations. Each of these is discussed below.

2. SCALES OF VARIATION

2.1 Within-Field variations

2.1.1 Intra-formation variations

Glasmann et al. (1989) documented variations in formation-water resistivity within the middle Jurassic Brent Group across the Heather Field. Reported resistivities range from 0.443 in B Block to 0.074 in west Heather, equivalent to a salinity variation of 14 000 - 100 000ppm. Although the formation water of the main Heather Field only varies by approximately one-half to one-third seawater salinity (Glasmann et al. 1989), the salinity in G Block of the main Heather Field is considerably higher. Glasmann et al. (1989) suggested that these present-day variations in formation water data might be comparable to variations in ancient formation waters indicated by stable isotopic data of authigenic clay minerals. If so, then this implies that variations in water chemistry within a single formation may have existed on a field-scale in the geologic past as well.

2.1.2 *Inter-formation variations*

Formation water chemistry variations in different stratigraphic intervals were observed within the Piper Field by Burley et al. (1989). Chemical and stable isotopic data were reported for the Eocene Sand, Upper Jurassic Piper Formation, Zechstein and Carboniferous sediments in the Piper Field together with Piper Formation water chemical compositions in the upthrown and downthrown blocks of the nearby Tartan field. Variations in the chemical composition of the formation waters in these different stratigraphic horizons is most obvious in terms of salinity (Figure 1) which clearly increases with depth. The chemical composition of the formation waters also varies, as revealed in changes in some major ion ratios (Figure 2). These suggest that the salinity profile in Figure 1 is unlikely to result from simple vertical diffusion.

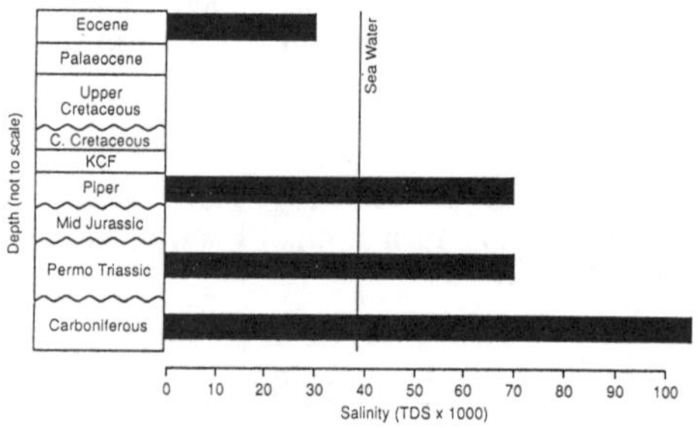

Figure 1. Salinity variations in different reservoir horizons in Piper Field (data from Burley et al. 1989).

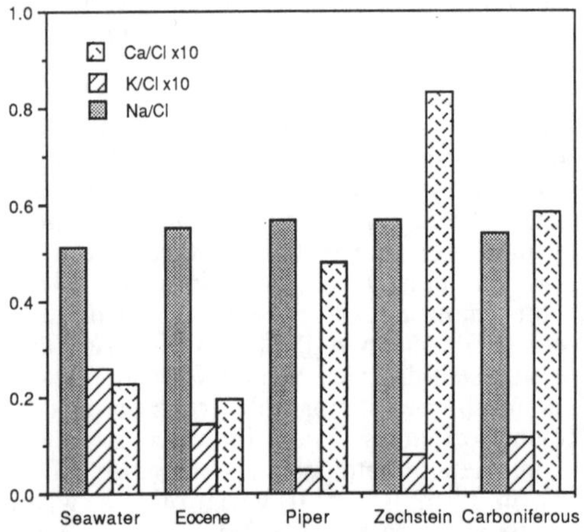

Figure 2. Formation water compositional variations in Piper Field. Note the variation in Ca/Cl and K/Cl (data from Burley et al. 1989).

Both the study on Heather Field and the Piper Field show that significant chemical variations in formation water compositions occur at a field-scale, both vertically and laterally, in the North Sea.

2.2 *Regional variations*

The magnitude of variations in the chemical compositions of formation waters is illustrated in the comprehensive data set from the Norwegian shelf published by Egeberg & Aagaard (1989). Major and minor element and stable isotopic compositions are reported for many fields in the Norwegian Continental Shelf (NOCS), although individual well-locations were not revealed. Such a comprehensive data set enables detailed examination of variations to be made. An example of such variation is that of stable isotopic composition (Figure 3). None of the formation waters from NOCS have the stable isotopic compositions of either unmodified meteoric water or seawater (Egeberg & Aagaard 1989), the two obvious sources of water. Furthermore, the stable isotopic composition of the formation waters in carbonate reservoirs appear to be different to those in clastic reservoirs. Aplin et al. (1992) plotted the same data for the sandstone reservoirs in terms of $\delta^{18}O$ with depth and observed that the deepest samples also had the most evolved oxygen isotopic compositions (Figure 4). Aplin et al. (1992) showed that a similar range and variation in $\delta^{18}O$ occurs in waters from sandstone reservoirs in the United Kingdom Continental Shelf (UKCS).

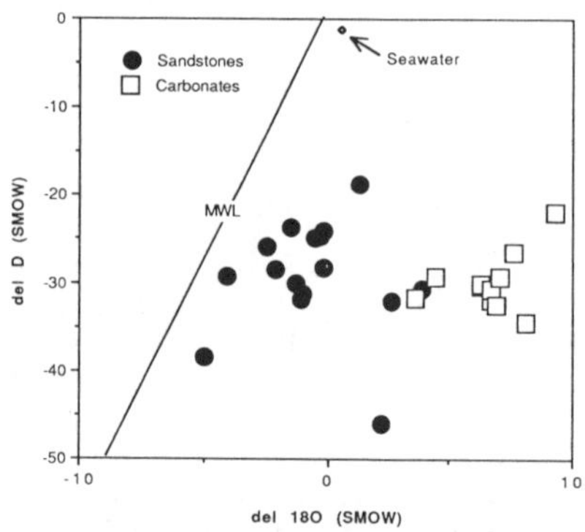

Figure 3. Stable isotopic compositions of NOCS formation waters (from Egeberg & Aagaard 1989). Present-day meteoric waters plot along the trajetory marked MWL.

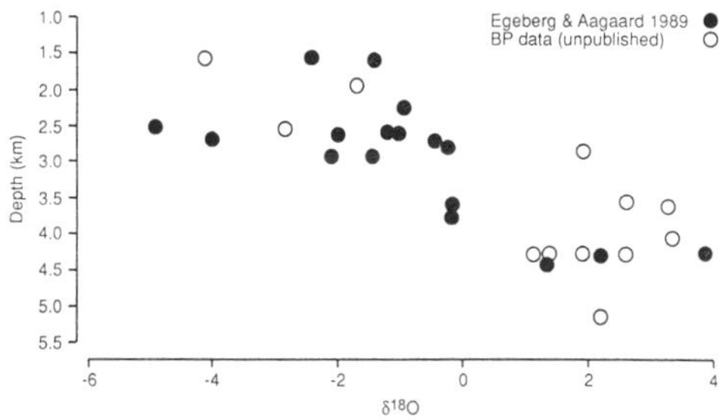

Figure 4. Variations in oxygen isotope ratios of formation waters from clastic reservoirs; full circles from NOCS, open circles from UKCS (Aplin et al. 1992).

Although the cause of these variations remains a subject of debate (e.g. Aplin et al. 1992), clearly, the isotopic chemistry of North Sea formation waters is both heterogeneous and complex. Variations in formation water composition occur on a regional scale and with depth.

2.3 *Basin-scale variations*

A basin-scale image of formation water compositional variations can be obtained from published salinity data. There are two main sources of data: salinity and density data for individual UKCS fields (Abbotts, 1991) and water-resistivity data for wells throughout the North Sea (SPWLA 1989). Converting water-resistivity into salinity using a simple conversion (Schlumberger 1989), the variations in salinity in the North Sea have been plotted for three stratigraphic intervals combined: Palaeocene, Middle Jurassic and Rotliegende (Figure 5).

Salinities are highest in the Southern and Central North Sea, and many exceed halite saturation (250 000 ppm). However, salinities are generally lower than seawater throughout the Northern North Sea, although the Heather Field study highlighted heterogeneity (Glasmann et al. 1989). Aplin et al. (1992) compared the distribution of formation water salinities with the distribution of the Zechstein Z3 polyhalite and found that the distribution of salt mimics the highest salinities (Figure 5). The authors argued that salt dissolution has influenced formation water chemistry in the underlying Rotliegende sandstone by simple diffusion or density-driven flow and in overlying strata by salt penetration through diapirism.

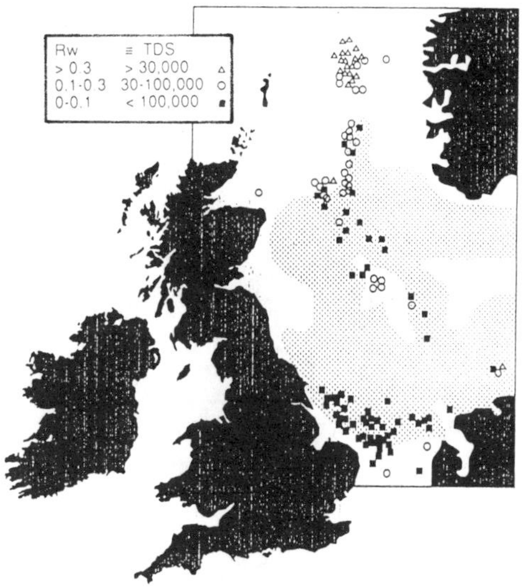

Figure 5. Water resistivity variations in North sea formation waters (data from SPWLA 1989). Lowest Rw equates with highest salinity.

3. LIMITATIONS OF THE PRESENT DATA

The results of all these studies show that there are considerable variations in North Sea formation water compositions. Heterogeneities occur both spatially and stratigraphically throughout the North Sea basin. Some simple controls on formation water composition can be deduced: the major control on salinity appears to be the distribution and diapirism of Zechstein salt. Significant variations in formation water composition are even observed within fields. Clearly, the North Sea basin cannot be regarded as a completely well-connected hydraulic system.

Although many heterogeneities have been revealed in North Sea formation waters, the processes responsible are not understood, those of stable isotopes, for example. At least one reason for this is the limited coverage, both spatially and stratigraphically afforded by the published data. Better data, both in terms of detail and geographical coverage, could enable the processes controlling formation water composition to be identified and understood, and hence impact prediction of formation water variations throughout the North Sea. Knowledge of the lateral and vertical variations in barium concentrations, for example, could greatly improve scale prediction, a costly

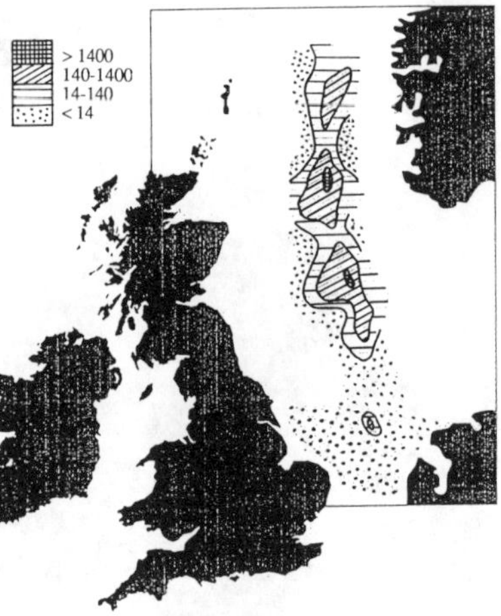

Figure 6. Barium concentration (mg/l) in North Sea formation waters (data from Warren et al. 1993, in prep.)

problem in many Central and Northern North Sea Fields. The ideal data set of North Sea formation waters would combine the comprehensive data of the NOCS data set (Egeberg & Aagaard 1989) with the basin-wide coverage of the water-resistivity catalogue (SPWLA 1989) and the lateral and vertical detail of the Heather and Piper studies (Glasmann et al. 1989; Burley et al. 1989) The Geological Society is now preparing such a volume (Warren et al. 1993).

The data compiled so far do give insight into the distributions of major solutes in formation waters throughout the North Sea basin. The distribution of barium is illustrated in Figure 6 and shows considerable variability. Barium concentrations are highest in the Central North Sea and can be in excess of 1000 ppm. However, barium is negligible in the Southern North Sea. Although the controls on barium concentration have not yet been established for the North Sea, it is clear that salinity (Figure 5) is not a major control.

4. CONCLUSIONS

Investigation of the limited published formation water data and new major element data contained in this study reveals significant heterogeneity in formation water composition from the field-scale to the basin scale. Variations are observed in salinity, major element chemistry and stable isotopic compositions both within formations and stratigraphically on a regional scale. However, the controls on these variations are not yet understood.

REFERENCES

Abbotts, I.L. (editor) 1991. *United Kingdom Oil and gas Fields 25 years commemorative volume*. Geol. Soc. Lon. Mem. 14, 565p.

Aplin, A.C., Warren, E.A. & Grant, S.M. 1992 (in press). Mechanisms of quartz cementation in North Sea reservoir sands: constraints from fluid inclusions. In: Robinson, A.G. & Horbury, A. (eds) AAPG Memoir.

Burley, S.D., Mullis, J. & Matter, A. 1989. Timing diagenesis in the Tartan Reservoir (UK North Sea): constraints from combined cathodoluminescence microscopy and fluid inclusion studies. *Mar. & Petrol. Geol.*, 6. 98-120.

Collins, A.G. 1975. Geochemistry of Oilfield Waters. *Developments in Petroleum Science,* 1, Elsevier, Amsterdam.

Egeberg, P.K. and Aagaard, P. 1989. Origin and evolution of formation waters from oil fields on the Norwegian shelf. *App. Geochem.*, 4, 131-142.

Glasmann, J.R., Lundegard, P.D., Clark, R.A., Penny, B.K. and Collins, I.D. 1989. Geochemical evidence for the history of diagenesis and fluid migration: Brent Sandstone, Heather Field, North Sea. *Clay Mins*, 24, 255-284.

Land, L.S., MacPherson, G.L. & Mack, L.E. 1988. Geochemistry of saline formation waters, Miocene, offshore La., *Trans. Gulf Coast Geol. Soc.,* 38, 503-511.

Schlumberger, 1989. *Log Interpretation Charts*. Schlumberger Educational Services, 150p.

SPWLA 1989. *North Sea Rw catalogue.* Society of Professional Well Log Analysts, London Chapter.

Warren, E.A., Smalley, P.C., Findlay, M. & Todd, A.C. 1993. *North Sea Formation Waters.* Geol. Soc. Lon. Mem. 15 (in prep.).

Water-Rock Interaction, Kharaka & Maest (eds) © 1992 Balkema, Rotterdam. ISBN 90 5410 075 3

Mixing of groundwater with isotopically distinct signatures: A multi-component mixing model for the Dundee Formation of southwestern Ontario, Canada

T.R.Weaver, S.K.Frape & J.A.Cherry
University of Waterloo, Ont., Canada

ABSTRACT: Groundwater from the oil-bearing Devonian Dundee Formation in southwestern Ontario has TDS contents ranging from 10 g/l to over 50 g/l. These saline waters occur at depths of less than 150 m in carbonate-dominated sequences. Stable-isotope and major-ion analyses indicate that large-scale fluid movement has influenced the groundwater composition in the shallow bedrock sequence since glaciation. Previously-mixed, low-δ^2H, high-salinity fluids derived from underlying Silurian salt formations have been emplaced into the Devonian sequence. Fluid migration between the Silurian salts and Devonian oil-bearing formations appears to be fracture controlled. Superimposed on the evidence of major, cross-formational fluid movement are effects resulting from petroleum production practices.

INTRODUCTION

Groundwater in the shallow oil-bearing formations of southwestern Ontario has total dissolved solids (TDS) contents significantly greater than those expected in shallow, carbonate-rich environments. Over the last century, much of the region has been exploited for petroleum production, liquid waste injection, brining, and gas storage at depth. Consequently, it was considered necessary to determine the source of salinity in the shallow system in order to evaluate the extent to which cross-formational fluid flow may have occurred in the study area.

The region lies on the eastern-most flanks of the Michigan Basin and is separated from the Appalachian Basin by the Algonquin and Findlay Arches to the east (Figure 1). Based on reconstructions from borehole data, Sanford et al. (1985) suggested that a major fracture network is present in the Silurian sequences of much of southern Ontario (Figure 1). The Palaeozoic sedimentary sequence extends from Cambrian sandstones through Ordovician shales and carbonates, Silurian salts, and Devonian carbonates and shales. This bedrock sequence is overlain by a thick blanket of glacial, clay-rich tills (Figure 2).

Evidence of major episodes of fluid movement since glaciation is found in the historical records. Petroleum products and saline fluids were observed in the overlying till in the mid-1800's. The most likely source of these fluids is the underlying Devonian carbonate oil-bearing Dundee Formation, present at depths of 50-100 m below the base of the till.

Other episodes of fluid movement involving more saline systems may have occurred within the bedrock sequence of southwestern Ontario. Although the oil-bearing Dundee Formation occurs at depths of only 150 m, the groundwater in this formation has solute concentrations that are much higher than those expected by simple water-rock interaction in a shallow carbonate-rock environment. Consequently, solutes and possibly aqueous components of the current groundwater must have been emplaced into the shallow Dundee Formation.

The goal of this research, therefore, was to determine the degree to which cross-formational fluid flow has occurred in the shallow bedrock sequence of southwestern Ontario, and in so doing, to determine the source of salinity and other groundwater components in the Dundee Formation of the region.

GROUNDWATER CHEMISTRY

Groundwater samples from 3 oil-producing regions (Figure 1) were taken from

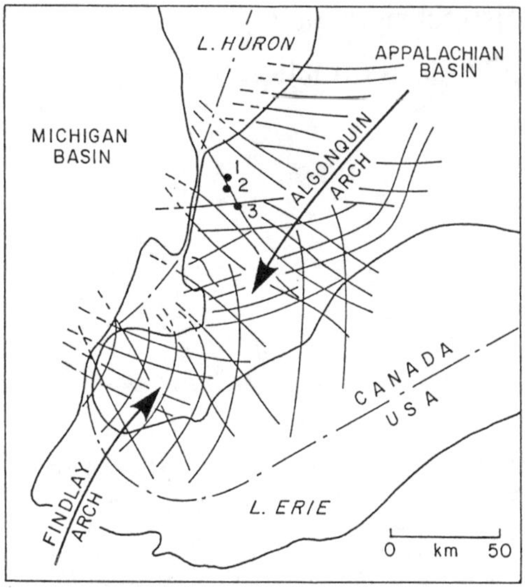

Figure 1: Major structural features (Ells et al., 1966): proposed fracture network (Sanford et al., 1985): sampling sites 1-Petrolia, 2-Oil Springs, 3-Bothwell.

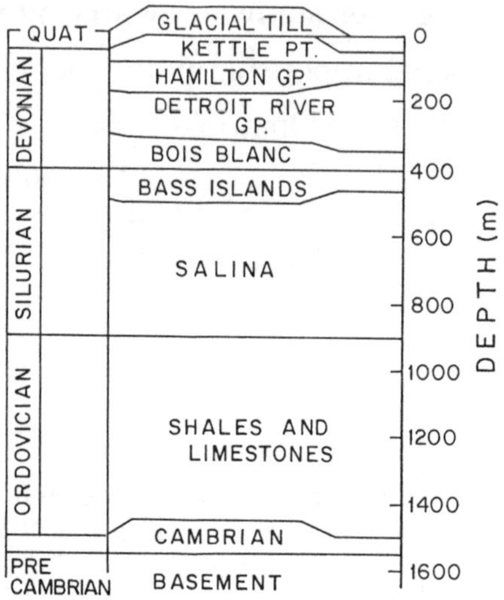

Figure 2: Simplified stratigraphy of the study area (after Sanford et al., 1985).

approximately 35 oil wells completed to the Dundee Formation. Samples were analysed for major ions, stable isotopes of water and sulphur species, tritium, and strontium isotopes.

In most cases, the groundwater chemistry within the Dundee Formation does not vary greatly between fields within the same oil-producing region. Chloride and sodium are the dominant solutes in all of the samples, and TDS concentrations range from 10-55 g/l.

DILUTION ALONG THE METEORIC WATER LINE

The stable-isotope compositions for most of the groundwater from these regions is such that it coincides with the local meteoric water line (Fritz et al., 1987). $\delta^{18}O$ values range from approximately -9 ‰ to -6 ‰ along the meteoric water line (Figure 3). The more negative value is within the range of $\delta^{18}O$ values for modern recharge in the region (Fritz et al., 1987).

The samples which lie on the meteoric water line, however, are characterized by a relatively high salinity, indicating that water-rock interaction must have occurred at some point during the recharge or flow history of groundwater within the Dundee Formation. This is a similar process to that observed by Banner et al. (1989) where solutes were contributed to meteoric water without altering the isotopic composition of the groundwater as it was recharged. In southwestern Ontario, as the $\delta^{18}O$ value of groundwater decreases along the meteoric water line, the salinity also decreases. This indicates that the primary process occurring along the meteoric water line is dilution by recently-recharged meteoric water. Additionally, this also suggests that the dominant groundwater present in the Dundee Formation prior to dilution by meteoric water lay in the region of $\delta^{18}O = -6$ ‰ and $\delta^{2}H = -40$ ‰.

The dilution of oil-field water with meteoric water recharged under current climatic conditions probably occurred as water was added to the formation during petroleum production, either as a result of oil-field flooding or leakage down poorly-sealed boreholes.

MULTI-COMPONENT MIXING MODEL

Although the majority of groundwater sampled from the Dundee Formation lies on the meteoric water line, samples from one well-defined area are characterized by almost constant $\delta^{18}O$ values of approximately -6 ‰. $\delta^{2}H$ values decrease by up to 50 ‰, concomitant with an increase in salinity. TDS values increase from 41 g/l to 55 g/l as $\delta^{2}H$ values decrease (Figure 3).

Based on the assumption that stable-isotopes of water act as a conservative tracer for the

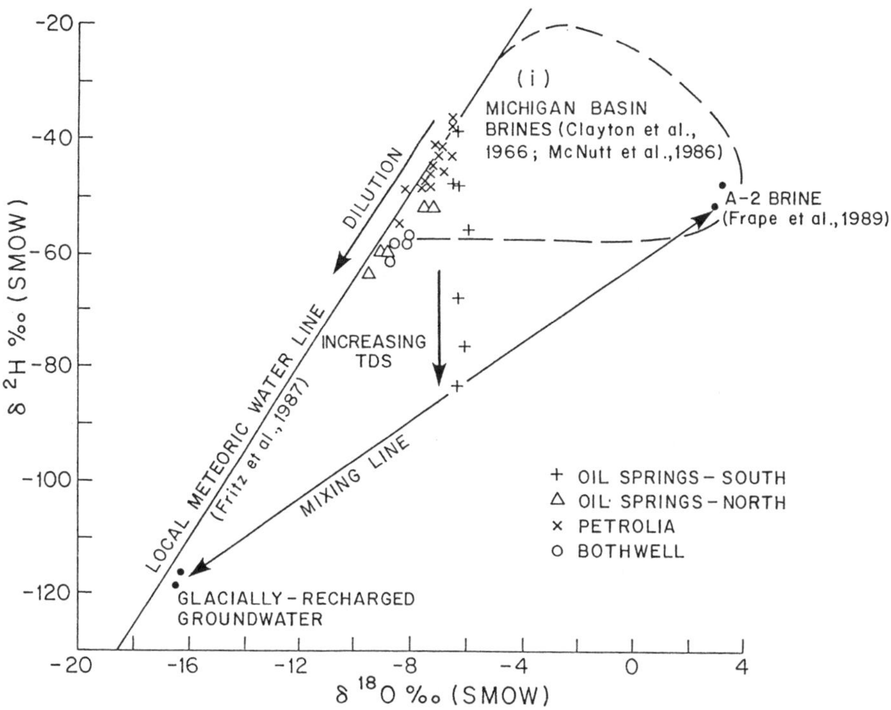

Figure 3: Hydrogen and oxygen isotopic compositions of groundwater from the Dundee Formation in the study area. Dilution along the meteoric water line. Mixing of glacially-recharged water and A-2 brine followed by mixing to produce a vertical mixing line.

aqueous component of groundwater in this system, a multi-stage, multi-component mixing model has been derived in order to explain the local occurrence of low-δ^2H, high-salinity groundwater in the Dundee Formation of southwestern Ontario. Stable-isotopic compositions of groundwater from surrounding bedrock formations in the eastern Michigan Basin (Clayton et al., 1966; Frape et al., 1989, McNutt et al., 1986) primarily lie above and to the right of the $\delta^{18}O = -6$ ‰ intersection with the meteoric water line (field (i); Figure 3).

If groundwater with an isotopic signature similar to that of the A-2 salt is diluted with a δ^2H-depleted groundwater similar to that which would result from recharge during glaciation (e.g. $\delta^2H = -120$ ‰, $\delta^{18}O = -17$ ‰, Desaulniers et al., 1981), the most saline, least ^{2}H-depleted sample from the study area can be duplicated (Figure 3). The tightly constrained δ^{18}O values observed for the large variation in δ^2H and salinity indicate that the line along which these samples lie results from mixing between only two endmembers: the previously mixed A-2/glacially recharged saline water and the groundwater present in the Dundee Formation prior to mixing.

Saturation indices calculated using PHRQPITZ (Plummer et al., 1988) indicate that fluids along this mixing line are significantly undersaturated with respect to the dominant host minerals in the Dundee Formation, calcite and dolomite. This further indicates that these fluids were derived from a different formation.

DISCUSSION

The current stable-isotope composition of groundwater in the oil-bearing Dundee Formation indicates that several episodes of mixing have occurred within this bedrock system.

In regions where significant petroleum production has occurred, groundwater samples are typically clustered along the meteoric water line. The spread along the meteoric water line probably results from dilution by recently-recharged meteoric water. The most likely mechanisms producing this dilution are; forced input of surface water or shallow groundwater to enhance petroleum production, and/or leakage from the surface or shallower formations through improperly-sealed abandoned wells. Tritium data indicate that leakage from the surface through abandoned wells is minimal (Weaver et al., 1990). Since production practices are involved, this dilution must have occurred recently, i.e. since the mid-1800's.

The earliest distinguishable mixing episode affecting groundwater chemistry within the Dundee Formation involved a saline fluid with an isotopic signature similar to that of brine from the Silurian A-2 salt and a low-δ^2H, low-$\delta^{18}O$ "fresh" water recharged under cold-climate conditions. The low-salinity component is likely to have been recharged during the last glaciation. Consequently the mixing between these two components must have occurred since glaciation. At this stage of mixing, fluids from the overlying Dundee Formation were not involved.

A later stage of mixing occurred when the previously-mixed Silurian/glacial fluid was input into the overlying Dundee Formation. The tight spatial distribution of groundwater samples along the mixing line indicates that the Silurian/glacial fluid was input into the Devonian system through a fracture network. The involvement of a glacially-derived component suggests that this fracture flow occurred after glaciation.

The large differences between groundwater chemistry and that expected from the surrounding matrix was the initial indicator that major inputs of water from different sources had occurred. The recent nature of these inputs as determined from the stable-isotope data is supported by the lack of equilibrium between the groundwater composition and the surrounding bedrock. If these inputs had occurred earlier, the system would be expected to have more closely approached equilibrium.

The evidence presented above indicates that recent, large-scale mixing episodes affecting the groundwater chemistry in the Dundee Formation have occurred in southwestern Ontario. Consequently, this hydrogeological regime must be viewed as a dynamic system in which cross-formational fluid movement could occur in response to structural and tectonic events in the area.

REFERENCES

Banner, J.L., G.J. Wasserburg, P.F. Dobson, A.B. Carpenter and C.H. Moore, 1989. Isotopic and trace element constraints on the origin and evolution of saline groundwaters from central Missouri. *Geochim. Cosmochim. Acta* 53: 383-398.

Clayton, R.N., I. Friedman, D.L. Graf, T.K. Mayeda, W.F. Meents, and N.F. Shimp, 1966. The origin of saline formation waters, 1. Isotopic Composition. *Jour. Geophys. Res.* 71: 3869-3882.

Desaulniers, D.E., J.A. Cherry, and P. Fritz, 1981. Origin, age and movement of pore water in argillaceous Quaternary deposits at four sites in southwestern Ontario. *Jour. Hydrology* 50: 231-257.

Ells, G.D., 1969. Architecture of the Michigan Basin. *Studies of the Precambrian of the Michigan Basin. Mich. Basin Geol. Soc. Ann. Field Excurs. Guidebook*: 60-80.

Frape, S.K., P. Dollar, B. Sherwood Lollar and R.H. McNutt, 1989. Mixing of saline basinal fluids in southern Ontario: Implications of rock-water interaction, hydrocarbon emplacement and Canadian Shield brines. *Proc. Water-Rock Interaction, 6*: 223-226. Rotterdam: Balkema.

Fritz, P., R.J. Drimmie, S.K. Frape and K. O'Shea, 1987. The isotopic composition of precipitation and groundwater in Canada. *IAEA Int. Symp. on the Use of Isotope Techniques in Water Resources Development IAEA-SM-299/017*: 14. Vienna.

McNutt, R.H., S.K. Frape and P. Dollar, 1987. A strontium, oxygen and hydrogen isotopic composition of brines, Michigan and Appalachian Basins, Ontario and Michigan. *App. Geochem.* 2: 495-505.

Plummer, L.N., D.L. Parkhurst, G.W. Fleming and S.A. Dunkle, 1988. A computer program incorporating Pitzer's equations for calculation of geochemical reactions in brines. *USGS WRI Report* 88-4153: 310.

Sanford, B.V., F.J. Thompson, and G.H. McFall, 1985. Plate tectonics - a possible controlling mechanism in the development of hydrocarbon traps in southwestern Ontario. *Bull. Can. Pet. Geol.* 33: 52-71.

Weaver, T.R., S.K. Frape and J.A. Cherry, 1990. The effects of production practices on groundwater chemistry of the Devonian oil fields of the Lambton County area, southwestern Ontario. *Proc. Ontario Petroleum Inst.* London, Ontario.

Water-Rock Interaction, Kharaka & Maest (eds) © 1992 Balkema, Rotterdam. ISBN 90 5410 075 3

Halogen compositions of fluid inclusions as tracers of crustal fluid behaviour

B.W.D.Yardley & D.A.Banks
University of Leeds, UK

I.A.Munz
Universitet i Oslo, Norway

ABSTRACT: Inclusion fluids have been extracted from their host quartz by a crush leach technique and analysed for Cl, Br and I by Ion Chromatography. Samples from different hydrothermal provinces produce discrete clusters of uniform fluid chemistry. Late-metamorphic veins from the Brusson gold district, Italian Alps, and from Precambrian basement in the Oslo district yield fluids with characteristics of surface waters, suggesting an origin from waters penetrating from above during uplift. High-T quartz veins from the SW England ore field yield homogeneous compositions of likely magmatic origin.

1 INTRODUCTION

The value of halogen analyses as conservative tracers of fluid masses has been recognised for some years in studies of contemporary near surface waters. More recently, attempts have been made to analyse halogen concentrations in fluid inclusions as tracers of ancient and deep seated fluid behaviour (Behr & Gehrler 1987, Bohlke & Irwin 1989, 1992, Banks et al. 1991). The seminal contribution of Bohlke & Irwin (1992) demonstrated that the fluids from a wide range of hydrothermal settings can be readily distinguished on the basis of I/Cl and Br/Cl ratios. Banks et al. (1991) demonstrated that these halogens continue to behave in a conservative manner at least to temperatures of 300°C, by finding that chemically variable saline fluids, included in quartz veins formed during thrusting in the central Pyrenees, had exchanged cations with adjacent wall rocks but all retained similar Cl and Br contents. The results from these Pyrenees samples are included here in Figure 1 for reference.

2 METHODS

Previous analyses of fluid inclusion halogen contents have largely been by neutron activation analysis (Behr & Gerler 1987) or noble gas mass spectrometry (Bohlke & Irwin 1992) of irradiated samples. In the case of the latter study, volatiles are sampled from selected small clusters of inclusions by laser ablation. However Cl and Br have also been determined by Ion Chromatography of bulk leached samples (Channer & Spooner 1992; Banks et al. 1991). This also allows the cation chemistry of the fluid to be determined (Bottrell et al. 1988). A single, average analysis is produced from around 4 g of purified quartz host, although partial analyses may be obtained from much smaller masses.

The results presented here were obtained by crushing c. 2 g of sand-sized, inclusion-bearing quartz, previously hand-picked for impurities, acid and electrolytically cleaned. The resulting powder was leached in c. 10 ml double distilled water and the solution analysed for Cl, Br and I with a Dionex 4500 Ion

Table 1. Analytical results for sample 2320, with comparison to the results of Bohlke & Irwin (1992)

	Cl	Br	I
Detection limit (ppb)	10	0.2	0.1
Analysed range (ppb)	8600-60200	13-21	0.2-2.7
log mol ratio to Cl	0	-3.17 to -3.23	-4.9 to -5.37
log ratio (B+I, 1992)	0	-2.89 to -3.15	-3.96 to -4.17

Chromatograph. Cl and Br were determined with a PAX-100 column and guard column, and carbonate-bicarbonate eluent. For I analysis a single short column only was utilised. In practice, Cl and Br were readily determined in all samples, but I was sometimes below detection.

Results on sample 2320 allow direct comparison between our results and the noble gas mass spectrometry results of Bohlke & Irwin (1992) on the same sample (Table 1). There is good agreement for Br/Cl, but a difference in I/Cl. To date we have been unable to identify the cause of this apparent discrepancy, however since it is reproducible and systematic it is unlikely to affect the interpretation attempted here.

3 RESULTS

3.1 SW England

A wide range of barren and mineralised veins, formed over a prolonged time span and a wide range of temperatures are present in this region (e.g. Jackson et al., 1989), and there have been many microthermometric studies of the fluid inclusions. This study is of relatively high-T quartz veins with tourmaline, cassiterite or wolframite, and of related hydrothermal rocks. Previously, halogen data on one sample (2320, see above) was given by Bohlke & Irwin (1992), who discussed its halogen geochemistry and concluded that it was in agreement with the magmatic origin proposed previously by Manning (1981) and Bottrell & Yardley (1989). However all our analyses to date cluster near our analysis of this sample (Fig. 1). Those hosted by graphitic slates around the intrusions tend to be enriched in I. This clustered data set, implying a magmatic origin for all the halogens, is surprising in view of existing models for the mineralisation. These commonly invoke a range of fluid types involved in the ore genesis (Jackson et al. 1989, Alderton & Harmon 1991). We have not yet analysed samples representative of the lower temperature, sulphide mineralisation in the region however, and this may account in part for the uniformity of the data.

3.2 Brusson, N Italy

Auriferous quartz veins occur at Brusson in Val d'Ayas, part of the Monte Rosa gold district of the western Alps. Steeply inclined veins traverse a wide range of rock types in the Monte Rosa and Piemonte Ophiolite Nappes. Gold formed during the main paragenetic stage with quartz and minor ferroan dolomite, muscovite and pyrite (Diamond (1990). Diamond demonstrated that the vein quartz is characterised by a main stage homogeneous fluid containing approximately 1.5 mol percent NaCl and 6 mol percent CO_2, which was trapped at 240-300°C. Immiscibility of H_2O and CO_2 developed in the final stages of mineralisation. Such mesothermal fluids are commonly ascribed to a metamorphic origin (e.g. Goldfarb et al. 1991). However there are clearly problems in accounting for the late formation of the veins relative to metamorphic dewatering, and their low salinity. Alternatively, Nesbitt (1991) has proposed that such veins form by the deep penetration of surface waters durng post-metamorphic uplift.

Halogen analyses from six vein quartz samples are plotted on Fig.1.

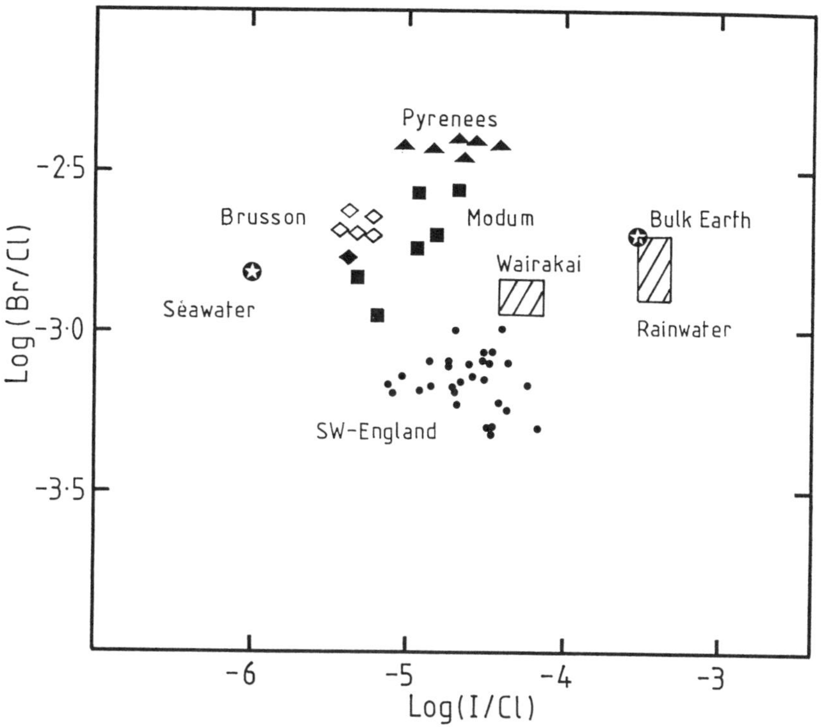

Fig. 1 Plot of halogen ratios determined for fluid inclusion samples, with some surface waters and bulk earth for reference. Triangular symbols are the Pyrenees fluids described by Banks et al. (1991). Other symbols are for each of the areas described in the text, as labelled. Open symbols denote I below detection, value plotted is an upper limit only for I/Cl.

Because of the very low yields of I from these samples, only one sample could be analysed for I, the other points plotted show an upper limit only. All the fluids fall in a tight cluster which lies on the array of surface water compositions between rain water and seawater. Until more extensive data are available on the behaviour of halogens in metamorphism it is not possible to rule out the possibility that metamorphic fluids do not exhibit any fractionation and also lie on the surface water array, but the results clearly show strong circumstantial support for an origin from deep penetrating surface waters.

3.3 Modum Complex, Norway

The Modum Complex is a region of high grade Precambrian basement rocks lying to the west of the Oslo graben. It contains metagabbros and amphibolites and enclosing supracrustal rocks. Late stage hydrothermal albitites occur within both metabasic and metasedimentary hosts. Vein or patchy quartz associated with the albitites contains fluid inclusions of two types: saline H_2O and CH_4-dominated carbonaceous inclusions. The salinity of the aqueous fluids is variable; some samples contain a populatoin of highly saline inclusions, c. 20 wt % NaCl equivalent, while most contain a less saline fluid, 1-12 wt % NaCl equivalent. Gas chromatography has revealed traces of hydrocarbons up to C5 in some samples, in addition to methane. Black, solid inclusions in some samples are tentatively interpreted as carbonaceous material.

The presence of the hydrocarbons suggests that there may be involvement of basinal fluids, possibly associated with the Oslo rift, in the formation of these basement veins. Halogen analyses appear to support such a suggestion (Fig. 1). There is a rather wide range in Br/Cl, with the highest values obtained from samples which were picked as enriched in solid carbonaceous inclusions. Overall however all the analyses lie in the array of surface-derived fluids, adjacent to the field of Brusson analyses.

4 CONCLUSIONS

Analyses of the Br/Cl and I/Cl ratios of inclusion fluids hosted by quartz yield distinct populations for different study areas, and appear to act as relatively conservative markers of fluid masses to mid crustal conditions. Our results suggest that post-metamorphic veining features in basement rocks are likely to result from the ingress of surface waters, and/or basinal fluids, into underlying crystalline rocks. In contrast, high to medium temperature hydrothermal veins from the SW England ore field all yield a similar signature of probable magmatic origin.

REFERENCES

Alderton, D.H.M. and Harmon, R.S. 1991. Fluid inclusion and stable isotope evidence for the origin of mineralising fluids in southwest England. Mineralogical Magazine 55: 605-612.

Banks, D.G., Davies, G.R., Yardley, B.W.D., McCaig, A.M. and Grant, N.T. 1991. The chemistry of brines from an Alpine thrust system in the central Pyrenees: an application of fluid inclusion analysis to the study of fluid behaviour in orogenesis. Geochimica et Cosmochimica Acta 55: 1021-1030.

Behr, H.-J. and Gerler, J. 1987. Inclusions of sedimentary brines in post-Variscan mineralisations in the Federal Republic of Germany - A study by neutron activation analysis. Chemical Geology 61: 65-77.

Bohlke, J.K. and Irwin, J.J. 1989. Halogen ratios in natural fluid inclusions analyzed by laser ablation and noble gas mass spectrometry. Geological Society of America Abstracts with Programs 21: 358.

Bohlke, J.K. and Irwin, J.J. 1992. Laser microprobe analyses of Cl, Br, I and K in fluid inclusions: implications for sources of salinity in some ancient hydrothermal fluids. Geochimica et Cosmochimica Acta 56: 203-225.

Bottrell, S.H., Yardley, B.W.D. and Buckley, F. 1988. A modified crush-leach method for the analysis of fluid inclusion electrolytes. Bulletin de Mineralogie 111: 279-290.

Bottrell, S.H. and Yardley, B.W.D. 1989. The composition of a primary granite-derived ore fluid from S.W England, determined by fluid inclusion analysis. Geochimica et Cosmochimica Acta 52: 585-588.

Channer, D.M.DeR. and Spooner, E.T.C. 1992. Analysis of fluid inclusion leachates from quartz by ion chromatography. Geochimica et Cosmochimica Acta 56: 249-259.

Diamond, L.W. 1990. Fluid inclusion evidence for P-V-T-X evolution of hydrothermal solutions in late Alpine gold-quartz veins at Brusson, Val d'Ayas, northwest Italian Alps. American Journal of Science 290: 912-958.

Goldfarb, R.J., Snee, L.W., Miller, L.D. and Newberry, R.J. 1991. Rapid dewatering of the crust deduced from ages of mesothermal gold deposits. Nature 354: 296-298.

Jackson, N.J., Willis Richards, J., Manning, D.A.C. and Sams, M.S. 1989. Evolution of the Cornubian ore field, southwest England: Part II. Mineral deposits and ore-forming processes. Economic Geology 84: 1101-1133.

Manning, D.A.C. 1981. The application of experimental studies in determining the origin of topaz-quartz-tourmaline rock and tourmaline-quartz rock. Proceedings of the Ussher Society 5: 121-127.

Nesbitt, B.E. 1991. Phanerozoic gold deposits in tectonically active continental margins. In R.P. Foster (ed.), Gold metallogeny and exploration: 104-132. Glasgow: Blackie.

Water-Rock Interaction, Kharaka & Maest (eds) © 1992 Balkema, Rotterdam. ISBN 90 5410 075 3

Hydrogeochemical patterns of major stages of the sedimentary process

V. P. Zverev
Scientific Center for Engineering Geology and Environment of the Russian Academy of Sciences, Moscow, Russia

ABSTRACT: Quantitative relations are calculated between the various stages of the sedimentary processes (weathering, transportation, sedimentation, diagenesis and katagenesis) and the mass and chemical compositions of natural waters interacting with rock, including duration of interaction, the dominant mass transfer mode, and the direction and intensity of mineral alterations. Using specific natural models, the author has evaluated the average rate of aluminosilicates alteration for major stages of the sedimentary process as ranging from 10^{-13} g^{Si}/sec-cm^3 during transportation to 10^{-10} g^{Si}/sec-cm^3 at the time of katagenesis.

1 INTRODUCTION

Major stages of the sedimentary process differ significantly in terms of the hydrodynamics and hydrogeochemical conditions. There is a distinct connection which has been validated between certain stages of the sedimentary process and types of natural waters. Mobilization and weathering are controlled by precipitation (rain and snow), soil solutions and underground waters in the zone of active water exchange, while transportation is dependent on streams and rivers. Sedimentation is active in river channels and on flood plains and in all topographic lows filled largely by water and forming surface reservoirs like swamps, lakes, etc., as well as in seas and oceans. Diagenesis is associated with pore water in the sediments during lithification. Finally, the next stages of lithogenesis, katagenesis and metagenesis, are controlled by ground waters which may be meteoric and flowing in the earth crust at certain rates or released under increasing pressure and temperature from the physically or chemically bound waters. Throughout the sedimentary process, minerals go through at least 7 or 8 transitions from one hydrogeochemical environment to another.

2 HYDROGEOCHEMICAL PATTERNS OF SEDIMENTARY PROCESSES

The major stages of the sedimentary process have their specific masses of natural waters and rocks and specific duration of interaction (Table 1). At the initial stages of sedimentogenesis, during mobilization and transportation, the ratio of interacting masses of water and rock is 2 to 3 thousand, while at the stage of katagenesis, it is a few hundredths, that is the ratio is reduced by more than five orders of magnitude. Interaction in the water-rock system may last as long as several minutes when atmospheric precipitation contacts the earth's surface, or up to tens and hundreds of millions of years during katagenesis, i.e., longer by 12 orders of magnitude. Duration of interaction is closely related to natural water velocities, which range from 10^2-10^3 cm/sec in surface streams, to 10^{-9}-10^{-11} cm/sec 2-4 km inside the crust, i.e. they range like the duration of interaction by the order of magnitude of 12. Natural water velocities exert a major control on the mass transfer regime. The transfer is realized through convection at the surface and in the upper crust, and through diffusion at depth.

An estimate of the rates of water-rock interaction at various stages of sedimentogenesis has been obtained using specific natural models that characterize weathering, transportation, sedimentation, and diagenesis and katagenesis (Table 1).

Table 1. Hydrogeological characteristics of the major stages of the sedimentary process.

Stages of the sedimentary process	Types of natural waters	Mass of rock (g)	Interacting mass of water (g)	Ratio Rock/water	Duration of inter action (sec)	Water velocity (cm/sec)	Predominant mass transfer mode
Surface weathering	Precipitation	$121x10^{19}$	$258x10^{21}$	$2x10^{3}$	$10^{3}-10^{4}$	10^{-1}	Convection
Underground	Ground water of active water exchange zones	$295x10^{21}$	$135x10^{24}$	$5x10^{2}$	$10^{5}-10^{7}$	10^{-7}	-"-
Transportation	Ephemeral streams rivers	$18.5x10^{15}$	$44.7x10^{21}$	$2.5x10^{3}$	$10^{4}-10^{5}$	$2-3x10^{2}$	-"-
	Inner seas oceans	$1.4x10^{19}$	$1.34x10^{24}$	10^{6}	$3x10^{6}$	$2-3x10^{2}$	-"-
Sedimentation	Rivers		$2.12x10^{19}$			10	
	Swamps, low-mineral lakes	$625x10^{15}$	$102.5x10^{19}$	$3x10^{-1}$	$10^{5}-10^{6}$	10^{-1}	
	Brackish and salt lakes		$85.4x10^{19}$			10^{-1}	
	Inner and marginal seas	$20.5x10^{15}$	$0.7x10^{21}$	$3x10^{4}$	$3x10^{7}$	10^{-1}	
	Oceans	$2x10^{15}$	$13x10^{21}$	$2x10^{6}$	$3x10^{7}$	10^{-1}	
Diagenesis	Continental reservoirs and streams	$13.3x10^{21}$	$1.2x10^{21}$	10^{-1}	$10^{6}-10^{7}$	$10^{-4}-10^{-5}$	diffusion
	Inner and marginal seas	$12.8x10^{15}$	$0.7x10^{21}$	10^{-1}	$10^{10}-10^{11}$	$10^{-4}-10^{-5}$	-"-
	Oceans	$190x10^{21}$	$1.9x10^{21}$	10^{-1}			
Infiltration	Underground waters of slow exchange zones	$38x10^{21}$	$40.9x10^{21}$	$3x10^{-1}$	$10^{11}-10^{12}$	$10^{-5}-10^{-7}$	-"-
Early Elision	Released physically bound waters	$64x10^{21}$	$11.1x10^{-2}$	$7x10^{-2}$	$10^{-9}-10^{-10}$	-"-	
Katagenesis Infiltration	Underground waters of extremely slow water exchange	$185x10^{21}$	$0.6x10^{21}$	$2x10^{-3}$	$10^{12}-10^{14}$	$10^{-9}-10^{-11}$	-"-
Deep Elison	zones of Released physically bound waters	$128x10^{21}$	$11.4x10^{21}$	$2x10^{-3}$	$10^{12}-10^{14}$		
	Released chemically bound waters		$6.2x10^{21}$	$5x10^{-2}$	$10^{14}-10^{16}$	$10^{-9}-10^{-11}$	-"-

Non-equilibrium controls on weathering at the initial stage of sedimentogenesis, along with high rates of water movements, leads to active discharge of solid components, and metal cations. The result is the high rate of mineral alterations (10^{-13}-10^{-14} g^{Si}/sec-cm^3) and the correspondingly large masses of altered rock.

Due to transportation, which is marked by strong yet short-term mineral alterations with a rate of ca. 10^{-13} - 10^{-14} g^{Si}/sec-cm^3, the mobilized material enters new hydrogeochemical environments that control sedimentation and diagenesis. These environments range from acidic swamp waters to alkaline salt lakes. Under diagenesis, mineral alteration rates average 10^{-15} g^{Si}/sec-cm^3.

With a longer time of interaction between water and rock and low rates of ground water movement, the rate of interaction is reduced, with values between 10^{-17} - 10^{-18} g^{Si}/sec-cm^3 during early filtration katagenesis. Still lower rate values (10^{-19} g^{-Si}/sec-cm^3 on average) characterize processes of elision katagenesis caused by the release of waters when sedimentary rocks are submerged deeper and subjected to higher temperatures and pressures.

Sedimentary basins

Mineral diagenesis

Water-Rock Interaction, Kharaka & Maest (eds) © 1992 Balkema, Rotterdam. ISBN 90 5410 075 3

North Sea clastic diagenesis and formation water constraints

P. Aagaard & J. S. Jahren
University of Oslo, Norway

P. Kr. Egeberg
Agder College, Kristiansand, Norway

ABSTRACT: Equilibrium calculations describing phase relations in North Sea Jurassic sandstones have been compared with formation water chemistry. The in situ activity of aqueous species were calculated by SOLMINEQ 88 by adopting the carbon-dioxide trend from North Sea and by assuming calcite saturation. Quartz, albite, illite and chlorite all appear to be in equilibrium or close to equilibrium with the pore water. Thus, the assumption of equilibrium among authigenic minerals and coexisting pore water, which frequently is made in diagenetic modeling, appears justified on geologic time scales.

1 INTRODUCTION

The diagenesis of the Jurassic major hydrocarbon reservoirs in the North Sea has been the focus of extensive investigations the last fifteen years (Blanche and Whitaker, 1978; Hancock and Taylor,1978; Burley et al.,1985; Jourdan et al.,1987; Glasman et al.,1989; Bjørlykke et al.,1992; Bjørlykke and Aagaard, 1992). The characteristics of the sediments and the main diagenetic mineral reactions are thus well documented. In addition, the formation water chemistry of a large number of producing oil fields and test samples from reservoirs under exploration is also known (Egeberg and Aagaard,1989, 1992). This offers a unique base to calculate dissolved species activities from the main diagenetic mineral reactions, by assuming equilibrium among authigenic phases and pore water, and compare the results with those derived from speciation/ activity calculations of corresponding formation waters.

The Jurassic clastic reservoirs are dominated by arkoses and subarkoses with minor occurrences of sublitharenites/ litharenites (Bjørlykke et al.,1992), e.g. the Brent Group sandstones normally have a quartz content between 40-90%, up to 15-20% kaolinite, a variable albite content, and a K-feldspar content decreasing from about 10% at shallow burial to below detection limit at 3.5-4 km depth. Calcite is present as cement in most sandstones, siderite is a frequent early cement, while the late carbonate cements are ankerites and ferroan calcite. The other late diagenetic minerals are quartz, albite, illite and minor chlorite, reflecting the original clastic mineral assemblage.

2 COMPUTATIONAL APPROACH

The calculation of equilibrium between minerals and their surrounding pore water, requires a consistent thermodynamic data set for the minerals and the aqueous species. The thermodynamic data base applied, is that of Helgeson and co-workers (Helgeson and Kirkham, 1974a,b, 1976; Helgeson et al., 1978, 1981), with its updates recently reported by Johnson et al. (1992). All calculations were done along an average North Sea P,T-gradient, taken from Smith and Ehrenberg (1989).

For estimation of in situ activities of

dissolved species we have used the geochemical computer code SOLMINEQ.88 (Kharaka et al., 1988) with the Pitzer formulation. The program is run at calcite saturation to the specified reservoir pressure, temperature and PCO_2. We here applied the carbon dioxide trend of Smith and Ehrenberg(1989) documented for the North Sea, and estimated concentrations of organic acid anions based on data from Barth (1986,1991). The details of the estimation scheme have been reported by Egeberg and Aagaard(1992), where also results of the calculations are given.

3 CONSTRAINTS IMPOSED BY AUTHIGENIC MINERALS

The carbonate minerals are normally considered to react rapidly with coexisting pore water (Kharaka et al., 1985) being able to buffer the water on short time scales. The equilibrium status of the carbonate minerals is not the subject of discussion in the present contribution, as we have based our pH calculation scheme on calcite saturation, and the iron analyses of formation water are either missing or of poor quality. However, our distribution/ activity calculations (Egeberg and Aagaard, 1992) have shown the pore waters to be undersaturated with respect to dolomite at temperatures above 140°C. This may reflect the increased iron substitution of the dolomite-ankerite solid solution towards higher temperatures.

3.1 Quartz

Quartz cement is unambiguously present in all North Sea sandstones studied, but the amount of authigenic quartz varies with burial. This feature is specially well documented for the Brent Group sandstones (Bjørlykke et al.,1992). At shallow to moderate burial conditions they contain only a few percentages of quartz cement, while the amount increases considerably at depths between 3 to 4 km. This pattern is mostly accounted for by pressure solution and stylolith formation, but other mineral reactions also produce quartz cement. These observations suggest that quartz is forming under most burial conditions from 65°C and upwards, even though the amount is minor at shallow burial. Fluid inclusion data (Walderhaug,1990; Glasman et al.,1989; Jourdan et al.,1987; Nedkvitne,1991) further support this hypothesis, as the range of homogenization temperatures reported from Jurassic sandstones vary from 90 to 155°C.

Formation water chemistry data reported by Egeberg and Aagaard (1992) follow quartz saturation closely, consistent with findings from other sedimentary basins as well (Kharaka et al., 1985). Under medium to upper diagenetic conditions, the precipitation rate of quartz is apparently fast enough to buffer the aqueous silica activity.

3.2 Illite

There is a strong depth related distribution of authigenic illite in the Jurassic sandstones offshore Norway (Bjørlykke et al., 1986; Ehrenberg and Nadeau, 1989). Illite is rather rare and occurs only in minor amounts at burial depths shallower than 3.5 km (Bjørlykke and Aagaard, 1992). However, between 3.5 and 4.0 km the illite content increases rapidly corresponding to a massive illitization of kaolinite. The process appears to be mainly isochemical and can be accounted for by the reaction

$$Al_2Si_2O_5(OH)_4 + KAlSi_3O_8 = KAl_3Si_3O_{10}(OH)_2 + 2\ SiO_2 + H_2O \quad (1)$$

The amount of illite formed is thus limited by available K-feldspar and/or kaolinite. In strongly kaolinized zones, the supply of K-feldspar is first exhausted and kaolinite will be preserved during further burial (Ehrenberg, 1991). Illite and kaolinite may thus constitute a stable pair in the upper diagenetic regime. In cases where the original sandstone contained K-feldspar in excess of kaolinite, illite will be stable together with K-feldspar instead. However, using the Brent Group as an example, the kaolinite/K-feldspar ratio is on the average between 1.5-2; thus formation waters from these sandstones are expected to be buffered by the illite-kaolinite pair below 4 km. This is also born out by the formation

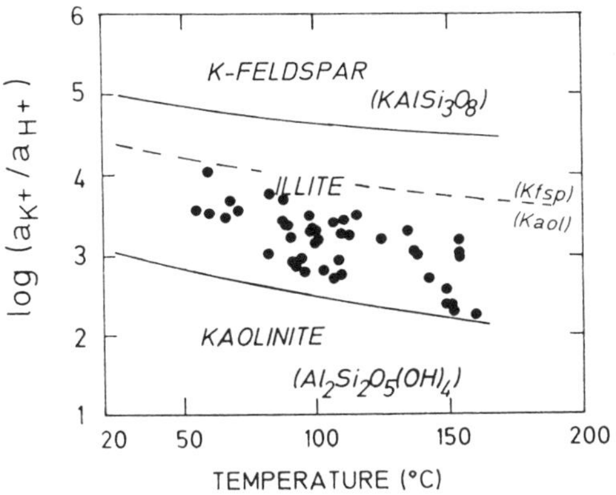

Fig.1 Stability field of illite at quartz saturation in terms of $\log(a_{K+}/a_{H+})$ versus temperature. The activity of the muscovite component is set equal to 0.5.

water chemistry.

Fig.1 depicts the stability relations among kaolinite, illite, K-feldspar and formation water at quartz saturation as a function of burial temperature. The activity of the muscovite component is set to an average value of 0.5 (Aagaard and Jahren, 1992). All formation water data plot in the stability field of illite, indicating that illite would be stable relative to K-feldspar + kaolinite at all diagenetic temperatures. In the lower temperature range, however, only a limited amount of illite is formed; possibly due to the slow dissolution rate of the reacting phases. At higher temperatures, where massive illite transformation has taken place, the (a_{K+}/a_{H+}) values approach the kaolinite illite boundary as predicted if K-feldspar is exhausted. The formation waters thus appear to be super-saturated or saturated with respect to illite.

3.3 Albite

Authigenic albite is found both as replacements of K-feldspar and separate cement (Saigal et al.,1988). Two processes appear to be responsible for the decrease of K-feldspar observed in the Jurassic clastic reservoirs on the Norwegian shelf (Bjørlykke et al.,1992). One is the albitization of K-feldspar observed at burial temperatures 65 - 125°C, which may be related with contemporaneous illitization (Pearson and Small, 1988) in adjoining shales through exchange of sodium and potassium (Aagaard et al.,1990). The other process is the illitization of kaolinite in the sandstones discussed above. The diagenetic formation of albite apparently may take place under most burial conditions, but is most intense before the major illitization of kaolinite, eg. the Brent sandstones reach a maximum level of albite cement at about 3.3 km depth.

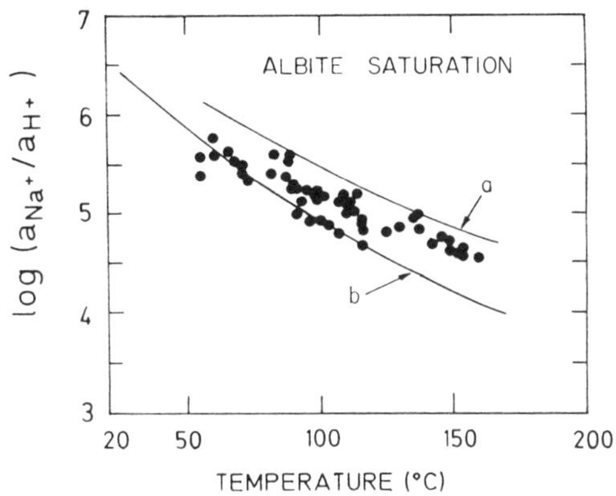

Fig.2 Saturation lines for albite. Line a) corresponds to the equilibrium assemblage albite, illite and quartz; line b) to the metastable assemblage albite, illite, kaolinite and K-feldspar.

The formation water chemistry relating to albite saturation also mirrors these overall diagenetic processes. Fig.2 depicts the saturation of albite expressed as a function of log (a_{Na+}/a_{H+}) versus temperature. Activity ratios calculated from chemical analyses of formation waters, define a narrow band bounded by two lines. The lower boundary corresponds to the metastable equilibrium between two major clastic components, K-feldspar and kaolinite, and the authigenic minerals, albite and illite. This is a metastable assemblage because the required aqueous silica activities will be above quartz saturation. The upper line is calculated by assuming equilibrium among albite, illite and quartz, and that the pore water will follow the (a_{K+}/a_{H+}) trend of Fig. 1. There appears to be a shift from a K-feldspar dominated system at lower temperatures to illite dominated at higher diagenetic grade. The spread in the formation water data can easily be accounted for by the uncertainty in analyses and pH estimates, and we will conclude that the formation waters are saturated or close to saturation with respect to albite.

3.4 Chlorite

While illite is a common late authigenic phase, smaller amounts of chlorite are also found (Kantorowicz, 1984; Jahren and Aagaard, 1989,1992; Bjørlykke and Aagaard,1992). They are commonly formed from kaolinite precursors. Jahren (1991) documented dynamic dissolution and growth behavior of these chlorites, with grain coarsening combined with chemical zonation, and argued that the Ostwald ripening process reflects a semiclosed system. The compositional changes were also consistent with internal equilibrium among coexisting illite chlorite pairs (Aagaard and Jahren, 1992).

The equilibrium among chlorite, illite and an aqueous phase, can be written in terms of their components clinochlore and muscovite:

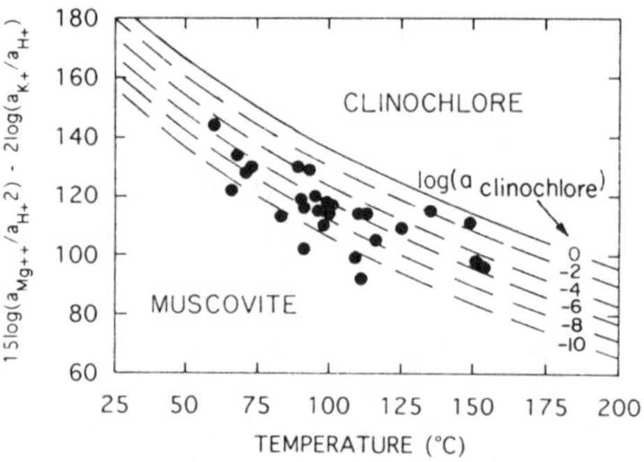

Fig.3 Chlorite-illite equilibrium written in terms of the clinochlore and muscovite components. The univariant boundary is drawn at a fixed muscovite activity and contoured at clinochlore activities between 1 and 10^{-10}.

$$3Mg_5AlAlSi_3O_{10}(OH)_8 + 2K^+ + 28H^+ = 2KAl_2AlSi_3O_{10}(OH)_2 + 15Mg^{++} + 3SiO_2(s) + 24H_2O \quad (2)$$

The equilibrium expression can be recast as

$$15\log(a_{Mg++}/a_{H+}^2) - 2\log(a_{K+}/a_{H+}) = \log K + 3\log(a_{clinochl}) - 2\log(a_{musc}) \quad (3)$$

suitable for graphical representation. The activity of the muscovite component is set equal to 0.5, while the clinochlore activity is allowed to vary. In this way, the right hand side of equation 3 is evaluated at different clinochlore activities and is represented in Fig.3 by a suit of equilibrium lines corresponding to clinochlore activities from 10^0 - 10^{-10}. The ion activity ratio expression on the left hand side of equation 3 was also evaluated for the different formation waters, and plotted onto Fig.3. Despite considerable uncertainties in ion activity and pH calculations, and the approximations involved in evaluating the clinochlore activity, the data point follow the equilibrium line trend surprisingly well. We will argue that also the pore water data support the hypothesis of equilibrium between authigenic chlorite illite pairs in the North Sea Jurassic sandstones (Aagaard and Jahren, 1992).

4 CONCLUDING REMARKS

The major late diagenetic silicate mineral reactions inferred from petrographic observations of North Sea Jurassic sandstones are clearly reflected in the formation water chemistry. Comparison of predicted ion activity ratios with the ones estimated from formation water analysis, reveal that authigenic minerals as quartz, albite, illite and chlorite are in equilibrium or close to equilibrium with the pore water. The buffering of silicate minerals (Nesbitt,1985; Hutcheon, 1990) thus appears to be effective in sandstone diagenesis on a geologic time scale.

REFERENCES

Aagaard, P., Egeberg, P.K., Saigal, G.C., Morad, S. and Bjørlykke, K., 1990.

Diagenetic albitization of detrital K-feldspars in Jurassic, Lower Cretaceous and Tertiary clastic reservoir rocks from offshore Norway. II. Formation water chemistry and kinetic considerations: J. Sediment. Petrol., in press.

Aagaard, P. and Jahren, J.S., 1992, Diagenetic illite-chlorite assemblages in arenites. II. Thermodynamic relations: Clays Clay Miner.(submitted).

Barth, T., 1987, Multivariate analysis of aqueous organic acid concentration and geological properties of North Sea reservoirs: Chemometrics and Intelligent Laboratory Systems, v. 2, p. 155-160.

Barth, T., 1991, Organic acids and inorganic ions in waters from petroleum reservoirs, Norwegian continental shelf: a multivariate statistical analysis and comparison with American reservoir formation waters: Applied Geochemistry, v. 6, p. 1-16.

Bjørlykke, K., and Aagaard, P., 1992, Clay minerals in North Sea sandstones: in Origin, Diagenesis, and Petrophysics of Clay Minerals in Sandstones, Soc. Econ. Paleontol. Mineral., Special Publication No 47

Bjørlykke, K., Aagaard, P., Dypvik, H., Hastings, D.S. and Harper, A.S., 1986, Diagenesis and reservoir properties of Jurassic sandstones from the Haltenbanken area, offshore mid Norway, in Spencer, A.M., ed., Habitat of hydrocarbons on the Norwegian Continental Shelf: (Norw. Petr. Soc.) p. 275-286.

Bjørlykke, K., Nedkvitne, T., Ramm, M., and Saigal, G.C., 1992, Diagenetic processes in the Brent Group (Middle Jurassic) reservoirs of the North Sea - An overview: Geol. Soc. London, Brent Conference Volume, in press.

Blanche, J.B., and Whitaker, J.H., 1978, Diagenesis of part of the Brent Sand Formation (Middle Jurassic) of the northern North Sea Basin: Geol. Soc. London, v. 135, p. 73-82.

Burley, S.D., Kantorowicz, J.D., and Waugh, B., 1985, Clastic diagenesis: in Brenchley, P.J. and Williams, B.P.J. (eds) Sedimentology, recent developments and applied aspects, Geol. Soc. London, Spec. Publ. No 18, p. 189-225.

Egeberg, P.K. and Aagaard, P., 1989, Origin and evolution of formation waters from oil fields on the Norwegian Shelf: Applied Geochemistry, v. 4, p. 131-142.

Egeberg, P.K. and Aagaard, P., 1992, Formation water chemistry in relation to the stability of detrital and authigenic minerals in clastic reservoirs from offshore Norway: submitted to Marine Petr. Geology

Ehrenberg, S.N., 1991, Kaolinized, potassium-leached zones at the contacts of the Garn Formation, Haltenbanken, mid-Norwegian continental shelf: Marine Petr. Geology, v.8, p. 250-269.

Ehrenberg, S.N., and Nadeau, P.H., 1989, Formation of diagenetic illite in sandstones of the Garn Formation, Haltenbanken, mid-Norwegian continental shelf: Clay Miner., v.24, p. 233-253.

Glasmann, J.R., Clark, R.A., Larter, S., Briedis, N.A., and Lundegard, P.D., 1989, Diagenesis and Hydrocarbon Accumulation, Brent Sandstones (Jurassic), Bergen High, North Sea: Amer. Ass. Petr. Geol., Bulletin, v.73, p. 1341-1360.

Hancock, N.J and Taylor, A.M., 1978, Clay mineral diagenesis and oil migration in the Middle Jurassic Brent Sand formation: J. Geol. Soc. Lond., v. 135, p. 69-72.

Hancock, N.J., 1978, Diagenetic modelling in the Middle Jurassic Brent sand of the northern North Sea: Proc. European Offshore Petroleum Conference & Exhibition, London, 1978, v .2, p. 275-280.

Helgeson, H. and Kirkham, D.H., 1974a: Theoretical prediction of the thermodynamic behaviour of aqueous electrolytes at high pressures and temperatures - I: Summary of the thermo-dynamic/electrostatic properties of the solvent. Amer. Jour. Sci. 274, 1089-1198.

Helgeson, H. and Kirkham, D.H., 1974b: Theoretical prediction of the thermodynamic behaviour of aqueous electrolytes at high pressures and temperatures: II. Debye-Huckel parameters for activity coefficients and relative partial molal properties: Amer. Jour. Sci., v. 274, p. 1199-1261.

Helgeson, H.C. and Kirkham, D.H., 1976: Theoretical predictions of the thermodynamic behaviour of aqueous electrolytes at high pressures and temperatures: III. Equation of state for aqueous species at infinite dilution. Amer. Jour. Sci., v. 276, 97-240.

Helgeson, H.C., Delany, J.M., Nesbitt, H.W. and Bird, D.K., 1978: Summary and citique of the thermodynamic properties of rock-forming minerals: Amer. Jour. Sci., v. 278A, 299 p.

Helgeson, H.C., Kirkham, D.H. and Flowers, G.C., 1981: Theoretical prediction of the thermodynamic behaviour of aqueous electorlytes at high pressures and temperatures: IV. Calculation of activity coefficients, osmotic coefficients, and apparent molal and standard and relative partial molal properties to 600°C and 5 kb: Amer. Jour. Sci., v. 281, p. 1249-1516.

Hutcheon, I., 1990, Clay-carbonate reactions, Scotian Shelf: in Spencer, R.J. and Chou, I.-Ming (eds) Fluid-Mineral Interactions: A Tribute to H.P. Eugster, The Geochemical Society, Spec. Publ. No 2, p. 199-212.

Jahren, J.S., 1991, Evidence of Ostwald ripening related recrystallization of chlorites from reservoir rocks offshore Norway: Clay Miner. 26, 169- 178.

Jahren, J.S. and Aagaard, P., 1989, Compositional vatiations in diagenetic chlorites and illites, and relationship with formation water chemistry: Clay Miner., v. 24, p. 157-170.

Jahren, J.S. and Aagaard, P., 1992, Diagenetic illite-chlorite assemblages in arenites. I. Chemical evolution: Clays Clay Miner. (submitted).

Johnson, J.W., Oelkers, E.H., and Helgeson, H.C., 1992, SUPCRT92: A software package for calculating the standard molal thermodynamic properties of minerals, gases, aqueous species, and rectionss from 1 to 5000 bars and 0° to 1000°C: Computers and Geosciences.

Jourdan, A., Thomas, M., Brevart, O., Sommer, F., and Sullivan, M., 1987, Diagenesis as the control of the Brent sandstone reservoir properties in the Greater Alwyn area (East Shetland Basin).: in Brooks,J. and Glennie, K.W. (eds) Petroleum Geology of North West Europe, Graham & Trotham, p.951-961.

Kantorowicz, J.D., 1984, Nature, origin and distribution of authigenic clay minerals from Middle Jurassic Ravenscar and Brent Group sandstones: Clay Miner., v.19, p. 359-375.

Kharaka, Y.K, Gunter, W., Aggarwal, P.K., Perkins, E.H. and DeBraal, J.D., 1988, SOLMINEQ.88: A computer program for geochemical modeling of water-rock interactions: U.S. Geol. Surv. Water-Resources Investigation Report 88-4227. 420 p.

Kharaka, Y.K., Hull, R.W., and Carothers, W.W., 1985, Water-rock interactions in sedimentary basins: in Gauthier, D.L. (ed) Relationship of Organic Matter and Mineral Diagenesis, Soc. Econ. Paleontol. Mineral., Short Course No 17.

Nedkvitne, T., Karlsen,D.A., Bjørlykke, K., and Larter, S.R., 1992, The relationship between reservoir diagenetic evolution and petroleum emplacement in the Ula Field, North Sea, Marine Petr. Geology

Nesbitt, H.W., 1985, A chemical equilibrium model for the Illinois Basin formation water: Am. Jour. Sci., v. 285, p. 436-458.

Pearson, M.J. and Small, J.S., 1988, Illite-Smectite diagenesis and paleotemperatures in Northern North Sea Quaternary to Mesozoic shale sequences: Clay Minerals, v. 23, p. 109-132.

Saigal, G.C., Morad, S., Bjørlykke, K., Egeberg, P.K. and Aagaard, P., 1988, Diagenetic albitization of detrital K-feldspars in Jurassic, Lower Cretaceous and Tertiary clastic reservoir rocks from offshore Norway. I. Textures and origin: J. Sediment. Petrol., in press.

Smith, J.T, and Ehrenberg, S.N., 1989, Correlation of carbon dioxide abundance with temperature in clastic hydrocarbon reservoirs: relationship to inorganic chemical equilibrium: Marine and Petroleum Geology, v. 6, p. 129-135.

Walderhaug, O., 1990, A fluid inclusion study of quartz cemented sandstones from offshore Mid-Norway - possible evidence for continued quartz cementation during oil emplacement: J. Sed. Petrology, v.60, p. 203-210.

Water-Rock Interaction, Kharaka & Maest (eds) © 1992 Balkema, Rotterdam. ISBN 90 5410 075 3

Cementation of North Sea sands: Constraints from fluid compositions

Andrew C.Aplin
NRG, The University, Newcastle upon Tyne, UK

Edward A.Warren & Shona M.Grant
BP Research, Sunbury-on-Thames, UK

ABSTRACT: Quartz is a widespread cement in pre-Tertiary North Sea sandstones. Fluid inclusion and oxygen isotope data show that much of the quartz formed from "basinal brines": ^{18}O-rich waters with variable, but often high salinities. These data limit potential mechanisms of silica transport. Transport could not have occurred in response to hydrodynamically-driven recharge of meteoric water. Compositional heterogeneities of the mineralising fluids suggest that large-scale convection or compaction-driven fluid flow are also unlikely transport mechanisms. Cementation appears to have been controlled by processes operating on a local scale. Possibilities allowed by the fluid data are (1) reservoir-scale convection and (2) diffusive transport.

1 INTRODUCTION

Quartz cement typically comprises 5-15 volume % of pre-Tertiary North Sea sandstones. Point-counting of optical thin sections generally suggests that much of the required silica could not have been sourced by local intergranular pressure solution, implying the import of significant masses of silica into the sandbody (e.g. Ehrenberg, 1990).

A fundamental but unresolved problem in diagenesis is to determine how large volumes of a sparingly soluble mineral like silica are transported on scales greater than an individual sandbody. In this paper we attempt to constrain the mass transport mechanisms which caused quartz cementation of pre-Tertiary sandstones from the North Sea. Our approach is to estimate the chemical and oxygen isotopic composition of the fluids from which quartz precipitated and to consider the implications of these compositions for possible mass transport mechanisms.

2 SAMPLES AND ANALYTICAL TECHNIQUES

Data was gathered from sands over a wide swathe of the North Sea (Fig. 1; including data from Hogg (1989) and Glasmann et al. (1989a,b)). Fluid inclusion homogenisation and final melting temperatures were determined to better than 1°C using a Linkam THM600 heating-cooling stage connected to a Linkam TMS90 controller. Oxygen isotopic analyses were made on quartz overgrowths which had been separated from detrital quartz using Lee and Savin's (1986) procedure. Purity checks using hot cathodoluminescence suggested that authigenic quartz was concentrated in finer grain sizes, which were separated and analysed for their $\delta^{18}O$ composition at SURRC, East Kilbride. We estimate that errors in the analyses are less than 1‰.

3 COMPOSITION OF MINERALISING FLUIDS

Homogenisation temperatures suggest that quartz precipitated between 70 and 150°C, at burial depths between 2 and 4 km. Combining these data with appropriate burial histories indicates that nowhere did cementation occur earlier than the late Cretaceous.

Precipitation temperatures were combined with

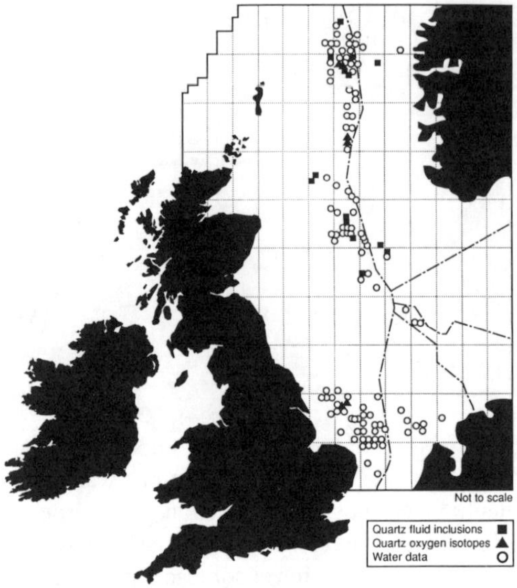

Figure 1 Sample location map.

the oxygen isotopic composition of authigenic quartz ($\delta^{18}O_q$) to estimate the oxygen isotopic composition of the mineralising fluids ($\delta^{18}O_w$). Values of $\delta^{18}O_w$ range between -3 and +3‰. Since, at least in coastal regions, $\delta^{18}O$ of meteoric water ($\delta^{18}O_{mw}$) varies as a function of latitude, it is possible to estimate $\delta^{18}O_{mw}$ at the time when quartz cementation occurred. These estimates range from -4 (late Cretaceous) to -7‰ (present-day) and indicate that the mineralising fluids were enriched in ^{18}O compared to local meteoric water (Fig. 2).

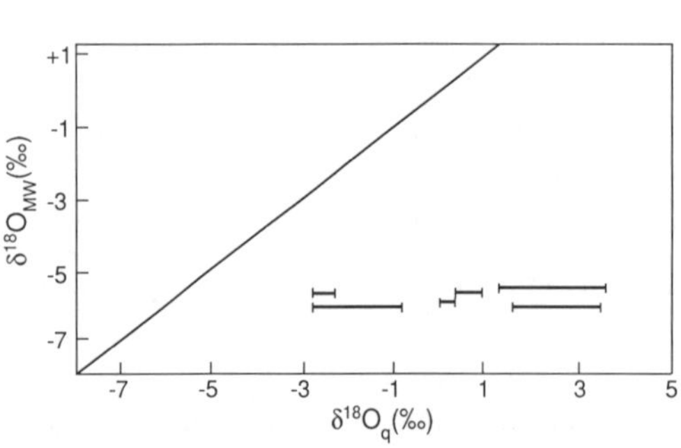

Figure 2 Comparison of the oxygen isotopic composition of quartz-precipitating fluids ($\delta^{18}O_q$) with that of local meteoric water ($\delta^{18}O_{mw}$) at the time of cementation. All mineralising fluids are enriched in ^{18}O compared to local meteoric water.

The salinity (weight % equivalent NaCl) of the mineralising fluids are plotted as a function of present-day burial depth in Fig. 3. Quartz precipitated from fluids with an extremely diverse compositional range. Many cements formed from fluids which were more saline than seawater but few formed from low salinity (< 0.5 % NaCl) fluids. Exceptionally fresh (i.e. potable) waters, which may be expected in situations where meteoric water is actively moving through the basin from exposed basin margins (e.g. Airey et al., 1979), are extremely rare.

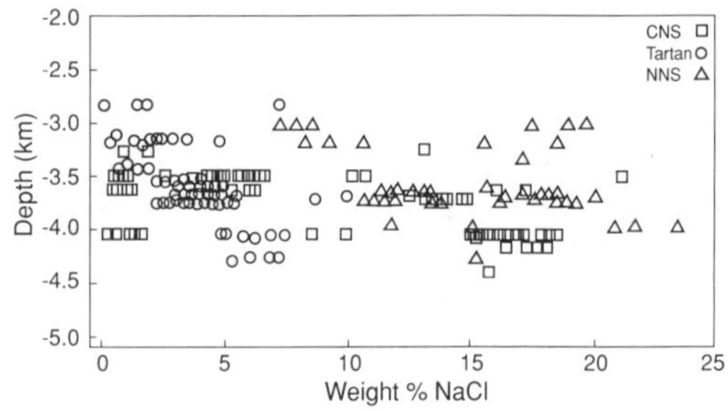

Figure 3 Salinity (weight % NaCl) of quartz-precipitating fluids, estimated from final melting temperatures of fluid inclusions. CNS = Central North Sea; NNS = Northern North Sea; Tartan Field is located 200 km due east of NE Scotland.

The mineralising fluids are also compositionally heterogeneous on a range of spatial scales. Figure 3 shows that salinities vary from 0 to 21% NaCl in the Central North Sea, and from 0 to 10% NaCl in Tartan Field (200 km east of NE Scotland). Such compositional heterogeneity indicates the local involvement of salt and suggests that the mineralising fluids were compartmentalised and not homogenised by large-scale fluid flow.

4 ORIGIN OF MINERALISING FLUIDS

Based on the few available data, it appears that both the isotopic and chemical composition of the mineralising fluids are similar to those of formation waters presently within the deeper parts of the North Sea sedimentary basin (Egeberg and Aagaard, 1989). Saline, ^{18}O-rich formation waters are common in the deeper parts of many sedimentary basins and generally result from (1) trapping of evolved seawater during periods of evaporite deposition and/or (2) isotopic exchange between water and recrystallising sedimentary minerals. Because (1) evaporites comprise a small fraction of basin fill, and (2) measurable enrichment of ^{18}O in waters by reaction with rocks implies a low water-rock ratio (Taylor, 1974), the absolute volumes of saline, ^{18}O-rich water in sedimentary basins must be small. A sensible approximation is that the amount of saline, ^{18}O-rich water in sedimentary basins is essentially that of the water contained within the interstices of the sediment pile.

5 MASS TRANSPORT MECHANISMS

Mass transport in sedimentary basins may occur by four basic processes:

1. Hydrodynamically-driven recharge of meteoric waters, especially at times when basin margins are exposed and/or uplifted.
2. Compaction-driven fluid flow.
3. Diffusion.
4. Fluid recycling, driven either by convection or stress-release mechanisms such as seismic pumping.

Since meteoric waters are fresh and ^{18}O-depleted, the saline, ^{18}O-rich fluids from which quartz precipitated in North Sea sandstones firmly rules out the possibility that silica transport was effected by meteoric water recharge. Although the fluid data cannot disprove the possibility that mass transport occurred in response to compaction-driven flow, mass balance arguments argue strongly against it. If quartz precipitates in response to the migration of a cooling fluid, then between 10^4 and 10^6 pore volumes of water must pass through each pore in order to generate 5 volume % quartz cement (Sibley and Blatt, 1976). For example, one can calculate how much water would be required to form 5 volume % quartz cement in a one billion barrel oilfield. The volume needed is the total volume of water expelled from a block of sediment 1 km x 160 km x 160 km during its burial from 2 to 3 km. Not only is it unlikely that such intense focussing of water might occur, but also this degree of focussing cannot explain the widespread occurrence of quartz cement in pre-Tertiary North Sea sands.

If quartz cementation had occurred in response to large-scale fluid flow, driven by compaction, convection or seismic activity, we believe that the composition of mineralising fluids would be much more homogeneous than is observed. In fact, the fluids are characterised by compositional heterogeneity on scales ranging from individual fields to sub-basins (Fig. 3). Silica transport and quartz cementation appear to have been controlled by processes operating on a local scale (intra-formation or between adjacent formations). In order to remain true to the fluid data, and in the absence of significant seismic activity, reservoir-scale convection, or localised diffusion, seem to be the only tenable transport mechanisms for silica in these sands. The fluid data cannot distinguish between these mechanisms.

6 CONCLUSIONS

Much of the widespread quartz cement in pre-Tertiary North Sea reservoir sands precipitated from ^{18}O-rich waters with variable, but often high salinities. The composition of the fluids rule out meteoric recharge and large-scale compaction- or convection-driven fluid flow as mechanisms by which silica may have been transported to cementation sites. Local (intra-formation or between adjacent formations) transport processes are implicated in quartz cementation. Reservoir-scale convection and diffusion are allowable transport mechanisms, but the fluid data cannot distinguish between them.

REFERENCES

Airey P.L., Calf G.E., Campbell B.L., Hartley P.E. and Habermehl M.A. 1979. Aspects of the isotope hydrology of the Great Artesian Basin. *IAEA-SM-228/12*: 205-212.

Egeberg P.K. and Aagaard P. 1989. Origin and evolution of formation waters from oil fields on the Norwegian shelf. *Appl. Geochem.* 4: 131-142.

Ehrenberg S.N. 1990. Relationship between diagenesis and reservoir quality in sandstones of the Garn Formation, Haltenbanken, Mid-Norwegian continental shelf. *AAPGB* 74: 1538-1558.

Glasmann J.R., Lundegard P.D., Clark R.A., Penny B.K. and Collins I.D. 1989a. Geochemical evidence for thc history of diagenesis and fluid migration: Brent Sandstone, Heather Field, North Sea. *Clay Mins.* 24: 255-284.

Glasmann J.R., Clark R.A., Larter S., Briedis N.A. and Lundegard P.D. 1989b. Diagenesis and hydrocarbon accumulation, Brent Sandstone (Jurassic), Bergen High Area, North Sea. *AAPGB* 73: 1341-1360.

Hogg A.J.C. 1989. *Petrographic and isotopic constraints on the diagenesis and reservoir properties of the Brent Group Sandstones, Alwyn South, northern North Sea.* Ph.D. Thesis, Univ. Aberdeen.

Lee M. and Savin S.M. 1986. Isolation of diagenetic overgrowths on quartz sand grains for oxygen isotope analysis. *Geochim. Cosmochim. Acta* 49: 497-501.

Sibley D.F. and Blatt H. 1976. Intergranular pressure solution and cementation of the Tuscarora orthoquartzite. *J. Sed. Petrol.* 46: 881-896.

Taylor H.P. 1974. The application of oxygen and hydrogen isotope studies to problems of hydrothermal alteration and ore deposition. *Econ. Geol.* 69: 843-883.

Water-Rock Interaction, Kharaka & Maest (eds) © 1992 Balkema, Rotterdam. ISBN 90 5410 075 3

Silica mass transport during shale diagenesis: Are shales a source or sink for silica?

J. Bloch
Geological Survey of Canada, Calgary, Alb., Canada

I. E. Hutcheon
The University of Calgary, Calgary, Alb., Canada

ABSTRACT: Detailed mineralogy of the illitic Lower Cretaceous Harmon shale has been evaluated from the bulk chemistry, clay mineralogy and mineral modes as input for linear programming to evaluate diagenetic mineralogy. Paleo pore-fluid composition with respect to silica is constrained by the mineralogy, paragenesis and the burial history. Silica dissolution and precipitation rates are evaluated using existing kinetic equations. The burial induced increase in temperature is inferred to be the primary control on silicate diagenesis in the Harmon shales. The mass of silica potentially available for transport in one pore-volume of fluid is calculated to be between 20 and 53 gm/m^3 shale using silica solubilitites consistent with modern pore-waters. Kinetic considerations suggest that as diagenesis proceeds, illitic shales become sinks for silica, not sources.

1. INTRODUCTION

Shales constitute the greatest volume of sedimentary basins yet their response to burial and diagenesis is poorly understood compared to coarser grained sandstones and carbonates. The source of material that constitutes the authigenic phases in coarser grained rocks is difficult to determine and shales frequently are invoked as sources of silica. Yet few studies have examined mass balance during the diagenesis of shales to determine if shales are the source of material attributed to them.

This study documents the aluminosilicate paragenesis of a Cretaceous marine mudstone, the Harmon Member (Peace River Formation) of Western Canada. Porosity loss data, determined from sonic velocity logs, are used to estimate the volume of fluid expelled. Thermal maturation studies (Kalkreuth and McMechan, 1988) provide a temperature history. These data are used to model the evolving pore-fluid composition with respect to dissolved silica to make an estimate of the mass of silica that was available to be expelled during diagenesis.

2. GEOLOGICAL BACKGROUND

The Harmon Member is an eastward-tapering wedge of transgressive marine siltstone and mudstone up to 120 m thick. It was deposited in a restricted basin about 102 Ma under dominantly anoxic to dysaerobic conditions (Stott, 1968; Stelck & Leckie, 1988; Yanagi et al, 1988). Harmon Member sediments comprise distal shelf mudstones to siltstones and hemipelagic mudstones. These sediments were progressively buried to a maximum of 4.0 km about 50-60 Ma during subsequent foreland basin development Estimated maximum burial temperatures are approximately 115 °C (Bloch, 1989).

3. PARAGENESIS AND PORE-FLUID EVOLUTION

The mineral modes and paragenesis are determined by synthesizing petrography, whole-rock chemistry (x-ray fluorescence, total carbon and sulphur analyses), x-ray diffraction data on bulk samples and clay separates as input data for linear programming (Slaughter, 1989). The paragenesis of the Harmon Member is determined to be kaolinite ⇒ illite ⇒ quartz ⇒ and possibly chlorite. Quartz and chert dissolution textures are pervasive but quartz overgrowths are observed. K-feldspar apparently was preserved during early diagenesis but underwent dissolution and replacement by illite during subsequent burial.

The mineral assemblage kaolinite, illite and K-feldspar can be represented on a phase diagram (Fig. 1) of log $aSiO_2$ vs. log $a(K^+/H^+)$. The paragenesis suggests that the stable assemblage during early diagenesis (low temperatures) is kaolinite-K-feldspar and that pore-fluids are therefore supersaturated with respect to quartz. This is consistent with the occurrence of authigenic quartz. With increasing temperature, the illite stability field expands. The observed paragenesis and associated dissolution of K-feldspar may be explained simply by an increase in temperature at a relatively constant fluid composition with respect to potassium. The inferred high activity of silica during Harmon Member

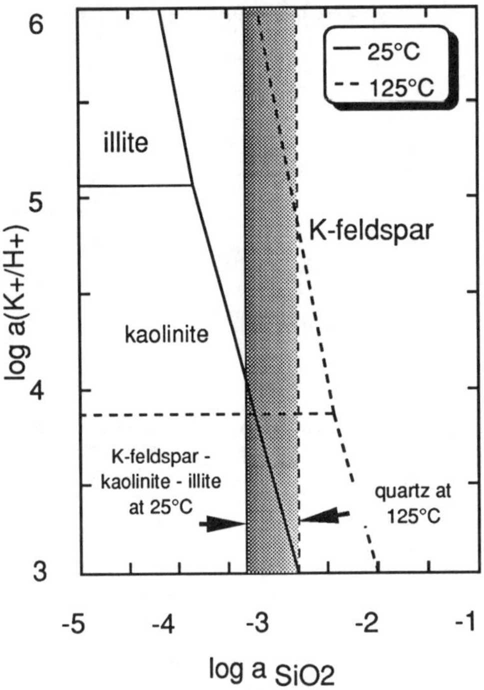

Figure 1. Log a_{SiO2} versus log $a_{(K+/H+)}$ showing the stability fields of illite (muscovite with component activity = 0.3), kaolinite and K-feldspar at 25 and 125 °C. Shaded area represents the range of SiO_2 activities during Harmon Member diagenesis inferred from the paragenesis and burial history

diagenesis is compatible with observed silica concentrations in modern marine shales (Fig. 2). At low temperatures, reactions between aluminosilicates and abundant biogenic silica appear to maintain high dissolved silica concentrations but somewhat less than amorphous silica saturation (Fig. 2). With increasing temperature, dissolved silica concentrations approach quartz saturation. These observations will be used to constrain silica solubilities and evaluate silica mass transport during diagenesis of the Harmon Member.

4. SILICA MASS TRANSPORT

To estimate the potential silica flux from a given body of shale it is necessary to know: 1) the volume of fluid expelled, 2) the timing of fluid expulsion, and 3) the dissolved silica content of the expelled fluid. Kinetic constraints provide information on the duration of quartz supersaturation of a pore-fluid in contact with a shale. The differences in dissolution and precipitation rates within the shale unit and between shales and sandstone must also be considered in the evaluation of silica mass transport between these two rock types as maintenance of supersaturation is necessary for significant silica transport.

The porosity represents a minimum volume of fluid (one pore volume) and compaction (porosity loss) curves therefore can be used to model fluid loss during burial. The equilibrium saturation state of the fluid can be estimated from temperature history and the paragenesis. Both chalcedony and the assemblage kaolinite-K-feldspar-illite fall within the wide range of possible silica saturation states (Fig 2) and jointly reflect the dissolved silica concentrations of many modern sedimentary pore fluids. Therefore, to model silica solubility during Harmon Member diagenesis, chalcedony and the mineral assemblage K-feldspar-kaolinite-muscovite are used to constrain $aSiO_2$.

4.1 Porosity Loss

Argillaceous sediments in particular show a wide range of porosity-depth values that reflect a complex relationship between porosity and depth. In comparison to coarser-grained sediments, shales undergo an initial rapid porosity loss within the first 500m of burial (Rieke & Chilingarian, 1974). This is due to the high initial water content and the orientation of dominantly platy grains under relatively small overburden stresses. After this initial stage of compaction, porosity decreases at a lower rate. Porosity reduction in shales may therefore be conceptualized as a two step process and approximated by two linear functions (Hamilton,1976; Issler,1990). The total porosity loss during burial (Fig 3) provides the minimum volume of fluid (one pore volume) expelled from the shale body.

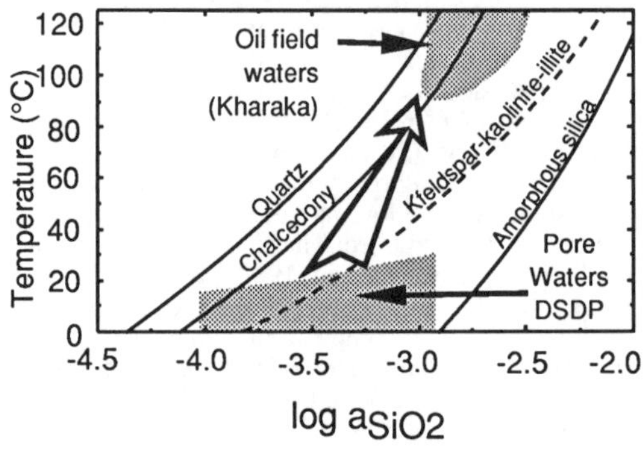

Figure 2. Log a_{SiO2} to 125 °C at quartz, chalcedony and amorphous silica saturation and at equilibrium with the assemblage K-feldspar - kaolinite - illite. Grey areas represent range of observed a_{SiO2} in marine shale pore waters (Sayles & Mannheim, 1975;. Zachos & Cederburg, 1989; Kastner et al., 1990; Swart & Burns, 1990) and overpressured shales of the Gulf Coast (Kharaka et al., 1977). Observed porewater a_{SiO2} indicate that a series of mineral reactions control SiO_2 solubility.

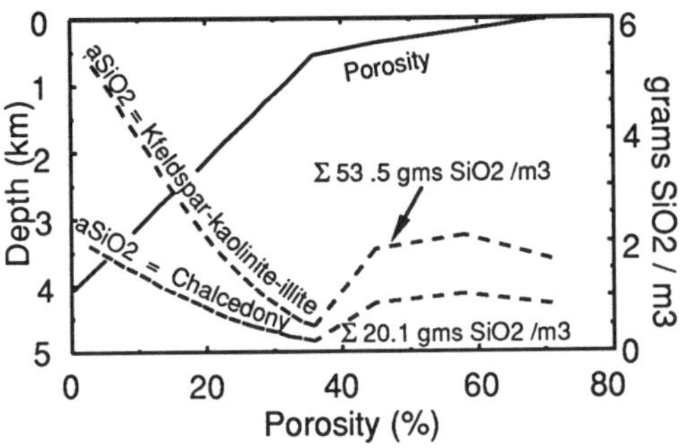

Figure 3. Porosity versus depth and equivalent mass of soluble SiO_2, normalized to 1 m^3 of shale, at the indicated a_{SiO2}. Σ's indicate the total amount of soluble a_{SiO2} potentially available for mass transport from 1 m^3 of shale in one pore volume of fluid. These totals do not consider authigenic SiO_2 precipitated within the shale.

4.2 Silica Flux

Temperature is the dominant control on silica solubility in sedimentary fluids under most naturally occurring pH conditions. Harmon Member porosity loss can be equated with thermal evolution. The univariant curves (Fig. 2) for chalcedony and muscovite-K-feldspar-kaolinite were used to calculate the dissolved silica concentration, assuming an activity coefficient of one for dissolved silica (Walther & Helgeson, 1977). Combined with fluid volume, concentrations of dissolved silica were converted to the mass of silica shown in Figure 3. The dissolved silica available for transport ranges from 20 g/m^3 at chalcedony saturation to 53 g/m^3 at K-feldspar-kaolinite-illite saturation.

4.3 Kinetic Considerations

The potential for silica mass transport may also be evaluated by comparing the rates of silica precipitation in shale and sandstone. A greater precipitation rate in shale would suggest that dissolved silica is preferentially retained within the shale body at constant flow rates.The kinetic approach used here is taken largely from Rimstidt and Barnes (1980). Assuming activities of quartz and aqueous silica to be one, the rate of change in the activity of dissolved silica (r_{H4SiO4}) may be expressed as:

$$r_{H4SiO4} = (A/V)\, k_{+}\, (1-Q/K) \quad (1)$$

where r = rate of change in the a_{H4SiO4} in mol/sec, A/V = surface area in m^2 of the mineral per unit volume of fluid, k_+ = dissolution rate constant in sec^{-1}, and Q/K = the saturation state of the fluid. The postulated range of silica activities for the Harmon Member indicate that typically, pore fluids would be undersaturated with respect to amorphous silica and supersaturated with respect to quartz. Therefore, the net rate of change for dissolved silica at a given time would be controlled by the dissolution of amorphous silica and the precipitation of quartz. The saturation state, Q/K, for amorphous silica would be less than one at deposition and decrease during burial as the saturation state approaches the quartz saturation curve (Fig. 2). In contrast, Q/K for quartz would be greater than one at deposition and approach equilibrium (Q/K = 1) at about 100 °C.

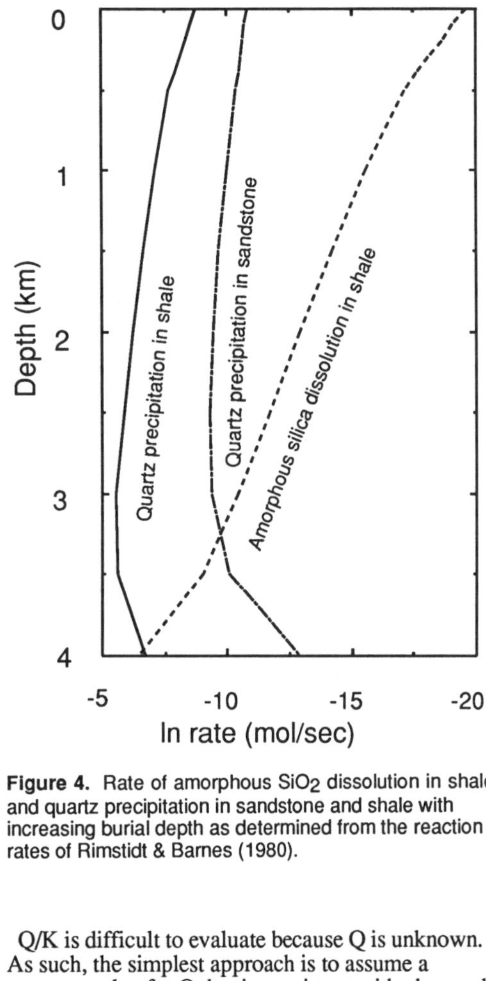

Figure 4. Rate of amorphous SiO_2 dissolution in shale and quartz precipitation in sandstone and shale with increasing burial depth as determined from the reaction rates of Rimstidt & Barnes (1980).

Q/K is difficult to evaluate because Q is unknown. As such, the simplest approach is to assume a constant value for Q that is consistent with observed pore-fluid values (Fig. 2). A value of 1.0 mm/l (log $aSiO_2$ = -3) was selected because it characterizes the range of observed silica concentrations in modern porewaters. A/V for shales and sandstones can be evaluated utilizing data from Lasaga (1984) and Wood (1989). Assuming constant grain size (neglecting the change in grain size due to overgrowths or dissolution), the change in the A/V term during burial is simply the porosity loss.

Equation 1 can now be evaluated for diagenetic conditions of the Harmon Member and the results are shown in Figure 4. The rate of quartz precipitation in shale is much greater than that of sandstone primarily because of grain size (A/V). With more rapid porosity loss in shales relative to sandstone during burial, this difference increases with increasing diagenesis and should result in an increase in authigenic quartz precipitation within the shale with time. The late diagenetic origin for much of the quartz in the Harmon Member is consistent with these calculations that suggest that as diagenesis proceeds, illitic shales become "sinks", not sources for dissolved silica.

REFERENCES

Bloch, J., 1989. Diagenesis and Rock-Fluid Interaction of Cretaceous Harmon Member (Fort St. John Group) Mudstones, Alberta and British Columbia. unp. Ph.D. Thesis, The University of Calgary, Calgary, Alberta. 165p.

Hamilton, E.L., 1976. Variations of density and porosity with depth in deep-sea sediments. Jour. Sed. Pet..46: 280-300.

Issler, D.R., 1990. A new approach to shale compaction and its geological significance. Bull. Can. Pet. Geol. 38: 168 (abs.).

Kalkreuth, W. & McMechan, M.E., 1988. Burial history and thermal maturity, Rocky Mountain front ranges, foothills, and foreland, east-central British Columbia and adjacent Alberta, Canada. AAPG Bull. 72: 1395-1410.

Kastner, M., Elderfield, H., Martin, J.B., Suess, E., Kvenvolden, K.A., & Garrison, R.E., 1990. Diagenesis and interstitial-water chemistry at the Peruvian continental margin - major constituents and strontium isotopes. In *Proceedings of the Ocean Drilling Program, Scientific Results.* Suess, E., von Huene, R., eds.112: 413-425.

Lasaga, A.C., 1984. Chemical kinetics of water-rock interactions. Journal of Geophysical Research. 89B: pp.4009-4025.

Rieke, H.H. & Chilingarian, G.V., 1974. *Compaction of Argillaceous Sediments.* Developments in Sedimentology #16. Elsevier Scientific, New York. p. 424.

Rimstidt, J.D., & Barnes, H.L., 1980. The kinetics of silica - water reactions. Geoch.et Cos. Acta. 44: 1683-1699.

Sayles, F.L., & Mannheim, F.T., 1975. Interstitial solutions and diagenesis in deeply buried marine sediments: results from the Deep Sea Drilling Project. Geoch.et Cos. Acta.39: 103-127.

Slaughter, M.1989. Quantitative determination of clays and other minerals in rocks. In *Quantitative mineral analysis of clays*. Pevear, D.R., Mumpton. F.A. (eds.). CMS Workshop Lectures. 1: 120-153

Stelck, C.R. & Leckie, D., 1988. Foraminiferal inventory and lithologic description of the Lower cretaceous (Albian) Hulcross shale, Monkman area, northeastern British Columbia. Can. Jour. Ear. Sci. 25: 793-798.

Stott, D.F. 1968. Lower Cretaceous Bullhead and Fort St. John Groups, between Smoky and Peace rivers, Rocky Mountain foothills, Alberta and British Columbia. GSC Bull. 152: p. 279.

Swart, P.K., & Burns, S.J., 1990. Pore-water chemistry and carbonate diagenesis in sediments from Leg 115: Indian Ocean. In *Proceedings of the Ocean Drilling Program, Scientific Results.* Duncan R.A., Backman, J., Peterson, L.C., et al., eds.115: 629-645.

Walther, J.V. & Helgeson, H.C., 1977. Calculation of the thermodynamic properties of aqueous silica and the solubility of quartz and its polymorphs at high pressures and temperatures. Am. Jour. Sci..277: 1315-1351.

Wood, J.R., 1989. Modelling the effect of compaction/dissolution on porosity. In *Mineralogical Association of Canada Short Course in Burial Diagenesis* . Hutcheon, I.E., ed. 15: 311-362.

Yanagi, T., Baadsgaard, H., Stelck, C.R., & McDougall, I., 1988. Radiometric dating of a tuff bed in the middle Albian Hulcross Formation at Hudson's Hope, British Columbia. Can. Jour. Ear. Sci. 25: 1123-1127.

Zachos, J.C., & Cederburg, T., 1989. Interstitial water chemistry, leg 105 sites 645, 646 & 647, Baffin Bay and Labrador Sea. In *Proceedings of the Ocean Drilling Program, Scientific Results.* Stewart, S., ed.115: 171-183.

Water-Rock Interaction, Kharaka & Maest (eds) © 1992 Balkema, Rotterdam. ISBN 90 5410 075 3

Overpressure release, cross-formational porewater flow and diagenesis

Patrice de Caritat
Department of Geology and Geophysics, University of Calgary, Alb., Canada

Julian C. Baker
Centre for Microscopy and Microanalysis, University of Queensland, Qld, Australia

ABSTRACT: Many thick, rapidly buried fine-grained sedimentary units become overpressured and retain relatively high volumes of depositional porewaters. These waters evolve chemically and isotopically by water-rock interaction during progressive burial. At some stage of the burial history of these units, overpressure may be released, allowing the evolved connate porewater to flow upward through overlying strata. There, it may become involved in further water-rock interaction, and leave behind a detectable signal of its passage.

Ankerite depleted in ^{18}O ($\delta^{18}O_{SMOW}$ +7.6 to +14.4‰) in the mid-Permian, marine Aldebaran Sandstone of the Denison Trough (Bowen Basin, Australia) is a mesogenetic phase which formed as the basin approached maximum burial (Late Triassic). Tectonic stress, which eventually resulted in the inversion of the trough, is the mechanism inferred to have triggered overpressure release in the underlying, rapidly buried, thick mudrocks and coals of the Early Permian *continental* Reids Dome beds. This allowed evolved, but still ^{18}O-depleted, "connate meteoric" porewater to flush the overlying units, where the isotopically-light ankerite precipitated.

1 INTRODUCTION

Porewater movement in sedimentary basins has the potential to redistribute mass and heat, and, therefore, is an important phenomenon to consider when diagenetic reconstruction and reservoir characterization studies are undertaken. Heat transport, mass transport, and mass balance are thus relevant considerations to take into account in the analysis of upward, cross-formational porewater flow. The importance of this type of subsurface fluid flow has been stressed in a few recent diagenetic studies (Jourdan *et al.* 1987, Burley *et al.* 1989, Sullivan *et al.* 1990, Taylor 1990, Wood & Boles 1991, Baker & Caritat *in press*, Caritat & Baker *in prep.*).

Porewater expulsion from overpressured units undergoing pressure relaxation is a potential mechanism for generating vertical, upward porewater flow. Many sedimentary basins have been noted for the development of overpressure (Fertl 1976, Hunt 1990), and it has been suggested that overpressured conditions may be the rule rather than the exception, at least in thick clayey units (Chapman 1972, Hunt 1979, Hanor 1987).

The development of overpressure in sedimentary basins can been attributed to a series of causes: disequilibrium compaction, (clay) mineral dehydration, hydrocarbon generation, pore fluid thermal expansion, and osmotic diffusion. The major cause for overpressure development in fine-grained sediments is generally attributed to disequilibrium compaction (Sharp 1976, Sharp & Domenico 1976, Chapman 1980, Keith & Rimstidt 1985, Hanor 1987, C.M. Bethke *pers. comm.* 1990), a process accompanied by undercompaction, or abnormally high porosity conservation and connate water retention.

In the present paper, we will examine, from a modeling point of view, some of the diagenetic consequences of upward, cross-formational porewater flow in sedimentary basins. Of particular interest is the capability to predict the oxygen isotopic composition of cements precipitating in sandstones that are subject to cross-formational porewater flow from below. This, in turn, requires a good understanding of the temperature conditions at the time of cementation, and hence the heat transport potential of this type of flow must be evaluated. Subsequently, we will apply these principles to a case study, where deep ankerite cementation is best explained to have taken place under conditions of upward, cross-formational flow of "connate meteoric" porewater (Caritat & Baker *in prep.*).

2 HEAT AND MASS TRANSPORT

When porewater moves in a permeable medium, such as porous or fractured sedimentary rocks, heat and mass are transported by diffusion (or conduction) through the solid framework and within the fluid phase, and by advection. The relative importance of diffusive *versus* advective transport is a function of the diffusion coefficients, the fluid flow rate and the partition coefficient of the species between solid and fluid (Bickle & McKenzie 1987).

Transport of heat and mass can be mathematically

expressed, and solved, in similar ways (Bickle & McKenzie 1987). Here, only the principles of heat transport are therefore developed, and a simple numerical solution is presented.

The conductive/advective heat transport equation, in its simplest form and in 1-D, is:

$$\frac{\partial T}{\partial t} = \frac{k}{\rho_s c_s} \frac{\partial^2 T}{\partial z^2} - \frac{\rho_f c_f}{\rho_s c_s} v_f \phi \frac{\partial T}{\partial z} \quad (1)$$

in which T is the temperature (K), t the time (sec), k the thermal conductivity of the solids ($W.m^{-1}.K^{-1}$), ρ_s and ρ_f are the solid and fluid specific density, respectively ($kg.m^{-3}$), c_s and c_f the solid and fluid specific heat capacity, respectively ($J.kg^{-1}.K^{-1}$), v_f is the fluid velocity, negative for upward flow ($m.sec^{-1}$), ϕ the porosity of the solid medium, and z the direction in which heat transport is considered, here vertical. The following substitutions:

$$\kappa = k / (\rho_s c_s) \ , \quad (2)$$

$$M = (\rho_f c_f) / (\rho_s c_s) \ , \quad (3)$$

$$W = v_f \phi \ , \quad (4)$$

where κ is called the thermal diffusivity of the solid medium ($m^2.sec^{-1}$), and W the effective fluid flux ($m.sec^{-1}$), result in:

$$\frac{\partial T}{\partial t} = \kappa \frac{\partial^2 T}{\partial z^2} - M W \frac{\partial T}{\partial z} \ . \quad (5)$$

This is the partial differential equation describing 1-D heat transport in terms of both conduction (first term of the right-hand side) and advection (second term of the right-hand side).

The solution for this type of equation can be very satisfactorily approximated by iteratively solving for the following implicit, finite-difference, Crank-Nicolson algorithm:

$$T_i^{n+1} = \frac{1}{1+r} \left\{ \frac{r}{2} (T_{i-1}^{n} + T_{i+1}^{n+1} + T_{i-1}^{n+1}) + \left(\frac{r}{2} - M W \frac{\Delta t}{\Delta z}\right) T_{i+1}^{n} + \left(1 - r + M W \frac{\Delta t}{\Delta z}\right) T_i^n \right\} \quad (6)$$

(after Gerald & Wheatley 1984). In this formulation, the space and time grids are represented by i (subscript) and n (superscript), respectively, Δz and Δt are the depth- and time-steps, respectively, where $\Delta t = r \, \Delta z^2 / \kappa$, and r is the stability factor. Choosing $r = 0.4$ results in an unconditionally stable solution.

Selected boundary conditions may be the surface temperature and the temperature at the top (or bottom) of the overpressured unit. Initial condition may be taken to be the conductive geothermal gradient between these two levels. Figure 1 illustrates, in non-dimensionalized terms, the effect of various conditions of upward, vertical porewater

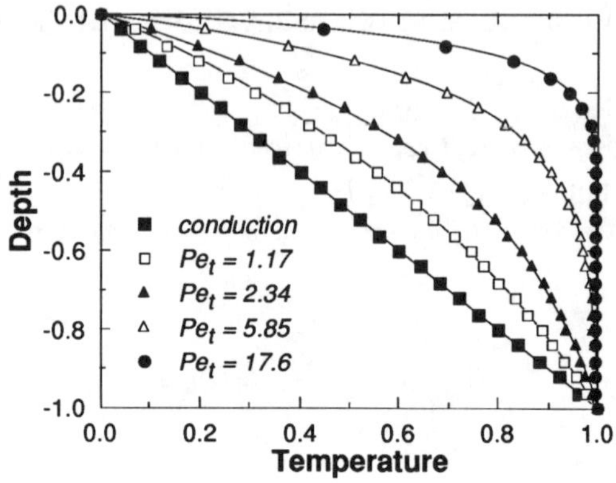

Figure 1. Effect of upward, vertical fluid flow upon temperature profile in sedimentary basins.

flow upon the equilibrium thermal profile ($t = t_\infty$) in sedimentary basins, with respect to the linear conductive geotherm (solid squares). The thermal Peclet number ($Pe_t = W h / \kappa$) used in Fig. 1 is a convenient way of expressing the relative importance of diffusion (conduction) and advection on the length-scale of interest, h (m). Pe_t varies from 0 to ∞; small *versus* large Pe_t indicates the predominance of conductive *versus* advective transport, respectively. For mass, rather than heat, transport, a chemical Peclet number can be calculated in a similar fashion.

3 MASS BALANCE

In modeling the oxygen isotopic composition of a cement, for instance a carbonate cement, precipitating from upward flowing porewater, one must take into account the mixing of the invading and *in situ* porewaters and the principle of mass balance. Once the oxygen isotopic composition of the mixed porewater has been calculated, the composition of the precipitating carbonate cement is computed according to the fractionation equation for that particular mineral, at the temperature of interest. The isotopic composition of the water after precipitation of the cement is derived from the law of mass conservation:

$$\delta_w^f - \delta_w^i = \frac{m}{w} (\delta_m^i - \delta_m^f) \ , \quad (7)$$

where δ_w and δ_m are the $\delta^{18}O$ composition (‰) of the water and precipitating mineral, respectively, and m and w the mole fractions of oxygen atoms in mineral and water, respectively, and the superscripts i or f indicate initial or final, respectively. Introducing the fractionation factor, α:

$$\alpha = (1000 + \delta_m^f) / (1000 + \delta_w^f) \ , \quad (8)$$

yields:

$$\delta^f_w = \frac{\delta^i_w + \frac{m}{w}(\delta^i_m - 1000\,\alpha + 1000)}{1 + \alpha\frac{m}{w}} . \qquad (9)$$

4 ANKERITE IN THE DENISON TROUGH

Authigenic ankerite in the gas-bearing, mid-Permian Aldebaran Sandstone of the Denison Trough (Bowen Basin, Australia) is a pore-filling and grain-replacing mesogenetic phase inferred to have precipitated at depths of around 3 km, when the sequence approached maximum burial during the Late Triassic. Despite having been partly dissolved as a result of meteoric water influx accompanying major uplift after the Late Triassic maximum burial, ankerite presently still accounts for up to 10% bulk rock volume.

This ankerite is extraordinarily depleted in ^{18}O ($\delta^{18}O_{SMOW}$ +7.6 to +14.4‰), so that temperatures in excess of 275°C would have to be inferred if one were to postulate that it precipitated from evolved connate porewaters of the *marine* Aldebaran Sandstone ($\delta^{18}O$ likely ≥ +5‰). Because depth, timing and temperature of ankerite precipitation in the Aldebaran Sandstone are well constrained by detailed petrological observations and burial, diagenetic and thermal reconstructions (Baker & Caritat *in press*), it is possible to rule out precipitation temperatures of the order of 275°C. Likewise, the possibility of meteoric influx from the surface at the time of ankerite cementation can be discredited on the basis that the sequence was then undergoing its first burial/compactional cycle, a time when porewaters are expelled upward and outward or are contained within overpressured units, but do not percolate downward from the surface.

An additional feature of interest is that the ^{18}O-composition of ankerite systematically increases from the bottom of the formation to its top (Fig. 2). Therefore, in seeking to understand the origin of this intriguing cement, one has not only to explain its very low ^{18}O-content, but also to provide an explanation for the observed isotopic trend.

The scenario which best explains these facts, and best fits the geology of the area, is as follows (Baker & Caritat *in press*, Caritat & Baker *in prep.*): the ankerite is suggested to have precipitated from porewater expelled upward, through the intervening Cattle Creek Formation aquitard, from the Early Permian Reids Dome beds. The latter unit, which exceeds 2.7 km in thickness and consists mainly of high latitude, *continental* sandstones, mudrocks and coals, was initially saturated with very ^{18}O-depleted meteoric water ($\delta^{18}O \approx$ -17‰; Bird & Chivas 1988), and is likely to have undergone overpressuring through disequilibrium compaction during rapid burial (at rates up to 1 km/Ma). Further contributions to overpressuring may have been provided by clay dewatering reactions, and gas generation from contained type III organic matter at that time. Indeed, XRD analysis of

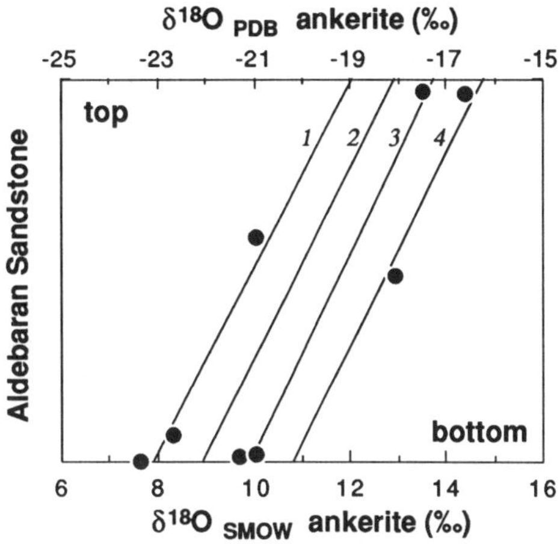

Figure 2. Ankerite $\delta^{18}O$ composition as a function of position in the Aldebaran Sandstone: observed values (solid dots) and modeling results (lines *1-4*, see text).

deep mudrock samples indicates that illite-rich interstratified illite-smectite (I/S) is present there, as opposed to more smectite-rich I/S higher in the section, and maturation modeling suggests that the Reids Dome beds had entered the gas window by the Mid Triassic. Incidentally, carbon isotopic values of the ankerite ($\delta^{13}C_{PDB}$ -9.3 to -4.9‰) indicate that part of the incorporated carbon was derived from CO_2 generated by thermal decarboxylation of organic matter.

Burial of the Reids Dome beds caused the $\delta^{18}O$ value of its connate porewater to increase by water-rock interaction, probably at least to ~ -12‰, a process well documented in many sedimentary basins. During the Late Triassic, tectonically-induced expulsion of evolved "connate meteoric" porewater out of the Reids Dome beds took place as the sequence approached maximum burial prior to major, regional uplift and folding. This water underwent mixing with heavier, evolved marine porewater as it was flushed upward through the Cattle Creek Formation, but evidently still retained a strongly meteoric isotopic signature ($\delta^{18}O$ ~ -9‰) as it entered the Aldebaran Sandstone.

Modeling of mass/heat transport, water mixing and isotopic mass balance, as detailed above, yields calculated ankerite isotopic compositions in excellent agreement with the data, both in terms of absolute values and in terms of depth-trends (Fig. 2).

A maximum vertical porewater velocity, v_f, of -6×10^{-10} m.sec^{-1} ($\approx -2\times10^{-2}$ m.a^{-1}) was estimated by assuming a maximum pressure gradient across the Cattle Creek Formation of 75 MPa and a vertical hydraulic conductivity of 10^{-11} m.sec^{-1} for this mudrock unit, in agreement with vertical conductivities of other aquitards (*e.g.*, Bredehoeft *et al.* 1983). Under these conditions, temperatures within the Aldebaran Sandstone are increased by about 5° to 10°C above the conductive geotherm values (initial condition is a geothermal gradient of 25° to 30°C/km). We have simulated the isotopic composition of ankerite using temperatures at the base of this unit ranging between 110° and 140°C. The hydraulic conductivity of the Cattle Creek Formation would, of course, have been significantly increased if the unit had been intensely fractured. On the other hand, the pressure gradient across this formation is likely to have been much less than 75 MPa, which represents the difference between lithostatic and hydrostatic pressures at a depth of ~ 6 km.

Lines *1* to *4* (Fig. 2) illustrate the output of several computer simulation runs, under the following conditions: thickness of the Aldebaran Sandstone, 730 m; temperature at the base of, and geothermal gradient through, the unit under advective heat transport conditions, 134°C and 27°C/km (lines *1-3*) or 113°C and 23°C/km (line *4*); porosity, 25%; volume of ankerite precipitating, 10%; ankerite composition, $Fe_{0.15}Ca_{0.55}Mg_{0.30}CO_3$; $\delta^{18}O$ composition of Aldebaran Sandstone porewater, +5‰; oxygen isotope fractionation equation, $1000 \ln \alpha_{(ankerite\text{-}water)} = 2.78 \times 10^6 T^{-2} + 0.11$ (Fisher & Land 1986); $\delta^{18}O$ composition of the "connate meteoric" porewater entering the base of the Aldebaran Sandstone, -9‰ (line *1*), -8‰ (lines *2, 4*), or -7‰ (line *3*); and volume of "connate meteoric" porewater, 1000 m^3 per m^2 of Aldebaran Sandstone basal area.

5 CONCLUSIONS

1. Overpressuring, which commonly occurs in rapidly buried, fine-grained sedimentary units, provides a mechanism by which anomalously high volumes of depositional connate porewater can be brought down into the subsurface;
2. Overpressure release, which can be triggered by tectonic or seismic stress, allows evolved connate porewater to flow upward through the overlying section, where it may leave a diagenetic imprint;
3. Extremely ^{18}O-depleted ankerite in the Aldebaran Sandstone is suggested to have precipitated from upward flowing "connate meteoric" porewater from the Reids Dome beds, as the latter underwent tectonically-triggered overpressure release;
4. Computer modeling can help constrain mass transport, heat transport and mass balance considerations of such porewater flow patterns.

REFERENCES

Baker, J.C. & P. de Caritat *in press*. Post-depositional history of the Permian sequence in the Denison Trough, eastern Australia. *AAPG Bull.*

Bickle, M.J. & D. McKenzie 1987. The transport of heat and matter by fluids during metamorphism. *Contrib. Mineral. Petrol.* 95: 384-392.

Bird, M.I. & A.R. Chivas 1988. Stable-isotope evidence for low-temperature kaolinitic weathering and post-formational hydrogen-isotope exchange in Permian kaolinites. *Chem. Geol.* 72: 249-265.

Bredehoeft, J.D., C.E. Neuzil & P.C.D. Milly 1983. Regional flow in the Dakota aquifer: a study of the role of confining layers. *U.S. Geol. Surv. Water Supply Paper* 2237.

Burley, S.D., J. Mullis & A. Matter 1989. Timing diagenesis in the Tartan Reservoir (UK North Sea): constraints from combined cathodoluminescence microscopy and fluid inclusion studies. *Marine Petrol. Geol.* 6: 98-120.

Caritat, P. de & J.C. Baker *in prep*. Oxygen-isotope evidence for upward, cross-formational porewater flow in a sedimentary basin near maximum burial.

Chapman, R.E. 1972. Primary migration of petroleum from clay source rock. *AAPG Bull.* 56: 2185-2191.

Chapman, R.E. 1980. Mechanical versus thermal cause of abnormally high pore pressures in shales. *AAPG Bull.* 64: 2179-2183.

Fertl, W.H. 1976. *Abnormal formation pressures*. New York: Elsevier.

Fisher, R.S. & L.S. Land 1986. Diagenetic history of Eocene Wilcox sandstones, south-central Texas. *Geochim. Cosmochim. Acta* 50: 551-561.

Gerald, C.F. & P.O Wheatley 1984. *Applied numerical analysis*. Reading: Addison-Wesley.

Hanor, J.S. 1987. Origin and migration of subsurface sedimentary brines. *SEPM Short Course* 21.

Hunt, J.M. 1979. *Petroleum geochemistry and geology*. San Francisco: Freeman.

Hunt, J.M. 1990. Generation and migration of petroleum from abnormally pressured fluid compartments. *AAPG Bull.* 74: 1-12.

Jourdan, A., M. Thomas, O. Brevart, P. Robson, F. Sommer & M. Sullivan 1987. Diagenesis as the control of the Brent sandstone reservoir properties in the Greater Alwyn area (East Shetland Basin). *Petroleum geology of north west Europe* 2: 951-961.

Keith, L.A. & J.D. Rimstidt 1985. A numerical compaction model of overpressuring in shales. *Math. Geol.* 17: 115-135.

Sharp, J.M. & P.A. Domenico 1976. Energy transport in thick sequences of compacting sediments. *Bull. Geol. Soc. Am.* 87: 390-400.

Sharp, J.M. 1976. Momentum and energy balance equations for compacting sediments. *Math. Geol.* 8: 305-322.

Sullivan, M.D., R.S. Haszeldine & A.E. Fallick 1990. Linear coupling of carbon and strontium isotopes in Rotliegend Sandstone, North Sea: evidence for cross-formational fluid flow: *Geology* 18: 1215-1218.

Taylor, T.R. 1990. The influence of calcite dissolution on reservoir porosity in Miocene sandstones, Picaroon Field, offshore Texas Gulf Coast. *Jour. Sed. Petrology* 60: 322-334.

Wood, J.R. & J.R. Boles 1991. Evidence for episodic cementation and diagenetic recording of seismic pumping events, North Coles Levee, California, U.S.A. *Appl. Geochem.* 6: 509-521.

Isotope analyses reported herein were carried out in the Isotope Laboratory, Department of Geology, University of Queensland

Water-Rock Interaction, Kharaka & Maest (eds) © 1992 Balkema, Rotterdam. ISBN 90 5410 075 3

Three zones for burial metamorphism

D. D. Eberl
US Geological Survey, Boulder, Colo., USA

ABSTRACT: A reconsideration of K/Ar measurements of shale cuttings from the Gulf of Mexico sedimentary basin reveals three zones for burial metamorphism: a shallow zone in which shales are unreactive, a middle zone in which illite is neoformed, and a deep zone in which illite recrystallizes. Age measurements from the shallow zone establish a time gradient for the entire sedimentary section. From this gradient, the amount of radiogenic Ar lost from the shales can be calculated. Radiogenic Ar loss then is used to calculate mass transfer for reactions in the deeper zones. The calculations reveal that the illitization reaction in the middle zone most likely is driven by hydrogen ions generated by the decomposition of organic matter. Illite recrystallization in response to increasing temperature in the deep zone may be driven by minimization of surface free energy (Ostwald ripening).

A reconsideration of published data (Hower et al., 1976; Aronson and Hower, 1976; Yeh and Savin, 1977) for shale cuttings from CWRU Well 6 reveals that there are at least three reaction zones for shales with increasing depth in the Gulf of Mexico sedimentary basin. Sharp boundaries between zones are defined by apparent K/Ar ages, and by the loss or gain of radiogenic Ar in the clay size fraction (<0.1 μm). In the first zone (depth <2750 m), the clay size fraction becomes progressively older with depth. The middle zone (2750 m to 4000 m) starts where the clays begin to accumulate radiogenic Ar at an accelerated rate. The clay in this zone becomes progressively younger with depth because it gains K (dissolved from the coarser size fraction) faster than it gains radiogenic Ar. The deepest zone (>4000 m) starts where the clay begins to lose radiogenic Ar.

The three zones correspond to changes in the types of chemical reactions occurring within the sediments. In the first zone inheritance is dominant. Both clay and whole rock become older because chemical reactions that transfer K between phases are minimal, and therefore radiogenic Ar accumulates. The clays become older with depth, rather than keep a constant age with depth (as would be expected from the burial of non-reacting detritus having a constant mean age inherited from the sediment source area), because the clay includes K-bearing material that formed near the time of sedimentation, either in the weathering environment (e.g. illite formed by the wetting and drying of smectite) or near the sediment-sea water interface (e.g. glauconite).

Apparent ages as a function of depth for the clays in this zone can be used to establish a time gradient for rocks in the section. Making the necessary assumptions, the time gradient for CWRU6 is 2.83 my/km, giving a total differential age of 16.44 my for the 5,810 m section drilled, a figure that is consistent with the section's middle Miocene to late Oligocene age (the top of the Miocene to the bottom of the Oligocene spans 23 my according to Kulp, 1961).

In the second zone, dissolution and precipitation reactions dominate. K-feldspar and muscovite react with hydrogen ions to produce illite, quartz, and K-ions. The K-ions thus generated react with smectite to produce additional illite and quartz, plus chlorite, Na-ions and Ca-ions. It is possible to calculate the quantity of K-bearing phases (K-feldspar and muscovite) that dissolved in this zone from the amount of radiogenic Ar that was lost, because K and radiogenic Ar occupy the same types of cation sites in the dissolving minerals, and therefore are released by dissolution in a constant proportion from rock of a given age. The moles of radiogenic Ar lost in the zone equals the moles of radiogenic Ar produced in the zone (calculated from the time gradient), minus the moles of radiogenic Ar retained (measured radiogenic Ar). The moles of K-bearing phases dissolved that are calculated by this method can be compared with the moles of feldspar and muscovite dissolved according to semi-quantitative X-ray mineralogy for the well (Hower et al., 1976). The radiogenic Ar calculation gives 0.035 moles of K-spar + muscovite dissolved in the middle zone/100 g shale, and the X-ray analysis gives a comparable value of 0.026 moles. The generation of 0.015 or 0.021 moles, respectively, of hydrogen ions/100 g shale is required to satisfy these illitization reactions. The source for these hydrogen ions presumably is shale organic matter that is decomposing in the same zone, a zone that also corresponds to the oil window.

The mass balance calculations indicate that 0.032 moles of illite/100 g shale are produced across the middle zone. In past studies the illitization reaction often has been followed by measuring the expandability of mixed-layer illite/smectite using X-ray diffraction. However, expandability measurements probably do not account for all of the illite produced, because additional illite also may be forming, for example, by epitaxial growth on detritial illite crystals, and may be crystallizing as illite particles that are not involved in swelling. Also, illite crystals may be growing larger than, and therefore out of, the clay size fraction that is used in expandability measurements.

In the deepest zone, feldspar and mica have dissolved completely, permitting the level of supersaturation with respect to illite in solution to fall to a point where illite crystals no longer can nucleate. Then illite crystal ripening reactions begin (Eberl et al., 1990): illite crystals in the finest fraction dissolve, supplying material to illite crystals growing in the coarser fraction. This situation is reversed from the middle zone, where the coarsest fraction dissolves, supplying material to the clay. Therefore in the deepest zone, radiogenic Ar is accumulating in the coarser fraction and is being lost from the finest fraction, whereas in the middle zone, radiogenic Ar is accumulating in the finest fraction and is being lost from the coarser fraction.

Expandability for illite/smectite in the deepest zone appears to be constant at about 20%, as though to suggest that reaction has ceased. But reaction has not ceased: the measurements indicate that the clays are continuing to lose radiogenic Ar with depth; the oxygen isotopes indicate that the clays are continuing to become lighter with depth; and particle size fraction analysis shows that the proportion of coarse to fine particles is increasing with depth. Reaction has not ceased, but it has slowed considerably because the reaction mechanism has changed from dissolution-precipitation to crystal ripening. Expandability measurements in the range of 20% or less are insensitive to small increases in particle thickness, and therefore could fail to indicate continued reaction.

Finally, another interesting feature of the deepest zone is that the clays do not appear to change age with depth. This observation has led to the idea of "punctuated diagenesis," in which illitization occurred in a single, short event (Morton, 1985; Ohr et al., 1991). This feature initially was discovered using Rb/Sr dating, but it also appears in the deepest zone in the K/Ar study of Aronson and Hower (1976). A more likely explanation for this feature is related to illite ripening.

The approximate age dating equation can be written:

Change in apparent age = constant x (radiogenic Ar gained - radiogenic Ar lost)/(%K_2O).

Apparent ages for the clays will not change in the deepest zone if the amount of radiogenic Ar produced by radioactive decay approximately equals the amount of radiogenic Ar lost by ripening.

REFERENCES

Aronson, J.L. and Hower, J. (1976) *Geol. Soc. America Bull.* **87**, 738-744.

Eberl, D.D., Srodon, J., Kralik, M., Taylor, B.E., and Peterman, Z.E. (1990) *Science* **248**, 474-477.

Hower, J., Eslinger, E.V., Hower, M., and Perry, E.A. (1976) *Geol. Soc. America Bull.* **87**, 725-737.

Kulp, J.L. (1961) *Science* **133**, 1105-1114.

Ohr, M., Halliday, A.N., and Peacor, D.R. (1991) *Earth and Planetary Sci. Lett.* **105**, 110-126.

Morton, J.P. (1985) *Geol. Soc. America Bull.* **96**, 114-122.

Yeh, H.W. and Savin, S.M. (1977) *Geol. Soc. America Bull.* **88**, 1321-1330.

Water-Rock Interaction, Kharaka & Maest (eds) © 1992 Balkema, Rotterdam. ISBN 90 5410 075 3

Secondary K-feldspar at the Precambrian-Paleozoic boundary, Ontario, Canada

D.A. Harper & F.J. Longstaffe
University of Western Ontario, Ont., Canada

M.A. Wadleigh
Memorial University, Canada

ABSTRACT: Secondary K-feldspar characterizes the Precambrian-Paleozoic boundary in the subsurface of southwestern Ontario. Both microcrystalline and rhombic varieties of secondary K-feldspar are present as veinlets in the weathered and altered Precambrian granitoid gneisses at the unconformity. Secondary K-feldspar is even more abundant as overgrowths on, and replacements of previously weathered grains. The K-feldspar is chemically very pure, has high $\delta^{18}O$ values (+18.9 to +21.4‰ SMOW) and shows no cathodoluminescence, all features suggestive of low temperature crystallization. K/Ar dates of 412±8 and 453±9 Ma have been obtained for the microcrystalline K-feldspar. Diagenetic K-feldspar also occurs in Cambro-Ordovician clastic and carbonate rocks that immediately overlie the Precambrian basement. Crystallization of secondary K-feldspar reflects interaction between the weathered zone at the unconformity and porewaters that travelled through this interval, early in the diagenesis of the Cambro-Ordovician rocks.

1 INTRODUCTION

In this paper, we describe development of secondary K-feldspar at the Precambrian/Paleozoic unconformity in the subsurface of southwestern Ontario, Canada. These observations extend further north-eastward the regional pattern of secondary K-feldspar occurrences known for similar settings in the Illinois, Michigan and Appalachian basins of mid-continental U.S.A. (e.g., Mensing and Faure, 1983; Hearn et al., 1987; Hay et al., 1988; Duffin, 1989; Duffin et al., 1989; Lidiak and Ceci, 1991).

Most of southwestern Ontario overlies the Algonquin Arch which divides the Appalachian and Michigan basins. The area examined in this study is located just southeast of this arch (Figure 1). Here, Cambro-Ordovician sandstones and intercalated carbonate rocks immediately overlie ≈1100 Ma old (Grenville) Precambrian basement rocks. The sedimentary rocks were deposited during a mid-continental, shallow marine transgression which blanketed the previously exposed and weathered Precambrian surface. Deposition continued throughout Paleozoic time. Presently, up to 1500 m of sedimentary rocks (Cambrian to Mississippian in age) cover the Precambrian basement in southwestern Ontario (Sanford et al., 1985). The sedimentary package thins to the northeast until the Precambrian unconformity is exposed along a topographic high known as the Frontenac Axis (Figure 1).

The Precambrian rocks have been divided into the central gneissic and central metasedimentary

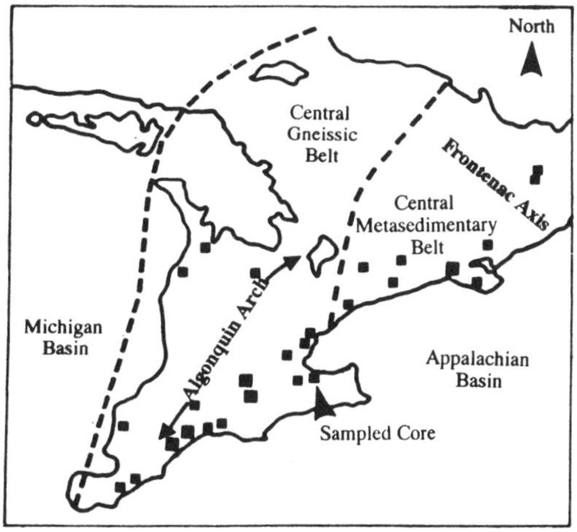

Fig. 1 Core locations. Larger squares indicate areas where numerous, closely spaced cores have been studied. The arrow indicates one core discussed in detail in this paper.

belts (Carter and Easton, 1990). Most of the fifty-nine cores examined (Figure 1) penetrate the granitoid gneisses of the central gneissic belt. Petrographic and geochemical results are reported here for a typical alteration zone, in this case a three metre thick profile encountered at the top of a tonalitic gneiss (Figure 1). The core was sampled at

35 cm intervals in the first 3 m below the unconformity and then at a 70 cm spacing.

2 FIELD AND MINERALOGICAL RELATIONSHIPS

The alteration zone that typifies the uppermost portions of Precambrian rocks in southwestern Ontario can be easily recognized by: (i) reddening and an increase in porosity, both probably the result of ancient weathering, (ii) loss of primary textures, and sericitization and chloritization of original minerals, and (iii) secondary veinlets of carbonate, chalcedony and K-feldspar. Salmon-coloured patches of K-feldspar are also commonly disseminated throughout the altered interval. Similar features have been described by Carter and Easton (1990). The extent of alteration varies on a centimetre to metre scale, and appears to be controlled by permeability and lithology. Alteration intensity is generally greatest at the contact and decreases with depth. Typically, a coarse-grained sandstone which contains clasts of Precambrian material immediately overlies the unconformity. The unconformable contact is generally gradational but can be sharp.

Secondary minerals in the alteration profile include K-feldspar, chalcedony, phyllosilicates (illite/smectite and chlorite), calcite and dolomite. Below 3 m, the weakly banded gneiss appears to be relatively unaltered. It is composed of equigranular, interlocking plagioclase and hornblende, biotite, interstitial quartz, very minor primary K-feldspar, and accessory apatite and magnetite. Some hornblende has undergone minor clay and carbonate alteration but biotite is visibly unaltered.

Three types of secondary K-feldspar, microcrystalline, rhombic (adularia habit) and alteration/replacement of weathered grains, which formed in at least two episodes, can be recognised in altered samples. Microcrystalline K-feldspar occurs as veinlets, 3.0 - 5.0 mm in width, which subsequently have been brecciated and cemented by later chalcedony and carbonate veinlets. K-Ar dates of 412 ±8 and 453±9 Ma have been obtained for the microcrystalline secondary K-feldspar. The rhombic K-feldspar, <0.2 mm in diameter, occurs scattered within these later silica and carbonate veinlets and in carbonate filled vugs. Most of the secondary K-feldspar, however, occurs as overgrowths on grains whose cores are clouded by clay and carbonate minerals. This variety of secondary K-feldspar is best developed closest to the unconformity, where porosity and permeability created during ancient weathering would have been most pronounced.

Clays present in the clouded grain cores may have first developed during weathering of the tonalitic gneiss prior to deposition of Paleozoic strata. Other clays that coat some secondary K-feldspar crystallized later. Both varieties consist of a mixture of illite/smectite (expandable layers <10 %) and Fe-rich chlorite. However, the generally finer grained (<1 μm) clay that coats K-feldspar is richer in ^{18}O by about +4 ‰ than the coarser (1-2 μm) clays typical of grain cores.

Diagenetic K-feldspar is also abundant in the immediately overlying Cambro/Ordovician mixed quartz arenites, dolomitic quartz sandstones, dolomitic arkoses and sandy dolostones (Longstaffe et al., 1990). It occurs as overgrowths, as euhedral, neoformed crystals and as fine-grained matrix in association with illitic clays. Other diagenetic minerals include chlorite, calcite, dolomite and quartz overgrowths; some lithologies also contain pelloidal glauconite. The diagenetic K-feldspar crystallized early, disrupting quartz overgrowths, coating some quartz grains and predating all but earliest grain-coating illite (Longstaffe et al., 1990).

3 GEOCHEMICAL CHARACTERISTICS OF SECONDARY K-FELDSPAR

Secondary versus primary K-feldspar can be distinguished in the tonalitic gneiss using cathodoluminescence, chemical composition and oxygen-isotope ratios. The bright blue luminescence of primary K-feldspar in unaltered or least altered samples sharply contrasts with the non-luminescent character of secondary K-feldspar in altered samples, reflecting the latter's much purer composition (e.g., absence of Ti^{4+}). The secondary K-feldspar has a near end-member orthoclase composition (Or_{99-100}), containing <0.2 wt. % Na_2O. Such compositions are very typical of low temperature precipitation (Kastner and Siever, 1979). In contrast, the primary K-feldspar is a mixture of orthoclase and albite end-members, typical of higher temperature crystallization. A low temperature origin for the secondary K-feldspar is further supported by its high $\delta^{18}O$ values (+18.9 to +21.4 ‰ SMOW). Much lower values generally characterize primary K-feldspar from granitic gneisses, including the Grenville province (e.g., +7 to +8 ‰; Shieh and Schwarcz, 1974). A hydrothermal or even lower temperature origin is also consistent with the adularia habit adopted by some crystals.

Whole rock and mineral oxygen-isotope data clearly demonstrate superposition of a low temperature event upon the tonalitic gneiss in the alteration zone. Unaltered samples have whole rock $\delta^{18}O$ values of ≈ +7 ‰, similar to results reported by Shieh and Schwarcz (1974) for the central gneissic belt. However, the whole rock $\delta^{18}O$ values rapidly increase to ≈ +19 ‰ within the alteration profile as the Precambrian unconformity is approached (Figure 2). The isotopic results for minerals show that the distribution and abundance of secondary K-feldspar largely controls the shift to higher $\delta^{18}O$ values (Figure 3). In the unaltered tonalitic gneiss, quartz, plagioclase and amphibole have oxygen-

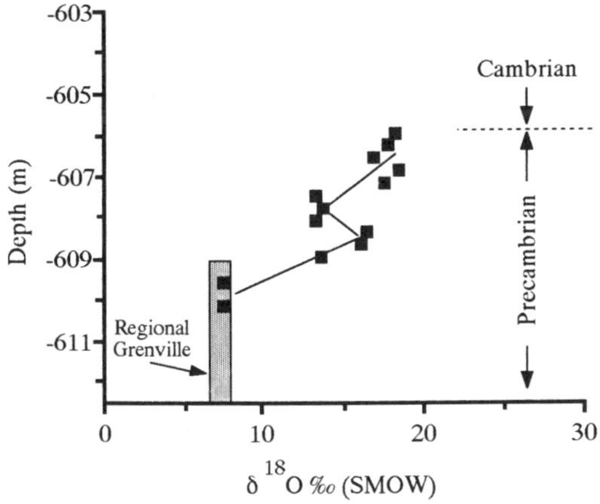

Fig. 2 Depth versus $\delta^{18}O$ values (solid squares) for whole rock samples of tonalitic gneiss. The shallowest sample was taken at the Precambrian unconformity. The shaded area represents whole rock $\delta^{18}O$ values reported by Shieh and Schwarcz (1974) for the central gneissic belt.

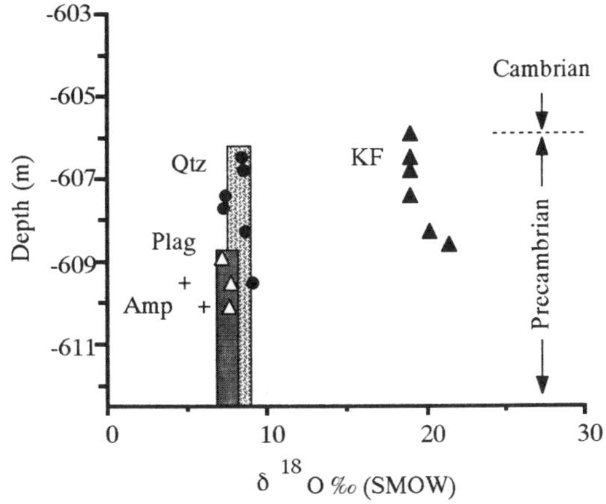

Fig. 3 Depth versus $\delta^{18}O$ values for minerals from the tonalitic gneiss: solid triangles - K-feldspar, open triangles - plagioclase, solid circles - quartz, crosses - amphibole. Light and dark areas represent quartz and feldspar, respectively, reported by Shieh and Schwarcz (1974) for the central gneissic belt.

isotope compositions typical of a high temperature mineral assemblage, and closely match analogous results given by Shieh and Schwarcz (1974) for the central gneissic belt. Amphibole and plagioclase are generally absent in the most altered parts of the core but quartz $\delta^{18}O$ values remain identical to those of unaltered samples. The secondary K-feldspar, in contrast, is out of high temperature oxygen-isotope equilibrium with quartz; the large negative $\Delta^{18}O$ values for quartz and feldspar are characteristic of low temperature, water-rock interaction.

Diagenetic K-feldspar in the immediately overlying Cambro-Ordovician sedimentary rocks is geochemically similar to its equivalent in the altered gneiss. This K-feldspar is rich in K and depleted in Na, is non-luminescent (Longstaffe et al., 1990), and has high $\delta^{18}O$ values (+17.5 to +17.9 ‰).

4 REGIONAL COMPARISON

Altered and unaltered basement granitoid lithologies throughout the mid-continental U.S.A. are similar to those of the central gneissic belt in southwestern Ontario. For example, Duffin (1989) and Duffin et al. (1989) reported secondary mineral assemblages in the Precambrian alteration profile (<1 to 8 m thick) from the Illinois basin that are very similar to those described here: secondary K-feldspar (discrete rhombs plus rims on, and replacement of pre-existing minerals), quartz, carbonates, iron oxides, Fe-rich chlorite, sericite and kaolinite. Likewise, Hay and Liu (1991) describe secondary K-feldspar from altered Precambrian granitoid rocks in Wisconsin. Here the thickness of the alteration profile ranges from 1 to 3 m. At this locality, kaolinite and secondary K-feldspar are intimately associated, suggesting perhaps that kaolinite (rather than illitic clay) was a precursor to secondary K-feldspar.

Like southwestern Ontario, secondary K-feldspar is also present in the overlying sedimentary rocks at most localities previously studied in the Illinois, Appalachian and Michigan basins. These K-feldspars are also chemically pure and have high $\delta^{18}O$ values (e.g., Hay et al., 1988; Duffin et al., 1989).

5 DISCUSSION

Our data suggest that secondary K-feldspar in southwestern Ontario formed during focusing of formation waters through the basal sandstones and weathered granitoid rocks at the Precambrian unconformity. Interaction between these porewaters and clays of weathering origin undoubtedly enhanced secondary K-feldspar precipitation, particularly if illitic clays dominated the weathering assemblage. Regardless, the occurrence of fracture-filling veinlets of microcrystalline K-feldspar suggests that the porewaters were highly saturated with K, Si and Al. Infiltration into sufficient pore space further allowed the adularia habit to be expressed and overgrowths to form.

Numerous Rb-Sr, K-Ar and Ar-Ar data in the literature provide dates for secondary K-feldspar formation throughout mid-continental North America. We discuss only a few of these results here. Our two K-Ar dates (453±9, 412±8 Ma) suggest a mid-Ordovician to late Silurian timing for micro-crystalline K-feldspar formation in southwestern Ontario. Hay and Liu (1991) report a similar K-Ar result (411±10 Ma) for K-feldspar in altered granitoid rocks from the Wisconsin locality, but a much older K-Ar date (549±18 Ma) is given by Duffin (1989) for similar rocks from the Illinois basin. Assuming that all of these samples contained only secondary K-feldspar, this disparity suggests that multiple episodes of potassic diagenesis or alteration affected the Precambrian basement rocks. Our textural data (microcrystalline versus rhombic K-feldspar varieties) support this conclusion.

The episodic nature of the potassic water-rock interaction is reinforced by the range of available K-Ar and Rb-Sr dates for secondary K-feldspar and illitic clays from the Cambro-Ordovician sedimentary rocks of mid-continental U.S.A. Diagenetic K-feldspar dates cluster around 400 Ma (like our results) whereas illitic clays range in age from 400 to 250 Ma (e.g., Hay et al., 1988; Duffin et al., 1989; Lee and Aronson, 1991).

Numerous researchers have speculated that tectonic overpressuring along the eastern continental margin of North America initiated large scale migration of warm brines which caused this ubiquitous potassic alteration throughout the mid-continental region (e.g., Hearn et al., 1987; Bethke and Marshak, 1990). If so, similar alteration events may have accompanied each of the Taconic, Acadian and Alleghanian Orogenies. In such a scenario, brines driven by the first event, the Taconic Orogeny, may have travelled at least as far north as southwestern Ontario.

6 CONCLUSIONS

Alteration, characterized by crystallization of abundant secondary K-feldspar, has severely affected the uppermost, weathered portions of Precambrian granitoid gneisses and immediately overlying Cambro-Ordovician sedimentary rocks in the subsurface of southwestern Ontario. This low temperature water-rock interaction was focused about the Precambrian unconformity, and was probably initiated early in the diagenesis of the sedimentary rocks, about mid-Ordovician time. The alteration may reflect the waning edge of large-scale, tectonically driven, episodic brine migrations proposed by others to explain similar features that are ubiquitous throughout mid-continental North America.

REFERENCES

Bethke, C.M. & S. Marshak 1990. Brine migrations across North America - The plate tectonics of groundwater. *Ann. Rev. Earth Planet. Sci* 18: 287-315.

Carter, T.R. & R.M. Easton 1990. Extension of Grenville basement beneath southwestern Ontario: lithology and tectonic subdivisions. In T.R. Carter (ed.), Subsurface geology of southwestern Ontario: a core workshop, p. 9-28. *Ont. Petrol. Inst., Amer. Assoc. Petrol. Geol. 1990 Eastern Sect. Meet., London, Canada.*

Duffin, M.E. 1989. Nature and origin of authigenic K-feldspar in Precambrian basement rocks of the North American midcontinent. *Geol.* 17: 765-768.

Duffin, M.E., M. Lee, G. deV. Klein & R.L. Hay 1989. Potassic diagenesis of Cambrian sandstones and Precambrian granitic basement in UPH-3 deep hole, Upper Mississippi Valley, U.S.A. *J. Sed. Petrol.* 59: 848-861.

Hay, R.L., M. Lee, D.R. Kolata, J.C. Matthews & J.P. Morton 1988. Episodic potassic diagenesis of Ordovician tuffs in the Mississippi Valley area. *Geol.* 16: 743-747.

Hay, R.L. & J. Liu 1991. Kaolinite was a precursor to diagenetic K-feldspar of Paleozoic age in Wisconsin. *Clay Minerals Society, 28th Ann. Meet.:* p. 70. Houston.

Hearn, P.P., J.F. Sutter & H.E. Belkin 1987. Evidence for Late-Paleozoic brine migration in Cambrian carbonate rocks of the central and southern Appalachians: Implications for Mississippi Valley-type sulfide mineralization. *Geochim. Cosmochim. Acta* 51: 1323-1334.

Kastner, M. & R. Siever 1979. Low temperature feldspars in sedimentary rocks. *Amer. J. Sci.* 279: 435-479.

Lee, M. & J.L. Aronson 1991. Repetitive occurrence of potassic diagenesis in the region of the Upper Mississippi Valley (UMV) mineral district: Implications for a persistent paleohydrological setting favorable for diagenesis. *Clay Minerals Society, 28th Ann. Meet.:* p. 98. Houston.

Lidiak, E.G. & V.M. Ceci 1991. Authigenic K-feldspar in the Precambrian basement of Ohio and its effects on tectonic discrimination of the granitic rocks. *Can. J. Earth Sci.* 28: 1624-1634.

Longstaffe, F.J., M.A. Wadleigh, S.K. Frape & R.H. McNutt 1990. Diagenetic evolution of Cambro-Ordovician clastic sedimentary rocks, southwestern Ontario. *Ont. Geol. Surv. Misc. Pap.* 150: 143-152.

Mensing, T.M. & G. Faure 1983. Identification and age of neoformed Paleozoic feldspar (adularia) in a Precambrian basement core from Scioto County, Ohio, U.S.A. *Contrib. Mineral. Petrol.* 82: 327-333.

Sanford, B.V., F.J. Thompson & G.H. McFall 1985. Plate tectonics - A possible controlling mechanism in the development of hydrocarbon traps in southwestern Ontario. *Bull. Can. Petrol. Geol.* 33: 52-71.

Shieh, Y-N. & H.P. Schwarcz 1974. Oxygen isotope studies of granite and migmatite, Grenville province of Ontario, Canada. *Geochim. Cosmochim. Acta* 38: 21-45.

Water-Rock Interaction, Kharaka & Maest (eds) © 1992 Balkema, Rotterdam. ISBN 90 5410 075 3

Illite/smectite diagenesis in overpressured and normally pressured environments, Beaufort-Mackenzie Basin, Arctic Canada

J. Ko & R. Hesse
McGill University, Montreal, Canada

ABSTRACT: The burial trends of (illite/smectite) I/S mixed-layer clays in the Beaufort-Mackenzie Basin on 215 core and drill-cutting samples from 22 wells onshore and offshore show increasing illite contents in a progressive sequence of mixed-layer types: random; mixuture of random and ordered; R1-ordered; and R>1-ordered I/S. Illitization starts at about 60°C. Illitization in overpressured shales is faster and more advanced than in normally compacted sections. Accelerated illitization in overpressure environments is probably due to higher water/particle ratios compared to normally compacted conditions (c.f. Whitney, 1990).

1 INTRODUCTION

The illitization of mixed-layer illite/smectite (I/S) clays in the Beaufort-Mackenzie Basin, one of the largest hydrocarbon habitats in Canada (estimated recoverable reserves: 1.5-2.0 billion bbl oil, 10.9-12.6 tcf gas; Dixon et al., 1988), has been scrutinized on the basis of mineralogical, geochemical and isotopic analyses of 215 core and drill-cutting samples from 22 wells onshore and offshore (Fig. 1). The Basin consists of thick, overpressured and normally pressured, progradational, deltaic sequences of Tertiary age (Fig. 2). The distribution of overpressure is interpreted to be facies-controlled (Hemingson and Carew, 1984; Hitchon et al., 1990).

2 GENERAL TRENDS

I/S clays in the Beaufort-Mackenzie Basin are characterized by four mixed-layer types: random, a mixture of random and ordered, R1-ordered, and R > 1-ordered I/S. The proportion of I-layers increases with depth, while the dominant type of I/S changes in that order (Fig. 3). The widespread occurrence of a mixture of random and ordered I/S in a transitional interval from random to R1-ordered I/S has important implications for the illitization mechanism, but has not yet been reported from other basins. A temperature of about 60°C appears to be the threshold temperature for activation of the illitization reaction based on geothermal gradients and logged temperatures in wells where the present depths may equal maximum burial depths. The increase in I-contents is fast and shows a parabolic trend in the interval of random I/S, whereas it is slow and linear in the range of ordered I/S (Fig. 4).

3 I/S DIAGENESIS IN OVERPRESSURED AND NORMALLY PRESSURED ENVIRONMENTS

Shale compaction appears to play an important role in I/S diagenesis. For instance, relatively low I-contents and composition gradients are observed in the normally pressured zone sandwiched between two overpressured zones in Reindeer D-27 (Fig. 3). Illitization in Taglu G-33 and Mallik A-06 is relatively advanced compared to Taglu C-42 where the overpressure top is about 350 m and 750 m deeper than in Taglu G-33 and Mallik A-06, respectively (Fig. 5). Accelerated illitization in overpressured environments appears to contradict Colten-Bradley's (1987) prediction based on

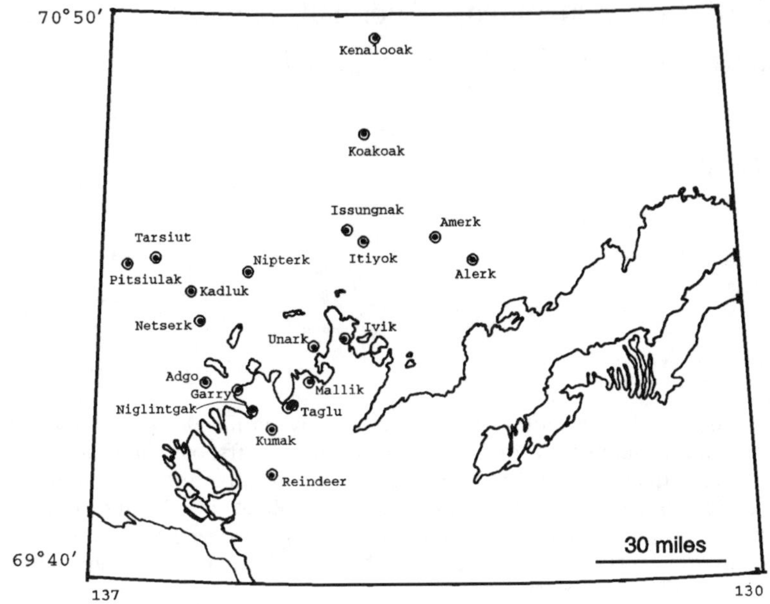

Figure 1. Well location map.

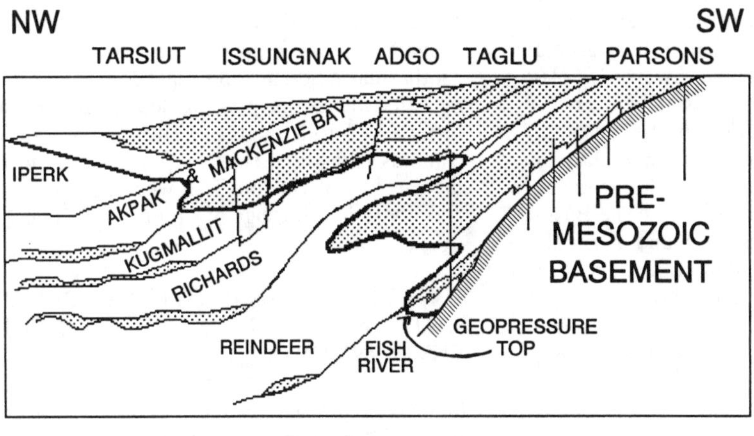

Figure 2. Distribution of overpressure (after Hemingson and Carew, 1984)

thermodynamic arguments that overpressure would inhibit the dehydration of smectite. However, dehydration of smectite in diagenetic environments is the result of chemical reactions involving layer-charge increase and absorption of cations with low dehydration energy (Ko, 1992). I/S in overpressured environments may have a higher probability of reacting with pore water solutes owing to higher water/particle ratios than under normally pressured conditions (Whitney, 1990). The latter author showed, on the basis of a series of hydrothermal experiments with water:particle ratios ranging from 1:1 to 1:20, that the higher the water content, the faster and the more advanced is the overall illitization reaction. Enhanced diagenesis is expected under these circumstances, in particular, if overpressured zones are associated

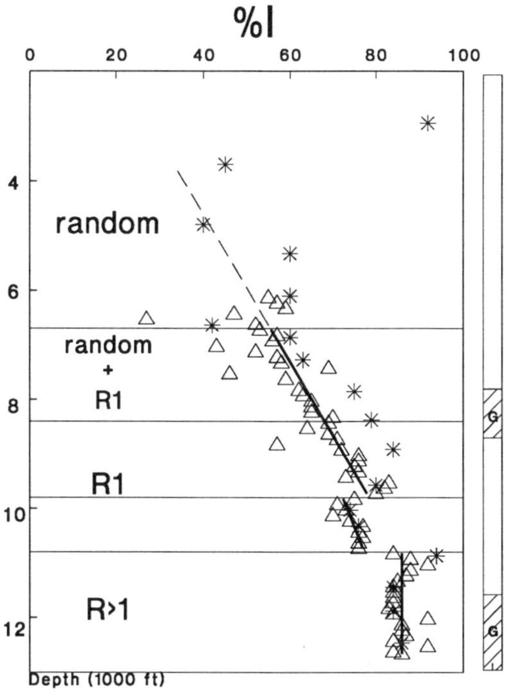

Figure 3. I/S burial trend for Reindeer D-27. * - core, Δ - drill-cutting. G - overpressured.

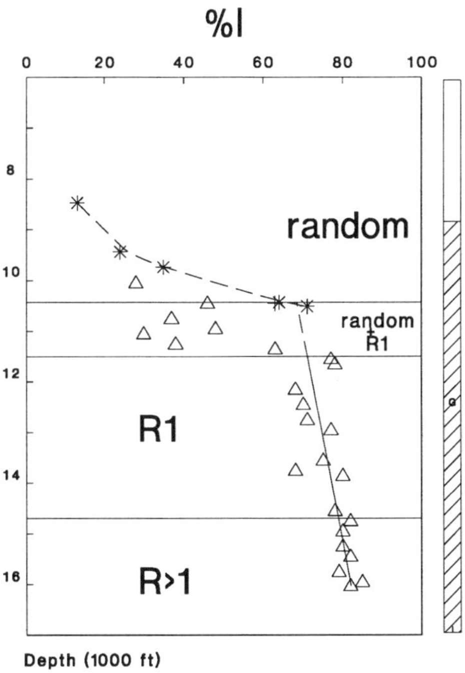

Figure 4. I/S burial trend for Taglu C-42. Symbols same as in Figure 3.

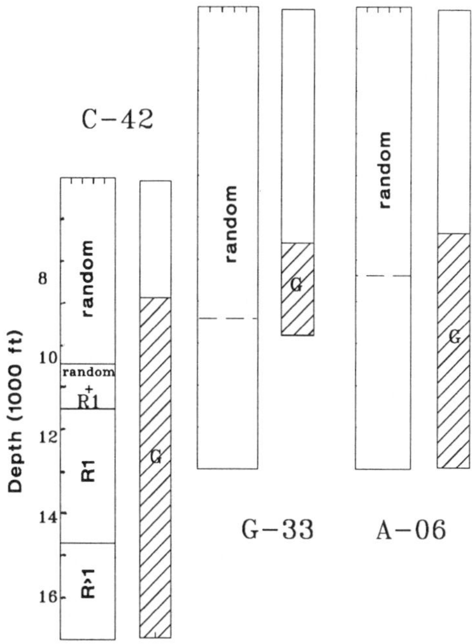

Figure 5. Relationship between overpressure and I/S diagenesis. G - overpressured.

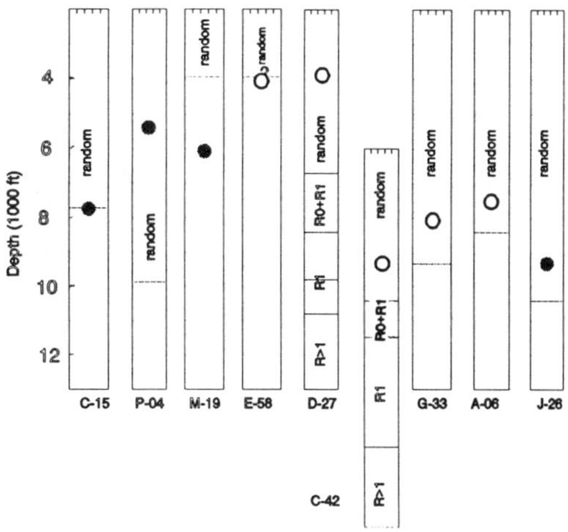

Figure 6. Relationship between hydrocarbon occurrence and I/S diagenesis. Filled circle: oil-dominant, empty circle: gas-dominant.

with elevated salinity compared to normally pressured zones as is the case in some Gulf-of-Mexico Tertiary sections (Kharaka et al., 1978), although many overpressure zones show lowered salinities.

There appears to be no obvious, direct relationship between overpressure and hydrocarbon occurrences (Hemingson and Carew, 1984), though most hydrocarbons were discovered in overpressure-related structures (growth faults and shale diapirs). Hydrocarbons occur above, within and below overpressured zones. However, the distribution of hydrocarbons seems to be related to I/S diagenesis. Hydrocarbons appear to occur always above or in the top portion of the active illitization zone, i.e. within the interval of random I/S, but rarely below the depth where ordered I/S appears (Fig. 6). There is a close relationship between hydrocarbon occurrence and I/S diagenesis. Deep hydrocarbons are found in wells where illitization starts at relatively great depth (Taglu C-42, G-33, Mallik A-06, Ivik J-26). In wells where the apparent depth of illitization is shallow due to uplift, caused either by growth faults, diapirs or tectonics, hydrocarbons also occur at shallow depths (Reindeer D-27, Kumak E-58, Niglintgak M-19, Garry P-04). In offshore wells where the geothermal gradients are 5-10°C/km higher (Judge and Bawden, 1987), and thus the depth of illitization is 1 to 1.5 km shallower than onshore, many hydrocarbon discoveries were made at depths of 1-2 km.

REFERENCES

Colten-Bradley, V.A. 1987. Role of pressure in smectite dehydration- Effects on geopressure and smectite-to-illite transformation. *American Assocication of Petroleum Geologists Bull.* 71: 1414-1427.

Dixon, J., Morrell, G.R., Dietrich, J.R., Proctor, R.M. & Taylor, G.C. 1988. Petroleum resources of the Mackenzie Delta-Beaufort Sea. *Geological Survey of Canada. Open File* 1926: 74p.

Hemingson, P. and Carew, W. 1984. *Hydrocarbons & geopressures in the Beaufort-Mackenzie Basin.* National Conference on Earth Sciences, Geopressures and Hydrocarbon Occurrences: Banff, Canada.

Hitchon, B., Underschultz, J.R, Bachu, S. & Sauveplane, C.M. 1990. Hydrogeology, geopressure and hydrocarbon occurrences, Beaufort-Mackenzie Basin. *Bull. Canadian Petroleum Geology* 38: 215-235.

Judge, A. & Bawden, W. 1987. Geothermal gradients. In Marine Science Atlas of the Beaufort Sea: Geology and Geophysics. *Geological Survey of Canada. Miscellaneous Report* 40, p.7.

Kharaka, Y.K., Carothers, W.W. & Brown, P.M. 1978. Origins of water and solutes in the geopressured zones of the Northern Gulf of Mexico Basin. *Society of Petroleum Eingineers of AIME* 7505, 8 pp.

Ko, J. 1992. *Illite/smectite diagenesis in the Beaufort-Mackenzie Basin, Arctic Canada.* Ph.D. thesis: McGill University, Montreal, Canada.

Whitney, G. 1990. Role of water in the smectite-to-illite reaction. *Clays & Clay Minerals* 38: 343-350.

Water-Rock Interaction, Kharaka & Maest (eds) © 1992 Balkema, Rotterdam. ISBN 90 5410 075 3

Isotopic, mineralogic and chemical evidence for multiple episodes of fluid movement during prograde and retrograde diagenesis in a Proterozoic basin

T.G. Kotzer & T.K. Kyser
Department of Geological Sciences, University of Saskatchewan, Saskatoon, Sask., Canada

ABSTRACT: In the Athabasca Basin, northern Saskatchewan, Canada, Proterozoic sandstones unconformably overlie Archean and Aphebian metasedimentary basement rocks. Multiple fluid events, involving isotopically and chemically distinct fluids, have produced clay and silicate mineral assemblages in permeable sandstone aquifers and along fault zones. $\delta^{18}O$ values of contemporaneous quartz-hematite and fluid inclusions in quartz overgrowths in sandstones indicate formation from a basinal fluid having moderate salinities (5-15 wt% NaCl) and temperatures (120 to 170°C) and a $\delta^{18}O$ value of 0±1°/oo. Altered sandstones and metasedimentary rocks proximal to the fault zones host high grade U, Ni, Au, Co, Cu, PGE, mineralization and contain an assemblage of 2M/3T illite, 1M kaolinite, hematite, carbonate, dravite, euhedral quartz and Al-Mg chlorite. $\delta^{18}O$ values of quartz-illite and illite-hematite, δD values of illite, kaolinite and dravite and fluid inclusions in the euhedral quartz indicate formation from saline (30 wt % NaCl), high temperature (180-240°C) fluids with $\delta^{18}O$ and δD values of 3+/-3 and -60+/-10, respectively. $^{18}O/^{16}O$ and D/H ratios in Al-Mg chlorite in the metasedimentary basement rocks in the alteration haloes indicate formation from a basement-derived fluid having $\delta^{18}O$ and δD values of about 7±2 and -30+/-10, respectively. $\delta^{18}O$ and δD values of late kaolinite in re-activated fault zones in the sandstones and basement rocks indicate formation from low-temperature (50 to 25°C) fluids having $\delta^{18}O$ and δD values of -16 and -130, repectively, similar to modern meteoric waters in the area. In the faults, previously formed illite, Al-Mg chlorite and dravite have aberrantly low δD values without concurrent shifts in their $\delta^{18}O$ values indicating substantial interaction and proton exchange with relatively modern meteoric waters at low temperatures and high integrated water/rock ratios. Clay and hydrous silicate minerals with aberrantly low δD values also show concurrent decreases in K and Mg contents, increases in H_2O contents and variably reset K-Ar and Rb-Sr systematics.

1 INTRODUCTION

Mature sandstone sequences in the Proterozoic Athabasca Basin (Fig.1), northern Saskatchewan, Canada have been affected by a number of isotopically and chemically distinct fluids which are recorded in the chemical and isotopic compositions of fluid inclusions, and clay, silicate and uranium minerals. A number of high-grade fault-controlled, unconformity-type U-Ni-Co-As-Cu-Au-PGE ore deposits have formed at the intersections of faults and the unconformity between Archean-Aphebian metasedimentary basement rocks and the overlying sandstones due to mixing of chemically and isotopically diverse fluids (Hoeve and Quirt, 1984; Wilson and Kyser, 1987; Kotzer and Kyser, 1990, 1991). Throughout the history of the basin, the paleolatitude and isotopic compositions of the fluids which have affected the basin have changed markedly through time, thus making it an ideal candidate for tracing fluids using stable isotope systematics (Wilson et al. 1987; Kotzer et al. 1991).

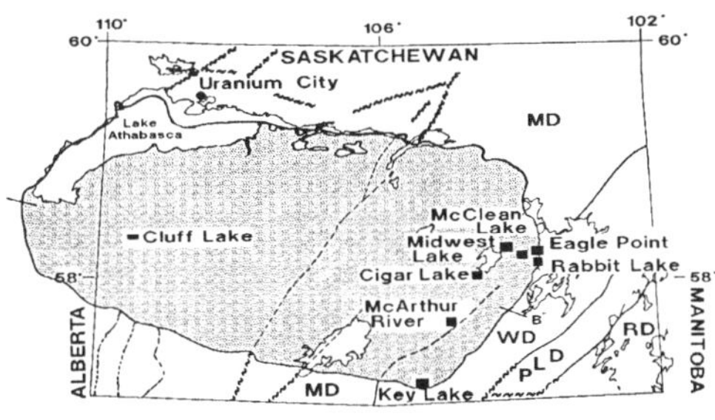

Fig.1 Map showing the extent of the Athabasca Basin, locations of U deposits and major lithostructural domains in the crystalline basement of northern Saskatchewan, Canada. (After Kotzer and Kyser, 1991).

2 ATHABASCA BASIN: GEOLOGY AND ALTERATION MINERAL PARAGENESIS

The Athabasca Basin in northern Saskatchewan, Canada, is a large intracratonic sandstone basin formed in the Mid-Early Proterozoic during the waning stages of the Hudsonian Orogeny (Hoeve and Quirt 1984; Kotzer et al. 1991). The basin fill consists of sequences of Helikian poly-cyclic, mature, fluvial to marine quartz-sandstones which were deposited in a near-shore, shallow shelf environment and are separated from the underlying basement gneisses by an unconformity (Ramaekers 1979). Large northeast-southwest trending faults rooted in the basement rocks and extending into the sandstones, represent boundaries between major crustal units (Hoeve and Quirt 1984). U-Pb and K-Ar ages and δD and $\delta^{18}O$ values on uranium minerals and illite indicate that these faults have remained active to Recent times (Kyser et al. 1990).

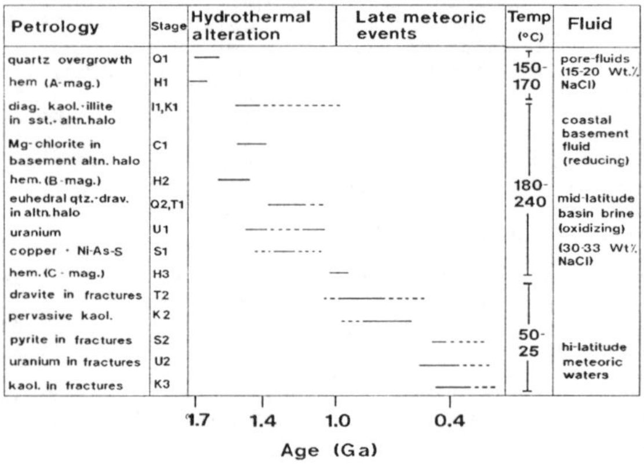

Fig. 2 Fluid-mineral-age relationships for various minerals within the Athabasca Basin determined from petrographic, fluid inclusion and stable and radiogenic isotope analyses. (After Kotzer and Kyser, 1991).

Field relationships, detailed XRD, SEM and petrographic and isotope studies indicate that an assemblage of kaolinite, 2M-3T illite, dravite, euhedral quartz, sudoite, clinochlore, hematite and various amounts of uranium (I1-K1, H1, C1, Q2-T1, U1 - Fig.2) is associated with prograde diagenesis, basin-basement fluid mixing, and hydrothermal alteration at 1200 to 1500 Ma (Wilson et al. 1987; Kotzer et al. 1991). Clinochlore (Fe-Mg-chlorite) occurs throughout the underlying metasedimentary rocks, whereas sudoite (Mg-Al chlorite) is mainly restricted to areas of hydrothermal alteration in the faulted basement rocks. Interstital 2M illite (I1) also occurs with kaolinite (K1) as a regional assemblage in diagenetically altered sandstones. Overprinting the diagenetic and hydrothermal minerals in reactivated faults is a late alteration assemblage consisting of kaolinite (K3) with various amounts of remobilized uranium (U2), and chemically and isotopically altered illite (I1) and chlorite (C1) (Fig. 2).

3 TEMPERATURES OF MINERAL FORMATION

Fluid inclusion microthermometry and stable isotope geothermometry on paragenetically equivalent quartz, silicate, oxide and clay minerals indicate that the Athabasca sediments have been subjected to high-temperature, prograde diagenesis from 1650 to 900 Ma involving moderate to high salinity basinal fluids (120 to 240°C, 10 to 36 wt.%NaCl; Q1, Q2 quartz - Figs.2,3; Kotzer and Kyser 1990), followed by retrograde alteration from late Precambrian to Recent, involving low-temperature and low-salinity fluids (20 to 50°C, <10 wt% NaCl; K3 - Figs.2,3; Kotzer and Kyser 1990). $\delta^{18}O$ values of intergrown quartz and illite in areas of early hydrothermal alteration yield similar temperatures of approximately 200°C (Fig. 3). In many parts of the basin, $\delta^{18}O$ values of coeval quartz, illite, kaolinite, dravite, and hematite minerals consistently yield oxygen isotope temperatures between 180 to 230°C suggesting that the highly saline, basinal fluids were homogeneous over large areas (Fig.3; Kotzer and Kyser 1990; Pagel et al. 1980). On the basis of their Al^{4+} contents, hydrothermal chlorites in sandstones and basement rocks formed at similar temperatures of approximately 180°C (Fig.3; Cathlineau and Niveau 1984).

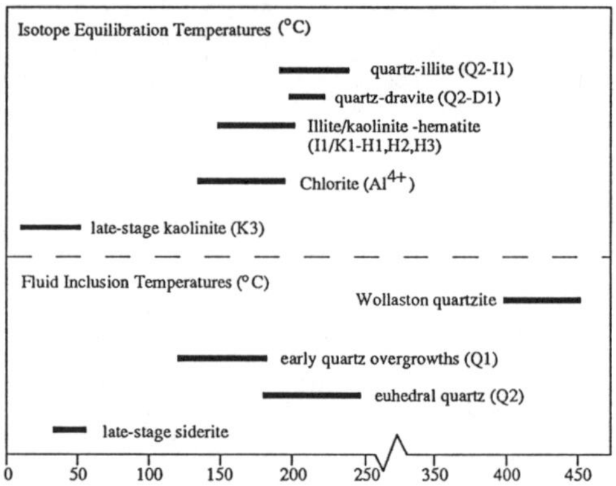

Fig.3 Temperatures of mineral formation determined by oxygen isotope geothermometry, fluid inclusion microthermometry and chemical compositions from electron microprobe analysis.

4 CALCULATED $\delta^{18}O$ AND δD VALUES OF THE EQUILIBRATION FLUIDS

$\delta^{18}O$ and δD values of fluids in equilibrium with paragentically (Fig. 2) and thermally (Fig. 3) characterised clay and silicate minerals plot in four distinct fields (Fig. 4).

4.1 200°C basinal fluids

Illite and kaolinite in hydrothermally and diagenetically altered sandstones have stable isotopic compositions indicating formation from a 200°C fluid having $\delta^{18}O$ and δD values of 4±2 and -60±20°/oo, respectively (#1 - Fig.4). Distinct Rb-Sr ages of 1477 and 957 Ma for the illites indicate the 200°C basinal fluid events were episodic over a time interval of approximately 600 million years (Kotzer and Kyser 1990) Isotopically, these fluids are comparable to formation fluids found in mature, mid-latitude basins of variable age such as the Alberta Basin (Clayton et al. 1966; Kharaka and Carouthers 1986; Wittrup and Kyser 1990).

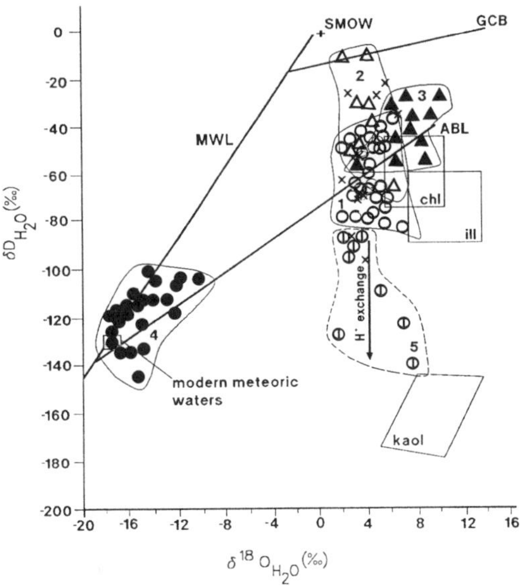

Fig. 4 Various fluids associated with the Athabasca Basin. Large boxes represent measured $\delta^{18}O$ and δD values for clay and silicate minerals as indicated. Numbers and symbols in fields represent calculated $\delta^{18}O$ and δD values of fluids in equilibrium with; (○#1) illite; (Φ#5) altered illite; (Δ #2) clinochlore; (▲ #3) sudoite and (x) dravite, all at temperatures of 200°C, and (● #5) kaolinite at temperatures of less than 50°C. Also shown are values for modern meteoric waters (MWL), low-latitude waters in Gulf Coast basins (GCB), and mid-latitude waters in the Alberta Basin (ABL).

4.2 Hydrothermal basement fluids

At temperatures of 180°C, fluids forming clinochlore have $\delta^{18}O$ and δD values of approximately +2 to +4 and -15 to -45°/oo, respectively (#2 - Fig.4). At similar temperatures, sudoite formed from fluids having δD values from -25 to -60 and were ^{18}O-enriched with $\delta^{18}O$ values between 7 to 9°/oo (#3- Fig.4). The $\delta^{18}O$ and δD values for these fluids are similar to those found low-latitude basins (Kharaka and Carouthers 1986) which differ in isotopic composition from the values of the contemporaneous basinal fluids. Petrographic relationships indicate that sudoite formed after clinochlore, therefore, $\delta^{18}O$ differences of up to 4 °/oo between the fluids forming the sudoite and clinochlore suggests ^{18}O-enrichments in the basement fluids as hydrothermal alteration of the siliceous metasedimentary rocks proceeded.

4.3 Basin-basement fluid mixing

Basin-basement fluid mixing, whereby ascending reducing basement fluids mixed with metal-bearing saline basinal fluids in hydrothermally altered faults hosting uranium mineralization, is evidenced by merging of the isotopic compositions of the basement and basin fluids (#1,2,3 - Fig. 4). The predominance of stable isotopic compositions of the basin fluids suggest high basin/basement fluid ratios prevailed in most cases.

Dravite in the faults is in textural equilibrium with both euhedral quartz and illite. Assuming the isotope fractionation properties of dravite are similar to illite (Kotzer et al. 1992), at temperatures of 200°C, dravite has formed from fluids having $\delta^{18}O$ values of 4±2°/oo and δD values which vary from -30 to -80°/oo (Fig.4). The δD values of the dravite-forming fluids overlap those of the fluids forming chlorite and illite and suggest that dravite has formed from mixing of variable amounts of the basement and basin fluids.

4.4 Late-stage meteoric waters

Late-formed kaolinite in reactivated faults formed from low-temperature (<50°C) fluids having δD and $\delta^{18}O$ values of -130±10 and -16±2°/oo, respectively, which are similar isotopically to post-Cretaceous Athabascan meteoric waters (#4 -Fig.4). Some of the late-formed kaolinites in the fracture systems have formed from fluids having isotopic compositions that lie along a line connecting the isotopic compositions of the 200°C saline, basinal fluids with those of modern-day meteoric fluids in the Athabasca area. This trend is similar to that found in mature sedimentary basins where the δD and $\delta^{18}O$ values of the formation fluids reflect mixing between a relatively modern meteoric water and an evolved basin fluid (Clayton et al. 1966).

Illites in the vicinity of fault zones hosting the late kaolinite have $^{18}O/^{16}O$ ratios which are identical to those of the unaltered illites distal to the fractures which have formed from the 200°C basinal fluids, but have D/H ratios indicating formation from a fluid that was substantially more deuterium depleted (#5 - Fig. 4). Abberantly low δD values for fluids in equilibrium with the illites in fractures can be attributed to extensive post-formational proton exchange with the same low-temperature, meteoric waters that percolated into the reactivated faults and precipitated the late kaolinite (Wilson et al. 1987; Kotzer and Kyser 1991).

5 WATER/ROCK RATIOS

Modelled water/rock (w/r) ratios in areas of hydrothermal alteration in the Athabasca Basin are based on the extent of oxygen isotope exchange. Due to mineral transformations and fracture-controlled fluid flow, these calculations represent minimum w/r ratios only (Taylor 1974). Calculated w/r ratios for basin and basement fluids are approximately 0.5 to 1.0 and 0.3 to 1.0, respectively, suggesting that the hydrothermal systems were rock-dominated. In reactivated faults, the presence of substantial amounts of kaolinite having δD and $\delta^{18}O$ values indicating formation from meteoric waters that have not been isotopically modified by interactions with the surrounding sandstones suggests the faults were fluid-dominated with w/r ratios as high as 10.

6 CHEMICAL AND ISOTOPIC EVIDENCE FOR RETROGRADE ALTERATION

Illite, dravite and chlorite minerals in and proximal to reactivated faults containing late kaolinite have anomalously low δD values in relation to their $\delta^{18}O$ values. Crystallinities of these minerals remain unchanged even at the XRD, SEM and TEM scale suggesting that hydrogen isotope exchange between the minerals and low-temperature meteoric waters has occurred without significant recrystallization. In addition to very low δD values, retrograde alteration of the minerals is further characterised by, 1) increased amounts of structural water, 2) increased FeO contents in the dravites and chlorites and, 3) decreased K_2O contents and reset K-Ar ages in illites (Wilson and Kyser, 1987; Kotzer and Kyser, 1991).

7 DISCUSSION AND CONCLUSIONS

Detailed XRD, SEM, petrographic, stable and radiogenic isotope analyses from clay and silicate minerals whose paragenesis is well-defined indicate that the Athabasca Basin has had a long and protracted fluid history involving at least three isotopically distinct fluids. Early diagenetic basinal fluids which evolved into high salinity basinal brines with increasing water/rock interactions and mixing between these saline basinal fluids and reducing basement fluids in faults have produced high temperature assemblages of illite, chlorite, dravite, kaolinite, quartz and high-grade uranium mineralization. Near reactivated fault zones, incursion of chemically and isotopically distinct meteoric waters has overprinted the early mineral assemblages, resulting in massive kaolinite formation and chemically and isotopically altered illite, dravite and chlorite. Pronounced post-formational decreases in the δD values of illite, chlorite and dravite without concurrent oxygen-isotope exchange and physical recrystallisation suggests that extensive hydrogen-isotope exchange has occurred in fault zones having high integrated water/rock ratios.

8 REFERENCES

Cathlelineau, M. and D.Nieva 1985. A chlorite solid solution geothermometer: the Los Azufres (Mexico) geothermal system. *Contrib. Mineral. Petrol.*, 91: 235-244.

Clayton, R., I. Friedman, D. Graf, T. Mayeda, W. Meents and N. Shimp, 1966. The origin of saline formation waters, 1. Isotopic composition. *J. Geophys. Res.*, 71: 3869-3882.

Hoeve, J., and D. Quirt 1984. Mineralization and host-rock alteration in relation to clay mineral diagenesis and evolution of the middle-Proterozoic Athabasca Basin, northern Saskatchewan, Canada. *Sask. Research Council Tech. Report #187.*

Kharaka, Y. and W. Carouthers 1986. Oxygen and hydrogen isotope geochemistry of deep basin brines. In (eds) P. Fritz and J. Fontes, *Handbook of Environ. Isotope Geochem.*, vol.2: 306-360. Elsevier: New York.

Kotzer, T. and T.K. Kyser 1990. Fluid history of the Athabasca Basin and its relation to uranium deposits. In *Summary of investigations 1990. Sask. Energy and Mines, Sask. Geol.Surv., Misc. Rep. 90-4*: pp. 153-157.

Kotzer, T. and T.K. Kyser 1991. Retrograde alteration of clay minerals in uranium deposits: radiation-catalyzed or simply low-temperature exchange? *Isotope Geosci.*, 86: 307-321.

Kotzer, T., T.K Kyser and E. Irving 1991. Use of stable isotopes and paleomagnetism to determine the isotopic and hydrologic evolution of fluids in a Proterozoic basin. *Geol. Soc. Amer., Pgm w. Abstr.*, A437.

Kotzer, T., R.W. King, T.K. Kyser and R. Kerrich 1992. An emperical oxygen and hydrogen isotope geothermometer for quartz-tourmaline and tourmaline-water. Submitted to *Geochim. Cosmochim. Acta.*

Kyser, T.K., T. Kotzer and T.I. Sibbald 1990. Oxygen, U-Pb and Pb-Pb isotope systematics in uraninite from complex U-Au-PGE vein-type and unconformity-type U deposits in northern Saskatchewan. In *Summary of investigations 1990. Sask. Energy and Mines, Sask. Geol. Surv., Misc. Rep. 90-4*: pp.64-70.

Pagel, M., B. Poty and S. Sheppard 1980. Contributions to some Saskatchewan uranium deposits mainly from fluid inclusion and isotopic data. In (eds) S.Ferguson and A. Goleby, *Uranium in the Pine Creek Geosyncline, IAEA: 639-654, Vienna.*

Ramaekers, P. 1979. Stratigraphy of the Athabasca Basin. In *Summary of investigations 1979. Sask. Energy and Mines, Sask. Geol. Surv., Misc. Rep. 79-10*: pp. 154-160.

Taylor, H.P. 1974. The applications of oxygen and hydrogen isotope studies to problems of hydrothermal alteration and ore deposition. *Econ. Geol.*, 69: 843-883.

Wilson, M.R. and T.K. Kyser 1987. Stable isotope geochemistry of alteration associated with the Key Lake uranium deposit. *Econ. Geol.*, 82: 1450-1557.

Wilson, M.R., T.K Kyser, H.H Mehnert, H.H. and J. Hoeve 1987. Changes in the H-O-Ar isotopic composition of clays during retrograde alteration. *Geochim. Cosmochim. Acta*, 51: 869-878.

Wittrup, M.B. and T.K. Kyser 1990. The petrogenesis of brines in Devonian potash deposits of western Canada. *Chem. Geol.*, 82: 103-128.

Water-Rock Interaction, Kharaka & Maest (eds) © 1992 Balkema, Rotterdam. ISBN 90 5410 075 3

The mechanism of porosity reduction, styolite development and quartz cementation in North Sea sandstones

E.H.Oelkers
Laboratoire de Géochimie, Université Paul Sabatier, Toulouse, France

P.A.Bjorkum
STATOIL, Stavanger, Norway

W.M.Murphy
Southwest Research Institute, San Antonio, Tex., USA

ABSTRACT: Petrographic investigation of the Jurassic North Sea sandstones indicates that porosity reduction at depths from 2.5 to 4.5 km is dominated by the chemical redistribution of silica. Although the driving force for this redistribution is commonly believed to be pressure solution, textural evidence indicates that silica transport in this system stems from the enhanced solubility of quartz adjoining mica surfaces, mainly at the edges of mica rich stylolites. This increased quartz solubility is likely the consequence of a local pH increase arising from the interaction of mica surfaces with the adjoining pore fluids. To better characterize this process, chemical mass transport calculations were carried out which simultaneously account for local equilibrium among solute species and diffusional mass transport coupled to kinetically controlled mineral dissolution/precipitation reactions. Results of these calculations indicate that a solution pH increase of 0.4 or less at the mica/quartz interface could account for all of the quartz cementation and porosity reduction in the Jurassic North Sea sandstones at depths from 2.5 to 4.5 km, and thus might also be the mechanism of stylolite development in many other sedimentary basins.

1. INTRODUCTION

Description of stylolite formation, porosity reduction, and quartz cementation in sandstones has been problematical. Calculations suggest that significant supply or removal of silica by advection from thick sedimentary sandstones would require orders of magnitude higher pore water fluxes than are likely in compactional regimes of sedimentary basins (Bjorlykke, 1979). Numerous attempts have been made to develop a non-advective mechanism to account for the chemical redistribution of quartz in the sandstones found in sedimentary basins. Most of these mechanisms rely on pressure solution to provide the driving force for silica dissolution and precipitation (Heald, 1955, Blatt, 1979). However luminescence petrography commonly indicates that much of what appears to be intergranular pressure solution in sandstones is in fact boundaries between precipitated silica cement (Bjorlykke et al., 1989). To better understand the mechanism of stylolite formation and quartz cementation in the Jurassic petroleum reservoir rocks of the North Sea, a combined petrographic/computational investigation has been carried out, which suggests that the driving force of this process is an increase of pH in the pore fluids adjoining quartz/mica interfaces. The purpose of the present communication is to summarize the results of this study, and to describe the development of stylolytes in deep sandstones of the North Sea sedimentary basin.

2. SUMMARY OF PETROGRAPHIC OBSERVATIONS

Porosity reduction in the Jurassic reservoir sandstones from the North Sea is dominated by mechanical processes at depths to 2.5 km, where most of this sand is loose due to lack of cementation. Significant amounts of silica cement and stylolite formation are apparent at depths from 2.5 to 3.5 km; and at a depth of 3.5 to 4.5 km, the stylolites are well developed and quartz cementation is extensive (Bjorlykke et al. 1989). The temperature of the pore fluids in these rocks is approximately 80° C at depths of 2.5 km, increasing to about 140° C at 4.5 km. The stylolites, which are generally parallel to the original bedding, consist primarily of muscovite with some interspersed clay fragments and minor heavy minerals. The distance between stylolites varies apparently randomly from 5 to 50 cm. The distribution of quartz cement in these sandstones is apparently independent of distance from the stylolite at core temperatures less than 120° C, but is strongly controlled by the distance from the stylolites at higher temperatures. For example, in a core taken from a depth corresponding to 170° C, the porosity was 3 to 4 percent within a centimeter of the stylolites, but 17 percent at a point equidistant between two stylolites 50 cm apart. There exists no textural evidence to support significant dissolution of quartz at quartz-quartz interfaces, though these surfaces are frequently flattened owing to the precipitation of quartz cement. The only place in this system which shows evidence of quartz dissolution is at mica/quartz interfaces. An example can be seen in Figure 1, which depicts the dissolution of a sand grain in contact with detrital muscovite. Note that although this sand grain exhibits a dramatic amount of dissolution at this interface, the mica shows no evidence of chemical alteration. Thus any mechanism describing the development of stylolites and quartz cement must account for the lack of dissolution of quartz at quartz/quartz interfaces, insubstantial silica cement at temperatures below 80° C, no muscovite alteration at mica/quartz interfaces, dramatic quartz dissolution at mica/quartz interfaces, and the dependence of silica cement on distance from the stylolite at high temperatures.

One possible mechanism stems from the interaction of mica surfaces with the adjoining pore fluids. An

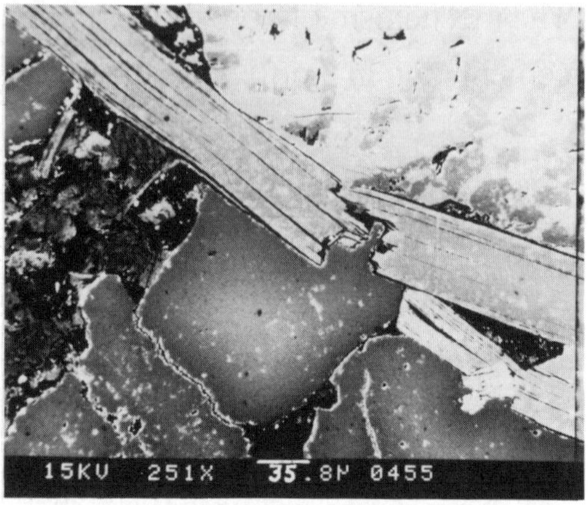

Fig. 1. Photomicrograph depicting interaction between mica and quartz grains from the Oseberg Formation of the North Sea at a depth of 2.8 km.

inherent positive charge on the muscovite surface would attract negatively charged ions, such as OH^-, creating an ionic concentration gradient from the bulk solution to this surface. Through the application of Guoy-Chapman double layer theory Boles and Johnson (1983) demonstrated that the charged surface of mica could result in a change in the pH of the adjoining saline pore waters by as much as 0.6 log units, without otherwise affecting the mica. A pH increase would then enhance the solubility of quartz adjacent to mica grains through the formation of the species $HSiO_3^-$ according to the reaction

$$SiO_2 + OH^- \rightleftharpoons HSiO_3^- \quad . (1)$$

This excess dissolved silica could then diffuse away and precipitate at some distance from this interface. The overall rate of silica cementation from this process would be simultaneously controlled by the amount of pH increase at the stylolite interface, the diffusion of dissolved silica away from the interface, and kinetically controlled dissolution/precipitation reactions occurring at the surfaces of the quartz grains.

3. MASS TRANSPORT CALCULATIONS

To assess the consistency of this quartz cementation mechanism with

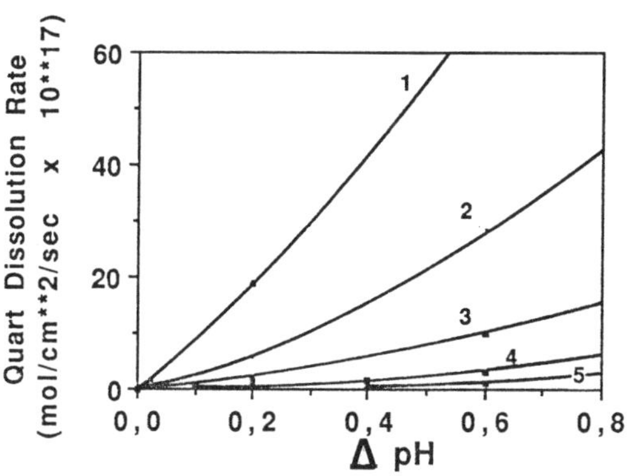

Fig. 2. Computed rate of quartz dissolution at the stylolite interface as a function of Δ pH. The curves labeled 1, 2, 3, 4, and 5, refer to the results of calculations corresponding to depths/temperatures of 4.5 km/140°C, 4.0 km/125°C, 3.5 km/110°C, 3.0 km/ 95°C, and 2.5 km/80°C, respectively (see text).

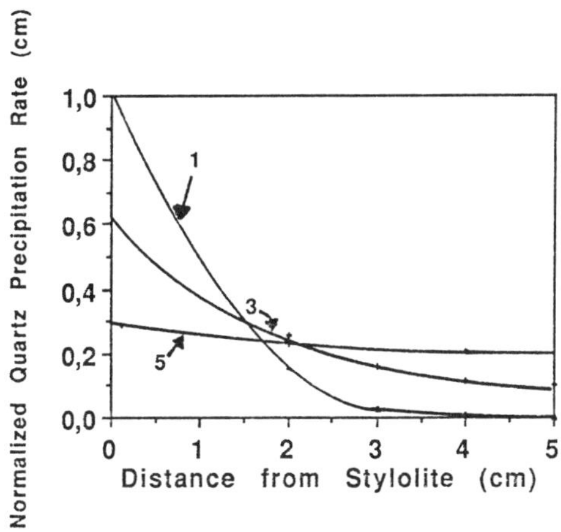

Fig. 3. Quartz cementation rate (mol /cm^3/sec) normalized by the quartz dissolution rate at the stylolite interface (mol/cm^2 /sec) as a function of distance from the stylolite. The curves labeled 1, 3, and 5 correspond to depths/temperatures of 4.5 km/ 140°C, 3.5 km/110°C, and 2.5 km/80°C, respectively.

experimental and theoretically generated kinetic, thermodynamic and transport parameters, chemical mass transfer calculations were carried out. These calculations are based on the approach described by Lichtner (1988). The combined mass transport/mass action/mass conservation equations are simultaneously solved for each point on a one dimensional finite difference grid in a sandstone oriented perpendicular to a mica rich stylolite. Taking account of symmetry, this system is assumed to be closed, bounded by a zero flux boundary at the midpoint between two stylolites, and the stylolite itself is assumed to be an impervious boundary. The sandstone is assumed to consist of pure quartz with a uniform porosity of 0.25 at a depth of 2.5 km, decreasing to 0.10 at a depth of 4.5 km. The pH of the bulk pore fluid is assumed to be neutral, but the pH at the mica rich stylolite was fixed at higher values. Standard state thermodynamic properties for the minerals and aqueous species were generated from the SUPCRT92 computer code (Johnson et al. 1992), activity coefficients for aqueous species were computed from the Hückel equation, kinetic rate constants for quartz dissolution/precipitation were taken from Murphy and Helgeson (1989), and the diffusion coeffients for the aqueous species were set to be equal to those of OH^- as reported by Oelkers and Helgeson (1988). The tortousities in the sandstone was taken to range from 2 to 4, increasing with depth. The surface area for quartz dissolution at the stylolite interface was taken to be 2 cm^2/cm^2, but the surface area in the sandstone itself was taken to be 62 cm^2/cm^3, consistent with an average grain size of 0.5-1.0 mm. Steady state solutions to the combined governing equations were generated at distinct depths in the sedimentary column for a characteristic 10 cm wide sandstone unit bounded by parallel stylolites.

The results of these calculations are summarized in Figures 2 and 3. Figure 2 depicts the computed rate of quartz dissolution at the sandstone/stylolite interface as a function of Δ pH, the pH increase at the stylolite interface. A quartz dissolution rate of 0.5 to $12.x10^{-17}$ mol/cm^2/sec at the stylolite interface is consistent with petrographic and field observations. It can be inferred from this figure that a pH increase of 0.4 log units or less at the stylolite/sandstone interface is sufficient to account for all of the observed quartz cementation and porosity reduction in the North Sea reservoir sandstones at depths from 2.5 to 4.5 km. Note that the computed rate of quartz dissolution at this interface is too slow to permit any significant quartz cement at temperatures of less than 80° C.

As soon as the excess dissolved silica is transported away from the mica surface, it is converted to the SiO_2 species in the neutral bulk solution. The resulting solution is supersaturated with respect to quartz, which precipitates as controlled by growth kinetics. A gradient in the SiO_2 concentration is generated which drives the excess silica towards the interstylolite zone. In accord with Murphy et al. (1989), the overall rate is dominated by the relatively slow growth kinetics at lower temperatures, where diffusion redistributes silica evenly between the stylolites. In contrast, at higher temperatures, the overall rate is dominated by relatively slow diffusion and more quartz is precipitated near the source of the excess silica at the stylolite interface. This effect is quantified for this system in Figure 3, which shows the distribution of quartz cement, represented by its relative precipitation rate, as a function of distance from the stylolite. It can be seen in this figure that the distribution of quartz cement is nearly uniform at temperatures/depths of 80° C and 2.5 kb, but tends to be controlled by distance from the stylolite at higher temperatures. It is likely that the major effect of the presence of hydrocarbons in the pore spaces is to increase the tortuosity, leading to a less homogeneous distribution of quartz cement.

4. CONCLUSIONS

The petrographic observations and results of chemical mass transport calculations presented above indicate that stylolite formation, porosity reduction, and quartz cementation in the North Sea reservior sandstones at depths from 2.5 to 4.5 km may be the consequence of a increase of pH in pore fluids at mica/quartz interfaces rather than pressure solution. Further efforts are planned to better characterize this mass transport mechanism.

ACKNOWLEDGEMENTS

We would like to thank Harold C. Helgeson, Knut Bjorlykke, and Jacques Schott for helpful discussions and suggestions during the course of this study.

REFERENCES

Bjorlykke,K. 1979. Cementation of sandstones. *J. Sed. Petrol.* 49:1358-1359.

Bjorlykke,K., M.Ramm, & G.C.Saigal 1989. Sandstone diagenesis and porosity modification during basin evolution. *Geol. Rund.* 78:243-268.

Blatt,H. 1979. Digenetic processes in sandstone. In: Aspects of diagenesis. *Soc. Econ. Paelo. Min. Spec. Pub.* 26:141-157.

Boles,J.R. & K.S.Johnson 1983. Influence of mica surfaces on pore water pH. *Chem. Geol.* 43:303-317.

Heald,M.T. 1955. Stylolites in sandstones *J. Geol.* 63:101-114.

Johnson,J.W., E.H.Oelkers & H.C. Helgeson 1992. SUPCRT92 A software package for calculating the standard molal properties of minerals gases, aqueous species, and reactions among them from 1 to 5000 bars and 0° to 1000°C.*Comp.Geosci.* (in press).

Lichtner,P.C. 1988. The quasi-stationary state approximation to coupled mass transport and fluid-rock interaction in porous media. *Geochim.Cosmochim.Acta* 52:143-165.

Murphy,W.M. & H.C.Helgeson 1989. Thermodynamic and kinetic constraints on reaction rates among minerals and aqueous solutions. IV. Retrieval of rate constants and activation parameters for the hydrolysis of pyroxene, wollastonite, olivine, andalusite, quartz, and nepheline. *Am.J.Sci.* 298:17-101.

Murphy,W.M., E.H.Oelkers & P.C. Lichtner 1989. Surface reaction versus diffusion control of mineral dissolution and growth rates in geochemical proceses.*Chem.Geol.* 78: 357-380.

Oelkers,E.H. & H.C.Helgeson 1988. Calculation of the thermodynamic and transport properties of aqueous species at high pressures and temperatures:Aqueous tracer diffusion coefficients of ions to 1000°C and 5 kb.*Geochim.Cosmochim. Acta* 52:63-85.

Water-Rock Interaction, Kharaka & Maest (eds) © 1992 Balkema, Rotterdam. ISBN 90 5410 075 3

Mathematical simulation of water-mineral interaction and fluid flow with respect to the diagenetic evolution of sandstones

R.Ondrak
Institute of Petroleum and Organic Geochemistry, Research Center Jülich, Germany

ABSTRACT: Porosity and permeability is changed by dissolution and precipitation of minerals. An important question is whether diagenetic processes take place under open or closed system conditions.

Fluid flow in an evolving basin can be caused by meteoric water inflow, thermal convection or compaction. Fluid flow causes the transport of solutes and implies diagenetic processes under open system conditions.

Calculation of mineral dissolution involves the solution of coupled, nonlinear equation systems for which numerical solutions have to be found. A silica-carbonate system is studied as an example. The numerical model developed for this system is then used to calculate mineral solubilities and the resulting changes in porosity and permeability.

Changes in permeability affect the flow rate through the sediment thus creating a complicated feedback between fluid flow and mineral reactions providing an interesting case for modeling diagenetic processes in sandstones.

However, it can also be shown that modeling of diagenetic evolution of sandstones may be limited due to the complicated interaction of the phases involved and the limitations regarding the input parameters. The results, therefore, represent an average statistical trend rather than a prediction of real distribution of porosity and permeability.

1. INTRODUCTION

Porosity is an important reservoir property. For this reason much attention has been paid to the diagenetic processes affecting this rock property. Several attempts have been made to quantify the process of cementation and dissolution in sandstones due to quartz redistribution (Giles, 1987; Bjorlykke, 1983; Wood & Hewett, 1982; Leder & Park 1986). However, none of the models includes a reaction term describing the interaction between the solid phase and the pore-fluid. It is usually assumed that the pore fluid is in chemical equilibrium with respect to temperature and pressure. Precipitation or dissolution of minerals is then induced by changes in the chemical equilibrium of the pore-fluid due to changed P-T conditions. In order to overcome this restriction of the models Ondrak and Bayer (1990) have developed an algorithm, which allows to model the changes of reservoir properties resulting from both mass transfer due to fluid flow and the reaction between the solid and the fluid phase.

2. QUARTZ SOLUBILITY

Quartz dissolves in water forming monosilicic acid (H_4SiO_4) (reaction <1>), which can dissociate into ($H_3SiO_4^-$) and ($H_2SiO_4^{2-}$) (reactions <2> and <3>).

$$SiO_2(s) + 2H_2O <==> H_4SiO_4 \quad <1>$$
$$H_4SiO_4 <==> H_3SiO_4^- + H^+ \quad <2>$$
$$H_3SiO_4^- <==> H_2SiO_4^{2-} + H+ \quad <3>$$

The total amount of silica in solution ($SiO_{2(aq)}$) is the sum of the concentrations of the different silica species, H_4SiO_4, $H_3SiO_4^-$, and $H_2SiO_4^{2-}$:

$$\Sigma\, SiO_{2(aq)} = [H_4SiO_4] + [H_3SiO_4^-] + [H_2SiO_4^{2-}] \quad (1)$$

The dissolution of quartz and the resulting amount of dissolved silica is therefore determined by reaction <1> which is independent of the pH of the solution and reactions <2> and <3> which depend on the pH of the solution. The dissociated species can be neglected for pH values less than 9 (Blatt et al., 1982). This is important, as most ground and formation waters fall between pH values of 5-8.5 (Collins, 1975).

3. CALCITE SOLUBILITY

The system $CaCO_3$-CO_2-H_2O, can be described by the following reactions:

$$CO_{2(g)} + H_2O \iff H_2CO_3 \quad <4>$$
$$H_2CO_3 \iff H^+ + HCO_3^- \quad <5>$$
$$HCO_3^- \iff H^+ + CO_3^{2-} \quad <6>$$
$$H_2O \iff H^+ + OH^- \quad <7>$$
$$CaCO_{3(S)} + H_2CO_3 \iff Ca^{2+} + 2HCO_3^- \quad <8>$$

In addition the requirement of electro-neutrality must be met. The charge balance equation (2) must be applied to assure that the number of charges carried by cations is equal to the number of charges carried by anions.

$$[H^+]+2[Ca^{2+}]= [HCO_3^-]+2[CO_3^{2-}]+[OH^-] \quad (2)$$

In the $CaCO_3$-CO_2-H_2O system two cases must be distinguished, depending on whether a CO_2 containing gaseous phase is present or not. If the fluid phase is in contact with a CO_2 containing atmosphere, it is called an open system and reaction <4> must be included in the calculations. If, however, the fluid phase is not in contact with a CO_2 containing atmosphere, the mass balance for the carbonate species must be satisfied and any change in the carbonate content can only be derived from the reaction of calcite with the fluid.

$$[Ca^{2+}]-[Ca^{2+}]_{(0)} = \Sigma CO_2 - \Sigma CO_{2(0)} \quad (3)$$

where

$$\Sigma CO_2 = [H_2CO_3]+[HCO_3^-]+[CO_3^{2-}] \quad (4)$$

The subscript (0) stands for the initial concentration before any changes took place.

Open system conditions consequently describe the setting for subsurface regions where CO_2 is released due to maturation of organic matter, whereas closed system conditions describe the environment of diagenetic processes where no source for CO_2 other than minerals is available.

4. COUPLED DIAGENETIC SYSTEMS

For the numerical solution of the system SiO_2-$CaCO_3$-CO_2-H_2O under closed system conditions with respect to CO_2, calculations have been split into reactions which take place in the fluid due to changes in temperature, mixing of fluids of different composition, or changes in composition due to mineral reactions, and reactions between fluids and solids. The different systems are solved separately ensuring exact mass balance of each species participating in the reaction. The coupling is achieved by iterative solution of the equations. This procedure allows the quantification of changes in concentration of all species, which is important for the correlation of porosity development and fluid composition.

The change in porosity with time due to dissolution or precipitation of minerals is proportional to the specific molar volume (Vsp) of the dissolved mineral. Porosity (Ø) evolution will be calculated from the product of the change in the absolute concentration (Ø C) in the fluid and the specific molar volume (Vsp) of the mineral (equation 5).

$$d\text{Ø}/dt = d(\text{Ø } C)/dt \; Vsp \quad (5)$$

5. GEOLOGICAL MODEL

A simple geological model is used for the simulations. A gently dipping, confined sandstone layer is bounded by two faults. The geothermal gradient is constant throughout the basin and time. Temperature increases with depth along the dip of the sandstone layer. Thus the temperature gradient along the flow path will vary depending on the slope of the layer. The thickness of the layer is small with respect to its lateral extension. The model is therefore reduced to a 2-D system. Fluid flow due to meteoric water inflow is moving up dip within the layer. Fluxes are further controlled by the permeability evolution due to dissolution or precipitation. The new permeability is calculated as a function of porosity using the Kozeny-Karman Equation (e.g. von Engelhardt, 1960).

6. COMPUTED EXAMPLES

The first example (fig. 1) shows the porosity evolution of a pure quartz sandstone with unevenly distributed porosity. The simulation shows how the initial porosity distribution determines the further porosity evolution. Starting from areas with relatively low

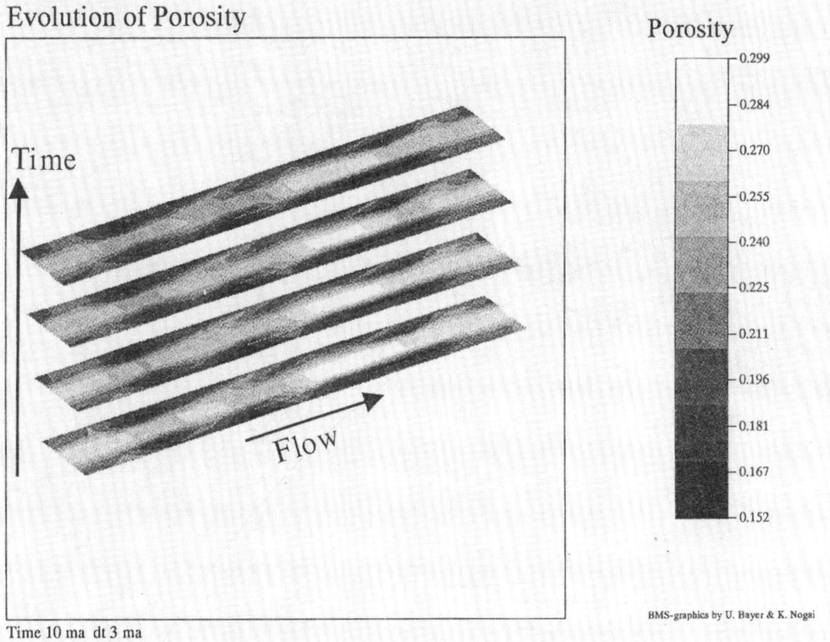

Figure 1: Porosity evolution of a sandstone with time. Fluid flow enters the sandstone layer form right precipitating quartz due to cooling as the fluid moves updip..

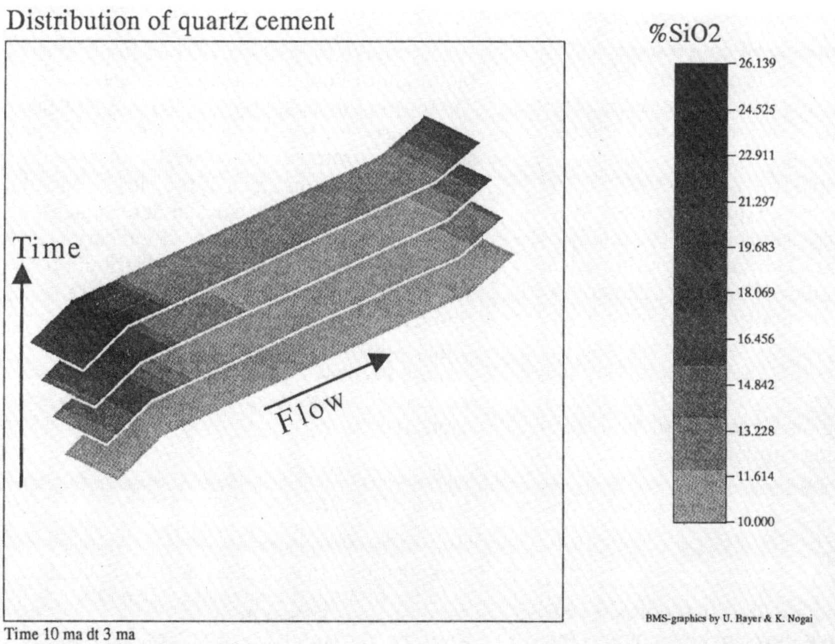

Figure 2: Temporal and spatial distribution of quartz cement. Fluid flow enters the sandstone layer form right precipitating quartz due to cooling as the fluid moves updip.

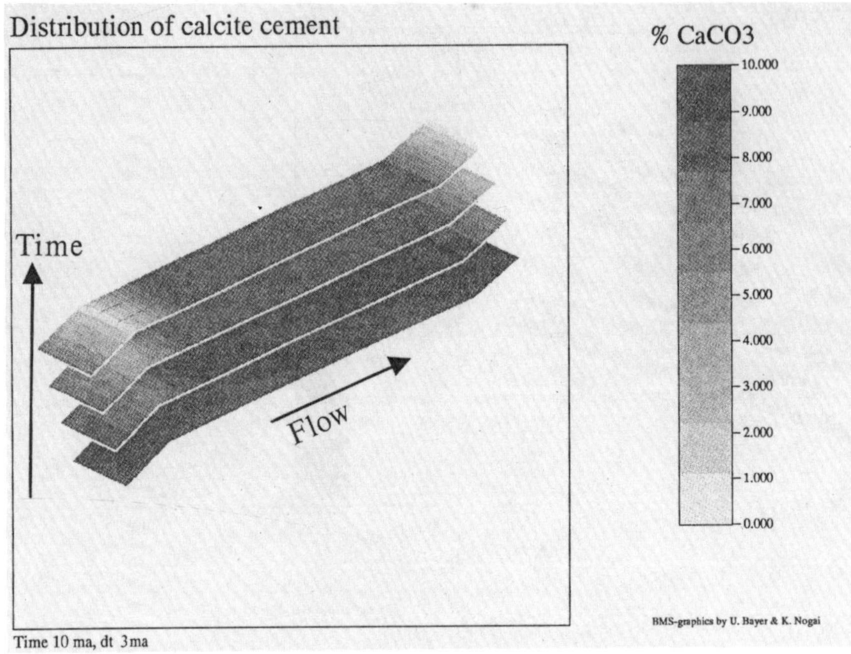

Figure 3: Temporal and spatial distribution of calcite cement. Fluid flow enters the sandstone layer form right dissolving calcite due to cooling as the fluid moves updip.

porosities, porosity is continuously reduced leading to the formation of concretions.
As these concretions are growing into areas of initially higher porosity they reduce the flow rate in these areas. This in turn results in a relative preservation of porosity.

The second example (fig. 2 & 3) shows the redistribution of calcite and quartz cements in a homogeneous quartz sandstone with initially 10% calcite cement. The simulation shows that calcite cement is replaced by quartz cement as the pore fluid moves updip and cools and, as a result dissolves calcite while quartz is precipitated. It also shows that these minerals are only redistributed in those areas where the temperature changes along the layer is relatively high, while almost nothing happens in areas with small temperature changes.

7. CONCLUSIONS

The model calculations show that porosity redistribution due to fluid flow can be an important factor in the diagenetic evolution of sandstones. While redistribution of quartz depends mainly on the temperature gradient along the layer, redistribution of calcite is also controlled by the chemical composition of the pore fluid. In this case a source for CO_2 is another important factor controlling the solubility of calcite. Therefore, the geometry and the temperature history used for the simulation, as well as a possible CO_2 source, must be well constrained. Another important factor is the facies depending porosity distribution because it determines the flow distribution in the layer. Great care must be put to describe the initial facies distribution. Above all, one should keep in mind that the simulation can only be interpreted as an average trend. Depending on the grid spacing, only general porosity trends can be accounted for while small scale changes must necessarily be neglected.

REFERENCES

Bjorlykke, K. 1983. Diagenetic reactions in sandstones. In A. Parker & B.W. Sellwood (eds), Sediment Diagenesis: 169-213, Dordrecht: Reidel Publ. Co.

Blatt, H., G. Middleton & R. Murray 1982. Origin of sedimentary rocks. Englewood Cliffs: Prentice Hall.

Collins, A.G. 1975. Geochemistry of oilfield waters. Amsterdam: Elsevier.

Engelhardt, W. v. 1960. Der Porenraum der Sedimente. Berlin: Springer Verlag.
Giles, M.R. 1987. Mass transfer and problems of secondary porosity creation in deeply buried hydrocarbon reservoirs. Marine & Petrol. Geol. 4: 188-204.
Leder, F. & W.C. Park 1986. Porosity reduction in sandstone by quartz overgrowth. AAPG Bull., 70: 1713-1728.
Ondrak, R. & U. Bayer 1990. Dissolution and cementation in basin simulation. (in publication)
Wood, J.R. & T.A. Hewett 1982. Fluid convection and masstransfer in porous sandstones - a theoretical model. GCA 46: 1707-1713.

Water-Rock Interaction, Kharaka & Maest (eds) © 1992 Balkema, Rotterdam. ISBN 90 5410 075 3

Experimental studies of mudstone diagenesis

Elspeth I.C. Rae & David A.C. Manning
Department of Geology, The University, Manchester, UK

ABSTRACT: Within sedimentary basins organic and base metal rich mudstones are regarded as potential sources of cement and ore forming solutes, organic acids and CO_2 (which may influence secondary porosity development) and petroleum fluids. However the behaviour of mudstones in detail during diagenesis is poorly understood. A series of experiments using fluid sampling vessels (150 - 310 °C, 50 MPa) is in progress to determine yields of solutes and organic matter for a range of mudstone source rocks, including samples from the North Sea petroleum province. Compositions of fluids sampled during experiments are controlled by mineral assemblages. Although a range of polar organic species has been identified (including ketones and short chain organic acids) no evidence has been found for their involvement in mineral solubility enhancement.

1 INTRODUCTION

A series of heating experiments has been carried out in Dickson-type rocking bombs to evaluate the potential yields of inorganic and organic solutes during mudstone diagenesis. Dickson-type rocking bombs use gold and treated titanium sample containment to minimise catalytic breakdown of products. These bombs allow the withdrawal of fluid samples during an experiment without quenching P and T so that the progress of reactions may be monitored.

2 RESULTS

Initial experiments were carried out using carbonate rich mudstones (Kimmeridge Blackstone Band, Jurassic Dorset, U.K. and Marl Slate, Permian Co. Durham) known to be rich in organic matter and trace metals. Carbonate reactions dominated fluid compositions in these experiments. In Marl Slate heating experiments at 310°C the assemblage dolomite + kaolinite + quartz reacted to form calcite and Mg-smectite. Na fixing from the NaCl brine used in some experiments resulted in the formation of feldspar and analcime. Na-fixing was also seen in the corresponding Kimmeridge Blackstone Band experiments, resulting in feldspar formation. Similar reactions have been noted as occurring during short term water-rock interactions (e.g. Hutcheon, 1984; Hutcheon et al., 1981; Milodowski et al., 1989), and during burial diagenesis (e.g. Hutcheon et al., 1980; Zen, 1959; Muffler and White, 1969)

Although polar organic species such as ketones and organic acids (see figure 1) were identified, no evidence was found for inorganic - organic interactions

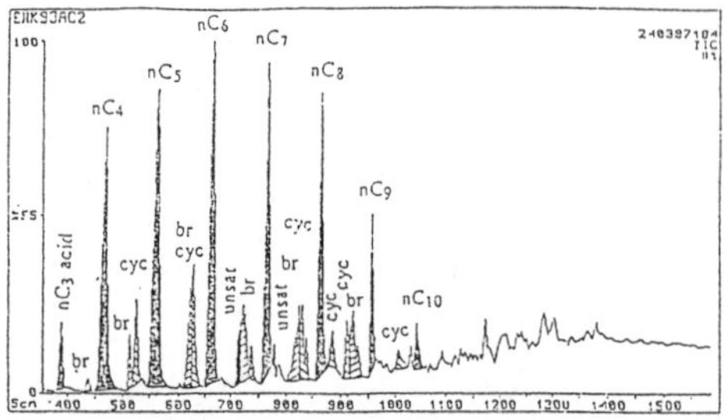

Figure 1 Short chain carboxylic acids tentatively identified by GCMS in samples taken from Kimmeridge Blackstone Band heating experiments. Peaks are labelled in the following manner nC_4 etc. = straight chain acid of designated chain length; unsat., cyc., br. = unsaturated, cyclic or branched carboxylic acids respectively.

which have been postulated as increasing aluminosilicate dissolution.

More recent experiments used mudstones with very low carbonate contents. Fluid compositions appear to be again buffered by mineral assemblages (see later section on geochemical modelling).

In early experiments oxygen fugacity was neither measured nor controlled and so attempts were made in later experiments to constrain the fugacity of dissolved gases in the system. In some experiments, deionised or distilled water was boiled prior to use in the bomb. The increase in measured pH after boiling was assumed to reflect the exsolution of some dissolved gases. As a matter of routine, the gold cell/titanium seal assembly is flushed with compressed gas (usually N_2) to ensure that the seal has been made. Excess gas is then squeezed out of the assembled components by deforming the gold cell.

By repeating the flushing process several times, it was hoped to displace dissolved CO_2 and O_2 in favour of N_2 which should be inert towards the reactants. Measured pH after flushing rose compared to that of the original DIW and as Figure 2 shows was higher than that measured on the boiled fluid. This was again assumed to reflect the exsolution of dissolved gases. An earlier experiment (abbreviated title 975) in which boiled, distilled water had been used was repeated (same P, T, mudstone, F/R ratio) using N_2 to flush the cell assembly for the purposes of comparison. This experiment is designated 9752. Data from the

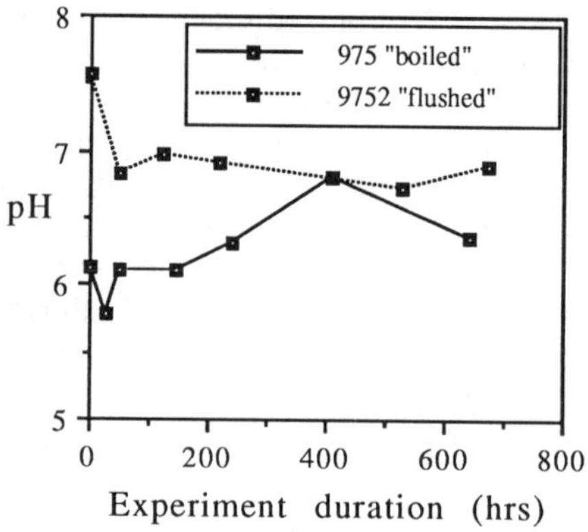

Figure 2 pH of experiments 975 and 9752 measured on samples taken throughout the experiments. 9752 is essentially a repeat of 975 (for details see text)

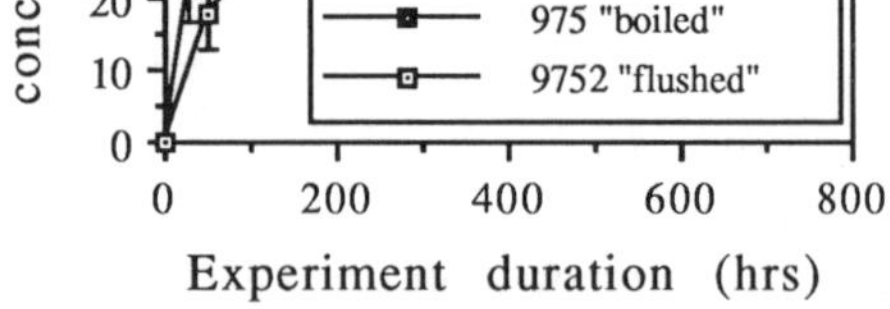

Figure 3 Si concentrations during repeat experiments. Experiment designations as for Figure 1.

Table 1 Comparison of solute yields from two experiments.

	975 (ppm)	975/2 (ppm)
Ca	26.7	15.4
Mg	3.04	0.2
citrate	24.6	8.92
oxalate	3.25	1.42
acetate	4.32	1.9
sulphate	103.8	46.6

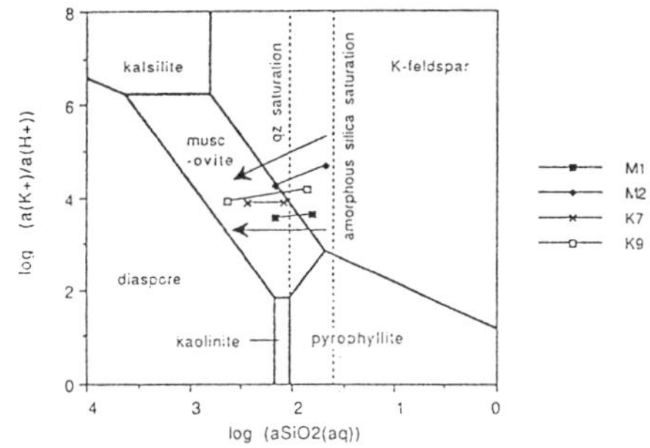

Figure 4

experiments are compared in Figures 2 and 3 and Table 1.

Si concentrations, as might be expected, are similar (see Figure 2) but in some cases solute yields differ (see Table 1). This is not likely to be a pH effect as Figure 2 shows that pH values measured during the two experiments converge.

The decrease in concentration of oxygenated species such as oxalate, citrate and sulphate might well have been predicted but the differing concentrations of Ca and Mg are harder to explain.

It is clear that more thought should be given to this aspect of the project if results are to be quantitative. Monitoring of oxygen fugacity during experiments similar to that described by Ulmer and Grandstaff (1990) may be necessary.

3 GEOCHEMICAL MODELLING

The geochemical code EQ3 has been used to calculate activities for dissolved species in these experiments, which can then be plotted on stability diagrams. Figure 4 shows activities for early carbonate rich mudstone heating experiments and as expected fluid compositions are buffered by clay mineral assemblages. Silica activities appear to be buffered to amorphous silica solubilities. Corresponding diagrams for carbonate poor mudstones are currently being produced.

4 CONCLUSIONS

The compositions of fluids derived from this series of mudstone heating experiments are buffered by mineral assemblages. Concentrations of silica have been shown to be buffered by amorphous silica and are thus oversaturated with respect to quartz. Carbonate rich mudstones produce fluids dominated by calcium, although changing carbonate solubility with pressure and temperature during sampling complicates interpretation of data. As these are experiments concerned only with source rocks, the effect of such fluids upon expulsion from mudstones, on adjacent lithologies cannot be stated with certainty. It seems likely, however, that there is potential for quartz and carbonate cement precipitation. Polar organic molecules (such as carboxylic acids) were generated in these experiments but did not appear to affect the solubility of inorganic species. Their potential for pH buffering could not be evaluated.

REFERENCES

Hutcheon, I., 1984 "A review of artificial diagenesis during thermally enhanced recovery" *Memoir of the American Association of Petroleum Geologists* 37, (eds. D.A. McDonald and R.C. Surdam), pp 413-429.

Hutcheon, I., Oldershaw, A. and Nahnybida, C., 1981 "The effects of in situ steam injection on Cold Lake Oil Sands-II. Petrophysical and petrological considerations" *Bulletin of Canadian Petroleum Geology* 29, pp 464-478. N.B. forms part of combined paper by Sedimentology Research Group, ibid., pp 447-478.

Milodowski, A.E., Savage, D., Bateman, K. and Hughes, C.R., 1989 "The precipitation of solids from potential heat exchange fluids for use in a HDR geothermal system in granite" *Geothermics* 18, pp 233-240.

Muffler, L.J.P. and White, D.E., 1969 "Active metamorphism of Upper Cenozoic sediments in the Salton Sea geothermal field and the Salton Trough, southeastern California" *Bulletin of Geological Society of America* 80, pp 157-182.

Ulmer, G.C., Grandstaff, D.E. and Myers, J., 1990 "A new hydrothermal technique for redox sensing using buffer capsules" *Geochemical Society Special Publication* 2, (eds. R.J. Spencer and I-Ming Chou) pp 17-21.

Zen, E-An, 1959 "Clay mineral-carbonate relations in sedimentary rocks" *American Journal of Science* 257, pp 29-43.

Water-Rock Interaction, Kharaka & Maest (eds) © 1992 Balkema, Rotterdam. ISBN 90 5410 075 3

Small scale trace element zonations in carbonates: An ion microprobe study of diagenesis in the Upper Devonian Nisku Formation, Western Canada Sedimentary Basin

Lee R. Riciputi & David R. Cole
Chemistry Division, Oak Ridge National Laboratory, Tenn., USA

Hans G. Machel
Department of Geology, University of Alberta, Edmonton, Alb., Canada

Warner H. Christie & Thomas M. Rosseel
Analytical Chemistry Division, Oak Ridge National Laboratory, Tenn., USA

ABSTRACT: Trace element contents of carbonates contain important information regarding the sample's diagenetic history. However, due to the fine scale and intergrown nature of many carbonate textures, separation of the different phases for bulk analysis is difficult or impossible. Analytical techniques have been developed using the ion microprobe that allow the analysis of a variety of trace elements in spots of 10 to 20 microns in diameter, at detection limits of 1 ppm or lower. The utility of this technique is demonstrated in a study of diagenetic carbonates from the Nisku Formation in the Western Canada Sedimentary Basin.

1 INTRODUCTION

Trace element concentrations and their variations provide potential insight into the diagenetic history and evolution of carbonate rocks. Unfortunately, small scale chemical zonations are impossible to analyze using bulk analytical methods, and useful information for concentrations <500 ppm are difficult to obtain using the electron microprobe. The ion microprobe has the potential to provide quantitative trace element analyses on spots of <20 microns in diameter at detection limits of 1 ppm or less for many elements, if proper standards and analytical techniques are utilized. Using the ion probe, trace element contents of diagenetic calcites have been measured in samples from the Upper Devonian Nisku Formation of the Western Canada Sedimentary Basin. The trace element variations may be related to changes in fluid composition during diagenesis of these carbonates, and the fine-scale zonations documented by the ion probe may provide clues as to the nature of fluid-rock interaction.

2 ANALYTICAL TECHNIQUES

Trace element contents in some Nisku Formation carbonates were determined using a Cameca ims-4f ion microprobe/microscope located at the Oak Ridge National Laboratory. This instrument is equipped with a computer controlled magnetic peak-switching system and scanning ion capabilities which allow trace element mapping of areas up to 500x500 microns in area. Au-coated samples were analyzed using a 2 to 5 nanoamp primary beam of $^{16}O^{-}$ accelerated at 12.5 kV, and the sample reference voltage was 4.5 kV. The size of the analyzed spot varied between 10 and 20 microns. Positive secondary ion counts were measured for B, C, Na, Mg, Al, Si, Ca, Fe, Mn, Sr, and Ba using an electron multiplier. Peak positions were calibrated each morning, and were checked before the start of each analysis. Ion signals for all elements but Al, Na, and B stabilized after ~2 minutes of sputtering. After ten minutes of sputtering, Al, B, and Na intensities had generally stabilized, varying less than 10% during the rest of the analytical period. In order to minimize the analytical uncertainties due to the volatility of Al, B, and Na, each spot was pre-sputtered for ten minutes before starting an analysis, during which the peak positions were checked and the magnet was cycled through the peak jumping routine to minimize hysteresis effects, and preliminary intensities were measured. Following the ten minute pre-sputter period, ten cycles of data were collected. During the analysis, the shape of the Ca energy distribution was automatically monitored for sample charging, which was usually insignificant. If any sample charging was present, the sample voltage was varied to compensate. Total analysis time was ~30 minutes for each spot.

Due to as yet poorly understood matrix effects, quantification of ion intensities requires the use of standards that are similar to the unknowns. Although internal precisions of ion probe analyses are generally quite good (<10% for Fe, Mn, Mg, Ca, Sr, and Ba, <20% for Al, B, and Na), the external accuracy is limited by the homogeneity of the standards and the accuracy of the bulk analyses (INAA, AA, ICP-MS) used to characterize each standard. Finding suitable homogeneous standards is one of the biggest challenges in obtaining accurate ion-probe analyses.

Ion intensities for elements of interest must be referenced to a measured ion intensity of a known concentration. Both C (concentrations assumed from stoichiometry) and Ca (concentrations measured on the electron probe) have been used as reference ions, and differences in the resulting analyses are within analytical error. Therefore, C is generally used as a reference species, assuming 11.8 wt% in dolomite and 12.2 wt% in calcite, thus avoiding the need to analyze each spot by electron probe as well as the ion probe. Absolute ion yields vary from day to day due to changes in the primary beam, instrument tuning, and electron multiplier performance. These variations are monitored by multiple analyses of a standard before and after each analytical session, and applying a correction factor to the measured intensities. Daily corrections are <10% for all elements except Na, Al, and B (<30%), and drift during any one day was <3%.

During the sputtering process, complex molecular ions are created, and these molecular ions present potential interference problems at the masses of interest. Previous ion microprobe studies of carbonates have used high mass resolution to resolve the interfering peaks from the masses of interest (Mason, 1987), or have tried to deconvolute the mass spectrum (Veizer et al., 1987). In contrast, we use energy filtering to eliminate most potential molecular interferences. Earlier studies have shown that molecular ion yields decrease much more rapidly than elemental ion yields as higher energy ions are sampled (Shimuzu et al., 1978). By shifting the sample offset voltage extraction field by 100V (relative to the upper edge of the molecular energy distribution), and avoiding peaks with obvious oxide interferences (ie, using ^{54}Fe rather than ^{56}Fe, which suffers from a $^{40}Ca^{16}O$ interference), we are able to obtain clean peaks for the elements of interest without the need for spectral deconvolution, and with much higher ion yields than obtainable using high mass resolution. The excellent vacuum conditions in the sputter chamber (typically 0.5 to 2×10^{-9} torr) also help to minimize the formation of some molecular species. Detection limits are <1 ppm for all the analyzed trace elements.

3 PETROLOGY OF THE NISKU SAMPLES

The Bigoray area of the Nisku Formation has been thoroughly studied by Machel (1986), who outlined a very complex limestone diagenetic history. Reefal portions of the Nisku Formation like the Bigoray C-Pool are generally a fossiliferous mud-mound facies, and many of these reefal areas have been the loci for fluid and hydrocarbon migration. Two samples from the Bigoray Pool were selected for ion probe analysis, each containing fine scale textures associated with early (<300m burial) limestone diagenesis of the carbonates. Sample N-19 consists primarily of two domains. One is a silty limestone matrix (MAT) which recrystallized early in the diagenetic sequence, and contains small dolomite rhombs. The second domain consists of a isopachous, white fibrous calcite cement (phase 5, WFCE, Machel, 1986) which is a primary high-Mg calcite that recrystallized to low-Mg calcite, fairly late in the limestone diagenetic sequence. These two domains are separated by a thin (<100 micron) calcite that is probably a syndepositional cement (CEM1). This cement is either a recrystallized primary aragonite or an atypically recrystallized phase 2 (Machel, 1986) high-Mg calcite. This zone is too fine to sample by drilling, so a series of spots traversing from the fibrous cement, across CEM1, and into the matrix calcite were

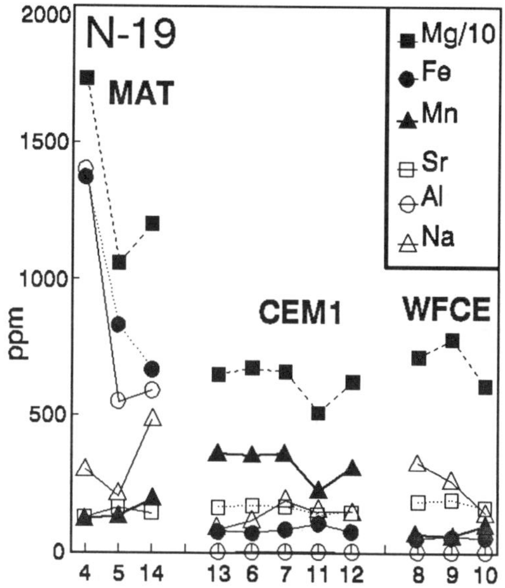

Figure 1a. Trace element contents in a profile across sample N-19. Analyses (x-axis) have been plotted sequentially with respect to position in the profile. Mg contents have been divided by 10. Abbreviations are in the text.

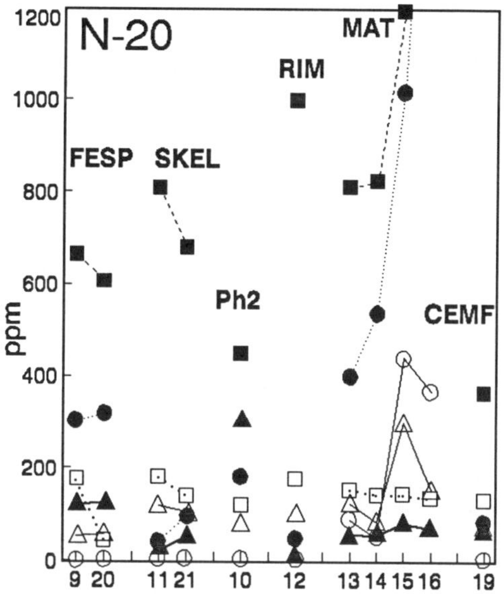

Figure 1b. Trace element contents in a profile across N-20. With the exception of CEMF, which is a fracture-filling calcite that crosscuts MAT and RIM, analyses are sequentially arranged. Symbols and abbreviations after Fig. 1a.

analyzed using the ion probe.

Sample N-20 contains a large coral with two types of internal cement, encrusted with at least one type of alga, and then embedded in the calcite and dolomite matrix. The coral contains both remnant skeletal calcite in the trabeculae (SKEL) and late-stage, blocky calcite spar cement (phase 7, FESP; Machel, 1986). Minor phase 2 scalenohedral cement (Ph2) occurs along the walls of the coral. The outer wall of the coral has a dark, Mg-rich lining (<2 microns wide). The coral is separated from the matrix calcite (MAT) and dolomite by a ~120 micron band of recrystallized calcite (RIM) that has an uncertain origin. A thin fracture filled by a clear calcite cement (CEMF) cuts through the matrix and RIM, and ends in the wall of the coral.

4 RESULTS AND DISCUSSION

Results of the N-19 analyses are shown in Fig. 1a, plotted in sequence from MAT, through the thin cement band (CEM1), and into the WFCE. Both the WFCE and CEM1 have similar Al and Mg contents (~1 ppm and ~6000-7500 ppm respectively), and there are no apparent trends relative to distance from the MAT border. Fe contents of WFCE and CEM1 are also similar, although contents are slightly higher in CEM1 (70-100 ppm as compared to 50-60 ppm). Al (500-1300 ppm), Mg (11000-18000), and Fe (700-1300) contents in MAT calcite are considerably higher than in CEM1 and WFCE. Although currently not quantified due to a lack of analyses for the standards, relative Ba and B distributions among the three phases are similar to those of Mg, Al, and Fe. Sr contents are relatively constant in all three types of calcite (130-200 ppm). Na concentrations are variable in all three phases (generally 100-300 ppm). The Mn profile is distinct from the other elements. WFCE and MAT have relatively low Mn contents (50-100 and 120-190 ppm, respectively). However, the thin cement zone (CEM1) that separates them has considerably higher Mn contents (220-360 ppm).

The high Al and B contents of MAT as compared to WFCE and CEM1 are consistent with the presence of very fine clay impurities inherited from the original silty lime mud that recrystallized to form MAT. WFCE recrystallized from a high-Mg calcite (Machel, 1986). The similar trace element characteristics of WFCE and CEM1 imply that CEM1 also formed from a high-Mg calcite, rather than from a primary aragonite. However, variations in Mn contents between WFCE and CEM1 suggest that redox-potentials were variable on a very fine scale in the sample. The coupled low Mn and Fe contents of WFCE are consistent with crystallization in an oxygenated environment with a high redox potential, as suggested for these cements by Machel (1986). The higher Mn contents of CEM1 suggest that Eh was much lower in this thin microenvironment. The low Fe contents of CEM1 do not necessarily conflict with this interpretation, but may instead reflect a lack of reducible Fe in the system. The inferred low Eh for CEM1 probably was caused by decaying films of organic matter during precipitation and/or recrystallization.

The elemental concentrations of a traverse in sample N-20 from the coral out into the matrix are shown in Fig. 1b. The phase 2 cement, fracture filling cement (CEMF), and the rim calcite (RIM) are all to small to be sampled for bulk techniques. The Al, Mg, and B contents of the different cements and the coral all are low. Coral and phase 7 cement recrystallized from low-Mg calcite, and the similar trace element contents of the other cements may indicate a similar low-Mg parental calcite. Al, Mg, and B contents in MAT are higher and variable, consistent with some primary high-Mg calcite and the presence of clay in the recrystallized matrix. Sr contents are relatively constant in all phases (120-170 ppm with the exception of one coral analysis). In general, Fe and Mn contents are positively correlated for the cements. RIM, CEMF, and SKEL all have low Mn and Fe contents (5-60 and 40-100 ppm respectively), which suggest more oxidizing conditions. The stage 7 spar that fills the coral (FESP) has higher Fe and Mn (~300 and 120 ppm), consistent with more reducing conditions. The matrix calcites (MAT) have more variable Fe and Mn (400-2000 and 55-80 ppm, respectively). The variability and lack of correlation between the Fe and Mn contents in MAT may reflect clay impurities.

In general, analyses obtained using the ion probe have concentrations similar to those obtained for phases that were bulk analyzed by ICP (Machel, 1986). Although the absolute concentrations do not always match, they are usually within 50%. Given the heterogeneity of the samples, this agreement is reasonable. Also, bulk samples are usually drilled from the center of large, "pure" areas with only one phase, but the ion probe analyses are from areas that are closely intergrown, so some of the differences may reflect real variability. The ion probe results support the conclusions of Machel (1986), who stated that the trace element compositions of the various calcites are variable within individual petrographic groups, and that most of the cements were formed in marine pore waters. The ion probe also revealed that compositional variability in the calcites is preserved on a very local scale (10-100 microns), as well as allowing analysis of samples that were previously uncharacterized.

Research sponsored by the Geosciences Research Program of BES/DOE under contract DE-AC05-84OR21400 with Martin Marietta Energy Systems, Inc. and NSERC.

REFERENCES

Machel, H.M. 1986. Limestone diagenesis of Upper Devonian Nisku Carbonates in the subsurface of central Alberta. Can. J. Earth Sci. 23:1804-1822.

Mason, R.A. 1987. Ion microprobe analysis of trace elements in calcite with an application to the cathodoluminescence zonation of limestone cements from the lower Carboniferous of South Wales, U.K. Chem. Geol. 64:209-224.

Shimuzu, N., Semet, M.P., & Allegre, C.J. 1978. Geochemical applications of quantitative ion-microprobe analysis. Geochem. Cosmochim. Acta. 42:1321-1334.

Veizer, J., Hinton, R.W., Clayton, R.N., & Lerman, A. 1987. Chemical diagenesis of carbonates in thin-sections: Ion microprobe as a trace element tool. Chem. Geol. 64:225-237.

Water-Rock Interaction, Kharaka & Maest (eds) © 1992 Balkema, Rotterdam. ISBN 90 5410 075 3

Spatial and temporal variations in formation water composition during diagenesis and petroleum charging of a chalk oilfield

P.C. Smalley & N.H. Oxtoby
BP Research Centre, Sunbury-on-Thames, Middlesex, UK

ABSTRACT: Formation waters were centrifuged from 5 preserved core samples of chalk reservoir rock from the Machar oilfield (UKCS) and analysed for salinity and $^{87}Sr/^{86}Sr$. A $^{87}Sr/^{86}Sr$-salinity relation was established, enabling pore water salinities to be estimated for unpreserved (dry) core by measuring the $^{87}Sr/^{86}Sr$ of residual pore salt extracts. The results show that, over a vertical distance of ~250 m, salinity rises with depth from 140,000 to 220,000 ppm, stays constant at this value for a while, then drops slightly. This is the first time that significant salinity variations have been reported within a single oil well; they probably relate to variable interaction between the waters and an underlying salt diapir during petroleum filling of the reservoir. Fluid inclusion and C,O,Sr isotopic analyses of fracture-filling carbonate cements reveal a wide range of water compositions during diagenesis, which overlapped with oil emplacement. Chalk fracturing and fracture cementation occurred repeatedly at various temperatures and with calcite cement sourced partly from outside the chalk. The overall picture is thus one of extreme fluid heterogeneity in both space and time.

1 INTRODUCTION

The Machar oilfield (UKCS block 23/26a) consists of a condensed raft of reservoir rocks (mainly Cretaceous chalk) developed over a Zechstein (Permian) salt diapir by a combination of salt movement, sediment downbuilding and extensional faulting (Fig. 1).

This study addresses two problems. Firstly, a knowledge of the composition (especially resistivity, R_W, related to salinity) of the formation water in the oil zone is needed in order to be able to deduce water saturation (from wireline resistivity logs), and thus to estimate the amount of oil in place. Secondly, predicting the degree to which fractures in the chalk reservoir have become cemented by calcite. This is important as calcite fracture fillings will reduce fracture permeability and act as barriers to fluid flow, reducing oil productivity. These problems are addressed by an integrated geochemical study of fluids and cements from well 23/26a-13 (Fig. 1).

2 A NOVEL APPROACH TO DETERMINING SALINITY VARIATIONS OF WATER-IN-OIL

2.1 Core centrifugation

In well 23/26a-13, preserved samples were available, i.e. core that had been encased in wax on the rig in order to preserve the fluids in the core. For 5 samples it was possible to extract formation water directly by ultra-centrifugation of core plugs. Additional extractions were unsuccessful due to difficulties caused by core fracturing and low matrix permeability. The extracted waters were analysed for ion concentrations using ICP-AES and ion chromatography, and for Sr isotopes using standard mass spectrometric methods. The results are presented in Table 1. Core contamination by drilling fluids is minimal in these samples due to the low permeability of the chalk.

With depth, over the sampled interval (~130m) concentrations of individual species increase (e.g. Cl from ~85,000 to ~130,000 ppm), total salinity increases from ~140,000 to ~220,000 ppm, and

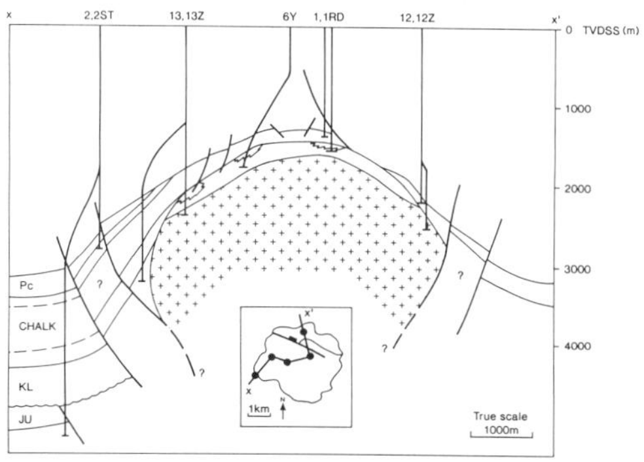

Figure 1. Cross section of the Machar Field, UKCS block 23/26a. Samples are from well 13.

Table 1. Analytical results for waters centrifuged from preserved core, well 23/26a-13

Sample Depth(m)	$^{87}Sr/^{86}Sr$@	Na	K	Mg	Ca	Sr	Ba	Cl	SO_4	TDS*	R_w+ (Ωm)
1889.10	0.70808	53600	400	483	1610	125	7	84600	1050	141875	0.0628
1934.25	0.70806	53800	1290	1090	5850	285	7	96500	1870	160629	0.0560
1977.10	0.70792	69900	710	791	3830	226	9	118100	1430	194996	0.0505
2016.10	0.70796	79600	785	663	1470	118	9	128000	1660	212305	0.0480
2016.25	0.70794	65000	1110	940	5280	218	13	115500	2280	190341	0.0517

* TDS = Total Dissolved Solids. + Rw measured at 25 °C. All concentrations in ppm. @ = ± 0.00003.

$^{87}Sr/^{86}Sr$ decreases from ~0.7081 to ~0.7079 (Fig. 2). This relation between salinity and $^{87}Sr/^{86}Sr$, albeit based on only 5 samples, can be used as a means of predicting salinity from $^{87}Sr/^{86}Sr$ (Fig. 2).

2.2 Residual salt analysis (RSA)

Strontium isotope ratios are not significantly fractionated by dissolution-precipitation processes. Hence, in conventional non-preserved core, the salts precipitated in the core from formation water as it evaporates during storage will have the same $^{87}Sr/^{86}Sr$ as the original formation water in the rock. These salts may be rehydrated and dissolved in the laboratory by treating coarsely crushed samples in ultrapure water (see Smalley et al. (1992) for RSA procedural details). These core extracts can then be analysed to determine the $^{87}Sr/^{86}Sr$ of the formation water. This allows Machar formation water $^{87}Sr/^{86}Sr$ data to be obtained for a much larger sample set. The RSA data confirm the depthwise fall in formation water $^{87}Sr/^{86}Sr$ that was identified using the centrifuged water samples, but also indicate that formation water $^{87}Sr/^{86}Sr$ begins to rise again below 2050 m.

2.3 Mapping salinity variations

Salinity data for the conventional core samples were calculated from the RSA $^{87}Sr/^{86}Sr$ data using the relation illustrated in Figure 2. These data support the rise in salinity from ~150,000 to ~250,000 ppm from ~1890 to 1960 m depth (Fig. 3). For the next ~100 m salinity remains roughly constant before dropping slightly again below 2050 m depth (Fig. 3).

3 FRACTURE CEMENTATION

The chalk in the Machar Field is extensively fractured, but many of the fractures have been infilled by calcite cement. Petrographically, the cements can be classified into two main types:

1. Blocky (type B): untwinned medium-coarse-grained calcite spar, filling <1 cm-wide fractures. They contain primary (i.e. trapped during crystal growth) aqueous fluid inclusions, but only secondary petroleum inclusions (trapped after crystal growth, along healed cracks).
2. Twinned (type T): very coarse drusy calcite spar occupying the thickest (≥2 cm) and petrographically latest fractures. The twinned cements contain both primary aqueous and petroleum fluid inclusions. The petroleum shows yellow-green(-blue) UV fluorescence typical of low API oil.

The overall picture is thus one of the blocky cement forming during fracturing but before oil emplacement, and the twinned calcite forming after fracturing and during oil emplacement.

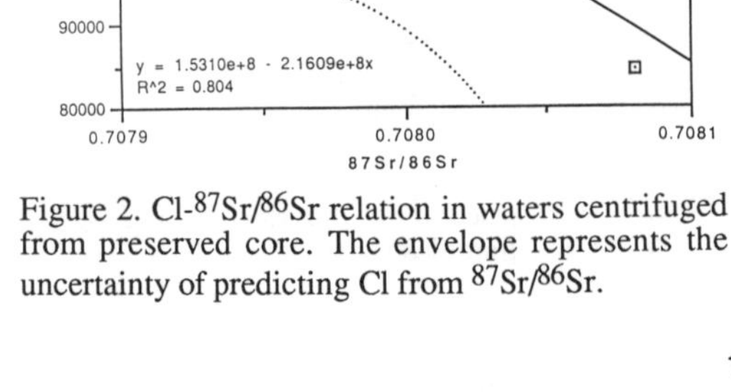

Figure 2. Cl-$^{87}Sr/^{86}Sr$ relation in waters centrifuged from preserved core. The envelope represents the uncertainty of predicting Cl from $^{87}Sr/^{86}Sr$.

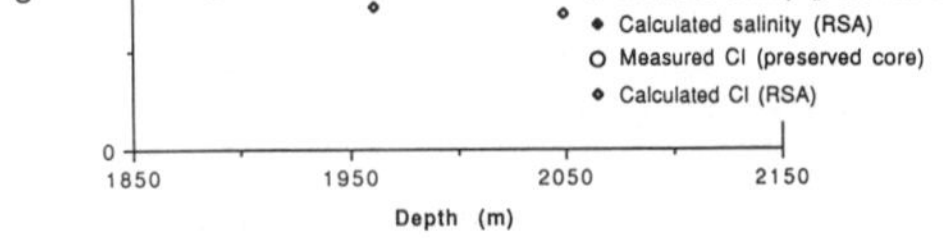

Figure 3. Measured (from preserved core) and calculated (from RSA) total salinities and Cl concentrations .

4 CARBONATE GEOCHEMISTRY

4.1 Chalk

Carbon, O and Sr isotopic analytical data for bulk chalk and fracture-filling calcite samples from Machar are presented in Figures 4-6.

The bulk chalk samples from Machar all have $\delta^{18}O$ values of ~-6 ‰ PDB, much lower than unaltered chalk from the North Sea basin (Fig. 4; cf. Jørgensen 1987). This indicates that the chalk has either partly "recrystallised" at temperatures of ≥40 °C, or contain a component of ≥40 °C microspar cement. However, $^{87}Sr/^{86}Sr$ and $\delta^{13}C$ values are identical to those of typical unaltered chalk (Fig. 5), indicating that any cement can only contain C and Sr derived internally from within the chalk.

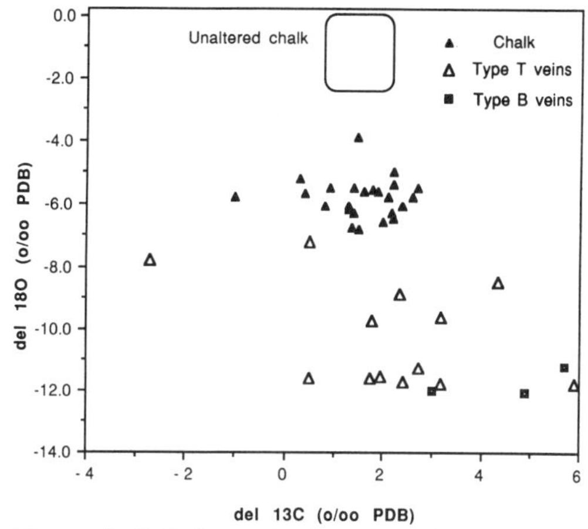

Figure 4. C-O isotopic data for twinned (T) and blocky (B) cements.

4.2 Fracture-filling calcite

The calcite fracture-fillings exhibit a much wider range of compositions (Figs. 4,5) than the chalk matrix, indicating some import of material into the chalk along the fractures.

Some fracture-filling cements, including all 3 blocky samples, have similar $^{87}Sr/^{86}Sr$ to the chalk, but much higher $\delta^{13}C$, up to +6 ‰ (Fig. 5). This heavy carbon signature has also been noted in fracture-filling calcites from other North Sea chalk oilfields, but its origin is problematical (Taylor and Lapre 1987, Smalley et al. 1992). Possible sources of the carbon include:

1. CO_2 produced by the bacterial fermentation of oil.
2. Pre-existing fermentation-derived carbonate (in e.g. the Kimmeridge Clay Formation, the regional source-rock) transported into the chalk by migrating waters (Taylor and Lapre 1987).
3. Permian dolomites, present at depth.

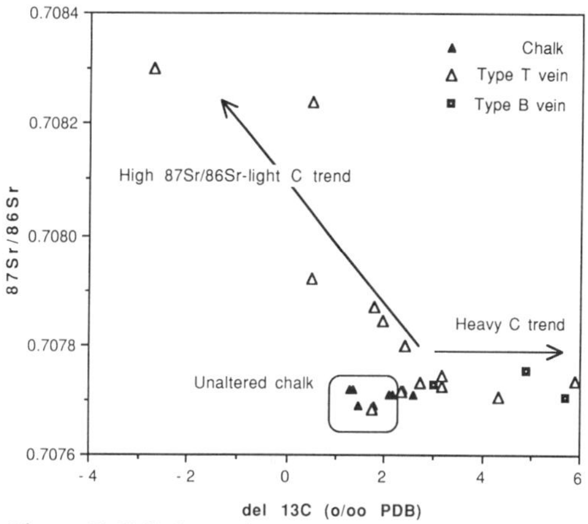

Figure 5. C-Sr isotopic data.

Another group of fracture-filling cements, made up of calcite of the twinned variety, trends towards higher $^{87}Sr/^{86}Sr$ (>0.7083) and lower $\delta^{13}C$ (<-2‰; Fig. 5). These must contain some organically-derived C (e.g. from source rock decarboxylation), and radiogenic Sr from, for example, detrital silicate minerals either internal or external to the chalk. Not only does this point to a completely different source of solutes for the two fracture-filling cement types; it requires fundamentally different styles of organic-inorganic and water-rock interactions.

5 FLUID INCLUSIONS IN CEMENTS

5.1 Temperatures of cement precipitation

The twinned calcite cement crystals contain primary aqueous and petroleum inclusions with identical homogenization temperature (T_{hom}) ranges (Fig. 6). It is very unlikely that this could result from inclusion leakage or stretching (a process that may lead to erroneous T_{hom} data in carbonates), but rather indicates that the inclusions are recording the actual temperature of mineral precipitation. The T_{hom} data fall in the range 65-120 °C, mostly between 80 and 100 °C (Fig. 6). The wide range of calcite T_{hom} data, from significantly below to significantly above present day temperatures (97 °C), suggests that fracture cementation has occurred either (a) at many different times, (b) from fluids flushing through the fractures out of thermal equilibrium with the host sediments. (cf. Jensenius and Munksgaard 1989), or (c) a combination of both.

5.2 Salinities of cement parent waters

The salinities of the waters from which the fracture-filling calcites precipitated, as deduced from fluid inclusion freezing point depression measurements, are illustrated in Figure 7. Salinity varies immensely (7,000-240,000 ppm), even within a single calcite sample, but most of the fluid inclusions are less saline than the present formation water (Fig. 7).

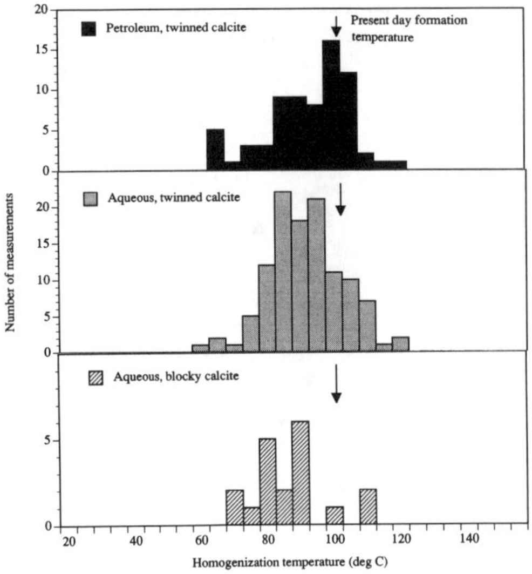

Figure 6. T_{hom} data for fracture-filling calcites.

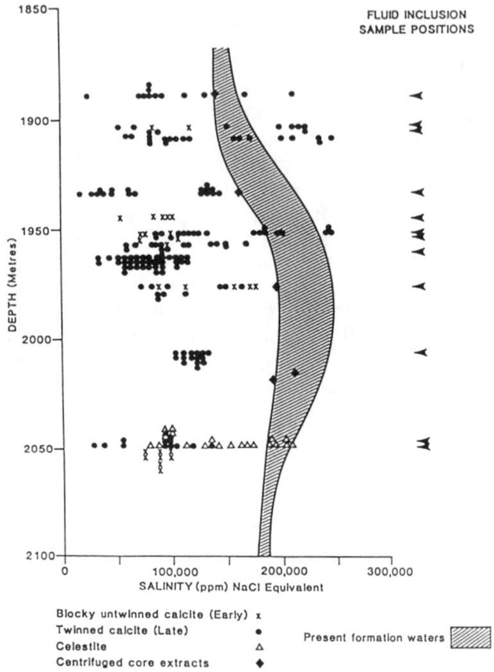

Figure 7. Fluid inclusion salinities in fracture-fills compared to present water salinity.

5.3 Isotopic composition of parent waters

Combining the fluid inclusion T_{hom} data with the calcite $\delta^{18}O$ values enables calculation of the $\delta^{18}O$ values of the parent waters to the fracture-filling calcite cements. Such calculated $\delta^{18}O$ values vary from -1 to +5 ‰ (SMOW). Similar values were deduced for fracture-filling calcites from chalk oilfields of the Danish sector of the North Sea (-1 to +7 ‰, Jensenius and Munksgaard 1989). This points to the involvement of a range of water types in the fracture cementation process.

6 WATER VARIATIONS IN SPACE AND TIME

Significant salinity variations have been identified in the Machar oil zone water. There is no obvious way that this could arise (a) through in situ water-rock interaction, or (b) as the result of a simple diffusive flux of dissolved NaCl from the underlying salt, since this could not explain the fall in salinity at the base of the sampled section (Fig. 3). Little is known about the ability of solutes to diffuse through the water in an oil column, but it is thought to be sufficiently slow that oil filling essentially "freezes in" the salinity of the water it traps (cf. Smalley et al. 1992). The fluid inclusions in fracture-filling calcite indicates that there were large changes in water salinity in the fractures, even within the relatively short time frame represented by oil filling, probably reflecting high fluid flow rates in the fractures. The salinity variations in the present Machar waters are thus interpreted as the result of a gradual rise and then slight fall in salinity of water in the chalk matrix during progressive oil filling, caused by variable interaction with the salt. Similar salinity variations across the oilfield may be predictable using a reservoir filling model.

REFERENCES

Jensenius, J. & Burruss, R.C. 1990. Hydrocarbon-water interactions during brine migration: Evidence from hydrocarbon inclusions in calcite cements from Danish North Sea oilfields. *Geochim. Cosmochim. Acta* 54: 705-713.

Jensenius, J. & Munksgaard, N.C. 1989. Large scale hot water migration systems around salt diapirs in the Danish Central Trough and their impact on diagenesis of chalk reservoirs. *Geochim. Cosmochim. Acta* 53: 79-88.

Jørgensen, N.O. 1987. Oxygen and carbon isotopic compositions of Upper Cretaceous chalk from the Danish sub-basin and the North Sea Central Graben. *Sedimentology* 34: 559-570.

Smalley, P.C., Lønøy, A. & Råheim, A. 1992. Spatial $^{87}Sr/^{86}Sr$ variations in formation water and calcite from the Ekofisk chalk oilfield: implications for reservoir connectivity and fluid composition. *Applied Geochem.*, in press.

Taylor, S.R. & Lapre, J.F. 1987. North Sea chalk diagenesis: its effect on reservoir location and properties. In: J. Brooks and K. Glennie (eds.), *Petroleum Geology of North West Europe*,Graham & Trotman, pp. 483-495.

Water-Rock Interaction, Kharaka & Maest (eds) © 1992 Balkema, Rotterdam. ISBN 90 5410 075 3

Coupled predictive models of diagenesis in sand-shale successions during progressive burial

R.C. Surdam & D.B. MacGowan
Department of Geology and Geophysics, University of Wyoming, Laramie, Wyo., USA

ABSTRACT: Petrographic, geologic and geochemical data are combined with pathway-type modeling to understand the sequence and evolution of diagenetic events. Thermal-burial history and kinetic models are combined to understand the timing of kinetically-controlled diagenetic events. The results of these models are exported to a thermodynamic, complex equilibria model to evaluate the magnitude of diagenetic events.

1. INTRODUCTION

The volumetrically important diagenetic reactions which result in significant porosity destruction or enhancement in clastic successions include framework grain dissolution and carbonate cement precipitation and dissolution. These reactions are mediated by biological, organic-inorganic and wholly inorganic interactions. The distribution in space and time of these reactions can be modeled by coupled burial-thermal history, diagenetic pathway, redox, kinetic and water-rock equilibria models. Burial-thermal models constrain the reactions in time-depth-temperature space. Diagenetic pathway models suggest which diagenetic reactions are likely to occur, and also constrain pore water chemical evolution through time. The kinetic models predict the position in space and time of kinetically-driven reactions (such as organic maturation and smectite-illite transformation). Water-rock equilibria models predict the timing of thermodynamically-driven reactions (e.g., carbonate mineral equilibria); in the zones of early and intermediate burial diagenesis, speciation of aqueous organic/metal complexes is of crucial importance to correct calculation of the thermodynamic equilibria in the rock-water system. Hydrocarbon migration not only alters the fluid-flow properties of the porous media which it invades, but also radically alters the redox equilibria within that reservoir. Redox reactions involving hydrocarbon invasion in some sandstones can significantly increase porosity, and thus must be critically evaluated. Modeling of sand-shale successions from basins world-wide suggest that this approach to diagenetic modeling neatly predicts porosity distributions and diagenetic sequences in many basins. These techniques appear to work best in basins where source and reservoir rocks are closely related (minimizing fluid migration distances), as well as basins characterized by sporadic hydrodynamic events, such as pressure chamber rupture or seismic pumping.

2. VOLUMETRICALLY IMPORTANT DIAGENETIC EVENTS

A typical chain, or series, of predictable progressive mineral reactions occur during continuous burial in sandstone/shale successions. Competition between key elements in this progressive reaction chain determines net porosity, and establishes positive or negative anomalies relative to compactional porosity loss. The most volumetrically significant of these competitions relative to porosity anomalies developing with increasing depth are as follows: 1) early mechanical deformation and grain rotation competing with early carbonate cementation for preservation intergranular volume (IGV), 2) dissolution of early carbonate cements and feldspar/lithic grains competing with precipitation of kaolinite and quartz for porosity enhancement, 3) formation of late carbonate cements and chlorite formation competing with quartz cementation and illite formation for preservation of IGV, and 4) dissolution of late carbonate and anhydrite competing with quartz cementation for deep porosity enhancement. The outcome of this series of overlapping diagenetic events determines the porosity/permeability characteristics relative to the compactional porosity loss through time and temperature. Positive porosity anomalies primarily are a function of carbonate, feldspar, chlorite, anhydrite and lithic material stability. We suggest that the parallel biological and organic reactions are serial and that they exert significant control on the inorganic reaction processes by at times controlling formation water alkalinity and aqueous speciation and thus the inorganic mineral stabilities. Process-wise, these biological and organic reactions are particularly significant to the mineral stabilities because they 1) determine the alkalinity of the fluid phase, 2) establish pH buffers, and 3) complex metals allowing key components in the system to become mobile. These three facts are critical in establishing the stability of both carbonates and aluminosilicates at any particular position in time temperature space.

3. THERMAL MODELING APPLIED TO DIAGENETIC STUDIES

The purpose of thermal modeling applied to diagenetic studies is to calculate a time temperature history (TTH) for the reservoir and source rocks of interest, incorporating as much constraining data as possible. The initial constraint on the thermal history of a reservoir rock is the temporal variation of terrestrial heat flow. On a large scale, many regional tectonic events can produce severe thermal perturbations, which may be modeled either analytically or with numerical approximations. Heat flow may also vary within a region due to factors other than tectonic history, such as differences in the insulative properties of sediment may cause heat to be refracted as a function of the relative contrast of insulative properties to the geometry of the layers.

Chrono-stratigraphic data are also very important in constraining burial history reconstructions and consequently, time-temperature histories. Without age, depth, and lithology data, it would be impossible to know the time-depth history of source-reservoir rock intervals. Lithologic data is important in calculations of compaction, or unit thickness and porosity, through time. Critical thermal properties of the rocks include thermal conductivity (for steady-state models) and thermal diffusivity (for time-dependent models). These parameters vary with changing lithology and porosity (depth) and directly affect the calculated (modeled) temperature. Thermal conductivity is also a function of the bulk mineralogy of the rock: it has been demonstrated that changes in the bulk mineralogy of sandstones due to progressive diagenesis cause major changes in the thermal conductivity and thus the thermal history of rock units.

4. KINETIC MODELING

The changes in the concentrations of reactive organic species through time must be modeled using measured kinetic parameters and time-temperature burial history models. The source of the reactive organic species (i.e., carboxylic acid anions, CO_2, phenols, etc.) affecting the inorganic mineral stabilities over this temperature interval is the same as for hydrocarbons in the source rocks. They are either expelled from the source rocks just prior to, or concomitant with, hydrocarbon generation, or form as a result of the interaction of liquid hydrocarbons and mineral matter in the reservoir rocks (i.e., redox reactions). It is possible to model the progress of these organic-inorganic interactions by calculating their transformation ratios through time. For the study of a specific basin, the source of the kinetic parameters are the hydrous pyrolysis experiments or natural mineral versus depth distributions (c.f., Surdam et al., 1989a and b). This allows maps of the distribution of carboxylic acids and anions (CAA), and CO_2 from decarboxylation in the formation waters in a reservoir with time to be constructed. From these data it is possible to qualitatively evaluate the relative stability of minerals (either precipitation or dissolution) at any time in the history of a targeted reservoir interval. In addition to the generation and destruction of CAA and oil generation, kinetic modeling can also be used to model the progress of other kinetically-controlled diagenetic reactions such as the smectite-illite transition and thermal reduction of sulfate by hydrocarbon oxidation (cf Surdam et al., 1989a and b).

5. REACTION PATHWAYS

Factors governing the evolutionary path of sandstones from the time of erosion to the onset of metamorphism include provenance and depositionally-controlled compositional and textural elements; near surface redox reactions at the time of early burial (sediment-water interface to burial temperatures of approximately 80°C), which control the composition of early carbonate cements and the fate of iron in the system; organic-inorganic interactions within the zone of intermediate burial diagenesis (80°C to 130°C ±10°C), which control carbonate and aluminosilicate mineral stability; deep diagenetic reactions, including abiotic thermal sulphate reduction, carbonate mineral alteration, and quartz cementation; and telogenetic alteration including meteoric influx and renewal of near-surface processes due to uplift, exposure, and/or base level fluctuations (cf. Surdam et al., 1989c).

These critical factors can be represented as is a system of heirarchically-arranged divides, similar to chemical divide models, but which also contain divides pertaining to non-geochemical, process-oriented factors (see Surdam et al., 1989c). Initial modeling of the evolution of rock-water systems begins with choosing appropriate initial pore water compositions. This is done by choosing a reasonable modern analog, based upon the depositional environment. The next step is to estimate the initial composition of the sandstone petrographically by examining very early calcite concretions, which essentially "lock-in" the initial sediment and do not allow further diagenetic reactions to take place. At this stage, pathway-type modeling (cf. Surdam et al., 1989c) is used to understand how the pore water composition evolves through time and to predict which diagenetic reactions may affect porosity and the water chemistry.

It is fairly straightforward to quantitatively model the chemical evolution of pore water, assuming that the water does not migrate, or does not mix with a water of a differing chemical composition. In the case of water mixing or migration, usually only qualitative calculations of water/mineral thermodynamic equilibria can be made. However, water mixing or migration commonly occurs in sandstones and can have a drastic affect on calculated water/mineral equilibria. For instance, in the case of calcite, the relationship of solubility with temperature, pressure and ionic strength is non-linear. This leads to the situation where two fluids in equilibrium with calcite can be mixed in equal proportion and produce a fluid which is undersaturated with respect to calcite (this is called "corrosive mixing" and has been proposed as mechanism for pervasive carbonate mineral dissolution [Harrison, 1989]). Also, because of the non-linearity of calcite/water equilibria with respect to these variables, mixing of two formation waters not in equilibrium with calcite, with differing chemistries and temperatures can lead to calcite precipitation. In addition, migration of a hot, deep formation water may cause that water to become undersaturated with respect to calcite. Pathway modeling can predict which diagenetic reactions are likely to take place, thermodynamic modeling predicts whether or not those reactions, in fact, go to completion.

6. THERMODYNAMIC ROCK/WATER EQUILIBRIA SIMULATION

Thermodynamic, complex equilibria type computer models with enhanced organic-inorganic reaction data bases and temperature-extrapolated reaction equilibria constants can be used to determine the direction and extent of water-rock interactions. In this way, the spatial stability of carbonate cements in a sandstone along a time-temperature profile can be quantified.

The results of this integrated burial-thermal-pathway-kinetic-thermodynamic approach to modeling the evolution of diagenetic systems yield quantitative information on the relative stabilities of mineral phases through time, the timing of the generation of reactive liquids, aqueous and gaseous organic phases, and the timing of diagenetic events such as cement precipitation and grain/cement dissolution. In this manner the evolution through time of primary, moldic and microporosity in a targeted clastic system can be more accurately evaluated.

7. EXAMPLE-STEVENS SANDSTONE, NORTH COLES LEVEE, SAN JOAQUIN BASIN, CALIFORNIA

7.1 Petrographically Observed Diagenesis

The diagenesis of the Stevens Sandstones of the San Joaquin basin has been extensively studied (see Fischer and Surdam, 1988; and Boles, 1987 for a review).

Early carbonate cements, their dissolution and the formation of late Fe, Mg-rich carbonate cements are recognized by the above mentioned investigators. Dissolution of feldspars, albitization of plagioclase, thin potassium feldspar overgrowths and the formation of kaolinite are also well documented. Quartz cementation is not reported. The compositions and isotopic signatures of the carbonate cements are related to various stages of formation water mixing and evolution (Boles and Ramseyer, in press). The early carbonates are recognized by the uncompacted nature of enclosed biotite grains (Boles, 1984). Carbonate dissolution is recognized by anhedral, relict forms within enlarged pores. Late stage Fe-dolomites occur, having formed after albite precipitation and replacing early-formed pore-filling calcites. Late stage dolomites also occur as minute rhombs within the abundant biotite flakes (Boles and Johnson, 1983).

The depth intervals over which the carbonate cements, the dissolution textures and the aluminosilicate authigenesis coincide with variations in the formation water chemistry greatly explain the sequence. Stable isotopic data of the carbonate cements also follow the geochemical sequence.

7.2 Thermal Modeling for the Stevens Formation

The time-temperature profile in this example is from Fischer et al., 1988 for the Stevens Sandstone at North Coles Levee in the San Joaquin Basin.

7.3 Diagenetic Modeling

The marine environment which characterizes the depositional environments (marine turbidites) of much of the Stevens indicate that the initial pore water composition would be marine, or slightly dilute marine water composition. During deposition, pH was likely buffered by aluminosilicate hydration (both the Stevens itself and the adjacent shales are rich in volcanogenic material) and the aqueous CO_2 system.

According to pathway modeling, in the zone of bacterial sulfate reduction, the SO_4^- in the marine water (initially about 2700 ppm) would be reduced to HS-; this reaction would also elevate the concentration of bicarbonate. It is likely that Fe concentrations exceeded sulfate concentrations, thus favoring the precipitation of early Fe-carbonate or dolomite cements. This would lower the concentrations of Fe, Ca, Mg, and bicarbonate in the pore water. In the zone of bacterial methanogenesis, the only significant reactions which took place likely were precipitation of carbonate minerals (Fe, Mn-rich). This would further lower the concentrations of Fe, Ca, Mg, and bicarbonate in the pore water. In both the zones of bacterial sulfate reduction and bacterial methogenesis the pH of the pore water would have been controlled by the aqueous CO_2 system. At some period during methanogenesis (e.g., before the sands reached about 80°C), it is likely that meteoric waters invaded the Stevens and mixed with the marine connate waters. Such mixing would promote the stabilization of kaolinite and calcite.

Aqueous carboxylic acids and anions (CAA) begin to become important as the Stevens enters the zone of intermediate burial diagenesis (according to kinetic modeling of a Type II kerogen, this occurs at ~6 Ma in the basin center, Figure 1). At this time, the pH is lowered, and, due to this and the ability of CAA to complex metals (such as Al, Si, Fe and Ca), Ca-feldspar and carbonate minerals become destabilized and begin to dissolve. The kinetic modeling suggests that malonate reaches its maximum concentration between about 6.1 Ma and that acetate reaches its maximum concentration at about 6 Ma. As the difunctional CAA begin to become thermally unstable and decarboxylate, the PCO_2 of the pore waters begins to increase, and some metals (such as Al, Si, Fe and Ca) which had been complexed by the difunctional CAA are released from the complexes as inorganic monomeric or hydroxylated aqueous species. During this period, the formation water alkalinity is still dominated by CAA and the increasing PCO_2 will actually stabilize carbonate minerals, this occurs from about 5.8 Ma to the present (Figure 2). Fe and Mg released from thermally-destroyed organic complexes and the smectite-illite transition would suggest that these late carbonates would be ferroan or dolomitic. Note that the kinetic modeling suggests that malonate concentrations at present are about 60% of their maximum value and that acetate has never undergone decarboxylation. It is in formation waters produced from the Stevens at North Coles Levee that some of the highest concentrations of CAA have been measured. Thermodynamic modeling of the stability of Ca-plagioclase suggests that it becomes destabilized at about 6 Ma, and is never stabilized again. The alteration of the Ca-feldspar would lead to albitization of feldspars and precipitation of laumontite.

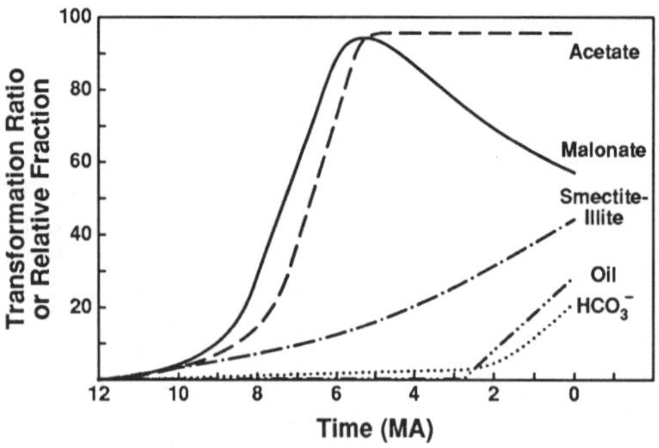

Figure 1. Summary diagram of kinetic modeling for the Stevens at North Coles Levee: bicarbonate, malonate and acetate shown in relative fraction vs time, oil and smectite-illite in transformation ratio vs time

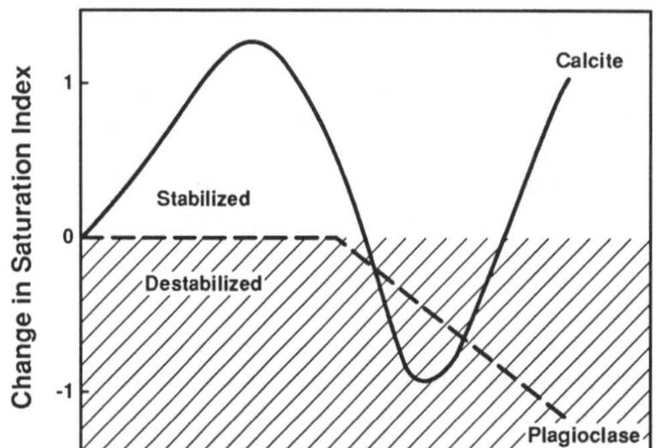

Figure 2. Summary diagram thermodynamic modeling of calcite and Ca-plagioclase in temperature vs log Q/K space. Values of log Q/K greater than zero indicate that the mineral is stable; values less than zero indicate destabilization

7.4 Summary

The San Joaquin basin modeling is an excellent example of the predictive abilities of this approach. This basin history is such that the straightforward approach used herein provides an excellent match of diagenetic sequence, carbonate mineral isotopic temperature gaps and the observed production intervals. The sequence of early carbonate cementation intermediate diagenetic dissolution of carbonates and feldspars, followed by the precipitation of the late stage Fe-bearing carbonates is simply classic.

8. REFERENCES

Boles, J.R., 1987, Six million year diagenetic history, North Coles Levee, San Joaquin Basin, California, in J. Marshall (ed.), Diagenesis of Sedimentary Sequences, Geological Society Spec. Publ. No. 36:191-200

Boles, J.R. and K.S. Johnson, 1983, Influence of mica surfaces on pore water pH, Chemical Geology, 43: 303-317.

Boles, J.R., and K. Ramseyer, in press, Diagenetic carbonate in Miocene sandstone reservoir, San Joaquin Basin, California, AAPG Bull.

Fischer, K.J. and R.C. Surdam, 1988, Contrasting diagenetic styles in a shelf turbidite sandstone sequence: The Santa Margarita and Stevens Sandstones, southern San Joaquin Basin, California, in S. Graham (ed.), Studies of the San Joaquin Basin, Pacific Section, SEPM, 60:233-247.

Fischer, K.J., R.C. Surdam, and H.P. Heasler, 1989, Hydrocarbon maturation modeling of the Southern San Joaquin Basin, California, in S. Graham (ed.), Studies of the San Joaquin Basin, Pacific Section, SEPM, 60:53-64.

Harrison, W.J., 1989, Modeling fluid/rock interaction in sedimentary basins, in T. Cross (ed.), Quantitative Dynamic Stratigraphy, Prentice Hall, 1-37.

Surdam, R.C., T.L. Dunn, H.P. Heasler,and D.B. MacGowan, 1989a, Porosity evolution in sandstone/shale systems, in I. Hutchinson (ed.), Short Course Notes: Burial Diagenesis, Mineral. Assoc. Canada, Toronto, pp. 61-134.

Surdam, R.C., T.L. Dunn, D.B. MacGowan, and H.P. Heasler, 1989b, Conceptual models for the prediction of porosity evolution: An example from the Bighorn Basin, Wyoming, in E. Coulson and S. Kaplan (eds.), Rocky Mountain Association of Geologists Guidebook: Petrogenesis and Petrophysics of Sandstone Reservoirs in the Rocky Mountain Region, RMAG, Denver, 7-28.

Surdam, R.C., D.B. MacGowan and T.L. Dunn, 1989c, Diagenetic pathways of sandstone and shale sequences, Univ. Wyoming Contrib. Geol, 27:21-32.

Water-Rock Interaction, Kharaka & Maest (eds) © 1992 Balkema, Rotterdam. ISBN 90 5410 075 3

A model for diagenesis in the Upper Wilcox reservoir sandstones at Fandango Field, south Texas, USA

Glenn A. Wilson, Charles E. Harvie & David T. Lawrence
Shell Development Co., Bellaire Research Center, Houston, Tex., USA

ABSTRACT: Diagenetic reactions in deep Gulf Coast Tertiary sandstones are controlled by the chemistry of the mineral/fluid system, and fluid flow which transports chemical components. Since fluids in the geopressured environment experience overall decreases in temperature and pressure as they migrate upward through the section, mineral/fluid reactions can occur along the flow path in response to the changing environment experienced by the fluid. These concepts can be applied to model the diagenetic evolution of the U. Wilcox sandstones at Fandango Field, S. TX. With the f_{CO_2} constrained according to gas compositions from the field, model calculations predict that feldspars dissolve, forming clay and quartz as products. With continued fluid flow, quartz continues to precipitate very gradually. The total amount of fluid which has passed through a specific location in the rock over geologic time can be inferred from the observed volume of quartz cement.

INTRODUCTION

Diagenesis of the Gulf Coast Tertiary sandstones typically involves the dissolution and alteration of feldspar and carbonate minerals, and the precipitation of carbonate, clay, and quartz cements. These open-system reactions are controlled by the chemistry of the minerals and the pore fluid, and by the physical mass-transport of chemical elements into and out of the system. The progress of diagenesis through time can be understood by establishing the relationship between the chemical phenomena and fluid flow, which is the dominant mass-transport mechanism.

In the geopressured environment, fluid potentials tend to increase with depth as does temperature so that formation fluids experience an overall decrease in temperature and pressure as they migrate through the section. This forms the basis for one possible model of how the reactions are driven. As the fluids flow through gradients in temperature, diagenetic reactions proceed in response to these changing conditions experienced by the fluid. At any location in the rock, the extent of diagenetic alteration is a direct function of the local time-integrated fluid flow rate. Thus, the volumes of dissolved and precipitated minerals can be directly related to the Darcy flux and geologic time. This model assumes that local equilibrium is maintained between the fluid and the minerals in the rock but it assumes nothing else about the initial fluid composition except P_{CO_2} and total NaCl concentration. These concepts have been applied in studies of diagenetic, metamorphic, and hydrothermal systems by Wood and Hewitt (1982), Baumgartner and Ferry (1991), Ferry and Dipple (1991), and Cathles (1991). Here we apply this model to gain some general insight into the relationship between diagenesis and fluid flow in the deep Upper Wilcox (Eocene) sandstone reservoirs at Fandango Field, Zapata County, south Texas.

Fandango Field is a series of stacked, fine-grained, gas-bearing sandstone reservoirs in a faulted roll-over anticline associated with a major listric growth fault. Down-dip from the crest of the anticline and away from the fault, the sandstones grade into shales which act as sources of fluids entering the sandstones. The sands experienced rapid burial to near their present depth approximately 50 million years ago. The reservoirs are geopressured at depths of approximately 4 to 5 km, and temperatures range from 185 to 235°C. The sands originally contained a detrital assemblage of quartz, plagioclase, K-feldspar, clay, and rock fragments, but have been extensively altered through diagenesis. The present-day mineralogy consists of quartz, plagioclase (>Ab_{98}), chlorite, illite, and ankerite, with local traces of siderite and pyrite.

REACTIONS FROM FLUID FLOW THROUGH TEMPERATURE GRADIENTS

Assuming only conductive heat flow and excluding mechanical compaction, the progress of diagenesis through geologic time at a specific point in the rock is given by,

$$\partial\phi_i/\partial t = (V_f^{-1}dV_i/dT)\,q\,\sin\theta\,(\partial T/\partial Z) \quad (1)$$

where ϕ_i is the volume fraction of mineral i in the rock, t is time, V_f is the unit volume of fluid, V_i is the volume of mineral i, q is the local Darcy flux, θ is the apparent dip in the flow direction, and $(\partial T/\partial Z)$ is the geothermal gradient.

The expression, $q\sin\theta\,(\partial T/\partial Z)$, is related to the change in temperature of the fluid with respect to time as the fluid migrates through the rock. The term, $(V_f^{-1}dV_i/dT)$, gives the change in the volume of mineral i with respect to temperature, per unit volume of fluid in equilibrium with a given buffer assemblage. The term is derived for each mineral from EQ6 (Wolery, 1984) calculations where equilibrium is maintained between the fluid and a specifically chosen mineral assemblage, while temperature is varied. Equation (1) transforms the fluid frame of reference of the EQ6 output into the frame of reference of the rock. For the present application, we have chosen a representative analogue sandstone assemblage of quartz, K-feldspar, plagioclase, calcite, and kaolinite from which comparisons will be made with the natural system.

CONSTRAINTS ON THE CO_2 FUGACITY

Although this assemblage should buffer the f_{CO_2} along with the activities of the other major components in the system, the buffer capacity for CO_2 is very small since the calcium-bearing silicate (typically, anorthite in plagioclase) is in low abundance in natural sandstones. This is reflected in the gas compositions at Fandango Field which, over a wide range in temperature, indicate that CO_2 is not internally buffered by the minerals in the sandstones. The mole fraction of CO_2 in the gas remains nearly constant over the entire range of reservoir depths, and varies only slightly due the changing phase relations in the H_2O–CH_4–CO_2 system with temperature and pressure. Variations in P_{CO_2}, then, are mainly due to variations in total pressure. These data are shown in Fig. 1 as $\log P_{CO_2}$ (back-calculated to *in situ* reservoir conditions) vs. reservoir temperature. A calculated curve which closely agrees with the data is shown for a system with a constant ratio of CO_2/CH_4, and temperature and pressure gradients that match those of the field.

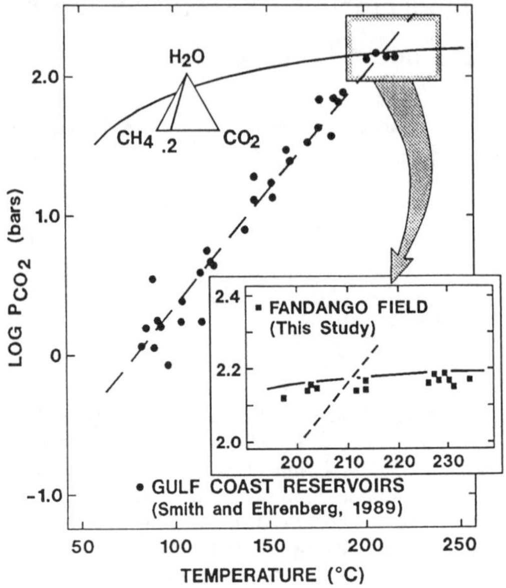

Figure 1. Log P_{CO_2} vs. *in situ* temperature from sandstone gas reservoirs. Square points (inset) are from Fandango Field, U. Wilcox, S. TX (this study). Round points are from Smith and Ehrenberg (1989) from a broad sampling of Gulf Coast sandstone reservoirs. The curve for $CO_2/(CO_2{+}CH_4){=}0.2$ (bulk system) was derived from an equation of state calculation where temperature and total pressure follow the gradients at Fandango Field.

Also shown in Fig. 1 are data from a large number of Gulf Coast sandstone reservoirs compiled by Smith and Ehrenberg (1989) who suggested that their data indicate mineral buffering of P_{CO_2}. Hutcheon (1990) showed that the buffer is probably an assemblage which includes the clay minerals smectite, illite, and kaolinite. These minerals are abundant in Gulf Coast shales but are of low abundance in the sandstones at Fandango Field. Since the data from the field overlap the general trend of the Gulf Coast data it appears likely that the P_{CO_2} was initially buffered in the down-dip shales. As CO_2-bearing fluids migrated into the sandstone reservoirs they retained their initial CO_2 content. It is possible that early in the diagenetic history, the CO_2 in these sandstones was controlled locally by the presence of a small amount of anorthite in detrital plagioclase together with kaolinite and calcite. The P_{CO_2} associated with that assemblage, however, is much lower than that which is associated with the shales and the buffer capacity would have been quickly exhausted by the irreversible removal of anorthite.

In this study, f_{CO_2} is constrained in the calculations of $(V_f^{-1}dV_i/dT)$ to mimic the calculated curve in Fig. 1 (converted to fugacity) by the addition of a fictitious mineral to the EQ6 data base.

MODEL RESULTS AND COMPARISON WITH THE NATURAL SYSTEM

The integrated form of equation (1), evaluated at a specific point in the sandstone, gives ϕ_i as a function of the volume of fluid which has passed through that location, $\int q\,dt$. As an approximation which is compatible with the rapid burial history at Fandango Field, $(V_f^{-1}dV_i/dT)$ is evaluated at a single rock temperature of 200°C (this greatly simplifies the model although it ignores any effects of near-surface early diagenesis). A geothermal gradient of -39°C/km and a dip of 15° will be assumed. To apply the integrated form of equation (1), an initial mineralogy must be assumed which for this illustration is loosely based on estimates from Fandango Field. Figure (2) shows the diagenetic evolution of a hypothetical sandstone with an initial composition given on the far left side of the figure.

According to Fig. 2, the feldspars begin to react as soon as fluid begins to flow through the rock, producing small amounts of calcite, kaolinite, and quartz cements as they dissolve. When anorthite runs out, calcite stops precipitating and its abundance remains nearly constant until K-feldspar runs out. At that point, calcite begins to dissolve. A few percent quartz cement forms in the early stages as a product of K-feldspar dissolution, and quartz continues to precipitate, but very gradually, after that due to its decreasing solubility with decreasing temperature. The discontinuities that occur where anorthite, K-feldspar, and calcite, respectively, are removed from the rock reflect changes in the buffering mineral assemblage where new values for $(V_f^{-1}dV_i/dT)$ apply.

At Fandango Field, petrographic evidence shows that compaction is negligible. K-feldspar which (presumably) once existed in the rock has been completely dissolved and replaced by chlorite and quartz. The plagioclase in the rock is presently composed of nearly pure albite with evidence of partial dissolution from the removal of the anorthite component. Small amounts of ankerite are present which in places are intergrown with plagioclase, suggesting replacement of anorthite. Rock fragments appear to have recrystallized in place to quartz and chlorite (±albite). Ankerite and chlorite here may have formed from the alteration of earlier calcite and kaolinite, by analogy to other nearby Wilcox sandstones (Boles, 1978; Boles and Franks, 1979). The most variable diagenetic feature is the amount of quartz cement which ranges from a few percent to as much as 25% of the bulk rock volume. Textural evidence indicates that most of the quartz formed after the other diagenetic features.

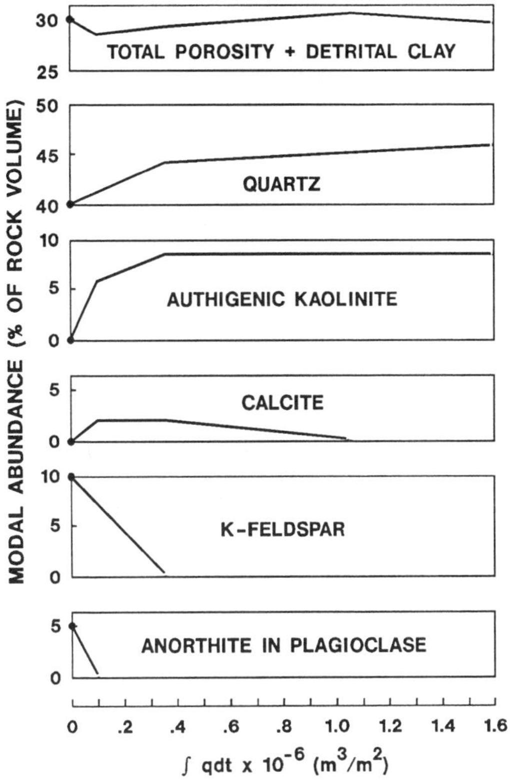

Figure 2. Diagenetic evolution of a sandstone as a function of the time-integrated fluid flow rate. The initial composition of the rock (at the far left) is loosely based on estimates of the initial composition of the sandstone at Fandango Field (with the balance of the rock being composed of rock fragments and albite). A constant reservoir temperature of 200°C, a geothermal gradient of -39°C/km, and a dip of 15° are assumed.

Comparing Fandango Field with the analogue system in Fig. 2 suggests that diagenesis in these rocks is well beyond the initial stages where the feldspars are reacting and well into a period of time when only quartz is precipitating. Figure 3 shows the point counted volume of quartz cement as an average through each cored interval at the field. Together with the quartz curve from Fig. 2, this is an indication of the local fluid flow rate at each cored location averaged over the lifetime of

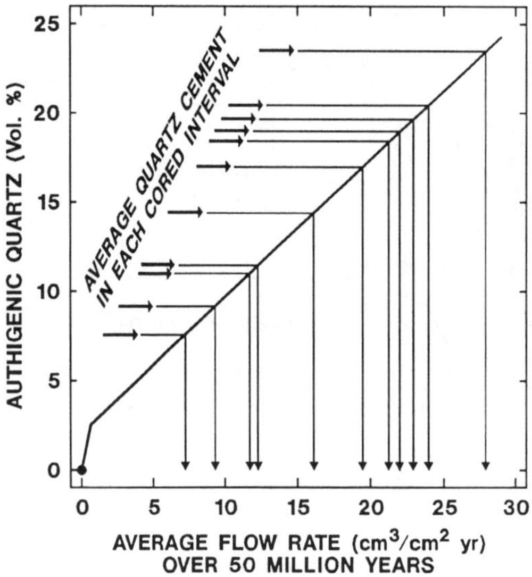

Figure 3. Average volume of quartz cement from point count data on each of 11 cored intervals from 6 sandstone horizons at Fandango Field, and the quartz reaction path from Fig. 2. Note that the scales are greatly expanded relative to Fig. 2, and the horizontal scale has been divided by 50×10^6 yr.

the sandstone. The averaged flow rate varies from 7 to 28 cm^3/cm^2 depending on the location. Given the permeability and likely pressure gradients, 28 cm^3/cm^2 is about the highest flow rate that could be expected at any time in these sandstones. Thus, quartz precipitation at Fandango Field must have occurred very gradually over nearly the entire 50 million year history rather than within a shorter period of time at higher flow rates.

CONCLUSIONS

The results of this study show that the observed diagenetic features at Fandango Field could be the products of the decreases in temperature experienced by fluids flowing through the rock, rather than the increases in temperature and pressure experienced by the rock during burial. The volumes of minerals dissolved and precipitated can be quantitatively related to the total amount of fluid which has passed through a given location in the rock. By this model, accumulations of quartz cement in excess of a few percent at Fandango Field must have been created as a very gradual process over tens of millions of years.

REFERENCES

Baumgartner, L.P. and Ferry, J.M. 1991. A model for coupled fluid-flow and mixed-volatile mineral reactions with applications to regional metamorphism. *Contributions to Mineralogy and Petrology* 106:273-285.

Boles, J.R. 1978. Active ankerite cementation in the subsurface Eocene of southwest Texas. *Contributions to Mineralogy and Petrology* 68:13-22.

Boles, J.R. and Franks, S.G. 1979. Clay diagenesis in Wilcox sandstones of southwest Texas: Implications of smectite diagenesis on sandstone cementation. *Journal of Sedimentary Petrology* 49:55-70.

Cathles, L.M. 1991. The importance of vein selvaging in controlling the intensity and character of subsurface alteration in hydrothermal systems. *Economic Geology* 86:466-471.

Ferry, J.M. and Dipple, G.M. 1991. Fluid flow, mineral reactions, and metasomatism. *Geology* 19:211-214.

Hutcheon, I. 1990. Clay-carbonate reactions in the Venture area, Scotian Shelf, Nova Scotia, Canada. In R.J. Spencer and I-Ming Chou (eds.), *Fluid-Mineral Interactions: A Tribute to H. P. Eugster. The Geochemical Society, Special Publication* 2:199-212.

Smith, J.T. and Ehrenberg, S.N. 1989. Correlation of carbon dioxide abundance with temperature in clastic hydrocarbon reservoirs: Relationship to inorganic chemical equilibrium. *Marine and Petroleum Geology* 6:129-135.

Wolery, T.J. (1984), EQ6, A Computer Program for Reaction-Path Modeling of Aqueous Geochemical Systems: User's Guide and Documentation, Lawrence Livermore National Laboratory, California, UCRL-51 Draft.

Wood, J.R. and Hewett, T.A. 1982. Fluid convection and mass transfer in porous sandstones - a theoretical model. *Geochimica et Cosmochimica Acta 46:1707-1713.*

Water-Rock Interaction, Kharaka & Maest (eds) © 1992 Balkema, Rotterdam. ISBN 90 5410 075 3

Evolution of $CaCl_2$ brine in Silurian aged formations of the Michigan basin, USA: The role of mineralogic reactions and evaporite diagenesis

T.P.Wilson
Kent State University, Ohio, USA

D.T.Long
Michigan State University, East Lansing, Mich., USA

ABSTRACT: Silurian aged carbonate and evaporite deposits in the Michigan basin contain highly concentrated $CaCl_2$ brine. Cl/Br ratios indicate that the brines originated from evapo-concentrated seawater, most likely residual from the deposition of the Salina evaporites. Dolomitization has been important in developing the Ca content of the brines; however, it cannot fully explain all of the measured $CaCl_2$. Reactions involving aluminosilicate minerals and fluids derived from the Salina A-1 potash salts are suggested to have been important in the evolution of these brines.

1 INTRODUCTION

The Silurian aged formations in the Michigan basin consist of interbedded carbonates, dolomites and a thick sequence of halite and potash. These formations contain extremely concentrated $CaCl_2$ brine.

For example, Case(1945) described a brine collected from the Silurian dolomites in Michigan having a total dissolved solids content >640g/l and a cation composition of >95% Ca. This is one of the most highly concentrated Ca-rich brines yet found in the crust. This paper suggests the diagenesis of potash salts has been important in the evolution of the $CaCl_2$ brines in Michigan.

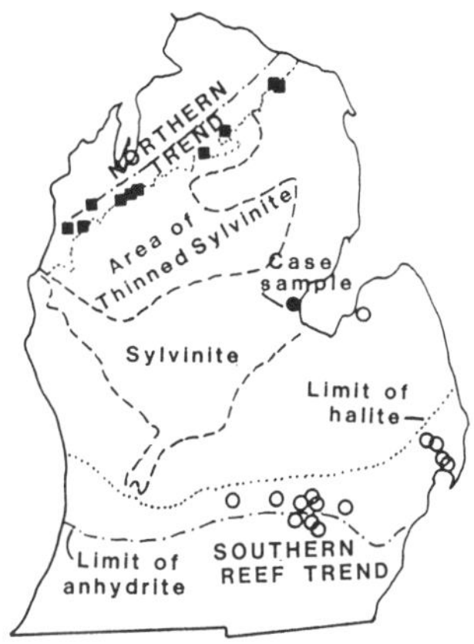

Fig. 1 Sample location map and Salina A-1 potash deposit.

1.1 *Study area*

The waters studied here are produced from the Silurian aged (Niagara Group) carbonate reefs in the northern and southern areas of the basin (Figure 1). The reefs are built upon the Niagaran dolomite and cut across the lower Salina salts (Figure 2). The reefs are from $400m^2$ to $20km^2$ in area and may reach >30m in height, and are highly dolomitized.

Also of interest is the Salina A-1 salt which contains potash (KCl) (Figure 1). The potash is sylvinite, a mixture of KCl (15-40%) and NaCl that occurs in 1m to 4m beds that coalesce into a 20m thick unit in the

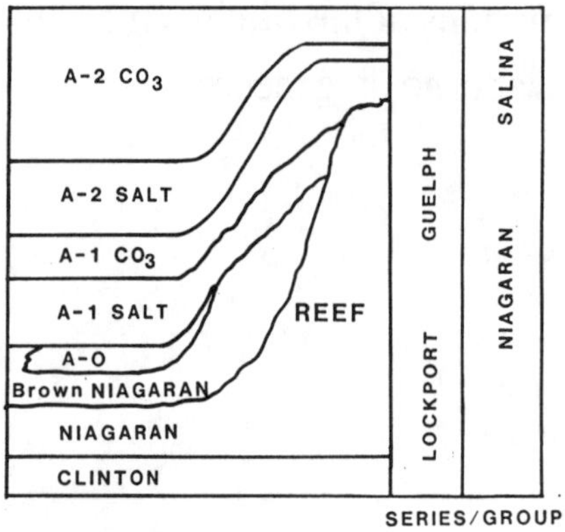

Fig. 2 Silurian stratigraphic section in Michigan.

Table 1. Composition of N/S waters in mg/l.

	mean (n=25)	Case	Seawater
Cl	215,000	403,207	185000
Br	2590	3500	2440
Ca	76,300	206,300	<126
Mg	11,800	7200	48200
Na	31,300	21	53800
K	9770	21,362	14500
ALK	91	1208	nm
SO_4	48	nm	64400
TDS	353,000	642,798	368340
MCl_2	4860	10866	2623

MCl2= Ca + Mg + Sr -HCO3 - SO4, in meq/l
n= number of analyses nm: not measurable
Seawater sample 36#1, McCaffery et al., (1987)

central basin. The sylvinite thins in the northwest and is truncated by an erosional surface with the A-1 carbonate unit, and grades into halite and anhydrite at the unit margins. The mineralogy of these units includes halite, sylvite, calcite and dolomite, along with polyhalite, Ca-borate nodules, authigenic quartz and chalcedony, hematite, pyrite, and illite-chlorite shales (Wilson and Long, 1991b).

2 METHODS

A description of the methods used is in Wilson and Long (1991a and b). 25 wells were sampled at the locations shown in Figure 1. Samples were diluted in-field to prevent salt precipitation, and were measured for temperature, pH, Eh and alkalinity. Major ions, selected trace elements and stable isotopes were measured using standard analytical techniques on suitable dilutions. Analysis was conducted at Michigan State University and at the University of Waterloo, Canada.

3 RESULTS

Complete results are presented in Wilson and Long (1991b). Table 1 lists the mean composition of the samples, the Case sample, and seawater concentrated to near the same degree as the average brine based on Br (McCaffrey et al., 1987).

The brines are extremely concentrated with TDS values averaging 353g/l. Ca is the dominant cation in the brines, followed by Na and Mg. Interestingly, the samples from the northern reef contain a greater percentage of Ca than do the southern trend samples, which contain equal proportions of Ca and Na (Figure 3). Cl is the dominant anion followed by Br with only minor amounts of SO_4 and alkalinity. The Case sample contains an anomalously high alkalinity, which was attributed to organic anions. No samples obtained in this study contained such a high alkalinity value.

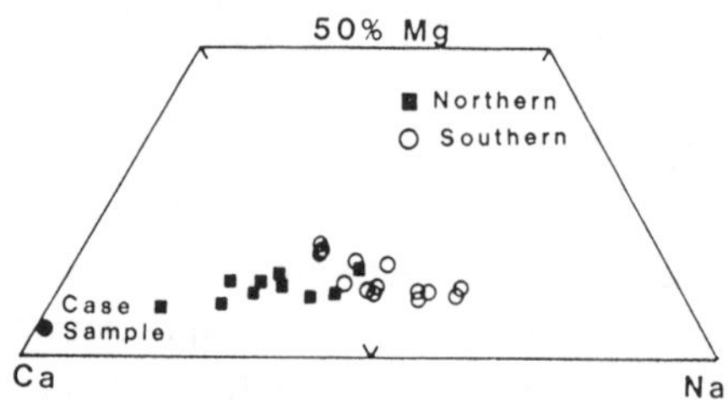

Fig. 3 Mole percentages of major cations in Silurian brines from Michigan.

3.1 Brine origin

The brine chemistry is compared with evapo-concentrated seawater to help determine the origin and evolution of the brine (Carpenter, 1978). Br is used as an indicator of degree of evapo-concentration. The generally good agreement between the brines and seawater in Cl and Br (Figure 4) supports that the brines originated from

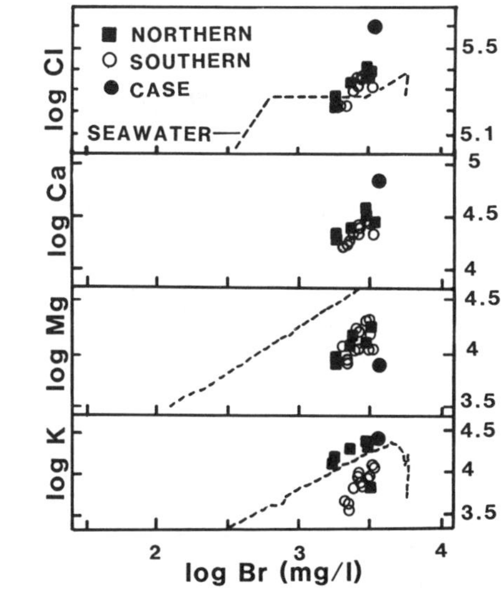

Fig. 4 Log concentration vs log Br (mg/l) in brines and seawater.

concentrated seawater. However, the samples plot in a linear trend that intersects the trajectory near the inflection signifying the start of $MgSO_4$-salt precipitation.

Samples having the highest Br and TDS, such as the Case sample, plot well above the seawater trajectory, indicating that they are enriched in Cl or possibly depleted in Br.

3.2 Ca-Mg-K chemistry

The brines are enriched in Ca and depleted in Mg from values in equally concentrated seawater (Figure 4). This is most likely the result of dolomitization, as it can be shown that the reaction; $Mg^{2+} + 2CaCO_3 = Ca^{2+} + CaMg(CO_3)_2$ explains a large percentage of the Ca in the brines and the difference in Mg between the brines and equally concentrated seawater (Wilson and Long, 1991b). However, many samples including the Case sample contain a large excess of Ca balanced by Cl that is not explained by dolomitization.

The K concentrations of the southern reef samples are lower than equally concentrated seawater, while most of the northern samples have values that match or exceed seawater concentrations. One explanation for the depletion is that K has been removed by reactions involving shales. The K depletion in the southern trend samples perhaps indicates a large amount of water-rock interaction has occurred in this area of the basin. The lack of depletion in the northern samples suggests a lack of mineral interaction, which is possible but not likely, or that the brines in this area were enriched in K by other processes.

3.3 Effects of salt diagenesis and mineral reactions.

Enrichment of $CaCl_2$ in formation waters has been attributed to reactions involving aluminosilicate minerals and carbonates (Carpenter, 1978) such as; $2KCl(aq) + CaCO_3 + Al_2Si_2O_3(OH)_4 + SiO_2 = CaCl_2 + 2KAlSi_3O_8 + H_2O + CO_2$. A similar reaction involving NaCl(aq) and authigenic albite can be proposed (Das et al., 1989). While these reactions are likely to have affected the brines, little supporting information exists on the types and abundance of authigenic silicates in these formations.

Another source for the CaCl enrichment is fluid derived from the diagenetic alteration of carnallite to sylvite; $KMgCl_3 \cdot (H_2O)_6 = KCl_{(s)} + MgCl_{2(aq)} + 6(H_2O)$. This exothermic reaction occurs when carnallite is buried or when it reacts with undersaturated seawater. Each 1 m^3 of carnallite produces 622 l of warm, $MgCl_2$-rich water, which would be altered to $CaCl_2$ by dolomitization. The water also has the potential to dissolve KCl and NaCl. Computer model results using PHRQPITZ (Plummer et al. 1989) support this and show that water in equilibrium with halite, sylvite, carnallite, dolomite and calcite has a composition that resembles the Case (1945) sample in Ca-K-Cl content (Wilson and Long, 1991b).

These results suggest that fluids liberated from the A-1 potash during basin compaction could have entered the Niagaran formations and enriched preexisting seawater brine with Ca-K-Cl. The history of these potash salts is not well known. Available Br data showing that the potash originated as sylvite can be interpreted to show they originated from carnallite. The erosional surface between the A-1 Salt and Carbonate units (Figure 1) is consistant with this origin.

4 CONCLUSIONS

Cl/Br ratios in brine produced from the Silurian formations of Michigan indicates a seawater origin. However, the concentrations of other major-ion components have been extensively modified by water-rock interactions, especially by dolomitization. The brines are highly enriched in $CaCl_2$ over equally concentrated seawater, which can be attributed to aluminosilicate mineral reactions and possibly to an input of fluids derived from the Salina A-1 potash.

5 ACKNOWLEDGEMENT

This research was funded by the A.C.S. Petroleum Research Fund (PRF #17606-AC2), Amoco Oil Company and the Shell Oil Company.

REFERENCES

Carpenter A.B. (1978) Origin and chemical evolution of brines in sedimentary basin. *Okl. Geol. Surv. Circ.* 79, 60-77.

Case L.C. (1945) Exceptional Silurian brine near Bay City Michigan. *Bull Am. Assoc. Petrol. Geol.* 68, 130-136.

Das N. and Holland H.D. (1989) Fluid inclusions in halite from the Salina group of the Michigan basin, and the chemistry of Late Silurian seawater. *Geochim. Cosmochim. Acta* 54,319-328.

Door J.A. and Eschmsn D.F (1984) *Geology of Michigan.* The Univ. Michigan Press.

McCaffrey M.A., Lazar B. and Holland H.D. (1987) The evaporation path of seawater and the coprecipitation of Br and K with halite. *J. Sediment. Petrol.* 57, 928-937.

Plummer L.N., Parkhurst D.L., Fleming G.W. and Dunkle S.A. (1988) A computer program incorporating Pitzer's equations for calculation of geochemical reactions in brines. U.S. Geol. Survey. Wat. resour. Invest. 88-4153. 1989

Wilson T.P. and Long D.T. (1991a) The geochemistry and isotopic chemistry of the Michigan basin brines: The Devonian formations. *Applied Geochemistry* (in press).

Wilson T.P. and Long D.T. (1991b) Origin and evolution of $CaCl_2$ brine in Silurian aged formations of the Michigan basin. In preparation.

Sedimentary basins

Mineral transformations and interactions

Water-Rock Interaction, Kharaka & Maest (eds) © 1992 Balkema, Rotterdam. ISBN 90 5410 075 3

Constraints for an evaporitic sedimentation modeling: Microanalysis of frozen fluid inclusions in halite

A.Canals, B.Carpentier, A.Y. Huc & B. Durand
Geology-Geochemistry Division, Institut Français du Pétrole, Rueil Malmaison, France

ABSTRACT: In the Mulhouse basin (France), part of the Tertiary Upper Rhine graben, the lithology and organic carbon content reveal an extended succession of cyclic sequences characterized by alternating halite, anhydrite and shale horizons which highlights the variations of depositional environment. A numerical model named Simsalt has been constructed (Carpentier et al., in press) in order to test hypothesis accounting for the sedimentogical sequence and organic matter distribution in an evaporitic basin permanently connected to the open sea. The Simsalt model is designed to provide a computed lithological column and associated organic content profile at a given location of the basin. Microanalysis of frozen primary fluid inclusions using Ayora & Fontarnau (1990) modified technique give a direct chemical composition of brine involved in halite precipitation to reconstruct a more accurate sedimentological environment. Halite fluid inclusions analysis have been carried out along one of the cored cycle of the Salt IV in the Mulhouse basin. Special care has been taken to analyze only fluid inclusions coming from halite primary textures. The brines chemical compositions obtained have been used as output constraint to estimate the initial fluid composition. This initial composition and the implementation of chemical approach in the Simsalt program provide a better understanding of the interplaying sedimentological factors (eustatism, fluids balance and evaporation) which control the evaporitic sedimentation in the Mulhouse basin.

1.INTRODUCTION

The numerical simulation of evaporitic sedimentation is one of the best ways for understanding the interplaying sedimentological factors that control sediment deposition (Harvie et al., 1980; Risacher and Fritz, 1984; Sanford and Wood, 1991 among others). Although the numerical approach allows us to take into account a lot of parameters, some of them being interdependent and others playing an independent role, the validity of the model solution could be easily critized due to the high degree of freedom that exists in the system.

The aim of this note is to report the result of a numerical evaporitic model which uses the lithological column and primary fluid inclusion data from the Tertiary evaporite sequence of Mulhouse basin (Upper Rhinegraben, France), to constraint some of the parameters. The physical parameters are derived from Simsalt model (Carpentier et al., in press) and the mass transfer module uses Pitzer equations .

2. GEOLOGICAL SETTING

The Rhine graben belongs to the Western-European intracontinental rift system that extends for more than 1,600 km, from the Mediterranean to the North Sea. The Upper Rhine graben is a Tertiary rift system 150 km in length and 10 to 25 km wide located along the French and German border. Within upper Eocene-lower Oligocene evaporitic deposits of the Upper Rhine graben, distinct marine influences are identified in the "Fossiliferous Zone" on the basis of paleontological evidence. The geographical location of the sea is still controversial (Perialpine Sea from the SE?, North Sea from the NW?) (Fontes et al., 1991). The marine conditions are definitively established at the end of the evaporitic episode. In the Mulhouse basin, a 1,600 m thick upper Eocene to lower Oligocene evaporitic series has been deposited. The underlying middle Eocene consists of limestone, clay and lignite deposits suggesting an environment of small lakes just prior to rifting. The evaporitic sequences can be divided in three parts from bottom to the top: Lower Salt (860 m of average thickness), Middle Salt (300 m of average thickness) and Upper Salt (530 m of average thickness). The upper part of this evaporitic series (Salt IV-V and Marls without Salt) shows no indication of sedimentary break nor of reworking, or extensive dissolution. It may thus be considered to have resulted from continuous sedimentation (for more details see Blanc-Valleron, 1991).

The evaporites from Salt IV are formed most likely as subaqueous deposits. The halite beds, 5-20 m thick, contain decimetric horizons of shale and/or anhydrite. The shaly-anhydritic beds (average 5 meters thick) are rhythmic interlayering of shales and anhydrite. Potash salts are virtually all found in two laterally continuous zones in the center of the basin. A lower potash interval (Couche Inférieure) of sylvite (up to 5 m thick) with sylvite and, in places carnallite, is separated by about 20 m from the upper zone (Couche Supérieure).

3.FLUID INCLUSION DATA

Fluid inclusions can provide more direct and precise information on the paleoenvironments where halite was deposited and the post-depositional diagenesis experienced by the halite (Roedder, 1984). Microanalysis of frozen primary fluid inclusions using Ayora & Fontarnau (1990) modified technique give a direct chemical composition of the brine involved in halite precipitation. Special care has been taken to analyze only fluid inclusions coming from halite primary textures. Chevron textures are significant because the inclusions were unequivocally trapped during crystal growth, and excellent preservation of fine fabrics suggests that the inclusions are unaltered.

Two scales of sampling have been done. One corresponds to an entire cored cycle. These cycles, which have been defined by Blanc-Valleron et al. (1991) as being halite-rich in the lower part of the interval and anhydrite-rich in the upper part (the organic-carbon profile, as displayed by the Carbolog method (Carpentier et al., 1991), shows a similar sedimentation pattern). The cored cycle corresponds to 23m of material from which the lower 15m has been sampled. In order to avoid dilution phenomena on the evaporation path (dilution represented by shaly-anhydritic beds), and to evaluate the meaning of the evolution trend found within one cycle, high frequency sampling was done within the Mi6 bed. Data reporting variations within one cycle are shown in Table 1. The overall accuracy of the analyses was cross-checked by calculating their saturation levels with respect to halite using the computer program from Zins-Pawlas (1988).

During the shale episode within halite beds, a high organic productivity, tentatively related to fresh water influx and a good preservation due to anoxic conditions established by the lack of mixing between the two water bodies (brine and fresh water), have been inferred to a maximum of TOC (Total Organic Carbon) in the shaly beds, as Blanc-Valleron et al., (1991) have already shown.

4. PHYSICO-CHEMICAL APPROACH

A numerical model named Simsalt has been constructed (Carpentier et al., in press) in order to test the hypothesis accounting for the sedimentological sequence and organic matter distribution in an evaporitic basin permanently connected to the open sea. The Simsalt model is designed to provide a computed lithological column and associated organic content profile at a given location of the basin. The chemical composition of the brine within a unit volume is computed according to the balance between

Table 1. molality of primary fluid inclusions. (SI saturation index)

sample	Cl	SO_4	Na	K
6,81	7,6	0,04	1,1	0,4
10,08	9,5	0,06	0,7	0,3
13,83	5,5	0,02	0,8	0,3
14,94	7,6	0,06	1,0	0,7
16,25	8,1	0,09	1,2	0,6
18,20	5,3	0,06	1,2	0,5
19,79	3,8	0,04	2,3	0,5
20,50	5,0	0,08	3,1	0,4

sample	Mg	Ca	Li	S.I.
6,81	3,3	0,2	0,2	0,02
10,08	4,1	-	0,4	-0,01
13,83	2,0	-	-	-0,68
14,94	2,7	0,1	0,2	-0,08
16,25	2,0	0,05	0,2	-0,18
18,20	1,2	-	-	-0,50
19,79	1,6	0,1	0,2	-0,15
20,50	1,2	0,04	0,1	-0,25

water influx (sea water, continental fresh water, eustatic variation) and regional evaporation using a simple chemical model. These variables are functions of time and are described by sinusoidal functions.

For the Mulhouse basin, the most representative environmental situation accounting for the observed geological and lithological data corresponds to a situation where sedimentation is controlled by the change of the balance between evaporation intensity, fresh water influx and seawater replenishment (function of sea level change). The evaporitic sedimentation is triggered when evaporation is greater than fresh water influx: halite is deposited when one brine body occurs at low sea level stand, anhydrite is deposited when two brine bodies occur at high sea level stand. Shales are deposited when fresh water influx exceeds evaporation. Simsalt results have been used as a base for a more elaborate model using a "true" chemical model constrained by primary fluid inclusion data.

Fluid inclusions analysis have been used as output data in a mass transfer program which gives us the number of moles of a given brine component that has been converted into evaporitic minerals as well as the change in brine composition as evaporation proceeds. Although the analytical data are "saturation sufficient" within the halite precipitation interval, according to Plummer et al., (1983) (the set of chemical analyses of initial and final brines are sufficient to define saturation indices for a given set of plausible phases in the system), outside the halite precipitation interval no chemical data are available. The reaction path simulation should estimate the chemical composition of the initial solution which was able to produce, via evaporation, the brine involved in halite precipitation.

5. CONCLUSIONS

In the tertiary evaporite sequence of Mulhouse basin (France), the observed lithological column and the chemical composition of brine involved in halite precipitation, could be used for constraining the Simsalt model to estimate the climatic and eustatic factors which control the sedimentation. The input in a mass transfer model, that uses Pitzer equations, has been constrained by primary halite fluid-inclusion data. This mass transfer model is used within Simsalt as a chemical approach in the evaporitic sedimentation simulation.

Fluid-inclusion data shows an upward confining system where the episodic shale within halite beds does not represent a dilution along the evaporitic sequence. The fresh water does not mix with the lower brine, but it can modify the system by increasing some of the "continental" brine constituents (Ca, Li, Fe, etc.). These constituents have been detected by ICP-EAS analysis of individual fluid inclusions.

The Simsalt approach is a helpful tool to understand the interplaying sedimentological factors (eustatism, fluids balance and evaporation) which control the evaporitic sedimentation and the organic matter content on the stratigraphic sequence in the Mulhouse basin. More elaborate coupling of the physico-chemical approach should be done in the future taking into account leakage of brine by ground water outflux, kinetic and isotopic data.

ACKNOWLEDGEMENTS

This work was performed under the auspices of ENOG (European Network of Organic Geochemistry laboratories) and has benefitted from scientific discussion within the group. ENOG comprises Laboratoire de Géochimie, IFP, Rueil-Malmaison, FRANCE; Institute of Petroleum and Organic Geochemistry, KFA, Jülich, GERMANY; Laboratoire de Chimie organique des Substances naturelles, ULP, Strasbourg, FRANCE; Organic Geochemistry Unit, Delft University of Technology, NETHERLANDS ; Organic Geochemistry Unit, University of Bristol, U.K., and receives funding from the European Economic Community (EEC contract N. SCI-0211-C(TT)).

We are also greatly indebted to C. Ayora, J. Garcia from Barcelona for their helpful comments, E. Rosenberg from I.F.P. for the laboratory help and enthusiasm.

REFERENCES

Ayora, C. & Fontarnau, R. 1990. X-ray microanalysis of frozen fluid inclusions, Chem. Geol. 89: 135-148.

Blanc-Valleron, M.M. 1991. Les formations paléogènes évaporitiques du bassin potassique de Mulhouse et des bassin plus septentrionaux d'Alsace. Doc. BRGM. 204.

Blanc-Valleron, M.M., Gély, J.P., Schuler, M., Dany, F. & Ansart, M. 1991. La matière organique associée aux evaporites de la base du Sel IV (Oligocène inférieur) du bassin de Mulhouse (Alsace, France). Bull. Soc. géol. France. 162: 113-122.

Carpentier, B., Huc, A.Y. and Bessereau, G. 1991 Wireline logging and source rocks. Estimation of organic carbon contents by Carbolog method. Log. Anal. 32(2).

Carpentier, B.; Gély, J.P.; Blanc-Valleron, M.M & Huc, A. in press. Simulation de la sedimentation dans un bassin évaporitique à niveau d'eau sous influence eustatique: application au bassin paléogène de Mulhouse (Alsace, France)., Rev. Inst. Fr. Pét.

Fontes, J.C., Filly, A., Gaudant, J. & Duringer, P. 1991. Origine continental des évaporites paléogégènes de Haute Alsace: arguments paléoécologiques, sédimentologiques et isotopiques. Bull. Soc. géol. France. 162: 725-737.

Harvie, C.E.; Weare, J.H.; Hardie, L.A. & Eugster, H.P. 1980. Evaporation of seawater: calculated mineral sequences. Science. 208: 498-500.

Plummer, L.N., Parkhust, D.L. & Thorstenson, D.C. 1983 Development of reaction models for ground-water systems. Geochim. Cosmochim. Acta. 47: 665-686.

Risacher, F. & Fritz, B. 1984. Simulation de l'évaporation sur micro-ordinateur. Sci. Géol. Bull. 37: 239-251.

Roedder, E., 1984. Fluids in salt., Am. Mineral., 69: 413-439.

Sanford, W.E. & Wood, W.W. 1991. Brine evolution and mineral deposition in hydrologically open evaporite basins. Am. Jour. Sci. 291: 687-710.

Zins-Pawlas, M.P., 1988: Géochimie de la silice dans les saumures et les milieux évaporitiques, Thèse Univ. Strasbourg, 179p.

Water-Rock Interaction, Kharaka & Maest (eds) © 1992 Balkema, Rotterdam. ISBN 90 5410 075 3

Computer modelling of steam flood experiments

Jon S. Dudley
Imperial Oil Limited, Calgary, Alb., Canada

Craig H. Moore
C.H. Moore Research & Applications, Bloomington, Ind., USA

ABSTRACT: The application of a numerical code coupling reactions, including kinetics and fluid transport, to the study of artificial diagenesis associated with heavy oil reservoir stimulation has been tested by simulating published results of high-temperature flow-through experiments conducted on the Lower Cretaceous Clearwater Formation of the Western Canada Sedimentary Basin. This formation is a feldspathic (volcanic) litharenite containing about 14 authigenic minerals including silicates and carbonates.The spatial distribution of mineralogical changes in the experiments is reasonably matched by numerical simulations.

The simulations indicate that coupled mineral reactions control the temporal and spatial distribution of neogenic minerals including clays, carbonates and zeolites. Understanding such couplings may be key to improving steam stimulation of oil sand deposits.

1 INTRODUCTION

Numerous studies have shown that water-rock interactions occur during steam-enhanced recovery of bitumen from oil sand deposits such as the Clearwater Formation, Cold Lake, east-central Alberta, Canada. Evidence from both experimental (e.g. Kirk et al., 1987; Gunter and Bird, 1988) and field studies (e.g. Hutcheon and Abercrombie, 1990; Fialka et al., 1990) has established that the reactions generally involve carbonates (siderite, dolomite, calcite), silica, and aluminosilicates (kaolinite, feldspars,) as reactants and smectite, CO2 and calcite as products.

A computer code coupling reactions and fluid transport promises to address the impact of these reactions upon reservoir quality and production by simulating the quantity and spatial distribution of neogenic minerals formed during steam-assisted recovery. This study tests the applicability of a coupled code to the Cold Lake oil sands by simulating published high-temperature flow-through experiments on the Clearwater Formation.

2 HIGH-TEMPERATURE FLOW-THROUGH EXPERIMENTS

As a preliminary test of the coupled code the published experimental results of Kirk et al. (1987) were selected for comparison to simulations. These results are some of the few on artificial diagenesis of oil sands to document the distribution of neogenic minerals.

The Kirk experiments were designed to investigate steam flood mineral reactions in an autoclave by passing fluids of variable pH and salinities through sand packs of Clearwater Formation sand under temperature and pressure conditions similar to those encountered in a reservoir during steam injection. One experimental run (run 3 of Kirk et al., 1987) demonstrating dramatic mineralogical changes was chosen for simulation. The general experimental setup and Kirk's run 3 conditions are shown in Figure 1. After a 21 day run, the sand pack was frozen, split longitudinally, and one half quartered for analysis of mineralogical changes. Pre- and post-flood mineralogy was determined using petrography of grain mounts, scanning electron microscopy and X-ray diffraction analysis.

2.1 *Experimental results*

The following is a summary of results of experimental run 3 as reported by Kirk et al. (1987). Numerous mineralogical changes

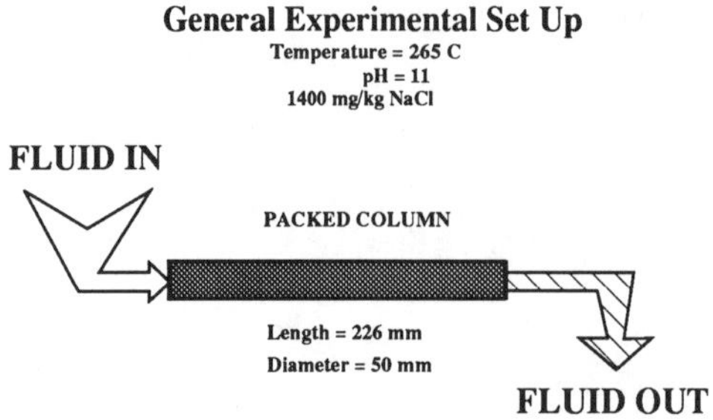

Fig. 1 Experimental setup and conditions for Kirk et al. (1987) run 3.

occurred involving carbonates, phyllosilicates, and framework silicates and are herein described as percent relative change. Dolomite decreased by 70% throughout the sandpack and more in the production quarter. Calcite increased by up to 200% with a minor decrease to 160% in the production quarter. Kaolinite decreased up to 50% throughout whereas smectite increased by up to 100%. Albite overgrowths were observed and K-feldspar overgrowths partly dissolved. Analcime, not present in the pre-flood sample, grew especially in the injection half with lesser amounts in the production half. Other zeolites tentatively identified as phillipsite were also observed to have grown.

A general increase in grains <38 μm was observed from 5% initial to 10-15% post-flood. This was accompanied by an 85% decrease in permeability from 6D to .9D.

3. THE NUMERICAL MODEL

The coupled numerical model used in this study is similar to REACTRAN (Ortoleva et al., 1987) in terms of its mathematics and algorithms and further details should be sought in the respective reference. Basic assumptions of the model can be found in Haskin et al. (1989). A linear one-dimensional flow version of the code was used in this study. Input to the code includes initial model mineralogy, inlet fluid composition, temperature, aqueous and mineral reactions, their equilibrium and rate constraints, the dimensions of the system, grid size, and the maximum time steps to be taken during the simulation. With each time step taken minerals are grown or dissolved according to their respective rate laws. Porosity and permeability are recalculated after changes in mineral amounts and new fluid velocities are determined for use in the next time step. New mineral surface areas are also determined.

4 DESCRIPTION OF MODEL INPUT

Inlet fluid velocity was held constant at 1.048 E-04 dm/sec and other conditions set to match those of the experiments (Figure 1). An initial grain size of .18 mm diameter was used for mineral grains other than clays and of .002 mm diameter for clays. Selecting the initial mineralogy was particularly problematic in that potentially reactive components of the Clearwater volcanic sediment such as volcanic rock fragments must be accounted for in the simulations yet are not readily characterized mineralogically. This combined with independent mineralogical descriptions of some of the experimental sand pack which differed from Kirk et al, (1987) made it necessary to construct a composite initial mineralogy based on all available data. For example, quartz, oligoclase and volcanic glass have been used to model the volcanic rock fragments and clays are modelled as individual clay components (Table 1). Thermodynamic and kinetic data were estimated by simple linear mixing formulae when data for end-members were available in the literature. When these data were not available, data used were those available for the mineral structurally closest to the one of interest, this being a first-pass simulation.

Table 1. Initial modal mineralogy (composite).

Silicates		Phyllosilicates	
QUARTZ*	22.9	ILLITE*	2.27
CHERT*	8.00	KAOLINITE*	0.70
MUSCOVITE*	1.00	SMECTITE(D)*	4.61
K-FELDSPAR*	5.75	SMECTITE(T)	0.03
OLIGOCLASE*	13.40	SEPTECHLORITE	0.09
"GLASS"*	10.50	CHLORITE*	0.15
Carbonates		**Secondary Minerals**	
CALCITE	0.70	ANALCIME	0.00
DOLOMITE**	2.20	K-CLINOPTILOLITE	0.00
SIDERITE**	0.70	AMORPHOUS SILICA	0.00
		ALBITE	0.00

vol % with 27% porosity

* occur in whole or part as rock fragments

** may not have been differentiated

5 RESULTS OF SIMULATION

Numerous simulations were performed until reasonable matches with the experimentally observed neogenic mineral changes were obtained. To obtain good matches refinements of rate constant values were required as were changes in initial modal mineralogy. For example, the initial volcanic glass content had to be reduced by 4 volume % to prevent excessive amounts of neogenic minerals being formed in the simulations. This may well reflect an armoring of the volcanic rock fragments by neogenic minerals as observed in the experiment. All changes, however, were constrained by observations of the system. Ultimately, reasonable matches between the observed and simulated mineral distributions were generally obtained. Figure 2, for example shows simulation results indicating the growth of some of the silicates observed to grow in the experiment. Note that the end of the experimental system occurs at 1.00 on the x-axis of the figures; simulations were run beyond this to detect any significant relationships missed by the experimental system.

The simulation shows a dolomite decrease especially toward the production end of the system and an increase in calcite with less calcite in the third quarter of the system (Figure 3). These trends are consistent with experimental observations but the per cent relative changes are less than those reported by Kirk et al. (1987). It is notable that a strong antipathetic relationship exists between the amount of dolomite and calcite and that per cent relative changes closer to those reported by Kirk et al. (1987) are predicted beyond the length of the experimental system.

Figure 4 depicts the distribution of trioctahedral smectite and ripidolite (chlorite) at two times during the simulation; time 2 is half the experimental time and time 4 equals the experimental time. At time 2, the two minerals show a moderate antipathetic relationship in the first and last quarters of the system and no strong relationship in the middle half. The nearly monotonic smectite distribution suggests that it is little affected by other reaction fronts at this time. The somewhat peaked ripidolite distribution, however, suggests control by other reactions and appears to be subtly

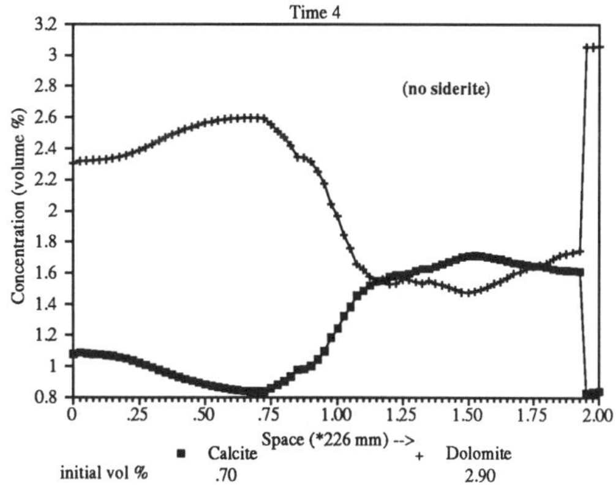

Fig. 3. Simulation at time 4 (21 days) showing relationship between calcite growth and dolomite dissolution.

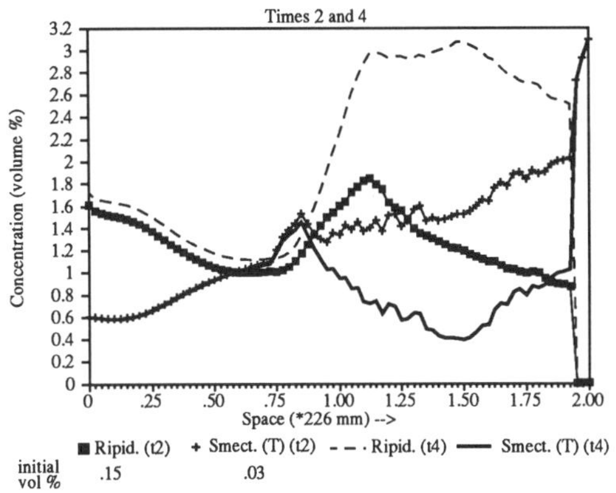

Fig. 4. The spatial distribution of smectite and ripidolite at two different times showing a change in system behaviour. Time 4 (21 days) is the same as in Figure 3.

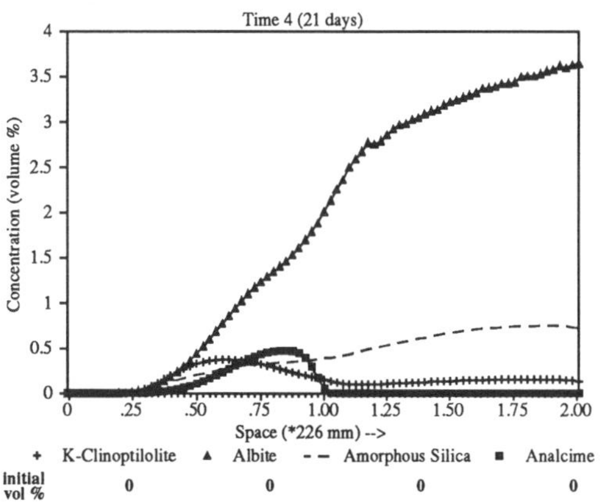

Fig. 2. Simulated spatial distribution of silicates at final time (21 days) is consistent with experimental results.

related to the calcite-dolomite-volcanic glass reaction interface. By time 4, chemical values critical to the coupling of the ripidolite and smectite reactions have been reached and we now see a strong antipathetic relationship in the region where the amounts of dolomite and calcite are nearly equal (compare with figure 3). This region of near equality of dolomite and calcite modes is probably a result of the ripidolite-smectite coupling.

6 DISCUSSION

The iterative process of modifying the simulation

input in order to get a match with observations can be instructive in itself and is to be expected since the initial mineralogy was only a best guess. The required reduction in volcanic glass content, for example, may be an indication of overestimation of this hard-to-quantify phase or that a significant portion of it is rendered non-reactive. This would be consistent with the dark brown rims observed on grains in Kirk's post-steam material. This kind of information will be necessary to determine which potentially reactive solid phases will in reality have a diminished impact on neogenesis due to such mechanisms as armouring. At this stage it is considered more beneficial to better match the general trends in modal mineralogy as opposed to absolute amounts for the purpose of identifying strong couplings that are controlling neogenic mineral distribution. Some of the couplings identified in this preliminary work may have application in the field. The coupling between calcite and dolomite illustrated in Figure 3, for instance, suggests that Clearwater zones rich in detrital dolomite (Longstaffe & Bird, 1988) should be avoided to minimize formation damage due to calcite cementation. Some subtle geological parameters controlling neogenesis may also be identified using coupled simulations. The coupling of calcite, dolomite, ripidolite and smectite reactions plays a significant role in the evolution of this system. This type of coupling may never be recognized through routine petrography, core description, or experimentation and is therefore one of the potentially profound contributions that coupled modelling could make to improving steam-assisted oil sands recovery by identifying new phenomena that are critical to the response of the system.

7 CONCLUSIONS

1) The Clearwater Formation - steam system is very complex and coupled reaction-transport modelling is a potential tool for understanding such complexity. 2) The reasonable match between experiment and simulation indicates that the numerical model is useful for identifying critical couplings which may have significant application in defining favourable reservoir quality in the Clearwater Formation. 3) Insights from this type of modelling may ultimately improve steam stimulation of oil sand deposits.

REFERENCES

Fialka, B.N., McClanahan, R.K., Robb, G.A., & Longstaffe, F.J. 1990. The evaluation of cyclic steam stimulation in an oilsand reservoir using post-steam core analysis. Paper 90-108 CIM-SPE International Technical Meeting, Calgary. Preprints 3.

Gunter, W.D. & Bird, G.W. 1988. CO2 production in tar sand reservoirs under in situ steam temperatures: reactive calcite dissolution. Chemical Geology 70: 301-311.

Haskin, H.K., Hart, T.J., Martir, W.K., Moss, R.M., Dudley, J.S., & Moore, C.H. 1989. Modelling acid stimulation of the Halfway Formation, Canada, using REACTRAN, a geochemical computer model. Proceedings of the Sixth International Symposium on Water-Rock Interaction, Malvern, U.K.:289-292.

Hutcheon I.E. & Abercrombie, H.J. 1990. Fluid-rock interactions in thermal recovery of bitumen, Tucker Lake pilot, Cold Lake, Alberta. In I.D. Meshri & P.J. Ortoleva (eds.), AAPG Memoir 49, Prediction of reservoir quality through chemical modelling: 1261-170.

Kirk, J.S., Bird, G.W., & Longstaffe, F.J. 1987. Laboratory study of the effects of steam-condensate flooding in the Clearwater Formation: high temperature flow experiments. Bulletin of Canadian Petroleum Geology 35, no.1: 34-47.

Longstaffe, F.J. & Bird, G.W. 1988. Data report: X-ray diffraction and stable isotope results for Clearwater Formation reservoirs in the Cold Lake area. Alberta Research Council - AOSTRA Open File Report A 8889-11, 82p.

Ortoleva, P., Merino, E., Moore, C., & Chadam, J. 1987. Geochemical self-organization I: reaction-transport feedbacks and modeling approach. American Journal of Science 287,: 979-1000.

Water-Rock Interaction, Kharaka & Maest (eds) © 1992 Balkema, Rotterdam. ISBN 90 5410 075 3

A stable isotope study of clay minerals from the Sancerre-Couy deep drill hole, Paris Basin, France

A.M.Fouillac
BRGM, Department of Geochemistry, Orleans, France

D.Beaufort
Université de Poitiers, Laboratoire Pétrologie des Altérations Hydrothermales, France

ABSTRACT: A mineralogical and isotopic study of authigenic minerals from veinlets in the metamorphic basement of the Paris Basin was performed on samples from cores of the deep drillhole at Sancerre-Couy. Two distinct stages of alteration affected the metamorphic rocks: a high temperature hydrothermal event dated at near 297 Ma and a low temperature alteration event. Oxygen, carbon and hydrogen isotopic analyses were carried out on calcite, albite and clay minerals. The present-day fluids, sampled during the drilling, have $\delta^{18}O$ values of 6.1 and 6.8 °/₀₀ and are not in equilibrium with the authigenic minerals. Large ranges of variation of $\delta^{18}O$ values of carbonates (from +7.9 to +23.5 °/₀₀ vs SMOW) and clay minerals (from +5.4 to +23.6 °/₀₀ vs SMOW) are interpreted as indicating the former existence of two hydrothermal systems separated in time. The first of these was probably related to cooling after the high temperature hydrothermal event.

1 INTRODUCTION

The Sancerre-Couy deep drill hole was initially drilled to investigate the magmatic anomaly of the Paris Basin. It provides the opportunity for a wider range of investigations such as water-rock interactions in the metamorphic basement and near the sediment basement interface. The drillhole reached 3500 m in depth, the unconformity between the sedimentary cover and the basement being at 942 m. Cores as well as fluids were collected during the drilling.

A mineralogical and stable isotope study was undertaken of samples of authigenic minerals in order to improve the understanding of the present day fluid circulation and of its interaction with rocks. The study showed the existence of fossil hydrothermal systems and has improved our knowledge of the history of this metamorphic basement.

2 GEOLOGY

The metamorphic basement of the southern part of the Paris Basin is essentially composed of interlayered hornblende-clinopyroxene-garnet amphibolite and biotite-garnet gneiss, dated at near 380 Ma, which have been affected by two distinct stages of alteration:

- A high temperature hydrothermal event, dated at near 297 Ma generated fracture-controlled crystallization of actinolite biotite, chlorite and phengite at different levels in the drillhole.
- A low temperature alteration event, structurally controlled by discontinuous late fractures related to a phase of NNE-SSW extension, affected the metamorphic basement down to the bottom of the hole. This event is characterized by successive mineral assemblages in open fractures and the adjacent wall-rocks. The same idealized sequence of mineral deposition, established from petrographic study of 100 thin sections, has been observed from top to bottom of the drilled basement. It can be summarized as follows:

1. albite + pyrite + adularia + quartz;
2. calcite + laumontite + anhydrite;
3. calcite + chlorite/smectite mixed-layer minerals + illite/smectite mixed-layer minerals.

3 ISOTOPE STUDY

3.1 Results

$\delta^{13}C$, $\delta^{18}O$ and δD analyses were carried out on calcite and clay mineral separates from the metamorphic basement, from the unconformity between the sedimentary cover and the basement at 942 m, up to 3478 m.

The $\delta^{13}C$ and $\delta^{18}O$ values of calcite vary from -15.1 to -5.9 ‰ versus PDB and from +7.9 to +23.5 ‰ versus SMOW, respectively. The δD and $\delta^{18}O$ values of clay minerals vary from -74 to -39 ‰ and +5.4 to 23.6 ‰ versus SMOW, respectively. Values are scattered near the unconformity and down to about 1700 m, the scatter then decreasing with depth. This could simply reflect a sampling effect, as the drilling was destructive between 2000 and 3200 m, few samples were collected in the lower part of the drillhole. The δD values of clay minerals are not so scattered as those of $\delta^{18}O$. The lowest values, between -74 and -64 ‰ are those of chlorite linked to the high temperature (around 350°C) hydrothermal event dated at 297 Ma and probably related to the intrusion of lamprophyres. For the other clay minerals (saponite, corrensite, chlorite/smectite mixed-layer minerals, illite/smectite mixed-layer minerals and laumontite) the range of δD values is -58 to -39 ‰.

Present-day fluids were collected during drilling at -558 m in the sedimentary cover and at -3272 m in the basement; the values for these are respectively $\delta^{18}O$=-6.8 and $\delta^{18}O$=-6.1 ‰ vs SMOW (Boulègue et al. 1990).

This communication focuses on the samples from the upper part of the drillhole down to 1883 m, i.e., on minerals related to the low temperature alteration event (chlorite related to the high temperature event at 297 Ma was sampled below 2000 m).

3.2 Present-day fluid-mineral equilibration

At the temperatures measured in situ in the drillhole, few of the calcite samples are in isotopic equilibrium with the present-day fluids. These samples could be the result of successive deposits, as shown in some cases by mineralogy. In this case the $\delta^{18}O$ of calcite is only a mean value of the different deposits in the same veinlet and the calculated isotopic temperature has no significance (Fouillac & Beaufort 1991).

The thermometric curves available for clay minerals are less precise than that for calcite. However, the same kind of calculations, using the available thermometric curves, indicate that few clay-minerals are likely to be close to isotopic equilibrium with the present-day fluids.

3.3 Fossil hydrothermal circulation systems

The distribution of $\delta^{18}O$ calcite values in two major groups, one with values lower than 13 ‰ (group L) and the other with values higher than 18 ‰ (group R), imply the existence of two fossil hydrothermal-circulation systems. Compared to present conditions of temperature and $\delta^{18}O$ of water, the hydrothermal circulation of group L was controlled by higher temperature and/or lower $\delta^{18}O$ values of water, whereas the hydrothermal circulation of group R was controlled by lower temperature and/or higher $\delta^{18}O$ values for water (Fouillac & Beaufort 1991). The $\delta^{18}O$ clay mineral values display the same distribution (group L with values from 5.4 to 13.7 ‰ and group R with values from 18.5 to 22.5 ‰, see Fig. 1) and confirm the previous result. However, results are more complex for clay minerals because of differences in mineralogy and in mineral water isotopic fractionation curves. Two secondary albites are also included in the results. The isotopic results show that the so called low temperature alteration, based on mineralogy, must in fact be attributed to two different fossil fluid circulations, separate in time.

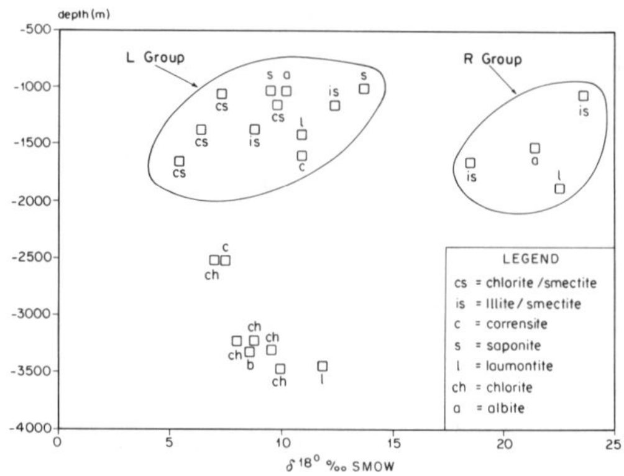

Figure 1 - Oxygen isotope values of albite and clay minerals versus depth

3.4 Thermometry

In some cases, minerals separated from the same veinlet were presumed to be deposited in the same conditions during the same event. In these cases, isotopic temperatures were calculated using the available thermometric curves. Three types of assemblages were obtained:

1. Fractures where minerals of group L have isotopic fractionations corresponding to a temperature around 200°C, in agreement with mineralogical data;

2. fractures where minerals of group L and of group R are both present and where minerals of group L show isotopic fractionation corresponding to lower temperatures or out of isotopic equilibria; and

3. fractures where clay mineral or albite of group R are present with calcite out of isotopic equilibrium.

In the two last types of fractures, partial isotopic reequilibration of minerals by the second fluid is suspected or more or less obvious, as in the case of two stage calcite samples.

3.5 Discussion

The distribution of isotopic values of group L and R minerals in the three different kinds of fractures indicates that the fluid circulation which deposited group L minerals was earlier than the fluid circulation which deposited group R minerals.

The first low temperature circulation, which deposited the group L minerals, took place at 200°C. The calculated oxygen value of the fluid is between 0 and +7 $^{0}/_{00}$ and is in the range of fluid values for sedimentary basins.

The second low temperature circulation, which deposited the group R minerals, probably took place at temperatures lower than 200°C, as indicated by the high ^{18}O values of the minerals.

The minerals of group L are albite, calcite, laumontite, corrensite, chlorite-smectite mixed-layer minerals and illite-smectite mixed-layer minerals. The same minerals composed group R, except for the trioctahedral minerals (saponite, corrensite and chlorite/smectite) which are not present.

This is the only mineralogical distinction between the two groups of minerals.

No distinction can be made between the δD values of minerals of the two groups which vary between -58 and -39 $^{0}/_{00}$. This means either that both fluids had similar δD values (the difference in temperature is too small to imply large difference in δD values of mineral), or that group L minerals suffered hydrogen isotopic exchange with the fluid of the second circulation. It could also be due to hydrogen isotopic reequilibration of fossil mineral deposits with the present-day fluid.

Referring to the geological setting, the first fossil fluid circulation at 200°C, can be related to cooling after the high temperature hydrothermal event linked to Stephanian volcanic events. The second fossil fluid circulation, at T<200°C, could be the result of fluid remobilization during Permo-Triassic tectonic events.

4 CONCLUSIONS

The mineralogical and isotopic study of authigenic minerals in the basement of the Paris Basin has revealed the existence of a high temperature hydrothermal event and of two different fossil fluid circulations at low temperature, at 200°C for the earlier circulation and at lower temperatures for the second. The fluids were both rich in ^{18}O, similar to fluids of sedimentary basins. The present-day fluid apparently precipitates few or no minerals.

REFERENCES

Boulegue J., M.Benedetti, B.Gauthier et B.Bosch 1990. Les fluides dans le socle du sondage GPF Sancerre-Couy. Bull. Soc. Géol. Fr. 8: 789-795.

Fouillac, A.M. & D.Beaufort 1991. Etude minéralogique et isotopique (^{13}C, ^{18}O) des carbonates néoformés dans le sondage profond de Sancerre-Couy (France. Bull. Soc. France, t. 162, 5: 939-945.

Water-Rock Interaction, Kharaka & Maest (eds) © 1992 Balkema, Rotterdam. ISBN 90 5410 075 3

Experimental study of reactions between selected clays and micas and aqueous fluid analogues of well-bore engineering fluids

C.R. Hughes & P.I. Hill
Clay Mineral Studies Unit, Department of Geology, University of Manchester, UK

ABSTRACT: Reactions between selected phyllosilicates and various aqueous fluids used during oil-field operations has been investigated in a matrix of batch and flow experiments. As expected, a very wide range of dissolution behaviour is exhibited. This varies according to the principal variables: mineral chemistry and structure, solution chemistry, temperature, pressure, flow rate and experiment duration. Whilst the precise mechanisms of phyllosilicate reaction have not been determined in any detail, the results demonstrate very rapid dissolution reactions occurring under a range of conditions relevant to oil-field engineering operations. Obviously, the results cannot be extrapolated simply to predict reactions in actual reservoirs. It is anticipated, nevertheless, that once the basic reactions are characterised, then experimental and numerical modelling approaches will lead to considerably improved predictive ability with respect to reservoir management during drilling, completion and workover operations.

1 INTRODUCTION

Authigenic clay minerals form an integral part of many sandstone hydrocarbon reservoirs. The clays precipitate as grain coating and pore bridging/occluding assemblages within the pore space. Prior to drilling the clays are broadly stable with respect to resident natural pore waters. Near well-bores, however, pore space is invariably invaded by a variety of complex artificial fluids introduced during drilling, completion and workover operations.

The reservoir clays vary considerably in structure and composition, and are usually extremely fine grained. Many are chemically reactive with high surface areas and ion exchange capacities, and as such there is potential for significant clay-fluid interaction with engineering fluids leading to near-well bore changes in permeability. These changes may have considerable impact upon reservoir management activities.

2 BACKGROUND AND EXPERIMENTAL RATIONALE

Previous experimental studies have used whole core or authigenic clays separated from cores (Hughes, Davey and Curtis, 1989a, 1989b). Although core flood tests are often carried out to determine reservoir sensitivity with respect to specific engineering fluids, the large number of chemical and physical variables involved mean that detailed interpretation of reaction paths, and thus predictive modelling, cannot be determined in any detail. Where authigenic clays have been separated from cores, isolation and evaluation of single clay population behaviour has proved very difficult because authigenic clay assemblages are usually mixtures. In an attempt to overcome these difficulties a selection of pure "end-member" clays and micas are currently being used to examine chemical reactivity of clays with respect to reagents representative of key components in drilling, completion and workover fluids.

Reported below are preliminary findings from the initial stages of an extended programme of clay-fluid reaction experiments. The overall programme is essentially a large matrix of batch and flow dissolution experiments with conditions varying from ambient up to "reservoir" temperatures and pressures. Batch experimental data only are reported.

3 EXPERIMENTAL METHODOLOGY

To date clays and micas used in experiments are commercially sourced pure minerals, used either as supplied or further refined to suitable size fractions (Table 1). Each mineral has been characterised in detail by XRD, SEM, ATEM and EMPA.

3.1 Batch experiments at 80° C ambient pressures

Selected clays and micas have been individually reacted in batch and flow experiments, with the following reagent solutions: KOH, NaOH, $Ca(OH)_2$, CH_3COOH, $C_6H_8O_7.H_2O$ and HCl all at Molar concentrations. Batch experiments were conducted using polymethlypentene flasks mounted in a thermostatically controlled agitating bath. 200 mg of clay or mica were suspended in 20 g of solution and agitated for periods of 8 and 144 hours at 80° C atmospheric pressure.

In addition, experiments using a fluid representative of a typical filtrate from a KCl/polymer water based drilling fluid system were carried out. The fluid was made up as KCl solution (1.5 Molar concentration) adjusted to pH 9.5 (measured at 22° C) by the addition of KOH. Experimental conditions were identical to above except durations were 6, 24 and 144 hours.

In all experiments leachates were filtered through 0.2 µm "Acrodisc" membrane filters and analysed by ICP-OES. Residual solids were freeze dried and analysed by XRD, SEM and ATEM.

Table 1. Summary data on starting materials

Starting Solid	Processing	Phases Present
kaolinite	used as supplied	kaolinite, tr.rutile
chamosite	refined to <8µm	trioctahedral Fe rich chlorite
clinochlore	crushed	trioctahedral Mg rich chlorite
muscovite	refined to <15µm	dioctahedral mica
ca-smectite	refined to >2<15µm	smectite, some cristobalite

3.2 Batch experiments at simulated reservoir temperatures

Experiments have been conducted in direct-sampling "Dickson type" autoclaves which allow extraction of fluids during experiments without disturbing P-T conditions.

To date experiments have been conducted using kaolinite with the KCl fluid as prepared above with pH adjusted to selected levels. 400 mg of each clay or mica have been reacted with 200 ml experimental fluid volume. Experiments were conducted at 120° C and 50 MPa.

Reactions were monitored by ICP-OES analysis of fluids extracted during each experiment run. Residual solids were freeze dried and then analysed by XRD, SEM and ATEM.

4. RESULTS

4.1 Batch experiments at 80° C ambient pressures

Analysis of all starting fluids show no significant concentrations of principal phyllosilicate cations.

In kaolinite experiments rapid stoichmetric dissolution was observed with NaOH and KOH and negligible reaction in other reagents after 8 hours. After 144 hours significant preferential leaching of Al was observed in KOH and $C_6H_8O_7.H_2O$ only. In the most aggressive solution, KOH, approximately 22% by weight dissolution of kaolinite occured.

In the case of muscovite reaction with HCl and $C_6H_8O_7.H_2O$, non-stoichmetric dissolution with preferential leaching of Al occurred within 8 hours and is continued up to 144 hours. Similar reaction is noted also with NaOH, KOH and CH_3COOH but with no marked changes in response between 8 and 144 hours.

Chamosite reacted markedly with HCl and $C_6H_8O_7.H_2O$ exhibiting rapid non-stoichmetric dissolution with preferential leaching of Al. In $C_6H_8O_7.H_2O$ the preferential dissolution of Al is further increased after 144 hours. Non-stoichmetric dissolution is evident also in NaOH and KOH with no changes in response from 8 to 144 hour duration experiments. Minor stoichmetric dissolution is observed in CH_3COOH.

Clinochlore behaved similarly to chamosite with HCl and $C_6H_8O_7.H_2O$ with non-stoichmetric preferential dissolution of Al, minor non-stoichmetric dissolution is evident in NaOH and KOH with minor stoichmetric dissolution in CH_3COOH. There were no marked changes in response for experiments of extended duration.

Fluid compositional data obtained from experiments with KCl/KOH solutions indicate minimal reaction for all the phases used except for Ca-smectite. Si concentrations are low, <10 ppm, for kaolinite, muscovite, chamosite and clinochlore. For Ca-smectite silicon levels increase to 26.4 ppm at 144 hours with no Al dissolution evident. pH monitoring of the experiments showed that for kaolinite, muscovite, chamosite and Ca-smectite final solution pH was between 6.0 and 7.0. Final solution pH for the reactions with clinochlore were significantly higher at pH 8.5.

XRD data from the Ca-smectite residue showed a decrease in 001 basal spacing suggesting partial ion exchange of both Ca and Mg for K.

4.2 Batch experiments at simulated reservoir temperatures

In the first experiment KCl solution was adjusted to pH 9.6 (at 20° C) using KOH and run for 350 hours. Only Si was detected in leachate samples, reaching a stable concentration of 13 ppm after 150 hours. pH was monitored with every fluid sample taken and a decline to pH 6.6 was noted after 16 hours, after which it remained constant at +/- 0.3 pH units.

The second experiment used KCl solution adjusted to pH 12.0 (at 20° C), run for 640 hours. Fluid data for Si and Al are presented in Figure 1. The first fluid sample was taken after 1 hour at 120° C, following a 6 hour heating period. Both Si and Al were significantly

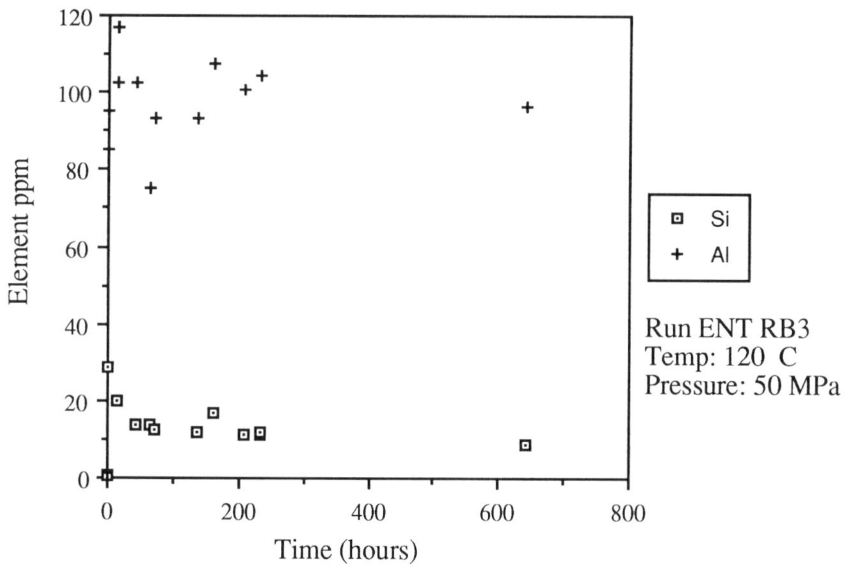

Fig. 1 Element concentration versus time for the reaction of kaolinite with KCl

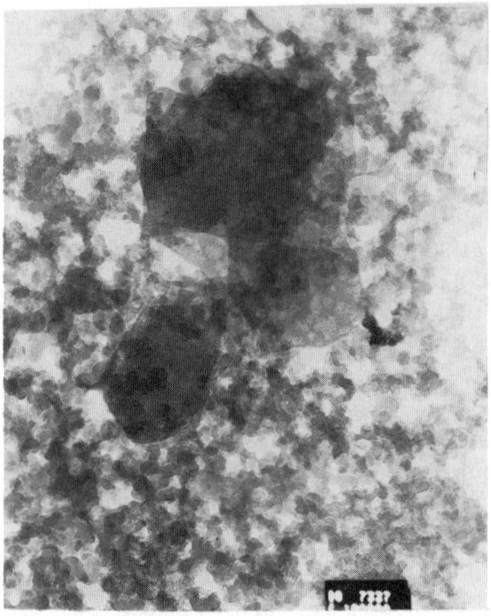

Fig. 2 Electron micrograph showing kaolinite with poorly ordered siliceous material Mag x 100,000

leached, Al at much higher concentrations than Si. The spread in Al data reflects poor ICP-OES precision for the fluid analysed. Si attains its highest concentration level after 1 hour, but progressively declines to 8.7 ppm at the end of the experiment. A pH drop of only 0.3 was observed during the first 16 hours. Lower and rather inconsistent values for pH were recorded for the rest of the run. This variance in pH may be due to sampling procedures (further investigations of the phenomenon are being undertaken).

ATEM examination of the solids retrieved post-experiment have identified a fine grained (0.05 μm) poorly ordered siliceous material (Figure 2) which on analysis contain up to 8 wt% CaO.

5 CONCLUSIONS

Whilst the precise mechanisms of phyllosilicate reaction have not been determined, it is clear that significant and very rapid dissolution reactions occur under a range of conditions. In some cases this may be followed by the precipitation of poorly ordered siliceous material. Where the siliceous material has incorporated significant Ca, some analogy can be drawn with calc-silicate-hydrate generation during cement setting.

Clearly, results from the experiments summarised above cannot be extrapolated simply to predict actual reactions around well-bores in real reservoirs. It is anticipated, however, that once basic reactions are more fully characterised, then experimental and numerical modelling approaches may lead to a considerably improved predictive ability with respect to reservoir management during drilling, completion and workover operations.

ACKNOWLEDGEMENTS

The research described is funded entirely by Enterprise Oil Plc. Special thanks are due to Dr K.P. Isaac, Senior Production Geologist with Enterprise, for his considerable contribution towards the development and running of the programme.

REFERENCES

Hughes, C.R., Davey, R.C., and Curtis C.D. 1989a. Aqueous solution reactions of hydrocrabon reservoir clays: Kaolinite, smectite, illite, chlorite. Proceedings of the 6th International Symposium on Water-Rock Interaction, 309-312.

Hughes, C.R., Davey, R.C., and Curtis C.D. 1989b. Chemical reactivity of some reservoir illites: Implications for petroleum production. Clay Minerals, 24, 445-458.

Water-Rock Interaction, Kharaka & Maest (eds) © 1992 Balkema, Rotterdam. ISBN 90 5410 075 3

Hydrodynamic and chemical controls of Paleozoic phosphorite and barite in the western Cordillera of North America

P.W. Jewell
University of Utah, USA

ABSTRACT: Organic-rich shales, bedded cherts, bedded barites, and phosphorites are common sediments in Paleozoic marine facies of the western Cordillera of North America. Field studies suggest that at least some of the phosphorite and bedded barite precipitated syngenetically from Paleozoic seawater. Elevated concentrations of nutrients (such as phosphorous) and nutrient-like elements (such as barium) is one plausible mechanism for this syngenetic precipitation. Recent field and geochemical work suggests that at least some phosphorites and bedded barites in deep sea sediments were deposited beneath coastal upwelling zones. In the modern ocean, surface biological productivity, mixing along isopycnal surfaces, and longshore transport of nutrients by bottom currents control the concentration of nutrients in coastal upwelling zones. A mass balance model which incorporates these processes suggests surface productivity and isopycnal mixing are the most important of these three factors. Although it is an extremely difficult parameter to quantify, isopycnal mixing appears to be related to distances over which isopycnal surfaces intersect the benthic boundary layer. The mass balance model and analysis of benthic boundary layer dynamics suggest that bathymetric configurations of the Paleozoic Cordilleran upwelling zones were critical in controlling the deep ocean chemistry and therefore precipitation of the barites and phosphorites.

1 INTRODUCTION

Upwelling represents a key link in the physical, chemical, and biological cycles of the ocean. Intermediate-depth, nutrient-rich seawater is brought to the surface via Ekman transport along the edges of ocean basins or in the open ocean due to wind divergence of surface wind stress. The resulting, enhanced biological activity consumes the nutrients and fluxes organic material back to the deep water and bottom sediments where remineralization processes consume oxygen. In addition to providing a connection between shallow and deep marine processes in the modern ocean, upwelling has produced a number of geochemical and sedimentological features which have been incorporated into the geologic record.

Among the most notable and important products of upwelling zones from the geological past are various types of chemical sediments which today constitute important economic and scientific artifacts of ancient oceans. Examples include bedded cherts, bedded barites, phosphorites, and organic-rich shales. Many of these rocks appear to have been deposited under conditions considerably different than those of the modern ocean. In addition to their importance in paleoceanography, many sediments below upwelling zones constitute economically important formations such as petroleum source beds, and as such they have been the subject of considerable research by both industry and government groups.

In this communication, I review the geologic and paleoceanographic setting of two Paleozoic formations which are believed to have been deposited in coastal upwelling settings. A simple, mass balance model of the controls of dissolved phosphorous and oxygen in coastal upwelling zones is then presented and applied to the Paleozoic settings.

2 PALEOZOIC PHOSPHORITES AND BEDDED BARITES

Phosphorites and bedded barites are found in a variety of sedimentary settings of various ages from throughout the world. Two Paleozoic phosphogenic and barite-bearing formations are discussed here on the basis of recent research suggesting that they were deposited beneath coastal upwelling zones. In both cases, precipitation of the chemical sediments is believed to have been the result of elevated nutrient or nutrient-like elements in the water column.

2.1 Phosphoria Formation (Permian)

The Phosphoria Formation outcrops extensively throughout the western United States. The formation consists of several members, with the most phosphogenic being the Retort phosphatic shale member and the Meade Peak phosphatic shale member. In the Meade Creek member, phosphatic oolite beds a few centimeters thick extend over thousands of square kilometers. Laminated, phosphatic shale is also found throughout the Meade Creek member, and is taken as evidence of inorganic apatite precipitation at the sediment-water interface (Sheldon, 1980). In addition to elevated amounts of phosphorite, chert, and organic-rich shales, cool water invertebrate faunal assemblages in the Phosphoria suggest the intrusion of cool water over a continental shelf (Wardlaw, 1980). Paleogeographic reconstructions suggest that the Phosphoria was deposited along a west facing, continental margin at a latitude appropriate for coastal upwelling (Parrish, 1982; Schopf, 1983). The general paleoceanographic setting of the Phosphoria formation is that of a broad continental shelf which may have been hundreds of kilometers wide, and was subjected to periodic intrusions of warm and cold (i.e. upwelling) waters.

2.2 Slaven Chert (Devonian)

The Late Devonian Slaven Chert outcrops throughout north-central Nevada. The formation is dominantly chert with lesser amounts of shale, siltstone, limestone, and bedded barite. The Slaven Chert forms a part of the Roberts Mountain allochthon which emplaced rocks of Ordovician through Devonian age over carbonate rocks of similar age during the Antler Orogeny in Late Devonian or Early Mississippian time. The Slaven contains economically important bedded barite which occurs as massive, stratiform lenses, concretions, and rosettes in the cherts and shales.

Recent work by Coles and Snyder (1985) and Coles and Varga (1988) suggests that the Slaven Chert and a number of other deep water Paleozoic formations in the western Cordillera were deposited beneath coastal upwelling zones. This conclusion is based largely on the ubiquitous presence of phosphorites in these units and paleolatitudes of the formations. Jewell and Stallard (1991) elaborated upon this idea to explain the presence of bedded barites in the Slaven Chert. This model explains the barite as the result of vigorous upwelling which leads to enhanced levels of barium and sulfate reduction in the underlying waters. Barite precipitates when the barium-rich, sulfate-depleted seawater interacts with oxygenated (i.e. sulfate-bearing) seawater.

3 MASS BALANCE MODELS OF COASTAL UPWELLING

A box model of coastal upwelling has been constructed in order to study the relative importance of specific physical and chemical boundary conditions of coastal upwelling zones. In this model, the deep water portion of the upwelling zone is considered to be a single box. Three basic mechanisms exist for transporting chemical components in or out: (1) particle fluxes from the surface, (2) offshore transport along isopycnal surfaces, or (3) transport by deep, longshore currents (Fig. 1). Analysis of geochemical data from modern upwelling zones demonstrates that in water beneath poleward undercurrents, the first two of these mechanisms are probably the most important (Jewell, in review). Mass balance equations highlight the critical

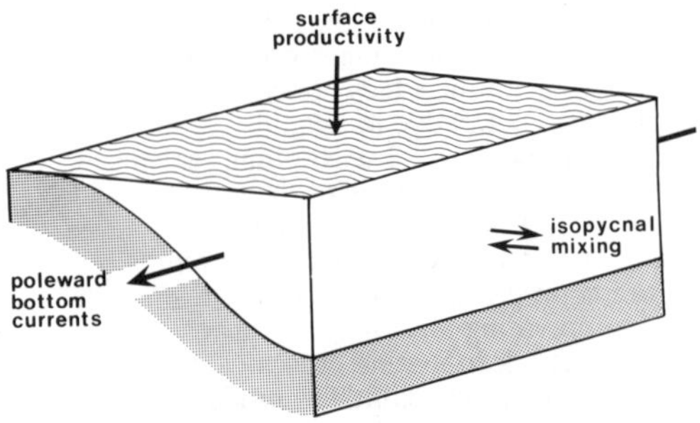

Figure 1. Diagrammatic representation of a coastal upwelling zone showing the three possible mass influxes of nutrients or oxygen: (1) surface productivity, (2) isopycnal mixing, and (3) bottom currents.

relationships between oxygen levels, nutrient levels, surface productivity, and isopycnal mixing. For oxygen, this relationship is:

$$O_{2,dc} = O_{2,do} - BOD/K_d$$

where $O_{2,dc}$ represents oxygen concentrations in deep coastal waters, $O_{2,do}$ represents open ocean water feeding into the upwelling zone, BOD is biological oxygen demand from sinking organic material, and K_d is cross-shelf isopycnal mixing rates. As K_d decreases, deep coastal oxygen decrease dramatically (Fig. 2). A similar equation and arguments can be constructed to explain elevated phosphate concentrations.

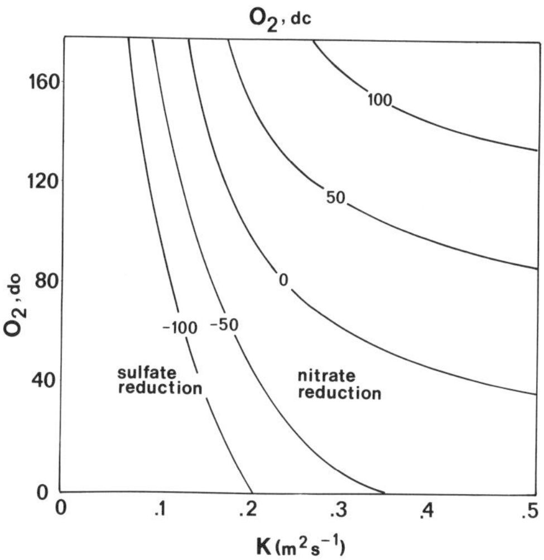

Figure 2. Dissolved oxygen concentrations in the interior of an upwelling zones ($O_{2,dc}$ in micromoles/L) as a function of deep ocean oxygen ($O_{2,do}$) entering the upwelling zone and isopycnal mixing rates. The mixing rates are a 2-dimensional mass flux, i.e. volume per longshore length per time. Oxygen concentrations from 0 to -30 in micromoles/L represent nitrate reduction and those below -30 in micromoles/L represent sulfate reduction. The diagram shows that nitrate reduction and sulfate reduction will be common at low isopycnal mixing rates, regardless of the dissolved oxygen concentration which enters the upwelling zone. Similar 2-5 fold increases of dissolved phosphate will also occur.

Unfortunately, there are very few ways to definitively predict isopycnal mixing rates. Geochemical tracer studies and physical considerations suggest that isopycnal diffusivity is related to the scale of the flow (Okubo, 1971; Lietzke and Lerman, 1975). Restricted settings or those close to bottom boundary layers tend to have lower isopycnal mixing than open settings far from boundary layers.

4 BATHYMETRIC AND DYNAMIC SETTINGS OF PALEOZOIC UPWELLING ZONES

Bathymetric configurations of coastal upwelling zones in the modern ocean are variable. For instance, near the well studied upwelling zone off Cape Blanc of northwest Africa (21°-22° N.), upwelling and the zone of maximum carbon accumulation occurs over the outer portion of a relatively wide continental shelf (Fig. 3). Between 15° and 16° S. off of Peru the continental shelf is much narrower, and carbon accumulation is greatest in the deeper waters of the shelf break. In all modern coastal upwelling, the high sea levels of the current interglacial period allow significant amounts of organic carbon to accumulate on the continental shelves.

The bathymetric slope of the Late Devonian ocean which bordered North American was probably much more extreme. The eastern margin of North America at this latitude was dominated by a thick sequence of bank carbonate deposits (Stewart, 1980). Under these conditions, the bathymetry of the continental margin may have resembled that of the modern day Bahamas (Fig. 3). Under these conditions, virtually all organic matter produced in the photic zone would be fluxed to very deep waters, rather than being sequestered in continental shelf sediments. In addition, the length scale of deep water, isopycnal mixing was probably restricted, leading to enhanced oxidant reduction and nutrient accumulation (Fig. 2).

The depositional setting of the Phosphoria Formation during Permian time may have resembled modern northwest Africa (Fig. 3) i.e. a wide continental shelf. Under these conditions, isopycnal mixing rates would be relatively large. The cause of inorganic apatite precipitation is speculative, but may have been the result of extremely high surface productivity, which in turn may have been caused by very high wind shear stress.

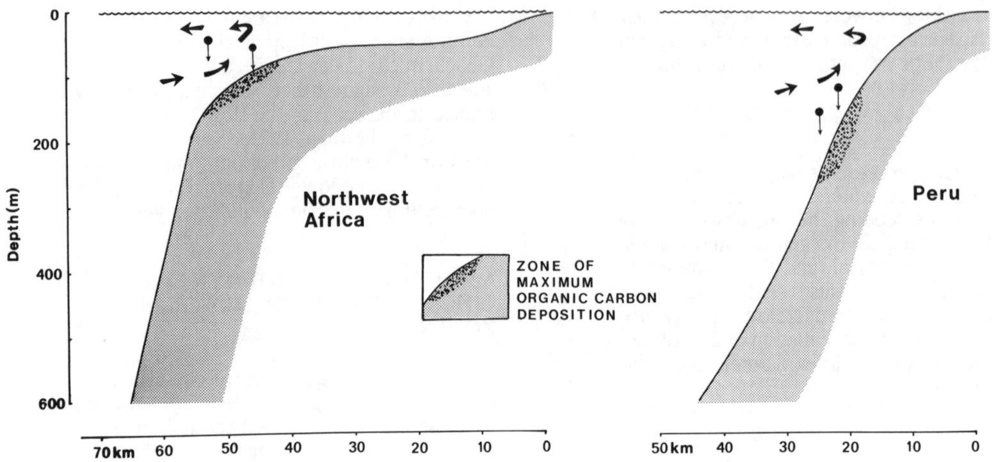

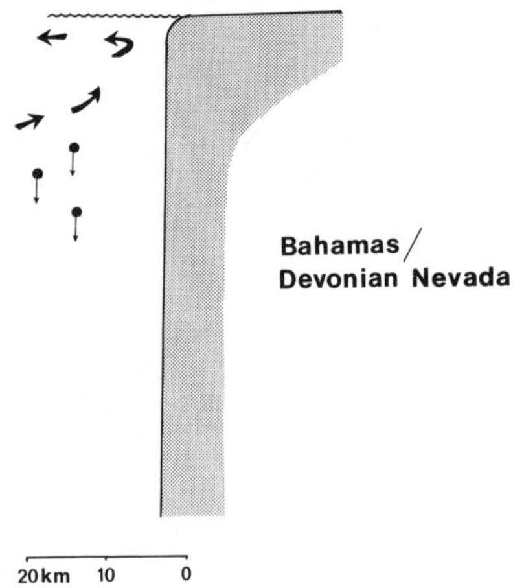

Figure 3. Generalized patterns of circulation and organic carbon accumulation for the modern northwest Africa upwelling system, the modern Peru system, and a proposed Devonian upwelling system in what is now Nevada. Bathymetry of the Nevada setting may have resembled that of the modern Bahamas bank carbonates.

REFERENCES

Coles, K. S., and Snyder, W. S., 1985, Significance of lower and middle Paleozoic phosphatic chert in the Toquima Range, central Nevada: *Geology*, v. 13, p. 573-576.

Coles, K. S., and Varga, R. J., 1988, Early to middle Paleozoic phosphogenic province in terranes of the southern Cordillera, western United States: *Am. J. Sci.*, v. 288, p. 891-924.

Jewell, P. W., in review, Mass balance constraints of modern coastal upwelling and application to the last glacial maximum: *Paleoceanography*.

Jewell, P. W., and Stallard, R. F., 1991, Geochemistry and paleoceanographic setting of central Nevada bedded barites: *J. Geol.*, v. 99, p. 151-170.

Lietzke, T. A., and Lerman, A., 1975, Effects of bottom relief in two-dimensional oceanic eddy diffusion models: *Earth Planet. Sci. Letters*, v. 24, p. 337-344

Okubo, A., 1971, Ocean diffusion diagrams: *Deep Sea Res.*, v. 18, p. 789-802.

Parrish, J. T., 1982, Upwelling and petroleum source beds, with reference to the Paleozoic, *Am. Assoc. Petrol. Geologists Bull.*, v. 66, p. 750-774.

Schopf, T., 1983, Paleozoic black shales in relation to continental margin upwelling, in, *Coastal upwelling: its sediment record. Part B: Sedimentary records of ancient coastal upwelling*, E. Suess and J. Thiede, eds., p. 579-596, Plenum Press, New York

Sheldon, R. P., 1980, Episodicity of phosphate deposition and deep ocean circulation -- a hypothesis, in *Phosporites*, Spec. Pub. 29, Y. Bentor, ed., pp. 239-247, Soc. Econ. Mineralogists and Paleontologists.

Stewart, J. H., 1980, Geology of Nevada: *Nevada Bur. Mines Geol. Spec. Pub. 4*

Wardlaw, B. R., 1980, Middle Permian paleoceanography of Idaho, Montan, Utah, and Wyoming, in, *Paleogeography of west-central United States*, T. D. Fouch and E. R. Magathan, eds., Soc. Econ. Paleontol. Mineral. Rocky Mountain Paleogeography Symposium 1, p. 353-361.

Water-Rock Interaction, Kharaka & Maest (eds) © 1992 Balkema, Rotterdam. ISBN 90 5410 075 3

Access of pore waters to carbonate precipitation sites during concretion growth

M.M.Jordan & C.D.Curtis
Department of Geology, University of Manchester, UK

A.C.Aplin
Newcastle Research Group in Fossil Fuel and Environmental Geochemistry, University of Newcastle Upon Tyne, UK

M.L.Coleman
BP Research International, Sunbury Research Centre, Sunbury upon Thames, Middlesex, UK

ABSTRACT: Carbonate concretions within mudstones of the upper Jurassic Heather Formation, Northern North Sea contain multiple generations of finely intergrown carbonate and silicate cements. Stable isotope and chemical analyses of separated carbonate minerals show that the bulk of each concretion comprises calcite which precipitated rapidly within the zone of active sulphate reduction. A rigid but porous structure formed at this early diagenetic stage. Most of the remaining pore space throughout the concretion was then infilled by authigenic siderite, kaolinite and illite. Finally, the outermost 10mm or so of the carbonate recrystallised with oxygen isotope exchange. The same recrystallisation also affected microconcretions within the body of the enclosing mudstone. The concretions in this study record processes to considerable depth during most of which time they remained chemically "open".

1 INTRODUCTION

Concretions are segregations of minerals which precipitated within the pores of sediments after deposition. Potentially, they preserve a record of the way in which the composition of pore-waters evolved in ancient mudrocks. This, in turn, records the sum of solid-solution reactions and solute migration. This is important information because, although the bulk of the pore waters in sedimentary basins are sourced from within mudstones, we have almost no direct evidence of their composition and evolution. In this paper we present chemical, isotopic and petrographic data for a series of carbonate concretions from the Upper Jurassic Heather Formation, (Magnus Field, North Sea). These samples are unusual in two ways. Firstly, they are from (mudstone) drill core, ensuring them to be pristine. Secondly, they were sampled at 3.2 km depth, the maximum depth to which the Heather Formation has been buried. Potentially, therefore, the concretions record a pore-fluid history spanning 150 Ma and 3.2 km of burial.

Concretions collected from between 3.22 km and 3.24 km depth are pale grey to buff in colour with distinctive lighter coloured, 4 mm to 8 mm wide margins that terminate sharply against black mudstone. There is no evidence of compaction within the concretion although differential compaction around them is very marked (compare Raiswell, 1971). The concretions are mostly sub-spherical to oblate, with minor uniaxial compression normal to bedding. The black mudstone which encloses the concretions is finely laminated with infrequent upward fining silty laminae. As well as "macroconcretions", small, < 1 mm diameter pale grey oblate carbonate "microconcretions" occur in large numbers in the mudstones, often concentrated within particular laminae. Mudstone laminae are also deformed around these structures indicating pre-compactional growth.

2 METHODS

5 mm slices were cut at right angles to the bedding, through each concretion and its enclosing mudstone. Subsamples were then taken regularly from all parts of the concretions, both for the preparation of polished thin sections (backscatter scanning electron microscopy, BSEM), and for isotopic analysis. BSEM examination (JEOL 6400) revealed that siderite and calcite are very finely-intergrown in the concretions. A chemical isolation technique was therefore developed which allowed the chemical and isotopic composition of both minerals to be determined both accurately and precisely. This is based on carefully characterised leaching in dilute acetic acid which removes calcite but not siderite. This procedure is outlined in figure 1. Isotope determinations, for both calcite and siderite, were made on a Finnegan MAT 251 mass spectrometer attached to an automatic carbonate reaction device.

3 CONCRETION PETROGRAPHY

Backscatter electron imagery reveals that calcite and siderite (identified by X-ray diffraction) are intergrown within both concretions and microconcretions. Calcite is volumetrically much

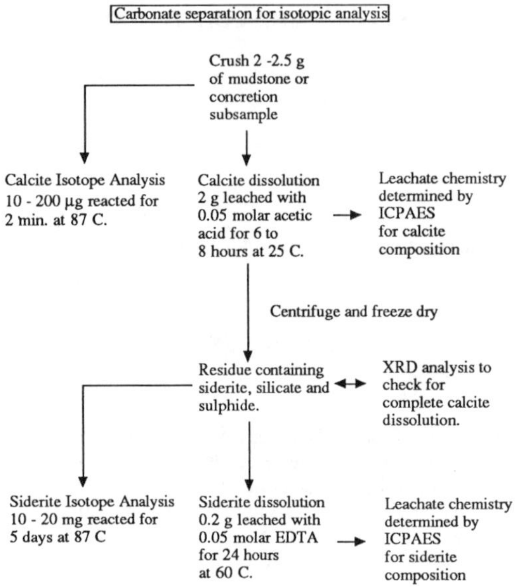

Figure 1

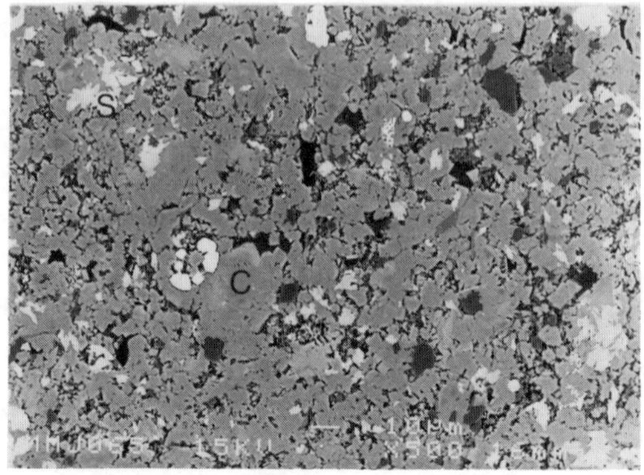

Electronmicrograph 1 Field of view 160µm BSEI of fine grained calcite, (C) and siderite, (S) from a typical concretion core area. Note intergrowth texture between calcite and siderite. Also present euhedral pyrite, detrital quartz, authigenic kaolinite and illite.

Electronmicrograph 2 Field of view 160µm BSEI of coarse grained calcite, (C) and siderite, (S) from the margin of the above concretion. Note carbonate grain size increase and increased proportion of siderite to calcite.

more important. The proportion of siderite to calcite increases from the core of the concretion to its margin, with a sharp rise in siderite concentration within the outer 3 - 9 mm of the concretions. Two distinct grain sizes ranges can be seen. The bulk of the concretion is extremely fine grained (2 to 5 µm) whilst the outer 3 to 9 mm is much coarser (10 to 15 µm). The microconcretions within the mudstone are coarse grained (10 to 15 µm) -see electronmicrographs 1 and 2. Detrital silicates present within the concretion and mudstone included k-feldspar, albite, kaolinite cemented mica and quartz. All of these have ragged or embayed margins, typical of silicate dissolution and replacement by carbonate. Organic matter preserved within both concretions and mudstone is dominated by terrestrial plant fragments. In most cases the cell structure has been preserved by the early precipitation of framboidal pyrite and non-ferroan calcite. This infilled cells prior to significant compaction of the sediment. Authigenic phyllosilicates within the concretions and mudstone include kaolinite booklets (2 to 0.2 µm) that precipitated within the pore spaces in both concretions and microconcretions. Hairy illite precipitated synchronously with (or possibly after) late kaolinite. Pore filling authigenic silicates are found in higher concentrations within the marginal zones of the concretion than in the main body and in all cases post -date siderite precipitation. Petrographic observation suggest a definite diagenetic sequence; early framboidal pyrite was enclosed by non-ferroan calcite. This apparently created a rigid but porous, sponge-like structure enclosing detrital organic matter and silicates. Pore spaces within the proto-concretions and microconcretions then became partially filled by siderite and the residual porosity was occluded by authigenic phyllosilicates.

4 PORE WATER EVOLUTION AS SEEN IN SEQUENTIAL CEMENT

4.1 CHEMICAL AND ISOTOPIC TRENDS WITHIN CARBONATE CEMENTS.

Quantitative analysis of calcite cements using the Joel 6400 SEM microprobe facility revealed single calcite crystals without zonation within both the bulk of the concretion and its margin. Leachate analysis of subsamples from the bulk of the concretion show relatively slight changes in cation chemistry. Only the manganese to iron ratio shows significant, systematic variation, falling away from a high within the core of the concretion towards lower values at the margin and into the mudstones.

Microprobe data from single siderite crystals show concentric iron zonation, both within the concretions and the microconcretions. Chemical data from the siderite leachate suggest the same trend as for calcite with manganese enrichment in the cores of carbonate concretions. Also evident from the siderite leachate data is an enrichment in calcium from the core of the concretion out into the microconcretions (table 1).

Table 1

Location and carbonate type	$\delta^{13}C_{PDB}$‰	$\delta^{18}O_{PDB}$‰	Cation mole faction Ca	Fe	Mg	Mn
Core calcite	-22 to-20	-0.5 to -1	0.90	0.01	0.03	0.06
Margin calcite	-18 to -12	-4 to -6	0.94	0.01	0.02	0.03
Microconcretion calcite	-4 to -2	-4 to -7	0.94	0.01	0.03	0.02
Core siderite	-5 to +1	-3 to -4	0.00	0.78	0.14	0.08
Margin siderite	-1 to +4	-5 to -7	0.07	0.75	0.15	0.03
Microconcretion siderite	+1 to +5	-10 to -13	0.14	0.74	0.13	0.02

$\delta^{13}C$ values for the non-ferroan calcite subsamples from the cores of the concretions range from -22‰ to -20‰, whereas the outer margins have more positive values (between -18‰ and -12‰). $\delta^{13}C$ values for the microconcretions within the mudrocks are even more positive (-4‰ to -2‰), perhaps indicating later precipitation from solutions with bicarbonate from a different source.

$\delta^{18}O$ values for calcite from the cores of the concretion are similar (-0.5‰ to -1.0‰) whereas the calcite subsample from the margins and microconcretions have values which are much more negative (-4‰ to -7‰).

$\delta^{13}C$ values for siderite display a similar trend to that for calcite, with values becoming more positive from the core of the concretion ($\delta^{13}C$ -5‰ to +1‰) towards the margin (-1‰ to +4‰) and out into the mudstone (+1‰ to +5‰). $\delta^{18}O$ values for siderite also show a similar trend to oxygen in calcites but the change is more marked with values becoming significantly more negative between the margin (-5‰ to -7‰) and the mudstone microconcretions (-10‰ to -13‰).

5 DISCUSSION

Non-ferroan calcite with its very negative $\delta^{13}C$ values probably formed as a result of organic matter oxidation by sulphate reducing bacteria. The lack of stable isotope fractionation associated with sulphate reduction leads to the formation of bicarbonate and ultimately carbonate which retains a negative $\delta^{13}C_{(PDB)}$ characteristic of the organic matter source (approximately -25‰). $\delta^{13}C$ values within the margin of the concretion and microconcretion are more positive. This trend must be due to the introduction of less negative $\delta^{13}C$ bicarbonate. Possible sources include dissolution of biogenic carbonate ($\delta^{13}C$ approximately zero ‰) or introduction of bicarbonate as one product of methanogenesis (Irwin *et al.*, 1977; Curtis, 1978). As biogenic carbonate in the form of coccolith fragments is still present within the mudrocks, this seems less likely as source of positive $\delta^{13}C$ bicarbonate, although local dissolution could have occurred. The simplest explanation of the trend is increasing importance of methanogenesis relative to sulphate reduction with time.

Siderite samples exhibit a similar carbon isotopic trend. Petrographic and geochemical evidence are consistent in indicating that siderite formed later than non-ferroan calcite and pyrite (figure 2). The "sponge-like" porous calcite concretion grew concentrically within sediments in the sulphate reduction zone. As sulphate reduction activity diminished with burial below the sediment-water interface, pyrite precipitation declined and the iron content of the pore waters increased. Siderite then started to precipitate in pore space within the core of the concretion and continued rather more slowly to cement outwards to the margins.

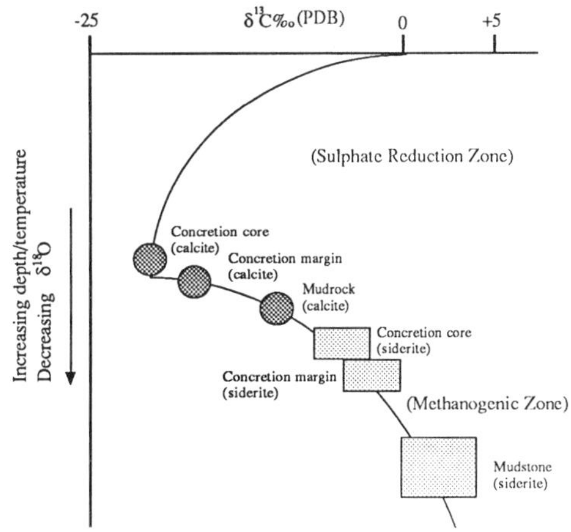

Figure 2

Microconcretion calcite and siderite display $\delta^{13}C$ values that are substantially more positive than either the bulk of the main carbonate concretion or its margin. This suggests that they grew somewhat later than the main concretion although they developed prior to significant compaction.

The oxygen isotope composition of carbonate minerals depends both on temperature and the isotopic composition of the water from which they precipitate. If precipitation of calcite is assumed to have been from pre-glacial sea water ($\delta^{18}O$ -1.2‰; Skackleton and Kennett, 1975), temperatures of 11 to 14°C are indicated for the core, with 35 to 45°C for the margin and microconcretions. We have no fluid inclusion data to support these figures but 11 to 14°C is a reasonable temperature for the sub-tropical shelf seas in which the Heather Formation is presumed to have accumulated. The much higher precipitation temperatures suggested for margins and microconcretions poses a problem. The suggested uniform initial porosity across the concretion and into the margin (i.e. simply assuming passive displacement of pore water by carbonate), implies development of the whole structure in sediments buried to no more than about 100 metres. Thus margins could only have precipitated at temperatures 3 or 4°C warmer than cores. It is possible that the water from which the carbonate precipitated was isotopically lighter than sea-water. This could result from either ingress of Jurassic meteoric water (for which there is no evidence) or by closed system evolution of sea water by reaction with sedimentary minerals (Lawrence and Gieskes, 1981). Alternatively, the margins of the concretions and the microconcretions recrystallised with oxygen exchange with pore waters at higher temperatures (although the cores apparently preserve original isotope ratios in spite of current burial temperatures in excess of 100°C. There is textural evidence to support this possibility in that concretion margin and microconcretion carbonates display a substantially coarser grain size range.

The precipitation temperature for siderite is not easy to estimate. If we make the following assumptions it is possible to estimate a pore water composition: 1) siderite precipitated immediately after calcite within the core of the concretion, 2) oxygen values within the core of the concretion were not affected by the late recrystallisation which has distorted the original values for the margins and microconcretions, 3) calcite at the core of the concretion precipitated at 15°C. Siderite $\delta^{18}O$ values from the core are -3‰, then the pore fluid from which it precipitated was approximately 3.5‰. Using this value for the pore fluids,temperatures at the margin range between 25 and 40°C and for the microconcretions, 45 to 60°C. The coarser grain size within the margin and the microconcretion, however, again suggests recrystallisation.

6 CONCLUSION

It is suggested that the growth and development of the Heather Formation concretions followed the sequence illustrated in figure 3.

The Heather Formation carbonate concretions behaved as chemically open systems for much of their growth history. Eventually, however, they became effectively closed so as to preserve isotopic signatures from early processes. Millimetre scale microconcretions and the outermost margins of the macroscopic concretions did exchange with porewater at higher temperature, thereby losing early process signals. Only in the cores of these macroconcretions is a record of a sequence of early diagenetic processes preserved.

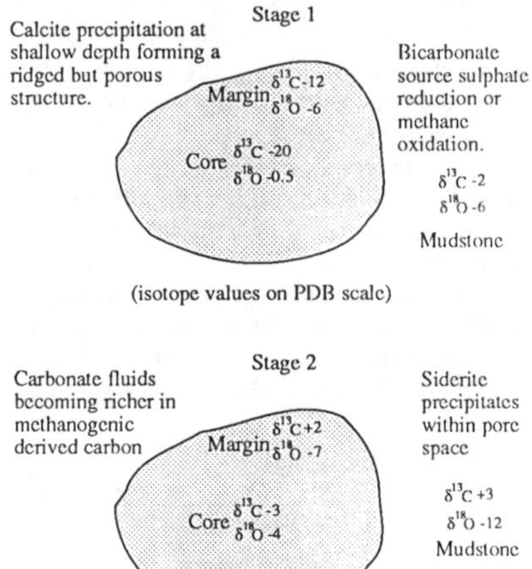

Stage 3

Precipitation of kaolinite and illite with remaining pore space resulting in reduced permeability and porosity.

Stage 4

Recrystallisation of carbonate at the margins of the concretion and within the mudstone with ambient pore fluids during burial.

Figure 3

ACKNOWLEDGEMENTS

This work was supported by a NERC/BP CASE studentship (MMJ) which is gratefully acknowledged. Stable isotope work was carried out at the BP Research Centre Sunbury and the guidance of Marge Savell and John Rouse was of great value.

REFERENCES

Curtis, C.D. (1977) Sedimentary geochemistry: environments and processes dominated by involvement of an aqueous phase. Phil. Trans. R. Soc. Lond. A. 286, 353-372.

Irwin, H., Curtis, C.D. and Coleman, M. (1977) Isotopic evidence for source of diagenetic carbonates formed during burial of organic-rich sediments.Nature, 269, 209-213.

Lawrence, J.R. and Gieskes, J.M. (1981) Constraints on water transport and alteration in the oceanic crust from the isotopic composition of pore water. J. Geophys. Res., 86, 7924-7934.

Raiswell, R.W.(1971). The growth of Cambrain and Liassic concretions. Sedimentology, 17, 147-171.

Shackleton, N.J and Kennett,J.P. (1975) Paleotemperture history of the Cenozoic and the initiation of Antartic glaciation: oxygen and carbon isotopic analysis in DSDP sites 277, 279 and 281. In: Init. Reports DSDP, 29, 743-755.

Water-Rock Interaction, Kharaka & Maest (eds) © 1992 Balkema, Rotterdam. ISBN 90 5410 075 3

Wettability of petroleum reservoir rock minerals

H. Lemke & K. Schwochau
Institute of Petroleum and Organic Geochemistry at the Research Centre (KFA) Juelich, Germany

ABSTRACT: The objective of this investigation is to quantify the rather limited wettability data for individual minerals by determining contact angles using the Dynamic Wilhelmy method. The mineral slices prepared from quartz, calcite, dolomite, kaolin, muscovite, albite and limestone are immersed in and removed from water and hexadecane as an idealized oil phase, while the forces vs. immersion depth are recorded. Without an ageing procedure the mineral slices reveal for a hexadecane/water system a strong water-wet behaviour, demonstrated by individual contact angles of $\Theta_A \leq 21°$ and $\Theta_R \leq 14°$ at 20°C. Ageing of the slices by keeping them 72 h at 20° in hexadecane results in oil-wet behaviour only for albite, while an ageing at 80°C for 72 h also induces oil wettability for calcite, limestone, and dolomite. Oil-wettability induced by ageing can be reversed to water-wet behaviour by exposing the slices to water at room temperature for 72 h. Monolayer sorptions of stearic acid, eicosylamine, α-lecithine or 1,2-hexadecylglycerol onto quartz slices each yield almost the same contact angle shifts of about 40° at the hexadecane/air interface.

1 INTRODUCTION

The wettability of petroleum reservoir rocks has proved to be a significant factor in oil reservoir engineering (Anderson 1986), since it has been established that reservoir rocks do not always exhibit a water-wet behaviour. It is often considered that some minerals possess an intrinsic hydrophobic nature (Leja 1982) or that sorption of crude oil surfactants onto the rock surface cause this non-water-wet behaviour (Collins et al. 1983). The aim of the experiments is to investigate the relevance of these considerations by quantifying wettability alterations on reservoir rock minerals.

2 EXPERIMENTAL

Fig. 1 illustrates the experimental set-up used for the contact angle measurements according to the Dynamic Wilhelmy method. The mineral slices hanging from the balance are immersed in and removed from the liquid phases, while the forces and the immersion depths are measured. The standard speed of immersion/emersion was 1.72 mm/min. According to the equation

$$F = P \cdot \gamma \cdot \cos\Theta_{A/R} - b \qquad (1)$$

the contact angle $\Theta_{A/R}$ (A=advancing, R=receding) can be calculated, were F=force

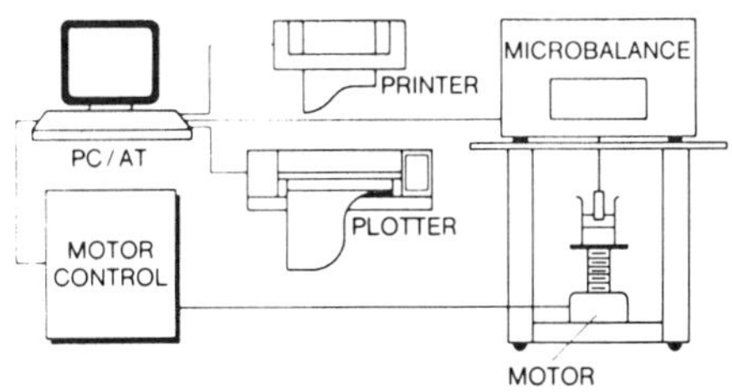

Fig. 1 Computerised Wilhelmy apparatus used for the measurements

(mN), P=perimeter (m), γ=interfacial tension (mN/m) and b=buoyancy (mN).

The mineral slices were prepared from quartz, calcite, dolomite, kaolin, muscovite, albite and limestone. Contact angle shifts caused by surface roughness were minimized by measuring polished slices with a roughness of about 1μm.

For contact angle measurements in surfactant free systems the clean, water soaked mineral slices were immersed in and withdrawn from the water/hexadecane phases in which hexadecane was used as an idealized oil phase. In order to study ageing effects, the slices were kept for 72 h in hexadecane at 20° and 80°C, respectively, before executing the measurements at 20°C.

To evaluate surfactant induced wettability alterations, contact angles were measured, when quartz slices were withdrawn from and immersed in the water/hexadecane phases containing 1 mM dodecyltrimethyl ammonium bromide (DTAB) and/or 1 mM stearic acid, respectively.

The sorption of crude oil surfactants onto reservoir rock minerals was simulated by coating mineral slices with a monolayer of an amphipatic molecule, e.g. long chain acids and bases, cholesterol and phospholipids, using the Langmuir-Blodgett-technique (Moebius 1975). Taking into account only influences which are induced by monolayer sorption, an internal standard was set by coating only half the slice and measuring the shift in contact angle, when the monolayer crossed the hexadecane/air interface. According to the equation

$$\Delta F = P \cdot \gamma \cdot \cos\Delta\Theta_{A/R} \qquad (2)$$

contact angle shifts can be calculated.

3 RESULTS

Advancing (Θ_A) and receding (Θ_R) contact angles of different mineral slices measured at the hexadecane/water interface, before and after ageing of the slices in hexadecane, are reported in Tables 1-3.

Table 1. Contact angles measured at 20°C without ageing the minerals.

Mineral	Θ_A	Θ_R
quartz	17°	0°
calcite (single crystal)	9°	0°
limestone	15°	0°
dolomite	9°	0°
kaolin	18°	13°
albite	21°	0°
muscovite	19°	14°

Table 2. Contact angles measured at 20°C after ageing the minerals for 72 h at 20°C.

Mineral	Θ_A	Θ_R
quartz	22°	10°
calcite (single crystal)	60°	8°
limestone	76°	17°
dolomite	25°	17°
kaolin	40°	28°
albite	94°	22°

Table 3. Contact angles measured at 20°C after ageing the minerals for 72 h at 80°C.

Mineral	Θ_A	Θ_R
quartz	57°	55°
calcite (single crystal)	138°	35°
limestone	118°	49°
dolomite	122°	48°
kaolin	83°	47°
albite	108°	52°

For the hexadecane/water system the mineral slices reveal a strong water-wet behaviour (Table 1), shown by individual contact angles of $\Theta_A \leq 21°$ and $\Theta_R \leq 14°$. Ageing of the slices by keeping them 72 h at 20°C in hexadecane (Table 2) results in oil-wet behaviour only for albite ($\Theta_A = 94°$), while an ageing at 80°C for 72 h (Table 3) also induces oil wettability for calcite, limestone, and dolomite. Oil-wettability induced by ageing can be reversed to water-wet behaviour by exposing the minerals to water at 20°C for 72 h.

Wettability alterations in the system hexadecane/water induced by the surfactants stearic acid and/or DTAB are shown in Table 4.

Table 4. Surfactant induced wettability alterations of quartz slices measured at 20°C.

oil phase	1mmol stearic acid		1mmol stearic acid
water phase		1mmol DTAB	1mmol DTAB
Θ_A	28°	71°	106°
Θ_R	25°	39°	54°

1 mM stearic acid in the hexadecane phase (idealized oil phase) increases the advancing contact angle Θ_A of quartz at the hexadecane/water interface from 17° (Table 1) to 28°, but quartz remains strongly water-wet. 1 mM DTAB in the water phase induces a distinctly higher Θ_A of 71°. However, the water-wet behaviour changes to oil wettability ($\Theta_A = 106°$) by synergetic effects, when each phase contains 1 mM of the respective surfactant.

Contact angle shifts $\Delta\Theta$, observed on quartz slices half-coated with monolayers of amphipatic molecules and described by equation (2), have been determined according to Fig. 2.

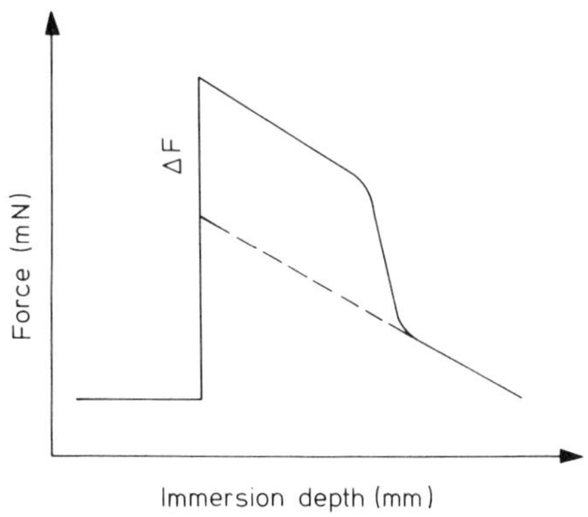

Fig. 2 Force/immersion depth diagram of a half-coated quartz slice.

The first attachment of hexadecane to the uncoated quartz slice yields a strong force (F) enhancement. With increasing immersion depth, F decreases linearly by buoyancy as long as the contact angle of the slice is constant. Additional reduction of F is induced by wettability alteration when the monolayer on the slice is crossing the hexadecane/air interface. ΔF can be evaluated by extrapolation. Table 5 shows some advancing contact angle shifts $\Delta\Theta_A$ for monolayers of amphipatic molecules.

Table 5. Monolayer induced advancing contact angle shifts $\Delta\Theta_A$ at the hexadecane/air interface measured at 20°C.

Monolayer	$\Delta\Theta_A$
eicosylamine	41°
stearic acid	42°
α-lecithine	42°
1,2,-hexadecylglycerol	44°
cholesterol	22°

Different monolayer sorptions of eicosylamine, stearic acid, α-lecithine or 1,2-hexadecylglycerol onto quartz slices each yield almost the same contact angle shifts of about 40°. These findings can be explained by a structural similarity of the outermost part of the monolayer molecules defining the surface energy. The entirely different molecular structure and the less pronounced amphipatic character of cholesterol appear to be responsible for its lower advancing contact angle shift.

4 CONCLUSION

The results demonstrate the variability of contact angles of petroleum reservoir rock minerals. The intrinsic water-wet behaviour of the minerals may be altered drastically by changing the experimental conditions. Thus the wettability of real reservoir rocks will be difficult to predict.

ACKNOWLEDGEMENT

We wish to thank Dr. H.S. Poelchau for reviewing the manuscript.

REFERENCES

Anderson, W.G. 1986. Wettability literature survey. Part 1: Rock/oil/brine interactions and the effects of core handling on wettability. J. Petr. Techn., Oct.: 1125-1144.

Collins, S.H. and Melrose, J.C. 1983. Adsorption of asphaltenes and water on reservoir rock minerals. Soc. Petr. Eng. AIME, SPE 11800: 249-256.

Leja, J. 1982. Surface chemistry of froth floatation. Plenum Press, New York City.

Moebius, D. 1975. Manipulieren in molekularen Dimensionen. Chemie in unserer Zeit 9: 173-182.

Water-Rock Interaction, Kharaka & Maest (eds) © 1992 Balkema, Rotterdam. ISBN 90 5410 075 3

Replacement of albite by K-feldspar and vice versa

Radomir Petrovich
Phillips Petroleum Co., Bartlesville, Okla., USA
Philip M.Orville†

ABSTRACT: Single crystals of low albite and adularia were reacted with 2M KCl and 2M NaCl aqueous solutions, respectively, at 200 MPa and 500-700°C for 1 to 256 hours in arc-welded gold capsules. Potassium feldspar was replaced by albite and *vice versa* by a dissolution-precipitation mechanism. The rate of advance of the replacement front is a function of the crystallographic orientation of the interface in accordance with Petrović's (1973) model of coherency-strain-controlled replacement of alkali feldspars by other alkali feldspars and with the orientational dependence of coherency-strain energies of alkali feldspar intergrowths calculated by Brown and Willaime (1974). However, in experiments that allowed determination of the rate of advance of the replacement front as a function of time t, this rate was proportional to $t^{1/2}$ and therefore was controlled by the diffusion of Na^+ and K^+ to and from the front.

INTRODUCTION

Reaction of an (Na,K)-feldspar with an aqueous solution that produces another (Na,K)-feldspar of a very different K/(Na+K) mole ratio proceeds by a dissolution-precipitation mechanism (O'Neil and Taylor, 1967; Christophe-Michel-Lévy, 1967; Heimann, 1979); in other words, it is a replacement reaction. However, when the K/(Na+K) ratio of the product (Na,K)-feldspar is relatively close to that of the original feldspar, the same reaction proceeds by cation exchange (Petrović, 1973). The reason for this difference in mechanisms is the coherency stress to which incipient thin films of cation-exchanged feldspar are exposed; if ΔK/(Na+K) is large, this stress results in a substantial strain energy that destabilizes the cation-exchanged feldspar film relative to reprecipitated feldspar of the same composition (Petrović, 1973). The present study was undertaken to elucidate the kinetics and mechanism of the said alkali feldspar → alkali feldspar replacement reaction.

EXPERIMENTAL TECHNIQUE

Single crystals of a low albite ($Or_1Ab_{98}An_1$) and single crystals, with cleaved {001} faces, of an adularia ($Or_{83}Ab_{14}An_1$ in the interior to $Or_{95}Ab_4An_1$ at {110} faces) were reacted with 2M KCl and 2M NaCl aqueous solutions, respectively, in arc-welded gold capsules. The solution/feldspar ratios were large enough to maintain approximate constancy of the $m_{Na}/(m_{Na}+m_K)$ ratio during the experiments but not so large as to cause marked net dissolution of the feldspar crystals. The experiments were carried out in cold-seal bombs, at 200 ± 2 MPa and 500 ± 5 or 700 ± 7°C, over times of 1 to 256 hours, and ended with quenching. The reacted crystals and their polished thin sections were first examined under a petrographic microscope, then back-scattered electron (BSE) images and some Na Kα X-ray maps of the polished thin sections were taken and some microimaging EDS analyses were done with a JEOL JXA-733 electron microprobe. Thicknesses of product layers were measured on BSE images of the sectioned crystals and on photomicrographs of crystal exteriors and calculated from microscope stage displacements corresponding to changes in focus from tops to bottoms of horizontally oriented product layers.

SPATIAL ASPECTS OF REPLACEMENT

The characteristic spatial patterns of replacement of albite by K-feldspar and *vice versa* are shown in Figure 1. The patterns are partially obscured by the extensive damage to the crystals that was caused by the differential thermal contraction during the quenching, but the qualitative picture is clear. Both the replacement of albite by K-feldspar (presumably sanidine) and the replacement of adularia by albite-twinned albite have occurred primarily by the rapid and irregular penetration of the replacement front in the [001] direction, accompanied by slow lateral spread. Our experiments show that this highly reproducible pattern, previously observed by Orville (1962), O'Neil and Taylor (1967), Christophe-Michel-Lévy (1967), and Heimann (1979) in experiments with albite and KCl aqueous solutions, also occurs in experiments with good K-feldspar crystals and NaCl solutions.

Uniformly thick replacement layers have also developed from a variety of faces, including {001}, but the advance of their flat replacement fronts was very much slower.

Coherency-strain energy of plane interfaces between albite-twinned albite and sanidine shows a very strong maximum for the interface that is perpendicular to [001] (Brown and Willaime, 1974). Therefore the strongly preferred advance of the replacement front in the [001] direction supports Petrović's (1973) model of destabilization of the cation-exchanged feldspar film by coherency strain energy. However, the irregularity of the actual albite/K-feldspar interfaces indicates that the advance involves either preferential transport along dislocations parallel to [001] or coupling between the replacement and localized mechanical failure along $(1\bar{1}0)$ and (110) cleavage planes.

Replacive K-feldspar also advanced rapidly into albite from some regions of {010} faces, forming relatively compact domains with irregular albite/K-feldspar interfaces that include many segments parallel to the $(1\bar{1}0)$ and (110) planes. Although filled with fluid inclusions, such K-feldspar shows no obvious oriented fabric.

THE MECHANISM OF REPLACEMENT

The partially replaced adularia crystals have yielded two sets of data on the kinetics of the K-feldspar → albite replacement reaction (Fig. 2). The slow increase in thickness of the uniform albite replacement film on the {110} faces of adularia definitely followed the $x \propto t^{1/2}$ rate law, which implies rate control by diffusion of Na^+ and K^+ ions through the product layer. Information on the fast replacement in the [001] direction is less good, but the replacement front certainly did not follow the $x \alpha t$ rate law and in view of the extensive loss of replaced material from {001} faces that occurred at the ends of the 16-h and 32-h experiments the agreement with the $x \propto t^{1/2}$ rate law is satisfactory.

Information on the kinetics of replacement of K-feldspar by albite was also obtained by Chelishchev and Nadzharyan (1971), who did their experiments with fragments of microcline, 25-50 μm across, and 1M NaCl aqueous solutions at 400, 500, and 600°C and 100 MPa. The increase in the amount of product albite with time in their experiments must have reflected mainly the fast advance of the replacement front in the [001] direction. Within the transformation range from 15 to 80 percent, this amount was proportional to $t^{1/2}$, which again indicates control by diffusion. The activation energy for this process was 54 kJ/mol, much too low for diffusion through the bulk feldspar but acceptable for diffusion along the subgrain boundaries of the replacive feldspar.

Acknowledgments: R.P. thanks Gary L. Lovell for the excellent electron microprobe images and analyses and the Phillips Petroleum Company for the permission to publish this work.

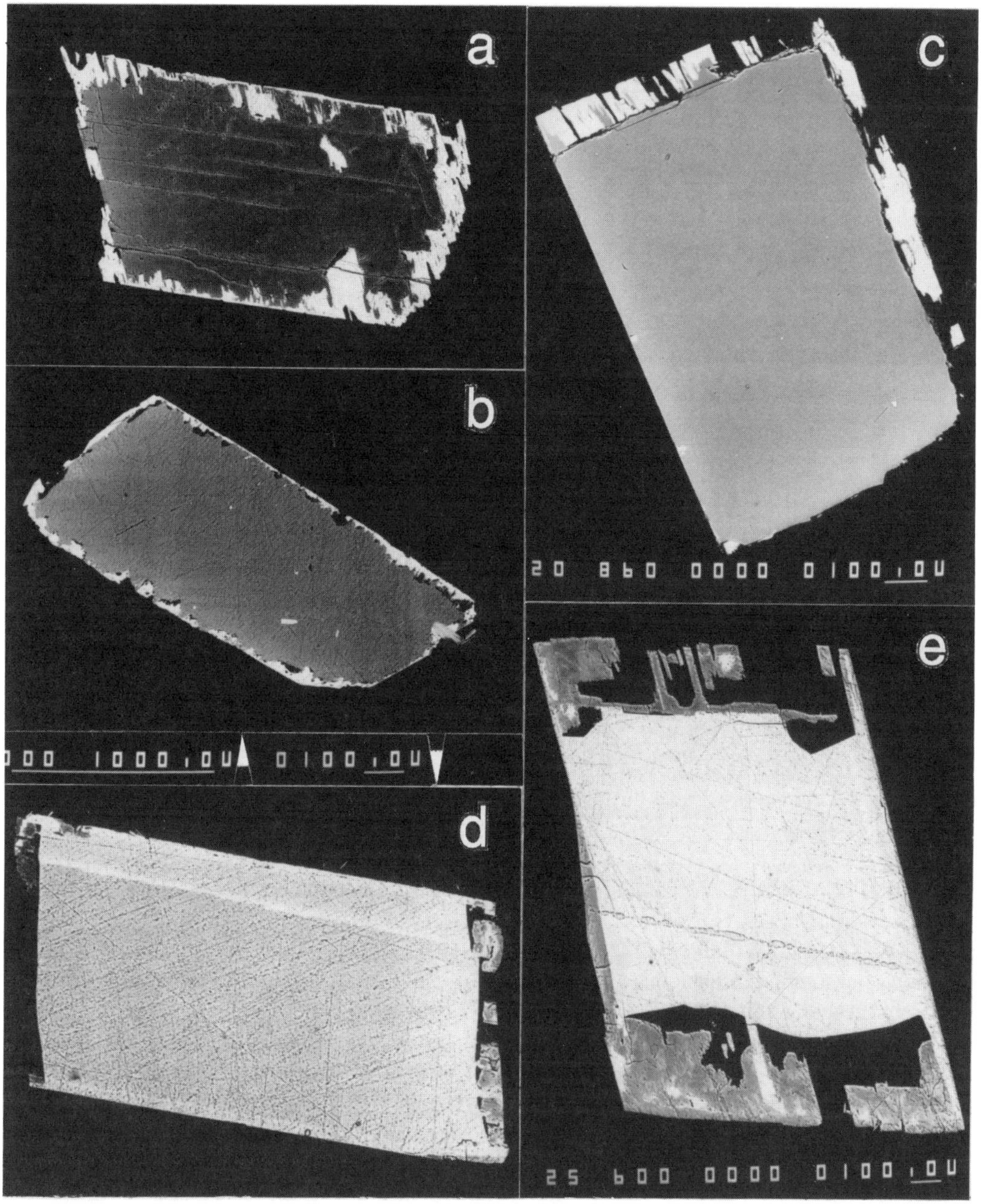

Figure 1. BSE images of sections of some of the partially replaced alkali feldspars; K-feldspar is bright, albite dark, scale-bar lengths are given in micrometers. The crystals are: (a) albite (700°C, 8 h), section $\|$(010) with (001) facing up; (b) albite (500°C, 8 h), section roughly parallel to (001) with (010) or $(0\bar{1}0)$ facing upper right; (c) albite (700°C, 1 h), section roughly perpendicular to [100] with (001) facing up and (010) facing right; (d) adularia (500°C, 8 h), section $\|$(110), and (e) adularia (500°C, 256 h), section $\|(1\bar{1}0)$, both elongated parallel to [001]. Note the preferential advance of the replacive feldspar parallel to [001] and the traces parallel to [001] within that feldspar.

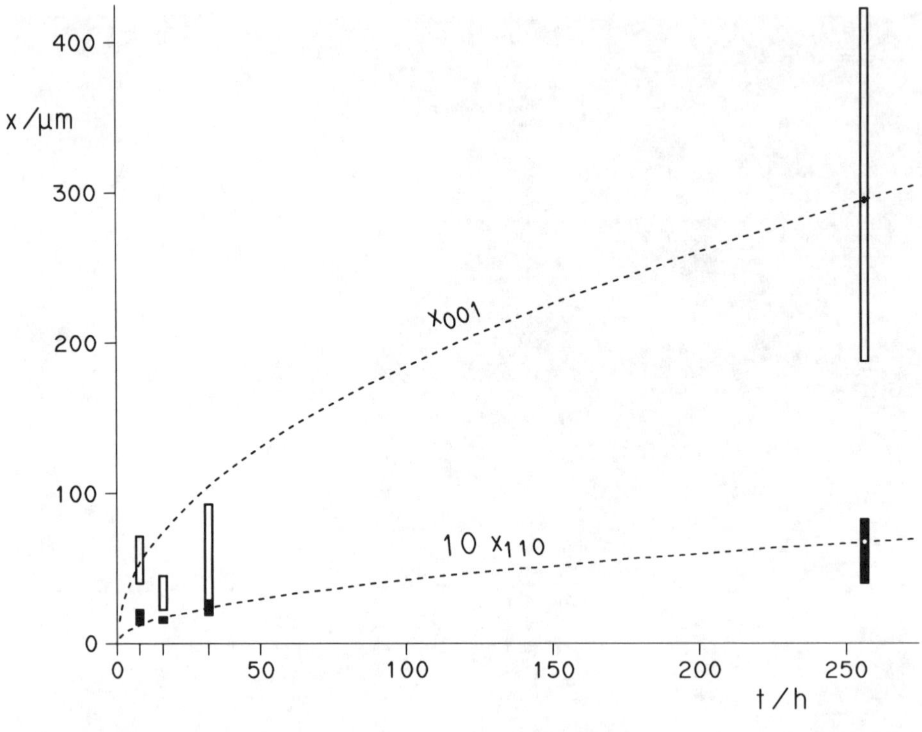

Figure 2. Thickness x of the replaced layer as a function of time t in experiments with adularia crystals at 500°C and 200 MPa. Rectangles indicate the ranges of measured thicknesses, the dashed lines are $x = k\, t^{1/2}$ curves drawn through the mean values (dots) for the 256-hour experiment.

REFERENCES

Brown, W. L. & C. Willaime 1974. An explanation of exsolution orientations and residual strain in cryptoperthites. *The Feldspars*: 440-459. Manchester Univ. Press.

Chelishchev, N. F. & V. N. Nadzharyan 1971. Kinetika reaktsii al'bitizatsii microklina v nadkriticheskikh usloviyakh. *Dokl. AN SSSR* 201: 658-660.

Christophe-Michel-Lévy M., 1967. Sur le mécanisme de "l'échange" Na-K par voie hydrothermale dans l'albite. *Bull. Soc. fr. Minéral. Cristallogr.* 90: 411-413.

Heimann, R. B. 1979. Hydrothermale Verdrängungsreaktionen an Alkalifeldspäten der Zusammensetzungen $Or_{93}Ab_7$ und $Ab_{96}Or_4$. *N. Jb. Miner. Abh.* 137: 1-19.

O'Neil, J. R. & H. P. Taylor, Jr. 1967. The oxygen isotope and cation exchange chemistry of feldspars. *Amer. Mineral.* 52: 1414-1437.

Orville, P. M. 1962. Alkali metasomatism and feldspars. *Norsk Geol. Tiddskr.* 42(II): 283-316.

Petrović, R. 1973. The effect of coherency stress on the mechanism of the reaction albite + K^+ ⇄ K-feldspar + Na^+ and on the mechanical state of the resulting feldspar. *Contrib. Mineral. Petrol.* 52: 17-28.

Water-Rock Interaction, Kharaka & Maest (eds) © 1992 Balkema, Rotterdam. ISBN 90 5410 075 3

Illite age analysis, a new tool for basin thermal history analysis

D.R.Pevear
Exxon Production Research Co., Houston, Tex., USA

ABSTRACT: Illite Age Analysis (IAA) is a method which provides both integrated age of diagenesis and mean age of illite in the source terrane from shale samples. IAA assumes that illite in shales is a mixture of diagenetic (young) illite, in mixed-layer illite / smectite (I /S) and discrete detrital (old) illite, derived by erosion. Laboratory processing gives size fractions with widely varying amounts of each end-member and therefore widely varying K/Ar ages. Modeling of the XRD patterns of each fraction permits determination of end-member proportions. Plots of K/Ar age vs. that percent of total illite which is diagenetic are linear confirming the two-component model; extrapolation to 0% and 100% detrital illite gives ages for the diagenetic and detrital end-members. Applications illustrate the validity and limitations of the method.

1 INTRODUCTION

Illite Age Analysis (IAA), our adaptation of the well-known K/Ar radiometric dating technique, provides both integrated age of diagenesis and mean age of illite in the source terrane from shale cuttings or core samples. Since the K/Ar method was introduced in the 1950's, results from the micaceous minerals in shales (illite and illite/smectite) have almost always shown the finer size fractions to be younger than the coarser fractions, in many cases even younger than the depositional age. This has lead to the assumption that shale is a mixture of detrital and diagenetic phases, the former older than the age of deposition, the latter younger. All attempts to physically separate the diagenetic material from the detrital, in order to obtain the age of diagenesis, have failed. Even the finest size fractions appear to be mixtures of detrital and diagenetic phases, therefore the dates obtained are not particularly useful. Illite Age Analysis has solved this important problem, and in so doing provides a new tool for thermal history analysis, maturation / migration / trap-fill timing, provenance studies and basin analysis.

Using IAA, the measured age of diagenetic illite can be compared with that forward-

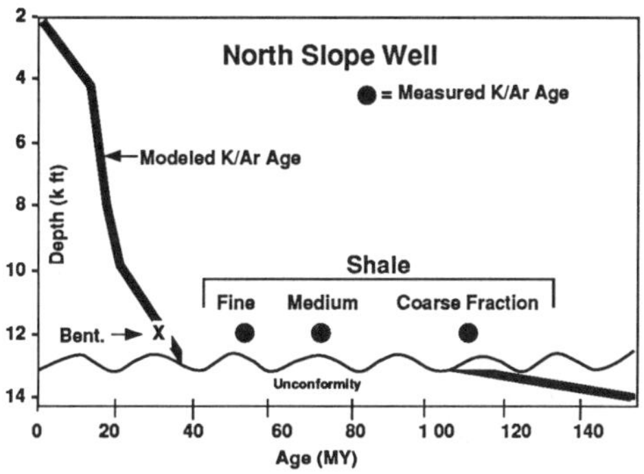

Figure 1 K/Ar ages of Turonian shale and bentonite

modeled from the burial/thermal history using W-L Huang's experimental kinetics for the smectite-to-illite reaction; any discrepancy between the calculated and measured date can then be used to adjust the thermal model. IAA assumes that illite in shales is a two-component mixture of diagenetic (young) illite, in mixed-layer illite / smectite (I / S) from the smectite-to-illite reaction, and discrete detrital (old) illite derived by erosion in the sediment source areas. Laboratory processing produces size fractions with widely varying amounts of each end-member and therefore widely varying K/Ar ages (black dots in Fig.1; from Turonian unit discussed below). Forward modeling of the

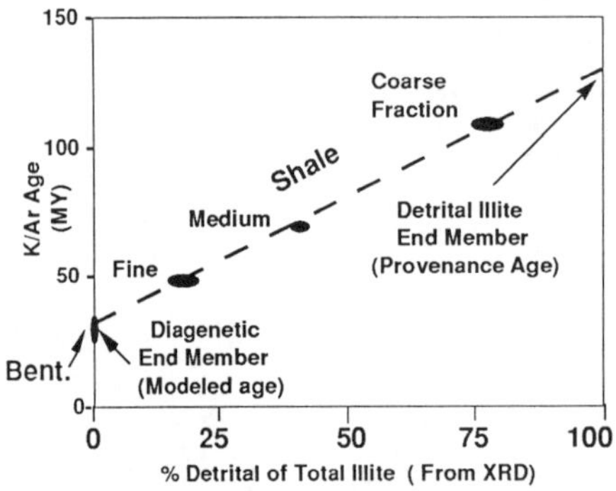

Figure 2 Illite Age Analysis plot of a single shale sample. Bar at left shows 3 bentonite samples

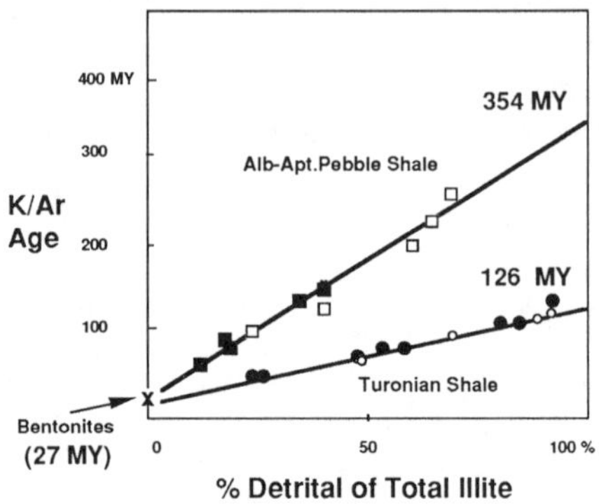

Figure 3 Illite Age Analysis Plot of fractions from four shale samples showing end-member ages.

XRD patterns of each fraction permits quantitative determination of end-member proportions. Plots of K/Ar age vs. that percent of total illite which is diagenetic are linear (Fig. 2; data points same as on Fig. 1), confirming the two-component model; extrapolation to 0% and 100% detrital illite gives ages for the diagenetic and detrital end-members, respectively (Fig. 2).

2 RESULTS

Initial testing of IAA was done using core from a well near Pt. Thomson, AK, containing shales interbedded with bentonites. The bentonites consist of I/S with 70% illite layers and lack detrital illite -- therefore they are entirely of one end-member and all fractions give the diagenetic (modeled) K/Ar age of ~30 my ("X" in Fig.1). This age provides an independent measure of the age of the diagenetic end-member as determined by IAA of adjacent shales. Because bentonite samples are seldom available, whereas shale cuttings are, the IAA technique is designed for shales. The 300'- thick cored interval overlies the Thomson Sand (depth ~13,000') and consists of the Albian-Aptian "Pebble Shale" (Ellesmerian provenance: Canadian Shield) and overlying bentonitic, Turonian Coleville Group of Brookian provenance. These organic-rich marine shales are the source for some of the hydrocarbons at Pt. Thomson. Figure 3 shows all data (including those in Fig. 2) for various fractions of two "Pebble Shale" samples (open and filled squares) and two Turonian samples (open and filled circles).

Statistically, the extrapolated diagenetic end-member value for each of the four samples in Fig. 3 has the same intercept within the 90% confidence limit of the least-squares regression. For all of these the IAA diagenetic age is ~30 my, confirmed by the bentonites ("X" on Fig. 3 at 27 my) and predicted by the kinetic model (Fig. 1). Although the details of the thermal model cannot be discussed here, IAA confirms both the absence of a Cretaceous heating event in this area, and the apparent requirement of a late Tertiary event, possibly associated with fluids driven by Brooks Range thrusting.

On the other hand, the age of the detrital end-member of the "Pebble Shale" (354 my; Fig. 3, right side) is much older than that of the overlying Brookian-sourced shale (126 my), confirming derivation of the former from the Canadian Shield. The numeric difference between the provenance age and the diagenetic age of a single sample (Illite Age Gap) is controlled by the regional rates of uplift, erosion and burial, and for these two shale units (324 vs. 96 my) indicates a major tectonic reorganization associated with opening of the Canada Basin, uplift of the Brooks Range, and establishment of a foreland basin-type setting. Illite Age Analysis has confirmed our previous finding based on other data that the HRZ (high-radioactivity zone on logs) consists of the Ellesmerian "Pebble Shale" and the overlying distal parts of the eastwardly prograding clinoforms derived from uplift of the Brooks Range.

3 CONCLUSIONS

Illite Age Analysis (IAA), our laboratory and data analysis application of the K/Ar radiometric dating technique, provides both integrated age of diagenesis and mean age of illite in the source terrane from shale samples. IAA assumes that illite in shales is a two-component mixture of diagenetic (young) illite from the smectite-to-illite reaction, and detrital (old) illite derived by erosion in the sediment source areas. Plots of K/Ar age vs. that percent of total illite which is diagenetic are linear, confirming the two-component model; extrapolation to 0% and 100% diagenetic illite gives ages for the detrital and diagenetic end-members, respectively. Testing of IAA was done using core from the Pt. Thomson, AK area containing shales interbedded with bentonites. The bentonites lack detrital illite -- therefore they are entirely of one end-member and all fractions give the diagenetic K/Ar age of ~30 my, the age predicted by the kinetic model. This age provides an independent measure of the age of the diagenetic end-member as determined by IAA of adjacent shales. The 300'-thick cored interval overlies the Thomson Sand (depth ~13,000') and consists of the "Pebble Shale" (Ellesmerian provenance) and overlying bentonitic, Albian-Aptian of Brooks Range provenance. For both of these the IAA-extrapolated diagenetic age is ~30 my, which is confirmed by the bentonites. However, the age of the detrital end-member of the Pebble Shale (354 my) is much older than that of the overlying Brookian-sourced shale (126 my), confirming derivation of the former from the Canadian Shield.

4 APPENDIX: METHODS

Details of the IAA procedure, an integrated package of laboratory and data analysis techniques, and of precision and accuracy are discussed below. The size fractions are prepared by chemical and ultrasonic disaggregation, high-speed centrifugation and high-gradient magnetic separation. We typically use 2.0-0.2μm, 0.2-0.02μm and <0.02μm size fractions (coarse, medium and fine). A shale sample weighing 100-500 gm is usually sufficient. The success of IAA depends upon four factors: 1) the degree to which a sample can be described by a two-component mixing model , 2) the standard errors of the linear regression, 3) the precision and accuracy of the K/Ar dates, and 4) the precision and accuracy of the quantitative XRD determinations of % detrital illite. The first of these, validity of a two-component model, is strongly supported if the data points from a single sample lie on a line (Figs. 2 & 3); this test requires at least three data points. Data from multiple samples (even from the same well) which have different end-member ages will plot on different lines; these lines may converge if they have a common end-member (Fig. 3). Analytical transmission electron microscopy may also be able to test the linear assumption; such work is in the beginning stages. The detrital end-member can be a mixture of illite and mica from multiple sources (and ages) yet still fit the linear model, providing that it has the same composition in all size fractions. The second factor, regression error, can be quite large if the points are scattered. In Fig.3 the standard error for the diagenetic age (y-intercept) is only 2 m.y. for the lower set of points, but is 7 m.y. for the upper points. The latter could reflect lack of homogeneity in the detrital end-member, although this would be expected to be more systematic than seen in Fig. 3.

The K/Ar dates typically reported for the small samples we use (~200 mg) have a precision of ± 1-2 % of the age; this is probably the most precise measurement within the IAA technique. The accuracy of the diagenetic end-member age depends largely on the material not having suffered argon loss after crystallization. The latter is typically caused by heating above ~200° C. Because the smallest crystals are expected to have the greatest diffusional argon loss and consequently younger ages, the absence of such loss is strongly suggested by the fact that all size-fractions of the bentonites in this study give the same age.

The XRD method for quantifying detrital (discrete) illite and diagenetic illite in mixed-layer I/S is new and will be discussed below; the errors of the determination are still being evaluated, but may be as high as ±10 %. The characterization and quantification of the K-bearing clay mineral phases are what make IAA possible; both depend on the forward-modeling techniques developed by R. C. Reynolds (Dartmouth) and sold commercially as NEWMOD, NEWMOD3C, and CLAYS. These programs, which model XRD patterns for mixed-layer and other clay minerals, are based on the first principles of X-ray crystallography and incorporate a description of the crystal structure and atomic site occupancies (chemical composition) as well as instrument factors.

Figure 4 illustrates the method applied to a difficult sample, the <0.02 μm (fine) fraction of the shale illustrated in Figs. 1 & 2. The difficulty is that the XRD peaks of the discrete illite (detrital end-member, 18% of this fraction) are nearly hidden by those of the much more abundant diagenetic I/S. The diagenetic phase (pattern C) is itself complex, consisting of a three-component mixed-layer I/S system: 70% illite layers, 20% smectite with 2 glycol layers and 10% smectite with one glycol layer. The illite and smectite are mixed with R-1 ordering, but the two types of smectite layers are randomly arranged with respect to each other. Only NEWMOD3C can model such a mineral. The 2Θ position of the superlattice peak on the extreme left side of Fig. 4 is very sensitive to small variations in sample position, therefore the mismatch from the modeled pattern is not a cause for concern.

The entire modeled system is remarkably responsive to small variations in input parameters; for example, no patterns produced by NEWMOD (2-components only) showed the 10 peaks in (C), all of which are visible in (A). Addition of 20% rather than 18% of discrete illite (D) to (C) gives the mixture (B) an unrealistically large bulge on the side of the large peak near 17°2Θ which does not match that labeled "ILL" on (A). Therefore the sample must have <20% detrital illite. For the coarser fractions of this sample the discrete illite peaks are, expectedly, much more prominent as in (E), the medium-size fraction. For routine application, a series of calculated patterns are produced to test the various possibilities and to obtain the final quantitative analysis. For a typical IAA plot (Figs. 2 & 3) the %illite in I/S and the % discrete illite (70 & 18, respectively, in Fig. 4) are normalized to 100% illite.

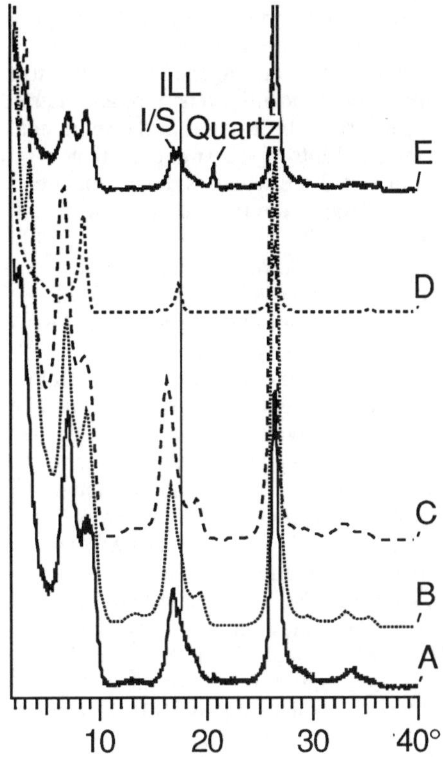

Figure 4 X-ray diffraction pattern of an ethylene glycol solvated oriented mount of the <0.02 μm (fine) fraction of the shale illustrated in Figs. 1 & 2.Scale is in 2Θ, Cu Kα radiation. XRD peaks of the discrete illite (detrital end-member, 18% of this fraction) are nearly hidden by those of the much more abundant diagenetic I/S. One such peak, a small shoulder on the side of a large I/S peak in (A), is identified by the line labeled "ILL" in the figure. XRD pattern (A) is the experimental pattern from a Na-saturated sample; pattern (B) is the final modeled pattern which consists of a mixture of 82% of (C), a modeled mixed-layer I/S, and 18% of (D), modeled discrete illite (the detrital component). For the coarser fractions of this sample the discrete illite peaks are, expectedly, much more prominent as in (E), the medium-size fraction.

Geothermal systems

Fluid chemistry

Water-Rock Interaction, Kharaka & Maest (eds) © 1992 Balkema, Rotterdam. ISBN 90 5410 075 3

The occurrence of acidic fluids in the Leirbotnar field, Krafla, Iceland

Halldór Ármannsson
Orkustofnun, Reykjavík, Iceland

Gestur Gíslason
Hitaveita Reykjavíkur, Reykjavík, Iceland

ABSTRACT: When magmatic gas entered the Krafla geothermal system during tectonic events that started in 1975 a low pH was observed in parts of the system, the northern Leirbotnar field being badly affected. The gas has mostly disappeared but acidity persists and is apparently confined to high-temperature aquifers at great depth.

1 INTRODUCTION

The Krafla high-temperature geothermal area, Iceland was troubled by magmatic activity from 1975 into the eighties (Ármannsson et al 1987).

It can be divided into at least four separate fields; the Leirhnúkur, Leirbotnar, Suðurhlíðar and Hvíthólar fields. The three latter ones have been drilled and revealed a complex geothermal system. In Leirbotnar (Fig.1) it consists of a cool (190-220°C) upper zone and a hotter (300-350°C) lower zone. On the basis of stable isotope concentrations it has been found that the recharge to it is essentially local in origin (Darling & Ármannsson 1989).

In early 1976 a large influx of gas was detected in wells and fumaroles in the Leirhnúkur and Leirbotnar fields. A gas excess was observed in the the lower zone, but hardly detected in the upper one. The gas concentrations of borehole fluids have shown a certain time-pattern, a maximum in 1977/1978, a secondary one in 1980 and a steady decline since, with some local variation.

2 THE NORTHERN LEIRBOTNAR FIELD

Of the wells in northern Leirbotnar, KW-2 and KG-8 draw from the upper zone only, KJ-15 the lower zone only, but KG-4, KG-10 and KG-25 a mixture from both zones. Acidic fluids are found at 1900-2100 m depth in an aquifer whose temperature is about 330°C near an acid intrusion in KG-4, KG-10 and KG-25 but not in KJ-15, the temperature of whose main inflow is about 310°C.

Well KG-4 was drilled in 1975 and went out of control. In early January 1976 its safety valves gave in and by the end of the month the drillsite had turned into a boiling pond from which a stream soon started to flow. Its initial flow was estimated 20-40 l/s, that in early spring 150-250 l/s but eventually stopped in December 1976.

Well KG-10 was drilled in 1976. Its wellhead pressure had reached >90 bar within 3 weeks and it discharged initially >50 kg/s at 20 bar a pressure but a decline set in after about 3 weeks

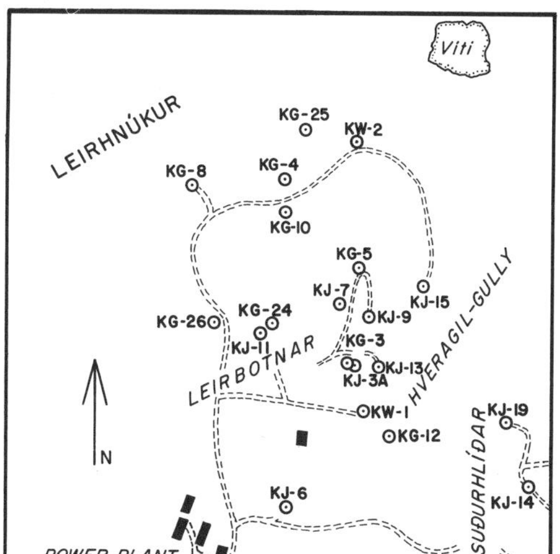

Figure 1. The Leirbotnar field wells

and at the end of January 1977 it was 15 kg/s at 4 bar a and did not change before it was drilled out in September 1977. The cause of the decline proved to be iron sulphide, iron oxide, iron silicate, silica and calcite deposits. The flow reached 50 kg/s at 25 bar a during early discharge after the work-over but declined and was 14 kg/s at 3.6 bar a when it was shut down in December 1979.

Well KJ-15 was drilled in 1980. Its flow has been steady at 4 kg/s and was dry until 1986 since when the enthalpy has fallen to 2400 kJ/kg. The high gas concentration of its fluid has declined recently.

Well KG-25 was drilled in 1990. Its initial flow was 40 - 50 kg/s at 15 - 22 bar a but had declined to 14.7 kg/s at 16 bar a when it was closed down in January 1991. During an attempted caliper survey a 1-1.5 m piece of the liner was discharged from the well. This apparently came from about 1400 m depth and was badly damaged by corrosion. Further caliper surveys and a work-over to 1550 m depth only revealed signs of minor deposition but very severe corrosion. Temperature logs from January and August 1991 showed that there was a flow from the bottom aquifer. The reduced flow is probably due to corrosion damage to the liner.

3 THE CHEMICAL COMPOSITION OF WELL FLUIDS

Analytical results for raw liquid samples from wells KG-4, KG-10 and KG-25 (Table 1) suggest changes in the chemistry of KG-4 fluid between November 1975 and March 1976. Magmatic activity started with earthquakes in the summer of 1975 and manifested itself with a volcanic eruption in Leirhnúkur in December 1975. A gas increase was observed in well KG-3 between January and March 1976 and in a nearby fumarole in January 1976 (Óskarsson 1978). As well KG-4 is closer to the centre of activity than both the fumarole and KG-3 a change may have occurred earlier there.

The chemical composition of the water in the stream from well KG-4 changed with time and decreasing flow and by October 1976 it had approached a typical upper zone fluid composition with pH 8.5.

The chemical composition of the fluid from KG-10 also approached an upper zone composition with reduced flow and as this was due to deposit formation in the lower zone the

Table 1. Sampling conditions and results of analyses for pH, major anions and silica in Krafla wells KG-4, KG-10 and KG-25.

Well	KG-4 well	KG-4 stream	KG-10	KG-10	KG-25
Year	1975	1976	1976	1977	1990
Date	11.24	03.22	12.13	10.21	11.15
H_0 kJ/kg	2350*	2350*	1362	1341	1741
P_0 bar a	8.8	1	5.5	21.3	23.5
pH	9.44	1.86	6.88	6.05	6.45
t°C	20	20	19	18	16
Cl^- mg/kg	45.2	151.4	40.5	44.6	71.8
SO_4^{-2} "	185.6	2200	177.4	131.9	175.9
CO_2 "	24.5	223	241	250	65.7
F^- "	5.32	0.83	1.46	1.97	4.31
SiO_2 "	447	519	524	446	261

* Estimated (see text)

acidity can be assumed to be a feature of the lower zone fluids. As wells KG-4, KG-10 and KG-25 draw a mixture of upper and lower zone fluids no direct analyses exist and calculation is needed to isolate lower zone properties.

The total enthalpy of the mixed fluid is known and so are the equilibrium temperatures of the upper and lower zone fluids. The upper zone fluid is assumed to be liquid only, all the silica to be present in liquid phase and its concentration to be a function of temperature. Thus there is enough information to calculate the vapour fraction of the lower zone fluid. The calculation method was described by Ármannsson et al (1982). The programme VDATA (Hauksson 1991) has been used for compositional and speciation calculations. Here we consider pH, alkalinity and the major anions, i.e. HCO_3^-, Cl^- and SO_4^{-2}.

The results for pH and speciation of the fluid in well KW-2 suggest that it was slightly affected by magmatic gas as shown by an increase in HCO_3^- (Fig. 2). Its pH came down at the same time.

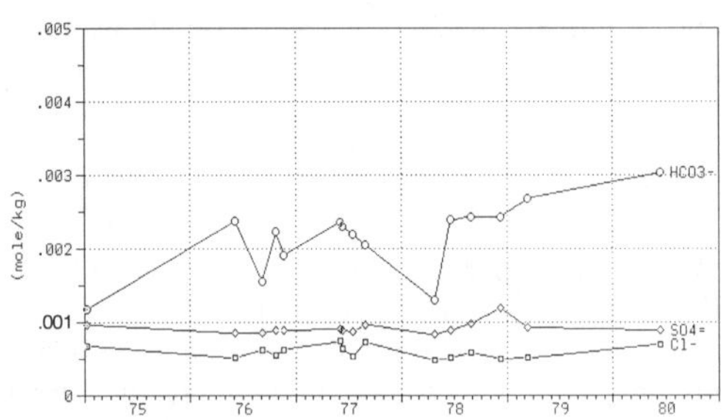

Fig. 2. KW-2 HCO_3^-, Cl^- and SO_4^{-2} in deep water

Compositional considerations, however, suggested that KW-2 fluid is representative for the upper zone and that changes with time in the upper zone after 1980 are negligible. The flow temperature in well KW-2 (195°C) is relatively close to that of the upper zone flow in wells KG-4, KG-10 and KG-25 (190°C). Therefore analytical results for its fluids are used as reference concentrations for upper zone fluids, accounting for closeness in time where possible.

The major upper zone aquifer in well KG-4 is at 700 m depth with a pressure of about 50 bar a, but during flow the wellhead pressure was about 40 bar a, so that the upper zone fluids could not have taken part in its discharge since the additional pressure due to the water column is at least 20 bar. This is supported by the fact that the initial chemical composition of the stream was typically lower zone (Table 1). Direct enthalpy and flow measurements could not be performed, but steam flow estimates from comparison with well KG-3, with the initial stream flow as liquid flow, gave an enthalpy value of 2350 kJ/kg.

4 RESULTS AND DISCUSSION

The alkalinity calculated for upper zone fluids from KW-2, lower zone fluids from KJ-13, and lower zone fluids from KG-4, KG-10 and KG-25, is shown in Fig. 3. The two former systems are close to neutral, but the fluid in KG-4 and KG-10 was apparently neutral to start with, with both becoming acidic after some time presumably due to volcanic effects. The KG-25 fluids were acidic from the start.

The values for major anions for KJ-15 and for the lower zone fluids of wells KG-4, KG-10 and KG-25 (Figs. 4 and 5) show that HCO_3^- is the dominant anion in KJ-15, Cl^- and HCO_3^- predominate initially in KG-4 and KG-10 fluids, but chloride after that and in KG-25 fluid. The HCO_3^- decrease in KJ-15 fluids is probably due to decrease of magmatic gas.

Table 2. Well KG-4 stream. Major anion concentrations (mole/kg) 1976.

Date	HCO_3^-	Cl^-	SO_4^{-2}
03.22	$2.2*10^{-8}$	$1.2*10^{-3}$	$3.1*10^{-6}$
10.17	$3.3*10^{-3}$	$1.0*10^{-3}$	$1.7*10^{-3}$

Calculations show that while the KG-4 steam was acidic chloride predominated but the concentration of all three anions was similar after the pH rise (Table 2).

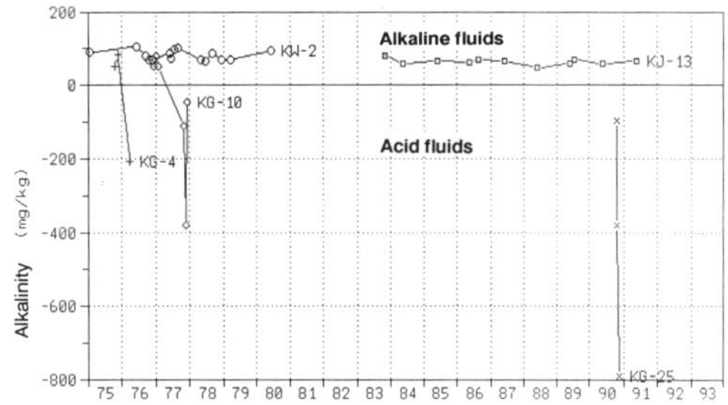

Fig. 3. Calculated alkalinity of fluids from different parts of the Leirbotnar field

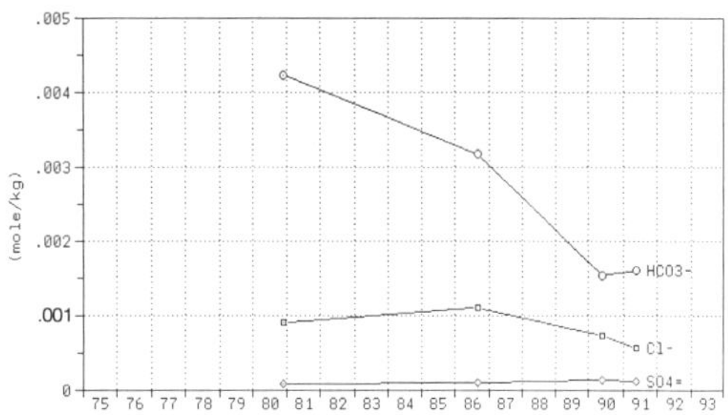

Fig. 4. Well KJ-15. The deep water concentration of major anions

Truesdell (1991) concludes that acid reservoir fluids in geothermal fields seem to originate from either the introduction of volcanic fluids or the volatilization and transport of HCl in superheated

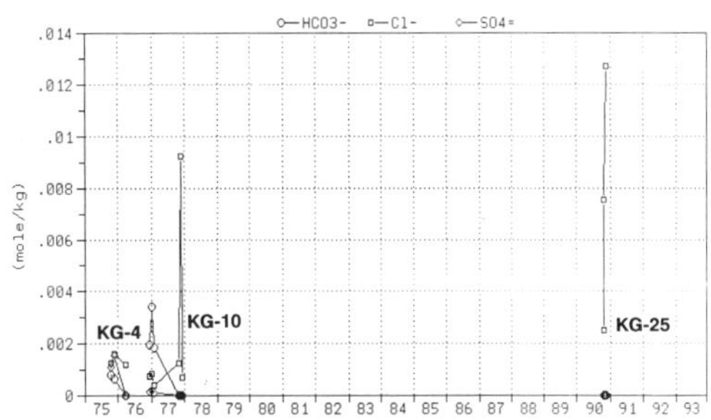

Fig. 5. The lower zone of wells KG-4, KG-10 and KG-25. The deep water concentrations of major anions

steam. At moderate temperatures HCO_3^- is the dominant anion in Leibotnar fluids, which are near neutral, but after the influx of magmatic gas, fluid in aquifers at temperatures ≥ 320°C became acidic and Cl^- the most important anion. The fluid enthalpy in KG-25 is very high and its inflow might be dry steam like the one in early KG-12 that produced acidic superheated steam from an aquifer at about 320°C (Truesdell et al. 1989). Evidence for chloride-rich acidic liquid has been found in well KJ-7 where a temperature of >340°C was logged. It is suggested that the acidic fluids originate from the introduction of volcanic fluids which locally may become superheated and transport HCl.

REFERENCES

Ármannsson, H. Gíslason, G. & Hauksson, T. 1982. Magmatic gases in well fluids aid the mapping of the flow pattern in a geothermal system. Geochim. Cosmochim. Acta 46: 167 - 177.

Ármannsson, H., Guðmundsson, Á, & Steingrímsson, B.S. 1987. Exploration and development of the Krafla geothermal area. Jökull 37: 13 - 29.

Darling, W.G. & Ármannsson, H. 1989. Stable isotopic aspects of fluid flow in the Krafla, Námafjall and Theistareykir geothermal systems of northeast Iceland. Chem. Geol. 76: 197 - 213.

Hauksson 1991. VDATA. A general programme for geothermal energy users. Kemía, Reykjavík, 1 pp.

Óskarsson, N. 1978. Effect of magmatic activity on fumarole gas composition in the Námafjall-Krafla volcanic center, N-Iceland. Nordic Volcanol. Inst. Publ. No. 78 03, 20 pp.

Truesdell, A.H. 1991. Origins of acid fluids in geothermal reservoirs. G.R.C. Trans. 15: 289 - 296.

Truesdell, A.H., Haizlip, J.R., Ármannsson, H. & D'Amore, F. 1989. Origin and transport of chloride in superheated steam. Geothermics 18: 295 - 304.

Water-Rock Interaction, Kharaka & Maest (eds) © 1992 Balkema, Rotterdam. ISBN 90 5410 075 3

Hydrochemistry of Olkaria and Eburru geothermal fields, Kenyan Rift Valley

M.K.Arusei
Olkaria Geothermal Project, Kenya

ABSTRACT: The chemical composition of waters from the Rift Valley is presented. The study of the chemical characteristics of these waters in used to suggest the movement, similarities and chemical processes affecting them. The study has shown that water from Lake Naivasha, which is the only fresh water lake in the Kenyan Rift Valley, outflows to recharge most of the nearby cold water boreholes and may recharge geothermal wells at Olkaria and Eburru. The waters from some boreholes to the north of Eburru and Elementeita spring may be outflowing from Eburru geothermal system. The waters from the escarpment springs are shallow waters which have been heated by shallow hot rock, probably below the location of the springs. The similarities of waters in the Rift make it difficult to determine which water is the main recharge to geothermal reservoirs, and thus the impact of geothermal exploitation at Olkaria on Lake Naivasha remains unknown.

1 INTRODUCTION

This report covers work done in the areas in the vicinity of Lake Naivasha extending from Eburru to Olkaria geothermal fields. A number of geothermal fields lie in the Kenyan Rift Valley. Preliminary surface investigations have been carried out in many of them (Clarke et al.,1990; Halldor, 1987), but drilling in just two, Eburru and Olkaria. Olkaria geothermal system consists of three fields, Olkaria West field, Olkaria Northeast field and Olkaria East field, the only one producing 45 Mwe. All the production wells for a 64 Mwe power station in Olkaria Northeast have been drilled. Six wells have been drilled in Eburru field and only one was successful producing 2.5 Mwe.

Lake Naivasha, the only freshwater lake in the Kenyan Rift Valley has no surface outlet. It gets its recharge from direct precipitation and the two rivers, Gilgil and Malewa (Figure 1). Its contribution to the hydrological regime of the area has given rise to a considerable amount of speculation when the water level fell drastically in 1987. The piezometric levels suggest that the water from the lake flows in both northerly and southerly directions (McCann, 1972; Allen et al., 1989; Ase et al., 1986; Darling et al., 1990).

A microseismic study of the Kenyan Rift Valley has shown that the grid faulting, unlike the escarpment faulting, is quite active, suggesting they are open (Allen et al, 1989). Thus, this grid faulting causes the groundwater from the escarpments to flow towards the center of the Rift and then follow longer flowpaths reaching greater depths.

Shallow groundwater can be characterized by some major and minor correlation diagrams. The correlation diagrams have been used worldwide to trace the movement and relative ages of groundwater. High evaporation rates in the Rift have caused high concentrations of chemical constituents, resulting in saline lakes such as Lake Nakuru and Lake Elementeita. Volcanic and magmatic activity such as the emanation of carbon dioxide from depth in some parts of the Rift Valley (Mt.Margaret) have also influenced the chemistry of the groundwater (Du Bois and Walsh 1970)

2 FIELD SAMPLING

Water samples were collected from the following selected sites: Musyoka's borehole (MYK), Munyua Railway Station borehole (MRS), Naivasha Town Water borehole (NWB), Marula Farm borehole 1-4 (MFB 1-4), Green Park borehole (GPB), boreholes C-2704, C-1287, C-3229, C-845, C-429 Elementeita spring (ELM), Kariandusi spring (KS), Kijabe spring (KJS), Mayer Farm spring (MS), Olkaria wells OW-23, OW-29, OW-714, OW-715 and OW-716,

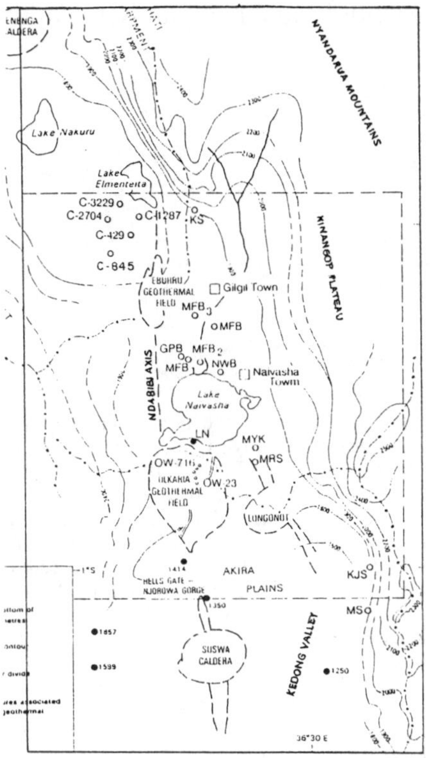

Figure 1. Simplified map of study area

Eburru wells, EW-01 and EW-04, Lake Naivasha (LN) and River Morendati (MR) in March 1991.

2.1 *Analysis and results*

The major constituents were analyzed in Olkaria laboratory while minor and some major constituents were analyzed in the Orkustofnun laboratory, Iceland. Ionic balances for all the samples were calculated to assess the data quality. All the results were below ± 10%.

3 DATA EVALUATION

The data were evaluated using various diagrams.

3.1 *Halide Diagrams*

The chloride and bromide concentrations in most waters from the area are generally low. There is, however, a gradual increase of both anions in the borehole waters around the lake. The lowest concentrations are found in Kariandusi spring and River Morendati, but the highest values are in the geothermal fluids. Figure 2 shows that a line can be drawn through water values from the lake, the cold borehole and the geothermal samples. Another line can be drawn through the values for the Green Park borehole, Marula Farm borehole 4, Musyoka's borehole, Mayer Farm spring and Munyua Railway station borehole waters. All of these boreholes are slightly to the east or the west of the lake.

Figure 3 shows a triangular diagram for Cl-F-Br. The diagram shows two groups of waters. The first group represents purely groundwater in the Rift floor, including the escarpment springs, lake and river waters. The points spread towards the fluoride corner. Kariandusi spring water is at the end of the spread with the highest fluoride ratio, while Munyua railway borehole and Mayer Farm spring waters are at the other end with the lowest fluoride ratio.

The second group is composed of waters from geothermal wells and boreholes and springs associated with geothermal activity. They plot almost parallel to Cl-F line with Olkaria wells OW-715 and OW-716 waters near the chloride corner and C-2704 the towards the fluoride corner.

3.2 *Cation Diagrams*

Figure 4 shows correlation between sodium and bromide. The values for borehole waters close to Lake Naivasha tend to plot close to sodium axis, but the values for borehole waters slightly east or west of Lake Naivasha plot close to bromide axis. The values for waters at Eburru and Elementeita boreholes plot on a different line from the one connecting values for Olkaria waters and Munyua Railway station borehole.

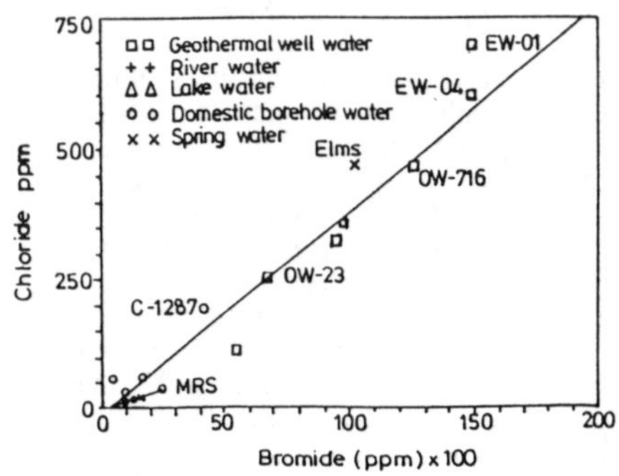

Figure 2. Cl/Br x 100 correlation diagram

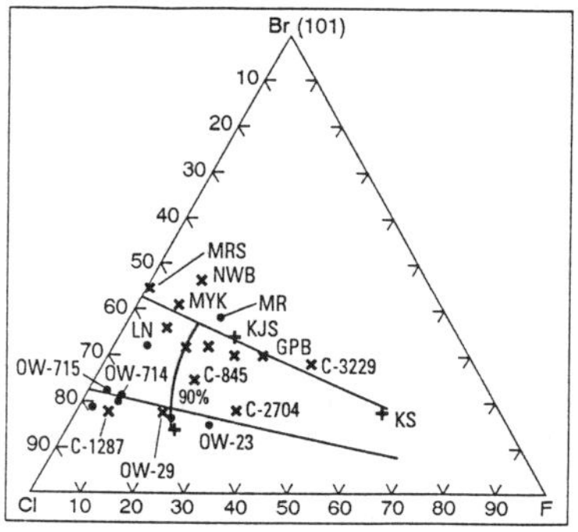

Figure 3. Cl-F-Br x 100 triangular diagram.

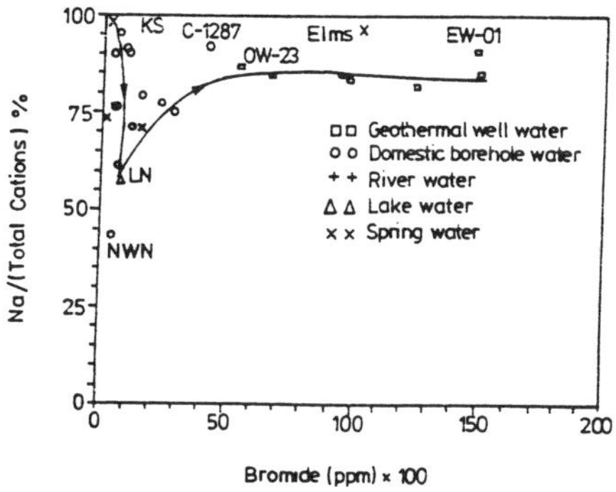

Figure 5 Na/Total cation % correlation diagram

The sodium concentration for waters from C-429, C-2704 and C-1287 are slightly higher than the rest of the values of the Rift floor waters. This may suggest influence of geothermal water in these boreholes. The water from boreholes east or west of Lake Naivasha could be recharging Olkaria geothermal water.

The sodium concentration was expressed in terms of its percentage of total cations to facilitate the observation of changes occurring as water moves from the escarpment to the Rift floor (Figure. 5).

The figure shows three types of water as follows;

a) high (Na/ Total cations)% and low bromide concentration

b) low (Na/ Total cations)% and low bromide concentration

c) high (Na/ Total cations)% and high bromide concentration

The first group consists of escarpment spring, river and all borehole waters except those from the Musyoka's, Munyua Railway station and C-1287 boreholes which fall within the third group. The second group consists of Lake Naivasha and Naivasha Town borehole waters, and the third group of geothermal well and Elementeita spring waters, as well as waters from boreholes around Lake Elementeita.

Figure 6 shows the correlation between log meq Na and log meq Cl to facilitate the cause of increment of concentrations of the constituents (Garrels and Christ, 1965). All the values for borehole, lake, river and escarpment spring waters plot in the general direction of a 45°. The geothermal well waters and Elementeita spring water plot slightly away from this line.

Figure 7 shows the Na/1000-K/100-√Mg values plotted on Giggenbach's diagram. Waters from the cold boreholes except C-2704 and C-1287 plot on the immature sector of the diagram. Water from Lake Naivasha plots away from the other immature waters. Waters from Kijabe spring, C-2704, and C-1287 are the only ones

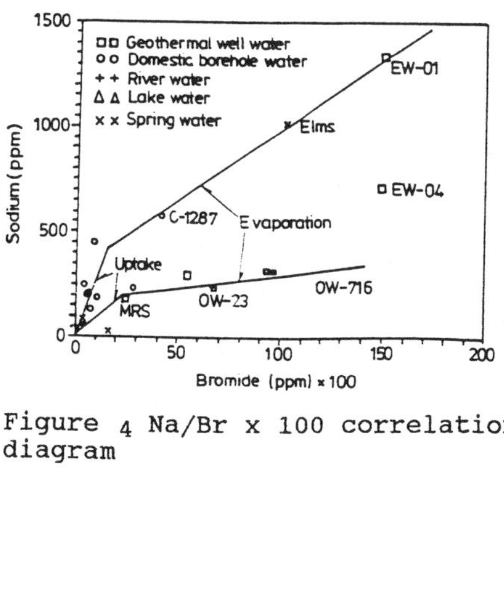

Figure 4 Na/Br x 100 correlation diagram

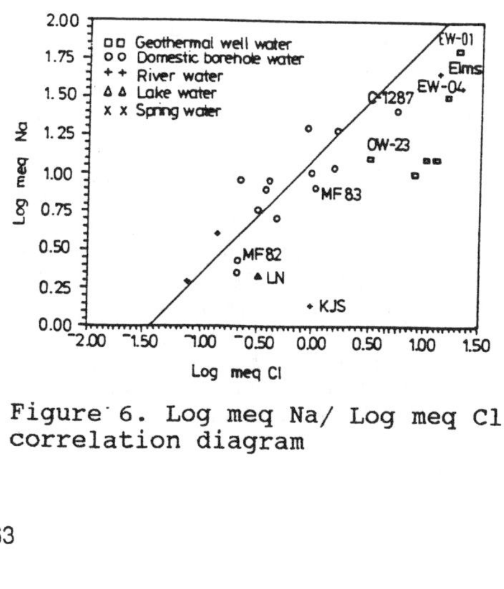

Figure 6. Log meq Na/ Log meq Cl correlation diagram

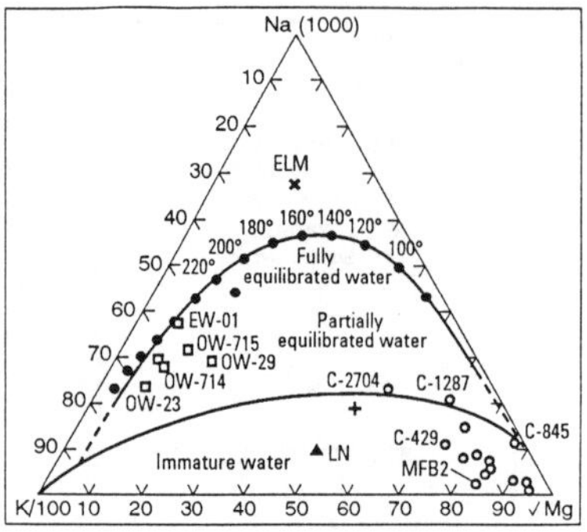

Figure 7. Na/1000-K/100-√Mg triangular diagram.

which are partially equilibrated. Waters from Olkaria wells are all partially equilibrated while Eburru well waters are fully equilibrated at 280°C and 250°C respectively. Water from Elementeita spring plots above the equilibration curve at 150°.

4 DISCUSSIONS

The various diagrams used in the study have shown that the escarpment waters, river and the groundwater in the Rift floor are related to the lake water. Since the lake is at the culmination of the Rift floor, its water recharges some of the boreholes to the north and south of it. The diagrams have provided chemical evidence for this suggestion. Since the concentrations of many constituents increase away from Lake Naivasha (Figure 6). This may suggest that the lake water flows out fairly fast to maintain the same concentration on an open lake where evaporation is high, to recharge the nearby boreholes. As the lake water moves away, evaporation causes the constituents to increase (Garrels and Christ, 1965).

Figure 2 shows that both Olkaria and Eburru well waters have almost the same Cl/Br ratio and that may mean both reservoirs get their waters from the same origin, which may be the Rift groundwater or escarpment water. The Cl/Br values for the waters from the Green Park borehole (GPB), Marula Farm borehole 4 (MFB 4), Musyoka's borehole (MYK) and Munyua railway station borehole (MRS), which are not directly north or south of the lake, seem to plot on a different line from those for the rest of the boreholes. This can be explained by, either the permeability between them and the lake is poorer, or they are not deriving their water from the lake. Bromide becomes more concentrated with respect to chloride upon intense evaporation.

Dilution factors between waters from Eburru well EW-01 and the waters from the boreholes with relatively high chloride and bromide concentrations and Elementeita spring were calculated (Table 1). The bromide and chloride dilution factors between EW-01 and EW-04, Elementeita spring, C-1287, C-429, and C-2704 show the conservative nature of the two ions. Water from C-845 has different dilution factors for bromide and chloride, suggesting that water from this borehole may not be derived from geothermal water from Eburru. The dilution factors for Na, K and F for waters from the same wells are not consistent, because of other processes that may affect them.

Table 1. Dilution constituents from EW-01 water to some other waters.

Dilution to	Cl	Br	Na	K	F
EW-04	1.1	1.0	1.9	3.4	2.0
Elm Spr.	1.4	1.4	1.7	7.8	1.2
C-1287	3.2	3.5	2.3	8.4	9.7
C-429	10.8	9.4	3.1	4.1	16.4
C-2704	20.9	16.7	2.9	6.3	9.7
C-845	41.1	25.0	6.7	9.9	3.6

The (Na/Total cations)% was calculated to check the weathering of feldspars with the water flow from the Rift escarpment to the Rift floor. Figure 5 shows that Marula Farm borehole 2 (MFB2), Naivasha Town borehole (NWB) and Lake Naivasha (LN) waters have the lowest percentage of sodium. The diagram may suggest that as the water flows from the Rift escarpment towards the Rift floor, sodium is dissolved more rapidly from feldspars and other sodium minerals than the other cations. As the water moves on, the other cations are also dissolved and the percentage drops. In the lake the percentage further drops because of other processes, such as ion-exchange and the dissolution of other cations. The lake water then flows to recharge the groundwater, and evaporation controls the sodium concentration at lower temperatures. As the residence time becomes shorter again when the water leaves the lake, sodium uptake can be expected to become faster than that of the other cations again, and the percentages rise. The geothermal activity both in Olkaria and Eburru further raises the percentage because although ion-exchange with potassium controls its concentration, the concentrations of the other cations are low. This can be seen also in figure 4, where sodium is extracted fast in the groundwater and controlled by evaporation and ion-exchange in the geothermal waters.

The two boreholes (NWB and MFB2) close to Lake Naivasha could derive water from a part of the lake which may be buried. The differences in the (Na/Total cations)% values in the lake water and the water from these two boreholes may be due to evaporation in the open lake. This may be supported by the fact that the concentrations of many species in the waters from these places are similar. The diagram may suggest that waters with high (Na/Total cations)% and low bromide concentration could be flowing towards the Rift floor, and the ones with high (Na/Total cations)% and high bromide concentration could be flowing along the Rift floor. If this is true, then Kijabe spring, C-3329, C-2704, Green Park borehole (GPB), River Morendati (MR), C-845 and Marula Farm borehole 1(MFB1) waters flow towards the Rift floor, while Marula Farm borehole 3 and 4, C-429, C-1287, Musyoka's borehole, Munyua railway station borehole, Elementeita spring and all the geothermal well waters flow along the Rift floor or away from the lake (Figure 5).

The Cl-F-Br diagram (Figure 3) was used to separate the local groundwater and the geothermal or mixed water. A mixing curve between water from Marula Farm borehole 3 and EW-01 was drawn. It showed that mixing up to 90% of Marula Farm borehole 3 (MFB3) water and 10% of Eburru well EW-01 water plotted in the geothermal water sector. According to this diagram then, C-429, C-1287, C-2704 waters may have resulted from the mixing of geothermal water from Eburru field and groundwater. This is supported by dilution values in Table 1. Although the water from C-3229 has an elevated temperature, it may not be derived from outflow from Eburru geothermal field because it plots with groundwater. Its fluoride value is closer to the value of Kariandusi spring which is on the Rift escarpment. The temperature of the water from Kariandusi spring and the Kijabe spring is also relatively high, but they do not appear to have the same source as geothermal water from Eburru and Olkaria.

Figure 7 (Na/1000-K/100-$\sqrt{Mg}$ diagram Giggenbach, 1981) shows that only C-2704, C-1287 and Kijabe spring waters have partially equilibrated geothermal waters. Borehole C-3229 water however, plots on the immature sector, although its temperature is elevated. Water from the Kijabe spring is partially equilibrated suggesting that it may have interacted with rocks for a long time. Water from Kariandusi spring may be derived from a shallow water which has been in contact with the hot rock for a short time.

5 CONCLUSIONS

In this study it has been possible to show that geothermal reservoirs at both Olkaria and Eburrumay are getting some recharge from the Rift groundwater which may be recharged by Lake Naivasha. It has also been shown that waters from boreholes C-1287, C-2704, C-429 and Elementeita spring may be outflows from Eburru geothermal field. Waters from C-845 and C-3229 may originate from Mau Escarpment and be heated shallowly in the same way as Kariandusi and Kijabe springs. It has not been possible to calculate the amount of outflow from the lake in this study and therefore it is difficult to determine the impact of geothermal exploitation on the lake.

REFERENCES

Allen, D.J., Darling, W.G., and Burgess, W.G., 1989. *Geothermics and hydrogeology of the southern part of the Kenyan Rift Valley with emphasis on the Magadi-Nakuru area*

Ase,L. Sernbo, K, and Syren, P., 1986. *Studies of Lake Naivasha, Kenya, and its Drainage Area.*

Clarke, M.G.G., Woodhall, D.G., Allen, D. and Darling, W.G., 1990. *Geological, volcanological and hydrogeological controls on the occurrence of geothermal activity in the area surrounding Lake Naivasha, Kenya.* Ministry of energy, Republic of Kenya. Nairobi, 138 pp + maps.

Darling, W.G., Allen, D.J. and Armannsson, H., 1990. Indirect detection of subsurface outflow from a Rift Valley lake. *J. Hydrol.*, 113, 297-305.

Du Bois, G.G.B. and Walsh, J., 1970. Minerals of Kenya. *Geol. surv. Kenya Bull*, 11, 102 pp.

Garrels, R.M. and Christ, C.L., 1965. *Solution, minerals, and equilibria.* Harper, Rowe, New York, 450 pp.

Giggenbach, W.F., 1981. Geothermal solute equilibria. Derivation of Na-K-Mg-Ca geoindicators. *Geochim. Cosmochim. Acta*, 52, p.2749-2765.

Halldor, A., 1987. /*Geochemistry of steam in the Suswa and Longonot geothermal areas.* DTCD report.

McCann, D.L; 1972. *Preliminary Hydrogeologic evaluation of the long-term yield of catchment related to the geothermal prospect areas in the Rift Valley of Kenya.* U.N. Report.

Water-Rock Interaction, Kharaka & Maest (eds) © 1992 Balkema, Rotterdam. ISBN 90 5410 075 3

Geochemistry of waters from three geothermal boreholes in the upper Rhein graben, Switzerland

L.Calmbach
Lausanne University, Switzerland

ABSTRACT: Three geothermal boreholes were drilled in the southernmost part of the Rhein valley near the city of Basel. Hot waters (65 °C) were encountered in Triassic carbonates (Upper Muschelkalk Formation) between 1400 - 1800 m. The waters are CO_2-rich (1000 mg/l) and highly saline (17-50 g/l). Their chemistry differs significantly from most other waters known from the same aquifer. Chemical and isotopic evidence has been found suggesting that these waters are composed of two fluids of a different origin and salinity.

1 INTRODUCTION

The Rhein valley has long been known for its hot and mineralized springs. The occurrence of most of these waters is related to the borders of the massifs enclosing the Rhein valley, the Black Forest to the east and the Vosges to the west (Fig. 1). This zone is intensively fractured by N-S striking normal faults, which allow the groundwater to penetrate deeply into the basement to heat up. The Rhein graben is also known for its thermal anomaly; geothermal gradients up to 100 °C/km have been measured in oil wells in the area of Pechelbronn north of Strasbourg. These facts led to the project of the first deep borehole for geothermal exploration and exploitation in Switzerland in the southernmost part of the Rhein graben. Two wells were drilled in Riehen and a third one in Reinach 10 km south of Basel. Hot water (60-68°C) was encountered in all three wells in the upper Muschelkalk Formation, but only the boreholes of Riehen supply a satisfactory discharge for a future use as a central heating station. As the geothermal fluid is too mineralized to be disposed into the sewage system, Riehen 2 will be used for reinjection of the geothermal fluids into the aquifer after heat exchange. The distance between the two boreholes is 1 km in order to prevent a hydraulic shortcut.

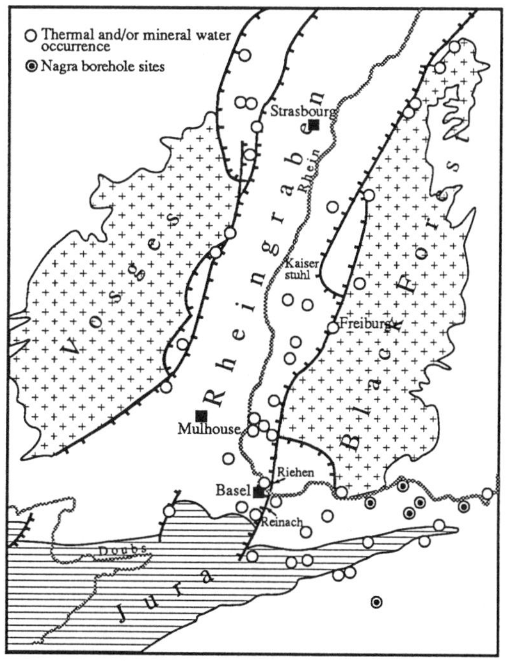

Fig 1: Thermal and mineral water occurrences along the faults of the Rhein graben borders. (after Carlé 1975)

2 GEOLOGICAL AND HYDROGEOLOGICAL SETTINGS

All boreholes are drilled into the Rhein valley flexure (Fig. 2). To the east, the basement rocks of the Black Forest are only covered by a thin layer of highly karstified Triassic limestones. Towards the graben to the west the sediment section becomes more complete including Jurassic and a thick layer of Tertiary sediments which filled the graben since its early beginning of spreading in the Eocene. Potassium rich evaporites occur in the Oligocene formations and build a major source for the high mineralization of many springs in the Rhein graben. The upper Muschelkalk which contains the aquifer of the studied cases is composed of dolomitic limestones at the base and limestones at the top. The underlying middle Muschelkalk frequently contains evaporitic minerals such as gypsum, anhydrite and rock salt.

3 HYDRAULIC CHARACTERISTICS OF THE BOREHOLES RIEHEN 1 AND 2

The hydraulic properties of the aquifer in Riehen 1 and 2 were extensively tested by means of pump- and reinjection tests and flowmeter logs. Both wells are artesian, wellhead pressures come to 2.5 (Riehen 1) and

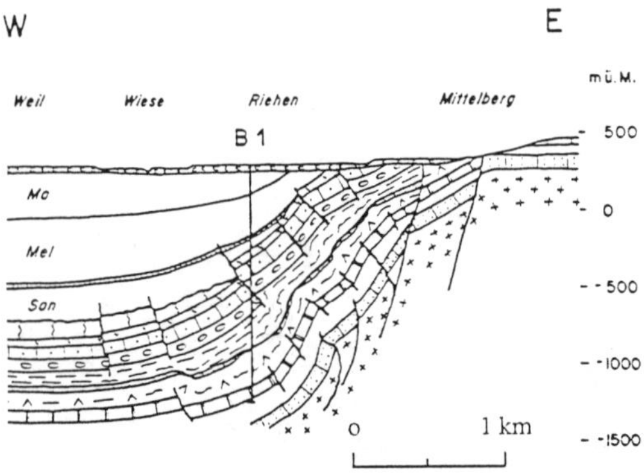

Fig. 2 Geological profile of the Rhein graben flexure at the Riehen 1 borehole site (from Hauber 1991).

1.5 bar (Riehen 2). The groundwater flows in through fissures in well defined zones. Flowmeter logging indicated one inflow zone in Riehen 1 (2 m thick) and two zones in Riehen 2 (2 m and 6 m thick). Paleokarstic openings occurring frequently in the studied formation do not participate in the active circulation systems as shown by comparing flowmeter- and microscanner-logs. Geothermal gradients of 37-40°C/km were determined using extrapolated bottomhole temperatures.

4 METHODS

25 water samples were taken during pump tests at wellhead conditions and at different inflow levels in the borehole. The standard field measurements included temperature, pH and alkalinity determination. Occasionally the Eh and the Liquid-Gas-Ratio (GLR) were measured at wellhead conditions. In the laboratory the samples were analyzed for chemical and isotopic composition of the liquid and gaseous phase.
Thermodynamic modelling was undertaken using the computer code PHREEQE (pH-Redox-equilibrium-equations, (Parkthurst et al. 1980). A second computer model was applied (TPDEGAZ: temperature-pressure degassing: Czernichowski-Lauriol, 1988) to reconstruct equilibrium bottomhole conditions and to study the evolution of the fluids during the pumping.
The composition was compared to other deep groundwaters of northern Switzerland sampled from boreholes drilled by Nagra, the Swiss National Cooperative for the Storage of Radioactive Waste. Some deep groundwater analyses were available from thermal stations.

5 CHEMICAL COMPOSITION OF THE STUDIED WATERS

Riehen 1 and 2 show very similar chemical and physical properites. Both waters belong to the Na-Cl type with high contents of sulfate. The conditions before degassing due to the pressure-drop during the upward water transport was simulated using the chemical composition, a gas analysis and the GLR of a water sample as raw data. These calculations indicate equilibrium with respect to calcite and dolomite, which are the rock forming minerals of the réservoir. The fluids are oversaturated with quartz, but in equilibrium with chalcedony. Saturation indexes around zero occur also for the minerals gypsum, anhydrite and fluorite which also occur frequently in the upper Muschelkalk. The GLR reaches up to 70 volume-percent at the wellhead. The gaseous phase is dominated by CO_2 (75-90 weight %). The carbonic acid is most likely of magmatic or thermometamorphic origin and was transported upwards along the deep reaching normal faults at the graben borders.

The waters sampled from Reinach show a similar chemical profile as Riehen, the overall concentration of dissolved solids is however significantly bigger (Fig 3). The main part of the dissolved solids weight difference is caused by the components Na and Cl. In Riehen very low pH values were observed, which increase the solubility of all carbonate species and leads to an aquifer fluid equilibrium in spite of the high calcium and magnesium concentrations. The reason for this low pH has not been explained yet.

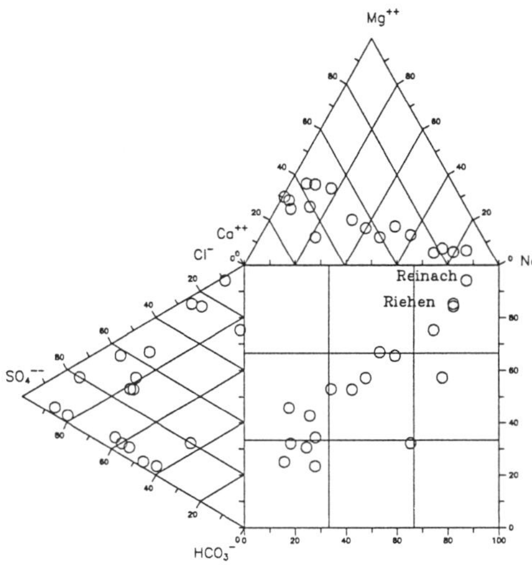

Fig 3: Durov representation of the thermal Muschelkalk-waters of northern Switzerland and the Rheingraben. The waters scatter around a straight line connecting low saline Ca-HCO_3 with highly saline Na-Cl type waters

Twelve of the most commonly used geothermometers were applied to the chemical analysis. The apparently most reliable data for the given formation were produced by the magnesium-corrected Na-K-Ca thermometer (Fournier & Truesdell, 1973), by the Mg/Li-thermometer (Kharaka and Mariner, 1989) and the chalcedony thermometer. They all indicate a reservoir temperature of 80-100°C close to the measured aquifer temperature.

6 ISOTOPES

As had been assumed, no traces of tritium could be detected in the studied waters. The nearly total absence of ^{14}C is surprising, if the karstified Triassic limestones covering the Black Forest high in the east is assumed to be the recharge area. Considering the measured permeability, an age of several ten thousands of years seems to be too high for the short flowpath. A significant amount of radiogenic ^{14}C must also have been removed during water-rock- interaction accelerating the radiogenic clock. As a minimum age of several thousands of years must still be admitted for these waters, an influence of modern recharge to the system can thus be excluded. The environmental isotopes ^2H and ^{18}O of all studied water samples from Riehen and Reinach lie slightly below the global meteoric waterline (Yurtsever & Gat, 1981) but indicate no clear evidence for isotopic exchange with the aquifer. The waters from Reinach are however significantly heavier than the waters from Riehen. This water or parts of it must have a different origin than the waters from Riehen. The local relation between infiltration altitude and isotopic composition (Pearson et al., 1991) permit to calculate a theoretical infiltration height for Riehen of 400 m which corresponds approximately to the altitude of the Triassic cover of the Black Forest.

The difference in isotopic and chemical composition is interpreted to be closely related to the different permeabilities met in the boreholes of Riehen and Reinach. Two circulation systems are assumed. Open fractures build an excellent circulation path for lower mineralized waters descending from the Black Forest. The rock volume contained between these "conducting tubes" have a homogeneous permeability provided by small fissures and fractures. It contains a highly saline, isotopically heavy water whose origin is still a matter of speculation. This water mixes with the water circulating in the larger fractures.

7 COMPARISON WITH OTHER DEEP GROUNDWATERS

Searching for the origin of the saline component assumed for Reinach, published data from other deep groundwaters in the adjacent area were compared to the studied waters (Fig 3). The natural occurrence of thermal waters is generally bound to two tectonic structures: the mentioned normal faults of the Rhein graben and the main Jura overthrust. A zone of deep groundwater upward flow, related to a Permo-Carboniferous trough was discovered by Nagra. Generally, most waters from the Muschelkalk contain little Na and Cl and differ strongly from the studied waters Nevertheless some ion ratios as F/(Ca+Mg) or Sr/(Ca+Mg) show a clear signature of the upper Muschelkalk Formation. The middle Muschelkalk, which contains important evaporite deposits can be excluded as a source for the dissolved solids. Its salt contains only little potassium, iodide and bromide while the studied waters are very rich in these elements. Brines of a similar composition were found in the Perm Formation of three Nagra boreholes. Some of these waters trend towards a heavy isotopic composition (Pearson, et al., 1991) as is the case in Reinach. The Rhein graben waters that were in contact with Tertiary evaporites also resemble strongly the fluids from Riehen. A hydraulic connection between these evaporites and the Muschelkalk aquifer can easily be established within the Rhein graben flexure. Nevertheless modelled groundwater circulation indicates an ascending flow direction making it difficult for Tertiary waters to reach the Muschelkalk aquifer against the hydraulic gradient.

8 CONCLUSIONS

The mineralisation-difference in the studied thermal waters of the upper Muschelkalk seems most likely to be related to the permeability, which can vary strongly on short distances. The more mineralized waters, flowing in the poorly fissured blocks cannot be explained by the longer residence time or the bigger rock-water contact surface alone. The isotopical compositions of the waters suggests that the water from Reinach is mixed with a fluid of a different, until now unknown origin. Further studies should be undertaken on Tertiary and Permian brines which seem most likely to be the source of the high mineralization.

REFERENCES

Carlé, W. (1975). Die Mineral- und Thermalwässer von Mitteleuropa. - Wiss. Verlagsges. Stuttgart.

Czernichowski-Lauriol, I. (1988). Modélisation de l'évolution des fluides géothermaux lors de leur exploitation par forages. ducuments du BRGM No. 159.

Fournier, R. O., & Truesdell, A. H. (1973). An Emperical Na-K-Ca Chemical geothermometer for natural waters. Geochim. Cosmochim. Acta, 37: 1255-1275.

Hauber, L. (1991). Ergebnisse der geothermiebohrungen Riehen 1 und 2 sowie reinach 1 im südosten des Rheingrabens Geol. Jb. 48: 167-184.

Kharaka, Y. K., & Mariner, R. H. (1989). Chemical geothermometers and their application to formation waters from sedimentary basins. In Naeser, N. D. and McCulloh, J. H. (eds.), Thermal History of sedimentary Basins, p. 99-118.

Parkhurst, D. L., Thorstenson, D. C., & Plummer, L. N. (1980). PHREEQE - A computer program for geochemical calculations. US Geol. Survey Water Res. Invest. 80-96.

Pearson, F. J., Balderer, W. Loosli, H. H., Lehmann, B. E., Matter, A., Peters, A., Peters, T., Schmassmann, H., & Gautschi, A. (1991). Applied Isotope Hydrogeology - A case study in northern Switzerland NTB No. 88-01.

Yurtsever, Y., & Gat, J. R. (1981). Atmospheric waters. In R. Gonfiantini (Ed.): Stable Isotope Hydrology, Deuterium and Oxygen-18 in the Water Cycle - Int. Atomic Energy Agency Vienna.

Water-Rock Interaction, Kharaka & Maest (eds) © 1992 Balkema, Rotterdam. ISBN 90 5410 075 3

Thermal springs in the Precambrian crystallines of the western continental margin of India: Field and experimental results

D.Chandrasekharam, A.Ramanathan & R.L.Selvakumar
Department of Earth Sciences, Indian Institute of Technology, Bombay, India

ABSTRACT: Puttur springs are located in the Precambrian crystallines bordering the western continental margin of India. Granite- meteoric water interaction experiment reveals that meteoric water is the main source for the springs. Albite, K-feldspar and biotite dissolution is controlling the chemical components in the springs. The reservoir temperature estimated using Na-K, Na-K-Ca and silica geochemical thermometers range between 210° and 250°C.

1.INTRODUCTION

The Irade and Banduru thermal springs collectively called the Puttur springs are located in the Precambrian crystallines of south India,along the western continental margin of India(Fig.1).These springs appear to be the southward extension of the Deccan thermal spring zone. Occurrence of Irade spring has been reported by Oldham(1882). Radhakrishna (1971)published chemical analysis of the Irade spring. The Bandru spring has been discovered during the present investigation. The springs have not been studied in detail with regard to their chemical evolution and reservoir temperature. The surface temperature of the springs recorded during 1990 field season are 37° and 39° C respectively.They are located along a shear zone.The discharge rates of the springs are about 20 l/m.The rock types recognised in this area include granite gneiss,biotite gneiss and garnet biotite gneiss.

As a part of our continuing work on thermal springs,chemical data obtained on the springs together with experimental results on granite-water interaction have been utilised to understand their evolution and to estimate the reservoir temperature.

2.MAJOR ION CHEMISTRY

The chemical analyses of water samples are given in table 1.The data together with the field of Deccan springs(Chandrasekharam et.al.,1989)are plotted in the Piper diagram(Piper,1944; Fig.2).The Puttur springs,unlike the Deccan springs,are of Na-HCO3 type.

3.GRANITE-WATER INTERACTION

Granite-water interaction experiments were conducted earlier by Savage(1986) and Savage et al.,(1987) to access the feasibility of geological disposal of high level radioactive waste(HLRW) by understanding the nature and magnitude of chemical interaction between the host rock and fluids surrounding it. The present investigation is aimed at understanding the chemical evolution and to estimate

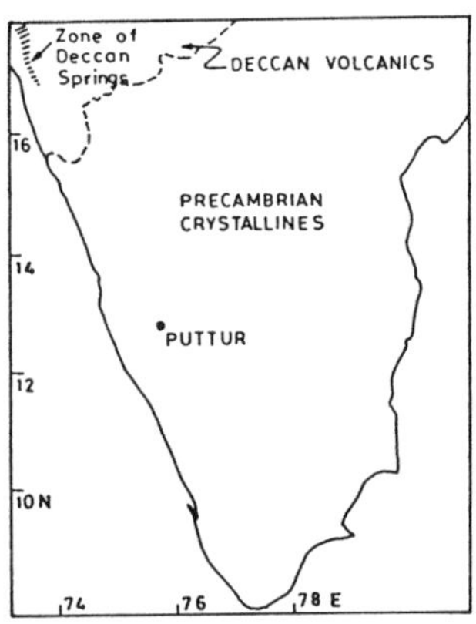

Fig.1 Location map of Puttur springs.

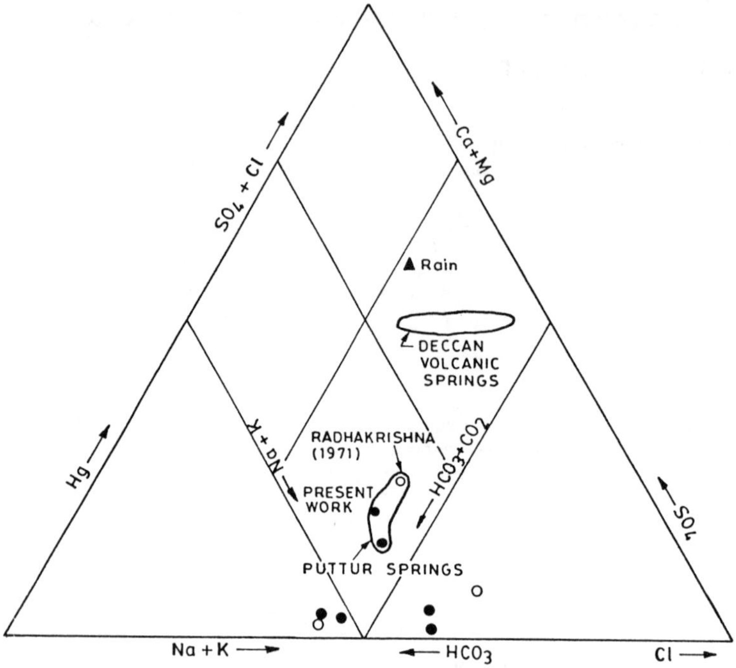

Fig.2 Trilinear variation diagram.

Table 1.Analytical data on water samples

	A	B*	C*	D*	E*
SiO2	470	300	270	nd	nd
Ca	14	11	3	0.8	nd
Mg	4	5	5	0.3	nd
Na	130	105	112	3.3	3.2
K	9	8	12	0.2	1.4
HCO3	260	189	149	4	nd
SO4	53	23	8	0.5	nd
Cl	57	28	35	5	14.2
pH	8.3	8.5	8.4	6.6	nd

Values in ppm.A:Irade(Radhakrishna,1971) B:Irade;C:Banduru;E:Experimental solution D:Rain. * Present work

the reservoir temperature of the springs. These results may help in future,if India wants to locate suitable repositories within the Precambrian rocks.

The experimental technique applied here is similar to that described by Crovisier et al. (1987) Water-to-rock ratio was kept at 10:1. Small chips of the host rock (granite gneiss,henceforth called granite)were interacted with rain water for 30 days at 80° C under atmospheric pressure. Rigaku DMAX 1C XRD unit has been utilised to obtain XRD pattern on the chips prior to and after the interaction experiment.The reacted water has been analysed for its chemical constituents.

4.RESULTS

Quartz,albite,K-felspar and biotite are the main phases present in the granite. Partial chemical analysis of water reacted with granite is shown in table 1.The chemical data of the springs,experimental liquid surface and groundwater occurring near the springs are plotted on the Na+K vs Cl diagram(Fig.3). X-ray diffractograms of the granite before and after interaction are shown in figure 4.

The experimental liquid falls on the tie line connecting the plots of the rain and the springs(Fig.3) thus indicating meteoric water as the main source for the thermal reservoir. Unlike the Deccan springs,the above springs have a single source and define a simple thermal circulatory system.

The XRD pattern(Fig.4)shows the presence

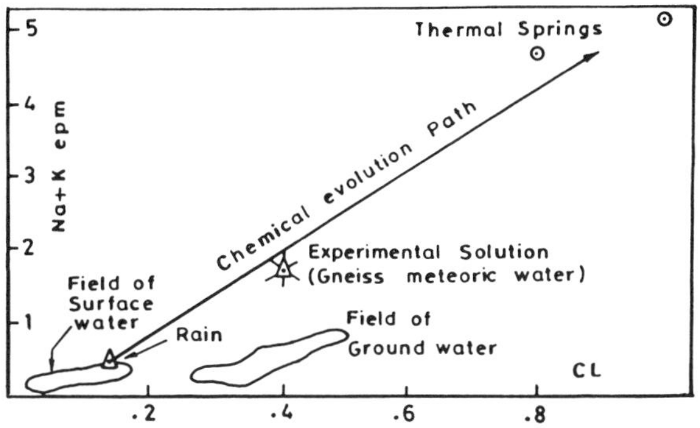

Fig.3 Na+K vs CL diagram.

of smectite(illite) and calcite as solid alteration products growing at the expense of albite,K-felspar and biotite.The dissolution rate of albite appears to be much faster relative to K-felspar.The formation of smectite as a reaction product within the reservoir and/or during the ascent of the fluids is indicated by the presence of kaolinite at the orifice of the springs.

5.DISCUSSION

Since the initial liquid interacted with granite is rain water,it is undersaturated with respect to all the starting minerals and posseses low cation/proton ratio. Even at low cation/proton ratio,the felspars have been altered to smectite and calcite (Fig.4).The Na+K and Cl concentrations in the reacted liquid show that the chemical components in the springs are controlled mainly by albite,K-felspar and biotite dissolution.The Na concentration in the initial and final liquids (D,E:Table 1) has not registered any change in spite of albite dissolution. The only means by which this could result is through conservation of Na by the solid phase smectite. Biotite dissolution (Fig. 4) has caused a three fold increase in Cl in the final liquid(Table 1). The main source of chloride in granite is biotite and 2 modal percent of biotite contains about 0.1 weight percent of chloride(Savage et.al.,1985).Since the experiment was conducted at low temperature,the concentrations of Na,K and Cl did not match with that of the springs but fall on the evolution line of the springs.The springs must have been equilibrated with the above phases at temperature above 80°C.

Occurrence of kaolinite at the orifice of the springs and low flow rates indicate that they must have cooled by conduction. As a result, at shallow levels low temperature alteration of felspars must have resulted in the formation of kaolinite (Giggenbach,1984).Besides low temperature alteration of felspars,sulphur gas bubbling from the springs indicates the presence of sulphide minerals at some depth,for, oxidation of sulphides promotes formation of kaolinite(Savage,et.al.,1987).Presence of sulphide ores or veins is expected in this area because of its proximity to copper mines at Kalyadi,a place which is about 150 km. east of Puttur.

The present investigation supports a simple thermal circulatory system with meteoric water as the source.In such cases Na-K(Fournier,1979) and Na-K-Ca(Fournier and Truesdell,1973)geochemical thermometers should indicate true reservoir temperature, unlike the Deccan springs,where mixing of saline water is causing a problem in estimating the true reservoir temperature. The reservoir temperatures thus calculated using the above method are given below:

	Irade	Banduru
Na-K	200	226°C
Na-K-Ca	236	237

Silica thermometry should also yield reliable reservoir temmperature for thermal systems with dilute solutions,low flow rate and conductive cooling(Fournier et.al. 1974;Fournier and Potter,1982).The enthalpy values of the springs with silica content between 270 and 470 ppm would fall between 860 and 1090 J/g.The reservoir temperature for these enthalpy values lie between 210° and 250° C which match well with Na-K and Na-K-Ca thermometry values.

Puttur falls within the zone III of heat flow value contours for the western continental margin of India.(Zone III

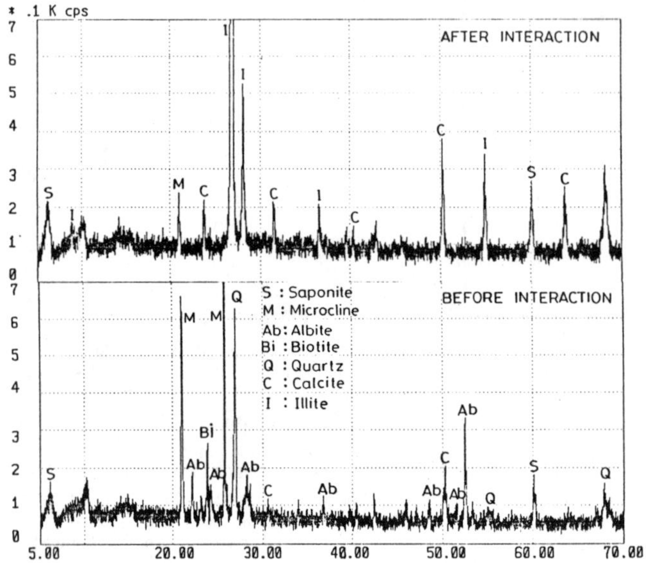

Fig.4. X-ray diffractograms.

h.f.v.range from 70 to 100 mW/sq.m.,Ravi Shankar,1988).A h.f.v. of 100 mW/sq.m can be assumed for Puttur area considering its proximity to Unhavre springs in Deccan volcanics located north of Puttur.The h.f.v. of Unhavre is 128 mW/sq.m. The temperature gradient of Puttur estimated using the above data would be around 50° C/km. Based on the reservoir temperature and temperature gradient, a depth of 4 to 5 km may be assumed for Puttur thermal reservoir.

REFERENCES

Chandrasekharam,D.,R.Ramesh and J.Balasubramanian. 1989. Geochemistry,oxygen and hydrogen isotope ratios of thermal springs,western continental margin of India. 6th Inter.Sym.Water-Rock:Ed.D.L. Miles, A.A.Balkema/Rotterdam/Bockfield 1989. 149-154.

Crovisier,J.L.,J.Honnorez and J.P.Eberhart. 1987. Dissolution of basaltic glass in seawater:Mechanism and rate. Geochim. Comochim.Acta.,51,2977-2990.

Fournier,R.O. 1979. A revised equation for the Na/K geothermometer.Geother. Resour.Council,3,221-224.

Fournier,R.O. and A.H.Truesdell. 1973. An empirical Na-K-Ca geothermometer for natural waters.Geochim.Cosmochim. Acta,37,1255-1276.

Fournier,R.O.,D.E.White and A.H.Truesdell. 1974. Geochemical indicators of subsurface temperature-Part 1,Basic assumptions. J.Res.U.S.Geol.Survey,2,259-262.

Fournier,R.O. and R.W.Potter. 1982. A revised and expanded silica(quartz) geothermometer.Geother.Resource Council. Bull.,3-9.

Giggenbach,W.F. 1984. Mass transfer in hydrothermal alteration systems-A conceptual approach.Geochim.Cosmochim.Acta. 48,2693-2711.

Oldham,T. 1982. Thermal springs of India. Geol.Survey India.Mem.,19,1-63.

Piper,A.M. 1944. A graphic procedure in the geochemical interpretation of water analysis. Am.Geophy.Union,Trans.,25,914-923.

Radhakrishna,B.P. 1971. On a little known thermal spring from near Puttur,Mysore State,India.J.Geol.Soc.India,12,399-401.

Ravi Shankar,D. 1988. Heat flow map of India and discussion on its geological and economic significance.Indian Miner., 42,89-110.

Savage,D.,M.R.Cave and A.E. Milodowski. 1985. The origin of saline groundwater in granitic rocks. Evidence from hydrothermal experiments.Met.Res.Soc.Sym., 50,121-128.

Savage,D. 1986. Granite-water interactions at 100 C,50Mpa:An experimental study. Chem.Geol.,54,81-95.

Savage,D.,M.R.Cave,A.E.Milodowski and I. George. 1987. Hydrothermal alteration of granite by meteoric fluid:an example from the Carnmellis granite,U.K.Con.Min. Petro.,96,391.

Water-Rock Interaction, Kharaka & Maest (eds) © 1992 Balkema, Rotterdam. ISBN 90 5410 075 3

Gas geochemistry for the Beppu hydrothermal system, Kyushu, Japan

F. D'Amore, G. Gianelli & A. Caprai
International Institute of Geothermal Research, Pisa, Italy
Y. Yusa, K. Kitaoka & K. Takemura
Beppu Geophysical Research Laboratory, Kyoto University, Japan

ABSTRACT: Surface gas chemical and isotopic compositions have permitted us to improve the conceptual model of the Beppu hydrothermal system. The results are consistent with a reservoir made up of a single liquid phase near the critical point and with a P(total) close to 45 MPa.

1 INTRODUCTION

The Beppu hydrothermal system is located on the island of Kyushu (Fig. 1) and is characterized by many surface manifestations such as hot springs, mud pools, fumaroles, steaming grounds and rock alteration over, an area of about 30 km2. The main features of the system are described in Allis & Yusa (1989). Studies based on water chemistry and tritium data available from many shallow wells (Kitaoka, 1990) suggest that the surface fluids could have originated from the mixing of meteoric fresh waters, steam heated acidic waters and geothermal chloride-rich fluids. According to Allis & Yusa (1989), the Beppu system is characterized by the upflow of a deep fluid from underneath the volcanic edifices of Garandake and Tsurumi located at about 10 km W of Beppu city. The fluid then flows eastwards with steam loss and mixing with recent meteoric waters. The deep fluid temperature was assumed to be of the order of 300°C at about 2 km below the steaming grounds. The aim of this paper is to reach a better definition of the conceptual model of the deep geothermal system. Chemical and isotopic analyses of selected gas samples from the area have been used to calculate the thermodynamic conditions at depth.

2 DATA PRESENTATION

Eight gas samples were collected from selected sites in the study area during October 1990. The dry

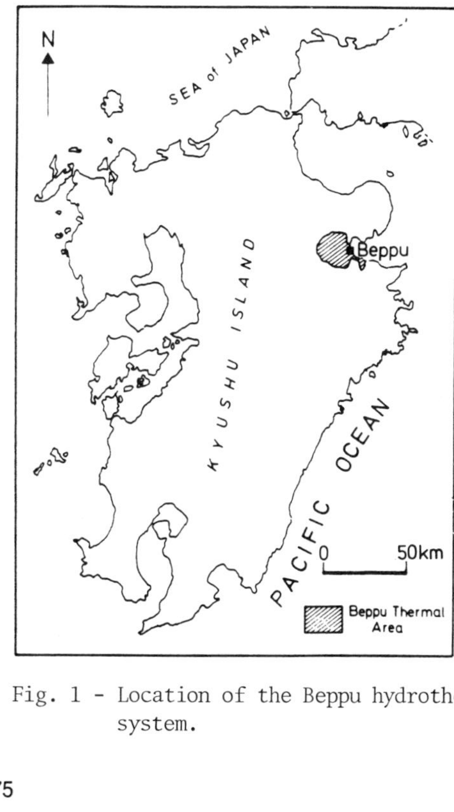

Fig. 1 - Location of the Beppu hydrothermal system.

gas compositions are reported in Table 1. The isotopic composition of the gases and associated spring waters is reported in Table 2. The maximum values of CO content, CO/H_2 and H_2S/H_2 ratios occurred close to the Garandake volcano. The CO contents and CO/H_2 ratios are both consistent with local maximum values of temperature and CO_2 partial pressure (see D' Amore et al., 1987 and Giggenbach and Goguel, 1989) The H_2S/H_2 measured ratios are anomalously high compared to the values found in fluids of exploited geothermal fields (e.g. D' Amore & Gianelli, 1984), and consistent with extremely high partial pressure of S_2.

In terms of the isotopic composition, the 13C content of CO_2 (analyzed after purification from H_2S) shows a mean value of -6.8 ± 1.2, which is consistent with a magmatic origin of this gas. The relative constancy of δ13C and δ2H of CH_4 also indicates a single source for CH_4. Different thermodynamic conditions of the deep H_2O as a source for hydrogen could however justify the observed variability in the values of δ2H(H_2). The stable isotopes of the local spring waters associated with the gas show a composition consistent with the local base meteoric line.

3 EVALUATION OF THE THERMODYNAMIC CONDITIONS OF THE DEEP FLUID

In our model we assumed local thermodynamic equilibrium in a deep reservoir in the system C-H-O-S-Ca-Fe. The structural geology and relative ratios between the reactive gases suggest that the gases in the samples from near to the Garandake volcano (e.g. B10 and B7) might be the most representative of the gas phase present in different layers of the deep fluid. The unknown thermodynamic parameters are temperature, partial pressure of CO_2, O_2 and S_2 and the amount of steam with respect to total water, in phase equilibrium in the reservoir (y = H_2O vapour/H_2O total). The following chemical equilibrium reactions are used:

1. $CH_4 + 3CO_2 = 4CO + 2H_2O$
2. $CH_4 + 2H_2O = CO_2 + 4H_2$
3. $H_2O = H_2 + 1/2O_2$
4. $H_2S = H_2 + 1/2S_2$
5. $2CaCO_3 + 1/3Fe_3O_4 + 2FeS + 2S_2 + 7/3O_2 = 2CaSO_4 + 3FeS_2 + 2CO_2$

This system is oversimplified. Only 5 unknowns have been considered: PS_2, PO_2, PCO_2, y and temperature when the mass action law is applied, neglecting salinity, solid solution effects and the presence of Ca-Fe-silicates. The partial pressure of the i-th gas is expressed by means of D'Amore and Truesdell's (1985) equation:

$$\log P(i) = \log[n(i)/n(CO_2)] - \log A(i) + \log A(CO_2) + \log P(CO_2)$$

where $n(i)/n(CO_2)$ is the molar ratio between the i-th gas and CO_2 at the sampling point, and $A(i) = y + (1-y)/B(i)$. In a two-phase system B(i) is the molar distribution coefficient of the i-th gas between the vapour and liquid phases. An iterative method can be used to solve the non-linear system.

The results obtained for samples B10 and B7 are, respectively:
temperature = 360 and 300°C (estimated uncertainty 5%)
PCO_2 = 24.5 and 3.3 MPa
y = 0 (pure liquid) and 0.04 (4% vapour)
$\log P(O_2)$ = -29.1 and -33.4
$\log P(S_2)$ = -5.8 and -7.8

A temperature value of about 360°C and a total fluid pressure of about 45 MPa can be inferred as the T-P conditions of the Beppu geothermal reservoir. The thermal value is confirmed by the isotope geothermometer CO_2-CH_4 and corroborated by the maximum temperature values computed from the isotope geothermometer CH_4-H_2. By applying this method to the other gas samples we obtain lower temperatures and $P(CO_2)$, with steam fraction values corresponding to a liquid phase for the manifestations B8 and B9 near the volcano (close to B10), and to a vapour-rich phase for the others.

4 DISCUSSION AND CONCLUSIONS

The results have led to a better definition of the conceptual model of the hydrothermal system of Beppu, which can be summarized as follows:

- upflow of deep, acidic magmatic fluid containing SO_2 and CO_2, from a magma chamber underneath Garandake volcano;
- recharge of recent meteoric waters into deep and shallow aquifers: the thermodynamic conditions of the deep geothermal reservoir are as reported above. Considering the computed total pressure of the fluid, the reservoir may be located at a maximum depth of 4.5 km below the top of Garandake volcano;
- mixing between magmatic fluid and recharge water produces an acidic, oxidant solution very rich in H_2S, CO_2 and sulphate, with dissolution of the volcanic rock and formation of calcite, Ca-sulphate, Fe-sulphides and magnetite. A possible irreversible reaction such as $2SO_2 + 2H_2O(l) >> SO_4^{=} + H_2S(l) + O_2 + 2H^+$ could justify the use of equation 5) as a P(S2) buffer and the high H_2S contents and H_2S/H_2 ratios. All the gas species equilibrate rapidly with the recharge water with high tritium content (Kitaoka, 1990) because of the very high temperature of the local geothermal reservoir;
- ascent to the surface of the geothermal fluid with production of steam through boiling and partial, local re-equilibration in a two-phase system at lower temperatures with respect to the original fluid;
- finally, mixing of the rising fluid with shallow fresh waters and oxidation of species such as H_2S and H_2 and dissolution of CO_2. Acidic waters may form at this stage.

Figure 2 is a simplified sketch of the conceptual model described above. The Beppu hydrothermal system deserves further study on fluid-rock interactions to improve our knowledge of the thermodynamic conditions of the deep system in space and time. Frequent monitoring of the gas composition is strongly recommended, in order to assess the risk of phreatic eruptions connected with the peculiar thermodynamic conditions of the deep fluid.

REFERENCES

Allis, R.G. & Yusa Y. 1989. Fluid flow processes in the Beppu geothermal system, Japan. Geothermics 18: 743-759.

D' Amore, F. & Gianelli, G. 1984. Mineral assemblages and oxygen and sulphur fugacities in natural water-rock interaction processes. Geochim. Cosmochim. Acta. 48: 847-857.

D' Amore, F. & Truesdell, A.H. 1985. Calculation of geothermal reservoir temperatures and steam fraction from gas compositions. Geotherm. Resources. Council. Trans. 9: 305-310.

D' Amore, F., Fancelli, R., Saracco, L. & Truesdell, A.H. 1987. Gas geothermometry based on CO content: application in Italian geothermal fields. Proc. 13th Workshop on Geothermal Reservoir Engng., Stanford, Calif.:247-252.

Giggenbach, W.F. & Goguel, R.L.

Table 1. Dry gas composition of the Beppu gases, in % vol. (CO_2, SO_2, H_2S, N_2) and ppm vol. (others). (*): f=fumarole, steaming ground; p=water pool, stream; w=shallow well

N	(*)	CO2	SO2	H2S	N2
B10	f	65.6	0.071	9.0	25.4
B9	f	91.1	1.42	5.8	1.19
B8	p	79.4	--	16.6	2.70
B7	p	81.6	--	17.6	0.93
B6	w	86.5	--	10.8	0.19
B5	f	97.9	0.37	1.62	0.49
B4	p	97.6	--	1.14	0.63
B3	p	35.7	--	<0.02	62.4

N	H2	CH4	CO	Ar	He
B10	404	429	24.0	3200	4.3
B9	624	397	0.71	45	9.0
B8	363	230	1.45	84	13.
B7	1760	369	16.0	58	10.
B6	22260	510	4.5	28	1.
B5	473	753	0.65	20	14.
B4	719	524	0.46	22	16.
B3	<1.5	8470	7.1	11100	43.

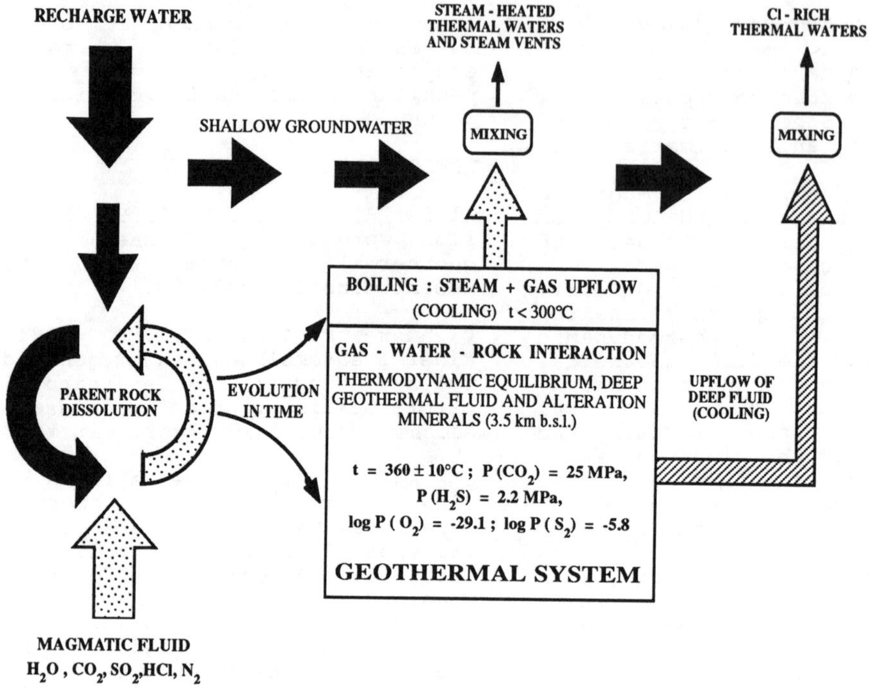

Fig. 2. Conceptual model of the Beppu hydrothermal system.

Table 2. Isotopic composition of Beppu gases and associated waters: δ13C vs PDB, δ18O and δ2H vs SMOW.

N	δ13C (CO2)	δ13C (CH4)	δ2H (CH4)	δ2H (H2)
B10	-7.0			
B9	-8.5	-28.7	-156	-359
B8	-6.9	-28.8	-170	
B7	-7.1	-29.5		-501
B6	-5.7			-737
B5	-5.6	-22.6	-156	-685
B4	-5.5	-26.3	-165	-382
B3	-8.4			

	δ18O (H2O)	δ2H (H2O)
B10	-12.3	-73.7
B9	-14.5	-83.5
B8	-8.3	-54.2
B7	-1.2	-32.3
B6	-8.2	-59.9
B5	-6.3	-53.2

1989. Collection and analysis of geothermal and volcanic water and gas discharges. New Zealand DSIR report n. CD 2387.
Kitaoka, K. 1990. Water circulation rates in a geothermal field: a study of tritium in the Beppu hydrothermal system, Japan. Geothermics 19: 515-539.

Water-Rock Interaction, Kharaka & Maest (eds) © 1992 Balkema, Rotterdam. ISBN 90 5410 075 3

Thermal waters and mineralization in the Geysers-Clear Lake area, California, USA

Julie M. Donnelly-Nolan
US Geological Survey, Menlo Park, Calif., USA

ABSTRACT: Geochemistry of evolved connate and evolved meteoric thermal waters in the Geysers-Clear Lake area of northern California is closely tied to host rock geology. Evolved connate waters are derived from ocean waters that were deposited with Mesozoic sedimentary rocks of the Great Valley sequence; these waters are associated with both gold and mercury mineralization. Evolved meteoric waters have developed locally in metamorphosed Mesozoic Franciscan rocks by repeated boiling and rock-water interaction; they are associated with mercury mineralization. The thermal waters of the region are heated by a large subjacent magma chamber under the 2.1 to 0.01 Ma Clear Lake Volcanics. The waters reach the surface along faults of the San Andreas transform fault system.

The Geysers-Clear Lake area in the northern Coast Ranges of California is the site of young volcanism, gold and mercury mineralization, abundant thermal springs, and numerous faults that belong to the San Andreas transform fault system. The Geysers vapor-dominated geothermal field, at the southwestern edge of the Clear Lake volcanic field, is the largest power-producing geothermal field in the world. The 2.1 to 0.01 Ma Clear Lake Volcanics are interpreted on the basis of geophysics to overlie a large subjacent magma chamber (Bufe et al., 1981; Chapman, 1966; Chapman, 1975; Isherwood, 1976, 1981; Iyer et al., 1979, 1981). Several water types are present in the Geysers-Clear Lake area of northern California (Donnelly-Nolan, 1983) including warm, isotopically meteoric springs that emerge in the main Clear Lake volcanic field, acid-sulfate thermal waters that issue in The Geysers geothermal field, and more saline, isotopically non-meteoric waters.

One important isotopically non-meteoric water type emerges from Great Valley sequence marine sedimentary rocks of Mesozoic age that are present in the eastern part of the area. This relatively saline water, which has about 12,000 mg/L of chloride, is interpreted to be evolved connate water that was originally ocean water (White et al, 1973). It has a B/Cl ratio <0.03. The other important isotopically non-meteoric water has about 1,000 mg/L of chloride and is found in metamorphosed and relatively impermeable rocks of the coeval Franciscan Complex. The water has a high B/Cl ratio, typically near 1.0, and is present at and near Sulphur Bank mercury mine (White et al., 1973; Beall, 1985). It is interpreted to have originated by Rayleigh distillation of an originally meteoric water within the host Franciscan graywackes. This evolved meteoric fluid is thought to be small in volume by the nature of its origin and localized to areas where faulting and thermal input from the magmatic system are focused. One such area is The Geysers geothermal field, where a deep boiling brine has been postulated to underlie the vapor-dominated reservoir (White et al., 1971), although none of the thermal water that boils to form steam has ever been sampled. If an evolved meteoric fluid is boiling to form the steam that powers electrical generation in The Geysers geothermal field, then reversal of the present decline in production (Barker et al., 1991) by discovery of large volumes of boiling water at depth is unlikely.

The isotopic composition of the two water

types is identical ($\delta^{18}O$=+5) and is thought to be a result of extended rock-water interaction with host sedimentary rocks with similar $\delta^{18}O$ values. Both types of isotopically non-meteoric water are associated with mercury mineralization, but only the evolved connate water is known to be associated with gold mineralization. Fluid inclusion studies (Thordsen, 1988; Peters, 1991) indicate that the mineralizing fluids associated with gold at the McLaughlin deposit and in the Wilbur Springs district are similar to evolved connate waters presently emerging at or near the areas of gold mineralization. These waters are presently depositing gold in the Wilbur Springs district (Peters, 1991).

Both the McLaughlin deposit and the Wilbur Springs gold mineralization are associated with early basaltic rocks of the Clear Lake Volcanics (Lehrman, 1986; McLaughlin et al., 1989). Although volcanism has been dominated in terms of volume by dacite and rhyolite, the early widespread phase consisted primarily of basalt and andesite (Donnelly-Nolan et al., 1981; Hearn et al., 1981). Mercury mineralization is strongly correlated with volcanism, and no mercury deposits are known in the Coast Ranges north of the Geysers-Clear Lake area which has been one of the most productive areas in the United States for mercury, primarily from the Wilbur Springs, Knoxville (location of the McLaughlin gold mine), and Mayacmas districts, and Sulphur Bank mine. Mercury deposits form a semi-circular array around the south and west side of The Geysers geothermal field, adjacent to the volcanic field.

Deep drilling in The Geysers field has revealed a major silicic intrusive body termed the "felsite" (Thompson, 1989) whose volume is estimated to be as large or larger than the total volume of erupted Clear Lake Volcanics, estimated at 100 km^3 (Donnelly-Nolan et al., 1981). Preliminary age data on the felsite (Schriener and Suemnicht, 1981; Pulka, 1991) indicate that it is too old to be providing heat for The Geysers field, but heat from the large magma body centered to the northeast under the volcanic field is probably tapped via faults. Other silicic intrusive bodies may be present elsewhere in the region.

CONCLUSIONS

The combination of long-term thermal input to the Geysers-Clear Lake area together with repeated fracturing of the area by faults of the San Andreas system has resulted in a focus for mineralization and for evolution of thermal fluids by rock-water interaction.

REFERENCES

Barker, B. J., Gulati, M. S., Bryan, M. A., and Riedel, K. L., 1991, Geysers reservoir performance: Geo-Heat Center Quarterly Bulletin, v. 13, no. 3, p. 1-14.

Beall, J. J., 1985, Exploration of a high temperature, fault localized, nonmeteoric geothermal system at the Sulphur Bank Mine, California: Geothermal Resources Council, Transactions, v. 9, p. 395-401.

Bufe, C. G., Marks, S. M., Lester, F. W., Ludwin, R. S., and Stickney, M. C., 1981, Seismicity of the Geysers-Clear Lake region: U. S. Geol. Survey Prof. Paper 1141, p. 129-137.

Chapman, R. H., 1966, Gravity map of Geysers area: Calif. Div. Mines and Geology, Mineral Information Service, v. 19, no. 9, p. 148-149.

Chapman, R. H., 1975, Geophysical study of the Clear Lake region, California: Calif. Div. Mines and Geology, Spec. Rept. 116, 23 p.

Donnelly-Nolan, J., 1983, Diverse Origins of Thermal Waters in the Geysers-Clear Lake Geothermal Area, Northern California: Extended Abstracts for Fourth International Symposium on Water-Rock Interaction, Misasa, Japan, p. 123-126.

Donnelly-Nolan, J. M., Hearn, B. C., Jr., Curtis, G. H., and Drake, R. E., 1981, Geochronology and evolution of the Clear Lake Volcanics: U. S. Geol. Survey Prof. Paper 1141, p. 47-60.

Hearn, B. C., Jr., Donnelly-Nolan, J. M., and Goff, F. E., 1981, The Clear Lake Volcanics: Tectonic setting and magma sources: U. S. Geol. Survey Prof. Paper 1141, p. 25-45.

Isherwood, W. F., 1976, Gravity and magnetic studies of The Geysers-Clear Lake geothermal region, California, U.S.A.: Proceedings of the Second United Nations Symposium on the Development and Use of Geothermal Resources, San Francisco, California, 20-29 May 1975, v. 2, p. 1065-1073.

Isherwood, W. F., 1981, Geophysical overview of The Geysers: U. S. Geol. Survey Prof. Paper 1141, p. 83-95.

Iyer, H. M., Oppenheimer, D. H., and Hitchcock, T., 1979, Abnormal P-wave delays in The Geysers-Clear Lake geothermal area, California: Science, v. 204, p. 495-497.

Iyer, H. M., Oppenheimer, D. H., Hitchcock, T., Roloff, J. N., and Coakley, J. M., 1981, Large teleseismic P-wave delays in the Geysers-Clear Lake geothermal area: U. S. Geol. Survey Prof. Paper 1141, p. 97-116.

Lehrman, N. J., 1986, The McLaughlin mine, Napa and Yolo counties, California: Nevada Bureau of Mines and Geology, Report 41, p. 85-89.

McLaughlin, R. J., Ohlin, H. N., Thormahlen, D. J., Jones, D. L., Miller, J. W., and Blome, C. D., 1989, Geologic map and structure sections of the Little Indian Valley-Wilbur Springs geothermal area, northern Coast Ranges, California: U. S. Geol. Survey Map I-1706, scale 1:24,000.

Peters, E.K., 1991, Gold bearing hot spring systems of the northern Coast Ranges, Calif.: Econ. Geol., v. 86, p. 1519-1528.

Pulka, F. S., 1991, Recent intrusives at Ford Flat and relationships to faulting and the geothermal heat source, Geysers geothermal field, California: Geol. Soc. Am. Abs. with Prog., v. 23, no. 2, p. 90.

Schriener, A., Jr., and Suemnicht, G. A., 1981, Subsurface intrusive rocks at The Geysers geothermal area, California: U. S. Geol. Survey Open-File Rept. 81-355, p. 294-303.

Thompson, R. C., 1989, Structural stratigraphy and intrusive rocks at the Geysers geothermal field: Geothermal Resources Council Transactions, v. 13, p. 481-485.

Thordsen, J. J., 1988, Fluid inclusion and geochemical study of eipthermal gold mineralization in the Wilbur Springs district, Colusa and Lake counties, California: Master's Thesis, Ohio State University, 229 p.

White, D., Barnes, I., and O'Neil, J. R., 1973, Thermal and mineral waters of nonmeteoric origin, California Coast Ranges: Geol. Soc. America Bulletin, v. 84, p. 547-560.

White, D. E., Muffler, L. J. P., and Truesdell, A. H., 1971, Vapor-dominated hydrothermal systems compared with hot-water systems: Econ. Geol., v. 66, p. 75-97.

Water-Rock Interaction, Kharaka & Maest (eds) © 1992 Balkema, Rotterdam. ISBN 90 5410 075 3

The Salton Sea Hydrothermal System, California, USA: A review

W.A. Elders, M.A. McKibben & A.E. Williams
University of California, Riverside, Calif., USA

ABSTRACT: The Salton Sea Geothermal System (SSGS) is well-known as an active, sediment-hosted, hydrothermal regime containing hot (365 °C), highly concentrated (25 wt% TDS), and metal-rich brines (1700 ppm Fe). This paper synthesizes studies made possible by the abundant new samples and data produced by the extensive commercial and scientific drilling in the SSGS in recent years. Our preferred model involves evaporites accumulated in a closed rift basin, being rapidly buried by deltaic and lacustrine sediments. Intrusion of a shallow, bimodal basalt-rhyolite complex generated a hot brine diapir. This rose to within 500 m of the surface, causing metamorphism of the sediments and ore mineralization, especially along a mixing front with the cooler, less saline, but more oxidizing, shallow groundwaters.

1 INTRODUCTION

For almost thirty years, since Donald White studied its first deep borehole, the SSGS has been a window to observe the interactions of Plio-Pleistocene sedimentary rocks with saline waters at temperatures up to 365 °C, at depths of only 2.2km (White, Anderson and Grubbs, 1963). These brines have high salinity, with up to 25 wt % of total dissolved solids, mostly sodium chloride (Helgeson, 1968) and are notable for their high metal contents, and for the ore minerals (Skinner et al., 1967; McKibben and Elders, 1985) and active metamorphism they produce (Muffler and White, 1969, McDowell and Elders, 1980, 1983).

Extensive drilling for the development of geothermal power plants and for scientific purposes has recently provided a wealth of new samples and data and shed new light on the nature and origin of the system (Elders, 1989; Elders and Sass, 1988; McKibben et al., 1987, 1988a, 1988b,: Williams, 1988; Williams and McKibben, 1989).

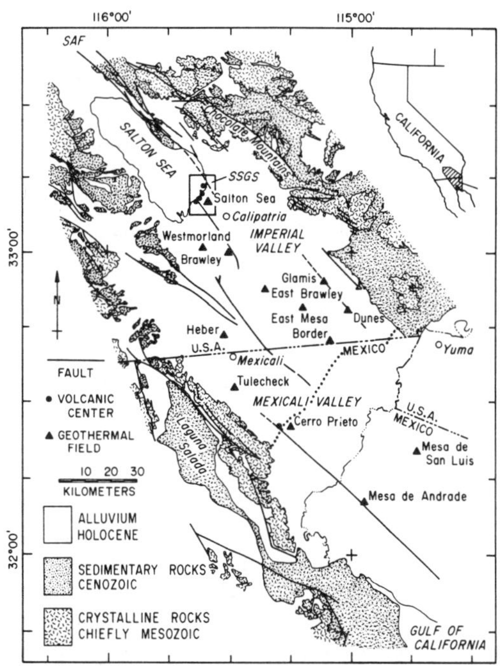

FIGURE 1. Geology of the Salton Trough and the location of its geothermal fields. SSGS is the Salton Sea Geothermal System.

1.1 *Geologic setting*

The Salton Sea occupies the lowest part of a closed depression, the landward extension of the Gulf of California, but isolated from the Gulf by the prograding delta of the Colorado River (Figure 1). It is a tectonic regime transitional between the divergent East Pacific Rise plate boundary, to the south, and the transform plate boundary of the San Andreas fault system, to the north. The SSGS, like other geothermal fields in the Trough, is situated in a "pull apart" structure, formed by right-stepping, right-lateral faults which are the dominant feature of this tectonic regime (Elders and others, 1972).

2 LITHOLOGY AND METAMORPHISM

The rocks so far penetrated by drilling in the SSGS are a sequence of Quaternary to Pliocene mudstones, siltstones and sandstones, deposited primarily in fluvatile, lacustrine, and deltaic environments (Herzig, et al., 1988). An important component of this sequence is a non-marine evaporite environment of bedded anhydrites, containing fluid inclusions with halite daughter crystals (McKibben et al. 1988a). Dikes of both rhyolite and diabase are also encountered. Similar basaltic rocks are found as xenoliths in five small Quaternary or Holocene rhyolite domes which occur near the center of the SSGS. These igneous rocks appear to be the surface manifestations of a composite intrusion, presumably the heat source for the hydrothermal system (Herzig and Elders, 1988).

With increasing temperatures all of these reservoir rocks become progressively metamorphosed to hornfelses. Dehydration and decarbonation reactions form abundant authigenic epidote, albite and quartz, with a chlorite-calcite zone at 180-320 ^{o}C, a biotite zone at 320-345 ^{o}C, and a clinopyroxene zone above 345 ^{o}C, and at the highest temperatures a garnet zone (Muffler and White, 1969; McDowell and Elders, 1980). In the clinopyroxene zone, the coexistence of authigenic actinolite, actinolitic hornblende and oligoclase demonstrates a transition from greenschist to amphibolite metamorphic facies assemblages (Cho et al., 1988).

3 BRINE CHEMISTRY

Some typical brine analyses are shown in Table 1. Isotopic and chemical studies indicate that the waters of the SSGS are meteoric in origin, but have undergone a great deal of near-surface evaporation (Williams and McKibben, 1989). The brine acquired its salts by dissolution of the non-marine evaporites at low temperature. Subsequent heating caused considerable modification of the brine chemistry by reaction with the host rocks, which were also the source of the dissolved metals (Shearer et al., 1988). The hypersaline fluids are Na-Ca-K-Cl brines containing 20-26 wt% TDS. Cl^- is the only major anion and the contents of H_2S and SO_4 are quite low. The high contents of Fe, Mn, Zn, Li and Pb are a notable feature of the hypersaline brine. Given the likely volume of this brine and the observed metal concentrations, the tonnages of metals in solution in and around the SSGS are comparable to those in a world class sediment-hosted base metal ore deposit. (McKibben, 1991). In addition to these components, the hypersaline brine also contains 1 mg/kg of Ag, 0.5 μg/kg of Au, and 1 μg/kg of Pt.

As will be seen from the table, at depths of less than 1km, a less saline brine occurs, containing less than 13 wt% TDS and lacking the high metal contents. However, even the hypersaline brines are relatively highly oxidized, corresponding to the hematite-pyrite and SO_4-H_2S redox couples (McKibben and Elders, 1985). Data on sulfur isotopes suggest that the source of sulfur in the brine is reduction of the ubiquitous anhydrite. The redox state of the brine is buffered by reactions between this sulfate and the Fe^{2+} in the brine (McKibben and Eldrige, 1989). The high activity of Ca^{2+} and the relatively high oxidation state keep the available sulfur in the brine to low levels.

Table 1. Representative flash-corrected brine analyses.

	HIGH TDS (a)				LOW TDS (a)		
	SSSDP Well, State 2-14 (b) 12-1985	SSSDP Well, State 2-14 (b) 3-1986 (c)	Commercial Well #11b	Commercial Well #10	Woolsey Well #1	Commercial Well #B1	Commercial Well #113
Temp (°C)	305	330	300	295	240	230	200
pH	5.4	5.1	5.2	5.3	?	6.9	7.6
Depth (m)	1,850-1,890	2,500-3,220	660-1,070	700-1,070	570-720	710-940	410-990
Na (wt ppm)	53,000	54,800	46,200	41,400	25,000	15,000	4,800
Ca	27,400	28,500	22,800	20,900	11,000	2,520	117
K	16,700	17,700	12,500	11,800	5,000	2,480	297
Fe	1,560	1,710	582	969	65	86	25
Mn	1,450	1,500	801	855	NA (a)	60	NA
SiO_2 (d)	>461	>588	>336	>404	NA	>255	>102
Zn	518	507	321	323	NA	11	NA
Sr	411	421	376	345	513	112	10
B	257	271	204	197	NA	92	32
Ba	203	353 (f)	183	156	NA	45	0.7
Li	194	209	157	152	93	55	9
Mg	33	49	19	33	NA	54	24
Pb	100	102	69	67	NA	2.6	NA
Cu	5.9	6.8	NA	2	NA	NA	NA
Cd	2.2	2.3	1.0	1.4	NA	NA	NA
Cs	NA (a)	NA	NA	NA	NA	NA	NA
NH_4	333	330	339	341	NA	103	NA
Cl	151,000	157,500	128,000	116,000	85,000	31,000	6,900
Br	99	111	95	78	NA	24	10
CO_2 (e)	1,600	580	1,100	5,500	NA	10,000	NA
H_2S	15	10	15	20	NA	NA	NA
SO_4	65	53	~100	53	NA	53	440
TDS (wt %)	~25.6	~26.5	~21.4	~20.0	~12.7	~6.2	~1.3

(a) TDS = Total dissolved solids; NA = Not available
(b) Salton Sea Scientific Drilling Project (Elders, 1989)
(c) Concentrations corrected for ≈5% dilution by drilling fluid
(d) Silica values low due to precipitation prior to sampling
(e) Volumetric measurement of total non-condensible gas

(Modified from McKibben, 1991, Table 1)

4 ORE MINERALIZATION

In contrast to the high concentrations in the brine, ore minerals are not particularly abundant in the SSGS and are mostly restricted to certain open mineralized fractures or veins (McKibben and Elders, 1985). Two kinds of ore veins occur. There is an earlier carbonate-sulfide vein system, with calcite, sphalerite, pyrite, chalcopyrite, galena, and pyrrhotite, and a later silicate-oxide-sulfide vein system, with epidote, quartz, chlorite and hematite, together with minor chalcopyrite, pyrite, sphalerite, and galena (McKibben et al., 1988a). Fluid inclusions in these veins suggest that the younger veins result from a cooler, more oxidizing fluid mixing with the hotter more reduced fluid.

5 THE BRINE DIAPIR MODEL

The reservoir of hypersaline brine (>20 wt% TDS) is everywhere overlain by the lower salinity (<13 wt% TDS) brines (Figure 2). A sharp, dome-shaped, interface exists between the two. The distribution of metamorphic mineral isograds and ore veins also follow the

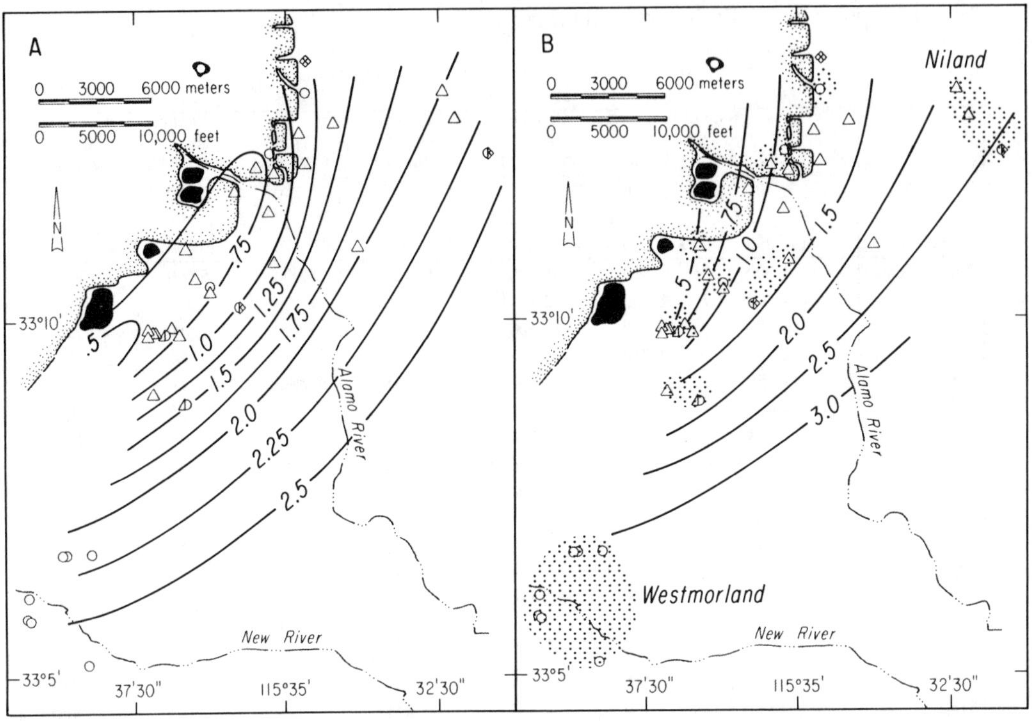

FIGURE 2. (a) Contour map of the SSGS showing depth (km) to the 250 °C isotherm as compiled from geothermal well temperature measurements from more than 30 wells. (b) Contour map of the SSGS showing depth (km) to the fluid interface constrained by salinity differences between closely spaced production zones. Stippled areas indicate locations of constraints on contours. Figure 7 of Williams & McKibben (1989).

form of the domal brine interface. This interface appears to be discordent to both lithological and structural features. Instead the interface follows the 250°C isotherm, reaching a depth of only 0.5 km near the center of the field (Figure 2). The top of this dome of brine is stabilized by the competing effects on its density of temperature and salinity gradients. At that temperature the density of the low TDS brine is approximately 0.85 g/cm^2 and that of the high TDS brine is approximately 1.0 g/cm^2 (Williams and McKibben, 1989). This density stratification is a major barrier to convective heat transfer and mixing. Thus the lithologic caprock inferred by previous workers is not the cause of the steep shallow temperature gradients observed. However, beneath the interface, even at the depth of the deepest wells so far drilled (3.2 km), the hypersaline brine seems well-mixed.

Thus, a model emerges of non-marine basinal brines accumulated in a closed basin, being heated by a sub-volcanic intrusion, dissolving evaporites, and rising advectively as a brine diapir, displacing the overlying cooler, less-saline brine, and rising to within a few hundred meters of the surface (Figure 3; Williams and McKibben, 1989). It appears that much of the metamorphism and ore genesis observed was produced in the zone where the two brines mixed. In fact, the evidence of fluid inclusions and the distribution of ore veins suggests that the brine diapir formerly reached an even higher level, that it has cooled and subsided slightly, and that the upper part of the system was then invaded by the cooler, lower TDS brine (McKibben et al., 1987, 1988a; Williams and McKibben, 1989).

6 CONCLUSIONS

Just as a single borehole in the SSGS in 1963 has become more than a hundred wells supplying steam to six electric power plants in 1992, the pioneering efforts of Don White, although they have not yet led to the discovery of "magmatic water", gave rise to a wealth of ideas on this interesting hydrothermal system.

ACKNOWLEDGEMENTS

This paper is largely based upon an extensive data base on the SSGS generated at the Center for Geothermal Resources Research at the University of California, Riverside (UCR). The authors are grateful for the funding support from various agencies including NSF grants EAR-8303557 and EAR-8502402 (Elders); NSF EAR-8502405 and EAR-8805426 (Williams); and DOE DE-FG03-85ER13408 (McKibben). This is report number 92/6 of the Institute of Geophysics and Planetary Physics (IGPP) at UCR.

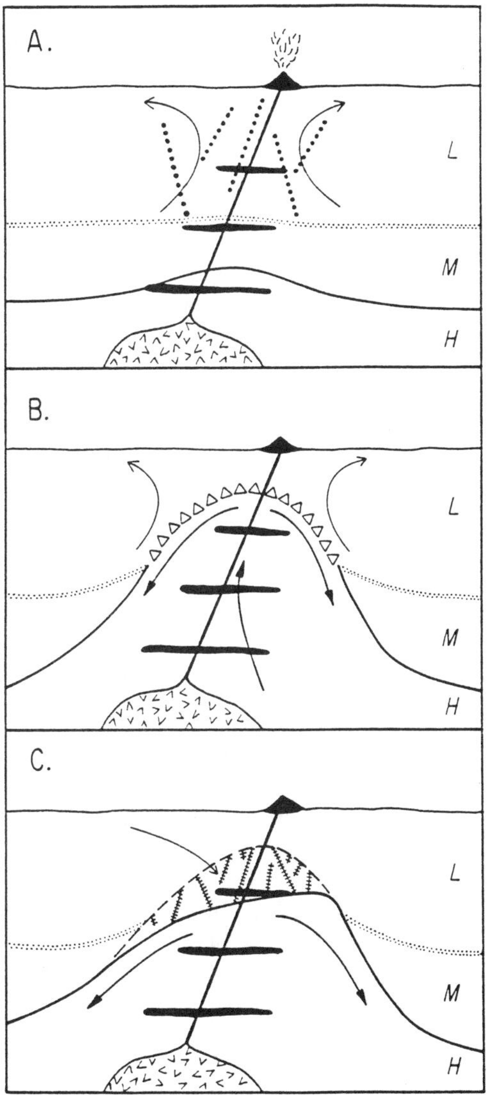

FIGURE 3. Schematic cross-sectional cartoon of our model for mineralization by brine upwelling and mixing in the SSGS. Arrows indicate direction of fluid movement. Prior to igneous intrusion, hypersaline brine (H) lies at depths greater than 5 km, overlain by moderate-salinity (M) and low-salinity (L) fluids.

A: Intrusion causes fracturing, heating, and the expulsion of reduced connate fluids, forming type 1 veins (large dotted lines).

B: An ascending diapir of hypersaline brine interacts with and displaces low-salinity fluids, forming type 2 mineralization in fractures that transect the rising fluid interface (triangles). Hypersaline brine rises to a maximum density-stabilized depth, pervasively altering the invaded host rocks to greenschist facies assemblages.

C: Upon cooling, the brine diapir begins to descend, drawing in cooler, oxidized waters and forming late-stage hematite in the type 2 veins (hatchured lines). From McKibben et al., 1987, 1988a.

REFERENCES

Cho, M., Liou, J.G., & Bird, D.K. 1988. Prograde phase relations in the State 2-14 well metasandstones, Salton Sea geothermal field, California. *Jour. Geophys. Res.* 93, B11: 13,081-13,103.

Elders, W.A. (ed.) 1989. A compendium of reports on the Salton Sea Scientific Drilling Project. Inst. Geophysics & Planetary Physics, Univ. Calif. Riverside: *UCR/IGPP* 89:14 1-320.

Elders, W.A., Rex, R.W., Meidav, T., Robinson, P.T., & Biehler, S. 1972. Crustal spreading in southern California: the Imperial Valley and Gulf of CA formed by rifting apart of a continental plate. *Science* 178, no 4056: 15-24.

Elders, W.A. & Sass, J.H. 1988. The Salton Sea Scientific Drilling Project: its scientific significance. *Jour. Geophysical Res.* 93: 12953-12968.

Helgeson, H.C. 1968. Geological and thermodynamic characteristics of the Salton Sea geothermal system. *Am. Jour. Science* 266: 129-166.

Herzig, C.T. & Elders, W.A. 1988. Nature and significance of igneous rocks cored in the State 2-14 Research Borehole: Salton Sea Scientific Drilling Project, California. *Jour. Geophys. Res.* 92, B11: 13,069-13.080.

Herzig, C.T., Mehegan, J.M. & Stelting, C.E. 1988. Lithostratigraphy of the State 2-14 borehole: Salton Sea Scientific Drilling Project. *Jour. Geophys. Res.* 93, B11: 12,969-12,980.

McDowell, S.D. & Elders, W.A. 1980. Authigenic layer silicate minerals in borehole Elmore #1, Salton Sea geothermal field, California, U.S.A. *Contrib. Mineral. Petrol.* 74: 293-310.

McDowell, S.D. & Elders, W.A. 1983. Allogenic layer silicates in Borehole Elmore #1, Salton Sea geothermal field, California, U.S.A. *Amer. Mineral.* 68: 1146-1159.

McKibben, M.A. 1991. The Salton Sea geothermal brines in M.A.McKibben (ed.) the Diversity of Mineral & Energy Resources of Southern California. *Society of Economic Geologists,* Guidebook Series 12: 127-138

McKibben, M.A., Andes, J.P., Jr., & Williams, A.E. 1988a. Active ore formation at a brine interface in metamorphosed deltaic lacustrine sediments: the Salton Sea geothermal system, California. Economic Geology 83: 511-523.

McKibben, M.A. & Elders, W.A. 1985. Fe-Zn-Cu-Pb mineralization in the Salton Sea geothermal system, Imperial Valley, California. *Economic Geology* 80: 539-559.

McKibben, M.A. & Eldrige, C.S. 1989. Sulfur isotopic variations among minerals and aqueous S species in the Salton Sea geothermal system: a SHRIMP ion microprobe and conventional study of active ore genesis in a sediment-hosted environment. *Amer. Jour. Sci.* 289: 661-707.

McKibben, M.A.,Williams, A.E., Elders, W.A., & Eldrige C.S. 1987. Saline brines and metallogenesis in a sediment-filled rift: the Salton Sea geothermal system, California. *Appl. Geochemistry* 2: 563-578.

McKibben, M.A., Williams, A.E., & Okubo S. 1988b. Metamorphosed Plio-Pleistocene evaporites and the origins of hypersaline brines in the Salton Sea geothermal system, California: fluid inclusion evidence. *Geochim. et Cosmochim. Acta* 52: 1047-1056.

Muffler , L.J.P. & White, D.E. 1969. Active metamorphism of upper Cenozoic sediments in the Salton Sea geothermal field and the Salton Trough, southeastern California. *Geol. Soc. America Bull.* 80: 157-182.

Shearer, C.K., Papike, S.B., Simon,S.B., & Davis, B.L. 1988. Mineral reactions in altered sediments from the California State 2-14 well: variations in the modal mineralogy, mineral chemistry, and bulk composition of the Salton Sea Scientific Drilling Project core. *Jour. Geophys. Res.:* 93, B11: 13,104-13,122.

Skinner, B.J., White, D.E., Rose, H.J., Jr., & Mays, R.E. 1967. Sulfides associated with the Salton Sea geothermal brine. *Economic Geology* 62: 316-330.

White, D.E., Anderson, E.T. & Grubbs, D.K. 1963. Geothermal Brine Well: mile deep hole may tap ore-bearing magmatic water and rocks undergoing meta-morphism. *Science* 139: 919-922.

Williams, A.E. 1988. Delineation of a brine interface in the Salton Sea geothermal system, California. *Geothermal Resources Council Transactions* 12: 151-157.

Williams, A.E. & McKibben, M.A. 1989. A brine interface in the Salton Sea geothermal system: fluid geochemical and isotopic characteristics. *Geochim. et Cosmochim. Acta* 53: 1905-1920.

Water-Rock Interaction, Kharaka & Maest (eds) © 1992 Balkema, Rotterdam. ISBN 90 5410 075 3

The geochemistry of hot spring waters at Norris Geyser Basin, Yellowstone National Park, USA

Robert O. Fournier & J. Michael Thompson
US Geological Survey, Menlo Park, Calif., USA

Roderick A. Hutchinson
US National Park Service, Yellowstone National Park, Wyo., USA

ABSTRACT: Relatively dilute alkali-chloride waters at Norris Geyser Basin attain chemical equilibrium with reservoir rocks underground in a series of reservoirs at temperatures ranging from about 200 to 325°C. Reservoirs of acid-sulfate-chloride waters at >190–200°C also exist underground at Norris and there is episodic mixing (usually in the fall) of acid-sulfate-chloride and alkali-chloride waters. The rate of discharge of sulfate by acid-sulfate-chloride water from deep in the system is large, and it is difficult to account for the generation of all of this type of water solely by present-day local oxidation of H_2S that is carried in steam to the earth's surface at Norris, followed by infiltration of acid-sulfate water back underground where it mixes with chloride-rich water. Additional sulfate may be generated by bacteria (*sulfolobus*) that consume native sulfur and liberate sulfuric acid in shallow underground environments. The source of native sulfur may be old solfataras in the Norris region, formed as a result of volcanic activity that produced silicic domes south of Norris, and that were subsequently covered either by later volcanic flows, or by glacial material deposited on hot ground. Another possibility, supported by $^3He/^4He$ data, is that a component of magmatic gas containing SO_2 is entering parts of the Norris system and reacting with water to form H_2SO_4 and H_2S.

1 INTRODUCTION

Fournier (1989) presented a general overview of the geochemistry of the thermal waters throughout Yellowstone National Park that categorized the Norris waters within a general model of the hydrothermal system, and White et al. (1988) focused on the geology and thermal activity of Norris Geyser Basin (Norris). Here we briefly outline the geochemical variations in the thermal waters found at Norris, where over 360 samples of thermal waters have been collected and analyzed for major dissolved constituents from about 70 hot springs and geysers over a period of 108 years.

2 CHEMICAL CHARACTERISTICS OF THERMAL WATERS

Representative chemical analyses of thermal waters at Norris are shown in Table 1. Waters from the hottest (275–323°C, based on silica and Na-K-Ca geothermometry, and on mixing models) and probably deepest reservoir at Norris flow to the surface at the side of Porcelain Terrace at the NE corner of the geyser basin (Figure 1). These waters generally are pH-neutral with relatively high chloride (550–800 mg/kg) and low sulfate (10–50 mg/kg). As a result of the high reservoir temperature and neutral pH of water, the rate of sinter deposition at the side of Porcelain Terrace is the fastest in Yellowstone. Anal. No. 1, Table 1 is typical of thermal waters issuing along a N-trending zone at the W side of the Terrace.

A USGS drill hole (Y–12) at the SE end of Porcelain Terrace attained a depth of 331.7 m and a temperature of 238°C (White et al., 1975). Fluid was collected from the bottom of this well using a downhole sampler. The chemical composition of that water is shown in Table 1 (No. 12). The most significant information provided by this sample is the chloride concentration of 528 mg/kg. There was no indication of steam separation before or after the sample was collected, so the measured chloride is probably close to the concentration of chloride in the deep reservoir beneath Norris that supplies water to the springs at Porcelain Terrace. The cation composition reflects reaction of hot water with glass-rich ash-flow tuff within the uncased portion of the drill hole where temperatures range from 190 to 238°C. Most of the sulfate in this water probably resulted from reaction of pyrite in

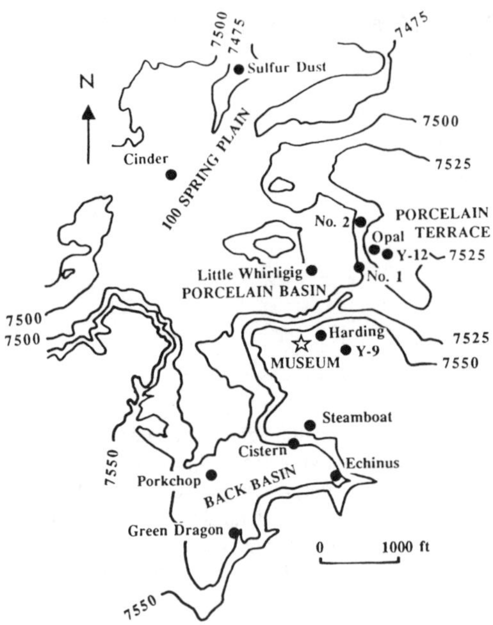

Fig. 1 Sketch map of Norris Geyser Basin showing topographic features and locations of selected hot springs and geysers. Contours showing equal altitude at 25-foot intervals are shown in the range 7475 feet to 7550 feet.

the wall rock with oxygenated water that filled the well after drilling was completed.

Progressing westerly from Porcelain Terrace through Porcelain Basin, and to the NW through 100 Spring Plain (Figure 1) there is a decrease in chloride, an increase in sulfate, and a decrease in pH (e.g., Little Whirligig and Sulfur Dust (Nos. 4 and 5, Table 1). Also, the estimated reservoir temperature decreases to about 200–230°C. Very little sinter deposits in Porcelain Basin and in the 100 Spring Plain because the water that is discharged there has a low pH.

In the Back Basin (Figure 1) the hydrothermal system is chemically and hydrologically complex. There, some waters issue from an acid-sulfate-chloride reservoir at about 195–215°C (e.g., Echinus, Table 1) and others issue from at least two pH-neutral, chloride-rich reservoirs, one at a temperature of about 210–230°C and the other at about 275–310°C. As an example, Porkchop Geyser (No. 7, Table 1) recently has discharged water mainly from a hotter (presumably deeper) neutral-chloride reservoir where dissolved silica concentrations and K/Na ratios are relatively high, while in the past it has discharged from an apparently different (presumably shallower) reservoir where pH-neutral water has reacted with rock at lower temperatures, yielding lower concentrations of dissolved silica and lower K/Na values (No. 6, Table 1). A gradual change in the composition of water discharged by Porkchop Geyser over the period 1960–1990 was documented by Fournier et al. (1991), who showed that the temperature of last water-rock equilibration of the discharged water increased by about 60–70°C during that same period of time, accompanied by an increase in eruptive activity which culminated in a hydrothermal explosion that destroyed the geyser in 1989.

Some springs that were sampled at different times show variable degrees of mixing of the acid-sulfate and neutral-chloride waters. This is most strikingly exhibited by Cistern Spring (Figure 2) that was sampled more than 100 times during a 25-year period. The variations shown in Figure 2 result partly from different amounts of mixing of different waters, and partly from different degrees of steam loss (adiabatic expansion) during upflow (Fournier et al., 1986).

3 EPISODIC INCREASED BOILING FOLLOWED BY INCREASED MIXING OF SULFATE-RICH AND CHLORIDE-RICH WATERS

Episodically (generally in the fall) there is a hydrothermal "disturbance" that affects waters in all of the Norris basins. The disturbance is characterized by increased turbidity of many hot springs, increased discharge of water and steam, extreme fluctuations in temperatures of pools, and, in some years, small hydrothermal explosions (White et al., 1988). In addition, underground mixing of alkali-chloride and acid-sulfate-chloride waters apparently coincides with the disturbance. Systematically increasing chloride in waters collected monthly from Cistern Spring in the Back Basin during 1976–1985 showed that there was increased boiling preceding these disturbances. Immediately following the onset of these disturbances decreased chloride and increased sulfate were found (Fournier et al., 1986). As a second example, one usually neutral-chloride spring at Opal Terrace was sampled on the first day of a disturbance (No. 2, Table 1); its pH was 2.82 and its sulfate concentration was 133 mg/kg.

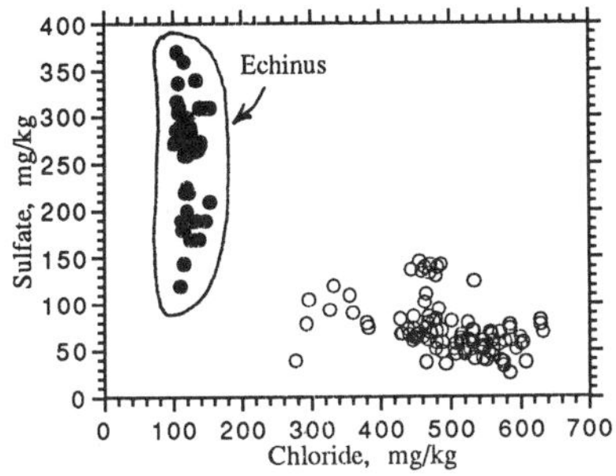

Fig. 2 Chloride-sulfate relations exhibited by waters collected at different times from Cistern Spring (circles) and Echinus Geyser (dots within the encircled area).

Four days later its pH had increased to 5.36 and its sulfate concentration had dropped to 99 mg/kg (see No. 3, Table 1). Previously waters from this and other nearby springs generally had been found to be very similar in composition to the analysis shown in No. 1, Table 1. The cause of the hydrothermal disturbance is uncertain, but may be related to seasonal fluctuations in the regional groundwater potentiometric surface.

4 CHEMICAL CHARACTERISTICS AND ORIGIN OF SULFATE-RICH ACID WATERS AT NORRIS

There are two types of sulfate-rich acid waters at Norris. One type is found in topographically high regions where nonoverflowing, boiling pools of low-chloride (<10 mg/kg), sulfate-rich waters originate by oxidation of H_2S to H_2SO_4 where steam and other gases bubble through surficial pools of perched groundwater and condensed steam. (e.g., Opal Spring, No. 10, Table 1). The second type generally contains >100 mg/kg chloride (acid-sulfate-chloride waters), and has relatively high rates of discharge, particularly from Echinus during its large geyser eruptions that currently occur almost every hour. The pH of the Echinus waters (3.0–3.5) appears to be buffered by the assemblage kaolinite-muscovite at about 195–215°C (the silica geothermometer temperature range).

The origin of the relatively high concentrations of sulfate in deep waters flowing from >200°C reservoirs at Norris is uncertain. Studies of sulfur isotopes are underway by others, but to date the results are very preliminary and inconclusive. Oxidation of H_2S at depth using oxygen initially dissolved in meteoric recharge water can account for only about 15 mg/kg dissolved sulfate in deep water. Present-day oxidation at and near the water table of H_2S initially carried in steam, followed by downward percolation may account for some of the sulfate and acidity. However, mass and thermal balance considerations indicate that the large rate of discharge of sulfate in the acid-sulfate-chloride water cannot be accounted for by the above mechanisms alone. Furthermore, the concentration of sulfate in Echinus water ranges widely while there is relatively little change in chloride (e.g., Nos. 8 and 9, Table 1). Figure 2 shows $Cl\text{-}SO_4$ relations for 53 samples from Echinus (all data points located within encircled area) collected from 1884 through 1986, mostly during 1980–1983. Variations in chloride in the Echinus waters are about as expected for different proportions of adiabatic and conductive cooling during upflow, while variations in sulfate are too large to be explained by different degrees of boiling during upflow from a 195–215°C reservoir.

One way to pick up sulfate with little change in chloride is for groundwater to come in contact with old native sulfur deposits that may be buried at relatively shallow depths in the cool margins of the Norris system. Sulfur-consuming bacteria (*sulfolobus*) are very effective at generating sulfuric acid using native sulfur and oxygen in the cool margins of hydrothermal systems (Brock, 1978). Native sulfur deposits could have formed in solfataras during early volcanic activity that produced silicic domes south of Norris, and these solfataras could later have been buried, either by younger volcanic flows that partly cover the domes, or by glacial material deposited from melting ice over hot ground.

Yet another mechanism for explaining the chloride-sulfate relations at Norris is by addition of a component of magmatic gas containing SO_2 that reacts with water to form H_2SO_4 and H_2S. This latter mechanism is consistent with recent observations that the highest $^3He/^4He$ ratios (>8–9 relative to atmospheric concentrations) in Norris waters are found in the acid-sulfate-chloride waters rather than alkaline-chloride waters (B.M. Kennedy, written communication, December 1991).

Table 1. **Chemical compositions (mg/kg) of selected thermal waters, Norris Geyser Basin**

No.	Name	Ref.	Date	°C	pH	SiO_2	Na	K	Ca	Mg	Li	Cl	SO_4	HCO_3
1	S. Base Porc. Terr.	(a)	07/03/75	92	7.00	589	400	93	2.60	0.01	7.18	675	24	60
2	N. Base Porc. Terr.	(b)	09/02/90	88	2.82	520	334	100	1.82	0.21	4.07	527	133	0
3	N. Base Porc. Terr.	(b)	09/06/90	89	5.36	665	400	112	2.82	0.08	5.36	634	99	32
4	Little Whirligig	(c)	09/27/71	91	3.20	420	349	83	2.50	0.50	5.30	607	113	0
5	Sulfur Dust	(d)	09/12/61	50	2.75	280	250	57	4.50	0.15	3.50	427	197	0
6	Porkchop	(d)	09/18/60	72	7.45	435	526	66	6.40	0.50	7.40	860	33	18
7	Porkchop	(e)	06/??/89	92	8.50	741	388	91	3.60	0.06	6.60	687	23	62
8	Echinus	(b)	11/15/81	90	3.14	257	151	45	3.70	0.52	1.00	125	170	0
9	Echinus	(b)	09/15/86		3.06	296	148	52	4.30	0.52	0.75	117	360	0
10	Opal	(d)	09/19/60	90	1.90	168	5	2	0.70		0.05	5	760	0
11	Harding	(c)	06/07/71	93	3.53	325	26	18	0.90	0.25	0.09	2	123	0
12	Y-12	(b)	09/13/69	238	7.86	352	377	21	1.90	0.20	1.43	528	34	79
13	Y-9	(b)	09/16/69	195	8.58	412	268	16	0.80	0.04	0.98	80	43	488

(a) Thompson & Yadav (1979); (b) Unpublished USGS data; (c) Thompson et. al. (1975); (d) Rowe et al. (1973); (e) Kharaka et al. (1990)

6 CONCLUSIONS

At the west side of Porcelain Terrace, Cl-rich and SO_4-poor water flows directly to the surface from the deepest and hottest reservoir at Norris (275–325°C). To the west and northwest of Porcelain Terrace, this Cl-rich water becomes progressively more mixed with an acid-sulfate water. A series of stacked reservoirs with different temperatures are present in the Back Basin so that waters of similar chloride content attain chemical equilibrium with rock at a variety of temperatures ranging from about 210 to 300°C before emerging at the earth's surface. Reservoirs of sulfate-rich waters also exist in the Back Basin region and discharge waters that show relatively large variations in sulfate, but little variation in chloride. Annual mixing of the sulfate- and chloride-rich waters commonly occurs in the fall immediately after a basinwide event characterized by an abnormally high degree of steam discharge (increased boiling).

REFERENCES

Brock, T.D. 1978. Thermophilic Micro-organisms and Life at High Temperatures. *Springer-Verlag New York,* 465 pp.

Fournier, R.O. 1989. Geochemistry and dynamics of the Yellowstone National Park hydrothermal system. *Ann. Rev. Earth Planetary Sci.* 17:13–53.

Fournier, R.O., J.M. Thompson, C.G. Cunningham & R.A.Hutchinson 1991. Conditions leading to a recent small hydrothermal explosion at Yellowstone National Park. *Geol. Soc. Am. Bull.* 103: 1114–1120.

Fournier, R.O., J.M. Thompson & R.A. Hutchinson 1986. Fluctuations in composition of Cistern Spring, Norris Geyser Basin, Yellowstone National Park, Wyoming—variable boiling and mixing 1962–1985. *Fifth Int. Symp. Water-Rock Interaction*, Reykjavík, Iceland: 206–209.

Kharaka, Y.K., R.H. Mariner, G. Ambats, W.C. Evans, L.D. White, T.D. Bullen & B.M. Kennedy 1990. Origins of water and solutes in and north of the Norris–Mammoth corridor, Yellowstone National Park. *Geother. Resourc. Counc. Trans.* 14(I): 705–714.

Rowe, J.J., R.O. Fournier & G.W. Morey 1973. Chemical analyses of thermal waters in Yellowstone National Park, Wyoming, 1960–1965. *US Geol. Surv. Bull.* 1303: 31 pp.

Thompson, J.M., T.S. Presser, R.B. Barnes & D.B. Bird 1975. Chemical analyses of the waters of Yellowstone National Park, from 1965–1973. *US Geol. Surv. Open-File Rep.* 75–25: 59 pp.

Thompson, J.M. & S. Yadav 1979. Chemical analyses of waters from geysers, hot springs and pools in Yellowstone National Park, 1974–1978. *US Geol. Surv. Open-File Rep.* 79–704: 49 pp.

White, D.E., R.O. Fournier, L.J.P. Muffler & A.H. Truesdell 1975. Physical results of research drilling in thermal areas of Yellowstone National Park. *US Geol. Surv. Prof. Paper* 892: 70 pp.

White, D.E., R.A. Hutchinson & T.E.C. Keith 1988. The geology and remarkable thermal activity of Norris Geyser Basin, Yellowstone National Park. *US Geol. Surv. Prof. Paper* 1456: 84 pp.

Water-Rock Interaction, Kharaka & Maest (eds) © 1992 Balkema, Rotterdam. ISBN 90 5410 075

Exploration progress of high enthalpy geothermal prospects in Indonesia

Sugiarto Ganda, Djoko Sunaryo & Tanda Tampubolon
Geothermal Pertamina, Indonesia

ABSTRACT: Geothermal exploration in Indonesia was officially started in 1972, although scientific drilling was recorded in 1926-1928 in Kamojang geothermal prospect, West Java. During the last 20 years, geoscientific field work has been carried out along the Indonesian volcanic belt to discover many geothermal features. The successful interpretation of the data obtained from those geoscientific surveys resulted in a conceptual model. More than 55 high enthalpy geothermal areas are spread along the islands of Sumatera, Java, Sulawesi, and East Indonesia. Some of them have been developed and utilized for electricity generation, such as Kamojang 140 MW, and Dieng 2 MW. Electricity production supplied from geothermal resources will be increased to 437 MW by the year 1995.

1 INTRODUCTION

It is known that Indonesia lies on the active volcanic arc, which has high enthalpy geothermal resources. Based on geoscientific investigation, total estimated geothermal potential is about 16,035 Megawatt (MW).

In Indonesia geothermal investigation has been carried out since the 1970's, and later detailed surveys pointed out several promising high enthalpy geothermal areas, as shown in Figure 1.

The exploration surveys were started in Java and Bali. In 1982 the survey was continued to Sulawesi, and in 1988 to Sumatera. East Indonesia will be investigated in the coming year.

The deep geothermal well has been drilled since 1974 in Kamojang. Development results and exploration activities can be seen on Tables 1 and 2.

2 SUMATERA ISLAND

The geothermal prospects in Sumatera are situated along the great Sumatera graben adjacent to the Bukit Barisan ridge. There are 15 prospects which have been investigated: 3 prospects are ready to be drilled; one just finished the reconnaissance stage, and the remainder are still being surveyed. Total estimated resource potential is 4885 MW. The first exploratory well is being drilled in Sibayak prospect.

3 JAVA ISLAND

All active geothermal prospects follow the volcanic chains in the center of Java island. There are 15 selected prospects which have been investigated in detail and have total resources of 8100 MW. Five prospects have been developed (Kamojang, Darajat, Dieng, Salak and Wayang Windu) with a proven potential of 895 MW. Seven prospects are being prepared for exploratory drilling and two prospects are being reevaluated.

4 SULAWESI ISLAND

The exploratory investigations have been carried out in North Sulawesi and South Sulawesi. In North Sulawesi, the Lahendong prospect will be developed for 20 MW in 1995/96. Two other prospects are still being surveyed. The only prospect that has been investigated in South Sulawesi is Bituang, which is in the reconnaissance stage. Total resource potential in Sulawesi is about 1500 MW.

5 BALI AND FLORES ISLANDS

In Bali, the Bratan geothermal prospect has

Table 1. The activities of geothermal investigation and drilling in Indonesia , status on 1 January 1992.

Geology:
A = Photogeology
B = Detail mapping
C = Alteration mapping
Geochemistry:
D = Cl seepage
E = Hg and CO_2
F = Isotope

Geophysics:
G = Resistivity
H = Gravity
I = Magnetic
J = MT-5 Ex
K = C.S.A.M.T.

* survey has been done

Drilling :#
L = Exploration
M = Production

\# numbers of well

	Locality	Geology			Geochemistry			Geophysics					Drilling	
		A	B	C	D	E	F	G	H	I	J	K	L	M
	Sumatra :													
1	Seulawah	*												
2	Sibayak	*	*	*	*	*	*	*	*	*				
3	Tarutung	*	*		*	*								
4	Sarulla	*			*	*		*	*					
5	Sibual-Buali	*	*	*	*	*	*	*	*	*				
6	Pusuk Bukit	*												
7	Sorik Merapi	*	*	*	*	*	*	*	*	*				
8	Muaralaboh	*	*		*	*		*	*					
9	Sungaipenuh	*	*		*	*		*	*	*				
10	Lempur	*	*					*	*	*			2	
11	Lumut Balai	*					*							
12	Ranau	*												
13	Suouh Sekincau	*	*	*	*	*	*	*	*	*				
14	Ulubelu	*	*	*	*	*	*	*	*	*				
15	Rajabasa	*	*	*	*	*	*	*	*	*				
	Java :													
16	Banten	*	*	*	*	*	*	*	*		*			
17	Cisolok	*	*		*	*	*	*	*		*			
18	Salak	*	*		*	*	*	*	*	*	*		12	
19	Tangkuban P.	*	*	*	*	*	*	*	*		*			
20	Patuha	*	*	*	*	*	*	*	*		*			
21	Wayang-Windu	*	*	*	*	*	*	*	*	*	*		1	
22	Darajat	*	*	*	*	*	*	*	*	*	*	*	7	
23	Kamojang	*	*	*	*	*	*	*	*	*	*	*	17	33
24	Karaha	*	*	*	*	*	*	*	*	*	*			
25	Telagabodas	*	*	*	*	*	*	*	*	*	*			
26	Dieng	*	*	*	*	*	*	*	*	*	*		10	13
27	Ungaran	*	*	*	*	*	*	*	*	*	*			
28	Telomoyo	*	*	*	*	*		*	*	*	*			
29	Ngebel-Wilis	*	*	*	*	*	*	*	*	*	*			
30	Ijen	*	*		*	*	*	*	*	*	*			
31	Bali/Bratan	*	*		*	*	*	*	*		*			
32	Flores/Ulumbu	*	*		*	*	*	*	*					
	Sulawesi :													
33	Lahendong	*	*	*	*	*	*	*	*		*		7	2
34	Tompaso	*	*	*	*	*	*	*	*		*			
35	Kotamubagu	*	*	*	*	*		*	*	*				
36	Bituang	*												

Table 2. List of geothermal performances, status on 1 January 1992.

Surface features:
A = Max. temp. of hot spring (C)
B = pH
C = Steam discharge
* presence

Reservoir:
D = Estimated temp. (C)
E = Measured temp. (C)
F = Depth (m)
G = Enthalpy

Potential (NW)
H = Installed
I = Proven
J = possible
K = Resources

	Locality	Features			Reservoir				Potential			
		A	B	C	D	E	F	G	H	I	J	K
	Sumatra :											
1	Seulawah	95	7	*	250							
2	Sibayak	66	6	*	250						140	240
3	Tarutung	65		*	180							
4	Sarulla				310						280	380
5	Sibual-Buali	53	6	*	260						600	750
6	Pusuk Bukit	63	1	*								
7	Sorik Merapi			*	290							
8	Muaralaboh	98	7	*	210							
9	Sungaipenuh	98	7									
10	Lempur			*	225						75	115
11	Lumut Balai	98	7	*	270							
12	Ranau	66	7		210							
13	Suouh Sekincau	99	6	*	300							
14	Ulubelu	99	6	*	250							
15	Rajabasa	74	7	*	200							
	Java :											
16	Banten			*								
17	Cisolok	94	7									
18	Salak			*		280		1842		280	370	600
19	Tangkuban P.	43	6	*	290							
20	Patuha	89	6	*	350						400	685
21	Wayang-Windu	50	6	*		270	1400			200	260	420
22	Darajat			*		245	1000	2783		120	250	420
23	Kamojang			*		252	900	2792	140	205	200	300
24	Karaha	49	7	*							200	250
25	Telagabodas	47	6	*	289						200	300
26	Dieng			*		368	1600	2560	2	285	575	1430
27	Ungaran	68	7	*	236							
28	Telomoyo											
29	Ngebel-Wilis	93	7								100	170
30	Ijen	39		*								
31	Bali/Bratan			*							215	325
32	Flores/Ulumbu	96	6	*							200	350
	Sulawesi :											
33	Lahendong			*		356	1300	2670		65	175	300
34	Tompaso			*							230	400
35	Kotamubagu	58	7	*	200						200	300
36	Bituang	47	7	*								

been investigated in detail, and has the possible potential of about 215 Megawatts. Some exploration methods have been applied, and the boundary of the prospect has been defined. Tourism facilities are the potential market in Bali.

Flores island is included in the program of rural electrification, and geothermal is the potential energy source in the area. Based on this aim, Ulumbu was investigated in 1989, and 200 Megawatts may be possible.

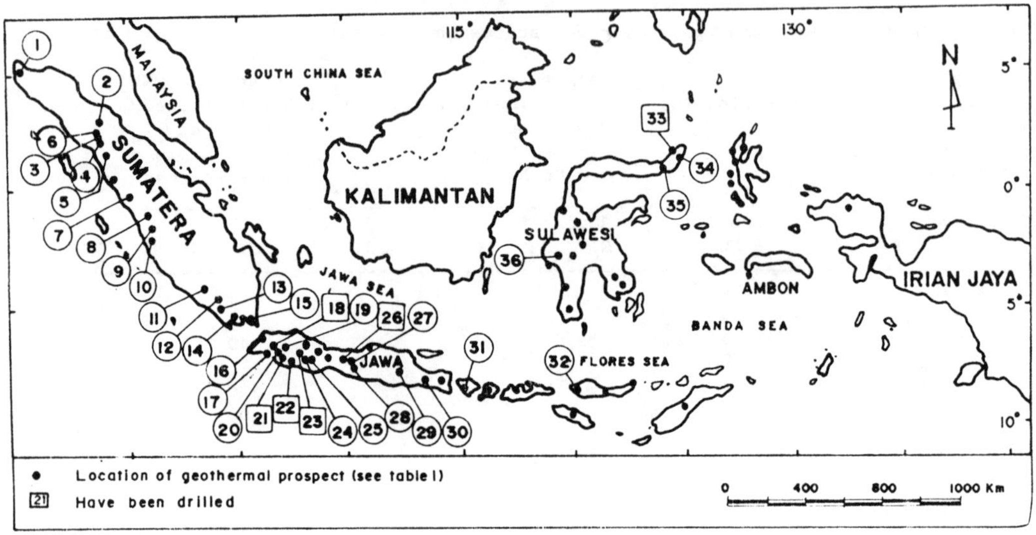

Figure 1. Location of the geothermal prospects in Indonesia, status 1 January 1992.

6 EAST INDONESIA

East Indonesia will be considered in the next round of geothermal investigations. This parallels the government's next Five-Year-Plan for developing East Indonesia. The estimated potential of about 1500 Megawatts is spread from Nusa Tenggara, Maluku to Irian Jaya. Forestry and fishery industries are the potential markets in the area.

ACKNOWLEDGEMENT

The authors would like to acknowledge the PERTAMINA Management for the opportunity to publish this paper, especially the Head of the Geothermal Division and the Head of the Planning Department.

REFERENCES

Andoyo; Pane, S.Z., A.Suardy, V.T. Radja & Widartomo 1989. Geothermal development in Indonesia present status and the future plan. 8th Meeting on Geothermal Power Development, Chiang Mai, Thailand.

Prijanto; Sinaga, D., S.Satibi, & S.Sitompul 1990. Geothermal steam pricing system in Indonesia, 9th Meeting on Geothermal Power Development, Baguio City, Philippines.

Pertamina, 1991. Coordination meeting Geothermal Division, Pertamina, Jakarta (unpublished).

Water-Rock Interaction, Kharaka & Maest (eds) © 1992 Balkema, Rotterdam. ISBN 90 5410 075 3

Hydrothermal systems of Hengduan Mountains, China

Guo Guoying
Department of Geology, Peking University, Beijing, People's Republic of China

ABSTRACT: There are 1032 hydrothermal features in Hengduan Mountains of southwestern part of China, which are more than one-third of the hydrothermal systems throughout China. Twenty-seven of these hydrothermal features are boiling springs, just behind Xizang Plateau in numbers. The distribution of these hydrothermal features is controlled by the old regional tectonics, the position of uplifts and troughs, neotectonism and Cenozoic volcanism. The water of these hydrothermal systems is generally Na-HCO_3 type water. According to the chemical model of Na-HCO_3 type water and position of relative contents of Na, K, Mg elements in triangular diagram, this kind of water is lixiviation water formed by meteoric water obtaining heat as it passes through fault systems to circulation depth of several kilometers. The chloride water discharged from the Hot Sea Field and Langpu Hot Pool in Tengchong county may obtain heat from magmatic pockets. Therefore, it could have more potential for geothermal energy.

1 INTRODUCTION

Since 1981, the geothermal survey of Hengduan Mountains in Qinghai-Xizang plateau has been carried out by Comprehensive Scientific Expedition. The survey area covers 10^5 km^2 in west part of Sichuam province and west part of Yunnan province. There are 1032 hydrothermal features in this area which are more than one-third of those hydrothermal systems throughout China. Twenty-seven of them are boiling springs, just behind Xizang Plateau in numbers. It really deserves to be called the concentration area of thermal springs. These hydrothermal areas are of various manifestations, such as thermal spring, thermal water pools, boiling springs, fumaroles and beautiful sinter deposits. In order to use these geothermal resources suitably, it is necessary to study their characteristics. This paper describes the hydrothermal systems of Hengduan Mountains and discusses their hydrogeochemistic characteristics.

2 THE DISTRIBUTION OF HYDROTHERMAL SYSTEMS IN HENGDUAN MOUNTAINS

The Hengduan Mountains are close to the collision zone between Eurasian and Indian plates. The regional tectonics are very complex. Systems of anticlinoriums and synclinoriums run almost parallel to north-south direction. Some famous large faults constitute the boundaries between uplifts and troughs. Late Cenozoic volcanic events mainly occur in the Tengchong-Gaoligongshan belt. The distribution of hydrothermal systems is controlled by the old regional tectonics, the position of uplifts and troughs, neotectonism and Cenozoic volcanism. Most of the hydrothermal systems concentrate in the uplift belts; in Changning-Lancang metamorphic belt, for example, there are 167 hydrothermal systems, eleven of which belong to high-temperature hydrothermal systems. Some boiling springs issue near the major faults or the fault intersections. In the Jinshajing fault belt, for example, there is a Chaluo geyser in Batang county, the only one in mainland China except on Xizang plateau. Ruili high-temperature hydrothermal system is located in the Nujing fault belt. Tengchong-Gaoligongshan belt is located at the convergence zone between continental plates where some magma pockets could still remain and the follow-up magmatism may exist in the shallow crust. This might serve as the local powerful heat source to the Hot Sea Field and Langpu Hot Pool in Tengchong county. Where depositional basins between the uplifts containing hydrothermal systems decrease in number, geothermal activity weakens and temperatures are lower. The wall rocks of most reservoirs are composed of crystalline rocks, such as granite, basalts, migmatites or gneisses and occasionally limestones (Fig. 1).

3 CHEMICAL COMPOSITION OF THE GEOTHERMAL WATERS

The chemical types of thermal spring waters are rather simple, mostly the neutral bicarbonate type; pH ranges from 6.5 to 10. It has a close relation with contents of H_2CO_3, HCO_3 and CO_3 in the water; when pH is more than 8, the thermal spring waters are HCO_3-CO_3-or CO_3-HCO_3-type. The thermal spring water usually contains HCO_3 of 100-300 mg/l (maximum HCO_3 content is 4560.78 mg/l), low chlorine (10-40 mg/l), and variable sulphate contents. When the anions in the water are mainly sodium, this kind of water is Na-HCO_3-type water, most of which are fresh waters with total dissolved solids lower than 1 g/l. These Na-HCO_3 type waters with total dissolved solids higher than 1 g/l are mainly located at the west part of Sichuan province. A few Ca-HCO_3 or Ca-Mg-HCO_3 type thermal spring waters exist in the depositional basins.

Another important kind of water is the chloride-type in this area. It is seen that Hot Sea Field in Tengchong county and Langpu Hot Pool are Na-Cl-HCO_3 or Na-HCO_3-Cl type of water which contains 57-372 mg/l of chlorine, 100-195 mg/l of silica and 1-2 g/l of total dissolved solids. Sodium exists mainly in anion form. Lithium, rubidium, cesium, and arsenic are of high concentrations. This chloride-type of water in Tengchong county is more or less similar to those thermal fields explored in the world and could have high potential for geothermal energy.

Other chemical contents in spring waters are lithium, rubidium, and cesium which have contents of no more than 0.5 mg/l, and low contents of boron. Besides this, silica content in high temperature springs has uniform contents of 100-200 mg/l and silica content of some geothermal waters is more than 200 mg/l.

Relative values of Na, K, and Mg in thermal spring waters over 45°C are plotted on a triangular diagram (Fig. 2 after Giggenbach 1986). The datum of Hot Sea Field is plotted near the full equilibrium line, the Langpu Hot Pool is shifted off the position of the full equilibrium line to a dilution line. Most spring waters are close to the Mg-corner marked "shallow water" and "partial equilibrium dilution or mixing" (Fig. 2). These positions of thermal spring waters on the triangular diagram are likely the Hot Sea Field discharge water reflecting deeper origin, and the Langpu Hot Pool waters reflecting dilution or mixing. Most bicarbonate waters reflect shallow or mixing waters.

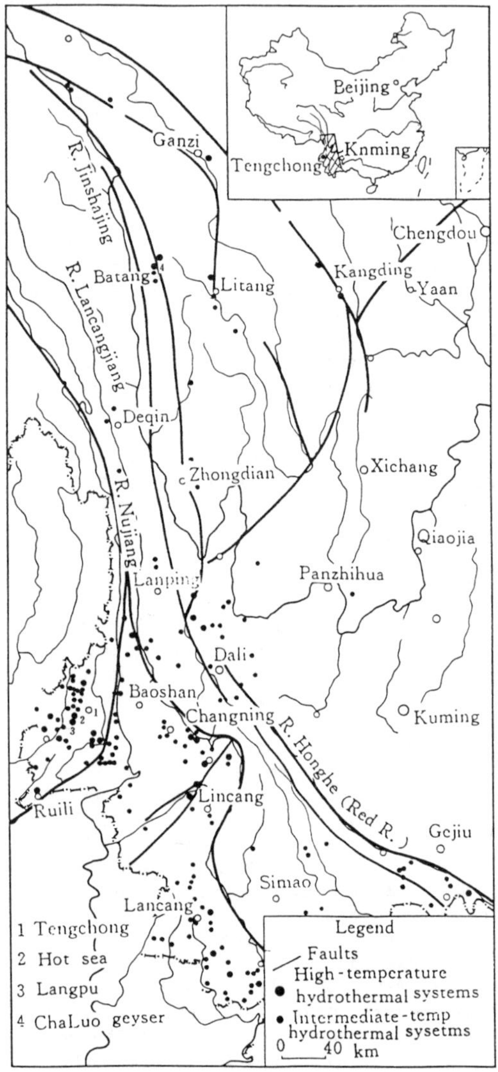

Figure 1 - The distribution of hydrothermal systems of Hengduan Mountains

4 THE CHEMICAL MODEL OF Na-HCO_3 TYPE OF GEOTHERMAL WATERS

The most common species in this area is the Na-HCO_3 type of geothermal water, the chemical model of which is discussed with the Na-HCO_3 type of thermal spring water in

Tengchong county.

The Na-HCO_3 type of thermal spring water in Tengchong county is divided into four groups (Table 1) based on the related logarithmic maps of Na versus HCO_3 contents and Na versus SiO_2 contents. The formation of the Na-HCO_3-CO_3 type of water and Na-CO_3-HCO_3 type of water in the fourth group is similar to that of the Na-HCO_3 type of water in the first group.

The major ions of the first and fourth groups result from the dissolution of albite and is strongly influenced by the

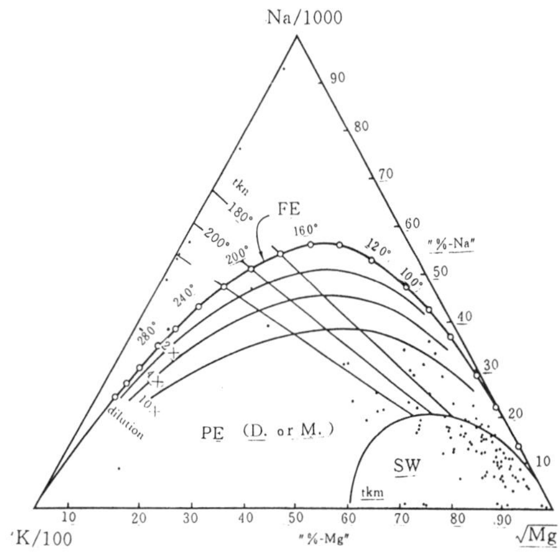

Figure 2 - Relative Na, K, Mg-contents of geothermal waters

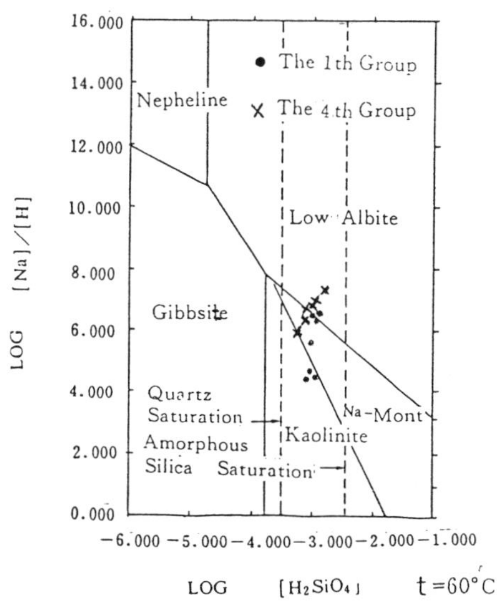

Figure 3 - Activity diagram

chemically aggressive nature of the water caused by dissolved CO_2 at 60°C. The chemical composition of the thermal spring water expressed in terms of Na, H and H_4SiO_4 is plotted in the kaolinite, Na-montmorillonite or low-albite stability field of a theoretical activity diagram (Helgeson et al. 1969) in Fig. 3, respectively. It is possible from a thermodynamic viewpoint that albite is incongruently dissolved early and kaolinite, H_4SiO_4 and Na are produced. As the dissolution continues, H_4SiO_4 and pH increase further and the water composition moves through the kaolinite field into the Na-montmorillonite stability field or moves directly toward low-albite. When the composition evolves to the boundary of the low-albite field, equilibrium with respect to the albite is attained, such as chemical reaction formula (1) (2):

$$\text{albite(s)} + CO_2(g) + 5\frac{1}{2}H_2O = Na^+ + HCO_3^- + 2H_4SiO_4 + \frac{1}{2}\text{Kaolinite(s)} \qquad k = 10^{-5.63} \quad (1)$$

$$3\frac{1}{2}\text{Kaolinite} + 4H_4SiO_4 + Na^+ \rightarrow Na_{0.33}Al_{2.33}Si_{3.67}O_{10}(OH)_2 \ \text{(Na-montmorillorite)} + H^+ + 11\frac{1}{2}H_2O \quad (2)$$

More Ca is characteristic of the thermal spring water in the second group than the first group. It results from the dissolution of plagioclase feldspar in the water at 100°C with abundant CO_2. The water composition containing H_4, SiO_4, Ca, Na, and H plots in the leonhardite field of the theoretical activity diagram (Helgeson et al. 1969). The leonhardite is in equilibrium or near equilibrium with the water composition (Fig. 4), following chemical reaction formula (3).

$$CaAl_2Si_2O_8 + 2H^+ + H_2O \rightleftharpoons Al_2Si_2O_5(OH) + Ca^{2+} \quad \Delta Gr < O$$

$$CO_2(g) + H_2O \rightleftharpoons H^+ + HCO_3$$

$$SiO_2(quartz) + H_2O \rightleftharpoons H_4SiO_4 \qquad pH 7.3\sim8.0$$

$$2Ca^{2+} + 4H_4SiO_4 + 2Al_2Si_2O_5(OH)_4$$

$$\rightleftharpoons Ca_2Al_4Si_8O_{24}7H_2O + 4H^+ + 3H_2O \quad (3)$$

In the Na-HCO_3-type of thermal spring water of the third group most of the wall rocks are gneiss, quartzite and some sedimentary rock. The thermal spring water contains high Na, K and Ca contents. Dissolution of biotite possibly results in increase of K and Cl contents. Moving

Table 1. Chemical composition of Na-HCO_3-type of geothermal water

Geothermal Area	The 1st group		The 2nd group		The 3rd group		The 4th group	
	Foot-rinsing Pond	Xiaodian Bath	Beidong Bath	Bali Big Bath	Sujiang Sp.	Black Stone Strean Sp.	Mayuwo Bath	Shang-beng-nai sp.
Na	43.8	42.0	100.0	112.0	420.0	290.0	42.0	44.0
K	0.9	1.0	2.0	4.0	42.0	21.0	<1	1.0
Ca	3.29	3.43	7.86	19.20	24.3	19.9	1.72	1.72
Mg	0.05	0.06	0.21	1.18	4.02	3.04	0.0	<0.01
HCO_3	56.0	78.0	209.0	264.0	1087.0	719.0	69.0	63.0
CO_3	0.0	0.0	0.0	0.0	0.0	0.0	37.0	37.0
SO_4	13.2	8.0	11.0	50.0	<1	21.00	2.0	4.0
Cl	1.39	2.00	7.0	39.9	82.0	59.0	10.0	13.0
SiO_2	64.3	50.0	75.0	50.0	45.0	115.0	150.0	46.0
pH	9.2	8.3	8.0	7.5	6.8	7.1	9.3	9.1
*P_{CO2}	$10^{-6.88}$	$10^{-6.30}$	$10^{-6.0}$	$10^{-4.45}$	$10^{3.15}$	$10^{-3.67}$	$10^{-7.00}$	$10^{-6.00}$

*P_{CO2} calculated value

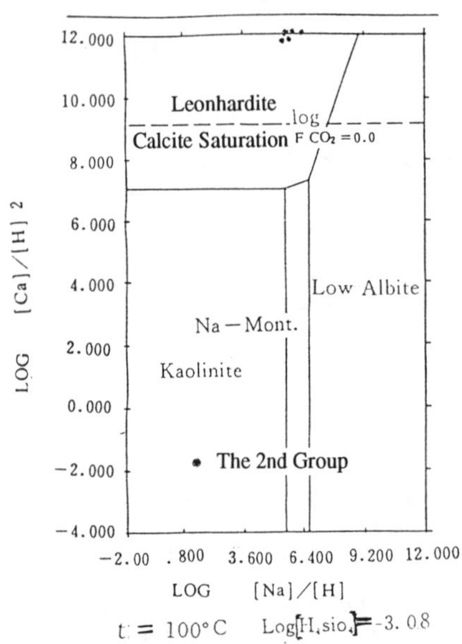

Figure - 4 Activity diagram

potassium ion from rocks expresses higher hydrothermal alteration temperature for these thermal springs which probably have higher geothermal reservoir temperatures.

All mentioned above show that the lixiviation of meteoric water is important for the formation of the Na-HCO_3-type of water in Tengchong county. The Na-HCO_3- type of water in the Hengduan Mountains is similar to that of Tengchong county in genesis which is lixiviation water without a magnetic heat source.

5 CONCLUSIONS

It is possible that the chloride water discharged from Hot Sea Field and Langpu Hot Pool is affected by the heat sources from a magmatic pocket; this kind of water has greater potential for geothermal energy. Most of the hydrothermal systems issuing bicarbonate water are lixiviation water without a magmatic heat source; their heat content comes from the movement of meteoric water through fault systems. These thermal spring waters containing higher Na, K, and Ca contents obtain their heat by passing through fault systems to circulation depths of several kilometers. They probably have higher geothermal reservoir temperatures.

6 ACKNOWLEDGEMENTS

This work was financially supported by the National Natural Science Foundation of China (NSFC). I especially appreciate the long time cooperation with the colleagues (Tong Wei, Liao Zhijie, Liu Shibin, Shen Minzi, Zhu Meixiang et al.) of our group for Hengduan Mountain's geothermal investigations.

REFERENCES

Tong, Wei, Zhang Mingtao (eds.) 1989. Geothermal in Hengduan Mountain's. Science Press, Beijing (In Chinese, in press).

Tong, Wei, Zhang Mingtao (eds.) 1989. Geothermal in Tengchong. Science Press, Beijing (In Chinese with English introduction and comprehensive summary).

Liao, Zhijie, Tong, Wei, Liu Shibin, & Zhao Fengsan 1986. High-temperature hydrothermal systems in west Yunnan province, China. Geothermics 15: 627-631.

Shen, Minzi, Liao, Zhijie 1990. Geology, geochemistry and isotopic composition of high-temperature geothermal systems in west Yunnan, China. Geothermal Resources Council Transactions 14: 975-977.

Guo, Guoying, Yin Chungu 1990. The chemical model of thermal spring water of HCO_3-Na type, Tengchong. Acta Scientiarun Naturalium Universi tatis Pekinensis 26: 700-710 (In Chinese with abstract).

Giggenbach, W.F. 1986. Graphical techniques for the evaluation of water/rock equilibration conditions by use of Na, K, Mg, and Ca-contents of discharge waters. Proc. of the New Zealand Geothermal Workshop: 37-43.

Helgeson, H.C., T.H.Brown, & R.H. Leeper 1969. Handbook of theoretical activity diagrams depicting chemical equilibria in geologic systems involving an aqueous phase at one atm and o° to 300°C, 1-115. Freeman, Cooper & Company, California 94133.

Water-Rock Interaction, Kharaka & Maest (eds) © 1992 Balkema, Rotterdam. ISBN 90 5410 075 3

Composition of gases from the Norris-Mammoth corridor, Yellowstone National Park, USA: Evidence for a magmatic source near Mammoth Hot Springs

Y.K. Kharaka, R.H. Mariner & W.C. Evans
US Geological Survey, Menlo Park, Calif., USA

B.M. Kennedy
Department of Physics, University of California, Berkeley, Calif., USA

ABSTRACT: Mammoth Hot Springs has one of the highest thermal water and gas discharges of the thermal areas outside the 0.6 Ma Yellowstone caldera. Thermal waters with surface temperatures of up to 73°C and calculated subsurface temperatures of about 100°C issue from nearly 100 hot springs scattered over a score of steplike travertine terraces that range in age from about 0.4 Ma to recent. Hydrologic and tracer tests conducted in 1989-1991 indicate that a very large flow, about 590 L/s, of thermal water from the Mammoth system enters the Gardner River. The heat flux from this system is comparable to that from the Norris Geyser Basin.

The isotopic and chemical compositions of thermal waters and solutes from the Norris-Mammoth corridor can be interpreted to indicate a common magmatic source for heat and volatile solutes located near Norris for the entire corridor. However, the chemical and isotopic compositions of gases in general, and the distribution of $^3He/^4He$ ratios in particular, provide strong evidence for a separate magmatic source for the Mammoth system. The highest $^3He/^4He$ ratios that could indicate proximity to a magmatic source in the corridor are obtained from Mammoth Hot Springs ($R/Ra \leq 8.4$) and Norris Geyser Basin ($R/Ra \leq 9.0$). A Norris to Mammoth flow path for the hydrothermal fluids at Mammoth, however, is likely precluded by the fact that fluids from all the six major thermal areas between them have R/Ra values that are appreciably lower than those from Mammoth Hot Springs.

1 INTRODUCTION

The Yellowstone Plateau volcanic field is one of the largest Quaternary silicic centers in the world. During the last 2.2 Ma an eruptive volume of about 6,000 km^3 of ash-flow tuffs and rhyolitic lavas have covered an area of about 17,000 km^2 (Christiansen, 1984: Hildreth et. al., 1991). A thermal mantle plume hypothesis is currently the preferred model for the origin of the eastern Snake River Plain-Yellowstone province (Morgan, 1973; Pierce and Morgan, 1990). In this model a hotspot, now centered beneath Yellowstone, was initiated 16 Ma. The track of this hotspot is defined by centers of caldera-forming volcanism that have migrated 700 km northeastward to the Park.

Mammoth Hot Springs, located in the northwest sector of Yellowstone National Park, is the world's most spectacular thermal area of active travertine precipitation. Massive amounts of travertine are being precipitated over a large area from nearly 100 hot springs scattered over a score of steplike terraces. The surface temperatures of the springs (≤73°C) are lower than those of boiling hot springs encountered at Norris and the other geyser basins in the Park. However, the total thermal-water discharge from the Mammoth system, calculated by flow measurements and extensive application of chemical tracers, is very large (590 ± 30 L/s), and the heat flux is comparable to that from the Norris Geyser Basin (Kharaka et al., 1990; Sorey et al., 1991). The gas flux from the Mammoth system is also very large and gas pressures >5 bar were calculated for samples from the Y-10 well located on the terraces at Mammoth (Kharaka et al., 1991).

The chemical and isotopic compositions of water and solutes, including gases, in Mammoth Hot Springs show a major component of thermal water that derives its heat and some solutes (e.g. Cl, B, ^3He and CO_2) from a magmatic source (Kharaka et al., 1991; Mariner et al., this volume). White et al. (1988) postulated that the magmatic source is located beneath Roaring Mountain and that this also is the source of heat and solutes for the Norris system (Fig. 1).

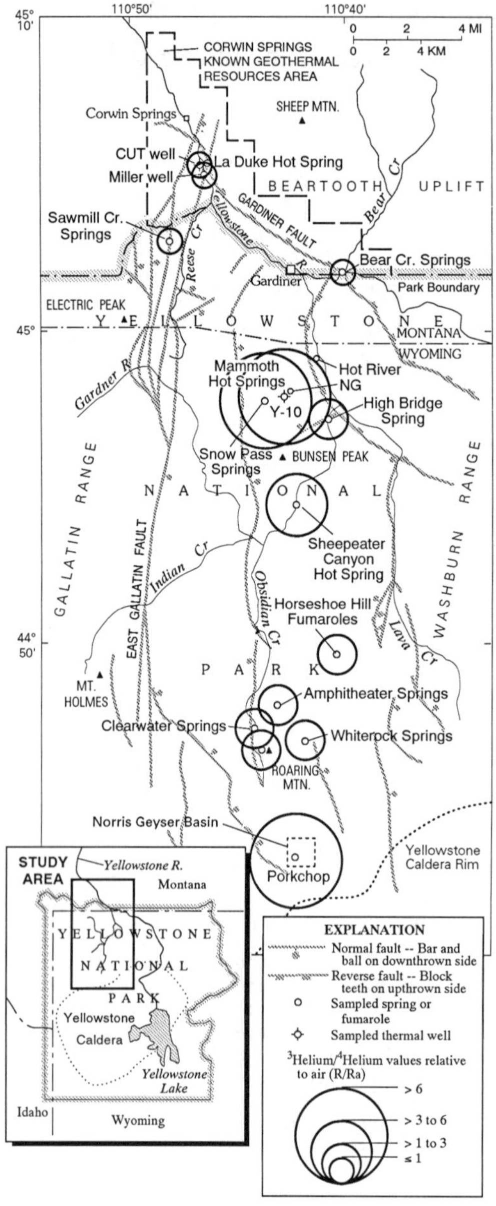

Figure 1. Location map showing the thermal areas sampled and the R/Ra values obtained.

The lower concentrations of Cl and other conservative solutes in the Mammoth water can be explained by mixing about 30-40% Norris-type water with 70-60% low-salinity meteoric water, the mixing taking place as the Norris-type water flows northward to Mammoth. Kharaka et al. (1990; 1991) presented an alternative model and postulated that about half of the volatiles, including He and CO_2 together with the heat for the Mammoth system are provided by a cooling magmatic body emplaced about 400 ka near the Mammoth area.

In this summary we will examine the chemical and isotopic data used by Kharaka et al. (1991). The chemical and isotopic compositions of water and gases from three additional thermal areas in the Norris-Mammoth corridor were sampled and analyzed for this study. The new results support the theory of a separate magmatic source for the Mammoth system.

2 CHEMICAL AND ISOTOPIC COMPOSITIONS OF GASES

The chemical and isotopic compositions of gases in general, and $^3He/^4He$ ratios in particular are the best geochemical indicators of mantle components (Clarke et al., 1969; Marty and Jambon, 1987). In Yellowstone National Park, an extensive data base on these gases has been obtained (Craig et al., 1978; Kennedy et al., 1985; Hearn et al., 1990; Kharaka et al., 1991). The data show that the chemical and isotopic composition of gases in the Park basins are extensively affected by repeated boiling and mixing with local meteoric water. The highest $^3He/^4He$ values from each basin or thermal area, however, reflect the efficiency with which magmatic volatiles are extracted and transported to the sampling site (Kennedy et al., 1985). The highest $^3He/^4He$ ratios obtained in thermal waters inside the 0.6 Ma Yellowstone caldera and at the Norris Geyser Basin give R/Ra values that range from 8 to 16 and clearly indicate the release of magmatic volatile components of "deep" mantle origin.

The chemical and isotopic compositions of gases from selected thermal springs, fumaroles and wells in the Norris-Mammoth corridor and adjacent areas are shown in Table 1; the sampling sites are shown in Fig. 1. In addition to the Y-10 well, five thermal springs in the Mammoth area were sampled together with at least one thermal site from all the major thermal areas in the corridor.

In the Norris-Mammoth corridor, CO_2 is the dominant gas, comprising 96% to more than 99% of H_2O-free total, with H_2S and N_2 being locally significant components,

Table 1. Chemical (mole %, H_2O-free basis) and isotopic compositions of gas phases from selected sites in the Norris-Mammoth corridor.

Site	Roaring Mountain	White Rock	Amphitheater	Horseshoe Hill	Sheepeater Canyon	Snow Pass	Y-10 (well)	High Bridge	Sawmill Creek
Sample#	89YNP-107	91YNP-28	91YNP-57	91YNP-30	89YNP-118	89YNP-253	89YNP-102	90YNP-503	90YNP-119
He	0.0094	0.0022	0.0033	0.0045	0.0002	0.0059	0.003	0.0053	0.0098
H_2	0.0787	0.550	0.103	0.0053	0.0001	<0.0002	0.004	0.0005	<0.0002
Ar	0.006	0.0106	0.0106	0.0051	0.0304	0.017	0.013	0.0169	0.966
O_2	<0.001	0.0865	0.104	0.0518	0.0024	0.04	0.003	0.0206	3.26
N_2	0.324	0.549	0.416	0.309	0.93	0.764	0.836	0.857	67.8
CH_4	0.0287	0.108	0.0692	0.0650	0.002	0.0462	0.026	0.0035	27.7
CO_2	96.0	96.6	96.0	99.6	99.5	98.6	99.1	99.1	0.0609
C_2H_6	<0.0002	0.0026	0.0011	0.0039	<0.0002	0.0008	----	<0.0002	0.0044
H_2S	1.26	2.24	3.08	1.22	0.209	0.0233	----	<0.0002	0.0113
Total	98.4	100.14	99.75	101.24	100.7	99.5	100.0	100.2	99.8
$\delta^{13}C(CO_2)$	-3.9	-3.2	-3.5	-3.5	-4.4	-3.4	-5.2	-5.1	----
R/Ra	2.2	3.0	2.0	2.8	5.8	7.5	8.3	1.2	0.05

especially in samples from fumaroles (Table 1). In the Mammoth area fluids, there is more than 50 m M/kg of excess CO_2 relative to Ca and Mg, indicating a source not related to dissolution of carbonate minerals. The molar total $C/^3He$ (5×10^9) and $\delta^{13}C$ values of CO_2 (Table 1) are compatible with the conclusion of a mantle origin for the bulk of the CO_2 (Des Marias, 1989; Kharaka et al., 1991). The composition of gas from Bear Creek-1 is generally similar to that of the Mammoth samples, and the pressure of CO_2 gas is moderately high at about 1.0 bar (Kharaka et al., 1991). The total gas content and the proportion of CO_2 (pressure of CO_2 gas is about 0.2 bar) are lower in samples from the La Duke Spring area of the Corwin Springs KGRA. The samples from Chico Hot Spring and Sawmill Creek Spring (Table 1) have N_2 concentrations expected in local ground water in equilibrium with air; the sample from Sawmill Creek, however, has a relatively high concentration of CH_4.

With the exception of He and in some cases ^{40}Ar, the concentrations of noble gases in the recharge and thermal waters of the study area are those expected from equilibrium with air (Mazor and Fournier, 1973; Kharaka et al., 1991). He concentrations in excess of air-saturated water reflect the addition of both magmatic and radiogenic He. The magnitude of excess He due to radiogenic sources will generally increase with increasing age of the water. The effect is most pronounced with He due to the paucity of He in air (~5.2 ppm) and air saturated water (0.002 μm/kg). The amount of radiogenic ^{40}Ar added, however, is minor relative to the atmospheric component in recharge water. The total Ar concentra- tion in thermal water, therefore, is expected to remain relatively constant at about 15-16 μm/kg; approximately the value obtained for water from the CUT and Miller wells, reflecting the dominance of the atmospheric component.

The calculated concentrations of He in the thermal waters of the Mammoth system are high, ranging from 4 to 10 μm/kg for four samples (two samples were obtained using downhole samplers) from the Y-10 well (Kharaka et al., 1991). The lower He value is calculated assuming gas loss by bulk stripping; the higher value is based on loss by Rayleigh-type distillation. The measured concentrations of He in the thermal waters from the La Duke Hot Springs area are also high, ranging from 9.8 to 11.3 μm/kg. Extensive degassing has affected the gas concentrations from Bear Creek-1; the calculated He concentrations assuming bulk stripping range from 0.3 to 1.0 μm/kg.

Noble gases in thermal fluids from the Norris-Mammoth corridor and the Yellowstone basins have characteristic isotopic compositions. Kennedy et al. (1985) showed that all the noble gases from the Yellowstone Park can be explained by mixing of three distinct components. The first component is of magmatic (mantle) origin with $^3He/^4He$ ~16 times the air value (R/Ra ~16). The second component is of radiogenic origin (crustal) with a theoretical R/Ra <0.05. The third component has the composition of air and originates from percolating meteoric watere saturated with atmospheric gases (R/Ra = 1). They also showed that all the Yellowstone gases plot in a relatively narrow planar field on a three dimensional He-Ar correlation diagram (axes being $^3He/^{40}Ar$, $^4He/^{40}Ar$, and $^{36}Ar/^{40}Ar$). The samples from the Mammoth area measured for this study as well as the sample from Bear Creek-1 also plot in or near this field. The samples from CUT and Miller wells, however, plot significantly outside the field for Yellowstone waters (Kharaka et al., 1991).

Results from this investigation show that Mammoth Hot Springs is an area of high gas discharge with up to 50% of He, CO_2, and possibly other gases being of mantle origin. The samples from the Y-10 well and three springs in its vicinity (Table 1) yield R/Ra values of 8.1 to 8.4. The $^3He/^4He$ ratios relative to air are still very high (R/Ra = 7.5) in the sample from Snow Pass located (Fig. 1) about 2 km west of Y-10, but the sample from High Bridge (~3 km S.E. of Y-10) has an R/Ra value of only 1.2. Gases from Sheepeater Springs yield a relatively high R/Ra value of 5.8, and because the water has many chemical and isotopic characteristics of Mammoth-type water, it is possible that the Mammoth hydrothermal system may extend as far south as Sheepeater canyon. Samples from CUT and Miller wells and Bear Creek-1 have R/Ra values of 0.13, 0.11, and 0.29, respectively). Based on a Yellowstone mantle value of R/Ra = 16 and local crustal values from 0.05 (Sawmill Creek) to 0.24 (Chico Hot Spring), it is unlikely that more than 2% of He from Bear Creek-1 and 1% from the CUT and Miller wells could be of mantle origin.

The $^3He/^4He$ ratios in fluids from Norris Geyser Basin yield R/Ra values that vary from about 2 to 9, with the high values being comparable to those at Mammoth (Welhan, 1981; Kennedy and others, 1985; Kharaka et al., 1991). The ratios decrease northward, to values of 2 to 3 (Table 1) in fluids from Roaring Mountain, White Rock Springs, Amphitheater Springs, Clearwater Springs, and Horseshoe Hill, before increasing backup to a value of 5.8 in fluids from Sheepeater Springs (Fig. 1). Thus, all the major thermal areas between Norris and Mammoth have R/Ra values that are lower than Mammoth. The overall pattern in He isotopic variations along the Norris-Mammoth corridor strongly favors a source of magmatic volatiles and heat for the Mammoth system which is independent of the main Yellowstone system defined by the fluids within the caldera and at Norris.

It is possible, but unlikely that the hydrothermal water in the Mammoth system could acquire its heat (temperatures up to 360°C) and solutes at or near the Norris Geyser Basin then flow north to Mammoth while bypassing all the major thermal areas in the corridor. The volume of the required hydrothermal water (~200 L/s) is too high to move in the corridor without leaving any He isotopic signature.

3 LOCATION OF A MAGMATIC SOURCE FOR THE MAMMOTH SYSTEM

The chemical and isotopic compositions of thermal waters from Mammoth and the adjoining areas indicate that Mammoth-type waters are issuing from springs and seeps located mainly within an area delineated by Sheepeater Canyon, Snow Pass, Mammoth Terraces, Hot River and High Bridge (Fig. 1); a small component of Mammoth-type water possibly is present in waters from Bear Creek Springs (Kharaka et al., 1991). Extensive travertine deposits occur in this area and indicate five periods of intermittent travertine precipitation from about 400 ka to present (Pierce et al., 1991; Kharaka et al., 1991). The $^{87}Sr/^{86}Sr$ ratios and ϵ_{Nd} values of most travertines including those from the Gardiner terraces are identical, indicating precipitation from thermal fluids derived from the same hydrothermal system (Kharaka et al., 1991); massive travertines at Bear Creek Springs (0 to 15 ka) and on Rattlesnake butte (50 to 60 ka) have different isotopic signature, and possibly different origins. The extensive travertine deposits with model U/Th ages that generally do not overlap from one location to another indicate that the Mammoth hydrothermal system has discharged fluids at different locations intermittently for about 400 ka, with the periods of activity possibly related to the thickness of glacial ice in the region (Pierce et al., 1991).

Geophysical surveys conducted by Stanley et al. (1991) in the Norris-Mammoth corridor are interpreted to indicate partial melt conditions (temperatures 500-600°C) at depths below about 6 km south of Bunsen Peak. Low resistivities and partial melt conditions may exist also at Mammoth, but at depths greater than about 10 km. Two extensive zones of low resistivities have tentatively been mapped (Stanley, personal communication, 1992) at relatively shallow depths in an area located between Mammoth and Claggett Butte (immediately north of Snow Pass) and also at the Reese Creek Fault (Fig. 1). These zones are interpreted to indicate intensive alteration possibly resulting from thermal water-rock interactions.

Additional evidence for the existence of magma in the Norris-Mammoth corridor comes from the fact that there are 13 rhyolite vents with ages ranging from 400 to 80 ka in the corridor. Detailed isotopic and chemical data show the extra-caldera rhyolites to be independent and from a different magmatic source/sources than the main caldera rhyolites (Hildreth et al., 1984; 1991). The corridor also contains many late Pleisticene basalts, at least two of which mixed syneruptively with rhyolitic magma. On the other hand, Hildreth et al. (1991) present strong evidence for the absence of an extensive magma chamber in the corridor and for the episodic generation of rhyolitic and basaltic magmas from partial crustal melts.

The geological, geophysical and geochemical data discussed in this paper cannot be used to conclusively prove that the Mammoth hydrothermal fluids do not flow from Roaring Mountain or the caldera as postulated by White et al. (1988) and Fournier (1989). The simplest interpretation of all the data, however, strongly favors an origin resulting from the heating of meteoric water recharged in the Gallatin Range to the west and northwest of Mammoth by a magmatic body in the Mammoth area, possibly at a depth of about 10 km. In this model, the magma was initially introduced into this area at least 400 ka, but episodic additions to this magma must have occurred to allow for the intermittent hydrothermal discharges and travertine deposition from 400 ka to present. Episodic thermal water discharges and travertine precipitation may reflect glacial influences (Pierce et al., 1991), or it may reflect a combination of glacial influences and magmatic intrusions.

REFERENCES

Christiansen, R.L., 1984. Yellowstone magmatic evolution: Its bearing on understanding large-volume explosive volcanism. In Explosive volcanism: Inception, evolution and hazards. Nat'l. Acad. of Sci., p. 84-95.

Craig, H., J.E. Lupton, J.A. Welhan & R. Poreda 1978. Helium isotope ratios in Yellowstone and Lassen Park volcanic gases: Geophys. Res. Letters: 5: 897-900.

Des Marais, D.J., 1989. Thermal CO_2 outgassing: Source signatures, sizes and significance: EOS, Trans. Am. Geophys. Union (abs): v. 43, p. 1377.

Fournier, R.O., 1989. Geochemistry and dynamics of the Yellowstone National Park hydrothermal system, Wyoming: An. Rev. Earth Planet. Sci. 17: 13-53.

Hearn, E.H., B.M. Kennedy & A.H. Tuesdell 1990. Coupled variations in helium isotopes and fluid chemistry: Shoshone Geyser Basin, Yellowstone National Park: Geochim. Cosmochim. Acta 54: 3103-3113.

Hildreth, W., A.N. Halliday & R.L. Christiansen 1991. Isotopic and chemical evidence concerning the genesis and contamination of basaltic and rhyolitic magma beneath the Yellowstone Plateau volcanic field: J. Petrology 32: 63-138.

Kennedy, B.M., M.A. Lynch, J.H. Reynolds & S.P. Smith 1985. Intensive sampling of noble gases in fluids at Yellowstone: I. Early overview of data; regional patterns: Geochim. Cosmochim. Acta 49: 1251-1261.

Kharaka, Y.K., R.H. Mariner, G. Ambats, W.C. Evans, L.D. White, T.D. Bullen & B.M. Kennedy 1990. Origin of water and solutes in and north of the Norris-Mammoth corridor, Yellowstone National Park. Trans. Geoth. Res. Counc. 14: 705-714.

Kharaka, Y.K., R.H. Mariner, T.D. Bullen, B.M. Kennedy & N.C. Sturchio 1991. Geochemical investigations of hydraulic connections between the Corwin Springs Known Geothermal Resources Area and adjacent parts of Yellowstone National Park. In M.L. Sorey (ed.) U.S.G.S. Water Res. Invest. Rept. 91-4052, F1-F38.

Marty, B., & A. Jambon 1987. C/^{3}He in volatile fluxes from the solid Earth: implications for carbon geodynamics. Geochim. Cosmochim. Acta 83: 16-26.

Mazor, E., & R.O. Fournier 1973. More on noble gases in Yellowstone National Park hot waters: Geochim. Cosmochim. Acta 37, 515-525.

Pierce, K.L., & L.A. Morgan 1990. The track of the Yellowstone hotspot: Volcanism, faulting, and uplift. U.S.G.S. Open-File Rept. 90-415, 50 p.

Pierce, K. L., K.D. Adams & N.C. Sturchio 1991. Geologic setting of the Corwin Springs Known Geothermal Resources Area Mammoth Hot Springs area in and adjacent to Yellowstone National Park, in M.L. Sorey (ed.) U.S.G.S. Water Res. Inves. Rept. 91-4052: C1-C37.

Sorey, M.M. (ed.) 1991. Effects of potential geothermal development in the Corwin Springs Known Geothermal Resources Area, Montana, on the thermal features of Yellowstone National Park. U.S.G.S. Water-Res. Invest. Rept. 91-4052, 148 p.

Stanley, W.D., D.B. Hoover, M.M. Sorey, B.D. Rodriques & W.D. Heran 1991. Electrical geophysical investigations in the Norris-Mammoth corridor, Yellowstone National Park, and adjacent Corwin Springs Known Geothermal Resources Area, in M.L. Sorey (ed.) U.S.G.S. Water Res. Inves. Rept. 91-4052: D1-D18.

Welhan, J.A., 1981. Carbon and hydrogen gases in hydrothermal systems: the search for a mantle source: Ph.D dissertation, University of California, San Diego, California, 194 p.

White, D.E., R. Hutchinson & T.E.C. Keith 1988. The geology and remarkable thermal activity of Norris Geyser Basin, Yellowstone National Park, Wyoming. U.S.G.S. Prof. Paper 1456, 84 p.

Water-Rock Interaction, Kharaka & Maest (eds) © 1992 Balkema, Rotterdam. ISBN 90 5410 075 3

The origin of thermal waters in crystalline rocks of the North China fault block

Li Guiru, Wang Jihua, Zhang Peiren & Jiang Fengliang
Institute of Geology, State Seismological Bureau, Beijing, People's Republic of China

ABSTRACT: The geochemical evolution of thermal waters in the Archaen gneiss and Mesozoic granite along the margin of North China fault block may be controlled by the hydrolysis of silicate minerals and the reaction of host rocks with circulating meteoric fluids in the fractured system.

1 INTRODUCTION

In the last decade great progress has been achieved in studying the origin of saline waters and brines in crystalline rocks. More attention has been paid to the importance of water-rock interaction in geochemical evolution of water chemistry (Frape et al. 1984; Edmunds et al. 1987; Fritz et al. 1990).

In China, hydrogeological and geochemical studies on sedimentary basins and oil fields have been carried out extensively (Li et al. 1988; Fan et al. 1988). Yet hydrogeochemical investigations of thermal waters and deep fluids in crystalline rocks are just at the beginning.

This paper evaluates hydrogeochemical data from the viewpoint of water-rock interaction as the process leading to the origin and geochemical features of thermal waters in crystalline rocks of the North China fault block.

2 GEOLOGY AND HYDROGEOLOGY

The investigated areas consist of the Hebei (HB) and Yishu rift zones (YS). HB is located about 120 km northwest of Beijing, and YS is situated in the Shandong province, about 700 km south of Beijing, as shown in Fig. 1.

Tectonically these two areas belong to the marginal zones of the North China fault block and have rather identical geological histories. The crystalline rocks in the study areas consist mainly of Archaen biotite-plagioclase gneiss and Mesozoic granite and rhyolite (in HB), and andesite and granite porphyry (in YS). The spatial

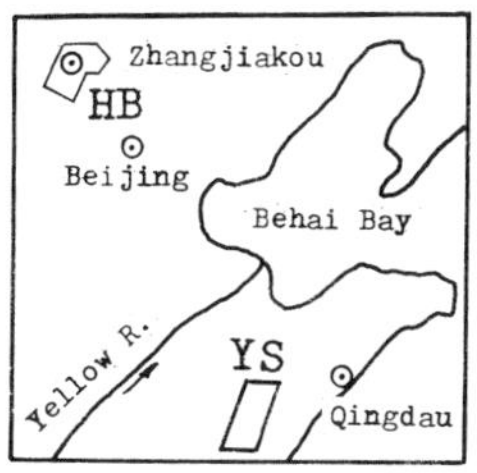

Figure 1 - The location of studied areas

distribution of these rocks is dominantly controlled by deep E-W mantle faults of Archaen age and the NNE fault systems of neotectonism. Plagioclase, hornblende, biotite and quartz are major rock-forming minerals with some accessory minerals, such as apatite and pegmatite. Chlorite, kaolinite, calcite and quartz are seen as secondary fillings of fissures and fractures. The HB igneous rocks are rich in multimetalliferous ore deposits; only a copper mine in Mz porphyry has been explored in YS area.

The thermal waters and hot springs associated with crystalline rocks dominantly occur along tectonic belts trending EW and NNE with temperatures of 35.5-87°C, pH 7.0-9.0 and TDS 0.5-6.9 g/l. Tritium contents for some surface and shallow groundwater range from 25 to 80 TU, and from 1-5 TU for thermal waters. According to Na/K geothermometer, the reservoir temperature is in the range of 95-100°C, giving a depth of circulation of about 2 km (the thermal gradient is 25°C/km). So_4-Na and Cl-Na type thermal water predominate in HB and YS, respectively. Nitrogen (N_2) usually is the major component of the dissolved gases with

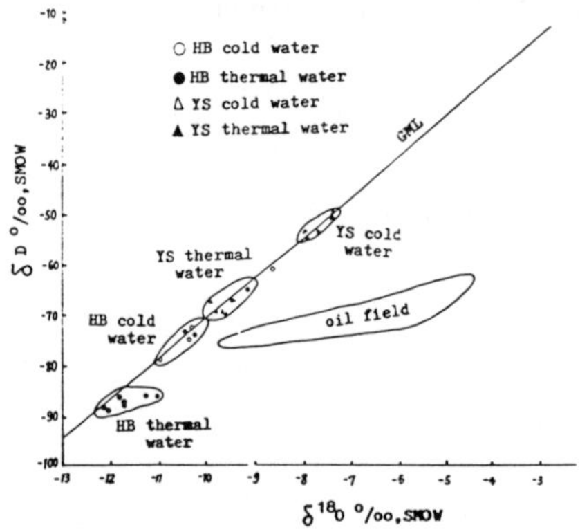

Figure 2 - Oxygen-18 and deuterium contents in gold water and thermal water from the crystalline rocks of N. China Fault Block.

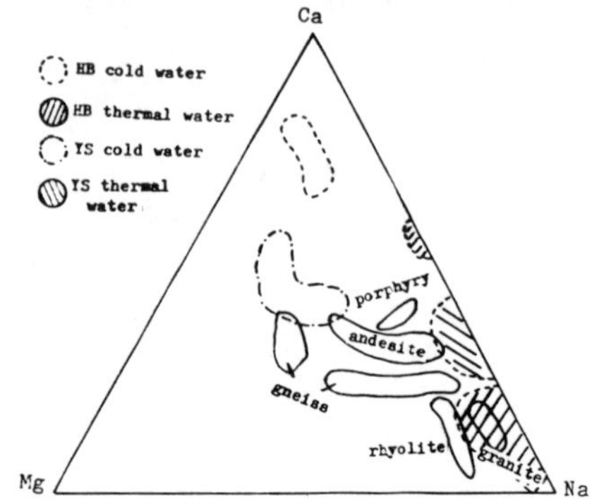

Figure 3 - Mole percentage of Ca, Na, and Mg concentrations in cold and thermal waters of crystalline rocks of N. China Block.

N_2/Ar = 50-55, favoring derivation of Ar from the radioactive decay of K of host rocks. H_2 and He contents can reach thousands of ppm.

3 RESULTS AND DISCUSSIONS

The analyses of light stable isotopes of waters, D/H and $^{18}O/^{16}O$, provide the most diagnostic tool for determining the origin of thermal waters. All the waters of HB and YS in Fig. 2 fall on or close to the global meteoric line (GML), suggesting that oceanic and oil field waters can be ruled out from the origin of these thermal waters. YS waters, both cold and hot, are richer in D and ^{18}O compared to those of HB waters. This results from the latitude effect and local geographical conditions. The insignificant oxygen shift may be due to either a large water-rock ratio or short contact time with rocks. Discharge of deep boreholes and hot springs is usually as low as 3-10 l/s.

3.1 Major ionic compositions of waters

To provide a plausible explanation for the major ionic compositions of waters related to mineralogy of host rocks, the mole percentage of Ca, Na and Mg concentrations are plotted in Fig. 3, with host rocks for comparison. This figure illustrates that thermal waters of HB and YS are remarkably similar in chemistry to their host rocks. HB thermal waters fall in granite and rhyolite areas, whereas YS waters are close to andesite and porphyry, suggesting that products of hydrolysis of plagioclase, biotite and other silicate minerals appear to be the main source of these constituents. This is in good agreement with the parallel increases in Ca^{2+}, Na^+, K^+ and Cl^-, while Mg decreases due to chloritization.

In the vicinity of multimetalliferous sulfide ore deposition the HB thermal waters are characterized by high sulfate contents as a result of oxidation of sulfide minerals at the ground surface. The produced H^+ ions can be utilized for subsequent acid hydrolysis, and SO^{2-}_4 (sulfuric acid) is neutralized by carbonate waters to form SO_4-Na waters. The interlayer exchange of OH^- ions may provide an additional source of Cl^- ions in the thermal waters (Edmunds et al. 1987).

3.2 Minor elements Li, B, F and Sr

The relationship of minor elements to chloride can provide a valuable index for geochemical evolution of thermal waters, as these elements are usually not incorporated in secondary minerals (Frape et al. 1984). Fig. 4 a, b, c and d show the relationships of Li^+, B, F^- and Sr^{2+} with cl^- in waters. The distribution patterns of these elements with Cl^- are similar, indicating that their concentrations in water may be regulated by water-rock interaction, which results in the dissolution of plagioclase, biotite and apatite. The reaction trends on the figures resemble HDR experimental results (see Edmunds et al. 1987), suggesting quick equilibration and similar processes occurring in the circulating system. In addition, dissolution of celestite in

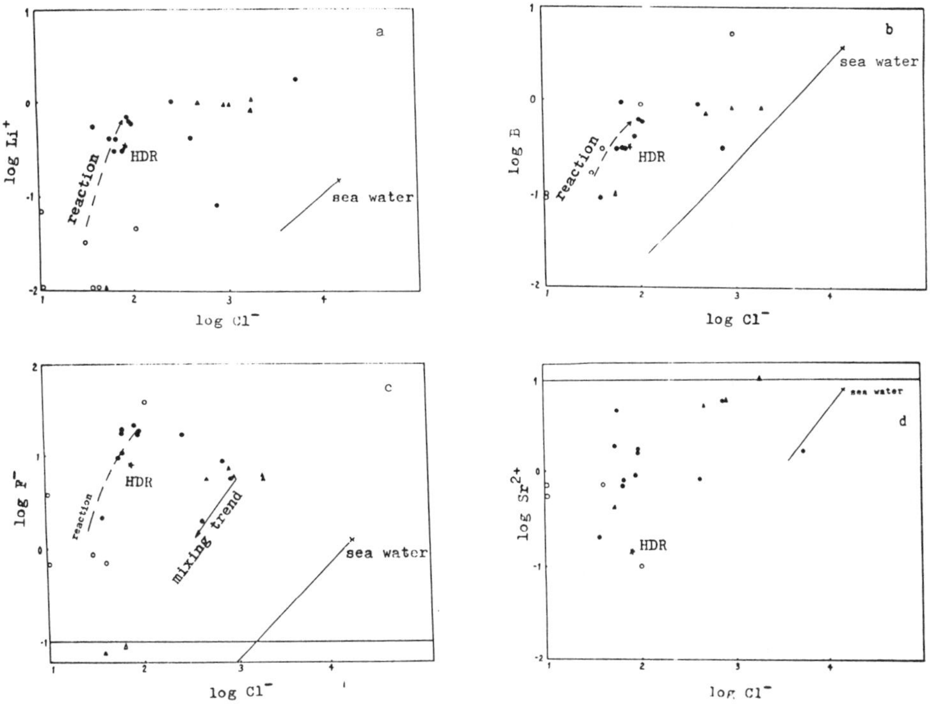

Figure 4 - Minor element relationships with chloride: a. lithium, b. boron, c. fluoride, d. strontium. The symbols in the Figures are the same as those in Fig. 2.

marine carbonate rocks by through-going water in the contact zone with crystalline rocks may offer another source of Sr^{2+} to thermal waters, as carbonate rocks of Proterozoic and Paleozoic ages are common in the investigated areas. The Sr^{2+} concentration increase with increasing SO^{2-}_4 supports this assumption (see Fig. 5).

It is seen in Fig. 4c that the relation of F^- and Cl^- shows a reaction trend similar to the HDR, which suggests that the under static conditions fluorite and/or topaz dissolutions, rather than biotite alteration, controls the concentration of F^- in the water (Savage et al. 1985).

The enrichment of thermal waters in minor elements compared to present sea water eliminates the possibility of marine contribution, but suggests that water-rock interaction occurred during geochemical evolution with relatively slow rates and short time frames, accounting for the general low mineralization of thermal waters. This may be attributed to the neotectonic structures and the active circulation condition.

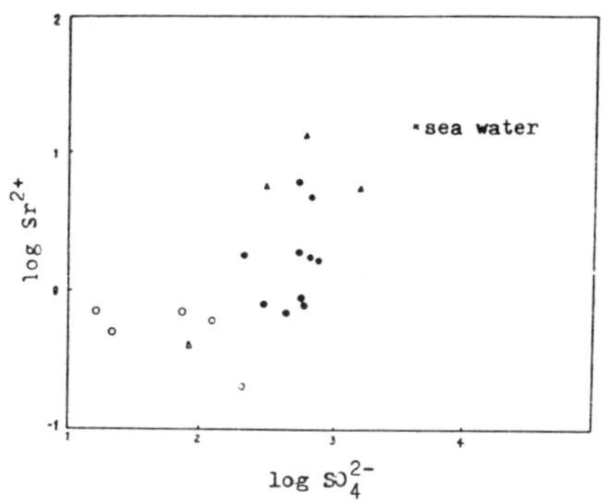

Figure 5 - The relationship between Sr^{2+} and So_4 of HB and YS waters.

4 CONCLUSIONS

The isotopic composition of waters and the relationships of major and minor constituents to chloride suggest that the thermal waters of HB and YS areas are mainly meteoric in origin with chemical contribution of primary and secondary minerals from host rocks. The presence of Mg^{2+} can best be explained as the result of chloritization of silicate minerals. The hydrolysis of plagioclase, biotite and apatite is the predominant mechanism controlling the composition of dissolved constituents in the waters, with additional contribution of F^-, Sr^{2+} and SO_4 from dissolution of fluorite and celestite.

The low mineralization and concentrations of minor elements in thermal waters suggest low rates and short term water/rock interaction, probably resulting from the neotectonic structure and active circulating system of the region.

Further study is required to determine the possible relationship between thermal waters and hydrothermal fluids, which may supply multimetalliferous ore deposits with Pb, Zn, Ag, Mo and other metals.

ACKNOWLEDGEMENTS

The project is supported by the Scientific Foundation of State Seismological Bureau and The National Committee of Sciences and Technology of China. Authors thank Mr. Zhang Minggang for his attentive reading of the manuscript and Ms. Tao Jingling for drawing the graphs.

REFERENCES

Edmunds, W.M., R.L.F. Kay, D.L. Mileo & J.M. Cook 1987. The origin of saline groundwaters in the Carnmenellis granite, Cornwall (UK): further evidence from minor and trace elements. Geol. Ass. Canada Special Paper 33: 197 210.

Fan, Zhicheng & Jiyang Wang. 1988. Review on data of oxygen and hydrogen isotope compositions in thermal waters in China. A Technical Report issued by Beijing Research Institute of Uranium Geology. Beijing.

Frape, S.K., P. Fritz & R.H. Mcnutt 1984. Water-rock interaction and chemistry of groundwaters from the Canadian Shield. Geochimica at Cosmochimica Acta 48: 1617-1627.

Fritz, P. & M. Lodemann 1990. Die salinaren Tiefenwasser der KTB-Vorbonrung. GEO WEN 8/9: 241-316.

Guiru, Li, Gao Wenxue, Zhang Peiren, Wang Jihua & Jiang Fengliang 1988. The characteristics of isotopic compositions of hydrogen and oxygen of geothermal waters in North China. Selection of Hydrogeology and Engineering Geology 4 (May 1988): 1-12.

Savage, D., M.R. Cave & A.E. Milodowski 1985. The origin of saline groundwaters in granitic rocks. Proc. 9th International Symposium on Scientific Basis for Nuclear Waste Management, Sept. 1985, Stockholm.

Water-Rock Interaction, Kharaka & Maest (eds) © 1992 Balkema, Rotterdam. ISBN 90 5410 075 3

Genesis of mineral waters of the Rhenish Massif, central Europe

Franz May, Stephan Hoernes & Horst J. Neugebauer
Bonn University, Germany

ABSTRACT: Based on a tectonic investigation, interpretation of hydrochemical data and analysis of altered rocks from mineral water wells, a genetic model for the mineral and thermal waters of the Rhenish Massif will be proposed. Two processes are of fundamental importance: 1. Dissolution of CO_2 ascending from the earth mantle in shallow groundwaters give rise to water rock reactions. 2. Opening of deep reaching fractures enabeling the ascent of basement brines and reactions at elevated temperatures of meteoric water with aquifer rocks.

1. INTRODUCTION

The Rhenish Massif is part of the Variscan Mountain belt of central Europe. Paleozoic greywacke and shale are the most common rock types in this area. In late Tertiary epirogenetic uplift of the Massif begun. Quaternary alkali basaltic volcanism and extensive tectonics indicate the present dynamic activity in this region.

Different mineral and thermal waters occur in numerous springs and wells of the massif and its vicinity. Among them are such famous locations as Selters and Spa.

Although scientists have been discussing various theories on the origin of these mineral waters for more than 300 years (Blondel 1688), many questions remain yet unanswered: Questions concerning the age of waters and the lifetime of springs; flowpaths, depths and driving forces of hydrothermal systems; origin of heat, chlorine and carbon dioxide; phases and physico-chemical conditions of subsurface reactions; influence of lithospheric structures and dynamics on the occurence and distribution of mineral springs.

To find the answers to some of the questions we are using a combination of different approaches to the problems:

1. Tectonic investigations,
2. Interpretation of the water chemistry,
3. Chemical Analysis of cuttings from mineral water wells,
4. Numerical modelling of fluid and heat transport and water-rock reactions.

2. RESULTS

This article presents some of the results obtained by the first three approaches mentioned above. They will be used as input data for the numerical models.

2.1 *Tectonical investigations*

Because of the low permeability of the paleozoic clastic rocks, the effective flow of groundwater is restricted to joints and fractures.

Strike direction data from joints, faults, ore veins, eruption fissures of Quaternary volcanoes, satellite lineations, in situ stress determinations, fault plane solutions and epicenter alignments of earthquakes, soil gas anomalies and river course directions from different subareas of the massif have been compiled and interpreted. They are shown in the rose diagram of figure 1. Since the Carboniferous varying lithospheric stress field orientations have generated a saturated fracture system. Springs occur preferably on N-S, E-W and NW-SE trending faults. These are the directions of active strike-slip and dip-slip movements which correspond to the recent extensional stress. Older fault zones are mostly sealed by mineralizations.

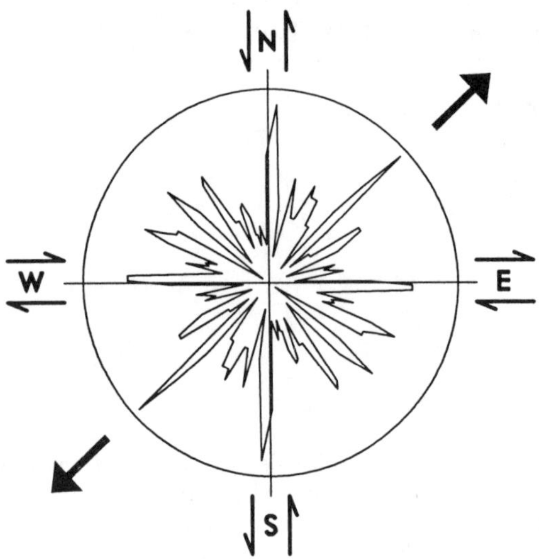

Figure 1. Rose diagram for tectonical elements of the Rhenish Massif. Bold arrows indicate the horizontal extension direction; light arrows indicate directions of shear movements.

Structural informations about the deeper crust were obtained by the deep seismic reflection profiles BELKORP, DEKORP1 and DEKORP2 (DEKORP research group 1991). The youngest features are north and south diping, conjugate reflectors in the middle and lower crust, which were interpreted as indicators of crustal dilatation. They are potential channels for deep reaching flow systems.

A laminated lower crust was observed in all profiles, which might be caused by magma underplating. The transition from laminated to transparent lower crust is correlated with a marked decrease in mineral spring frequency and present day seismicity.

A compositional profile through the crust was deduced by Mengel et al. (1991) from a petrological examination of xenolihs in the Cenozoic volcanics of the Eifel. They expect a crystalline basement of metamorphic rocks in amphibolite facies below 6 to 8 Km depth.

In summarizing the above observations, we can state that the active extensional stress field is one important factor for the genesis of mineral waters, enabling deep circulation of surface waters (Langguth & Plum 1984) on permeable fault zones, in some places down to the crystalline basement. The fault pattern is also of importance for the spatial distribution and the locations of mineral springs.

2.2 *Chemistry of mineral waters*

A compilation of about 800 published analyses from 240 sites was available for evaluation.

To establish a classification for the different mineral waters we used different types of diagrams to compare the mineral water analyses. The graphics depicted a continous cloud of points, or curves, but no individual point or line clusters. Nevertheless we delimited five groups to which most of the samples could be assigned, these were based on their main components:

Type I = Ca-Mg-HCO_3
Type II = Ca-Mg-Na-HCO_3
Type III = Na-HCO_3
Type IV = Na-HCO_3-Cl
Type V = Na-Cl

In figure 2 the main components of the water types are depicted. Other waters could mostly be explained by local aquifer petrology (evaporates e.g.) or mixing.

Systematic variations exist in physical and chemical properties between waters of type I and type V. Thermal waters mainly belong to types IV and V. The maximum temperature is reached in Aachen with 75 °C. Average wellhead and spring temperatures in all five groups are slightly above the mean annual surface temperature. The maximum wellhead temperatures in each group are in the order of average estimated aquifer temperatures. Both increase from type I to V. K-Mg, Mg-Na, Mg-Li and Na-K-Ca-Mg geothermometers were used for temperature estimations. They produced the most consistent results of the 19 different geothermometers applied. Only waters of type V show estimated aquifer temperatures above 100 °C. Mean and maximum depth of wells and average salinities also increase from type I to V. The contents of free CO_2 in the sampled waters are often in a range of 1.5 to 2.5 g/l. About 85 % of the waters from types I, II and III, but only about 50 % of the samples from types IV and V, contain more than 1 g/l CO_2. The waters of type I with concentrations below 0.5 g/l CO_2 are mixed with fresh water.

Table 1 summarizes mean values and maximum depths (above M.S.L.) for the mineral water groups.

Table 1. Properties of water types I to V.

Type	I	II	III	IV	V
Depth (m)	35	-287	-650	-790	-976
$T_{surface}$ (°C)	11.2	15.1	20.3	22.8	22.5
$T_{aquifer}$ (°C)	29.8	43.4	51.4	66.5	93.4
Sal. (mval/Kg)	40.3	69.2	102	107	344
CO_2 (g/Kg)	1.92	2.18	1.71	1.17	1.04

2.3 *Distribution of mineral waters*

The map (fig. 3) displays the mineral and thermal water occurences of the Rhenish Massif. The dotted line encircles outcrops of Paleozoic rocks. The Lower Rhine embayment, around Cologne, is a rifted basin, whose subsidence began in the Middle Oligocene.

Waters of types I and II are predominant in the Rhenish Massif. There they are concentrated in the Quaternary volcanic fields of the eastern and western Eifel. Waters of type III are frequent at the southern end of the Lower Rhine embayment and in the adjacent Middle Rhine valley. Waters of type IV are found in different places of the eastern part of the Rheinisches Schiefergebirge. Waters of type V are only abundant at deep reaching faults at the margins of the Rhenish Massif, in the Lower Rhine embayment and the Upper Rhine graben (south of Frankfurt).

In general no unequivocal correlation exists between the distribution of present day seismicity, Quaternary volcanism and the occurrence of mineral waters.

2.4 *Alteration of aquifer rocks*

We hoped to find aquifer rocks in exchange with mineral waters in deep wells producing bicarbonate waters, because this type appears to spring from shallow depths. Cuttings from four wells in the Eifel (Wehr 600 m, Neuenahr 358 m, Birresborn 140 m, Daun 82 m) were sampled for chemical and phase analysis.

Replacement of chlorite, and partly albite, by newly formed kaolinite and siderite are common in all cases. Some samples contain dolomite, smectite and alunite as secondary minerals. The phases probably limiting the chemistry of the mineral waters were calculated using the thermodynamical computer programs WATEQ, and PHREEQE; these were carbonates, SiO_2-modifications, clay minerals, amorphous sesquioxides, alunite and zeolites.

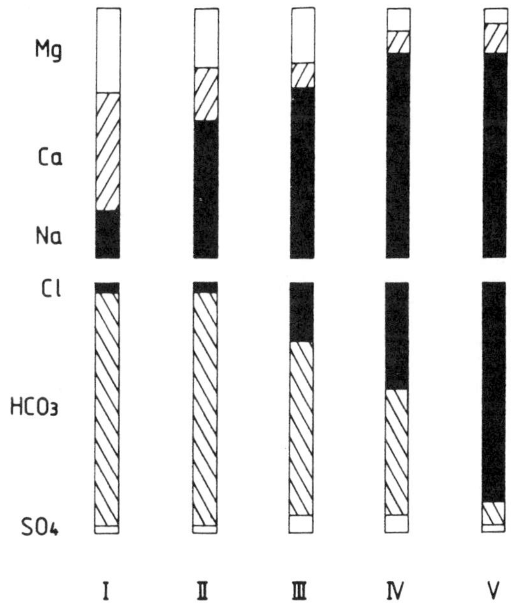

Figure 2. Average composition of the five types of mineral waters in equivalent portions.

The comparison of altered rocks and protoliths revealed losses of Ca, Na and Si and gains of Mn, Fe, Pb, Zn, Cu, As and Th. The average compositions of clastic rocks of the region were assumed to resemble the protoliths (Schulz-Dobrick & Wedepohl 1983).

The leaching of clastic rocks at shallow depths produces some of the components of the mineral waters. Other components come from different sources. Percipitation of carbonates and the enrichment of metals in the altered rocks show that even low temperature hydrothermal systems contribute to a significant redistribution of elements and to the formation of ore deposits.

3. GENETIC MODEL

The formation of the majority of the mineral waters of the Rhenish Massif is controled by two basic processes:

1. Ascent of mantle derived CO_2 (indicated by C and He isotopes (Griesshaber et al. 1991) and mass balances) and its dissolution in ground waters, which causes reactions between the water and the aquifer rocks.

2. Opening of deep reaching fractures into

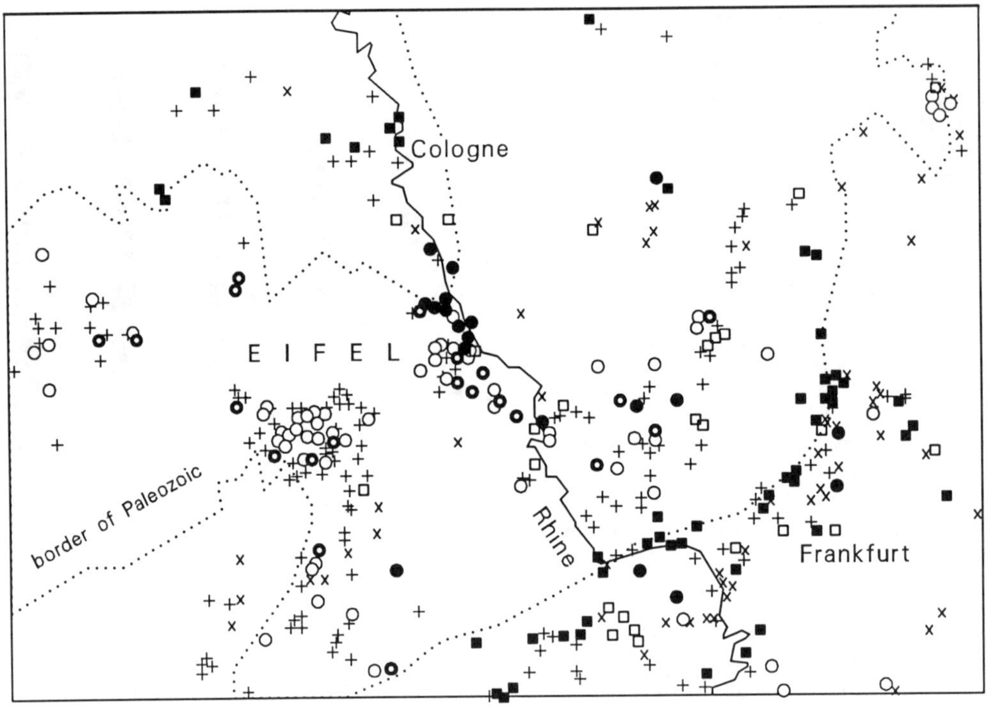

Figure 3. Mineral waters of the Rhenish Massif: open circles = type I, rings = type II, solid circles = type III, open squares = type IV, solid squares = type V, x = other types, + = no data.

the crystalline basement enabeling high saline fluids under overpressure to rise into higher groundwater systems and deep circulation of meteoric waters at elevated temperatures.

The first process results in the formation of bicarbonate waters. Their composition is further determined by secondary processes: low temperatures in shallow depths and mixing with fresh waters (type I), higher temperatures and young fractures with unaltered protolith (type III) and ion exchange (all bicarbonate waters). The second process produces waters with Cl^- anions. Mixing with fresh or mineralized ground waters contributes to the chemical variability of the resulting waters.

Further investigations on the extent of fluid-rock exchange will include stable isotope analysis of hydrothermaly altered rocks, and numerical modeling of reactions and hydrodynamics.

4. REFERENCES

Blondel, F. 1688. *Auszführliche Erklärung und augenscheinliche Wunderwirckung deren heylsamen Badt- und Trinckwässern zu Aach.* Aachen.

DEKORP research group 1991. Results of deep seismic reflection investigations in the Rhenish Massif. *Tectonophysics* 173: 507-513.

Griesshaber, E., R.K. O'Nions & E.R. Oxburgh 1991. Helium and carbon isotope systematics in the crustal fluids from the Eifel, the Rhine Graben and the Black Forest, F.R.G.. Submitted to *Chemical Geology*.

Langguth, H.-R. & H. Plum 1984. Untersuchung der Mineral- und Thermalquellen der Eifel auf geothermische Indikationen. *Forschungsbericht T 84-019 Bundesministerium für Forschung und Technologie*. Bonn.

Mengel, K., P.M. Sachs, H.G. Stosch, G. Wörner & G. Loock 1991. Crustal xenoliths from cenozoic volcanic fields of West Germany; implications for structure and composition of the continental crust. *Tectonophysics* 195: 271-289.

Schulz-Dobrick, B. & K.H. Wedepohl 1983. The chemical composition of sedimentary deposits in the Rhenohercynian belt of central Europe. *Intracontinental fold belts:* 211-229. Berlin, Heidelberg: Springer.

Water-Rock Interaction, Kharaka & Maest (eds) © 1992 Balkema, Rotterdam. ISBN 90 5410 075 3

Environmental isotope study of saline geothermal systems in western Anatolia, Canakkale, Turkey

Stefan Mützenberg & Werner Balderer
Ingenieurgeologie ETH-Hönggerberg, Zürich, Switzerland

Werner Rauert
Institut für Hydrologie, Gesellschaft für Strahlen- und Umweltforschung mbh, Neuherberg, Germany

ABSTRACT: From the observed stratigraphic and tectonic features it results that the position of the three sites of saline thermal springs in the western Biga Peninsula coincides with the intersection of three fault systems. The outflowing hot water of these saline springs are of Na-Cl type and contain up to 65000 ppm of total dissolved solids. The ratios of the most soluble constituents indicate a common origin from a hot saline fluid. The actual composition of the present day outflowing hot waters of the observed thermal springs can be explained by mixing processes with heated fresh water of regional aquifers and locally with near surface groundwater of meteoric origin up to 85%.

1 GEOLOGICAL- SETTING

The studied area is located in the northwestern part of Anatolia on the Biga Penninsula, known in the antiquity as the Troas. Along the western coast, three centers with thermal water discharges occur. Since classical times these thermal springs were used in spas.

Within the investigated area the following main rock types are present:

Metamorphic rocks are exposed in several outcrops around an intrusive body and consist mainly of diopside-plagioclase-actinolite-biotite schists, quartzites and also marbles (Mützenberg 1990).

Folded slates and sandstones of probably Carboniferous age are outcropping in a small area south of Akçekeçili, followed by a 400 to 500 m thick sequence of Permian limestones.

Clastic rocks of the Upper Miocene are the next following sedimentary deposits present in a narrow stripe along the coast line only.

Intrusive rocks of a quartz-monzonite type are found in the Kestanbol area, where an intrusive body was formed during the Oligocene. The exposed intrusion covers an area of about 80 km.

Products of calcalcaline volcanism of the Lower and Middle Miocene are also present in western Anatolia and consist mainly of lava flows, lava domes and volcanic agglomerates.

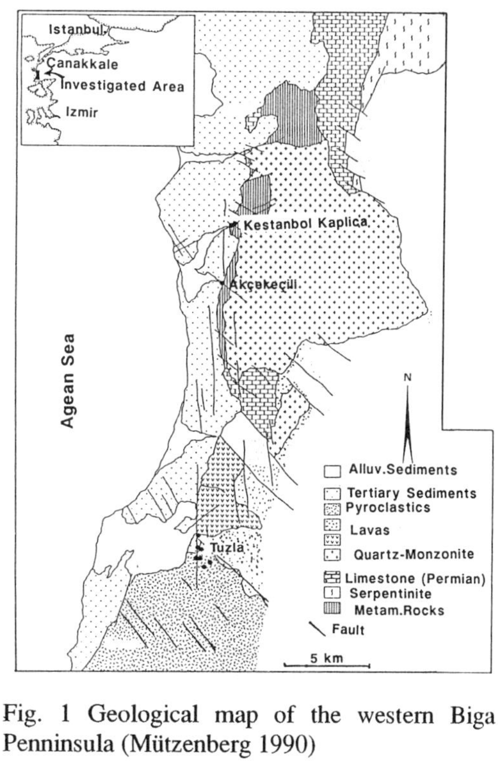

Fig. 1 Geological map of the western Biga Penninsula (Mützenberg 1990)

2. HYDROGEOLOGICAL SITUATION

The small thermal spring of Akçekeçili appears on the western limit of a complex of about 100 m thick metamorphic series just before they dip beneath Upper Miocene sediments.

Near Kestanbol Kaplica (Kaplica meaning spa) a larger metamorphic complex is exposed along a NE-SW striking normal-dextral fault-zone. The two main thermal springs lie on the northern flank of the Ilica valley (Figure 2).

The intrusive quartz-monzonite type rocks on both sides of the valley are intensively cleft and show salty exhalations. A 290m deep drill hole (Yürür 1985) completed in 1975 provides the thermal water used for the spa.

The Tuzla area is dominated by calcalcaline volcanics and ignimbrites. Below the about 600 m thick volcanic sequence Permian limestones and metamorphic rocks have been encountered in two drill holes (Karmanderesi 1987).

The thermal activities around Tuzla cover a surface area of about 1 km^2. Since Upper Miocene to the Present evidences of at least four thermal phases can be recognized in this area:

1. Red and black ferromanganese crusts intercalated in the Upper Miocene sediments with feeders along joint sets;
2. Silicified zones in sedimentary and volcanic rocks, probably associated with N-S striking faults;
3. Travertine terraces;
4. Outflowing Na-Cl rich hot brines, which characterize the present hydrothermal activity.

Based on the structural pattern of major fault systems (Figure 1) it can be recognized that the origins of the thermal springs within the three investigated areas coincide with the intersection of the following three systems:

i) NNE-trending dip-slip fault zones,

ii) major NE-SW striking, right lateral strike-slip faults,

iii) sets of NW-SE trending joints/faults cutting through all three thermal areas.

The resulting increase in vertical permeability along the intersection of these three major fault systems provides a path for the rise of the thermal water from the deep reservoir through the nearly impermeable cover of fine grained Miocene sediments to the springs at the surface (Figure 3).

The outflow zones of the thermal brines are situated about 20 to 30 m above sea level in the Kestanbol and 20 to 80 m in the Tuzla area.

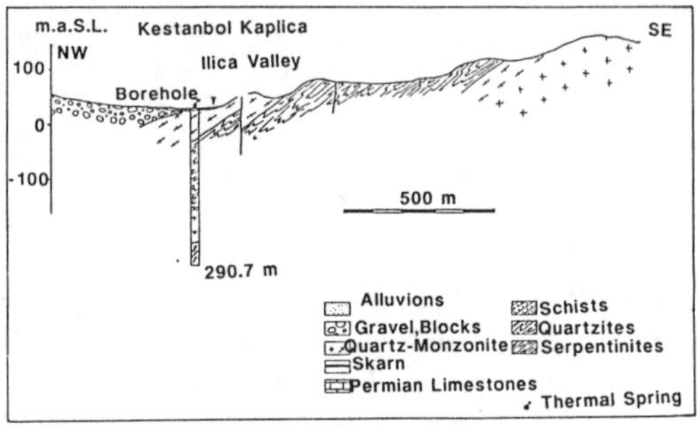

Fig. 2 Geological cross section of the Kestanbol Kaplica thermal area

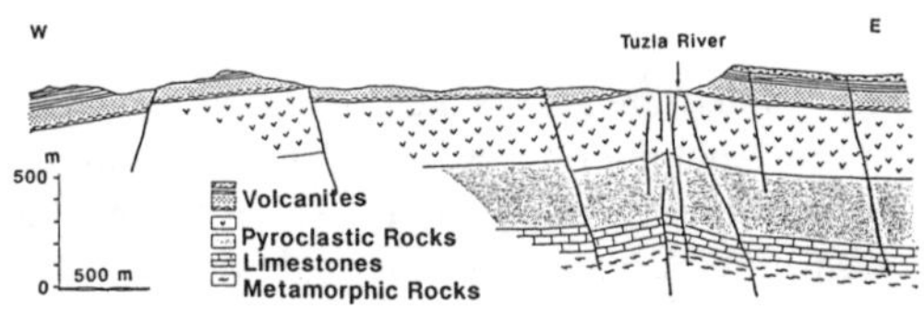

Fig. 3 Geological cross section of the Tuzla area

In both areas springs of low mineralized fresh water at lower temperatures are observed at the same level or at slightly higher altitudes. They represent the outflowing local groundwater recharged and circulating in the fissured rocks of the intrusion of the quartz-monzonite in the Kestanbol, or in layers of increased permeability within the volcanic sediments in the Tuzla area respectively.

3 CHEMICAL CHARACTERISATION OF THE THERMAL WATERS (BRINES)

During the present study the outflow of 14 thermal springs and one artesian borehole were sampled in different seasons from 1987 to 1988.

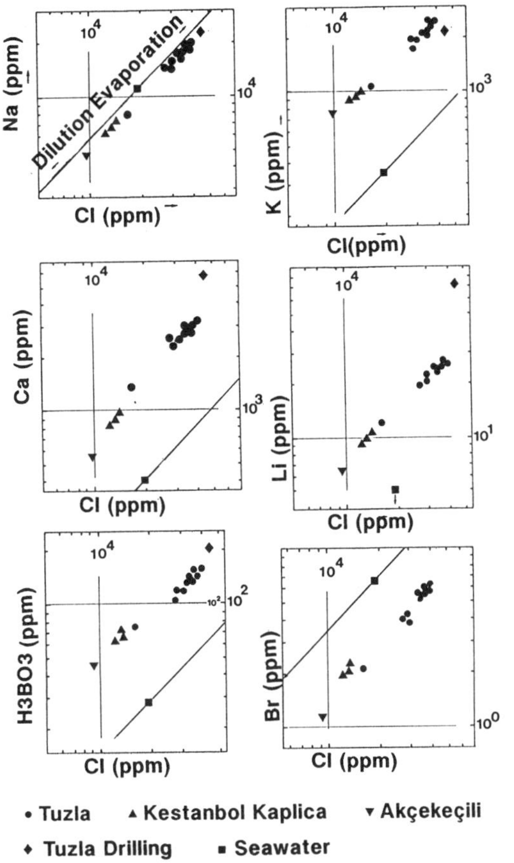

Fig. 4 Plots of the dissolved ions versus chloride (in logarithmic scale)

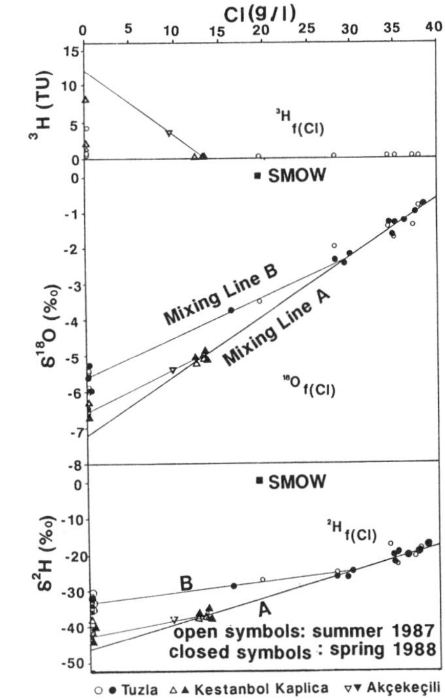

Fig. 5 Diagram of isotope contents(^3H,^2H,^{18}O) versus Chloride content.

The temperature and total mineralisation of the outflowing thermal water ranges from 33.5 °C and 16 g/l for the spring of Akçekeçili to the boiling point temperature and a maximum value of 65 g/l of total dissolved solids for the highest mineralized spring at Tuzla.

Based on the main constituents (with Na about 80 mval % of total cations and Cl greater than 99 mval % of total anions), the brine can be characterized as Na-Cl type, with Ca greater than K and low concentrations of Mg, SO_4 and HCO_3.

The ratio Na/Cl is similar to seawater with a slight depletion of Na.

The Ca/Cl, K/Cl, and Li/Cl ratios are markedly increased with respect to seawater (Figure 4).

4. ISOTOPIC SIGNATURES

Samples of the thermal waters and of adjacent fresh water springs have been analyzed for their oxygen-18, deuterium and tritium contents.

The tritium data as represented in Figure 5 versus the chloride concentration reveal that:
i) only some of the low mineralized ground water of freshwater springs (with very low chloride concentration) contain anthropogenic tritium; ii) whereas all heavy mineralized thermal waters of the Tuzla and Kestanbol area do not contain any measurable tritium.

Therefore a mixing process between local low mineralized fresh ground water and the deep thermal brine is evidenced for the thermal spring of Akcekecili by the measuered tritium content (3.5+0.7 TU).

The stable isotope contents δ^2H and $\delta^{18}O$ are also represented in Figure 5. By this diagrame a clear distinction between low mineralized waters of low chloride content and the hot thermal brines is possible.

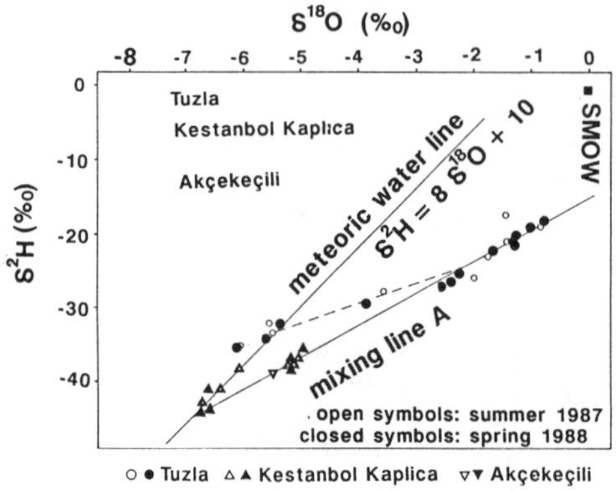

Fig. 6 δ^2H versus δ^{18}O diagram

On the δ^2H versus δ^{18}O diagram in Figure 6 the corresponding values of the low mineralized shallow groundwaters with low chloride concentration are situated along the the meteoric water line (with values in the range of:-45 to-30 ‰ for δ^2H and -7 to -5 ‰ for δ^{18}O).

On the same diagram in Figure 6 the values of the thermal brines are situated below the meteoric water line on a line with lower slope (line A) showing a marked oxygen-18 enrichment.

This line A could be interpreted as mixing line between the low mineralized ground waters and the hot thermal brines, where the most concentrated brine of the Tuzla area represents also the most enriched endmember with respect to its stable isotope contents.

5. CONCLUSIONS

Based on this presented study the following conclusions can be made:

As the ratios of non-reactive species of the sampled thermal waters differ very little it can be concluded that they evolved from a common saline end-member and represent mixing series with low mineralized fresh groundwater.

The last mentioned correspond either to shallow waters (containing anthropogenic tritium) or heated regional ground waters of meteoric origin.

For the saline endmember an evolution from a fossil altered seawater (connate water) is suggested based on the similar Na/Cl and only slightly differing Br/Cl and B/Cl ratios (Figure 5).

The high salinity and the observed isotopic signatures can be explained by diagenetic reactions in the sediments (Clayton et al. 1966) and water rock interactions at higher temperatures during circulation in volcanic/ crystalline and metamorphic rocks (also involving oxygen isotope exchange reactions, Savin, 1980, Kharaka and Carothers 1986, Balderer et al. 1991).

Therefore the actual composition of the present day outflowing hot waters of the observed springs can be explained by mixing processes with heated fresh water of regional aquifers and locally with near surface groundwater of meteoric origin up to 85%.

The ascent of the saline thermal water results of the combined effect of (1) deep regional circulation along major fault zones, (2) the high heat flow in this region of active tectonic movements and (3) the dilution effect by mixing with low mineralized hot freshwater of low density which led to the observed upwelling of the hot brines due to the buoyancy effect along zones of high vertical permeability .

REFERENCES.

Balderer, W., Greber, E., Imbach, Th., Rauert W., Trimborn, P., Güler, S. 1991. Isotope Study of Thermal, Mineral and Normal Groundwater in the Bursa/Kuzuluk Area. IAEA-SM-319/52 Vienna.

Clayton R.N. et al.1966 J. Geophys. Res. 71/6

Karamanderesi M. 1987. Hydrothermal alteration in well Tuzla T-2, Canakkale, Turkey. MTA Internal Report, Izmir.

Kharaka Y.K. and Carothers W.W. 1986.In: Handbook of Environmental Isotope Geochemistry vol 2 (Fritz, P.J.Ch., Eds.), Elsevier, Amsterdam.

Mützenberg, St. 1990. Westliche Biga - Halbinsel Geologie, Tektonik und Thermalquellen (Canakkale, Türkei). Mitt. Geol. Inst. der ETH und der Universität Zürich Nr.287.

Savin S. 1980.In: Handbook of Environmental Isotope Geochemistry vol.1 (Fritz, P., Fontes, J.Ch., Eds.), Elsevier, Amsterdam.,

Yürür, T. 1985. Diploma thesis ITÜ, Istanbul.

Water-Rock Interaction, Kharaka & Maest (eds) © 1992 Balkema, Rotterdam. ISBN 90 5410 075 3

Trace element geochemistry of the system 'rock – thermal water – suspended matter – deposits' in a granitic environment

Elka N. Pentcheva
Bulgarian Academy of Sciences, Geological Institute, Sofia, Bulgaria
Els Veldeman, Luc Van't dack & Renaat Gijbels
Department of Chemistry, University of Antwerp (UIA), Antwerp-Wilrijk, Belgium

ABSTRACT: In order to elucidate the processes leading to the definitive chemical composition (in its complexity and dynamics) of thermal waters from granites and metamorphosed silicate rocks and to understand their geochemical evolution, an experimental study of water-rock interaction at elevated temperatures and pressures was set up, in juxtaposition with the investigations on the geochemistry of the natural hydrothermal system: solution, suspended matter (< 0.4 μm) and possible deposits. The results on the complex dynamic system "rock - water - suspended phase - deposits" and the interphase hydrogeochemical processes are discussed in a primary and secondary genetical respect, yielding possibilities for a mastering and prognosis of the evolution of the hydrothermal system.

1 INTRODUCTION

The understanding and the mastering of water-rock interaction processes and the obtainment of an as complete as possible view on the chemistry of the whole resulting system (in dynamic balance): rock - thermal fluid - suspended matter - possibly associated concretions, are essential for the proper exploitation and management of geothermal systems from different aspects and for various uses. The present large scale study attempts to elucidate the processes leading to the definitive complex and dynamic trace and major element composition of hydrothermae from granites and metamorphosed silicate rocks, and to understand their chemical evolution. Therefore, experimental studies of water-rock interactions at elevated temperatures and pressures are performed and the trace element composition of the experimental fluids is compared with those of the natural hydrothermal solutions, observed as a dynamic association with the suspension (< 0.4 μm) in them and the solid phases deposited at the outflow. The results are based on the study of more than 80 N_2-alkaline hydrothermae in highly fissured crystalline rocks in Southwest Bulgaria (low TDS, high SiO_2 and F, Na-HCO_3-SO_4 (or Na-SO_4-HCO_3), very low Mg), very rich in trace elements, with a selective and specific distribution (Pentcheva 1967; 1984).

2 METHODS

Trace elements in all kinds of phases were analyzed by means of instrumental neutron activation analysis (INAA) and by inductively-coupled plasma mass spectrometry (ICP-MS), which is a powerful technique for low TDS waters (Veldeman 1991). The experimental simulation of the primary and secondary water-rock interaction processes is based on the experience of Pentcheva (1973; 1984) and on the recent investigations of Veldeman (1991). Principal Component Analysis (PCA) and chemical geothermometry were also applied. Furthermore, the suspended matter and solid deposits were studied with an electron microprobe.

3 RESULTS AND DISCUSSION

1. Experiments, reacting granite rock samples

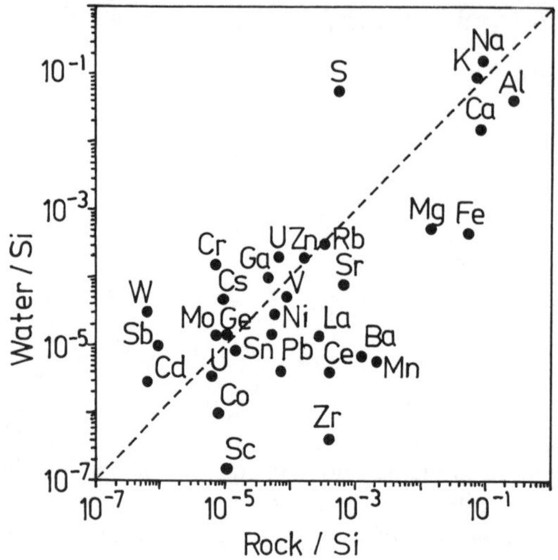

Fig. 1. C_i/C_{Si} ratios for the 200 ^{o}C deionized water reaction fluid versus the corresponding ratio in the granite rock.

with dilute Na-HCO_3-SO_4 solutions (simulating natural thermal waters) and deionized water produce alkaline solutions of low salinity and trace- (and major-) components present in the natural hydrothermal systems are also demonstrated in the experimental fluids (with increased pH during the interactions) (Fig. 1). At thermodynamic equilibrium conditions, the solubility of the elements leached into the Na-HCO_3-SO_4 solution (in comparison with deionized water) may be affected by the precipitation of carbonate, sulphate and/or hydroxide compounds and, on the other hand, by complexation with HCO_3^-, SO_4^{2-} and/or OH^- ligands. When equilibrium is not attained, the element contents in solution will be controlled by kinetic factors, i.e. dissolution and/or precipitation rates of the minerals involved limit the dissolved element contents. It is evident from the results (Fig. 1) that at these conditions the specific elements, with their anomaly and selective distribution in alkaline thermal waters from granites (Li, Rb, Cs, Ga, Ge, W, Mo, V, S, etc.) (Pentcheva 1967; 1984) are enriched in the fluid phase, while the other ones (e.g. Mn, Fe, Sr, Ba, Mg, etc.) show more affinity for the solid phase, also at 200 ^{o}C. This non-stoichiometric dissolution behaviour could be explained by preferential leaching from the rock or by selective removal of elements from the solution by secondary reactions in the alkaline aqueous medium (Pentcheva 1984).

2. The trace element concentrations for both types of reaction fluids tend to increase with temperature (factor 5 to 20 between 52 and 200 ^{o}C; higher in the Na-HCO_3-SO_4 solutions) (Fig. 2), except in the case of the more difficultly leached Cs, Ge and Sb (see also Pentcheva 1973; 1984) (Si/Ge ratios in the experimental fluids vary between 10000 and 70000, consistent with the relative rock abundance, (Si/Ge = 87000). Comparing the hydrothermal experimental fluids, at 150 ^{o}C after 7 and 14 days reaction time respectively, it is evident that the reaction time is not so significant for the experimental fluid composition (in spite of the higher silica content, the 14 days solution is still undersaturated with respect to quartz). For the unmixed natural waters, the quartz geothermometer yields the most reliable reservoir temperature and the data indicate a deep temperature up to about 150 ^{o}C. An attempt is also made to extract deep reservoir temperature information from the PCA derived components (Fig. 3), where component 1 is assumed to represent the influence of temperature at depth on the included variables. The correlation of each trace element with temperature at depth is assessed, yielding a kind of classification of the various trace elements. Those typical for granite associated thermal waters (Li, Rb, Cs, Ge, W, B, F, etc.) show a strong positive correlation with temperature at depth, while with increasing influence of surface conditions Ca, Mg, Eh, U, Ba, ΣCO_3, etc., increase. These results are of essential primary and secondary genetic importance and give prognostic possibilities.

3. The comparison of experimental (reactions at 125 ^{o}C) and natural fluids is presented in Fig. 4. It is evident that only some atypical elements are enriched in the experimental fluids (Ba, Al, Mn, Mg); in the natural thermal water they are controlled by the precipitation of a slightly soluble secondary phase while in the reaction solution they are undersaturated with respect to most studied minerals. A whole suite of elements, and especially the specific ones, stay enriched in the natural waters, controlled by coprecipitation and / or sorption processes (and

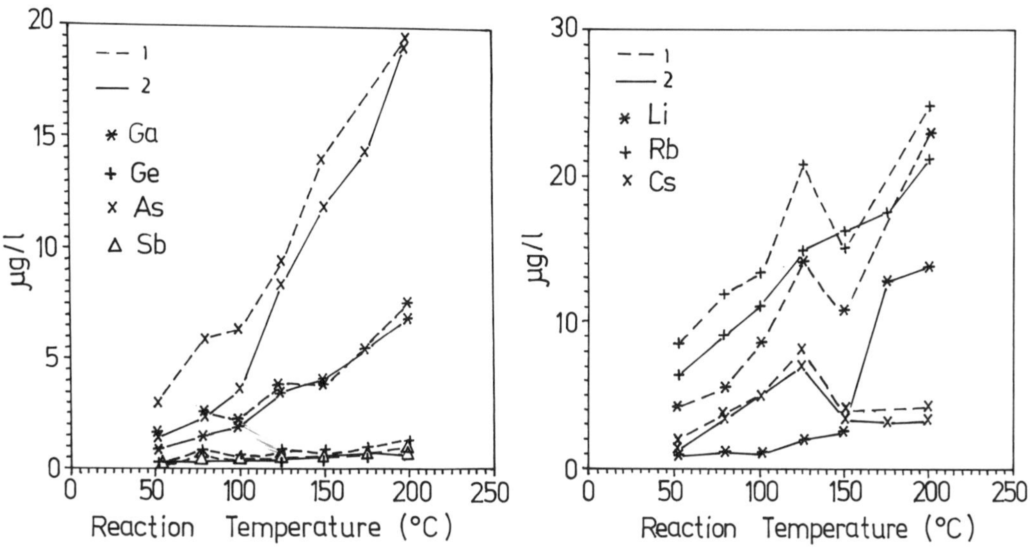

Fig. 2. Trace element concentrations versus the reaction temperature for deionized water and Na-HCO_3-SO_4 reaction fluids.

with specific values of the typical element ratios: Rb/Cs, Li/Cs, Sr/Ba, etc.). The concentrations of most major and trace elements here are probably controlled by kinetic factors rather than by equilibrium thermodynamics. This implies that the content of a range of elements is controlled by the rate of dissolution of the corresponding minerals, which increase with increasing reaction temperature. Some elements in the reaction fluids (W, As, S) are rapidly leached from the rock and their content in solution may be limited by their presence in the rock sample. F and B, present in relatively high amounts in the natural fluids, could not be demonstrated in the experimental solutions. This is probably due to a high resistance to weathering of the corresponding minerals. The chemical thermodynamic modeling of the natural thermal waters, calculation of element speciation and saturation degree with respect to selected solid phases (WATEQ4F), is treated in the light of element affinity for sulphur (both in the suspended phase and in solution) and classified in 3 groups (Veldeman 1991).

4. Suspended matter in thermal waters (non-visible with the naked eye) isolated on membrane filter and characterized by an exclusive sorption, coprecipitation and ion-exchange capacity, originates from 2 distinct sources, namely physical weathering, erosion of the rock during the ascent of the water to the surface, and chemical weathering (i.e., precipitation due to changes in temperature or other parameter). The suspended matter studied is enriched in Ba, Sr, Zn, Cu, Ni, Cr, Cs, W, As, U, REE with respect to the host rock (Fig. 5). It is typically controlled by the precipitation of secondary minerals (e.g. sulphides, carbonates, hydroxides) having a low solubility. It originates, at least partially, from the secondary precipitation and/or adsorption of material from solution. This may suggest that the physical weathering products are not so significant for the origin of the suspended matter. Electron probe microanalysis (EPMA) reveals the presence of, e.g., calcite, silica, aluminosilicates, S-containing particles, Fe-S, Fe-Zn-S, Cu-S (and mixed Fe-Si-Al material). $CaCO_3$ and SiO_2 aggregates are formed in the suspended material of hydrothermal waters (similar with observations made for Eastern Pyrenean thermal waters; Vandelannoote 1984). Cold waters are mainly characterized by aluminosilicate particles, with various contributions of, e.g., Ca, Fe, K. The relative

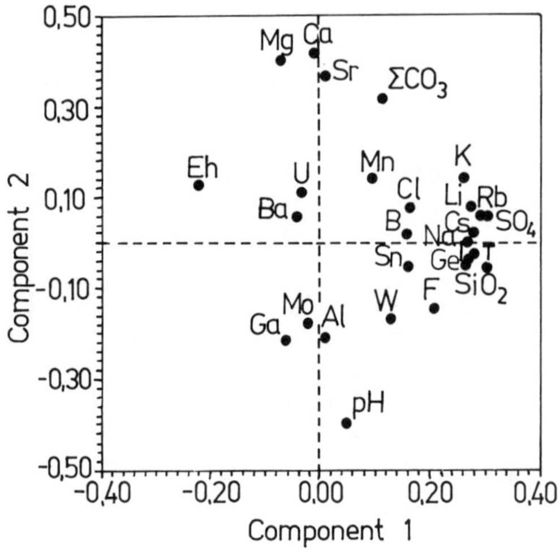

Fig. 3. Loadings of the variables on the first two components extracted from PCA.

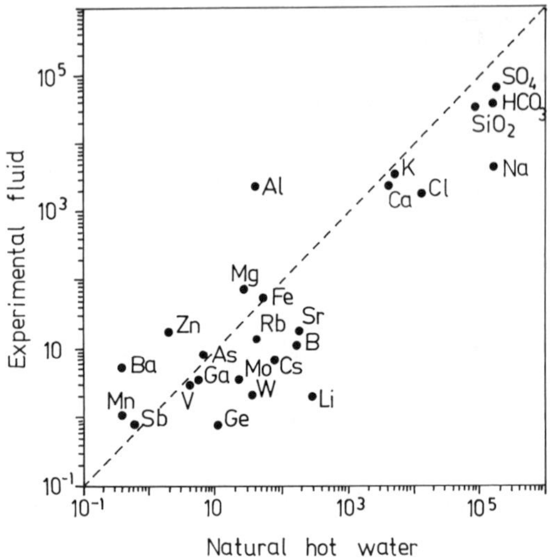

Fig. 4. Chemical composition of the deionized water reaction fluid versus that of the natural thermal water (in $\mu g.l^{-1}$).

amount of elements transported in the dissolved and suspended phases (Fig. 5) shows: a) typochemical elements (W, Cs, Rb, Sr, As, also Ge, Ga) mainly transported in the dissolved phase; b) for elements like Mn, Cr, Al, Zn, Cu, also Fe and especially REE, however, a significant fraction of the total concentration (dissolved + suspended) is contained in the suspended phase. Consequently, it is evident that suspended particles play an important role during redistribution processes in hydrothermal systems. This is demonstrated by means of the $R_{solution}/R_{suspension}$ coefficient. These values, always above 100 for trace elements typical for alkaline hydrothermal waters, show clearly their existence and transport in soluble form (ionic, complexed). Less than 1% of their content (As and V included) is transported as suspended matter. It manifests the low vulnerability of these elements with respect to metamorphization processes due to their transport in dissolved forms. Another important proof for the virulence of the mobile trace elements during the evolution of the water is their stability in the alkaline reducing medium towards migration barriers (R_{Cs} max = 88000 and R_W max = 9000, even though they could be detected in the suspended matter). The less specific elements (like V and As) display a more balanced hydrogeochemical behaviour between the two phases (R_V max = 280 and R_{As} max = 583) whereas the REE show a tendency towards fixation in the solid phase. The results for the whole dynamic system can be interpreted as to the metamorphization probability of the migration forms during the evolution of the fluid, and gives the possibility of mastering the hydrothermal system and of making hydrogeochemical prognoses.

5. Deposits at the outflow of the waters are believed to originate from precipitation after partial evaporation of the water. Results show that the deposits reflect the major chemical composition of the waters, and are mainly composed of Na- and/or Ca-sulphates and -carbonates, as well as silica aggregates (and sometimes F-rich deposits). Concretions studied reflect also the water's microchemical composition, illustrated in Fig. 5. Concentration data for water and deposits seem to be near one another, which suggests a direct chemical link between them. The tendency for the highest mobility of the components typical for alkaline N_2-waters (W, Rb, As, Sr, V, etc.) is, however, always real, as opposed to the trend for Al, Fe, Mn, Ba, Zn, Sb, Co, Th, Hf, Sc, U, Ni, REE, and also Ca, which are depleted from solution,

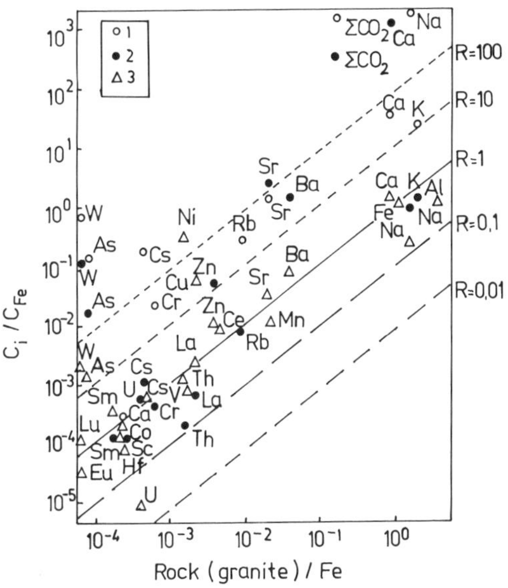

Fig. 5. C_i/C_{Fe} ratio for 1) natural thermal water; 2) suspended matter; 3) concretions, versus the corresponding ratio in the granite rock.

due to secondary immobilization in a solid phase, probably related to the reduced mobility in alkaline media.

REFERENCES

Pentcheva, E.N. 1967. Sur les particularités hydrochimiques de certains oligoéléments des eaux naturelles. Ann. Idrologia 5: 90-112.

Pentcheva, E.N. 1973. On the hydrogeochemical behaviour of gallium and germanium in the dynamic system "thermal water - rock" and its genetic significance. *Proc. Intern. Symp. Hydrogeochemistry - Biogeochemistry*: 140-155. Washington: Clarke & Co.

Pentcheva, E.N. 1984. Formation et évolution de la spécificité microgéochimique indicatrice des hydrothermes azotés de roches silicatées. Doctor Habil. Thesis, Geol. Inst. Bulgarian Academy of Sciences.

Vandelannoote, R. 1984. Trace element geochemistry in thermal waters (Eastern Pyrenees) and in waters related to hidden Zn-Pb-S ore deposits. Ph. D. Thesis, University of Antwerp.

Veldeman, E. 1991. Geochemical characterization of hydrothermal systems in a granitic environment. Ph. D. Thesis, University of Antwerp.

Water-Rock Interaction, Kharaka & Maest (eds) © 1992 Balkema, Rotterdam. ISBN 90 5410 075 3

Thermal springs in granitic rocks from the Grimsel Pass (Swiss Alps): The late stage of a hydrothermal system related to Alpine Orogeny

H.-R. Pfeifer & A. Sanchez
Earth Sciences, Lausanne University, Switzerland

C. Degueldre
Paul Scherrer Institute, Villigen, Switzerland

ABSTRACT: The isotopic composition of warm Na-SO4-type spring waters issuing from granitic rocks with temperatures around 28°C indicates that they are of meteoric origin. Their silica content of up to 100 mg/liter suggests a reservoir temperature of at least 100°C and a corresponding penetration depth of several thousand meters. This present day warm water circulation system is interpreted as the low temperature stage of a hydrothermal alteration system related to the rapid uplifting of this part of the Alps that started towards the end of the Tertiary metamorphic event around 20Ma ago.

1 INTRODUCTION

1.1 General remarks

Most of thermal springs of the Swiss Alps are related to the steep southern border of the basement of the Aar massif and issue at low altitudes (Lavey, Leuk, Brig, Pfäfers; Vuataz, 1982). This project deals with a series of warm springs which occurs in a tunnel, located at 2000 m above sea level, that was completed in 1973 to house a gas-pipeline going from the Netherlands to Italy (Fig. 1). During the summer and fall of 1991 eighteen low output springs were collected, of which six have temperatures between 18 and 28°C. These occur in a 300 m long section of the tunnel in which the air has a temperature of 15°C, instead of 10°C as in the other parts. In this zone the normally light coarse metagranite has large yellowish- brown spots and is obviously hydrothermally altered.

1.2 Geological setting

The tunnel is located within the carboniferous granitic basement of the Aar massif which has been deformed and metamorphosed to greenschist facies during the Tertiary (Fig. 1). Within the tunnel the different lithological units have a subvertical orientiation, dipping to the southeast. From north to south the tunnel traverses

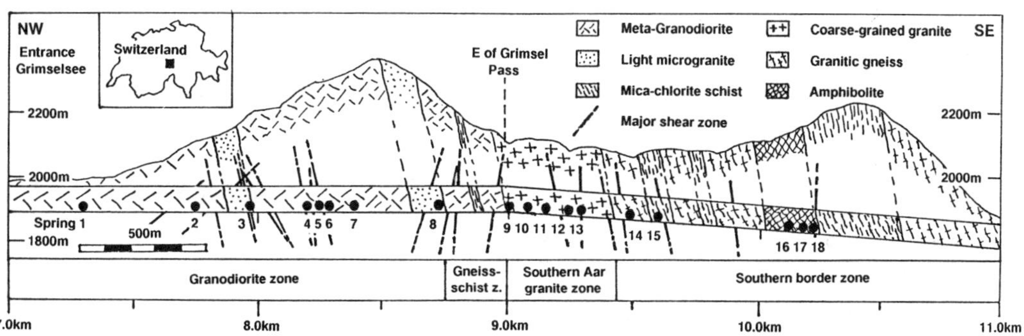

Fig. 1: Location and geology of the tunnel section containing the warm springs (no.9 to 13).

the following rock types (Fig. 1; Lüthi 1961, Stalder 1964, Boerner et al.1973): (1) mesocratic augen- gneisses, leucocratic meta-aplites and -microgranites belonging to the Grimsel granodiorite unit, (2) granodioritic gneisses and micaschists belonging to an old migmatic intercalation separating the central part of the Aar massif from its southern part, (3) light coarse grained less deformed metagranites belonging to the southern Aar granite unit, (4) various meso- and melanocratic gneisses and schists with mafic intercalations belonging to the pre-Carboniferous southern border zone of the Aar massif.

The whole Grimsel area, at the surface as well as in the different tunnels, is characterized by a complex joint system which has been extensively studied by many authors (cf. NAGRA 1985, Dollinger 1989). Some of the joint systems have been attributed to the carboniferous magmatic cooling but most seem to be Tertiary. The dominant joint directions in the tunnel are subparallel to the SW-NE strike of the main lithological units and the alpine schistosity. Some important cataclastic and tectonic breccia zones have the same orientation (Fig.1).

2 SPRING WATERS

2.1 Introduction

Sampling dates were July, August-September and December 1991. Since the tunnel is situated only 100 to 200 m below the surface which has an average altitude of about 2100 m, a maximum of springs were found during the snow melt in July and a minimum in December. Of the 18 springs 6 have temperatures between 18 and 28°C and the others between 3.5 and 10°C (mean annual surface temperatures around 4°C). For simplicity we only report data from 4 cold and 2 warm springs which persisted year-round and which could be sampled under ideal conditions (no contamination from concrete or metal of the tunnel construction, Tab.1, Figs.1 and 2).

2.2 Chemistry, colloids, secondary minerals and stable isotopes

The warm spring waters have between 300 and 500 mg/L total dissoved solids (TDS) whereas the cold ones contain only between 20 and 100 mg/L (Table 1, Fig. 2). The warm springs are of the Na-SO4-HCO3-type with very high SiO2 contents of up to 100 mg/L. They are also rich in B, F, As and W. The cold waters (except S.3) are of the Ca-HCO3-type as is typical for cold shallow ground waters. S.3 is most likely a mixture of cold and warm water. It is of Na-Ca-SO4-HCO3-type and contains about a fifth of the TDS contained in the warm springs.

Two of the five warm springs (S.12, S.13) contain between 10 and 20 mg/liter suspended particles smaller than 5 µm and colloids larger than 100nm, comparable to the waters found by Degueldre et al.(1990). They have a composition and an X-ray diffraction pattern close to average illite (Degueldre 1992). Argillaceous fissure fillings at the site of S.3 are composed of illite, mixed layer clays, chlorite and zeolites.

The typical yellowish-brown domains of the warm zone have also been found at the surface by Dollinger (1989) who identified them as rich in smectite and Fe-hydroxide (about 5 km to the SW of the tunnel in question, locality: SW of Grosses Sidelhorn). The small grey crystals that cover the rocks in the warm zone of the tunnel turned out to be Na_2SO_4 (thenardite) and its hydrated counterpart (mirabilite).

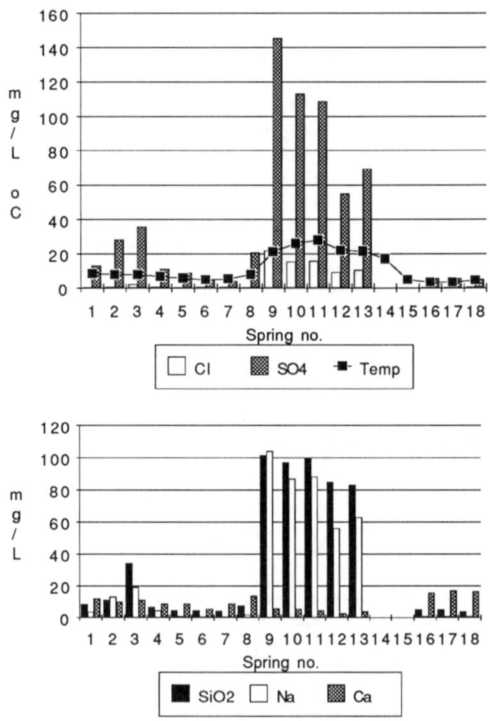

Fig. 2: Major chemical composition and temperature of the different springs

Table 1: Selected water compositions of the gas tunnel at Grimsel Pass

	warm springs			cold springs			
	S.11	S.12nf	S.12f	S.3	S.5	S.7	S.17
Date (1991)	Aug.27	Aug.27	Aug.27	Aug.27	Aug.27	Aug.27	Aug.27
Locat. km	9.124	9.145	9.145	7.960	8.240	8.380	10.125
Flow l/min	0.80	1.50	--	0.07	0.40	0.25	50.0
Filtered[1]	no	no	yes	no	no	no	no
Temp. °C	27.00	22.50	--	7.50	5.50	5.00	3.50
pH	9.20	9.30	--	9.40	9.30	8.50	8.50
Eh [2] mV	-250.00	-80.00	--	400.00	200.00	450.00	n.m
TDS mg/L	468.00	316.00	232.00	98.00	56.00	26.00	40.00
Li μg/L	134.00	108.00	96.00	12.00	1.00	1.00	1.00
Na mg/L	83.90	65.70	58.70	17.80	0.98	0.68	0.85
K mg/L	6.79	5.73	5.51	1.06	1.73	1.33	1.67
Rb μg/L	52.00	47.00	31.00	1.00	0.40	0.30	1.00
Cs μg/L	9.00	12.00	5.00	<0.10	<0.10	<0.10	0.10
Mg μg/L	140.00	370.00	137.00	81.00	487.00	288.00	420.00
Ca mg/L	4.10	3.44	3.85	8.89	10.00	8.81	16.80
Sr μg/L	31.00	26.00	32.0	41.00	73.00	29.00	104.00
Ba μg/L	0.70	6.30	2.60	0.70	0.90	0.40	5.80
Ti μg/L	12.90	47.50	0.80	1.20	1.50	4.60	<0.10
Mn μg/L	10.10	23.30	0.9	0.50	0.80	1.30	<0.10
Fe μg/L	56.00	964.00	17.50	21.80	32.30	70.10	7.70
Co μg/L	<0.10	0.20	0.10	<0.10	<0.10	<0.10	<0.1
Ni μg/L	0.10	0.10	1.00	0.20	0.10	0.30	0.30
La μg/L	<0.10	0.40	0.10	<0.10	<0.10	<0.10	<0.10
Th μg/L	<0.10	0.40	0.10	<0.10	<0.10	<0.10	0.01
U μg/L	1.00	9.00	7.00	4.00	2.00	6.00	6.00
W μg/L	48.00	83.00	70.00	46.00	0.30	0.50	0.50
F mg/L	9.30	7.90	n.m	2.80	0.10	0.20	0.10
Cl mg/L	16.20	12.50	n.m	2.00	0.20	2.00	0.20
Br μg/L	5.00	4.00	4.00	1.00	0.30	0.40	0.20
I μg/L	0.4	1.00	2.00	<0.10	0.10	0.20	0.10
SO_4 mg/L	101.00	75.30	65.10	27.40	9.60	4.00	5.80
Se μg/L	0.9	4.0	5.0	0.9	1.0	2.0	5.0
NO_3 mg/L	0.2	1.1	n.m	0.2	0.8	1.0	0.8
P μg/L	<2.0	73.00	65.00	4.00	3.00	7.00	<2.00
As μg/L	3.00	80.00	7.00	0.10	<0.10	<0.10	<0.10
IC mg/L	12.63	9.36	n.m.	5.00	3.20	4.68	10.55
OC μg/L	121.00	185.00	n.m.	165.00	n.m.	128.00	157.00
Al μg/L	29.20	1010.00	18.50	21.00	32.10	51.60	7.00
SiO_2 mg/L	96.30	82.82	67.41	10.27	5.20	4.34	4.94
H_3BO_3mg/L	0.50	0.40	0.378	0.20	0.023	0.034	0.029
Coll.mg/L	0.50	10.00	-.-	2.00	1.00	0.50	0.10
Part./mL[3]	$1.9x10^7$	$5.7x10^8$	-.-	$9.2x10^6$	$1.8x10^7$	$6x10^5$	$4.4x10^6$
T-calcedony[4]	108.0	98.0	93.0	7.0	-.-	-.-	-.-
T-Na-K-Ca[4]	124.0	116.0	112.0	38.0	-.-	-.-	-.-
T-Na/Li[5]	103.0	99.0	101.0	61.0	-.-	-.-	-.-
T-Mg/Li[4]	98.0	76.0	94.0	43.0	-.-	-.-	-.-
$\delta^{18}O$ SMOW [6]	-13.97	-14.6	-.-	-13.45	-13.5	-13.63	-14.43
δD SMOW [6]	-107.27	-109.7	-.-	-97.2	-98.28	-99.0	-106.08

Abbreviations: Locat.: location, measured from the main tunnel entrance at Handegg, nf: not filtered, f, [1]: filtered <0.1μ, [2]: measured in the field, indicative, n.m. not measured, IC: total inorganic carbon, OC: total organic carbon (laboratory measurement), Coll.: colloids, Part.:colloidal particles > 100nm, 4) calculated after Kharaka (1987), [5]: Fouillac and Michard, 1981 in Vuataz (1982), [6] for comparison: Creek S of Grimsel Pass: $\delta^{18}O$: -14.56, δD: -105.5, temp. 7°C.

A test with the program code PHREEQUE indicated that the warm waters are close to saturation with fluorite, Ca- and Mn-carbonate, but supersaturated with respect to quartz, K-feldspar, smectite and illite.

The stable isotope values (Tab.1) show that all waters are of clear meteoric origin. The warm waters are the lightest and if one assumes infiltration of these waters under recent recharge conditions, the altitude of infiltration is around 2200m (using equations proposed by Kullin and Schmassmann 1991).

3 CONCLUSIONS

As expected, the warm spring waters are of meteoric origin and their Na-SO4-character is comparable to similar waters from granitic basement areas in Switzerland (Vuataz 1982, Pearson et al. 1989, Dubois et Parriaux 1990). Both spring types have small flow rates and are related to the many joints and young shear zones present. However, the warm springs occur in a zone which is relatively little deformed but shows clear signs of recent hydrothermal alteration (southern Aar granite). What is striking is the high SiO2- contents of the warm springs, for which the calcedony geothermometer and other emprirical water thermometers indicate temperatures close to 100°C (cf.Tab.1). The geothermal gradient of around 25°C/km of the western Aar massif (Bodmer 1982) implies that meteoric waters percolate down to several thousand meters where they are heated and move up again interacting hydrothermally with the adjacent rocks. The high sulfur content (possibly as HS^- at depth and $SO4^{2-}$ only in the upper parts) seem to be typical for granitic basement in Switzerland, but its exact origin is not known.

Having cold waters percolating down to considerable depth and warm water moving up in a relatively small area can only be explained with the following hydraulic situation: small amounts of warm waters move up in the whole area, but due to the omnipresent fissure zones that go up to the surface, the major flow is cold water downwards. Only in selected small areas, with small fissures that do not correspond with the surface, can the warm fluids move up close to surface without being to much cooled and diluted.

Comparing these sodium-silica and sulphate rich waters with the alteration products in the hydrothermally altered rocks at the surface of the Grimsel area (presence of calcedony, pyrite, smectite in a tectonic breccia: Stalder 1964 and Dollinger 1989, locality N of Kleines Sidelhorn) strongly suggests that these warm waters are the low temperature end stage of a long living hydrothermal system that is related to the rapid uplift of this part of the alps and the subsequent thermal disequilibrium. Fission track data for apatite, corresponding to a temperature of 120°C, are as young as 10 Ma (Hurford 1990; Hunziker, pers.comm.).

AKNOWLEDGEMENTS

The company Transit Gas AG provided technical support during sampling. R. Brütsch, M.Cordisco, S.Granacher, R.Keil, P.Thélin and B.Wernli have done the chemical and X-ray analyses. Critical comments and reviews were made by L.Calmbach, L.Eisenlohr, A. Gautschi, J. Hunziker, D.Marshall, A. Steck, A. Scholtis and Z.Sharp. We thank them all very much.

REFERENCES

Bodmer, P. 1982. Geothermal map of Switzerland. Schweiz. geophys. Komm. Kümmerly & Frey, Bern.

Boerner, C., Jenny, P. & Klemenz, W. 1973: Geologische und Technische Aufnahme Grimselstollen, Transitgas AG. Unpubl. manuscr. Dr.T.Schneider, consult. geol., Männedorf.

Degueldre, C. et al. 1990. Grimsel colloid exercice: an international intercomparison exercise on the sampling and characterisation of ground water colloids. Nagra techn. Ber. 90-01. Nagra, Wettingen, Switzerland.

Degueldre, C. 1992. Colloid properties and behavior in granitic groundwater systems with emphasis on the safety asessement of radioactive waste depositories. Nagra Techn. Ber.(in prep.).

Dollinger, H. 1989. Hydrothermale Veränderung granitischer Gesteine in der Grimselregion.Diploma thesis, Univ.Bern.

Dubois, J.-D. & Parriaux, A. 1990. Hydrogeological charateristics of crystalline aquifers. Example of the Mont Blanc and Aiguilles Rouges massifs. First results. In Parriaux, A. ed. Water in mountaineous regions. Mem. 22nd Congress IAH, Lausanne.

Hurford, A. 1990. Spaltspurdatierungen. In: Soom, M.A. Abkühlungs- und Hebungsgeschichte der Externmassive und der Penninsichen Decken beidseits der Simplon- Rhone -Linie seit dem Oligozän. Ph.D.thesis, Univ. Bern.

Kharaka, Y.& Mariner R.H. 1989. Chemical geothermometers and their application to formation waters from sedimentary basins. In: Naeser N.D. & McCulloh, T.H. eds. Thermal history of sedimentary basins. methods and case studies. Springer, Berlin.

Kullin M. & Schmassmann, H. 1991. Infiltration conditions. In: Pearson et al.,Applied isotope hydrogeology- a case study in Switzerland. Nagra techn.Ber. 88-01. Nagra, Wettingen, Switzerland.

Lüthi, H. 1961. Querprofil durch das zentrale Aarmassiv (Grimselroute). Führer Schweiz. Alpenposten, PTT-Bern.

NAGRA 1985. Grimsel test site: overview and test programs.Nagra Techn.Bericht 85-46. Nagra, Wettingen, Switzerland.

Pearson, F. Localma, J. & Scholtis,A. 1989. Chemistry of waters in the Böttstein, Weiach, Riniken, Schafisheim, Kaisten and Leuggern boreholes. Nagra techn. Ber. 86-19. Nagra, Wettingen, Switzerland.

Stalder, H.A. 1964. Petrographische und mineralogische Untersuchungen im Grimsel-gebiet (mittleres Aarmassiv). Schweiz. Mineral.Petrogr. Mitt.44: 187-399.

Vuataz, F. 1982. Hydrologie et géochimie des eaux minerales thermales de suisse et régions limitrophes. Matér. Géol. Suisse, Série hydrol. 29. Kümmerly & Frey, Bern.

Water-Rock Interaction, Kharaka & Maest (eds) © 1992 Balkema, Rotterdam. ISBN 90 5410 075 3

Thermal waters in México

R. M. Prol-Ledesma
Instituto de Geofisica, Mexican National University, Mexico

ABSTRACT: Due to the active regional tectonic history, hot and warm springs are frequently found throughout the continental part of Mexico as well as in its islands and the sea bottom. Here, chemical analyses of most continental springs are presented and discussed. Most geothermal discharge features are located within the limits of the Mexican Volcanic Belt (MVB) and the Sierra Madre Occidental (SMO), related with the high terrestrial heat flow observed in both geological provinces. However, outside those two provinces warm bicarbonate springs are common. Chemical analyses of springs show they are immature and partially equilibrated waters. Chemistry of the water discharged by hydrothermal systems in the central part of Mexico indicate partial equilibrium conditions and in the case of Los Humeros geothermal field the analyses plot in the immature waters area. Conditions close to equilibrium were observed only for deep reservoir water in the large hydrothermal system of Cerro Prieto.

1 INTRODUCTION

Numerous surface thermal manifestations in Mexico have been subjected to chemical studies to determine their potential as energy sources (Medina, 1986; Tello, 1986a, 1986b; Quijano, 1986; Tello, 1987), and also to establish their relation with the regional tectonics. There are published works on the silica concentration of thermal waters (Marvin, 1984; Prol and Juárez, 1985; Prol-Ledesma and Juárez, 1986), and also on the dissolved gases composition (Polak et al., 1982; Polak et al., 1985). These works attempt an interpretation of the chemical variations of the fluid discharged by surface thermal manifestations and their relation with geologic structures and tectonic activity. However, the attainment of fluid-rock equilibrium has not been yet assessed for the water discharged by thermal springs and geothermal wells, except for the Cerro Prieto geothermal field (Giggenbach, 1988). The condition of full equilibrium is critical for the application of the Na-K-Mg-Ca geothermometers, which are commonly used for the preliminary assessment of geothermal fields; therefore determination of the equilibrium condition of thermal waters in Mexico is necessary to identify those waters that can provide reliable information about deep temperatures.

2 TECTONICS AND GEOLOGICAL SETTING

Morán-Zenteno (1986) has classified the geological and tectonic setting of Mexico into 13 geologic provinces (Figure 1). Since the Tertiary, most structural features in Mexico have been closely related to the evolution of the plate boundaries in the Pacific Coast. In particular, many hydrothermal manifestations may be associated with processes related to the subduction in the Eastern Pacific. The subduction process along the western coast of Mexico started about 100 m.y.b.p. and the

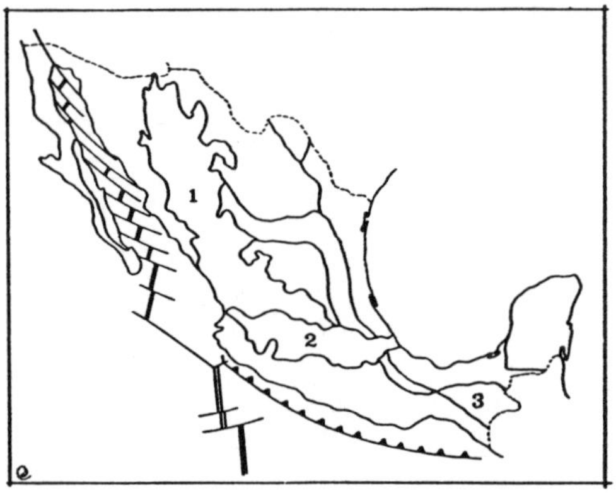

Fig. 1 - Main tectonic structures in Mexico: 1-Sierra Madre Occidental; 2-Mexican Volcanic Belt; 3-Sierra de Chiapas (After Moran-Zenteno, 1986).

formation of the Sierra Madre Occidental, as well as the activity in the Mexican Volcanic Belt (TMVB) have generally been related to this process (Demant and Robin, 1975).

The main consequences of the changes in the ocean plate tectonics have been the diversity in the types of volcanic activity that have taken place in Mexico since the Tertiary up to the present. This volcanic activity has formed the extensive ignimbrite fields in the western part of Mexico (Sierra Madre Occidental) (Figure 1), the alkaline volcanics of the Sierra Madre Oriental, the igneous rocks in the Baja California Peninsula, and the TMVB. The recent volcanic activity in Chiapas, the Southernmost state of Mexico, represented by the activity of El Chichon and El Tacana is more related to the Central America Volcanic Chain than to the TMVB.

3 CHEMICAL COMPOSITION OF THERMAL WATERS

The chemical analyses studied here (approximately 600) correspond to waters from geothermal well discharge and hot and warm springs with temperatures that range from slightly above 30°C up to the boiling point. They have been taken from the literature (Henley et al., 1984; Medina, 1986; Quijano, 1986; Tello, 1986a, 1986b, 1987) and from unpublished reports by the Geothermal Division of the Federal Electricity Commission. The pH measurements were made in the laboratory; most values correspond to neutral and alkaline-pH waters. Acid waters were sampled only in the southeastern part of Mexico, in the vicinity of El Chichon and Tacana volcanoes, where mixed sulphate-type waters predominate.

Concentration contours show that the main positive anomalies in chloride and silica are associated with the TMVB and the tectonically active areas in the northern and southern parts of México (Prol-Ledesma and Quijano, 1988). Chloride anomalies on the Gulf of Mexico coast are due to salt water intrusion. High sulphate concentrations are located in areas where two different phenomena take place: anhydrite dissolution from buried marine sediments, and H_2S oxidation from hydrothermal or volcanic gas discharge.

A chemical-type zonation is clearly defined for the thermal waters (Figure 2). Springs in the TMVB and SMO areas have a composition that resulted from the mixing of $Cl-SO_4-HCO_3$ waters; this marks the areas where higher temperatures and more intense hydrothermal activity is observed.

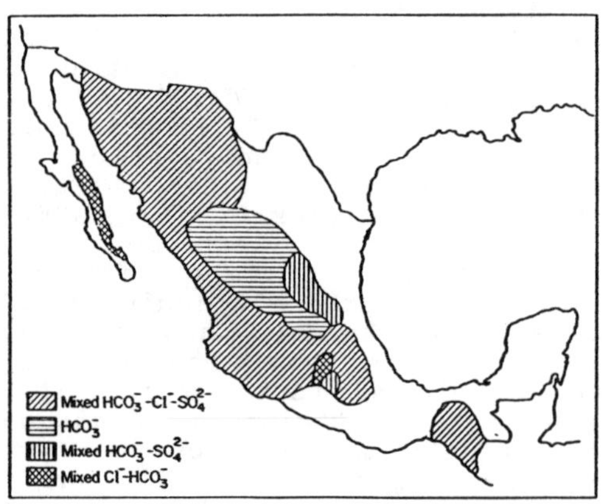

Fig.2 Chemical zonation of thermal waters in Mexico.

SO_4-HCO_3 mixing is restricted to areas with evaporite-bearing lithologies in the central part of México. Mixing of bicarbonate and chloride waters is recognized in two areas: one within the TMVB, in the proximity of the Los Azufres geothermal field, and in the Baja California peninsula. Many bicarbonate springs in the central part of Mexico are depositing calcite at the surface, and some have formed large travertine deposits.

4 ASSESSMENT OF WATER-ROCK EQUILIBRIUM

Here, the assessment of the degree of attainment of fluid-rock equilibrium follows the criteria proposed by Giggenbach (1988) for thermal waters on the basis of Na-K-Mg-Ca concentrations.

In a triangular diagram (Na/1000, K/100, ´Mg) spring waters plot in the lower right-hand corner in the immature waters area in the early stages of rock dissolution (Figure 3). Some springs attain low temperature partial equilibrium. Geothermal wells in the Los Humeros field also discharge immature waters. Los Azufres and Ixtlán geothermal fields are characterized by partially equilibrated waters and only in the Cerro Prieto geothermal field is fluid-rock equilibrium attained.

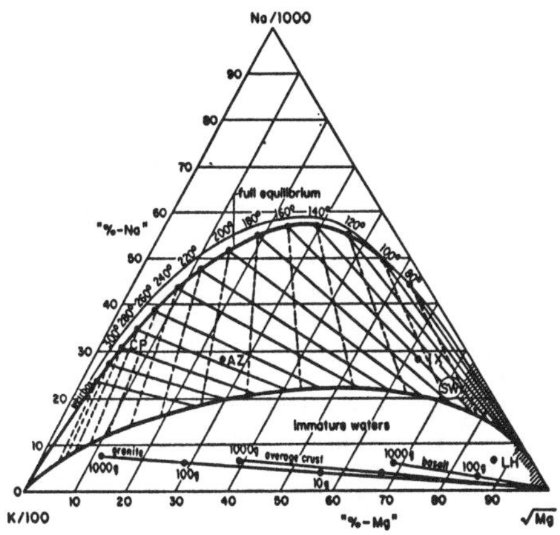

Fig.3 Relative contents of Na-K-Mg. Full and partial equilibrium curves from Giggenbach (1988).

Evaluation of equilibrium for the K-Mg-Ca system (Figure 4) also shows low temperature K-Mg equilibration. The chemical composition of most spring waters corresponds to the early stages of rock dissolution and calcite formation. The waters from deep geothermal wells plot above the equilibrium line (rock-dominated

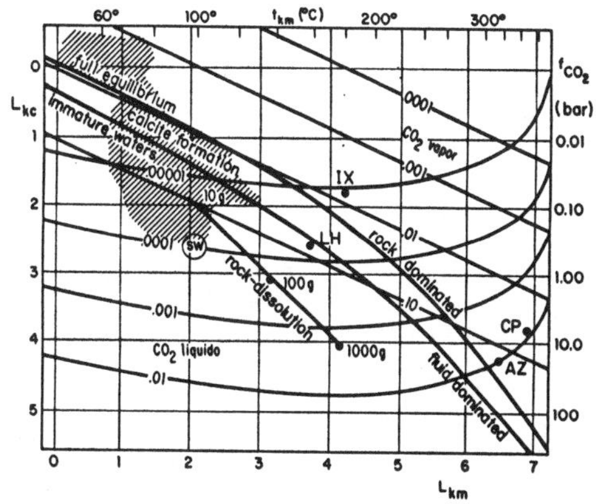

Fig.4 Logarithms of the K-Mg-Ca concentrations of thermal waters to estimate temperature and CO_2 fugacity (after Giggenbach, 1988).

systems) except for Los Humeros geothermal field. According to the conclusions reached by Giggenbach (1988), such a system should be dominated by acid fluid, thus it is not surprising that acid fluids corrosion has been reported in Los Humeros.

5 DISCUSSION

From the presented chemical data, the lack of fluid-rock equilibrium in the thermal waters from surface discharges and also from deep reservoirs is evident for any of the Na-K geothermometers reported in the literature. This phenomenon may be the result of intensive mixing of the thermal component with cold Mg-rich groundwater at shallow depths. The fact that waters from wells deeper than 2000 m in Los Humeros geothermal field are clasified as immature indicates that other parameters should be taken into account in addition to mixing. In particular, inferences about the reservoir permeability may be made in the Los Humeros geothermal field, where a decrease in the permeability in the shallow parts of the field has been attributed to self-sealing. This may contribute to a poor fluid-rock interaction. In Los Azufres geothermal field the existence of a zonation within the reservoir has been proposed: a shallow zone (steam-dominated) and a deep zone (liquid-dominated). Some degree of mixing between both zones may account for the partial equilibrium observed there.

6 CONCLUSIONS.

Thermal waters from springs and some deep wells have not attained full equilibrium with the rocks, therefore the use of geothermometers based on the cation concentrations must be used with great care when the goal is the deep temperature estimation. In most cases the obtained value of temperature will be affected by the lack of equilibrium. Due to its rapid equilibration, the K-Mg geothermometer is the most suitable to be applied to the spring waters; however the temperature obtained will reflect shallow conditions.

REFERENCES

Demant, A. and Robin, C. 1975. Las fases del volcanismo en Mexico, una sintesis en relación con la evolución geodinámica desde el Cretácico. Rev. Inst. Geol. UNAM 1: 70-82.

Giggenbach, W.F. 1988. Geothermal solute equilibria. Derivation of Na-K-Mg-Ca geoindicators. Geochim. Cosmoquim. Acta 52: 2749-2765.

Henley, R.W., Truesdell, A.H., Barton, P.B. and Whitney, J.A. 1984. Fluid-mineral equilibria in hydrothermal systems. Rev. in Econ. Geol. 1: 267 pp.

Marvin, P.R. 1984. Regional heat flow based on the silica content of ground waters from northcentral Mexico. M.Sc. Thesis. New Mex. St. Univ. 107 pp.

Medina, A. 1986. Geoquímica de aguas y gases del Volcán Tacaná. Chis. Geotermia 2: 95-110.

Moran-Zenteno, D.J. 1986. Breve revisión sobre la evolución tectónica de México. Geofis. Int. 25: 9-38.

Polak, B.G., Prasolov, E.M., Kononov, V.I., Verkhovskhiy, A.B., Gonzalez, A., Templos, L.A., Espindola, J.M., Arellano, J.M., Mañón, A. 1982. Isotopic composition and concentration of inert gases in Mexican hydrothermal systems. Geofis. Int. 21: 193-227.

Polak, B.G., Kononov, V.I., Prasolov, E.M., Sharkov, I.V., Prol, R.M., Gonzalez, A., Razo, A. and Molina, R. 1985. First estimations of terrestrial heat flow in the Transmexican Volcanic Belt and adjacent areas based on isotopic composition of natural helium. Geofis. Int. 24: 465-476.

Prol, R.M. and Juárez, G. 1985. Silica geotemperatures and thermal regime in the Mexican Volcanic Belt. Geofis. Int. 24: 609-621.

Prol-Ledesma, R.M. and Juarez, G. 1986. Geothermal Map of Mexico. J. Volcanol. Geoth. Res. 28: 351-362.

Prol-Ledesma, R.M. and Quijano, J.L. 1988. Geochemical characterization of thermal waters in Mexico. Proc. 10th New Zealand Geothermal Workshop 245-249.

Quijano, J.L. 1986. Geoquímica de los manantiales termales de La Soledad y Santa Cruz Atistique, Jalisco. Geotermia 2: 195-200.

Tello, E. 1986a. Geoquímica del agua de manantiales termales de la zona geotérmica de San Antonio el Bravo, Chih. Geotermia 2: 11-123.

Tello, E. 1986b. Características químicas e isotópicas del agua de manantiales termales de las zonas geotérmicas de Ixtlán de los Hervores y Los Negritos, Michoacán. Geotermia 2: 43-53.

Tello, E. 1987. Características geoquímicas de las descargas de los pozos de Los Humeros, Puebla. Comisión Federal de Electricidad Informe 12-87.

Water-Rock Interaction, Kharaka & Maest (eds) © 1992 Balkema, Rotterdam. ISBN 90 5410 075 3

REE distribution in calcites from the Lisbon volcanic complex, Portugal

M.I. Prudêncio & J.M.P.Cabral
Departamento de Química, ICEN-LNETI, Sacavém, Portugal

M.A. Sequeira Braga
Universidade do Minho, Largo do Paço, Braga, Portugal

ABSTRACT: REE concentrations were determined in calcites occurring in the Lisbon volcanic complex (Portugal) filling fissures and vacuoles, and compared with their abundances in the adjacent basaltic rocks. The results mainly show that: (1) in some cases a small positive Eu anomaly appears indicating temperatures of the fluids large enough to reduce part of the Eu during fluid-rock interaction and (2) Nd and Sm seem to be the less enriched REE in calcites when compared with the adjacent basaltic rocks.

1. INTRODUCTION

The distribution of the rare earth elements (REE) in hydrothermal fluids and minerals has been a subject of several studies in order to know the behavior of those elements during interaction of hydrothermal solutions with hot rocks (Michard & Albarede (1986); Alderton, Pearce & Potts (1980); Morgan & Wandless (1980)).

Among the REE, Eu may play an important role in determining fluid conditions, specially the temperature, due to its ability to exist in either divalent or trivalent states. As a matter of fact, and as already pointed out by Sverjensky (1984), at surface conditions (weathering), the aqueous geochemistry of Eu should be dominated by the trivalent state. However, as temperature increases, the Eu^{2+}/Eu^{3+} ratio also increases - around 100°C significant activities of both Eu^{3+} and Eu^{2+} can occur in solutions and at temperatures greater than 250°C and elevated pressures, Eu^{2+} should predominate.

Calcites, sometimes associated with palygorskite and/or smectites, occur filling fissures and vacuoles in basaltic rocks of the Lisbon volcanic complex (~72 m.y.). The origin - hydrothermal or meteoric - of these minerals has not yet been sufficiently clarified (Galopim de Carvalho (1969-1970)).

In this work an attempt is made to contribute to a better understanding of the origin of those calcites by using the rare earth element distribution.

2. EXPERIMENTAL

Samples of calcites were collected at Restelo (1) from veins (1-2cm thick) in a lava flow (RF1) and (2) as crystals filling vacuoles (up to 5cm of diameter) of a very altered basaltic rock (RC); and at Carnaxide from veins parallel and crossing tuff beds (CH and CV, respectively).

All these samples, as well as samples of the adjacent basaltic rocks, were analysed by the instrumental neutron activation analysis method. Details of the technique used have been described elsewhere (Cabral, Gouveia & Prudêncio (1979); Prudêncio, Gouveia & Cabral (1986)).

3. RESULTS AND DISCUSSION

The REE concentrations obtained in the analyses are presented in Table 1.

Table1. REE concentrations (ppm) in calcites and respective adjacent basaltic rocks.

	RF1	Basalt	RC	Basalt	CH	CV	Tuffs
La	15	53	5.4	50	16	26	40
Ce	23	105	8.5	99.0	18	33	87
Nd	11	58	4.1	54	5.8	11	46
Sm	1.99	10.6	0.80	9.90	0.89	1.85	8.04
Eu	0.71	3.04	0.33	2.93	0.30	0.58	2.23
Tb	0.26	1.4	0.18	1.3	0.22	0.26	1.00
Yb	0.42	2.1	0.38	1.9	1.0	0.92	1.5
Lu	0.06	0.30	0.06	0.32	0.19	0.15	0.24

The chondrite-normalized REE profiles of the calcite samples (Fig. 1) mainly show an enrichment of the light REE relative to the heavy REE.

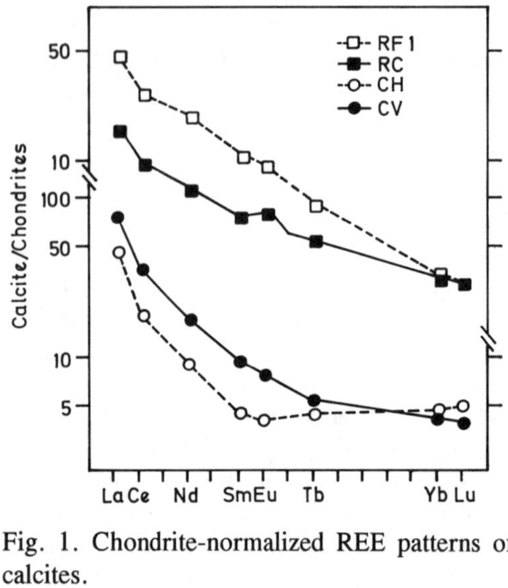

Fig. 1. Chondrite-normalized REE patterns of calcites.

A small positive Eu anomaly occurs in the calcite samples from Restelo (RF1 and RC). This anomaly becomes more evident when one compares the REE contents in these samples with those in samples of the adjacent basaltic rocks (Fig. 2).

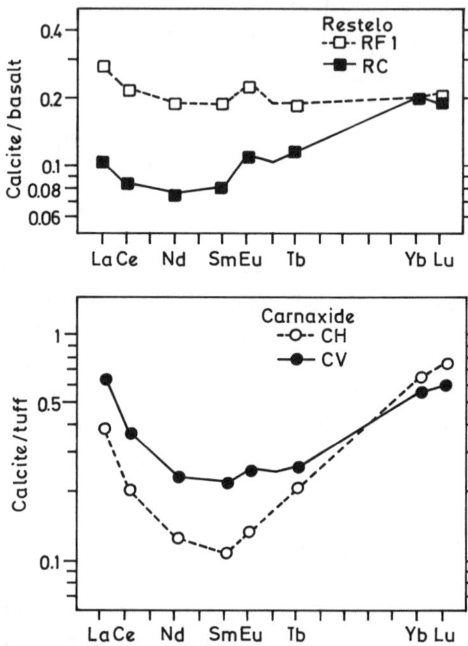

Fig. 2. REE concentrations in calcites normalized relative to the respective adjacent basaltic rocks.

The existence of this small anomaly indicates the presence of Eu^{2+} in the fluids from which calcite precipitated since this mineral can accomodate preferentially Eu in the divalent state by replacing Ca^{2+}. One may conclude therefore that the temperature of the solutions should have been sufficiently high to reduce some of the Eu during the interaction with the basaltic rocks (Sverjensky (1984)).

On the other hand, the comparison of the REE contents in the calcites and in the adjacent basaltic rocks shows that Nd and Sm appear to be the less enriched REE in calcites. This is in agreement with the observed enrichment of these two elements, specially Nd, during the alteration of the basaltic rocks of the Lisbon volcanic complex (Prudêncio, Sequeira Braga & Cabral (1991)).

REFERENCES

Alderton, D.H.M., Pearce, J.A. & Potts, P.J. 1980. Rare earth element mobility during granite alteration: evidence from southwest England. *Earth Planet. Sc. Letters* 49: 149-165.

Cabral, J.M.P., Gouveia, M.A. & Prudêncio, M.I. 1979. Determinação de terras raras e doutros oligoelementos em rochas por activação com neutrões térmicos.*Bol. Museu Lab. Min. e Geol.* Faculdade de Ciências 16: 205-218.

Galopim de Carvalho, A.M., Ubaldo, M.L. & Quadrado, R. 1969-1970. Notas de mineralogia e petrografia portuguesas. XV - Paligorsquite do complexo vulcânico de Lisboa. *Bol. Museu Lab. Min. e Geol.* Faculdade de Ciências 11: 233-239.

Michard, A. & Albarède, F. 1986. The REE Content of some hydrothermal fluids. *Chemical Geology* 55: 51-60.

Morgan, J.W. & Wandless, G.A. 1980. Rare earth distribution in some hydrothermal minerals: evidence for crystallographic control. *Geoch. Cosmoch. Acta* 44: 973-980.

Prudêncio, M.I., Gouveia, M.A. & Cabral, J.M.P., 1986. Instrumental neutron activation analysis of two French geochemical reference samples - Basalt BR and Biotite Mica-Fe.*Geost. Newsletters* 10: 29-31.

Prudêncio, M.I., Sequeira Braga, M.A. & Cabral, J.M.P. 1991. REE behaviour during continental weathering of basalts - their distribution in paleoweathering, sub-actual weathering profiles and alluvial deposits. *Proc. 7th Euroclay Conf.*: 863. Dresden, Germany.

Sverjensky, D.A. 1984. Europium redox equilibria in aqueous solution. *Earth Plant. Sc. Letters* 67: 70-78.

Water-Rock Interaction, Kharaka & Maest (eds) © 1992 Balkema, Rotterdam. ISBN 90 5410 075 3

Petrology and fluid chemistry of magmatic-hydrothermal systems in the Philippines

A.G. Reyes
Geothermal Division, Philippine National Oil Company, Energy Development Corporation, Fort Bonifacio, Metro Manila, The Philippines

W.F. Giggenbach
DSIR Chemistry, Petone, New Zealand

ABSTRACT: Early warning signs of the likely presence of volcanic-magmatic fluids, within hydrothermal systems closely associated with andesitic magmatism, consist of the discharge of highly "immature", predominantly Cl, but SO_4-rich, frequently acid waters, enriched in both ^{18}O and D with respect to local groundwater. Relative N_2 contents of associated vapors are much higher than those in air-saturated groundwater, $N_2/Ar > 200$. More detailed information on the underground distribution of such fluids, whether relict or active, is obtained from the alteration mineralogy and fluid inclusion data. Contact minerals are present, pneumatolytic-type minerals may be associated with acid alteration assemblages. Fluid inclusions homogenize to the liquid and/or vapor phase at $t > 350^{o}C$, or sometimes to a critical point. Equivalent salinities are >10% NaCl and daughter minerals, including salts and anisotropic minerals, are abundant; some contain elemental S or exhibit two or three immiscible fluids at low temperatures. The comparative rarity of well developed magmatic-hydrothermal systems suggests that their temporal and spatial extents are limited to comparatively short periods and to small rock volumes.

1 INTRODUCTION

With 894 MWe installed, geothermal energy provides 21 to 26% of the power needs of the Philippines. An additional 1010 MWe of steam-field capacity is planned to be developed by 1996 (Javellana, 1992). Most of the explored and developed geothermal systems are associated with Late Pliocene to Recent andesitic volcanic complexes and/or dacite to silicic andesite domes (Reyes 1990). A typical well, from top to bottom, penetrates: a sequence of andesitic, basaltic and dacitic lava flows and pyroclastics; andesitic hyaloclastites grading into a sedimentary sequence of limestone, breccias, sandstones, siltstones, claystones and shales; an intrusive complex that predates the sedimentary formations; and a series of young dikes that cross-cut the older intrusives and the sedimentary rocks. The sedimentary formations are usually late Miocene to Pliocene in age. Aquifers are largely associated with faults and fractures, occasionally with lithological contacts and primary porosity.

2 MAGMATIC HYDROTHERMAL SYSTEMS

Based on fluid discharge chemistry and isotopic compositions, alteration mineralogy and fluid inclusion characteristics, about ten of the more than 400 geothermal wells drilled in the Philippines show strong indications for the presence of an active magmatic component. They are, or were encountered at Biliran, Mt Pinatubo, Mt Cagua, Alto Peak, Mahagnao, Palinpinon, and Bacon-Manito. Alteration mineralogy and fluid inclusion studies suggest that ten other wells may have harbored magmatic-hydrothermal fluids in the past, as at Palinpinon, Bacon-Manito, Daklan, Natib and Alto Peak.

3 SURFACE INDICATIONS FOR POSSIBLE ACTIVE MAGMATIC INPUT

In most cases there are good indications that wells may encounter active magmatic input at depth, even during the early surface exploration stages. Such wells are usually found close to Pleistocene-Recent silicic andesite to dacite domes, often associated with an active solfatara, substantial sulfur flows, and/or kaipohans. Some of these volcanoes have erupted within the last 1 ka., e. g. Pinatubo (1991), Cagua (1860), Canlaon (1988), Biliran (1939?) and Mahagnao (1895). Carbonized wood in sulfur flows from Biliran was dated 500 a b.p. (NZ Geol. Survey). Active solfataras in such areas sometimes discharge high Cl vapors suach as at Biliran (Ruaya, 1981), Pinatubo (Clemente 1984) and Mt Apo (Delfin et al. 1984). The presence of highly acid Cl-SO_4 springs at high elevations at Pinatubo and Cagua were used as indications for possible influx of magmatic volatiles (Clemente, 1984; and pers. commun. 1992). Wells are sited to avoid faults that may act as conduits for magmatic fluids. Otherwise the wells are drilled away from the upflow of magmatic fluids.

4 EVIDENCE OF MAGMATIC INPUT FROM DRILLHOLES

The main thrust of this study is to establish, in addition to chemical, petrological criteria useful as an "early warning device" for the occurrence of active magmatic components at depth while a well is being drilled. Such information is required for safety reasons, to assist in the assessment of the suitability af a system for power production, and in the location of future wells.

4.1 Fluid discharges

Strong evidence for the input of an active magmatic input is gleaned from the chemical and isotopic composition of well discharges. All the discharges have $^3He/^4He$-ratios >7 R_A. (R. J. Poreda, pers. comm.). The waters, associated with andesitic magmatism, generally have elevated SO_4 contents and show, in addition to the more common ^{18}O-shift, a distinct deuterium shift (Matsuhisa 1991). It may be ascribed to the addition of varying proportions of D-enriched, "andesitic" water (Giggenbach 1992a). Fig. 1 shows the total discharge isotopic compositions of wells drilled recently into four "magmatic-hydrothermal" systems.

According to this evaluation, the proportions of andesitic water ranges from 45% at Alto Peak to 85% at Mahagnao. Another good indicator for the presence of magmatic components are relative N_2, He and Ar contents (Giggenbach 1992b), as shown in Fig. 2. All indicators point to the presence of considerable proportions of magmatic fluids.

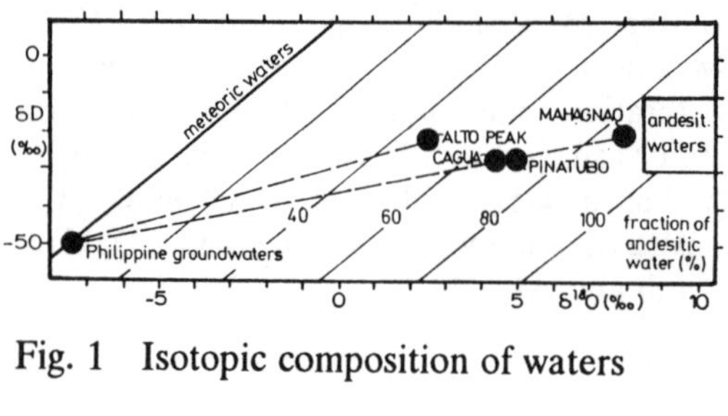

Fig. 1 Isotopic composition of waters

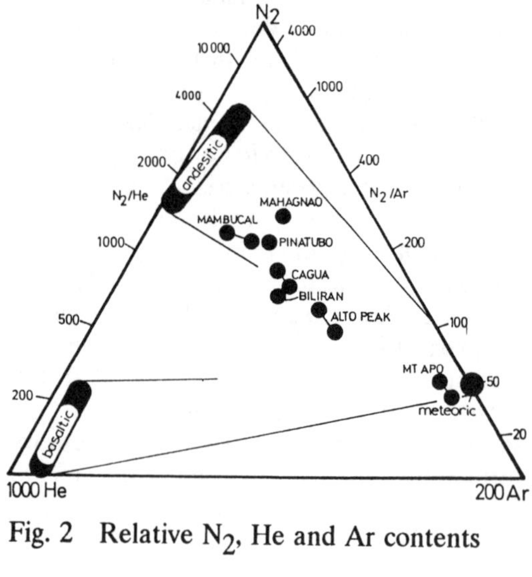

Fig. 2 Relative N_2, He and Ar contents

Other common features of vapors produced in volcanic-magmatic-hydrothermal systems are their exceptionally high H_2 and CO contents, pointing to the existence of high temperature, possibly magmatic, environments "nearby".

Fluid discharges are acid in wells from Pinatubo, Biliran, and initially, from Cagua. Cagua like Mahagnao subsequently produced only steam. A well at Palinpinon, OK-11D, did not sustain discharge, but eventually produced molten sulfur inside the casing.

4.2 Reservoir characteristics

Except for well AP-2D at Alto Peak, most wells with high magmatic contributions have low permeabilities and eventually produce super-heated steam, some show large pressure drops. Measured downhole temperatures range from 330° to >400°C.

4.3 Petrological characteristics

Most "magmatic" wells are drilled into andesitic volcanics or derived clastics. The wells at Mahagnao encountered serpentinites and basic rocks and thus also a different set of alteration minerals. In all wells, young dikes induced contact metamorphism grading into high temperature hydrothermal alteration. The degree of alteration depends on permeability and "water/rock" ratios at the time of intrusion. Fig. 3 provides a list of minerals found in these magmatic-hydrothermal wells. The temperatures indicated are based on measured and fluid inclusion homogenization temperatures. The upper limit of 700°C is imposed by the capabilities of the heating stage used. The alteration suite encompasses both neutral Al-silicate and acid sulfate mineral associations grading into contact metamorphism.

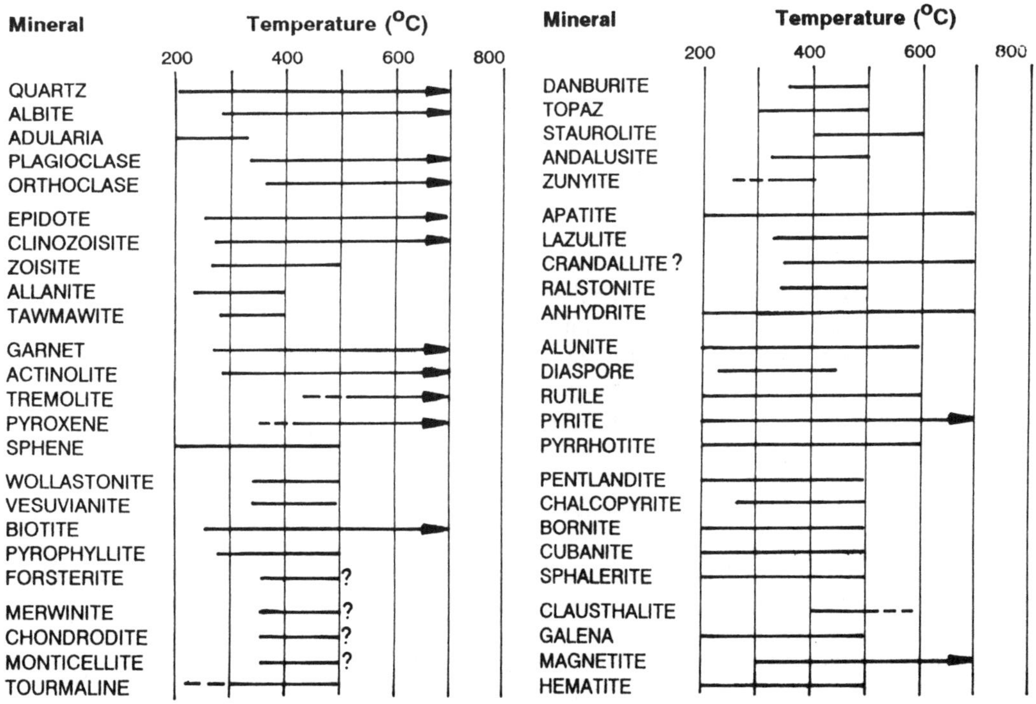

Fig. 3 Distribution of alteration minerals

Albeit present in only small quantities, Cl, B, PO_4 and F bearing minerals are relatively common in wells that encounter magmatic-hydrothermal fluids. Opaque minerals are mainly sulfides and oxides such as pyrite, chalcopyrite, bornite, cubanite, pyrrhotite, pentlandite, sphalerite, magnetite and rare enargite and clausthalite. Clay deterioration, especially of muscovite and illite, to an interlayered type clay is common when temperatures exceed 350°C.

4.4 Fluid inclusions

Fluid inclusion homogenization and melting temperatures were measured in quartz, orthoclase, anhydrite and plagioclase. Homogenization temperatures range from 350° to >700°C, with the upper limit for most between 450 to 500°C. The inclusions may homogenize to a critical point or to the liquid and/or vapor phase at >350°C. Salt crystals and anisotropic daughter minerals are common, equivalent salinities are frequently >10% NaCl. Some inclusions show the presence of elemental S; two or three immiscible fluids may appear at low temperatures.

5 DISCUSSION AND CONCLUSIONS

Since most of the geothermal systems in the Philippines are closely associated with Recent volcanism, there is always the danger of encountering highly immature fluids, unsuitable for geothermal power production with existing technology. In combination, pre-, syn- and post- drilling observations can provide unequivocal evidence for the presence of sub-magmatic fluids within or close to a geothermal well, allowing evasive action to be taken.

The comparatively rare occurrence of wells with distinct magmatic inputs, however, suggests that the extents of the primary magmatic environments are spatially quite limited, as compared to those of the "geothermal" zones proper, or that imprints left by the incursion of "immature" magmatic fluids are quickly erased. The major reasons for the ephemeral nature of imprints left by magmatic fluids are probably the comparatively short duration of volcanic-magmatic episodes and the low fluid/rock ratios accompanying alteration during a "magmatic" stage, leaving a considerable proportion of the wall rock unaltered. Later mixing of invading groundwaters with residual fluids and interaction with the unaltered rock, at lower temperatures, is likely to give rise to the formation of the neutral Cl waters typical of "mature" hydrothermal systems.

REFERENCES

Clemente, V. C. 1984. A re-evaluation of the Mt Pinatubo geothermal system. PNOC-EDC Internal Report.

Delfin, F. G., Castro, C. C., Zaide, M. C., Clemente, V. C., DeJesus, R. G., Seatres, J. S., Layugan, D. B and Rodriquez R. C. 1984. Mt Apo Geothermal Project, First stage exploration evaluation. PNOC-EDC Internal Report.

Giggenbach, W. F. 1992a. Isotopic shifts in waters from geothermal and volcanic systems along convergent plate boundaries and the origin of "andesitic" waters. Earth Planet. Sci. Let. (in review).

Giggenbach W. F. 1992b. The composition of gases in geothermal and volcanic systems as a function of tectonic setting. Proc. 7th Intl Symp. on Water/Rock Interaction, Park City, Utah, this volume.

Javellana, S. P. 1992. Country Update: Geothermal development in the Philippines. Proc. Intl Workshp on Geoth. Energy Training, Nov. 11-15, Dumaguete, The Philippines.

Matsuhisa Y. 1991. Origin of magmatic waters in subduction zones: Stable isotope constraints. Extended Abstr. Japan-U.S. Seminar on Magmatic Contrib. to Hydrothermal Systems. Kagoshima-Ebino: 91-94.

Reyes A. G. 1990. Petrology of Philippine geothermal systems and the application of alteration mineralogy to their assessment. J Volcanol. Geotherm. Res. 43: 279-309.

Ruaya, J. R. 1981. Geochemistry of Biliran Island: Water chemistry and mineral equilibria. PNOC-EDC Internal Report.

Water-Rock Interaction, Kharaka & Maest (eds) © 1992 Balkema, Rotterdam. ISBN 90 5410 075 3

The question of recharge to the geysers and hot springs of Yellowstone National Park, Wyoming, USA

Robert O. Rye
US Geological Survey, Denver, Colo., USA
Alfred H. Truesdell
Menlo Park, Calif., USA

ABSTRACT: Isotopic studies of Yellowstone National Park geyser and hot-spring waters indicate that recharge occurs at shallow and deep levels. Shallow recharge is local to thermal areas but deep recharge (from drillholes and modeling) is lighter isotopically (-148 to -150 per mil δD) than all park cold waters sampled except high-altitude waters in the extreme north and northwest parts of the park. The geology of this area is favorable for recharge to deep reservoirs in the western part of the caldera. Volumetric models suggest that a 0.5 to 5 km deep reservoir would refill in less than 1800 years. The amount of isotopically light water infiltrating now may not be enough to provide the present outflow so that a contribution of water from a cooler time such as the Little Ice Age (15th century) may be required.

1 INTRODUCTION

The extraordinary number, size and unspoiled beauty of the geysers and hot springs of Yellowstone National Park make them a national treasure. In 1967 as part of a study of the geochemistry and hydrology of the thermal areas we began isotopic studies of cold waters around the various geyser basins. Because of the oxygen-18 shift we used deuterium to trace water origins. We expected at that time that the hot-spring water of each basin represented outflow from an individual hydrothermal system with local cold recharge. We found, however, that water samples from research drillholes were significantly lighter isotopically than local cold waters. This difference caused us to broaden the area of our cold-water sample collections to include the entire park and surrounding areas. We discuss here the evidence that deep recharge waters were isotopically light and our quest for the source of those waters.

2 δD OF DEEP THERMAL WATER

Research holes drilled in 1967-68 in the geyser basins of the park provided samples of water from relatively shallow reservoirs from 65 to 332 m depth with maximum temperatures ranging from 73 to 237.5°C. In most holes maximum temperatures were 170 to 205°C. The measured temperatures were similar to average geothermometer temperatures of surface thermal waters in these basins of 179 to 210°C (silica) and 162 to 251°C (Na-K-Ca). These reservoirs were clearly feeding the thermal features. The δD values of drillhole waters were significantly lower than those of surface thermal waters (Table 1), and modeling by Truesdell et al. (1977) showed that these differences resulted from boiling and mixing processes occurring between the reservoirs and the surface. This model indicated that if the maximum temperature of deep water in the system was 360° C, as indicated by isotope geothermometers and mixing calculations, the δD of this water would be -148 to -150, in the same range as the drillhole samples and in all cases quite different from local cold waters.

These results indicated that recharge to the Yellowstone hydrothermal system was only in part local (water mixing at shallow levels) and that a source of isotopically light water recharging the deep, high-temperature part of the system must be sought. We rejected an igneous source of water which could have been no lighter than -80 δD, based on numerous earlier studies. We looked instead for infiltrating ground waters in the park

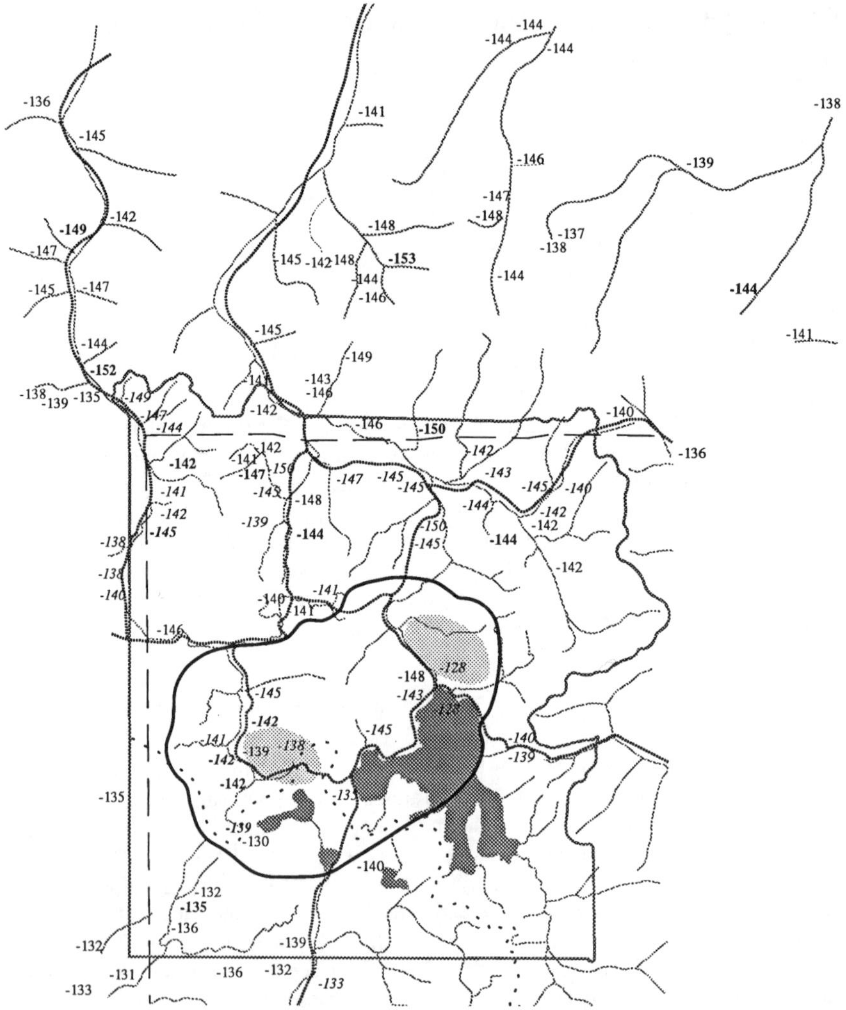

Figure 1. δD of cold waters of Yellowstone. Shown with lines: caldera (heavy), roads (medium), streams and rivers (light), park (dotted), and state (dashed) boundaries ,and patterns: resurgent domes (light), and lakes (dark)

Table 1. δD of drillhole, hot-spring and cold water samples from Yellowstone Nat. Park geyser basins

Basin	Drillholes	Hot Springs	Cold Waters
Norris	-148 to -151	-139 to -149	-140 to -144
Lower	-146 to -149	-140 to -146	-140 to -145
Shoshone	--	-135 to -139	-130 to -135

area that had δD similar to the calculated deep water. The location of this recharge area is of political as well as scientific interest because of the importance of protecting all parts of the Yellowstone thermal system from environmental and hydrologic damage.

3 δD OF COLD WATERS

We collected samples of cold springs and cold rivers and creeks yearly and at different times of the year since 1967, with the greatest number collected in 1978. Although our search was for infiltrating meteoric waters, our collections were of cold springs and cold surface waters. We also analyzed precipitation samples collected almost

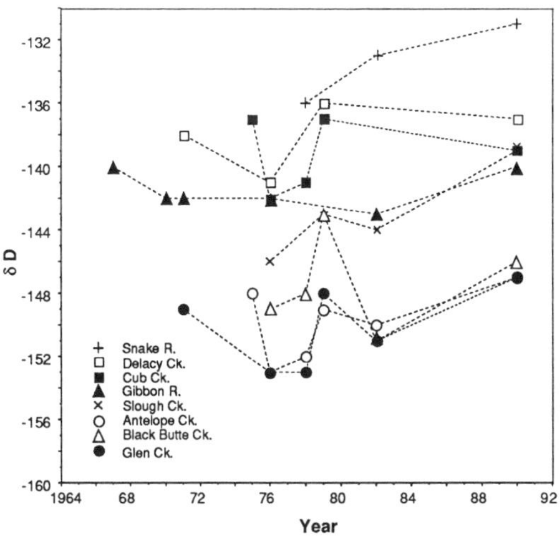

Figure 2. δD in early Fall discharge of eight Yellowstone streams sampled repeatedly.

monthly at Old Faithful and (less frequently) at Mammoth from 1978 to 1981. Comparison of these cold waters showed that although the precipitation samples varied widely in δD (from -200 to -60 observed), the weighted average of winter samples was close to yearly average cold spring and ground-water (cold wellwater) δD in the same area. Spring runoff (from snowmelt) was similar to cold spring discharge in early fall. Late spring and summer rainwater was different isotopically, so we chose early fall cold waters as most representative of snowmelt. We also compared δD in cold spring waters (which are scarce) with that in waters from small creeks and rivers and found that cold, high-altitude surface waters did not significantly differ from spring waters or change along drainages. We avoided sampling waters from lakes (or marshes) that might have evaporated and become isotopically heavier.

δD values of cold water for Yellowstone National Park and areas to the north are shown in Figure 1. Average values over several years are in italics; spring and well values, in boldface. A number of points are evident from the data: 1) δD values of the surface water become less negative from north to south and from west to east north of the park and to some extent in the northern part of the park, 2) δD values of surface water more negative than -144 are not found in the southern half of the park, 3) Surface waters with δD values of -147 to -150 are not found near the geyser basins but only at higher altitudes, mainly north of the caldera. Figure 2 shows δD of small streams in fall sample collections over more than 20 years. The variation in δD of a stream was much less than the overall range of all streams, and in most years the order of δD values was similar. Thus the δD values of possible infiltrating cold waters vary but have consistent patterns in both time and space.

4 AREAS OF RECHARGE TO THE DEEP RESERVOIR

From comparison of δD values in cold surface waters (Figure 1) with the calculated δD values of water in the deep Yellowstone thermal aquifer (Truesdell et al., 1977), we judge that the most likely recharge areas to lie to the north and northwest of the caldera. These areas include: 1) the Gallatin Range in the northwestern part of the park and north of the park between the Gallatin and Yellowstone Rivers, 2) the Washburn Range, 3) the Buffalo Plateau and 4) the Absaroka Mountains to the north of the park. From geologic considerations the most likely recharge area for the geyser basins on the western side of the park is the Gallatin Range. This area is bounded by several major north-south trending faults which parallel regional faults known from earthquake data to be active. These faults cross the caldera and may provide channels for access of water into the caldera. The Gallatin Range also has large areas of Precambrian metamorphic rocks and Paleozoic, Mesozoic and Eocene sedimentary and volcanic rocks which may have permeable contacts. The Washburn Range and Buffalo Plateau are also close to north-south faults connecting with the caldera and may recharge fumarolic areas in the eastern part of the caldera. Little is known about the isotopic composition of deep water in this area.

If infiltration on the Gallatin Range is the source of recharge to the deep reservoir feeding the western thermal areas of the park, then the average amount of infiltration should equal the average outflow of deep water. The outflow of 360°C water calculated from data of Fournier et al.(1976) from chloride flux measurements is 7×10^{10} kg /yr. The area of the Gallatin Range within the park is 1000 km^2 with 400 km^2 above 2560-m elevation. Precipitation averages 75 cm/yr at West Yellowstone and probably exceeds 100 cm/yr above 2560 m in the Gallatin Range. Thus precipitation on high areas of these mountains within the park probably exceeds 4×10^{11} kg/yr, but not all of this water has δD <-148. Infiltration of about 20% of this water could provide recharge to the western caldera thermal areas to equal the amount of deep water outflow from these areas. However, modern infiltrating waters, although adequate in volume, may not be light enough isotopically to match the present outflow.

5 AGE OF THE DEEP THERMAL WATER

The age of the recharging water must be considered for several reasons. We assumed in our search for recharge that the isotopic composition and the rate of precipitation were about the same at the time of recharge as now. These quantities must have varied significantly between glacial times and the present. In cooler periods the amount of precipitation and probably infiltration was greater, but during maximum glaciation, with high-altitude ground water frozen, infiltration was probably greatly reduced. Dating of deep Yellowstone waters with tritium and ^{14}C has been tried. Tritium analyses showed that most thermal waters were mixtures of dilute modern (1 to 100 year old) water and high-chloride water with undetectable tritium (Pearson and Truesdell, 1978). Carbon-14 was not detected in CO_2 from fumarolic steam as is to be expected for CO_2 of igneous or sedimentary origin (F.J. Pearson, Jr., unpublished data, 1978).

The age of the water may, however, be estimated from the volumes involved. Outflow of the Yellowstone hydrothermal system is so great that the reservoir and recharge conduit(s) are probably flushed reasonably frequently. The volume of the deep reservoir in the caldera (using 5% porosity and assuming that the reservoir extends from 0.5 to 5 km depth) can be estimated as between 25 and 125 x 10^{12} kg. The outflow of 360°C water is 7×10^{11} kg/yr calculated from Cl flux as mentioned earlier. This calculation suggests that hydraulic exchange (by piston flow) occurs in 360 to 1800 years. This time span allows an effect on the volume of flow from the Little Ice Age (15th century) but not from glacial or near glacial times. The isotopic effect of any residual glacial water would be very small. The contribution of water from the Little Ice Age is unknown but it probably would have resulted in recharge water of lower δD.

6 SUMMARY

An extensive park-wide collection and analysis of cold water samples was undertaken to determine the sources of recharge to the geyser and hot spring basins of Yellowstone National Park. Earlier chemical and isotopic modeling of boiling and mixing processes, along with compositions of waters from 200 - 300 m deep drillholes, show that δD of deep thermal water (-148 to -150) is significantly lower than that of local cold waters which enter thermal reservoirs only near the surface. We found that cold waters in the central, southern, and eastern parts of the park are isotopically too heavy (generally -130 to -144) to provide deep recharge. Only in the Gallatin Range in the northwestern park (and outside the park) are cold water δD values sufficiently low and geological structures appropriate to carry recharging water into the deep caldera reservoir. The quantity of high-altitude infiltration with δD <-148 in this area may not be sufficient at present to match measured thermal outflow. The age of the deep water estimated from volume and flow is less than 1800 years. δD values of infiltrating waters during this period may at times have been somewhat lower than today, indicative of a cooler climate .

REFERENCES

Fournier, R.O., D.E. White and A.H. Truesdell 1976. Convective heat flow in Yellowstone National Park, Proc. 2nd United Nations Symp. on the Development and Use of Geothermal Resources, San Francisco, 1: 731-739.

Pearson, F.J., Jr. and A.H. Truesdell 1978. Tritium in the waters of Yellowstone National Park: in Zartman, R.E. (ed.), Short Papers of the 4th Int. Conf., Geochron., Cosmochron., Isotope Geology: U.S. Geol. Surv. Open-File Report 78-701: 327-329.

Truesdell, A.H., M. Nathenson, and R.O. Rye 1977. The effects of subsurface boiling and dilution on the isotopic compositions of Yellowstone thermal waters: J. Geophys. Res. 82: 3694-3704.

Water-Rock Interaction, Kharaka & Maest (eds) © 1992 Balkema, Rotterdam. ISBN 90 5410 075 3

Secondary effects of reinjection on the Larderello, Italy geothermal system evaluated by the study of the isotopic and chemical composition of the gas mixtures

G. Scandiffio, F. Baccarin & M. Valenti
ENEL-VG Geothermal Activity, Pisa, Italy

C. Panichi, P. Noto & L. Bellucci
CNR-International Institute for Geothermal Researches, Pisa, Italy

ABSTRACT: isotopic and chemical analyses of steam and gas delivered from producing wells of the Larderello were used to study the effects of the reinjection of waste fluids. Chemical variations of Ar appear to be available tools for evaluating a recovery factor of a reinjected water. Isotopic variations in H_2, CH_4 and CO_2 are used to describe disequilibrium conditions between gas components induced by the establishment of a liquid plume at the producing level of the reservoir.

1. INTRODUCTION

In a vapor-dominated geothermal field such as that of Larderello, where the fracture-derived permeability of the reservoir is extremely variable (Baldi et al. 1980; Celati et al.. 1975) it is not easy to predict the effect of the reinjection of the industrial waste waters at the same level of the producing reservoir. Theoretically, a "secondary recovery" of heat seems to be possible from the exploited reservoir volume where the temperature is still within the 240°-260°C range, while pressure largely decreases throughout the field (Giovannoni et al.. 1981). Favourable conditions for obtaining significant production increases can be found in the horizons most exploited nowadays, over the wide zones of Larderello characterized by high permeability and low pressure (O'Sullivan and Pruess, 1980; Schroeder et al. 1980). At the moment, however, we have not a sufficient knowledge of the spatial distribution of the fractures, nor, consequently, of the path taken by the injected fluid in the reservoir and the sweep efficiency attainable.

In order to obtain valuable information from a reinjection study of vapor-dominated systems, one must use the methodology most suited to the type of fluid being produced, i.e. gas and steam. Tracers are, in this respect, of considerable importance as they can provide information on the processes occurring in the reservoir and on their evolution with time. The isotopic composition of the condensate leaving the cooling towers at the power-plants was successfully used as a "natural" tracer during monitoring of reinjection tests at Larderello (Nuti et al. 1981; Bertrami et al. 1985; D'Amore et al. 1987). The close correlation found between the hydrogen and oxygen stable isotopes of the steam and gas/steam ratios in the fluid collected in the monitored wells indicates that a mixing occurs at depth between reinjected fluids and deep components, accompanied by the formation of a liquid plume undergoing different degrees of evaporation (D'Amore et al. 1987).

Variations of isotopic temperatures, derived from H and C isotopic exchanges between gas molecules such as CO_2, CH_4, H_2 and H_2O, before and after the reinjection at Larderello, were interpreted by Bolognesi et al. (1990) as a consequence of the developing of the liquid phase in the producing steam reservoir.

Additional geochemical studies were carried out on a transect in the inner part of the Larderello field, collecting gases both for chemical and isotopic analysis. The aim of this work is to present the preliminary results obtained by the comparison of the chemical and isotopic data with those collected before the beginning of the reinjection process, started in 1979.

2 PRIMARY EFFECTS OF THE REINJECTION

Figures 1 and 2 show the effects on stable isotopes and gas contents of steam produced at Larderello field as a consequence of reinjection of the waste fluids of the power-plants. Data refers to 20 productive wells of the inner part of the field, where additional processes such as infiltration of local cold water or steam lateral condensation are little or absent. Data collected after reinjection of Fig. 1 refer either to the wells where the fraction of delivered steam from reinjected water approaches 100% or where this contribution is less than 10%. Fig. 2 shows the close relationship existing between the gas content and oxygen-18 of the steam deliveredby three selected wells, which had undergone different interactions with the reinjected waters.

From Fig.2, the correlation between gas and stable isotope contents appears to persist for any proportion

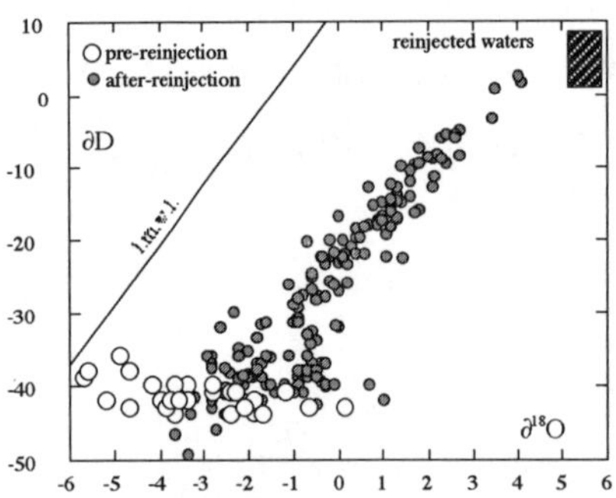

Figue 1.Stable isotopic composition of the steam produced at Larderello before and after the reinjection.Delta values are expressed in per mil vs V-SMOW.

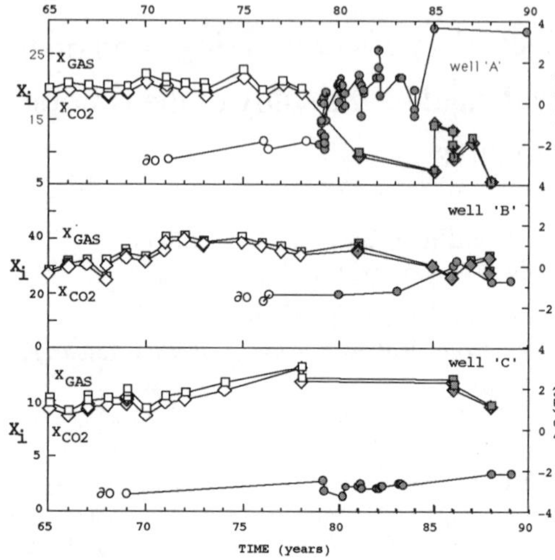

Figure 2. Variations with time of gas and oxygen- 18 contents of selected producted wells. Xi are molar fractions refer to the total fluid; ∂ values are in per mill vs V-SMOW.

of deep and reinjected components. Secondarily, the relative amount of the CO_2 seems to decrease at a lesser extent with respect to the total gas content, after the beginning of the reinjection.

In general ∂D, $\partial^{18}O$, gas/steam ratio and CO_2 contents may be successfully used in order to evaluate the amount of the reinjected water which has been recovered by each productive well in the Larderello area using the methodology proposed by D'Amore et al. (1987).

D'Amore et al. (1987) reported small variations in the composition of the H_2 and H_2S components during the 1979-82 reinjection tests. This suggests that the variationsof other gas components may be used for the evaluation of the equilibrium conditions at depth before and after the reinjection process.

3 SECONDARY EFFECTS OF REINJECTION ON DISCHARGED FLUIDS

Whereas the isotopic composition of the steam is clearly affected by mixing with injected waters in the vicinity of the reinjection sites only, the isotopic composition of gas appears to be modified over much wider zones (Bolognesi et al. 1990).

3.1 Chemical variations of the gas components

The inter-relationship among the gas constituents are shown in triangular diagrams where each species is plotted with respect the two main constituents of the geothermal fluids: water and carbon dioxide. Figure 3

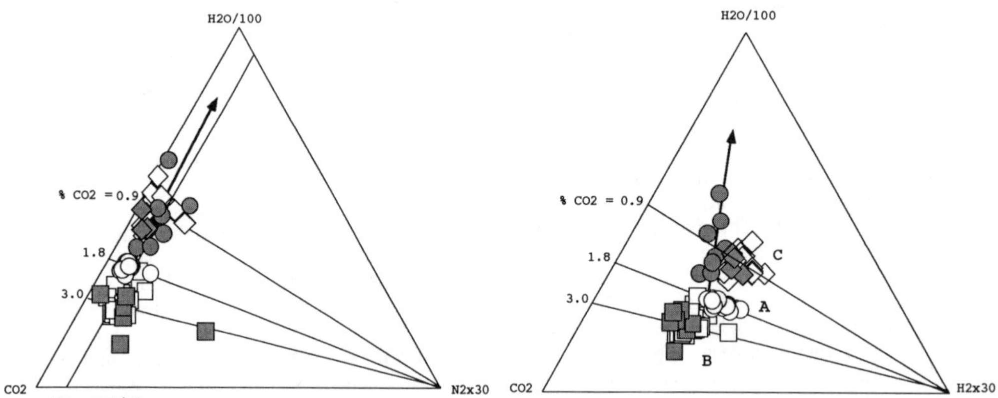

Figure 3. Relative concentration of H2O,CO2,N2 and H2 (mmol/mol of total fluid) obsrved in the selected well of the Fig.2 before (open symbols) and after (full symbols) reinjection.

shows the compositional variations between H_2O-CO_2-N_2 and H_2O-CO_2-H_2 species, of the gas mixtures discharged by the same three wells of Fig. 2. The variations observed for CH_4 and H_2S components are similar to those observed for N_2 and H_2 respectively. The comparison between samples collected before and after-reinjection (empty and full symbols) indicates that only the well "A" is strongly affected by the injected water. The distribution of data in Fig. 2 can be explained by a mixing between a deep component with a CO_2 content of about 1.8% and injected gas-depleted water.

In the case of wells B and C the composition of gas is very little changed in terms of H_2 content, while it remains constant with respect to N_2 content.

Hydrogen variations imply that the dilution effects of reinjected water are still evident even if the gas/steam ratios are not sensibly affected. On the other hand, the distribution of N_2 suggests that the injected water contributes largely to define the content of nitrogen in the geothermal fluid. Well "A" trends, in fact, towards the point where the N_2/H_2O ratio corresponds to that of the air saturated groundwater at 20°C.

The contribution of meteoric gases to the fluids discharged after reinjection is well documented by Fig. 4 where the relative content of N_2, He and Ar components is reported for gas mixtures collected before and after reinjection.

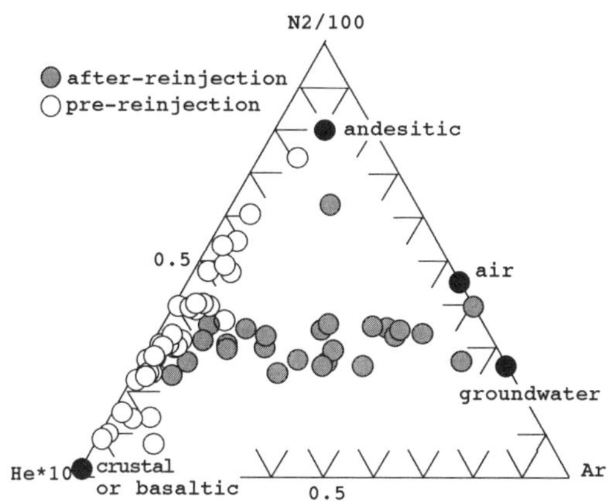

Figure 4. Relative concentrations (m mol/mol on water free basis) of N_2, He and Ar components for steam samples of the Larderello field. Compositions of crustal and magmatic fluids, together with N_2/Ar ratio for the atmosphere and air saturated groundwaters are reported for comparison.

Empty circles refer to gas emerging everywhere in the Larderello geothermal systems, while filled circles represent the situation recorded in the inner part of the field. In general, we can see that pre-reinjection data are able to define a mixing process between an almost pure magmatic component and a crustal one, and post-reinjection data reveal Ar as a very sensitive gas component with respect to the evaluation of the amount of the reinjected water recovered in a single well. The usefulness of this tracer will be of the same order as the isotopic composition of steam.

3.2 Variations of physical conditions of the fluid at depth

Preliminary evaluations of isotopic disequilibrium between gaseous components of the geothermal fluid induced by reinjection was described by Bolognesi et al. (1990). Additional data on O, H and C isotope variations of the fluid discharged by the 20 wells of the above described transect, confirm that after-reinjection samples show isotopic temperatures always lower than those obtained during previous periods (Fig. 5).

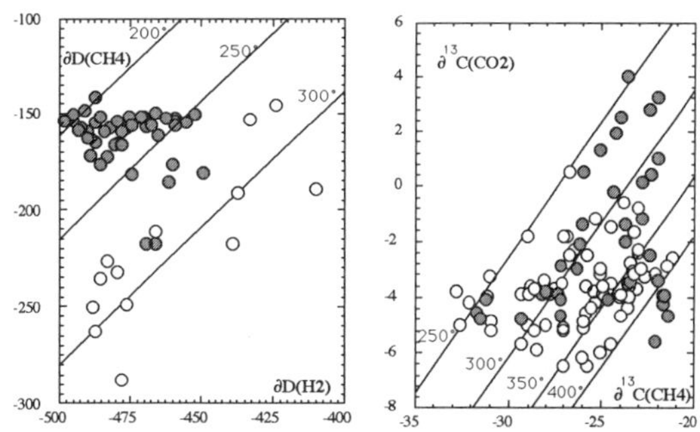

Figure 5. Isotopic distribution of the hydrogen in the pair (CH_4-H_2) and of carbon in the pair (CH_4-CO_2) observed in the steam samples collected after-reinjection (full circles). Data from previous surveys (empty circles) are reported for comparison. ∂D are vs V-SMOW, and ∂C refers to PDB standard.

In the case of (CH_4-H_2) and (CO_2-CH_4) pairs these variations appear to be essentially due to significant changes in the isotopic composition of each component. Excluding $\partial^{13}C(CH_4)$, which shows the same range, other ∂ values suggest that heavier CO_2 and lighter H_2 are newly produced in the geothermal system and that these species do not have enough time to reach new equilibrium conditions. Isotopic temperatures for the (CH_2-H_2) pair are indeed often lower than the measured ones.

Disequilibrium conditions at depth are also supported by the relationship between H_2, CO_2 and Ar components of the gas of the transect (after-reinjection period) compared with same "historical" data from Sborgi (1935) and other unpublished data of a 1963 survey. From an original composition representing near-vapor equilibrium conditions at about 250°C, an alignment along the 200°C isotherm is shown in Fig. 6, as a consequence of a progressive influence of the injected water on the deep component. However, evaluated chemical temperatures from Fig.6 may be only apparent, because the observed variations may be ascribed to the addition of the Ar componet to the geothermal system.

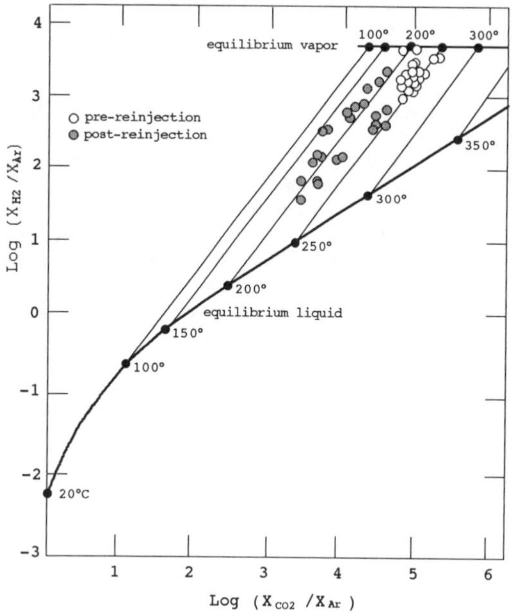

Figure 6. Evaluation of H_2-Ar, CO_2-Ar equilibrium conditions. The concentration of the gases is in mmol/mol on water free bases.

Approaching the equilibrium liquid curve, data of Fig.6 support the hypothesis relative to the formation of a liquid phase inside the vapor-dominated system.

4 CONCLUSIONS

The evaluation of the amount of injected water which is recovered by the producing wells may be improved by the use of Ar data together with those relative to ∂ values of steam and gas/steam ratios. In particular, better results may be obtained in the peripheral areas where some infiltration of recent meteoric water (depleted in oxygen-18 content with respect to the deep steam) is concomitant with the mixing process between deep and injected fluids (strongly enriched in heavy isotopes).

Further investigations are needed in order to follow the developing of the liquid plume inside the productive volume of rocks using, in particular, the isotopic composition of the gas components of the geothermal fluids.

REFERENCES

Baldi, P., Bertini, G., Calore, C., Cappetti, G., Cataldi, C., Celati, R., & Squarci, P. 1980. Selection of dry wells in Tuscany for stimulation tests. Proc. 2nd DOE-Enel Workshop for cooperative research in geothermal energy, Berkeley, 98-115.

Bertrami, R., Calore, C., Cappetti, G., Celati. R.& D'Amore, F. 1985 .A three years recharge test by reinjection in the central area of Larderello field: analysisi of production rate. Geothermal Resources Council Trans. 9/2:293-298.

Bolognesi, L., Noto, P. &Panichi C. 1990. More mileage for the isotope gas geothermometers: the case of the Larderello geothermal field. Geothermal Resources Council Trans. 14/2: 857-861.

Celati, R., Neri,G., Perusini, P. & Squarci, P. 1975. An attempt to correlate Kh distribution with geological structure of Larderello geothermal field.Proc. 1st Stanford Geothermal Reservoir Engineering Workshop, 37-41.

Giovannoni, A., Allegrini, G., Cappetti, G. & Celati, R.1981. First result of a reinjection experiment at Larderello.Proc. 7st Stanford Geothermal Reservoir Engineering Workshop, 77-83 .

Nuti, S., Calore, C. & Noto, P. 1981. Use of environmental Giovannoni, A., Allegrini, G., Cappetti, G. & Celati, R.1981. First result of a reinjection experiment at Larderello.Proc. 7st Stanford Geothermal Reservoir Engineering Workshop, 77-83 .

Nuti, S., Calore, C. & Noto, P. 1981. Use of environmental isotopes as natutal trecers in a reinjection experiment at Larderello. Proc.7st Stanford Geothermal Reservoir Engineering Workshop, 85-89.

O'Sullivan , M.J. & Pruess, K. 1980. Numerical studies of energy sweep in five-spot geothermal production-injection system. Proc. 6st Stanford Geothermal Reservoir Engineering Workshop, 214-214.

Sborgi, U. 1934. Studi e ricerche sui gas dei soffioni boraciferi con particolare riguardo al loro contenuto in elio ed altri gas nobili. Reale Accademia d'Italia, 5:667-713.

Schroeder, R.C., O'Sullivan, M.J., Pruess, K., Celati, R. & Ruffilli, C. 1980. Reinjection studies of vapor-dominated systems Proc. 2nd DOE-Enel Work shop for cooperative research in geothermal energy, Berkeley, 381-433.

Water-Rock Interaction, Kharaka & Maest (eds) © 1992 Balkema, Rotterdam. ISBN 90 5410 075 3

Inactive gas ratios of geothermal gases from the Sengan field, Northeast Japan, and their implications in origins of hydrothermal systems

H. Shigeno
Geological Survey of Japan, Hokkaido Branch, Sapporo, Japan

M. Takahashi & T. Noda
Geological Survey of Japan, Headquarters, Tsukuba, Japan

ABSTRACT: Gas compositions including principal inactive gases (He, Ar, N2, CH4 and C2H6) were analyzed for geothermal fluids collected from the Sengan field, where three active andesite volcanoes and three geothermal power plants (total capacity, 82MWe) are located. The gases were classified into N2-He and C2H6 types based on N2/Ar, He/Ar and C2H6/CH4 ratios. The N2-He type gases discharge mainly at or near the summit areas of the active volcanoes. Hydrothermal systems of these areas are expected to be heated mainly by fluid injection from magma chambers under the volcanoes. In contrast, the C2H6 type gases discharge mainly in the areas surrounded by both the active and inactive Quaternary volcanoes. This result and high B/Cl ratios of hot waters from these areas indicate that the hydrothermal systems are distributed, at least partly, in the Neogene to pre-Tertiary detrital marine sedimentary rocks, and are heated mainly by thermal conduction from relatively old magmas of the volcanoes.

1 INTRODUCTION

According to essential heating mechanisms, hydrothermal systems related to magma systems (igneous systems) at depth could be classified into two types: type A, heated by fluid convection from magma systems (magmatic fluid injection); type B, heated by thermal conduction from magma systems (Fig. 1).

Understanding the heating mechanism of each geothermal reservoir is very important for its exploitation toward electric power generation, especially for development of the deeper part of the hydrothermal system. Namely, the question has to be answered is whether expensive drilling to the depth will encounter acidic saline fluids from magma (bad condition), hot dry rock, or neutral dilute fluids of meteoric water origin (good condition), even if the temperature at the deeper part is conveniently high.

In order to predict the environments of the deeper parts of hydrothermal systems, inactive (soluble, conservative, non-reactive) components can play a very important role (refer to Fig. 1). In this context, one of us reported geochemistry of soluble components, especially of Cl, B, Li and Cs, for Japanese hydrothermal systems (Shigeno & Abe, 1983, 1986). Concerning the above objective, this paper shows results of geochemical study for major inactive gases at the Sengan geothermal field (Shigeno et al., 1989; Shigeno, 1992).

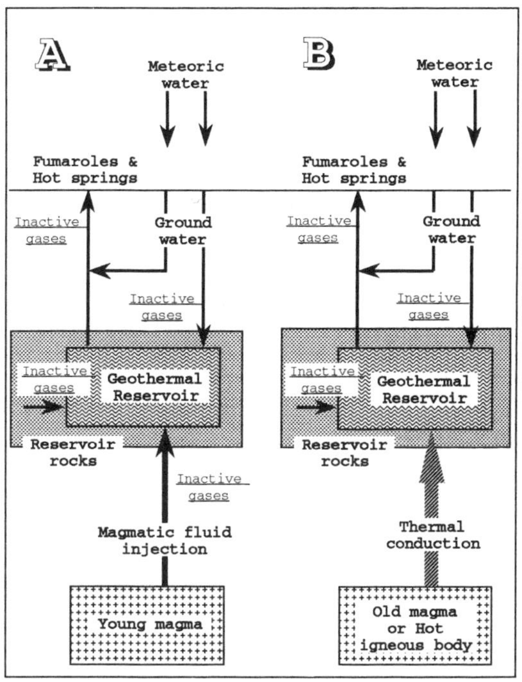

Fig. 1. Heating mechanisms of magma-related (igneous-related) hydrothermal systems with relations to origins of inactive gases.
(A) Magmatic fluid injection type;
(B) Thermal conduction type.

2 OUTLINE OF THE SENGAN GEOTHERMAL FIELD

The Sengan geothermal field (ca. 40x30km) is located at the northern part of Honshu Island, Japan. In this field, three active andesite composite volcanoes, i.e. Akita-Yakeyama, Iwatesan and Akita-Komagatake, and many inactive Quaternary andesite volcanoes are distributed. Numerous fumaroles and hot springs are located generally along the distributions of the volcanoes (Shigeno & Abe, 1987).

Three geothermal power plants, Onuma (10MWe), Matsukawa (22MWe), and Kakkonda (50MWe), have been operating, and well drilling and testing for new power plant construction have been in progress at the Sumikawa and Kakkonda II areas. At these areas, exploited geothermal reservoirs are in the Neogene formations which are composed mainly of volcanic rocks and partly of marine sedimentary rocks.

3 GAS SAMPLES AND ANALYTICAL METHODS

About sixty gas samples were collected from fumaroles and geothermal wells at the Sengan field in 1987. Most fumaroles showed boiling temperature. Ratios of residual gases to condensed water and acidic gases were determined using alkaline solutions at field. CO_2 concentrations were analyzed using alkaline solution acidification method in vacuo. H_2S concentrations were measured by conversion of H_2S through CdS to $BaSO_4$. Residual gas compositions were analyzed using three gas chromatographies for (1) He, Ne, H_2 and Ar (Molecular Sieve 5A column, O_2 carrier gas, TCD); (2) H_2, O_2, N_2 and CH_4 (Molecular Sieve 5A column, Ar carrier gas, TCD); (3) CH_4, C_2H_6 and other hydrocarbons (Porapak Q column, H_2 carrier gas, FID).

Gas samples were collected by Mitsubishi Materials Co., Ltd. (MMC) and Japan Metals and Chemicals Co., Ltd. (JMC), and were analyzed by MMC.

4 RESULTS AND DISCUSSION

4.1 Classification of the geothermal gases from the Sengan field

Concerning inactive gas ratios of the geothermal fluids from the Sengan field, Figs. 2 and 3 show the correlations between He/Ar and N_2/Ar ratios, and He/Ar and C_2H_6/CH_4 ratios, respectively. Fig. 4 shows the most dominant gas components, and He/Ar and C_2H_6/CH_4 ratios of the geothermal fluids from the Sengan field on a map.

Figs. 2 and 3 indicate that the N_2/Ar and He/Ar ratios have a positive correlation, but the He/Ar and C_2H_6/CH_4 ratios have a negative correlation. Based on these correlations and the areal distributions of the characteristics of the gases (Fig. 4), both the fumarolic and geothermal well gases can roughly be grouped into two types, namely N_2-He and C_2H_6 types, by the relative dominance of these gases. The boundary values for the two types are roughly 70, 0.010 and 0.007 for mole ratios of N_2/Ar, He/Ar and C_2H_6/CH_4, respectively.

4.2 Indications of N_2-He type gases

Along the volcanic front of the Northeast Japan arc, high temperature (100 to 800℃) fumaroles located mainly at the summit areas of andesite composite volcanoes have high N_2/Ar and He/Ar ratios up to about 40 and 4000 times those of the atmosphere, respectively (Matsuo et al., 1978; Kiyosu, 1986). These high N_2/Ar ratios are probably caused by addition of N_2 to magmas from sedimentary materials which were brought to depth by subduction of the Pacific plate (Matsuo et al., 1978). The high He/Ar ratios are associated with high $^3He/^4He$ ratios up to about ten times the atmospheric value. These volcanic gases should be derived fairly directly from magma chambers under the volcanoes. Yoshida (1986) indicated that geothermal well gases from Northeast Japan, especially from the Okuaizu and Mori areas, have such high N_2/Ar and He/Ar ratios, and estimated that the above magmatic fluids contribute to the geothermal fluids.

In the Sengan field, the N_2-He type fumarolic gases discharge on and near the active volcanoes, i.e. at the summit areas of Akita-Yakeyama and Iwatesan, and at the west slop of the Nyutosan-Akita-Komagatame volcano complex (Fig. 4). This result indicates that the N_2-He type fumarolic gases are derived fairly directly from magmas under the active volcanos also in the Sengan field.

The geothermal well gases from the Matsukawa geothermal power plant, where a vapor-dominated reservoir has been exploited, have relatively high N_2/Ar and He/Ar ratios as Yoshida (1986) indicated (Fig. 2; refer to the mixing line and the caption). In the Matsukawa area, the wells drilled at the eastern part, along the Sumikawa fault of NW-SE trend, show clearly higher gas fractions, and N_2/Ar and He/Ar ratios than the wells of the western part (Yoshida & Ishizaki, 1988). These results suggest that the Matsukawa reservoir is heated, at least partly, by magmatic fluid injection from igneous system distributed southeast to the area under Iwatesan Volcano.

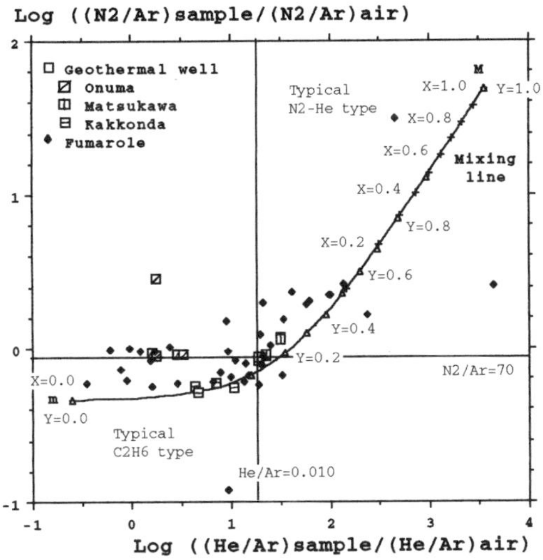

Fig. 2. Correlation between He/Ar and N2/Ar ratios for the geothermal fluids from the Sengan field. Mixing line is based on Kiyosu (1986). M: essential magmatic fluid; m: meteoric water saturated with air. For the calculated mixed fluids (symbol +), X is the fraction of the essential magmatic fluid assumed to contain relatively high N2, Ar and He fraction of 4E-5, 1E-8 and 2E-8, respectively. Also, for the calculated mixed fluids (symbol Δ), Y is the fraction of the essential magmatic fluid assumed to contain relatively low N2, Ar and He fraction of 4E-6, 1E-9 and 2E-9, respectively.

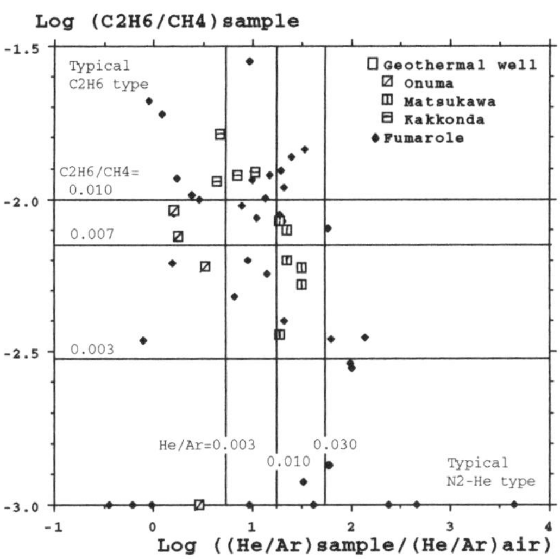

Fig. 3. Correlation between He/Ar and C2H6/CH4 ratios for the geothermal fluids from the Sengan field.

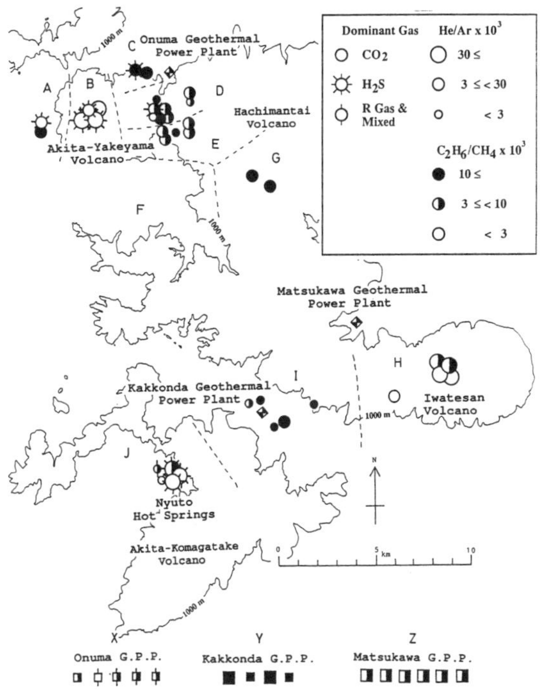

Fig. 4. Most dominant gas components, and He/Ar and C2H6/CH4 ratios of fumarolic gases from the Sengan field. Fumarole areas: A, Tamagawa; B, Akita-Yakeyama; C, Sumikawa; D, Goshogake; E, Shinzanzawa; F, Yudamatazawa; G, Toshichi; H, Iwatesan; I, Takinoue; J, Nyuto. Geothermal well gas compositions are shown at the bottom with square symbols. (Modified from Shigeno et al. (1989), and Shigeno (1992))

4.3 Indications of C2H6 type gases

Nehring & Truesdell (1978) indicated that geothermal well gases from sedimentary rock reservoirs have clearly higher concentrations of C2H6 and other hydrocarbons than fumarolic gases from active volcanoes, suggesting that these components are produced mostly by thermal decomposition of organic matter in reservoir rocks. Average C2H6/CH4 mole ratios are, for example, about 0.001 to 0.006 for the fumarolic gases from the geyser areas in Yellowstone (Gunter, 1978; reservoirs are probably in volcanic rocks), and 0.006 to 0.030 for the geothermal well gases from Cerro Prieto (DesMarais et al., 1982; reservoir is in marine sedimentary rocks).

In the Sengan field, the C2H6 type fumarolic gases discharge at the Onuma-Goshogake, Toshichi and Kakkonda-Takinoue areas (Fig. 4). These areas are surrounded by the active and inactive volcanoes or distributed on one of the inactive Quaternary volcanoes. The geothermal well gases from the Onuma and Kakkonda areas are also mainly of the C2H6 type. These distributions and the low N2/Ar and He/Ar ratios of the C2H6 type gases (Fig. 2; refer to the mixing line and the caption) indicate that the hydrothermal systems in these areas are not supplied with large amounts of magmatic fluids by igneous systems, but are probably heated mainly by thermal conduction from slightly older molten or solidified magma chambers of the Quaternary volcanoes.

The Onuma-Goshogake geothermal area is located in the Hanawa graben of N-S trend. Neutral Na-Cl type dilute waters from hot springs and geothermal wells in the area have abnormally high B/Cl mole ratio of about 1 (Shigeno & Abe, 1983). Based on these data, Shigeno & Abe (1983, 1987) estimated that the exploited hydrothermal system has a deep root in the basement rocks composed mainly of matured Paleozoic-Mesozoic detrital marine sedimentary rocks, and the deep hydrothermal system is heated mainly by thermal conduction from magmas distributed both east and west to the area under Hachimantai and Akita-Yakeyama Volcanoes. Dominance of the C2H6 type fumarolic gases at the Onuma-Goshogake area supports the above conceptual hydrothermal system model. Concerning the geothermal well gases from the Onuma area, however, the N2/Ar and He/Ar ratios which are close to those of air (Fig. 2) are probably caused, at least partly, by continuous reinjection of all the produced hot waters, which are open to the atmosphere, to the reservoir.

For the Kakkonda-Takinoue area, the geothermal reservoir is distributed partly in the Neogene detrital marine sedimentary rocks, and the hot waters are dilute neutral Na-Cl type with relatively high B/Cl mole ratio of around 0.1 (Shigeno & Abe, 1987). The dominance of the C2H6 type gases at the Kakkonda-Takinoue area is concordant with the above conditions, suggesting that the hydrothermal system is probably heated mainly by thermal conduction from relatively old magmas of the surrounding active and inactive volcanoes.

ACKNOWLEDGEMENTS: We are grateful to the people of Mitsubishi Materials Co., Ltd. (MMC) and Japan Metals and Chemicals Co., Ltd. (JMC) for their contributions to this study, especially to Drs. Akira Ueda (MMC) and Yutaka Yoshida (JMC). We thank to Dr. Yasuhiro Kiyosu, Nagoya University, for his kindly commenting on chemistry of essential magmatic fluids.

REFERENCES

DesMarais, D.J., M.L. Stallard, N.L. Nehring & A.H. Truesdell 1982. Hydrocarbon production in the Cerro Prieto geothermal field. Proc. 4th Symp. on the Cerro Prieto Geothermal Field, Baja California, Mexico, Guadalajara 1982: 7 p.

Gunter, B.D. 1978. C1-C4 hydrocarbons in hydrothermal gases. Geochim. Cosmochim. Acta. 42: 137-139.

Kiyosu, Y. 1986. Variations in N2/Ar and He/Ar ratios of gases from some volcanic areas in Northeastern Japan, Geochem. Jour. 19: 275-281.

Matsuo, S., M. Suzuki & Y. Mizutani 1978. Nitrogen to argon ratio in volcanic gases. in Alexander, E.C.Jr. and M. Ozima ed., Terrestrial Rare Gases: 17-25. Tokyo: Japan Scientific Societies Press.

Nehring, N.L. & A.H. Truesdell 1978. Hydrocarbon gases in some volcanic and geothermal systems. Geothermal Resources Council, Trans. 2: 483-486.

Shigeno, H. & K. Abe 1983. B-Cl geochemistry applied to geothermal fluids in Japan, especially as an indicator for deep-rooted hydrothermal systems. Extended Abstr. of 4th Internat. Symp. on Water-Rock Interaction, Misasa, Japan 1983: 437-440.

Shigeno, H. & K. Abe 1986. Some aspects of 'soluble elements', especially on behavior of Li and Cs in hydrothermal systems in Japan. Extended Abstr. of 5th Internat. Symp. on Water-Rock Interaction, Reykjavik 1986: 517-520.

Shigeno, H. & K. Abe 1987. Conceptual hydrothermal system model for the Sengan area based on geochemistry of hot springs and fumaroles. Rept. Geol. Surv. Japan, 266: 251-283 (in Japn. with Engl. abstr.).

Shigeno, H., M. Takahashi & T. Noda 1989. Chemical compositions of fumarole and geothermal-well gases from the Sengan field, and their implications. Abstr. Annual Meeting of Geothermal Res. Soc. Japan, Tokyo 1989: B34 (in Japn.)

Shigeno, H. 1992. Applications of geochemistry in geothermal-field development of Japan. in D'Amore, F. ed., Applications of Geochemistry in Geothermal Reservoir Development: 365-382. Rome: U.N. Centre on Small Energy Resources.

Yoshida, Y. 1986. He, Ar and N2 concentrations in geothermal steams from Northeast Japan. Extended Abstr. of 5th Internat. Symp. on Water-Rock Interaction, Reykjavik 1986: 648-651.

Yoshida, Y. & H. Ishizaki, 1988. Geochemical model of the Matsukawa geothermal field. Proc. Internat. Symp. on Geothermal Energy: Exploration and Development of Geothermal Resources, Kumamoto and Beppu, Japan 1988: 128-131.

Water-Rock Interaction, Kharaka & Maest (eds) © 1992 Balkema, Rotterdam. ISBN 90 5410 075 3

A conceptual model of the hydrothermal system in Long Valley caldera, California, USA

Michael L. Sorey
US Geological Survey, Menlo Park, Calif., USA

Christopher D. Farrar
US Geological Survey, Santa Rosa, Calif., USA

ABSTRACT: A synthesis of data from studies of Long Valley caldera, California by public and private agencies conducted since the 1960's, including the results of test drilling in the 1986-1990 period, indicates that the roots of the present-day hydrothermal system in Long Valley caldera, California, lie beneath the caldera's west moat. Geochemical characteristics of thermal fluids and reservoir rocks suggests that the source reservoir occurs in metamorphic rocks of the Sierran roof pendant that underlie the volcanic fill in the west moat, although an older, but no longer active, system involved deep circulation beneath the resurgent dome. Zones of hot water upflow could exist beneath the plateau of moat rhyolite located west of the resurgent dome or beneath Mammoth Mountain along the caldera's southwestern rim. Lateral flow of thermal water through reservoirs in the Bishop Tuff and early rhyolite accounts for temperature reversals in wells completed within the volcanic fill. Shallow magma beneath the Mono-Inyo Craters volcanic chain could provide a potent heat source for this hydrothermal system.

1. INTRODUCTION

Geothermal investigations began in the Long Valley caldera in eastern California in the 1960's with the drilling of several exploration holes. Since that time many studies by public and private agencies have been conducted to delineate the details of the present-day hydrothermal system and its evolution through time. The Long Valley hydrothermal system is characterized by magmatic heating from upper crustal sources, thick deposits of volcanic fill cut by active faults, and abundant recharge from the Sierra Nevada that borders the caldera on the west and south (fig. 1). A wealth of data is available from investigations of surface features and drilling to depths of about 2 km. Geothermal power production from a shallow 170°C reservoir currently totals about 30 MW_e from three binary-electric plants located at Casa Diablo. This short paper summarizes the current state of our understanding of this system and the techniques used to investigate it. Additional details are provided by Sorey et al. (1991), Goff et al. (1991), White and Peterson (1991), Flexser (1991), and Smith and Suemnicht (1991).

2. GEOCHEMICAL STUDIES

Geochemical techniques have been particularly useful in delineating the general circulation patterns within the hydrothermal system, including the locations of recharge areas around the caldera rim and zones of lateral outflow from upflow areas in the west moat to hot-springs around the resurgent dome and in the east moat. Isotopic evidence (carbon, strontium, and boron) and temperature reversals in wells argue for a high-temperature source reservoir in the metamorphic rocks of the Sierran roof pendant. However, chemical mass balance considerations

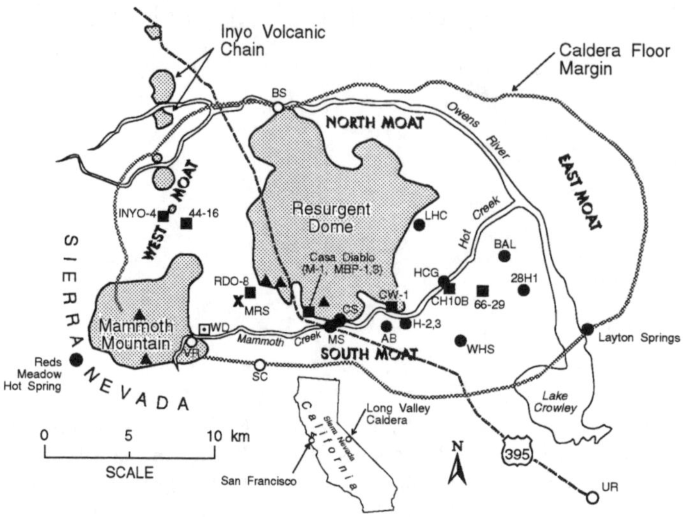

Fig. 1. Map of Long Valley caldera showing principal thermal springs (filled circles), fumaroles (triangles), selected cold springs (open circles), selected thermal (filled squares) and nonthermal wells (open squares). *MRS* denotes Mammoth Ranger Station, *LHC* Little Hot Creek, and *HCG* Hot Creek Gorge.

indicate that thermal water re-equilibrates within rhyolitic rocks of the volcanic fill, which includes the Bishop Tuff and post-caldera early rhyolite. Maximum measured temperatures in west-moat wells are near 214°C; chemical geothermometers indicate maximum reservoir temperatures as high as 248°C. Ratios of conservative ions (Cl, B, Li) in thermal fluids encountered in wells and springs in the caldera are relatively constant (fig. 2), indicating a common fluid source. Deuterium and oxygen-isotope data show that the hydrothermal system is recharged by meteoric water originating around the western rim and on Mammoth Mountain and that thermal waters flowing eastward from the west moat are progressively diluted by ground water derived from the mountainous areas adjacent the south and east moats (fig. 3).

Elevated helium-isotope ratios in thermal fluids and the volcanic history of the caldera indicate that magmatic heat input to the present-day system is most likely derived from the same magma source that drove the 500-600 year-old volcanic and phreatic eruptions along the Inyo volcanic chain in the west moat. Hydrothermal alteration from an earlier (~300,000 year) flow system, involving deeper and hotter circulation beneath the resurgent dome than occurs at present, accounts for oxygen-isotope and Cl depletion in rocks in this region. A record of the ages of both this older system and the present-day system can be developed from U/Th dating of hot-spring deposits in Long Valley and from quantities and ages of borate and other saline deposits in Searles Lake, located 200 km south of Long Valley along the ancestral Owens River. The present-day hydrothermal system appears to have been active for about 40,000 years; its initiation could be related to the onset of volcanic activity along the Mono-Inyo Craters volcanic chain.

3. GEOPHYSICAL STUDIES

Temperature profiles in selected wells show consistent evidence, in the form of reversals in temperature, of lateral flows of thermal water within the volcanic and sedimentary fill in the southern part of the caldera (fig. 4). The temperatures in these flow zones decrease

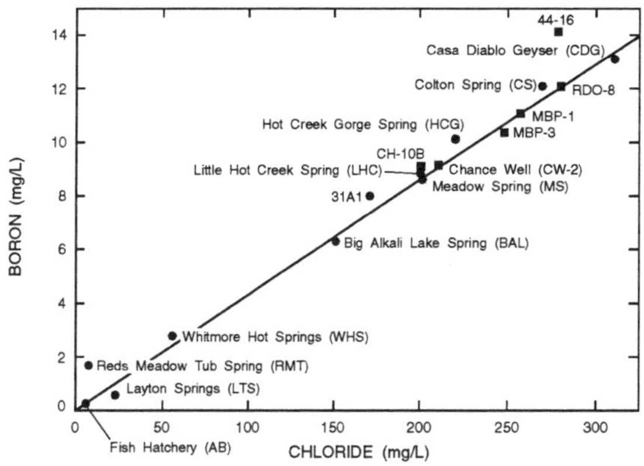

Fig. 2. Chloride and boron concentrations for thermal springs (circles) and wells (squares) in the Long Valley caldera region. Solid line represents a Cl/B ratio of 23. Locations of thermal features shown in Fig. 1.

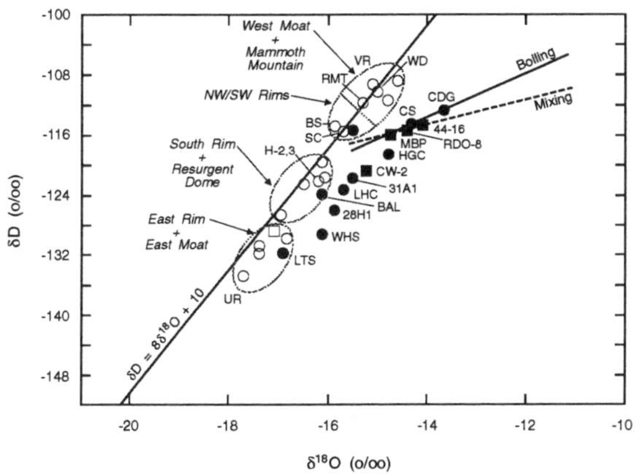

Fig. 3. Stable-isotope values for cold springs (open circles) and nonthermal wells (open squares), and thermal springs (filled circles) and thermal wells (filled squares). Value labeled *MBP* is the average of several values for wells MBP-1 and MBP-3. Locations and names of features shown in Figs. 1 and 2. Meteoric water line (labeled with equation) is that of Craig (1961). Ellipses enclose values for cold waters within general geographic regions of the Long Valley caldera. Isotope values for cold waters from the western part of the caldera are divided into Northwest (NW)/Southwest (SW) Rims and West Moat + Mammoth Mountain subareas.

monotonically from 214°C in the west moat (well 44-16) to 70°C in the east moat (well 66-29). Two such zones are in evidence in well M-1 at Casa Diablo; fluid production from the shallower zone and fluid injection into the deeper zone accompanies the binary-electric geothermal power development in this area. These temperature data, along with water-level and fluid chemistry data indicate that thermal water at temperatures of 214°-248°C flows upward into the volcanic fill in the west moat and then moves laterally eastward toward hot springs around the

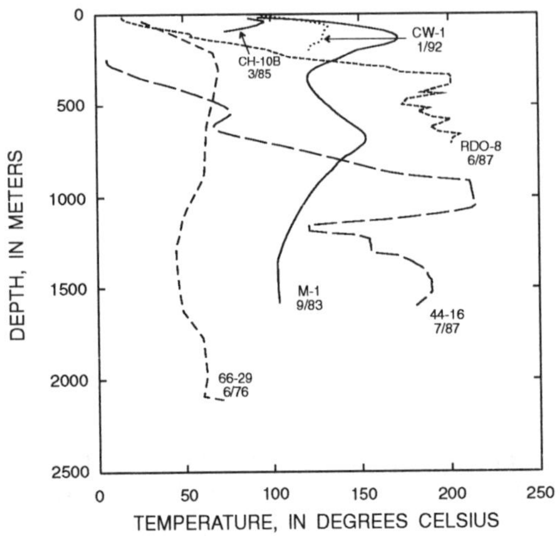

Fig. 4. Temperature profiles in selected wells in Long Valley caldera, with month/year of each profile.

south side of the resurgent dome and in the east moat. Two possible locations for the upflow zone are indicated in fig. 5 - one near the northern end of the plateau of 100 ka-moat rhyolite (RP) and one beneath Mammoth Mountain. The locations were delineated in large part from the temperature data and from electrical geophysical data noted below.

Electrical geophysical studies (magnetotelluric and time-domain electromagnetic surveys) yield a map of apparent resistivity that shows a broad region with less than 50 ohm-m resistivity and four regions with less than 20 ohm-m resistivity (fig. 6). As discussed by Sorey et al. (1991) and Nordquist (1987), the region of deep-seated 20 ohm-m resistivity on the moat-rhyolite plateau east of well 44-16 is most likely to be associated with a zone of active hydrothermal fluid upflow. Owing to sparse data coverage, the resistivity structure beneath Mammoth Mountain is not adequately delineated, and the presence of such an upflow zone in that area cannot be ruled out. Regions of low resistivity beneath the resurgent dome appear to reflect hydrothermal alteration that occurred during previous periods of hydrothermal activity.

4. GEOHYDROLOGIC STUDIES

Possible locations of deep upflow zones correspond with intersections of significant structural features. The dominant structural trend in the Long Valley caldera is NW/SE, parallel with the Sierran frontal faults that intersect the caldera along the northwestern and southeastern rims. These faults show post-caldera activation in response to regional extensional strain. The postulated upflow zone beneath the rhyolite plateau occurs near the intersection of the extensions of the Hartley Springs and Laurel-Convict faults with the northeast-trending Discovery fault zone (fig. 5), discussed by Suemnicht and Varga (1988). Similarly, fluid upflow beneath the north side of Mammoth Mountain may be associated with the intersection of the caldera ring fracture zone, the Discovery fault zone, and north-trending faults along the Inyo volcanic chain. The available information on hydraulic head and temperature are consistent with upflow at either location. Upflow of hot water along the ring fracture zone west of well 44-16 is not considered a viable alternative because of the deep water level (300 m) and relatively low temperatures (75°C) encountered in well INYO-4, drilled beneath the Inyo Craters (Sorey et al., 1991). Significant increases in $^{3}He/^{4}He$ in fumarolic gas discharged from the north side of Mammoth Mountain following seismic swarm activity in 1989 may be indicative of shallow magmatic intrusions that could provide potent heat sources for fluid circulating beneath Mammoth Mountain and out into the caldera.

5. PLANS FOR DEEP DRILLING

A scientific exploration hole (51-20 in fig. 5) is currently being drilled on the resurgent dome, northeast of Casa Diablo, to investigate thermal conditions and stress regimes in the region of recent crustal unrest. Well LVEW 51-20, being drilled under the supervision of Sandia National Laboratory, reached a depth of 2.3 km in October 1991. Temperatures and thermal gradients are relatively low (maximum temperature 104°C) thus far, but both may

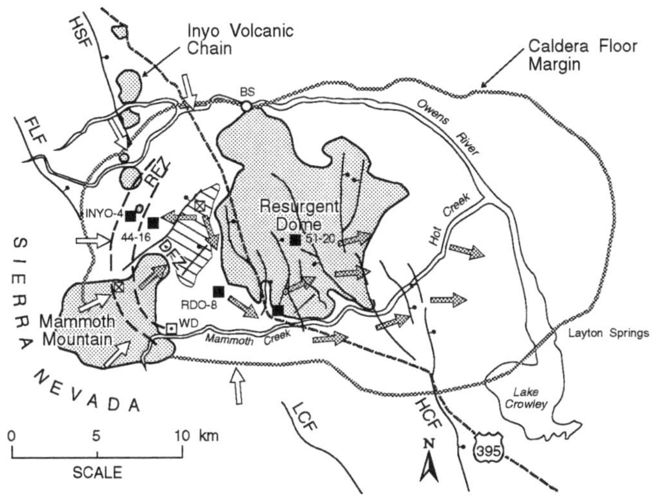

Fig. 5. Postulated directions of groundwater flow within the post-caldera volcanic and sedimentary fill in the present-day hydrothermal system in Long Valley caldera. Open arrows denote cold-water recharge; solid arrows denote thermal-water flow. Thermal wells in the western moat and at Casa Diablo shown as squares with well numbers; principal normal faults shown with bar and ball on downthrown side. *RFZ* denotes ring fault zone; *HSF* Hartley Springs fault; *FLF* Fern Lake fault; *HCF* Hilton Creek fault; *LCF* Laurel Convict fault; *DFZ* Discovery fault zone. Also shown are target areas (squares with cross) for future drilling beneath the rhyolite plateau (*RP*) and Mammoth Mountain.

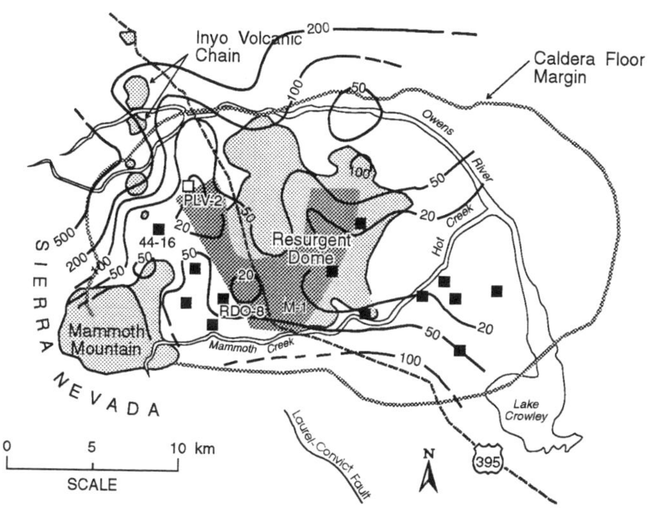

Fig. 6. Invariant apparent resistivity at 1 Hz, based on data collected from MT stations by Unocal Geothermal Division (Nordquist, 1987). Resistivity contours in ohm-meters. Also shown are wells with reservoir temperatures above 50°C (filled squares), well PLV-2 (open square), and the U-shaped region of unusually deep low resistivity interpreted from 2-D and 1-D MT interpretations (darkest pattern).

increase significantly if plans for drilling the next phase to a depth of 4.3 km are carried out in 1993. Sections of this hole are being continuous cored to provide a detailed record of lithology and fracture distribution and access for in-hole measurements of stress and hydrologic parameters.

Scientific drilling to a depth of 2 km is being proposed for the west moat at sites on the rhyolite plateau and Mammoth Mountain (fig. 5). Drilling and scientific experiments in these wells would be coordinated by the U.S. Geological Survey, Lawrence Berkeley Laboratory, and Unocal Geothermal Division. The goal of this drilling effort is to delineate the roots of the present-day hydrothermal system, including the principal upflow zone, the source reservoir, and the heat source. Both holes would be continuously cored, and completed so as to allow various types of post-drilling borehole testing to be accomplished.

6. CONCLUSIONS

The roots of the present-day hydrothermal system in Long Valley caldera appear to lie to the west of the resurgent dome. Although geochemical and well-temperature information indicate that a source reservoir exists in metamorphic basement rocks beneath the volcanic fill, its location and extent are as yet uncertain. Thermal water at temperatures of 70°-214°C encountered in reservoirs in the Bishop Tuff and early rhyolite are supplied by lateral flow from deep-seated zones of upflow in the west moat. Additional drilling to depths of about 2 km is needed to evaluate possible locations for the upflow zone. Such a delineation would facilitate future exploration and development of geothermal resources of this region, and may provide useful information on the nature of the heat source for the present-day hydrothermal system.

REFERENCES

Flexser, S. 1991. Hydrothermal alteration and past and present thermal regimes in the western moat of Long Valley caldera. *Journal of Volcanology and Geothermal Research* 48:303-318.

Goff, F., H.A. Wollenberg, D.G. Brookins, & R.W. Kistler 1991. A Sr-isotope comparison between thermal waters, rocks, and hydrothermal calcites, Long Valley caldera, California. *Journal of Volcanology and Geothermal Research* 48:265-281.

Nordquist, G. 1987. Mapping the hydrothermal system beneath the western moat of Long Valley caldera using magnetotelluric and time-domain electromagnetic measurements. In N.E. Goldstein (ed.), *Proceedings of the Symposium on the Long Valley Caldera: a pre-drilling data review*. Lawrence Berkeley Laboratory Report LBL-23940:24-31.

Smith, B.M. & G.A. Suemnicht 1991. Oxygen isotope evidence for past and present hydrothermal regimes of Long Valley caldera, California. *Journal of Volcanology and Geothermal Research* 48:319-339.

Sorey, M.L., G.A. Suemnicht, N.C. Sturchio, & G.A. Nordquist 1991. New evidence on the hydrothermal system in Long Valley caldera, California, from wells, fluid sampling, electrical geophysics, and age determinations of hot-spring deposits. *Journal of Volcanology and Geothermal Research* 48:229-263.

Suemnicht, G.A. & R.J. Varga 1988. Basement structure and implications for hydrothermal circulation patterns in the western moat of Long Valley caldera, California. *Journal of Geothermal Research* 93(B11):13,191-13,207.

White, A.F. & M.L. Peterson 1991. Chemical equilibrium and mass balance relationships associated with Long Valley hydrothermal system, California, U.S.A. *Journal of Volcanology and Geothermal Research* 48:283-302.

Water-Rock Interaction, Kharaka & Maest (eds) © 1992 Balkema, Rotterdam. ISBN 90 5410 075 3

Temporal changes in mud volcanos, Salton Sea Geothermal Area

A. A. Sturz, R. L. Kamps & P. J. Earley
University of San Diego, Calif., USA

ABSTRACT: The Salton Sea Geothermal Area is a well-documented site of hydrothermal activity in southeastern California. Thermal leakage and localized surface expression of subsurface fluid flow have formed mud volcanos in an otherwise uniformly impermeable cap confining the geothermal reservoir. Chemical compositions of surface fluids and interstitial waters (sampled December 1985, June 1991, October 1991, January 1992) associated with nine mud volcanos suggest a complex plumbing system which results in various degrees of mixing between the brines of the deep-seated geothermal reservoir and surface waters. Preliminary interpretation of the fluid flow rates and chemical compositions suggests that fluid flow is greater in winter than summer, and was greater in 1985 than 1991-92. The mud volcano edifices were larger in 1985 than in 1991-92. Cl/Br ratios higher in winter than in summer suggest that mud volcano fluids are influenced by a greater proportion of hypersaline brine in the winter. This indicates that local basin recharge influences mud volcano edifice formation, fluid flow volume, and degree of mixing between surface runoff and deep reservoir brine.

1 INTRODUCTION

The Salton Trough is located at the northern end of the Gulf of California, at the transition from extensional tectonics of the East Pacific Rise to transform tectonics of the San Andreas Fault System. It is a seismically active region of high heat flow, >100 mW/m^2 (Lachenbruch et al., 1985). Within the Salton Trough, the Salton Sea Geothermal Area is one of several well-documented sites of hydrothermal activity associated with magma intrusion at depth (e.g. Elders, 1984; Elders and Sass, 1988). Newmark et al. (1988) describe the Salton Sea Geothermal Area as an arcuate region about 4 km wide and 12 km long located near the southern shore of the Salton Sea. Thermal gradients rise across the area from near regional (0.090°C/m) to extreme (0.816°C/m) (Newmark et al., 1988).

The Salton Trough is a below-sea-level closed basin isolated from the Gulf of California by Colorado River delta deposits. Local precipitation and irrigation runoff accumulate in the lowest portion of the basin, forming the Salton Sea. The trough is a tectonic depression which is partially filled by alluvial fan materials, Colorado River deltaic deposits, and lacustrine sediments (Muffler and Doe, 1968; Mufffler and White, 1969; Randall, 1974). The varied depositional environments in which these sediments accumulated impose a heterogeneous distribution of primary permeability. Generally in this area, transitional terrestrial-shallow marine sediments underlie lacustrine sediments. The transitional terrestrial-shallow marine sediments are dominated by more permeable fluvial-deltaic siltstones, sandstones, and gravels deposited in the Gulf of California. The lacustrine units are dominated by less permeable evaporites, poorly consolidated muds, and shales deposited in the Salton Trough after it was isolated from the Gulf of California in the mid-Pleistocene (Randall, 1974).

The overall permeability structure in this region has been divided into two sections, an impermeable upper cap overlying a more permeable sequence. Based on geophysical well logs and petrology of well cores and cuttings, Randall (1974) interprets this permeability distribution to be original, a depositional permeability distribution. Younker et al. (1982) interpret the permeability distribution in this region as original depositional permeability that has been enhanced by hydrothermal alteration associated with the high thermal gradient. This interpretation is based on geophysical well log data and on heat transfer profiles which suggest that in the upper cap unit vertical heat transfer is by conduction and the thermal gradient is steep. Underlying the thermal cap is a more permeable zone which is isothermal and heat transfer is by convection. Younker et al. (1982) define a lower boundary for the upper cap unit which varies in depth from 750 to 1000 meters, and does not coincide the lithologic permeability boundary of Randall (1974).

Leakage in the otherwise impermeable thermal cap may be observed in localized areas. A line of mud volcanos and CO_2 vents runs northeasterly from the southeast corner of the Salton Sea. Based on anecdotal information from local residents, the mud volcanos and CO_2 vents have remained active at least since the early 1920's. Local residents say that the intensity of CO_2 degassing, fluid flow volume, and temperature of some vents has varied through time. Some say that the variation is seasonal, some say the variation is decadal, some say that variation follows earthquake activity. To date, there is has been no consistent scientific monitoring of these vents.

The South mud volcano field is located at the southeast corner of the Salton Sea, along Davis Road, between MacDonald and Schrimpf Roads. South field consists of four large mud mounds, several smaller mud mounds, and more than 20 circular depressions. Some of the depressions are filled with warm water, some with cool water, and some are filled with soft, water saturated mud. Both the mud mounds and circular depressions exhale CO_2.

North field, a second region of water and mud-filled depressions, is on Davis Road about 1 km north of South field. North field is privately owned and has been developed for hot mineral baths. Development has been abandoned, but an open-ended, vertically oriented pipe remains in place. Warm water fountains constantly from the pipe. It is unknown how deep the pipe extends.

Several CO_2 vents have been observed off shore in the Salton Sea, between Mullet Island and the southeast shoreline. Based on thermal gradient measurements off shore near the CO_2 vents (Newmark et al., 1988) and near-surface (76 meters subbottom) sediment interstitial water chemistry (Sturz, 1989), localized leakage disrupts the impermeable cap.

2 THIS STUDY

This study investigates the source of fluids and muds that build the mud volcanos at North and South fields. We hope to continue this study for several more years to establish a baseline data set for comparison to future fluid chemical composition, fluid flow volume, and temperature changes.

3 METHODS

Interstitial water was separated from mud by centrifugation and analyzed according to the methods of Gieskes and Peretsman (1986). Dissolved chloride, calcium, and magnesium analyses were by titration. Dissolved bromide analysis was by colorimetry. Mineralogy of the muds, using the $<2\mu$ sediment size fraction, was determined by X-ray diffraction analyses of the air dried, glycolated, and 525°c heated (001) and (060) reflections. Table 1 gives the chemical compositions of the fluids.

4 RESULTS AND DISCUSSION

One site was sampled at South field in December 1985. At that time, the largest mud volcano was a conical structure greater than 2 meters tall and greater than 5 meters diameter at the base. Three smaller mud volcanos (between 1 - 1.5 meters tall) were also active. All were expelling water saturated mud from their apices as bursts of gas and mud bombs, at 3-10 second intervals. Standing water about 15 cm deep surrounded the bases. Outside the rim of standing water, were more than 20 circular depressions filled to about 3 cm below the surface with constantly bubbling mud. Temperature of the mud in these depressions varied from 18 to 27 °C. Mud was collected from the 27 °C mud pot (site #3) and from near the apex of the largest mud volcano. (Table 1.)

In July 1991, South field was revisited. Mud mounds were much smaller than in 1985, about 1 meter high, and no longer exhaling mud and gas from apices. Small, shallow (2-3 cm deep) pools of standing water surrounded the mud mounds. Many circular depressions were still active, including the depression sampled in 1985. As in 1985, soft water saturated, bubbling mud filled the depressions, but standing water or mud level was about 15 cm below the surrounding surface. Seven of the mud-filled depressions were sampled, sites 1-7. Site 3 of this series is the same depression that was sampled in 1985. Temperature of mud in the depressions varied from 28 to 39°C.

In October 1991, water from the abandoned pipe at North field was sampled, site 9. Flow rate was about 30 liters per minute and water temperature was 43°C.

In January 1992, South field sites 1-7, and water from the North field pipe, site 9, were resampled. Also at this time, muds from two additional actively bubbling, mud-filled depressions, sites 8 and 10, were collected. Site 8 is about 20 meters southwest of sites 1-7. Site 10 is a newly built vent on the side of one of the smaller mud volcanos that was expelling mud and gas from the apex in 1985, but not in July, 1991. The new vent is a bubbling, mud filled crater, about 0.5 meters up the side of the original mud mound. Site 10 is about 5 meters northwest of site 3. At this time, temperatures of the bubbling muds at sites 1-7 ranged from

18 to 32°C. Temperature of waterflowing from the North field pipe, site 9, was 43°C and flow rate was about 30 liters per minute, unchanged from the previous October.

Muds from sites 1-7 are dominated by trioctahedral smectite (45-70% smectite; 20-35% illite; 10-20% kaolinite).

Table 1. Chemical composition of mud volcano fluids. Tem = temperature.

	Date	Tem °C	pH	Mg mM	Ca mM	Cl mM	Br mM	Cl/Br
North field								
#9	10/91	43	6.8	5.9	15.4	313	0.14	2236
	1/92	43	6.8	7.8	16.0	317	0.16	1918
South field								
#3	12/85	27	7.7	8.7	9.1	460	0.0	2300
	7/91	33	7.9	11.2	10.9	431	1.29	334
	1/92	25	7.5	12.4	9.2	450	0.2	2250
#1	7/91	36	7.4	19.7	14.5	3200	2.13	1502
	1/92	18	6.6	14.4	24.4	1890	0.95	1989
#2	7/91		7.5	24.0	7.6	953	0.18	5294
	1/92	24	7.6	10.1	9.3	371	0.21	1767
#4	7/91	39	7.8	16.4	10.3	693	2.20	315
	1/92	22	7.2	19.8	11.5	782	0.29	2697
#5	7/91	32	7.4	15.0	11.3	1644	0.90	1827
	1/92		7.4	44.7	16.2	1372	0.52	2638
#6	7/91	28	7.6	72.6	11.5	3724	2.75	1354
	1/92	20	7.5	10.8	14.8	948	0.34	2788
#7	7/91	32	8.1	33.7	11.0	1323	0.40	3308
	1/92	31	7.6	5.4	11.6	326	0.18	1811
#8	1/92	24	7.6	1.0	22.9	2304	0.67	3439
#10	1/92	21	7.2	4.3	11.8	193	0.15	1287

Hydrothermal reactions associated with emplacement of magma at depth in the Salton Sea Geothermal Area have resulted in sediment alteration to the greenschist facies (e.g. Elders, 1984; McKibben and Elders, 1985; Cho et al., 1988; Caruso et al., 1988) and the formation of base metals enriched interstitial fluids in the geothermal reservoir at depths greater than 1000 meters (Rex, 1983; Thompson and Fournier, 1988). Temperature in the geothermal reservoir exceeds 300°C. Hypersaline brines associated with the geothermal reservoir originated from dissolution by cold groundwater of halite associated with lacustrine sediments (Rex, 1983). The brine has been hydrothermally altered by interaction at elevated temperature with reservoir rocks (Rex, 1983).

Blending of the geothermal reservoir brine with fossil lake waters and desert runoff gives rise to variation in subsurface interstitial waters. Rex (1983) suggests that the various subsurface and surface fluids can be characterized by Cl/Br. The geothermal reservoir brine is characterized by Cl/Br >20,000. Salton Sea surface water, originated from local precipitation and irrigation runoff, has Cl/Br <300. Mixtures of the geothermal reservoir brine with surface runoff and fossil lake waters produces intermediate Cl/Br fluids.

We utilize the Rex (1983) Cl/Br classification, and other geochemical signals, to infer the source of fluids associated with mud volcanos of North and South fields. The source of the fluids can give some indication of the extent to which the otherwise impermeable thermal cap is breached. If the fluids are derived directly from the deep seated geothermal brine (Cl/Br >20,000), this suggests that the conduits through which the fluids flow rupture the entire cap sequence. If the fluids are entirely surface runoff (Cl/Br <300), this suggests that fluid circulation is near surface and the thermal cap is sealed at depth. Intermediate Cl/Br, indicating mixtures of fluids, are more difficult to interpret but suggest that the plumbing system is complex and may or may not entirely breach the thermal cap. By monitoring changes in the fluid chemistry through time, perhaps we can say something about seasonal and longer response time to mixing of fluids from different sources. Such a study may give some insight into physical and chemical processes associated with localized rupture of the otherwise impermeable cap.

South field site 3 is the only site sampled in 1985, 1991, and 1992. Temporal temperature fluctuation of the mud pot at site 3 suggests a seasonal response. In December 1985, and January 1992, the muds were higher than ambient air temperature, but cooler than temperatures measured in July 1991. Dissolved chloride at site 3 has not varied considerably (460-431 mM) between 1985, 1991, and 1992, suggesting little dilution from direct rainfall. Dissolved bromide was much greater (1.29 mM Br) in July, 1991, than in the two winter time samples (0.20 mM Br). Summer time Cl/Br is similar to that of surface runoff. Winter time Cl/Br is higher. This suggests that site 3 responds seasonally to mixing with the high Cl/Br brine of the geothermal reservoir. Based on the differences in mud volcano size and fluid flow rates between 1985 and 1991, these vents also seem to respond to differences in total annual runoff.

Dissolved chloride concentrations for South field sites 1, 2, 6, and 7 are significantly lower in the winter than summer. Such dissolved Cl differences might indicate direct rainfall dilution. However, other parameters, such as dissolved Br, Mg, and Ca do not corroborate simple dilution. At sites 1, 2, 4 and 7, dissolved Mg is lower and dissolved Ca is higher in the winter time. Mud temperatures for these sites were above ambient air temperature during both summer and winter, but were lower in the winter than in the summer. Similar to site 3, dissolved bromide, and Cl/Br ratios for sites 4 and 6 suggest that these sites are dominated by surface waters in the summer months and are mixed with some component of geothermal brine in winter.

North field site 9, water flowing from the pipe, has remained constant from October 1991 to January 1992. This site seems to be unaffected by changes in ambient air temperature and local rainfall. Dissolved chloride at site 9 is relatively low (313-317 mM Cl) and Cl/Br is intermediate (2236-1981 Cl/Br), suggesting some mixture between surface precipitation and irrigation runoff and the deep-seated geothermal brine.

5 CONCLUSIONS

It is obvious that much more work needs to be done to unravel the complexities of surface and subsurface fluid flow feeding the Salton Sea Geothermal Area mud volcanos. Preliminary interpretation of the data from mud volcano interstitial waters collected in 1985, 1991, and 1992, suggest that these volcanos are fed by a complex plumbing system that responds to variations in seasonal surface runoff and total annual rainfall, and includes some component of hypersaline brine. Conduits that feed fluid to the vents locally disrupt the otherwise impermeable thermal cap. We find no compelling evidence that the plumbing system completely breaches the thermal cap to provide a direct link to the deep-seated geothermal reservoir.

REFERENCES

Caruso, L. J., D. K. Bird, M. Cho, and J. G. Liou, 1988. Epidote-bearing veins in the State 2-14 drill hole. *Jour. Geophys. Res.*, 93: 13123-13134.

Cho, M., L. G. Liou, and D. K. Bird, 1988. Prograde phase relationships in State 2-14 well metasandstones, Salton Sea Geothermal Area. *Jour. Geophys. Res.*, 93: 13081-13103.

Elders, W. A. 1984. Notes on geology, geothermics and ore genesis in the Salton Trough. Soc. Econ. Geologists Field Trip, Feb. 24-26, *Rep. UCR/IGPP 85/4*, pp. 1-37.

Elders, W. A., and J. H. Sass, 1988. The Salton Sea Scientific Drilling Project. *Jour. Geophys. Res.*, 93: 12953-12968.

Gieskes, J. M., and G. Peretsman, 1986. Water-chemistry procedures aboard the JOIDES Resolution. *Ocean Drilling Program Technical Note No. 5.*

Lachenbruch, A. H., J. H. Sass, and S. P Galanis, Jr., 1985. Heat flow in southernmost California and the origin of the Salton Trough, *Jour. Geophys. Res.*, 90: 6709-6736.

McKibben, M. A., and W. A. Elders, 1985. Fe-Zn-Pb mineralization in the Salton Sea Geothermal System, Imperial Valley, California. *Econ. Geol.*, 80: 539-559.

Muffler, J. L. P., and B. R. Doe, 1968. Composition and mean age of detritus of the Colorado River delta in the Salton Trough, southern California. *Jour. Sedimentary Petrology*, 38: 384-399.

Muffler, J. L. P., and D. E. White, 1969. Active metamorphism of upper Cenozoic sediments in the Salton Sea Geothermal Field and the Salton Trough, southern California. *Geol. Soc.. Amer. Bull.*, 80: 157-182.

Newmark, R. L., P. W. Kasameyer, and L. W. Younker, 1988. Shallow drilling in the Salton Sea region: The thermal anomaly. *Jour. Geophys. Res.*, 93: 13005-13024.

Randall, W., 1974. An analysis of the subsurface structure and stratigraphy of the Salton Sea geothermal anomaly, Imperial Valley, California. Ph.D. Thesis, Univ. California, Riverside.

Rex, R. W., 1983. The origin of the brines of the Imperial Valley, California. *Geothermal Resources Council Transactions,* 7: 321-324.

Sturz, A. A., 1989. Low- temperature hydrothermal alteration in near-surface, Salton Sea Geothermal Area. *Jour. Geophys. Res.*, 94: 4015-4024.

Thompson, J. M., and R. O. Fournier, 1988. Chemistry and geothermometry of brine produced from the Salton Sea Scientific Drill Hole, Imperial Valley, California. *Jour. Geophys. Res.*, 93: 13165-13173.

Younker, L. W., P. W. Kasameyer, and J. D. Tewey, 1982. Geological, geophysical, and thermal characteristics of the Salton Sea Geothermal Field, California. *Jour. Volcanol. Geotherm. Res.*, 12: 221-258.

Water-Rock Interaction, Kharaka & Maest (eds) © 1992 Balkema, Rotterdam. ISBN 90 5410 075 3

Carbon stable isotope geochemistry of a geothermal system: Los Azufres, Mexico

Mahendra P.Verma, Enrique Portugal & David Nieva
Geotermia, Instituto de Investigaciones Electricas, Cuernavaca, Mexico

ABSTRACT: The fractionation of carbon-13 between two phases, vapor and liquid, of geothermal fluid is a function of temperature as well as pH of the fluid. An algorithm has been developped to obtain the isotopic composition of carbon-13 in the two phases. The isotopic analysis of carbon dioxide gas has been performed on samples from 11 geothermal wells of Los Azufres geothermal system, collected in Feb., 1989 and Sept., 1989. The $\delta^{13}C$ values lie in the range -11 to -7 ‰ and are related to magmatic carbon dioxide. A positive correlation between $\delta^{13}C$ and elevation of well production zone confirms the existence of boiling and condensation in the reservoir, which is explained according to a model of reservoir stratification.

1 INTRODUCTION

A knowledge of isotopic composition of light stable isotopes in geothermal fluids, which are the consequence of rock-water interaction at high temperature, may help to elucidate fluid flow-patterns, the process of mixing of different types of subterranean waters, possible geochemical origin of the species, and existence of boiling and condensation of geothermal fluids during their ascent. Carbon dioxide is the second most prominent species in geothermal fluid which influences to a major extent the chemical and physical characteristics and properties of the system. Within a geothermal system concentrations of CO_2 are, generally, highest at shallow depths, but they may vary widely with depth and position in the system. Mahon et al (1980) have considered that CO_2 is a critical component in determining the discharge behaviour of wells, whether the system produces steam or a steam/water mixture. The chemical composition and pH of the fluid together with hydrothermal rock alteration and deposition of secondary minerals are controlled by the concentration of CO_2. The carbon dioxide may originate from magma, decarbonation, organic matter, or dissolved CO_2 in meteoric waters. The isotopic data of carbon-13 may provide a clue to understand the origin of carbon dioxide; it is also required to know the effects of the mechanism of fractionation associated with the processes of boiling and partial condensation of geothermal fluid in the system (Tabaco et al, 1991).

The article presents $\delta^{13}C$ data on Los Azufres geothermal system and a stratification model of the reservoir to explain the behaviour of the isotopic data. The geochemical and isotopic (δD and $\delta^{18}O$) studies of the system have been reported elsewhere (Verma et al, 1989).

2 THEORETICAL ASPECTS ON ISOTOPE GEOCHEMISTRY OF CARBON

Carbon dioxide (or total carbon) is present in various forms, $CO_{2(g)}$, $CO_{2(aq)}$, HCO_3^- ($CO_{2(aq)}$), and CO_3^{2-}, in geothermal fluids. The proportion of the species is a function of temperature and pH of the fluid. Thus, the fractionation of carbon-13 between liquid (including all the carbonic aqueous species) and CO_2 in gas phase is a function of temperature and pH of the fluid.

2.1 Carbonic system

The geothermal fluid in a reservoir is generally present in two phases, vapor and liquid. The distribution of CO_2 between the two phases is governed by Henry's law. The carbon dioxide undergoes

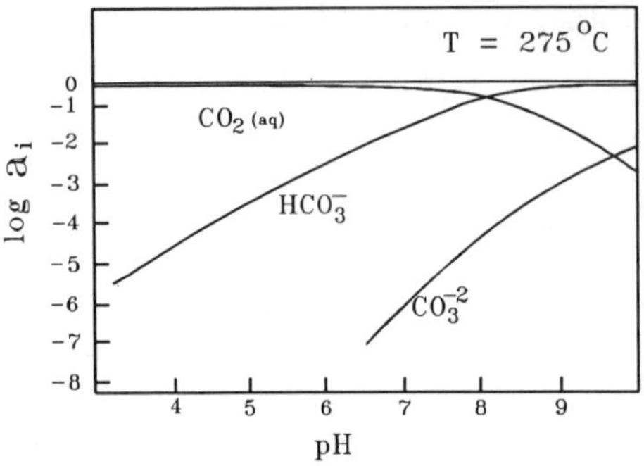

Fig. 1: Variation of carbonic species with pH

the reactions of hydration and ionization. The hydration of CO_2 forms H_2CO_3:

$$CO_2 + H_2O \overset{K_{CO_2}}{\rightleftharpoons} H_2CO_3 \qquad (1)$$

The H_2CO_3 ionizes according to the following reactions:

$$H_2CO_3 \overset{K_{H_2CO_3}}{\rightleftharpoons} H^+ + HCO_3^- \qquad (2)$$

$$HCO_3^- \overset{K_{HCO_3^-}}{\rightleftharpoons} H^+ + CO_3^{2-} \qquad (3)$$

The concentration of the carbonic species in a geothermal reservoir depends on pH and temperature. Figure 1 depicts the variation of carbonate species with pH of the fluid at 275°C. The variation of the species has been calculated with the values of constants of the reactions.

2.2 Isotopic fractionation mechanism

The fractionation of carbon-13 between the vapor and liquid phases may be expressed in the form of the following isotopic exchange reactions (Tabaco et al, 1991, and ref. cited therein):

$$^{13}CO_{2(g)} + {}^{12}CO_{2(aq)} \overset{\alpha_1}{\rightleftharpoons} {}^{12}CO_{2(g)} + {}^{13}CO_{2(aq)}$$
$$^{13}CO_{2(g)} + H^{12}CO_3^- \overset{\alpha_2}{\rightleftharpoons} {}^{12}CO_{2(g)} + H^{13}CO_3^-$$
$$^{13}CO_{2(g)} + {}^{12}CO_3^{2-} \overset{\alpha_3}{\rightleftharpoons} {}^{12}CO_{2(g)} + {}^{13}CO_3^{2-} \qquad (4)$$

where α_1, α_2, and α_3 are the respective coefficients of fractionation. The gross fractionation (α_T) may be represented in a composite reaction:

$$^{13}CO_{2(g)} + \begin{pmatrix} X_{CO_{2(aq)}} \cdot {}^{12}CO_{2(aq)} \\ X_{HCO_3^-} \cdot {}^{12}HCO_3^- \\ X_{CO_3^{2-}} \cdot {}^{12}CO_3^{2-} \end{pmatrix} \overset{\alpha_T}{\rightleftharpoons} {}^{12}CO_{2(g)} +$$

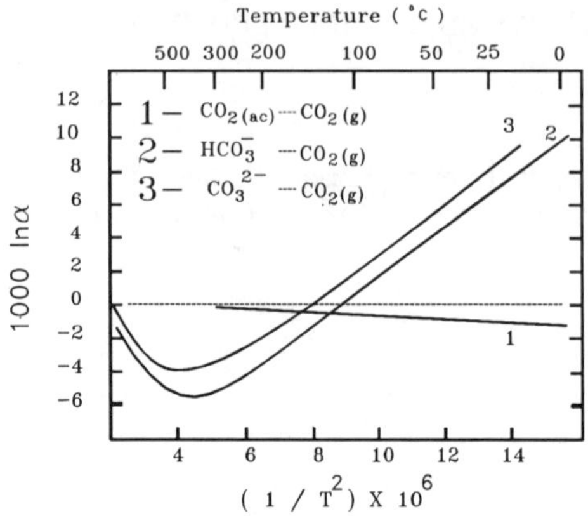

Fig. 2: The fractionation curves of aqueous carbonic species with CO_2 in vapor phase

$$\begin{pmatrix} X_{CO_{2(aq)}} \cdot {}^{13}CO_{2(aq)} \\ X_{HCO_3^-} \cdot {}^{13}HCO_3^- \\ X_{CO_3^{2-}} \cdot {}^{13}CO_3^{2-} \end{pmatrix} \qquad (5)$$

where the coefficient X_i represents corresponding molar fraction of each carbonate species in the liquid phase. The gross fractionation coefficient may be expressed as:

$$\alpha_T = \alpha_1 X_{CO_{2(aq)}} + \alpha_2 X_{HCO_3^-} + \alpha_3 X_{CO_3^{2-}} \qquad (6)$$

where

$$X_{CO_{2(aq)}} + X_{HCO_3^-} + X_{CO_3^{2-}} = 1$$

Figure 2 shows the fractionation coefficients of the carbonic species. The curves for the pairs HCO_3^-–$CO_{2(g)}$ and CO_3^{2-}–$CO_{2(g)}$- shows the positive values for the fractionation coefficients upto $130°C$. This indicates that there is an enrichment of carbon-13 in liquid phase, but the behaviour is reversed at high temperature. The curve for $CO_{2(aq)}$–$CO_{2(g)}$ indicates an enrichment of carbon-13 in gas phase at all temperatures. This means carbon-13 is enriched in the gas phase at temperatures above $130°C$ (i.e. in all the conventionally called high-temperature geothermal reservoirs).

2.3 Isotopic mass balance

The geothermal fluid, produced by the drilled wells, consists of a mixture of vapor and liquid, which separates in the two phases at the sur-

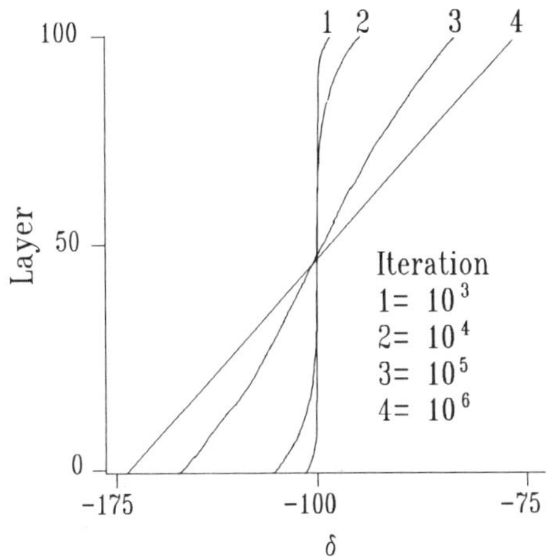

Fig. 3: Theorical behaviour of δ-value acoording to stratification model of reservoir

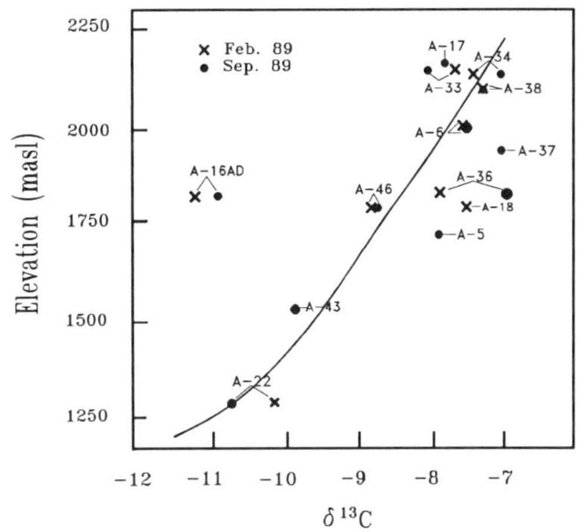

Fig. 4: Behaviuor of $\delta^{13}C$ with well production zone elevation of Los Azufres geothermal system

face. The vapor contains the non-condensable gases such as $CO_{2(g)}$, whereas the liquid phase has aqueous carbonate sepecies such as $CO_{2(aq)}$, HCO_3^- and CO_3^{2-}. The fraction (proportion) of the carbonic species present in the liquid phase is much less than that of CO_2 in the vapor phase under the fluid separation conditions generally implied during sample collection.

3 EXPERIMENTAL

The vapor phase samples of geothermal fluids were collected in evacuated glass sample bottles containing a 4N *Na OH* solution. The carbon dioxide is precipitated in the form of $SrCO_3$ by mixing a supersaturated solution of $SrCl_2$-NH_4OH in an inert atmosphere to avoid contamination with atmospheric CO_2. The reproducibility of isotopic measurement is within $\pm 0.2‰$.

4 STRATIFICATION MODEL OF A RESERVOIR

A simplified stratification model of a geothermal reservoir is proposed to comprehend the behaviour of isotopic data of geothermal fluid obtained from drilled wells. In this model the geothermal reservoir is considered as a two-dimensional closed system and fluid moves only vertically. Thus the problem is tackled one-dimensionally. To make the model simple, we assume a uniform pressure and temperature throughout the reservoir. The reservoir is considered initially homogeneous and divided in uniform layers (0 to 100) with constant isotopic ratio, $R(0,n)$. Let us consider a fraction x (say 0.1 % of fluid M in each layer) evaporates from layer n and, condenses and mixes up in layer $n+1$. And, then, the same fraction x from layer $n+1$ mixes back to layer n. Using the Rayleigh fractionation equation, the average isotopic ratio of vapor, which separates from layer n and mixes in layer $n+1$ at i^{th} iteration, may be expressed as

$$TM(i,n) = R(i,n)\frac{1-f^{\alpha}}{1-f} \tag{7}$$

Where fractionation coefficient α is defined as vapor to liquid (R_v/R_l). The modified isotopic composition of $n+1$ th layer at i^{th} iteration

$$R(i,n+1) = \frac{R(i-1,n+1)\,M + TM(i,n)\,x}{m+x} \tag{8}$$

The composition of nth layer after i^{th} iteration

$$R(i,n) = \frac{R(i,n)\,M - TM(i,n)\,x + R(i,n+1)}{M} \tag{9}$$

A restriction is also imposed in the algorithm that the fluid does not leave the boundary layers (i.e. fluid is not going down to layer 0 and not going up to the layer 100). If one considers the behaviour of the species dissolved in water, it is required to know the distribution of the species in liquid and vapor phase at evaporation according to Henry's law. Figure 3 represents the theoretical curves which were constructed with using the following parameter values: $\alpha = 1.0005$, $\delta_0 = -100‰$ and $x = 0.1M$.

5 CARBON ISOTOPE GEOCHEMISTRY OF LOS AZUFRES GEOTHERMAL SYSTEM

The carbon isotope data of 11 wells from Los Azufres geothermal system which were sampled two times in 1989 are plotted against elevation (masl) of production zone of the well in figure 4. The $\delta^{13}C$ values lie in the range -11 to -7 ‰ and show a positive correlation with elevation, although the behaviour is slightly parabolic.

Verma et al (1989) have shown with isotopic data ($\delta^{16}O$ and δD data) the existence of a possible recharge zone in far south-east area of the main geothermal field. The elevation of the east-south zone is about 1000 m below than that of the main field. Giggenbach and Quijano (1981) have demonstrated the lack of direct infiltration into the reservoir as local meteoric water is lighter in deuterium value. Nieva et al (1987) have presented a model of steam upflow and partial condensation in which the reservoir and the upper part of the reservoir contain super-heated steam. It is also in favour of the hypothesis of lack of direct infiltation or there may exist a small amount of direct recharge in the reservoir. Thus, it is a good approximation to consider as a closed reservoir for Los Azufres geothermal field.

Nieva et al (1987) have observed a positive correlation between the concentration of volatiles and elevation of the wells. Deuterium is distributed preferentially in vapor phase with evaporation above 220°C, it can be observed with the variation curve of fractionation coefficient with temperature. The Los Azufres reservoir temperature, which is an average of temperatures measured in the wells and estimated with geochemical thermometers, is about 290°C (Verma et al 1989). The δD values measured in geothermal fluid of Los Azufres geothermal system also show a positive correlation with the elevation of production zone of the well. This behaviour may be explained with the above proposed stratification model.

As mentioned above, the concentration of carbon dioxide, sampled in vapor phase samples at superficies, may be considered as concentration of total dissolved carbon in geothermal fluids. It can be observed in figure 2 that the carbon-13 distributed preferentially in vapor phase above 130°C. In case of Los Azufres, the reservoir temperature is 290°C. Thus, the tendency of $\delta^{13}C$ values to vary with elevation is related to the model of evaporation and partial condensation (Nieva et al, 1987) and can be explained with the stratification model of Los Azufres geothermal reservoir.

The $\delta^{13}C$ values gets modified due to process of evaporation and condensation in the reservoir, but the average value still represents original $\delta^{13}C$ value of the reservoir. Thus, an average of all the $\delta^{13}C$ value is close to the reservoir $\delta^{13}C$ value. The average of all the $\delta^{13}C$ values of 11 wells at Los Azufres is about -8.3 ‰) and it lies in the range of magmatic origin carbon (-8 to -5 ‰). This supports the prediction of magmatic origin carbon for Los Azufres geothermal system which was made by Quijano (1983).

CONCLUSION

The $\delta^{13}C$ values of Los Azufres geothermal field support the evidence of existence of evaporation and condensation (Nieva et al, 1987). The positive correlation between $\delta^{13}C$ values and elevation of production zone of the well may be explained with a one dimensional stratification model which considers the process evaporation and mixing using Rayleigh fractionation mechanism. The model can be improved by including the variation of temperature and pressure within the reservoir. The possible source of origin of CO_2 is magmatic origin, although there may be contribution from other sources.

REFERENCES

Giggenbach, W. and J. L. Quijano 1981. Estudio isotopico de las aguas del campo geotermico de Los Azufres. Internal report, CFE, Mexico.

Mahon, W. A. J., G. D. MacDowell and J. B. Finlayson 1980. Carbon dioxide: Its role in geothermal systems. N. Z. J. Sci., 323: 133-148.

Nieva, D., M.P. Verma, E. Santoyo, R. M. Barragan , E. Portugal, J. Ortiz and J. L. Quijano 1987. Chemical and isotopic evidence of steam up flow and partial condensation in Los Azufres reservoir. XII Work. Geo. Reserv. Eng., Stan. Univer. 253-259.

Quijano, J. L. 1983. Consideraciones geoquimicas de los pozos A6, A13, A17 Y A19 del campo geotermico de Los Azufres, Michoacan. Memorias de la 1ra reunion interdisciplinaria de actividades sobre el campo geotermico de Los azufres, CFE/IIE.

Tabaco, F., M.P. Verma, D., Nieva and E. Portugal 1991. Caracteristicas geoquimicas e isotopicas del carbono en el sistema geotermico de Los Azufres, Michoacan. Geofisica Internacional, 30: 173-182.

Verma, M.V., D. Nieva, E. Santoyo, R. M. Barragan and E. Portugal 1989. Water Rock Interaction WRI-6, D.L. Miles (Ed.): 723-726. Rotterdan: Balkema.

Water-Rock Interaction, Kharaka & Maest (eds) © 1992 Balkema, Rotterdam. ISBN 90 5410 075 3

Geothermal systems in continental area of China

Wang Ji-yang, Chen Moxing, Xiong Liang-ping & Pang Zhong-he
Institute of Geology, Academia Sinica, Beijing, People's Republic of China

ABSTRACT: Geothermal systems are widely distributed in the contintental area of China. High temperature geothermal systems of the convective type are mainly located in southern Tibet and the Tengchong volcanic zone of Yunan Province, SW China. Low-medium temperature geothermal systems of convective type are spread over a vast area of continental China, especially concentrated in SE coast. Low temperature geothermal systems of the conductive type are mainly found in large-scale sedimentary basins. The specifics and characteristics for each type of geothermal system in China are given in this paper.

1 INTRODUCTION

Geothermal systems are widely distributed in continental area of China. According to geological settings and heat transfer mechanism (Rybach and Muffler, 1981), the geothermal systems can be classified into three groups: (1) high temperature geothermal systems of convective type related to young magma body at shallow depth of the earth's crust, such as Yangbajing geothermal system in Tibet and Tengchong geothermal system in Yunan Province, SW China; (2) Low-medium temperature geothermal systems of convective type, not related to young magma body but heated by normal to relatively high regional heat flow, such as the Zhangzhou and Fuzhou geothermal systems in SE China; (3) Low temperature geothermal systems mainly occurring in large-scale sedimentary basins, such as North China Basin, Sichuan Basin and Talimu Basin.

2 HIGH TEMPERATURE GEOTHERMAL SYSTEMS RELATED TO YOUNG MAGMA BODY AT SHALLOW DEPTH IN THE EARTH'S CRUST

This type of geothermal system usually occurs in areas of plate convergence, continental rift and/or subduction zones in the world. In the continental area of China, high temperature geothermal systems of this type are concentrated in southern Tibet and Tengchong volcanic zone of Yunan Province, SW china, which is the eastern extension of the Mediterranean Geothermal Belt and has been named in China as Himalayan Geothermal Belt (Figure 1; Tong et al., 1981). Tectonically, this geothermal belt is the product of continent-continent collision of Indian and Eurasian Plates which occurred 40 Ma ago. High temperature surface manifestations, such as geysers, hydrothermal explosion, fumaroles, boiling springs, etc. are widespread in this belt (Tong et al., 1981; Tong & Zhang, 1989). Heat flow studies indicate that the geothermal background for this belt is also very high. Average heat flow for geothermal systems in southern Tibet appears around 80-100 mW/m^2 with the highest up to 364 mW/m^2 (Shen et al., 1989), and for Tengchong volcanic zone, >80 mW/m^2 (Wang, 1990). It is believed that the granitic magma body at 10-12 km depth in the earth's crust is the heat source for the high tem-

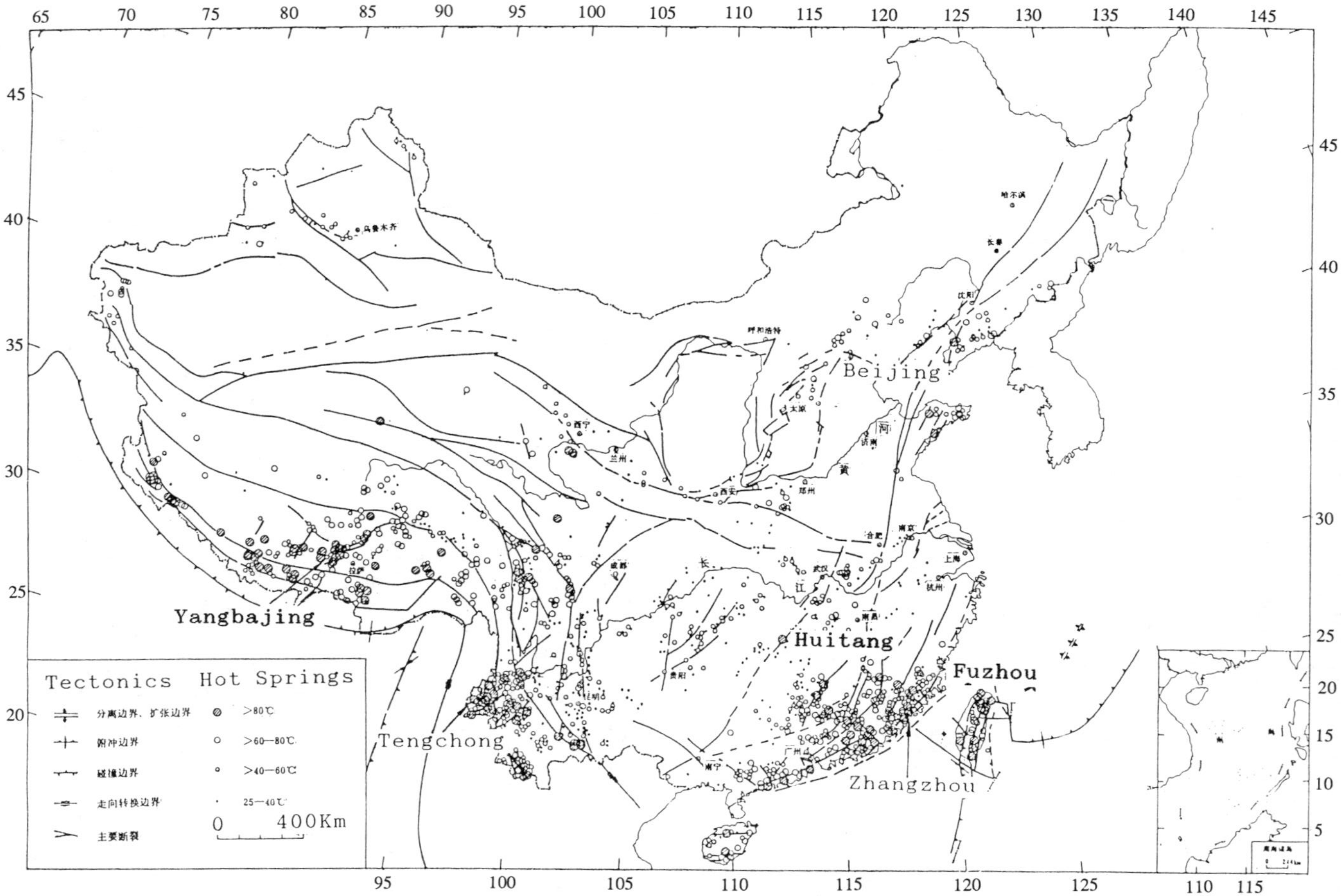

Fig. 1 Distribution of geothermal systems of convective type in the continental area of China

perature geothermal systems in this area (Shen, 1991).

Isotope and geochemical studies demonstrate that the thermal water in the high temperature geothermal systems seems to originate from meteoric water but with high contents of Li, Rb, Cs (Tong et al., 1981). Thermal fluid in this geothermal belt has been used for power generation since 1977. At present, the installed capacity of the Yangbajing geothermal power plant (largest in China) has reached 25 MW. It is estimated that the power generation potential from the Tibetan section of Himalayan Geothermal Belt is nearly 1000MW, and the Tengchong section alone provides about 450 MW (Shen & Chen, 1992; Tong & Zhang, 1989).

3 LOW-MEDIUM TEMPERATURE GEOTHERMAL SYSTEMS OF CONVECTIVE TYPE, NOT RELATED TO YOUNG MAGMA BODY

Unlike the above mentioned type, these geothermal systems are heated by normal to relatively high heat flow from depth. There is no particular heat source-magma body-underneath the systems. The temperature of thermal water in this type of systems depends mainly upon the circulation depth of the water. The deeper the water penetrates, the higher its temperature will be and vice versa.

It is clear from Figure 1 that this type of geothermal systems spreads over a vast area of continental China, especially concentrated along the SE coast. Statistics show that the background heat flow for these systems varies between 40-75 mW/m^2 depending upon tectonogeological settings in different regions (Wang & Huang, 1990). For instance, regional heat flow for Zhangzhou and Fuzhou Geothermal Systems in Fujian Province is about 70-75 mW/m^2 and for Huitang Geothermal System in Hunan Province, 40-60 mW/m^2 (Wang et al.,1992). The circulation depth of thermal water in Zhangzhou geothermal system is believed to be 3.5-4.0 km and the geotemperature pattern reflects the water flow path in the system (Xiong et al., 1990).

Isotope and geochemical studies show that the thermal water in these systems also originates from meteoric water (Pang & Wang, 1990). Along the coastal area, thermal water in some systems was revealed to be a mixture of meteoric water with sea water. In Zhangzhou Geothermal System, the mixing ratio of thermal water to seawater is 2:1 (Pang et al., 1990a). The reservoir temperature calculated by using a mixing model for Zhangzhou geothermal system is about 140 °C (Pang et al., 1990b). This result is consistent with the regional geothermal background and circulation depth of thermal water in the system (Wang et al., 1989). Fuzhou geothermal system has similar parameters as the Zhangzhou, and these two systems may be considered to be the representative ones of this type (Wang et al., 1992).

4 LOW TEMPERATURE GEOTHERMAL SYSTEMS OF CONDUCTIVE TYPE

Geothermal systems of this type in China are mainly located in the large-scale sedimentary basins, such as North China Basin, Sichuan Basin and Talimu Basin. North China Basin is a Meso-Cenozoic basin of faulting-depression origin and is viewed as a rift and/or graben. For this reason, the regional geothermal background is higher than Sichuan and Talimu Basins, which are Mesozoic and Meso-Cenozoic basins of depression origin. Heat flow measurements indicate that the average heat flow for the North China Basin, Sichuan Basin, Talimu Basin is of 62, 55, 44 mW/m^2, respectively (Chen et al., 1988, Wei, 1992). It is obvious that the systems of this type are supported by normal to relatively high regional heat flow. Thermal water in these systems is again of meteoric origin as evidenced by isotope and geochemical studies. However, in the deeper parts of the sedimentary basin, some sort of connate water has been found. As the water flow in large-scale sedimentary basins is quite low, thermal equilibrium between water and its surrounding rocks has been established for a long time. In this context, thermal water resources in these systems may be regarded as unrecoverable. Nevertheless, due to the large volume of the reservoirs, the heat content and energy potential of thermal water in theses systems turn out to be very large, too. Calculation shows that the thermal energy contained in the thermal water only from the northern part of the North China Basin is around 0.9 Billion tons of standard coal equivalent (Deng et al., 1992)

Low-medium temperature thermal waters are mainly used for non-electrical direct use. In Beijing, Tianjin, Fuzhou, Zhangzhou, thermal water is widely used for space heating, dying, drying and industrial processing. In the vast rural areas throughout the country, thermal water is extensively used for agriculture, aquaculture etc.

REFERENCES

Chen, M., Yang, S., Shen, X. and Lu, X., 1988, Terrestrial heat flow, Chapter 2 in: Geothermics of North China (Chen, M., ed.), Science Press, Beijing, 58-88.

Deng, X., Chen, M. and Zhang, J., 1992, North China Basin, Chapter 6 of Part 3, in: Geothermal Resources in China (Chen, M., Wang, J. and Deng, X., eds.), Science Press, Beijing (in Press).

Pang, Z. and Wang, J., 1990, Oxygen and hydrogen isotope study on Zhangzhou Basin Hydrothermal System SE China, GRC TRANSACTIONS, V. 14, Part II, 945-951.

Pang, Z., Fan, Z. and Wang, J., 1990a, Isotopic evidence for the genesis and sea water involvement of geothermal water in Zhangzhou Basin, Geochimica, 4, 296-302.

Pang, Z., Fan, Z. and Wang, J., 1990b, Calculation of the reservoir temperature of the Zhangzhou Geothermal Field using a SiO_2 mixing model, Chinese Science Bulletin, 35(16):1360-1363.

Rybach, L. and Muffler, P., 1981, Geothermal systems, principles and case histories, John Wiley & Sons, Ltd., New York, pp. 359.

Shen, X., 1991, Crustal and upper mantle thermal structure of Xiang (Tibet) inferred from the mechanism of high heat flow observed in S. Tibet, in: Terrestrial heat flow and the lithosphere structure (Cermak, V. and Rybach, L., eds.), Springer-Verlag, Berlin, pp. 293-307.

Shen, X., Zhang, W., Yang, S., Shen, J., Zhang, R., Yang, Z. and Xia, S., 1989, Updated report on heat flow data in Xizang (Tibet), Chinese Science Bulletin, 34 (20):1728-1732.

Shen, S. and Chen, M., 1992, Xizang (Tibet) autonomous region, Chapter 3 of Part 2, in: Geothermal Resources in China (Chen, M., Wang, J. and Deng, X., eds.), Science Press, Beijing (in press).

Tong, W. and Zhang, M., 1989, Geothermics in Tengchong, Science Press, Beijing, pp. 262.

Tong, W., Zhang, M., Zhang, Z., Liao, Z., You, M., Zhu, M., Guo, G. and Liu, S., 1981, Geothermics beneath Xizang (Tibetan) Plateau, Science Press, Beijing, pp. 170.

Wang, J., Xu, Q. and Zhang, W., 1990, Heat flow data and some geologic-geothermal problems in Yunan Province, Seismology and Geology, 12(4):367-377.

Wang, J. and Huang, S., 1990, Compilation of heat flow data in the China continental area (Second edition), Seismology and Geology, 12(4):351-366.

Wang, J., Pang, Z., Xiong, L., Jiang, Z. and Zuang, Q., 1989, Genesis analysis of Zhangzhou Basin Geothermal system, in: Advance in Geosciences (1), IGAS, China Ocean Press, Beijing, 409-417.

Wang, J., Xiong, L. and Chen, M., 1992, Nature of hydrothermal system and geothermal resources type in coastal area of SE China, Chapter 14 of Part 4, in: Geothermal Resources in China (Chen, M., Wang, J. and Deng, X., eds.), Science Press, Beijing (in press).

Xiong, L., Wang, J. and Pang, Z., 1990, Circulation depth of thermal water in Zhangzhou Geothermal Field, Scientia Geologica Sinica, 4, 377-384.

Wei, D., 1992, Terrestrial heat flow on the northern side of Tarim Basin, Scientia Geologica Sinica, 1, 93-96.

Water-Rock Interaction, Kharaka & Maest (eds) © 1992 Balkema, Rotterdam. ISBN 90 5410 075 3

Determination of cation exchange parameters for a hot dry rock geothermal energy reservoir

Luke Wilkinson, John H.Tellam & John R.Ashworth
School of Earth Sciences, University of Birmingham, Edgbaston, UK

Hugh G. Richards
Camborne School of Mines Geothermal Energy Project, Rosemanowes Quarry, Penryn, Cornwall, UK
(Presently: Environmental Protection Group, BNFL Ltd, Sellafield, Cumbria, UK)

ABSTRACT: The Camborne School of Mines Hot Dry Rock (HDR) Geothermal Energy Project has sited an HDR reservoir at a depth of 2-2.8 km in the Carnmenellis Granite, Cornwall, U.K. Geochemical experiments by previous workers have determined that mineral dissolution and precipitation are important processes. However certain experiments indicate that cation exchange is also a significant process. Laboratory experiments using rock from the production well, measured exchange capacities of 12-24 meq/m^2 for natural joint surfaces likely to form the surfaces swept by the flow during circulation, and 3.5 meq/100g for granite drilling chips. XRD and SEM analyses show that mixed-layer clay minerals are the probable exchange sites. Subsequent laboratory experiments were aimed at establishing whether existing models of ion exchange equilibria suitably described exchange in the granite of the HDR reservoir. Applying a Vanselow type equilibrium model to the results, produces selectivity coefficients which show a strong dependance on exchange site occupation, and site occupation history.

1. INTRODUCTION

1.1 The Hot Dry Rock experimental system.

Experimental investigation into the viability of Hot Dry Rock (HDR) as an energy resource for the UK began in Rosemanowes Quarry in the Carnmenellis Granite in 1977, the location of the site is shown in Figure 1.. The objective of the research has been to demonstrate that it is possible to create reservoirs of economic size at shallow depths, and develop the technology required to construct and operate economically viable HDR reservoirs at depths of 6 km.

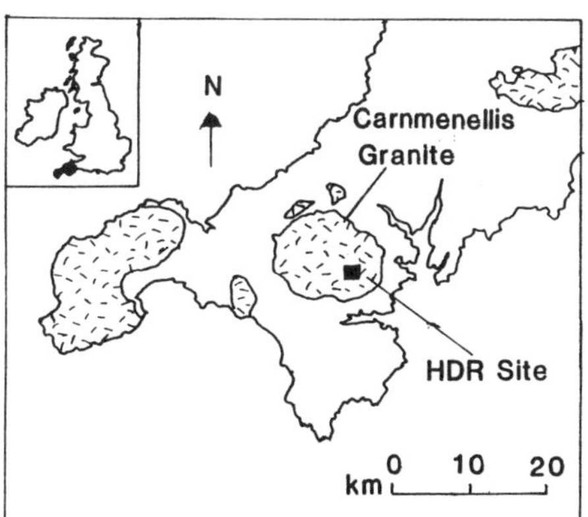

Figure 1. Location of the HDR experimental site.

Since 1980 a HDR reservoir has been developed at a depth in excess of 2 km, where the rockmass consists of medium grained biotite-muscovite granite, and contains an extensive bi-directional natural joint system. The injection and production boreholes (RH12 and RH15 respectively) have been successfully connected through hydraulic stimulation of this joint system, which contains the main flowpaths. Successive stimulations have reduced water losses to <15%. The injected circulation fluid is abstracted from a local stream, and has a $Na\text{-}Ca\text{-}K\text{-}Mg\text{-}Cl\text{-}HCO_3\text{-}SO_4$ based composition with total dissolved solids (TDS) of 100 mg/l, and a pH of 6.5-7. Injection rates under normal circulation conditions are around 15 ls^{-1}; however due to a fast flowpath in the system temperatures of production fluid have declined rapidly with circulation time from 80 °C in 1985 to 42 °C in 1990.

1.2 Previous studies of water-rock reaction processes in the HDR reservoir.

Geochemical investigations have been a major area of research, focusing on two areas of HDR operation: reservoir diagnostics, and long term operations. Diagnostic applications have included using inert tracers to evaluate reservoir hydraulics, and radon emanation to estimate fracture apertures. One major concern of the long term operational study is to determine the effects of water-rock interactions on reservoir permeability, and potential corrosion of borehole casing and surface plant.

Chemical results from reservoir circulation experiments indicate that mineral dissolution and precipitation are major influences on circulation fluid chemistry. The characteristics of these results are a net increase of TDS in production fluids, and uptake of in particular, K and Mg by the reservoir Laboratory studies by Savage et al. (1987) provided information on likely reactants and precipitates in a series of batch autoclave experiments, using granite drilling chips from 1800m in RH12 and waters of injection fluid compositions, at temperatures ranging from 80° to 250°C. Savage et al. concluded that plagioclase and K-feldspar were the major dissolving phases at 80° C, and inferred that smectite and calcite were likely precipitates. The rates of reaction and amounts of precipitate indicated that an injection fluid of the above composition would not produce a chemical environment that could significantly impede reservoir performance even at high temperatures.

2. EVIDENCE OF CATION EXCHANGE IN THE HDR SYSTEM

2.1 Recirculation production run chemistry

Some heat production experiments in the reservoir took place under closed-loop circulation conditions i.e. the production fluid was recirculated for considerable time periods. At the end of these experiments, fresh stream water was injected, producing rapid changes in reservoir fluid chemistry (Richards et al. 1989). The change depended on the chemistry of the displaced solution, and could only be explained by dissolution and/or precipitation phenomena if unrealistic reaction rates wre invoked. An alternative process which could more satisfactorily explain the observed chemical changes was cation exchange.

Exchange reactions will affect the inter-element ratios in the production fluids and estimates of dissolution rates derived from circulation experiments.

2.2 The magnesium injection experiment

The cation exchange which occurred during the fresh water flush experiments appeared to affect magnesium to a greater degree than the other major ions. Therefore, a field experiment was designed to monitor the reservoir response to a step increase in magnesium concentrations. The experiment was conducted when the reservoir was in closed loop circulation, under conditions which would minimise the effects of other reactions. Magnesium, in the form $MgSO_4.7H_2O$, was injected for 7 days, with total Mg content of the injection stream maintained at 1 meq/l. Fluid chemistry was monitored for 20 days, and ICP analyses of injection and production fluids, shown in Figure 2, indicate that Mg exchanged for sorbed Ca, and this has been tested with mass balance estimates. The results of the magnesium injection experiment confirmed that ion exchange was likely to be a significant process in the HDR reservoir, but the potential exchange capacity of the reservoir was unknown.

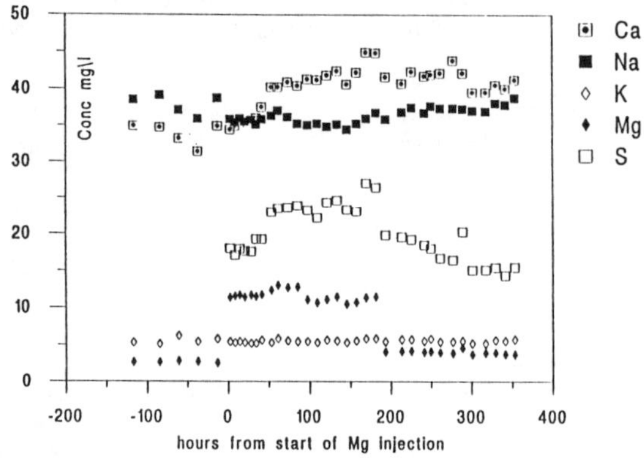

Fig 2a: RH12 fluid compn during Mg injection expt.

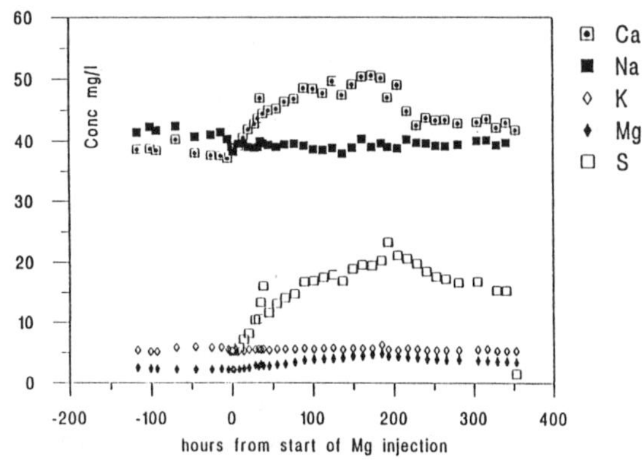

Fig 2b: RH15 fluid compn. during Mg injection expt.

Figures 2a & 2b: Chemistry of the circulation system during the Mg injection experiment.

2.3 Measuring Cation Exchange Capcitities (C.E.C's) for reservoir materials.

Following the field studies, a series of laboratory experiments were carried out to measure C.E.C's for two materials representing reservoir mineralogy:

1. Drilling chips from two sample heights in RH15, and a mixed sample from 8 sample levels in RH15. The chips are analagous to those used by Savage et. al. (1987).
2. Surfaces of joints from RH15 core lightly coated with alteration minerals.

The drill chippings were placed in stoppered conical flasks with 1M LiCl, and allowed to react for 2 hours. The C.E.C.'s of the joints were measured by pooling a known volume of LiCl on the surfaces, for 2 hours. The resulting solutions were analysed by ICP, and exchange capacities calculated; the results are given in Table 1.

Table 1. C.E.C.'s of granite drill chippings and joint surfaces.

Exchanger	Expt	C.E.C.
Granite chips	batch	
1785m RH15		4.30 meq/100g
Granite chips	batch	
2243m RH15		3.50 meq/100g
Granite chips	batch #1	
mixed RH15		3.47 meq/100g
Granite chips	batch #2	
mixed RH15		3.40 meq/100g
Granite chips	batch #3	
mixed RH15		3.50 meq/100g
RH15 2100m	pool	
Joint surfaces		12-24 meq/m^2

These results have been corrected for the presence of precipitated salts by measuring the concentrations of salts released in distilled water batches. EQ3NR/6 (Wolery 1983, 1984) was employed to check that dissolution reactions did not contribute to the measured exchange capacities.

3. DETERMINING CATION EXCHANGE EQUILIBRIUM MODELS FOR THE HDR SYSTEM

If cation exchange is a significant process in the HDR system, then a suitable model of exchange equilibrium, incorporating selectivity coefficients, is necessary for use in long term operational prediction of fluid chemistry, and in re-interpreting past field experiments. Laboratory experiments were designed to study exchange equilibria, and measure selectivity coefficients for granite chips in a range of solution compositions containing the four main cations present in the reservoir.

3.1 Laboratory Method

Granite drilling chips, taken from the openhole section of RH15, were sampled and thoroughly mixed to produce a stock of chips representing the mineralogy of the reservoir. Four 500g. aliquots of this stock sample were placed in columns and repeatedly flushed with 0.1N monocationic chloride solutions of Na, K, Ca or Mg. Samples of eluent were filtered and analysed by ICP. This tested that the exchange sites were thoroughly saturated, and checked their exchange capacities. A methanol flush of the chips removed any salt precipitated during the pretreatment flushes.

Varying amounts of treated chips were placed in flasks and mixed with 50 or 100 ml of 0.1N chloride solutions of different cation compositions. For example samples of Na-saturated chips were reacted with solutions containing 100% KCl, or 75% KCl + 25% NaCl, or 50% KCL + 50% NaCl, or finally, 25% KCl + 75% NaCl.

The contents of the reactors were mixed frequently by shaking, and allowed to react for up to 7 days. Sub-samples of solution were taken at four and twenty four hours, to assess the degree of equilibration in each system. The main series of experiments were carried out at laboratory temperatures for ease of operation. Limited experiments at 60°C suggested that similar phenomena were occuring, though this is yet to be confirmed.

3.2 Results and discussion

The solutions were analysed by ICP, all systems reached equilibrium in less than 24 hours. The equilibrium exchange site chemistry was calculated from the changes in solution chemistry, and a Vanselow-type equilibrium model was used to calculate selectivity coefficients K_V (Vanselow 1932). Mass balances were checked in each case. For each pair of exchange reactions, e.g. Ca-Mg, the experiments measured the approach to equilibrium from both directions i.e using Ca-saturated chips, and Mg-saturated chips and reacting both batches with the same solutions as described above. The results indicated that different equilibrium chemistries were attained dependant on the original saturated ion, both in the Na-K and Ca-Mg systems. Figure 3 is a plot of ln K_V against exchange site composition for; Mg replacing sorbed Ca, and Ca replacing sorbed Mg.

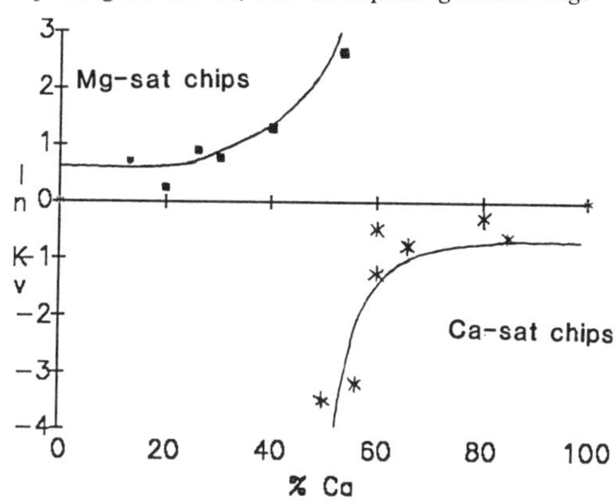

Figure 3. Plot of ln K_V against sorbed Ca for the Ca-Mg exchange system.

These results strongly suggest that K_V's are a function of exchange site occupation, and previous site occupation history. dependance of K_V on site occupation has not infrequently been found in soil chemistry work (e.g. Jensen and Babcock 1973), and arises at least in part because the activities of the sorbed ions is not adequately described by equivalent fractions. The dependance of K_V on site occupation history is not normally found to be so marked as it is here. However even Vanselow (1932) found some degree of dependance in the (relatively) simple exchangers he used in his experiments. In the crushed granite system examined here more complex behaviour might not entirely be unexpected considering the potential exchangers.

4. EXCHANGE SITE MINERALOGY

It can be inferred that ion exchange in the HDR reservoir is most likely to be due to the presence of sheet silicate minerals with large surface area to volume ratios, and unsatisfied lattice charges. Although investigation is in an early stage, minerals of this type have been identified in this and previous studies. In thin sections of the borehole cores, apparently fresh granite is seen to contain several neoformed phases:

1. Green chlorites and colourless sheet silicates replace biotite micas along cleavages and crystal edges
2. Feldspars are replaced by small crystals of clay minerals.

The clay fractions of natural joint surfaces from RH15 cores have been studied using XRD by Thomas and Milodowski (in Edmunds et al. 1987), a trace for a pale green alteration mineral assemblage, similair to one used in the laboratory measurements of C.E.C.'s, are shown in Figure 4.

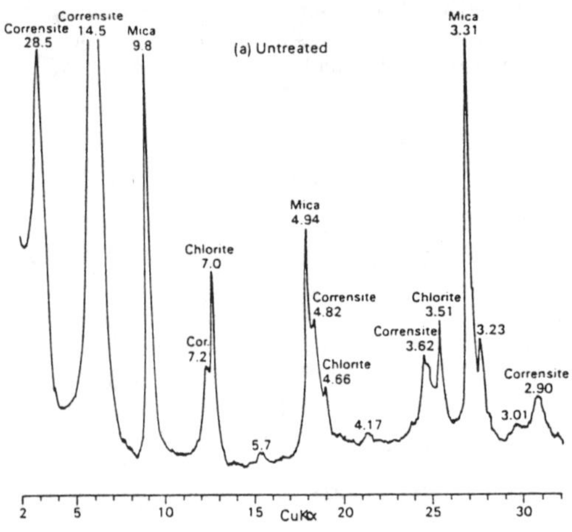

Figure 4. XRD trace of clay minerals from joint surface from 2050m in the HDR reservoir.

Corrensite has an estimated C.E.C. of 30meq/100g (Richards 1991), and chlorite of 10-40 meq/100g. The drill chippings are from the same boreholes as the cores, and XRD analysis has detected mixed layered clay minerals in the <2 m fraction. Further studies currently in progress, using a JEOL SEM with EDAX, have located Ca-smectites on the chip surfaces.

It is likely that the smectites, chlorites and micas on the joint surfaces are analogous to those in the bulk granite. These minerals contain the exchange sites responsible for the phenomena observed in the laboratory and field experiments.

5. CONCLUSIONS

1. Production run chemistry, the field experiment and laboratory experiment indicate that cation exchange is influencing fluid chemistry in the HDR system.
2. Laboratory experiments on drilling chips indicate, as often with ion exchange work, that the exchange mechanisms are complex. For example applying a Vanselow equation gives selectivity coefficients strongly dependant on site occupation and site occupation history. The dependance on site occupation history appears to be particularly strong in the present instance.
3. The exchange sites are located in chlorites, smectites and micas in the granite and the fissures.
4. Future work will be aimed at confirming these conclusions for the drilling chips, and fissure material, and investigating the implications of exchange using a mixing cell model incorporating an ion exchange algorithm based on the experimental results.

REFERENCES

Jensen H.E and K.L. Babcock 1973. Cation exchange equilibria on a Yolo loam. Hilgardia Vol 41 pp 475-487.

Richards H.G., C. Wilkins, R.L.F. Kay, and D. Savage, 1989. Geochemical results from the Rosemanowes HDR system 1986-88. Geochemistry in relation to Hot Dry Rock Geothermal Development in Cornwall Vol.2 Research Report SD/89/2 Brit. Geol Surv. Keyworth 240pp.

Richards H.G., D. Savage, J.N. Andrews 1992. Granite-meteoric water reactions in an experimental hot dry rock geothermal reservoir at the Rosemanowes test site, Cornwall, UK. Applied Geochemistry (In Press).

Savage D., M.R. Cave, A.E. Milodowski and I. George 1987. Hydrothermal alteration of granite by meteoric fluid: an example from the Carnmenellis Granite, UK. Contrib. Mineral. Petrol. 96 pp 391-405.

Thomas L.J. and A.E. Milodowski 1988. Alteration mineralogy. in Edmunds W.M et Al., Granite water interactions in relation to hot dry rock geothermal development. Investigations of the Geothermal potential of the UK. British Geological Survey, Keyworth. pp 57-80.

Vanselow A.P. 1932. Equilibria of the base exchange reactions of bentonites, permutites, soil colloids and zeolites. Soil Science 33, pp 95-113.

Wolery T.J. 1983. EQ3NR. A computer program for geochemical aquaeous speciation solubility calculations, users guide and documentation. LLNL, UCRL No. 53414 Cal.

Wolery T.J. 1984. EQ6. A computer program for reaction path modelling of aquaeous geochemical systems: users guide and documentation. LLNL manuscript. UCRL Cal.

Water-Rock Interaction, Kharaka & Maest (eds) © 1992 Balkema, Rotterdam. ISBN 90 5410 075 3

The rare-alkali metal characteristics of Tibetan geothermal systems and the potential for exploitation and utilization

Zheng Yaxin
Commission for Integrated Survey of Natural Resources, The Chinese Academy of Sciences, People's Republic of China

ABSTRACT: The world's highest values of geothermal fluid rare-alkali metal contents were discovered in Tibetan geothermal systems. Rare-alkali metal springs were discovered in the Yangbajain geothermal field, the largest hydrothermal system in the Himalayan geothermal belt. The unique geochemistry offers information on the origin of heat for the Himalayan geothermal belt, which is located on the global continental crust convergence zone with the greatest crust thickness. The highest rare-alkali metal contents in Tibetan geothermal fluids are Li 65.4 ppm, Rb 13.5 ppm and Cs 99.8 ppm. The recovery of minerals from the discharged fluids from the Yangbajain geothermal power plant creates a rare opportunity and a good alternative for comprehensive utilization of geothermal fluids, elimination of environmental pollution and enhancement of economic benefits.

1 GENERAL DESCRIPTION OF HIGH RARE-ALKALI METAL CHARACTERISTICS

In the south Tibetan plateau, geothermal manifestations along Yarlung Zangbo River and Senge Zangbo River basins are world-famous for their high temperature and high rare-alkali metal content characteristics. The "Geothermals under Tibetan Plateau" published in 1981 was the summation of field work and laboratory analyses in the 1970's (Tong et al. 1981). In the book, the highest Li value in the Sodo No. 1 spring of Gar county is 50 ppm. At that time, the Li value in Tibet was the second highest value in the world. The highest Rb value in Tibetan hot springs (Zedowa hot spring, Gerze county) was 11.04 ppm, the Rb value was also the highest value in world literatures. The highest Cs value in the Rabrong hot spring, Ngamring county, Tibet, was 58 ppm, the second highest Cs value in the Semi boiling spring, Ngamring county, was 50 ppm. These two values were also the highest and the second highest world records in the literature. Some Cs value records abroad are lower than these. The Cs value in the Puge Hot Spring, Kashmir, is 11-12 ppm. The Cs value in the Ninakevu Hot Spring, Hokkaido, Japan, is only 6.2 ppm.

In 1989, new analytical data from the Zhumosha hot spring were reported. The Zhumosha Hot Spring is located at the middle reaches of Yarlung Zangbo River, just at the entrance of Sangri-Gyaca Gorge, within Qusum county. The Zhumosha Hot Spring analytical data are 65.4 ppm Li; 13.5 ppm Rb; and 99.8 ppm Cs. These values create new world records for Li, Rb and Cs.

On June 27, 1989, we did the field work at the riverside of the Yarlung Zangbo River to get the Zhumosha Hot Spring water sample. Tibetan locals told us that the water level was rather high, and some parts of the spring outlets were inundated by the river at that time. The flowrate was estimated at approximately 2 l/sec. The annual amount of transported cesium is almost 6294.59 kg. The amount is about 100 times the current world demand. The Zhumosha Hot Spring is really a cesium spring on the Roof of the World.

The Tibetan plateau covers a vast area. Most of geological resources, including geothermal resources on the plateau have only been sparsely surveyed. The large-scale and extensive survey work is still at its initial stage. The newly found content values at the end of the 1980's exceed those published at the beginning of the 1980's. The possibility of finding more springs with higher rare-alkali metal values in Tibetan geothermal systems is not excluded. However, the rare-alkali values now found on the Tibetan Plateau might be recognized as some important decipherable codes of the Tibetan plateau, known as the golden key to the Geoscience--within scientific circles.

2 YANGBAJAIN GEOTHERMAL FIELD

There are limited geothermal fields in Tibet under trial exploitation. These fields include Yangbajain, Yangyixiang, Langjiu, etc., have not only been surveyed, but also have undergone drilling and logging. The Yangbajain geothermal field is one of the largest geothermal systems, located in a basin faulted at the Nyainqentanglha piedmont benchland.

The main strata outcropped in the Nyainqentanglha Mountain are granite-gneiss of Palaeozoic era. The K-Ar absolute age determined from biotite is 21 Ma. It is the nearest metamorphic age. The Nyainqentanglha fault zone, located on the southeastern side of Nyainqentanglha Mountain and the northwestern side of Yangbajain basin, with a strike of N-E (40-50 degrees) is characteristic of compression and torsion, while the fault zone is intersected by N-W tension fault.

The Yangbajain fault basin is located on the upper wall of the Nyainqentanglha fault. The oldest outcropping strata in the basin are of Carboniferous age with overlying volcanic rock of early Cretaceous period and moraine debris of Pleistocene age. There are multiple Himalayan movements affecting Yangbajain granite intrusions in slates of Pangduo group of Carboniferous age and volcanic rock of early Cretaceous period. The isotope age of the Yangbajain granite is 60 Ma. These strata are distributed in a strike line of E-W on eastern side of the basin, but these strata turn to strike N-E, parallel with the fault in the basin. The drag strikes are formed by the Nyainqentanglha fault. The fault was a left-lateral parallel displacement fault of compression and torsion at its early development. When the Tibetan plateau uplifted in the Cenozoic era, the fault was a gravity fault along the weak faces which emerged at the earlier stage, and fault basins were formed.

In the southeastern fringe, there is the Maruguo-danguoguo fault, with a strike of NNE and overlying Quaternary drifted and talus materials. There is also a NW local uplifted block within the Yangbajain basin. There exist many hidden faults on these uplifted bedrock massifs. The Yangbajain geothermal field is located on the local uplifted block and is closely related to the faults, which are cutting the uplifted block. The most important faults are perhaps the NE hidden sulfur mine fault and the NW fault in parallel with the uplifted crest.

To investigate the relationship between the plane of Yangbajain geothermal field and the two-dimensional distribution of rare-alkali metals, the lithium anomaly area with 80 ppm isopleth (Fig. 2), the Rb anomaly area with 200 ppm isopleth (Fig. 3) and the cesium anomaly area with 250 ppm isopleth (Fig. 4) are drawn. Fig. 2, Fig. 3, and Fig. 4 show no difference between rare-alkali metal anomaly scopes and the geothermal field limits. However, the rare-

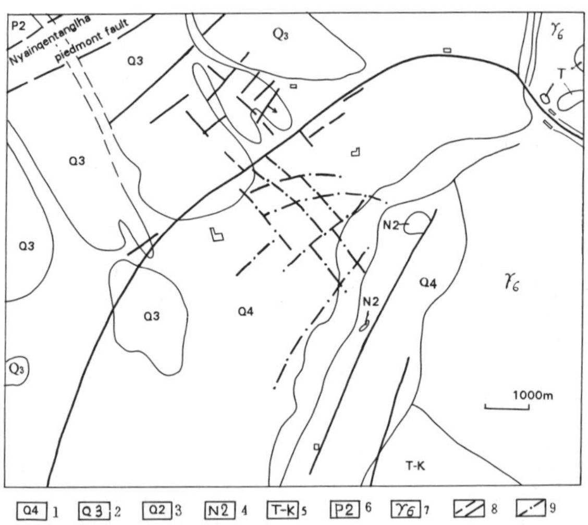

Figure 1 - Geology of the Yangbajain geothermal field, Tibet. 1 Holocene; 2 Upper Pleistocene series; 3 Middle Pleistocene series; 4 Andesite in Late Tertiary; 5 Metamorphic rocks in Trias-Cretaceous; 6 Metamorphic rocks in Palaeozoic era; 7 Granite in Himalayan movement; 8 Faults surveyed on-the-spot or inferred; 9 Faults inferred by geophysical survey

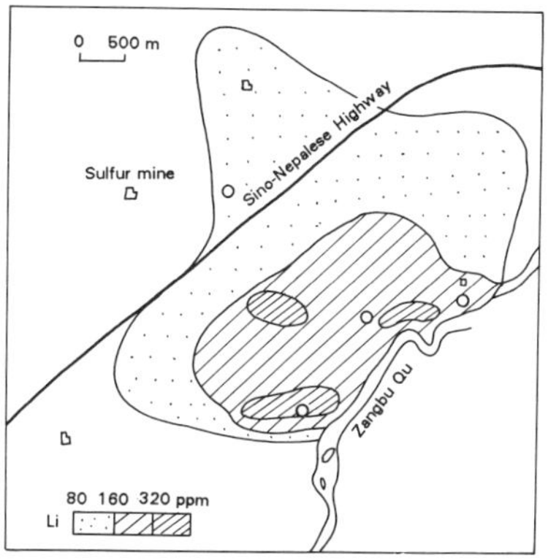

Figure 2 - Lithium anomalies in soil, Yangbajain geothermal field, Tibet.

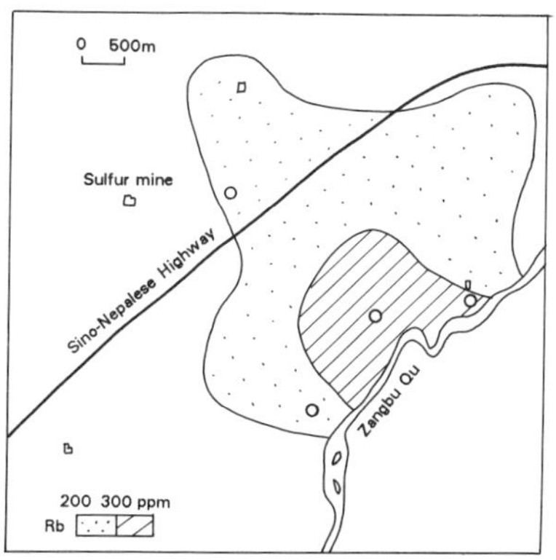

Figure 3 - Rubidium anomalies in soil, Yangbajain geothermal field, Tibet.

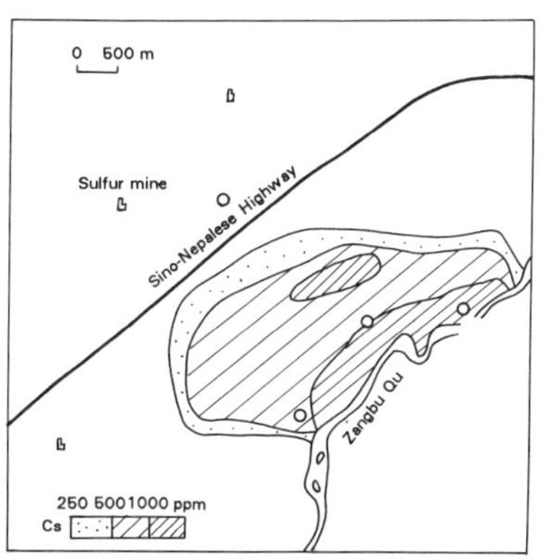

Figure 4 - Cesium anomalies in soil, Yangbajain geothermal field, Tibet.

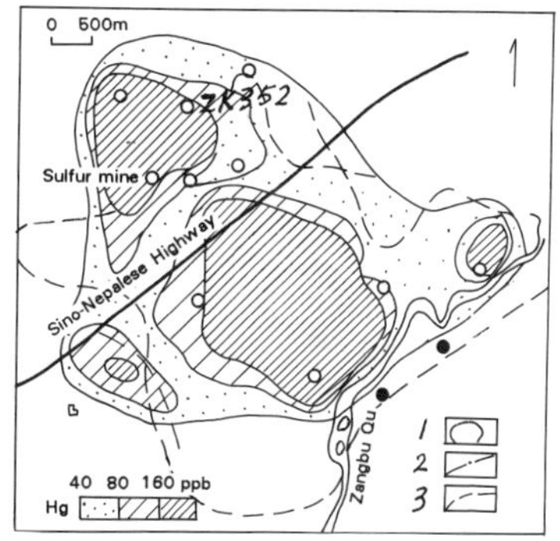

1 Hg anomaly 40-80-160 ppb; 2 Ground temperature anomaly 10°C isotherm; 3 Electric log anomaly (30 m) wells drilled before geochemical exploration; O wells drilled after geochemical exploration

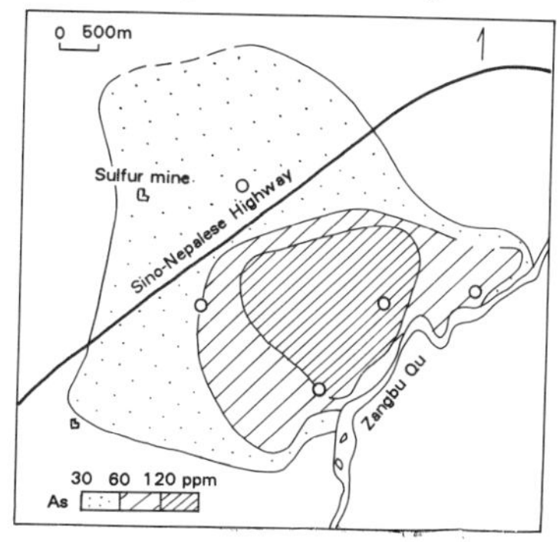

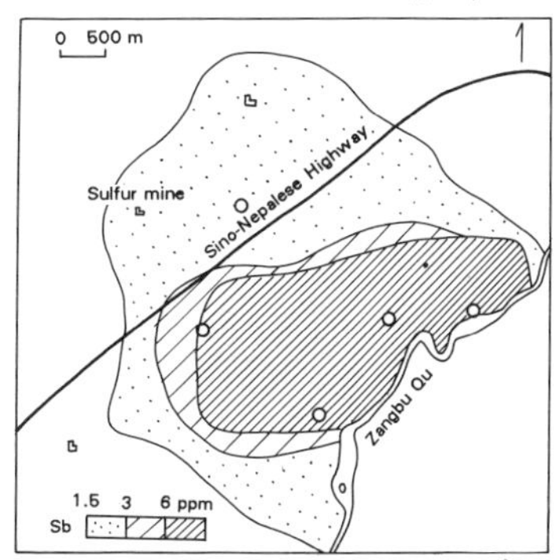

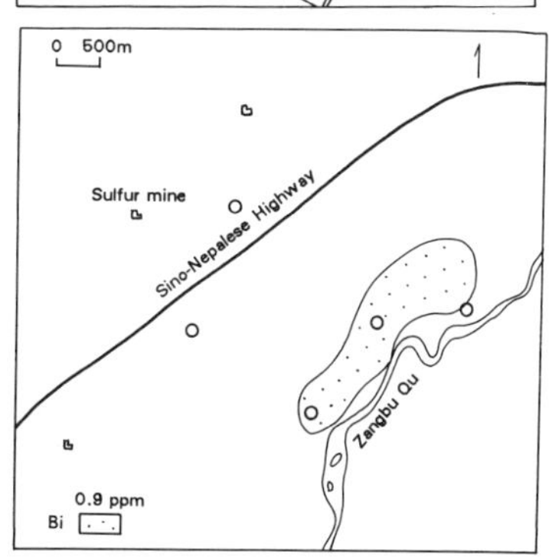

Figure 5 - Mercury, arsenic, antimony and bismuth anomalies in soil, Yangbajain geothermal field, Tibet.

Table 1. Characteristic parameters of indicator elements in anomaly area and background area, Yangbajain geothermal field, Tibet.

scope	element / parameter	Li	Rb	Cs	Hg	As	Sb	Bi
anomaly area	Maximum	452	470	3113	> 1600	198	243	2.1
	Minimum	80	196	296	40	30	1.1	1.0
	Average	167	264	1035	265	105	16.7	1.4
	Standard Deviation	83	57	569	326	136	36.2	0.3
	Contrast	6	2	29	14	6.5	21.8	2.8
	Maximum Contrast	17	4	88	> 84	60.8	317.2	4.1
	Number of Samples	76	57	45	117	80	105	11
back-ground area	Maximum	39.5	141	97.5	38	20	1	0.7
	Minimum	8	72.5	7.5	6	10	0.4	0.3
	Average	26	114	35	19	16.1	0.77	0.5
	Standard Deviation	9	18	17	7.6	2.7	0.18	0.12
	Number of Samples	58	65	187	80	48	47	60

(Li,Rb,Cs,As,Sb,Bi,ppm;Hg, ppb)

Table 2. Correlation coefficient matrix for 50 sample data of Li, Rb, Cs, Hg, As, Sb, Be.

element	Hg	As	Sb	Bi	Li	Rb	Cs	Be
Hg	1.000	0.168	0.168	0.099	0.168	0.168	0.168	0.168
As	0.168	1.000	0.328	0.099	0.342	0.342	0.342	0.342
Sb	0.168	0.328	1.000	0.099	0.328	0.328	0.328	0.328
Bi	0.099	0.099	0.099	1.000	0.099	0.099	0.099	0.099
Li	0.168	0.342	0.328	0.099	1.000	0.712	0.712	0.712
Rb	0.168	0.342	0.328	0.099	0.712	1.000	0.712	0.712
Cs	0.168	0.342	0.328	0.099	0.712	0.721	1.000	0.775
Be	0.168	0.342	0.328	0.099	0.712	0.721	0.775	1.000

$n = 50 \quad r_{0.05} = 0.273$

alkali metal anomalies are located on the southern part of the geothermal field, with comparatively lower elevation. In particular, the cesium anomaly area is located entirely at the south of Sino-Nepalese Highway. The Cs anomaly seems to represent mainly the hydrothermal fluid discharged area from the shallow reservoir. The very successful geochemical survey work had been done in the Yangbajain basin, using traditional element indicators, Hg, As, Sb and Bi. The mercury anomaly data combined with other related geological characters fixed the ZK 352 well position. The temperature of the well fluid is over 200°C. The determined temperature at the well bottom is 202°C. The highest temperature ever recorded in the exploitation history of the Yangbajain geothermal field effects a breakthrough in the exploring and drilling of the deep reservoir. For comparison, Fig. 5 shows the anomalies of Hg, As, Sb and Bi of the geochemical exploration in the Yangbajain basin. Table 1 shows the characteristic parameters of geochemical exploration. The correlation analyses between the rare-alkali metals and other indicator elements in geochemical exploration (e.g., Hg, As, Sb and Bi) were made. Table 2 shows the results. The correlation coefficient matrix shows that close correlations exist not only among rare-alkali metals (e.g., Li, Rb, and Cs) but also between rare-alkali metals and beryllium.

However, the beryllium in the discharged thermal water from the Yangbajain Power Plant is far in excess of the standards set by the State Environment Agency. The causes, therefore, can be reasoned out. There exists a general correlation between rare-alkali metals and As and Sb.

The anomalies of rare-alkali metals in the Yangbajain geothermal field are not only perfectly confining, but also show extremely high contrast values. Average lithium contrast value is 6, much larger than the lithium contrast value (2.1) in hot sea of Tengchong, Yunnan Province, where lithium anomalies were first determined in the global geothermal areas. The average cesium contrast reaches a remarkable value of 29, with the highest value of 88. The highest cesium content in the samples is 3,113 ppm. Such high cesium value is not only the value first found among world geothermal fields, but also almost as same as the cesium content of pollucite pegmatite (4,000 ppm), which contains cesium almost 400-fold increase of cesium in granite. The data provide some convincing geochemical clues for the stages of magmatic differentiation in magmatic pockets, in melting or half melting state, as heat sources for the Yangbajain geothermal field. The deductions in the "Geothermals under Tibet Plateau" assume that there are cooling granite magmatic pockets, which are blocking channels in shallow parts of crust as heat sources for Himalayan geothermal belt. From the cesium anomalies in the largest geothermal field, the Yangbajain geothermal field, in the geothermal belt, we may now more accurately describe the cooling pocket of magmatic body as melting pegmatite with a high initial cesium concentration. This is rare in various geothermal belts around the world and formed by specific mechanism of continental crust convergency, with the greatest crust thickness.

3 RARE-ALKALI METAL GEOCHEMISTRY IN TIBETAN GEOTHERMAL SYSTEMS

Himalayan geothermal belt on the Tibetan Plateau is a quite specific active belt on the earth. The signals of general active belts on the earth are earthquakes and eruptions of volcanoes. At the present there are about 800 active volcanoes around the globe, among them about 500 volcanoes are located in the Pacific Ring Zone area. In the Alps-Asia active zone, volcanoes only exist in Mediterranean Sea area and Indonesia and there are very few volcanoes in other areas. Until now, no active volcano in the Himalayan geothermal belt on the southern Tibetan Plateau was found. However, there are frequent earthquakes on the belt and there exists an epicentral intensive zone, which extends to a broad area in the North, crossing Tibet to the interior of China.

The crust thickness on both sides of the Yarlung Zangbo fossil suture line in the central Tibet is about 70 km, almost twice the depth of normal Mohorovicic discontinuity, called by some scholars as double layer crust structure. The zone is also the area where the most intensive geothermal manifestations are mostly located. Based on the hot spring distribution, some scholars advocate that the zone should be called the Yarlung Zangbo geothermal belt instead of Himalayan geothermal belt. The stress state inside the crust with great thickness in the sole continental crust convergence area is interesting to world geoscience circles and an important topic to be further investigated.

There only exists fossil descending oceanic crust remains in the Yarlung Zangbo fossil suture line, where the mechanism of modern ocean crust subduction no longer exists. It is very difficult for magmatic activities to occur in forms of active volcanoes in the belt, but quite easy to occur in forms of shallow magmatic intrusions or crypto-volcanism. The Bouguer anomalies of geophysical survey show that the upper crust in this area is in a melting or half-melting stage, which infers that remelting granite permeates throughout the upper crust in this area. Two heat-flow values (91 mw/m^2, 146 mw/m^2) were determined in the south of the Yarlung Zangbo suture line. Heat-flow values determined in Western US Basin and Range Area is 2 hfu (83.74 mw/m^2). Heat flow value determined in Soviet Kamchatka Peninsula is 2.4 hfu (100.48 mw/m^2). All these heat flow values are on the high side. Some scientists divide the geothermal resources in the Yarlung Zangbo geothermal belt into recent magmatic subtype of magmatic activity type, making it different from the recent volcanic subtype of magmatic type.

The world highest rare-alkali metal values, repeatedly found in geothermal fluids of the geothermal zone, offer strong evidences for existence of some kind of recent magmatic activities in crust shallow parts and relevant clues for geochemistry of magmatic differentiation. The cesium content in the geothermal fluid is remarkably high. The cesium content is even larger than the lithium and rubidium contents. Therefore, the magmatic activities in shallow crustal parts of the Yarlung

Zangbo zone might be in the form of melting pigmatite with high cesium concentration, similar to the magmatic evolutionary process of tourmaline muscovite granite outcropped in the Higher Himalaya of Southern Tibet.

There are the highest rare-alkali metal (Li, Rb, Cs) contents in the Zhumosha hot spring, Qusum county. This area is located in an ophiolite suite, i.e., zonal ultrabasic rocks with marine-face basalt and siliceous euabyssite. The ophiolite suite is interpreted as remains of descending fossil Tethys oceanic crust. The largest proven chromite deposit in China, the famous Luobusha chromite deposit on ultrabasic rocks is located on the southern mountain of Zhumosha. The Zhumosha hot spring is also located on the right bank of the entrance of the famous Sangri-Gyaca Gorge. There were landslides recorded on the Wuoka of the left bank, opposite of Zhumosha, caused by frequent and strong earthquakes. The most serious one once temporarily blocked the Yarlung Zangbo River. The ultrabasic rocks, outcropped in the Zhumosha area, contain less rare-alkali metal contents, not only lower than intermediate and acid rocks contain, but also lower than the average abundance of the rare-alkali metals of the crust. Therefore it is impossible for the hot springs to get such high rare-alkali metal concentrations from leaching the ultrabasic rocks. There must exist sources of rare-alkali metals in the deep crust. Strong and frequent earthquakes give evidences of the fossil suture line to be turned from compression to tension in stress analyses and earthquake mechanism solutions. Deep geological sources might be pegmatitic phase magma with initial high cesium concentration character. While geothermal fluids rise along abyssal tension faults, some rare-alkali metal components might be also brought up to the surface.

4 THE FEASIBILITY AND ITS SIGNIFICANCE OF TIBETAN ECONOMIC DEVELOPMENT AND RECOVERY OF RARE-ALKALI METALS FROM TIBETAN GEOTHERMAL SYSTEMS

Recovery of mineral products from geothermal systems is one important aspect of comprehensively utilizing geothermal fluids. Based on the feasibility study on recovery of mineral products from geothermal brine in Salton Sea, California, U.S.A., the output values of recovered mineral products from the hot brine with a recovery rate of 60 percent are twice the thermal heat values or 40 times the value of the electricity produced. Among the mineral product varieties, lithium chloride yields the highest value. Lithium chloride occupies only 5.25 per mille of the total weight of recovered mineral products, but accounts for almost 41 percent of the total output values.

Tibet is one of the regions with the most abundant rare-alkali metal reserves in China. The potential lithium chloride reserve in Tibet is not only in the best in China, but also among the best around the world. In Tibet, the salt lakes in Northern Tibet are the main reserves of lithium and other rare-alkali metals, in the form of chloride. Restricted by necessary conditions for developing metallurgical industry, e.g., transportation, power industry, etc., large scale exploitation of rare-alkali metal resources in salt lakes can only be listed in intermediate-range or long-range plans, while putting intensive geological survey and other earlier stage capital construction in advance.

Recovery of rare-alkali metals from Tibetan geothermal systems is a scheme with economic, social and environmental benefits, warranting careful consideration and may be effective in the near future. The Yangbajain geothermal power plant has run for more than ten years, with total capacity of 19 MW. The power plant discharges more than 800 tons of hot water per hour. Based on monitoring discharged water from Yang No. 23 well (ZK313), lithium content is 13 ppm. Estimated from these data, annual discharged lithium from Yangbajain geothermal plant is about 91 tons, almost one percent of the world lithium supply (9,000 short ton) in 1977, and is a valuable mineral resource.

However, the discharged water is a pollution source for local hydrographic net at present. Some harmful element contents exceed the state drinking water standards many times: As 79 to 199-fold; Be, 13- to 15-fold; fluoride 12 to 14-fold. Therefore, discharged thermal water from the Yangbajain power plant makes the discharged rivers: Dueilong Qu, Lhasa River and Yarlung Zangbo River polluted in varying degrees. In 1978, famous hydrogeologist Prof. Wang Dachun stressed at the early stage of geothermal development in our country, "There are such high boron and fluoride content in Tibetan high-temperature waters, very seldom seen in other places around the world. If we did our best to extract these components, we can increase output of very valuable mineral materials. If the waters were discharged freely after generating electricty, the discharged waters may become sources of pollution. Making a comparison between advantages and disadvantages, the two are far apart. So it is necessary to have comprehensive utilization."

Reinjection is one of the ways to solve the problems of polluted hydrographic net. However, the refilling techniques of waste water are rather complicated and closely related to local stratigraphical structure. In 1981, the concrete plans were already made, but the construction schemes had never been realized. Even if the plans are put into practice in the future, all the schemes have to be followed in order, gradually phased-in and advanced step by step.

Comprehensive utilization of discharged thermal water from the power plant may be divided into two parts. The first half uses the thermal energy of the discharged water. At almost 800 tons of discharged thermal water per hour, generally 80°C, Yangbajain power plant is sufficient to supply the heat demand of a 100,000 m^2 vegetable greenhouse and use of wool washing and artificial incubation. Now, about 3,000 m^2 greenhouses have been built in the Yangbajain geothermal field, with annual output of 15 kg vegetables/m^2. If all the discharged thermal water from the power plant were fully utilized, the total vegetable supply for Lhasa city will be more than 1.5 million kg per year. The greenhouses will be one of the major vegetable production bases for Lhasa city, with a population of 100,000.

As the discharged water temperature decreases, the second half of the comprehensive utilization is recovery of mineral products. The key project could be extraction of a single mineral product, e.g., lithium chloride. Abundant solar energy and large evaporation in Tibet provide advantages to design the process to concentrate discharged thermal water into brine with little cost. TDS (total dissolved solid) of the discharged thermal water (determined in No. 23 well) is 1.912 g/l. TDS of Zhacang Caka, one famous northern Tibetan salt lake, is 340.58 g/l. If the discharged thermal water of Yangbajain is concentrated to the Zhacang Caka TDS, the Yangbajain concentrated brine will have a lithium content of 2315.66 ppm. The value is even larger than the lithium content (2120 ppm) in the Zhacang Caka, known as a brine lithium deposit. Extraction of metals in brine state is known as wet-metallurgy, commonly used in modern industry. A functional ion-exchanger (e.g., crown ether) will selectively remove certain metallic ions such as potassium, cesium, sodium and lithium into an organic phase. If some parameters such as concentration and temperature are chosen to optimize the conditions, the processes of absorption, extraction, and stripping might cost very little.

Development of the metallurgical industry is restricted by infrastructure, such as transportation and power as well as techniques in Tibet at present. It is difficult to compete against the interior of China in metallurgical industry over a short period. However, "the outline of socioeconomic development plans of Tibet Autonomous Region for 2000," edited by Tsinhua University and Beijing Economical Institute holds that "developing nonferrous metallurgical industry is the bridge to future Tibetan economic development." The recovery of mineral products from discharged thermal water of the Yangbajain power plant is an alternative not only with little investment and quick returns, but also with economic, environmental and social benefits, such as recycling waste materials. And the recovery of rare-alkali metals may be regarded as a breakthrough in developing non-ferrous metallurgical industry in Tibet. According to the current amount of discharged thermal water, annual recovery of only one mineral product (e.g., lithium chloride with a recovery rate of 60 percent from the discharged thermal water of the power plant) is about 331.5 tons. One ton of lithium chloride priced in international market is 1,900 US dollars. The recovered lithium chloride is valued 0.62985 million US dollars, converted into 2.9729 million Yuan (RMB) by official rate (one US dollar = 4.72 Yuan RMB), equivalent to more than 85 percent of the power output value in 1984 (3.492 million Yuan RMB). At present, the efficiency of Tibetan power industry is poor. Total investment in ten years is 1.92 billion yuan (RMB), the increased power output value is only 41 million yuan (RMB), i.e., the input/output ratio is about 47. The ratio is 30X higher than that of other general industries. Comprehensive utilization of discharged thermal water of the power plant will greatly improve the input/output efficiency of the Yangbajain power plant. As long as the recovery processes of mineral products gradually improves, reinjection with complicated techniques over a short time, can be conducted without damage to the environment.

At present, lithium is mainly used in aluminum, lubricant and cosmetic industries. The world lithium demand is not large. Recently lithium rechargeable batteries have made a major breakthrough. Button batteries consume more and more lithium. If large capacity rechargeable battery makes some progress, lithium rechargeable battery can be used in the electric car. Lithium, known as a new metal noble, will distinguish itself in the 1990's. The lithium world demand will rise quickly and the world market quotation on lithium is also expected to rise. The recovery of lithium can be a starting point for developing Tibetan non-ferrous metallurgical industry, and a bridge for future Tibetan economic development.

REFERENCES

Huang, S. et al. 1986. Volcanoes, hot springs and geothermal energy. Geology Press. Beijing (in Chinese).

Li, C. 1981. From the point of view of the plate theory to study geothermals in China. Symposium on geothermals in China. Science Press (in Chinese).

Liu, Y. et al. 1984. Elemental geochemistry. Science Press (in Chinese).

Si, D. et al. 1985, The chemical pollution impacts on the hydrographical net by the exploitation of the Yangbajain geothermal field, Study on the Yangbajain geothermal power plant. Science and Technology Literature Press, Chongqing Branch (in Chinese).

Tong, W. et al. 1981. Geothermals under Tibet. Science Press (in Chinese).

Wahl, E.F. 1977. Geothermal energy utilization. A Wiley-Interscience Publication.

Wang, D. 1981. Some problems concerning geothermal exploitation in China. Symposium on geothermals in China. Science Press (in Chinese).

Zheng, X. 1988. Salt lakes in Tibet. Science Press (in Chinese).

Geothermal systems

Mineral interactions

Water-Rock Interaction, Kharaka & Maest (eds) © 1992 Balkema, Rotterdam. ISBN 90 5410 075 3

Calcite deposition in geothermal wells caused by infiltrating seawater

Stefán Arnórsson
Science Institute, University of Iceland, Reykjavík, Iceland

Abstract: Water salinity has increased with time in a few production wells in Iceland located near the coast due to infiltration of seawater or seawater-freshwater mixtures into the geothermal reservoir. This increase in salinity has been associated with calcite deposition in the wells and in downhole pumps causing some operational problems. Mixing of saline water with dilute geothermal water leads to substantial increase in its Ca concentrations but other chemical parameters affecting calcite saturation are little affected. Thus, H_2CO_3 concentrations change little and pH is maintained about constant through the buffering action of aqueous silica. It is the increase in Ca that makes the initially calcite saturated geothermal water supersaturated and causes calcite to deposit.

INTRODUCTION

Many hot water geothermal wells in Iceland are located near the coast because villages and towns are mostly coastal and the purpose of the drillings is to provide hot water for domestic heating for these communities. Initially, the water discharged/pumped from these wells was in many instances quite low in dissolved solids and similar to that of many geothermal systems located some distance inland. It is, thus, evident that the water was meteoric in origin, with little or no seawater component. During prolonged exploitation the water salinity has, however, increased, evidently as a result of inflow of seawater or seawater-freshwater mixtures into the geothermal system. Examples of such salinity increase are provided by few wells in the Laugarnes field within Reykjavík and a well at Sudureyri in the NW-Peninsula. Wells at Seltjarnarnes, which is close to Reykjavík, have also shown an increase in water salinity, but initially the water in these wells contained a small seawater component.

Before exploitation was initiated in the coastal geothermal systems under discussion reservoir pressures were such that their water rest level was considerably above sealevel. The reason must be connection to aquifers inland that are associated with a higher groundwater table. The hydrostatic head in the geothermal system may have been further elevated by the difference in density of the hot water column within the system and a cold water column outside it. Pressure drawdown resulting from exploitation may change the flow pattern of water recharging the geothermal system. Initially, it was dominated by fresh groundwater flow from inland but recharge from other directions were stimulated by the pressure drawdown in the reservoir.

In this contribution an attempt will be made to elucidate the chemical changes that occur when seawater or seawater-freshwater mixtures mix with dilute geothermal water and cause calcite deposition.

RESULTS

Geothermal waters are generally very close to equilibrium with calcite (Arnórsson, 1978, 1989). The reaction involving calcite and aqueous species may be expressed as

$$CaCO_3 + 2H^+ = Ca^{+2} + H_2CO_3 \qquad (1)$$

It is seen from the above equation that the factors affecting the state of saturation with respect to calcite are pH and the activities of calcium ions and dissolved carbon dioxide. Note that dissolved carbon dioxide is expressed as H_2CO_3 in equation (1) above. H_2CO_3 is taken to be equivalent to CO_2 + H_2O. Increase in Ca^{+2}, H_2CO_3 and decrease in H^+ (increase in pH) all change a calcite saturated solution into a supersaturated one and, thus, favour precipitation.

An assessment has been made of the changes of these factors that occur when saline water mixes with typical dilute water from a low-temperature field in Iceland (Table 1). The saline waters selected are seawater and water from Seltjarnarnes which is located on the tip of the Seltjarnarnes Peninsula a few kilometres west from Reykjavík (Table 1). The WATCH programme (Arnórsson et al., 1982) was used to calculate the aqueous speciation of various mixtures of the saline and dilute waters. The calcite solubility data used are those given by Arnórsson et al. (1982). The results are shown in Figs. 1 and 2. The fraction of the salt water component in the mixture is expressed in terms of Cl concentrations. 200 ppm Cl in Fig. 1 corresponds to 11.7% of Seltjarnarnes water in the mixture and 500 ppm Cl in Fig. 2 corresponds to 2.35% seawater in the mixture.

Mixing of both Seltjarnarnes- and seawater with the Reykjavík low-temperature water has very little effect on pH (Figs. 1 and 2). The reason is that aqueous silica buffers quite stongly the relatively high pH of the Reykjavík water but the buffering capacity of both the Seltjarnarnes- and seawater is weak.

Seawater contains 102.7 total carbonate and the Seltjarnarnes water 7.1 ppm which is to be compared with 18.8 ppm in the Reykjavík water. Although the differences between these values are considrable, the fraction of saline water taken to mix with the dilute water is not sufficiently high to change the total carbonate content significantly and since pH stays also about constant, so do H_2CO_3 concentrations.

In contrast to H_2CO_3, Ca concentrations are much higher in the saline waters than in the dilute Reykjavík water, or 150-200 times. For Ca^{+2} activity this difference is somewhat smaller due to more complexing and lower activity cofficients in the saline waters as compared with the dilute water, or 36-80. It is, therefore, clear that adding limited amount of saline water to the dilute Reykjavík water will increase the calcium activity of the latter substantially and lead to calcite supersaturation since the water was initially saturated and H_2CO_3 concentrations and pH are little affected by the mixing (see equation (1)). The extent of this supersaturation is shown on the bottom diagrams of Figs. 1 and 2. It is considerably greater when Seltjarnarnes water is involved rather than seawater, the reason being the higher Ca content of the former relative to Cl. The degree of supersaturation is comparable with that often produced by boiling of geothermal waters (Arnórsson, 1989). Observations show that calcite deposits instantaneously when such supersaturation is produced by boiling. By comparison one would also except calcite to deposit instantaneously when comparable supersaturation is produced by mixing saline and dilute waters.

DISCUSSION

The degree of calcite supersaturation produced by mixing seawater or seawater-freshwater mixtures with dilute geothermal water increases with increasing salt water component in the mixture and its calcium content. The Ca/Cl ratio of saltine groundwaters relative to seawater depends basically on two factors, at least if equilibration with secondary minerals has been closely approached. They are temperature and water salinity. At equilibrium the ratio Ca/Na^2 is constant and it decreases with increasing temperature (Arnórsson et al., 1983). Since Na is the dominant cation in

Table 1. The chemical composition of dilute and saline waters used in the mixing calculations Concentrations in ppm.

	Seawater[a]	Seltjarnarnes well 4	Reykjavík well 15
Temp.°C	5	114	124
pH/°C	8.15/25[b]	8.37/24	9.63/22
SiO_2	1.0	112.3	146.9
B	4.5	0.24	0.04
Na	11050	473.9	56.1
K	416	11.2	2.13
Ca	422	560.0	2.79
Mg	1326	0.204	0.072
Fe	0.003	0.052	0.018
Al	0.005	0.103	0.337
ΣCO_2	102.7	7.1	18.8
SO_4	2784	256.0	14.8
H_2S	-	0.11	0.30
Cl	19870	1458	33.5
F	1.4	0.68	0.64

[a]Riley and Chester (1971). [b]Garrels and Christ (1965)

seawater and saline groundwaters and Cl the dominant anion it follows that Ca concentration, or more accurately Ca^{+2} activity increases in a quadradic manner with increase in aqueous Cl concentrations. It is the higher Ca/Cl ratio of the Seltjarnarnes water relative to seawater that explains why mixing of this water with Reykjavík water produces greater calcite supersaturation than mixing involving seawater.

It is well known that boiling of initially calcite saturated water leads to supersaturation (Arnórsson, 1989). When geothermal waters begin to boil within discharging wells calcite deposition may be substantial and it has been observed that it is most intense at the first level of boiling (Thórhallsson, 1988). This indicates that calcite deposits readily from supersaturated solution under geothermal conditions.

An attempt has not been made to predict the rate of calcite precipitation which occurs as a result of mixing of saline water and dilute water. If this precipitation involves removal of calcium and carbonate from solution by cumulation on calcite crystals the rate is not predictable because the surface area of the calcite crystals in a geothermal system cannot be evaluated with any confidence. However, the ease with which calcite appears to precipitate suggests that mixing of dilute geothermal water and saline water that leads to calcite deposition in downhole pumps and wells must either occur within the well or in its immediate vicinity. If mixing occurred at some distance from the well the water would have re-equilibrated by depositing calcite in the aquifer before entering the well.

The geothermal water selected for the present study

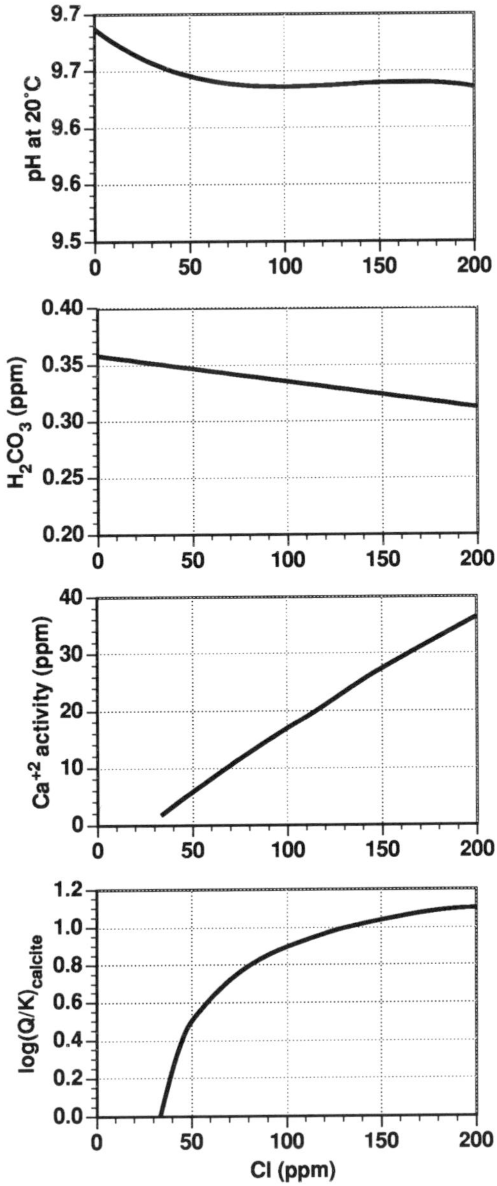

Fig. 1. Calculated changes in the chemical parameters that affect calcite saturation when water from well 15 in Reykjavík is mixed with saline water from well 4 at Seltjarnarnes. The saline water component in the mixture is expressed in terms of Cl, 200 ppm Cl corresponding to 11.7% of Seltjarnarnes water.

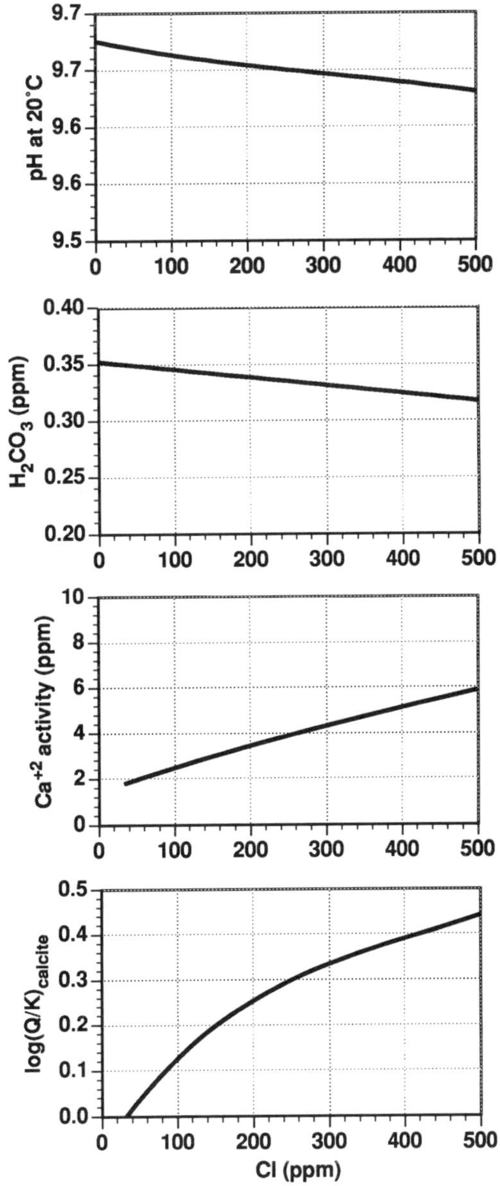

Fig. 2. Calculated changes in the chemical parameters that affect calcite saturation when water from well 15 in Reykjavík is mixed with seawater. The seawater component in the mixture is expressed in terms of Cl, 500 ppm Cl corresponding to 2.35% of seawater.

to assess theoretically the effect of mixing it with saline water upon calcite supersaturation possesses a high pH. Geothermal water with a substantially lower pH, low enough to prevent significant buffering of aqueous silica, would not become so calcite supersaturated upon mixing with saline water. Lowering of pH would, at least, compensate to some extent for the increase in Ca^{+2} activity caused by the mixing.

There appears to be several ways of solving the problem of calcite scaling induced by infiltrating seawater into exploited geothermal systems. One would be to use inhibitors. Another possibility would be to reduce the production to such an extent that the water level in the reservoir would be maintained above sealevel. Alternatively reservoir pressures could be maintained high by injecting fresh water. By keeping the groundwater table within the reservoir above sealevel inflow of ocean water would be impeded, at least at shallow levels.

If the inflow of seawater occurred at a shallow level, the problem could be solved, at least temporarily, by reworking the wells and sinking the production casing to a deeper level. In order to be able to verify if the seawater inflow occurred through shallow aquifers, it would be necessary to collect downhole samples while water was being pumped from the well. This would make it possible to locate the depth of inflow of the saline water.

REFERENCES

Arnórsson, S. 1978. Precipitation of calcite from flashed geothermal waters in Iceland. Contrib. Mineral. Petrol. 66: 21-28.

Arnórsson, S. 1989. Deposition of calcium carbonate minerals from geothermal waters - theoretical considerations. Geothermics 18: 33-39.

Arnórsson, S., Sigurðsson, S. and Svavarsson, H. 1982. The chemistry of geothermal waters in Iceland. Calculation of aqueous speciation from 0° to 370°C. Geochim. Cosmochim. Acta 46: 1513-1532.

Arnórsson, S., Gunnlaugsson, E. and Svavarsson, H. 1983. The chemistry of geothermal waters in Iceland. II. Mineral equilibria and independent variables controlling water compositions. Geochim. Cosmochim. Acta 47: 547-566.

Garrels, R. M. and Christ, C. L. 1965. Solutions, Minerals and Equilibria. Freeman, Cooper & Company, San Francisco, 450p.

Riley, J. P. and Chester, R. 1971. Introduction to Marine Chemistry. Academic Press, London. 465p.

Thórhallsson, Sv. 1988. Experience in developing and utilizing geothermal resources in Iceland. Geothermics 17: 205-223.

Water-Rock Interaction, Kharaka & Maest (eds) © 1992 Balkema, Rotterdam. ISBN 90 5410 075 3

An autoclave reinjection study of a vitreous rhyolite tuff

Cheng Yan, A.C. Lee, V.J. Grassi, D.E. Grandstaff & Gene C. Ulmer
Geology Department, Temple University, Philadelphia, Pa., USA

ABSTRACT: The Bannock Rhyolite tuff (69 wt% SiO_2) was used in an autoclave reinjection study at 300°C and 30 MPa. This tuff has 95 wt % glass and is a textural and chemical 'end member' in our continuing studies of glassy rhyolites and basalts. Data are presented for pre- and post- injection solution-time profiles, for solution chemistry, for pH and for log fO_2. Comparisons were then made to reinjection experiments involving glassy (32%) Cohassett basalt (Gardiner *et al.*, 1988) conducted at similar P and T in our lab. The comparison of rock type in both pre-injection and post-injection data allow the conclusion that rock type does not greatly affect most water-rock parameters (differences range from a factor of 5 to zero), but does strongly influence pH (differences of as much as one log unit) and fO_2 (differences as much as four log units).

INTRODUCTION

Knowledge of the hydrothermal reactions of rhyolite is critical to existing projects such as the Valles Caldera Geothermal Project (NM) and the Yucca Mountain Nuclear Waste Repository (NV). Accordingly, in this manuscript we report data for hydrothermal autoclave studies of the Bannock Rhyolite Tuff (BRT) with 93% glass from Montana. Re-injection studies of the glass-rich Bannock Tuff continued, however, beyond the initial work of Grassi *et al.* (1991) and is the subject of this paper. Furthermore, results from the Bannock Tuff reinjection experiment may be compared with a similar reinjection study of the Cohassett Basalt (Gardiner *et al.*, 1988). Thus, the point of the comparisons, particularly after re-injections, is to examine the influence of these two diverse rock types on the development of geochemical solution parameters.

TECHNIQUES

The experiment (BRT-1) was performed at 300°C and 30 MPa in a Dickson autoclave with flexible gold cell following the methods of Gardiner *et al.* (1988). High-temperature speciation parameters were calculated using computer program HIPH/FOLLY (see Grassi *et al.* this volume). The rock sample used is from the Bannock Rhyolite Tuff; the chemical composition and modal mineralogy are given in Table I; grain size was 125-250 μm. The initial solution was I= 0.0034 NaCl, and the initial water:rock mass ratio was 10. BRT was reinjected after stable fluid compositions were achieved (3531 hours- see Figure 1). This re-injection produced: (1) replacement of > 90 vol% of the initial, but well reacted, fluid to produce a cumulative water/rock ratio of 19; and (2) a new oxidative stress of *ca.* 9 ppm dissolved oxygen in the 120 ml of new solution. Secondary minerals were characterized by XRD.

Results from this study are compared with a similar reinjection experiment (BCI-1: 300°C, 30 MPa, 50:1 initial W:R) conducted using basalt (125-250 μm and simulated groundwater [I=0.015](Gardiner *et al.*, 1988). The composition and mineralogy of the basalt are also shown in Table I.

RESULTS

Major secondary minerals identified in BRT-1 are: high sanidine, a mordenite group zeolite (ferrierite), and dioctahedral illite. The total amount of secondary phases produced, as determined gravimetrically, was about 48 percent.

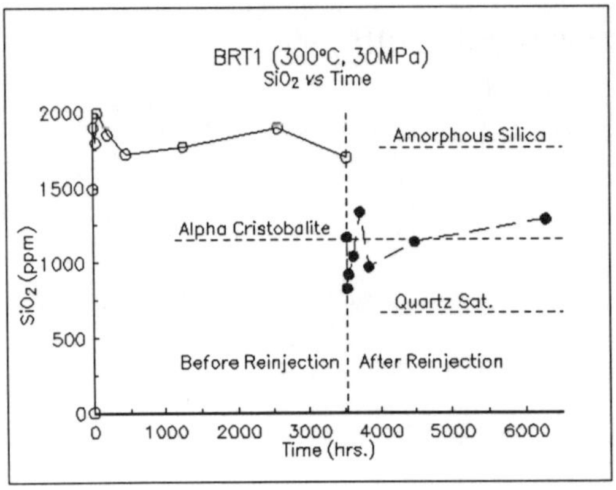

Figure 1. BRT SiO_2 solution-time profile. See text for discussion.

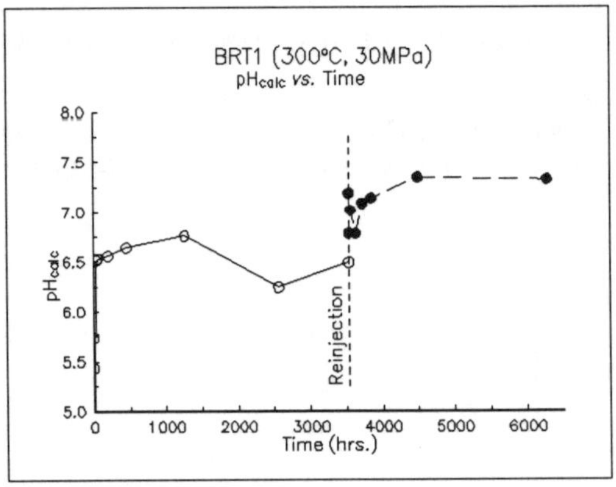

Figure 2. BRT solution pH *vs.* time. See text for discussion.

Concen-trations	BRT	BCI	Composi-tion	BRT	BCI
SiO_2	69.53	52.81	Glass	94.7	31.5
Al_2O_3	14.32	13.70	Qtz	1.8	-
Fe_2O_3	2.03	total Fe as FeO	Biot	0.4	-
FeO	0.43	11.62	Plag	0.8	36.9
MgO	0.25	5.04	Sanid	1.7	-
CaO	1.35	8.79	Calc	tr.	-
Na_2O	2.01	2.66	Crist	tr.	-
K_2O	5.13	1.11	Fayal	tr.	-
H_2O^+	3.02	NA	Magn	-	4.1
H_2O^-	0.47	NA	Pyx	-	20.8
TiO_2	0.27	NA	Clphae-ite	-	6.6
MnO	0.03	0.19			
P_2O_5	0.01	0.27			
TOTAL %	98.85	97.98	TOTAL %	100.00	100.00

Table I. Chemical and Modal Analysis of Bannock Rhyolite Tuff (BRT) and Cohassett Basalt (BCI) used in reinjection studies at 300°C and 30 MPa.

GROUP	Concen-trations (ppm)	BRT-1		BCI-1	
		Pre-reinjection (3531 hrs.)	Post-reinjection (6266 hrs)	Pre-reinjection (2137 hrs)	Post-reinjection (4657 hrs.)
I	Ca	0.39	0.3 (4489 hrs.)	0.22	0.12
	K	94.7	17.2	30	34.3
	Mg	0.03	0.03	0.01	0.01
	Al	5.01	6.8 (4489 hrs.)	3.5	3.1
	SO_4	12.69	7.50	20.1	9.2
II	SiO_2	1699	1283	1189	622
	HCO_3	64.0	72.7	157	132
	Na	74.9	161.3	322	352
	Cl	145.9	210	403	412
III	pH_{calc}	6.49	7.32	7.41	7.89
	log fO_2	-27.32	-27.5	-31.1 (1871 hr)	-31.8

Table II. Note three groupings stand out in Table II. Group I and II are parameters that differ by less than an order of magnitude either when comparing rock type or comparing pre- and post-reinjection data. Group II are parameters that have differences that are essentially anticipated from initial conditions (see text). Group III are parameters whose values differ by log units when examined by rock type.

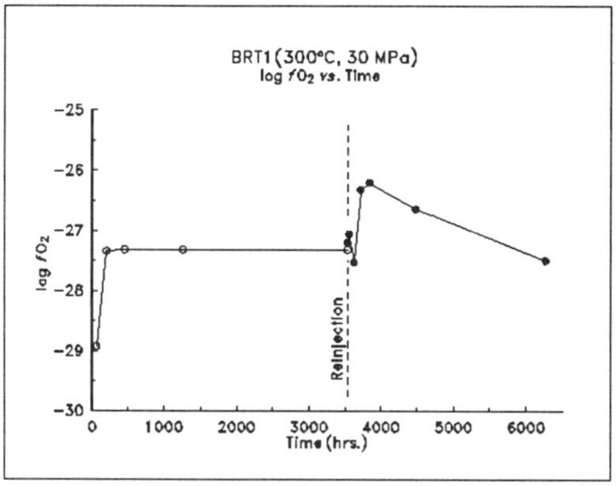

Figure 3. BRT solution fO_2 *vs.* time. See text for discussion.

Concentration-time data for some dissolved species in BRT-1 and BCI-1 are given in Figures 1-3 and Table II.

During the initial period of BRT-1 (Fig. 1) dissolved SiO_2 concentrations quickly rose to values equivalent to saturation with amorphous silica as might be expected from its initial 93% glass content (Table I). In contrast, concentrations after re-injection appear consistent with control by precipitation of secondary α-cristobalite. Concentrations of dissolved species are controlled by equilibrium with secondary minerals and rates of dissolution and precipitation reactions. In BRT-1, during the initial period, the high proportion of glass and rapid rate of glass dissolution was apparently much faster than precipitation of quartz or any other crystalline silica polymorph, or other silicate secondary phases. Thus SiO_2 rose to amorphous silica saturation. After reinjection the rate of glass dissolution was apparently slower and the rate of secondary mineral precipitation faster leading to decreased concentrations (see also discussion in Gardiner *et al.*, 1988); maturation and increased grain size of secondary minerals may have also contributed to decreased SiO_2 concentrations. In the comparable (BCI) study, the pre-injection SiO_2 concentrations appeared controlled by α-cristobalite and the late post-reinjection SiO_2 matched quartz saturation (Figure 1, and Table I). The amount of glass and bulk SiO_2 content in the basalt was less than in BRT-1 (Table I); the difference was apparently sufficient to result in lower dissolved SiO_2 concentrations in that experiment, however, the difference is less than a factor of two.

Before reinjection the calculated high-temperature pH of BRT ranged between 6.7 and 6.3 in contrast with the initial pH range of 8.0 - 7.6 for BCI (Gardiner *et al*, 1988). At P and T of these studies, neutral pH is *ca.* 5.5. For both studies, the immediate post-reinjection pH values abruptly increased at least half a pH unit (more alkaline). After thousands of hours both the BRT and BCI re-injection pH values come closer together with the basalt pH being about 8.05 and the tuff pH being about 7.6. Comparison of cation/proton ratios (to eliminate effect of differences in ionic strength) from the two experiments, (see Na^+ and Cl^- in Table II), also show that both the pre-reinjection and the post reinjection solution pH values are lower in the tuff than in the basalt. The increase in alkalinity (decrease in proton) after both reinjections is primarily explained by equilibria with secondary minerals and the change in dissolved silica. For example, for equilibria between potassium feldspar, muscovite and solution:

$$1.5KAlSi_3O_8 + H^+ = 0.5KAl_3Si_3O_{10}(OH)_2 + K^+ + 3\ SiO_{2(aq)}$$

with

$$pH = \log K_{eq} - \log [K^+] - 3 \log[SiO_{2(aq)}]$$

At constant potassium concentration the pH will increase with decreasing dissolved silica concentration. In this study the pH varies directly with log $SiO_{2(aq)}$ (rather than 3log SiO_2,); this relationship was also found in the BCI experiment. The direct relationship between pH and log SiO_2 reflects the stoichiometry of the secondary and clay minerals in equilibrium with the solution. Changes in dissolved carbonate, aluminum, and other species with reinjection will also affect the pH. The difference in pH values between the granite and basalt experiments may result from the difference in octahedral character of the clays (Grassi *et al.*, 1991).

The fO_2 data for the BRT study show a very early attainment of nearly constant -log fO_2 of 27.3, *i.e.*, about four orders of magnitude more oxidized than the (MH) buffer. With the oxidative stress of added reinjection fresh solution, the post-reinjection fO_2 increase recovers only over 2500 hours as shown in Figure 3). This very sluggish recovery presumably has to do with both the low ferrous iron in the BRT (see Table I., 0.43 wt% FeO) and the slow release of the ferrous iron. By comparison, much harsher oxidative stresses (initial water rock ratio = 50:1 and after first re-injection = 90/1) put upon the BCI (Gardiner et al, 1988) show faster recovery (within 600 hours) to the lower fO_2 value of -log fO_2 = 30 $\pm$ 1.0 [essentially the (MH) buffer] presumably because of the higher total ferrous iron (typical basalts have about 8 wt%) and faster release rate of the basaltic iron to solution. Hence to generalize, the re-injection data strongly suggest that rhyolite tuffs have far less buffering capacity to oxidative stress than do typical basalts.

The reinjection data of both BCI and BRT are presented in Table II. Note that the results are grouped according to the general behavior of the species upon reinjection.

Group I includes Ca, K, Mg, Al, and SO_4. All values change by less than an order of magnitude and usually by less than even a factor of five, whether compared by rock type or by reinjection history.

Group II includes SiO_2, HCO_3, Na, and Cl. Note that either rock type or reinjection still change these parameters by less than an order of magnitude.

In Group II, rock texture/rock type may both control silica in that the BCI basalt has been reported to have *ca.* 32% glass (65% SiO_2). By comparison the BRT tuff (see Table I) has 69.53% SiO_2 and 93% glass. With the higher glass and silica activity in BRT, a higher solution level of SiO_2 is expectable with BRT than with BCI. The SiO_2 data of Table II summarize the comparison to BCI. The Na and Cl differences between BRT and BCI vary sympathetically and expectably with the two different salinity values of the appropriate synthetic groundwaters involved, *i.e.*, higher molality NaCl in the BCI experiments. The higher HCO_3 in the basalt originates in the same way, *i.e.*, a higher HCO_3 in the BCI-1 synthetic groundwater.

Finally the Group III parameters, *i.e.*, fO_2 and pH are, indeed, significantly different, given that they are log based values. When correlated with rock type they are seen to vary by orders of magnitude. Fig. 2 details the exact behavior of solution pH in the BRT experiment; a similar graph exist in Gardiner *et al.* (1988) for pH in the BCI experiment. The data given in Table II summarize the comparison of BRT and BCI pH values. It may add understanding to Table II to realize that in both reinjections, the fO_2 becomes more oxidized with the oxidative stress of the oxygen dissolved in the newly injected groundwater (about 9 ppm O_2). Fig. 3 shows the exact behavior of log fO_2 for BRT. A similar graph again exists in Gardiner *et al.* (1988), but the data are summarized in Table II herein.

CONCLUSIONS

At 300°C and 30 MPa, the results of this autoclave study show that long term dissolved SiO_2 varies only slightly despite differences in glass and SiO_2 in the two types of bulk rock. After correcting for the ionic strength differences of the groundwaters used, the pH values measured by the rhyolite glass-rich tuff, both before and after re-injection, are as much as 0.5 pH units more alkaline than the solution pH that develops in basalt-water interactions at the same P and T. This difference may result from differences in the octahedral character of the clay minerals (Grassi *et al.*, 1991). Again, the redox value measured for the BRT is as much as 4 orders of magnitude more oxidized, both before and after reinjection than the fO_2 in experiments at the same P and T on BCI. Furthermore, after reinjection the recovery of the BRT with a total water rock ratio of 19/1 is still much more sluggish than the recovery for BCI, with water/rock cumulative ratio of 90/1.

These comparative data on solubility, pH and fO_2 should be of interest in planning not only geothermal well projects but also to nuclear waste management decisions and conceivably even to hydrothermal ore deposit modeling in various host bedrock types. Rock type seems to produce large differences at the P and T of this study only in pH and log fO_2. All other solution geochemistry parameters are much less (same to $<$5-fold) influenced by rock type.

REFERENCES

Gardiner, M.A., G.H. Kacandes, G.C. Ulmer, and D.E. Grandstaff, 1988. Application and results of a reinjection technique for hydrothermal experiments in Dickson autoclaves. Mat. Res. Soc. Symp. Proc. Vol. 112, 383-392.

Grassi, V.J., G.C. Ulmer and D.E. Grandstaff, 1991. The effect of igneous rock type on nuclear waste siting, in NUCLEAR WASTE MANAGEMENT IV, (Amer. Ceram. Soc., Westerville, OH), 199-706.

Grassi, V.J., G.C. Ulmer, and D.E. Grandstaff, 1992. This volume, Water-rock hydrothermal experiments: Influence of rock type on solution pH and Redox.

Lee, A.C., 1990. Hydrothermal experiments (300°C) conducted using Bannock rhyolite tuff (Montana) and Conway Granite (New Hampshire): Implications for high-level nuclear waste disposal and comparisons with geothermal solution data. M.A. Thesis at Geology Dept., Temple University, Philadelphia, PA 19122, 207 pp.

Water-Rock Interaction, Kharaka & Maest (eds) © 1992 Balkema, Rotterdam. ISBN 90 5410 075 3

Experimental approach to the self-sealing of rock-water systems

Masahiro Chigira
Central Research Institute of Electric Power Industry, Japan
Masakazu Watanabe
Central Research Service Corporation, Japan

ABSTRACT: A new apparatus was developed that simulates slow hydrothermal flow in a narrow rock fracture. The apparatus enables hydrothermal solutions up to 200 °C to flow along a fracture as narrow as 0.05 mm and 1.4 m long with or without temperature gradient along the flow path. By using this apparatus, hydrothermal solutions that were saturated with silica at 200 °C and 100 °C were pushed through a granite fracture with a temperature gradient of 1 and 0.5 °C/cm, respectively. The hydrothermal solutions deposited silica particles on the fracture surfaces, quartz, plagioclase, K-feldspar, and also on biotite, narrowing or plugging the fractures.

1 INTRODUCTION

Some geothermal fields have structures made by self-sealing, which is the phenomenon that convecting geothermal fluids deposit hydrothermal minerals along the flow paths and they hydraulically isolate themselves (Facca and Tonani (1967); White et al. (1975); Keith et al. (1978); Keith and Muffler (1978); White et al. (1988)). Geysers, Wairakei, Waiotapu are considered to have such structures. Among the hydrothermal minerals that cause self-sealing, silica minerals are most important. Silica deposition from flowing or stagnant hydrothermal solutions has been studied by many researchers whose concern, however, was in prolonging the life time of geothermal power plants or some other artificial plants (Bohlmann et al. (1980); Garibaldi and Freeston (1981); Weres et al. (1981, 1982); Itoi et al. (1985)). Silica deposition for the self-sealing, however, has somehow different aspects from such silica scaling; self-sealing is considered to occur by slow hydrothermal flow along flow paths that have negative temperature gradients, so that the hydrothermal solutions supersaturate with silica minerals and would deposit them along the flow paths.

Water-rock interactions in open systems have been challenged by experiments since 1950s (Morey and Chen (1955)), and experiments incorporating the hydraulic properties of rocks started 1970s (Summers et al. (1978); Morrow et al. (1981); Moore et al. (1983, 1986), Potter et al. (1985); Hochella et al. (1988)). Water-rock interactions in a field with temperature gradients were studied for granite or tuff by some of the previous studies, but temperature gradients used were as high as 30-70° C/cm and flow paths were as short as several centimeters.

2 NEW HYDROTHERMAL FRACTURE FLOW APPARATUS

We developed a new hydrothermal fracture flow apparatus, which consists of an autoclave with 2 m length, a pump, accumulators, a capillary tube system for the solution sampling and the measurement of differential fluid pressure along the flow path, and a temperature controller (Fig. 1). The autoclave installs 7 samples connected in series, through which hydrothermal solutions flow; each sample, 20 cm long, may be a rock cylinder or granular sample in a cylindrical container. The temperature in the autoclave is controlled by seven furnaces surrounding the autoclave, and they can be measured with a thermocouple inserted into a well within the autoclave. Almost all of the metal used are SUS 316.

The fractures in rock samples are made by cutting rock cylinders along the axes and polishing the cut surfaces by alundum powder (#1,500 in this case). The aperture of the fractures are set by inserting stainless steel

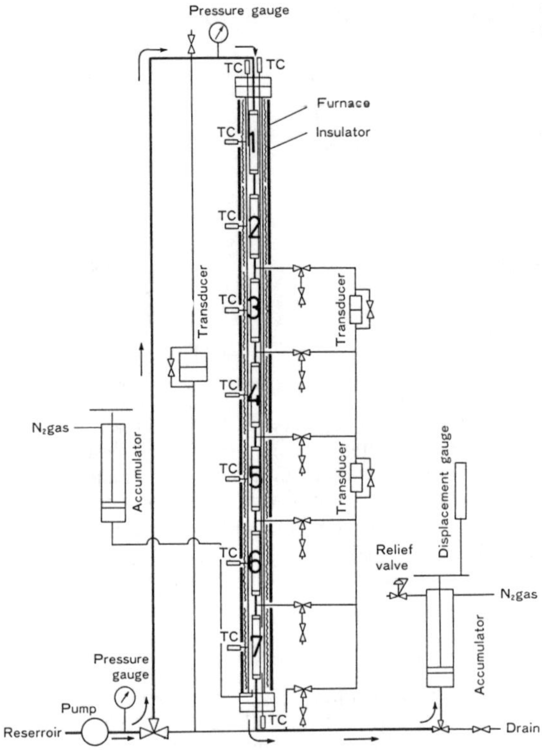

Fig. 1 Hydrothermal fracture flow apparatus TC: thermocouple

plates between the two polished surfaces. The rock samples with fractures are wrapped with Teflon sleeves together with the end caps.

3 SELF-SEALING OF GRANITE FRACTURES BY SILICA

3.1 *Experimental condition*

By using the apparatus, two runs of self-sealing experiment were performed. Temperature gradients of 1°/cm or 0.5^{o}/cm were given to the autoclave, and distilled water was pushed through silica gel first at high temperatures (200 or 100 ^{o}C) and then five rock fractures with decreasing temperatures (Figs. 2 & 3). The samples were named 1 to 7 from the top (inlet) to the bottom (outlet) of the autoclave; sample 1 in RUN 0 was absent. The silica gel (Silica gel 40 of MERCK in Germany) ranged from 0.2 to 0.5 mm in diameter. The rock sample used was Inada Granite consisting of quartz, plagioclase, orthoclase, and biotite partially chloritized. The experimental conditions are summarized in Table 1. The flow rate of the solution was 0.039 cm^3/minute; the residence time of the solution within a single fracture was 3 or 6

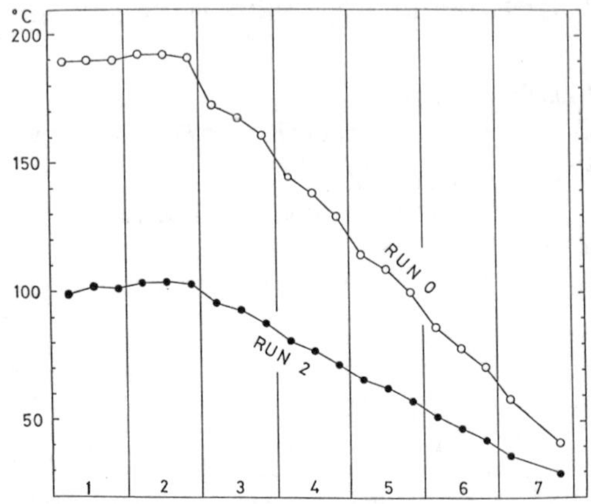

Fig. 2 Temperature gradients given to the two runs

Table 1 Experimental condition

RUN number	RUN 0	RUN 2
Fluid pressure (kgf/cm^2)	18	17.5
Confining pressure (kgf/cm^2)	28	28
Flow rate (cm^3/min)	0.039	0.039
Maximum temperature (^{o}C)	192	102
Temperature gradient (^{o}C/cm)	1	0.5
Duration time (days)	10	29
Sample Silica gel Rock sample Fracture aperture Fracture width Fracture length	 1 container (90 cm^3) 5 pieces 0.05 mm 10 mm 200 mm	 2 containers (180 cm^3) 5 pieces 0.10 mm 10mm 200 mm

minutes when the aperture was 0.05 mm or 0.1 mm, respectively, while the residence time within the joint tubes connecting the samples was approximately 2 hours.

The differential pressure measurement and the solution sampling were performed for RUN 2 but not for RUN 0. For the solution sampling, the first 1 cm^3 of the solution (dead volume of the capillary tube) was discarded, and the next 2 cm^3 was used for the pH measurement by a glass electrode and the analysis of silica by ICP; 1 cm^3 of the 2 cm^3 was dropped into distilled water and diluted to 25 cm^3 , then 3 drops of 30%-sodium hydroxide solution were added. The hydrothermal solutions were sampled 2 times, 10 and 20 days after the outset of the run. All solution was sampled at the 10th day without filters, while at the 20th day, the solution was sampled through filters with the mesh size of 0.2 micrometer.

3.2 *Results and discussion*

The fracture surfaces of the samples of both the runs were covered by a film of silica that consisted of silica particles, which was observed with a scanning electron microscope (Fig. 4). Silica particles, however, did not adhere to the fractures of the samples just beneath the silica gel container, but adhered to the downstream fractures (sample 4 to 7 in Fig. 3). The existence of silica-free zone may be due to low cation contents in the hydrothermal solutions; cations play the role of bridges between the surfaces of silica particles and minerals (Iler, 1979). To acquire the cations, the hydrothermal solutions must flow for some distance after leaving the silica-gel container, and this interval may correspond to the silica-free zone. As described later, colloidal particles of silica seems to have formed enough before entering sample 3 for RUN 2.

Optical microscope observations of the fracture surfaces revealed that the silica particles adhered to all the main minerals almost evenly, quartz, plagioclase, K-feldspar, and biotite. Silica particles may adhere to preferable minerals at an incipient stage, but after all the surfaces are covered by silica the preferential adherence may disappear as was suggested by Bohlmann et al. (1980).

The size of the silica particles was very uniform at a given location. It became as large as 1.5 micrometers in the middle of the flow path for RUN 2 (sample 5), decreasing upstream and downstream. This suggests that silica particles of this size easily adhere to fracture surfaces within a fracture as narrow as 0.05 mm. The silica particles were larger for RUN 2 than for RUN 0, suggesting that the growth rate is much faster under higher temperatures.

The silica contents and pH values of the solutions are shown in Fig. 5 for RUN 2. Silica concentrations including all colloids or colloids smaller than 0.2 micrometers did not decrease systematically downstream, but are almost constant. This fact suggests that the hydrothermal solutions became more and more supersaturated downstream.

Differential fluid pressure along the rock fracture (between the inlet of sample 3 and the outlet of sample 7), that was 5mm H_2O initially, gradually increased from 10th day and reached to 60 m H_2O at 29th day. This increase was due to the plugging of the front ends of rock fracture 6 and 7 by silica that consisted of silica particles smaller than 0.1 micrometer (shown in Fig. 3). Silica thus precipitated in a larger amount in a stagnant zone than in a fast-flowing zone along a flow path with temperature gradients.

4 CONCLUSIONS

The experiments in this paper confirmed that hydrothermal solutions saturated with silica precipitates silica particles when they flow along a fracture with decreasing temperatures. The silica particles decrease

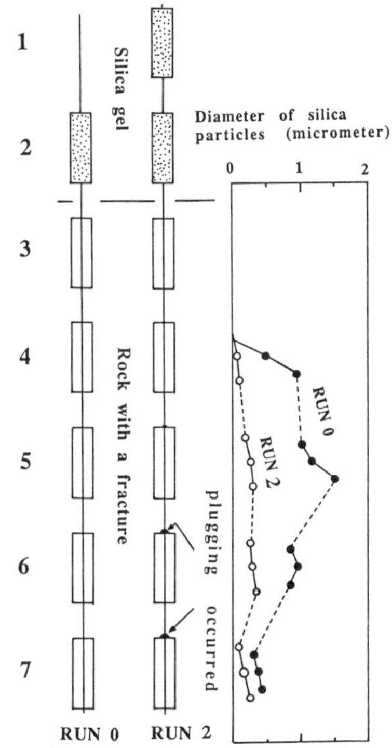

Fig. 3 Diameter of silica particles

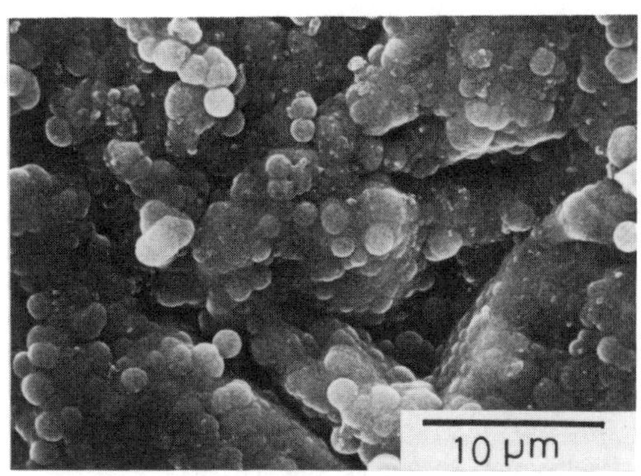

Fig. 4 Scanning electron microscope image of the silica particles on sample 5 of RUN 2

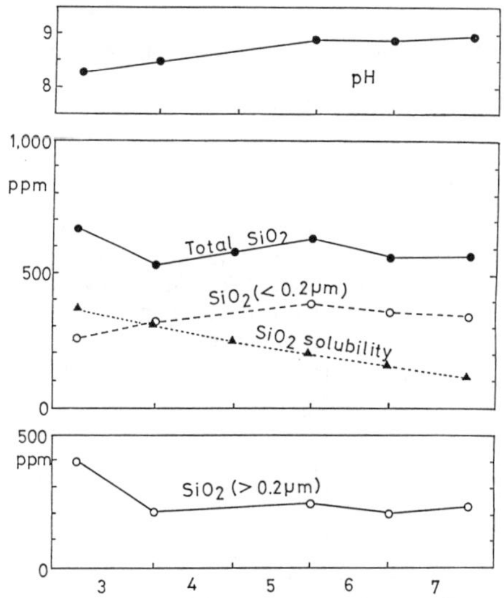

Fig. 5 pH and silica concentration profiles along the flow path from left to rihght of the solutions of RUN 2. Numbers show samples.

their size under lower temperatures, but the silica precipitation processes may not change under the temperatures from 150° C to 30° C. Silica particles can adhere to quartz, plagioclase, K-feldspar, and biotite almost evenly.

REFERENCES

Bohlmann, E. G. , Mesmer, R. G., & Berlinski, P. 1980. Kinetics of silica deposition from simulated geothermal brines. Soc. Petrol. Eng. J. August, 239-248.

Facca, G. & Tonani, F. 1967. The self-sealing geothermal field. Bull. Volcanol. 30: 271-273.

Garibaldi, F. & Freeston, D. H. 1981. Effect of hydrodynamic flow conditions on silica deposition. New Zealand Geothermal Workshop, 3rd, 203-208.

Hochella, M. F. Jr., Ponder, H. B., Turner, A. M. & Morris, D. W. 1988. The complexity of mineral dissolution as viewed by high resolution scanning Auger microscopy: Labradorite under hydrothermal conditions. Geochim. Cosmochim. Acta. 52: 385-394.

Iler, R. K. 1979. The chemistry of silica: Solubility, polymerization, colloid and surface properties, and biochemistry. John Wiley and Sons Inc., 866p.

Itoi, R., Maekawa, H., Tatsuta, K., Fukuda, M., Jinno, K., Hatanaka, K., Nozaki, S., Yokoyama, T., & Shimizu, S. 1985. A mathematical model of silica deposition in a porous medium. Geothermal Resources Council, Trans. 9: 337-340.

Keith, T. E. C., White, D. E., & Besson, M. H. 1978. Hydrothermal alteration and self-sealing in Y-7 and Y-8 drill holes in northern part of upper Geyser Basin, Yellowstone National Park, Wyoming. U.S.G.S. Prof. Paper, 1054-A.

Keith, T. E. C., & Muffler, L. J. P. 1978. Minerals produced during cooling and hydrothermal alteration of ash flow tuff from Yellowstone drill hole Y-5. J. Volcanol. Geotherm. Res. 3: 373-402.

Moore, D. E., Morrow, C. A., & Byerlee, J. D. 1983. Chemical reactions accompanying fluid flow through granite held in a temperature gradient. Geochim. Cosmochim. Acta, 47: 445-453.

Moore, D. E., Morrow, C. A., & Byerlee, J. D. 1986. High-temperature permeability and groundwater chemistry of some Nevada Test Site Tuffs. J. Geophys. Res. 91: B2, 2163-2171.

Morey, G. W. & Chen, W. T. 1955. The action of hot water on some feldspars. Am. Mineral. 40: 996-1000.

Morrow, C. A., Lockner, D., Moore, D. E., & Byerlee, J. D. 1981. Permeability of granite in a temperature gradient. J. Geophys. Res. 86: 3002-3008.

Potter, J. M., Pohl, D. C., Guillemette, R. N. Ponder, H. B., & Liou, J. B. 1985. A system for flow through experimental studies at hydrothermal conditions. N. Jb. Miner. Mh. H. 7: 329-335.

Summers, R., Winkler, K., & Byerlee, J. 1978. Permeability change during the flow of water through Westerly granite at temperatures of 100 °C-400 °C. J. Geophys. Res. 83: 339-344.

Weres, O., Yee, A., & Tsao, L. 1981. Kinetics of silica polymerization. J. Colloid Interface Sci. 84: 379-402.

Weres, O., Yee, A., & Tsao, L. 1982. Equations and type curves for predicting the polymerization of amorphous silica in geothermal brines. Soc. Petrol. Eng. J. February, 9-16.

White, D. E., Brannoch, W. W., & Murata, K. J. 1956. Silica in hot springs. Geochim. Cosmochim. Acta, 10: 27-59.

White, D. E., Hutchinson, R. A., & Keith, T. E. C. 1988. The geology and remarkable thermal activity of Norris Geyser Basin, Yellowstone National Park. U. S. G. S. Prof. Paper, 1456, 84p.

White, D. E., Fournier, R. O., Muffler, L. J. P., Truesdell, A. H. 1975. Physical results of research drilling in thermal areas of Yellowstone National Park, Wyoming. U. S. G. S. Prof. Paper, 892, 70p.

Water-Rock Interaction, Kharaka & Maest (eds) © 1992 Balkema, Rotterdam. ISBN 90 5410 075 3

Travertines deposited by CO_2-rich springs in the geothermal province of northern Latium, Italy

G.Chiodini & A.R.Zanzari
Dipartimento di Scienze della Terra, Universita' di Perugia, Italy

L.Fanfani
Dipartimento di Scienze della Terra, Universita' di Cagliari, Italy

ABSTRACT: The amounts of minor and trace elements in travertine are related to the dissolved amounts of the same elements in the depositing water. This linkage may represent a guide in understanding the nature of fossil discharges of fluids.

1 INTRODUCTION

Numerous deposits of travertines, up to some tens of square kilometers in extent and hundreds of metres in thickness, are located in central Italy. They were mostly deposited by thermal or cold waters during the last uplift stages of the Appenninic chain (middle to late Pleistocene).

The geothermal area of northern Latium, where travertines are more frequently found, is characterized by a wide anomaly of CO_2 flux. Isotopic studies have shown that the carbon of the travertines and of the gas emissions in this area originates principally from the decarbonitization of the deep Mesozoic carbonates (Gonfiantini et al., 1968, Manfra et al., 1976).

The aim of this paper is to investigate the chemistry of the travertines and the relationships between rock and depositing water compositions in an area where these carbonate deposits are originated by different thermal and cold waters. Since most travertines are the direct or collateral manifestation of geothermal systems, such a study may be suitable in identifying fossil discharges of geothermal fluids and evaluating the extent of these systems in space and time.

Chemical data on five deposits in the northern Latium and of their parent waters will be compared with those of deposits from localities in Umbria and Sardinia (Caboi 1991) lacking any geothermal manifestation.

2 HYDROGEOCHEMISTRY

Many known geothermal systems (Larderello, Mt Amiata, Latera, Torre Alfina, Cesano) are located in central-western Italy; they are linked to the presence in depth of the permeable Mesozoic carbonate and evaporitic formations of the Tuscan and Umbrian-Marches sedimentary series. Two main types of hydrothermal systems have been recognized within the carbonate and/or evaporitic reservoirs: (a) deep medium-high enthalpy systems characterized by Na-Cl waters as at Torre Alfina and Latera (Buonasorte et al., 1988; Gianelli and Scandiffio, 1989); (b) shallower low enthalpy systems characterized by $Ca-SO_4$ waters (Checcucci et al., 1988). Large amounts of thermal waters, principally of type b, depositing travertine are discharged in areas generally near the outcrops of the reservoir formations. Their chemical characteristics reveal a complex hydrogeochemistry due to numerous processes affecting the waters, namely mixing with shallow meteoric waters from volcanic reservoirs or from sedimentary formations, degassing and consequent precipitation of travertine, interaction with rising gasses rich in CO_2 and H_2S. The Langelier-Ludwig diagram (Fig. 1) shows the general picture of the hydrogeochemistry of central-western Italy based on data of springs, wells and shallow waters, from DST-PG and from Principe et al. (1990).

Travertine deposits are principally associated with $Ca-SO_4$ and Na-Cl thermal water discharges, but are also precipitated from shallow cold waters of sedimentary formations.

The springs considered in this paper (Tab. 1) give us a complete selection of the thermal hydrotypes : the springs of Poggio Olivastro (PO) and Bagnarello (BA) show a $Ca-SO_4$ composition, while the

Table 1. Analytical data of waters related to travertines.

Name		pH	T °C	Ca ppm	Mg ppm	Na ppm	K ppm	HCO_3 ppm	Cl ppm	SO_4 ppm	SIO_2 ppm	F ppm	Li ppm	Sr ppm	Mn ppb	Zn ppb
P.Olivastro	(PO)	6.2	45	620	115	113	45	692	75	1539	56	2.2	0.4	14	22	40
Monterozzi	(MR)	6.7	22	578	113	1081	285	2342	1453	1006	28	1.5	5.2	21	350	50
B.S.Agostino	(BS)	6.2	49	624	82	431	60	1708	548	828	27	1.6	4.0	14	-	30
Bagnarello	(BA)	6.5	45	578	120	35	9	604	29	1554	51	3.0	0.1	13	-	5
Rota	(RO)	6.8	18	146	55	1850	67	5139	272	229	23	1.3	1.5	1.0	-	-
Triponzo	(TR)	6.7	29	292	72	137	20	194	212	974	-	1.7	0.2	6.2	19	12
Biselli	(BI)	7.2	11	63	8	2.2	0.7	104	3.5	18	-	0.7	0.002	0.2	5	64
Velino	(MA)	8.3	10	87	12	5.1	2.1	317	6.4	50	-	-	0.003	0.7	11	25

bicarbonate waters from Monterozzi (MR), Bagni Sant'Agostino (BS) and Rota (RO) are characterized by the presence of an important Na-Cl component. All waters show PCO_2 values approaching 1 bar and are saturated with respect to calcite, quartz, fluorite and, except Rota, to gypsum.

The amounts of Ca, Mg, SO_4, F and SiO_2 are almost constant. On the contrary the mobile species Cl, B and alkalies, show different values related to the type of aquifer. Using their content as tracers of the deep Na-Cl waters (Giggenbach and Goguel, 1989; Chiodini et al., 1991), different proportions of geothermal brine, for instance similar to that of Latera, are recognized in the five springs. In particular, B.S.Agostino and Monterozzi are the most representative samples of the deep Na-Cl component, Bagnarello is a pure $Ca-SO_4$ water type and Poggio Olivastro spring represents a mixing of the two hydrotypes with the $Ca-SO_4$ component prevailing (Chiodini et al., 1991).

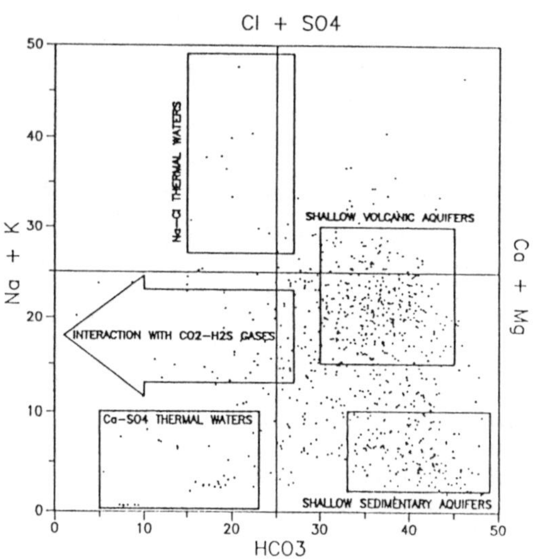

Fig. 1 Hydrochemical characters in central Italy.

The last three samples, reported in Table 1 for comparison (DST-PG unpublished data), refer to waters depositing travertine located in the region of Umbria which borders on Latium and which, unlike this, is not thermally anomalous. Triponzo (TR) is a hypothermal $Ca-SO_4$ water whose aquifer is located in the deep evaporitic formation. The Biselli (BI) spring and Velino river (MA) are typical examples of cold $Ca-HCO_3$ waters from shallow carbonate aquifers (Chiodini et al.1982). The Velino river is still forming travertine at the Marmore fall.

3 CHEMISTRY OF TRAVERTINES

Travertine deposits related to the waters considered in this study are characterized by different extension and morphology.

Samples collected not far from the springs of Poggio Olivastro come from Canino quarry where sub-horizontal layers of both lithoid and friable porous travertine make up a bank about 100 m thick. Smaller deposits characterize the other areas sampled. Samples from several travertine cones formed by the flowing out of high-CO_2 waters from scattered springs were collected at Bagni Sant' Agostino and Monte Rozzi. At Bagnarello travertine specimens were taken from the wall of a partially exploited quarry. Few samples represent the carbonatic coating on the outcroping rocks deposited by Rota spring.

The mean values of the chemical composition of 32 samples of travertine are shown in Tab. 2, together with those of 75 samples from the three deposits from Umbria reported for comparison (DST-PG, unpublished data).

High CaO and LOI values indicate that travertine from all sites is essentially built up of calcium carbonate. The low amounts of SiO_2, TiO_2, Al_2O_3 and K_2O represent the contribution of silicate materials to most of the insoluble fraction of these travertines; the observed constant ratios among these species indicate, furthermore,the noteworthy composi-

Table 2. Mean composition of travertines.

Name	CaO	SiO_2	TiO_2	Al_2O_3	Fe_2O_3	K_2O	SO_4	F	Mg	Sr	Mn	Zn	Na	Li
	%w	%w	%w	%w	%w	%w	ppm	ppm	ppm	ppm	ppm	ppm	ppm	ppm
CANINO	55.23	0.33	0.003	0.09	0.05	0.004	3490	504	1220	886	98	<10	264	4
M.ROZZI	55.25	0.07	0.001	0.03	0.01	0.002	4010	700	1650	1413	626	<10	431	7
B.S.AGOSTINO	52.89	2.08	0.016	0.47	0.11	0.026	16400	726	2252	2414	1082	33	794	16
BAGNARELLO	55.05	0.63	0.005	0.14	0.32	0.007	2116	307	839	759	72	106	137	3
ROTA	50.48	5.71	0.045	1.47	0.72	0.050	2960	359	3701	754	692	74	1247	8
TRIPONZO	54.33	1.65	0.012	0.31	0.18	0.020	3165	114	1318	363	282	37	296	4
BISELLI	55.57	0.20	0.002	0.00	0.09	0.003	1053	-	1097	156	<50	180	314	2
MARMORE	54.48	1.74	0.027	0.49	0.54	0.028	2463	-	2088	363	262	68	134	5

tional homogeneity of this fraction in all sampled areas. Iron does not belong to the silicate minerals to any great extent but is mainly present in the travertine as finely scattered oxide. Sr and Mg are to be considered as replacing Ca in the calcite lattice, while Na, Li, F, and SO_4 can be associated with the carbonate fraction as dissolved by dilute hydrochloric acid. Mn, though dissolvable by dilute acid, belongs to the carbonate portion only to a minor extent. Zn is probably adsorbed on Fe and Mn oxides.

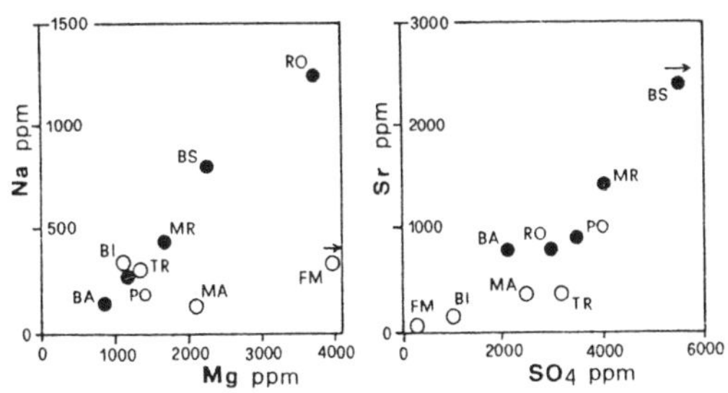

Fig. 2 Minor elements in travertines.

A very similar Sr/Mg ratio is observed in all areas of northern Latium, except Rota where travertine is enriched in Mg, while the Umbrian deposits, because of their lower amounts of Sr, can be distinguished by this ratio. All elements associated with the carbonate fraction exhibit approximately proportional amounts, again with the exception of Rota which is also enriched in Na (Fig.2). The linear relations, among the elements associated to the carbonate fraction, signify their common origin. The different absolute amounts of the trace elements allow a chemical distinction of the travertines from different localities.

4 RELATIONSHIP WITH DEPOSITING WATER

As deep Na-Cl geothermal water is richer in all mobile elements than shallow Ca-SO_4 waters, the increase in Na, Li and SO_4 from Bagnarello to Bagni Sant' Agostino through Poggio Olivastro and Monte Rozzi, indicates that the amount of these elements in depositing waters affects the chemical composition of the travertines more than other factors such as temperature, deposition rate and diagenesis. Rota travertine also maintains a memory of the depositing water as it is shown in the graphs of fig 3.

Sr and Mg, as distinctive chemical parameters in the Latium travertines, apparently work less than the mobile elements because of their relatively constant

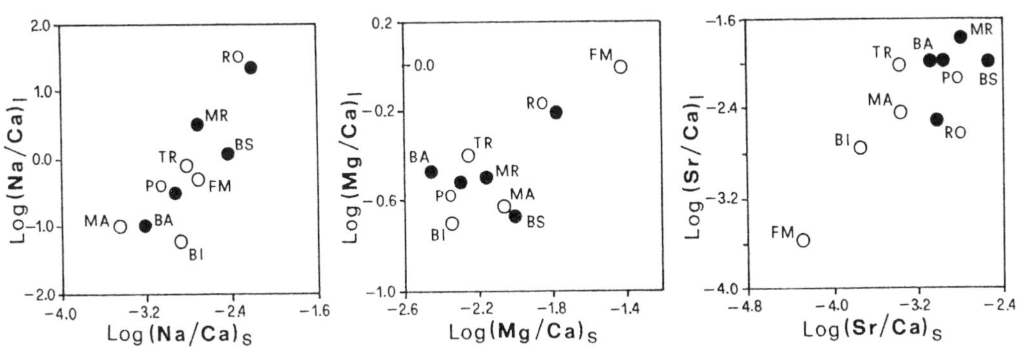

Fig. 3

amounts in these waters; but if we consider examples from other areas the relation between the abundance of these elements in travertine calcites and their concentration in depositing waters is clearly shown (fig. 3) and can be approximately quantified using apparent partition coefficients between rock and water. Their values are in agreement with those from laboratory experiments on calcite precipitation as reported by Veizer (1983).

In the graph Na-Mg of fig. 2, Funtana Maore (FM) clearly departs from the linear trend. This is easily explained on taking into account that the FM aquifer is essentially dolomitic while all others are carbonate or evaporitic aquifers.

5 CONCLUSIONS

Notwithstanding the influence of temperature, rate of precipitation and diagenesis on the geochemistry of travertine, the chemical composition of waters from different aquifers gives an imprinting to the rock recognizable not by a single marker but by a set of geochemical data, considered in its entirety. However, further data are needed to understand how the partition of minor and trace elements between water and travertine can be employed in the identification of fossil discharges of geothermal fluids.

REFERENCES

Buonasorte, G., Cataldi, R., Ceccarelli, A., Costantini, A., D'Offizi, S., Lazzarotto, A., Ridolfi, A., Baldi, P., Barelli, A., Bertini, G., Bertrami, R., Calamai, A., Cameli, G.M., Corsi, R., D'Aquino, C., Fiordelisi, A., Ghezzi, C. and Lovari, F. 1988. *Ricerca ed esplorazione nell'area geotermica di Torre Alfina (Lazio-Umbria)*. Boll. Soc. Geol. Ital. 107.

Caboi, R., Cidu, R., Fanfani, L., Zuddas, P. and Zuddas, P.P. 1991. Geochemistry of Funtana Maore travertines (Central Sardinia, Italy). *Miner. Petrogr. Acta*. 34:77-95.

Checcucci, R., Chiodini, G. and Giaquinto, S. 1988. Idrogeochimica dei Monti Vulsini: sintesi e rielaborazione dei dati esistenti. *CNR-PFE* SI5:269-290 Roma.

Chiodini, G., Giaquinto, S. and Zanzari, A.R. 1982. Relazione tra il chimismo delle sorgenti umbre e le caratteristiche litologiche degli acquiferi. *CNR- PFE-SPEG 3*:30-47. Roma.

Chiodini, G., Giaquinto, S., Frondini, F. and Santucci, A. 1991. Hydrogeochemistry and Hydrogeology of the Canino Hydrothermal System(Italy). *Geothermics* 20:329-342.

Gianelli, G. and Scandiffio, G. 1989. The Latera geothermal system (Italy): chemical composition of the geothermal fluid and hypotheses on its origin. *Geothermics* 18:447-463

Giggenbach, W. F. and Goguel, R.F. 1989.Collection and analyses of geothermal and volcanic water and gas discharges. Chemistry Division, Dept. of Scientific and Industrial Research, Petone, New Zeland. Internal report.

Gonfiantini, R., Panichi, C. and Tongiorgi, E. 1968. Isotopic disequilibrium in travertine deposition. *Earth Plan. Sci. Letters* 5:55-58.

Manfra, L., Masi, U. and Turi, B. 1976. La composizione isotopica dei travertini del Lazio. *Geol. Rom.*15:127-174.

Principe, C. and Romano, G.A. 1990. GEOCH: Geochemical data of natural fluids from surveilled italian active volcanoes. *EARNT/BITNET* ASTRA service.

Veizer, J. 1983. *Carbonates: mineralogy and chemistry*. Reviews in Mineralogy 11: 265-299 Reeder R.J..

Water-Rock Interaction, Kharaka & Maest (eds) © 1992 Balkema, Rotterdam. ISBN 90 5410 075 3

Alteration mineralogy of the Wayang Windu geothermal field, West Java, Indonesia

Sugiarto Ganda & Djoko Hantono
Geothermal Pertamina, Jakarta, Indonesia

ABSTRACT: Wayang Windu is an active geothermal field in west Java that has been explored since 1985. The first successful exploration well was drilled in 1991 and produced two-phase fluids during the production test. This interesting field is included in the next five-year plan of the Geothermal Development in Indonesia and should be developed soon. The alteration mineralogy of the Wayang Windu field indicates that the rocks have undergone hydrothermal alteration in two main phases. High temperature alteration took place at temperatures in excess of 325°C, when the area was subjected to magmatic activity. Biotite, sphene and galena were formed during this period along with calcite and quartz. Moderate temperature alteration took place as the hydrothermal system cooled, and a second main phase of alteration occurred at temperature of 240 - 270°C. The moderate temperature activity resulted in the precipitation of chlorite, epidote, laumontite, and wairakite associated with calcite and quartz. These minerals seem to have been formed in active hot water aquifers.

1 INTRODUCTION

Two high temperature geothermal fields have been explored within the Wayang Windu Contract area (Figure 1).

Those fields lie at the Pangalengan district, south of Bandung, the capital city of West Java Province. The Wayang Windu field, drilled in 1991, has a maximum temperature of 270°C and potential energy of about 200 Megawatt. The Malabar field has a predicted potential of about 220 Megawatt.

The first exploration well drilling took place in the Wayang Windu field in 1991 with the drilling of well WWD-1. Two additional wells will be drilled in the next fiscal year to improve the geothermal modelling and to confirm the total energy potential.

This paper deals with the alteration mineralogy of the Wayang Windu field as it appears in well WWD-1. The mineral distribution is mostly based on data from petrographic analysis and X-ray diffraction examination.

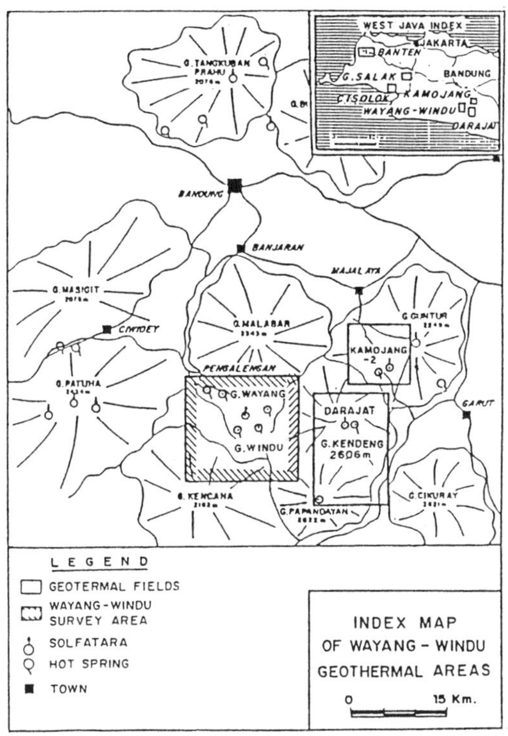

Figure 1 - Location map of the Wayang-Windu geothermal area, Indonesia

2 GEOLOGICAL SETTING AND WATER CHEMISTRY

The Wayang Windu geothermal field lies just

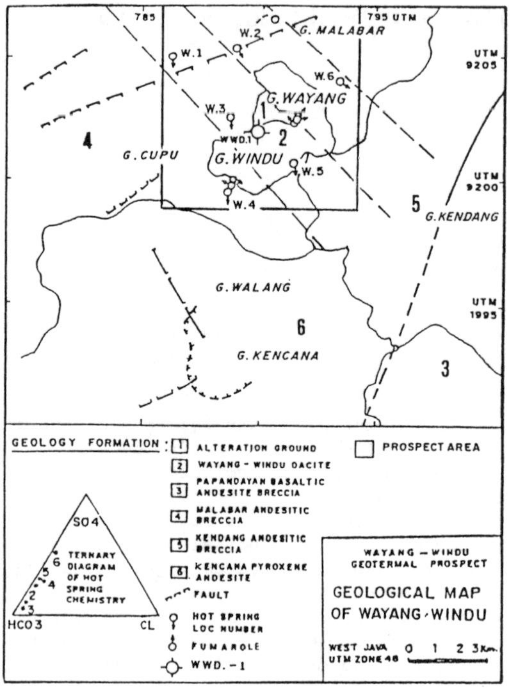

Figure 2 - Geological map and location of wells and springs Wayang-Windu geothermal prospect

inside the active volcanic zone in the center of west Java (Figure 2).

The active volcanic zone is flanked symmetrically by Quaternary volcanics which in turn are flanked by Tertiary volcanics.

The Wayang Windu geothermal system is situated in Pleistocene rocks ranging in age from about 1.0 to 0.147 million years. Two central volcanoes were active in this region during this period, the Malabar Volcanic Complex and the Wayang Windu lava domes, which produced relatively young lavas in part of the strata at this field dating back to about 147,000 years ago. These are overlain by younger sediments and epiclastic volcanics in the Wayang Windu Valley, which are younger than 50,000 years (Table 1).

Pleistocene strata are characterized by an interlayered series of tuff breccia and lava flows. The surface rocks were subsequently cut by shallow level dykes and sheets, which were related to growing of the Wayang Windu lava dome (Dewi and Ganda 1985).

Thermal springs in this region are mainly bicarbonate in character, with varying amounts of sulfate. Some hot springs

Table 1. General stratigraphy of the G. Wayang Windu, West Java.

Age		Rock Unit	Volcano Activity	
Recent		Alluvium (Qal)		
PLEISTOCENE	Middle	Wayang-Windu Dacite (QWynl)		0.147 m.y.a
		Papandayan Basatt (QPndl)	G.Papandayan	0.226m.y.a
		Papandayan Basaltic-Andesite Breccia (QPndl)		
		Malabar Andesittic Breccia (QMlgv)	G.Malabar	0.252 m.y.a
		Cupu Tuff Breccia (QCpV)	?	
		Waspada Tuff Breccia (QWspV)	?	
	Lower	Kendang Andesitic Breccia (QKndV)	G. Kendang	1.00 m.y.a
PLIOCENE	Upper	Limo Andesitic Breccia (QLmV)		2.63 m.y.a
	Lower	Talun Tuff Breccia (QTlnV)		4.36 m.y.a
		Walang Pyroxene-Andesite (QWlkL)	G.Kencana	
		Hanjawar Andesite (QHjl)		
		Kencana Andesitic Breccia (QKncV)	G.Kencana	
			?	
		Kencana Andesite (QKncL)	G.Kencana	
Middle-Upper Miocene		Pojok Andesite	?	12.1 m.y.a

associated with fumarole manifestation show rather high sulfate concentrations at the higher elevation. Water pH values range from 6 to 7, HCO_3 ranges from 115 to 560 ppm, SO_4 from 45 to 214 ppm and Cl from 1 to 20 ppm (Table 2).

Sometimes neutral-pH hot springs are associated with carbonate deposition due to steam loss at the surface.

Table 2. Chemical analyses of hot springs at the G. Wayang Windu (ppm).

SAMPLE NUMBERS	T °C	pH 25°C	Na	K	Ca	Mg	HCO_3	Cl	SO_4	SiO_3
1	49	6.45	61.3	32	33.6	22.4	306	2	85	42
2	50	6.79	80	28	36.2	27.2	418	5	45	145
3	36	6.69	90.7	9.7	50	46.8	560	1	50	52
4	61	6.07	66.7	16.7	56.8	37.6	325	20	145	75
5	46	6.12	50.3	11.3	70.8	30.6	220	6	214	52
6	48	6.41	18.6	8.7	22.4	10.6	115	1	50	60
7	92	1	10.9	6.8	16	8.6	–	1	2400	85

3 GENERAL STRATIGRAPHY

Table 1 shows the simplified stratigraphy through drillhole WWD-1. The stratigraphic section is composed of lava flows with mostly thin interlayers of tuff and tuff breccia from several volcanic centers. Several thick sediment beds are encountered in the drillholes, but their horizontal extent appears to be small.

In the upper part of the profile dacitic lavas are significant, making up 25-30% of the profile. Below 500 m depth andesitic breccia is dominant (Figure 3).

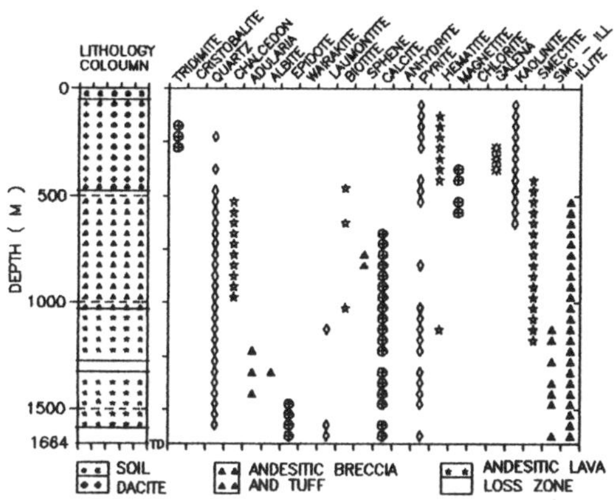

Figure 3 - Hydrothermal minerals in well WWD-1.

The porphyritic holocrystalline dacitic andesite seems to be an intrusive rock that varies from a coarse grain to fine grain dyke and which has sheet interlayering of andesitic lavas at the lower part of the lithologic section in well WWD-1 (Emmy 1991).

The main aquifer, situated in the andesitic lava and dacitic dykes, could be a potential heat source in this field. Contact metamorphic minerals grew along the contact with the dyke.

4 AQUIFERS AND TEMPERATURE

The main aquifers of the Wayang Windu field lie at depth intervals of 1400-1450 m and 1600-1664 m (bottom) in well WWD-1. The maximum temperature of the field is about 278°C in the andesitic formation, which may have intersected the fractures that control flow into the present hydrothermal system (Figure 4).

The pressure gradient results in no water in the well. Maximum pressure at the reservoir is 75 kg/cm^2; if it is related to the temperature of 278°C, this zone represents two-phase fluids. This prediction is confirmed by the appearance of anhydrite at the lower part of the well. Anhydrite is precipitated during the rise of water temperature and mixing with saline or non-saline fluids (Kristmannsdottir 1979). The fractures trending NW-SE may presently control flow within the hydrothermal system of the Wayang Windu field.

5 ALTERATION AND MINERALOGY

The alteration mineralogy of the Wayang Windu geothermal field has been described

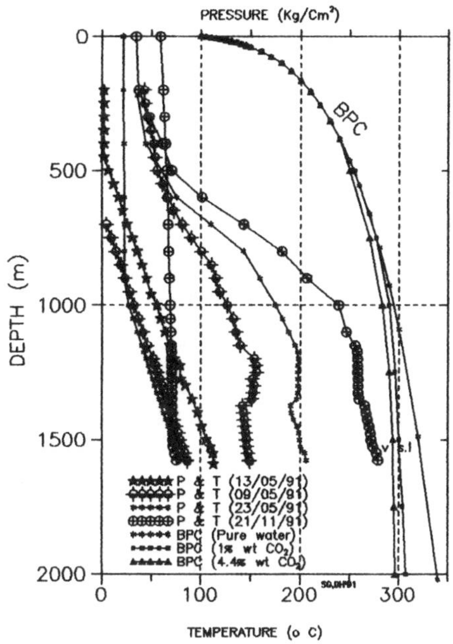

Figure 4 - Temperature and pressure in well WWD-1.

in some detail by Hantono and Masdjuk (1991). In this contribution we focus on well WWD-1 in the Wayang Windu field.

The distribution of the alteration minerals in well WWD-1 is shown in Figure 3 along with the alteration zones. Five mineral zones are identified: I. Chalcedony (500-1000 m); II. Biotite-Sphene (500-1000 m); III. Epidote-Chlorite (1400-1664 m/TD); IV. Smectite-Illite (600-1400 m); and V. Kaolinite (50-600 m).

The Chalcedony Zone (I) and Biotite-Sphene Zone (II) represented the fossil alteration system, which was overlain by the younger alteration system in the present active geothermal system.

The younger alteration system is represented by Epidote-Chlorite Zone (III), Smectite-Illite Zone (IV) and Kaolinite Zone (V).

The Kaolinite Zone is found at 50-600 m in well WWD-1. The zone ends when illite appears at 600 m. Other minerals found in this zone are anhydrite, cristobalite, hematite, and galena.

The Smectite-Illite Zone occupies the depth interval of 600-1400 m in well WWD-1. Quartz, calcite, and anhydrite are found throughout this zone along with smectite and illite. This zone ends with the disappearance of smectite and appearance of lower epidote.

The Epidote-Chlorite Zone occupies the lowest part of well WWD-1, below 1400 m. It is first identified with certainty where the well intersects a 220°C hot aquifer at 1400 m. Epidote and lower-chlorite laths seem to be growing in this aquifer, judging from its fresh appearance. Other minerals found in the Epidote-Chlorite (270°C) Zone are calcite and zeolite (laumontite and wairakite).

Minerals such as sphene, biotite, upper epidote, upper-chlorite, and galena belong to a high-temperature system (325°C) that prevailed during the active eruption phase of the Wayang Windu lava domes. Chalcedony was formed soon after the Wayang Windu volcanic activity when the system cooled and reheated at the beginning of the real hydrothermal system in this field.

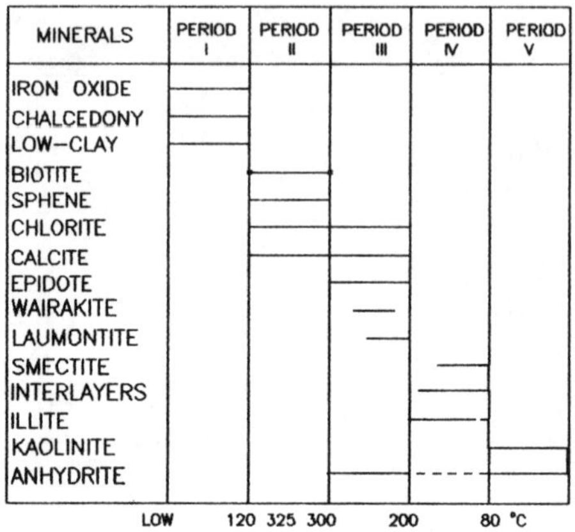

Figure 5 - The alteration history of well WWD-1

6. ALTERATION HISTORY

In the Wayang Windu field, five distinct stages of alteration and mineralization are recognized (Figure 5).

The first one of these represents a time period of a low temperature geothermal system, characterized by the deposition of chalcedony, iron oxide and low temperature clay minerals (<120°C). The minerals precipitating during this period indicated a system which was formed soon after volcanic emplacement upon rising in the geothermal gradient.

The second time period involves prograde alteration of primary minerals and open-space mineralization, with the extent of alteration increasing with depth. This period represents temperatures up to 325°C and reactions with the extinct earlier system. Some relict minerals from the first period are seen within mineral assemblages of the second period.

When the system cooled down, the third period of mineralization appears, forming an Epidote-Chlorite Zone. The two-phase fluid system cooled from 300°C to 270°C with the precipitation of anhydrite. Recent active geothermal fluids are supplied from this zone at depth intervals of 1400-1660 m.

Because of lower permeability and partial flashing, the present day system continues with the fourth period, which is represented by the Smectite-Illite Zone in the middle section of the well (600-1400 m). This is confirmed by the conductive layer shown by the temperature gradient (Figure 4). This zone is superimposed on the first and second period of old alteration. Average temperature is from 80°C-200°C.

The last zone is kaolinite, which is produced by steam or gas activity. This zone is seen to be a cap of the present active system which has temperatures of less than 80°C. The abundance of anhydrite within this zone indicated that condensation is taking place, and cold fluid might be in contact with geothermal fluid.

In the reservoir, the present fluid is saturated with respect to epidote, chlorite, and anhydrite, which are formed at temperatures of 270°C. High shut-in pressure is a result of the high gas content in the water-dominated system. The geothermal system in this field tends to cool slowly, and the upper part of the system has already been sealed by fossil chlorite-calcite in vesicles.

7 DISCUSSION AND CONCLUSION

The hydrothermal system forms a mushroom-shaped reservoir that follows the root of Wayang Windu lava domes at 600-1400 m in well WWD-1. The fluid is saturated with respect to epidote-chlorite. The alteration minerals in the Wayang Windu field can be divided into five alteration zones:

I. Chalcedony Zone (500-1000 m).
II. Biotite-Sphene Zone (500-1000 m).
III. Epidote-Chlorite Zone (1400-1650 m).
IV. Smectite-Illite Zone (600-1400 m).
V. Kaolinite Zone (50-600 m).

The alteration history can be divided into two main generations of geothermal activity. The first generation resulted in the formation of the Biotite-Sphene Zone during a phase of very high temperature activity (300-325°C) in association with the Wayang Windu volcanic activity.

The last three zones represent the present active geothermal system (after volcanic activity) when the system is slowly cooling down. The main reservoir is dominated by near neutral hot water at a depth of 1400-1600 m.

8 ACKNOWLEDGMENT

The authors would like to acknowledge the Pertamina Management for the opportunity to publish this paper, especially the Head of Geothermal Division and Head of Planning Department.

REFERENCES

Dewi, G.M. & S.Ganda 1985. Geology of Wayang Windu area, west Java. Geothermal Pertamina, Jakarta (unpublished).

Emmy, S. 1991. Report on petrographical and XRD analysis well WWD-1, Wayang Windu. PT. Citra Andana, Jakarta (unpublished).

Hantono, D. & M.Masdjuk. 1991. Final report on Drilling Well WWD-1, Wayang Windu, west Java. Geothermal Pertamina, Jakarta (unpublished).

Kristmannsdottir, H. 1979. Alteration of basaltic rocks by hydrothermal activity at 100-300°. Elsevier Scientific Publishing Company, Amsterdam: 359-367.

Water-Rock Interaction, Kharaka & Maest (eds) © 1992 Balkema, Rotterdam. ISBN 90 5410 075 3

Augustine volcano fumarole wall rock alteration: Mineralogy, zoning and numerical models of its formation process

Aberra Getahun & Mark H. Reed
University of Oregon, Oreg., USA

Robert B. Symonds
USGS Cascade Volcano Observatory, Wash., USA

ABSTRACT: Altered fumarole wall rocks on Augustine volcano exhibit distinct mineral zoning differentiated by color bands. Color bands correspond to the following mineral assemblage zones (in order of decreasing temperature): a) tridymite-anhydrite (white to grey), b) tridymite-anhydrite-hematite-iron hydroxide-molysite (pink to red), and c) tridymite-halite-sylvite-soda and potash alum and other sodium and potassium sulfates (dark green to green). Thermochemical modeling suggests that the mineral zoning results from cooling of volcanic gases and reaction of the wall rock with volcanic gas mixed with air.

1 INTRODUCTION

Mount St. Augustine is an andesitic-dacitic island volcano, located 285 Km southwest of Anchorage, Alaska, in the lower Cook Inlet. It is one of the most active volcanoes on the Aleutian arc, last erupting in 1986. It is also well known for its Cl-rich volatiles (Johnston, 1980). Following the 1986 eruption, the most active fumaroles were located at the base of the spine on the 1986 lava dome. Due to intense HCl emission, the dome fumaroles were inaccessible in 1986. In 1987, they were discharging 870 °C gases that were exceptionally enriched in HCl, S, and metals (Symonds et al., 1990). Other fumarole fields were located on the moat, a remnant on the 1976 lava dome, and on the 1986 pyroclastic flows (Kodosky and Keskinen, 1990; Symonds, et al., 1991).

Samples of intensely altered wall rock were collected from the dome fumaroles in 1989, when the maximum temperature had decreased to 640 °C. Incrustations around the low temperature (< 250 °C) rootless fumaroles on the 1986 pyroclastic flows are documented by Kodosky and Keskinen (1990). In this study, we document the high temperature alteration mineral assemblages and use thermochemical modeling to help understand the origin of the wall rock alteration in Augustine's fumaroles.

The numerical modeling was carried out using program GASWORKS, which is designed for calculating heterogeneous equilibria among minerals, liquids, and gases during processes of temperature change, pressure change, gas-gas mixing, and gas-rock reaction (Reed and Symonds, 1992). The program computes the distribution of gases, liquids, and solids in systems of up to 42 components as a function of temperature and pressure using the basic formulation of equilibrium calculation of Reed (1982).

2 MINERALOGY AND ZONING

In the high temperature (640 °C) fumaroles (spine-1), distinct zoning of alteration minerals in the interiors of the wall rocks is differentiated by color bands ranging from white to grey, pink to red, and dark green to green in order of decreasing temperature. The color boundaries are sharp except for the colder end, where the change is gradual. The thickness of each individual zone varies from less than a centimeter to few centimeters. X-ray diffraction (XRD) analysis on eight zones from spine-1 fumarole wall rock indicates the presence of ten crystalline phases, dominantly oxides, chlorides, and sulfates. Tridymite is present in all zones, and it is a dominant phase at the high temperature end (white to grey). Oxides, hydroxides, and chlorides of iron occupy the middle zone (pink to red) followed by Na and K chlorides and sulfates at the relatively colder end (dark green to green).

SEM analysis on broken surfaces and microprobe analysis of polished surfaces of spine-1 wall rock confirm the presence of all the mineral phases identified by XRD and reveal a number of additional crystalline and amorphous phases that were not otherwise positively identified by XRD. At the high temperature end of spine-1 samples, tridymite is the dominant phase together with subordinate anhydrite and an Al-Si phase. The middle color band (pink to red) is constituted by tridymite, iron-chlorides (probably molysite – $FeCl_3$), iron-hydroxide and(or) hematite. The iron-chloride occurs as small clusters of hexagonal platelets. Other minor phases identified in this zone are small cubes of halite, an amorphous Fe-Zn phase and unidentified phases containing Fe-Al-Si, and Fe-Mg-Al-Cl-Si. Most of the amorphous phases have apparent dehydration cracks that run across the aggregates. The relatively colder ends of the samples are composed of halite, sylvite, anhydrite, and gypsum accompanied by unidentified phases containing Ca-Mg-Si, Ca-Al-Si, Ca-Mg-Fe-Si, Ca-Al, slender crystals of K-Fe-Ca-Ti-Al-S-Cl, and amorphous aggregates of Al-hydroxide (probably $Al(OH)_3$).

In the low temperature (350 °C, spine-3) fumarole, color zones in the altered wall rock change gradually from white to grey. These samples exhibit cristobalite surrounded by amorphous forms of SiO_2. The other major alteration phases are sulfate of K-Al-S (potash alum), Na-Al-S (soda alum), K-Na-S (aphthitalite), and Al-S (alunogen).

3 NUMERICAL MODELING

We hypothesize that the minerals in the altered wall rock form by direct reaction of volcanic gases with the wall rock, cooling of the volcanic gases, mixing of the volcanic gases with air, or a combination of these processes. We explored these processes with the following calculations: a) reaction of the dome gases with wall rock at 870 °C and 640 °C, b) cooling of the dome gases from 870 °C to less than 100 °C, and c) mixing of air with the dome gases. The necessary inputs for the numerical simulations are the volcanic gas composition, the composition of air, the chemical composition of unaltered wall rock, and the temperature and pressure at which reactions occurred. We used the 1987 restored dome-3 sample, including trace metals, from Symonds, et al. (1990) for volcanic gas composition.

The 870 °C gas-rock reaction produces abundant tridymite at high gas/rock (g/r) ratios corresponding to the high temperature mineral zone. As the g/r ratio decreases, anorthite, orthoenstitite, albite, diopside, and sanidine precipitate. At low g/r ratios magnetite and tephroite (Mn_2SiO_4) precipitate. The 640 °C gas-rock reaction produces a larger assemblage than the 870 °C case, consisting of sulfates (anhydrite and thenardite), halides (halite and sylvite), hematite, andalusite, orthoenstatite, and rhodonite (Fig. 1). This result closely resembles much of the observed alteration. The cooling calculation produces anhydrite, halite, sylvite, molybdenite, bornite, and pyrrhotite between 700 and 600 °C followed by fluorides, chlorides, and sulfates of Mg, Mn, Fe, Al, Cu, and Zn from 600 to 200 °C and sulfates of Na and K, hydrated sulfates of Zn, Fe, and Mn, liquid sulfur and hygroscopic H_2SO_4 at temperatures lower than 200 °C (Fig. 2).

The air mixing calculation (with simultaneous cooling) produces entirely sulfides, fluorides, and chlorides without sulfates. At high gas/air (g/a) ratios molybdenite and bornite form, followed by halite, sylvite and pyrrhotite. As the g/a ratio decreases, pyrrhotite is replaced by iron chloride (Fig. 3). Iron chloride, which is computed only in the air mixing calculation is observed with hematite in the altered fumarole rocks. One of the distinctive results from the cooling and air mixing simulations is that most of the incrustation and sublimate mineral assemblages were formed below 700 °C.

4 CONCLUSIONS

A comparison of model results to the observed alteration assemblages suggests the following:

1) Direct volcanic gas-rock reaction at 870 °C produces abundant tridymite and silicate minerals. The silicates that precipitate at this temperature resemble most of the unidentified high-temperature minerals. This indicates that the gas-rock reaction might have started at a higher temperature (such as observed in 1987) than 640 °C, the temperature of spine-1 when wall rock was collected. The gas-rock reaction at 640 °C produces tridymite along with anhydrite, halite and sylvite at higher g/r ratios and Na and K sulfates at low g/r ratios. Hence the sulfates that are produced at low g/r ratios, and observed around relatively lower temperature zones might have been formed by either secondary processes of alteration of anhydrite, halite, and

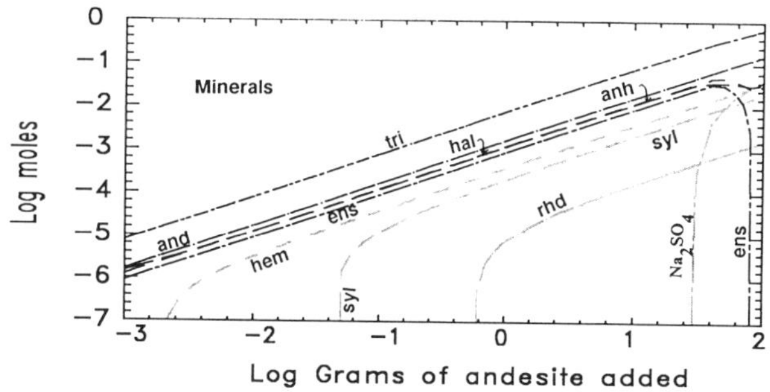

Fig.1 Reaction of andesitic rock with Mt. Augustine volcanic gas at 640 C. Mineral abbreviation used here and in Figures 2 and 3 are as follows: tri, tridymite; hal, halite; anh, anhydrite; and, andalusite; ens, enstitite; hem, hematite; syl, sylvite; rhd, rhodonite; pyr, pyrrhotite; py, pyrite; moly, molybdenite; sl, sphalerite; bn, bornite; cv, covellite; cor, cordierite; S(L), liquid sulfur; S(M), monoclinic crystalline sulfur.

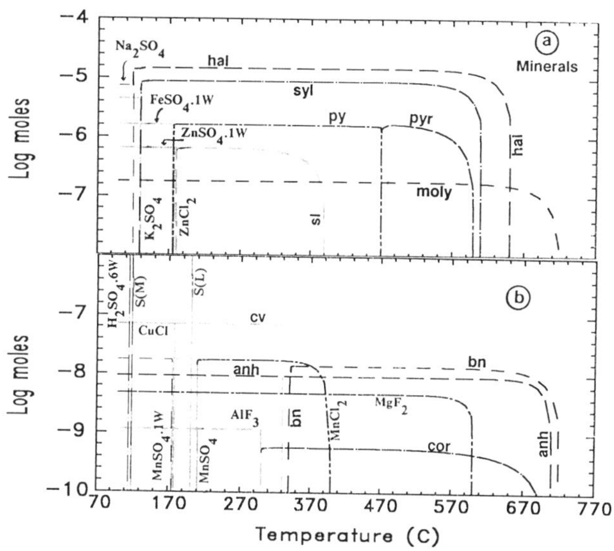

Fig.2 Results showing cooling of Augustine volcanic gas from 870 C to 100 C without mineral fractionation.

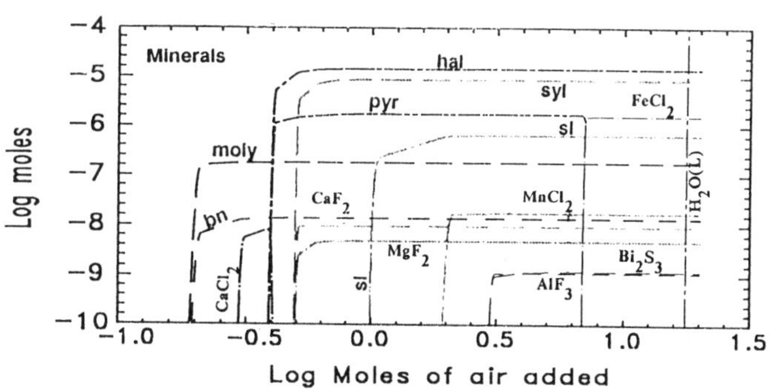

Fig.3 Results showing the mixing of Augustine volcanic gas with air. The starting mixing temperature of the gas and air was 870 and 25 C respectively.

sylvite or by direct reaction of original rock forming minerals (e.g. feldspars) with the volcanic gas.

2) Cooling and air-gas mixing produce dominantly halite and sylvite over a large range of temperature and g/a ratios. Most of the sulfides, chlorides, fluorides, sulfates and hydrates are produced by these two formation processes. Iron-chloride which is observed with hematite and iron-hydroxide in the intermediate (pink to red) zone is computed only in the air-mixing calculation. This suggests that it formed by an episode of mixing of volcanic gas and air.

Even though these calculations might not entirely duplicate the fumarolic alteration processes, they put significant constraints on the various processes that might have formed the zoned alteration sequences in the wall rock of Augustine's fumaroles. In particular, the modeling has successfully demonstrated that the fumarole wall rock alteration may have been caused by the interplay of a continuous gas-rock reaction, cooling of volcanic gases and mixing of volcanic gases with air.

REFERENCES

Johnston, D.A. 1980. Volcanic contribution of chlorine to the stratosphere: more significant to ozone than previously estimated? Science 209, 491-492.

Kodosky, L. and M. Keskinen 1990. Fumarole distribution, morphology, and encrustation mineralogy associated with the 1986 eruptive deposits of Mount St. Augustine, Alaska. Bull Vol. 52: 175-185.

Reed, M. H. 1982. Calculation of multicomponent chemical equilibria and reaction processes in systems involving minerals, gases and aqueous phase. Geochim. et Cosmochim. Acta, 46: 513-528.

Reed, M. H. and R.B. Symonds 1992. Calculation of multicomponent equilibria in gaseous-solid-liquid systems, Part I: Numerical methods: Am. Jour. Sci., in review.

Symonds, R.B., W.I. Rose, T.M. Gerlach, P.H. Briggs, R.S. Harmon 1990. Evaluation of gases, condensates, and SO_2 emissions from Augustine Volcano, Alaska: the degassing of a Cl-rich volcanic system. Bull. Volcanology 52: 355-374.

Symonds, R.B., M.H. Reed, W.I. Rose 1991. Speciation and fluxes of gases at Augustine Volcano, Alaska: metal transport in a Cl-rich gas phase. (in press).

Symonds, R.B. and M.H. Reed 1992. Calculation of multicomponent chemical equilibria in gas-solid-liquid systems, Part II: Thermochemical data and applications to studies of high temperature volcanic gases with examples from Mount St. Helens. Am. Jour. Sci., in review.

Water-Rock Interaction, Kharaka & Maest (eds) © 1992 Balkema, Rotterdam. ISBN 90 5410 075 3

Water-rock hydrothermal experiments: Influence of rock type on solution pH and oxygen fugacity

V.J.Grassi, G.C.Ulmer & D.E.Grandstaff
Geology Department, Temple University, Philadelphia, Pa., USA

ABSTRACT: Systematic differences in pH, cation/proton ion activity ratios, and redox state have been observed between solutions produced in rock-water hydrothermal experiments with tuff, granite, and basalt. Stable pH values in tuff-water experiments may be as much as two pH units more acidic than basalt-water experiments at the same temperature and ionic strength. Oxygen fugacity values in 300°C tuff experiments are 4-7 orders of magnitude more oxidizing than basalt experiments and *ca.* 4 log units more oxidizing than the magnetite-hematite buffer. Such systematic differences are often obscured in natural systems.

1 INTRODUCTION

Examination of data from various hydrothermal experiments reveals some small, systematic variations between fluids which may result from different rock types used in experiments. This paper will concentrate on systematic differences in pH and fO_2 observed between solutions reacting with granite, tuff, and basalt. Previous studies (e.g. Grandstaff *et al.*, 1991) have suggested that solutions produced during rock-water hydrothermal experiments are similar to geothermal fluids in most respects. Thus, variations observed in experiments should also be seen in natural systems.

2 RESULTS

Parameters for experiments are compared in Table 1. All hydrothermal experiments were conducted using Dickson-design rocking autoclaves with gold reaction cells. Internally consistent pressure-corrected high-temperature solution speciation parameters and fO_2 values were calculated from analytical data using the computer program HIPH/FOLLY (Kacandes and Grandstaff, 1990). Values of fO_2 were calculated from measured H_2 concentrations following the method of Kishima and Sakai (1984).

2.1 pH

Figure 1 shows calculated high temperature pH values *vs.* time for several 300°C rock-water experiments conducted in low-salinity fluids of similar ionic strength. Stabilized high-temperature pH values for the experiments range from neutral to 2 pH units basic. Solution pH values for granitic rocks tend to be lower (more acidic) than those for basaltic rock experiments.

Detailed pH comparisons cannot be made between fluids of greatly different ionic strength because pH values are a function of ionic strength. However, direct comparisons may be made using cation/proton ion activity ratios, whose values are not affected by ionic strength. An example using the Na/H ion activity ratio is shown in Figure 2. Sodium was the dominant cation in fluids produced during most of these experiments. Differences in the ratio are caused, primarily, by differences in hydrogen ion activity (pH) rather than differences in sodium ion activity. This confirms the relations shown in Figure 1 and shows that for temperatures between *ca.* 90° and 300°C, the Na/H ion activity ratio (and, at given sodium ion activity, the solution pH) is lowest for experiments conducted with tuff, intermediate for granite, and highest for basalt.

Differences in pH (and in concentrations of other ions) could be produced if different secondary phase assemblages formed during the experiments. However, although proportions of identified

TABLE 1. HYDROTHERMAL EXPERIMENTS

Starting Material	Starting Solution	Temperature (°C)	P (MPa)	References
Devitrified Topopah Tuff (NV)	J-13	200,300°	15	Grassi *et al.*, 1991
Devitrified Topopah Tuff (NV)	J-13	90, 50,250°	5-10	Knauss *et al.*, 1985
Vitric Topopah Tuff (NV)	J-13	90, 50,250°	10	Knauss and Peifer, 1986
Vitric Bannock Tuff (WY)	NaCl	300°	30	Lee *et al.*, 1990
Conway Granite (NH)	NaCl	300°	30	Lee *et al.*, 1990
Carnmenellis Granite (UK)	meteoric	80°, 50°,250°	50	Savage *et al.*, 1987
Monzogranite (UK)	DDW	100°	50	Savage., 1986
Umtanum /Cohasset Basalt (WA)	GR-3/4	100, 200, 300°	30	Moore *et al.*, 1985, Korn *et al.*, 1986
Krafla Basalt, Iceland	meteoric	100, 200, 300°	30	Kacandes and Grandstaff, 1989
Bentonite/Umtanum Basalt (WA)	GR-3	200°, 300°	30	Wood *et al.*, 1982

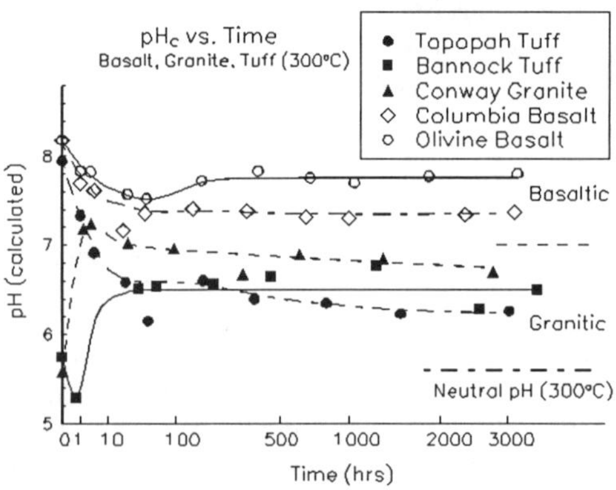

Figure 1. Calculated pH *vs.* time for representative tuff, granite, and basalt experiments at 300°C.

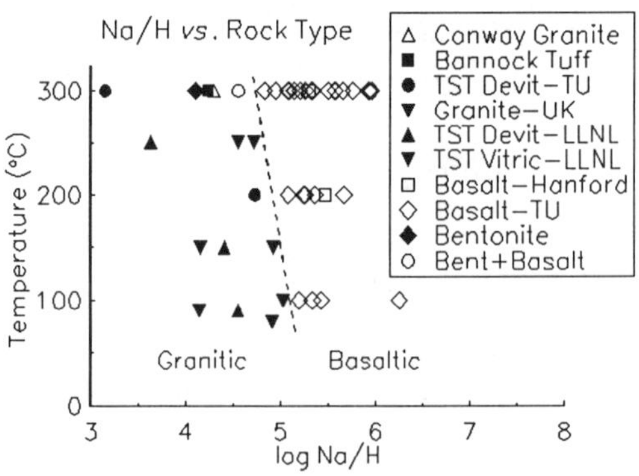

Figure 2. Calculated high-temperature log (Na/H) cation/proton ratios for experiments conducted using tuff, granite, and basalt.

minerals varied with rock type, it appears that all of these experiments produced a broadly similar secondary phase assemblage: smectite, illite, silica polymorph, zeolite, plus accessories (Grassi *et al.* 1991). Different zeolites were identified in various experiments. Solubility and phase relationships of zeolites are virtually unknown and it is uncertain whether this difference could produce the pH variations observed. Substitution of various silica polymorphs also could not explain the total variation observed (Kacandes and Grandstaff, 1989).

However, the composition of smectites formed during the experiments differed considerably. Smectites formed in basalt experiments were iron- and magnesium-rich saponites with (060) near 1.535Å, indicating a trioctahedral structure. In contrast, because iron and magnesium concentrations in granite and tuffs were very low, smectites formed during the granite and tuff experiments were quite aluminous (beidellitic) with (060) near 1.50Å, in the dioctahedral range. This difference in composition and structure might produce the pH difference observed (see further discussion in Grandstaff *et al.*, 1990; and Grassi *et al.*, 1991).

2.2 Redox

Values of fO_2 in geothermal fields lie between the magnetite-hematite (MH) and fayalite-magnetite-quartz (FMQ) buffers and are not obviously dependent on rock type (Figure 3). In

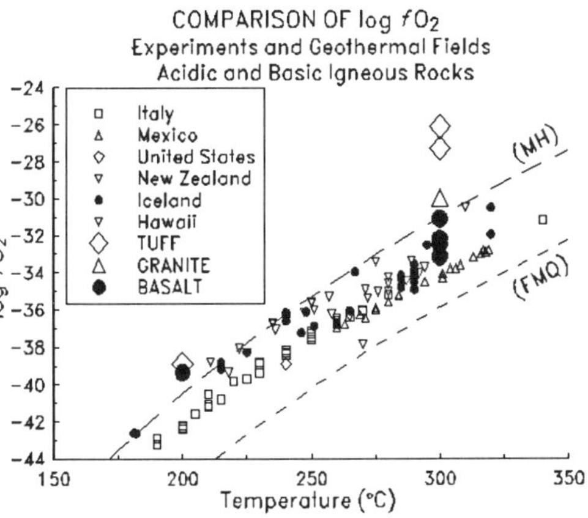

Figure 3. Comparison of redox values from geothermal fields (small symbols) and experiments (large) involving granitic (open) and basaltic (closed) rocks. Data from D'Amore and Gianelli (1984) and sources in Kacandes and Grandstaff (1989).

200°C experiments long-term fO_2 values for tuff (log fO_2 = -39.1 $\pm$ 0.1) are similar to values from basalt experiments (log $fO_2 \approx$ -39.5), near the magnetite-hematite (MH) buffer (log fO_2 = -40.5), and close to those of natural geothermal fields (Figure 3). However differences are observed in 300°C experiments. Calculated log fO_2 values as a function of time for representative tuff, granite, and basalt experiments conducted at 300°C are shown in Figure 4. In the basalt experiment log fO_2 stabilized near -32.5, comparable to values in other 300°C basalt experiments (Figure 3), and near the envelope of geothermal data. In the granite experiment (Figure 4) log fO_2 values decreased to *ca.* -30, the inferred maghemite-magnetite buffer value, at the end of the experiment. The fO_2 value was decreasing, even after 2800 hours, thus the eventual redox value would have been more reduced and nearer the (MH) value. However, log fO_2 values for tuff experiments stabilized near -26 to -27, *ca.* 4 orders of magnitude more oxidized than (MH) buffer values and 4 to 7 orders of magnitude more oxidized than basalt and geothermal values (Figure 4).

The reason for the discrepancy in 300°C tuff experiments is not known. Kinetics, rates of oxygen consumption, are unlikely to cause the discrepancy because the presumably slower 200°C experiments show no major differences. Long-term redox conditions are controlled by the composition and thermodynamic characteristics of the secondary minerals. The secondary phase assemblage contains not only simple oxides and sulfides but also silicates. In basalt experiments the redox state may have been controlled by equilibrium between iron-saponite and magnetite (Kacandes and Grandstaff, 1989). The projected long-term granite fO_2 value is consistent with (MH) buffering. The very oxidizing tuff fO_2 may reflect buffering by magnetite and Fe^{3+}-bearing clays. It is difficult to characterize secondary minerals in these experiments, however, it appears that the value of log fO_2 correlates with the whole rock values of total iron content and Fe^{3+}/Fe^{2+} ratio. Further work is required to confirm the oxidizing value and determine phases responsible.

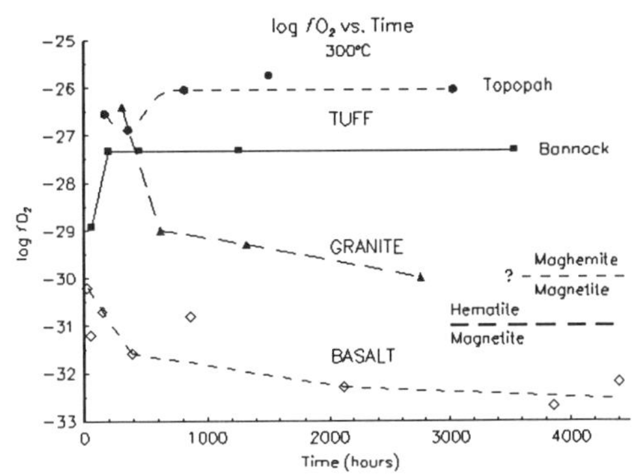

Figure 4. Values of log fO_2 (H_2O/H_2) *vs.* time for some tuff, granite, and basalt hydrothermal experiments at 300°C.

3 CONCLUSIONS

Systematic differences in pH, cation/proton ion activity ratios and fO_2 have been observed between solutions produced in rock-water hydrothermal experiments with tuff, granite, and basalt. In general at a given ionic strength $pH_{(tuff)} < pH_{(granite)} < pH_{(basalt)}$.

Stable pH values in tuff-water experiments may be as much as two pH units more acidic than basalt-water experiments at the same temperature and ionic strength. Although fO_2 values in 200°C experiments are quite comparable, 300°C tuff experiments are 4-7 orders of magnitude more oxidizing than basalt experiments and ca. 4 log units more oxidizing than (MH) buffer.

Such differences in pH and cation/proton ratios cannot be easily demonstrated in geothermal systems. The heterogeneity of rock types in most geothermal areas, the input of gases and fluids from magma crystallization and country rock devolatilization, mixing of fluids from different aquifers, uncertainties in temperature, are some factors that might obscure or minimize variations which occur in natural situations. However, isothermal comparisons by Hutcheon (1990) demonstrated that CO_2 pressures in fluids in kaolinitic sediments were higher than those from Icelandic geothermal fields at the same temperature. Further work must be done to document these type of variations in natural systems.

The phase assemblage produced in autoclave experiments often contains metastable minerals, for example, clay minerals and low temperature zeolites are usually formed in 300°C experiments, although they do not form at that temperature in natural environments. The inclusion of these metastable minerals may account for differences between experiments and higher temperature natural systems and may limit direct application of these results to lower temperature geothermal regimes where clay minerals do form.

REFERENCES

D'Amore F., and G. Gianelli, 1984. Mineral assemblages and oxygen and sulphur fugacities in natural water-rock interaction processes. Geochim. Cosmochim. Acta, **48**, 847-857.

Grandstaff D.E., V.J. Grassi, A. Lee, and G.C. Ulmer, 1991. Comparison of Granite, Tuff, and Basalt as geologic media for long-term storage of high-level nuclear waste, in Scientific Basis for Nuclear Waste Management XIV, (North-Holland, New York), pp 849-854.

Grassi V.J., G.C. Ulmer, and D.E. Grandstaff, 1991. The effect of igneous rock type on nuclear waste siting, in Nuclear Waste Management IV, (Amer. Ceram. Soc., Westerville, OH), pp. 695-706.

Hutcheon I. 1990. Clay-carbonate reactions in the Venture area, Scotian Shelf, Nova Scotia, Canada. in Fluid Mineral Interactions, a Tribute to H.P. Eugster, (Geochemical Soc, Sp. Pub #2), 199-212.

Kacandes G.H. and D. E. Grandstaff, 1989. Differences between geothermal and experimentally derived fluids. Geochim. Cosmochim. Acta **53**, 343-358.

Kishima N., and H. Sakai. (1984). A simple gas analytical technique for the Dickson-type hydrothermal apparatus and its application to the calibration of MH, NNO, and FMQ oxygen buffers. Geochem J., **18**, 19.

Knauss K.G., W.J. Beiriger, and D.W. Peifer, 1985. Hydrothermal interaction of crushed Topopah Spring Tuff and J-13 water. UCRL-53630.

K.G. Knauss and D.W. Peifer, 1986. UCRL-53795, 39p.

Korn R., G.C. Ulmer, and D.E. Grandstaff, 1986. Hydrothermal reactions of Columbia Plateau basalts with simulated Hanford Groundwaters, in International Symposium on Water-Rock Interactions 5th (Orkustofnun, Reykjavik 1986) , pp. 333-336.

Lee A.C., G.C. Ulmer, D.E. Grandstaff, 1990. The effect of rhyolitic glass on solution parameters in autoclave experiments. Trans. Am. Geophys. Union **71**, 663.

Moore E.L., G.C. Ulmer, and D.E. Grandstaff, 1985. Hydrothermal interaction of Columbia basalt from the Umtanum flow with its coexisting groundwater. Chem. Geol., **49**, 53-71.

Savage, D., 1986. Granite-water interactions at 100°C, 50 MPa. Chem. Geol., **54**, 81-95.

Savage D, M.R. Cave, A.E. Milodowski, and I. George, 1987. Hydrothermal alteration of granite by meteoric fluid. Contr. Min. Pet., **96**, 391-405.

Wood M I., G.D. Aden, D.L. Lane, 1982. Evaluation of Sodium Bentonite and Crushed Basalt as Waste Package Backfill Materials, RHO-BW-ST-21P, 40 pp, Rockwell-Hanford Operations.

Water-Rock Interaction, Kharaka & Maest (eds) © *1992 Balkema, Rotterdam. ISBN 90 5410 075 3*

The travertine totem forest of Yellowstone National Park, USA: Geologic controls and geochemistry

R.A. Hutchinson
National Park Service, Yellowstone National Park, Wyo., USA

J. Michael Thompson
US Geological Survey, Menlo Park, Calif., USA

ABSTRACT: At the confluence of Broad and Shallow Creeks in a remote portion of Yellowstone National Park is a "forest" or collection of 59 hot spring cones. These narrow, totem pole-like travertine features stand as tall as 4.8 m on the hot spring apron and 7.9 m above the two tributaries which drain the area. Thermal activity of the unnamed area is associated with the ring fracture zone arcing into the Hot Springs Basin Group, immediately outside the inferred boundary of the 0.6 Ma. Yellowstone Caldera. The bicarbonate rich springs' chemistry suggest Eocene volcanoclastic rock sediments are reacting with fluids rising up deep ring fractures, with the source area below the extensive Hot Spring Basin Group. Their unusual narrow columnar shapes and densely lined hollow central tube serving as vents indicate each has a long history of minimal discharge from point sources.

1 INTRODUCTION

Active travertine deposition on and around the Yellowstone Quaternary volcanic plateau is relatively rare. Less than 3 percent of the mapped thermal areas within the national park have recent or current travertine deposition. The largest and best known are the Mammoth Hot Spring terraces, Terrace Spring near Madison Junction, and the Firehole-Black Warrior Lake complex. Other smaller bicarbonate-rich spring groups are scattered about the park, such as Soda Butte, Hillside Springs, Boundary Creek thermal areas, and Snake River Hot Springs.

One of the most unusual and least known is an unnamed thermal area covering about 2,000 m^2 at the confluence of Broad and Shallow Creeks. This hot spring area contains 59 active and inactive travertine cones that range in size from 0.14 to 4.80 m high, (or up to ~7.90 m high when measured from the adjacent stream bed) and are 0.42 to 2.97 m in diameter when measured at the base. These totem or pillar-like cones are predominately scattered on and along a triangular bench of bedded hot spring deposits between the confluence of the two creeks. Access to the thermal area is exceptionally difficult because it requires a day's hike to the canyon and a steep drop of 316 m down a poorly defined game trail.

Isolated hot spring travertine spring cones and mounds are present in numerous areas of the world. In Yellowstone, three well known examples are Liberty Cap (14 m high), Devil's Thumb, and Soda Butte. However, large assemblages of tall, narrow travertine cones are extremely rare. The best examples of similar thermal systems are in the People's Republic of China. Banglia travertine hydrothermal area is located in Rongging Village on the upper reaches of the Bangalong River of Jiangmian District, Baqen County, Xizang Autonomous Region. Hundreds of travertine bodies of all shapes lie on both sides of Bangalong River, creating a "stone forest" (Dunzhujiacan 1985). A second collection of narrow travertine cones is on the Longma' er River Valley, Shuanghu County, Northern Xizang Autonomous Region (Dunzhujiacan 1985).

2 GEOLOGIC SETTING

Volcanic stratigraphic units are well exposed above the confluence of Broad and Shallow Creeks in the 316 m deep canyon. Most of the vertical cliff faces consist of early to middle Eocene Lamar River Formation volcaniclastic rocks (U.S. Geological Survey 1972). The Lamar River Formation consists predominately of dark-colored andesitic bedded rocks and is the basal volcanic unit for much of northeastern Yellowstone National Park (Smedes and

Table 1. Chemical compositions of springs at the confluence of Broad and Shallow Creeks and other selected areas in Yellowstone National Park.

(Compositions are in mg/L.)

Sample No.	Y-83-C12	Y-83-C41	Y-83-C57	YJ-7503	YJ-7626	YJ-8026
Broad & Shallow Creek Thermal Area				Hillside Springs[1]	Terrace Spring[1]	Silver Scarf[2]
SiO_2	236	226	218	229	193	212
Fe	0.22	0.21	0.20	--	--	--
Mn	0.15	0.07	0.18	--	--	--
Rb	--	0.14	0.17	--	--	0.06
Cs	0.11	0.07	0.22	--	--	0.05
Zn	0.04	--	--	--	--	--
Ca	55.5	39.3	35.6	6.6	23.2	4.0
Mg	1.32	1.24	0.90	0.13	0.92	0.20
Na	182	168	177	130	319	179
K	26.6	26.6	27.7	10.9	38.5	7.0
Li	0.50	0.51	0.53	0.5	0.8	1.4
HCO_3	626	599	493	256	617	300
SO_4	34	29	38	13	28	9
Cl	67	68	71	43.6	75	99
F	3.4	3.5	4.0	10.5	7.3	22
B	8.9	8.7	9.0	0.75	--	2.5
Field pH	6.8	6.9	--	7.3	5.9	--
Lab pH	8.34	8.42	8.29	7.48	9.21	7.77
Temp. °C	91	90	--	68	59	90
Flow l/min.	>230	22	--	17-20	>80	12

[1] Thompson and Yadav (1979).
[2] Thompson and Hutchinson (1980).

Prostka 1972). At the upper end of the thermal area along the Broad Creek drainage, a large unmapped dike is exposed in the west canyon wall that may be associated with the Lamar River Formation's Sulpfur Creek stock (U.S. Geological Survey 1972).

Higher up in the canyon walls are exposures of the Mount Jackson Rhyolite. Situated unconformably above the Eocene volcaniclastics, the rhyolite is a product of the third eruptive cycle of the Yellowstone calderas. It occurs in areas adjacent to the inner scarp of the 75 by 45 km crater and is essentially conformable with the Lava Creek Tuff on top of the canyon rims which erupted 0.6 Ma. However, above the thermal area, mainly on the east side of Broad Creek, a relatively thin unit of Undine Falls Basalt is present between the Mount Jackson Rhyolite and Lava Creek Tuff and is assumed to be less than 700,000 years old (Christiansen and Blank 1972).

The inferred boundary of the 600,000 year old Yellowstone Caldera meets Broad Creek at Josephs Coat Hot Springs only 2 km south of the confluence of Broad and Shallow Creeks, and then curves south to parallel all of Broad Creek's upper length. A complex series of ring fracture type faults arc northwest into and along Shallow Creek less than 1 km upstream of the confluence from Hot Spring Basin Group and Coffee Pot Hot Springs.

3 GEOCHEMISTRY

Three chemical analyses of thermal waters from the confluence of Broad and Shallow Creeks are shown in Table 1. For comparison, bicarbonate-rich water analyses from Hillside Springs between Biscuit and Black Sand Basins, Terrace Spring east of Madison Junction, and Silver Scarf thermal area near Boundary Creek are also listed.

Using Fournier's (1990) modification of Giggenbach's (1988) diagram, the water from the Broad and Shallow Creeks thermal area plot in "the partly equilibrated or mixed" portion. Because the measured temperatures are just slightly below boiling, mixing of thermal and nonthermal waters in the immediate vicinity does not seem appropriate. Indeed, if mixing and no boiling occurred, no solution is found for the quantity of hot and cold water there. If the waters are mixed and boiling occurred at 140 °C, then the hot water component was near 300 °C. If Na-K geothermometers are correct and they did preserve an imprint of high temperature, then that water is slightly above 300 °C. The Na-K-Ca (Fournier and Truesdell 1973) geothermometer temperatures are near 265 °C. No fumaroles are located nearby suggesting that the hot water may have boiled elsewhere, perhaps in the vicinity of Hot Springs Basin Group, and traveled NW 6.5 km to discharge in the Broad and Shallow Creeks thermal area. If so, then this thermal water most likely was cooled and diluted in transit. High Ca concentrations may be attributed to the partial pressure of CO_2 being at saturation with travertine and a slow flow rate. This explains the tall columnar cone shapes present as opposed to massive terraces typical of the Mammoth Hot Spring area.

Fournier (written communication, 1992) has suggested that mixing is occurring to produce these waters. The cooler diluting water is relatively unusual because it is a calcium bicarbonate water at about 100 °C and contains most of the chloride in the mixed water. The hot water is essentially a steam condensate from the Hot Spring Basin Group. However, this water is almost equilibrated with quartz at 200 °C. The hot and cooler components mix in approximately equal ratios, perhaps boiling to produce the waters seen in the Broad and Shallow Creek thermal area.

If no mixing is assumed, then SiO_2 (Fournier and Potter 1982) quartz adiabatic and conductive geothermometers yield temperature estimates near 190 °C. No cation geothermometer except Tonani's (1980) Ca/Na are close. The cation geothermometers using Mg concentrations, Mg-Li (Kharaka and Mariner 1990) and K-Mg (Giggenbach et al. 1983), are not in agreement, 124 and 210 °C respectively. Those using Li concentrations, Mg-Li and Na-Li (Kharaka and Mariner 1990) are in better agreement, 124 and 139 °C respectively. (Boiling at 140 °C was suggested from these geothermometers.) Obviously the increased Ca is affecting geothermometer calculations. Thus, no comments other than that these waters are not steady-state or equilibrium water seem appropriate, and that further work (more detailed chemical and isotopic sampling and analysis) is required to understand the origin of this thermal water.

REFERENCES

Christiansen, R.L. and H.R. Blank, Jr. 1972. Volcanic stratigraphy of the Quaternary rhyolite plateau in Yellowstone National Park. U.S. Geological Survey Professional Paper 729-B: 18 pp.

Dunzhujiacan 1985. A pearl on the roof of the world, the Yangbajain geothermal demonstration power plant. China Pictorial Publishing Co., Beijing, P.R.C.: 152 pp.

Fournier, R.O. 1981. Application of water geochemistry to geothermal exploration and reservoir engineering. In Rybach, L. & L.J.P. Muffler (eds.), Geothermal Systems: Principles and Case Histories. John Wiley & Sons, Chichester: 109-143.

Fournier, R.O. 1990. The interpretation of Na-K-Mg relations in geothermal waters. Geothermal Resources Council Transactions, 14, Part II: 1421-1425.

Fournier, R.O. and R.W. Potter II 1982. A revised and expanded silica (quartz) geothermometer. Geothermal Resources Council Bulletin, 11(10): 3-12.

Fournier, R.O. and Truesdell, A.H. 1973. An empirical Na-K-Ca geothermometer for natural waters. Geochim. Cosmochim. Acta, 37: 515-525.

Giggenbach, W.F. 1988. Geothermal solute equilibria, derivation of Na-K-Mg-Ca geoindicators. Geochim. Cosmochim. Acta, 52: 2749-2765.

Giggenbach, W.F., R. Gonflantini, B.L. Jangi, and A.H. Truesdell 1983. Isotopic and chemical composition of Parbati Valley geothermal discharges, NW-Himalaya, India. Geotherm. 12: 199-222.

Kharaka, Y.K. and R.H. Mariner 1990. Chemical geothermometers and their application to formation waters from sedimentary basins. In Naeser & McCollon (eds.), Thermal History of Sedimentary Basins. Sprg.-Verlag, New York: 99-117.

Smedes, H.W. and H.J. Prostka 1972. Stratigraphic framework of the Absaroka Volcanic Supergroup in the Yellowstone National Park region. U.S. Geological Survey Professional Paper 729-C: 33 pp.

Thompson, J.M. and R.A. Hutchinson 1980. Boundary Creek thermal areas of Yellowstone National Park II: Thermal water analysis. Geothermal Resources Council, Transactions 4: 189-192.

Thompson, J.M. and S. Yadav 1979. Chemical analyses of waters from geysers, hot springs, and pools in Yellowstone National Park, Wyoming, from 1974-1978. U.S. Geological Survey Open-File Report 79-704: 49 pp.

Tonani, F. 1980. Some remarks on the application of geochemical techniques in geothermal exploration, in Proc. Adv. Eur. Geoth. Res. Symp., Strasbourg: 428-443.

U.S. Geological Survey 1972. Geologic map of Yellowstone National Park. Misc. Geologic Investigations Map I-711.

Water-Rock Interaction, Kharaka & Maest (eds) © 1992 Balkema, Rotterdam. ISBN 90 5410 075 3

A look at silica phases in evolving hydrothermal systems

T.E.C. Keith
US Geological Survey, Menlo Park, Calif., USA

ABSTRACT: Details of silica phases deposited in evolving hydrothermal systems can aid in interpreting the dynamics of the system. Inclusions and deposits of halite in and on opal-AG deposits from near-neutral, NaCl-rich Porkchop Geyser waters provide evidence that colloidal silica forms a sponge-like gel that includes the NaCl-rich waters in the pore spaces; the gel then dehydrates and solidifies to opal-AG with residual halite deposited from the waters. In acid fumarolic systems, much of the silica at the surface is residual from acid leaching of other chemical components from primary host rock material.

1 INTRODUCTION

Various studies of silica crystallinity and polymorphs show that, in detail, silica is a very complex topic and that silica structures vary on a very fine scale. Transformations from noncrystalline to crystalline silica and degree of crystallinity in various environments can be discerned but only with difficulty (e.g., Betterman and Leibau, 1975; Flörke et al., 1971, 1990; Heaney and Post, 1992; Jackson et al., 1977; Jones and Segnit, 1971a, 1971b; Kano and Taguchi, 1982; Oehler, 1976). Fournier (1983, 1985) has emphasized the significance of observing details of silica phases in hydrothermal systems and epithermal mineralization in order to interpret the conditions of formation. Donald E. White (in many papers, e.g., 1981) has considered the importance of silica phases and crystallinity in interpreting fossil and active hydrothermal systems.

X-ray diffraction and powder camera, petrographic, and scanning electron microscope (SEM) equipped with x-ray energy dispersive analytical capability (EDS) techniques have been used in this study, but microstructural techniques such as transmission electron microscopy would also be useful. Terminology for silica phases in this paper generally follows that of Flörke et al. (1990, 1991) except that the term β-cristobalite is used for weakly to moderately disordered cristobalite (C_{Dw} and C_{Dm} of Flörke et al., 1990) and α-cristobalite includes both opal-C and cristobalite of Flörke et al. (1991).

Three groups of silica phases are discussed: (1) layered, noncrystalline silica deposited from thermal water on the walls of the Porkchop Geyser conduit, Yellowstone National Park (YNP), Wyoming, (2) the significance of chalcedony vs microcrystalline quartz and temperature effects on cristobalites vs chalcedony structural ordering with regard to YNP samples in research drill cores and at Canyon where chalcedonic sinter occurs, and (3) noncrystalline to poorly ordered silica from fossil fumarolic deposits in the 1912 ash-flow sheet and overlying fallout deposits at the Valley of Ten Thousand Smokes (VTTS), Katmai National Park, Alaska.

2 SILICA IN PORKCHOP GEYSER

Porkchop Geyser, located in the Back Basin of Norris Geyser Basin, was a quiet hot spring until 1971 when small geyser eruptions began, followed by its becoming a perpetual spouter in 1985 (White et al., 1988; Fournier et al., 1991). On September 5, 1989, a small hydrothermal explosion occurred at Porkchop Geyser, and now a boiling pool fills the reamed geyser vent (Fournier et al., 1991). The waters of Porkchop have been notably rich in SiO_2, Cl, and Na and temperatures have ranged from 68° to 95°C throughout the years Porkchop has been sampled (Fournier et al., 1991).

During the 1989 explosion, angular blocks (some larger than one meter on a side) of banded and brecciated, recemented siliceous sinter were thrown out of Porkchop (Fournier et al., 1991). The latest deposit of the sinter was clear to translucent, botryoidal silica, as thick as one cm, most of which was very hard but some of which was still pliable gel 6 days after the explosion. Fournier et al. (1991) interpreted this clear, noncrystalline silica (both hard and soft) as having been deposited from highly polymerized and colloidal silica in the hot waters as a gelatinous silica on the subsurface walls of Porkchop Geyser. The conversion of the gelatinous silica to a hard deposit probably took place on a time scale of days to weeks (Fournier et al., 1991).

One softball-sized piece of the layered sinter ejecta was preserved by R.A. Hutchinson and R.O. Fournier for study. The clear, botryoidal silica discussed by Fournier et al. (1991) is the latest deposit of the hand specimen. X-ray diffraction traces show that this phase is noncrystalline and, since its formation from gel has been demonstrated, it is truly the opal-AG of Flörke et al. (1991). Layers of brown to dark caramel-colored opal-AG underlie the clear, botryoidal silica; spherical surfaces in micropore spaces in these layers are smooth. The oldest silica layer of the hand specimen is light tan and X-ray amorphous (opal-AG); however, SEM micrographs show tiny pore spaces in which spheres have developed, some of which are smooth (at a scale of 4 um = 1 cm) indicating no structure, and some of which have a blocky surface (at a scale of 5 um = 1 cm) suggesting local crystallization to α–cristobalite. Brown algae (White et al., 1988) may be the cause of the tan and brown color in the silica bands.

SEM micrographs show patches of abundant halite crystals on the surface of the clear, botryoidal surface silica layer and in cracks within the layer. The halite appears to have crystallized following the explosion of Porkchop Geyser. The likely process resulting in halite precipitation in the hand specimen is the soft silica gel, initially precipitated in NaCl-rich hot water, dehydrated and hardened to opal-AG in air leaving the residual NaCl from the water as halite crystals.

Using the SEM, tiny (1-4 um long) crystals of NaCl are seen in micropore spaces of the middle silica layers, but none have been positively identified in the oldest layer. Halite included in the middle silica layers suggests that the Na- and Cl-rich waters became included in pore spaces in the subsurface silica gel, and halite precipitated as inclusions in the gel as it dehydrated and became solid opal-AG.

The siliceous sinter deposits of Porkchop Geyser indicate strongly that they formed from a gel in the subsurface as concluded by Fournier et al. (1991). The accompanying halite crystals are further evidence that as the colloidal silica precipitated in the hot waters, the silica first became a sponge-like gel with the NaCl-rich waters becoming isolated to pore spaces of the sponge as colloidal silica particles became more compact (by addition of silica from the waters to the gel or dehydration of the gel or both?) and the gel hardened. This process may have occurred more slowly in the subsurface water of Porkchop Geyser, so that much of the NaCl-rich water was squeezed out of the gel back into the hot waters of Porkchop as the gel dehydrated and hardened to opal-AG. A small proportion of the NaCl from the waters in the gel sponge was preserved as tiny inclusions of halite in the solid opal-AG deposits. The silica gel that was thrown into the air during the explosion of Porkchop dehydrated and hardened more rapidly than the subsurface gel so that as the NaCl-rich waters were expelled from the spongy silica gel and evaporated in air, abundant halite crystallized on the clear, botryoidal opal-AG as it hardened.

3 MICROCRYSTALLINE QUARTZ AND CHALCEDONY

During studies of drill cores and chalcedonic sinter in YNP, several samples having X-ray diffraction patterns of well-crystallized quartz appear to be chalcedony in thin section, but often SEM micrographs clearly show microcrystalline euhedral quartz ±3 um long and in bundles forming crude dipyramids about 10 um long and wide. In thin section, massive chalcedony appears as banded spherulitic layers or as a cryptocrystalline mosaic. In SEM micrographs, the chalcedony appears massive and sometimes has both irregularly- or regularly-spaced, tiny holes. When crystal forms are seen in chalcedony deposits with the SEM, the forms are always that of quartz.

Chalcedonic silica has been differentiated into several forms by Flörke et al. (1991) and one of the forms, moganite, has most recently been described in additional detail by Heaney and Post (1992). The differences in the forms are subtle; moganite has not yet been identified in either thin section or X-ray diffraction traces in YNP samples.

4 SILICA IN LEACHED PHENOCRYSTS FROM VTTS FUMAROLES

Fossil fumaroles in the 1912 deposits of the VTTS formed when high-temperature gases vigorously escaped from the ash-flow sheet (Keith, 1991). In the exposed fumarolic conduits, phenocrysts of plagioclase and pyroxene have been completely leached of most chemical constituents except for most of the SiO_2 (Keith, 1991). Depending upon location relative to the conduit in a fumarolically altered deposit, all gradations of alteration of phenocrysts occur. Totally replaced feldspar phenocrysts are impossible to distinguish from fresh or partly altered phenocrysts in hand specimen. Fresh and altered mafic phenocrysts can be identified by color, but altered mafic and plagioclase phenocrysts cannot be easily separated. In thin section, an unusual patchy to vermicular replacement texture is seen in altered phenocrysts and the silica replacing the phenocrysts is isotropic or very weakly birefringent with refractive index usually below 1.45. In SEM, phenocryst surfaces are usually smooth, cleavage is seen, but analysis by EDS shows that the material is all silica. X-ray powder camera diffraction was used to look at the structure of small parts of altered phenocrysts. The X-ray patterns show a variety of silica phases from opal-A through disordered silica with peaks representing partial ordering to β- and/or α-cristobalite, opal-CT, and, rarely tridymite.

Acid leaching by fumarolic gases removes the critical structural components (alkalis and Al) of the phenocrysts leaving only noncrystalline silica. Noncrystalline silica may remain in the leached phenocrysts that have structurally collapsed and/or the silica may be locally dissolved into steam condensate. Textures suggest that the leaching of the alkalis and Al, and deposition of silica in the phenocrysts are very closely related and that the silica mobilized as a result of leaching of other chemical constituents is immediately redeposited forming a pseudomorph after the origin phenocryst. If fumarolic alteration persists, the residual silica from the leached phenocrysts begins ordering to some level, depending largely on fumarolic temperatures.

5 IMPLICATIONS

In areas of active hydrothermal systems, evidence of silica being both dissolved and deposited exists. Any identification of silica structural phases and use of silica in bulk chemical calculations should be done with appropriate considerations of the system.

n the case of hot spring silica deposits, silica is dissolved into thermal water at depth, brought to the surface, and is deposited as various silica phases reflecting physical and chemical conditions of fluids in the dynamic hydrothermal system. In the case of fumarolic alteration at the VTTS, silicification is only apparent where acid leaching has removed most chemical constituents, including the main mineral structural components, except most of the silica.

In the case of fumarolic silica produced in an acid environment, opal-CT is prevalent, with the conspicuous tridymite peak at 4.11Å, but in the hot spring systems penetrated by Yellowstone drill holes, the silica phases are truly β- or α-cristobalite with not a hint of the tridymite peak (Keith et al., 1978; White et al., 1988; Keith, unpub. data). In crystalline devitrified ash-flow tuffs and rhyolite lava flows in the Yellowstone caldera, tridymite X-ray diffraction peaks show clearly along with α-cristobalite peaks defining two distinct mineral phases that can also be seen in thin sections. Tridymite occurs as high-temperature, vapor-phase crystals and α-cristobalite is a product of both lava flow cooling and hydrothermal devitrification in the ash-flow tuffs and rhyolite flows of YNP. In the 1912 ash-flow sheet of the VTTS, tridymite is uncommon and has not been identified as a vapor-phase mineral produced by cooling of the ash-flow sheet, but it does occur rarely where high-temperature fumaroles have leached primary minerals and glass and redeposited residual silica.

Increases in temperature and age will result in structural ordering of noncrystalline to crystalline silica phases. The possibility must be considered that original deposition of inconsistently layered silica phases in hydrothermal systems occurred where changing silica saturation of fluids was caused by changes deep within the system. Rapid changes result in depositional layers of various silica structures from noncrystalline through cristobalite, or chalcedony and quartz, being deposited seemingly randomly without a specific orderly sequence of crystallization. This is evidence, as in several of the Yellowstone drill cores, of varying silica saturation of the waters, as well as temperature variations, in the hydrothermal fluids.

Study of contrasting silica phase occurrences indicates the importance of looking carefully at silica phases and textures when studying evolving hydrothermal systems which are commonly in disequilibrium.

The assistance of Robert Oscarson with the SEM studies and of Elizabeth Bailey with some of the x-ray diffraction studies is gratefully acknowledged.

REFERENCES

Betterman, P. and Liebau, F., 1975. The transformation of amorphous silica to crystalline silica under hydrothermal conditions. *Contrib. Mineral. Petrol.*, 53: 25-36.

Flörke, O.W., Graetsch, H., Martin, B., Roller, K. and Wirth, R., 1991. Nomenclature of micro- and non-crystalline silica minerals, based on structure and microstructure. *N. Jb. Min. Abh.*, 163: 19-42.

Flörke, O.W., Graetsch, H. and Jones, J.B., 1990. Hydrothermal deposition of cristobalite. *N. Jb. Min. Mh.*, H. 2: 81-95.

Flörke, O.W., Jones, J.B. and Segnit, E.R., 1971. Opal-CT crystals. *N. Jb. Mineral. Mh.*, H. 8: 369-377.

Fournier, R.O., 1983. Active hydrothermal systems as analogues of fossil systems. In The Role of Heat in the Development of Energy and Mineral Resources in the Northern Basin and Range Province, *Geothermal Resources Council Special Rept. 13*, 263-284.

Fournier, R.O., 1985. The behavior of silica in hydrothermal solutions. In B.R. Berger and P.M. Bethke (eds.), Geology and Geochemistry of Epithermal Systems, *Reviews in Economic Geology*, v. 2, 45-61.

Fournier, R.O., Thompson, J.M., Cunningham, C.G. and Hutchinson, R.A., 1991. Conditions leading to a recent small hydrothermal explosion at Yellowstone National Park. *Geol. Soc. Amer. Bull.*, 103: 1114-1120.

Heaney, P.J. and Post, J.E., 1992. The widespread distribution of a novel silica polymorph in microcrystalline quartz varieties. *Science*, 255: 441-443.

Jackson, M.L., Clayton, R.N., Fujii, N. and Henderson, J., 1977, Cristobalite morphology and oxygen isotopic composition variation under hydrothermal alteration. *Clays & Clay Minerals*, 25: 31-38.

Jones, J.B. and Segnit, E.R., 1971a. The nature of opal I. Nomenclature and constituent phases. J. Geol. Soc. Austral., 18: 57-68.

Jones, J.B. and Segnit, E.R., 1971b. Genesis of cristobalite and tridymite at low temperatures. *J. Geol. Soc. Austral.*, 18: 419-422.

Kano, K. and Taguchi, K., 1982. Experimental study on the ordering of opal-CT. *Geochem. J.*, 16: 33-41.

Keith, T.E.C., 1991. Argillic alteration in the Novarupta vent region, Katmai National Park, Alaska. *Geophys. Res. Letters*, 18: 1549-1552.

Oehler, J.H., 1976. Hydrothermal crystallization of silica gel. *Geol. Soc. Amer. Bull.*, 87: 1143-1152.

White, D.E., 1981. Active geothermal systems and hydrothermal ore deposits. *Econ. Geol. 75th Ann. Vol.*, 392-423.

White, D.E., Hutchinson, R.A. and Keith, T.E.C., 1988. The geology and remarkable thermal activity of Norris Geyser Basin, Yellowstone National Park, Wyoming. *U.S. Geol. Surv. Prof. Pap. 1456*, 84 pp.

Water-Rock Interaction, Kharaka & Maest (eds) © 1992 Balkema, Rotterdam. ISBN 90 5410 075 3

Yellowstone travertines: U-series ages and isotope ratios (C, O, Sr, U)

N.C. Sturchio
Argonne National Laboratory, Ill., USA

M.T. Murrell
Los Alamos National Laboratory, N.Mex., USA

K.L. Pierce
US Geological Survey, Denver, Colo., USA

M.L. Sorey
US Geological Survey, Menlo Park, Calif., USA

ABSTRACT: Geochronologic and isotopic investigations of travertines yield new insights into late Quaternary hydrothermal activity and glacial chronology in the northern Yellowstone National Park area. Periodic travertine deposition has occurred from ~375 ka to present. Location and timing of deposition were modulated by climate, topography, and geologic structure. Known periods of deposition were ~375 ka, ~134 ka, ~50 ka, ~20 ka, and ~15 ka to present. Isotope ratios of C, O, Sr, and U in travertines and modern thermal waters permit inferences regarding deposition temperatures, isotope ratios of paleowaters and their solutes, and travertine diagenesis. Full Pinedale glacial conditions occurred between 47 and 22 ka, and may have caused relatively low $\delta^{18}O$ and $(^{234}U/^{238}U)_i$ values in travertines from 22 to 15 ka.

1 INTRODUCTION

There are extensive hot-spring travertine deposits in the northern Yellowstone National Park area, Wyoming-Montana, U.S.A. (Fig. 1). We have determined U-series ages and isotope ratios of C, O, Sr, and U for a representative set of these travertines, and have used these data to gain insight into the evolution of the underlying hydrothermal system, with emphasis on the effects of glaciation. Here we present a synopsis of the data and some salient observations. More detailed discussions of objectives, methods, results, and implications will be published elsewhere.

2 BACKGROUND

The reader is referred to Fournier (1989) for a recent review of the general geology, geochemistry, and dynamics of the Yellowstone hydrothermal system, and to Pierce (1979) for a comprehensive review of the glacial geology of the northern Yellowstone area. Recent detailed studies of the geology, geochemistry, and hydrogeology of the study area are presented by Sorey (1991).

Large masses of ice covered the area during the Bull Lake and Pinedale glaciations of late Quaternary time (Pierce, 1979). Full glacial conditions occurred at 150 to 140 ka (Bull Lake) and 40 to 30 ka (Pinedale) according to Pierce et al. (1976). Evidence for pre-Bull Lake glaciations in the area is fragmentary (Pierce, 1979). The northern Yellowstone glacier during Pinedale time, defined as the 60-km long outlet glacier in the Yellowstone valley and its accumulation area, covered ~3400 km^2, had an average thickness of >700 m, and included a flow line 146 km long (Pierce, 1979).

Travertine deposits sampled for this study are located mostly within the former extent of the outflow zone of the northern Yellowstone glacier (Fig. 1). The sampled deposits occur over an elevation range from about 1700 to 2400 m, and resemble those now being deposited subaerially at Mammoth Hot Springs and other locations within Yellowstone National Park (relatively pure, >98% $CaCO_3$). Most of the travertine deposits in the area were

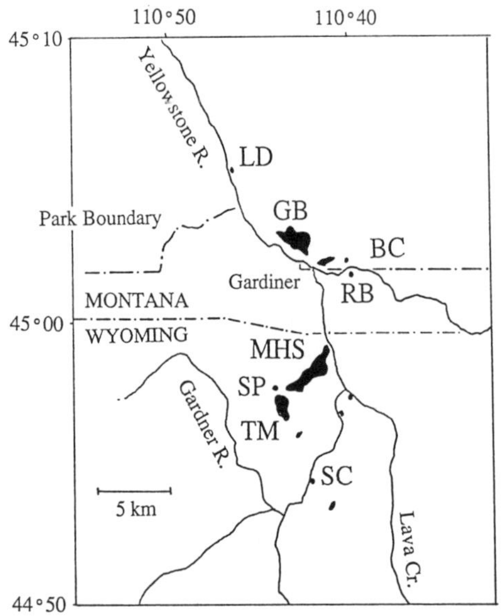

Fig. 1. Location map for travertines (black). LD, LaDuke; GB, Gardiner bench; Bear Creek; RB, Rattlesnake Butte mound; MHS, Mammoth Hot Spr.; SP, Snow Pass; TM, Terrace Mtn.; SC, Sheepeater Canyon.

sampled, from travertine outcrops where stratigraphic relations with glacial till deposits were unambiguous.

3 U-SERIES AGE DETERMINATIONS

Ages determined for travertines are based either on isochrons generated from leach-residue analyses (Ku and Liang, 1984), or are single-leach ages that use an assumed initial ($^{230}Th/^{232}Th$) ratio of 0.9±0.3 (the range of six initial Th ratios calculated from isochron intercepts and one measured Th ratio for zero-age travertine from Mammoth Hot Springs).

The oldest travertine in the area (389±26, 361±23 ka) caps Terrace Mountain. The horizontal layering of this travertine at the eastern cliff edge of the mountain indicates that it was deposited on a landscape without the present deep valley to the east.

Travertine within glacio-fluvial kame deposits of probable Bull Lake age at the north end of Terrace Mtn. (Snow Pass) has an age of 134±3 ka.

The next known period of travertine deposition is represented by three deposits: Rattlesnake Butte mound, a large, half-eroded travertine mound across from the confluence of Bear Creek and the Yellowstone River, 57.1±0.6 ka; the southernmost known travertine accumulation in the study area, south of Bunsen peak and Sheepeater Canyon, mantled by Pinedale glacial erratics, 54.5±1.0 ka; and basal travertine on the travertine bench at Gardiner, overlying Bull Lake till and overlain by Pinedale erratics, 46.9±2.7 ka.

A younger period of deposition was identified SE of the 46.9 ka deposit on the travertine bench at Gardiner, represented by ages of 22.3±0.2 and 19.5±0.1 ka for bedded travertine samples, 20.2±0.2 ka for the outer white portion of a layered carbonate vein within the basal travertine near the sample dated at 46.9 ka, and 22.5±0.4 ka for a vent carbonate vein at the top of the 46.9 ka deposit. These deposits also are mantled by Pinedale till and may represent travertine deposition during a time of glacial recession between Pinedale full glacial time and a later readvance called the Deckard Flats readjustment (Pierce, 1979). Within bedded travertine near the sample dated at 22.3 ka there are structures resembling casts of grass-like plants.

The next period of travertine deposition is represented by ages ranging from 15.5 ka to present. An age pf 15.5±0.1 ka was obtained for the inner laminated portion of vein carbonate (center of layered vein mentioned above) on the Gardiner bench, and may represent the resumption of travertine deposition upon deglaciation following the Deckard Flats readjustment. Travertine at the top of a post-glacial mound 50 m thick, partly tree-covered, near the outlet of Bear Creek has an age of 9.87±0.06 ka, and may have built up over a period of thousands of years. Samples of similar post-glacial travertine from outcrops on Pinyon Terrace, uphill from Mammoth Hot Springs, have ages of 8.8±0.4 and 7.3±1.2 ka. These ages are approximately equal to that determined for basal travertine from Y-10 drill core from the Highland Terrace area at Mammoth Hot Springs, 7.8±0.9 ka, indicating that travertine deposition apparently migrated downhill with time following deglaciation.

Travertine deposition is currently very active at Mammoth Hot Springs, having been continuous since at least ~9 ka and probably thousands of years earlier. Travertine from the top of a travertine mound about 5 m high at the bottom of Sheepeater Canyon has an age of 2.7±0.2 ka, next to a spring depositing minor travertine. Minor amounts of travertine also are being deposited from springs near the outlet of Bear Creek and at LaDuke hot spring, but none has been deposited on the travertine bench at Gardiner since about 20 ka.

The alternation of travertine deposition between the Gardiner/Bear Creek and Mammoth Hot Springs areas through time indicates the influence of two major factors: hydrogeologic connections afforded by geologic structures linking the two areas (several intersecting faults and the Gardiner syncline that includes the Mission Canyon Limestone aquifer) and by the variable presence of glacial ice in the two areas at different times (discussed in detail by Pierce et al. in Chapter C of Sorey, 1991).

4 ISOTOPE RATIOS OF TRAVERTINES

Figure 2 shows calculated values of $(^{234}U/^{238}U)_i$ vs. age for the travertines from 70 ka to present. Travertines from the Gardiner bench having ages of 47 ka and in the 22 to 15 ka range have lower $(^{234}U/^{238}U)_i$ values than older and younger travertines elsewhere and present thermal waters throughout the area. Pre-50 ka and post-15 ka travertines have $(^{234}U/^{238}U)_i$ values generally in the range of present thermal waters. This may reflect increased dissolution of carbonate aquifer rock in response to increased hydrostatic pressure at depth during glacial times; aquifer rock generally has $(^{234}U/^{238}U) = 1.00$.

Figure 3 shows $\delta^{13}C$ vs. $\delta^{18}O$ for 51 travertine samples. There is a positive

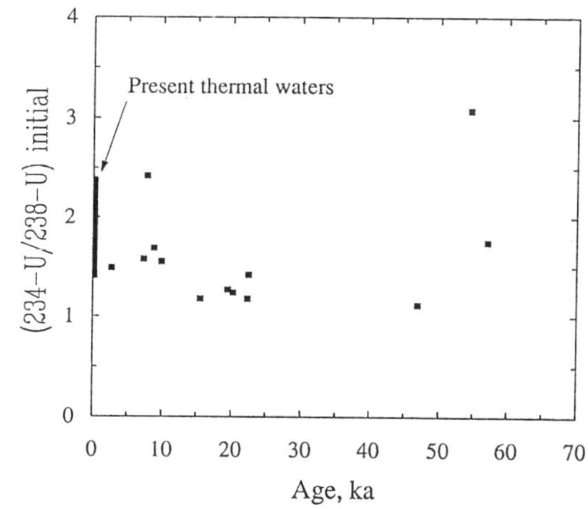

Fig. 2. $(^{234}U/^{238}U)_i$ vs. age (ka) for travertines, and range of $(^{234}U/^{238}U)$ in thermal waters.

correlation between these two values that was discussed by Friedman (1970). Fields outlined on Figure 3 are for: (1) presently-depositing travertines at Mammoth Hot Springs; (2) Y-10 drill core samples from 0.7 to 67.3 m depth; (3) outcrop samples from inactive deposits in the Mammoth area (including Terrace Mtn., Pinyon Terrace, and Hot River); (4) outcrop samples from the Gardiner bench area (samples having ages in the 22 to 15 ka range are in the low end of this field); and (5) outcrop samples from inactive deposits in the Bear Creek area.

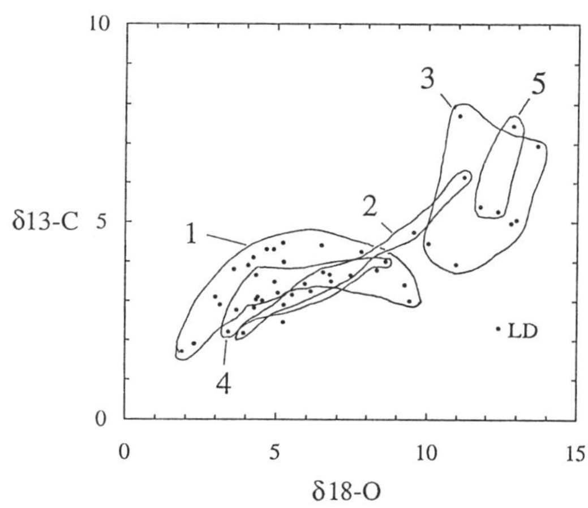

Fig. 3. $\delta^{13}O$ vs. $\delta^{18}O$ for travertines. See text for definition of outlined fields.

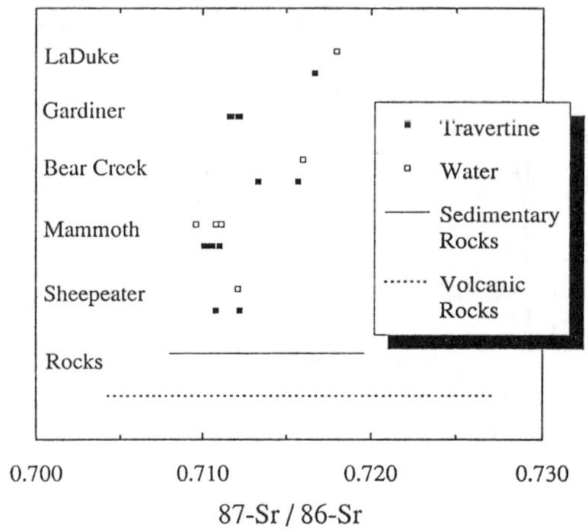

Fig. 4. $^{87}Sr/^{86}Sr$ for travertines, thermal waters, and rocks.

Travertines from the Gardiner bench define a field that is distinctly lower in $\delta^{13}C$ and $\delta^{18}O$ than those of inactive Mammoth and Bear Creek areas. The Gardiner travertines may have either experienced less post-depositional low-temperature diagenetic isotope exchange than those from inactive Mammoth and Bear Creek area deposits, or they may have had lower initial $\delta^{18}O$ values. Further data are required to clarify this.

Figure 4 shows $^{87}Sr/^{86}Sr$ ratios for travertines, thermal waters, and rocks. Travertines have $^{87}Sr/^{86}Sr$ ratios identical to those of thermal waters in each location. The range in $^{87}Sr/^{86}Sr$ for travertines at Sheepeater, Mammoth area, and Gardiner is fairly narrow (0.710-0.712), and has been nearly invariant with time since at least 375 ka. This may indicate that the deep hydrothermal flow system feeding the surface discharges in these areas was established prior to 375 ka, which is consistent with the structural controls postulated by Pierce et al. (Chapter C in Sorey, 1991). However, the entire range of Sr isotope ratios for travertines and thermal waters is encompassed by those of both Paleozoic/Mesozoic sedimentary rocks and Quaternary volcanic rocks in the area, and thus Sr isotopes are not very diagnostic of the source of Sr in this system.

REFERENCES

Fournier, R.O. 1989. Geochemistry and dynamics of the Yellowstone National Park hydrothermal system. Ann. Rev. Earth Planet. Sci. 17: 13-53.

Friedman, I. 1970. Some investigations of the deposition of travertine from hot springs--I. The isotopic chemistry of a travertine-depositing spring. Geoch. Cosmochim. Acta 34: 1303-1315.

Ku, T.L. and Liang, Z.C. 1984. The dating of impure carbonates with decay-series isotopes. Nucl. Instr. Meth. Phys. Res. 223: 563-571.

Pierce, K.L. 1979. Histroy and dynamics of glaciation in the northern Yellowstone National Park area. U.S. Geol. Surv. Paper 729F: 90 p.

Pierce, K.L., Obradovich, J.D., and Friedman, I. 1976. Obsidian-hydration dating and correlation of Bull Lake and Pinedale glaciations near West Yellowstone, Montana. Geol. Soc. Amer. Bull. 87:703-710.

Sorey, M.L. (ed.) 1991. Effects of potential geothermal development in the Corwin Springs Known Geothermal Resources Area, Montana, on the thermal features of Yellowstone National Park. U.S. Geol. Surv. Water Res. Inv. Rept. 91-4052.

ACKNOWLEDGEMENTS

This work was supported by the US. Department of Energy, Office of Basic Energy Sciences, Division of Engineering and Geosciences.

Water-Rock Interaction, Kharaka & Maest (eds) © 1992 Balkema, Rotterdam. ISBN 90 5410 075 3

Magnesium silicate scales in geothermal utilization

Guðrún Sverrisdóttir, Hrefna Kristmannsdóttir & Magnús Ólafsson
Orkustofnun, Reykjavík, Iceland

ABSTRACT: Magnesium silicate scale has been found in two kinds of Icelandic district heating systems: those in which fresh water is heated, and those where there is mixing of geothermal and fresh water. The fluids precipitating magnesium silicates have a large compositional range, and their Mg/Si ratios span two orders of magnitude. These silicates are poorly crystalline, but show fairly regular microstructures and have fixed stoichiometry. Their structures resemble some known crystalline sheet silicates, mainly belonging to the serpentine group, but talc-like structures occur where the fluid Mg/Si ratio is low.

1 INTRODUCTION

High temperature geothermal fluids are not suitable for direct space heating, due to their chemical composition. Where high temperature geothermal fields are exploited for this purpose in Iceland, heat exchangers are employed to heat fresh water for use in the distribution systems. In all such cases, there have been problems due to magnesium silicate scaling. Similar precipitation occurs where low temperature geothermal water is mixed with fresh water.

The main aim of this work was to study the magnesium silicate mineral species formed by precipitation from heated fresh water, and how they differ from those formed by precipitation from mixtures of geothermal and fresh water. Another objective was to determine whether the character of the minerals depends on the proportions of fresh and geothermal water in the mixture. Data from four heating systems, *hitaveitas*, were used in this study.

1) Hitaveita Suðurnesja. Samples of scales and fluids provide an example of heated fresh water with a high Mg/Si ratio.

2) Hitaveita Reykjahlíðar uses heated fresh water with a moderate Mg/Si ratio. Since large amounts of good scale samples were available, these were chosen for detailed examination.

3) Hitaveita Laugaráss was used as an example of low temperature geothermal water with a low Mg/Si ratio, which had been "contaminated" by natural inflow of fresh water, causing precipitation.

4) Hitaveita Reykjavíkur. Samples of precipitate from heated, unmixed Nesjavellir fresh water were examined as were samples precipitated from that water after mixing with the geothermal water from the Reykir geothermal field, which has been used by the district heating system of Reykjavík for a number of years.

2 PRECIPITATION OF Mg-SILICATES

Conditions for magnesium silicate precipitation and the consequent problems that Icelandic district heating systems are dealing with have been described in detail by Kristmannsdóttir (1980), Kristmannsdóttir et al. (1983,1989), and Gunnlaugsson and Einarsson (1989).

Magnesium concentrations in Icelandic fresh waters range from 1 to 10 mg/l, whereas the magnesium content of geothermal water is generally less than 0.001 mg/l (Fig. 1). Due to the retrograde solubility of magnesium silicates there is a great risk of their precipitation when the fresh water is heated. Table 1 shows the composition of fresh water and heated water in three district heating systems in Iceland: Hitaveita Suðurnesja, Hitaveita Reykjahlíðar, and Hitaveita Reykjavíkur. Also shown is the

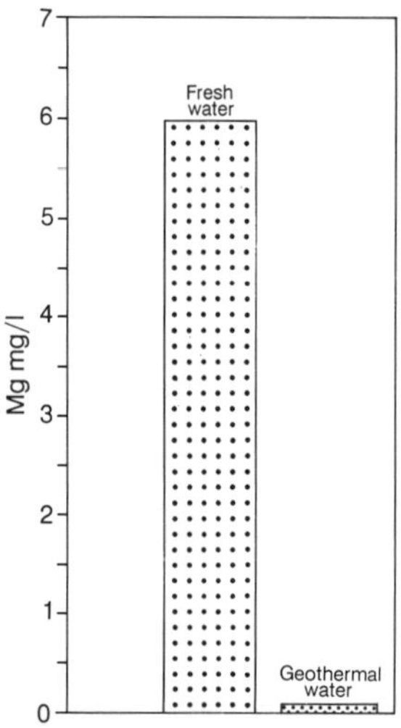

Fig. 1 Concentrations of Mg in fresh water and geothermal water in Iceland.

composition of mixed water from two systems: Hitaveita Reykjavíkur and Hitaveita Laugaráss.

Several water-rock experiments have been undertaken and they have all led to similar results. Mottl and Seyfried (1980), and others, have shown that magnesium reacts with, and is retained by, the rock phase. The magnesium silicate sepiolite precipitates from seawater even at temperatures as low as 25°C, if the concentration of dissolved silica is increased (Wollast et al. 1968). Various magnesium silicate minerals have been shown to form in these experiments. Talc seems to be the most easily formed, frequently outside its stability field (Whitney and Eberl, 1982). In addition, stevensite, saponite, and chrysotile are all formed at hydrothermal conditions at relatively low pressure and temperature.

The solubility of chrysotile is shown in Fig. 2, and the respective activity products for some fresh, geothermal, and mixed waters are plotted for comparison. The solubility is represented by the log K for the precipitation of the mineral. The figure demonstrates clearly the inverse temperature dependence of the magnesium silicate solubility. The same figure also displays curves of simulated heating of fresh water, followed by boiling at 108°C with 30% and 100% deaeration, respectively, and subsequent cooling to 20°C. The curves show that the boiling temperature and the degree of deaeration, together with the pH, are the controlling factors for supersaturation.

3 MAGNESIUM SILICATE MINERALS

Table 2 shows chemical analyses, made by a scanning electron microscope, of samples of precipitate from the four systems studied. The analyses are semiquantitative, but give a good measure of the Mg/Si ratio. One wet chemical analysis is shown, of a sample named Barmahlíð, which is a precipitate from the heating system in Reykjavík after the Nesjavellir fresh water had been mixed with the geothermal water of Reykir. The chemical

Table 1. Chemical composition of water (mg/l).

	Hitaveita Suðurnesja		Hitaveita Reykjahlíðar		Hitaveita Laugaráss		Hitaveita Reykjavíkur		
	fresh water	heated water	fresh water	heated water	geothermal water	contaminated geothermal water	Nesjavellir fresh water	Reykir geothermal water	mixed water N+R 1:1
Temp °C	5	100	4	96	98	98	3.8	93.6	80
pH/°C	7.6/21	9.1/22	7.5/18	8.8/18	9.7/21	9.7/20	8.0/23	9.8/22	9.2/22
SiO_2	14.6	12.5	26.4	25.9	112	112	16.5	96.9	59.0
Na^+	35.5	30	8.5	8.6	7.9	81	8.1	50.0	29.2
K^+	1.7	1.7	1.3	1.2	1.9	2.1	0.9	1.4	1.1
Ca^{++}	8.8	7.1	8.8	9.0	3.8	4.8	7.3	1.8	4.0
Mg^{++}	8.0	6.3	5.4	4.9	0.002	0.22	4.5	0.003	2.05
CO_2(tot.)	23	10	51	45	17	19	29	25	30
SO_4^{2-}	10.5	9.7	5.6	5.6	53	61	7.8	19.0	12.9
H_2S	<0.03	<0.03	0.0	1.03	0.64	0.63	0.0	0.9	0.9
Cl^-	71	64	3.3	3.3	47	48	8.0	12.6	12.2
F^-	0.09	0.06	0.11	0.12	1.88	1.88	0.2	0.92	0.51
TDS	159	154	87	87	340	349	80	-	-
Mg/Si (atomic ratio)	1.35	1.25	0.51	0.47	0.00004	0.005	0.67	0.00008	0.09

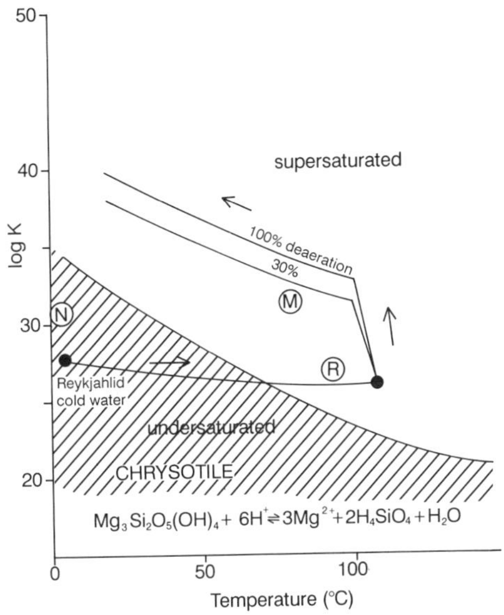

N: Nesjavellir, cold water
R: Reykir, geothermal water
M: Mixed water, Nesjavellir + Reykir 1:1

Fig. 2 Solubility of chrysotile and activity products for various fluids.

composition of the waters forming the precipitates is shown in Table 1. The Mg/Si ratios in the precipitate from the heated, unmixed water range between 1.1 and 1.3, which is typical for serpentine minerals. The waters have rather high, but variable, Mg/Si ratios (Table 1), which are clearly not proportional to the ratios in the precipitate. The precipitate from the Laugarás "contaminated" geothermal water and the Barmahlíð sample have considerably lower Mg/Si ratios of 0.7 to 0.8. These are common ratios for sheet silicates such as talc, stevensite, and saponite. The mixed water has a Mg/Si ratio nearly two orders of magnitude lower than the unmixed water.

Table 2. Chemical composition of precipitates (wt.%).

	Hitaveita Suðurnesja SEM	Hitaveita Reykjahlíðar SEM	Hitaveita Laugaráss SEM	Hitaveita Reykjavíkur	
				Nesjavellir SEM	Barmahlíð Wet chem.
SiO_2	50.07	56.21	64.85	52.31	41.39
Al_2O_3	2.31	0.29	3.24	0.51	-
FeO(tot.)	0.66	0.16	0.71	0.10	14.83
MgO	41.20	42.81	29.68	46.62	22.80
CaO	0.27	0.49	1.50	0.41	0.96
	94.51	99.96	99.98	99.95	79.98
Mg/Si (atomic ratio)	1.22	1.14	0.68	1.33	0.82

The crystal structure of several samples was first analysed by the X-ray diffraction method (XRD). The patterns obtained were broad reflections, revealing the extremely fine-grained habit of the precipitate. The patterns for precipitates from unmixed water were all similar, resembling closely those of minerals belonging to the serpentine group, primarily chrysotile. But there were also some patterns indicating the presence of the mineral talc. The sample from Laugarás had only patterns resembling talc, and no indication of a serpentine mineral was present. Precipitates from the mixtures of geothermal and fresh water from the Reykjavík system show more complicated XRD-patterns. Varieties of talc, saponite, sepiolite, and chrysotile are probably present.

Some of the precipitates, especially the serpentine-like ones, show characteristic infrared spectra (Kristmannsdóttir et al. 1989.) Other precipitates have poorly defined IR spectra, resembling those of poorly crystalline smectites.

The high melting point (>1450°C) of the sample from Hitaveita Reykjahlíðar verifies that the mineral belongs to the serpentine group. The differential thermal curve (DTA) from the same sample resembles a reference sample of antigorite. The XRD-pattern, on the other hand, resembles more that of chrysotile. Examination of this sample in a scanning electron microscope reveals the morphology of the sample as compact plates connected by dendritic or columnar growths (Fig. 3). The columns seem to be composed of small vesicles. A closer view at higher magnification reveals that the plates are composed of the same columnar structures as the connections, only more closely packed (Fig. 4). It can be clearly seen from the spheres on the surfaces of the plates that the columns are also constructed of spherical microstructures or vesicles.

The precipitates may be divided into two distinct groups on the basis of their Mg/Si ratios, in contrast to the waters they precipitate from, which exhibit a wide range of Mg/Si ratios. This leads us to the conclusion that the precipitation is governed by the equilibrium and supersaturation of two well-defined magnesium silicate species, rather than by some kind of polymerization and subsequent annealing with time. The equilibrium with the species with

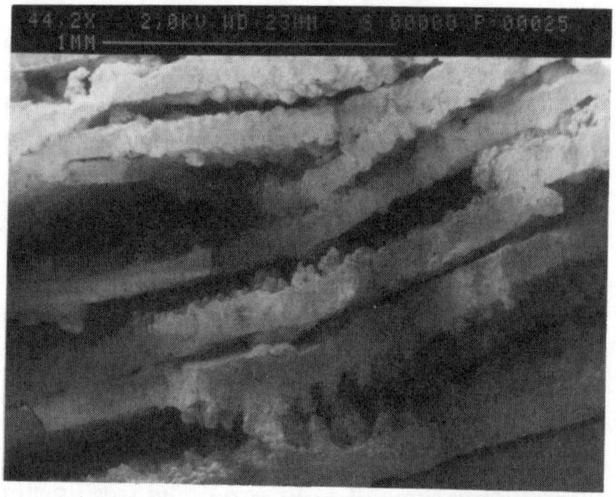

Fig. 3 An electron-micrograph. Platy structure of a precipitate from hitaveita Reykjahlíðar.

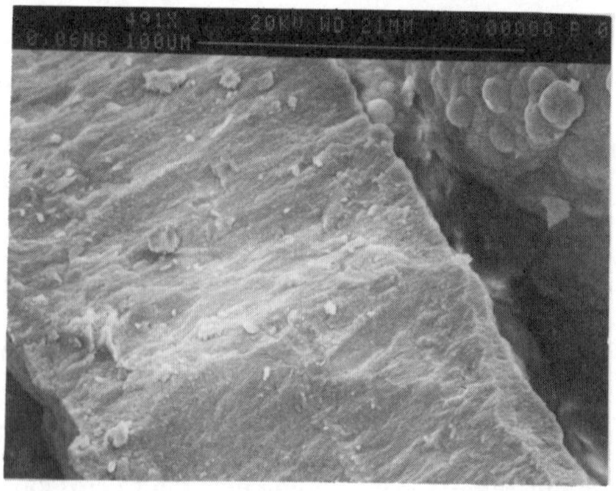

Fig. 4 An electron-micrograph. A closer view of the same sample as in Fig. 3

Mg/Si ratios of about 1.1 is maintained over a considerable range of fluid Mg/Si ratios. When the Mg/Si ratio in the water falls below 0.1, the magnesium-rich mineral species are replaced by silica-rich species, with Mg/Si ratios of approximately 0.8, as the equilibrium species.

All this led to the conclusion that the mineral formed in the unmixed heated fresh water is a serpentine mineral, probably a variety intermediate between chrysotile and antigorite. Minerals formed in the mixed water are richer in silica. Various disordered stacking arrangements may occur among layered minerals, and, in addition to substitution of ions in place of Si and Mg, the existence of curvatures in the composite sheet structure of serpentine minerals is well known.

4 CONCLUSION

The magnesium silicate mineraloids formed by precipitation from heated fresh water in Icelandic district heating systems are most clearly related to the serpentine mineral group. Mixtures of geothermal and heated fresh water precipitate more complicated mineral assemblages, composed of talc, sepiolite, saponite, and chrysotile. The Mg/Si ratios of the scales indicate that the precipitation is controlled by equilibrium between water and the two mineral species described, rather than by polymerization and subsequent annealing.

ACKNOWLEDGEMENTS

Thanks are due to all the *hitaveitas* for providing samples, and for cooperation and support. Dr. Per Arne Melkerud and Birgitta Hultén are thanked for assistance with the DTA analyses and heating experiments. Our colleagues in the Geochemistry Department have cooperated on this project for many years.

REFERENCES

Gunnlaugsson, E. and Á. Einarsson, 1989. Magnesium-silicate scaling in mixture of geothermal water and deaerated fresh water in a district heating system. *Geothermics*, Vol 18, No. 1/2: pp. 113-120.

Kristmannsdóttir, H. 1980. Magnesium silicate scaling in Icelandic district heating systems. *Proc. 3rd Int. Symp. Water-Rock Interaction*, 110-111.

Kristmannsdóttir, H., S. Thórhallsson and K. Ragnars, 1983. Magnesium silicate scaling in district heating systems (in Icelandic). *Orkustofnun report OS-83051/JHD-10*, June 1983. 28 pp.

Kristmannsdóttir, H., M. Ólafsson and S. Thórhallsson, 1989. Magnesium silicate scaling in district heating systems in Iceland. *Geothermics*, Vol. 18, No.1/2: pp. 191-198.

Mottl, M. and W. E. Seyfried, 1980. Sub-sea-floor hydrothermal systems: rock vs seawater-dominated. In *Seafloor Spreading Centers: Hydrothermal Systems*. (Eds. Rona, P.A. and R.P. Lowell) Stroudsbury, Pennsylvania: Dowden Hutchinson and Ross, Inc. 66-82.

Whitney, G. and D. Eberl, 1982. Mineral paragenesis in a talc-water experimental hydrothermal system. *Amer. Miner.*, 67, 944-949.

Wollast, R., F.T. Mackenzie and O.P. Bricker, 1968. Experimental precipitation and genesis of sepiolite at earth-surface condition. *Amer. Miner.*,53, 1645-1662.

Geothermal systems

Geothermometry

Water-Rock Interaction, Kharaka & Maest (eds) © 1992 Balkema, Rotterdam. ISBN 90 5410 075 3

Correlation of the Na/K ratio in geothermal well waters with the thermodynamic properties of low albite and potash feldspar

John A. Apps & Gee Minn Chang
Lawrence Berkeley Laboratory, University of California, Calif., USA

ABSTRACT: The Na/K ratio in geothermal well waters provides a better estimate of the relative stability of low albite and potash feldspar than do predictions from calorimetry and high temperature phase equilibria. The calculated saturation indices from field data for low albite, potash feldspar suggest that $\Delta G^{o}_{f,298}$ for the latter should be revised to -3748.6 ± 3.7 kJ.mol^{-1}.

INTRODUCTION

For many years, the element concentration ratio, Na/K, in the liquid aqueous phase has been used to estimate the source region temperature of hot springs and geothermal effluents. As a concentration ratio, it possesses the advantage that it is unaffected by transient boiling or condensation, and is affected only slowly by conductive cooling. Fournier and Truesdell (1973) recommended its use only where source region temperatures are greater than 150°C.

The thermochemical basis for the Na/K geothermometer is not well defined. Usually, discussion centers on the "exchange" reaction between plagioclase (low albite) and potash feldspar (adularia or microcline);

$$NaAlSi_3O_8 + K^+ = KAlSi_3O_8 + Na^+ \quad (1)$$

where

$$\log K(T) = \frac{[Na^+]}{[K^+]} \quad (2)$$

Attempts to reconcile field observations with calculations using published thermodynamic properties of the participating species, have not been particularly successful.

The coexistence of secondary low albite and potash feldspar in geothermal systems is frequently mentioned in the literature. Tomasson and Kristmannsdottir (1972), in discussing mineral alteration in the geothermal area of Reykjanes, Iceland, refer to the occasional presence of albitized plagioclase as well as the sporadic occurrence of newly formed potash feldspar in all geothermal holes. Mehegan and Robinson (1982) and Viereck et al. (1982) describe secondary hydrothermal alteration in the Reydarfjordur drill core from Eastern Iceland. Low albite and adularia coexist in most of the intersected volcaniclastic rocks, whereas authigenic albite replaces primary plagioclase crystals in some of the interdigitated basaltic flows. Browne and Ellis (1970) noted the presence of adularia and secondary albite in hydrothermally altered intermediate and acid lavas and volcaniclastic rocks intersected by boreholes in the Ohaki-Broadlands hydrothermal area in New Zealand. On the Kamchatka Peninsula, Naboko, et al. (1965) refer to secondary adularia and albite in dacitic and andestic volcaniclastics of the Pauzhetka geothermal field, whereas Trukhin and Petrova (1974) describe alteration zones containing secondary albite and adularia in andesitic lavas and andesitic and dacitic tuffs of the Bolshe-Bann geothermal field. In contrast, mineralogical studies of cores penetrating sea floor basalts, between ~4°C and 70°C, never mention the coexistence of low albite with potash feldspar, although frequent reference is made to the presence of the latter, sometimes replacing plagioclase. Plagioclase is usually replaced by secondary clays instead of low albite. Analcime is commonly observed as an authigenic phase, although it is rarely observed in basalts saturated with meteoric waters at similar temperatures.

Because of extensive mineralogical observations confirming the coexistence of secondary low albite and potash feldspar in geothermal fields, it is reasonable to assume that the Na/K ratio in the aqueous phase reflects the equilibrium state between these two minerals.

PREDICTION OF Na/K FROM THERMODYNAMIC DATA

Several comprehensive internally consistent compilations of thermodynamic data of minerals have been published in which $\Delta H^o_{f,298}$ or $\Delta G^o_{f,298}$, of both low albite and potash feldspar (adularia or microcline) are included (Helgeson et al., 1978, Robie et al., 1979; Berman, 1988; Holland and Powell, 1990). Their properties are based on calorimetry and high temperature, i.e. > 400°C, phase equilibria with coexisting minerals. Helgeson et al. also used the Na/K ratio from gas field brines to correlate the properties of the two feldspars. Recently, Kiseleva et al. (1990) redetermined $\Delta H^o_{f,298}$ for microline using high temperature lead borate melt calorimetry.

In Figure 1, the calculated values of log K(T) for Eqn. (1) from all five sources are compared with the analytical determinations of the Na/K ratio in well waters from geothermal fields and deep water-saturated formations from around the world. Field temperatures were measured down hole or computed from well discharge data. The log K(T) values are calculated along the water saturation curve, employing the entropies cited by Robie et al. (1979) and Maier-Kelley heat capacity functions, cited by Helgeson et al. (1978), for low albite and potash feldspar. The thermodynamic properties for the ionic species, Na^+ and K^+, are those given by Shock and Helgeson (1988). The uncertainty is estimated from those given in the cited references.

Although mineralogical studies from most sites selected for the correlation describe the presence of authigenic low albite and potash feldspar, there is no assurance that the coexisting groundwaters were in equilibrium or that the recovered groundwaters originated where low albite and potash feldspar coexisted. Some sampled well waters may have been mixtures from several producing zones. In spite of these uncertainties the field data are generally consistent and can be fitted by a univariant curve, positioned as indicated on Figure 1 to within ±0.2 $\log(Na^+)/(K^+)$, or 1 kJ in ΔG^o_r for reaction (1) at 25° and 2 kJ at 300°C. This contrasts with an uncertainty of nearly 3.5 kJ for the log K(T) predicted from calorimetry or phase equilibrium measurements. The fitted curve follows closely the equation proposed by Fournier (1981) for the Na/K geothermometer and the data reported by Il'in et al. (1979), for the Pauzhetka geothermal field, but it does not fit their data from the lower temperature Paratunka field in the same region.

The precision of Na/K predictions using either calorimetric and/or high temperature phase equilibrium data do not do justice to the quality of routinely obtainable field measurements. Groundwater analyses may therefore have the potential for calibrating thermodynamic data, provided that rigorous control is maintained over sampling and analysis procedures.

The chemical analyses of geothermal wells below 200°C reported by Arnorsson et al. (1983) include Al^{+++}. The saturation indices of low albite and potash feldspar may therefore be computed for those well waters and their solubility products compared with those predicted from calorimetry and high temperature phase equilibria. The results, illustrated in Figure 2, show the saturation indices referenced to the data by Berman (1988), who used the thermodynamic properties of low albite taken from Hemingway and Robie (1977). The saturation indices of both feldspars show trends towards supersaturation below 100°C, which may be indicative of the sampled waters originating at temperatures higher than those measured. In spite of the scatter, the saturation indices suggest that $\Delta G^o_{f,298}$ of low albite may be more nearly correct than that of potash feldspar in Berman's database.

CONCLUSIONS

A correction to $\Delta G^o_{f,298}$ for potash feldspar alone using the field correlation of Na/K yields a revised value of −3748.6±3.7 $kJ.mol^{-1}$, which is more negative than that of Hemingway et al. (1979), −3742.3±3.4 $kJ.mol^{-1}$, or that of Berman (1988), −3745.4 $kJ.mol^{-1}$, but is much less negative than −3754.2±3.7 $kJ.mol^{-1}$ calculated from $\Delta H^o_{f,298}$ of microcline determined by Kiseleva et al. (1990).

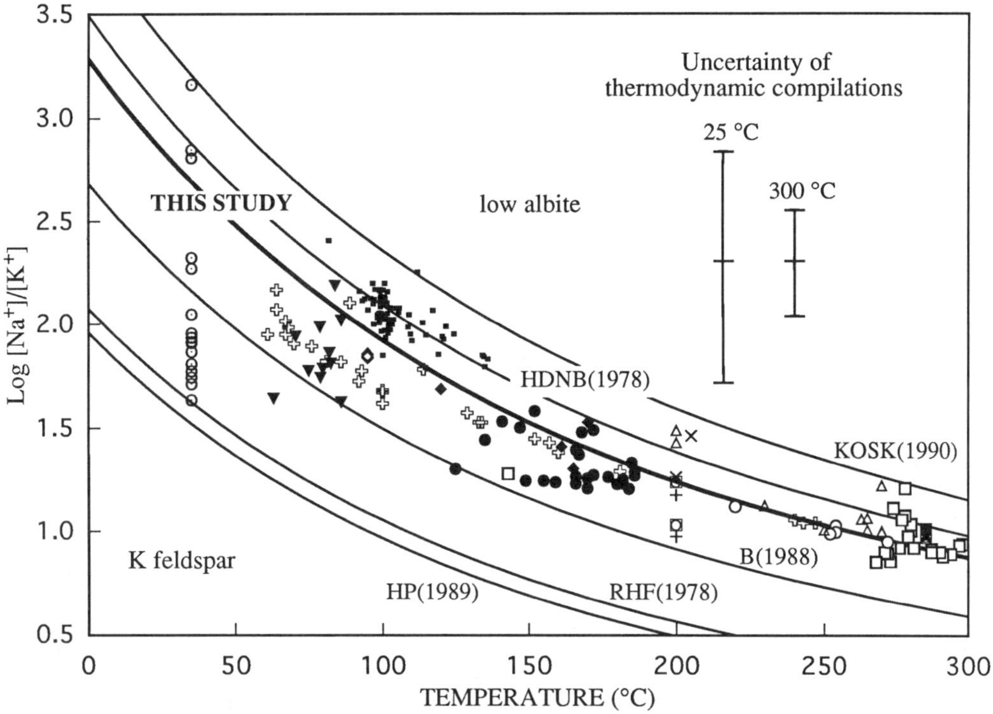

Fig.1. Log[Na+]/[K+], as a function of temperature, calculated from thermodynamic compilations (labeled by authors' first letters and date) and well waters (symbols). ✥ Iceland ; New Zealand, ⊠ Kawerau, □ Ohaki-Broadlands, ⧈ Orakeikorako, × Rotorua, ○ Tauhara, + Waiotapu, △ Wairakei; USA, ◇ Cajon Pass, ▪ Kettleman Dome, ○ Savannah River; Russia, ◆ Bolshe Bann, ▼ Paratunka, ● Pauzhetka.

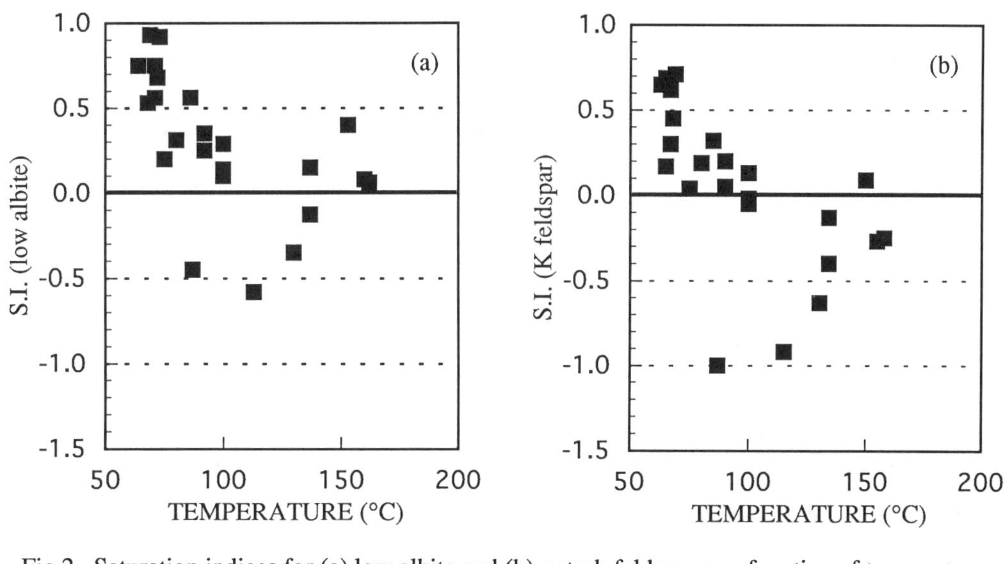

Fig.2. Saturation indices for (a) low albite and (b) potash feldspar as a function of temperature, calculated from well waters in Iceland from data by Arnorsson et al. (1983)

ACKNOWLEDGEMENT

This work was supported by the U.S. Nuclear Regulatory Commission through the U.S. Department of Energy, Contract No. DE-AC03-76SF00098.

REFERENCES

Arnorsson, S., E. Gunnlaugsson & H. Svavarsson 1983. The chemistry of geothermal waters in Iceland, II. Mineral equilibria and independent variables controlling water compositions. *Geochim. Cosmochim. Acta* 47:547–566.

Berman, R.G. 1988. Internally-consistent thermodynamic data for minerals in the system Na_2O-K_2O-CaO-MgO-FeO-Fe_2O_3-Al_2O_3-SiO_2-H_2O-CO_2. *J. Petrol.* 29:445-522.

Browne, P.R.L. & A.J. Ellis 1970. The Ohaki-Broadlands hydrothermal area, New Zealand: mineralogy and related geochemistry. *Am. J. Sci.* 269:97–131.

Fournier, R.O. 1981. Application of water geochemistry to geothermal exploration and reservoir engineering. Chapter 4 in *Geothermal Systems: Principles and Case Histories* (L. Rybach and L.J.P. Muller, Eds.). John Wiley and Sons, Ltd.

Fournier, R.O. & A.H. Truesdell 1973. An empirical Na-K-Ca geothermometer for natural waters. *Geochim. Cosmochim. Acta* 37:1255-1275.

Helgeson, H.C., J.M. Delany, H.W. Nesbitt & D.K. Bird 1978. Summary and critique of the thermodynamic properties of rock forming minerals. *Am. J. Sci.* 278-A:1–229.

Hemingway, B.S. & R.A. Robie 1977. Enthalpies of formation of low albite ($NaAlSi_3O_8$), gibbsite ($Al(OH)_3$) and $NaAlO_2$: revised values for $\Delta G^o_{f,298}$ and $\Delta G^o_{f,298}$ of some alumino-silicate minerals. *J. Res. U.S. Geol. Surv.* 5:413–429.

Holland, T.J.B. & R. Powell 1990. An enlarged and updated internally consistent thermodynamic dataset with uncertainties and correlations: The system K_2O-Na_2O-CaO-MgO-MnO-FeO-Fe_2O_3-Al_2O_3-TiO_2-SiO_2-C-H_2-O_2. *J. Metamorphic Geol.* 8:89-124.

Il'in, V.A., B.G. Kononov, B.G. Polyak & S.V. Kozlovtseva 1979. Hydrochemical estimation of deep rock temperatures. *Geochem. Int. 1979* 119–132.

Kiseleva, I.A., L.P. Ogorodova, Yu.I. Siderov & I.L. Khodakovskiy 1990. Thermodynamic properties of alkali feldspars. *Geochem. Int. 1990* 93–100.

Mehegan, J.M. & P.T. Robinson 1982. Secondary mineralization and hydrothermal alteration in the Reydarfjordur drill core, Eastern Iceland. *J. Geophys. Res.* 87:6511–6524.

Naboko, S.I., G.A. Karpov & A.P. Roznikova 1965. Gidrotermal'nyi metamorphizm porod i mineralo-obrazovanie [Hydrothermal metamorphism of rocks and mineralization. In *Pauzhetskie Goryachie Vody na Kamchatka.* Nauka, Moscow, 76–118.

Robie, R.A., B.S. Hemingway & J.R. Fisher 1979. Thermodynamic properties of minerals and related substances at 298.15 K and 1 bar (10^5 Pascals) pressure and at higher temperature. *U.S. Geol. Surv. Bull.* 1452.

Shock, E. & H.C. Helgeson, 1988. Calculation of thermodynamic and transport properties of aqueous species at high pressures and temperatures: Correlation algorithms for ionic species and equation of state prediction to 5 kb and 1000°C. *Geochim. Cosmochim. Acta* 52:2009–2036.

Tomasson, J. & H. Kristmannsdottir 1972. High temperature alteration minerals and thermal brines, Reykjanes, Iceland. *Contrib. Mineral. Petrol.* 36:123-134.

Trukhin, Yu., P. & V.V. Petrova, 1974. Geochemical effect of hydrothermal metasomatism and variation in the composition of thermal solutions with time, (based on Bolshe-Banny geothermal deposit of Kamchatka). In *Hydrothermal Mineral-Forming Solutions in the Areas of Active Volcanism.* [Gidrotermal'nye Mineraloobrazuyushchio Rastvory Oblastei Aktivnogo Vulkanizma] (S.I. Naboko, Ed.) Nauka Publishers, Siberian Branch Novosibirisk, 1974. Translated from Russian, and published for the U.S. Dep. Inter. and Nat. Sci. Found., Amerind Publishing Co. Pvt. Ltd., New Delhi, 1982 274–285.

Viereck, L.G., B.J. Griffin, H.-U. Schmincke & R.G. Pritchard 1982. Volcaniclastic rocks of the Reydarfjordur drill hole, Eastern Iceland 2. Alteration. *J. Geophys. Res.* 87:6459–6476.

Water-Rock Interaction, Kharaka & Maest (eds) © 1992 Balkema, Rotterdam. ISBN 90 5410 075 3

Hydrocarbon gas ratio geothermometry in the East African Rift System

W.G. Darling & J.C. Talbot
British Geological Survey, Wallingford, UK

ABSTRACT: A gas geothermometer based on the ratio of methane to ethane (C_1/C_2) has been developed for use in geothermal exploration in the East African Rift System. The geothermometer was calibrated using data from geothermal wells in Kenya, Ethiopia and Djibouti. It appears to be robust, apparently not suffering to the same extent the problems (condensation, diffusion, reactivity) which beset the conventional ratio and concentration geothermometers using CO_2, H_2 and H_2S. It is postulated that a mixture of hydrocarbons is produced at $\leq 100°C$ from dissolved or sedimentary organic matter. When these gases enter the hotter parts of hydrothermal plumes, breakdown to CH_4 increases almost logarithmically with temperature. Independent evidence that C_1/C_2 is predictably temperature-dependent comes from a good correlation with $^3He/^4He$. Both ratios are likely to reach a maximum where temperature conditions are hottest. Sparse C_1/C_2 data from hydrothermal systems elsewhere in the world suggest that the methane/ethane geothermometer may have a wider application.

1 INTRODUCTION

The East African Rift System (Fig 1) contains many areas of geothermal activity, usually found in association with late-Quaternary volcanic centres. The generally rather arid nature of Rift climates has resulted in groundwater levels often a considerable distance below surface, and consequently few areas of natural discharge. The use of conventional cation and silica geothermometry is therefore seldom possible. Fumarolic discharges of steam and gas are however fairly common, and in most areas offer the only key to underground temperatures. Various gas geothermometers have been proposed, but their application to the comparatively weak fumaroles of the Rift is frequently inappropriate because the more reactive gases tend to disappear while traversing the often large distance from reservoir boil-off to the surface, and because subsurface condensation of steam changes the gas-water ratio (Armannsson, 1987).

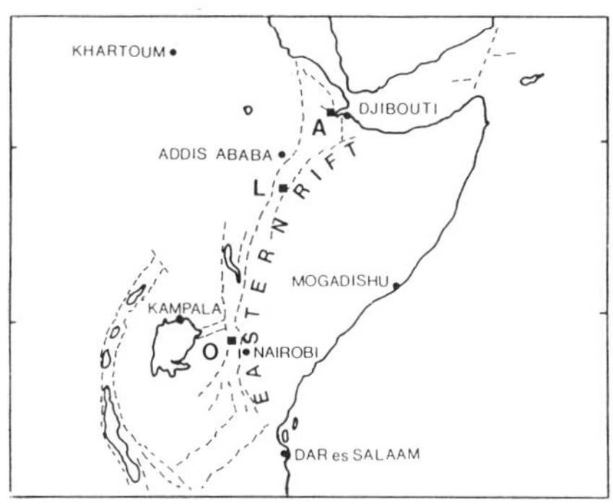

Fig 1. The East African Rift System. O, L and A refer to the Olkaria-Eburru, Langano and Asal geothermal fields respectively.

The ideal gas geothermometer would: (a) be based on a ratio, which would be much less affected by condensation than a concentration geothermometer, (b) use similar gases, which would be relatively immune to differentiation by diffusion, (c) use comparatively unreactive gases, which would not be consumed during passage to the surface, and (d) use gases which can be detected down to low concentrations that could be measured even in weak or air-contaminated fumaroles.

During geothermal exploration in the Kenya Rift, it became apparent that the ratio between methane and ethane might provide a basis for geothermometry (Darling and Talbot, 1991).

2 RESULTS

To calibrate the putative C_1/C_2 geothermometer, gases were collected from geothermal wells at Olkaria and Eburru (Kenya), Langano (Ethiopia) and Asal (Djibouti). Results were supplemented by sampling from two localities where fumaroles could be sampled close to related boiling springs, and one outgassing hot spring.

Values of log CH_4/C_2H_6 are plotted against reservoir solute temperatures in Fig 2. A curve with a correlation coefficient of $r^2 = 0.64$ can be fitted to the data according to the following equation where R represents log C_1/C_2:

$$t°C = 454.8 - (586.4/R) + (281.4/R^2)$$

The plotted points fit the curve with a standard deviation of 27°C. The preponderance of data points from geothermal wells means that the curve is adequately constrained in the 200-300°C range, but less well so at lower temperatures. The curve is intended to be used with values of C_1/C_2 from 10^1 to 10^5, which is the approximate limit of detection, and which correspond to a temperature range of 150-350°C.

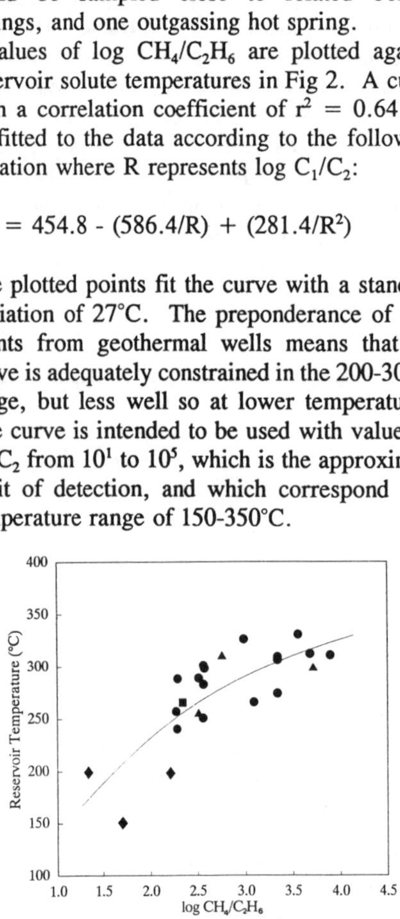

Fig 2. Log CH_4/C_2H_6 plotted versus reservoir solute temperature for geothermal wells at Olkaria (circles), Langano (triangles) and Asal (square). Diamonds represent springs.

3 APPLICATION

The histograms in Fig 3 show the C_1/C_2 geothermometer applied to fumaroles (some within the wellfields and some in unrelated areas) in Kenya, Ethiopia and Djibouti. In addition other geothermometers are shown for the same data set: the empirical equation of D'Amore and Panichi (1980), and the CO_2, H_2S, H_2 and CO_2/H_2 equations of Arnorsson and Gunnlaugsson (1985). These latter equations come in two different versions according to salinity and temperature of the underlying reservoir, and the appropriate version has been used for each fumarole plotted (details in Darling and Talbot, 1991). Some of the geothermometers give results above or below any feasible temperature range and these are omitted from the figures. Given that geothermal reservoir temperatures are commonly in the range 200-300°C, the methane/ethane geothermometer gives a likely distribution of results. CO_2 temperatures are almost uniformly high, and the others low or very low.

4 DISCUSSION

Unlike most other gas geothermometers, the methane/ethane geothermometer seems unlikely to be dependent on gas-mineral equilibria. Carbon stable isotope values for methane in the Rift are nearly all within the -30 to -20‰ range (Darling and Griesshaber, 1990), which suggests that the hydrocarbon gases are produced thermogenically. Abundant petroleum industry literature testifies that hydrocarbon production can begin at temperatures <100°C, and it is considered here that mixtures relatively rich in C_{2+} alkanes are produced from dissolved or sedimentary organic material in the warm outer regions of hydrothermal plumes. If these mixtures are drawn into the hotter parts of the plume, progressive breakdown of the C_{2+} alkanes occurs resulting in an almost logarithmic increase in C_1/C_2 with temperature.

Evidence that breakdown to CH_4 will increase with temperature is shown in Fig 4. This is based on the lignite pyrolysis experiments of Des Marais et al (1988) and clearly indicates a rise in log C_1/C_2 with temperature. Conditions in geothermal systems are very different from laboratory experiments, but independent evidence that there is a relationship between

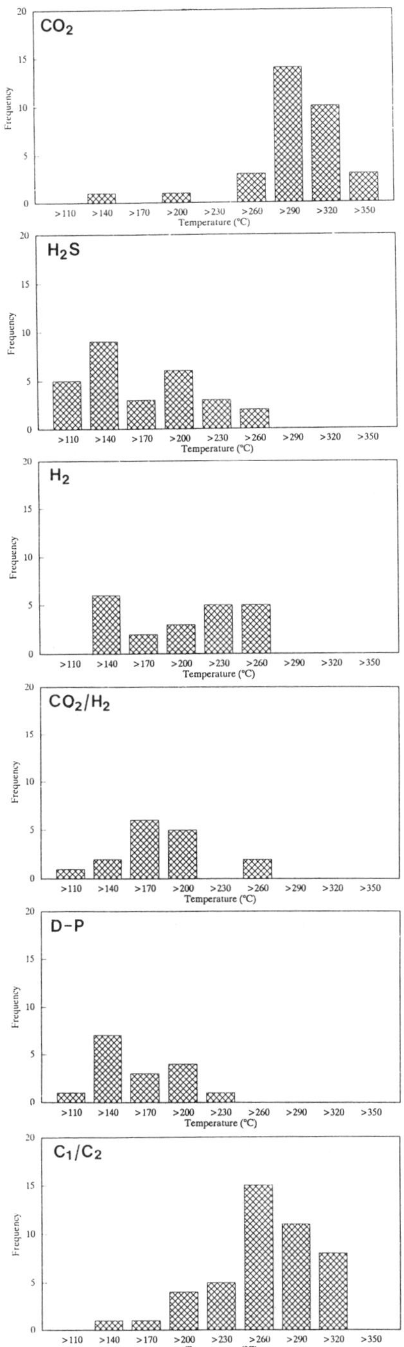

Fig 3. Histograms of the C_1/C_2 geothermometer and other published geothermometers on the same data set. (D-P is the D'Amore-Panichi geothermometer).

C_1/C_2 and temperature is provided by helium isotope ratios. Fig 5 shows the correlation between C_1/C_2 and $^3He/^4He$ for various fumaroles and wells from Olkaria in the southern Rift to Silali in the northern Rift of Kenya. In the absence of complicating factors, the highest helium R_A values will be found overlying the hottest parts of hydrothermal plumes where magmatic influences are largest. Clearly these are also the zones where greatest thermal breakdown could be expected.

5 APPLICABILITY ELSEWHERE

There appear to be few localities worldwide where both C_1/C_2 values and reservoir temperatures are known and therefore the

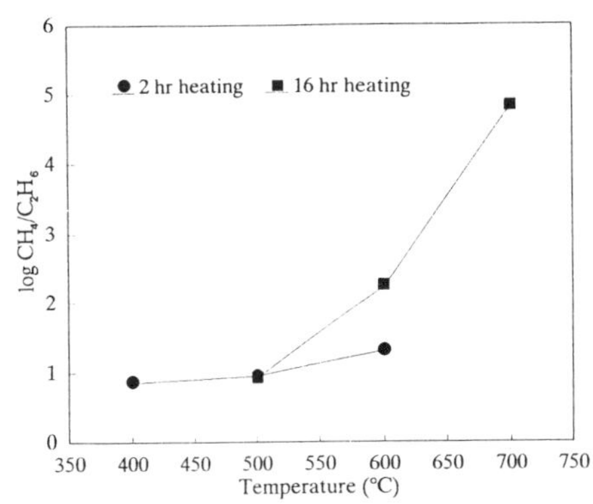

Fig 4. The lignite pyrolysis experiments of Des Marais et al (1988) plotted as log CH_4/C_2H_6 versus heating temperature.

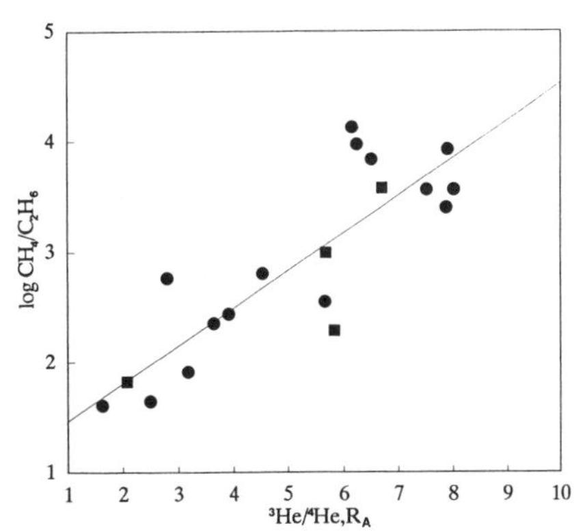

Fig 5. Plot of log CH_4/C_2H_6 versus $^3He/^4He$ (units R_A) for fumarole (circles) and well (squares) samples in the Kenya Rift.

performance of the methane-ethane geothermometer cannot be widely evaluated. Fig 6 shows the available data (extracted from Gunter, 1978; Welhan and Lupton, 1987; Des Marais et al, 1988; Evans et al, 1988; Whiticar and Suess, 1990).

They show that for hydrothermal fluids from ocean floor sources and from Yellowstone geyser basins the C_1/C_2 geothermometer works well. In the Cerro Prieto geothermal field, however, a different relationship has developed. A correlation between log C_1/C_2 with temperature is still observed, but the trend is offset from the 1:1 line, while remaining almost parallel to it.

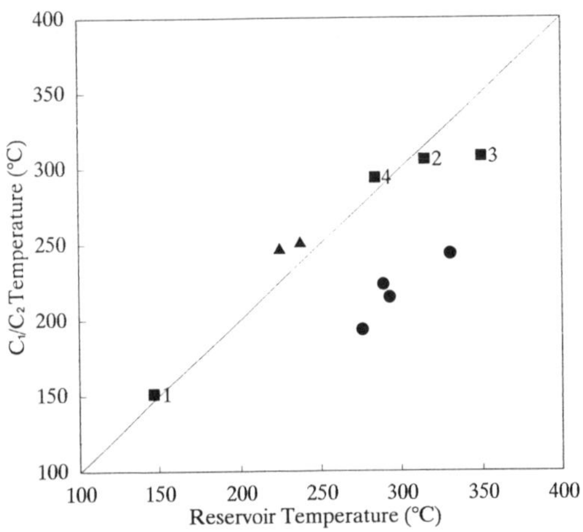

Fig 6. Plot of C_1/C_2 temperatures versus measured or solute temperatures for Cerro Prieto geothermal wells (circles), Yellowstone geyser basins (triangles) and ocean hydrothermal systems (squares). 1 - Bransfield Strait, Antarctica, 2 - Guyamas basin, 3 - East Pacific Rise, 4 - Juan de Fuca ridge.

6 CONCLUSIONS

The methane/ethane geothermometer tends to outperform other geothermometers on the relatively weak fumaroles of the Rift Valley and is easy to apply, requiring no knowledge of steam/gas ratio or reservoir chemistry. Although there is only sparse evidence from outside East Africa, there are indications that the geothermometer will work elsewhere. More observations from geothermal fields worldwide are required to test the relationship further.

REFERENCES

Armannsson, H. 1987. Studies on the geochemistry of steam in the Suswa and Longonot areas and water in the Lake Magadi, Kedong Valley and Lake Turkana Areas, Rift Valley, Kenya. UNDP Tech Rep.

Arnorsson, S. & E. Gunnlaugsson 1985. New gas geothermometers for geothermal exploration - calibration and application. Geochim et Cosmochim Acta 49: 1307-1325.

D'Amore, F. & C. Panichi 1980. Evaluation of deep temperatures in hydrothermal systems by a new gas geothermometer. Geochim et Cosmochim Acta 44: 549-556.

Darling, W.G. & E. Griesshaber 1990. Helium and carbon isotope ratios in crustal fluids of the Eastern Rift System in Kenya. Geological Society of Australia Abstracts 27: 24.

Darling, W.G. & J.C. Talbot 1991. Evaluation and development of gas geothermometry for geothermal exploration in the East African Rift System. Brit Geol Surv Rep WD/91/72.

Des Marais, D.J., M.L. Stallard, N.L. Nehring & A.H. Truesdell 1988. Carbon isotope geochemistry of hydrocarbons in the Cerro Prieto geothermal field, Baja California Norte, Mexico. Chem Geol 71: 159-167.

Evans, W.C., L.D. White & J.B. Rapp 1988. Geochemistry of some gases in hydrothermal fluids from the southern Juan de Fuca ridge. J Geophys Res 93: 15305-15313.

Gunter, B.D. 1978. C_1-C_4 hydrocarbons in hydrothermal gases. Geochim et Cosmochim Acta 42: 137-139.

Welhan, J.A. & J.E. Lupton 1987. Light hydrocarbon gases in Guyamas Basin hydrothermal fluids: thermogenic versus abiogenic origin. AAPG Bull 71: 215-223.

Whiticar, M.J. & E. Suess 1990. Hydrothermal hydrocarbon gases in the sediments of the King George Basin, Bransfield Strait, Antarctica. Appl Geochem 5: 135-147.

ACKNOWLEDGEMENTS

This research was supported by the British Overseas Development Administration and carried out with the cooperation of the Kenya Power and Light Company, the Ethiopian Institute of Geological Surveys and the Institut Superieur d'Etudes et de Recherches Scientifiques et Techniques of Djibouti.

Water-Rock Interaction, Kharaka & Maest (eds) © 1992 Balkema, Rotterdam. ISBN 90 5410 075 3

Past and present subsurface geothermal structure revealed by fluid inclusion thermometry of borehole SN-1, Sengan geothermal area, Japan

B. De Vivo & F. Raia
Dipartimento di Geofisica e Vulcanologia, Università di Napoli Federico II, Italy
M. Sasada
Geothermal Energy Department, NEDO, Tokyo, Japan
T. Sawaki
Geothermal Resources Department, Geological Survey Japan, Tsukuba, Japan

ABSTRACT: The SN-1 drill hole is located in the southwestern part of Sengan geothermal area, northeast Honshu, where three geothermal power stations are operating (Ohnuma, Matsukawa and Kakkonda). Thick dacitic welded tuffs of Teriary to Quaternary ages are distributed around SN-1. The fluid inclusions (in core samples from 582, 664, 816, 1487-89 and 1969 m.) occur mainly along fracture trails, clearly denoting their secondary origin in igneous quartz crystals of the tonalite and volcanic rocks, as well as in late hydrothermal veins of calcite and quartz. The widespread coexistence of liquid-rich and vapor-rich inclusions indicates extensive boiling in the hot-water dominated system. The fluids are very dilute in 582, 664, 816, 1487-1489 levels (0.1 wt % NaCl equiv.), with a salinity ranging from 4.3 to 40 wt % NaCl equiv. in the level 1969. Comparing the Th values with in-hole measured temperatures, cooling of at least a portion of the Sengan geothermal area is evident. The total cooling is in excess of 200°C in this portion of the system. The present water table is 300 m below the surface (about 200 m a.s.l.), but, based on the fluid inclusion temperatures, the water table in the past must have been at least 500 m (700 m a.s.l.) higher than the present level. Fluid inclusion data available for other wells, although indicating the existence of previous, hotter thermal events in the system, show that other portions of the Sengan geothermal area have not cooled as much. The pattern of fluid inclusion Th values recording cooling is probably the result of a falling water table, with consequent cooling related to changes in permeability of the area. The cooling might also be caused by incursion of cold water and/or shifts in the heat source.

1 INTRODUCTION

The SN-1 drill hole is located in the southwestern part of the Sengan geothermal area, northeast Honshu (Fig. 1), where three geothermal power stations are operating (Ohnuma, Matsukawa and Kakkonda).

The SN-1, drilled as a cooperative project between Geological Survey of Japan and the New Energy Development Organization (NEDO, 1989), is one of the wells in the Sengan geothermal area currently being investigated by fluid inclusion studies in a cooperative research project between Italy and Japan. The goal was to determine the hydrothermal history and the present subsurface thermal structure of the geothermal field and to elucidate the characteristics of cooling processes in the area. The cooperative Italian-Japanese research project, completed the study of drill hole SN-7D in the Sumikawa geothermal field, on the northern flank of the andesitic strata volcano Akita-Yakeyama (De Vivo & Sasada, in press).

2 GEOLOGICAL SETTING

The Sengan geothermal area (Fig. 1), located in the Hachimantai volcanic region at the border between Akita and Iwate Prefectures, is one of the largest-scale active geothermal fields in Japan. In the Sengan geothermal area, Neogene formations and Quaternary volcanic rocks are widely distributed on the pre-Tertiary basement rocks, whereas Paleozoic sediments and Cretaceous granitic rocks crop out at the peripheral zone of the field.

The Neogene formations, defined by drilling, reach more than 2000 m in the central portion of Sengan area (Suto, 1987a). The lower portion is mainly composed of altered volcanic rocks ("Green Tuff"), whereas the upper portion is made up of rhythmic, alternating siltstones and tuffs.

Geothermal drilling revealed the presence of two cauldrons in the central part of Sengan area. One is located just below the Akita-Yakeyama volcano, and the other one is in the southern neighboring area. The former one is called Pre-Yakeyama caldera and is thought to have collapsed during the eruption of Tamagawa Welded Tuffs, a widespread felsic flow deposit, made up of two units: Rhyolite Welded Tuff and Dacite Welded Tuff. The eruption age and the volume of the deposits of the two tuff units are estimated as 2 Ma and 130 km^3, and 1 Ma and 50 km^3, respectively (Suto, 1987b). The latter caldera is called Old-Tamagawa caldera and is thought to have collapsed repeatedly during the eruption of Old-Tamagawa Welded Tuffs, probably Upper Miocene or Early Pliocene in age, whose thickness exceeds 2500 m.

The volcanic activity has continued during the last 3 Ma in the Sengan area (Suto & Mokoyama, 1987).

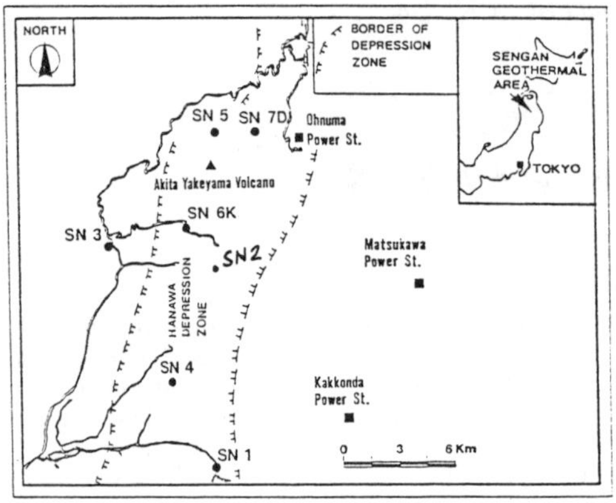

Figure 1 - Location map of Sengan geothermal area and of borehole SN-1.

The latest activity took place in 1719 at Iwate volcano and 1970 at Akita-Komugatake volcano, where effusions of andesite flows occurred. Heat energy from repeated magma intrusions in the latest 3 Ma may play an important role as heat source for the geothermal reservoirs of Sengan area (Suto, 1988).

3 SN-1 DRILL HOLE

The SN-1 penetrates the Rhyolite Welded Tuffs unit (belonging to the Tamagawa Welded Tuffs of the pre-Yakeyama caldera). These tuffs include (from 30 to 70 vol. %) large quartz and plagioclase crystals, showing little lithofacies changes with depth (Suto, 1987b). At about 1250 m the hole penetrates as well layers of altered andesites and siltstones.

The well head is 500 m a.s.l. The minerals, in order of appearance from top to bottom of the borehole are: montmorillonite, sericite/montmorillonite, chlorite, sericite, laumontite, wairakite, albite, calcite, prehnite, epidote, pyrite, pyrrhotite, Ca-amphibole, biotite, sphalerite and chalcopyrite.

Geochemical analyses on core samples of the exploration well show that the remarkable leaching of Na and Ca, and enrichment of K during alteration, was caused by geothermal fluids flowing into the system, equilibrated at high temperatures (Takeno & Noda, 1987). Three thermal events occurred in the geothermal system: contact metamorphism, ore mineralization and geothermal alteration. These thermal events are mineralogically characterized by the presence of biotite-actinolite, sphalerite, galena, Ca-Al silicates and zeolites-clay minerals, respectively (Takeno & Noda, 1987).

Core samples were recovered from levels 582, 664, 816, 1487-89 and 1969 m. Veins of hydrothermal minerals are present in samples from levels 582 and 664 m. The borehole temperatures were measured several times after drilling. The highest temperature recorded is 130°C at 2000 m.

4 NATURE OF FLUID INCLUSIONS

The fluid inclusions are distributed both in late hydrothermal veins of calcite and in igneous quartz crystals of rhyolite. The fluid inclusions in late hydrothermal veins of calcite from the levels 582 m and 664 m occur always along fracture trails, denoting their secondary origin. They are generally flat, with various shapes (from subspherical to ellipsoidal to irregular) and range in size from 1 um up to 30 um. All of them are two-phase (L+V), L-rich, except for rare monophase inclusions (L), occurring along fracture planes in samples from 582 m. The vapor/liquid ratio is almost constant.

The igneous quartz in core samples from 816 m, 1487-89 m, and 1969 m hosts both secondary and primary (?) inclusions. Both types tend to be regular in shape (cylindrical and sub-spherical); their size ranges from 2 um to 25 um.

Three types of inclusions have been recognized: (i) two-phase (L+V), liquid-rich; (ii) two-phase (L+V), vapor-rich; (iii) polyphase (L+V+Halite). Types (i) and (ii) are ubiquitous in igneous quartz from 816 m, 1487-89 m and 1969 m, whereas type (iii) occurs only at 1969 m.

5 FLUID INCLUSION DATA

Measurements of fluid inclusion homogenization temperature (Th) and final melting point temperature (Tm-ice) were made using a CHAIX-MECA heating/freezing stage. Successive calibration runs, using both synthetic fluid inclusions (Bodnar & Sterner, 1984) and chemical compounds of known melting points (Roedder, 1984), suggest that the temperature uncertainty in the range of Tm-ice is $\pm$ 0.1°C, whereas in the heating mode it is $\pm$ 2°C.

5.1 Freezing Temperatures

For most of the experiments on calcite and quartz samples from 582, 664, 816, 1487-89 m, it was not possible to see the ice on warming previously frozen inclusions; the melting of the last ice crystal (Tm-ice) has been determined from the last visible bubble movement or by a sudden appearance of the bubble in the inclusions. The Tm-ice on 350 freezing experiments estimated in this way, ranges from -0.1 to +5.5°C. In some cases in samples from 582 m, the reappearance of the bubble occurred up to +25°C. When frozen, the inclusions, both in calcite and

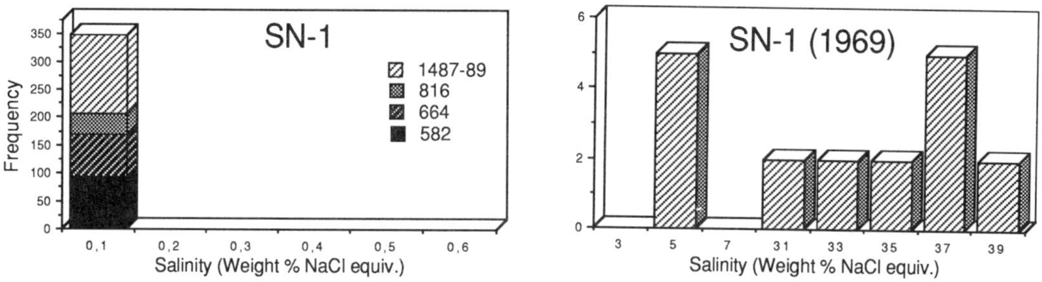

Figure 2 - Histograms showing the 582, 664, 816, 1487-89 and 1969 levels salinity values of borehole SN-1.

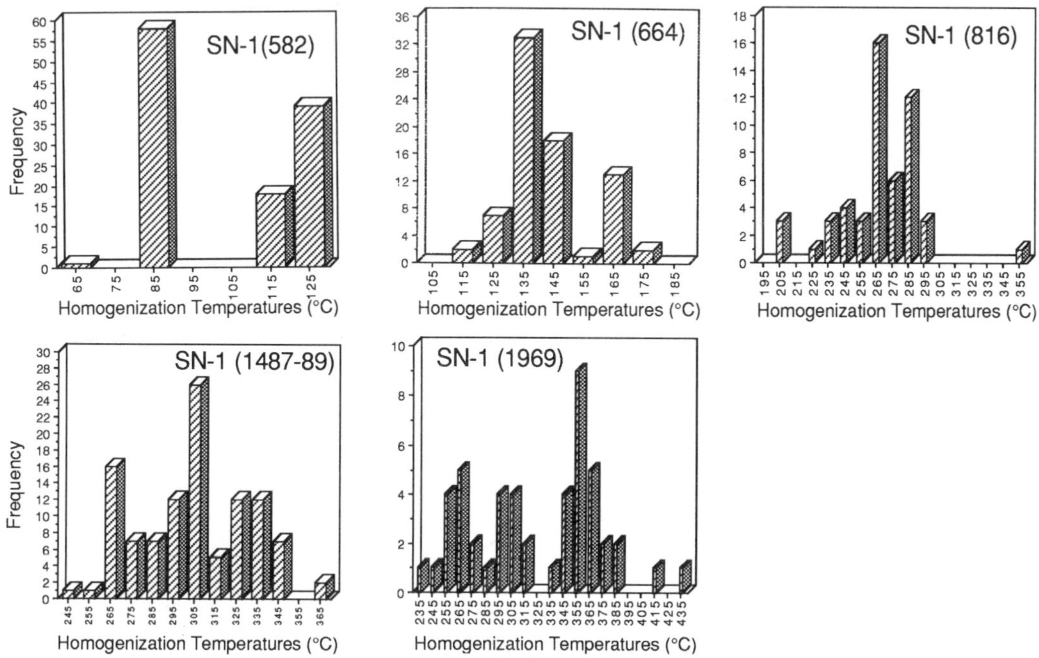

Figure 3 - Histograms showing the 582, 664, 816, 1487-89 and 1969 levels homogenization temperatures of borehole SN-1.

quartz, because of the volume expansion of the ice, eliminated the small vapor bubble, and on subsequent warming, the ice persisted metastably in this vapor-free system to temperatures as high as +5.5°C. This phenomenon occurs when the liquid, without a bubble, is maintened in a stretched condition, occupying a larger volume than the equilibrium volume, so that the pressure on the liquid is much less than the vapor pressure, hence yielding a negative pressure (Roedder, 1971). Superheated (metastable) ice precludes freezing point determinations, hence salinity estimation. Because this phenomenon is most common in very dilute (approaching fresh water) inclusions, we assumed that the inclusions are very dilute, i.e., 0.1 wt % NaCl equivalent. When visible and then successfully detected the Tm-ice, in a few experiments, indicates a very dilute solution (0.1 wt % NaCl equiv.).

The eutectic temperature (Te) was not determined in this type of inclusion. Although Te was not determined in the samples, the prevalence of calcium minerals in this system suggests a Ca component in solution. Nevertheless, we have recast the few Tm-ice data determined, into wt % NaCl equivalent, assuming the dominance of NaCl as the dissolved species (Potter et al., 1978). No liquid CO_2 or evidence of clathrate formation was detected in any sample upon freezing.

The eutectic temperature, measured in igneous quartz from 1969 m, gives evidence that the inclusion fluids contained other species in solution. Few measured Te show values in the range from -76°C to <-50°C; this

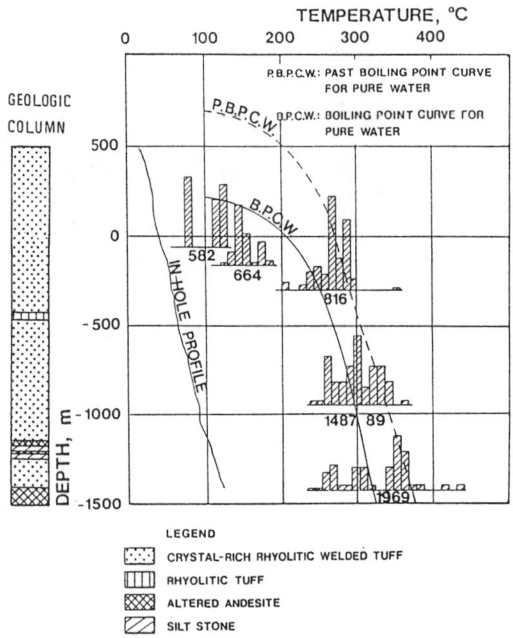

Figure 4 - Diagram showing the geologic column, the in-hole temperaures profile, the homogenization temperature histograms and the boiling point curve for pure water of borehole SN-1.

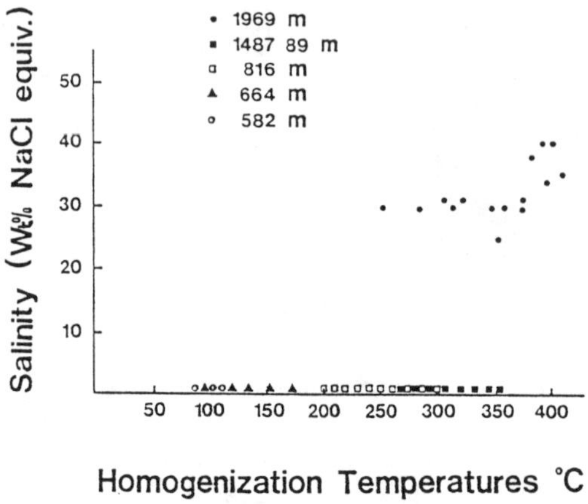

Figure 5 - Diagram showing salinity values versus homogenization temperatures for borehole SN-1.

demands the presence in solution of Ca^{2+} and possibly Li^{+} (Crawford, 1981; Linke, 1965). In some of the inclusions from this level Tm-ice (from -38.7°C to -24.4°C), and the melting temperature of hydrohalite (T m-hydr)(from -37.8°C to -25.5°C) has been determined. The salinity of the fluids referred to the system H_2O-NaCl-$CaCl_2$ (Yanatieva, 1946; Vanko et al., 1988) have values of about 30 % NaCl and 10 % $CaCl_2$.

Many inclusions in the quartz from 1969 m contained cubic, isotropic daughter crystals, assumed to be halite. The halite crystals dissolved at temperatures below the V/L homogenization; only in a few cases the dissolution was slightly higher than V/L homogenization. This might be explained by slow dissolution kinetics of the halite or by necking-down after initial halite precipitation. The salinity of the 13 inclusions with halite was estimated by using vapor present NaCl daughter dissolution temperature (Tm-NaCl) assuming a NaCl-H_2O system (Potter et al., 1977). Salinity ranges from 31 to 40 wt % NaCl equivalent.

The salinity of the inclusion fluids at the five measured levels is shown in figure 2. The figure shows that the salinity of levels 582 m, 664 m, 816 m and 1487-89 m is very homogeneous, never exceeding 0.1 wt % NaCl equivalent, whereas the deeper level (1969 m) shows fluids, from moderate (4.3 wt % NaCl equiv.) to very high salinity (40 wt % NaCl equiv.).

5.2 Homogenization Temperatures

Homogenization temperatures were measured for 408 liquid-rich (i) and polyphase (iii) inclusions; none of the vapor-rich (ii) have been measured because optical conditions and the size of inclusions did not permit accurate determination. Particular care was used while measuring inclusions in calcite to avoid leakage and stretching phenomena.

The range of homogenization temperature at any given depth is quite variable: it is 40°C at 582 m, 60°C at 664 m, 90°C at 816 m, 120°C at 1487-89 m and 160°C at 1969 m (Fig.3). The largest spread of values occurs mostly in secondary inclusions of igneous quartz, especially from the deepest levels 1487-89 m and 1969 m.

The diffuse presence of liquid-rich (i) and vapor-rich (ii) inclusions suggests the very strong possibility that the range of homogenization temperatures may be explained by heterogeneous trapping (and hence boiling) of fluids (Roedder & Bodnar, 1980; Sasada et al., 1986). This is confirmed by the agreement of the homogenization temperature profile with the boiling point curves originating from the present water level at about 200 m a.s.l. and from a paleowater level at 700 m a.s.l. (Fig. 4). An additional possibility is that the spread of homogenization temperatures represents different trapping fluid events, recording the cooling down of the system at various times.

6 THE GEOTHERMAL SYSTEM EVOLUTION

The salinity - Th diagram (Fig. 5) indicates that it is possible to distinguish three groups of inclusions in SN-1:

i) Liquid-rich inclusions, trapped in calcite of hydrothermal origin (Suto, 1988) from 582 m and 664 m levels, have a salinity of 0.1 wt % NaCl equiv. and relatively low homogenization temperatures (modal value of 135°C). These fluids are of meteoric origin, as supported also by isotopic data on the fluids currently circulating in the Sengan geothermal field Takahashi & Noda, 1987). According to Takeno and Noda (1987) these meteoric fluids are the source of the hot, shallow water in the Sengan geothermal field.

ii) Liquid-rich and vapor-rich inclusions in quartz from 816 m and 1487-89 m levels, have very low salinity (0.1 wt % NaCl equiv.) and homogenization temperatures ranging from 205°C to 356°C at 816 m and from 245°C to 365°C at 1487-89 m. The inclusions, both at 816 m and 1487-89 m show a wide range of values: 90°C and 120°C, respectively. The widespread coexistence of liquid-rich and vapor-rich inclusions indicates extensive boiling in the hot-water dominated system. This is consistent with the fact that Th values match the boiling point curve for water adjusted to the present water table (200 m a.s.l.).

iii) Liquid-rich and vapor-rich polyphase fluid inclusions in igneous quartz from 1969 m, have homogenization temperatures from 250°C to 437°C and salinity from moderate (4.3 wt % NaCl equiv.) to very high (up to 40 wt % NaCl equiv.). The inclusions show a wide range of values (160°C), and the possible presence in the fluids of other components (Li^+, Ca^{2+}) in addition to Na^+. Fluids with these characteristics might be explained as the result of a boiling and condensation processes of an initially dilute homogeneous fluid, which in addition to producing two-phase liquid-rich and vapor-rich inclusions, produced as well hypersaline brines as well. This occurrence has been found in the SN-7D drill hole (De Vivo & Sasada, in press) and in other geothermal fields of Italy (Belkin et al., 1985; De Vivo et al., 1989). These types of fluids, as already suggested for SN-7D (De Vivo & Sasada, in press), could be genetically magmatic, related to Tertiary granodioritic and tonalitic intrusives, which are found at the deepest levels of the boreholes in the Sengan geothermal area (Takeno & Noda, 1987).

Because pumping tests reveal no evidence of hypersaline fluids circulating in the present geothermal system, it has to be concluded that the hot hypersaline fluids were sealed by mineral crystallization, no longer circulate through the system and represent a fossil geothermal episode no longer present in the area. This hypothesis is realistic when comparing the Th thermal profile with in-hole measured temperatures profile (Fig. 4). From this figure it is evident that at least the Sengan geothermal area drilled by SN-1 is cooling down. Cooling of portions of active geothermal systems has been recognized in other studies in Japan (Taguchi & Hayashi, 1983; Taguchi et al., 1984), in Costa Rica (Bargar & Fournier, 1988), in Italy (De Vivo et al., 1989) and in Philippine (Reyes, 1990).

The cooling generally occurs on the margins of the system, because of the incursion of cool fluid or of a falling water table caused by uplift and erosion or because of the waning of the thermal activity (Hedenquist et al., in press).

In our case, fluid inclusion data for SN-1 show that the cooling in this portion of the system is in excess of 200°C and occurred from boiling conditions during inclusion trapping to the present thermal regime indicated by the in-hole temperaures profile.

The above reported cooling is, anyway, a minimum estimate, because, based on fluid inclusion temperatures, the water table in the past must have been at least 500 m (700 m a.s.l.) higher than the present level. In this case the cooling from boiling conditions originating in the ancient water table to the present in-hole temperature profile is almost 300 °C. Fluid inclusion data available for other wells (De Vivo & Sasada, in press), although confirming the existence of previous, hotter thermal events in the system, show that other portions of the Sengan geothermal area have not cooled as much.

Although cooling is often deduced to be the result of a falling water table, this does not seem to apply completely to our case. The cooling in SN-1 system is caused more probably by incursion of cold water and/or shifts in the heat source.

The fall in the water table and the consequent cooling related to changes in the permeability of the area, is registered by different inclusions trapped at different times. In other words, the cooling history since the falling of water table from 700 m altitude to 200 m altitude (present water table) is registered by the range of homogenization temperatures found at levels 816 m, 1487-89 m and 1969 m.

The homogenization temperatures profile of the upper levels (582 m and 664 m), departs more markedly from the boiling point curve and falls much more near the in-hole temperature profile, because it has been more influenced by the influx of meteoric cold water from surface.

7 CONCLUSIONS

1 - The very low salinity fluids occurring in inclusions in calcite from 582 m and 664 m levels are very similar to the fluids of meteoric origin presently circulating at the surface in the Sengan geothermal area, and hence are of meteoric origin.

2 - In the deepest levels of the borehole, the widespread coexistence of liquid-rich and vapor-rich inclusions indicates boiling in the hot-water dominated system; the presence of Li^+ and Ca^{2+} is indicated in the system in addition to Na^+. This type of fluid is most probably magmatic, genetically related to intrusive granodiorite and tonalite bodies.

3 - The pattern of fluid inclusion homogenization temperatures compared to the in-hole temperature profile records cooling of at least 200°C of a portion of the Sengan geothermal area. The cooling might be the result of a combination of factors such as falling water table (with consequent cooling related to change in permeability), incursion of cold water and shifts in the heat source.

ACKNOWLEDGEMENTS

This research has been supported by a CNR (Italy) grant to B.D.V (Bilateral Cooperation Project between Italy and Japan). The authors are grateful to H. E. Belkin (USGS) and M. Martini (Univ. of Firenze) for reviewing the manuscript

REFERENCES

Bargar, K.E. & R.O. Fournier 1988. Fluid-inclusion evidence for previous higher temperatures in Miravalles geothermal field, Costa Rica. Geothermics 17 : 681-693.

Belkin, H.E., B. De Vivo, G. Gianelli & P. Lattanzi 1985.Fluid inclusions in minerals from geothermal fields of Tuscany, Italy. Geothermics 14 : 59-72.

Bodnar, R. J. & S. M. Sterner 1984. Synthetic fluid inclusions in natural quartz: I: Compositional types synthesized and applications to experimental geochemistry. Geochim. Cosmochim. Acta 48 : 2659-2668.

Crawford, M.L. 1981. Phase equilibria in aqueous fluid inclusions.In L.S.Hollister & M.L.Crawford (eds), Fluid inclusions, Applications to petrology, Mineral. Assoc. Canada Short Course, 6 : 75-100.

De Vivo, B., H.E. Belkin, M. Barbieri, W. Chelini, P. Lattanzi, A. Lima & L. Tolomeo 1989. The Campi Flegrei (Italy) geothermal system: a fluid inclusion study of the Mofete and San Vito fields. J. Volcanol. Geotherm. Res., 36 : 303-326.

De Vivo, B. & M. Sasada. Fluid inclusions from deep borehole SN-7D, Sumikawa geothermal field, Sengan area, northeast Honshu, Japan. Geother. Res. Soc. Japan (in press).

Hedenquist, J.W., A.G. Reyes, S.F. Simmons & S. Taguchi. The thermal and geochemical structure of geothermal and epithermal systems: a framework for interpreting fluid inclusion data. Eur. J. Mineral. (in press).

Linke, W.F. 1965. Solubilities. Washington: Amer. Chem. Soc. 1914 pp.

New Energy Development Organization (NEDO) 1988. Abstract of Showa 61 FY "Confirmation study of the effectiveness of prospecting techniques for deep geothermal resources", 187 pp.

New Energy Development Organization (NEDO) 1989. Abstract of Showa G1FY "Confirmation study of the effectiveness of prospecting techniques for deep geothermal resources", 143 pp.

Potter, R.W.II, S. Babcock & D.L. Brown 1977. A new method for detrmining the solubility of salts in aqueous solutions at elevated temperatures. J. Res. U. S. Geol. Survey 5 : 389-395.

Potter, R.W.II., M.A. Clynne & D.L. Brown 1978. Freezing point depression of aqueous sodium chloride solutions. Econ. Geol. 73 : 284-285.

Reyes, A.G. 1990. Petrology of Philippine geothermal systems and the application of alteration mineralogy in their assessment. J. Volcanol. Geotherm. Res. 43 : 279-309.

Roedder, E. 1971. Metastability in fluid inclusions. Soc. Mining Geol. Japan, Sp. Issue 3 : 327-334.

Roedder, E. & R. J. Bodnar 1980. Geologic pressure determinations from fluid inclusion studies. Ann. Rev. Earth Planet. Sci. 8 : 263-301.

Roedder, E. 1984. Fluid inclusions. In Review in mineralogy 12, Washington : Mineral. Soc. America, 644 pp.

Sasada, M., E. Roedder & H.E. Belkin 1986. Fluid inclusions from drill hole DW-5, Hohi geothermal area, Japan: Evidence of boiling and procedure for estimating CO_2 content. J. Volcanol. Geotherm. Res. 30 : 231-251.

Suto, S. 1987a. Geological structure of the central part of the Sengan geothermal area, northeast Japan. Rept. Geol. Survey Japan 266 : 43-76.

Suto, S. 1987b. Large scale felsic pyroclastic flow deposit in the Sengan geothermal area, northeast Japan - Tamagawa and Old Tamagawa Welded Tuffs. Rept. Geol. Survey Japan 266 : 77-142.

Suto, S. & S. Mukoyama 1987. Volcanic history at northern part of the Sengan geothermal area, northeast Japan, on the basis of paleomagnetic survey. Rept. Geol. Survey Japan 266 : 143-158.

Suto, S. 1988. Geothermal fields in Tohoku. The Geother. Res. Soc. Japan. The Intern. Heat Flow Comm. of Jaspei, IUGC 27-34.

Taguchi, S. & M. Hayashi, 1983. Past and present subsurface structures of the Kirishima geothermal area, Japan. Trans. Geother. Res. Council 7 : 199-203.

Taguchi, S., M. Hayashi, T. Mimura, Y. Kinoshita, K. Gokou & I. Abe 1984. Fluid inclusion temperature of hydrothermal minerals from the Kirishima geothermal area, Kyushu, Japan. Chinetsu 21 : 119-129 (in Japanese).

Takahashi, M. & T. Noda 1987. Isotopic investigation of waters taken from boreholes, hot springs and surface waters in the Sengan geothermal area. Rept. Geol. Survey Japan, 266 : 285-293.

Takeno, N. & T. Noda 1987. Alteration of the core samples from the geothermal exploration wells in the Sengan geothermal area, northeast Japan. Rept. Geol. Survey Japan 266 : 233-249.

Vanko, D.A., R.J. Bodnar & S.M. Sterner 1988. Synthetic fluid inclusions: VIII. Vapor-saturated halite solubility in part of the system $NaCl-CaCl_2-H_2O$, with application to fluid inclusions from oceanic hydrothermal systems. Geochim. Cosmochim. Acta 52 : 2451-2456.

Yanatieva, O.K. 1946. Polithermal solubilities in the systems $CaCl_2-MgCl_2-H_2O$ and $CaCl_2-NaCl-H_2O$. Zhur. Priklad. Khim. 19 : 709-722.

Water-Rock Interaction, Kharaka & Maest (eds) © 1992 Balkema, Rotterdam. ISBN 90 5410 075 3

Silica geothermometers for use in predicting temperatures in Alberta Oil Sand Reservoirs during steam-assisted thermal recovery

W. D. Gunter, Brian Wiwchar, Larry Holloway & E. H. Perkins
AOSTRA / ARC / CANMET Industry Research Program, Oil Sands and Hydrocarbon Recovery Department, Alberta Research Council, Edmonton, Alb., Canada

ABSTRACT: The concentrations of dissolved silica in the produced water from cyclic steam-assisted oil recovery from the Alberta Oil Sands vary in a regular fashion reflecting temperature changes in the reservoir. A series of autoclave runs were made to calibrate a silica geothermometer for these reservoirs. All of the Oil Sands data lie between the quartz and chalcedony geothermal thermometers, regardless of sand mineralogy. Consequently approximate reservoir temperatures can be estimated using the quartz or chalcedony geothermometers if a specific calibration for the Oil Sands Reservoir of interest does not exist. A more rigorous treatment should incorporate the kinetics of dissolution and precipitation of the silica phases (which are buffering the dissolved silica concentration in the water) into a reservoir model to accurately "position" the geothermometer temperature.

1. INTRODUCTION

Recently geothermometry has been used by a number of investigators (e.g. Saltuklaroglu, 1985; Abercrombie and Hutcheon, 1986; Gunter et al, 1988) to estimate reservoir temperatures in steam-assisted thermal recovery. For the most part, these geothermometers were developed for geothermal reservoirs. They are based on the principal that concentrations of dissolved constituents in geothermal fluids are controlled by the temperature of the reservoir and the mineral assemblage present in the reservoir. Any constituent whose concentration is controlled by a temperature-dependant reaction could, in theory, be used as a geothermometer. Consequently, as long as the composition of the water reflects temperature-dependent mineral solubilities in the reservoir, geothermometers can be used to augment downhole thermocouple measurements.

Applying geothermometers to oil reservoirs introduces several additional assumptions. Oil is assumed to have no effect on the geothermometer calibration. A packet of water produced from an oil reservoir by cyclic steam or steam drive has travelled a more complex path compared to a geothermal reservoir. The residence time of the packet is shorter, the packet is composed of both injected and connate water, and the temperature gradients in the oil reservoir are much higher. In spite of all these uncertainties, several of the geothermometers generate reasonable temperature estimates when compared to direct field measurements.

The families of geothermometers which have proved most successful are based on the concentration of SiO_2 in the produced waters (Henley et al., 1984). The SiO_2 family of geothermometers depends on the solubility of one of the solid silica polymorphs (quartz, chalcedony, opal-CT, cristobalite or amorphous silica - Fournier and Rowe, 1966; and Fournier, 1981) controlling the amount of dissolved SiO_2. Each polymorph has a different solubility relationship with temperature. In the oil sands, both quartz and chalcedony are commonly present; opal is rarer. Either mineral may control the concentration of SiO_2 in the produced water. Kharaka and Mariner (1988) recommend that the chalcedony thermometer be used for temperatures between 40 and 70°C and quartz from 70 to 250°C. However both thermometers must be used with caution above 200°C because the waters become supersaturated with amorphous silica at surface temperatures which can readily precipitate during their ascent to the surface; and low temperatures would be calculated because of the loss of dissolved SiO_2. On the other hand, if there is any boiling during the passage of the water to the surface, the dissolved silica will become more concentrated. If this is suspected, an adiabatic calculation should be made as opposed to the conductive calculation where boiling is assumed not to occur.

The purpose of this communication is to report on the calibration of silica geothermometers for use in Oil Sand Reservoirs in Alberta. The final water compositions from a series of autoclave tests were examined and the estimated geothermometer temperatures are compared to the set temperatures of the experiments. This allows the correct form of the silica geothermometer to be chosen for use in the Oil Sands Reservoirs.

2. SiO_2 THERMOMETER CALIBRATION IN THE ALBERTA OIL SANDS

Boon (1982) and Boon et al (1983) statistically examined mineral-water reactions in an experimental study of oil sands from the major Alberta Oil Sands deposits. In the factorial design, they tested the effect of temperature, salinity, bitumen and pH in batch autoclave runs on sands from the McMurray Formation in the Athabasca Reservoir, from the Grand Rapids and Wabiskaw Formations in the Wabasca Reservoir, from the Gething Formation in the Peace River Reservoir and from the Clearwater Formation in the Cold Lake Reservoir.

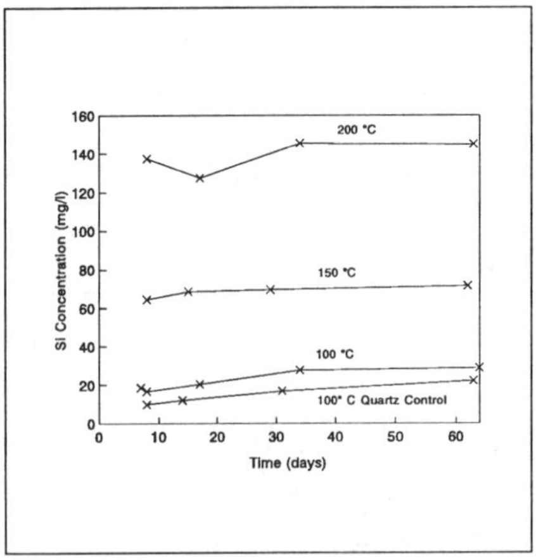

Figure 1: Release of silica from Clearwater and quartz control autoclave experiments.

In Boon's experiments, the temperatures were 200 and 250°C. Experiments were run both with oil present and absent in distilled water and 0.1 molal NaCl for 26 days with a water:sand ratio of 20:1. No regular offset in dissolved silica concentration between saline and distilled water experiments was found, and the same symbol was used to plot both sets of data. The twenty-six day run time for Boon's experiments should be enough time to have reached equilibrium based on the rate law for quartz dissolution proposed by Bird et al. (1986). Produced water from oil sand operations normally ranges in pH from 6 to 8; the experiments were performed in the neutral pH range. Under these conditions, the only dominant silica aqueous species is H_4SiO_4; thus the geothermometer calibration was not affected.

We have extended the data on the Cold Lake Clearwater sand to lower temperature through a series of batch autoclave runs. The sands were placed in small volume stainless steel autoclaves with deionized water in a water/sand weight ratio of 4.33 to 1 and subjected to temperatures of 100, 150 and 200°C. Run times to reach equilibrium were calculated using the Bird et al. rate law for quartz. Run times were chosen of 8, 16, 32 and 64 days for the dissolution experiments to ensure equilibrium was reached.

2.1 Cold Lake Clearwater Oil Sands

The Cold Lake Clearwater Reservoir sand mineralogy is the most complex of all the tar sand reservoirs and has been well described (Putnam and Pedskalny, 1983; Kirk et al, 1987; Harrison et al, 1981, Visser et al, 1985). Detrital quartz and feldspar (includes plagioclase and alkali feldspar) are the dominant minerals, being present in amounts between 15 and 30%. Lithic fragments range from 20 to 50%. Up to 15% of the rock fragments can be attributed to chert (i.e. massive chalcedony). Amphiboles, pyroxenes, biotites and other forms of silica, such as opal CT, may be present in small quantities. The clay content normally lies between 3 and 10%, and consists of smectite, illite, chlorite and kaolinite. Carbonates are either siderite, calcite or dolomite and are quite variable in quantity, ranging from 0 to 25%. Given the diversity of the minerals present; what mineral or mineral assemblage is controlling the amount of dissolved silica in the water?

In order to decide if the quartz or chalcedony thermometers are appropriate, three samples of extracted oil sand from the Clearwater Formation at Husky Oil's Tucker Lake Pilot were obtained from Dr. Ian Hutcheon of the University of Calgary. Equal weight fractions of the three samples were homogenized to make up enough material for the planned experiments. A control sample of pure quartz sand used previously by Dr. G.W. Bird in his kinetic experiments (Bird et al., 1986) was also obtained.

The results of experiments at 100, 150 and 200°C on this extracted Clearwater sand are presented in Figure 1.

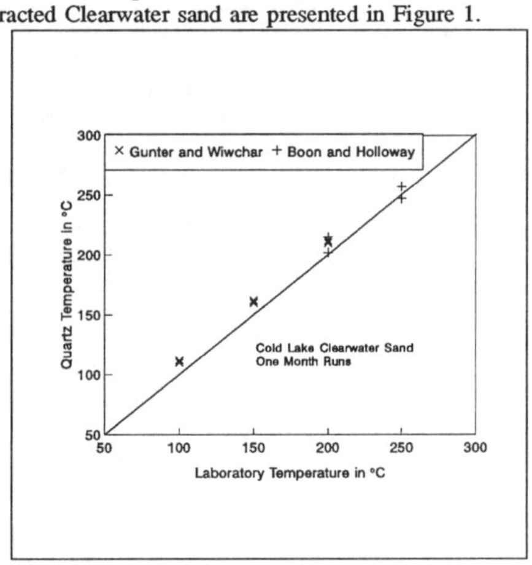

Figure 2: Quartz temperature prediction for autoclave experiments using extracted Clearwater sand.

Except for the 17 day data point at 200°C, the dissolved silica concentrations increase gradually with time approaching a maximum limiting steady state value at each temperature. This maximum value is the calibration point for the silica geothermometer; and presumably represents saturation with a silica phase which is a constituent of the sand (We think the position of the anomalous data point at 200°C on Figure 1 is in error). Boon's data for extracted Cold Lake Clearwater sand extend the geothermometer calibration to 250°C (Figure 2).

These steady state or equilibrium dissolved silica concentrations are significantly displaced from that expected from a simple quartz control. Geothermometer temperature estimates would be 12°C too high if the Fournier's quartz thermometer was used and approximately 8 to 17°C too low if Fournier's chalcedony thermometer was used.

2.2 Silica Data For the Alberta Oil Sands

The autoclave experimental data for the Cold Lake Clearwater, Athabasca McMurray, Peace River Gething, Wabasca Grand Rapids and Wabasca Wabiskaw oil sands collected by the Alberta Research Council have been plotted in Figure 3.

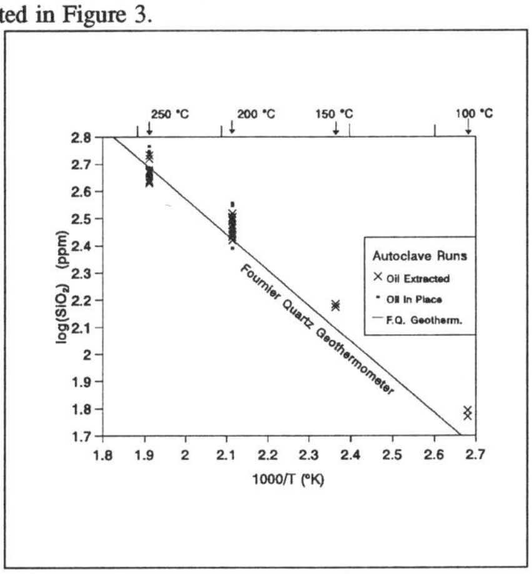

Figure 3: Autoclave silica release data plotted against temperature. Solid line is trace of Fournier's quartz geothermometer calibration.

The data is concentrated at two temperatures, 200 and 250°C. Only the Clearwater is available for lower temperatures. At 200°C and below the dissolved silica concentrations are concentrated slightly above the quartz geothermometer calibration of Fournier (1981); at 250°C it straddles Fournier's curve. The oil saturated experiments have slightly more scatter in measured dissolved silica but still do not differ significantly from those values measured for the experiments in which the oil had been extracted beforehand.

3. DISCUSSION AND CONCLUSIONS

Silica geothermometers have been calibrated for use in predicting reservoir temperatures during thermal recovery in Oil Sand Reservoirs in Alberta. Approximate temperatures can be estimated by using either the quartz or chalcedony thermometers if a specific calibration for the reservoir of interest does not exist. In such a case maximum errors would be limited to the gap between the quartz and chalcedony thermometers. This maximum error is 30°C at 100°C, 16°C at 200°C and 7°C at 250°C; going to zero at 275°C. The silica geothermometer is insensitive to moderate variation in TDS. The silica geothermometer does not appear to be affected by the presence of dissolved organics, even though the solubility of silica has been shown to increase due to the formation of aqueous complexes with certain organic ligands (Bennett and Siegel, 1989). At high pHs, which occurs during alkaline flooding, the solubility of the silica minerals increases tremendously and anonomously high temperatures would be determined unless this is taken into account. Computer programs such as SOLMINEQ.88 (Kharaka et al, 1988) which are designed to calculate and evaluate mineral saturations can handle these calculations easily, and we recommend their routine use in the field to evaluate the produced water chemistry (Perkins et al, 1988; Perkins and Gunter, 1988).

In the field, Abercrombie and Hutcheon (1986) have successfully used the chalcedony geothermometer calibration of Arnorsson et al. (1983) in Cold Lake while Gunter et al (1988) used the quartz geothermometer calibration of Fournier (1981). It is encouraging that the pure silica phases appear to control the concentration of dissolved silica in these reservoirs, even though other complex reactions are occurring.

Consequently simple applications of the silica geothermometer, developed for geothermal reservoirs, to thermal oil recovery for in situ cyclic steam operations should prove to be quite successful, in spite of the differences between the two types of reservoirs. A more rigorous treatment incorporating the kinetics of dissolution and precipitation of the silica phases (which are buffering the dissolved silica concentration in the water) into a reservoir model to accurately "position" the geothermometer temperature is underway. To test the silica geothermometer, quartz kinetics has been inserted into our numerical reservoir simulator to model dissolved silica production (Perkins, in preparation). We find that in cyclic steam, the predicted temperatures exist approximately 3 to 20 meters radially from the sandface, depending on flow rate. However the accumulation of silica present in each produced water packet comes from a much larger portion of the reservoir.

ACKNOWLEDGEMENTS

Thanks to J. Boon for permission to use his data from his factorial designed set of autoclave experiments. B. Young and N. Khiamal helped with the calculations.

REFERENCES

Abercrombie, H.J. and Hutcheon, I. (1986). Remote monitoring of water-rock-bitumen interactions during steam-assisted heavy oil recovery. Fifth Int. Symp. on Water-Rock Interaction, Ext. Abs., 1-4.

Arnorsson, S., Gunnlaugsson, E., and Svavarsson, H., (1983). The chemistry of geothermal waters in Iceland. III. Chemical geothermometry in geothermal investigation. Geochim. Cosmochim. Acta, 47, 567-577.

Bennett, P.C. and Siegel, D.I., (1989) Silica-organic complexes and enhanced quartz dissolution in water by organic acids. Sixth Int. Symp. on Water-Rock Interaction, Ext. Abs., 69-72.

Bird, G.W., Boon, J. and Stone, T., (1986a). Silica transport during steam injection into oil sands. I. Dissolution and precipitation kinetics of quartz; new results and review of existing data. Chem. Geol., 54, 69-80.

Boon, J.A., (1982). An experimental - statistical study of mineral transformations during in situ recovery of bitumen from various oil sand deposits in Alberta, Canada, Third Int. Symp. on Water-Rock Interaction. Ext. Abs., 97-99.

Boon, J.A., Hamilton, T., Holloway, L., and Wiwchar, B., (1983). Reaction between rock matrix and injected fluids in Cold Lake oil sand - Potential for formation damage. J. Can. Petr. Tech. 22, 55-66.

Fournier, R.O., (1981). Application of water chemistry to geothermal exploration and reservoir engineering, in Rybach, and Muffler, L.J.P., eds., Geothermal Systems: Principles and Case Histories: New York, Wiley, 109-143.

Fournier, R.O and Rowe, J.J., (1966). Estimation of underground temperatures from the silica content of water from hot water from hot springs and wet-steam wells: Am. J. of Sc., 264, 685-672.

Gunter, W.D., Perkins, E.H., Young, B., and Bird, G.W., (1988). Geochemical Monitoring of Oil Field Fluids. Fourth UNITAR/UNDP Int. Conf. on Heavy Crude and Tar Sands. Held August 7 to 12 in Edmonton, Canada, Paper #12, 18 pages.

Harrison, D.B., Glaister, R.P. and Nelson, H.W., (1981). Reservoir description of the Clearwater oil sand, Cold Lake, Alberta. in: The Future of Heavy Crude Oils and Tar Sands, R.F. Meyer and C.T. Steele (eds.) McGraw-Hill, 264-280.

Henley, R.W., Truesdell, A.H., and P.B. Barton Jr. (1984). Fluid-Mineral equilibria in hydrothermal systems: Reviews in Economic Geology, v.1, Soc. of Econ. Geol., Book Crafters, Inc., Chelsea, MI, 267 pages.

Hutcheon, I., Abercrombie, H.J. Putnam, P., Garner, R. and Krouse, H.R., (1989). Diagenesis and sedimentology of the Clearwater Formation at Tucker Lake. Bull. of Can. Pet. Geol. 37, 83-97.

Kharaka, Y.K., Gunter, W.D., Aggarwal, P.K., Hull, R.W., and Perkins, E.H., (1988). SOLMINEQ.88: A Computer Program Code for Geochemical Modeling of Water-Rock Interactions, U.S. Geological Survey Water Resources Investigations Report 88-4227, Menlo Park, California, 420 pages.

Kharaka, Y.K., and Mariner, R.H., (1988). Chemical geothermometers and their application to formation waters from sedimentary basins, in Thermal History of Sedimentary Basins. Naeser, N.D. and McCulloh, T., eds., Springer Verlag, New York, 99-117.

Kirk, J.S., Bird, G.W. and Longstaffe, F.J., (1987). Laboratory study of the effects of steam-condensate flooding in the Clearwater Formation: High temperature flow experiments. Bull. Can. Pet. Geol., 35, 34-47.

Perkins, E.H., Gunter, W.D. and Bird, G.W., (1989). Applications of SOLMINEQ.88 and SOLMINEQ.88 PC Shell. Proc. Sixth Int. Symp. on Water-Rock Interaction. August 3-8, Malvern, UK., D.L. Miles (editor), 553-556.

Perkins, E.H. and Gunter, W.D., (1988). Applications of SOLMINEQ/88 and SOLMINEQ.88 PC/SHELL to Thermally Enhanced Oil Recovery. Fourth UNITAR/UNDP Int. Conf. on Heavy Crude and Tar Sands. held August 7-12 in Edmonton, Canada, Paper #11, 13 pages.

Putnam, P.E., and Pedskalny, M.A., (1983). Provenance of Clearwater Formation sandstones, Cold Lake, Alberta, with comments on feldspar compositions. Bull. Can. Pet. Geol., 31, 148-160.

Saltuklaraglu, M., (1985). Monitoring reservoir temperature by means of produced water chemistry in steam based recovery processes. Pet. Soc. Can. Inst. Min., Paper No. 85-36-31.

Visser, K., Dankers, P.H.M., Leckie, D. and van Der Marel, A.G.P. (1985). Mineralogy and Geology of the Clearwater Reservoir Sands in the Wolf Lake Area, Cold Lake, Alberta. in The Third UNITAR/UNDP Int. Conf. on Heavy Crude and Tar Sands, Chapter 14, 119-133.

Water-Rock Interaction, Kharaka & Maest (eds) © 1992 Balkema, Rotterdam. ISBN 90 5410 075 3

Calibration and reequilibration kinetics of cation geothermometers: An experimental study

Houssein Ibrahim, Pierpaolo Zuddas, Gil Michard & Daniel Grimaud
Laboratoire de Géochimie des Eaux, Université Paris 7, France

ABSTRACT: A calibration study of both Na/K and Na-K-Ca geothermometers and the kinetics of their reequilibration by relaxation experiments was carried out. Starting from a calculated equilibrium composition (for a given concentration of Cl at 180°C), we prepared solutions at equilibrium, and enriched or depleted in K or Ca. In all experiments, the concentration of K increased, then decreased and reached the same steady state value after 2-3 months. In contrast, Ca exhibited oscillating variation and did not reach a steady state value. This can be related to reactions involving CO_2 and $CaCO_3$.

1 INTRODUCTION

Geothermometers are applicable when an equilibrium is established between elements of the solution and minerals of the rock. The calibration of Na/K and Na-K-Ca geothermometers is improved with studies of natural fluids and experiments in the laboratory. The study of hydrothermal and geothermal solutions has allowed, since the first observations of White (1957) who saw the decrease in Na/K ratio when temperature increased, the elaboration of formula which associate concentration of elements (Na, K and Ca) and temperature (White 1970; Ellis and Mahon 1964, 1967; Fournier and Truesdell 1973; Fouillac and Michard 1977; Arnorsson et al. 1983; and Giggenbach 1988).

The first experimental studies are from Ellis and Mahon (1967) who noticed that concentration of some elements are controlled by mineral solubility or ion exchange equilibria. The experimental works of Orville (1963) and Hemley (1967) on alkali feldspars helped the calibration of the Na/K geothermometer. Later on, Benjamin et al. (1983) pointed out the prominent part played by secondary phases instead of feldspar-quartz in the Na-K-Ca geothermometer. Pope et al. (1987) tested the possible effect of different fluid compositions on Na/K, SiO_2 and Na-K-Ca geothermometers. However, these experimental studies were conducted using as starting solution deionized water or other solutions far from equilibrium composition with rock. It is known that fluid equilibrium composition in contact with rock depends on the temperature and the concentration of mobile ion (Cl^-).

In order to study the calibration of Na/K and Na-K-Ca geothermometers and the kinetics of their reequilibration by relaxation experiments, we used a starting solution calculated to be in equilibrium with rock at 180°C and a fixed content of (Cl^-). In these solutions, the concentration of one element (K or Ca) was perturbed. Three kinds of starting solutions were prepared as follows:
-- the first with the concentration of the elements at equilibrium with rock,
-- the second with only the concentration of K or Ca enriched with regard to equilibrium conditions,

-- the third with only the concentration of K or Ca lowered with regard to equilibrium conditions.

The Na/K ratio is 25 and Na/Ca is 218, so changes in K or Ca contents are balanced by small changes in Na concentration to keep the equality of negative and positive electric charges of the initial solution. In some experiments, initial K (or Sr for Ca experiment) in solution was enriched in one isotope (^{39}K, ^{84}K) to follow processes. These solutions react at 180°C with powdered granite, (0.08< size <0.2 mm). These experiments were carried out for a range of Cl^- concentrations (0.5 x 10^{-3} M, 1.5 x 10^{-3} M and 5 x 10^{-3} M). The aims of this study were:

-- the calibration of Na/K and Na-K-Ca geothermometers;

-- an estimation of the time needed for their reequilibrations after a temperature change;

-- an understanding of the processes.

2 RESULTS AND DISCUSSION

2.1 Behaviour of potassium

In all experiments the K concentration begins to increase (indicating that the dissolution phenomena is very dominant) during the two first weeks, then the concentration of K decreases (meaning that precipitation is dominant). The content of K shows a stable value after 2 months, when the dissolution compensates for precipitation. After 2 months, concentrations of K converge to the same value, whatever initial conditions, so we can invoke the same mineral phase buffering this element (Fig. 1).

The experimental ratio of Na/K is 38 $\pm$ 8. It is far from the natural Na/K ratio given by Truesdell 1976 (18), Arnorsson 1983 (20) and Michard 1989 (20), but it is close to Na/K ratio given by Giggenbach 1989 (35). These results agree with experimental data of Pope et al. 1986 at 200°C. He calculated, using Truesdell (1976) Na/K geothermometer, temperatures 50°C to 100°C lower than the experimental one.

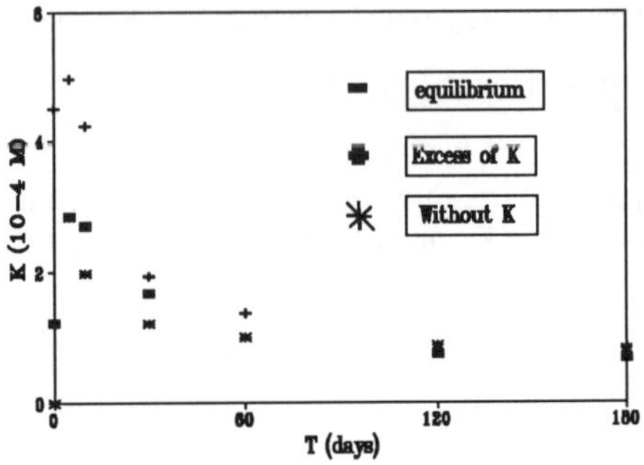

Figure 1 - K versus Time (days).

The use of K isotopes points out three steps in the interaction processes:

-- a very rapid exchange of K between rock and solution;

-- a dissolution of a small part of the rock (0.4%) with a characteristic time of 3 days;

-- a precipitation step with a characteristic time of 10 days (Figures 2, 3).

2.2 Calcium behaviour

In the first step of the interaction, calcium concentration increases and after 1 hour decreases, suggesting the precipitation of neogenic calcic phase. The Ca/Sr ratio of 1000 suggests that calcite is the first calcic mineral. In contrast to potassium, calcium exhibits

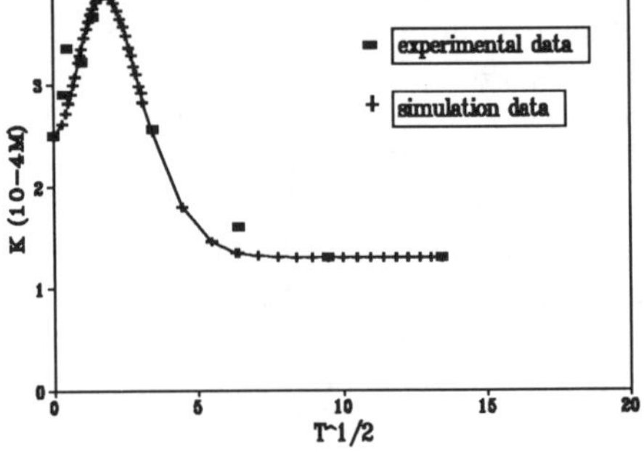

Figure 2 - Experimental and simulated data for K. K concentration versus square root of time.

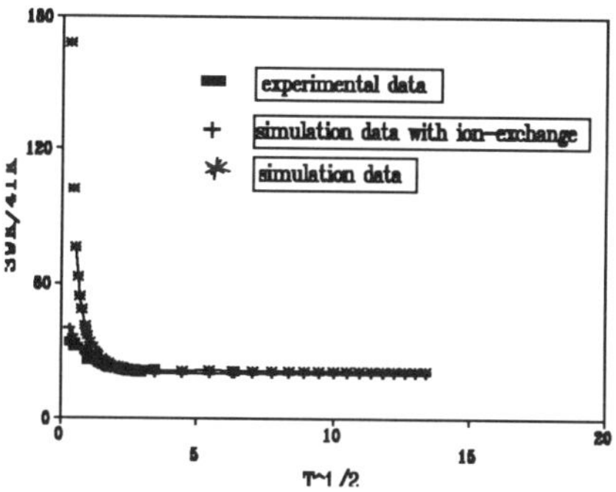

Figure 3 - Experimental, simulated, and simulated ion-exchange data. K concentration versus square root of time.

oscillating variations and does not reach steady state values for all experiments. The Ca/Sr ratio variations suggest precipitation of silicate (Ca/Sr = 100) only in the later step. The Fournier and Truesdell (1973) relationships can only be used if calcium is buffered by aluminosilicate minerals. The kinetics of carbon have an influence on the calcium behaviour because of reactions involving CO_2 and $CaCO_3$.

CONCLUSIONS

According to these experimental results, potassium is controlled, at 180°C, at a steady state value significantly lower than that of natural data characterized by Arnorsson, Truesdell, and Michard Na/K geothermometers.

This discrepancy would be interpreted as:
-- a lack of equilibrium of either experimental solutions or natural solutions,
-- a lack of reaction between solution and rock,
-- controls on elements (K, Na) that are different in experimental and natural systems at 180°C.

In experimental conditions, T = 180°C and water/rock ratio of 10, characteristic time for reequilibration is about 10 days.

Experiments at 120°C are in progress and we hope to clarify these observations because thermometric relation Na/K = f(T°) is better defined at 120°C.

REFERENCES

Arnorsson, S., E. Gunnlaugsson & H. Svarvarsson 1983.The chemistry of geothermal waters in Iceland III. Chemical geothermometry in geothermal investigations. Geochim. Cosmochim. Acta 47: 567-577.

Benjamin, T., R. Chales & R. Vidale 1983. Thermodynamic parameters and experimental data for the Na-K-Ca-geothermometer. J. Volcanol. Geoth. Res. 15: 167-186.

Ellis, A.J., & W.A.J. Mahon 1964. Natural hydrothermal systems and experimental hot water/rock interactions. Geochim. Cosmochim. Acta 28: 1323-1357.

Ellis, A.J., & W.A.J. Mahon 1967. Natural hydrothermal system and experimental hot water/rock interaction (II). Geochim. Cosmochim. Acta 31: 519 538.

Fouillac, C. & G. Michard 1977. Sodium, potassium, calcium relationships in hot springs of Massif Central. Proc. 2nd Intl. Symp. on Water-Rock Interaction, Strasbourg 3: 109-116.

Fournier, R.O. & A.H. Truesdell 1973. An empirical Na-K-Ca geothermometer for natural waters. Geochim. Cosmochim. Acta 37: 1255-1275.

Giggenbach, W.F. 1988. Geothermal solute equilibria. Derivation of Na-K-Mg-Ca geoindicators. Geochim. Cosmochim. Acta 52: 2749-2765.

Hemley, J.J. 1967. Aqueous Na/K ratios in system K_2O-Na_2O-Al_2O_3-SiO_2-H_2O. Program, Ann. Meeting, Geol. Soc. Amer., New Orleans: 94-95.

Michard, G. 1989. Equilibres chimiques dans les eaux naturelles. Publisud, 357 pp.

Orville, P.M. 1963. Alkali ion exchange between vapor and feldspar phases. Amer. J. Sci. 261: 201-237.

Pope, L.A., A. Hajash & R.K. Popp 1987. An experimental investigation of quartz, Na/K, Na-K-Ca geothermometers and the effects of fluid composition. J. Volcanol. Geoth. Res. 31: 151-161.

Truesdell, A.H. 1976. Summary of Section III Geochimical Techniques in exploration. In Proceedings of the Second United Nation

Symposium on the Development and Use of Geothermal Resources, San Francisco, 1, liii-lxxix.

White, D.E. 1957. Magmatic connate and metamorphic waters. Geol. Soc. Amer. Bull. 69: 1659-1682.

White, D.E. 1970. Geochemistry applied to the discovery, evaluation, and exploitation of geothermal energy. Resources: Rapporteur's Report. Geothermics 2(1): 58-80.

Water-Rock Interaction, Kharaka & Maest (eds) © 1992 Balkema, Rotterdam. ISBN 90 5410 075 3

Chemical reactions in upflow zones of geothermal systems affecting subsurface temperature estimates by chemical geothermometry

G.Th. Magnúsdóttir, S. Arnórsson & Audur Andrésdóttir
Science Institute, University of Iceland, Reykjavík, Iceland

Abstract: Mixing geothermal water with cold groundwaters in upflow zones of geothermal systems tends to cause lowering of pH. Conductive cooling has the same effect, at least if the hot water component in the mixture had temperature in excess of some 150°C. The result of pH lowering is enhanced dissolution of primary igneous minerals like olivine and orthopyroxene and sometimes of plagioclase. Consequently aqueous Mg and Fe concentration increase stongly and sometimes also Ca and K concentrations causing faulty geothermometry results for spring waters of mixed origin.

INTRODUCTION

One of the major tasks of geochemistry in the exploration of geothermal resources is to estimate subsurface temperatures in the reservoir from the chemical composition of thermal springs or shallow drillholes.

The fundamental assumption involved in the application of chemical geothermometry is that temperature dependent chemical equilibria are attained, or better closely approached, between particular minerals and solutes in the reservoir, at least for the aqueous components or component ratios used as geothermometers, and that reactions in the upflow where cooling tends to occur are insignificant. In other words the system is essentially taken to be isochemical between the reservoir and the surface. Comparable assumptions are made for mixing models. The system is taken to be isochemical after mixing has occurred.

Numerous studies of geothermal well fluids support the validity of the assumption of close approach to chemical equilibrium between hydrothermal minerals and solution, at least at temperatures above about 100°C (e.g. Giggenbach, 1981, 1988; Arnórsson et al., 1983a). Yet, there is always some uncertainty inherent to this assumption, if not in establishing which minerals are involved, then in reactions in the upflow or mixing with shallow water, all of which change the water composition and, therefore, the geothermometry temperature. Data from Iceland on the chemical composition of water from wells and nearby thermal springs show that reaction and mixing processes in upflow zones sometimes modify water compositions much but in other instances they are negligible.

In the present paper an attempt has been made to assess changes in the saturation state of selected geothermal waters with respect to selected hydrothermal minerals and primary minerals in basalt when they mix with cold water or cool conductively. The changes are subsequently used to predict changes in water compositions that tend to occur in upflow zones of geothermal systems as a result of the cooling and mixing processes. Data are presented (Table 1) from selected wells and springs to demonstrate differences in water compositions that must have resulted from reactions in upflow zones.

DISSOLUTION AND PRECIPITATION REACTIONS ASSOCIATED WITH BASALT-WATER INTERACTION

Geothermal waters, which are generally accepted to be of meteoric origin, represent groundwaters that have percolated deep into the ground and gained heat through contact with hot rock. The initial parent meteoric water will contain some salt, derived from seawater spray. The salt content of this precipitation depends mostly on wind speed, wind direction, altitude, distance from the ocean and whether this precipitation falls as rain or snow. Rain water contacting soil or bedrock is undersaturated with most if not all weathering minerals as well as the primary minerals of basaltic rocks (Gíslason and Arnórsson, 1990). Consequently it tends to dissolve these minerals and in this way its dissolved solids content increases. Continued dissolution does not, however, lead to saturation with the primary basalt minerals. The reason is that weathering and hydrothermal minerals forming in basalts are less soluble, i.e. more stable, than the primary minerals (Gíslason and Arnórsson, 1990). When the water has become somewhat supersaturated with a hydrothermal mineral, this mineral will start precipitating from solution at a rate determined by the degree of supersaturation and kinetic factors.

The hydrothermal alteration process may be viewed as two sets of non-reversible reactions, one set involving primary mineral dissolution and the second hydrothermal mineral precipitation. The alteration process will continue as long as there are any primary minerals available to dissolve regardless of whether the system is open or closed to water flow.

THE CHEMICAL THERMODYNAMIC DATA BASE AND CALCULATION OF MINERAL SATURATION

The WATCH programme (Arnórsson et al., 1982) was used to calculate the aqueous speciation of the mixed cold and thermal waters, and of conductively cooled geothermal waters. This permits evaluation of the reaction quotient (Q) for each mineral. The data base in the WATCH programme is that given in Table 5 of Arnórsson et al. (1982) except for aqueous aluminium hydroxide species where the data of Kuyunko et al. (1983) were used. The solubility data for hydrothermal minerals and end-member compositions of solid solutions in the primary minerals of basalt are based on Berman (1988) correcting the albite Gibbs free energy data are proposed by Sverjensky (1991). However, the Gibbs free energy data for microcline has been modified to be consistent with low-albite and the Na-K geothermometer equation of Arnórsson et al. (1983a).

The standard molal Gibbs free energy of formation of the aqueous species was calculated using the revised equation of state for the standard molal properties of ions and electrolytes (Tanger and Helgeson, 1988 using equation of state parameters for charged species from Shock and Helgeson (1988) and for neutral species from Shock et al. (1989) except for Al^{+3}. A value of -489,578 J/mole was selected for the standard Gibbs free energy of formation of this ion at 25°C and 1 bar which is consistent with the data of Kuyunko et al. (1983) on aluminium hydroxide species.

The igneous minerals in basalt represent solid solutions. Equilibrium constant for the stoichiometric (congruent) dissolution of the actual igneous minerals were obtained by assuming that they represented ideal solid solutions and choosing the standard state to be a solid solution of a fixed composition at the temperature of interest (Gíslason and Arnórsson, 1990). The compositions selected are considered to correspond well with the average groundmass composition of olivine ($Fo_{0.43}$) and plagioclase ($An_{0.29}$) in basalt and of orthopyroxene ($En_{0.38}$) in icelandite (andesite).

Having derived the solubility constants (K) for the minerals comparison with the reaction quotients (Q) permits evaluation of the degree of super- or undersaturation of a water with a particular composition. We have

$$\Delta G_r = -2.303RTlogK + 2.303RTlogQ$$

At equilibrium $\Delta G_r = 0$ so Q = K. For supersaturated solution Q>K and Q<K for undersaturated solution.

FIELD DATA

The first four columns in Table 1 provide data on the chemical composition of water discharged from wells and nearby springs in two localities in the Southern Lowlands low-temperature area in Iceland.The springs emerge in peat soil. Their water contains considerably higher concentrations of total carbonate, Mg and Fe than the well water but their pH is much lower. Calcium and potassium concentrations are also high in the spring water at Laugarbakkar compared with the well water but similar at Sölvholt. As the spring water at Laugarbakkar contains somewhat lower chloride than the well water, it is concluded that the rising geothermal water has mixed somewhat with dilute, cold water in the upflow. The cold water component is the spring water is about 25%. At Sölvholt dilution in the upflow is insignificant. The increase in carbonate, and the corresponding decrease in pH, is probably due to a supply from an organic source.

The differences in the composition of spring and well waters cause considerable differences in the calculated geothermometry temperatures (Table 2). Thus, Na-K-Ca-, and in particular K-Mg temperatures are significantly lower for springs than wells which can be attributed to increases in aqueous Ca and Mg concentrations, respectively. Chalcedony equilibrium temperatures show opposite changes. The increase for

Table 1. Chemical analyses of selected thermal and non-thermal waters in Iceland. Concentrations in ppm.

	Laugarbakkar		Sölvholt		Reykir	Hveragerdi	Hólsá	Eyvindart.	Vatnskot
	well	spring	well	spring	well 5	well 4	river	spring	bogwater
Temp.°C	53	41	51	35	81	181	6	5.5	10
pH/°C	10.10/22	6.83/22	9.03/22	7.34/22	9.42/20	6.96/181	6.68/20	9.68/20	6.56/22
SiO_2	56.2	60.8	60.6	41.0	71.2	270.8	11.2	20.0	40.2
B	0.28	0.14	0.65	0.50	0.04	147.7	0.0023	0.0026	0.012
Na	105.4	78.9	471.5	414.9	42.1	0.60	7.6	15.4	23.4
K	0.93	1.67	9.6	8.0	1.08	12.9	0.46	0.55	1.98
Ca	7.48	14.7	202.0	221.3	2.08	1.67	2.31	1.61	21.7
Mg	0.043	3.81	0.096	1.37	0.06	0.002	1.01	0.17	15.3
Fe	0.015	5.89	0.089	0.156	0.010	0.008	0.013	0.066	18.7
Al	0.069	0.028	0.045	0.254	0.13	0.13	0.022	0.063	0.046
ΣCO_2	4.4	96.4	1.7	16.7	26.8	154.0	8.8	16.2	150.4
SO_4	57.3	27.0	120.2	114.6	17.1	42.1	2.4	2.6	15.5
H_2S	0.02	0.26	0.06	0.03	0.034	23.2	<0.01	<0.01	<0.01
Cl	102.6	78.8	1045.0	999.0	10.5	105.5	10.1	5.1	24.1
F	0.82	0.33	0.22	0.18	0.69	1.75	0.03	0.11	0.26

Laugarbakkar can be attributed to pH lowering whereas at Sölvholt some silica precipitation from solution must be postulated. The Na-K geothermometer of Giggenbach (1988) yields unrealistically high values, probably due to faulty calibration. The Na-K temperature of the Laugarbakkar spring is considerably higher than that of the well because of the higher K content of the spring water. The Na-K-Ca geothermometer definitely gives the most reliable result, i.e. Na-K-Ca temperatures for springs match best aquifer temperatures encountered by drilling.

RESULTS

According to the thermodynamic data used in the present study the 181°C water from Hveragerdi is undersaturated with low-albite and microcline whereas the 81°C Reykir water is somewhat supersaturated. The cold waters are strongly supersaturated. Mixing causes a progressive change towards supersaturation as the cold water component increases (Fig. 1). Thus, mixing favours precipitation of low-albite and microcline, but these minerals are considered to control aqueous Na/K ratios. Chalcedony also tends to precipitate upon mixing. The mixing process produces calcite undersaturation despite an increase in the Ca content of the solution, the cause being the lowering of pH.

The effects of mixing processes upon the saturation state of the primary igneous minerals is rather complicated (Fig. 2). The stability of these minerals depends much on their composition, decreasing with increasing Mg content of the olivine and orthopyroxene and with increasing anorthite content of the plagioclase. Thus, phenocrysts which are more "basic" than groundmass minerals are less stable. The following discussion is confined to groundmass mineral compositions.

The surface waters (near neutral pH) selected for the present study are strongly olivine and orthopyroxene undersaturated whereas the cold groundwater (alkaline) is rather close to saturation with these minerals. All the cold waters are significantly supersaturated with groundmass plagioclase The 81°C Reykir water is close to saturation but Hveragerdi strongly undersaturated.

Mixing of the geothermal water with surface water leads initially to decreasing undersaturation but subsequently to progressive increase in undersaturation with respect to both olivine and orthopyroxene (Fig. 2) and, thus, favours their dissolution. On the other hand mixing of the 81°C Reykir water with alkaline groundwater stabilizes these minerals For the 181°C Hveragerdi water undersaturation is maintained for all mixing ratios except those with almost a pure cold water component. Mixing always increases plagioclase stability.

The solubility of both olivine and orthopyroxene is strongly pH dependent whereas plagioclase solubility is practically pH independent in the range (6.5-10) occurring the natural waters in Iceland. Mixing involving relatively low pH waters, such as surface waters will, therefore, favour dissolution of both olivine and orthopyroxene. pH lowering by addition of CO_2 from an organic source has, of course, the same effect. Such seems to be the case for the waters at Laugabakkar and Sölvholt presented in Table 1. If pH is lowered sufficiently to cause a significant portion of the aqueous Al to occur as $Al(OH)_3$ together with $Al(OH)_4^-$ this may lead to plagioclase undersaturation and through its dissolution the Ca and K content of the water tends to increase.

Table 2. Geothermometry results for springs and wells at Laugarbakkar and Sölvholt

Geothermometer	Laugarbakkar well	Laugarbakkar spring	Sölvholt well	Sölvholt spring
Discharge	53	41	51	35
SiO_2 (chalcedony)[1]	37	82	76	62
Na-K[2]	33	77	74	71
Na-K[3]	92	133	131	128
Na-K-Ca[4]	58	47	64	56
K-Mg[3]	74	39	127	84

[1]Fournier (1977). [2]Arnórsson et al. (1983a). [3]Giggenbach (1988). [4]Fournier and Truesdell (1973).

Conductive cooling of waters relatively rich in carbonate has generally the same effect as mixing as it causes lowering og pH.

REFERENCES

Arnórsson, S., Sigurdsson, S. and Svavarsson, H. 1982. The chemistry of geothermal waters in Iceland. I. Calculation of aqueous speciation from 0° to 370°C. Geochim. Cosmochim. Acta 46: 1513-1532.

Arnórsson, S., Gunnlaugsson, E. and Svavarsson, H. 1983a. The chemistry of geothermal waters in Iceland. II. Mineral equilibria and independent variables controlling water compositions. Geochim. Cosmochim. Acta 47: 547-566.

Arnórsson, S., Gunnlaugsson, E. and Svavarsson, H. 1983b. The chemistry of geothermal waters in Iceland. III. Chemical geothermometry in geotherm-al investigations. Geochim. Cosmochim. Acta 47: 567-577.

Berman, R. G. 1988. Internally consitent thermodynamic data for minerals in the system Na_2O-K_2O-CaO-MgO-FeO-Fe_2O_3-Al_2O_3-SiO_2-TiO_2-H_2O-CO_2. J. Petrol. 29: 445-522.

Fournier, R. O. 1977. Chemical geothermometers and mixing models for geothermal systems. Geothermics 5: 41-50.

Fournier, R. O. and Truesdell, A. H. 1973. An empirical Na-K-Ca geothermometer for natural waters. Geochim. Cosmochim Acta. 37: 1255-1275.

Giggenbach, W. F. 1981. Geothermal mineral equilibria. Geochim. Cosmochim. Acta 45: 393-410.

Giggenbach, W. F. 1988. Geothermal solute equilibria. Derivation of Na-K-Ca-Mg geoindicators. Geochim. Cosmochim. Acta 52: 2749-2765.

Gíslason, S. R. and Arnórsson, S. 1990. Saturation state of natural waters in Iceland relative to primary and secondary minerals in basalt. In: R. J. Spencer and I-Ming Chou (Eds.), Fluid-Mineral Interactions: A tribute to H. P. Eugster. The Geochemical Society, Sp. Publ. 2: 373-393.

Kuyunko, N. S., Malinin, S. D. and Khodakovskiy, I. L. 1983. An experimental study of aluminium hydrolysis at 150, 200 and 250°C. Geochem. Intl. 19: 76-86.

Shock, E. L. and Helgseon, H. C. 1988. Calculation of the thermo-dynamic transport properties of aqueous species at high pressures and temperatures. Geochim. Cosmochim. Acta 52: 2009-2036.

Shock, E. L., Helgeson, H. C. and Sverjensky, D. A. 1989.

Calculation of the thermodynamic and transport properties of aqueous species at high pressures and temperatures: Standard partial molal properties of inorganic neutral species. Geochim. Cosmochim. Acta 53: 2157-2183.

Sverjensky, D. A., Hemley, J. J. and D'Angelo, W. M. 1991. Thermodynamic assessment of hydrothermal alkali feldspar-mica-aluminosilicate equilibria. Geochim. Cosmochim. Acta 55: 989-1004.

Tanger, J. C. and Helgeson, H. C. 1988. Calculation of the thermodynamic and transport properties of aqueous species. Am. J. Sci. 288: 19-98.

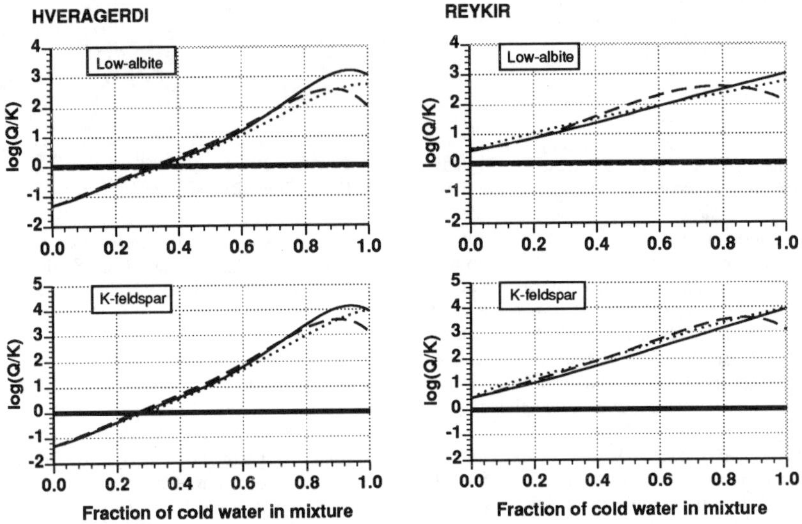

Fig. 1. Changes in the saturation state of low-albite and microcline (K-feldspar) as geothermal water from Hveragerdi and Reykir mix with different types of cold water in Iceland. Solid, dashed and dotted curves designate cold waters from Eyvindartunga, Hólsá and Vatnskot, respectively (see Table 1 for chemical composition). The solid horizontal line corresponds to saturation.

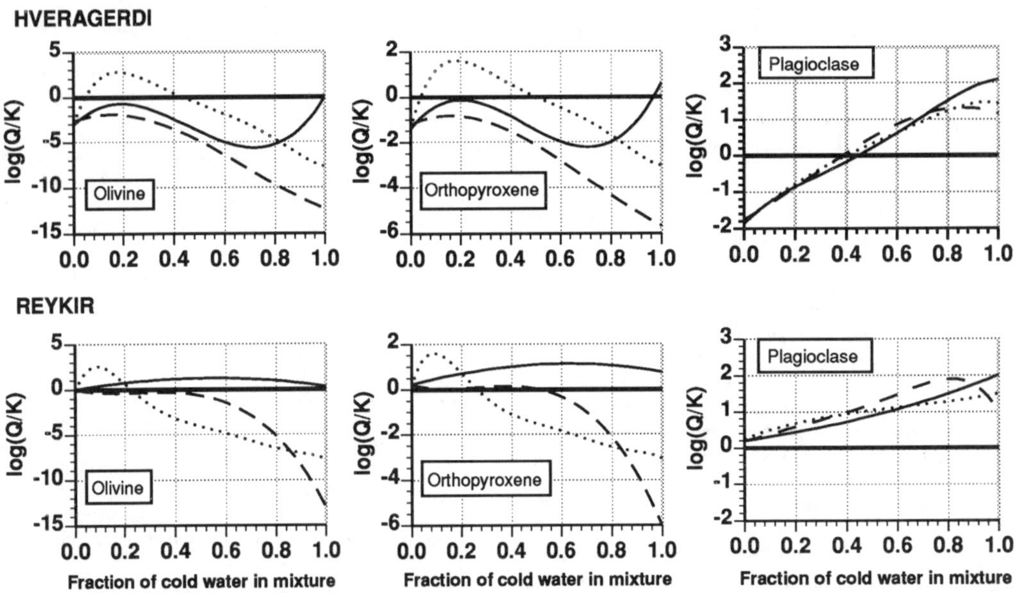

Fig. 2. Changes in the saturation state of groundmass olivine ($Fe_{0.43}$) and plagioclas ($An_{0.29}$) in basalt and groundmass orthopyroxene ($En_{0.38}$) in icelandite as geothermal waters from Hveragerdi and Reykir mix with different types of cold water in Iceland. Curves and line have same signature as in Fig. 1.

Water-Rock Interaction, Kharaka & Maest (eds) © 1992 Balkema, Rotterdam. ISBN 90 5410 075 3

Theoretical calibration of chemical geothermometers and its application to the granitic geothermal areas of SE China

Pang Zhong-he
Institute of Geology, Academia Sinica, Beijing, People's Republic of China

ABSTRACT: Conventional geothermometers are reviewed and the significant uncertainties discussed. Methodology for the theoretical calibration of the chemical geothermometers is proposed with emphasis on the understanding of the fluid-mineral equilibrium state of the studied systems. A case study is presented of the application of the theoretical calibration to the granitic geothermal areas of southeastern China.

1 INTRODUCTION

Conventional geothermometers (Fournier, 1989; Truesdell, 1984 and D'amore & Panichi, 1987) have some assumptions in common, i.e., fluid-mineral equilibrium has to be attained in the studied system with respect to the specific mineral or mineral assemblage and no re-equilibration takes place as the thermal water flows to the surface.

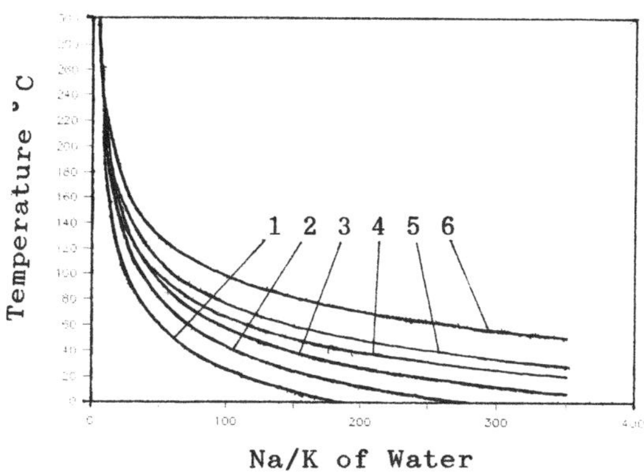

Fig. 1 Comparison of temperature dependent Na/K values for different Na-K geothermometers. 1. Truesdell (1976); 2. Tonani (1980); 3. Arnorsson (1983); 4. Nieva (1987); 5. Fournier(1983); 6.Giggenbach (1988)

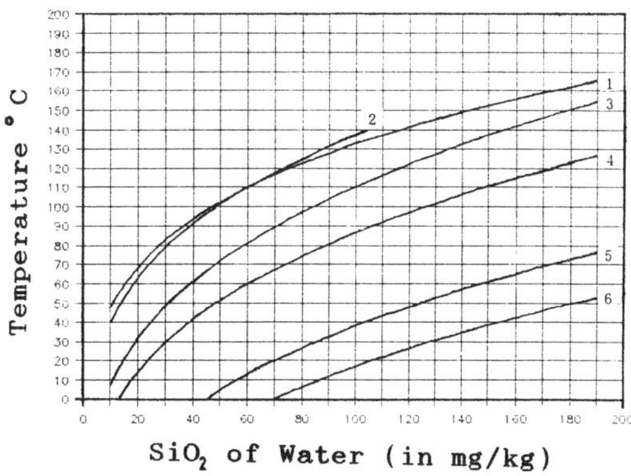

Fig. 2 Comparison of temperature dependent silica solubilities for different silica mineral phases. 1. Quartz-nsl; 2. Quartz-msl; 3. Chalcedony; 4. α-Cristobalite; 5. β-Cristobalite; 6. Amorphous silica

However, the geothermometers themselves can not facilitate the judgement on these pre-conditions. Different geothermometers give different results for the same data they very rarely yield the same result. A particular geothermometer may work well in one place but not in another.

This is very clear for the cation geothermometers. Taking the Na-K geothermometers as an example, for the same Na/K ratio, different equations give very different temperatures, as much as 50-70 °C (Figure 1). And the Na-K-Mg-Ca geoindicators (Giggenbach, 1988) have the similar problems as that of the cation geothermometers were pointed out by Fournier (1990).

Silica geothermometer is the one with a relatively more solid experimental basis, but the identification of appropriate mineral phase is essential to the correctness of the results, for the temperature difference can be as much as 100 °C if a different phase is assumed (Figure 2). For the above reasons, calibration and optimization are a necessary step before a geothermometer can be used with confidence.

2 METHODOLOGY FOR CALIBRATION

The purposes of the calibration are to identify the appropriate mineral phase and to chose suitable geothermometer equations. To achieve these goals, representative geothermal systems with the geological settings and geochemical environment identical to the target area should be selected. The measured reservoir temperature of the geothermal reservoir is necessary.

Thermodynamic calculations of the equilibrium between fluid and minerals have been discussed by many authors (Michard & Roekens, 1983; Reed & Spycher, 1984; Pang & Armannsson, 1989). This modelling will provide a judgement on the overall fluid-mineral equilibrium state in the studied system and comparison of temperatures by geothermometers with that by down-hole temperature logging can be regarded as the final resolution on the correctness of a particular geothermometer temperature, and the comparison of the result with those by conventional geothermometers should make a clear distinguishment of the suitability for each of them.

3 A CASE STUDY FROM SE CHINA

Calibration of geothermometers was carried out for the granitic geothermal areas in the southeast of China, including Fujian, Guangdong and Hainan Provinces, SE China (Figure 1 in Wang et al., this volume).

Nanjin hot spring system is selected as a representative system for the calibration. The reservoir rock is granodiorite. The total natural flow rate of hot water from the hot springs in the area is greater than 20 l/s. The highest temperature measured at the bottom of a 628-meter-deep hole is 90 °C which can be used to represent the reservoir temperature of the system.

Modelling work of this geothermal system was presented by Pang & Armannsson (1989), the temperature obtained by the equilibrium cal-

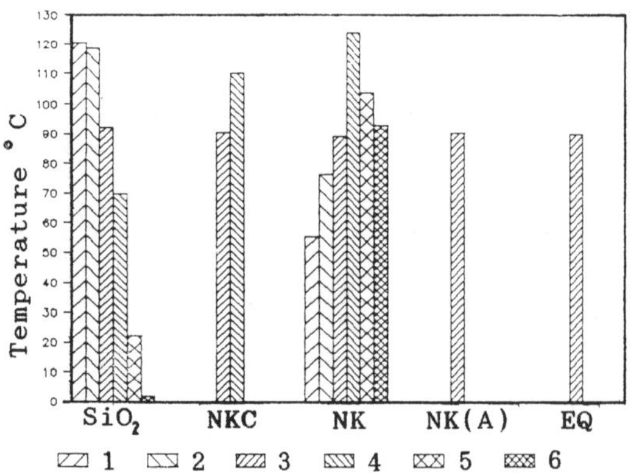

Figure 3. Comparison of results for reservoir temperature of the Nanjin hot spring by different geothermometry methods. The order of the legends coincides with the left to right order of the bars in the diagram. The order for Na-K geothermometers is: 1. Truesdell (1976); 2. Tonani (1980); 3. Arnorsson (1983); 4. Giggenbach (1988); 5. Fournier(1983); 6. Nieva (1987). For silica geothermometers : 1. Quartz-nsl; 2. Quartz-msl; 3. Chalcedony; 4. α-Cristobalite; 5. β-Cristobalite; 6. Amorphous silica. For Na-K-Ca geothermometers is: 1. Na-K-Ca(β=4/3); 2. Na-K-Ca (β=1/3). NK(A) is the averaged temperature by the six Na-K geothermometers. EQ is the temperature derived from the fluid-mineral equilibrium calculations.

culations is in good agreement with the measured temperature. Moreover, the reservoir temperature has been estimated by using different geothermometry methods for a comparison. As is shown in Figure 3, geothermometers that are in good accordance with the measured temperature are: Na-K-Ca ($\beta=4/3$) by Fournier & Truesdell (1973); Chalcedony by Fournier (1989); the Na-K geothermometer by Arnorsson (1983) and by Nieva (1987); and result from equilibrium calculations. But Quartz, Na-K-Ca ($\beta=1/3$) and some other Na-K geother-mometers indicate higher temperatures.

Result of the calibration indicates that for silica geothermometers, chalcedony controls that solubility of silica under this circumstance; and for the cation geothermometers, the Ca correction has increased that suitability of the Na-K geothermometer to low temperature situations. Some other cation geothermometers may also be applicable to low temperature environments but evidently the priority should be given to the former ones.

Reservoir temperature estimations were made for many geothermal systems from the southeast of China. Taking the geothermal systems in southern Fujian Province as an example, chalcedony and Na-K-Ca geothermometers show quite good agreement in the reservoir temperature estimations while most of the Na-K geothermometers tend to give higher temperature predictions (Pang, 1991).

There are many saline hot water systems along the coastal areas of SE China and the high salinity comes from the mixing with seawater (Pang & Wang, 1990). Applicability of conventional geothermometers to this type of systems is different from the fresh water systems. Modelling work on the fluid-mineral equilibrium in these systems was done and it was found that no overall equilibrium exists in them. All cation geothermometers fail to work in this situation. However, the influence of the salinity on the silica geothermometers is neglectable, so they are still applicable. The reservoir temperature of the Zhangzhou geothermal field was calculated by using chalcedony geothermometer with consideration of the mixing process of deep saline thermal water with shallow groundwater (Pang, et al., 1990).

4 DISCUSSIONS AND CONCLUSIONS

Conventional geothermometers have significant uncertainties, so they must be calibrated theoretically to the practical situation.

Representative geothermal systems with measured reservoir temperature can be selected for the calibration. These systems should be identical in geological settings and geochemical environment to the target areas.

Calibration should be based not only on the temperature agreement between the measured and the geothermometers' but also on the fluid-mineral equilibrium state of the systems.

Calibration of geothermometers for the low temperature granitic geothermal areas in the southeast of China indicates that the chalcedony and Na-K-Ca geothermometers are more suitable to this type of geothermal systems.

For the saline hot water systems in the coastal areas. all cation geothermometers fail, but silica geothermometers are more applicable.

ACKNOWLEDGEMENTS

National Natural Science Foundation of China (project 49000034) and the International Atomic Energy Agency (contract 4850/RB) financially supported this study. Sincere thanks are due to Wang Ji-yang for his encouragement on the work and helpful discussions with the author. The author is also grateful to Chen Shun-gui for providing with background information of some of the studied areas.

REFERENCES

Arnorsson, S., Gunnlaugsson, E. and Svavarsson, H., 1983, The chemistry of geothermal waters in Iceland II. Mineral equilibria and independent variables controlling water compositions, Geochim. Coschim. Acta, 47, 547-566.

D'amore, F. and Panichi, C., 1987, Geochemistry in geothermal exploration, in: Applied Geothermics, John Wiley & Sons Ltd., 69-90.

Fournier, R., 1977, Chemical geothermometers and mixing models for geothermal systems, Geothermics, 5, 31-40.

Fournier, R., 1989, Lectures on geochemical interpretation of hydrothermal waters, UNU Geothermal Training Programme, Report 10, pp 73.

Fournier, R., 1990, The interpretation of Na-K-Mg relations in geothermal waters, GRC TRANSACTIONS, Vol. 14, Part II, 1421-1425.

Giggenbach, W., 1988, Geothermal solute equilibria, derivation of Na-K-Mg-Ca geoindicators, Geochim. Cosmochim. Acta, 52, 2749--2765.

Michard, G. and Reokens, E., 1983, Modelling of the chemical composition of alkaline hot waters, Geothermics, 12, 161-169.

Nieva, D. and Nieva, R., 1987, Developments in geothermal energy in Mexico, part 12 -- A cation composition geothermometer for prospection of geothermal resources: Heat recovery systems and CHP, 7, 243-258.

Pang, Z. and Armmannson, H., 1989, Modelling chemical equilibrium in hydrothermal systems: with examples from Iceland and China, in: Water-Rock Interaction (Miles ed.), Balkema, Rotterdam, 541-545.

Pang, Z. and Wang, J., 1990, Oxygen and hydrogen isotope study on Zhangzhou Basin hydrothermal system, southeast of China, GRC TRANSACTIONS, Vol. 14, Part II, 945-951.

Pang, Z., 1991, Calibration of chemical geothermometers based on fluid-mineral equilibrium calculations with application to the hot spring areas in the sout of Fujian Province, SE China, GRC TRANSACTIONS, V.15, 273-277.

Reed, M. and Spycher, 1984, Calculation of pH and mineral equilibria in hydrothermal waters with application to geothermometry and studies of boiling and dilution, Geochim. Cosmochim. Acta, 48, 1479-1492.

Tonani, F., 1980, Some remarks on the application of geochemical techniques in geothermal exploration, in: Proc. Adv. Eur. Geoth. Res., Second Symposium, Strasbourg, pp 428-443.

Truesdell, A., 1976, Summry of Section III geochemical techniques in exploration, in: Truesdell, A., 1984, Chemical geothermometers for geothermal exploration, Chapter 3 in Henley, R. W., Truesdell, A.H. and Barton, P. B., Fluid-mineral equilibria in hydrothermal systems, Reviews in Economic Geology, Vol. 1 (Robertson, J. M., series editor), Soc. Econ. Geol., 31-43.

Water-Rock Interaction, Kharaka & Maest (eds) © 1992 Balkema, Rotterdam. ISBN 90 5410 075 3

Alteration and temperature distribution within and at the margin of the volcanic zone on the Reykjanes peninsula, SW-Iceland

Jens Tómasson & Hjalti Franzson
Orkustofnun, Reykjavík, Iceland

ABSTRACT: A study of the hydrology and alteration within the Reykjanes volcanic rift zone shows that the upper boundary of the zeolite deposition coincides with a transition from a zone of high fracture permeability and a near zero thermal gradient and a zone below with a lower permeability and a relatively high thermal conductive gradient. A distinct cooling episode in the scolecite/mesolite zone has occurred within the Krísuvík swarm at the outer margin of the volcanic zone where cold groundwater percolates down to over 1000 m depth.

1 INTRODUCTION

The aim of this paper is to compare temperature and the onset of mineral deposition in a volcanic rifting environment. The example chosen is the alteration and temperature distribution in drillholes outside the high-temperature areas on the Reykjanes peninsula. The western part of the volcanic rift zone which extends through the Reykjanes peninsula is represented by fissure swarms. The drillholes of this research are situated near or inside two of these; the Reykjanes and the Krísuvík fissure swarms. Both of these swarms propagate out of the volcanic zone. The Reykjanes swarm is near the western tip of the peninsula but the Krísuvík one is further inland (figure 1). The succession of the volcanic zone and its margins consists of subaerial lavas and hyaloclastite formations. Glacial erosion becomes more notable at the margin of the zone. The geological data is derived chiefly from drillcutting analysis, except in well H-37 which was continuously cored. The cutting analysis was done by binocular stereomicroscope and supplemented by petrographic and XRD-analyses in all wells except STG-1, where the first method was solely used.

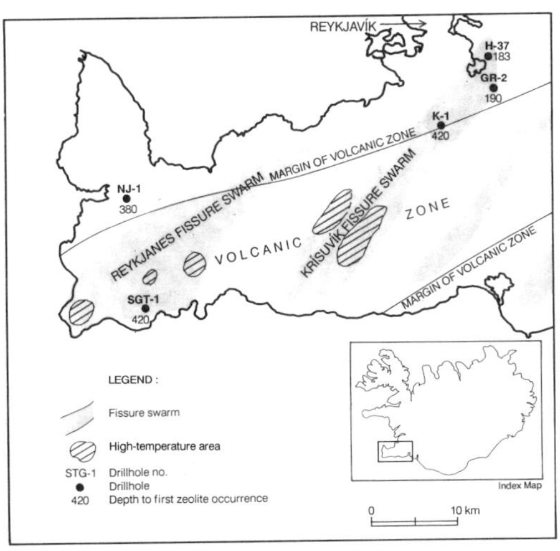

Figure 1. The Reykjanes volcanic zone, fissure swarms, high-temperature areas and location of drillholes.

2. ALTERATION

2.1 Alteration of primary phases.

Figures 2 and 3 shows the geological sections and the alteration along with the measured temperature curves in two drillholes, one sunk in the Krísuvík fissure swarm (K-1) and the other (STG-1) within the Reykjanes fissure swarm. The location of these wells is shown in figure 1.

The rock sequence consists, as mentioned above, of lavas and hyaloclastites. The composition is believed to be largely olivine tholeiite. No indication of marine sediments were seen. It is probable that there may be some variation in alteration susceptibility, as a result of variable cristallinity of the rock. Sideromelane glass is the constituent most sensitive to alteration and alters readily to palogonite or is oxidized. These processes are related more to weathering than to temperature. The next stage is the alteration of palagonite into smectite, which generally coincides with the first appearance of zeolite. The smectite found in drillhole K-1 is relatively poorly crystallized, where the crystal structure collapsed upon heating. The first sign of olivine alteration also coincides with the first occurrences of zeolites. In well K-1 olivine alteration is incomplete above 860 m depth. There are also indications that the groundmass pyroxenes start to alter at depths similar to the first occurrences of zeolites. Pyroxene phenocrysts, however, show no sign of alteration. Plagioclase shows a considerable resistance to alteration, and only minor indications of its alterations along microcracks were observed in phenocrysts.

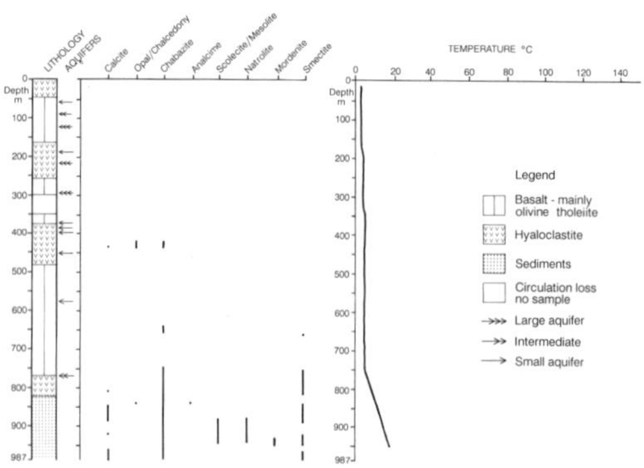

Figure 2. Drillhole K-1. Geological section, aquifers, secondary minerals, and measured temperature. Cemented casing down to 430 m depth.

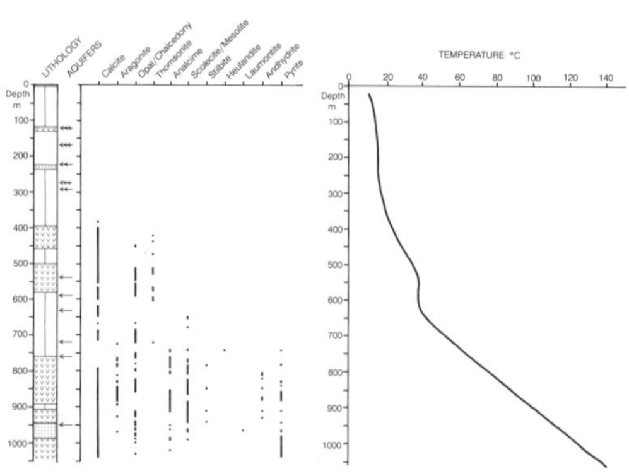

Figure 3. Drillhole STG-1. Same legend as in figure 2. Casing cemented to 488 m depth.

2.2 Mineral deposition

Both wells STG-1 and K-1 are around 1000 m deep. The bottom temperature of K-1 is 19°C while 130°C is measured on the bottom of STG-1. The distribution of the alteration minerals is, despite the temperature difference, quite similar in both wells. The first mineral deposition occurs at 420 m depth in both wells, thomsonite and calcite in STG-1 and chabazite and calcite in well K-1. Chabazite and thomsonite appear in the same zeolite zones within a succession dominated by olivine tholeiites (Walker 1960). Mesolite/scolecite occurs first at 650 m depth in STG-1 and becomes frequent at about 750 m depth. Analcime occurs at similar depth, but stilbite is first found sporadically below 790 m depth. Only one occurrence of laumontite is reported at 980 m depth. In well K-1 analcime occurs at 840 m depth, mesolite/scolecite at 870 m and mordenite at 920 m depth. There are therefore similarities in zeolite distribution in both drillholes down to the mesolite/scolecite zone. Below that laumontite at 980 m depth and mordenite in K-1 at 920 m may further imply a similar type of mineral temperature curve at that depth interval. The occurrence of anhydrite in well STG-1 is probably attributed to the high salinity of the groundwater.

Figure 1 shows the location of well NJ-1 drilled at the margin of the volcanic zone. The first zeolite occurrence (phillipsite) at 380 m depth coincides with a temperature of about 30°C. Thomsonite is found at a slightly greater depth (Tómasson *et al.* 1977). All the above mentioned wells show a relatively constant depth interval where no mineral deposition occurs, which coincides with a zone of cold groundwater system. Walker (1960) concluded from his regional study of the zeolite distribution in eastern Iceland that the top of the chabazite-thomsonite zone in the region of olivine tholeiite lava succession was at least 250 m below the original surface of the basalt succession. This compares well with the Reykjanes volcanic zone, and further implies that a cold groundwater system penetrating down to about 400 m depth may be a general feature of a volcanic rift zone in Iceland. The wells GR-2 and H-37 are situated further outside the volcanic zone. The shallower occurrences of the zeolites there is probably related to glacial erosion and/or a higher thermal gradient. There is an overall concordance of the zeolite distribution in wells STG-1 and K-1, where both show a progressive downward intensity in zeolitization. If a lower formation temperature of chabazite-thomsonite is assumed to be in the range of about 30-50°C and that of laumontite being slightly over 100°C (e.g. Kristmannsdóttir & Tómasson 1978), then it conforms relatively well with the measured temperature in well STG-1, though having slightly lower temperatures at 400-600 m depth than expected. A similar comparison in drillhole K-1, however, shows that the measured temperature is drastically out of phase with the alteration assemblage, where only 5°C is measured at 750 m depth and 19°C at 920 m depth. Similarily temperature gradients in wells GR-2 and H-37 are near to zero where bottom temperatures are 3-6°C (180-190 m depth), in spite of the alteration indicating a location well within the chabazite-thomsonite zone.

3 DISCUSSION AND CONCLUSION

Volcanic rift zones within an oceanic crust is a very dynamic environment. The tectonism related to the Reykjanes volcanism is very intense and represented on surface as dense fissure and fault swarms. The fluid permeability accordingly is extremely high in this environment. The wells STG-1 and NJ-1 show a low temperature down to some 400 m depth succeeded below by a partly linear conductive gradient. The top of the zeolite zone conforms roughly to the boundary between the zero gradient and the conductive one. Hydrologically it is also a transition from a very permeable groundwater system and a less permeable formation below. The close similarity of the alteration with alteration in other areas in Iceland (Walker 1960, Kristmannsdóttir & Tómasson 1978, Kristmannsdóttir 1982, Pálmason *et al.* 1979, Mehegan *et al.* 1982) suggests strongly that this is a dominant permeability/gradient structure within the volcanic rift zones of Iceland. If that is so it also implies that the high fracture permeability decreases in general at 400-500 m depth below the surface of the volcanic zone. The very low temperatures measured in K-1, GR-2 and H-37 show, however, that the cold groundwater system may extend further down if permeable fractures penetrate deeper. Such conditions are, however, considered semistable.

REFERENCES

Kristmannsdóttir, H. and J. Tómasson 1978. Zeolite zones in geothermal areas in Iceland. Natural zeolites occurrance, properties, use: 277-284. Oxford and New York. Pergamon press.

Kristmannsdóttir, H. 1982. Alteration in the IRDP Drillhole compared with other drillholes in Iceland: 6525-6532.

Mehegan, J.M., P.T. Robinson and J.R. Delandy 1982. Secondary Mineralization and Hydrothermal Alteration in the Reydarfjordur Drill Core, Eastern Iceland. Jour. Geophys. Research, vol. 87: 6511-6525.

Pálmason, G., S. Arnórsson, L. Kristjánsson, I.B. Friðleifsson, H. Kristmannsdóttir, K. Sæmundsson, V. Stefánsson, B. Steingrímsson & J. Tómasson 1979. The Icelandic crust: Evidence from drillhole data on structure and processes. In: Deep drilling result in the Atlantic ocean crust. Maurice Ewing Series, 2. Am. Geophys. Union: 43-66.

Tómasson, J., H. Kristmannsdóttir, M. Arnórsson 1977. The interaction of seawater with bassaltic volcanic rocks on the Reykjanes peninsula. Proceedings of the second international Symposiun on the Water-rock interaction Strasbourg august 17 to 25, 1977.

Walker, G.P.L. 1960: Zeolite zones and dyke distribution in relation to the structure of basalts in eastern Iceland. J. Geol., vol.68, p.515-528.

Metamorphic environments

Water-Rock Interaction, Kharaka & Maest (eds) © 1992 Balkema, Rotterdam. ISBN 90 5410 075 3

Formation of zoned epidote in hydrothermal systems

John G. Arnason & Dennis K. Bird
Stanford University, USA

ABSTRACT: Epidote minerals in active and fossil hydrothermal systems exhibit a wide range in Fe^{+3}-Al subsitution and commonly form chemically zoned crystals. Petrologic observations of epidotes in basalt-hosted systems demonstrate that epidote commonly replaces plagioclase and is zoned with Al-rich rims. In contrast, observations from calcareous sediment-hosted hydrothermal systems reveal that epidote replacing calcite cement is zoned with Fe-rich rims. These zoning patterns are replicated in a series of irreversible mass transfer computer experiments using the program EQ3/6 (Wolery, 1983). Epidote with Al-rich rims is precipitated during the irreversible dissolution of anorthite into aqueous chloride solutions of variable initial pH and oxygen fugacities. Epidote with Fe-rich rims can be formed during dissolution of calcite into the same starting solutions, but zoning depends on the initial Fe/Al ratio in the solution. These results suggest that bulk rock composition is one of the controlling variables for epidote zoning patterns in hydrothermal systems.

1. INTRODUCTION

Epidote $Ca_2Fe_xAl_{3-x}Si_3O_{12}(OH)$ is formed in a wide variety of geological environments by reactions of hot aqueous solutions with many different rock types. Compositional zoning, caused by variable Fe^{3+}-Al substitution, is a common characteristic of hydrothermal epidotes. Zoning records a complex sequence of open-system, irreversible chemical reactions under changing conditions of temperature, pressure, and fluid composition. In some cases epidote zoning has been used to infer the evolution of physical and chemical conditions during regional metamorphism (e.g. Raith, 1976) and hydrothermal metasomatism (e.g., Kitamura, 1975; Exley, 1982).

The present study focuses on irreversible processes that produce zoned epidote in low-temperature (200-350°C) and low-pressure (liquid-vapor equilibrium for H_2O) conditions characteristic of geothermal systems in subvolcanic environments and low-pressure greenschist or prehnite-actinolite facies regional metamorphism. The purpose of this study is to 1) characterize the compositional zoning characteristics of epidotes in different rock types from active and fossil hydrothermal systems; and 2) through computer modeling, to establish cause and effect relations governing the formation of zoned epidotes by irreversible reactions of aqueous solutions with common rock-forming minerals.

2. ZONED EPIDOTE IN HYDROTHERMAL SYSTEMS

Epidote occurs in most geothermal systems in which measured or calculated temperatures exceed 200 to 250°C (Bird *et al.*, 1984). Host-rock lithology and mineralogy is strongly correlated with the type of zoning observed: in basaltic rocks, epidote replaces primary igneous plagioclase and is zoned with Fe-rich cores and Al-rich rims, whereas in calcareous sediments, epidote replaces matrix calcite and has the opposite zoning pattern: Al-rich cores and Fe-rich rims (Arnason, 1991).

2.1 Basalt-hosted systems

Compositional data are summarized below for epidotes in two basalt-hosted hydrothermal systems: the Nesjavellir geothermal system of Iceland, and

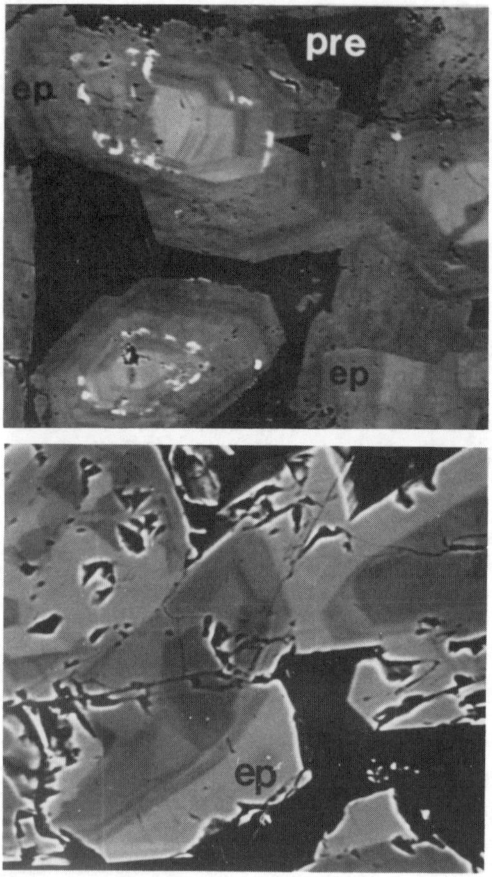

Figure 1. Backscatter electron images of zoned epidote. Light areas are Fe-rich, dark areas Al-rich.
A. (top): Epidote (ep) and prehnite (pre) in basalt vug from the Nesjavellir geothermal system. Epidote has Fe-rich cores, Al-rich rims, and oscillatory zoning. Bright spot indicated by arrow is grandite garnet inclusion. Field is 175 μm wide.
B. (bottom): Epidote with Al-rich cores and Fe-rich rims in the State 2-14 well of the Salton Sea geothermal system. Field is 110 μm wide.

the Lower Basalts, a fossil hydrothermal system in East Greenland.

The Nesjavellir geothermal system is located in an active spreading zone that is part of the Mid-Atlantic Ridge in central Iceland. Drill hole samples reveal active metamorphism of tholeiitic lavas, hyaloclastites, dikes, and sills from greenschist to amphibolite facies at temperatures from 350 to 380°C (Hreggvidsdottir, 1987). Epidotes (± albite ± K-feldspar) typically replace plagioclase and have Fe-rich cores (Figure 1A). Epidote compositions range continuously from X_{Fe} = Fe/(Fe + Al) = 0.14 to 0.38 with a mean value of X_{Fe} = 0.25 and an average core-rim difference of X_{Fe} = 0.11. Some crystals exhibit micron-scale oscillatory zoning superimposed on a monotonic pattern.

The Lower Basalts are a sequence of early Tertiary basaltic lavas, tuffs, hyaloclastites, and breccias that were intruded by the Skaergaard gabbro in East Greenland. During emplacement of the Skaergaard, basalt pore waters heated up and reacted with the basalts to form wide-spread prehnite-actinolite to greenschist facies mineral assemblages (Manning & Bird, 1991). Epidote is one of the most common alteration minerals in the basalt matrix, vesicles, vugs, and veins. The majority of epidotes have Fe-rich cores and Al-rich rims (Arnason, 1991). Some are unzoned, but epidotes with Al-rich cores and Fe-rich rims were not found. Compositions range continuously from X_{Fe} = 0.17 to 0.45 with an average value of 0.29. The average core composition is X_{Fe} = 0.34 and the average rim composition is X_{Fe} = 0.28. The extent of compositional variation within a single crystal ranges from 0.01 mole fraction Fe to 0.16, with the average being 0.09.

2.2 Calcareous Sediment-Hosted Systems

Compositional and zoning data are summarized below for epidotes from two hydrothermal systems hosted by calcareous sediments: the Salton Sea geothermal system and the Cerro Prieto geothermal system. Both are active systems located in the Salton Trough of southern California and northern Mexico. Although similar in host rock lithology, these two systems differ dramatically in the chemistry of their hydrothermal solutions. The Salton Sea is characterized by hypersaline brines (>300,000 ppm total dissolved solids), whereas the Cerro Prieto system contains more dilute alkali chloride solutions (<10,000 ppm total dissolved solids). In addition, the Cerro Prieto sediments contain abundant organic material.

In the Salton Sea geothermal system, epidote is found in veins and replaces matrix calcite cement. Most matrix epidotes have Al-rich cores and Fe-rich rims (Figure 1B), some have the opposite pattern, others show oscillatory zoning, and some show no zoning at all (Cho *et al.*, 1987; this study). They range in composition from X_{Fe} =

0.11 to X_{Fe} = 0.42. Vein epidotes are also zoned with Al-rich cores and Fe-rich rims, and are overall more Fe-rich than matrix epidotes (Caruso *et al.*, 1987; this study).

Hydrothermal mineralogy at Cerro Prieto is summarized by Schiffman *et al.* (1985). Epidote is the most abundant calc-silicate mineral and occurs at temperatures exceeding 230°C where it replaces calcite and wairakite. Epidote compositions define a normal distribution with X_{Fe} ranging from 0.11 to 0.31. Backscatter electron images of epidotes reveal complex zoning patterns. Matrix epidotes display complex oscillatory zoning superimposed on a pattern of Fe-rich cores, Al-rich mantles, and Fe-rich outer rims.

3. FORMATION OF ZONED EPIDOTE BY IRREVERSIBLE MASS TRANSFER

To better understand the mechanisms for the formation of zoned epidote, we examined in detail irreversible chemical processes between common rock-forming minerals and aqueous chloride solutions under hydrothermal conditions. With the aid of the computer program package EQ3/6 and its associated database (Wolery, 1983), we calculated irreversible mass transfer and the changes in mineral and fluid compositions accompanying these processes. We present a series of models representing isothermal, isobaric reaction of (1) anorthite and (2) calcite with epidote-saturated aqueous solutions over a range of initial fluid compositions in order to clarify the role of pH, oxygen fugacity, and mass transfer in determining epidote zoning patterns in hydrothermal systems. Theory of irreversible mass transfer between minerals and aqueous solutions is reviewed by Helgeson (1979). Details of the models and thermodynamic data used are given in Arnason (1991). The results of the computer experiments are summarized on the activity-activity phase diagram of Figure 2.

3.1 Anorthite dissolution

In anorthite dissolution reactions (1–4 in Table 1 and Figure 2) X_{Fe} in epidote decreases from 0.17 to ~ 0.06, due to the increasing concentration of aqueous Al^{3+} relative to that of Fe^{3+}. The pH decreases during the reactions as OH^- ion is consumed to form Al hydroxide species. The mass of

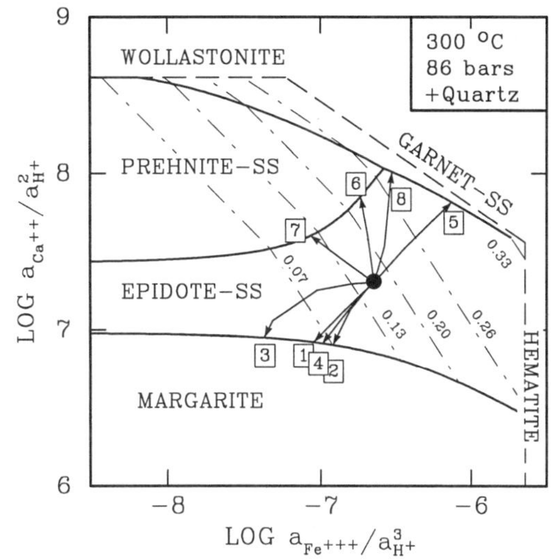

Figure 2. Activity-activity phase diagram showing the stabilities of minerals in the system CaO-Fe_2O_3-Al_2O_3-SiO_2-H_2O-HCl at 300°C and 86 bars in the presence of quartz and an aqueous solution (a_{H_2O}=1). Solid lines represent mineral phase boundaries. Dashed lines represent saturation surfaces of the fluid with respect to the indicated phase. Dash-dot lines denote isopleths of constant epidote, prehnite, and garnet composition (numbers for epidote are X_{Fe}). The solid circle represents the initial fluid and epidote composition. Irreversible reaction paths are indicated by solid lines with arrows and numbers in boxes corresponding to the reactions discussed in the text.

Table 1. Summary of initial solution compositions used in mass transfer calculations. All solutions are saturated with respect to quartz and epidote (X_{Fe} = 0.17). For reactions 1-4, anorthite is the reactant. For reactions 5-8, calcite.

Reaction numbers	pH	Log oxygen fugacity	total Cl^- (log molal)
1 & 5	5	-31.0	-1.00
2 & 6	6	-31.0	-2.00
3 & 7	7	-31.0	-3.00
4 & 8	6	-35.6	-2.00

epidote produced is approximately equal to the mass of anorthite dissolved, and the total volume of solids decreases by < 1%. During these reactions, oxygen fugacity decreases slightly or remains constant. Variations in initial solution pH, chloride concentration, and oxygen fugacity have a minor effect on the epidote zoning patterns.

3.2 Calcite dissolution

In calcite dissolution reactions (reactions 5–8 in Table 1 and Figure 2), epidote zoning patterns are functions of the inital Fe^{3+}/Al^{3+} concentration ratio in aqueous solution. At pH=5 to 6 (reactions 5, 6, and 8) the initial Fe^{3+}/Al^{3+} ratio is > 1 and the epidote produced has Fe-rich rims. At pH=7 (reaction 7), the initial Fe^{3+}/Al^{3+} ratio is < 1 and epidote with Al-rich rims is produced. The mass of epidote formed per gram of calcite dissolved is approximately 2 orders of magnitude less than that for anorthite. As a result, the volume of solid phases decreases ~ 96%. In all cases, pH increases due to calcite dissolution, and oxygen fugacity decreases slightly or remains constant. Comparing reaction paths 6 and 8 in Figure 2, it is apparent that the initial oxygen fugacity has a small effect on epidote zoning.

4. CONCLUSION

Our observations from active and fossil hydrothermal systems indicate that epidote zoning patterns are largely controlled by the host-rock mineralogy. In basalt-hosted systems, epidote replaces plagiclase and has Al-rich rims. In calcareous sediment-hosted systems it replaces calcite cement and has Fe-rich rims. Zoning patterns observed in hydrothermal systems are consistent with the formation of epidote by irreversible dissolution of primary minerals in these systems. Contrasting zoning patterns may be caused by differences in the relative degree of mass transfer of Fe and Al among reactant minerals, the aqueous solution, and the growing epidote. The effects of pH on the stabilities of hydroxide complexes of ferric Fe and Al play an important role in this process. The models demonstrate that closed system precipitation of epidote from an aqueous solution will result in a decrease in oxygen fugacity by the removal of ferric Fe from the fluid. In natural systems, however, oxygen fugacity may be buffered by other phases. Because of the complex dependence of epidote composition on fluid chemistry, determination of the evolution of physical and chemical conditions from epidote growth zoning patterns will often be non-unique.

REFERENCES

Arnason, J. G., 1991. Formation of zoned epidote by irreversible mass transfer in hydrothermal systems. *M. S. Thesis, Stanford University.*

Caruso, L. J., Bird, D. K., Cho, M., & Liou, J. G., 1988. Epidote-bearing veins in the State 2-14 Drill Hole: Implications for hydrothermal fluid composition. *J. Geophys. Res.*, 93, 13123-13133.

Cho, M., Liou., J. G., & Bird, D. K., 1988. Prograde phase relations in the State 2-14 well metasandstones, Salton Sea geothermal field, California. *J. Geophys. Res.*, 93, 13081-13103.

Exley, R. A., 1982. Electron microprobe studies of Iceland research drilling project high-temperature hydrothermal mineral geochemistry. *J. Geophys. Res.*, 87, 6547-6557.

Helgeson, H. C., 1979. Mass Transfer among minerals and hydrothermal solutions. *in* Barnes, H. L., ed., *Geochemistry of Hydrothermal Ore Deposits, v. II.* New York: Holt, Rinehart, and Winston, Inc., 1979.

Hreggvidsdottir, H., 1987. The Greenschist to Amphibolite Facies Transition in the Nesjavellir Hydrothermal System, Southwest Iceland. *M. S. Thesis, Stanford University.*

Kitamura, K., 1975. Al-Fe partitioning between garnet and epidote from the contact metasomatic copper deposits of the Chichibu mine, Japan. *Econ. Geol.*, 70, 725-38.

Manning, C. E., & Bird, D. K., 1991. Porosity evolution and fluid flow in the basalts of the Skaergaard magma-hydrothermal system, East Greenland. *Am. J. Sci*, 291, 201-257.

Raith, M., 1976. The Al-Fe(III) epidote miscibility gap in a metamorphic profile through the Penninic series of the Tauern window, Austria. *Contr. Miner. Petrol.*, 57, 99-117.

Schiffman, P., Bird, D. K., & Elders, W. A., 1985. Hydrothermal mineralogy of calcareous sandstones from the Colorado River delta in the Cerro Prieto geothermal system, Baja California, Mexico. *Miner. Mag.*, 49, 435-449.

Wolery, T. J., 1983. EQ3NR, a computer program for geochemical aqueous speciation-solubility calculations: User's guide and documentation. *Lawrence Livermore Laboratory*, UCRL-53414.

Water-Rock Interaction, Kharaka & Maest (eds) © 1992 Balkema, Rotterdam. ISBN 90 5410 075 3

Dependence of hydrothermal alteration on rock texture and microporosity in mafic crystalline rocks

Mark E. Brandriss & Dennis K. Bird
Stanford University, Calif., USA

James R. O'Neil
University of Michigan, Mich., USA

ABSTRACT: Layered gabbros of the Kap Edvard Holm Intrusive Complex (East Greenland) contain xenoliths of metabasalt country rock. The xenoliths are fine-grained equigranular granoblastic hornfelses with sparse traces of secondary alteration minerals; the gabbros are medium-grained, heterogranular and sub-idiomorphic, with secondary minerals common along grain boundaries and microfractures. Oxygen isotope fractionations between coexisting plagioclase and pyroxene in the xenoliths reflect equilibration at temperatures of approximately 1000°C, indicating that the xenoliths experienced very little retrograde alteration during subsolidus cooling of the intrusion. In contrast, gabbros surrounding the xenoliths have disequilibrium isotopic fractionations, reflecting subsolidus reaction with a low-^{18}O meteoric-hydrothermal fluid. These mineralogic and isotopic data suggest that fluid infiltration at the scale of individual mineral grains was more pervasive in the gabbro than in the xenoliths, probably as a consequence of rock texture and its effects on microporosity.

1 INTRODUCTION

Emplacement of plutons into the Earth's crust is accompanied by an input of thermal and mechanical energy that often results in the generation of large-scale hydrothermal systems (Norton, 1984; Criss and Taylor, 1986). These systems enhance metamorphic recrystallization and metasomatic alteration of country rocks and plutons, with complex fracture systems providing the main conduits for fluid flow. The sizes, spacings and geometries of interconnected fractures will therefore largely control the spatial distribution of mineralogic alteration and hydrothermal metasomatism in contact aureoles and solidified plutons. Although much progress has been made in modeling large-scale permeability and fluid flow in fractured crystalline rocks, the nature of permeability and fluid flow at the small scale, and particularly at the grain-size scale, is poorly understood (Brenan, 1991). This study demonstrates that rock textures can control the nature of small-scale fracturing and hydrothermal alteration in mineralogically similar but texturally distinct granular mafic rocks from a single locality in a cooling pluton.

2 GEOLOGY AND PETROGRAPHY

The Early Tertiary Kap Edvard Holm intrusive complex of East Greenland consists predominantly of layered gabbros emplaced near a regional unconformity between Precambrian gneisses and overlying Tertiary basalts (Elsdon, 1969). Near the western margin of the complex, the gabbros contain large tabular xenoliths of hornfelsed metabasaltic country rock up to five meters thick and >100 meters long. The gabbros contain plagioclase, clinopyroxene, olivine and ilmenite, are weakly foliated and modally layered, and have medium-grained, subidiomorphic heterogranular textures (Fig. 1a). The xenoliths are fine-grained granoblastic hornfelses consisting of clinopyroxene, olivine, plagioclase, orthopyroxene and ilmenite (Fig. 1b).

All gabbros and xenoliths from the study area contain secondary alteration minerals associated with hydrothermal veins. Macroscopic veins are thin (<0.5mm), are continuous from tens of centimeters to at least several meters, and are typically spaced from centimeters to tens of centimeters apart. In veins and vein selvedges, the common secondary minerals are chlorite, actinolite, anthophyllite, hornblende, magnetite and talc.

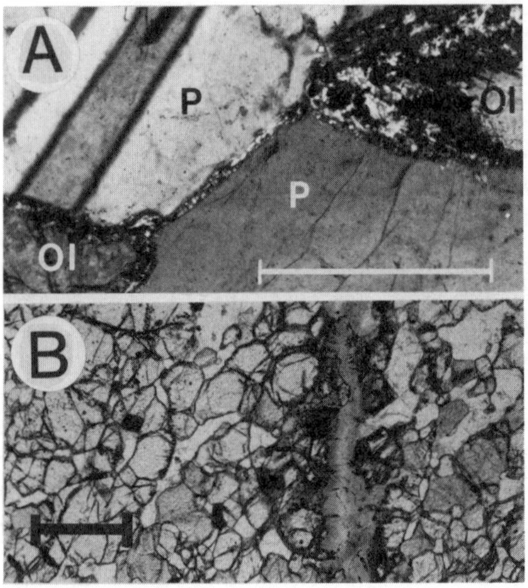

FIGURE 1: Photos from a single thin section, on opposite sides of a contact between metabasalt xenolith and gabbro. Scale bars = 0.4mm.
A) Gabbro; two olivine grains (ol) are partially replaced by magnetite + anthophyllite, and are linked by a grain–boundary veinlet containing amphibole + chlorite. P=plagioclase.
B) Macroscopic chlorite vein in xenolith.

These minerals formed over a temperature range of approximately 400–800°C (Bird et al., 1988).

Within the xenoliths, secondary minerals are confined mostly to macroscopic veins and their narrow selvedges (Fig. 1b). In contrast, secondary minerals within the gabbros are found not only within macroscopic veins and their distinct selvedges, but also as patches and as fillings in irregular networks of microcracks; secondary minerals are ubiquitous (though volumetrically minor) in the rock. Patches of alteration include cloudy feldspars and pyroxenes as well as igneous grains partially or completely replaced by secondary minerals. Zones of patchy alteration are commonly linked to one another or to macroscopic veins by veinlets filled with chlorite and actinolite (Fig. 1a). These veinlets are common along grain boundaries, and also along twin planes and cleavages in feldspar. The distribution of veinlets throughout the gabbro indicates that microcracks previously formed an interconnected, fluid–filled porous network at the scale of individual mineral grains. In the hornfelsed xenoliths, in contrast, the distribution of secondary minerals indicates that hydrothermal fluids were mostly confined to widely–spaced throughgoing fractures.

3 STABLE ISOTOPE DATA

Oxygen and hydrogen isotope analyses indicate that the mineralogic alteration involved meteoric–hydrothermal fluids with low $\delta^{18}O$ and low δD. Vein–filling actinolite has δD of −120 per mil (relative to SMOW), consistent with formation in the presence of an early Tertiary meteoric–hydrothermal fluid with $\delta D \approx -100$ (Taylor and Forester, 1979). This δD requires a meteoric water with $\delta^{18}O \approx -14$ (Craig, 1961). Hydrothermal alteration by low–$\delta^{18}O$ meteoric waters resulted in substantial ^{18}O depletion of many gabbros and basalts in the North Atlantic Tertiary igneous province (Criss and Taylor, 1986), and similar ^{18}O depletions are characteristic of the Kap Edvard Holm gabbros and their basaltic country rocks (Fehlhaber and Bird, 1991). New data, summarized below, indicate that retrograde hydrothermal alteration during cooling of the intrusive complex was more pervasive in the gabbros than in adjacent hornfelsed basaltic xenoliths.

3.1 Oxygen isotopic fractionations

Oxygen isotopic analyses of plagioclase and pyroxene separates from basaltic xenoliths and gabbros of Kap Edvard Holm are shown in Figure 2. Point A in this diagram is assumed to represent the isotopic composition of unaltered basalts and gabbros, in accordance with the observation that the great majority of mantle–derived mafic magmas have $\delta^{18}O = 5.0 - 7.0$ (Kyser, 1986). Because water exchanges oxygen with plagioclase much more readily than with pyroxene at hydrothermal temperatures (e.g. Taylor and Forester, 1979), subsolidus hydrothermal alteration of the gabbros resulted in disequilibrium ^{18}O fractionations between coexisting plagioclase and pyroxene (i.e. path A–B, Fig. 2).

The basalt xenoliths, in contrast, have plagioclase–pyroxene ^{18}O fractionations that are very near to equilibrium values for a temperature of about 1000°C (Friedman and O'Neil, 1977). However, the xenoliths are substantially depleted in ^{18}O relative to fresh basalts. This indicates that the basalts were hydrothermally altered prior to stoping: if hydrothermally altered, ^{18}O–depleted

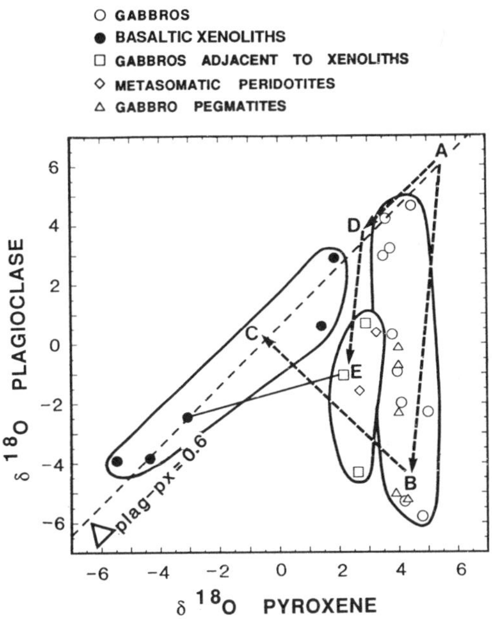

FIGURE 2: Measured $\delta^{18}O$ of coexisting plagioclase and pyroxene. The thin dashed line represents equilibrium fractionation at $\sim 1000°C$. The thin solid line joins samples taken within 2 centimeters on either side of a xenolith/gabbro contact. Gabbros <0.5 meter from xenoliths, and peridotites of metasomatic origin, appear to have assimilated small amounts of low-^{18}O xenolithic material (i.e. along path A–D) before being hydrothermally altered during subsolidus cooling (i.e. path D–E). Some gabbro and gabbro pegmatite data are from Fehlhaber and Bird (1991).

basalts were stoped into a gabbroic magma and reheated to magmatic temperature, they would be expected to recrystallize such that equilibrium between plagioclase and pyroxene would be restored but at lower whole-rock $\delta^{18}O$ than the original fresh basalts. The data therefore indicate that the xenoliths were hydrothermally altered prior to stoping, and followed exchange paths similar to A–B–C in Figure 2 (with plagioclase and pyroxene probably altered to albite–epidote hornfels assemblages somewhere along path A–B).

Unlike the gabbros, most xenoliths have positive plagioclase–pyroxene fractionations indicating that they underwent little or no isotopic exchange with hydrothermal fluids during subsolidus cooling of the intrusion. The isotopic data are therefore consistent with the very limited amount of secondary mineralogic alteration observed in the xenolith hornfelses. In contrast, the substantial ^{18}O depletions in many gabbro plagioclases are consistent with the pervasive grain-scale mineralogic alteration described earlier.

4 EFFECTS OF MICROFRACTURES

The differences in alteration between xenoliths and gabbros appear to be related to different distributions of small-scale tensional fractures that formed during subsolidus cooling and thermal contraction of the plutonic complex. Microfracturing occurs when the local stress exceeds the local strength of the rock; cracks therefore tend to form along planes of weakness such as grain boundaries, cleavage planes, and twin lamellae (Kranz, 1983). With the exception of macroscopic veins, the distributions of hydrous secondary minerals indicate that fracturing within the gabbros and xenoliths was controlled by these grain-sized planes of weakness. However, only within the coarser-grained gabbro did these microcracks coalesce to form interconnected porous networks capable of being infiltrated by hydrothermal fluids.

The link between small structural weaknesses and hydrothermal alteration patterns may be supplied by the Griffith model for tensile fracturing, which proposes that the stress concentration at the tip of an elliptical opening-mode crack increases with crack length (Griffith, 1921). This relationship suggests that microcracks along the surfaces of large grains would have greater tendencies to lengthen and interconnect with other cracks than would microcracks at the surfaces of smaller grains (Brace, 1961). Grain shape could also be a factor: the equant, equigranular texture of the hornfels could foster an isotropic distribution of grain boundary fractures, with little tendency for individual fractures to preferentially lengthen and interconnect with one another. These factors would tend to increase the connectivity of pores in the gabbro more than in the hornfels, thus accounting for the more pervasive grain-scale alteration in the former.

In addition to providing paths for diffusive transport of elemental and isotopic components, interconnected microfractures may have formed permeable networks that hosted intergranular fluid flow through the gabbro between macroscopic fractures. Rock texture could significantly affect the nature of this permeability. If we assume that grain boundary cracks are distributed randomly among grains during thermal contraction, then the average crack apertures should be greater in a coarse-grained rock than in an otherwise equivalent fine-grained rock. This is because the surface-to-volume ratio of grains decreases as grain size increases; a given amount of crack dilation would therefore be accomodated on a smaller crack surface area in the coarser-grained rock. This increase in average fracture aperture would tend to increase permeability, because the permeability of a planar fracture is roughly proportional to the cube of the aperture (Norton and Knapp, 1977). The lower tortuosity of flow paths in the coarser-grained rock would also favor higher permeability and provide shorter diffusion pathways.

5 CONCLUSIONS

Mineralogic, petrographic and isotopic evidence indicates that hydrothermal fluid infiltration was more pervasive in the Kap Edvard Holm gabbros than in hornfelsed basalt xenoliths within the gabbros. Xenoliths underwent less mineralogic and isotopic alteration than did gabbros only centimeters away. Macroscopic hydrothermal veins are common in both lithologies; only in the gabbros, however, is there significant alteration within the rock between these veins. Therefore, although outcrop-scale permeabilities were probably controlled by large open fractures (Norton and Knapp, 1977), the extents of bulk rock alteration were largely dependent on the geometries of smaller-scale networks of interconnected micropores. The geometries of these networks depended, in turn, on properties such as grain size, grain shape and textural isotropy. This demonstrates that rock textures and microstructures can substantially affect relationships between large-scale permeability and chemical alteration during hydrothermal circulation in fractured crystalline rocks.

REFERENCES

Bird, D.K., C.E. Manning & N.M. Rose 1988. Hydrothermal alteration of Tertiary layered gabbros, East Greenland. *Am. J. Sci.* 25: 405–457.

Brace, W.F. 1961. Dependence of fracture strength of rocks on grain size. *Bulletin of the Mineral Industries Experiment Station* no. 76: 99–103.

Brenan, J.M. 1991. Development and maintenance of metamorphic permeability: implications for fluid transport. *In* Kerrick, D.M. (ed.), *Contact metamorphism*: 291–319. Washington D.C.: Mineralogical Society of America.

Craig, H. 1961. Isotopic variations in meteoric waters. *Science* 133: 1702–1703.

Criss, R.E. & H.P. Taylor Jr. 1986. Meteoric-hydrothermal systems. *In* Valley, J.W., H.P. Taylor Jr. & J.R. O'Neil (eds.), *Stable isotopes in high temperature geological processes*: 373–424. Washington D.C.: Mineralogical Society of America.

Elsdon, R. 1969. The structure and intrusive mechanism of the Kap Edvard Holm layered gabbro complex, East Greenland. *Geol. Mag.* 106: 46–56.

Fehlhaber, K. & D.K. Bird 1991. Oxygen -isotope exchange and mineral alteration in gabbros of the Lower Layered Series, Kap Edvard Holm Complex, East Greenland. *Geology* 19: 819–822.

Friedman, I. & J.R. O'Neil 1977. Data of geochemistry, sixth edition: Chapter KK. Compilation of stable isotope fractionation factors of geochemical interest. *Geological Survey Professional Paper 440-KK*. Washington D.C.: United States Government Printing Office.

Griffith, A.A. 1921. The phenomena of rupture and flow in solids. *Phil. Trans. Royal Soc. Lon., Ser. A*, 221: 163–198.

Kranz, R.L. 1983. Microcracks in rocks: a review. *Tectonophysics* 100: 449–480.

Kyser, T.K. 1986. Stable isotope variations in the mantle. *In* Valley, J.W., H.P. Taylor Jr. & J.R. O'Neil (eds.), *Stable isotopes in high temperature geological processes*: 141–164. Washington D.C.: Mineralogical Society of America.

Norton, D.L. 1984. Theory of hydrothermal systems. *An. Rev. Earth Planet. Sci.* 12: 155–177.

Norton, D.L. & R.B. Knapp 1977. Transport phenomena in hydrothermal systems: nature of porosity. *Am. J. Sci.* 277: 913–936.

Taylor, H.P. Jr. & R.W. Forester 1979. An oxygen and hydrogen isotope study of the Skaergaard intrusion and its country rocks: a description of a 55 m.y. old fossil hydrothermal system. *J. Petrol.* 20: 355–419.

Water-Rock Interaction, Kharaka & Maest (eds) © 1992 Balkema, Rotterdam. ISBN 90 5410 075 3

A revised model for the formation of dequartzified and alkalinized granites (episyenites)

C.Casquet, J.M.Caballero & C.Galindo
Departamento de Petrología, Facultade Ciencias Geológicas, Universidad Complutense Madrid, Spain

F.Tornos
Instituto Tecnológico Geominero de España, Madrid, Spain

ABSTRACT: Dequartzified granites (episyenites) are interpreted as resulting from reaction of different types of granitiods with a simple low-saline fluid equilibrated at depth with regional solid adamellites. Sudden decompression, a P-T evolution into the region of retrograde solubility of silica and a high fluid/rock ratio are considered the main variables involved in the genesis of these rocks. Different amounts of alkali exchange along the channelway walls could explain the observed differences in the relative contents of albite and K-feldspar.

1.INTRODUCTION

Episyenites are dequartzified and alkalinized rocks that form by reaction of crystalline rocks, granitoids namely, with hydrothermal fluids. They are relatively common in the Sierra del Guadarrama, to the north of Madrid, which is a part of the Iberian Hercynian Basement.

Most of the episyenites found here are albite rich, although a few microcline-rich varieties can be found. Three types of the albite-rich variety can be distinguished on the basis of mafic minerals: 1) Hedenbergitic, 2) Epidote-Amphibolitic (hastingsite to hornblende), and 3) Chloritic (oxidized Fe-rich chlorite). Petrographic continuity exists between them. Temperature seems to be the main variable controlling the petrographic diversity, ranging from about 600°C in the first type, down to 350-400°C in the third. Microcline-rich episyenites are mineralogically simpler, with microcline and biotite as the dominant minerals (Caballero et al., 1992); their temperature of formation is close to 450-400°C. On the basis of fluid inclusion studies, pressures of formation (P_{fluid}) were mostly in the range 0.5-1Kb.

Episyenites are found as swarms of lens-shaped bodies that follow two structural, almost orthogonal directions: an E-W main one and a N-S secondary one. Although they preserve the original textures of the host granitoid to the naked eye, a tectonic fabric with cataclastic and recrystallization textures is fairly common. Structural analysis suggest that episyenites formed along bands of pervasive fracturing during an extensional phase in agreement with conclusions by Lepinasse and Pecher (1986). Episyenitization is a late phenomenon, as it affects all the igneous rocks exposed in the area.

Zoning within individual episyenites, is very poor or absent. However, sometimes an intermediate zone with quartz enrichment is found between the episyenite s.s. and the host granite. The outer fronts of the alteration band are always sharp.

2. GEOCHEMICAL AND CHRONOLOGICAL CONSTRAINTS

The most remarkable features of episyenites are a small content of SiO_2 (58-67% with a maximum between 60-63%), a high content in Al_2O_3 (15-23%, with a maximum between 19-21%) owed to the increase in feldspars and formation of peraluminous minerals like chlorite, and a high content of alkalies ($Na_2O + K_2O = 10\text{-}12\%$). REE, Y and Nb display a very uniform behavior. All the REE increase in most cases by a factor of 1 to 2 with regard to the protoliths (values of 2.5 - 4 have locally been found). La/Lu varies slightly while the Eu anomaly of the granite is preserved. The significant increase in the REE contents suggest that high fluid/rock ratios were involved; in fact, in moderately alkaline fluids like the ones expected here (see below) concentrations of REE are very low (10^{-6} to 10^{-5} times chondritic, Michard, 1989).

K-Ar dating carried out on minerals (amphibole and biotite) vary from 260 to 220 Ma.(Caballero et al., 1992). These ages are younger than the youngest granite dated in the area (275 Ma., Ibarrola et al. 1988). Rb-Sr internal isochrons (minerals + rock, table 1) give ages of 275 to 270 Ma. suggesting that some modification of the K-Ar system took place. As it can be seen, the initial $^{87}Sr/^{86}Sr$ values for the internal isochrons, are close to the calculated ones for the regional adamellites for the reported age (0.7104- 0.7113); therefore, the possibility of fluids being in equilibrium with adamellites at their sources is strengthened.

Stable isotopes (δD and $\delta^{18}O$) geochemistry (work in progress) points also to the participation

Table 1. Rb-Sr internal isochrons of three episyenites.

Age M.A.	MSWD	$^{87}Sr/^{86}Sr_i$
274±7	7.80	0.7121
275±3	0.02	0.7107
270±3	0.00	0.7125

of fluids with an important meteoric component, with no or little magmatic component. Such data agree with the conventional geochemistry and geological evidence.

3. A MODEL FOR THE GENESIS OF EPISYENITES

Any genetic model for the episyenites must take into account the following geological constraints: 1) they formed along pervasive fracturing zones, and 2) fluids were simple low-saline meteoric brines with "peak" temperatures ranging from 600°C to 350-400°C. The model must account for the two basic features of these rocks, i.e. the leaching of quartz and the introduction of alkalies, with precipitation of albite and/or microcline.

An interpretation is thus envisaged in which fluids resident at different depths and temperatures, in equilibrium with regional adamellites move quickly upwards due to the sudden opening of deep microfracture zones. Changes in the chemical properties of this fluid as a consequence of P-T changes induced by fracturing and movement along the conduit, with little exchange with the traversed rocks, are thought to be responsible for the formation of episyenites.

In order to check this possibility a chemical model of the fluid-rock interaction in the simple system SiO_2 - Al_2O_3 - Na_2O - K_2O - H_2O - HCl, has been envisaged. Chemical codes SOLVEQ and CHILLER (Reed and Spycher, 1988) have been employed. As the interaction took place at temperatures and pressures higher than those allowed for in the SOLTHERM database, new expressions for solubility and dissociation constants up to 2 Kb and 500°C were calculated by regression of corresponding values obtained with SUPCRT92 program (Johnson et al., 1992) or as described in Tornos et al. (1991). Activity coefficients of charged and neutral species were calculated as described in Helgeson et al. (1981) and Oelkers and Helgeson (1991), respectively.

A low-saline fluid ($m\Sigma Cl=0.4$) is first reacted with a muscovite bearing adamellite (to initially buffer the pH), with plagioclase, K-feldspar and quartz ($a_{albite} \approx a_{orthose} = 0.7$ and $a_{muscovite} = 0.8$). Initial conditions, i.e. before fracturing, are estimated at $P_f = 2$ Kb and T = 500°C; calculated pH is 5.2. The fluid is then "separated" from the rock and allowed to re-equilibrate alone at $P_f = 0.5$ Kb and T = 450°C, simulating a first step of adiabatic thermal expansion by the sudden pressure release. A strong pH increase takes place (≈ 7) because of the expected decrease of the ionic strength. Subsequently, the fluid is allowed to cool under an almost isobaric (P_f) trend to 420°C and reacted again with an adamellite. The pH values decrease again because of the increase of HCl dissociation towards lower temperatures.

Fig. 1, shows the composition of the reaction products in terms of grams of adamellite involved in the reaction. It can be seen that under high fluid/rock ratios alkali feldspar precipitates from the fluid, simultaneously the other minerals are dissolved from the rock. In fact, mass balance considerations suggest that high fluid/rock ratios had to be involved in the formation of episyenites. As the fluid/rock ratio decreases the fluid becomes saturated first with muscovite followed by quartz and albite. At this stage, the pH of the fluid varies from about 6 at high fluid/rock ratios to a buffered value of 5.8 when quartz joins muscovite and K-feldspar as stable phases.

This chemical simulation although very simplistic, confirms that no unusual, alkaline and silica undersaturated fluids must be involved at the source. Precipitation of an alkali feldspar is the consequence of the increase of pH following decompression, other minerals being dissolved under high fluid/rock ratios as is expected in this case. Quartz dissolution is also favored by P_{fluid} - T values laying in the field of retrograde solubility of silica as suggested by Leroy (1978) and Tornos (1990). This strongly backs the geological interpretation.

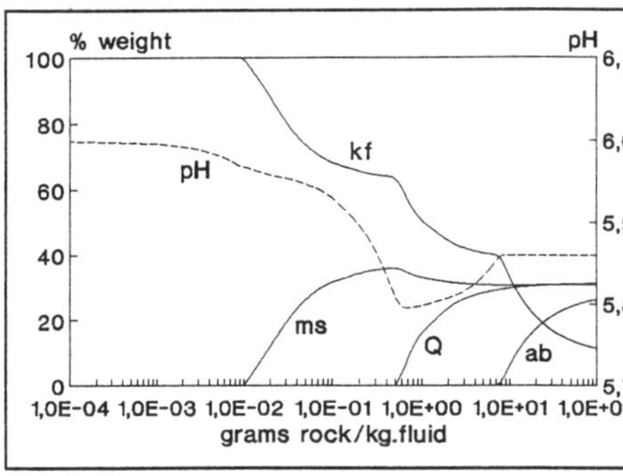

Fig.1. Hydrothermal minerals (wt%) precipitated during fluid-granite reaction at 420°C and 500 bars fluid pressure.

4. PITFALLS OF THE MODEL

The model above fails at explaining two important points: 1) the precipitation of K-feldspar instead of albite, and 2) the presence of muscovite in the chemical model which is not found in the investigated episyenites. It is also noteworthy that under ≈410°C the fluid is saturated with muscovite at any fluid/rock ratio in disagreement with geothermometry that suggests that episyenites form down to 350°C.

The first problem can be solved if we accept that reaction of the fluid with the rocks has taken place to a greater or lesser extent on its way to the surface. As a consequence, the initial Na:K ratio in the fluid (6.8) should evolve to higher values in agreement with experimental results by Orville (1963). A similar model with an Na:K ratio greater than 10 leads to the formation of albite-rich episyenites (Tornos et al., 1991).

Formation of muscovite is probably a consequence of the simplicity of the chemical model system. Minerals like epidote and chlorite are abundant in intermediate to low temperature episyenites. Thus, significant amounts of Al must be withdrawn from the fluid to form them, probably hindering the formation of muscovite.

5. CONCLUSIONS

Chemical modelling strengthens the idea, based on geological and geochemical arguments, of episyenites resulting from reaction of simple low-saline brines with granitoids. The main factors involved seem to be a sudden decompression of a confined deep meteoric fluid and its ascent to the surface; formation of episyenites needs fluid/rock ratios to be necessarily high. Fluid is quartz saturated during most of the P-T path to the surface, episyenites forming in the field of retrograde solubility of quartz with a pH increased by the decompression.

Acknowledgements: This work has been partially financiated by the Spanish CICYT project num.-PB88-0124.

REFERENCES

Caballero, J.M., Casquet, C., Tornos, F. & Pelli cer, M.J. 1991. Caracterizaci6n petrográfica de las episienitas de la Sierra del Guadarrama, Sistema Central Español. Bol. Soc. Esp. Min. 14: 273-284.2

Caballero, J.M., Casquet, C., Galindo, C., Gonzalez Casado, J.M., Snelling, N. & Tornos,F. 1992. Dating of hydrothermal events in the Sierra del Guadarrama, Iberian Hercynian Belt. Geogaceta, 11, (in press).

Helgeson, H.C., Kirkham, D.H. & Flowers, G.C. 1981. Theoretical prediction of the thermodynamic behavior of aqueous electrolytes at high pressure and temperature. IV: Calculation of activity coefficients, osmotic coefficients and apparent molal properties to 5 kb and 600. Amer. Jour. Sci. 281: 1241-1516

Ibarrola, E., Villaseca, C., Vialette, Y., Fuster, J.M., Navidad, M., Peinado, M., Casquet,C. 1988. Dating of Hercynian granites in the Sierra de Guadarrama (Spanish Central System). In Geología de los granitoides y rocas asociadas del Macizo Hespérico, ed.Rueda, Madrid, p.377-383

Johnson, J.W., Oelkers, E.H., Helgeson, H.C. 1992. SUPCRT92: A software package for calculating the standart molal thermodynamic properties of minerals, gases, aqueous species and reactions from 1 to 5000 bars and 0° to 1000°C. Comp.& Geosc. (in press)

Lepinache, M. & Pecher, A. 1986. Microfractu ring and regional stress field: a study of the preferred orientations of fluid-inclusion planes in a granite from the Massif Central, France. Jour. Struct. Geol. 8-2: 169-180.

Leroy, J. 1978. The Margnac and Fanay U depo sits of the La Crouzielle district (Western Massif Central, France): Geologic and fluid inclusions studies. Econ. Geol. 73: 1611-1634

Michard, A. 1989. Rare earth element systematics in hydrothermal fluids. Geochim. Cosmochim. Acta. 53: 745-750.

Oelkers, E.H. & Helgeson, H.C. 1991. Calculation of activity coefficients and degrees of formation of neutral ion pairs in supercritical electrolyte solutions. Geochim. Cosmochim. Acta. 55:1235-1251

Orville, P.M. 1963. Alkali ion exchange between vapor and feldspar phases. Amer. Jour. Sci. 261: 201-237.

Reed, M.H. & Spycher, N.F. (1988). Chemical modeling of boiling, condensation, fluid-fluid mixing and water rock reaction using programs CHILLER and SOLVEQ. Abs. Amer. Chem. Soc. Sym., Chem. Mod. Aqueous systems: 25-30.

Tornos, F. 1990. Los skarns y mineralizaciones asociadas del Sistema Central Español. Modelo de caracterizaci6n petrol6gica, geoquímica y metalogénica. Doctoral Thesis, Universidad Complutense de Madrid, 487 pp.

Tornos, F., Casquet, C., Caballero, J.M., Galindo, C. 1991. A chemical model for the genesis of episyenites and superimposed ores linked with phyllic alteration. In M.Pagel & J.L.Leroy (eds.), Source, Transport and Deposition of Metals, p.143-146.

Water-Rock Interaction, Kharaka & Maest (eds) © 1992 Balkema, Rotterdam. ISBN 90 5410 075 3

Origin of the Maqarin, Jordan hyperalkaline groundwaters: Isotopic and geochemical evidence for in situ combustion, calcination and recarbonation of bituminous marls

Ian D. Clark – *Ottawa-Carleton Geoscience Centre, University of Ottawa, Ont., Canada*

Hani N. Khoury & Elias Salameh – *Department of Geology and Mineralogy, University of Jordan, Amman, Jordan*

Peter Fritz & Harald K. Seidlitz – *GSF-Institute für Hydrologie, Neuherberg, Germany*

Tony E. Milodowski – *British Geological Survey, Keyworth, Nottingham, UK*

ABSTRACT: Spontaneous combustion of bituminous marls is implicated in the origin of high temperature "contact" metamorphic minerals found in brecciated zones at Maqarin, northern Jordan, although only retrograde alteration and dissolution persist today. Hyperalkaline groundwaters (pH 12.5) discharging from the brecciated zones show two distinct geochemical facies, representing early and later stage dissolution within two discrete flow systems.

The high temperature history of the system is recorded by isotopes in secondary carbonate and sulphate. Values for $\delta^{34}S$ fall within the range for Mesozoic hydrocarbon, and the positive $\delta^{18}O_{SO_4}$ values suggest oxidation of reduced sulphur from the bitumen, with the participation of atmospheric O_2. Mineralogical evidence and stable isotope information acquired through Nd-YAG laser micro-sampling and standard methods suggest that calcination with subsequent recarbonation has been extensive. Partially calcined "marble" shows ^{13}C and ^{18}O depletions characteristic of high-temperature metamorphic carbonates. Secondary carbonates are further depleted in ^{13}C, documenting recarbonation in a cooling environment with the participation of organic C from combusted bitumen.

1 INTRODUCTION

Hyperalkaline groundwaters occur in a variety of geological settings around the world, and make fascinating studies of extremities in rock-water interaction (e.g. Barnes and O'Neil, 1969; Barnes et al., 1978; Kraynov et al., 1969). However, in the Maqarin region of northern Jordan occur possibly the most highly alkaline natural groundwaters reported in the literature, discharging with pH's exceeding 12.5.

First described during site investigations for the Jordan-Syria Unity Dam project (Barnes et al., 1982; Khoury et al., 1985; Khoury et al., 1992), these solutions discharge from high temperature metamorphic zones within Maestrichtian-age bituminous marls (Fig. 1).

In the metamorphic zone, high temperature assemblages include spurrite [$Ca_5CO_3(SiO_4)_2$], with wollastonite [$CaSiO_3$] and calcite, plus minor diopside, anorthite, brownmillerite, ellestadite, fluoroapatite, graphite, Cu-sulfide, Cu-selenide, and Ca and Fe oxides.

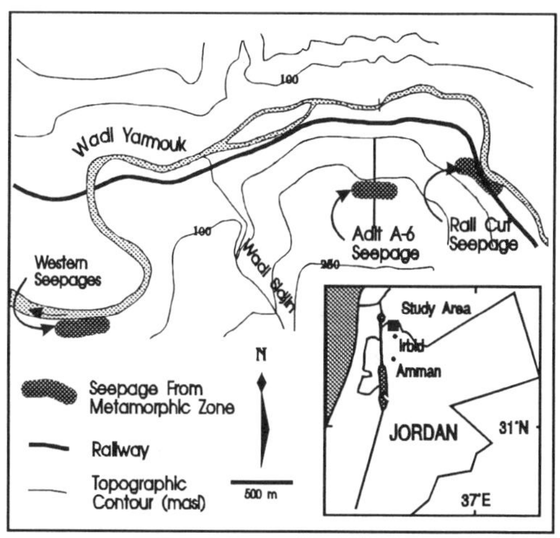

Fig. 1 The Maqarin area of northern Jordan, showing discharge sites of hyperalkaline waters.

Low temperature, retrograde mineralization occurs within the brecciated zone matrix, fractures and cavities. Dominant minerals in this suite include secondary calcite, portlandite [$Ca(OH)_2$], thaumasite [$Ca_6H_4(SiO_4)_2SO_4(CO_3)_2 26H_2O$], ettringite [$Ca_6(Al(OH)_6)_2(SO_4)_3 \cdot 26H_2O$], plus tobermorite and a host of pure and solid-solution sulphate phases including gypsum, bassanite, ellestadite, barite and celestite (Khoury and Nassir, 1982; Milodowski et al., 1991).

Similar high and low temperature mineral assemblages have been observed in stratigraphic equivalents in central Jordan (Khoury and Nassir, 1982) and at other localities in Israel and the West Bank (Bentor, 1963). In all cases, the metamorphic event has been attributed to the spontaneous *in situ* combustion of bitumen in fracture-dominated zones. That spontaneous combustion can occur is documented by a dump of burning waste rock excavated from a dam diversion tunnel at Maqarin.

Sites in central Jordan have been dated to early Pleistocene (Clark et al., 1991). At Maqarin, the persistence of retrograde alteration to present time suggests that metamorphism at this site is recent.

2 GROUNDWATERS

Seepages occur within Adit A-6 constructed in 1980, along a railway rock face cut around 1910 for the Damascus-Palestinian railway, and naturally along local wadis (Fig. 1). Two distinctly different geochemical facies exist, which relate to early and more advanced dissolution of secondary minerals (Fig. 2). The western seeps have the highest alkalinities (pH 12.5) and much higher mineralization. Their chemistry is controlled largely by equilibrium with portlandite and sulphate minerals in the alteration zone (Table 1). The eastern seeps, by contrast, are less mineralized and have lower pH values (12.3). These groundwaters are undersaturated with $Ca(OH)_2$ and sulphates, suggesting more advanced dissolution in the alteration zone. Minor and trace element concentrations follow a similar pattern between the eastern and western groundwater systems (Fig. 2). High alkali and selenium concentrations are attributed to alteration of CuKNa-selenides found in the primary metamorphic suite. Dissolved carbonate contents (1.2 to 2.2 mg/l) are controlled by calcite solubility. The P_{CO_2} of these waters (10^{-11}) is nearly eight orders of magnitude less than that of the atmosphere. Rapid uptake of $CO_{2(atm)}$ during discharge has led to the precipitation of impressive travertine formations.

Stable isotope data provide some insights to the origin of and processes affecting these highly

Table 1 Saturation Indices for portlandite and sulphate minerals.

	MQ-2 (Eastern)	MQ-5 (Western)
logSI[$Ca(OH)_2$]	-0.46	-0.04
logSI[$CaSO_4 2H_2O$]	-0.71	-0.01
logSI[$BaSO_4$]	-0.88	0.13
logSI[$SrSO_4$]	-0.07	-0.09

alkaline waters. In Fig. 3, it is clear that they are considerably more positive than local rainfall, demonstrating that either evaporation occurs during recharge or that subsurface processes are important. Evaporation effects are observed in shallow groundwaters in potential recharge areas between Irbid and the Yarmouk river valley (Bajjali, 1990). However, in highly alkaline groundwaters, strong H_2O -hydroxide fractionation during hydration reactions may impart ^{18}O enrichments on H_2O. Dakin et al. (1983) noted an ^{18}O enrichment of 1 to 2‰ in water during hydration of cement (pH >10). Although the controlling reactions need clarification this may account for the enrichments observed in Fig. 3.

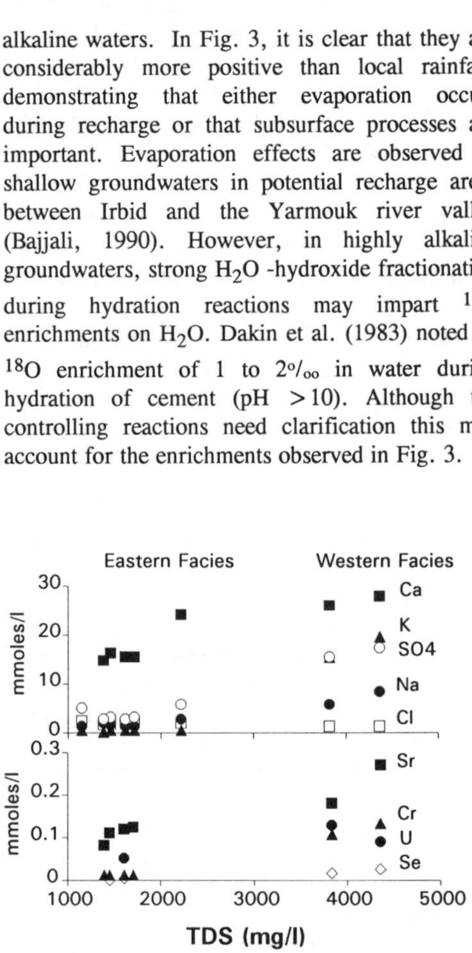

Fig. 2 Correlation of major and minor species with total dissolved solids. Differences between eastern and western facies geochemistry reflect early and later stage dissolution of secondary minerals.

A second observation emerges from the isotope data, which supports geochemical evidence for two separate flow systems. A clear distinction is observed in Fig. 3 between the samples collected from the eastern discharges versus the western springs along the Yarmouk River. With the exception of one sample (3 TU), the hyperalkaline groundwaters are tritium free, indicating that they have mean residence times of at least several decades.

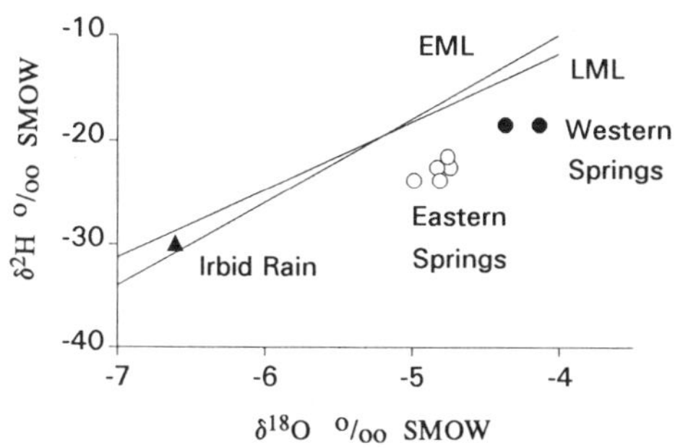

Fig. 3 $\delta^{18}O$ vs δ^2H relationship for meteoric waters and groundwaters at Maqarin. The local meteoric water line is defined as $\delta^2H = 6.5\ \delta^{18}O + 14$ from precipitation at Irbid (Bajjali, 1990). The Eastern Mediterranean Line (EML) is defined by Gat and Carmi (1970).

3 ISOTOPES IN CARBONATE

The primary calcite in the host biomicritic marl has $\delta^{13}C$ values in the order of 0.0 ± 1‰. Secondary phases are considerably depleted from this value (Fig. 4) due to two principal mechanisms: (i) high temperature decarbonation, and (ii) secondary (*in situ*) recarbonation reactions. Isotope data have been acquired using a Nd-YAG laser microprobe. Details on methodology are described by Clark et al. (1992).

High temperature metamorphism of limestone yields CO_2 gas enriched in both ^{13}C and ^{18}O, and imparts a reciprocal depletion on the carbonate phase (e.g. Valley, 1986; Bottinga, 1968). This trend is observed in samples collected from the burning waste-rock at Maqarin (Fig. 4). In the subsurface, laser microprobe data of marble from the high temperature mineral suite fall on the same "decarbonation" trend line. However, these data contrast with the considerably more ^{13}C-depleted secondary carbonates formed by recarbonation of primary metamorphic phases.

Early recarbonated phases include secondary calcite precipitated within the shrinkage cracks of the high temperature marble and a matrix of calcite, portlandite and ettringite with vestiges of primary metamorphic minerals. Calcite in shrinkage cracks has highly depleted $\delta^{13}C$ values due to incorporation of CO_2 released during combustion of bitumen ($\delta^{13}C$ = -26‰), possibly mixed with a minor component of isotopically enriched metamorphic CO_2. The low $\delta^{18}O$ values for these secondary calcites indicate that they formed at a relatively high temperature (>100°C), probably from condensing vapour phases following combustion (based on calculations using fractionation factors of Bottinga, 1968 and $\delta^{18}O_{H_2O} \sim 4$‰).

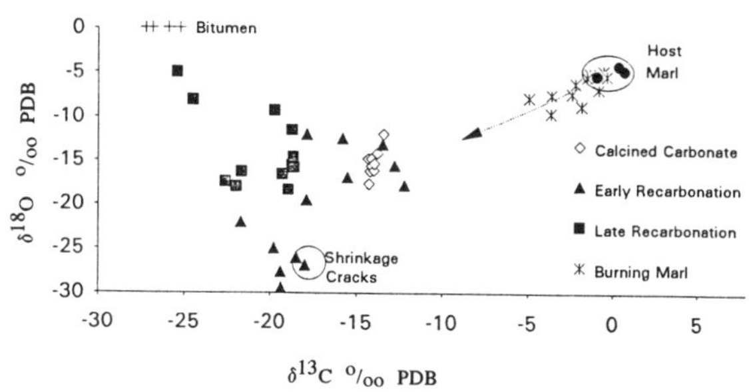

Fig. 4 Stable isotope contents for carbonate samples by laser microprobe. $\delta^{13}C$ values were within the analytical error ($\pm$ 0.1‰) of comparative samples analysed by acidification. $\delta^{18}O$ values are complicated by the complex mineralogy and are less accurate ($\pm$ 1‰).

Early matrix carbonate has evolved from high temperature calcite and spurrite precursors ($\delta^{13}C \sim -14$ and $\delta^{18}O \sim -16$‰) by incorporation of CO_2 released during combustion of bitumen. Variations in the degree of alteration, and thus of the contributions from both CO_2 and marble precursor, would account for the strong variations in both isotopes. The highly depleted $\delta^{18}O$ values in much of this early carbonate represents growth when temperatures were highest.

Late recarbonation samples represent the final phase of retrograde alteration. The matrix of this material contains fine-grained calcite and ettringite/thaumasite solid solution with younger ettringite/thaumasite veinlets. The ^{13}C content of this material is considerably more depleted than the early recarbonation phases, demonstrating essentially complete recarbonation by the CO_2 (from bitumen) dominated solution. $\delta^{18}O$ values, however, are similar to or enriched above values from the early recarbonation phases, suggesting that temperatures were decreasing as retrograde alteration proceeded.

4 ISOTOPES IN SULPHATE

Both aqueous sulphate and sulphate minerals from the alteration zones originate as organic sulphur in bitumen, as clearly shown by their $\delta^{34}S$ values (Fig. 5) which fall into the narrow -5 to 0‰ range for Mesozoic petroleum and bitumen (Krouse, 1980). The enriched $\delta^{18}O$ values are typical for sulphate which has formed through the oxidation of reduced sulphur with the participation of atmospheric oxygen. Thus, SO_4^{2-} and sulphate minerals likely originated from combustion of

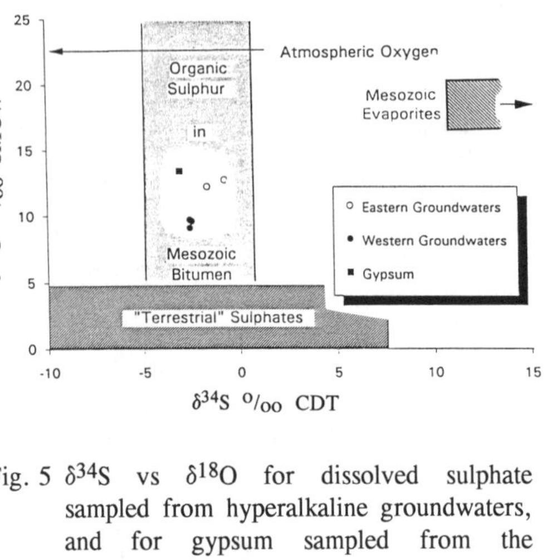

Fig. 5 $\delta^{34}S$ vs $\delta^{18}O$ for dissolved sulphate sampled from hyperalkaline groundwaters, and for gypsum sampled from the secondary alteration zone.

organic sulphur, which can reach 12% or more in the bitumen. The exact reactions which took place are difficult to establish, although a series of intermediary sulphur species, including SO_2, sulfite [SO_3^{2-}] and thiosulphate [$S_2O_3^{2-}$], are likely to have formed. Subsequent oxidation would then have occured under aqueous conditions with the participation of H_2O to produce the range of values observed.

5 SUMMARY

The stable isotopes in the hyperalkaline groundwaters and metamorphic suites serve to clarify the process of *in situ* combustion and secondary alteration of the bituminous biomicritic marls at Maqarin. The high temperatures generated in these reactions caused extensive decarbonation, leaving a prograde metamorphic assemblage dominated by isotopically depleted carbonate minerals. Organic sulphur combusted with the bitumen in this highly oxidizing environment.

Retrograde alteration followed, with a solution probably characterized by elevated concentrations of dissolved carbonate originating as CO_2 from combusted bitumen, high dissolved sulphate, and T ~ 100°C. Ensuing hydration, carbonation and sulphatization reactions led to the formation of secondary phases including the thaumasite - ettringite solid solution series and secondary calcite. Secondary carbonates have $\delta^{13}C$ values varying between organic carbon ($\delta^{13}C_{bitumen}$ ~ -26‰) and the primary metamorphic phases (~ -14‰).

The circulating groundwaters have since evolved towards low P_{CO_2} values and high pHs, representing a change from alteration/precipitation reactions to dissolution of secondary phases.

These extensive recarbonation reactions and high trace element concentrations provide an interesting natural analogue for the attenuation in a cementitious barrier of safety-relevant radionuclides and $^{14}CO_2$ from nuclear waste.

ACKNOWLEDGEMENTS

This work was funded by Ontario Hydro in collaboration with Nagra (Switzerland), NIREX (U.K.) and the British Geological Survey. Dr. Benno Reuter, Fachhochschule, Munich, put the laser used in this study at our disposal. Mr. Werner Weindorf was responsible for its testing and operation.

REFERENCES

Bajjali, W, 1990. Isotopic and hydrochemical characteristics of precipitation in Jordan. Unpublished M.Sc. Thesis, University of Jordan, Amman, Jordan, 99 pp.

Barnes, I. and O'Neil, J.R., 1969. The relationship between fluids in some fresh alpine-type ultramafics and some possible modern serpentinization, western United States. Geol. Soc. Amer. Bull. 80: 1947-1960.

Barnes I., O'Neil J.R. and Trescases J.J. 1978. Present day serpentinization in New Caledonia, Oman and Yugoslavia. Geochim. Cosmochim. Acta 42: 144-145.

Barnes, I., Presser, T., Saines, M., Dickson, P. and Koster Van Gross, A., 1982. Geochemistry of highly basic calcium hydroxide groundwater in Jordan. Chemical Geology, v. 35, pp. 147-154.

Bentor, Y.K., Gross, S. and Heller, L.,1963. High temperature minerals in non-metamorphosed sediments. Nature 199: 478.

Bottinga Y., 1968. Calculation of fractionation factors for carbon and oxygen in the system calcite - carbon dioxide - water. Journal of Physical Chemistry 72: 800-808.

Clark, I.D., H.N. Khoury, E. Salameh, P. Fritz, H.Y. Göksu, A. Wieser, C. Causse and J.-Ch. Fontes, 1991. Travertines in central Jordan: Implications for palaeohydrology and dating. Proc., Int. Atomic Energy Agency, Symp.312, Vienna.

Clark, I.D., Seidlitz, H.K., Fritz, P. Trimborn, P, Milodowski, A.E., Pearce, J.M. and Khoury, H.N., 1992. Metamorphic sequences in spontaneously combusted bituminous marls from Jordan: ^{13}C and ^{18}O systematics examined by laser microsampling. Submitted, Geology.

Dakin, R.A., Farvolden, R.N., Cherry, J.A. and Fritz, P., 1983. Origin of dissolved solids in groundwaters of Mayne Island, British Columbia, Canada. J. Hydrology 63: 233-270.

Gat, J. and Carmi, I., 1970. Evaluation of the isotopic composition of atmospheric waters in the Mediterranean Sea area. J. Geophys. Res. 75: 3039-3078.

Khoury, H.N. and Nassir, S., 1982. High temperature mineralization in Maqarin area North Jordan. N. Jb. Miner. Abh., (Schwizerbart'sche Verlagsbuchhandlung, Stuttgart),144: 197-213.

Khoury, H.N., Salameh, E. and Abdul-Jaber, Q., 1985. Characteristics of an unusual highly alkaline water from the Maqarin Area, northern Jordan. Journal of Hydrology, v. 81, pp. 79-91.

Khoury, H.N., Salameh, E., Clark, I.D., Fritz, P., Bajjali, W., Milowdowski, A.E., Cave, M.R. and Alexander, W.R., 1992. A natural analogue of high pH cement pore waters from the Maqarin area of northern Jordan I: introduction to the site. *in press*, Journal of Geochemical Exploration.

Kraynov, S.R., Mer'kov, A.N., Petrova, N.G., Baturinskaya, I.V. and Zharikova, V.M., 1969. Highly alkaline (pH 12) fluorosilicate waters in the deeper zones of the Lovozero massif. Geokhimiya 7: 791-796.

Krouse, H.R., 1980. Sulphur isotopes in our environment. In P. Fritz and J-Ch. Fontes (eds.), Handbook of Environmental Isotope Geochemistry, 1A, Elsevier.

Milodowski, A.E., Khoury, H.N., Pearce, J.M., Hyslop, E.K. and Clark, I.D., 1992. Discussion of the mineralogy, petrography and geochemistry of the Maqarin source-term rocks and their secondary alteration products. Section 4.1 in: A natural analogue study of cement-buffered hyperalkaline groundwaters and their interaction with a sedimentary host rock. Nagra NTB 91-10.

Valley, J.W., 1986. Stable isotope geochemistry of metamorphic rocks. *in* Stable Isotopes in High Temperature Geologic Processes, *ed* Paul H. Ribbe, Min. Soc. Amer. Reviews in Mineral., 16: 445-487.

Water-Rock Interaction, Kharaka & Maest (eds) © 1992 Balkema, Rotterdam. ISBN 90 5410 075 3

Hydrothermal alteration in two Palaeoproterozoic volcanic associations in northern Finland

P. Eilu
Geological Survey of Finland, Rovaniemi, Finland

ABSTRACT: Four alteration phases are recognized at the Palaeoproterozoic metavolcanic associations of Myllyvaara and Eksymäselkä: 1) metamorphic recrystallization in greenschist facies, 2) albitization, 3) carbonation, and 4) sericitization. The alteration during the first and second phases took place in subaqueous hydrothermal systems where the temperature was 300 - 380 °C, the pressure < 2 kb, and water/rock (w/r) ratio < 10. In the albitized zones, the w/r ratio has been about 10. The carbonation at Eksymäselkä took place at about 380 °C and at Myllyvaara, where it was the final alteration phase, at about 240 °C. At Eksymäselkä, carbonation was followed by sericitization. The temperature of the hydrothermal system during sericitization probably was less than 200 °C, and the pH and the a_{K+} of the fluid were low. These conditions are comparable to argillic alteration of subaerial hydrothermal systems and submarine off-axis systems.

1. INTRODUCTION

Greenstone belts composed of Palaeoproterozoic metavolcanics, metadiabases, and metasediments, are significant formations of the bedrock in northern Finland and have probably a rift related origin (Lehtonen 1989). Within these greenstone belts, there are lithological units which have often been called adinoles or albitites. They are pale in colour and mostly felsic or intermediate in their present chemical composition. These albite rich rocks are variably hydrothermally altered and can be either volcanic, subvolcanic or sedimentary in origin. The size of the altered rock units varies considerably between greenstone belts and within rock formations. The regional metamorphosis, which postdated hydrothermal alterations, has not remarkably changed the mineral composition and has not overprinted the textures of the alteration zones. Within the Lapland Palaeoproterozoic greenstone belts, two sites of albite rich rocks, Myllyvaara and Eksymäselkä (Figure 1), have been studied in detail. The general lithology of these study sites is shown in Figures 2 and 3.

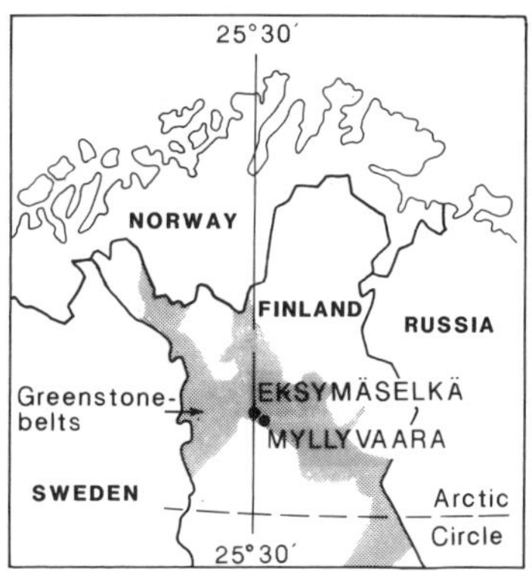

Fig. 1. Location map showing the study sites.

2. DESCRIPTION OF THE ALTERATION ZONES

2.1. *Myllyvaara*

At Myllyvaara, three zones of alteration can be distinguished in the tuffites and in the lamprophyre dykes:

1. The least altered rocks have a dark colour

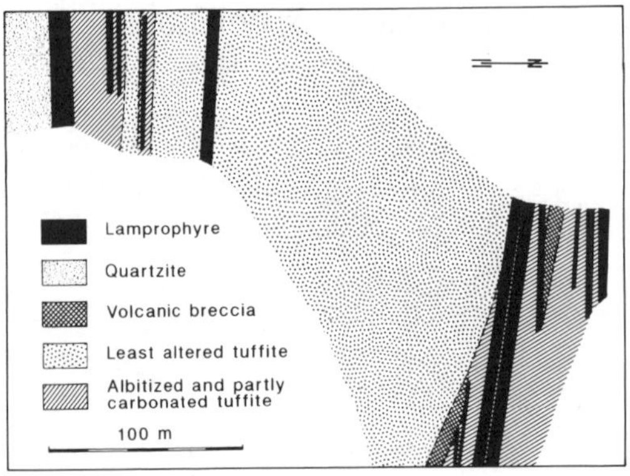

Fig. 2. Geological map of Myllyvaara.

and have no or very little brecciation (recrystallization in greenschist facies with hydration).

2. Brecciated, bleached, and albitized zones.

3. Carbonated zones which overlap the bleached zones.

Typical mineralogy and textures in Myllyvaara alteration zones are:

Zone 1: Albite + K-feldspar (only in lamprophyres) + biotite or actinolite + magnetite + sphene or rutile ± epidote. The primary magmatic textures in lamprophyres and granoblastic textures in tuffites are typically still visible.

Zone 2: Rocks gradually lose their primary textures. Lamprophyres become granoblastic. In tuffites, the bleaching begins with albitized haloes (2 to 50 cm wide) surrounding the breccia fractures. Biotite, actinolite, K-feldspar, sphene, and magnetite are replaced by albite + muscovite + quartz + rutile. Finally, muscovite is replaced by albite and quartz. The mineral association in most altered tuffites and dykes is albite + quartz + rutile.

Zone 3: Calcite and dolomite replace all the other minerals except quartz, rutile, and zircon.

2.2. *Eksymäselkä*

At Eksymäselkä four different alteration zones can be distinguished:

1. Least altered rocks with a greenschist facies mineralogy.

2. Albitized (bleached) and more or less brecciated zones.

3. Carbonated and brecciated zones.

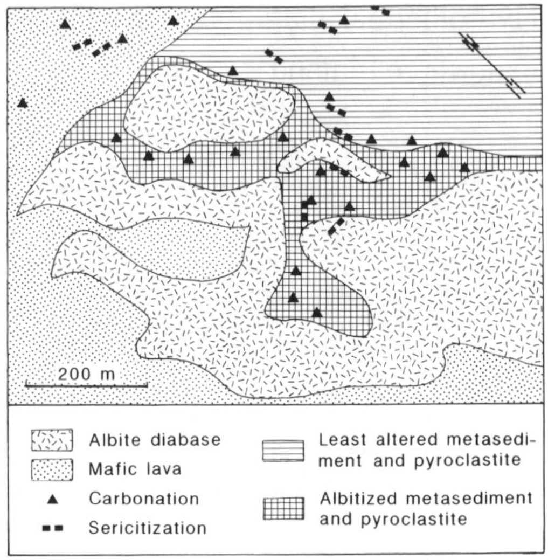

Fig. 3. Geological map of Eksymäselkä.

4. Sericitized and further brecciated types.

Alteration zones 3 and 4 often occur together and overlap the first and second alteration zone.

The mineralogy and textures in Eksymäselkä alteration zones are:

Zone 1: Albite + biotite ± actinolite or chlorite ± epidote ± magnetite + sphene or rutile. The mineral association and texture varies with respect to primary rock type. Textures are primary magmatic in the dykes and lavas and granoblastic in the metapyroclastites and metasediments.

Zone 2: Rocks lose their primary textures and become granoblastic. The typical mineral association is albite + quartz + rutile + haematite. Albite ± quartz and rutile totally replace the mafic minerals.

Zone 3: Calcite and dolomite replace all the other minerals except quartz, zircon, and rutile.

Zone 4: White mica replaces carbonates, albite, biotite, and chlorite.

3. THE CHEMICAL CHANGES

In order to determine the chemical changes which have occurred between the different alteration zones, the mass balance calculations were carried out following Gresens' (1967) method. The calculations of the Myllyvaara rocks are

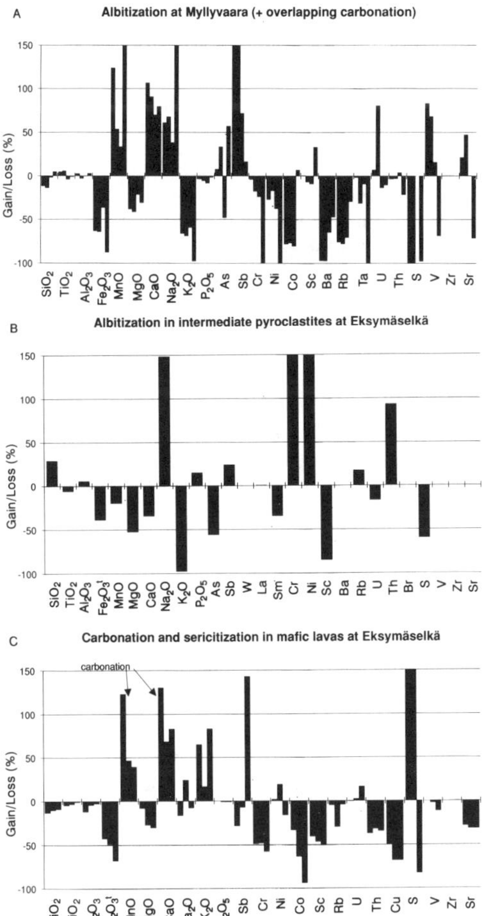

Figs. 4A - C. The chemical changes in different alteration phases at Myllyvaara and Eksymäselkä.

based on the immobility of Al, V, and Sc and suggest a minor volume increase during brecciation and bleaching. The mass balance calculations of the Eksymäselkä rocks were based on the immobility of Ti, Al, Zr, and V. They indicate a minor volume increase during brecciation and albitization for felsic metasediments and metapyroclastites but indicate an over 20 % volume increase for intermediate metasediments and metapyroclastites. The most intensively carbonated rocks have a volume increase of 30 - 50 % while sericitized rocks have a minor volume decrease. The chemical changes for different stages of alteration are shown in Figure 4.

4. THE CONDITIONS OF ALTERATION

4.1. *Recrystallization and albitization*

Albitization took place contemporaneously with or slightly postdated the greenschist facies recrystallization (spilitization). Typical features are mineral associations of greenschist facies recrystallization and albitization as well as most of the chemical changes related to albitization - especially the Na gain and loss of Fe, Mn, Mg, Ca, K, Co, Ba, Rb, and S (Fig. 4). These changes are typical for rocks altered in the upflow zone of a submarine hydrothermal system (eg. Mottl 1983; Reed 1983; Alt et al. 1986). Especially at Eksymäselkä, a submarine hydrothermal system is suggested because of the existence of tholeiitic metavolcanites and dykes, primary volcanic structures like pillow lavas, and the chemical and mineralogical changes related to greenschist facies alteration and albitization. In submarine volcanism, the alteration conditions in the greenschist facies and albitization zones are: temperature from 300 to 380 °C, pressure less than 1 - 2 kb, and w/r ratio low. In albitized zones, the conditions differed to some extent from the less altered greenschist zones: The albitized zones are more brecciated, so the effective w/r ratio was probably higher, approximately 10 (eg. Reed 1983). It is apparent that the a_{Na+} in the hydrothermal fluid was higher in albitized zones than in the less altered zones.

4.2. *Carbonation*

Carbonation at both study sites is concentrated in most albitized and brecciated zones which, due to the zones' high permeability, formed suitable channelways for CO_2-rich fluids. For temperature determination of carbonation, the mineral pair calcite - dolomite was used. Accordingly, the carbonation temperature at Eksymäselkä was about 380 °C and at Myllyvaara about 240 °C (Figure 5). At Eksymäselkä, the carbonation temperature suggests no remarkable temperature change between carbonation and albitization. At Myllyvaara, there may have been a significant temperature decrease from the albitization stage to carbonation. Because of the

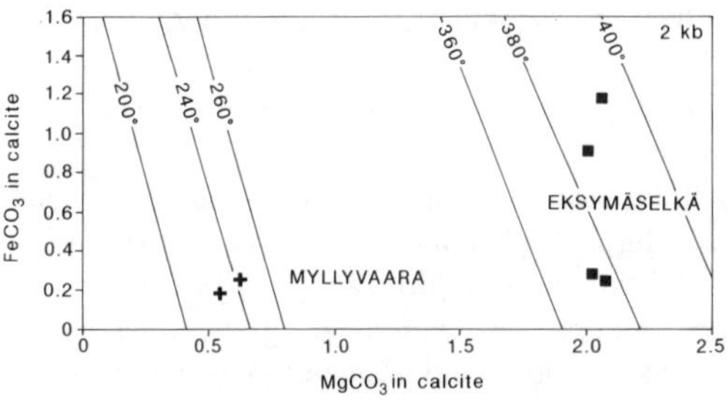

Fig. 5. Carbonate geothermometry at Myllyvaara and Eksymäselkä. The diagram is after Bickle and Powell (1977).

overprinting of the alteration zones, only some estimates about the carbonating fluids can be made: CO_2 content was moderate, not high, because no complete carbonation of the rocks existed; Ca and Mn were added to the rocks; and the S content of the fluid was low because very few sulphides and no sulphates exist in association with the carbonates.

4.3. *Sericitization*

The mobility of K is primarily a function of temperature: The K-content of a rock in a submarine hydrothermal system only rises in relative low temperatures (Mottl 1983; de Ronde & Blattner 1988). In K-metasomatosis, low pH and low a_{K+} in the fluid favor the formation of K-mica instead of K-feldspar (Reed & Spycher 1985). Hence it appears that during sericitization at Eksymäselkä, the temperature of the hydrothermal system was lower than during the earlier stages, probably less than +200 °C. The pH of the fluid was acid and the a_{K+} in the fluid was low. The alteration conditions were near those of argillic alteration in subaerial hydrothermal systems (Reed & Spycher 1985) or those of low temperature off-axis zones in submarine hydrothermal systems (Mottl 1983).

5. CONCLUSIONS

The petrographical and chemical data show that recrystallization in greenschist facies conditions and albitization took place in a submarine hydrothermal system with w/r ratio of ≤ 10. With time, these alteration conditions changed and were followed by carbonation and then by sericitization. Before sericitization, the hydrothermal system in Eksymäselkä may have changed from submarine to subaerial. Regional greenschist facies metamorphosis followed these alterations but did not change the chemical and mineralogical relationships within the rocks.

REFERENCES

Alt, J.A., Honnorez, J., Laverne, C. and Emmermann, R. 1986. Hydrothermal alteration of a 1 km section through the upper oceanic crust, Deep Sea Drilling Project Hole 504B: mineralogy, chemistry, and evolution of seawater-basalt interactions. *J. Geophys. Res.* 91: 10309-10335.

Bickle, M.J. and Powell, R. 1977. Calcite-dolomite geothermometry for iron-bearing carbonates. *Contr. Miner. Petrol.* 59: 281-292.

Gresens, R.L. 1967. Composition - volume relationships of metasomatism. *Chem. Geol.* 2: 47-65.

Lehtonen, M. (1989) The Lapland Volcanite Project. In: *S. Autio (ed.) Geological Survey of Finland, Current Resesearch 1988. Geol. Surv. Finl. Spec. Paper* 10: 19-23.

Mottl, M.J. 1983. Metabasalts, axial hot springs, and the structure of hydrothermal systems at mid-ocean ridges. *Geol. Soc. Am. Bull.* 94: 161-180.

Reed, M.H. 1983. Seawater-basalt reaction and the origin of greenstones and related ore deposits. *Econ. Geol.* 78: 466-485.

Reed, M.H. and Spycher, N.F. 1985. Boiling, cooling and oxidation in epithermal systems: a numerical modelling approach. In: *B.R. Berger & P.M. Bethke (eds.). Geology and geochemistry of epithermal systems. Rev. Econ. Geol.* 2: 249-272.

de Ronde, C.E.J. and Blattner, P. 1988. Hydrothermal alteration, stable isotopes, and fluid inclusions of the Golden Cross epithermal gold-silver deposit, Waihi, New Zealand. *Econ. Geol.* 83: 895-91.

Water-Rock Interaction, Kharaka & Maest (eds) © 1992 Balkema, Rotterdam. ISBN 90 5410 075 3

Chemical variations accompanying detachment faulting in the Whipple Mountains, California, USA

Harold Ekstrom & Jean Morrison
University of Southern California, Los Angeles, Calif., USA

ABSTRACT: Systematic major element and isotopic variations in a suite of brittly deformed mylonitic rocks show a sharp increase in SiO_2, a decrease in Al_2O_3, and a decrease in $\delta^{18}O$ from ~10‰ to 8‰ toward the Whipple detachment fault. Chemical variations in the suite correspond to the replacement of plagioclase by epidote and quartz, accompanied by increasing brecciation. It is likely that the 1.5 to 2‰ decrease in $\delta^{18}O$ with proximity to the fault may be the result of interaction with meteoric fluids at low temperatures.

1 INTRODUCTION

Lower-plate crystalline rocks in the Whipple Mountains metamorphic core complex of southeastern California (USA) have experienced deformation in both the ductile and brittle regimes. Metamorphic and plutonic rocks developed mylonitic fabrics at depths of 16±5 km (Anderson and others 1988), then were deformed in the brittle regime at higher crustal levels during decompression along the regionally extensive Whipple detachment fault. The presence of rocks deformed at both ductile and brittle crustal levels provides an opportunity to evaluate the contrasting role of fluids in the two regimes. This study evaluates the elemental and isotopic changes accompanying brittle detachment faulting in lower-plate mylonitic rocks.

2 GEOLOGY

Several compositionally distinct igneous suites have been recognized intruding Proterozoic crystalline rocks in the lower-plate of the Whipple Mountains metamorphic core complex (Anderson and Cullers 1990). Crystallization ages, in conjunction with thermobarometry, have been used to determine the uplift history of the region by Anderson and others (1988). Two stages of decompression have been recognized: (1) an early stage from ~89 to 26 Ma with uplift rates of ~0.3 mm/yr, and (2) a later accelerated stage from ~26 Ma to ~19 Ma with uplift rates of ~1.3 mm/yr. Mylonitic deformation in lower-plate rocks occurred from before 26±5 Ma to 19±2 Ma (Wright and others 1986; DeWitt and others 1986). Davis and others (1980; 1982; 1986) conclude that detachment faulting occurred between 18 and 14 Ma.

This investigation focuses on chemical changes accompanying brecciation and epidotization associated with detachment faulting in a leucocratic biotite-tonalite. The intrusion of the 89±3 Ma Whipple Wash suite (Anderson and Cullers 1990; Wright and others 1986) is the oldest Phanerozoic igneous event recognized in the lower-plate. The biotite-tonalite is one of the structurally highest members of the Whipple Wash suite, located in the upper 1.3 km of the >3.9 km-thick lower-plate mylonitic section (Anderson and Cullers 1990).

The geometry of present outcrops of the biotite-tonalite suggest that it was originally intruded as a sill (Thurn 1983). Mylonitic fabrics are not as strong as in gneissic country rocks. The upper contact of the sill has been truncated by the Whipple detachment fault. Rocks are variably brecciated down to ~60 m below the fault in this area. Linear regions of strong brecciation and epidotization extend from the fault into underlying mylonites. The material surrounding these zones is bleached white. Numerous small epidote-cored veins showing similar bleaching in host rocks are scattered throughout.

3 PETROGRAPHY

The most prominent feature of the suite is an increase in brecciation approaching the detachment fault. Furthest from the fault, rocks are coarse-grained and exhibit the weakest brecciation. Plagioclase (~An_{32}) is euhedral to subhedral with minor seritization in grain margins. Quartz occurs in elongate mosaics with individual grains displaying undulatory extinction. Chlorite pseudomorphs biotite in aggregates bending around large plagioclase crystals and quartz mosaics. Granular epidote embays plagioclase. Minor constituents include medium-grained length-parallel twinned sphene, euhedral apatite, anhedral opaque minerals, and secondary calcite in thin veins.

Closer to the fault, rocks show increasing brecciation with a finer-grained matrix and coarse plagioclase clasts. Plagioclase crystals are fractured, and display bent twin lamellae and weak polysynthetic twinning. Sericitization of plagioclase in grain margins is greatly advanced. Chlorite completely replaces biotite and decreases in abundance relative to epidote.

Within a few meters from the fault, rocks are considerably more brecciated. Strongly deformed and highly altered plagioclase clasts are set in a fine-grained matrix of quartz and angular feldspar. Coarse-grained quartz mosaics are rimmed by fine-grained quartz matrix. Chlorite is absent and epidote is minor in abundance. Fluid inclusions are abundant in quartz.

At the fault (90WP9) an aphanitic cataclasite forms the detachment surface. Brecciated material is present as clasts in a dark-brown microcrystalline matrix. Minor, small euhedral pyrite grains are present.

4 GEOCHEMISTRY

Major and trace element compositions and values of $\delta^{18}O_{SMOW}$ are listed in Table 1. The

Table 1: Chemical and isotopic compositions

	90WP 9	90WP 10	90WP 11	90WP 12	90WP 13	90WP 14
Dist. from detachment (meters)	0	3	12	25	40	50
SiO_2	81.24	75.70	60.99	61.11	64.28	62.11
Al_2O_3	5.50	12.89	19.55	19.29	18.06	17.22
TiO_2	0.31	0.21	0.37	0.49	0.40	0.58
FeO_t	6.36	1.79	3.40	4.12	3.32	4.72
MgO	0.79	0.78	0.91	1.33	1.20	1.48
MnO	0.03	0.02	0.06	0.07	0.05	0.11
CaO	0.34	0.66	4.80	5.20	4.94	5.83
Na_2O	0.02	6.42	6.56	4.47	4.63	3.86
K_2O	2.26	0.12	1.02	1.22	0.72	1.07
P_2O_5	0.10	0.07	0.17	0.19	0.17	0.22
LOI	1.55	1.02	1.64	1.75	1.50	2.87
Total	98.49	99.68	99.45	99.23	99.26	100.06
$\delta^{18}O$ (‰)	8.3[a]	7.9	8.9	10.2	9.3	9.7
Rb	51.8	2.2	30.5	38.0	20.0	31.1
Sr	90.5	224	915	997	846	1066
Ba	3032	n.d.	n.d.	n.d.	n.d.	n.d.
Zr	65.8	94.1	147	203	186	242
Hf	1.7	2.9	5.2	6.2	5.8	7.5
Th	2.3	2.2	1.9	3.8	4.0	3.5
U	2.8	0.1	2.2	1.1	2.6	1.7
Pb	5.8	1.0	4.7	8.0	8.2	7.4
Ga	8.4	8.6	21.7	22.5	21.0	19.8
Ni	16.2	0.9	0.5	2.3	n.d.	1.8
Cr	71.0	5.5	5.4	6.7	5.3	5.7
Nb	5.06	13.2	13.1	15.1	17.1	8.9
Y	10.9	8.4	23.4	39.6	36.1	26.1
La	n.d.	5.8	7.3	38.4	34.7	47.3

[a] Whole rock. Remaining values are from quartz separates.
Major elements in weight percent. Trace elements in ppm. All determinations by XRF. Total iron as FeO. LOI = loss on ignition, n.d. = not detected.
Values for NCSU-QTZ standard analyzed during collection: +11.71‰ and +11.65‰. Accepted value is +11.7‰.

variation of major element composition and $\delta^{18}O$ of quartz is shown in Fig. 1 as a function of distance from the detachment fault. Distances are nearly perpendicular to the detachment surface. The $\delta^{18}O_{qtz}$ shows an erratic but steady decrease with proximity to detachment fault. SiO_2 increases sharply within 10 m of the fault. Al_2O_3 shows behavior opposite to SiO_2, with large decreases from 3 to 0 m. Generally systematic decreases are observed in TiO_2, MgO and MnO contents towards the detachment fault. FeO_t and K_2O also slightly decrease with approach to the fault, but show large enrichments at the fault surface. Na_2O displays opposite relations.

Major element composition is observed to correlate with mineralogy. Toward the detachment fault, SiO_2 shows a sharp increase. Over the same distance, relative amounts of

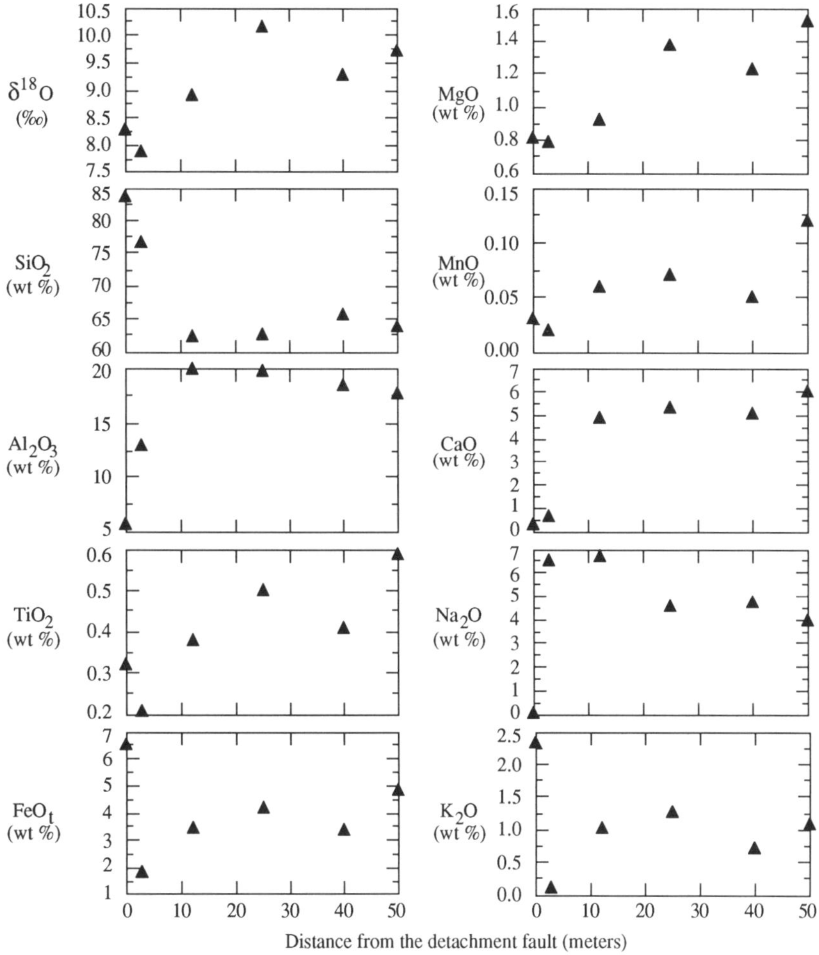

Figure 1: Variation of major element composition and $\delta^{18}O$ with distance from the detachment fault. The sample at zero distance is the detachment surface and was analyzed for whole-rock $\delta^{18}O$. Major elements are recalculated to 100 weight % on an anhydrous basis.

epidote increase and chlorite decrease. The amount of quartz is greatly increased in 90WP10 and chlorite and epidote are reduced to trace amounts. The high SiO_2 and FeO_t, contents of 90WP9 suggest this specimen is micro-crystalline quartz and magnetite; the rock powder is magnetic. Cr and Ni contents are 10 and 7 times higher than in less brecciated rocks.

5 DISCUSSION

Major element variations are interpreted to reflect the replacement of plagioclase by epidote and quartz during progressive brecciation and hydrothermal alteration. Steady depletions in Sr and Ga contents also correspond to the breakdown of feldspar. These chemical variations are opposite to those observed by Halfkenney and others (1989) for chlorite breccias near the Copper Penny mining prospect in the Buckskin Mountains. It is probable that mineralogic, elemental and isotopic shifts produced by brittle deformation are strongly dependent on protolith composition and P-T conditions.

The general decrease in $\delta^{18}O$ toward the detachment fault may be the result of interaction with meteoric fluids with low $\delta^{18}O$.

If pure water with a $\delta^{18}O$ of 0‰ exchanged with quartz in a fluid dominated system (infinite fluid flux), equilibrium temperatures are in the range 280 to 300°C. Under similar conditions, interaction with H_2O-fluids with a $\delta^{18}O$ of -10‰, would indicate temperatures between 145° and 163°C. The large fractionation between quartz and water requires either low-temperature interaction or incomplete exchange at lower fluid fluxes.

The steadily depleted profile may suggest pervasive infiltration of fluids and increasing fluid fluxes toward the detachment fault. Alternatively, infiltrating fluids may have been channelized along epidote-cored veins within bleached host rocks. A quantitative assessment of fluid flow and the effects of net-transfer reactions will require the $\delta^{18}O$ of coexisting minerals.

6 CONCLUSION

Systematic major and trace element variations in brittly deformed mylonites reflect the breakdown of plagioclase to epidote and quartz with increasing brecciation and proximity to the Whipple detachment fault. The shifts correspond to a decrease in $\delta^{18}O_{qtz}$ toward the fault. It is likely that large fluid fluxes have accompanied the chemical changes. Further, the depletion in ^{18}O may suggest interaction with a low-^{18}O reservoir, such as meteoric water at low temperatures.

ACKNOWLEDGMENTS

This work has been supported by GSA Student Research Grant 4676-91 to Ekstrom and NSF grant EAR-9105924 to Morrison.

REFERENCES

Anderson, J.L., Barth, A.P. & E.D. Young 1988. Mid-crustal Cretaceous roots of Cordilleran metamorphic core complexes. *Geology* 16: 366-369.

Anderson, J.L. & R.L. Cullers 1990. Middle to upper crustal plutonic construction of a magmatic arc; An example from the Whipple Mountains metamorphic core complex. In: Anderson, J.L. (ed.) *The Nature and Origin of Cordilleran Magmatism, Geol. Soc. Am. Mem.* 174: 47-69.

Davis, G.A., Anderson, J.L., Frost, E.G. & T.J. Shackelford 1980. Mylonitization and detachment faulting in the Whipple-Buckskin-Rawhide Mountain terrane, southeastern California and western Arizona. In: Crittenden, M.D., Coney, P.J. & G. H. Davis (eds.), *Cordilleran metamorphic core complexes. Geol. Soc. Am. Mem.* 153: 79-130.

Davis, G.A., Anderson ,J.L., Krummenacher, D., Frost, E.G. & R.L. Armstrong 1982. Geologic and geochronologic relations in the lower plate of the Whipple detachment fault, Whipple Mountains, southeastern California; A progress report. In: Frost, E.G. & E.L. Martin (eds.), *Mesozoic-Cenozoic tectonic evolution of the Colorado River region, California, Arizona, and Nevada*, Cordilleran Publishers, San Diego, California: 408-432.

Davis, G.A., Lister, G.S., & S.J. Reynolds 1986. Structural evolution of the Whipple and South Mountains shear zones, southwestern United States. *Geology* 14: 7-10.

DeWitt, E., Sutter, J.F., Davis, G.A. & J.L. Anderson 1986. $^{40}Ar/^{39}Ar$ age-spectrum dating of Miocene mylonitic rocks, Whipple Mountains, southeastern California. *Geol. Soc. Am. Abs. Prog.* 18: 584.

Halfkenny, R.D., Kerrich, R. & W.A. Rehrig 1989. Fluid regimes and geochemical mass transport in the development of mylonites and chlorite breccias at Copper Penny, Buckskin Mountains, In: Spencer, J.E. & S.J. Reynolds (eds.),*Geology and mineral resources of the Buckskin and Rawhide Mountains, west-central Arizona, Arizona Geol. Surv. Bull.* 198: 190-202.

Thurn, L. C. 1983. Geology of a portion of the southeastern Whipple Mountains, San Bernadino county, California, M. S. Thesis, University of Southern California, 186 p.

Wright, J.E., Anderson, J.L. & G.A. Davis 1986. Timing of plutonism, mylonitization, and decompression in a metamorphic core complex, Whipple Mountains, California. *Geol. Soc. Am. Abs. Prog.* 18: 201.

Water-Rock Interaction, Kharaka & Maest (eds) © 1992 Balkema, Rotterdam. ISBN 90 5410 075 3

Evidence for the controlling effect of the high-pressure metamorphic PTt path on the mass transfer of major elements

B.Goffé & O.Vidal
Ecole Normale Supérieure Paris, France

ABSTRACT: New experimental and natural data suggest that large amounts of mass transfer in metapelites metamorphosed under high-pressure, low temperature conditions are controlled by local equilibrium among aqueous mineral phases during the P-T-t history. Increase of pressure and decrease of temperature will enhance the formation of pure aluminosilicates at the expense of ferromagnesian or alkaline aluminosilicates. Under conditions of increasing temperature or decreasing pressure, the inverse will be true.

INTRODUCTION

Goffé et al. (1987) in a study of hydrothermal vein mineralization in meta-argilites subjected to high-pressure, low-temperature (HP-LT) metamorphism revealed that ferromagnesian (e.g. chlorite) and pure aluminosilicate (e.g., pyrophyllite) mineralization can be correlated with regimes of increasing and decreasing temperature, respectively. These authors also performed an experimental study of the transport of silica, aluminum and magnesium in hydrothermal solutions to simulate variations in the physical conditions during metamorphism and the accompanying mass transport in a closed system. At last they concluded that local equilibrium among aqueous and mineral phases controls the distribution and composition of hydrothermal vein mineralization and that vein mineralogy can be used to infer the sense of variation of pressure and temperature during metamorphism. These results were applied to a relatively small and particular rock system which can be reasonably considered as closed. New natural and experimental data presented here indicate that metamorphic mineralization of very large pelitic systems of common composition can be also controlled by the form of P-T-t paths.

1 THE NATURAL ROCK SYSTEM: THE METAPELITES

In recent years studies of low-grade metapelites metamorphosed at high-pressure, low-temperature conditions (blueschist facies, 6-12 kbar, 300-400 °C) have demonstrated extensive transformations of these rocks during the HP-LT metamorphic evolution (Betic chain, Corsica, Western and Central Alps, Egean Sea, Oman; Chopin and Schreyer, 1983; Goffé and Chopin, 1986; Goffé et al., 1988; Goffé et al., 1989; Fournier et al., 1991; Theye et al., 1991; Goffé and Oberhänsli, 1992). These transformations are revealed by the massive (up to 40% of the rock volume) crystallisation of the high pressure mineral Fe, Mg carpholite ($(Fe,Mg)Al_2Si_2O_6(OH)_4$) in mesoscopic quartz ($\pm$carbonates) segregations. The surrounding schist, sandwiched at the mesoscopic scale between the quartz seggregations, is composed of Fe Mg carpholite, chloritoid or/and lawsonite porphyroblasts in a foliated matrix of organic matter, Fe and Ti oxydes, chlorite or pyrophyllite and minor white micas. Because the protoliths are, generally, common marine or continental shales in which illite is an abundant mineral, the scarcity of the white micas in units where carpholite and lawsonite are fresh is an intriguing observation. On the other hand, in units or part of units where carpholite or lawsonite are partially or totally retrograded, the white micas (muscovite, paragonite) are widespread minerals. These white micas, associated with chlorite, systematically replace the carpholite and lawsonite prisms. In some cases albite appears. Finally, when carpholite or lawsonite are totally retrograded to chlorite and white micas, the rock assemblage is that typical of the greenschist facies. The bulk composition of these greenschist rocks is not significantly different from that of the protoliths (Brosse, 1982). These transformations affect a very large part of the chain implicated in the HP metamorphic evolution (10000 to 20000 km^3 for the Western Alps). Large scale mapping of the preservation or retrogression of carpholite and lawsonite shows that these preserved minerals are distributed along small zones (some kilometers in size) located at the external or the upper parts of the thrusted units (Goffé and Chopin, 1986; Goffé et al., 1988). These parts, where aragonite or liquid hydrocarbons are also preserved, correspond to rock volumes which have undergone a retrograde metamorphism characterised by an anticlockwise P-T-t path (Gillet and Goffé et al. (1987) experimentally demons-partially retrograded carpholite or lawsonite are found in all the other parts of the HP metapelitic units, sometimes along very large zones (250 km long, 10 to 30 km wide in the "schistes lustrés" nappe of the Western Alps). This widespread distribution of carpholite or lawsonite relics shows: 1) the independence of this mineralogy to the initial protolithe composition (in the range of the pelitic composition variations from marl to shale and uncommonly quartzite), 2) that the whole volume of the metasediment units have undergone a unique prograde history. The intensity of the carpholite and lawsonite replacement by white micas increases toward the deeper or the more internal part of the units. Albite appears in the more intensively retrograded assemblages. In associated metabasites, blue amphiboles, jadeite and lawsonite are replaced by low-pressure, medium-temperature assemblages characteristic of the greenschist facies. These rocks have enjoyed the clockwise P-T-t path classically described in the HP metamorphic chains (Thompson and Ridley, 1987). The surprising correlation between metamorphic evolution and either the absence or the development of white micas (and albite) suggests that HP anticlockwise P-T-t paths favour disapearance of white micas while clockwise paths favour the contrary (Fig 1).

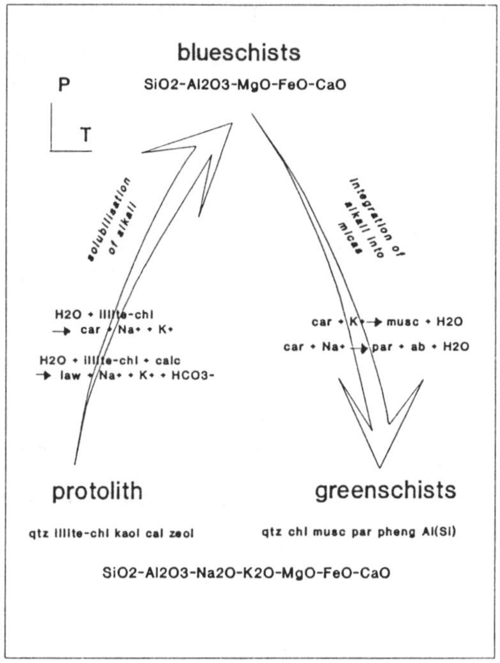

Figure 1: Classical prograde- retrograde mineralogical evolution in HP metapelites.

2 EXPERIMENTAL INVESTIGATION

Goffé et al. (1987) experimentally demonstrated the feasibility of the mass transfer process in response to the variations of temperature with time in a perfectly closed system: In order to extend the results to elements other than Si, Al and Mg and to determine the influence of pressure on their fluid-solid transfer ability, same type of experiments have been undertaken in considering a mixture of paragonite, pyrophyllite and quartz. The experimental apparatus consisted of a small (2 cm long, 35 mm in diameter) partially open gold capsule containing the mixture (50 mg), inside a larger (5 cm x 4 mm) sealed tube containing water (300 μl). The inner capsule should simulate the rock and the external capsule the vein. Five sets of experiment were performed : 1) at constant pressure, increasing of temperature (from 300 to 500°C at 2 kbar and 400 to 500°C at 4 kbar by steps of 5°C/day), decreasing of temperature (400 to 300°C at 4 kbar and 500 to 250°C at 2.5 kbar by steps of 5°C/day); 2) at constant temperature, increasing of pressure (4 to6 kbar and 2 to 5.5 kbar at 400°C by step of 130 bar/day), decreasing of pressure (6 to 4 kbar and 5 to 1.5 kbar at 400°C by step of 130 bar/day); 3) at constant pressure and temperature (400 °C at 4 kbar and 2kbar, 30 days). After experiments (quenched by air jet in 5mn), the run products growing on the wall of the external capsule (which were without ambiguity newly formed minerals) were observed by SEM. The increase of temperature and the decrease of pressure favoured the formation of paragonite. On the other hand, the decrease of temperature and the increase of pressure favours pyrophyllite crystallization. In the inner capsule, the same reaction were observed, the new products growing on the destabilized one. In the experiments were P and T were held constant with time, no reaction were observed (only a dissolution of both minerals paragonite + pyrophyllite and a growth of large grains at the expense of smaller one).

Results of these experiments seem to lend credence to a mass transfer process involving Na, Fe, Mg, Si and Al in closed systems, controlled exclusively by the change of the thermobarometric conditions with time. These results are consistent with the thermodynamic predictions proposed by Goffé et al. (1987): in hydrothermal system where fluid and solid phases are in equilibrium, increasing of pressure and decreasing of temperature will enhance the formation of pure aluminosilicates at the expense of ferromagnesian or alkaline aluminosilicates. Under conditions of increasing temperature or decreasing pressure, the inverse will be true.

Using the thermodynamic data for aqueous species proposed by Helgeson et al. (1981) and those of Berman (1988) for paragonite, quartz and pyrophyllite, it is posssible to calculate the compositional evolution of a fluid phase with a variation of the P-T conditions. Results are in agreement with the experimental observations and the general rule stated above.

Although well-calibrated thermodynamic data for aqueous species are not available for T > 600°C, and P > 5 kbar, an analysis can been attempted since metamorphic conditions are within the T calibrated range. Extrapolation to higher P will invalidate numerical results but not relative values. It can be seen in figure 2 that the results are in agreement with the field observations: considering $aH_2O = 1$

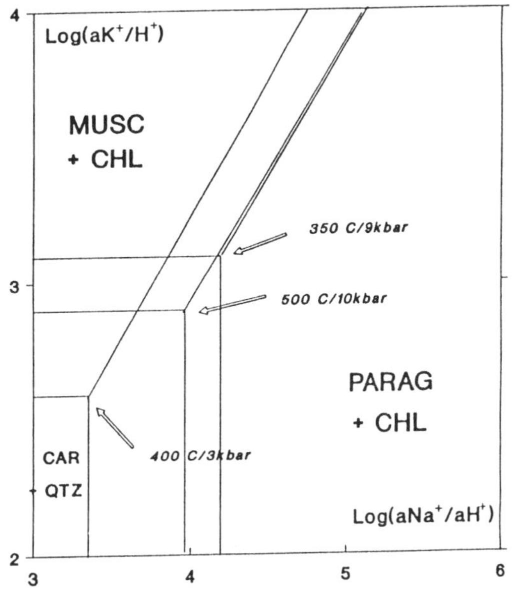

Figure 2: Compositional change of a fluid phase buffered by the assemblages car + qtz + chl + micas

and a complete ionic dissociation, the ratios aNa^+/aH^+ and aK^+/H^+ in the fluid phase in equilibrium with the paragenesis carpholite-paragonite-muscovite-chlorite-quartz decrease with a decrease of the P-T conditions (thermodynamic data for carpholite are from Vidal et al., 1992). Moreover, assuming the only possible variations of the pH to be controlled by the changes in the dissociation constant of water, the variations of aNa^+/aH^+ and aK^+/aH^+ are directly reliable to the variations of aNa^+ and aK^+: the pH increases from 4.08 at 500 °C, 10 kbar to 5.7 at 400 °C, 3 kbar and 5 at 250 °C, 2.5 kbar. That tends consequently, to increase the values of aNa^+/aH^+ and aK^+/H^+. As a result, a decrease of aNa^+/aH^+ and aK^+/aH^+ can only be due to a decrease of the sodium and potassium concentrations in the fluid phase. It results in growth of phyllosilicates at the expense of carpholite. Synmetamorphic and retrograde vein mineralizations of the metapelites metamorphosed in the green- to blueschist facies can be used as a tool to determine the tendancy of the thermobarometric evolution. The same reasoning has been also applied by Jamieson and

O'Beirne-Ryan (1991), in order to explain the regional decompression-induced growth of albite in the Fleur de Lys supergroup (Western Newfoundland).

CONCLUSION:

Independent of other hydrothermal process, like fluid percolation, which can effect the HP-LT metapelites of the blueschist facies, the control of major elements mass transfer by the form of specific P-T-t paths seems to be important. In this hypothesis, the HP-LT metamorphism could be considered as a process which enriches the fluids in alkaline elements. Under the HP conditions, a possible opening of the system induces the migration of saline fluids toward the upper parts of the metamorphic structure. This mechanism may induce the enrichment of the upper parts of the metamorphic domains in alkaline elements and the coeval depletion of the deeper ones with the subsequent crystallization of aluminum silicates (kyanite, sillimanite), these minerals being common in high grade metamorphic rocks.

REFERENCES

Berman, R.G. 1988. Internally-Consistent Thermodynamic Data for Minerals in the system Na_2O-K_2O-CaO-MgO-FeO-Fe_2O_3-Al_2O_3-SiO_2-TiO_2-H_2O-CO_2. J. Petr., 29, part 2, 445-22.

Brosse, E. 1982. Contribution à la minéralogie et à la géochimie des sédiments pélagiques profonds. Comparaison des "black-shales" du Crétacé dans l'atlantique central Nord et des dépôts du Malm et du Crétacé en Briançonnais. Thèse de docteur Ingénieur de l'E.N.S.M. Paris, 474 p.

Chopin, C. & W. Schreyer 1983. Magnesiocarpholite and Magnesiochloritoid: two index minerals of pelitic blueschists and their preliminary phase relations in the model system MgO-Al_2O_3-SiO_2-H_2O. Am. J. Sci., 283-A, 72-96.

Fournier, M., L. Jolivet, B. Goffé, & R. Dubois 1991. The alpine Corsica metamorphic core complex. Tectonics, 10, 1173-1186.

Gillet, P.& B. Goffé 1988. On the significance of aragonite occurrence in the Western Alps. Contrib. Mineral. Petrol., 99, 70-81.

Goffé, B., A. Michard, J.R. Kienast, O.& Le Mer 1988. A case of obduction-related high pressure, low temperature metamorphism in upper crustal nappes, Arabian continental margin, Oman: P-T paths and kinematic interpretation. Tectonophysics, 151, 363-386.

Goffé, B., W.M. Murphy, & M. Lagache 1987. Experimental transport of Si, Al and Mg in hydrothermal solutions: an application to vein mineralization during high-temperature, low-pressure metamorphism in French Alps. Contrib. Mineral. Petrol., 97, 438-450.

Goffé, B. & C. Chopin 1986. HP metamorphism in Western Alps : zoneography of metapelites, chronology and consequences. Schweiz. Mineral. Petrogr. Mitt., 66, 41-52.

Goffé, B., A. Michard, V. Garcia-Duenas, F. Gonzalez-Lodeiro, P. Monié, J. Campos, J. Galindo-Zaldivar, A. Jabaloy, J.M. Martinez-Martinez & J.F. Simancas 1989. First evidence of high-pressure, low-temperature metamorphism in the Alpujarride nappes, Betic Cord-illeras (S.E. Spain). Eur. J. Mineral., 1, 139-142.

Goffé, B. & R. Oberhänsli 1992. (Fe,Mg)-carpholite occurrences in the "Bündnerschiefer" of the eastern Central Alps (Grisons and Engadine window). Eur. J. MineraL (in press).

Helgeson, H.C., D.M. Kirkham & G.C. Flowers 1981. Theoritical prediction of the thermodynamic behavior of aqueous electrolytes at high pressures and temperatures: IV. Calculation of activity coefficients, osmotic coefficients, and apparent and relative partieal molal properties to 600 °C and 5 kbar. Amer. J. Sci., 2381, 1249-1516.

Jamieson, R.A. & A.M. O'beirne-ryan 1991. Decompression-induced growth of albite porphyroblasts, Fleur de Lys Supergroup, western New-found-land. J. Metam. Geol., 9, 433-439.

Theye, T., E. Seidel, O. Vidal 1992. Petrology of carpholite bearing metapelites and related rocks from the high pressure metamorphic phyllite-quartzite unit of Crete and the Peloponnese (Greece). Eur. J. Mineral, in press.

Thompson, A.B.& J.R. Ridley 1987. Pressure-temperature-time (P-T-t) histories of orogenic belts. In E., Oxburgh, B. Yardley & P. England (eds), The royal Society London, 27-45.

Vidal, O., B. Goffé, Theye T. 1992. Experimental study of the stability of sudoite and magnesiocarpholite and calculation of a new petrogenetic grid for the system FeO-MgO-Al_2O_3-SiO_2-H_2O. J. metamorphic Geol., 10, in press.

Water-Rock Interaction, Kharaka & Maest (eds) © 1992 Balkema, Rotterdam. ISBN 90 5410 075 3

A different approach to fluid-rock-interactions applied to microtextures in metacarbonates

S. Heuss-Aßbichler & C. Schneider
Institut für Mineralogie und Petrographie, München, Germany

ABSTRACT: An alternative conception to the fluid infiltration model is presented, based on the principle of internal buffering of the fluid phase. According to this conception low Xco_2-values can be maintained by the precipitation of calcite due to a change in calcite-solubility in dependence of Xco_2.

1 INTRODUCTION

Calculations of fluid-rock-ratios for metamorphic reactions (Ferry, 1991) in carbonate rocks show that generally large quantities of external fluid are necessary to maintain the equilibrium-Xco_2 during reaction progress. Investigations in the thermoaureole of the Ballachulish Intrusion (Scotland), focused on one decarbonation reaction, show no evidence for external fluid exchange. In this paper we aim to present a model able to maintain even H_2O-rich fluid compositions during reaction progress without external influx of H_2O-rich fluids, based on the fact that the volume of the fluid phase produced by a decarbonation reaction prevents the infiltration of external fluid. We assume a buffering mechanism taking the calcite-solubility into account for the CO_2-reduction within the fluid phase.

2 FIELD OBSERVATIONS IN THE BALLACHULISH-AUREOLE

The Ballachulish Intrusion Complex (W-Scotland) has intruded in Dalradian metasediments. The pressure deduced for the thermometamorphism is 3 kbar (Voll et al., 1991). In the northeastern section of the aureole thin carbonate layers are exposed from a distance of 700 m up to the contact (Fig.1) (Masch & Heuss-Aßbichler, 1991). In this section a marked difference in Xco_2 between the central part on the one hand and the inner part and the outer part of the aureole on the other hand is striking. At the Fo-Cal-isograd (9), in the outer part, an $Xco_2 < 0.2$ was deduced, the assemblage Fo-Cal-Dol-Per/Bru at the contact suggests an Xco_2 less than 0.1. For the isobaric invariant assemblage Tr-Di-Fo-Cal-Dol (CIII) an $Xco_2 > 0.5$ was estimated. The change in Xco_2 from the isograd

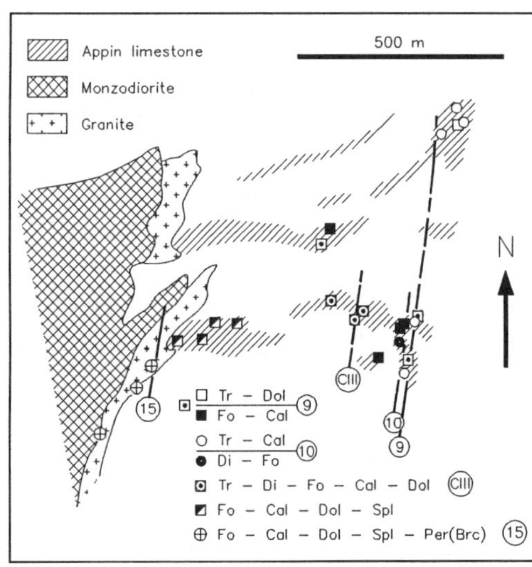

Fig. 1: Isograd map of siliceous dolomites in the NE-section of the Ballachulish-Aureole.

up to the isobaric invariant point (CIII) can be explained by the conception of fluid buffering (Greenwood, 1975). The H_2O-rich conditions needed near the contact in order to produce periclase requires a dilution of the fluid phase during reaction progress. However, the isotopic data (O^{18}/O^{16}) (Hoernes, pers. comm.) show no evidence for external fluid exchange for these samples.

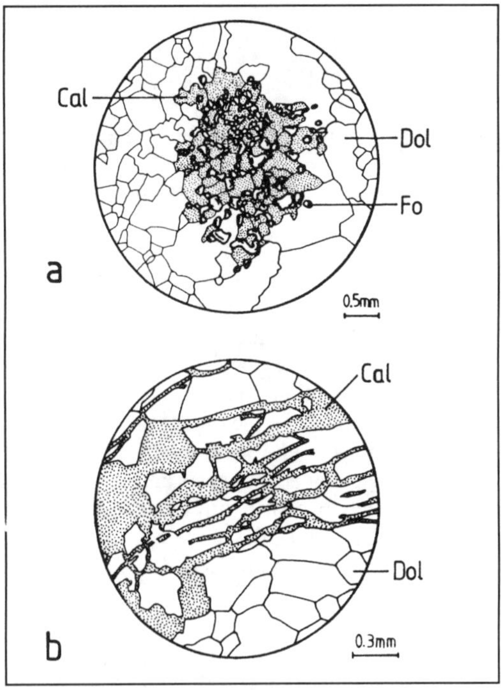

Fig. 2: Microtextures in siliceous dolomites of the Ballachulish Aureole:
a Intergrowth of forsterite and calcite within a domain surrounded by a dolomite matrix.
b Calcite veinlets in dolomite matrix.

A detailed study of thin sections indicates a systematic change in microtextures (Heuss-Aßbichler and Masch, 1991) reflecting a change in distribution and size of reaction sites. In the specimens from the Fo-Cal-isograd up to the isobaric invariant assemblage intergrowth textures are exposed (Fig. 2a), whereas the samples near the contact show granoblastic textures. Remarkable is the intergrowth texture occurring in samples for which high Xco_2-values (> 0.5) were deduced: the product phases are limited to isolated sites which stand out against the more fine-grained dolomite matrix. The distinct reaction sites may be connected by calcite veinlets (Fig. 2b). This texture implies a limited exchange of the fluid phase generated by the reaction with the initial fluid present along the interconnected grain boundary system. We suggest that the higher values of Xco_2 within the locally restricted reaction sites are caused by this fact. From these observations we inferred the following conception.

3 A DIFFERENT APPROACH TO THE CALCULATION OF FLUID-ROCK-RATIOS

Temperature and pressure during reaction progress are considered as constant; P_{fluid} is equal to P_{lith}. The composition of the fluid phase is restricted to the system H_2O-CO_2. The reaction progress variable ξ is defined from 0 to 1; reaction progress 1 signifying reaction completion.

In order to demonstrate the computation procedure the wollastonite-producing reaction was selected, as this decarbonation reaction liberates CO_2 only. The reaction conditions chosen are 600°C, 1 kbar, Xco_2-equilibrium (Xco_2^{Eq}) = 0.35, pore space 1 %.

The considered carbonate rock is thought to consist of two subsystems (see fig. 3a). The volume of subsystems I and II is equal to 100 Vol-%. Subsystem I is defined by the volume of the reacting solid phases needed to produce the observed amount of the considered product phases, it is standardized to 100 cm^3. The reaction progress ξ is 1, when this subsystem contains the product phases only. The remaining volume of the considered carbonate rock is defined as subsystem II. Subsystem III comprises the country rock and/or igneous rock. All three subsystems are thought to be in local equilibrium. At the beginning of the reaction the pore space is filled with a H_2O-rich fluid.

With the onset of the decarbonation reaction in subsystem I a CO_2-rich fluid phase is liberated, which shifts the Xco_2 in subsystem I rapidly to Xco_2^{Eq} (see fig. 3b, step 1; see tab. 1, Xco_2^{I}). In order to maintain reaction progress, a decrease of Xco_2 in subsystem I is required. For the calculation procedure we assume an immediate mixing of the fluids of subsystems I and II at this step. This causes a decrease of Xco_2 in

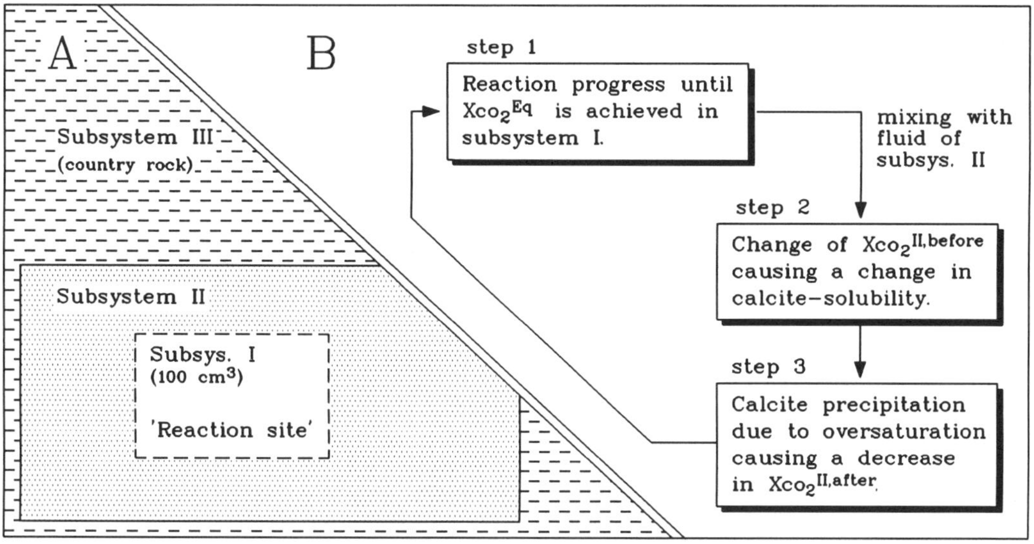

Fig. 3: A: Schematic illustration of the subsystems B: Reaction steps, describing one iteration step of the calculation procedure.

Table 1: Calculated iteration steps for the wollastonite-producing reaction for an initial rock-composition of 80 Vol-% Cal and 20 % Qtz.

pressure	100 MPa			pore space		1.00 Vol.-%	
temperature	600 °C			equilibrium Xco_2		0.35	
volume subsys. I	100 cm^3			initial Xco_2		0.01	
volume subsys. II	276 cm^3						

selected iteration steps	0	1	4	7	10	13	15	17
Xco_2^{I}	0.01	0.35	0.35	0.35	0.35	0.35	0.35	0.35
$Xco_2^{II,before}$	0.01	0.01	0.16	0.21	0.26	0.29	0.31	0.32
$Xco_2^{II,after}$	-	-	0.00	0.00	0.00	0.00	0.00	0.00
Calcite precipitated (cm^3)	-	-	1.94	4.21	13.61	29.09	46.94	74.58
Reaction progress ξ	0.00	0.01	0.04	0.10	0.21	0.41	0.64	1.00

subsystem I, and a shift in Xco_2 from H_2O-rich conditions to higher Xco_2 values in subsystem II ($Xco_2^{II} < Xco_2^{Eq}$)(see fig. 3b, step 2; see tab. 1, $Xco_2^{II,before}$). The increasing CO_2-content of subsystem II, however, affects a change in the calcite solubility. According to the thermodynamic calculations of Bowers et al. (1984) and the experimental studies of Fein and Walther (1987) an increase of the CO_2 concentration in the fluid phase results in a decrease of the calcite solubility ($Xco_2 > 0.05$), and thus induces the precipitation of calcite from the fluid phase - and a decrease in the CO_2-content (see fig. 3b, step 3). This enables the reaction to proceed (see fig. 3b, step 1). The sequence of the three reaction steps form one iteration step in the calculation procedure. The reaction progress ξ for the wollastonite-producing reaction obtained at different iteration steps is listed in Table 1. The thermodynamic data of Bowers et al. (1984) were used to evaluate the amount of calcite, which will precipitate due to Ca^{++}-oversaturation in the fluid phase of subsystem II.

Fugacities of CO_2 and H_2O were calculated following Kerrick & Jacobs (1981). In the example above the change in calcite-solubility is sufficient to reduce the Xco_2 of subsystem II, respectively subsystem I, to 0.0 (see tab. 1, $Xco_2^{II,after}$).

This reaction cycle stops:

1. if one of the reacting phases is exhausted (reaction progress $\xi = 1$),
2. if the Xco_2-condition in subsystem II is equal to the equilibrium-Xco_2 ($Xco_2^{II} = Xco_2^{Eq}$).

A further reaction progress in case 2 is possible by removing the excess-CO_2 not buffered by the interaction between subsystem I and II to subsystem III. Changes in the reaction conditions resulting from an increase in temperature induces a shift of Xco_2^{Eq} to higher values, and so effects a new cycle of interaction between the subsystems at the new reaction conditions.

4 DISCUSSION

The above described example (Tab. 1) demonstrates that calcite-precipitation is a powerful mechanism to buffer the CO_2-content of the fluid phase at low values. Assuming that the calcite-solubility values calculated by Bowers et al. (1984) are correct, the Xco_2 in subsystem I is buffered at the equilibrium-Xco_2, but in subsystem II the Xco_2 after calcite-precipitation does not change considerably from the initial value. Note that this mechanism produces large amounts of calcite (Tab. 1: 74 cm^3), which would normally fill the pore space rapidly, thus preventing further reaction progress. The decarbonation reaction, however, is accompanied by an increase of the fluid volume, which means that P_{fluid} may be locally higher than P_{lith}. This is thought to induce hydraulic fracturing and the formation of veins in subsystem II, so providing an additional space for the calcite precipitation. It is debatable whether all the calcite is precipitated in the pore space and/or in additional produced veins. But it should be considered that this model yields maximum values for precipitated calcite as it does not consider the escape of CO_2 to the country rock. The assumption of hydraulic fracturing being effective is supported by field observations in the Ballachulish Aureole (Fig. 3b).

The application of this model to natural conditions is limited by the following facts:

- the model is restricted to the fluid system H_2O-CO_2.
- only few experimental data on calcite solubility at high P/T-conditions are available. They deviate from the calculated values from Bowers et al. (1984).

Nevertheless, this conception can explain easily the intergrowth textures with interconnected calcite veinlets of Ballachulish aureole described above (fig. 2a, b).

REFERENCES

Bowers, T.S.; Jackson, K.J.; Helgeson, H.C. 1984. *Equilibrium activity diagrams*. Springer, Berlin.

Fein, J.B.; Walther, J.V. 1987. Calcite solubility in supercritical CO_2-H_2O-fluids. *Geochim. Cosmochim. Acta*, 51: 1665 - 1673.

Ferry, J.M. 1991. Dehydration and decarbonation reactions as a record of fluid infiltration. In: Kerrick, D.M. (ed.). *Contact metamorphism*. Reviews in mineralogy, 26: 353 - 394.

Greenwood, H.J. 1975. Buffering of pore fluids by metamorphic reactions. *Am. J. Sci.*, 275: 573 - 593.

Heuss-Aßbichler, S; Masch, L. 1991. Microtextures and reaction mechanisms of carbonate rocks: A comparison between the thermoaureoles of Ballachulish and Monzoni (N. Italy). In: Voll, G.; Töpel, J.; Pattison, D.R.M.; Seifert, F. (eds.). *Equilibrium and kinetics in contact metamorphism*. Springer, Berlin. 229 - 249.

Kerrick, D.M.; Jacobs, G.K. 1981. A modified Redlich-Kwong equation for H_2O, CO_2, and H_2O-CO_2 mixtures at elevated pressures and temperatures. *Am. J. Sci.*, 281: 735 - 767.

Masch, L.; Heuss-Aßbichler, S. 1991. Decarbonation reactions in siliceous dolomites and impure limestones. In: Voll, G.; Töpel, J.; Pattison, D.R.M.; Seifert, F. (eds.). *Equilibrium and kinetics in contact metamorphism*. Springer, Berlin. 211 - 227.

Voll, G.; Töpel, J.; Pattison, D.R.M.; Seifert, F. (eds.) 1991. *Equilibrium and kinetics in contact metamorphism*. Springer, Berlin.

Water-Rock Interaction, Kharaka & Maest (eds) © 1992 Balkema, Rotterdam. ISBN 90 5410 075 3

Kinetic aspects of formation of Japanese bentonite deposits

Daizo Ishiyama & Osamu Matsubaya
Mining College, Akita University, Japan

Masakazu Ito
Research Laboratory, Kunimine Industries Co., Ltd, Tochigi, Japan

ABSTRACT: Na- and Ca-montmorillonites of many Japanese bentonite deposits were formed by the interaction of fossil seawater and volcanic glass. The estimated range of formation temperature of these montmorillonites is 51 to 72°C. The time of formation for the Na-bentonite deposit and the Ca-bentonite deposits is about 8 Ma and 1 Ma, respectively. Na-bentonite deposits were formed by the interaction of volcanic glass and fossil seawater under the diagenetic condition, while Ca-bentonite deposits were formed by the interaction of volcanic glass and fossil seawater with supply of heat in a short-time.

1 INTRODUCTION

Many bentonite deposits occur in Tertiary strata of Japan which suffered intense submarine volcanism. They are divided into Na-bentonite deposits and Ca-bentonite deposits. There are many geologic and mineralogical studies for the mode of occurrence of Japanese bentonite deposits (e.g., Takeshi & Uno 1979; Ossaka et al. 1981). Takeshi & Uno (1979) suggested that Japanese Ca-bentonite deposits were formed under a hydrothermal condition. Ossaka et al. (1981) proposed that bentonite was derived from volcanic glass of fresh pumice tuff through the formation of montmorillonite and cristobalite. However, formation temperature and duration have not been investigated; they were determined in the present study.

2. OUTLINE OF BENTONITE DEPOSITS

The Na- and Ca-bentonite deposits respectively from Tsukinuno, the Dobuyama, Kawasaki and Akita were examined (Table 1). These four bentonite deposits are located in northeast Japan. The Tsukinuno Na-bentonite deposit is a stratiform type whereas the Kawasaki, Dobuyama and Akita Ca-bentonite deposits occur as lenticular or massive body (Takeshi & Uno 1979). Their protoliths are Miocene vitric tuff and pumice tuff. Fission track age of the tuffaceous rocks of the Tsukinuno, Dobuyama and Kawasaki deposits is 13.9, 12.1, 17.0 Ma, respectively. Potassium-argon age of the initial rock of the Akita bentonite deposit is 10.7 Ma. The vitric tuff of the Tsukinuno deposit is intercalated with indurated mudstone. The radiolarian age of authigenic dolomite nodule in the mudstone ranges from 10 to 8.5 Ma, and it is younger than the fission-track age of vitric tuff of the Tsukinuno deposit.

Table 1. Characteristic features of the Tsukinuno, Dobuyama, Kawasaki and Akita bentonite deposits.

Deposits	Tsukinuno	Dobuyama	Kawasaki	Akita bentonite
Initial rock	Vitric tuff	Pumiceous tuff	Pumiceous tuff	Pumiceous tuff
Age (Ma)	13.9±0.9 (F.T.) 10-8.5 (radiolaria*)	12.2±0.9 (F.T.)	17.0±1.3 (F.T.)	10.7±2.6 (K-Ar)
Wall rock	Hard mudstone	Black mudstone	Black mudstone	Hard mudstone
Depositional environment	Marine	Marine	Marine	Marine
Smectite	Na-Montmorillonite	Ca-Montmorillonite	Ca-Montmorillonite	Ca-Montmorillonite
Relative amounts of quartz and cristobalite	Quartz	Cristobalite >> quartz	Cristobalite >> quartz	Cristobalite >> quartz
Formation temperature (°C)	55 - 70	63 - 65	53 - 69	51 - 72

F.T.: Fission track method; K-Ar: Potassium-argon method (M.M.A.J., 1986);
*: *Lychnocanoma nipponica magnacornuta Sakai*

Vitric and pumice tuffs of the Dobuyama, Kawasaki and Akita bentonite deposits are closely associated with black mudstone and mudstone. Their intimate association suggests that the acidic tuff and indurated mudstone were formed at submarine environment.

3 MINERALOGY OF BENTONITE

Mineral assemblage and content of illite layers in the <2 micrometers fraction of the bentonite and mudstone of the four deposits were examined by X-ray diffraction method. The bentonite samples from the Tsukinuno, Dobuyama and Kawasaki deposits are not weathered, whereas those of the Akita bentonite deposit are weakly weathered. The bentonite from Tsukinuno contains abundant Na-montmorillonite and quartz with minor plagioclase, biotite, analcime, calcite, dolomite and pyrite. Indurated mudstone of the Tsukinuno deposit consists of large amounts of quartz with lesser amounts of illite/montmorillonite mixed-layer mineral, and minor amounts of plagioclase, biotite and adularia. Quartz of volcaniclastic material and authigenic quartz occur in the bentonite and indurated mudstone. Authigenic minerals of the Tsukinuno Na-bentonite deposit are Na-montmorillonite, quartz, analcime, calcite, dolomite and pyrite. The bentonite of the Dobuyama deposit is composed of abundant cristobalite and Ca-montmorillonite with small amounts of quartz, plagioclase and calcite. Those from the Kawasaki deposit are abundant cristobalite and Ca-montmorillonite with minor quartz, clinoptilolite and calcite. The bentonite from Akita consists of large amounts of cristobalite Ca-montmorillonite with small amounts of quartz and plagioclase. Plagioclase and some quartz in the Dobuyama, Kawasaki and Akita Ca-bentonite deposits are fragments of volcaniclastic material, whereas authigenic minerals of these deposits are Ca-montmorillonite, cristobalite, quartz and calcite.

Na- and Ca-montmorillonite of the bentonite deposits replaces glass shards of acidic tuff, not phenocrysts of the acidic tuff. The replacement of volcanic glass by Na- and Ca-montmorillonite occurs at marine environment according to the mode of occurrence of montmorillonite and the geologic setting of these bentonite deposits.

The Na-montmorillonite of the Tsukinuno bentonite deposit contains less than 5 percent illite layer and has randomly interstratified structure. The Ca-montmorillonite of the Dobuyama, Kawasaki and Akita bentonite deposit contains 10 percent illite layer and has also randomly interstratified structure.

4 HYDROGEN AND OXYGEN ISOTOPIC RATIO OF BENTONITE

Hydrogen and oxygen isotopic ratios of the <2 micrometers size fraction of montmorillonite were measured. The hydrogen and oxygen isotopic ratios of the Na-montmorillonite of the Tsukinuno deposit range from -82 to -60 per mil and +18.0 to +19.6 per mil, respectively. The hydrogen and oxygen isotopic ratios of the Ca-montmorillonite of the Dobuyama deposit varies from -68 to -63 per mil and +18 to +20.3 per mil, respectively. The range of hydrogen and oxygen isotopic ratios of the Ca-montmorillonite of the Kawasaki deposit are from -75 to -66 per mil and +18.6 to +18.8 per mil, respectively. The hydrogen and oxygen isotopic ratios of the Ca-montmorillonite of the Akita deposit range from -43 to -34 per mil and +17.6 to +20.6 per mil, respectively (Fig. 1).

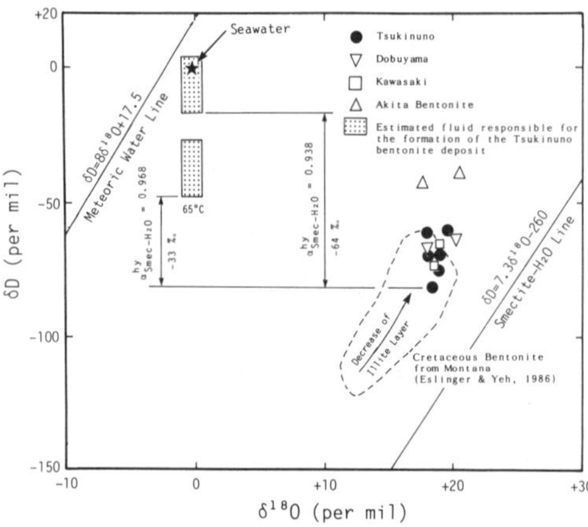

Figure 1 - δD versus $\delta^{18}O$ relationships showing the positions of δD and $\delta^{18}O$ values of bentonite samples. Dotted areas show the estimated fluid based on the δD and $\delta^{18}O$ values of montmorillonite and hydrogen and oxygen isotopic fractionation factors (Savin & Epstein 1970; Yeh 1980; Yeh & Savin 1977).

The hydrogen and oxygen isotopic ratios of the Tsukinuno Na-montmorillonite are similar to those of Ca-montmorillonite of the Dobuyama and Kawasaki deposits. The oxygen isotopic ratio of Akita Ca-montmorillonite is similar to that of Na- and Ca-montmorillonite of the Tsukinuno, Dobuyama and Kawasaki deposits, whereas the hydrogen isotopic ratio of Ca-montmorillonite of the Akita bentonite deposit is higher than that of Na- and Ca-montmorillonite of the Tsukinuno, Dobuyama and Kawasaki deposits by 30 per mil. The hydrogen and oxygen isotopic ratios of montmorillonite of the Tsukinuno, Dobuyama and Kawasaki deposits also resembles those (Eslinger & Yeh 1986) of montmorillonite, which has the lowest content of illite layer, of Cretaceous bentonite in Montana, U.S.A. (Fig. 1).

5 FORMATION TEMPERATURE AND DURATION OF FORMATION OF BENTONITE

The bentonite of the Tsukinuno, Dobuyama, Kawasaki and Akita deposits resulted from the interaction of volcanic glass of acidic tuff and seawater on the basis of the mode of occurrence of Na- and Ca-montmorillonite. According to the isotopic equilibrium between montmorillonite and seawater, oxygen isotopic fractionation factor between smectite and water (Yeh & Savin 1977) and the oxygen isotopic ratios of montmorillonite, the estimated formation temperature of the Tsukinuno Na-bentonite deposit ranges from 55 to 70°C (Table 1). This estimate is in good agreement with the temperature estimated (Matsubaya & Honda 1991) from the oxygen isotopic ratio of authigenic calcite coexisting with Na-montmorillonite of the Tsukinuno deposit. Hence, Na-montmorillonite and calcite of the Tsukinuno deposit were formed under the oxygen isotopic equilibrium condition. The estimated formation temperature of the Dobuyama, Kawasaki and Akita Ca-bentonite deposits ranges from 63 to 65°C, from 53 to 69°C and from 51 to 72°C, respectively. There is no large different between the formation temperature of Na-bentonite deposits and that of Ca-bentonite deposits.

The estimated hydrogen isotopic ratio of fluid forming the Tsukinuno Na-bentonite deposit varies from -49 to -25 per mil on the basis of the isotopic equilibrium between montmorillonite and seawater at the temperature range from 55 to 70°C, hydrogen isotopic fractionation factor between smectite and water (Yeh 1980) and the hydrogen isotopic ratios of montmorillonite of the Tsukinuno deposit. The estimated hydrogen isotopic ratios are lower than that of seawater by -30 per mil. On the other hand, the estimated hydrogen isotopic ratio of fluid forming the Tsukinuno Na-bentonite deposit varies from -18 to +4 per mil on the basis of the isotopic equilibrium between montmorillonite and seawater, hydrogen isotopic fractionation factor between smectite and water at the sedimentary temperature (Savin & Epstein 1970) and the hydrogen isotopic ratios of montmorillonite of the Tsukinuno deposit. Its estimated hydrogen isotopic ratios include that of seawater.

Large amounts of amorphous silica and cristobalite were not recognized in the bentonite and indurated mudstone of the Tsukinuno deposits, while large amounts of amorphous silica and quartz are not identified in the bentonite of the Dobuyama, Kawasaki and Akita. Transformation process of authigenic silica minerals of Miocene mudstones under diagenetic conditions is amorphous silica --> low-cristobalite --> low-quartz (Murata & Larsen 1975; Utada 1965). The Na-bentonite of the Tsukinuno deposit is at the low-quartz formation stage, whereas the Ca-bentonite of the Dobuyama, Kawasaki and Akita bentonite deposits is at the low-cristobalite formation stage. According to kinetic experimental work for silica diagenesis under hydrothermal conditions (Mizutani 1967), the transformation of silica minerals is consecutive reactions involving two irreversible first-order steps; relative amounts of amorphous silica (A), cristobalite (C) and quartz (Q) are shown by the following equations:

$$\frac{dA}{dt} = -k_1A,$$

$$\frac{dC}{dt} = k_1A - k_2C,$$

$$\frac{dQ}{dt} = k_2C$$

where k_1 and k_2 are the rate constants for reactions from amorphous silica to cristobalite and from cristobalite to quartz, respectively.

The mineral assemblage of silica minerals and montmorillonite, the estimated temperatures for these bentonite deposits, the transformation of silica minerals and the kinetic data (Mizutani 1967) suggest that the duration for formation of the Tsukinuno Na-bentonite deposit is estimated about 8 Ma, and for the Dobuyama, Kawasaki and Akita Ca-bentonite deposits about 1 Ma (Fig. 2). These time estimates are shorter than fission-track, potassium-argon and radiolarian ages of the host rocks of these bentonite deposits. Therefore Na-bentonite deposits must have formed by interaction of volcanic glass and fossil seawater in the

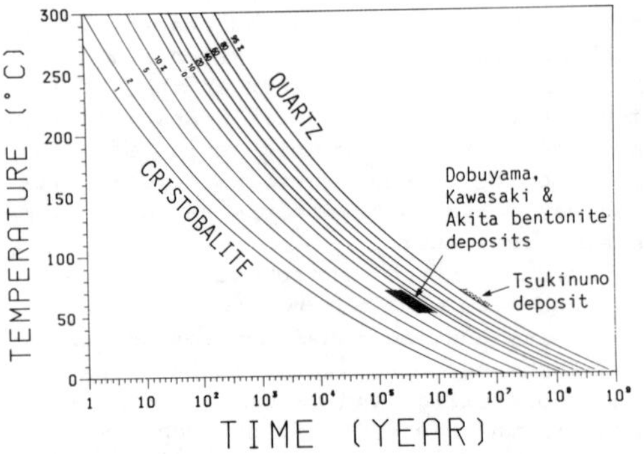

Figure 2 - Relationships among the mineralogical variation for silica minerals, formation temperature and duration of the formation of the Tsukinuno, Dobuyama, awasaki and Akita bentonite deposits.

pores of host rock under diagenetic condition and Ca-bentonite deposits were formed by similar process at higher temperature but shorter duration.

REFERENCES

Eslinger, E.V. & H.W.Yeh 1986. Oxygen and hydrogen isotope geochemistry of Cretaceous bentonites and shales form the Disturbed Belt, Montana. Geochim. Cosmochim. Acta 50: 59-68.

Matsubaya, O. & S.Honda 1991. Carbon and oxygen isotopic ratios of carbonate nodules from Onnagawa Formation in Oga peninsula and some bentonite deposits in Yamagata and Gunma prefecture. Rep. Res. Inst. Nat. Res. Mining Coll. Akita Univ. 56: 131-142.

Metal Mining Agency of Japan 1986. Reports on the regional geology of the Tazawa district. M.I.T.I.: 151 p.

Mizutani, S. 1967. Kinetic aspects of diagenesis of silica in sediments. J. Earth Sci. Nagoya Univ. 15: 99-111.

Murata, K.J. & R.R.Larsen 1975. Diagenesis of Miocene siliceous shale, Tembler Range, California. J. Res. U.S. Geol. Surv. 3: 553-566.

Ossaka, J., J.Hirabayashi, K.Okada & M.Kato 1981. The occurrence and its chemical composition changes in the Dobuyama montmorillonite deposit, Miyagi prefecture. J. Mineral. Soc. Japan 15 Spec. Issue: 170-186.

Savin, S. & S.Epstein 1970. The oxygen and hydrogen isotope geochemistry of clay minerals. Geochim. Cosmochim. Acta 43: 25-42.

Takeshi, H. & Y.Uno 1979. Notes on the formation and transformation of montmorillonites in Japan. J. Mineral. Soc. Japan 14 Spec. Issue: 70-77.

Utada, M. 1965. Zonal distribution of authigenic zeolites in the Tertiary pyroclastic rocks in Mogami district, Yamagata prefecture. Sci. Papers Coll. General Educ. Univ. Tokyo 15: 173-216.

Yeh, H.W. 1980. D/H ratios and late-stage dehydration of shales during burial. Geochim. Cosmochim. Acta 44: 341-352.

Yeh, H.W. & S.M. Savin 1977. Mechanism of burial metamorphism of argillaceous sediments: 3. O-isotope evidence. Geol. Soc. Am. Bull. 88: 1321-1330.

Water-Rock Interaction, Kharaka & Maest (eds) © 1992 Balkema, Rotterdam. ISBN 90 5410 075 3

Composition and temperature effects on mass transfer during serpentinization of ultramafic rocks

C.F.Jove
Department of Geology, Stanford University, Calif., USA

ABSTRACT: Changes in fluid composition and in parageneses of secondary minerals during irreversible hydration of ultramafic rocks were modeled using the EQ3/6 computer program (Wolery, 1983). Rocks composed of pure forsterite (fo), pure enstatite (en), and fo-en mixtures were reacted with dilute solutions of pH=5.8 and 8 at 25 °C and 1 bar, and at 150 °C and 4.8 bars. The pH increased in all cases due to hydrolysis reactions, but the increase was smaller at high temperature due to the higher degree of Mg-hydroxy complexation. Pure forsterite and pure enstatite produced chrysotile + brucite, and chrysotile + talc, respectively. Dissolution of en/fo=0.33 and 1 produced chrysotile + brucite, and chrysotile, respectively. Initial variations in pH within the range of 5.8-8 had no effect in secondary mineral assemblage. Dissolution of fo-en mixture into CO_2-rich fluids did not predict quartz as an alteration product, unlike the models of Pfeifer (1977). Theoretical evidence from the present study supports a low-temperature regime for the process of serpentinization.

1. INTRODUCTION

Serpentinization of ultramafic rocks has received a great deal of attention due to the complexity of the process and the broad range of physico-chemical conditions in which it may occur. Most of the descriptions of hydration reactions during serpentinization are based on detailed petrographic studies (Page, 1968; Whittaker and Wicks, 1977), mineral-stability relations (Martin and Fyfe, 1970), and stoichiometric mineral solubility experiments (Hemley et al., 1977a,b) which are based to a limited extent on thermodynamic requirements. Pfeifer (1977) evaluated irreversible chemical reactions involving minerals and aqueous solutions at 20 °C and 1 bar to explain the occurrence of high-pH springs issuing from ultramafic rocks. The present study focuses on modeling the process of low-temperature alteration of ultramafics and how it compares to higher temperatures using a modified version of the algorithms used by Pfeifer (1977) with updated, internally-consistent thermodynamic data.

2. PROCEDURES

The computer program EQ3/6 (Wolery, 1983) was used to calculate the changes in the concentration of aqueous and mineral species during isothermal-isobaric hydration reactions of ultramafic rocks. Modes of the modeled ultramafic rocks were : (1) 100% enstatite (en); (2) 100% forsterite (fo); (3) harzburgite (en/fo=0.33); and (4) harzburgite (en/fo=1) and (5) peridotite (90%fo:8%en:2%diop). These rocks were reacted at 25 °C and 1 bar and at 150 °C and 4.8 bars with a dilute aqueous solution with initial pH=5.8 and 8. Model peridotite (5) was reacted only at 25 °C and 1 bar with a CO_2-rich (0.033 molal) fluid with pH of 5.8. The latter is a duplication of a previous run made by Pfeifer (1977). Another run using a CO_2 fluid involves a fluid in equilibrium with atmospheric CO_2 at 25 °C and 1 bar with pH=5.8. To interpret the metasomatic reactions and the compositional evolution of the fluid phase along the reaction paths the results are plotted on an activity-activity phase diagram (figure 1). Changes in pH and total molal amounts of Mg and Si are plotted as a function of the reaction progress variable ξ (zi) which is the number of moles of reactant dissolved (See Helgeson, 1979). In the following section, only results from runs at a pH value of 5.8 are shown unless otherwise specified.

3. RESULTS

The results of the computer models are presented in figure 1 as reaction paths describing dissolution trends for the reacted rocks at 25 °C and 1 bar.

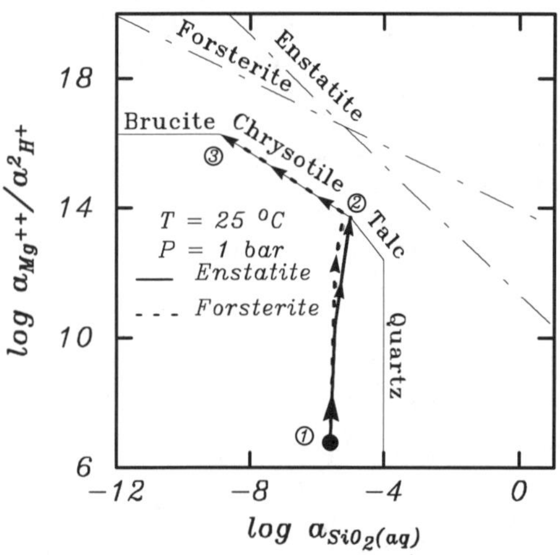

Figure 1 - Log activity phase diagram for the system MgO-SiO_2-H_2O at 25 °C and 1 bar showing the reaction path for the dissolution of forsterite and enstatite. Dot-dash-dot line represent the metastable lines of the reacted minerals. Arrows indicate directions of reaction path. Numbers 1, 2, and 3 indicate initial fluid, either talc or chrysotile precipitation, and brucite-chrysotile equilibrium point respectively (Data from EQ3NR(Wolery,1983)

3.1 Low-temperature reactions

Forsterite dissolution into a solution composition at (1) (figure 1) produced chrysotile at (2), and then brucite at (3). The pH increased by more than 4 log units (figure 2) due to forsterite hydrolysis which consumes H^+. Also, further removal of H^+ is caused by precipitation of chrysotile at (2) which increased the pH by an additional log unit. Si concentration increased with forsterite hydrolysis and brucite precipitation, and decreased rapidly with chrysotile precipitation (figure 3). Mg concentration increased constantly through the whole reaction (figure 3). At (3), brucite precipitates and pH remain constant (figures 1 and 2).

Enstatite hydrolysis produced a 3 log-unit increase in pH (figure 2). In this hydrolysis reaction more silica is released and the fluid evolved towards talc, a more silica-rich phase (figure 1).

Modeled harzburgites of two different en/fo ratios of 0.33 and 1 produced two different reaction paths. Rock of en/fo ratio=0.33 shows almost the same reaction path as pure forsterite at both temperatures and pressures (Figure 1). The other rock with an en/fo ratio of unity only produced chrysotile as a secondary mineral. Therefore, the reaction path went from the initial fluid at (1) to chrysotile saturation at (2).

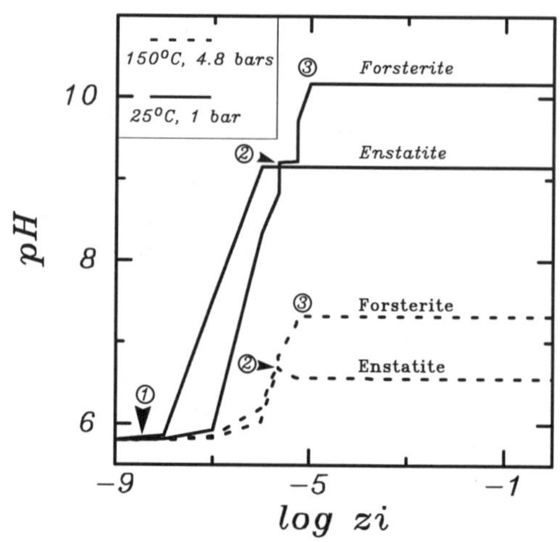

Figure 2- pH vs log zi (Helgeson, 1979) for the dissolution of the minerals forsterite and enstatite at 25 °C and 1 bar (solid line) and 150 °C and 4.8 bars (dotted line) - numbers as figure 1.

3.2 High-temperature reactions

Reaction paths at 150 °C and 4.8 bars show the same relative variations in Mg and Si as those at 25 °C and 1 bar but to a lesser degree. pH only increases by 1 and 0.5 log units above the initial value of 5.8 for forsterite and enstatite dissolution respectively (figure 2). This increase is even less in reactions involving initial alkaline solutions(pH=8) because of the relatively high-pH equilibria of serpentinization reactions.

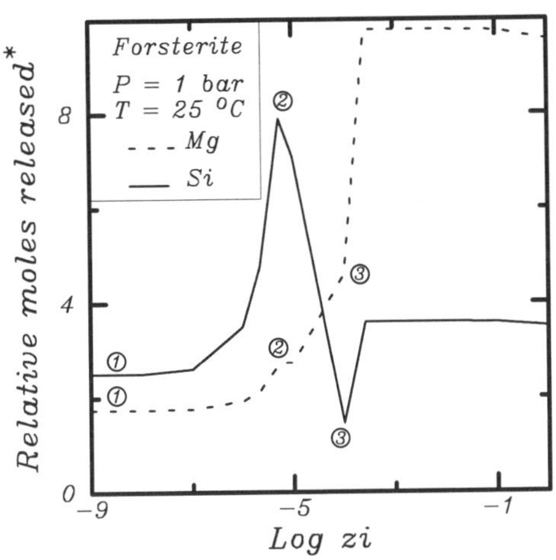

Figure 3 - Concentration of Mg and Si released during forsterite dissolution *(m_{Mg} X 10^5, m_{Si} X 10^6) vs Log zi (Helgeson, 1979) - numbers as figure 1.

3.3 Reactions involving CO_2

The resulted reaction paths using a fluid in equilibrium with atmospheric CO_2 are essentially the same as the CO_2-free reaction for forsterite (figure 1) with no saturation of Mg-carbonates such as magnesite and dolomite. An exact duplication of the run made by Pfeifer(1977) using his model peridotite (90%Fo:8%En:2%Diop) and a CO_2-rich fluid (R5 and F6 in Pfeifer(1977)) was done to test the reaction paths published in his work and to check for quartz saturation which his model predicts at 20 °C and 1 bar. The present model did not predict quartz saturation but does predict calcite, dolomite, and magnesite saturation in the early stages of the reaction path calculation.

4. DISCUSSION

In general, reaction paths for pure forsterite and harzburgite (en/fo=.33) both produced chrysotile and brucite as secondary mineral assemblages. However, very small differences exist between their reaction paths. The harzburgite reaction path lies between the pure enstatite and pure forsterite reaction paths. This predicted behavior for harzburgite dissolution having a reaction path that is intermediate between the one's for the pure minerals is in agreement with the observations of Hemley et al.(1977b).

The absolute increase in pH was lower at higher temperatures because of increasing Mg^{++} complexation with OH^- ion with increasing temperature. Thus, less H^+ is consumed in the hydrolysis reactions at higher temperatures.

In contrast to the results of Pfeifer (1977), reactions involving CO_2-rich solutions did not result in quartz saturation. However, his observations about non-saturation of carbonate minerals in low-CO_2 fluids is consistent with the present work.

5. CONCLUSIONS

The most distinctive characteristic of all reactions was the large increase in pH during mineral hydrolysis which located the equilibria of hydration reactions in the high-pH range. This pH increase was larger at lower pressures and temperatures, because of a greater degree of Mg-OH complexation with increasing pressures and temperatures resulting in less H^+ consumption. This result agrees with present day serpentinization at low temperature where high pH waters have been observed (Barnes and O'Neil, 1969).

The relative proportions of forsterite and enstatite had little effect on the reaction paths during harzburgite dissolution reactions but it has some effects on the presence and relative amount of product minerals.

Similarly, the presence of CO_2 in low concentrations didn't have a significant effect on the dissolution reactions. However, in high concentrations the presence of CO_2 will allow precipitation of carbonate minerals and hence changes in the resulting reaction path.

REFERENCES

Barnes, I and O'Neil, J.R., 1969. The relationship between fluids in some fresh alpine-type ultramafics and possible modern serpentinization, western United States. Geol. Soc. Am. Bull. V.80.

Helgeson, H.C., 1979. Mass transfer among minerals and hydrothermal solutions: in geochemistry of hydrothermal ore deposits. H.L. Barnes ed., New York: John Wiley & Sons. 568-610.

Hemley, J.J., Montoya, J.W., Christ, C.L., and Hostetler, P.B., 1977a. Mineral equilibria in the MgO-SiO_2-H_2O system: I Talc-Chrysotile-Forsterite-Brucite stability relations. Am J. Sci. 277 322-351.

Hemley, J.J., Montoya,J.W., Shaw, D.R., and Luce R.W., 1977b. Mineral equilibria in the MgO-SiO_2-H_2O system: II Talc-Antigorite-Forsterite-Anthophyllite-Enstatite stability relations and some geologic implications in the system. Am J. Sci. 277 353-383.

Martin, B. and Fyfe, W.S., 1970. Some experimental and theoretical observation on the kinetics of hydration reaction with particular reference to serpentinization. Chem. Geol., 6 185-202.

Whittaker, E.J.W. and Wicks, F.J., 1970. Chemical differences among serpentine "polymorphs": A discussion. Am. Mineral. Vol.55 1025-1047.

Wolery, T.J., 1979. Calculation of chemical equilibrium between aqueous solution and minerals: The EQ3/6 software package. Lawrence Livermore Laboratory URCL-52658.

Wolery, T.J., 1983. EQ3NR, a computer program for geochemical aqueous speciation-solubility calculations: Users guide and documentation. Lawrence Livermore Laboratory URCL-53414.

Water-Rock Interaction, Kharaka & Maest (eds) © 1992 Balkema, Rotterdam. ISBN 90 5410 075 3

High temperature metamorphism and low temperature retrograde alteration of spontaneously combusted marls: The Maqarin cement analogue, Jordan

Hani N. Khoury
Department of Geology and Mineralogy, University of Jordan, Amman, Jordan

Tony E. Milodowski
British Geological Survey, Keyworth, Nottingham, UK

ABSTRACT: An unusual assemblage of high- and low-temperature metamorphic minerals occurs in the Maqarin area of northern Jordan. The bituminous host "Chalk-Marl" unit and its stratigraphic equivalents have been thermally metamorphosed by in situ spontaneous combustion of the bitumen, which represents up to 20 wt% of the unaltered rock. The minerals in the Maqarin area lie in two assemblages:

(i) A univariant progressive reaction series was established at high temperature, low presure conditions (isochemical decarbonation-rehydration), normally characteristic of contact metamorphic rocks. The mineral suite includes larnite, wollastonite, diopside, and graphite, plus garnet, anorthite, ferrites, and recrystallized calcite, consistent with in situ combustion of impure limestones and marls. Textural and structural analyses suggest a rapid heating where maximum temperatures were achieved early.

(ii) A retrograde recrystallization accompanied with hydration, carbonation, and sulphatization. The metamorphic assemblage includes calcium silicate hydrates (tobermorite, jennite, afwillite, and apophyllite), sulphates (ettringite, hashemite, barite, thaumasite and gypsum), and stable and metastable carbonates (calcite, vaterite, aragonite, and kutanahorite) and oxides and hydroxides (hematite, portlandite, and gibbsite).

In spite of high- and low-temperature metamorphism, the metamorphic fluid is still present at Maqarin. Highly alkaline groundwaters discharged from recrystallized rocks have pH values of up to 12.5, consistent with the solubility of portlandite and calcium silica hydroxide (CSH) phases. The Maqarin site is being studied as a natural analogue for a nuclear waste repository in a sedimentary environment.

1 INTRODUCTION

A high-temperature mineralization zone occurs within the bituminous limestone beds of Upper Cretaceous-Lower Paleocene age in Maqarin area, northern Jordan (Figs. 1 and 2). The metamorphic rocks are restricted to the left flank of the Yarmouk River and occur in an unusual brecciated unit, which extends from Wadi es Sidjin eastward to the dam site towards Wadi Shallala to the southeast for about 1900m. The rocks are highly heterogeneous in texture, color and composition and have been investigated by Khoury and Nassir (1982).

Thermally metamorphosed rocks by spontaneous combustion of the bituminous limestones and marls are present in other parts of Jordan and its stratigraphic equivalent of Palestine, the Hartrurim Formation (Mottled Zone). The concept of organic matter combustion as responsible for thermal metamorphism has been discussed (Lee 1928; Bentor et al. 1963, 1972; Gross et al. 1967; Kolodny et al. 1971; Kolodny and

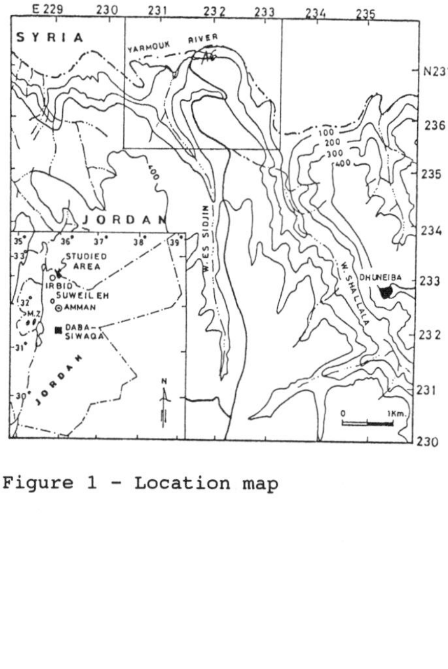

Figure 1 - Location map

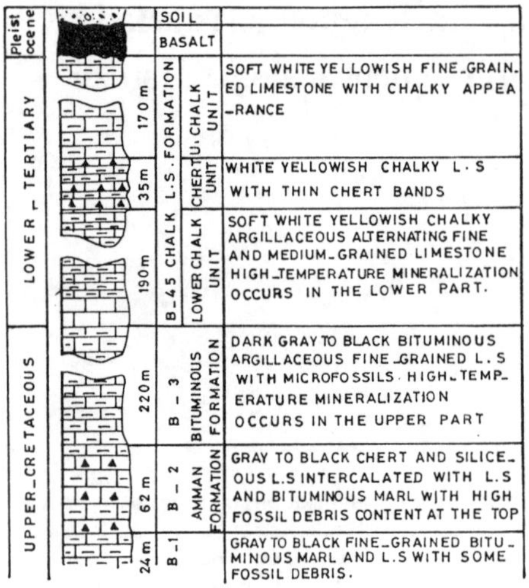

Figure 2 - A columnar section

Gross 1974; Gross 1977; Mathews and Kolodny 1978; Kolodny 1979). The metamorphic rocks in Jordan belong to the sanidinite and pyroxene hornfels facies and show strong coloration in all shades of brown, red, pink, violet, grey, green and black. The rocks are highly fractured and secondary mineralization is very common in voids, fissures and fractures. In Jordan, three marble areas were recognized and have been studied by Heimbach and Rösch (1980), Nassir and Khoury (1982), Khoury and Nassir (1982 a & b), and Khoury and Salameh (1986). The largest marble area occurs in Daba-Siwaqa, 50km south of Amman. The rocks in Daba-Siwaqa area belong stratigraphically to the Chalk Marl and Chert Limestone Units of Maestrichtian-Danjan-Lower Paleocene age. The second and the third metamorphic areas are in Suweileh, 16km NW of Amman, and in Maqarin, 16km north of Irbid. The marbles in Daba-Siwaqa, Suweileh, Maqarin and the Mottled Zone have similar stratigraphy, lithology, mineralogy and origin. In the past isotopic evidence was used to confirm the thermal metamorphism by combustion of organic matter in the Mottled Zone, Daba-Siwaqa and Suweileh (Kolodny and Gross 1974; Heimbach and Rösch 1980; and Khoury and Salameh 1986).

2 RESULTS AND DISCUSSION

2.1 The bituminous limestone unit

The bituminous limestone rocks are the protoliths for the metamorphic rocks and are essentially biomicrites made up of calcite. Oil droplets and authigenic pyrite and quartz were noticed in some foraminiferal tests (Abed and Amireh 1983). The principle minerals of the unit are calcite, dolomite, quartz, clay minerals, apatite and pyrite. The average composition of calcite, quartz and clay minerals (kaolinite and illite) is 69%, 6% and 5% respectively. Organic matter constitutes 15% in the outcrops and 20% in the core samples. The average sulfur content is 1.15% and the trace elements Cr, Cu, Mo, Ni, V and Zn are associated with the organic matter (Abed and Amireh 1983).

2.2 Highly alkaline waters

An unusual highly alkaline calcium hydroxide-sulfate waters rich in base metal concentrations (pH 12.5) are recorded in the Maqarin area (Khoury et al. 1985). The high pH is due to the interaction with natural cement minerals. The alkaline waters precipitate soft travertine and extract base metals from the original bituminous rocks. The waters are however, supersaturated with respect to wollastonite and are of meteoric origin ($\delta^{18}O=-4.4°/_{oo}$ SMOW). All the travertines in the area are recent and are precipitating as a result of the reaction of the hydroxide waters with atmospheric CO_2.

2.3 The mineralogy of the metamorphic rocks

Two groups of minerals are recognized and are almost similar to those in the Mottled Zone; the identified minerals are listed in Tables 1 and 2. High temperature assemblage reflects an isochemical decarbonation-dehydration metamorphism of impure limestones and marls.

Low-temperature minerals fill fissures, cracks, voids and cavities with the exception of the massive portlandite. Portlandite shows a common sheet-like structure where stacking of the sheets gives an acicular and fibrous habit. The assemblage indicates a later multistage retrograde reaction.

The mineralogy is comparable to that of cement clinker, and to hydrated cement products.

2.4 The origin of the unusual mineralization

The origin of high and low temperature mineral assemblages is explained below. The bituminous limestone and marl calacines were subjected to spontaneous combustion. Textures and structures (porphyroblastic, poikiloblastic, granulated, broken and cracked crystals, columnar structure cavity, vein and fracture fillings and concretionary growth), suggest a rapid heating followed by rapid cooling. The lime hydrates were introduced by groundwater to form portlandite. Portlandite reacts with the SO_4^{-2}, $H_3SiO_4^-$, $Al(OH)_4^-$ and CO_3^{-2} in the groundwater (at pH values of 12.4 and higher) to form ettringite, thaumasite and other minerals. Such reaction is similar to the hydration mechanism of cementing material. If the marls are siliceous, spurrite, wollastonite and other silicates may be expected. The variation in mineralogy is a function of strating material, temperature, P_{CO2} and pH.

During the crystallization of the unusual low-temperature minerals in cavities, voides and fissures, the composition of the circulating alkaline water and the availability of CO_2 were the important factors. The occurrence of portlandite and single crystals of different calcium silicate hydrates, sulfates and calcium aluminum hydroxisalts is due to the precipitation from the alkaline circulating water and/or to the water-rock interaction.

Table 1. Primary metamorphic minerals in the Maqarin metamorphic zone.

Mineral species	Ideal formula
flourapatite	$Ca_{10}(PO_4)_6F_2$
francolite	$Ca_{10-x-y}(Na,K)_xMg_x(PO_4)_{6-z}(CO_3)_zF_{0.4z}F_2$
ellestadite	$Ca_{10}(SiO_4)_3(SO_4)_3O_{24}(Cl,OH,F)_2$
spurrite	$Ca_5(SiO_4)_2(CO_3)$
wollastonite	$CaSiO_3$
larnite	$-Ca_2SiO_4$
diopside-hedenbergite	$Ca(Al,Fe)Si_2O_6$
anorthite	$CaAl_2Si_2O_8$
brownmillerite	$Ca_2(Al,Fe)_2O_5$
Ca-ferrite	$CaFe_2O_3$
(?)ferrites	undefined Ba,Cr,Al,Ti,Mg,Zn,Mn-bearing
hematite or ferric oxide	Fe_2O_3
Ca-aluminate	undefined
calcite	$CaCO_3$
graphite	C
lime	CaO
Ba,Ca,S-silicate	undefined
Ba,Ca,Zr,Mo,silicate	undefined
oldhamite	CaS to $CaS_{0.9}Se_{0.1}$
Cu,K,Na-selenide	$Cu_{10.2}K_3Na_{0.2}Se_{7.7}S_{2.3}$(approx)
UCa-oxycarbonate (?)	Ca : U = 2

Table 2. Secondary minerals in the Maqarin metamorphic zone.

Mineral species	Ideal formula
calcite	$CaCO_3$
aragonite	$CaCO_3$
vaterite	$CaCO_3$
kutnahorite	$Ca_{0.75}(Mn,Mg)_{0.26}CO_3$
strontianite	$SrCO_3$
hematite	$-Fe_2O_3$
maghemite	$-Fe_2O_3$
gibbsite	$-Al(OH)_3$
brucite	$Mg(OH)_2$
portlandite	$Ca(OH)_2$
quartz	SiO_2
opal-CT	SiO_2
opal-A	$SiO_2.nH_2O$
baryte	$BaSO_4$
barytocelestite	$(SrBa)SO_4$
calcian barytocelestite	$(SrBaCa)SO_4$
hashemite	$BaCrO_4$ to $BaSO_4$ complete solid solution
Cd-sulphate	undefined
Pb-sulphate	undefined
gypsum	$CaSO_4$
bassanite	$CaSO_4.0.5H_2O$
ettrinigte	$Ca_6Al_2(SO_4)_3(OH)_{12}.25H_2O$
thaumasite	$Ca_6Si_2(SO_4)_2(CO_3)_2(OH)_{12}.24H_2O$
Cu,Zn-sulphate	undefined
hydroxyapatite	$Ca_{10}(PO_4)_6(OH)_2$
flourapatite	$Ca_{10}(PO_4)_6F_2$
francolite	$Ca_{10-x-y}(Na,K)_xMg_x(PO_4)_{6-z}(CO_3)_zF_{0.4z}F_2$
ellestadite	$Ca_{10}(SiO_4)_3(PO_4)_3O_{24}(Cl,OH,F)_2$
afwillite	$Ca_3Si_2O_4(OH)_6$
tobermorites	$Ca_5Si_6O_{16}(OH)_2.2\text{-}8H_2O$
jennite	$Ca_9H_2Si_6O_{18}(OH)_8.6h_2O$
apophyllite	$KCa_4Si_8O_{20}(OH,F)8H_2O$
birunite	$Ca_{15}(CO_3)_{5.5}(SiO_3)_{8.5}SO_4.15H_2O$
CSH gel	amorphous, undefined
U,Ca-silicate	undefined

The availability of oxygen is the most important factor for combustion. Folding in Maqarin area lead to cracking and created channels for supplies of oxygen. Upon cooling, contraction and collapse took place to form new channels of oxygen supply. The release of carbon dioxide combined with other volatiles such as water would result in pressures sufficient to brecciate or fracture rocks overlying the area of combustion. The circulating meteoric waters dissolve portlandite which is 125 times more soluble than calcite in distilled water. The high alkalinity of the waters is maintained by saturation of portlandite. Trace elements could be introduced into the circulating waters from the calcined bituminous limestone, which originally contains heavy metals. Recarbonation and hydration processes accompanying the cooling stages might have increased the temperature (exothermal reactions) and could explain the occurrence of hydrothermal minerals like apophyllite. The hydrated minerals often contain high trace elements. The amorphous calcium silicate contains

1.78 CaO; SiO_2 contains 0.34% Zn; CrO_4^{-2} replaces SO_4^{-2} in these minerals. Solid solution exists between ettringite and thaumasite and between the different types of apatites and carbonates. Additional minerals however, are expected to be found in the studied area.

3 CONCLUSIONS

The highly alkaline water precipitates low-temperature minerals in fissures and voids; such low-temperature process is still active as a result of reaction of these waters with the rocks. A complex mineralogy however, is expected in Maqarin area as a direct action of the highly alkaline circulating water. The bonding of the different trace elements to the hydrated mineral assemblage is of great importance in preventing contamination of the groundwater system.

The Maqarin site offers a good area to investigate the interactions of the high pH water with the metamorphic and non-metamorphic rocks. A similar situation is likely to occur if a low and intermediate level waste repository was constructed in a sedimentary site.

REFERENCES

Abed, A. and B.Amireh 1983. Petrography and geochemistry of some Jordanian oil shales from north Jordan. Jour. of Petroleum Geology 5.3: 261-274.

Barnes, I., T.Presser, M.Saines, P.Dickson & A.Koster Van Gross 1982. Geochemistry of highly basic calcium hydroxide groundwater in Jordan. Chem. Geol. 35: 147-154.

Bentor, Y.K., S.Gross & L. Heller 1963. Some unusual minerals from "Mottled Zone" complex, Israel. Amer. Min. 48: 924-930.

Bentor, Y.K., S.Gross & Y.Kolodny 1972. New evidence on the origin of high temperature assemblage of the "Mottled Zone" Israel. International Geol. Congress 2: 265-275.

Gross, S., E.Mazor, S.Sass & I.Zak 1967. The "Mottled Zone" complex of Nahal Ayalon (Central Israel). Israel J. Earth-Sci. 16: 84-96.

Gross, S. 1977. The mineralogy of Hartrurim Formation. Israel. Geol. Survey of Israel Bull. 70: 80 p.

Heimbach, W. & H.Rösch 1980. Die Mottled Zone in Zentral-Jordanien. Geol. Jb. 340: 3-17.

Khoury, H. & S.Nassir 1982a. High temperature mineralization in Maqarin area, north Jordan. N. Jb. Miner. Abh. 144: 197-213.

Khoury, H. & S. Nassir 1982b. A discussion on the origin of Daba-Siwaqa marble. Dirasat 9: 55-66.

Khoury, H., E.Salameh & Q.Abdul-Jaber 1985. Characteristics of an unusual highly alkaline water from the Maqarin area, northern Jordan. Journal of Hydrology 81: 79-91.

Khoury, H. 1985. The origin of highly alkaline waters from the Maqarin area, north Jordan. Dirasat. 12: 125-131.

Khoury, H. & E.Salameh 1986. The origin of high temperature minerals from Suweileh area, Jordan. Dirasat. 13: 261-269.

Kolodny, Y., M.Bar & E.Sass 1971. Fission track age on the "Mottled Zone" in Israel. Earth and Planet Sci. Lett. 2: 269-272.

Kolodny, Y. & S.Gross 1974. Thermal metamorphism by combustion of organic matter; isotopic and petrological evidence. J. Geol. 82: 489-506.

Kolodny, Y. 1979. Natural cement factory: A geological story. Engineering Foundation Conference: 1-15.

Lee, G. 1928. The chert beds of Palestine. Proc. Geol. Ass. London 39: 445-463.

Matthews, A. & Y. Kolodny 1978. Oxygen isotope fraction in decarbonation metamorphism; The Mottled Zone event. Earth and Planet. Sci. Lett. 39: 179-192.

Nassir, S. & H.Khoury 1982. Geology, Mineralogy and petrology of Daba marble, Jordan. Dirasat. 9: 107-130.

Water-Rock Interaction, Kharaka & Maest (eds) © 1992 Balkema, Rotterdam. ISBN 90 5410 075 3

Fluid regime of glaucophane metamorphism

G.G. Likhoidov & L.P. Plyusnina
Far East Geological Institute, Vladivostok, Russia

ABSTRACT: To quantify the blueschist facies metamorphism the glaucophane (Gl) stability field was experimentally studied within a temperature range 220-500°C and P = 4-6 kbar as functions of fluid chemistry and oxygen fugacity. The Gl stability field expands with increasing activity of Na in solutions and reaches its maximum ranges of 220-410°C at a given P. In the mixed 1 m NaCl with 0.1 m carbon dioxide solution, the equilibrium Gl with Gr solid solutions was defined at 370±10°C (P = 4 kbar, Fe-Mt buffer). With increasing pH of the Na_2SiO_3 solutions, the Gl + Gr association is replaced by the Gl + Cpx at the same P-T, and further pH increases cause Gl replacement by Mg-riebeckite. Variations in oxygen fugacity defined by WM, NNO, and HM buffers at constant P-T results in the crystallization of ferroglaucophane, crossite, and riebeckite, respectively. Thus the Gl stability field is highly dependent on fluid chemistry, and is restricted to the environments with reducing Na-bearing metamorphic solutions.

1 INTRODUCTION

Glaucophane-bearing metamorphic rocks occupy a somewhat unusual tectonic setting. Commonly they are restricted to the ophiolitic belts and were described in terrains that suffered different-grade metamorphism from the zeolite to eclogite facies. This is why the blueschist facies metamorphism has always been difficult to quantify, and there are contradictory ideas about the glaucophane origin and its intrinsic fluid regime (Ernst 1963; Dobretsov 1974; Marakushev 1979; Wood 1970; Okay 1980; etc.). The purpose of our report is, first, to study experimentally an interaction of the glaucophane-bearing assemblages with solutions of different composition, and second, to discuss the petrologic significance of inferred results for best understanding the peculiarities of Gl metamorphism as a special case of a more general interaction between fluid and minerals in the course of metamorphism.

2 EXPERIMENTAL RESULTS

The low limit of the Gl stability field in the Fe-free system may be described by a reaction of hydrolysis

$$3Chl+4Ab+19Q+6Na^+ = 5Gl4H_2O+6H^+. \quad (1)$$

It was studied using 1m NaCl, 0.25m Na_2SiO_3, and 0.5m Na_2SiO_3 solutions at P = 4 and 6 kbar; preliminary runs using pure water revealed entire Gl breakdown into the Ab+Chl assemblage at T below 400°C. Experiments were performed in hydrothermal cold seal pressure vessels. The accuracy of temperature measurements was to ±5°C, and pressure to ±50 bar. The natural Ab of stoichiometric composition, Mg-Chl - $Mg_{4.65}Fe_{0.05}Al_{2.17}Si_{3.10}(OH)_8$, and Mg-Gl from blueschist of the Maksiutov metamorphic complex were used as starting materials. Experimental procedure was described in detail before (Likhoidov & Plyusnina 1988).

Data obtained are plotted in Fig. 1 where the equilibrium T is a function of log $([Na^+]/[H^+])$-value of reacting solutions. At the increase of the latter from 5.20 to 6.00, the equilibrium T decreases from 360 to 220°C (P = 4 kbar). Meanwhile, increase of P from 4 to 6 kbar shifts slightly the equilibrium curve towards lower a_{Na^+}/a_{H^+} ratio. The comparison of isobars 4 and 6 kbar depicts a negligible pressure effect on equilibrium T presumably induced by small delta V solids of reaction (1).

In the more alkali 0.5m Na_2SiO_3 solutions (pH=9.0), Mg-riebeckite (Mrb) forms without seeding along with Gl within the range 340-400°C. To evaluate the stability limits of

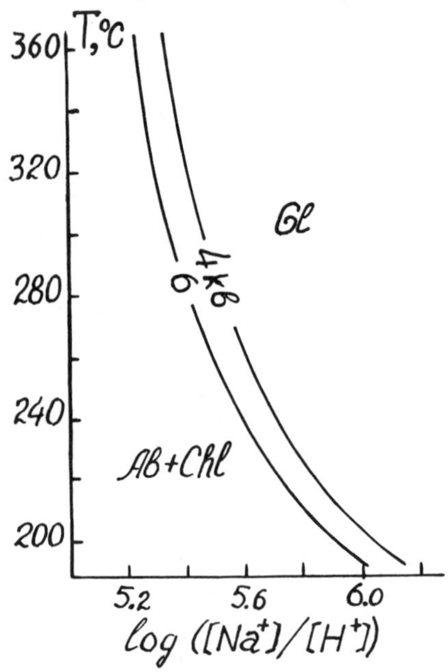

Figure 1 - Temperature of the equilibrium for reaction 3CHl + 4Ab + 19Q + $6Na^+$ = 5Gl + $4H_2O$ + $6H^+$ versus $\log([Na^+]/[H^+])$

sodium amphiboles, the reaction

$$3Mg\text{-}Gl + 5Mt + 25Q + 5H_2O + 6Na^+ = 3Rb + 2Ab + 3En + O_2 + 6H^+ \qquad (2)$$

was calculated using known Rb thermodynamic constants (Makarova et al. 1984). The thermodynamic data on Gl were taken from Likhoidov et al. (1982), and data on the other minerals from Helgeson et al. (1978). Under the conditions of the HM buffer, the equilibrium curve of reaction (2) was defined at 300°C, and $\log([Na^+]/[H^+])$ = 10.90; 400°C and 10.23; and 500°C and 9.87 (P = 4 kbar). Thus, the equilibria (1) and (2) limit the Gl stability field in the Na-bearing solutions.

Then, conditions for the stable coexistence of the higher temperature association Gl_{ss}+ Gr_{ss}+ Ab + Zo were studied in the model NCMFAS system. Its equilibrium may be described by the

$$Gl_{ss}+Zo+CO_2=Gr_{ss}+Ab+Cc+Mt+H_2O. \qquad (3)$$

The technique and experimental procedure were discussed in detail (Likhoidov & Plyusnina 1989); we briefly report the results. At P = P_{H2O}= 4 and 6 kbar, garnet with essentially almandine composition is replaced by Chl+Mt assemblage at T below 500°C which agrees with the experimental data of Hsu (1968). As a result, Gl-Gr association is stable under the conditions of $P_{H2O} < P_{fl}$, achieved in our runs using the complex (1m NaCl + 0.1m CO_2) fluid. The equilibrium of (3) is defined at 370±10°C, P = 4 kbar, and at 380 ± 10°C, 6 kbar at P_{O2} defined by the Fe-magnetite buffer. Equilibrium compositions of coexisting Gl and Gr solid solutions were determined by reversal experiments using starting Gl and Gr with different contrasting Fe/Mg ratio (fig. 2). The equilibrium composition was approached from both sides. It is noteworthy that at T above 400°C, Gl is not observed in run products, but Gr modal amount increases with rising T.

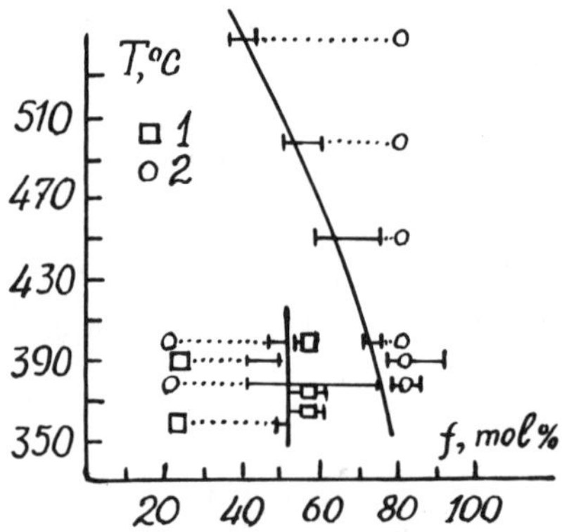

Figure 2 - The equilibrium compositions of coexisting Gl and Gr versus T at P = 6 kbar (f = Fe/(Fe+Mg)x100; 1, and 2 - starting compositions of Gl and Gr, respectively; bars indicate the uncertainty)

Low content of the grossularite end-member in the latter was measured by EMA (0.03 per units at 380°C, and 0.043 at 400°C). The coexistence of Gl with Gr of higher Fe/Mg ratio is confirmed by natural assemblages (Graham & Powell 1984).

The other set of runs was performed in the more alkali 0.5m Na_2SiO_3 solutions. To avoid the high activity of Fe in this solution that resulted from the noticeable solubility of the Fe-bearing buffers, we used the Mo-MoO mixture, providing identical oxygen fugacity. In those runs, clinopyroxene (Cpx_{ss}) formed at the expense of Gr_{ss}, the first appearance of Cpx was recorded at 410°C (4 kbar). The modal Cpx amount increases with increasing T. Newly formed Cpx contains the jadeite end-member; positive correlation between Na and Al contents was ascertained by EMA. For example, at 550°C, Na and Al contents achieved 0.5 per f.unit for each. Increasing T at constant P is also accompanied by the shift of the Gl composition towards Ca-Na amphiboles with lower Al content, beginning from 410°C. An exchange reaction between Cpx and Ca-Na Amph can be studied separately. Hence, the increase in alkalinity of reacting solutions leads to transition reaction (3) into

$$Gl_{ss} + Zo + Mt = Cpx_{ss} + Ab + Cc. \quad (4).$$

A stable coexistence of 3-phase assemblage Gl+Cpx+Gr is realized in the solutions with a fixed value of $\log([Na^+]/[H^+])$. The calculation of the hydrolysis reactions of almandine $Fe_3Al_2Si_3O_{12} + Q + H_2O + 2.5O_2 + 2Na^+ = 2Ab + Mt + 2H^+$, Na-Cpx (aegirine)-$2NaFeSi_2O_6 + 2H^+ = Hem + 4Q + H_2O + 2Na^+$, and the Gl via (1) indicates that the 3-phase compatible assemblage coexist at a restricted narrow range of $\log([Na^+]/[H^+])$ = 6.5 - 6.6 for a given P-T (Likhoidov & Plyusnina 1989).

The studied equilibria (1), (2), and (3) illustrate the enlargement of the Gl stability field with the increase of alkalinity of reacting solutions in the range 220-410°C at P = 4-6 kbar. However, a further increase of the alkalinity leads to the Gl replacement by Na-amphiboles of Rb-Mrb composition which is typical for alkali metasomatites (Miyashiro 1957).

Major mineral species of Na-amphiboles are also highly sensitive to oxygen fugacity during metamorphism (Okay 1980). The lack of reliable thermodynamic and experimental data on these had encouraged us to study experimentally the effect of fO_2 on the compositions of Na-amphibole solid solutions. The Fe-Gl-Rb series was synthesized at 500°C, 5 kbar as function of fO_2, which is defined by the WM, NNO, and HM buffers. The starting mixtures consist of $Al(OH)_3$, Na_2CO_3, $Fe_2C_2O_4.2H_2O$, and amorphous SiO_2, taken in the stoichiometric proportion corresponding to the Fe-Gl composition - $Na_2Fe_3Al_2Si_8O_{22}(OH)_2$. Runs were performed in the 1m NaCl solution, and run products were studied by X-ray and EMA (Plyusnina & Likhoidov 1986). At more oxidizing conditions (HM buffer), the content of the Rb end-member in newly formed Na-amphiboles increased, along with albite in run products. Na-amphiboles of the Fe-Gl composition were synthesized in the runs with WM oxygen buffer only (Fig. 3). The composition of newly formed amphiboles (buffered by NNO) ranges Al_2O_3 content from 2 to 8.5 wt%, and FeO (total) content from 36.2 to 41.0 wt%. It corresponds to the variation in content of Fe-Gl end member from 0.13 to 0.73 mole%. At the maximum oxygen fugacity defined by the HM-buffer, the systematic

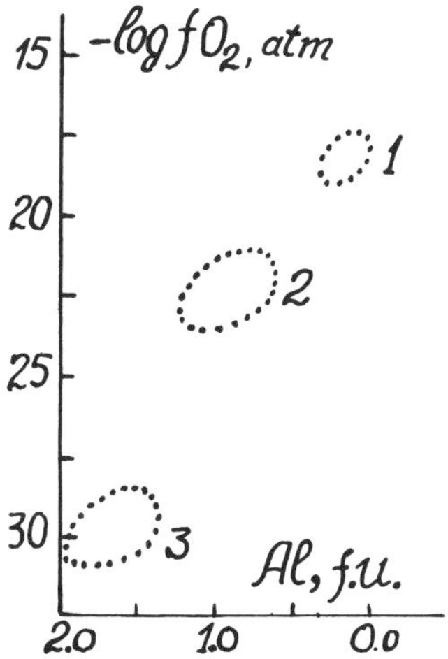

Figure 3 - Compositions of synthesized sodic amphiboles versus oxygen fugacity defined by: 1 - WM, 2 - NNO, 3 - HM (500°C / 5 kbar).

Na-amphiboles contain insignificant amount of the Fe-Gl end-member (0.02-0.15 mole%, see Fig. 3). Thus, the sodic amphiboles of the general composition

$Na_2(Mg,Fe)_3(Al,Fe)_2Si_8O_{22}(OH)_2$ are stable at the specific value of oxygen fugacity at a given P-T.

3 CONCLUSION

Although Gl has been assumed to be a high-pressure indicator following Ernst (1963), our new data demonstrate that the Gl stability field may be also affected by solution chemistry. Increasing oxygen as well as sodium activities in metamorphic fluids at a constant P-T promote the formation of Na-amphiboles with decreasing Al content. Variations of these fluid parameters during their interaction with permeable metamorphic rocks may occur within a restricted area, thus causing unusual heterogeneity, diversification, and complexity of the mineralogy often observed within a single metamorphic complex and even in a single sample (Black et al. 1973; Brown & Ghent 1983; Wood 1979, 1980). This explanation seems more reliable than the idea about possible variations in pressure within a narrow rock volume (Dobretsov 1974; and others). A compositional divergency of Na-amphiboles from the graphite-bearing metapelites intercalated with metabasites (Wood

1980) illustrates a different buffering capacity of these rocks in relation to oxygen fugacity in the course of amphibole crystallization.

In summary, the fluid regime of glaucophane metamorphism is defined by the low oxygen potential and enhanced Na activity. This explains an observed areal connection of blueschists with deep fault zones in the ophiolite belts (Marakushev 1979). Study of chemical compositions of blue and green schists from the Penginskii metamorphic complex, Kamchatka detected an addition of soda by reducing fluids in the fault zone (Likhoidov & Plyusnina 1979).

It is, however, important that the so-called glaucophane metamorphism is really not isochemical features, and the influence of metasomatism is much greater within this facies than within other facies (Wood 1979). This suggestion is confirmed by the Gl formation in such contrasting rock as metabasite, metachert, metapelite, serpentinite, and graphitic slate, by the Gl appearance in rocks of different metamorphic grade, and by a common Gl association with alkali metasomatic minerals. Hence, we propose that the concept of blueschist facies metamorphism under high pressure, low temperature conditions disregarding the specific fluid regime should not be accepted. In conclusion, it should be noted that a water-rock interaction during initial low-temperature metamorphic events presents some examples of particular significance of modal fluid chemistry; it likely exceeds the significance of P-T. The role of fluid in blueschist facies metamorphism is more complex than that of the normal regional metamorphism. The zonation and overgrowth often observed in the sodic amphiboles may then be interpreted as a record of changing fluid conditions.

REFERENCES

Black, Ph.M. 1973. Mineralogy of New Caledonian metamorphic rocks 2. Amphiboles from the Oregon district. Contrib. Mineral. Petrology 39: 55-64.

Brown, E.N. & J.D.Gent 1983. Mineralogy and phase relations in the blueschist of the Black Butte and Ball Rock areas, northern California Coast Ranges. Amer. Mineralogist 68: 365-372.

Dobretsov, N.L. 1974. Glaucophane-schist and glaucophane-schist eclogite complexses of the USSR. Novosibirsk: Nauka (in Russian).

Ernst, W.G. 1963. Polymorphism in alkaly amphiboles. Amer. Mineralog. 48: 241-260.

Graham, C.M. & R.A.Powell 1984. A garnet-hornblende geothermometer: calibration, testing and application to the Pelone schists, southern California. Journ. Metamorph. Petrology 2: 13-31.

Helgeson, H.C., J.M.Delany, H.W.Nesbitt & D.K.Bird 1978. Summary and critique of the thermodynamic properties of rock-forming minerals. Amer. J. Sci. 278A: 1-230.

Hsu, L.C. 1968. Selected phase relationships in the system Al-Mn-Fe-Si-O-H: a model of garnet equilibria. J. Petrology 9: 40-43.

Likhoidov, G.G. & L.P.Plyusnina 1979. About petrogenesis of glaucophane-bearing rocks for an example of SW Kamchatka region. Dokl. Acad. Nauk USSR 246: 1203-1205 (in Russian).

Likhoidov, G.G. & L.P.Plyusnina 1988. Experimental study of lower stability limit of magnesian glaucophane. Geochimia 7: 973-980 (in Russian).

Likhoidov, G.G. & L.P.Plyusnina 1989. Experimental study of glaucophane stability in the presence of garnet and clinopyroxene. Isvestia Acad. Nauk USSR 7: 68-76 (in Russian).

Likhoidov, G.G., Ju.I.Sidorov, J.M.Gurevich et al. 1982. Thermodynamic properties of glaucophane $Na_2Mg_3Al_2Si_8O_{22}(OH)_2$ and some petrologic consequences. Geochimia 7: 1002-1013 (in Russian).

Makarova, T.I., Ju.I.Sidorov & V.B.Naumov 1984. Conditions of formation of ferruginous minerals in uranium-bearing alkaly metasomatites. Geochimia 1: 70-83 (in Russian).

Marakushev, A.A. 1979. Petrogenesis and ore formation. Moscow: Nauka (in Russian).

Miyashiro, A. 1957. The chemistry, optics, and genesis of alkali-amphiboles. J. Fas. Sci. Univer. Tokyo ser. 2, pt. 11: 57-83.

Okay, A.J. 1980. Sodic amphiboles as oxygen fugacity indicators in metamorphism. Journ. Geology 88: 225-232.

Plyusnina, L.P. &G.G.Likhoidov 1986. Composition of Na-amphiboles of ferro-glaucophane-riebeckite series as a function of oxygen fugacity after experimental data. Mineralogicheskii Zhurnal 8: 26-31 (in Russian).

Wood, R.M. 1979. A re-evaluation of the blueschist facies. 1. The pressure-temperature hysteresis cycle. Geol. Mag. 116: 21-33.

Wood, R.M. 1980. Compositional zonning in sodic amphiboles from the blueschist facies. Miner. Mag. 43: 741-752.

Water-Rock Interaction, Kharaka & Maest (eds) © 1992 Balkema, Rotterdam. ISBN 90 5410 075 3

Thermochemical determination of water activities relevant to eclogitic rocks

H.-J. Massonne
'High-Pressure' Research Group, Institut für Mineralogie, Ruhr-Universität Bochum, Germany

ABSTRACT: Thermodynamic properties of end-members hedenbergite and Mg-Al-celadonite as well as mixing properties for components in phengite and P2/n and C2/c omphacite were obtained by least-sqare fits to experimental data. P-T paths of eclogitic rocks from the European Variszides and the Alps were determined by thermochemical calculations of equilibria involving components in garnet, omphacite and phengite. Water activities were calculated considering additional reactions with H_2O, quartz and the tremolite component in amphibole. This geohydrometric method yielded water activities above 0.5 only for a limited section of the exhumation path of the rocks investigated, which is explained by the influx of an external hydrous fluid. This event is accompanied by growth of large amphibole porphyroblasts or by partial melting indicated by quartz-feldspar inclusions in omphacite and garnet rims.

1 THERMOCHEMICAL METHOD

As reported recently by Massonne (1991a), equilibria involving muscovite and Mg-Al-celadonite components in phengite, diopside and hedenbergite components in omphacite and grossular, pyrope and almandine components in garnet define a single P-T point and, thus, allow to determine P-T paths of phengite-bearing eclogites in chemically-zoned minerals. If such eclogites also contain quartz and amphibole, additional equilibria with H_2O, quartz and tremolite component can be calculated. The water activity for these equilibria is defined by the P-T point of the water-absent equilibria.

The ΔG_R=0 relation of mineral equilibria was solved by thermochemical calculations using an advanced version of the GE0-CALC software package (Brown et al. 1988) distributed by Berman. Activity formulations were applied as given in Table 1 . Standard state thermodynamic properties of diopside, grossular and tremolite are those of Berman (1988). The data for almandine were taken from Berman (1990). Muscovite and pyrope data from Berman (1988) were slightly changed (Table 2) according to the author's recent experimental results. Mixing properties of components in Al-garnet were broadly taken from Berman (1990). Those of components in phengite and P2/n and C2/c omphacite (Table 3) are based either on experimental investigations reported in the literature or on estimates considering mixing data of similar solid solutions and solvus relations as observed from natural parageneses. The latter concerns, for example, properties of the paragonite - Mg-Al-celadonite solid solution. Among the experimental studies, the work of Gasparik (1985) on the jadeite-diopside join is of particular importance, because of the interrelation of P2/n and C2/c omphacites, which was later on also transferred to natural omphacites with unknown space group.

The mixing properties of the muscovite - Mg-Al-celadonite solid solution as well as the thermodynamic data of end-member Mg-Al-celadonite in Table 2 were obtained by a least-square fit to experimental data of Massonne & Schreyer (1987, 1989) and Massonne (unpubl.). The α_P and β_T functions of Mg-Al-celadonite were assumed to be identical to muscovite. The method of least-squares was also applied to experimental results of Ellis & Green (1979) and Pattison & Newton (1989) to derive the following data simultaneously: thermo-

Table 1 . End-member components and activity models. Tr = Tremolite, ❒ = vacancy (A-site).

	Omphacite	Phengite	Al-Garnet	Amphibole (molar fractions)
1	Diopside (Dp)	Muscovite (Ms)	Grossular (Gr)	Ca/2 = XCa
2	Hedenbergite	$K(Mg,Fe^{2+})_3AlSi_3O_{10}(OH)_2$	Pyrope	$Na^{M4}\cdot(Al^{VI}/(Al^{VI}+Fe^{3+}))/2$
3	Jadeite	Mg-Al-celadonite	Almandine	$Na^{M4}\cdot(Fe^{3+}/(Al^{VI}+Fe^{3+}))/2$
4	Opx:$(Mg,Fe^{2+})_2Si_2O_6$	Fe^{2+}-Al-celadonite	Spessartine	
5	$CaAlAlSiO_6$	Paragonite		$a_{Tr} = XCa^2\cdot\gamma_1 \cdot OH/(OH+F)$
6	rest (acmite, etc.)	rest (e.g. $KTi(MgFe^{2+})AlSi_3O_{10}(OH)_2$)		$\cdot❒\cdot\gamma_2 \cdot Mg/(5-Na^{M4})$
	$a_{Dp} = X_{Dp}\cdot\gamma_{Dp}$	$a_{Ms} = X_{Ms}\cdot\gamma_{Ms}\cdot XOH$	$a_{Gr} = XCa^3\cdot\gamma_{Gr}\cdot XAl^2$	

Table 2. Thermodynamic standard state properties. * = Berman (1988). ** = following Berman & Brown (1985)

	H_f J/mol	S J/K/mol	V J/bar/mol	c_P
Muscovite	-5976740*	293.157*	14.080	*
Mg-Al-celadonite	-5832415	288.527	13.870	**
Hedenbergite	-2845280	170.7	6.788	**
Pyrope	-6281700	270.8	11.316*	*

Table 3. Mixing properties (Margules parameters) to be used with the general form of Wohl's asymmetric solution model (e.g. Berman 1990). Units: J/mol (WH), J/K/mol (WS), and J/bar/mol (WV). * = estimated

	P2/n / C2/c Omphacite				Phengite			
Parameter	WH	WS	WV		WH	WS	WV	
12	-14550	-10	0		5000	-10	-.135	
21	36500	15	0		5000	-10	-.135	
13	-4340/17600	0/12.1	0		0	58.598	.735	
31	13800/8600	0/7.6	0		0	15.92	.187	
14	26125	0	-.0384		0	58.598	.735	*
41	32301	0	-.0067		0	15.92	.187	*
15	0	-12.26	-.17		38500	39	.32	
51	41242	30.92	0		9500	-9	.152	
16,61	10000	5	0	*	0	0	0	*
23	15090	8	0		15000	-30	-.20	
32	3670	9	0		15000	-30	-.20	
24	18682	0	-.0864		15000	-30	-.20	*
42	20001	0	.0271		15000	-30	-.20	*
25	0	-12.26	-.17	*	30000	0	0	*
52	41242	30.92	0	*	30000	0	0	*
26,62	10000	5	0	*	5000	-10	-.135	*
34	22000	9	-.09	*	0	0	0	*
43	35000	8	.03	*	0	0	0	*
35	9740	7.15	0		30000	0	0	*
53	19800	7.15	0		30000	0	0	*
36,63	0	0	0	*		as 31,13		*
45,54	24000	0	-.026	*	30000	0	0	*
46,64	28000	8	0	*		as 41,14		*
56,65	15000	7	0	*		as 51,15		*
xyz	≈0.25·(xy+yx+xz+zx+yz+zy)*				0	0	0	*
125	16785	12.685	-.085					
145	-36500	-58.5	-.067					

	Al-Garnet				Amphibole (γ_1 only)
Parameter	WH	WS	WV		WH
12	69200	29.829	.10		27000 *
21	21560	-31.718	.10		27000 *
13	2620	5.08	.09		15000 *
31	20320	5.08	.17		15000 *
14	0	0	0	*	
41	0	0	0	*	
23	3720	0	.06		15000 *
32	230	0	.01		15000 *
24	13500	0	0		
42	13500	0	0		
34	1757	0	0		
43	1757	0	0		
xyz	0.5·(xy+yx+xz+zx+yz+zy)				as garnet *

dynamic data for end-member hedenbergite (Table 2, α_P and β_T as diopside), mixing properties (Table 3) of diopside-hedenbergite (WH and WS), grossular-pyrope (WS) and diopside-hedenbergite-$CaAlAlSiO_6$ (ternary interaction only: WH and WS). Compositional data of garnet and omphacite from numerous eclogites metamorphosed at fairly well-known P-T conditions were also used for that purpose.

The thermobarometric method described above was applied to several eclogites reported in the literature using the newly derived thermodynamic data. For instance, metamorphic pressures for phengite-plagioclase-quartz-bearing eclogites were found to be nearly identical to those obtained from the equilibrium jadeite + quartz = albite.

The activity formulation for tremolite component in amphibole (Table 1) is only a preliminary one. It considers substituting Na for Ca in the M4-site coupled with Al or Fe^{3+} exchange for Mg and thus mixing of tremolite, glaucophane and Mg-riebeckite components. The corresponding mixing properties (Table 3) were estimated from solvus relations of natural amphiboles. The same was undertaken to derive a Margules parameter for a symmetric solution taking the occupation of the A-site through the edenite substitution into account. A value of 13500 Joule/mol was estimated leading to γ_2 of the tremolite activity model. The OH-F exchange and the further substitution of Al (Tschermaks substitution) for Mg as well as Fe^{2+} and other cations for Mg is simply considered by the introduction of X_{OH} and X_{Mg} in the activity formulation.

2 APPLICATION

Numerous phengite-amphibole-quartz-bearing eclogites of the European Variszides and the Alps were studied with the aid of the electron microprobe. The compositional variation of the eclogitic minerals were then used to derive P-T paths applying the thermochemical method described above. The example used for demonstration here comes from the Saxonian Erzgebirge.

Garnet in eclogite sample E25d shows a rather complex zonation pattern (Table 4) which can be subdivided into two areas yet by means of the polarizing microscope. Only the rim region is nearly free of mineral inclusions whereas most of the garnet is inclusion-rich. For instance, the grossular content decreases from the core region with about 25 mol% (stage I) to below 20 mol% (stage II). In the rim region it increases to nearly 23 mol% (stage IIIa)

Table 4 . Representative structural formulae (double unit, simplified) of garnet from eclogite sample E25d

	core		rim	
	I	II	IIIa	IIIb
Al^{VI}	3.912	3.977	3.971	3.984
Fe^{3+}+Cr	.088	.023	.029	.016
Ca	1.519	1.120	1.376	1.320
Mg	1.096	1.371	1.130	1.092
Fe^{2+}	3.310	3.447	3.415	3.502
Mn	.076	.062	.079	.086

Table 5 . Representative structural formulae of minerals from sample E25d and molar fractions of components

	Omphacite (4 cat., O=6)				Phengite (42val.+ Na+K+Ba/2)				Amphibole (15+A cat., O=23)		
	I	II	IIIa	IIIb	I	II	IIIa	IIIb	I-II	II-IIIa	IIIa-b
Al^{IV}	.0194	.0203	.0242	.0174	1.207	1.290	1.485	1.717	1.832	1.249	1.619
Ti	.0014	.0032	.0027	.0036	.078	.083	.085	.093	.046	.061	.073
Al^{VI}	.4110	.4475	.4341	.4218	2.983	3.023	3.132	3.280	1.265	.930	1.060
Fe^{3+}+Cr	.0254	.0021	.0105	.0342	.003	.005	.009	.011	.110	.244	.284
Fe^{2+}	.0702	.1367	.1179	.0988	.178	.178	.188	.174	1.557	.981	1.199
Mn	.0012	.0018	-	.0018	-	.003	-	.004	.014	.003	-
Mg	.5124	.4376	.4534	.4544	.790	.755	.674	.558	2.008	2.781	2.384
Ca	.5585	.5356	.5557	.5394	-	-	-	-	1.605	1.424	1.461
Na	.4198	.4356	.4257	.4459	.105	.114	.169	.138	.935	.980	1.041
K	-	-	-	-	1.716	1.729	1.703	1.672	.214	.127	.166
Ba	-	-	-	-	.014	.027	.021	.021	.002	-	-
F	-	-	-	-	-	-	-	-	-	.086	.093
X1	.4743	.3938	.4240	.4304	.4753	.4878	.5426	.6507	.8027	.7117	.7304
X2	.0650	.1230	.1102	.0936	.0162	.0239	.0437	.0598	.1830	.2296	.2131
X3	.3930	.4304	.4126	.4081	.3298	.2993	.2101	.1169	.0143	.0586	.0565
X4	.0217	.0288	.0185	.0146	.0744	.0706	.0586	.0364			
X5	.0180	.0171	.0215	.0137	.0569	.0611	.0893	.0753			
X6	.0281	.0070	.0132	.0396	.0473	.0573	.0557	.0609			

and decreases again somewhat (stage IIIb). The variable chemical composition of omphacite and phengite (Table 5) was related to the four stages I to IIIb considering e.g. the relations of mineral inclusions and host.

The P-T conditions for the four eclogite stages of sample E25d were calculated. The resulting metamorphic path is shown in Fig. 1 and interpreted according to Massonne et al. (1991). Eclogites of stage I formed by subduction of mafic rocks. After cessation of the subduction process due to continent-continent collision, eclogites were heated up by uprising isotherms and were already somewhat uplifted. Once in contact with the relatively cold continental crust (stage II), that has been thickened by collision, the eclogites were chilled somewhat during immediate tectonic uplift (stages III).

Large amphibole porphyroblasts (Fig 2a) grew subsequent to stage II. Some contain an older amphibole generation, relatively rich in Al, which is related to eclogite stages I to II (Table 5). The rim of the amphibole porphyroblasts is again chemically different (Fig. 2b, Table 5) due to the changes of P, T and a_{H2O} along the exhumation path (stages III).

Calculated water activities were found to be low for the early and the late eclogite stages (Fig. 1). They are probably due to water-absent conditions. However, water activities were as high as 0.7 during the growth of the amphibole porphyroblasts between the eclogite stages II and IIIa. This is interpreted to be caused by the influx of external H_2O into the eclogites. The source for that might be micaschists and gneisses at or near the base of the thickened crust, which were heated by uprising isotherms after the collisional event. In addition, these crustal rocks were tectonically uplifted as the eclogites. Both temperature increase and pressure decrease lead to dehydration reactions in common quartzo-feldspathic gneisses and metapelites (e.g. Heinrich 1982).

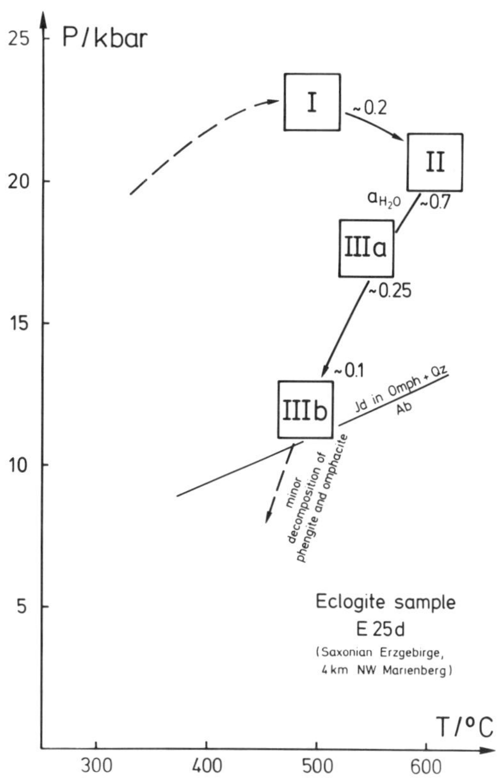

Figure 1. P-T path derived for sample E25d from mineral analyses of Tables 4 and 5. The curve of the reaction jadeite + quartz = albite was calculated for a_{Ab} = 0.9 and omphacite composition IIIb .

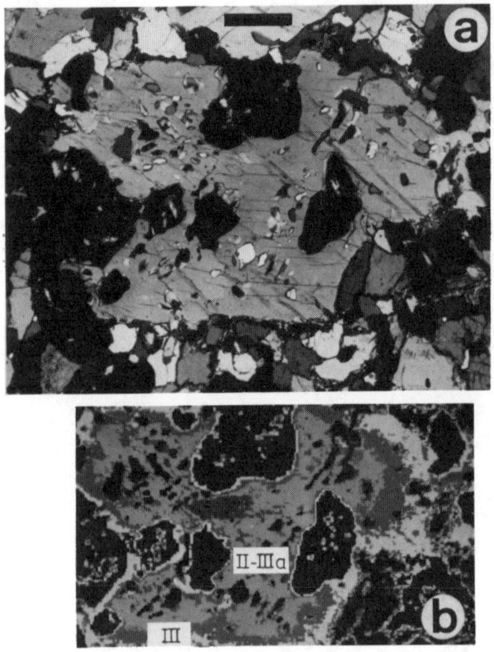

Figure 2. Amphibole porphyroblast in eclogite E25d. Scale bar is 300 µm. a) microphotograph, crossed nicols b) Al-distribution map obtained with the electron microprobe: various grey tones represent fixed ranges of counting rates that were related to amphibole compositions of Table 5 .

3 CONCLUDING REMARKS

Low water activities at early and late eclogite stages as well as high water activities prevailing along a limited section of the exhumation path corresponding to pressures close to 20 kbar are a general result of the investigations on eclogites from the European Variszides and the Alps. Growth of large amphibole porphyroblasts seems to be a good indicator for the influx of H_2O at a certain stage of eclogite metamorphism. However, some eclogites yielding calculated water activities of 0.7 and above, do not show amphibole as porphyroblasts. For example, this holds true for eclogites of the Saxonian Erzgebirge and the nearby Münchberg Massif (that are not related to ultrabasic rocks) showing peak metamorphic temperatures between 650° and 800°C. However in these eclogites, aggregates consisting mainly of feldspars and quartz were found as inclusions in omphacite and garnet particularly in rim regions. These features are interpreted as melt inclusions (Massonne 1991b). Partial melting during high-pressure metamorphism has probably been caused by the influx of H_2O as well.

ACKNOWLEDGMENTS

This work was financially supported by Deutsche Forschungsgemeinschaft.

REFERENCES

Berman, R.G. 1988. Internally-consistent thermodynamic data for minerals in the system Na_2O-K_2O-CaO-MgO-FeO-Fe_2O_3-Al_2O_3-SiO_2-TiO_2-H_2O-CO_2. J. Petrol. 29: 445-522.

Berman, R.G. 1990. Mixing properties of Ca-Mg-Fe-Mn garnets. Am. Mineralogist 75: 328-344.

Berman, R.G. & T.H. Brown 1985. Heat capacity in the system Na_2O-K_2O-CaO-MgO-FeO-Fe_2O_3-Al_2O_3-SiO_2-TiO_2-H_2O-CO_2: representation, estimation, and high temperature extrapolation. Contrib. Mineral. Petrol. 89: 168-183.

Brown, T.H., R.G. Berman & E.H. Perkins 1988. Ge0-Calc: Software package for calculation and display of pressure-temperature-composition phase diagrams using an IBM or compatible Personal Computer. Computer & Geosci. 14: 279-289.

Ellis, D.J. & D.H. Green 1979. An experimental study of the effect of Ca upon garnet-clinopyroxene Fe-Mg exchange equilibria. Contrib. Mineral. Petrol. 71: 13-22.

Gasparik, T. 1985. Experimentally determined compositions of diopside-jadeite pyroxene in equilibrium with albite and quartz at 1200-1350°C and 15-34 kbar. Geochim. Cosmochim. Acta 49: 865-870.

Heinrich, C.A. 1982. Kyanite-eclogite to amphibolite facies evolution of hydrous mafic and pelitic rocks, Adula nappe, Central Alps. Contrib. Mineral. Petrol. 81: 30-38.

Massonne, H.-J. 1991a. Druck-Temperatur-Entwicklung phengitführender Eklogite. Mitt. Österr. Mineral. Ges. 136: in press.

Massonne, H.-J. 1991b. High-pressure, low-temperature metamorphism of pelitic and other protoliths based on experiments in the system K_2O-MgO-Al_2O_3-SiO_2-H_2O. Habilitation thesis, Ruhr-Universität Bochum: 172 pp.

Massonne, H.-J. & W. Schreyer 1987. Phengite geobarometry based on the limiting assemblage with K-feldspar, phlogopite, and quartz. Contrib. Mineral. Petrol. 96: 212-224.

Massonne, H.-J. & W. Schreyer 1989. Stability field of the high-pressure assemblage talc + phengite and two new phengite barometers. Eur. J. Mineral. 1: 391-410.

Massonne, H.-J., J. Kopp & K. Rötzler 1991. P-T-Entwicklung der Eklogite des Sächsischen Erzgebirges und ihre geodynamische Bedeutung. Zentralbl. Geol. Paläont., Teil I, 1991: 7. Rundgespräch "Geodynamik des europäischen Variszikums", Kurzfassungen, 30.

Pattison, D.R.M. & R.C. Newton 1989. Reversed experimental calibration of the garnet-clinopyroxene Fe-Mg exchange thermometer. Contrib. Mineral. Petrol. 101: 87-103.

Water-Rock Interaction, Kharaka & Maest (eds) © 1992 Balkema, Rotterdam. ISBN 90 5410 075 3

Chemical and oxygen isotopic variations in upper-plate rocks of the Whipple Mountains detachment system, California, USA

D. Mayo, J. Morrison & J. L. Anderson
University of Southern California, Los Angeles, Calif., USA

ABSTRACT: A suite of K-metasomatized upper-plate volcanic rocks from the Whipple Mountains in southeastern California displays distinctive chemical and oxygen isotopic characteristics that differ from unmetasomatized counterparts in the neighboring Mopah Range. The metasomatized rocks have relatively high K_2O (5 to 10 wt. %), low Na_2O (0.1 to 3.5 wt. %) and CaO (6.5 to 0.4 wt. %), and contain plagioclase partially replaced by secondary K-feldspar. Whole-rock silicate $\delta^{18}O$ in the metasomatized rocks varies from 5.7 to 12.4‰. Secondary calcite has a wide range of $\delta^{18}O$ (10.6 to 24.2 ‰), consistent with precipitation over a range of temperatures or from a fluid with variable $\delta^{18}O$ or both.

1 INTRODUCTION

The Whipple Mountains are one of a number of highly extended terranes exposed in a northwest-trending belt from southeastern Arizona to southeastern California. The detached upper plate in the Whipple Mountains contains a thick sequence of Miocene volcanic and sedimentary rocks deposited nonconformably on pre-Tertiary basement rocks. Blocks of upper-plate rocks dip steeply to the southwest and are separated by northeast-dipping, high-angle normal faults rooted to the regionally extensive Whipple detachment fault. As in many areas of regional extension, the volcanic rocks in the upper plate have been overprinted by potassium metasomatism and secondary mineralization. To explain the origin of this distinctive type of alteration, numerous hypotheses regarding (1) the relationship between metasomatism and extensional deformation and (2) the mechanisms responsible for the metasomatism have been proposed (Kerrich et al., 1989; Roddy et al., 1988; Glazner, 1988; Kerrich and Rehrig, 1987; Chapin and Lindley, 1986; Wilkins et al., 1986; Brooks, 1986). The present study compares the chemical and stable isotopic compositions of metasomatized volcanic rocks and secondary calcite from the Whipple Mountains with unmetasomatized counterparts from the neighboring Mopah Range, approximately 30 km to the west. Tertiary volcanic rocks from the Mopah Range, which occupy a higher structural level in the Whipple Mountains detachment system (Hazlett, 1986), serve as unaltered references with which to compare metasomatized samples.

Samples from the Whipple Mountains are from a 2 km-long traverse within the upper plate, parallel to the present exposure of the Whipple Mountains detachment fault in the southernmost part of the range. All sample localities are within ≈ 200 m of the detachment fault (distance approximately perpendicular to the fault). The samples from the Mopah Range were collected and described in detail by Hazlett (1986).

2 PETROGRAPHY

Three of the four Mopah Range samples are basaltic in composition, containing plagioclase, clinopyroxene, orthopyroxene, and olivine phenocrysts in a groundmass composed predominantly of plagioclase and pyroxene. Mineralogic alteration is minor, consisting of hematization along fractures and iddingsitization of olivine phenocrysts. Secondary calcite is not visible in thin section, but likely exists along minute fractures. Sample RH122-2 is a porphyritic dacite with plagioclase, biotite, hornblende, and opaque oxide phenocrysts in a flow-banded matrix. Minor alteration consists of hematization and perlitization of groundmass oxides and glass, respectively.

Secondary alteration is much more evident in the samples from the Whipple Mountains. Highly fritted plagioclase and oxidized biotite are the only distinguishable primary phenocryst phases. Plagioclase is partially replaced by secondary potassium feldspar. The groundmass in these rocks is dominated by fine-grained potassium feldspar, opaque oxides, and hematite. Vesicles, irregular microscopic voids, and fractures are filled with polycrystalline quartz and secondary calcite.

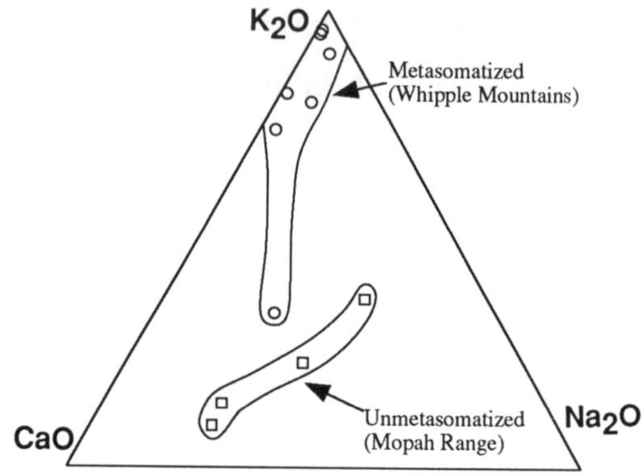

Figure 1. Ternary illustration of the Na_2O and CaO depletion of metasomatized volcanic rocks from the Whipple Mountains. Oxides are in wt. %, normalized to 100 %.

3 CHEMICAL AND ISOTOPIC COMPOSITIONS

Chemical and isotopic compositions of samples from both mountain ranges are listed in Table 1, and pertinent correlations discussed below are displayed graphically in Figures 1 and 2. The metasomatized rocks from the Whipple Mountains contain much higher K_2O and lower Na_2O and CaO than unmetasomatized rocks from the Mopah Range.(Figure 1). Concentrations of FeO and MgO are also lower in the metasomatized rocks at comparable SiO_2 values.

The ratios K_2O/Na_2O (K/N) and K_2O/Al_2O_3 (K/A) serve as indexes of K-metasomatism. K/N is the more sensitive of the two, because of concomitant increase in K_2O, decrease in Na_2O, and relatively small change in Al_2O_3. K/N ratios in the unmetasomatized samples from the Mopah Range vary from 0.40 to 0.95, while K/N ratios in the metasomatized samples from the Whipple Mountains have a wide range from 1.5 to ≈ 100. K/A in the unmetasomatized samples correlates positively with SiO_2, with a value of about 0.3 at 65 wt. % SiO_2 (Figure 2a). The metasomatized samples also display a positive K/A vs. SiO_2 correlation, but have elevated K/A ratios at similar SiO_2 contents (≈ 0.4 at 65 wt. % SiO_2). The apparent K_2O enrichment and Na_2O + CaO depletion in the metasomatized samples is consistent with the replacement of plagioclase by K-feldspar evident in thin sections.

The whole-rock silicate oxygen isotopic compositions of metasomatized and unmetasomatized samples over the range of SiO_2 contents for which the two sample sets overlap (< 65 wt. %) are very similar, ranging from 7.1 to 8.7 ‰ (Figure 2b). These isotopic compositions are within the range (5 to 10‰) of typical basaltic, andesitic, and dacitic rocks that have not undergone $\delta^{18}O$-altering processes during their formation or subsequent history (Taylor and Sheppard, 1986). Thus, the oxygen isotopic composition of the two metasomatized basaltic andesites (W80-217 and W80-292), in spite of chemical and petrographic evidence for post-eruption alteration, is similar to that of unaltered volcanic rocks.

Of the five remaining metasomatized samples, three (W80-222, W80-239, and W80-237) are high-SiO_2 rhyolites with $\delta^{18}O$ values (5.7 to 7.2) which are slightly lower than the two metasomatized basaltic andesites. If the basaltic andesite values are primary and the rhyolites are related by fractional crystallization, then the rhyolite $\delta^{18}O$ values should be slightly higher, because of the tendency for quartz to concentrate ^{18}O (Taylor and Sheppard, 1986). Thus, the slightly lower $\delta^{18}O$ observed in these three high-SiO_2 rocks could be attributed to subsolidus reequilibration with isotopically light fluid. Alternatively, the rhyolites may not be comagmatic with the less siliceous rocks, and their whole-rock $\delta^{18}O$ may be primary.

Table 1. Chemical and stable isotopic compositions.

	Metasomatized (Whipple Mountains)							Unmetasomatized (Mopah Range)			
	W80 217	W80 292	W80 232	W80 287	W80 222	W80 239	W80 237	RH 1120-3	RH 49-3	RH 1213-4	RH 122-2
SiO2	51.6	51.7	65.3	69.8	74.3	75.0	75.5	46.3	49.9	55.7	62.2
TiO2	1.6	1.3	0.7	0.4	0.2	0.2	0.3	1.7	1.3	1.2	0.7
Al2O3	17.9	16.2	15.0	14.4	11.1	10.7	11.7	17.0	17.5	18.2	17.0
FeO	6.4	6.9	2.8	1.7	0.8	4.9	1.2	10.4	9.1	6.6	3.6
MgO	2.3	3.3	1.3	0.1	0.2	0.3	nd	7.4	5.2	3.2	0.9
MnO	0.1	0.2	0.1	nd	nd	nd	nd	0.1	0.1	0.1	nd
CaO	6.5	2.8	1.4	0.3	0.3	1.6	0.4	10.2	9.8	6.2	3.0
Na2O	3.5	0.2	0.7	0.5	0.1	0.1	0.1	3.6	3.6	4.9	5.0
K2O	5.0	8.3	8.3	8.4	9.3	8.0	9.8	1.4	2.2	3.3	4.8
LOI	3.0	4.2	1.9	0.6	0.6	1.7	0.6	0.2	0.7	0.5	0.4
Total	97.9	95.0	97.5	96.2	96.8	102.4	99.6	98.0	98.7	99.4	97.0
K/A	0.28	0.51	0.55	0.59	0.83	0.75	0.84	0.08	0.12	0.18	0.28
K/N	1.46	48.71	11.24	16.49	103	61.46	75.46	0.40	0.61	0.66	0.95
Rb	127	230	232	235	267	238	259	24	31	58	108
Sr	1021	137	260	94	58	144	87	456	971	626	486
$\delta^{18}O$, wr	7.4	7.8	9.5	12.4	7.2	6.7	5.7	8.7	7.1	8.2	7.4
$\delta^{18}O$, cc	10.6	12.6	13.4	24.2	21.9	17.0	23.3	25.3	22.4	23.1	26.0
$\delta^{13}C$, cc	-4.0	-3.0	-5.0	-1.8	-0.8	-4.0	-4.5	-2.4	-0.4	-2.7	-1.7

Whole-rock chemistry by AA (Whipple Mountains samples) and XRF (Mopah Range samples); nd = not detected; major elements in wt. %; trace elements in ppm.
K/A = K_2O/Al_2O_3; K/N = K_2O/Na_2O.
wr = whole rock leached of calcite; cc = calcite extracted from whole rock.
NCSU-QTZ standard (accepted value +11.7 ‰) analyzed with whole rocks yielded +11.5 ± 0.1 ‰.
$\delta^{18}O$,wr for RH1213-4 is an average of three determinations with a 1σ std dev. of 0.08 ‰.
Isotopic ratios are reported in standard delta notation relative to SMOW.

Sample W80-287, with a $\delta^{18}O$ of 12.4 (and perhaps sample W80-232, with a $\delta^{18}O$ of 9.5) is enriched in ^{18}O beyond that due to magmatic differentiation. Assuming rock-water fractionation approaching that of quartz and K-feldspar (because of their high modal abundance), sample W80-287 could have equilibrated with a reservoir having $\delta^{18}O$ between 5 and 10 ‰ at temperatures below about 300°C, or with a reservoir having $\delta^{18}O < 0$ at temperatures below about 100°C.

The oxygen isotopic composition of secondary calcite from the unmetasomatized rocks falls in a narrow range between 22 and 26 ‰ (Figure 2c). These rocks lack mineralogic evidence of high-temperature, post-eruption alteration. Assuming a low temperature of equilibration (< 50 °C), this calcite precipitated from a reservoir with $\delta^{18}O < 0$, such as meteoric water.

Secondary calcite from the metasomatized samples, in contrast, displays a wide range of $\delta^{18}O$ (from 10.6 to 24.2) that correlates positively with host rock SiO_2 (Figure 2c). The highest $\delta^{18}O_{cc}$ values in the metasomatized rocks are similar to $\delta^{18}O_{cc}$ values in the unmetasomatized rocks.

4 DISCUSSION

In the absence of temperature constraints, Figure

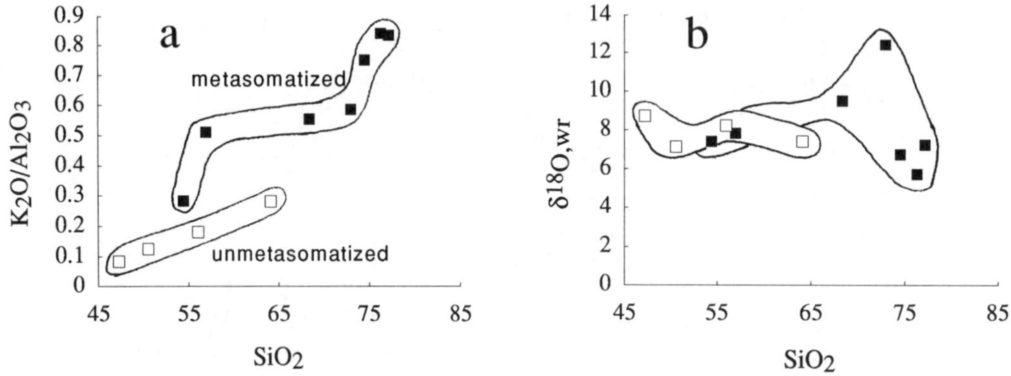

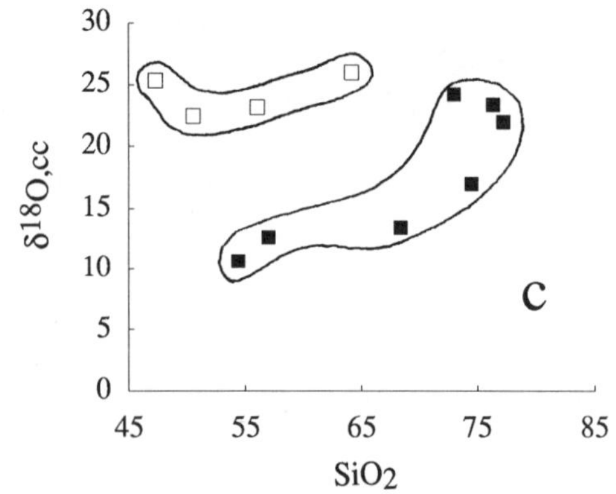

Figure 2. Chemical and isotopic covariations. Abbreviations as in Table 1. Open symbols = unmetasomatized samples from the Mopah Range. Filled symbols = metasomatized samples from the Whipple Mountains.

3 presents allowable calcite-fluid models to account for the observed range of calcite isotopic compositions. The models assume that calcite was precipitated in equilibrium with H_2O-rich fluid and that host rock composition did not affect the calcite. Model 'a' assumes roughly constant $\delta^{18}O_{fluid}$ of ≈ 7.5, in the range of isotopically heavy formation waters (Hoefs, 1987). The temperature range (100 to 400°C) assumed in model 'a' corresponds approximately to the range of homogenization temperatures determined for fluid inclusions in secondary minerals from the upper plates of several detachment systems (e.g. Wilkins et. al., 1986), including the Copper Basin area in the northeastern Whipple Mountains. Model 'b' assumes a roughly constant $\delta^{18}O_{fluid}$ of 0, the approximate upper limit for meteoric waters (Hoefs, 1987). This model requires significantly lower equilibration temperatures than model 'a'. Model 'c' illustrates the fact that a variation of $\delta^{18}O_{fluid}$ requires a much smaller range of equilibration temperatures to generate the observed range of calcite isotopic composition.

Because of the positive $\delta^{18}O_{cc}$ vs. SiO_2 correlation, models such as 'a' and 'b' above would suggest that progressively higher-temperature calcite equilibration occurred in progressively more mafic rocks. If progressively more mafic rocks happened to be exposed nearer to the detachment fault, then the $\delta^{18}O_{cc}$ vs. SiO_2 correlation might be fortuitous, simply reflecting a fault-related temperature gradient during calcite precipitation. There is, however, no apparent correlation between SiO_2 content (or $\delta^{18}O_{cc}$) and distance to the detachment fault.

Model 'c' requires a relatively small range of equilibration temperatures. It also allows for a wider range of fluid isotopic compositions, and thus a more variable range of sources, including mixtures of isotopically different reservoirs. Some of the required fluid $\delta^{18}O$ variation could also reflect exchange with the host rock.

No one model is presently favored over another, because the equilibration temperature range is unconstrained. An important goal of future work is to extract other secondary

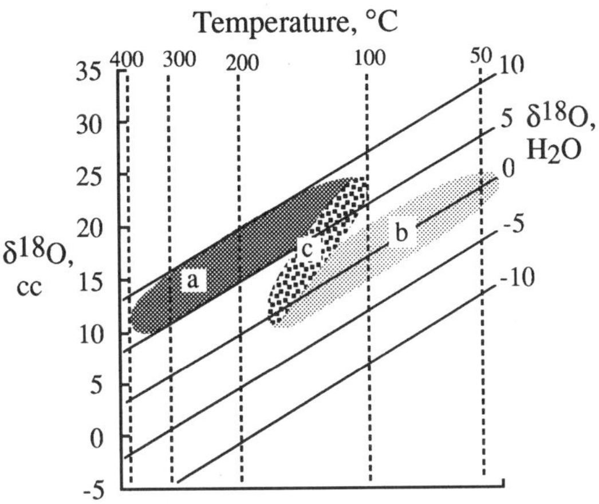

Figure 3. T vs. fluid composition models that account for the observed range of $\delta^{18}O_{cc}$. Model 'a' assumes a temperature range similar to Wilkins et al. (1986). Model 'b' assumes a fluid composition of about 0 ‰, to illustrate that equilibrium precipitation from a fluid reservoir more similar to meteoric water requires significantly lower temperatures. Model 'c' is a representative composite of 'a' and 'b', in which both temperature and fluid composition vary. Calcite-water fractionation equation from O'Neil et al. (1969).

minerals, in addition to calcite, to test for isotopic equilibrium and constrain the secondary mineral equilibration temperature range with appropriate stable isotopic thermometry.

Wilkins et al. (1986) concluded, on the basis of fluid inclusion determinations, that mineralizing solutions responsible for detachment fault-related deposits were hot (200 to 325°C) brines derived from synorogenic basins overlying detachment systems. Their conclusion corresponds to model 'a' in Figure 3 above. Chapin and Lindley (1986) also concluded that the mineralizing fluid in the detached terrane near Socorro, New Mexico, was a synorogenic basinal brine, but favor a low-temperature (< 100°C), diagenetic metasomatic process. Such a low temperature is not consistent with the calcite isotopic data presented here, requiring an unreasonably low $\delta^{18}O$ value (-5 ‰) for the brines.

5 CONCLUSIONS

The upper plate volcanic rocks from the Whipple Mountains detachment system considered in this investigation display abundant petrographic and major element chemical evidence of potassic metasomatism, and yet whole rock oxygen isotopic values, with one exception, are broadly similar to typical igneous values. Secondary calcite extracted from the metasomatized rocks displays a wide range of $\delta^{18}O$, consistent with either (1) precipitation from a reservoir with $\delta^{18}O$ between 5 and 10 ‰ at temperatures between 100 and 400°C or (2) precipitation from a reservoir with $\delta^{18}O \approx 0$ ‰ at temperatures between 50 and 150°C. The variable precipitation temperatures implied by the calcite data do not correlate with distance to the Whipple detachment fault, but do correlate with SiO_2 of the host rock. Work is continuing to assess the significance of this correlation.

ACKNOWLEDGMENTS

This research was supported in part by Geological Society of America Research Grant #4746-91 to Mayo and NSF grant EAR-9105924 to Morrison.

REFERENCES

Brooks, W.E. 1986. Distribution of anomalously high K_2O volcanic rocks in Arizona: metasomatism at the Picacho Peak detachment fault. *Geology* 14: 339-342.

Chapin, C.E., and Lindley, J.I. 1986. Potassium metasomatism of igneous and sedimentary rocks in detachment terranes and other sedimentary basins: economic implications. *Arizona Geological Society Digest* 16: 118-126.

Glazner, A.F. 1988. Stratigraphy, structure, and potassic alteration of Miocene volcanic rocks in the Sleeping Beauty area, central Mojave Desert, California. *Geological Society of America Bulletin* 100: 424-435.

Hazlett, R.W. 1986. Geology of a Tertiary volcanic center, Mopah Range, San Bernardino County, California. Ph.D. dissertation, University of Southern California: 303 p.

Hoefs, J. 1987. *Stable isotope geochemistry.* Springer Verlag.

Kerrich, R., Rehrig, W.A., and McLarty, E. 1989. Volcanic rocks in the suprastructure of metamorphic core complexes, southwest Arizona: geochemical and isotopic evidence for primary magma types and secondary hydrothermal regimes. *Arizona Geological Survey Bulletin* 198: 203-214.

Kerrich, R., and Rehrig, W. 1987. Fluid motion associated with Tertiary mylonitization and detachment faulting: $^{18}O/^{16}O$ evidence from the Picacho metamorphic core complex, Arizona. *Geology* 15: 58-62.

O'Neil, J.R., Clayton, R.N., and Mayeda, T.K. 1969. Oxygen isotope fractionation in divalent metal carbonates. *Journal of Chemistry and Physics* 51: 5547-5558.

Roddy, M.S., Reynolds, S.J., Smith, B.M., and Ruiz, J. 1988. K-metasomatism and detachment-related mineralization, Harcuvar Mountains, Arizona. *Geological Society of America Bulletin* 100: 1627-1639.

Taylor, H.P., and Sheppard, S.M.F., 1986. Igneous rocks: I. Processes of isotopic fractionation and isotope systematics. *Stable isotopes in high temperature geological processes*: 227-271. Mineralogical Society of America.

Wilkins, J.Jr., Beane, R.E., and Heidrick, T.L. 1986. Mineralization related to detachment faults: a model. *Arizona Geological Society Digest* 16: 108-117.

Water-Rock Interaction, Kharaka & Maest (eds) © 1992 Balkema, Rotterdam. ISBN 90 5410 075 3

Enhancement of porosity in gabbros by dike intrusion

Richard J. Nevle & Dennis K. Bird
Stanford University, Calif., USA

ABSTRACT: The distribution of vein systems and hydrothermal alteration in the Lower Layered Series gabbros of the Kap Edvard Holm complex, East Greenland, suggests that hydrothermal fluid flow was concentrated in and around mafic dikes discordant to the gabbro layering. Fracturing caused by the emplacement of mafic dikes increased porosity in dike-adjacent gabbros to values 3 times as high as gabbros >1 meter from dikes. Within these porous zones, fracture systems transported hydrothermal fluids that reacted with their mafic host rocks to produce numerous vein-filling assemblages and extensive mineralogic and isotopic alteration (up to 50% secondary minerals by volume in gabbro and $\delta^{18}O$ of plagioclase as low as -5.8‰). These results demonstrate the extent to which dike emplacement can enhance hydraulic conductivity in magma-hydrothermal systems.

1. INTRODUCTION

Hydrothermal convection systems near cooling plutons redistribute heat and mass in the Earth's crust. For example, the Skaergaard magma-hydrothermal system produced systematic depletions in the ^{18}O of primary minerals ($\delta^{18}O$ of plagioclase typically ranges from 2 to 6.3‰) and convected heat away from the cooling gabbro pluton (Taylor and Forester, 1979; Norton and Taylor, 1979). The investigations of Norton et al. (1984); Bird et al. (1986); and Manning and Bird (1991) demonstrate that hydrothermal fluids transported through fractures reacted with the Skaergaard gabbro and its host rocks, resulting in the present-day distribution of oxygen isotope depletion, mineral-filled pores, and mineralogic alteration. Norton and Taylor (1979) attributed the limited extent of mineralogic alteration to the relatively high temperatures (>480° C) maintained during most of the hydrothermal circulation.

In contrast to the Skaergaard intrusion, the Lower Layered Series gabbros of the Kap Edvard Holm complex (located ~20 km SW of the Skaergaard intrusion) record both extensive mineralogic alteration and ^{18}O depletion ($\delta^{18}O$ of plagioclase ranges from 0.5 to -5.8‰; Fehlhaber and Bird, 1991). Whereas a single magma-hydrothermal system produced alteration in the Skaergaard intrusion, multiple magma-hydrothermal systems that overlapped in time and space caused the extensive alteration of the Lower Layered Series (Bird et al., 1988; Nevle and Bird, 1990).

The present-day distribution of vein systems and hydrothermal alteration results from the combined tectonic, intrusive, and hydrothermal history of the region during early Tertiary continental rifting. Because of similarities of the Kap Edvard Holm complex to ophiolites (Bernstein et al., 1992), the Lower Layered Series (LLS) records hydrothermal processes similar to those active in oceanic gabbros.

2. GEOLOGIC BACKGROUND

The Kap Edvard Holm complex (KEHC) consists of ultramafic, gabbroic, and syenitic intrusions exposed over an area of 360 km^2 (Wager, 1947). The largest lithologic unit is a <4 km thick composite sequence of layered gabbros emplaced during Tertiary continental rifting and formation of the North Atlantic ocean basin (Brooks and Nielsen, 1982). Like the Skaergaard intrusion, the layered gabbros of the KEHC were emplaced near the unconformity between the Precambrian crystal-

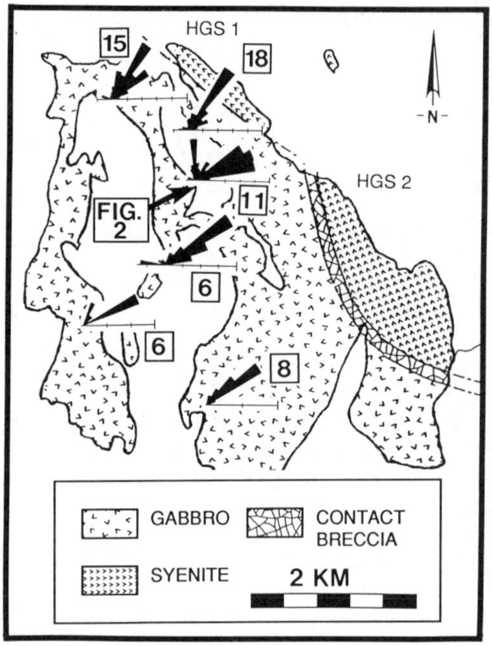

Fig. 1. Rose histograms showing strikes and abundances of veins in LLS gabbros. Numbers in boxes represent veins per meter. HGS 1 and HGS 2 represent Hutchinson Glacier Syenites 1 and 2, respectively. See text for explanation.

line basement and an overlying 2 to 4 km thick Tertiary mafic volcanic pile.

Multiple pulses of basaltic magma intruded and crystallized to form the LLS (Abbott and Deer, 1972; Bernstein et al. 1992). Field observations suggest that numerous mafic dikes intruded and fractured the LLS before emplacement of the Middle and Upper Layered Series gabbros, exposed to the south. Following gabbro emplacement, intrusion of syenitic plugs, ring complexes and complex breccias caused brittle deformation and extensive hydrothermal metasomatism of the gabbros and mafic dikes. Faulting and tilting were followed by uplift and glacial erosion.

3. VEIN SYSTEMS

The distribution and orientations of hydrothermal veins in the LLS gabbros reflect a regional vein system formed during gabbro cooling and mafic dike emplacement, and veins formed during syenite emplacement. As shown in Fig. 1, the most abundant veins <400 meters from the Hutchinson Glacier Syenite 1 (HGS1) strike ~030°, approximately perpendicular to the intrusive contact; whereas veins more distant from the contact strike ~060°, parallel to a regional mafic dike swarm.

Similarly, the dominant vein-filling mineral assemblages change with distance from the HGS1 contact. Hornblende + bytownite + clinopyroxene, and albite + epidote ± chlorite ± calcic amphibole occur as the most abundant vein-filling assemblages within gabbros and dikes <400 meters from the HGS1 contact; >400 meters, calcic amphibole ± chlorite and chlorite assemblages fill most veins.

The gabbros <400 meters of the syenites were extensively veined and altered during the circulation of syenite-heated hydrothermal fluids. The distinct chemical alteration produced during syenite emplacement and cooling allows the effects of overlapping magma-hydrothermal systems in the LLS to be distinguished. Near the syenites, older veins within and near dikes were reopened during syenite emplacement and exploited by silicic magmas and hydrothermal fluids which precipitated hornblende + bytownite + pyroxene and albite + epidote ± chlorite ± calcic amphibole assemblages.

In general, vein abundances in the gabbros increase near contacts with mafic dikes and are highest within mafic dikes (Fig. 2). Cross-cutting relations between mafic dikes and calcic amphibole ± chlorite and chlorite veins indicate vein formation during dike emplacement (Rogers and Bird, 1987). These veins emanate from dike terminations and occur as the most abundant vein type within mafic dikes. Closely spaced (~20 veins/cm), narrow (~20 μm wide), calcic amphibole ± chlorite veins not visible in outcrop occur in gabbros collected <1-2 meters from mafic dikes.

4. ALTERATION

Hydrothermal systems associated with emplacement and cooling of the gabbros, mafic dike intrusion, and syenite emplacement produced extensive hydrothermal alteration throughout the region shown in Fig. 1. In Fig. 2, the volume percent of secondary minerals in gabbros collected from a ~ 50 meter wide outcrop is plotted as a function of distance along the outcrop (see Fig. 1 for outcrop location). Several 0.5 meter wide altered dolerite dikes cut the gabbros in this outcrop.

The gabbros are olivine-plagioclase adcumulates; augite and Fe-Ti oxides occur as interstitial phases. Plagioclase occupies ~60% of the rock by

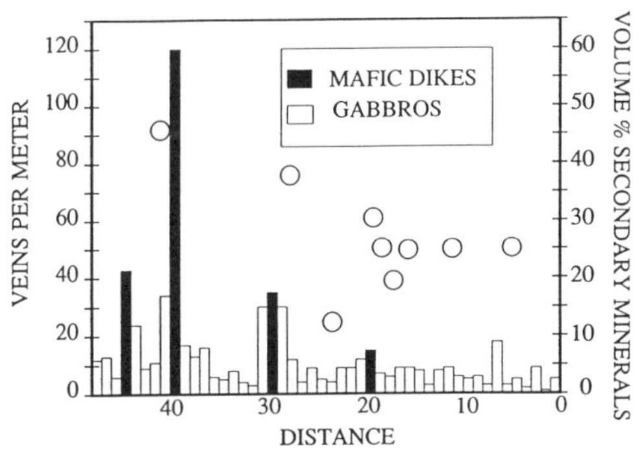

Fig. 2. Variation of vein abundances and mineralogic alteration in a ~50 meter wide outcrop in the LLS. Abscissa represents distance (in meters) along the outcrop. Bars represent measurements of veins per meter. Open circles represent the volume percent of secondary minerals in gabbros determined by optical point counting. See text for explanation and Fig. 1 for location of outcrop.

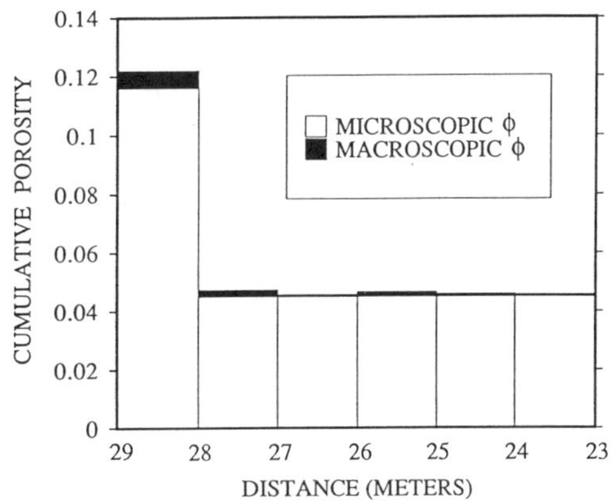

Fig. 3. Variation of cumulative porosity in gabbros near a mafic dike. Distance units along abscissa correspond to Fig. 2. See text for explanation.

volume and occurs as granular, equidimensional crystals typically <0.2 mm in diameter. Olivine occurs as anhedral, equigranular to acicular grains that occasionally enclose plagioclase oikocrysts. Augite poikilitically encloses olivine and plagioclase. Fe-Ti oxides occur as small (<0.5 mm wide) interstitial grains. Primary phlogopite and hornblende are rare (<0.5% by volume).

In gabbros, alteration occurs predominantly along grain boundaries and is most extensive between plagioclase and olivine. Grain boundaries between plagioclase grains are marked by assemblages of calcic amphibole and chlorite. Olivine is partially to completely replaced by metasomatic mineral assemblages. Outward from olivine, the assemblages proceed from anthophyllite + magnetite ± talc to chlorite.

Along vein margins, hydrothermal pyroxene and actinolite replace augite, and plagioclase contains numerous fine-grained fluid and mineral inclusions. Inclusion-filled plagioclase appears bleached in outcrop and is particularly noticeable near contacts with mafic dikes. The extent of mineralogic alteration in the gabbros increases with the number of veins (Fig. 2).

Mafic dikes are typically 0.5 to 2 meters wide, have chilled margins, and are extremely hydrothermally altered. Altered, inclusion filled plagioclase suggests a doleritic texture. Dikes consist of over 60% calcic amphibole and chlorite by volume; magnetite and titanite occur as accessory phases. Vugs formed during alteration of the dikes contain actinolite, epidote, chlorite, and calcite.

5. DISCUSSION

The geometric and cross-cutting relations among mafic dikes, calcic amphibole ± chlorite veins, and chlorite veins suggest synchronous formation of these features, perhaps during regional extension associated with continental rifting of Greenland from Eurasia. Calcic amphibole ± chlorite veins formed at temperatures >500-600°C as suggested by phase and compositional relations among the alteration products of olivine adjacent to the veins (Bird et al., 1988). The high abundances of both macro- and microscopic veins near mafic dikes, as well as the presence of these veins emanating from dike terminations, suggest that dike propagation induced localized fracturing of the gabbros (Rogers and Bird, 1987). Geometric relations among veins within mafic dikes indicate fracture propagation during cooling and in response to regional stresses.

The cumulative porosity, ϕ^*, in gabbros adjacent to the dikes is almost three times that of gabbros >1 meter from dikes due to the presence of numerous microscopic veins (Fig. 3). Microscopic porosity is estimated using the relation:

$$\phi^* = D\ (N/L),$$

where D is the mean microscopic vein aperture, and N is the number of veins measured in a traverse of length L (Manning and Bird, 1991). The quotient N/L represents vein density in cm^{-1}. The estimates for microscopic porosity illustrated in Fig. 3 were determined from several randomly oriented traverses in thin sections. The width of mineral-filled microcracks along grain boundaries typically varies from 1.0×10^{-3} to 2.5×10^{-4} cm, and N/L is ~ 60 cm^{-1}. Through-going microscopic veins in gabbros adjacent to dikes have widths of ~2.0×10^{-3} cm and densities of 20 cm^{-1}. Macroscopic porosity can be estimated by measuring N/L as the number of veins counted along a section oriented perpendicular to strike. Values for macroscopic porosity shown in Fig. 3 were calculated using the vein densities reported in Fig. 2 and an average vein width of 2.0×10^{-2} cm; macroscopic veins are typically 3 to 5 meters in length. Although macroscopic porosity increases near the dikes, it is two orders of magnitude lower than the estimated microscopic porosity.

6. CONCLUSIONS

The distribution of vein systems and hydrothermal alteration in the LLS suggests that zones of high porosity within and near mafic dikes channelized circulating hydrothermal fluids. Highly fractured, structurally weak zones formed during dike emplacement and cooling were later exploited by syenite-heated hydrothermal fluids. The distribution of hydrothermal features in the LLS suggests that in regions characterized by long-lived plutonism, fractures formed during dike emplacement and cooling repeatedly focus the flow of hydrothermal fluids. Because of the structural and petrologic similarities of the LLS to ophiolites, the distribution of veins and mineralogic alteration in the LLS demonstrates the extent to which dike intrusion can effect hydrothermal alteration in oceanic crust.

7. REFERENCES

Abbott, D. and Deer, W.A. 1972. Geological investigations in East Greenland. IX. The gabbro cumulates of the Kap Edvard Holm Lower Layered Series: Medd. om Gronland 190, 6: 42 pp.

Bernstein, S., Rosing , M.T., Brooks, C.K., and Bird, D.K. 1992. An ocean-ridge type magma chamber at a passive volcanic continental margin: The Kap Edvard Holm Layered Gabbro Complex, East Greenland, Geological Magazine, in review.

Bird, D.K., Rogers, R.D., and Manning, C.E. 1988. Mineralized fracture systems of the Skaergaard intrusion, East Greenland. Medd. om Gronland, Geosci. 16: 68 pp.

Bird, D.K., Manning, C.E., and Rose, N.M. 1988. Hydrothermal alteration of Tertiary layered gabbros, East Greenland: Am. Jour. Sci. 288: 405–457.

Brooks, C.K., and Nielsen, T.F.D. 1982. The East Greenland continental margin: A transition between oceanic and continental magmatism. Geol. Soc. London Jour. 139: 265–275.

Fehlhaber, K., and Bird, D.K. 1991. Oxygen isotope exchange and mineralogic alteration in gabbros of the Lower Layered Series, Kap Edvard Holm Complex, East Greenland. Geology. 19: 819–822.

Manning, C.E., and Bird, D.K. 1991. Porosity evolution and fluid flow in the basalts of the Skaergaard magma-hydrothermal system, East Greenland 291: 201–257.

Nevle, R.J. and Bird, D.K. 1990. The effect of dikes on hydrothermal alteration of layered gabbros: The Kap Edvard Holm complex, East Greenland. Geol. Soc. Amer. Abstr. Prog. 22: 213.

Norton, D., and Taylor, H.P. 1979. Quantitative simulation of the hydrothermal systems of crystallizing magmas on the basis of transport theory and oxygen isotope data: an analysis of the Skaergaard intrusion. J. Petrol. 20: 421–486.

Norton, D., Taylor, H.P., and Bird, D.K. 1984. The geometry and high temperature brittle deformation of the Skaergaard intrusion. J. Geophys. Res. 89: 10,178–10,192.

Rogers, R.D., and Bird, D.K. 1987. Fracture propagation associated with dike emplacement at the Skaergaard intrusion, East Greenland. Jour. Struct. Geol. 9: 71-86.

Taylor, H.P. and Forester, R.W. 1979. An oxygen isotope study of the Skaergaard intrusion and its country rocks: a description of a 55 m. y. old fossil hydrothermal system. - J. Petrol. 20: 335–419.

Wager, L.R. 1947. Geological investigations in East Greenland, Pt. IV. The stratigraphy and tectonics of Knud Rasmussens Land and the Kangerdlugssuaq region. Medd. om Gronland 134: 5, 64 pp.

Water-Rock Interaction, Kharaka & Maest (eds) © 1992 Balkema, Rotterdam. ISBN 90 5410 075 3

Constraining the temperature, pressure, and fluid compositional evolution of an accretionary complex by an integrated study of cement stratigraphy and tectonics

David L.Orange, Duncan Geddes & J.Casey Moore
Earth Sciences Board and Institute for Tectonics, University of California, Santa Cruz, Calif., USA
Randy Cox & Michael Underwood
Department of Geology, University of Missouri, Columbia, Mo., USA

ABSTRACT: The Hoh rock assemblage of the western Olympic Peninsula, Washington provides an exceptional area to study the processes of accretion and dewatering. These rocks were deposited in Eocene to Miocene time, and accreted, deformed, uplifted and eroded by the late Miocene, resulting in a punctuated deformational history with little or no structural or thermal overprint. The structural history of the Hoh includes two post-peak thermal events, an early generation of folding and a late phase of thrust faulting. The Hoh contains abundant evidence for fluid involvement throughout its deformational history, including an early micritic carbonate cement, six generations of carbonate (+laumontite) veins and a final phase of laumontite veining. By combining field mapping and careful sample collection we are able to document the progression of fluid precipitates within the relative structural evolution. Carbonate veins progress from micritic to sparitic in crystal habit, dirty to clear in plane light, and largely increasing brightness in luminescence. These changes in habit, number of inclusions, and luminescence reflect the changing physical and chemical conditions, and together with field mapping and vitrinite reflectance constrain the thermal and hydrologic evolution of this young accretionary complex.

1. Introduction and Regional Geology

Accretionary complexes are regions where sediments are scraped off subducting plates and accreted to the overriding crust. During accretion sediments are compacted, lithified, and deformed, leading to a dramatic reduction in pore space and the expulsion of large amounts of fluid (Moore et al., 1991). In order to understand the mechanisms that control fluid flow (intergranular or fracture permeability), the changes that occur in the fluid, and where in the deformation-pressure-temperature history these changes occur we have focused our studies on the Hoh accretionary complex of the western Olympic Peninsula, Washington (**Figure 1**). In this paper we will first present the structural setting, and then place the progression of fluid precipitates in this tectonic framework. We will then place thermal constraints on the tectonic and fluid evolution by using vitrinite reflectance data.

Upper Miocene to Pliocene oceanic crust is currently being underthrust beneath coastal Washington and Oregon at 3.9 cm/yr along an azimuth of 70° (DeMets et al., 1990); reconstructions indicate that the adjacent spreading center has been in close proximity to the continental margin for the last 30 my. Off Washington the incoming oceanic crust is covered by at least several kilometers of sediment deposited by a prograding submarine fan that has filled the trench with predominantly terrigenous sediments (Kulm et al., 1984). These sediments are quite porous, and therefore provide an abundant source of water as they are accreted to the margin and compacted.

The Olympic Peninsula exposes the products of Tertiary accretion in the Cascadia subduction zone (**Figure 1**; Stewart, 1970; Rau, 1973; Tabor and Cady, 1978). In the western Olympic Peninsula the

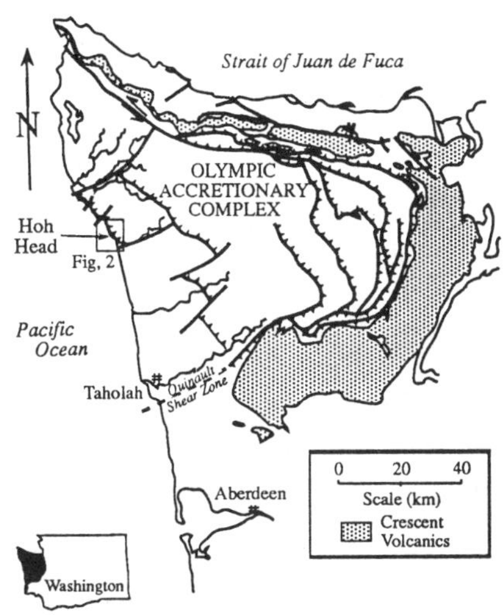

Figure 1: Generalized geology of the Olympic Peninsula (after Tabor and Cady, 1978).

Hoh rock assemblage is composed of Eocene to Miocene age sediments that occur as broken formation and mélange, and reach maximum metamorphic conditions of laumontite grade (Stewart, 1974). The highly deformed Eocene to Miocene rocks of the western Olympic peninsula are unconformably overlain by shallow water Miocene to Pliocene marine sequences (Rau, 1979).

In order to understand the relationship of fluid events to the tectonic evolution the general structural framework must first be presented. The Hoh rock assemblage is characterized by thick mud matrix mélange fault zones and wall rocks composed of relatively coherent Eocene to Miocene siliciclastic sections (Rau, 1979). The hanging walls are composed of conglomeritic to massive sandstone and footwalls are composed of finer grained and thinner bedded sandstones and siltstones. A fault origin for the mélange is inferred based upon the intense matrix deformation, consistent matrix orientations, millimeter scale scaly fabric throughout the entire mélange zone, and the similar orientation of clasts within the mélange (see Orange, 1990). Units within the wall rocks are broad to tightly folded and are faulted as well. In the present study we focus on the Hoh-Goodman mélange zone exposed at Hoh Head, although studies of other fault zones demonstrate similar results (Orange et al., 1991).

Hanging wall bedding orientations indicate a moderate to steep northerly dip. The contact between the hanging wall and the mélange fault zone trends north and is steeply dipping, varying from vertical to 60° east. The fault zone mélange has a consistent foliation that strikes north-northwest and dips steeply to the northeast (**Figure 2**). Asymmetric folds in the mélange matrix indicate thrust vergence to the west, with a small component of left lateral shear, and slickenlines associated with hanging wall and footwall contacts suggest thrust asymmetry as well. In cross-section, the Hoh Head fault zone mélange varies from 200m to 500m thick. The contact between the fault zone and the footwall is well exposed, and dips from 51 to 70° to the northeast. The Hoh Head footwall syncline has a steep east-west trending axial plane and a shallowly plunging fold axis. The fault zone mélange is unaffected by and truncates this fold, indicating that folding occurred prior to faulting (**Figure 2**).

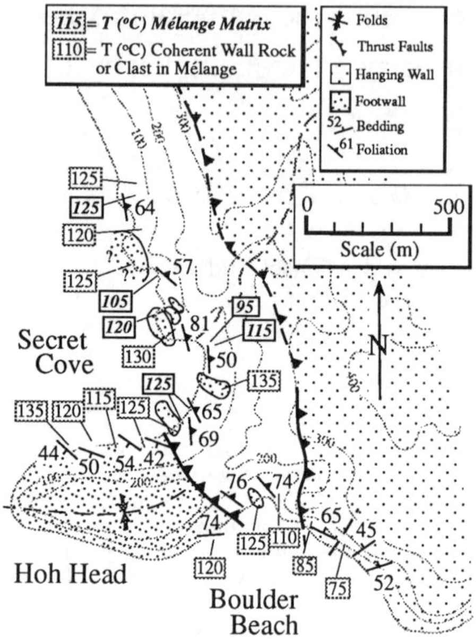

Figure 2: Structural and thermal data from Hoh Head. Note that the fault zone truncates (post-dates) folding in the footwall. The mixed temperatures indicate that faulting was post-peak thermal and took place at T<75-85°C.

2. Fluid Evolution

Field observations in the Hoh rock assemblage indicate abundant fluid involvement in many parts of the complex. "Smell muds", petroliferous matrix mélange and clasts, are common in many of the fault zones on the western Olympic Peninsula (Rau, 1979; Kvenvolden et al., 1989). Veins are also common throughout the Hoh rock assemblage, and include carbonate as well as laumontite veins.

In order to assemble accurate and meaningful data regarding the evolution of fluid precipitates, samples must be examined in the field for the relationships between fluid and deformational events. In this manner, mesoscopic observations of faulting, discrete cataclastic grain breakage, veins and cements can be documented. All samples were slabbed in order to investigate the relationship between deformation and fluid events, and thin-section size targets were selected from these slabs. Targets were selected based upon the presence of multiple fluid events and/or a clear relationship between a fluid precipitate and a deformational feature.

In the following section, the terms "micritic" (<30μ), "fine spar" (30-80μ), and "sparitic" (>80μ) will be used to describe the size of carbonate crystals, while "dirty" and "clear" are used to describe the clarity of a vein in plane light. "Dirty" veins are those where the presence of included particulate matter results in a dusty appearance under low power magnification (a function of the number and size of inclusions). The terms "dark", "bright", etc. will be used to describe the intensity of carbonate cathodoluminescence (CL).

2.1 Observations

Samples from the Hoh rock assemblage show a consistent progression of fluid precipitates regardless of structural setting. Early cements contain undeformed mica grains and also separate adjacent broken grains, indicating that the cement post-dates a phase of distributed grain breakage. In addition, discrete cataclastic shear zones post-date cementation. The progression from distributed grain breakage to cementation to localized zones of grain breakage is consistent with deformation occurring during an increase in effective stress and lithification (Atkinson and Bramsby, 1978).The early carbonate cement is micritic to fine-spar, occurs as both an alteration product and replacement of an earlier non-carbonate cement, and pre-dates localized cataclastic shear zones (**Figure 3**). Other authigenic relationships include clay rinds on some grains and chlorite replacement of tuffaceous and mafic grains and cement. Grains with clay rinds are found within sparitic carbonate veins, while the timing of chlorite with respect to veining is unclear.

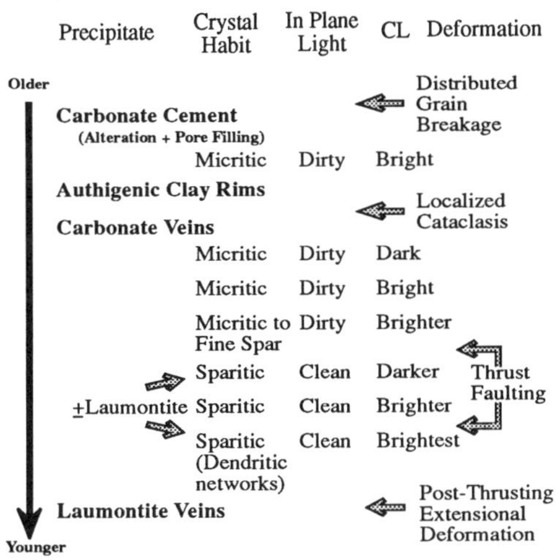

Figure 3: Evolution of fluid precipitates from the western Olympic Peninsula.

The first vein event occurs as a diffuse carbonate vein that is micritic, brightly luminescent, and frequently contains inclusions of disaggregated grain and matrix. The disaggregation gives this early vein a "dirty" appearance in plane light, and suggests that this vein formed early in the history of the sample prior to much of the cementation. Carbonate veins progress from micritic to sparitic in crystal habit, dirty to clear in plane light, and largely show increasing luminescence intensity throughout the progression. Two of the carbonate veins, however, show relatively low cathodoluminescence ("dark" on **Figure 3**). Early veins follow sediment grain boundaries, while later veins truncate grains. Equilibrium carbonate+ laumontite veins occur in the later parts of the fluid evolution (are syn- or post-deformational), and pure laumontite veins characterize the last fluid event.

Veins commonly show multiple phases of fluid emplacement. These observations, and the cross-cutting relationships of vein phases and deformation, determine their relative timing. Most of the thrust-related deformation occurs after the micritic dirty carbonate veins, and before the last phase of laumontite veining. The last laumontite veins, in fact, occur along extensional fractures that cut across the mélange deformational fabric.

2.2 Interpretation of Fluid Precipitates

A remarkable aspect of the Hoh rock assemblage samples is that the fluid stratigraphies for all samples can be placed within the same overall fluid evolution (a similar progression is also seen in nearby autochthonous basins; Galloway, 1974). This progression reflects the changing physical and chemical conditions within the accretionary complex.

The large amount of entrained particulate matter in the early carbonate cement and veins is most likely due to precipitation before significant cementation. If fractures open before the host rock is well cemented then these fractures could be filled with particulate matter from the matrix. As host rock cementation increases, any fracture that opens up will be less likely to pull particulate matter from the matrix, and the resulting precipitate will be clearer in plane light.

The progression from early micritic to sparitic veins may be due a number of factors. Tucker (1981, p. 135-136) and Folk (1974) attribute this transition to a freshening of the fluid phase. This hypothesis is supported by observations of negative chloride anomalies along fault zones in active accretionary complexes (Kastner et al., 1991), and of increasingly fresh of fluids landward of the toe (Vrolijk et al., 1991). Alternatively, the presence of entrained particulate matter common in these early veins may provide a large number of nucleation sites for crystal formation (e.g.: Kingery, 1967, p. 297-298), and the resultant vein could have numerous very small crystals, giving the overall micritic appearance. In this interpretation, sparitic veins would be indicative of a higher degree of wall-rock cementation and less entrained matrix within the vein. Both processes, consolidation and fluid freshening, may be active during the increasing burial and diagenesis of these sediments.

The changes in carbonate luminescence throughout the fluid evolution are more difficult to interpret. Regardless of the interpretation, the dark luminescent carbonate veins are regionally significant and are consistently found as an early micritic event and as a later, syn-deformational sparitic event. Although a wide variety of elements and chemical changes can affect cathodoluminescence (e.g. Pb^{+2}, Ce^{+2}, Ni^{+2}, rare earth elements, clay diagenesis, variations in trace element supply: Machel et al., 1991), we will base our preliminary interpretation upon the $Mn:Fe^{+2}$ ratio alone (Machel et al., 1991; Tucker, 1981, p. 135). The brightness of carbonate luminescence is strongly affected by the ratio of $Mn:Fe^{+2}$ in the carbonate, with an increase in this ratio leading to an increase in the luminescence. Preliminary EDS analyses of several samples from the Hoh rock assemblage support a relationship between luminescence and $Mn:Fe^{+2}$ ratio. Reducing conditions can lead to an increase in Fe^{+2} at the expense of Fe^{+3} and thus a decrease in carbonate luminescence. Since the first dark carbonate vein is micritic and rich in included particulate matter, we suggest that it formed in the sulphate reduction zone near the sediment-water interface.

Progressively brighter carbonates could signify less reducing conditions during burial and diagenesis, but this progression is interrupted by the regionally common dark sparitic carbonate vein. This vein may be due to precipitation under reducing conditions related to the presence of hydrocarbons (Barker and Dalziel, 1991). The darkly luminescent sparitic carbonate vein has clear evidence of petroleum fluid inclusions (Orange et al., 1990), yet successive generations of bright sparitic veins also have petroleum inclusions. The dark sparitic vein may indicate the onset of petroleum migration, but the progression to brighter cements remains enigmatic.

The final fluid event documented in the Hoh rock assemblage is the precipitation of laumontite. The Hoh rocks also show a progression from carbonate -> carbonate+laumontite -> laumontite veins, with equilibrium carbonate+laumontite occurring in numerous samples. The coexistence of laumontite and carbonate is indicative of a high calcium activity and a low P_{CO_2} (Thompson, 1971). The progression of carbonate- to laumontite-dominated precipitation could be due to an increase in the calcium activity in the fluid phase, a decrease in the P_{CO_2}, or a change in the P-T conditions of the sample (Thompson, 1971).

3. Integrating Fluid Precipitates, Deformational Events, and Thermal Evolution

In addition to the overall progression of fluid events, the cross-cutting relationships of deformational features and fluid precipitates make it possible to constrain their relative timing, structural, and thermal history (e.g.: Fisher and Byrne, 1990). The transition from distributed grain breakage (independent particulate flow) to localized cataclastic shear zones requires an increase in the effective stress and/or an increase in the degree of lithification (Atkinson and Bramsby, 1978). In the Hoh rocks the presence of carbonate cement indicates that increased lithification (cementation, consolidation) was an important part of this transition; if cementation occurred during burial then effective stress may also have played a role.

It is also possible to constrain the fluid evolution within the thermal framework of the Hoh accretionary complex by using vitrinite reflectance data (paleotemperatures based upon Barker, 1988; internal reproducibility ±5°C, total error ±30°C; although absolute temperatures may be different, relative changes in paleotemperature and spatial gradients are robust). Paleotemperatures within the Hoh Head fault zone, for example, range from 105-130°C (**Figure 2**). If faulting were pre-peak thermal then all of the material within the fault zone would have experienced the same peak temperature. Folds in the footwall also indicate that they formed after peak thermal conditions. The syncline at Hoh Head does not have any values within the core of the fold (limb T=120-130°C), but synclines at Toleak Point indicate cooler temperatures in their cores (Orange et al., 1991). Note that paleotemperatures within the hanging wall at Hoh Head are substantially lower than in the fault zone, in the neighborhood of 80°C (**Figure 2**). In order to thrust colder rocks over hot rocks there needs to be a pre-faulting event that deforms isograds; in the case of Hoh Head, we suggest that the early phase of post-peak thermal folding is responsible for this.

The combination of structural data and vitrinite reflectance from the Hoh rocks indicates that both folding and thrust faulting are post-peak thermal. The maximum regional thermal conditions during fault zone emplacement, therefore, could not be higher than ~80°C, or else the vitrinite would have been reset. This provides an important constraint for analyzing future stable isotope and fluid inclusion data, as it implies an upper limit to the regional thermal conditions during deformation. Assuming a geothermal gradient of 25°C/km for the Hoh rocks (assumed valid from deposition through deformation; gradient from well data; after Snavely and Kvenvolden, 1989), this implies maximum burial depths of 3±1km and maximum pressures of ~80±30 MPa (includes absolute error from vitrinite method).

The earliest stages of the fluid evolution involve the precipitation of cements and diffuse, micritic carbonate veins. Obviously intergranular permeability is the controlling mechanism for such precipitates. Later phases of fluid, however, precipitate as clear, sparitic carbonate filling discrete fractures, or veins. These veins indicate formation after some degree of cementation, cementation that also occludes the intergranular permeability. The later veins that form along discrete fractures imply that fracture permeability becomes increasingly dominant in controlling fluid migration. Since these veins are syn-deformational, and fault zone emplacement occurred at temperatures less than ~80°C, then the transition to fracture-controlled dewatering in an accretionary complex must occur below this maximum condition. The combination of structural and thermal data, and the observations of fluid precipitates, places important constraints on the physical, chemical, and hydrologic evolution of accretionary complexes.

ACKNOWLEDGEMENTS

This research supported by NSF grants EAR 88-10789 and EAR 90-05376, and by the Donors of the Petroleum Research Fund, administered by the American Chemical Society (ACS PRF 23472-AC2). We also wish to thank the Quinault and Quillayute Indian Nations, and the Olympic National Park, for making this research possible.

REFERENCES

Atkinson, J.H. and Bramsby, P.L. 1978, The Mechanics of Soils - An Introduction to Critical State Soil Mechanics: McGraw-Hill, Maidenhead, England, 375 p.

Barker, C.E. 1988, Geothermics of petroleum systems: Implications of the stabilization of kerogen thermal maturation after a geologically brief heating duration at peak temperature; in Magoon, L., ed., Petroleum Systems in the United States, USGS Bulletin 1870, 26-29.

Barker, C.E. and Dalziel, M.C. 1991, Using cathodoluminescence to map regionally zoned carbonate cements occurring in diagenetic aureoles above oil reservoirs, Velma oil Field, Oklahoma (abs.): Amer. Assoc. Pet. Geol. Bull. 75, 538.

Byrne, T. 1984, Early deformation in mélange terranes of the Ghost Rocks Formation, Kodiak Islands, Alaska: in Raymond, L.A., ed., Mélanges: Their nature, origin, and significance: GSA Special Paper 198, 21-52.

DeMets, C., Gordon, R.G., Argus, D.F., and Stein, S. 1990, Current plate motions, Geophys. Jour. I. 101, 425-478.

Fisher, D. and Byrne, T. 1990, The character and distribution of mineralized fractures in the Kodiak Formation, Alaska: Implications for fluid flow in an underthrust sequence: J. Geophys. Res. 95, 9069-9080.

Folk, R.L. 1974, The natural history of crystalline calcium carbonate: Effect of magnesium content and salinity: J. Sed. Pet. 44, 40-53.

Galloway, W.E. 1974, Deposition and diagenetic alteration of sandstone in Northeast Pacific arc-related basins: Implications for graywacke genesis: GSA Bull. 85, 379-390.

Kastner, M., Elderfield, H. and Martin, J.B. 1991, Fluids in convergent margins: What do we know about their composition, origin, role in diagenesis and importance for oceanic chemical fluxes?: Phil. Trans. Royal Soc. London, A. 335, 275-288.

Kingery, W.D. 1967, *Introduction to Ceramics*: John Wiley and Sons, New York, 781 p.

Kulm, L.D., Peterson, C.P., Connard, G.G., Johnson, S.Y., Niem, W.A., McClain, K.J., Loubere, P.W., Lewis, B., Couch, R.W., Niem, A.R., Scheidegger, K.F., Crosson, R.S. and Stewart, R.J., 1984, Western North American continental margin and adjacent ocean floor off Oregon and Washington: Atlas 1 Ocean Margin Drilling Program, Regional Atlas Series, Marine Sciences International, Woods Hole, MA, 32 sheets.

Kvenvolden, K.A., Rapp, J.B., Hostettler, F.D. and Snavely, P.D.Jr. 1989, Preliminary Evaluation of the Petroleum Potential of the Tertiary Accretionary Terrane, West Side of the Olympic Peninsula, Washington: Part 2 - Comparison of Molecular Markers in Oil and Rock Extracts: USGS Bull. 1892, 19-36.

Machel, H.-G., Mason, R.A., Mariano, A.N. and Mucci, A. 1991, Causes and emission of luminescence in calcite and dolomite: in Barker, C.E. and O.C. Kopp, eds., Luminescence Microscopy and Spectroscopy: Qualitative and Quantitative Applications: SEPM Short Course Notes 25, 9-25.

Moore, J.C., Brown, K.M., Horath, F., Cochrane, G., MacKay, M. and Moore, G. 1991, Plumbing accretionary prisms: Effects of permeability variations: Phil. Trans. Royal Soc. London, A. 335, 275-288.

Orange, D.L. 1990, Criteria helpful in recognizing shear zone and diapiric mélanges: Examples from the Hoh Accretionary Complex, Olympic Peninsula, Washington: GSA Bull. 102, 935-951.

Orange, D.L., Knittle, E., Farber, D. and Williams, Q. 1990, Spectroscopic studies of hydrocarbon fluid inclusions in calcite: Implications for fluid evolution of the Hoh accretionary complex, Olympic Peninsula, Washington: PACROFI III: Third Biennial Pan-American Conference on Research on Fluid Inclusions, May 20-22, 1990, Toronto, Ontario, Canada, p. 67.

Orange, D.L, Geddes, D., Cox, R. and Underwood, M. 1991, Out-of-sequence thrust faulting in a young accretionary complex: evidence from fluid inclusion and vitrinite reflectance data, Trans., AGU. 72, 549.

Rau, W.W. 1973, Geology of the Washington Coast between Point Grenville and the Hoh River: Wash. Dept. of Natl. Res., Geol. and Earth Res. Div., Bulletin 66, 58 p.

Rau, W.W. 1979, Geologic map in the vicinity of the Lower Bogachiel and Hoh rivers, and the Washington Coast: Wash. Dept. of Natl. Res., Geol. -and Earth Res. Div., Geologic Map GM-24.

Snavely, P.D. Jr. and Kvenvolden, K.A. 1989, Preliminary Evaluation of the Petroleum Potential of the Tertiary Accretionary Terrane, West Side of the Olympic Peninsula, Washington: Part 1 - Geology and Hydrocarbon Potential: USGS Bull. 1892, 1-18.

Stewart, R.L. 1970, Geologic Map of the Hoh River - Queets River Area, Olympic Peninsula, Washington: U. Wash., M.S. thesis, 74 p.

Stewart, R.J., 1974, Zeolite metamorphism of sandstone in the western Olympic Peninsula, Washington: GSA Bull. 85, 1139-1142.

Tabor, R.W. and Cady, W.M. 1978, The structure of the Olympic Mountains, Washington- Analysis of a subduction zone: USGS Prof. Paper 1033, 38 p.

Thompson, A.B. 1971, P CO2 in low-grade metamorphism: Zeolite, carbonate, clay mineral, prehnite relations in the system CaO-Al2O3-SiO2-CO2-H2O. Cont. Min. Pet. 33, 145-161.

Tucker, M.E. 1981, *Sedimentary Petrology: An Introduction*. Halsted Press, John Wiley and Sons, New York, 252 p.

Vrolijk, P., Fisher, A. and Gieskes, J. 1991, Geochemical and geothermal evidence for fluid migration in the Barbados accretionary prism (ODP Leg 110): Geophys. Res. Let. 18, 947-950.

Water-Rock Interaction, Kharaka & Maest (eds) © 1992 Balkema, Rotterdam. ISBN 90 5410 075 3

Structural controls of fluid flow during regional metamorphism: Application of reaction front advection theory to greenschist facies metabasites in the Scottish Dalradian

A.D.L. Skelton & C.M. Graham
Department of Geology and Geophysics, University of Edinburgh, UK
M.J. Bickle
Department of Earth Sciences, University of Cambridge, UK

ABSTRACT: Infiltration of H_2O-CO_2 fluids during greenschist facies regional metamorphism in Dalradian rocks of the SW Scottish Highlands has driven calcite- and dolomite- forming reactions at the margins of metabasic sills. A simple geochemical technique has enabled reaction fronts to be identified in the field. Time integrated fluid fluxes have been calculated from the distances which these fronts have been advected into sill margins. Fluid flow directions have been constrained from the asymmetry of front profiles. Regional scale mapping of directionally-constrained time-integrated fluid fluxes has revealed large-scale fluid flow patterns. Fluid flow was channelised along both regional and outcrop-scale antiformal hinge zones. Time-integrated fluid fluxes vary reciprocally with distance from the axial plane of a regional recumbent anticline, and average fluxes across the sampled area of about 75 m^3/m^2 are consistent with derivation of the fluid from devolatilisation of underlying, higher grade lithologies, without recourse to extraneous fluid sources.

1. INTRODUCTION

Crustal metamorphism must be accompanied by movement of substantial volumes of H_2O-CO_2 fluids in order to remove fluid lost during prograde devolatilisation reactions (eg Walther and Orville 1982). The movement of such fluids through metamorphic rocks may result in systematic changes in the stable isotopic compositions, mineral assemblages and mobile element chemistry of rocks along the flowpath. Isotopic, petrological and trace element data may thus provide important constraints on fluid sources, fluid pathways, fluid volumes and transport mechanisms, but results of such studies have remained controversial (eg Wood and Graham 1986). Estimates of fluid volumes in metamorphic terrains vary from small volumes consistent with fluid derived purely through devolatilisation reactions (Baker et al 1989; Bickle and Baker 1990a) to large volumes sufficient to modify the geochemical and thermal structure of metamorphic terrains (eg Wickham and Taylor 1985, Ferry 1988). Such large fluid volumes could only be derived from an external source such as surface waters convected to depths of metamorphism which is physically implausible (eg Wood and Walther 1986), or if fluid is channelised.

Quantities of fluid interacting with rocks undergoing metamorphism have commonly been calculated from reaction progress or isotopic mass balance equations in terms of volumetric fluid:rock ratios. The basis and significance of such calculations have been criticised by several authors (eg Wood and Graham 1986, Wood and Walther 1986). Calculations based purely on mass-balance equations provide no constraints on the direction of fluid flow, and imply significant and implausible departures from local chemical equilibrium along the (unconstrained) flowpath.

Where a potential tracer of fluid flow such as oxygen isotopes has been homogenised on a regional scale (eg Wickham and Taylor 1985), information on fluid flow paths and fluid fluxes cannot be deduced. However, where advective displacement of isotopic or petrological discontinuities (eg lithological boundaries) from their original, pre-metamorphic location may be identified and quantified, advective movement of fluid may be unambiguously demonstrated, and fluid fluxes in the direction of front advection computed (eg Bickle and Baker 1990a,b). Such studies yield directionally-constrained, time-integrated fluid fluxes which can be used as meaningful indicators of fluid volumes and transport directions, thus aiding in the identification of fluid sources.

Here we outline the application of reaction front theory to the infiltration of CO_2-bearing, H_2O-rich fluid into greenschist facies metabasic rocks in the Dalradian of the SW Scottish Highlands. Directionally constrained, time-integrated fluid fluxes have been mapped on a regional scale for the first time, the relationship of local and regional structure to fluid flow pathways investigated, and fluid sources constrained.

2. GEOLOGY AND INFILTRATION HISTORY

The rocks of the SW Scottish Highlands comprise a series of psammites, phyllites, calc-pelites, marbles and metabasic sills, dykes, lavas and tuffs, all metamorphosed in the greenschist and epidote amphibolite facies and isoclinally folded on a regional scale during the Caledonian Orogeny. NW and SE facing recumbent antiforms (Fig 1) are separated by a synformal root-zone (the Loch Awe Syncline). These rocks show evidence of three chronologically distinct episodes of infiltration (Graham et al 1983): (i) Metabasites proximal to the Loch Awe Syncline show mineralogical and geochemical evidence of pre-metamorphic spilitic alteration, probably resulting from sub-seafloor hydrothermal activity during emplacement of sills and volcanics; (ii) The majority of metabasites show marginal alteration resulting from infiltration of CO_2-bearing hydrous fluid during metamorphism, documented by the metamorphic carbonation/hydration reactions: (1) Amph + Epid + CO_2 + H_2O = Chl + Cc + Qz, or, more rarely and at higher X(CO2) (2) Chl + Cc + Ab + CO_2 = Dol + Parag + Qz + H_2O (Fig 2); (iii) Many lithologies are locally retrogressed and altered by vein-controlled H_2O-CO_2 fluid. The results of a study of the peak-metamorphic infiltration event are now presented and discussed.

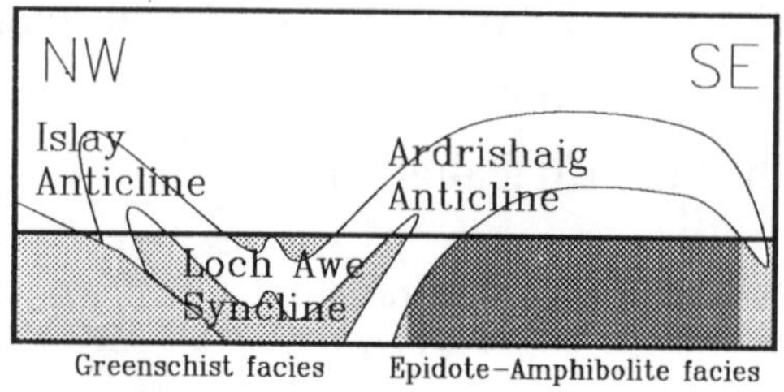

Figure 1: Structural and Metamorphic Section of the S.W. Scottish Highlands.

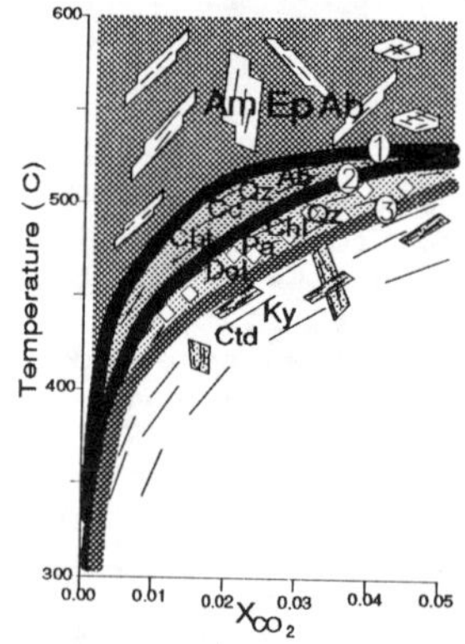

Reaction 1 is the calcite-forming reaction: Am+Ep+CO2+H2O =Chl+Cc+Qz, Reaction 2 is the dolomite-forming reaction: Chl+Ab+Cc+CO2 =Dol+Pa+Qz+H2O, Reactions 2 and/or 3 buffer the infiltrating fluid composition within the host phyllites.

Figure 2: T-X(CO2) phase relations (adapted from Holland and Powell 1988)

3. METHODOLOGY

The advection distances of carbonation reaction fronts from metabasite lithological boundaries were measured using a simple field-based geochemical technique. Samples taken at intervals across metabasites were crushed, and a known volume was reacted in a portable, sealed reaction vessel with an excess of 10% HCl. The volume of CO_2 evolved was measured from the pressure recorded on a built-in manometer. The reaction vessel was previously calibrated in the laboratory for samples containing calcite, dolomite, and calcite + dolomite, the relative amounts of coexisting calcite and dolomite being estimated from reaction rate differences.

Using these data, the progress of calcite and dolomite generating carbonation reactions (1) and (2) was calculated. Reaction progress was plotted against distance from metabasite contacts, and steps in reaction progress constrained the location of reaction fronts (Fig 3). Results of this approach were found to be in good agreement with location of reaction fronts from point counting. The ease and rapidity of this technique, without recourse to laboratory work, allows a comprehensive regional picture of fluid flux in relation to structure and lithology to be constructed for the first time.

The advection distance of a reaction front ($V_f t$) measured as above is given by Baker and Bickle (1990a) as:

$$V_f t = \frac{w\phi t(X_{CO2,2} - X_{CO2,1})}{v_{CO2}(1 - X_{CO2,2}) - v_{H2O} X_{CO2,2}} \qquad (3)$$

where v_{CO2} and v_{H2O} are the stoichiometric coefficients of CO_2 and H_2O, respectively, absorbed by reaction (1) or (2) at the reaction front, and $X_{CO2,2}$ and $X_{CO2,1}$ are the volume fractions of CO_2 in the fluid downstream and upstream of the reaction front respectively. V_f is the velocity of the front, t is time, and therefore $V_f t$ is the (time-integrated) distance which the front has travelled. w is the velocity of the fluid, ø is the porosity (instantaneous fluid volume), and thus wø is the fluid flux and wøt the time-integrated fluid flux. v_{CO2} and v_{H2O} can be estimated from the reaction stoichiometry. $X_{CO2,2}$ is the equilibrium fluid composition for the reaction (1) or (2), computed from calculated T-X(CO_2) phase relations (Fig 2; Holland and Powell 1988). $X_{CO2,1}$ is the composition of the infiltrating fluid and has been shown to be that which is buffered by the host phyllite assemblage and is computed from similar calculated T-X(CO_2) phase relations.

The measured advection distance, $V_f t$, must be corrected for diffusive broadening of the front, which should be symmetrical on each side of infiltrated metabasic sills, and can readily be deduced for example from Fig 3. If suitable assemblages are not available to constrain the X(CO_2) of infiltrating fluid, values can be obtained by simultaneously solving equation (3) for two reaction fronts ((1) and (2)) in the same sill. From equation (3), we calculate $V_f t$ = 0.2wøt for reaction (1) and $V_f t$ = 0.1wøt for reaction (2). Calculated fluxes are directionally constrained by the asymmetry of reaction zones, the thicker margin being on the infiltration side and the thinner margin resulting from diffusive front broadening (Fig 3).

4. DISCUSSION

The fluid flux map constructed for the abundant metabasic sills in the greenschist facies of the SW Scottish Highlands (Fig 4) shows a remarkable correlation between flux direction and geological structure over the whole region. Fluid flow direction has a major component outwards from antiformal axial surfaces and inwards towards synformal axial surfaces. Of particular interest is the observation that the magnitude of the time-integrated fluid flux shows a strong linear correlation with reciprocal distance from the axis of the major regional antiformal structure - the Ardrishaig Anticline (Fig. 5). Fluid is thus strongly channelised along both local and regional antiforms.

The time-integrated fluid flux as constrained in a NW-SE direction and averaged across the area sampled in the SW Highlands (18km x 12km) is about 75 m^3/m^2. A flux of this magnitude could be generated, for example, by metamorphic devolatilisation of about 2-3km thickness of underlying metabasites and metasediments along the fluid flow path undergoing prograde dehydration at a greenschist to epidote-amphibolite transition comparable to that exposed at the garnet isograd to the SE of the study area, where breakdown of chlorite liberates about 3-4 vol.% water. The volumes of fluid required are consistent with the known geological and

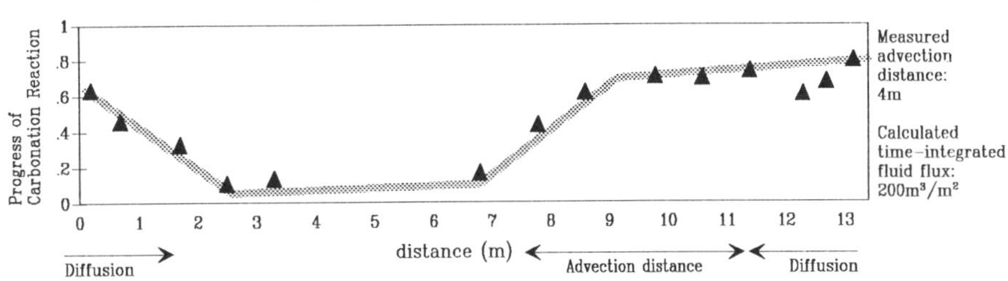

Figure 3: Profile of Field Estimates of Reaction Progress across a Metabasite Sill at Port Cill Maluaig in Knapdale S.W. Scottish Highlands

lithological structure of the area, and no external sources of fluid are required.

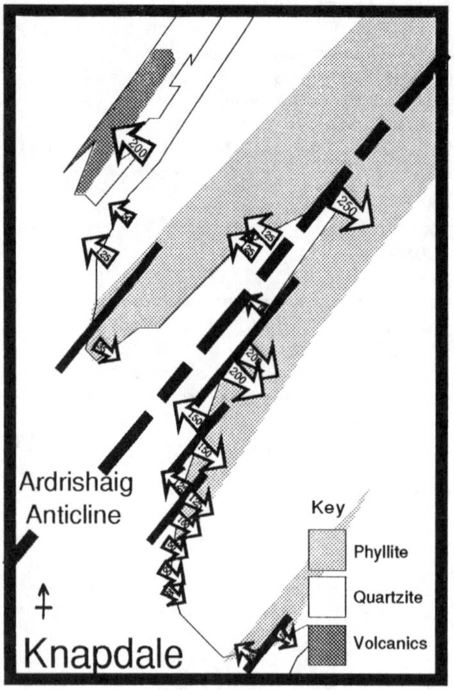

Figure 4; Map showing directionally constrained Time-integrated fluid fluxes

Arrows indicate the direction and magnitude of fluid flow. Dashed Lines are anticlinal axes. Fluid is seen to flow outwards from anticlinal axes suggesting that they act as fluid channels. The magnitude of the time-integrated fluid flux decreases with distance from the Ardrishaig anticline implying that it acts as a regional fluid channel.

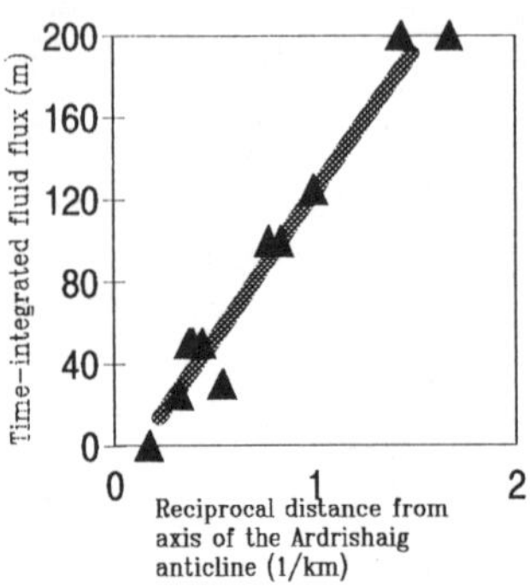

The above plot shows that time-integrated fluid flux is inversely proportional to the distance from the Ardrishaig anticline, thus implying that fluid was channelised along this regional-scale structure.

Figure 5 ; Fluid-channelisation along the Ardrishaig Anticline

5. REFERENCES

Baker J, Bickle M J, Buick I S, Holland T J B and Matthews A (1989) Isotopic and petrological evidence for the infiltration of water-rich fluids during the Miocene M2 metamorphism on Naxos, Greece. Geochim Cosmochim Acta 53, 2037-2050.

Bickle M J and Baker J (1990a) Migration of reaction and isotopic fronts in infiltration zones: assessments of fluid flux in metamorphic terrains. Earth Planet Sci Letters 98, 1-13.

Bickle M J and Baker J (1990b) Advective-diffusive transport of isotopic fronts: an example from Naxos, Greece. Earth Planet Sci Letters 97, 78-93.

Ferry J M (1988) Infiltration-driven metamorphism in Northern New England. J Petrol 29, 1121-1159.

Graham C M , Greig K M, Sheppard S M F and Turi B (1983) Genesis and mobility of the mobility of the H_2O-CO_2 fluid phase during regional greenschist and epidote amphibolite facies metamorphism: a petrological and stable isotope study in the Scottish Dalradian. J geol. Soc. London 140, 577-599.

Powell R and Holland T J B (1988) An internally consistent thermodynamic dataset with uncertainties and correlations: 3. Applications to geobarometry, worked examples and a computer program. J Metam Geol 6, 173-204.

Walther J V and Orville P M (1982) Rates of metamorphism and volatile production and transport in regional metamorphism. Contrib Miner Petrol 79, 252-257.

Wickham S M and Taylor H P (1985) Stable isotope evidence for large-scale seawater infiltration in a regional metamorphic terrain: the Trois Seigneurs massif, Pyrenees, France. Contrib Miner Petrol 91, 122-137.

Wood B J and Graham C M (1986) Infiltration of aqueous fluid and high fluid:rock ratios during greenschist facies metamorphism: a discussion. J Petrol 27, 751-761.

Wood B J and Walther J V (1986) Fluid flow during metamorphism and its implications for fluid-rock ratios. In J V Walther and B J Wood (eds). Fluid-rock interactions during metamorphism, Advances in Physical Geochemistry 5, 89-108, Springer-Verlag: New York.

Water-Rock Interaction, Kharaka & Maest (eds) © 1992 Balkema, Rotterdam. ISBN 90 5410 075 3

Fluid-rock interaction in skarns from La Cerdanya, Central Pyrenees, Spain: Evidence from carbon and oxygen isotopes

A. Soler & J. Delgado
Departamento de Mineralogia, Universitat de Barcelona, Spain
E. Cardellach
Departamento de Geologia, Universitat Autonoma de Barcelona, Spain
C. Ayora
Instituto Ciències de la Terra, CSIC, Barcelona, Spain

ABSTRACT: C and O isotope data of calcites in skarns of La Cerdanya (Central Pyrenees) range from $\delta^{18}O$=25.2 to 10.3‰ and $\delta^{13}C$=0.2 to -12.2‰. The variation has been modeled by isotopic exchange between Devonian limestones and a fluid of magmatic origin at temperatures between 600 and 350°C and X_{CO_2} from 0.1 to 0.01. Calculated fluid to rock ratios indicate that the metasomatic fluid circulated from the granite contacts towards the country rocks preferentially through interstrata porosity (channelized flow).

INTRODUCTION

Carbon and oxygen isotope compositions of skarn calcites have been demostrated to be useful tracers of the interaction between metasomatic fluids and the regional carbonates (Bowman et al., 1985a, b; Taylor and O'Neil, 1977, among others). These studies show that stable isotope data provide important information on progresive water-rock interaction, fluid sources and ore forming conditions (X_{CO_2}, T, pH, etc).

Detailed petrologic and C-O isotope studies have been carried out on the calcic skarns of the La Cerdanya area (Central Pyrenees, Spain). The geology of the studied area and the petrology, zoning and physico-chemical conditions of skarn formation have been presented previously (Soler et al., 1990; Soler & Ayora, 1991; Cardellach et al., 1992). The main objective of this study is to define the extent of metasomatic fluid/wallrock interaction and to trace the path and direction of fluid migration during skarn formation. The study is based on 78 carbon and oxygen isotope analyses of calcite samples from different types of skarns.

GEOLOGICAL SETTING

The study area is situated in a fragment of the Hercynian orogen outcropping in the axial zone of the Pyrenees. The area is constituted of Paleozoic metasediments ranging from pre-Caradocian to Carboniferous in age. The materials are siliciclastic (clays, sandstones, black shales) except the Devonian which is mainly made up of limestones. The Hercynian tectonics caused the formation of E-W folds, thrusts and low grade metamorphism (chlorite). The Andorra-Mont Lluis granite intruded these materials during late Hercynian times, causing the development of a thermic aureola of more than 300 m wide. Andalusite and/or sillimanite are the main silicates formed in the pelitic lithologies close to the contact, whereas cordierite and biotite are formed further away. The crystallization of idocrase, garnet, epidote, diopside and biotite occurred in the impure limestone horizons. Mineral equilibria in the KFMASH system suggest that the P-T conditions of the contact metamorphism were about 650° C and 2 Kbar. The contact between the granite and the Devonian limestones is sharp and parallel to bedding.

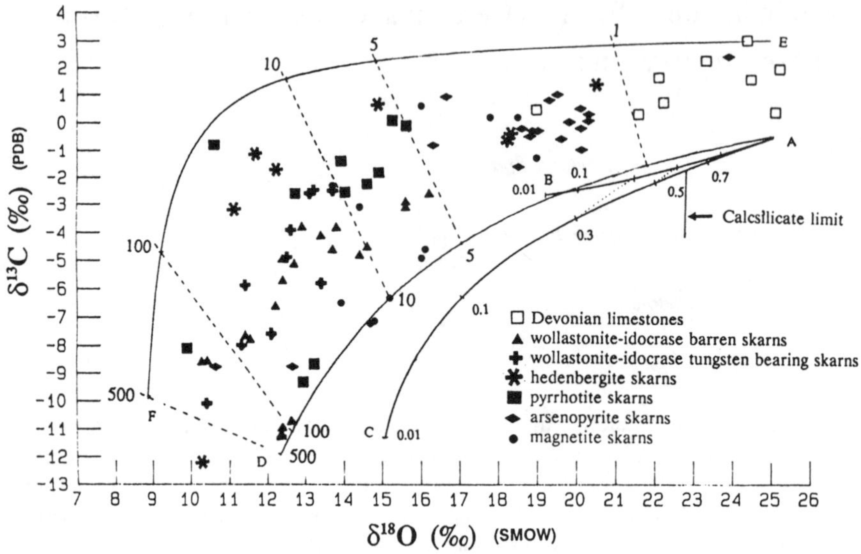

Fig. 1.- Plot of $\delta^{13}C$ versus $\delta^{18}O$ values of calcites from Devonian limestones and the studied skarns. The lines A-B and A-C are the trends of $\delta^{13}C$ and $\delta^{18}O$ depletion resulting from single-stage and Rayleigh decarbonation respectively. The lines are marked with the fraction of residual carbonate. Lines A-D and E-F are the plot of the $\delta^{13}C$ and $\delta^{18}O$ depletion in calcite resulting from the isotopic exchange between the average regional limestone and a magmatic fluid ($\delta^{13}C$=-10‰ and $\delta^{18}O$=+8‰) at temperatures of 350 and 600°C, and X_{CO_2} of 0.1 and 0.01, respectively. These lines are marked with the water/rock ratio (closed system).

Only in places where the intrusive contact cuts across the carbonate bedding, meter to tens-of-meters-sized skarns developed.

Based on mineralogical and geochemical criteria, six types of skarns are distinguished: 1. wollastonite-idocrase skarns, 2. wollastonite-idocrase tungsten bearing skarns, 3. hedenbergite skarns, 4. pyrrhotite skarns, 5. arșenopyrite skarns, 6. magnetite skarns. All skarns types have undergone the same metasomatic process, in which three stages of formation can be distinguished: a first, barren stage, including the formation of wollastonite, idocrase, diopside and granditic garnet; a second, iron-rich stage, with hedenbergite, stanniferous andradite and epidote; and third, sulfide rich stage, made up of a quartz, calcite, chlorite, actinolite and grossular associated with arsenopyrite, pyrrhotite, löllingite, native gold, galena, sphalerite, chalcopyrite, STeBi phases, hessite, native bismuth and occasionally magnetite and hematite.

Temperatures of skarn formation deduced from fluid inclusion microthermometry and silicate and sulfide equilibria, range from about 600° C for the first stage to 350° C for the third stage.

C and O ISOTOPE DATA

The $\delta^{18}O$ and $\delta^{13}C$ values of calcites from regional limestones vary from 22.1 to 25.2 ‰ and from 0.2 to 3 ‰ respectively. All these values are typical from marine carbonates of Devonian age (Veizer & Hoefs, 1976). On the other hand, calcites from skarns and veins show a depletion in both ^{18}O and ^{13}C values (up to ($\delta^{18}O$= +10.3‰ and δ^{13}= -12.2‰). This depletion is comparable to those described in other localities, where decarbonation reactions and/or exchange processes have taken place (Taylor & O'Neil, 1977; Bowman et al., 1985a, 1985b; Brown et al, 1985; Valley, 1986). Some of our ^{18}O and ^{13}C depleted values can be interpreted as the result of decarbonation processes, either by single-stage (batch) or continuous fluid removal (Rayleigh distillation).

At 600°C the progressive decrease in $\delta^{18}O$ and $\delta^{13}C$ values of the wallrock limestone by a single-stage (batch) decarbonation process cannot produce sufficient ^{18}O depletion (Fig.1). But at this temperature the lowest $\delta^{18}O$ and $\delta^{13}C$ values obtained could be explained by a continuous (Rayleigh) distillation process. However, this process would imply values of F for oxygen below 0.6 (the calcsilicate limit for wollastonite formation of Valley, 1986), which are very unlikely in the studied skarns because calcsilicates (wollastonite, diopside, idocrase) coexist with the remaining carbonate and also because of the stoichiometry of the volatilization reactions. Consequently, volatilization processes do not seem to have played a substantial role during the skarn formation. The large isotopic shifts seen in Fig. 1 must then result largely from an isotopic exchange between the wallrock limestone and an infiltrating fluid. Taking the initial composition of the regional limestone and $\delta^{18}O$= +8.0 and δ^{13}= -10.0 for the infiltrating fluid (magmatic origin), the isotopic exchange can be modeled at temperatures between 600 and 350°C and X_{CO_2} from 0.1 to 0.01. The depletions observed in the skarn calcites are due to different water/rock ratios. The obtained temperature and X_{CO_2} are in agreement with those obtained from silicate equilibria and fluid inclusion microthermometry (Soler et al., 1990; Soler and Ayora, 1991). The isotopic composition of the metasomatic fluid has been considered to be that of water in equilibrium with the vein calcite showing the lightest $\delta^{18}O$ and $\delta^{13}C$ values at the temperature of calcite.

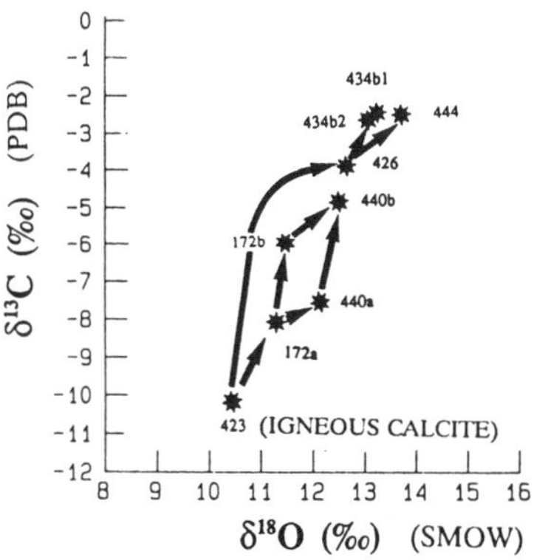

Fig. 2.- Plot of $\delta^{13}C$ versus $\delta^{18}O$ values of calcites from wollastonite-idocrase tungsten skarn, showing the decrease in the water/rock ratio between the successive samples (see Fig. 3).

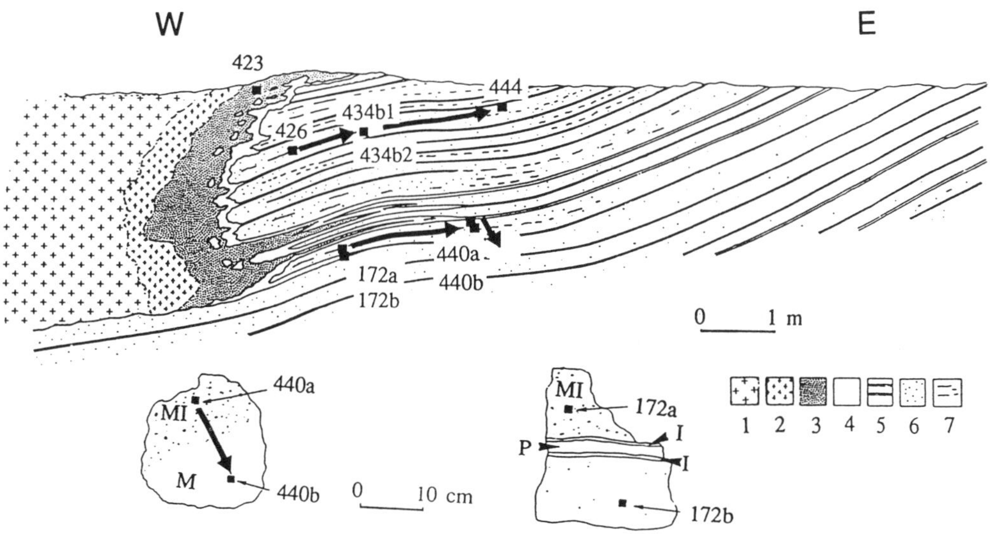

Fig. 3.- Location of samples in the wollastonite-idocrase tungsten-bearing skarn. From the decrease in the water rock ratio (see Fig. 2). The arrows indicate the fluid flow path. Legend: 1= granodiorite; 2= leucogranite; 3= massive skarn; 4= marble; 5= calcsilicate veins; 6= disseminated idocrase; 7= millimeter-scale idocrase veins; M= limestone; MI= limestone with disseminated idocrase; I= idocrase; P= prehnite vein.

formation (independently estimated as 450ºC). Taking into account the C-O isotopic composition of the calcites from the six types of skarn, it is possible to relate all of them to the same metasomatic event.

Coupling detailed sampling with calculation of the corresponding fluid-rock ratios, the path and direction of fluid migration has been determined. For the estimated X_{CO_2} and temperature conditions the exchange process took place at water/rock ratios either between 1 and 500 (closed system) or 0.7 to 6 (open system). Calcites sampled close to the skarn are more ^{18}O and ^{13}C depleted compared to calcites sampled further away (Fig. 2 and 3), suggesting an exchange progressively dominated by the rock as the fluid progressed and indicating the sense of the fluid flow. The changes in the water/rock ratios are strongly dependent on the sampling direction. Thus, as observed in Fig. 3, calcites sampled perpendicularly to the regional limestone bedding show a sharp decrease in the estimated water/rock ratio over very short distances (centimeters), whereas an equivalent decrease in calcites sampled along the interstrata (from the skarn to the regional limestone) are observed over longer distances (meters). Therefore, the metasomatic fluid circulated from the granite towards the Devonian limestones was channelized through the interbedding. Minor flow took place perpendicular to the interstrata.

CONCLUSIONS

The C and O isotope analysis of calcites in skarns from La Cerdanya (Central Pyrenees), coupled with a detailed and systematic sampling, has allowed to evaluate the evolving water/rock ratios during the metasomatic process and to determine the fluid path and the type of flow. The depletion in the skarn calcites resulted from an isotopic exchange between an igneous fluid and the Devonian regional limestone, at temperatures from 600 to 350ºC and X_{CO_2} from 0.1 to 0.01. The infiltrating fluid circulated from the intrusive body to the enclosing limestone mainly along the interstrata (channelized flow), and to a lesser extent through microporosity perpendicular to the bedding. Since the C and O stable isotope composition of carbonates have shown to be very sensitive to fluid-rock interactions, this technique, coupled with careful sampling, can be used to prospect the source of mineralizating fluids infiltrating carbonate rocks.

REFERENCES

Bowman,J.R., Covert,J.J., Clark,A.H. & Mathieson, G.A. 1985a. The Cantung E zone Scheelite Orebody, Tungsten, NW territories: O,H and C isotope studies. *Economic Geology*, 80: 1872-1985.

Bowman, J.R.; O'Neil, J.R. & Essene, E.J. 1985b. Contact skarn formation at Elkhorn, Montana II: origin and evolution of C-O-H skarn fluids. *American Journal Science*, 285, 621-660.

Brown, P.E.; Bowman, J.R. & Kelly, W.C. 1985. Petrologic and stable isotope constraints on the source and evolution of skarn-forming fluids at Pine Creek, California, *Economic Geology*, 80: 72-95.

Cardellach,E., Ayora,C., Soler,A. & Delgado,J. 1992. The origin of the fluids involved in the formation of gold-bearing skarns of the Andorra granite (Central Pyrenees, Spain): sulphur isotope data. *Mineralogy & Petrology*, in press.

Soler,A., Ayora,C., Cardellach,E. & Delgado,J. 1990. Gold-bearing hedenbergite skarns from the SW contact of the Andorra granite (Central Pyrenees, Spain). *Mineral Deposita*, 25S: 59-68.

Soler,A. & Ayora,C. 1991. Gold-bearing arsenopyrite bodies in skarns from l'Alt Urgell, Central Pyrenees, Spain. In: S.S.Augustithis (Ed.). *Skarns their genesis and* Metallogeny: 227-251. Athens: Teophrastus.

Taylor,B.E. & O'Neil,J.R. 1977. Stable isotope studies of metasomatic Ca-Fe-Al-Si skarns and associated metamorphic and igneous rocks, Osgood Mts., Nevada. *Contributions Mineralogy Petrology*, 63: 1-49.

Valley,J.W. 1986. Stable isotope geochemistry of metamorphic rocks. In Valley, J.W., Taylor,H.P. & O'Neil,J.R. (Ed): *Stable isotopes in high temperature geological processes*. Reviews in Mineralogy, 16: 445-486.

Veizer,J. & Hoefs,J. 1971. The nature of $^{18}O/^{16}O$ and $^{13}C/^{12}C$ secular trends in sedimentary carbonate rocks. *Geochimica Cosmochimica Acta*, 40: 1387-1395.

Water-Rock Interaction, Kharaka & Maest (eds) © 1992 Balkema, Rotterdam. ISBN 90 5410 075 3

Fluid interaction and deformation, the Brevard Shear Zone, Rosman, North Carolina, USA

Jane R. Stockey
DOE, Office of Civilian Radioactive Waste Management, Washington, D.C., USA
Eileen L. McLellan
Department of Geology, University of Maryland, College Park, Md., USA

ABSTRACT: A combination of petrographic analysis, fluid inclusion studies, $\delta^{18}O$ and $^{40}Ar/^{39}Ar$ geochronology was used to constrain deformation mechanisms, strain, and fluid distributions and the timing and conditions of deformation. The study was conducted in the Brevard Shear Zone, Rosman, North Carolina. Based on data from these analyses, the proposed P-T conditions of ultramylonitization were around 380°C and 3 kbar. High water-rock ratios were a necessity for the deformation mechanism to operate, if not the controlling factor, and it occurred in the Late Paleozoic, most probably during the late stages of the Alleghenian orogeny.

1 INTRODUCTION

The Brevard Shear Zone is the major structural boundary between the Blue Ridge and Piedmont in the central and southern Appalachians. It records a complex Paleozoic poly-meatmorphic and poly-deformational history which has been much studied but is still not fully understood. This study examines the role of fluid in deformation mechanism control in the Brevard Zone where it cuts the henderson Augen Gneiss at the type section at Rosman, North Carolina.section at Rosman, North Carolina.

At this locality, the Brevard Zone cuts the moderately sheared gneiss in a wide zone of highly sheared gneiss and mylonite and several smaller zones of ultramylonite. In the ultramylonite, the protolith gneiss has been so extremely deformed no augen are visible in hand sample, the rock is extremely fine-grained and a concurrent re-crystallization has occurred. Preliminary studies (Sinha et al., 1986, Stockey et al., 1987) supported the open-system interaction of an exotic fluid in the ultramylonite to accommodate the open-system geochemical behavior involved in the development of the hydrated mineralogy of the re-crystallized zone. This study documented the difference in deformation mechanism between the sheared mylonite and the ultramylonite and attempted to associate the change in mechanism with a change in fluid regime.

2 DISCUSSION

Petrographic data show localization of strain in the Brevard Shear Zone during two periods of deformation. The first was a high temperature event with the feldspar augen deforming ductilely and no evidence of open-system fluid interaction. The second was a low temperature event where the feldspar augen deformed brittlely in a fine-grained matrix of feldspar and quartz in which the dominant deformation mechanism was dynamic re-crystallization. This formed the main body of the shear zone, the mylonite. At the same time, there was highly localized strain forming the ultramylonite in which the dominant deformation mechanism was grain boundary sliding accommodated by pressure solution. This type of deformation and accommodation mechanism requires a high fluid-rock ratio. This fluid presence in the ultramylonite, originally indicated by the geochemical studies (Sinha et al, 1986, Stockey et al., 1987), and required by the dominant deformation mechanism active in the zone, is supported by both the fluid inclusion data, the $\delta^{18}O$ data and the $^{40}Ar/^{39}Ar$ data.

The original fluid inclusion population in the Henderson Augen Gneiss was modified and partially destroyed during the low temperature deformation throughout the Brevard Zone, but was totally destroyed and replaced by a new population only within the recrystallized ultramylonite. This population of very small (average diameter one micron), rounded inclusions containing tiny bubbles, occurs in planes of various orientations. The inclusions are not visibly necked or decrepitated and yielded on analysis an average homogenization temperature of 237.5°C. We believe these data represent conditions prevailing during the geologic event responsible for the formation of the ultramylonite, and the isochore defined by these data defines the P-T space occupied by these rocks during this low-temperature deformation event. The fluid inclusion data also indicate a second low temperature fluid event which occurred post-ultramylonitization, as shown by an inclusion population associated with brittle fractures which cross-cut the mineralogy and foliation of the ultramylonite.

The $\delta^{18}O$ mineral separate data document a fluid interaction with an exotic, non-igneous fluid that lowered $\delta^{18}O$ values, not only of the whole rock, but of all the constituent minerals including quartz. These data show increasing fluid interaction from the protolith and through the shear zone with the largest drop in value in the ultramylonite. Thus, the change from matrix crystal plastic deformation in the mylonite to grain boundary sliding in the ultramylonite cannot be attributed solely to presence or absence of a fluid flux. It does correlate with the greatest volume of flux, however. This fluid interaction took place just before or during a long period of middle to upper greenschist metamorphic conditions with a temperature range of 370-480°C as shown by the quartz-K-feldspar mineral pair data. The mineral pair fractionation values from these rocks also support the exotic nature of the interacting fluid as they are substantially higher than those seen in igneous rocks and are characteristic of metamorphic rocks which have reached equilibrium over a long period of time.

The $^{40}Ar/^{39}Ar$ data show that hornblende and probably muscovite contain extraneous argon. Using argon isotope correlation plots, the data indicate the region passed through the hornblende closure temperature of 480-530°C at about 330 Ma. The muscovite sample from the ultramylonite contains two populations. The fine-grained, foliation-defining population was shown by petrographc evidence to have formed during the fluid-influenced deformation event (ultramylonitization). Argon data show this population passed through muscovite closure temperature at about 315 Ma. Thus, the ultramylonitization took place after the closure of hornblende at about 330 Ma, but prior to or at the muscovite closure date of about 315 Ma. Data from the low temperature end of the K-feldspar spectra show a second, regional fluid interaction that occurred, post 315 Ma, but prior to the K-feldspar closure (150°C) date of 250 Ma. This is consistent with the fluid inclusion data which also indicated a second low-temperature event in the population associated with cross-cutting brittle fractures.

Various techniques were used to independently define temperature ranges for the deformation event. The best supported temperature values were then used to define a pressure range on the fluid inclusion isochore in order to define the P-T space in which the fluid/deformation event occurred.

Using the brittle/ductile threshold for feldspar of 350-450°C as a temperature maximum, and the $^{40}Ar/^{39}Ar$ closure temperature of muscovite (320-350°C) which grew during the deformation, as a temperature minimum, and applying these temperatures to the isochore defined by the fluid inclusion population from the ultramylonite, gives a pressure range of 1.25-3.2 kbar. These temperatures are also consistent with the $\delta^{18}O$ data which give a temperature range of 370-480°C. Based on this P-T range, we propose a probable P-T regime of 380°C and 2.5 kbar for the ultramylonitization in the Brevard Shear Zone.

3 CONCLUSION

The whole region around the Brevard Zone at Rosman, North Carolina, acted as a single event during the metamorphic/deformation event when the ultramylonite was formed in association with a fluid infiltration event. This event reached temperatures of at least 480-530°C during its maximum; deformation continued during cooling to at least 320-350°C and the event took place during the Late Paleozoic, presumably during the Alleghenian orogeny.

Detailed fluid characterization is difficult due to the complexity of the multiple events. The data show at least two fluid events in the region. One fluid interaction occurred during the ultramylonitization event as shown by the geochemical data of Sinha et al., (1986) and the $\delta^{18}O$, argon and fluid inclusion data of this study. This fluid varied at least by volume between the area of ultramylonitization, extreme strain localization and re-crystallization of the mineral assemblage, and both the rest of the shear zone and the rest of the region. A second fluid interaction, very low temperature and probably also Alleghenian, is documented by the argon and fluid inclusion data. The occurrence of at least two fluid events which affected these rocks precludes any characterization and source modeling of the earlier fluid.

Thus, while the data support a change in deformation mechanism associated with both extreme strain localization by grain boundary sliding and high water-rock ratios required for the accommodating mechanism of pressure solution, it is also apparent that the variable controlling the switch in deformation mechanisms from crystal plastic deformation to grain boundary sliding was not simple presence of a fluid flux. The change in deformation mechanisms does, however, correlate with both a fluid volume change and fluid controlled, open-system behavior geochemically, either of which, or both, could have been a controlling variable in the change in deformation mechanism.

REFERENCES

Sinha, A. K., Hewitt, D.A., and Rimstadt, J.D., 1986, Fluid interaction and element mobility in the development of ultramylonites. Geology, 14:10:883.

Stockey, J. R., Criss, R.E. and Sinha, A. K., 1987, $\delta^{18}O$ and geochemical evidence for fluid/rock interaction during mylonitization, Brevard Fault Zone, Rosman, N.C., Geological Society of America Abs with Progs., 19:7:856.

Water-Rock Interaction, Kharaka & Maest (eds) © 1992 Balkema, Rotterdam. ISBN 90 5410 075 3

Deposition from skarn-forming solution: The genesis of some vein-like skarns (with an example from the Longjiaoshan Cu-W deposit: southeastern Hubei, P.R. China)

Yanxin Wang, Zhaoli Shen & Shaozhuo Xu
China University of Geosciences, People's Republic of China

ABSTRACT: The geologic characteristics of garnet skarns in the Longjiaoshan ore deposit are not related either to hydrothermal metasomatism of carbonate rocks or to filling of "skarn magma" in fractures. The concept "skarn-forming solution" is introduced, defined and discussed to explain the genesis of these skarns.

1 INTRODUCTION

Skarns are normally related to magmatic or hydrothermal metasomatism. Metasomatic skarns have been thoroughly studied (Einaudi et al. 1981).

However, it had been noticed in the 1950s that the metasomatism theory of Korzinskii can hardly be applied to explain the genesis of vein-like skarns located inside an intrusive body or in its sedimentary hostrocks. In recent years, vein-like skarns have been recognized in some skarn-type mineral deposits of the polymetallic orebelt of the middle and lower reaches of the Yangtze River, one of the well-known orebelts in China. Their specific geologic features can be summarized as follows:

(1) The vein-like skarns are found inside intrusive masses, or at contact zones, or even in carbonate strata far from the contact zones. The contact between the skarns and their hostrocks is sharp.

(2) The mineral assemblage of the skarns varies from monomineralic to more than 20 minerals, consisting of skarn minerals, ore minerals, calcite, quartz, etc. Without cross-cutting in replacement relation, these minerals were formed at the same stage, following the crystallization differentiation sequence. Besides, a mineral assemblage of feldspar, quartz, apatite and zircon, similar to that of granite, is occasionally observed in the vein-like skarns.

(3) Vesicular structures occur in the skarns.

(4) Flow banding made up of different minerals and flow structure displayed by the orientation of vesicles in the vein-like skarns.

Thus, it has been suggested that the skarns result from cooling and consolidation of "skarn magma" in fractures which differentiated from intermediate-siliceous magmas during their upward intrusion into shallower crust (Lin et al. 1989).

The geologic features of some garnet skarns from the Longjiaoshan Cu-W deposit and the concept "skarn-forming solution" are discussed in this paper, which represents some results of our paleohydrogeologic study on the mineral genesis of the ore deposits in southeastern Hubei, the western part of the polymetallic orebelt of the middle and lower reaches of the Yangtze River.

2 A BRIEF SUMMARY OF THE ORE GEOLOGY OF THE LONGJIAOSHAN DEPOSIT

The Longjiaoshan deposit is located on the northwestern flank of the Longjiaoshan anticline, the core of which consists of Silurian shales and sandstones and the northwestern flank of which Carboniferous-Permian carbonates.

The faults of the ore body trend in two directions -- SWW and NNE. The intrusion and subsequent emplacement of the Longjiaoshan granodioritic porphyry stock occurred along the NNE fault zones in late Jurassic. Three groups of SWW-trending faults are the host structures for orebodies with the following basic characteristics.

(1) The Longjiaoshan fault zone. It

extends throughout the deposit and was active prior to, during and after ore genesis. The faults cut an unconformity contact between the middle Carboniferous Huanglong Formation (C_{2h}) and the middle Silurian Fentou Group (S_{2f}). The No. 320 orebody was found in this fault zone; the ores are mostly brecciated with weak alterations. And ore minerals (chalcopyrite, scheelite, etc.) are observed only in the cement of the mineralized breccias.

(2) The intra-stratal faults. They are located within the Huanglong Formation and composed of a series of parallel, dislocated and small-scale fractured blocks of rocks. Recrystallized paleokarst breccias, formed prior to or during the intrusion of the Longjiaoshan stock, were identified in the faults and the carbonate rocks of the Huanglong Formation (Wang 1990). The No. 420 orebody, consisting of intermittently connected ore veins, is found in the faults. The occurrence of the orebody is basically congruent with that of the strata. The skarn-type ores are massive and brecciated. Unmetasomatized marble breccias are often included in the orebody (Fig. 1), which is in sharp contact with the host carbonate rocks.

(3) The inter-stratal faults. Most are bedding plane faults separating the Huanglong Formation and the Chuanshan Formation (upper Carboniferous, C_{3c}, and they host No. 520 orebody whose ore types are similar to those of No. 420 orebody. Unmetasomatized marble breccias are also common in the orebody. The contact between the orebody and its hostrocks is also sharp.

3 THE CHARACTERISTICS OF THE GARNET SKARNS OF THE LONGJIAOSHAN DEPOSIT

(1) The occurrence of the skarns, like the

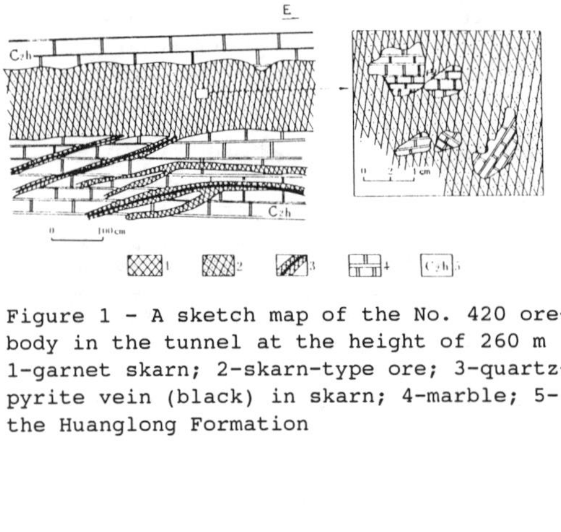

Figure 1 - A sketch map of the No. 420 orebody in the tunnel at the height of 260 m 1-garnet skarn; 2-skarn-type ore; 3-quartz-pyrite vein (black) in skarn; 4-marble; 5-the Huanglong Formation

orebodies, is also controlled by the SWW-trending faults.

(2) The skarns are found only in the intra-stratal or inner-stratal fractures located at a certain distance away from the main contact between the intrusive body and the host carbonate rocks.

(3) The skarns associated with orebodies are either massive or vein-like (Fig. 1) and have sharp contact with the orebody; those vein-like skarns usually cut or include the hostrocks. The included carbonate rocks do not exhibit any trace of alteration or mineralization.

(4) Separate garnet skarn veins (5-10 cm wide) have been observed in fractures of the hostrocks, especially in the Huanglong Formation. For instance, in a tunnel at the height of 260 m, near the contact between the Fentou Group and the Huanglong Formation, a garnet-pyrite vein 10 cm wide extends from hornfels of the Fentou Group to marble of the Huanglong Formation. Within the Huanglong Formation, several separate garnet skarn veins have also been observed.

(5) In the garnet skarns, garnets have andradite composition and contain primary fluid inclusions with high salinities (15.7-17.5 wt% NaCl) and homogenization temperatures of 337-423°C. Primary gas-liquid inclusions, multiphase inclusions and liquid inclusions often coexist in one andradite crystal, indicating that there are boiling phenomena during the formation of andradite skarns.

4 DISCUSSION ON THE GENESIS OF THE SKARNS

The above-mentioned characteristics of the garnet skarns of the Longjiaoshan ore deposit are not related to hydrothermal metasomatism and they probably result from filling of skarns in fractures. Since the garnet skarns are closely associated with the orebodies, a model for their genesis would be helpful for explaining the genesis of the ore deposit; this subject is discussed elsewhere.

The skarns do not exhibit the specific geologic characteristics typical for magmatic skarns, such as a mineral assemblage similar to that of granite, gas vesicle structure, flow banding, etc. Moreover, the studied region has not undergone any other metamorphism except for contact metamorphism. We introduced the concept "skarn-forming solution" to explain the genesis of these skarns.

The concept "skarn-forming solution" was first proposed by Abdullaev (1954) without specific explanation of its connotation.

The "skarn-forming solution" refers to fluid containing fugitive constituents and elements making up skarn minerals. It primarily originates from degassing and dewatering of shallowly emplaced magmas and their hostrocks which are intensely heated by the intruding magmas. During its upward migration to shallower crust, the solution assimilates groundwater as well as chemical components (Ca, Mg, Fe, O, etc.) of the hostrocks and gradually becomes saturated. When the solution flows into a low-pressure region or an open space, skarn minerals are deposited.

In subvolcanic environments such as around the Longjiaoshan porphyry stock, it is possible to form hydrothermal fluid which may contain element assemblage of Si-Fe-Ca-O-H. Study of geothermal waters proves that volcanic hydrothermal fluids contain fluid species such as H_2O, H_2S, CO_2, HCl, etc., and elements such as Si, Al, Fe, Na, Ca, Mg, etc.

Laboratory mineral synthesis experiments, on the other had, have shown that andradite possesses a wide fO_2-T oxygen fugacity stability field (Gustafson 1974). A relatively oxidized environment is favorable for the stable occurrence of andradite; pure andradite becomes unstable only when its environment is moderately reduced. Andradite has been synthesized, using oxide mixtures $CaO+SiO_2+Fe_2O_3$, with a low limit of formation temperature around 400°C. Therefore, andradite may directly crystallize from oxidized hydrothermal solution if there are suitable conditions and constituents such as CaO, SiO_2, Fe_2O_3, etc.

Geologic and paleohydrogeologic study shows that, at the time of the intrusion and emplacement of the Longjiaoshan porphyry stock in late Jurassic, the Longjiaoshan region was subjected to active extension and faulting in NNE direction. The carbonate rocks of the Huanglong Formation have strongly developed paleokarst. All these correspond to a relatively open and oxidized geologic environment. When high-temperature, high-pressure, and high-concentration skarn-forming solution flowed into such a low-pressure zone, boiling would have taken place. The skarn-forming solution would be further concentrated and skarn minerals be deposited to fill the fractures. The direct evidence of the boiling process is the coexistence of primary gas-liquid inclusions, multiphase inclusions and liquid inclusions in one single andradite host mineral, which is a quite common phenomenon for the Longjiaoshan Cu-W deposit.

ACKNOWLEDGEMENTS

This work was supported by National Science Foundation of China (NSFC-49102034). We also thank Prof. Lin Xinduo for his valuable suggestions.

REFERENCES

Abdullaev, Kh.M. 1954. Genetic relationship between mineralization and granitoid intrusions. Moscow: Kosgeoltechnizdat (in Russian).

Einaudi, M.T. et al. 1981. Skarn deposits. Econ. Geol. 57 Ann. Vol.: 317-391.

Gustafson, W.T. 1974. The stability of andradite, hedenbergite and related minerals in the system Ca-Fe-SiO-O-H. Jour. Petrol. 15: 455-486.

Korotaev, M.U. 1987. Vestnik Moscow University. Series Geol. 2: 49-57 (in Russian).

Lin, X. et al. 1989. Geoscience 3: 351-358 (in Chinese, with English abstract).

Wang, Y. 1990. Paleohydrogeology of the ore deposits of the southeastern Hubei ore-belt, P.R. China. Doctoral Dissertation of China University of Geosciences. 138 Pages (in Chinese, with English abstract).

Zhen, Y. 1987. Experimental Geochemistry. Beijing: Beijing University Press (in Chinese).

Water-Rock Interaction, Kharaka & Maest (eds) © 1992 Balkema, Rotterdam. ISBN 90 5410 075 3

Fluid flow during thrusting: A Sr isotopic study

D. M. Wayne & A. M. McCaig
The University, Leeds, UK

ABSTRACT

Sr isotopic data indicate that fluids carrying highly radiogenic Sr moved along fault zones in Cretaceous limestones of the central Pyrenees for distances of several kilometers. The patterns of isotopic alteration show that deformed carbonates proximal to the thrust plane were more permeable than the adjacent undeformed carbonates, and that fluid movement was primarily southward and parallel to the thrust plane.

1 INTRODUCTION AND GEOLOGIC SETTING

One of the central aims of water-rock interaction research is to quantify total fluid fluxes during tectonic events. This may be acheived by observing the displacement of pre-existing chemical or isotopic inhomogeneities in the direction of fluid flow (e.g. Baumgartner and Rumble 1988; Bickle and McKenzie 1987). However, the direction of fluid flow is rarely known and the perceived "front" may, in fact, be perpendicular to the actual fluid movement direction. Faults and shear zones provide a well-constrained setting for such studies since fluid flow is generally confined to the zone of deformation (e.g. McCaig et al. 1990).

The Gavarnie thrust plane is marked by several meters of mylonite which occur in footwall Cretaceous carbonates, and in the Cretaceous, Silurian and Devonian metapelitic rocks of the hangingwall. Syntectonic quartz and carbonate veins are abundant in all lithologies. The main movement on the Gavarnie thrust occurred approximately 42 Ma ago (Farrell et al. 1987), with a minimum southward displacement of 12 to 16 km (Seguret, 1972). Samples have been collected from three main areas along a N-S traverse across the thrust belt. At Pic de Port Vieux (area 1, Fig. 1), intense thrust imbrication has duplicated Cretaceous carbonates and Triassic redbeds in the footwall, and the Gavarnie thrust itself is breached by lower thrusts (Grant 1990). At the Plan de Larri (area 2) footwall imbrication is minor, and the thrust cuts up section so that Cretaceous carbonates form both the footwall and the hangingwall. Area 3 comprises a traverse through cover rocks of the traverse through cover rocks of the Gavarnie-Mont Perdu thrust sheet, 1 to 4 km above the thrust plane.

2 RESULTS AND INTERPRETATION

Sr concentrations and isotopic ratios were obtained from whole rock samples and carbonate vein-matrix pairs (Fig. 2). CL and microtextural observations were used to detect the presence of chemically and/or texturally distinct vein material. Coexisting fibrous and coarsely crystalline calcite from these veins show little systematic variation in terms of Sr content or isotopic composition.

Cretaceous carbonates from area 3 are slightly more radiogenic than contemporaneous seawater (Table 1), and 87Sr/86Sr ratios of calcite veins are generally similar to those of the host rocks (Fig. 2). In area 2, 87Sr/86Sr ratios of Cretaceous carbonates >20 m away from the thrust plane in both the footwall and the hangingwall are similar to those from area 3 (Table 1). Cataclastic and mylonitic Cretaceous carbonates <20 m from the thrust plane are markedly more radiogenic. Contrasts between vein and matrix 87Sr/86Sr are generally greater in mildly-deformed Cretaceous carbonates than in mylonitic or cataclastic carbonates (Fig. 2). Carbonate veins in the Devonian metapelites of the hangingwall are

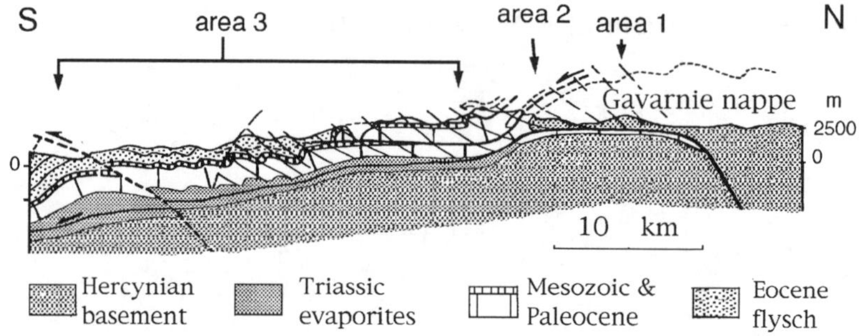

Figure 1: Geologic cross section of the study area after Seguret (1972).

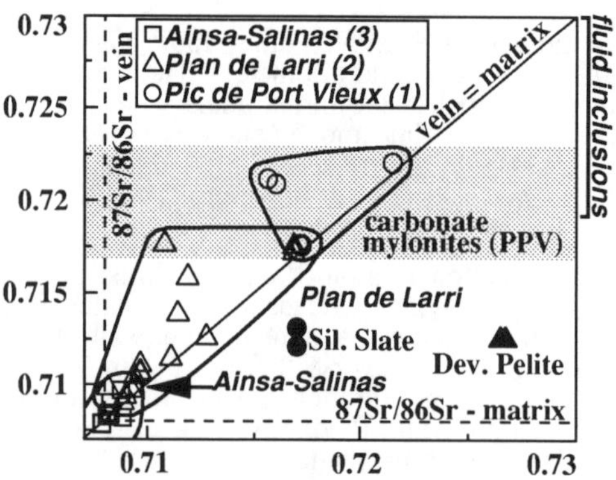

Figure 2: 87Sr/86Sr of vein-matrix pairs from areas 1, 2 and 3. Dashed line denotes 87Sr/86Sr of Cretaceous seawater (Veizer 1989). Sr87/Sr86 of Cretaceous carbonates and Alpine veins increases from Area 3 to Area 1. Shaded region is the range in 87Sr/86Sr of mylonitic Cretaceous carbonates at Area 1. The range in 87Sr/86Sr of fluid inclusions (Banks et al. 1991) from Area 1 appears on the right.

less radiogenic than the enclosing rock (Fig. 2). Cleaved to mylonitic Cretaceous carbonates from area 1 have highly radiogenic 87Sr/86Sr, in excess of mylonitized Cretaceous carbonates sampled at area 2 (Fig. 2). Banks et al. (1991) showed that 87Sr/86Sr of fluid inclusions in quartz veins from area 1 (0.7330-0.7179) fell between those of the Triassic (0.7460-0.7548) and those of the deformed Cretaceous carbonates (Table 1). Carbonate veins in the Cretaceous fall in the same range of values (Table 1), and tend to be slightly more radiogenic than the matrix. Carbonate veins from other lithologies at area 1 tend to be much less radiogenic than the adjacent matrix (Fig. 2).

The systematic decrease in 87Sr/86Sr values of carbonate mylonites from area 1 southward towards area 2 can be explained if radiogenic Sr-rich Triassic-derived fluids entered the thrust zone at area 1 and were expelled southwards along the Gavarnie Thrust (Banks et al. 1991). Fluid flow within the Cretaceous rocks was confined largely to a thin mylonitic/cataclastic layer, and fluid-related Sr transport extended over at least several km. Negative vein-matrix shifts in 87Sr/86Sr seen in the Devonian and Silurian (Fig. 2) suggest that limited amounts of fluid also escaped upwards through vein systems in

Table 1. Summary of Sr isotope data from Cretaceous carbonates, Areas 1-3.

Sample	Rb ppm	Sr ppm	87Sr/86Sr @ 42 Ma
AREA 1			
Cretaceous	3-78	107-400	0.7157-0.7216
CC veins:			
in Cret.	< 1	158-561	0.7176-0.7221
in Trias.	< 1	279-418	0.7181-0.7198
in Sil.	< 1	213-253	0.7121-0.7131
AREA 2			
Cretaceous:			
undef.	<107	78-1019	0.7082-0.7108
myl.	2-43	132-403	0.7094-0.7169
CC veins:			
in Cret.	< 1	126-633	0.7090-0.7177
in Dev.	< 1	556-649	0.7125-0.7128
AREA 3			
Cretac.	<31	435-2526	0.7078-0.7088
CC Veins	< 1	583-7091	0.7079-0.7097

the hangingwall. If any Triassic-derived fluid reached area 3, it had already lost its anomalous Sr signature.

2.1 Fluid flux and fluid-rock Sr exchange

Equilibrium fluid-rock exchange of Sr requires the transport of sharply stepped isotopic "fronts" in both the fluid and the rock in the direction of flow (e.g. Bickle and McKenzie 1987), and allows only small-scale degradation of step sharpness due to diffusion and/or hydrodynamic dispersion. Where fluid-rock equilibrium is not acheived, isotope transport is best described in terms of the Damkohler number:

$$N_D = Kh/\omega \qquad (1)$$

where K is the rate constant for fluid-rock exchange, h is the length scale for fluid-rock exchange, and ω is the pore flow velocity (Lassey and Blattner 1988; Bickle 1992). At high N_D (i.e. near-equilibrium), front sharpness is preserved and the isotopic composition of fluid and rock are similar at all points along the flow path. At low N_D, front sharpness is diminished in the direction of flow, and fluid-rock isotopic equilibrium is not approximated. In our study area, deformed Cretaceous carbonates show a gradual southward decrease in 87Sr/86Sr, indicative of south-directed fluid flow at low N_D. While our data do not permit close constraints on profile shape or step location, they do suggest transport of a Sr isotopic front over a distance of about 8 km.

The transport distance (d) of a tracer is related to the volumetric fluid-solid partition coefficient:

$$K_v = K_d(\rho_f/\rho_s) \qquad (2)$$

and integrated fluid flux (J) by:

$$J = d/K_v \qquad (3)$$

(Bickle and McKenzie 1987). At area 1, fluids contain 1000 to 2000 ppm Sr (Banks et al. 1991) and coexisting vein calcite contains 100 to 200 ppm Sr, giving K_v=5. This yields a minimum time-integrated fluid flux of 1600 m, which is modest compared to previous estimates of fluid flux in shear zones and hydrothermal systems (e.g. McCaig et al. 1990; Norton and Taylor 1979).

Data from vein-matrix pairs indicate that the degree of Sr isotopic re-equilibration in the Cretaceous carbonates was high relative to Devonian, Silurian and Triassic rocks, where extremely large vein-matrix contrasts are seen (Fig. 2). This suggests that fluid-silicate Sr exchange was more sluggish than fluid-carbonate Sr exchange, although some of the apparent isotopic re-equilibration in mylonitized carbonates may be deformation-assisted. The maximum vein-matrix contrasts in the Cretaceous carbonates do not exceed those observed in other rock types, and may reflect unusually rapid fluid movement (e.g. during seismic events), and/or interaction with a low temperature fluid.

3 CONCLUSIONS

The chief conclusions of this study are:

1. Thrust faults and mylonites channeled hypersaline, radiogenic fluids expelled from Triassic redbeds during thrusting.
2. Fluid-rock Sr exchange was generally rapid relative to fluid velocity in the Cretaceous carbonates, but slow in the Silurian and Devonian metapelites.
3. Experimental determinations of Sr exchange kinetics may provide further constraints on fluid flow rates from estimates of the Damkohler number.

ACKNOWLEDGMENTS

Funding for this project was provided by NERC Grant GR3-7495. We thank Dr. B.W.D. Yardley for his review, and Dr. R.A. Cliff and Mr. R. Green for their assistance in the isotope lab at Leeds.

REFERENCES

Banks, D.A., G.R. Davies, B.W.D. Yardley, A.M. McCaig & N.T. Grant 1991. The chemistry of brines from an Alpine thrust system in the Central Pyrenees: An application of fluid inclusion analysis to the study of fluid behavior in orogenesis. Geochim. Cosmochim. Acta. 55: 1021-1030.

Baumgartner, L.P. & D. Rumble III 1988. Transport of stable isotopes I: Development of a kinetic continuum theory for stable isotope transport. Contrib. Mineral. Petrol. 98: 417-430.

Bickle, M.J & D. MacKenzie 1987. The transport of heat and matter by fluids during metamorphism. Contrib. Mineral. Petrol. 95: 384-392.

Bickle, M.J. 1992. Transport mechanisms by fluid-flow in metamorphic rocks: Oxygen and strontium decoupling in the Trois Seigneurs Massif - a consequence of kinetic dispersion? J. Petrol., in press.

Farrell, S.G., G.D. Williams & C.D. Atkinson 1987. Constraints on the age of movement of the Montsech and Cotiella Thrusts, south central Pyrenees, Spain. J. Geol. Soc. London. 144: 907-914.

Grant, N.T. 1990. Episodic discrete and distributed deformation: consequences and controls in a thrust culmination from the central Pyrenees. J. Struct. Geol. 12: 835-850.

Lassey, K.R. & P. Blattner 1988. Kinetically controlled oxygen isotope exchange between fluid and rock in one-dimensional flow. Geochim. Cosmochim. Acta 52: 2169-2175.

McCaig, A.M., S.M. Wickham & H.P. Taylor Jr. 1990. Deep fluid circulation in Alpine shear zones, Pyrenees, France: field and oxygen isotope studies. Contrib. Mineral. Petrol. 106: 41-60.

Norton, D. & H.P. Taylor Jr. 1979. Quantitative simulation of the hydrothermal systems of crystallizing magmas on the basis of transport theory and oxygen isotope data: an analysis of the Skaergaard intrusion. J. Petrol. 20: 421-486.

Seguret, M. 1972. Etude tectonique des nappes et series decollees de la partie centrale du versant sud des Pyrenees. Pub. Ustela Ser. Geol. Struct. Nr. 2, Montpellier.

Veizer, J. 1989 Strontium isotopes in seawater through time. Ann. Rev. Earth Planet. Sci. 17: 141-167.

Hydrothermal ore deposits

Water-Rock Interaction, Kharaka & Maest (eds) © 1992 Balkema, Rotterdam. ISBN 90 5410 075 3

Hydrothermal replacement of carbonates by sulfides

H.L. Barnes
Ore Deposits Research Section, Pennsylvania State University, University Park, Pa., USA
W.W.Gould
Ore Deposits Research Section, Pennsylvania State University, University Park, Pa., USA
(Presently: Earth Sciences Consultants, Pa., USA)

ABSTRACT: True replacement requires that the dissolution of a carbonate and precipitation of a sulfide proceed by microscopically interdependent reactions which can retain initial rock textures. Textural evidence from hydrothermal deposits which formed above about 250°C, such as skarn ores, shows that these processes are especially common in ore deposits

We have identified, and tested experimentally three processes by which carbonate replacement can occur: (1) flow of sulfide-carrying, weakly acidic brines from silicate rock into a carbonate where metal sulfides are deposited by neutralizing reactions at the contact; (2) mixing of bisulfide-carrying solutions with circulating, oxidizing solutions within carbonates where redox reactions cause sulfide precipitation; and (3) flow of a near neutral, sulfidic brine carrying chloride-complexed metals down a thermal gradient in a carbonate host where cooling causes carbonate dissolution and simultaneous sulfide deposition. This third, thermal gradient process, better fits typical field criteria.

There are several processes that might link in time and place the dissolution of carbonates and of deposition of sulfides to cause the replacement common in many types of hydrothermal ore deposits. These include reactions with interdependent stoichiometries (Dobrovolsky and Lyalko, 1979), diffusional kinetics (Frantz and Mao, 1979), pressures of crystallization (Maliva and Siever, 1988), and molar volumes (Gould and Barnes, in press). Characteristically, there is little change in the volumes between the host carbonates and the replacing sulfides and the textures of the host are at least partially preserved. Carbonates are replaced both ar their contacts with other rock types and often also within these rocks through which the ore solution must have flowed for some distance before reaction.

Following these criteria, three replacement processes have been simulated experimentally: (1) reaction of acidic solutions at a carbonate contact, (2) redox reactions between solutions mixing within carbonates, and (3) reactions along thermal gradients within carbonates.

1. REPLACEMENT BY ACIDIC SOLUTION

Hydrothermal solutions of 1.0*m* NaCl were equilibrated first with kaolinite + montmorillonite + quartz, pyrite + pyrrhotite, graphite + CO_2, and sphalerite at 250 - 300°C to control the weak acidity and redox state. Second, the vessel was tilted to bring the solution in contact with a marble core for a few days (D'Andrea, 1976).

The surface of the core was quickly replaced by a coating of fine grained sphalerite, pyrite and pyrrhotite. The volume of sulfides were about equal to that of the original marble in the surface so that the outer core diameter remained nearly constant. However, there was additional porosity developed under the sulfide coating where an excess volume of calcite was dissolved. The dominant reactions of replacement are linked here by the decrease in acidity due to

(1) $CaCO_3\ (s) + 2H^+ = Ca^{2+} + H_2O\ (l) + CO_2\ (aq)$

(2) $ZnCl_2\ (aq) + H_2S\ (aq) = ZnS\ (s) + 2H^+ + 2Cl^-$

The molar volumes of the solids in the net reaction decrease from calcite (36.9 cc/mole) to sphalerite (23.8 cc/mole) by 36%, accounting in part for the increase in porosity.

This replacement process, where neutralization occurs at first contact of a weakly acidic ore solution with a carbonate, could be typical, for example, of some scarn environments. However, it cannot account for replacement where an ore solution passes for some distance through carbonates before the process takes place. Because the neutralization reactions are very rapid at hydrothermal temperatures, any repalcement by neutralization must be immediately at the carbonate contact. For these reasons, there must be other important processes that can cause replacement within carbonates, as described in sections 2 and 3 below.

2. REPLACEMENT BY OXIDATION OF BISULFIDE SOLUTIONS

Hydrothermal sulfide solutions may carry the metals with soft electron shells (including Ag, Au, and Hg, for examples) at concentrations adequate for ore deposition. Their bisulfide complexes, when they are oxidized by suitable solutes on mixing with groundwater or by reaction with oxide minerals, precipitate their metals or metal sulfides. This process was tested experimentally by Howd and Barnes (1975).

A hollow marble core filled with various solid, soluble, oxidants was immersed in, and reacted near 400°C with, bisulfide solutions saturated in sphalerite and pyrrhotite. Within a few days, both minerals replaced calcite within the core near the interface between oxidizing and reducing solutions. In addition, there was a coating of the same minerals on the surface of the core at the liquid - gas interface.

The reaction with limited oxidants available,

(3) $2HS^- + 2O_2(aq) = H_2S\,(aq) + SO_4^{2-}$

causes depletion of the bisulfide ion and metal sulfide deposition but no dissolution of the carbonate host. Consequently, the process can only fill available porosity and cannot cause replacement. However, where sufficient oxidants are present, the reactions

(4) $HS^- + 2O_2\,(aq) = SO_4^{2-} + H^+$

(5) $H_2S\,(aq) + 2O_2\,(aq) = SO_4^{2-} + 2H^+$

release acidity to also dissolve the host carbonates as in reaction (1). In these experiments, there was dissolution of the calcite both along grain boundaries, to form a sanded carbonate host, as well as around the sulfides allowing their replacement of the marble.

At higher hydrothermal temperatures, near 400°C or above, there is a transition from bisulfide to chloride complex transport for the metals with soft electron shells and below that for the transition metals. Consequently, this oxidation mechanism cannot cause replacement under such conditions because bisulfide complexing is important only at the lower temperatures. Therefore, another process must also be important for typical replacement.

3. THERMAL GRADIENT REPLACEMENT

Experiments were carried out to test the possibility that changing the aqueous speciation of a metal - carrying hydrothermal solution in response to a thermal gradient could cause replacement within a carbonate host rock (Gould, 1989). Solutions of $NaCl + CaCl_2 + H_2S + CO_2$, saturated with sphalerite + pyrite + pyrrhotite, were circulated upward along a decreasing temperature gradient about 15 cm long in either a core of, or crushed, marble. Circulation rates were varied over 2 - 11 kg/day.

The pH was controlled by the reaction

(6) $CaCl^+ + H_2O\,(l) + CO_2\,(aq) = CaCO_3\,(s) + 2H^+ + Cl^-$

at neutral to very weakly acidic conditions, near those of natural hydrothermal fluids (Barnes, 1979).

Thermal gradients, dependent on flow rates and wattage settings of the dual heaters, were calculated for the various run combinations and then checked by thermocouple measurements. Gradients of 29 - 90C° through about 15 cm were used at temperatures up to 420°C.

3.1. Reactions during cooling

When cooling, ion - pairing decreases

(7) $HCl\,(aq) = H^+ + Cl^-$

(8) $CaCl^+ = Ca^{2+} + Cl^-$

and calcite dissolves

(9) $2H^+ + CaCO_3\,(s) = Ca^{2+} + H_2O\,(l) + CO_2\,(aq)$

while sphalerite precipitates

(10) $ZnCl_2\,(aq) + H_2S\,(aq) = ZnS + 2H^+ + 2Cl^-$

thereby causing replacement. The detailed speciation of this process has been calculated, and the results confimed by sampling of the circulating fluid during experiments, followed by analysis for solute concentrations

3.2. Products of thermal replacement

Microscopic and chemical analyses showed euhedral sphalerite deposited in greater quantities in the hotter zone of the marble host. Pyrrhotite and, especially in the cooler zone, siderite were also deposited. The marble host was etched near sulfides and also often along calcite grain boundaries to form a sandy texture.

The quantity of sphalerite deposited in each segment of the marble was calculated using

$$(11) \quad \Delta C_r = \frac{dC_s}{dT} \cdot \frac{dT}{dX} \quad (Qt\,dt)$$

where C_r is the sphalerite concentration in the rock, C_s is the concentratioin of zinc in the solution, and the bracket is the integrated flux through time. The calculation of the quantity deposited agreed with the amount measured within a factor of two.

4. IMPLICATIONS

The results from the experiments and calculations on replacement due to temperature gradients indicate that this process is more likel to produce disseminated ores. Concentrated ores can precipitate by this process only where there are probably geologically exceptionally high flow rates, steep temperature gradients, or high solution concentrations. In contrast, both the acid neutralization and the oxidation processes are easily able to confine sulfide precipitation to form high grade ore bodies. However, these require hydrothermal fluids with relatively low or high pH conditions, respectively, compared to that of the thermal gradient process.

5. VOLUME CHANGES

In all of these experiments, there has been consistently a decrease in volume of the replacing sulfides compared to that of the initial carbonate host rock. This observation is consistent with most detailed description of replacement ores. Characteristically, there is an increase in porosity, sometimes to the extent of the collapse of the initial host rock. This decrease in the volume of the mineralogy is to be expected. If there were an increase in volume on forming the secondary, ore sulfides, then the surface of the carbonate would become armored and replacement could only continue by solid state diffusion, a very slow process. Consequently, the probability of replacement of one mineral by a second can be ascertained by the sign of the change in the molar volumes of the secondary versus host minerals, in the pertinent stoichiometric ratio. Because calcite and dolomite have exceptionally large molar volumes, most sulfides can replace them and realize a decrease in net molar volumes. This discussion implies that the "Law of Constant Volumes" for replacement processes is not well obeyed.

REFERENCES

Barnes, H. L. 1979. Solubilities of the ore min--erals. In H. L. Barnes (ed.), Geochemistry of Hydrothermal Ore Deposits, Sec. Ed.: 404-460. Wiley - Interscience, New York.

D'Andrea, R. F. 1976. Replacement of marble by ZnS in chloride solutioins. MSc Paper, Dept. of Geosciences, Pennsylvania State University, 33 p.

Dobrovolsky, E. V. & V. I. Lyalko 1979. Dynamic model of fluorite mineralization of carbonate - bearing rocks. Am. J. Sci. 279: 01022 - 1032.

Frantz, J. D. & H. K. Mao 1979. Bimetasomatism resulting from intergranular diffusion: II. Prediction of multimineralic zone sequences. Am. J. Sci. 279: 302 - 323.

Gould, W. W. 1989. Hydrothermal replacement of calcite by sphalerite in a temperature gradient. Doctoral Dissertation, Dept. Geosciences, Pennsylvania State University, 115 p.

Gould, W. W. & H. L. Barnes 1992. Hydrothermal replacement of calcite by sulfides in a temperature gradient. Econ. Geol., in press.

Howd, F. H. & H. L. Barnes 1975. Ore Solution Chemistry IV. Replacement of marble by sulfides at $450^{o}C$. Econ. Geol. 70: 698 - 981.

Maliva, R. G. & R. Siever 1988. Diagenetic replacement controlled by force of crystallization. Geology 16: 688 - 691.

Water-Rock Interaction, Kharaka & Maest (eds) © 1992 Balkema, Rotterdam. ISBN 90 5410 075 3

Reaction path models for hydrothermal Au-PGE mineralization at Coronation Hill and similar deposits of the South Alligator Mineral Field, Australia

M.S. Bloom & C.H.Gammons
Department of Earth Sciences, Monash University, Clayton, Vic., Australia

D.J.Gilbert & A.R.Wilde
BHP-Minerals Exploration, Asia-Pacific Division, Hawthorn, Vic., Australia

ABSTRACT: The South Alligator Valley Mineral Field of Australia's Northern Territory contains Au-PGE deposits of low-temperature hydrothermal origin. Mineralization is associated with an unconformity separating felsic volcanics and volcaniclastic sediments from deformed Lower Proterozoic metasediments, and is comprized of electrum/porpezite, stibiopalladinite and native Pd together with less common palladseite, clausthalite and other rare Pd and Pt selenides. Reaction path models of Au-PGE mobilization and redistribution indicate that oxidized saline solutions are capable of transporting concentrations of Au, PGE, As, Sb and Se sufficient to account for the observed mineralization at Coronation Hill and similar deposits. Precious metals were mobilized as chloride complexes and co-transported with arsenite, selenite, antimonite and sulfate, subsequent to oxidation by reaction with evaporites. PGE deposition was a complex function of intermetallic and metal/ligand ratios and conversion of oxidized to reduced ligand species at or near the site of deposition.

INTRODUCTION

The Coronation Hill Au-Pt-Pd deposit is located in the valley of the South Alligator River (latitude 13°35'S, longitude 132°36'E) in the Northern Territory, Australia. Coronation Hill is one of thirteen sites in the South Alligator Valley at which uranium was mined prior to 1964, and which has been the site of renewed exploration interest for precious metals since 1984. The resource at Coronation Hill is estimated to contain 4.85 million tons of ore grading 4.31 g/t Au, 0.65 g/t Pd and 0.19 g/t Pt (Carville *et al.*, 1991). A distinct combinaton of host and source lithologies, dilational structures within a reginal strike-slip fault system and the potential for generating oxidized brines is present. As part of a continuing effort to understand the processes of hydrothermal transport and redistribution of the platinum-group elements (PGE), we have used models of progressive advective fluid-rock interaction to examine possible processes leading to Au-PGE mineralization in the South Alligator Valley.

GEOLOGICAL CONTROLS ON U-Au-PGE

The geology of the South Alligator River Mineral Field has been described in detail by Needham and Stewart-Smith (1987), Carville *et al.* (1990, 1991) and Wyborn *et al.* (1991). The South Alligator Valley is a NW-trending fault-bound depression in which Lower Proterozoic sediments are unconformably overlain by Middle Proterozoic platform sediments. In the Coronation Hill area, the Koolpin Fm is stratigraphically the oldest unit. It is dominated by siliceous and carbonaceous pelites, but stromatolitic dolomite is also known to form thick lenses at several locations. The Koolpin Fm is overlain by a sequence dominated by volcaniclastic sediments which Needham and Stuart-Smith (1987) assigned to the Coronation Sandstone, but which Carville *et al.* (1990; 1991) include with the upper Koolpin Fm. Both the Koolpin Fm and the volcaniclastic sequence are unconformably overlain by a distinctive sedimentary breccia (Type C of Carville *et al.*, 1990, 1991), and by two additional breccias collectively referred to as debris-flow conglomerates (Types A and B of Needham and Stuart-Smith, 1987). All of these rock types are unconformably overlain by the Coronation Sandstone, comprised of feldspathic sandstone lenses with massive polymictic or boulder conglomerates near the base. Two prominent intrusive rock suites have also been encountered in core and surface exposures: quartz-feldspar porphyries unrelated to mineralization and a group of fractionated, fine to coarse-grained, intermediate to basic rocks essentially of quartz diorite composition which form stocks and finer grained dikes in the mineralized zone.

The locus of mineralization in the Coronation Hill and related deposits is given by Wyborn (1990) and Wyborn *et al.* (1991). The deposits lie close to the unconfornity between little-deformed 1860 Ma felsic volcanics and sandstones belonging to the Coronation Sandstone and the underlying highly folded and deformed Koolpin Fm. There are two styles of unconformity-related mineralization (Wyborn *et al.*, 1991). U-Au-PGE mineralization generally occurs in cherty ferruginous units of the carbonaceous Koolpin Fm below and within 100m of the unconformity. Au-PGE mineralization without U is more widely distributed and occurs both above and below the unconformity. Mineralization is contained within a major NW-trending dextral strike-slip fault system (the Rockhole-El Sherana-Palette fault). Almost all mineralization is located around fault zones that are clearly associated with dilatancy (Wyborn, 1990; Wyborn *et al.*, 1991). Two different structural sites are recognized, one related to contractional jogs and the other to dilational jogs along the fault system.

ALTERATION

Carville *et al.* (1990,1991) have described a series of alteration mineralogies which vary from rock type to rock type and which are generally overprinted by a late and areally extensive earthy hematite alteration. Early pale green chloritic alteration occurs in dioritic intrusions and tuffaceous shale. In feldspathic rocks this alteration is characterized by pervasive sericitization. This alteration type is clearly pre-mineralization and is probably associated with metamorphism of basement lithologies (Wyborn *et al.*, 1991). Tuffaceous shale and diorite locally also show a red hematitic alteration which pseudomorphously overprints early chloritic alteration, and is clearly replaced by chloritic alteration envelopes associated with mineralized veins which indicates that it pre-dates mineralization. Chloritic alteration envelopes occur in association with some of the mineralized veins, particularly those associated with the U-bearing style of mineralization. Late veins of earthy and(or) specular hematite occur in mineralized host rocks.

MINERALIZATION

Mineralization at Coronation Hill has been outlined by Carville *et al.* (1990,1991), based on the detailed studies of Gilbert (1987). The principal Au and Pd minerals are electrum of high fineness, stibiopalladinite (Pd_5Sb_2), sudburyite (PdSb) and native Pd together with less common PdSbSe, palladseite ($Pd_{17}Se_{15}$) and porpezite (Au-Pd). Platinum is present as $(Pt,Pd)Se_2$, Pt-Pd-Fe and rare native Pt. Rare bismuthian stibiopaladinite and Bi-Pd-Se are present. Other selenide minerals include clausthalite (PbSe), tiemannite (HgSe) and Ni-Co selenide. Volumetrically insignificant base metal sulfides and arsenides include galena, chalcopyrite, sphalerite, cobaltnickelpyrite, gersdorffite and rare pentlandite, bornite, pyrrhotite and chalcocite.

Visible gold occurs in association with late veins and breccias containing varying porportions of quartz, carbonate, chlorite and pitchblende (Wyborn *et al.*, 1991). Gold also commonly forms intergrowths with clausthalite and stibiopaladinite. The $Au+PbSe+Pd_5Sb_2$ intergrowths are generally confined to quartz+dolomite±hematite micro-veinlets in altered metapelites. $(Pt,Pd)Se_2$ occurs in narrow wispy microveinlets and fine-grained porous patches closely associated with florencite and clausthalite. In some cases, there is significant chloritization of the groundmass associated with mineralization. Veinlets postdate most deformation features and vein fill is generally undeformed, suggesting a late timing for mineralization.

Veinlet paragenesis is complex and is not completely described. A simplified paragenetic sequence comprizes early Fe-rich chlorite → earthy and specular hematite → pyrite+uraninite → $Au+PbSe+Pd_5Sb_2 \pm (Pt,Pd)Se_2$. Clausthalite, PGE phases and minor gold are often found interstitial to specular hematite where they are clearly later than hematite in the paragenesis. In rare cases, Au is closely associated with chalcopyrite+clausthalite in dolomite-bearing veinlets. Uraninite is closely associated with pyrite, galena and other base metal sulfides. Carboniferous material/graphite of suspected hydrothermal origin occurs in some veinlets and within the alteration matrix.

COMPOSITION OF ORE-FORMING FLUIDS

Mernaugh *et al.* (1992) have investigated fluid inclusions in quartz and carbonate veinlets from Coronation Hill. The majority of primary inclusions homogenize at temperatures of approximately 140°C and are characterized by high salinities with an average of 26.3 equivalent weight % $CaCl_2$. These data are in excellent agreement with the earlier estimates of Wilde *et al.* (1989a,b) for the composition of fluids associated with unconformity-related mineralization in the Alligator River uranium field. They are also in reasonable agreement with fluid inclusion evidence for the chemistry of fluids associated with intermediate-temperature PGE mineralization at the New Rambler deposit, SE Wyoming (Nyman *et al.*, 1990). We have used an initial fluid composition enriched in $CaCl_2$ which incorporates these observations and the calculations of Sverjensky (1984) for $CaCl_2$-rich oil field brines. The temperature chosen for our calculations is 150°C, which is within the uncertainties of the fluid inclusion data of both Mernaugh *et al.* (1992) and Wilde *et al.* (1989b).

PROGRESSIVE FLUID-ROCK INTERACTION

Preliminary reaction-path models of possible processes leading to Au-PGE mineralization have been calculated using the "stage" directive of program *CHEMIX* (Turnbull and Wadsley, 1986; Bloom *et al.*, 1992). Thermochemical data for minerals and aqueous species have been taken from the sources reported in Bloom *et al.* (1992) and references therein. Stability constants for Pt and Pd chloride and bisulfide complexes are from Gammons *et al.*(1992) and Gammons and Bloom (1992).

A one-dimensional advection model assuming local equilibrium between successive small batches of fluid and segments of a column of reactive rock along a flow line has been used (see Heinrich, 1990, for a comparison of progresive fluid-rock interaction at an alteration front with sequential single-step water-rock titrations). A fixed mass of rock was subdivided into segments along a linear array to form a reaction cell, and one kg of fluid was partitioned into allotments which were sequentially equilibrated with each segment of the rock column in successive cells.

The model presented here consists of four stacked cells, each comprizing a distinctive lithology, which refer to successive hypothetical stages during the evolution of mineralizing fluids and deposition of Au-PGE mineralization at Coronation Hill. In the first cell, the reaction of a Ca-enriched brine with a feldspathic and hematitic sandstone such as the Coronation or Kombalgie Sandstones is modeled. The fluid composition from the first cell is then permitted to react with hematitic and sulfate-rich evaporites in the second cell. In the third cell, the effluent solution is permitted to acquire Au and PGE by reaction with basic intrusive rocks. The final set of calculations in cell four simulates Au-PGE deposition by reactions between the fluid from cell three and a bulk composition approximating debris-flow breccias and other host rocks at Coronation Hill, but also containing open space filled by an aqueous fluid of lower oxidation state.

Oxidation and aqueous sulfide depletion of the inital solution occurs when the fluid reacts with the dolomite+anhydrite+hematite assemblage. Similar results are obtained if the initial solution reacts first with the assemblage quartz+K-feldspar+muscovite±chlorite±hematite±pyrite, followed by reaction with the evaporite assemblage. This result is in agreement with the calculations of Sverjensky (1987) for migration of oxidized fluids through evaporitic and red-bed sequences. An important result obtained from both studies is that reaction of a brine with a quartzo-feldspathic assemblage containing hematite but not anhydrite is inadequate for oxidizing the initial fluid. Trace elements exclusive of the precious metals (As, Sb, Se, etc.) are acquired in the first cell by reaction of brines of inter-mediate oxidation state with pyrite, and are advected into the next cell as aqueous antimonates, arsenates and selenates.

Au and the PGE are well above average crustal abundance in a greater proportion of the country rocks in the South Alligator Valley Mineral Field, particularly carbonaceous shales and mafic igneous rocks of the Koolpin Fm and Zamu Dolerite (Warren and Kamprad, 1990; Wyborn *et al.*, 1991). The third cell consists of basic intrusive rocks presumed to be the source of Au and the PGE. Au and Pt concentrations are given by Warren (1990) and Warren and Kamprad (1990), but no analytical data are available for Pd. We have estimated this using Pd concentrations from diabase (dolerite) sills elsewhere having similar mineralogies, fractionation histories and Au and Pt abundances (Gottfried *et al.*, 1990). Mobilization of PGE in the third cell takes place along a chromatographic front, i.e., there is incremental redistribution of PGE with dissolution at the rear and precipitation on the leading edge of the front, in response to local variation in the redox capacity of the hydrothermal solution. Very large fluid/rock ratios are necessary in order to mobilize substantial quantities of the contained PGE, and that this can be achieved only when the intrinsic buffering capacity of the source rock (related to the abundance of ferromagnesian minerals) is exhausted by oxidation.

In the final cell, the PGE-enriched solution from the previous cells is allowed to react with altered lithologies known to host ore. Unlike the U-Au-PGE deposit at Jabiluka (Wilde *et al.*, 1989a) Au-PGE mineralization at Coronation Hill does not show a clear association with a specific lithology which localizes mineralization. Both mineralization styles occur in a variety of rock types, suggesting that the chemical cause(s) of mineralization are not lithologically controlled. Carville *et al.* (1991), Wyborn *et al.* (1991) and Mernaugh *et al.* (1992) have invoked increase in fluid pH as a depostional mechanism. The results of our numerical experiments suggest that this option is unlikely, because the solution composition is constrained in our model to a mildly acidic pH by the presence of co-existing pre-mineral muscovite (sericite) and chlorite. Also, because the pH of a brine which has undergone reaction with a dolomite+ anhydrite+hematite assemblage does not differ significantly from the alumino-silicate-buffered pH (see also Sverjensky, 1987), there is unlikely to be a sufficient pH contrast for mixing with a fluid derived from carbonate-bearing lithologies to be effective.

Possible alternatives to neutralization include reduction by mixing with methane-bearing fluids (Wilde *et al.*, 1989a) or by interaction between fluids and chloritic lithologies. Both alternatives are consistent with the geological constraints. PGE deposition in either of these scenarios differs largely in the rate of precipitation, and is a complex function of intermetallic and metal/ligand ratios, and the rate of conversion of oxidized ligand species to reduced ligand species at or near the site of ore deposition. PGE chloride complexes are transported together with oxyanions (selenite, arsenite, sulfate) in a single oxidized brine. The minerals precipitated, their sequence of saturation and the masses deposited are extremely sensitive to Se/Sb, Se/As and As/Sb ratios and to Pd/Se and Pt/Se etc. ratios in the fluid, and to the relative redox conditions at which the respective oxyanions are reduced and become available for reactions precipitating the observed selenide and native metal phases.

REFERENCES

Bloom, M.S., Wadsley, M.W. & Gu, Y. 1992. Thermodynamic modeling of geochemical processes using the CSIRO-Monash THERMOCHEMISTRY System. *7th Int. Symp. Water-Rock Interaction, 9-23 July, Park City, Utah. (this volume)*

Carville, D.P., Leckie, J.F., Moorhead, C.F., Rayner, J.G. & Durbin, A.A. 1990. Coronation Hill gold-platinum-palladium deposit, *in* Hughes, F.E. (ed.) *Geology of the Mineral Deposits of Australia and Papua New Guinea*, Aus. I.M.M., Melbourne, 759-762.

Carville, D.P., Leckie, J.F., Moorhead, C.F. & Rayner, J.G. 1991. Coronation Hill unconformity related gold platinum palladium prospect, Northern Territory. *World Gold '91, Cairns, April 21-25, 1991*, 287-293.

Gammons, C.H., Bloom, M.S. & Yu, Y., 1992. Experimental investigations of the hydrothermal geochemistry of platinum and palladium. I. Solubility of Pd and Pt sulfide minerals in $NaCl/H_2SO_4$ solutions at 300°C. *Geochim. Cosmochim. Acta (in press)*.

Gammons, C.H. & Bloom, M.S. 1992. Experimental investigations of the hydrothermal geochemistry of platinum and palladium. IV. Solubility of PdS and PtS in aqueous sulphide solutions to 300°C. (in prep.)

Gottfried, D., Froelich, A.J., Rait, N. & Aruscavage, P.J. 1990. Fractionation of palladium and platinum in a Mesozoic diabase sheet, Gettysburg basin, Pennsylvania: implications for mineral exploration. *Jour. Geochem. Exploration 37: 75-89.*

Heinrich, C.A. 1990. The chemistry of hydrothermal tin(-tungsten) ore deposition. *Econ. Geol.* 85: 457-481.

Mernaugh, T.P., Leckie, J.F., Carville, D.P., Valenta, R.K. & Wyborn, L.A.I. 1992. A fluid inclusion study of mineralization at Coronation Hill, Northern Territory, Australia. *Geol. Soc. Australia, Abstracts 32:72.*

Needham, R.S. & Stuart-Smith, P.G. 1987. Coronation Hill U-Au mine, South Alligator Valley, Northern Territory: an epigenetic sandstone type deposit hosted by debris flow conglomerate. *BMR Journal of Australian Geology and Geophysics 10:121-131.*

Nyman, M.W., Sheets, R.W. & Bodnar, R.J. 1990. Fluid inclusion evidence for the physical and chemical conditions associated with intermediate-temperature PGE mineralization at the New Rambler deposit, southeastern Wyoming. *Can. Mineral. 28: 629-638.*

Sverjensky, D.A. 1984. Oil field brines as ore-forming solutions. *Econ. Geol.* 79:23-37.

Sverjensky, D.A. 1987. The role of migrating oil-field brines in the formation of sediment-hosted Cu deposits. *Econ. Geol.* 82:1130-1141.

Turnbull, A. G. & Wadsley, M.W. 1986. The CSIRO-SGTE THERMODATA System, version V: Melbourne, CSIRO Div. Mineral Chemistry Commun., v. 1-7, 413 p.

Warren, R.G. 1990. Geochemical discrimination of mafic layered units, eastern Pine Creek Inlier. *BMR Research Newsletter, 13*, October, 1990.

Warren, R.G. & Kamprad, J.L. 1990. Mineralogical, petrographic and geochemical studies in the South Alligator Region, Pine Creek Inlier, N.T. *BMR Geology and Geophysics, Record 1990/54:1-114.*

Wilde, A.R., Bloom, M.S. & Wall, V.J., 1989a. Transport and deposition of gold, uranium and platinum-group elements in unconformity-related uranium deposits. *Economic Geology, Monograph 6, 637-650.*

Wilde, A.R., Mernaugh, T.P., Bloom, M.S. & Hoffmann, C.F. 1989b. Fluid inclusion evidence on the origin of some Australian unconformity-related uranium deposits. *Econ. Geol. 84: 1627-1641.*

Wyborn, L.A.I. 1990. Localization of mineralization in the Coronation Hill and related deposits, South Alligator Mineral Field, N.T. *BMR Research Newsletter, 12*, April, 1990.

Wyborn, L.A.I., Valenta, R.K., Jagodzinski, E.A., Whitaker, A., Morse, M.P. & Needham, R.S, 1991. A review of the geological, geophysical and geochemical data of the Kakadu Conservation Zone as a basis for estimating the resource potential. *BMR, Geology and Geophysics, Australia, Report to Resource Assessment Commission (in press).*

Water-Rock Interaction, Kharaka & Maest (eds) © 1992 Balkema, Rotterdam. ISBN 90 5410 075 3

Computer models of formation of hydrothermal ore bodies at the expense of metals of the country rocks

M.V. Borisov
Moscow State University, Russia

Vict. L. Barsukov
Vernadsky Institute of Geochemistry & Analytical Chemistry, Moscow, Russia

ABSTRACT: The results of thermodynamic computer modeling of the process of formation of uraninite-galena veins and impregnated ore bodies at the expense of metals of the country rocks are presented. Dependence of mineral composition of the ore bodies and metasomatic rocks on the composition of the initial solution, of the country rocks and on the regime of filtration have been analyzed.

1 INTRODUCTION

The reaction of hydrothermal solution and country rock gives rise to two products: a static one - altered rocks and a dynamic one - solutions that have reacted with the rocks in the infiltration aureoles and have come into relative equilibrium with the medium (pore solutions). This paper is dedicated to the analysis of compositions and to the elucidation of the role of pore solutions. All then "solution+rock" and "solution+solution" equilibria calculations were carried out using the GIBBS computer program (Shvarov 1982) for 150°C in the 14-component system (H,O,K,Na,Ca, Mg,Al,Si, Fe,C,Cl,S,Pb,U), described by 69 particles of aqueous solution and 36 possible minerals. The dynamics of process was constructed in the form of a multistage flow reactors.

The production of ore bodies at the expense of metals of the country rocks becomes possible, if there are mechanisms performing three obligatory functions: 1. collection (leaching-out) of disseminated metals from large volumes of rocks; 2. their transportation to the limited volumes of the forming ore bodies (sites in the earth crust); 3. their selective deposition in these limited volumes providing the increasing concentration of metals as compared to their initial content in the rocks. Two of such ore-forming mechanisms: selfmixing (or automixing) of hydrothermal solutions and mixing solutions that have come through different media, are considered.

2 ORE - FORMATION BY SELFMIXING OF HYDROTHERMAL SOLUTIONS

The selfmixing phenomenon takes place due to hydrodynamic effects of divergence and convergence of the flow as a result the differences in the permeability of rocks and (or) of fracture structures (Barsukov et al. 1977). The generalized geological section of ore deposit is represented in Figure 1.

The selfmixing is caused by transition of a rising flow controlled by vertical fracture zone (a) from low permeable liparite lavas (b) into highly permeable tuffs (c) of the same composition and coming back from tuffs into lavas. The liparite contains nE-4% U and nE-3% Pb. In the diverging parts, the rising flow splits into two branches (d - solution current lines), represented by fracture stream (A,C,E,G) and pore stream (B,D,F); in the converging parts, these branches of the flow reunify in the fracture channel (Barsukov et al. 1980, 1984).

The three rhythms of the solution selfmixing have been analyzed in the computer models. The rhythm of selfmixing starts with the divergence of the part of the hydrothermal flow aside from the controlling fracture channels.

It was shown that this process leads to the production of metasomatic aureoles surrounding the fractures (e) by acid leaching liparite (argillization-M(I), seritization - M(II),M(III)); to the formation of negative and positive (up to impregnated ore) aureoles of lead and uranium redistribution (Figure 2a), to saturation of the pore solutions (solutions B,D,F in Figure 1) with this metals up to nE-5 mole for U and nE-3 mole for Pb (Figure 2b and Table 1). Thickness of the argillization aureoles and degree of the redistribution of metals increased as a result of divergence of a new portions solution.

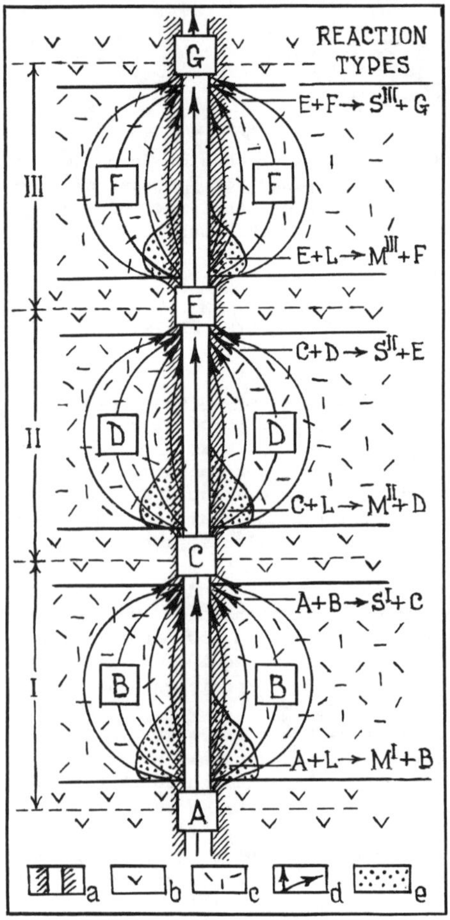

Fig.1 The three-rhythm model of hydrothermal flow selfmixing

The mixing of ore-bearing pore solutions (B,D,F) and barren fracture solutions (A,C,E) at the hydrodynamic barriers within a limited space of fracture cavities at the upper boundaries of each rhythm provokes precipitation of galena, uraninite, calcite and other minerals, and correlation between them depends on the initial composition of the model fluid (Tables 1 and 2). The content of galena in the vein paragenesis for different models varies from 0 to 98%, for uraninite - to 17% and so on (Barsukov and Borisov 1989a, 1989b, 1990).

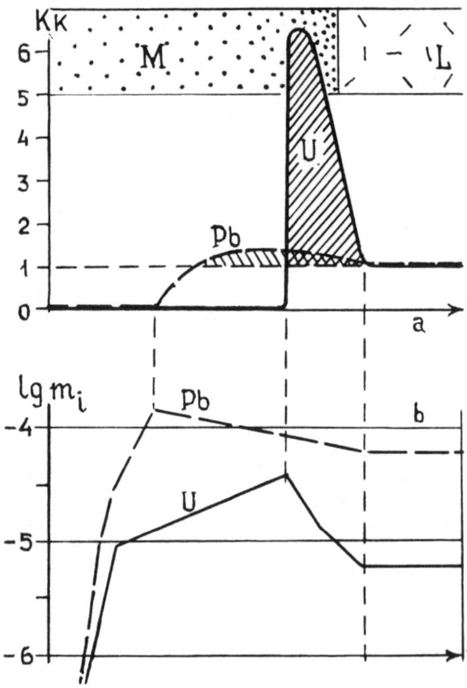

Fig.2 Redistribution of metals of the liparite: a - degree of concentration (Kk) of metals (ratio of metals concentration in the solid phase to their abundance taken as 1), b - concentration of U and Pb in the pore filtrating solutions

The hydrothermal systems with a pronounced selfmixing of the solutions are characterized by a high specific outputs of ore minerals due to the rhythmical self-rehabilitation of their productivity. For example, in one of the considered models 1 kg H_2O of initial fluid redeposits 8 mg of uraninite in the form of disseminated ore at the external front of metasomatic aureole (Figure 2a) and 1 mg of uraninite in vein-ores (Tables 1 and 2) during only one rhythm of selfmixing. As a whole this 1 kg H_2O redeposits 27 mg UO_2 in 3 rhythms and retains the ability for redeposition further quantity of metals.

Table 1. U,Pb,S and CO_2 behaviour in the processes of rocks alteration and formation of the ore veins.

	Type of solution				
	A	B(D,F)	C	...	G
CO_2	1.0*	0.11	0.53	...	0.21
S(II)	1.0E-2	nE-32	4.5E-3	...	1.1E-3
S(VI)	0	3.7E-4	1.04E-4	...	3.1E-4
Pb	0	7.3E-5	9E-11	...	1E-11
U	0	7.6E-6	1E-12	...	5E-13
pH	3.38	7.22	4.88	...	5.95
	Minerals of vein (mg/1 kg H_2O)				
$CaCO_3$			1671		
PbS			8.75		
UO_2			1.05		

*Concentration in mole/1 kg H_2O.

Table 2. Dependence of composition of ore vein from initial solution composition.

	Initial solution (mole/1 kg H_2O)			
	NaCl=0.1; KCl=0.01			
CO_2	1.0	0.1	1.0	1.0
H_2S	0.01	0.01	3E-5	0.01
$CaCl_2$	0.1	0.1	0.1	0.001
	Vein mineralization (mg/1 kg H_2O)			
$CaCO_3$	1671	873	1687	-
SiO_2	2.1	2.3	2.2	3.1
PbS	8.75	8.44	-	10.16
UO_2	1.05	0.15	0.98	2.72

3 ORE-FORMATION BY MEANS OF MIXING SOLUTIONS THAT HAVE COME THROUGH THE CONTRAST-DIFFERENT GEOCHEMICAL MEDIA

The interaction of the solutions with rocks of contrast-different composition that are simultaneously present in the ore-fields are of high interest. The two consecutive stages of the unique process were reconstructed in the models. On the first stage, "solution+rock" reactions (I and II in Figure 3) describe phenomenon that happens during the filtration of initial solution through carbonaceous and separately through leucocratic rocks. On the second stage "solution+solution" reactions (III in Figure 3) are used for the investigation of the results of the interaction of the equilibrium solutions with various rocks for different ratios of mixing in the fracture channel.

The initial chloride-carbonate solution lose the part of the CO_2 and are saturated

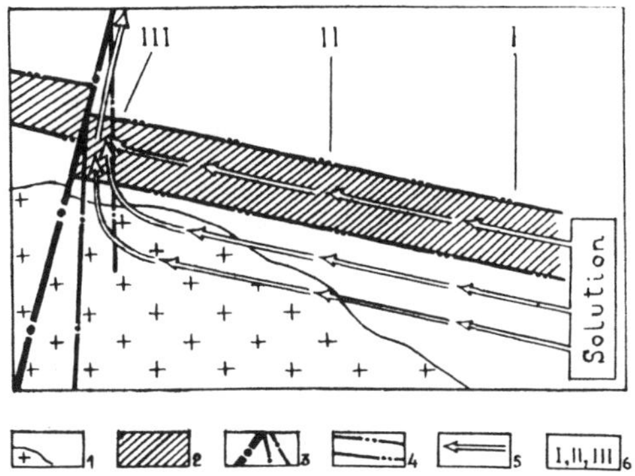

Fig.3 The schematic geological section: 1 - leucocratic metamorphic rocks of granitoid composition or granite; 2 - metamorphic black schits; 3 - fault zone; 4 - dislocation by shifting; 5 - solution flow; 6 - consecutive stages in the models

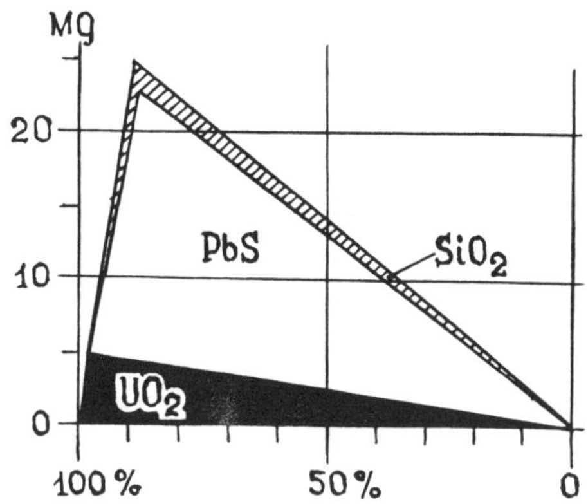

Fig.4 The vein mineral assemblages formed by the mixing of pore solutions (abscissa - percentage of pore solution from leucocratic rock; ordinate - minerals, mg/1 kg H_2O)

with H_2S to nE-2 - nE-3 mole, H_2 to nE-7 mole, CH_4 to 4E-5 mole, but they do not leache uranium and lead (<nE-10 mole) due to the reactions with black schists. The pore solutions are saturated with U and Pb, and becomes ore-bearing as a result of the reactions with leucocratic rocks (example see in Table 1).

The mixing of these different pore solutions in the fracture channel leads to deposition of galena and uraninite and to formation of vein ores. The monomineral uraninite bodies have been formed, if the flow coming from leucocratic rocks predominated (Figure 4).

Change of the filtration regime (increasing one or another flow of solutions) generates the different mineral assemblages in the fracture space of the drained structures: calcite vein with ores, mono- or polyore and barren veins (example see in Figure 4). The maximum output of the ore minerals by the mixing reaches 5.6 mg of uraninite and 19.2 mg of galena from 1 kg solution. It proves the possible high productivity of hydrothermal systems which are close to the considered models.

REFERENCES

Barsukov, Vict.L. and Borisov, M.V. 1989a. Heavy-element behaviour in automixing of hydrothermal solutions. Part 1. A computer model for U-Pb ore vein formation from country-rock metals. Geochemistry International, Vol.26, No.5:46-60.

Barsukov, Vict.L. and Borisov, M.V. 1989b. .... Part 2. Influence of various factors on the formation of U-Pb veins by leaching of metals from country rocks. Geochemistry International, Vol. 26, No.8: 59-75.

Barsukov, Vict.L. and Borisov, M.V. 1990. .... Part 3. Redistribution of U and Pb in the by the pore component of the solution. Geochemistry International, Vol.27, No.11: 72-83.

Barsukov, Vict.L., Borisov, M.V. and Pek, A.A. 1984. Hydrothermal solution selfmixing in the production of ore veins and wall-rock alterations aureoles. 27th International Geological Congress, Vol.11: 521-535.

Barsukov, Vict.L., Laverov, N.P. and Pek, A.A. 1977. The structure of the ore-bearing solutions flow as a factor of sonality in hydrothermal deposits. Geochronology and Problems in Ore Formation. Moscow: Nauka (in Russian).

Barsukov, Vikt.L. and Pek, A.A. 1980. Solution selfmixing in the production of hydrothermal vein deposits. Geochemistry International, Vol.17, No.6: 100-113.

Shvarov, Yu.V. 1982. Algorithms for determining the equilibrium compositions of multicomponent heterogeneous systems. PhD thesis Moscow State University, USSR

Water-Rock Interaction, Kharaka & Maest (eds) © 1992 Balkema, Rotterdam. ISBN 90 5410 075 3

Numerical models of mineral deposition for the Acupan vein system, Philippines

David R.Cooke
Department of Earth Sciences, Monash University, Clayton, Vic., Australia
(Currently: CODES Key Centre, University of Tasmania, Tas., Hobart, Australia)

ABSTRACT: Boiling was the principal depositional mechanism for carbonates, base metal sulfides and precious metals in the Acupan low sulfidation vein system. Some gold mineralisation may have formed during acidification of boiling chloride brines. Conductive cooling or fluid mixing caused the precipitation of barren quartz-pyrite-K-mica bands in the higher mine levels.

1 INTRODUCTION

Prolonged and varied geothermal activity during the Pliocene/Pleistocene at the Acupan gold deposit resulted in the formation of an extensive series of auriferous low sulfidation veins and mineralised breccias. Meteoric fluids of 300°C or more ascended along a a series of interconnected steeply-dipping faults and joints to the palaeosurface. Interaction of ascending chloride brines with granodioritic wallrocks produced intense, symmetrically zoned metre-wide K-mica alteration halos that give way to diffuse propylitic alteration zones.

The fluids responsible for epithermal mineralisation were analogous to fluids in certain modern-day subareal geothermal systems. Salinities were low (0 - 5.1 eq. wt % NaCl) and dissolved CO_2 concentrations high (≈ 0.18 molal). Temperatures varied with depth and time (≈ 200° to 300°C) and the fluid pH was near-neutral (Cooke and Bloom, 1990).

Spatial and temporal variations in depositional conditions formed a complex series of 'paragenetic' bands characterised by: (A) fine-grained chalcedony; (B) fine-grained grey quartz + pyrite ± K-mica ± calcite ± adularia ± sphalerite ± galena ± gold; (C) coarse-grained white quartz ± adularia ± calcite/rhodochrosite ± K-mica ± base metal sulfides ± tellurides ± gold; (D) coarse-grained calcite + adularia ± quartz ± base metal sulfides ± gold ± tellurides; and (E) coarse-grained anhydrite ± quartz ± pyrite. Several of these mineral assemblages occur at any given locality in the vein structures, yet no single band is traceable over the full vertical or lateral extent of the deposit. Quartz + K-mica + pyrite vein assemblages in type B and C bands are prevalent in the upper 500 metres of the system, whereas type B, C and D quartz + carbonate + adularia bands occur at greater depths. Type A bands are restricted to the upper mine levels, whereas type E bands occupy the centre of deeper veins. Repetitive cycles of mineral deposition and spatially equivalent vein assemblages prevent the compilation of a detailed paragenetic scheme for the entire deposit.

Likely mechanisms of ore and gangue deposition were determined in the first instance using textural, mineralogical and fluid inclusion criteria. Temperatures waned from the onset of mineralisation to the final deposition of type C and D bands, and mineralising fluids cooled as they ascended along the veins. Fluid inclusion evidence for boiling is present in some type B and C bands. However, on mineralogical and textural criteria, boiling may have also caused deposition of type D bands. No evidence for fluid mixing was detected.

Further insights into possible depositional processes have been gained by applying the computer program CHILLER (Reed and Spycher, 1985; Spycher and Reed, 1989) to a hypothetical high temperature (300°C) Acupan-type fluid constructed using fluid inclusion data and forced mineral equilibria. (Table 1).

Table 1: Composition of the 300°C Acupan-type water (relative to 1 kg H_2O). This fluid is saturated with respect to quartz, pyrite, muscovite and K-feldspar.

Component	ppm	Component	ppb
H^+	180.6	Al^{3+}	66.0
Cl^-	6704.0	Mg^{2+}	27.0
HCO_3^-	10968.0	Fe^{2+}	14.0
HS^-	66.1	Mn^{2+}	39.0
SiO_2	533.7	Zn^{2+}	12.0
Ca^{2+}	32.0	Cu^+	1.4
K^+	719.5	Pb^{2+}	6.6
Na^+	3926.0	Au^+	7.1

2 COOLING

The effects of cooling were examined by cooling the Acupan-type water from 300°C to 100°C without boiling (Figure 1). Quartz formed throughout the simulation, muscovite was deposited from 300° to 180°C, and pyrite precipitated from 300° to 270°C and from 210° to 100°C. Pyrite was not deposited between 270° and 210°C due to the low iron and HS^- concentrations in the initial fluid (Table 1). Increasing the iron concentration caused pyrite to deposit from 300° to 100°C; however, it also caused magnetite to precipitate during boiling.

Sphalerite and galena were precipitated at temperatures below 175° and 115°C; no copper sulfides were deposited. Additional cooling simulations revealed that if the initial fluid was saturated with respect to base metal sulfides, sphalerite and galena were precipitated from 300°C, chalcocite between 260°C and 125°C, and covellite below 125°C. The temperature at which base metal sulfides deposited during cooling is therefore primarily an artifact of the concentration of zinc, lead and copper in the initial fluid. Sphalerite and galena could have been deposited by cooling at temperatures of 200° to 300°C, if the initial fluid contained sufficient lead and zinc.

Gold did not precipitate during the cooling simulation, even when the initial fluid was saturated with respect to gold. This suggests that cooling alone is unlikely to have resulted in gold deposition in the Acupan vein system.

The mineral assemblages precipitated during the cooling simulation are similar to those recognised in barren type B and C bands in high level veins at Acupan. Conductive cooling may therefore have been an active depositional process in the upper levels of the system.

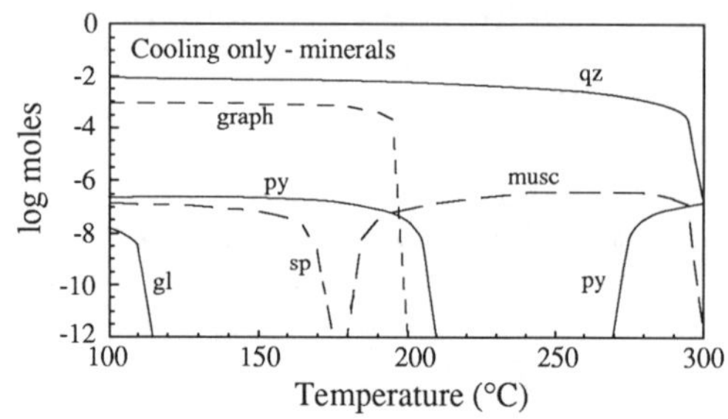

Figure 1: Minerals in equilibrium with the aqueous phase during cooling of the Acupan-type fluid from 300° to 100°C. Rate of mineral precipitation is moles per degree of temperature change per kilogram of solution. Minerals were not fractionated during this run. Boiling was prevented by fixing the pressure at 99 bars (saturation pressure = 98.4 bars at 300°C). The appearance of graphite is an artifact of the suppression of phase separation during cooling (Reed and Spycher, 1985). Mineral abbreviations used here and in the following diagrams are: ca = calcite; cc = chalcocite; cl = clinochlore; cpy = chalcopyrite; cv = covellite; dap = daphnite; gl = galena; graph = graphite; kf = K-feldspar; musc = muscovite; py = pyrite; qz = quartz; sp = sphalerite.

3 BOILING

Figure 2 illustrates the effects of isoenthalpic boiling with mineral fractionation from 300° to 100°C. Gold, K-feldspar and chlorite all commenced deposition within the first 5°C of boiling-induced cooling, and calcite began to precipitate at 285°C. Quartz was deposited throughout the boiling run, while pyrite precipitated at temperatures below 175°C.

Continuous isoenthalpic boiling converted 40.6% of the original fluid to gas by the completion of the run at 100°C. Volatile release from the aqueous phase was greatest during the first increment of boiling, where the concentration of all exsolved gases in the vapour phase increased by over two orders of magnitude. CO_2 loss resulted in an increase in pH and calcite deposition (as discussed by Reed and Spycher, 1985). This contrasts with the cooling simulation, where the dissociation of weak acids caused pH to decrease (Figure 2*C*).

Isoenthalpic boiling proved to be a more effective mechanism than cooling for inducing base metal sulfide and gold precipitation. Sphalerite and galena began precipitating at 275° and 270°C respectively (Figure 2*A*), chalcocite was deposited between 235° and 220°C, and chalcopyrite at temperatures below 225°C (Figure 2*B*). Gold precipitation began during the first increment of boiling, even

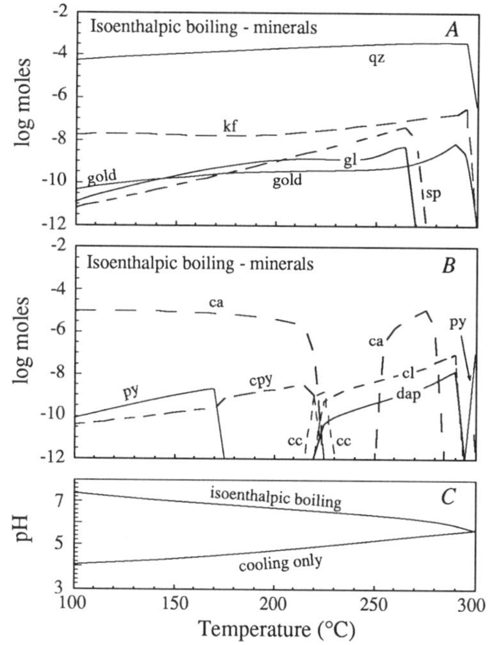

Figure 2: Isoenthalpic boiling with mineral fractionation of the hypothetical fluid tabulated in Table 1. Gases were not fractionated during this simulation. Rate of mineral precipitation is in moles per degree of temperature change per kilogram of liquid + gas solution. *A* and *B*: Minerals in equilibrium with the aqueous phase. *C*: Comparison of pH from the cooling (Figure 1) and isoenthalpic boiling simulations. See Figure 1 for mineral abbreviations.

though the initial fluid was undersaturated with respect to gold by one order of magnitude.

The most complex mineral assemblages of the Acupan vein system are recognised in type C bands. The results of this simulation indicate that such vein assemblages are a likely product of boiling. Similarly, carbonate and gold-bearing assemblages in type D and deep level type B bands are likely products of boiling.

4 COOLING OF A BOILED FLUID

To test how the transition from deep carbonate + adularia-bearing veins to high-level quartz + K-mica + pyrite bands occurred, a simulation was developed whereby a boiling Acupan-type fluid ceased boiling during its ascent. This situation may arise if fluid pressures increased in response to self-sealing, or if mixing with groundwaters occurred. The results of this simulation (isobaric cooling a boiled fluid from 250°C to 100°C) are presented on Figure 3.

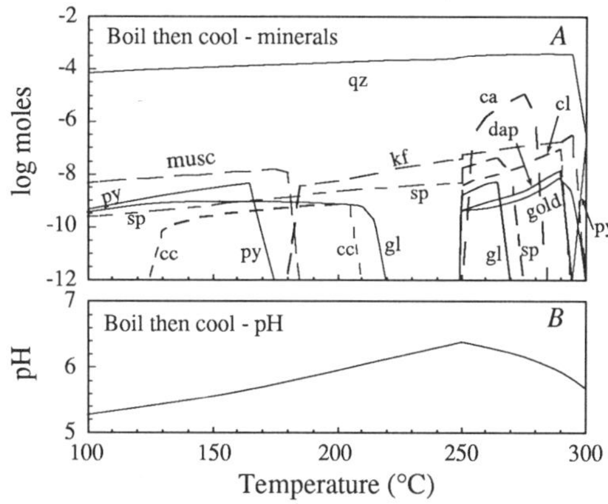

Figure 3: Isoenthalpic boiling (300° to 250°C) followed by cooling (250° to 100°C). Minerals were fractionated throughout the simulation. Gases were fractionated at 250°C to prevent their resorption into the aqueous phase. Rate of mineral precipitation is in moles per degree of temperature change per kilogram of liquid + gas solution. *A*: Minerals in equilibrium with the aqueous phase. *B*: pH vs. temperature. See Figure 1 for mineral abbreviations.

Of particular interest is the behaviour of K-feldspar below 250°C (Figure 3*A*). After boiling ceased, K-feldspar continued to precipitate until the temperature dropped to 185°C, whereupon muscovite became the stable aluminosilicate phase. This is a consequence of the decrease in pH caused by weak acid dissociation during cooling, which resulted in the fluids passing from K-feldspar-stable to muscovite-stable conditions at 185°C (Figure 3*B*). These results indicate that adularia need not be a product of boiling in low sulfidation systems, although it may indicate that boiling has occurred at greater depths. Adularia-quartz assemblages deposited from cooling solutions are likely to be free of carbonate and precious metal mineralisation, as these phases will not precipitate from the cooling fluid. A transition from adularia to high-level K-mica-bearing vein assemblages could be induced by this process, although the transition would occur somewhere above the final boiling zone.

5 FLUID MIXING

Fluid mixing is a common process in the near-surface environment of geothermal systems. By analogy, fluid mixing could have taken place in the near-surface environment at Acupan. However, the mineralising fluids were too dilute to allow the detection of any systematic changes in salinity with elevation by microthermometric techniques. CHILLER can be used to predict what might have happened during fluid mixing of an Acupan-type fluid, provided some assumptions are made concerning the nature of the reactant solution.

Mixing of cold groundwater (represented by 25°C pure water) with a 300°C non-boiling chloride brine yielded a sequence of minerals almost identical to those predicted by the cooling simulation, suggesting that temperature decrease was the fundamental effect of this mixing process. The initial salinity of the Acupan-type fluid (≈ 1 eq. wt. % NaCl) is probably too low for dilution to have any effect on aqueous species behaviour and on the mineral species precipitated; a higher initial salinity for the deep chloride brine could have produced a more unique precipitation sequence.

Mixing of acid sulfate water (represented by 0.005 molal 25°C sulfuric acid) with a 300°C non-boiling chloride brine caused quartz, gold and covellite to precipitate below 200°C. This vein assemblage has no counterpart at Acupan. Also, because it formed at low pH values (<3), this assemblage should be related to advanced argillic alteration (also unknown at Acupan).

The third mixing simulation involved acidification of a boiling chloride brine. Increments of cold 0.005 molal sulfuric acid were titrated into the fluid from the isoenthalpic boiling experiment at an intermediate point in its boiling trajectory (200°C), causing resorption of the gas phase at constant temperature. Minerals were not fractionated during this simulation. Figure 4 shows the sequence of mineral precipitation from the commencement of mixing (pH = 6.9) to the end of boiling (pH = 3.2) at constant temperature (200°C). During the initial mixing increments, acidification resulted in the initial dissolution of all mineral phases that were previously precipitating from the boiling solution. Gold recommenced precipitation as the pH dropped below 6. No other minerals were deposited prior to the cessation of boiling. Textural evidence for gold deposition along quartz grain boundary margins in type C bands suggests that at least some gold mineralisation at Acupan formed at a time when no silicate, carbonate or sulfide minerals were being deposited. Such relationships may be accounted for by the acidification of boiling chloride waters via mixing with acid sulfate waters. If such a process occurred during the deposition

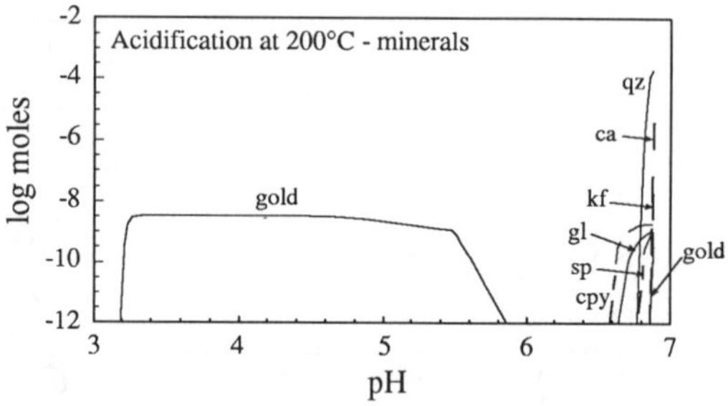

Figure 4: Minerals in equilibrium with the aqueous and gas phases during closed system acidification of a boiling chloride water at 200°C with 0.005 molal sulfuric acid. Results are presented as a function of pH. Abundances of minerals in moles per kilogram of liquid + gas solution. See Figure 1 for mineral abbreviations.

of type C bands, it did not proceed to the point where advanced argillic alteration would form.

6 CONCLUSIONS

The results of the numerical simulations suggest that boiling as a depositional mechanism is consistent with many of the vein assemblages observed at Acupan. This process resulted in mineral deposition in type D and deep-level type B and C bands. Mixing of high temperature boiling chloride brines with cool groundwaters or conductive cooling may have been responsible for a transition from carbonate + adularia-dominated gangue below 500 m to K-mica + pyrite-bearing vein assemblages at higher elevations. Acidification of a boiling deep chloride brine may have induced some gold deposition in type C bands. Further work is required to establish the likely depositional processes for type A and E bands.

REFERENCES

Cooke, D.R. & Bloom, M.S., 1990. Epithermal and subjacent porphyry mineralization, Acupan, Baguio District, Philippines: a fluid-inclusion and paragenetic study. *In*: J.W. Hedenquist, N.C. White & G. Siddeley (Eds.) Epithermal Gold Mineralisation of the Circum-Pacific; Geology, Geochemistry; Origin and Exploration, I. J. Geochem. Explor. 35, 297-340.

Reed, M.H. & Spycher, N.F., 1985. Boiling, cooling and oxidation in epithermal systems: a numerical approach. *In*: B.R. Berger & P.M. Bethke (Eds.) Geology and geochemistry of epithermal systems. Rev. Econ. Geol., 2: 249-272.

Spycher, N.F. & Reed, M.H., 1989. Evolution of a Broadlands-type epithermal ore fluid along alternative P-T paths: Implications for the transport and deposition of base, precious and volatile metals. Econ. Geol., 84: 328-359.

Water-Rock Interaction, Kharaka & Maest (eds) © 1992 Balkema, Rotterdam. ISBN 90 5410 075 3

Mineral paragenesis and regional zonation of granite-related Sn-As-Cu-Pb-Zn deposits: A chemical model for the Mole Granite district (Australia) based on PIXE fluid inclusion analyses

Christoph A. Heinrich
BMR Geology and Geophysics, Canberra, A.C.T., Australia

Christopher G. Ryan
CSIRO Division of Exploration Geoscience, North Ryde, N.S.W., Australia

ABSTRACT: Quantitative trace element analysis of individual fluid inclusions by proton induced X-ray emission (PIXE) spectrometry are combined with thermodynamic modelling, to determine the chemical mechanisms controlling oxide and sulfide mineral precipitation in the Sn – W – base metal – silver mining district centred on the Mole Granite (northeastern New South Wales, Australia). Published data on regional metal zonation around the granite, the precipitation sequence indicated by mineral textures in single deposits, fluid inclusion microthermometry and H, O and S isotopes can be explained by a simple model of fluid mixing: a hot magmatic brine is injected into and progressively diluted by a cooler meteoric fluid containing minor CO_2 and H_2S derived from the intruded metasediments. The PIXE data show that the magmatic brine contains chalcogenic metals in great excess over sulfur, so that an external sulfur source is required for efficient precipitation of metals; Fe and Zn, in particular. An interesting complication, not included in the modelling for lack of thermodynamic data, is the fluid inclusion evidence for high-temperature brine – vapor separation, and the unexpected PIXE result indicating that copper preferentially partitions into the coexisting sulfur-rich vapor, whereas all other base metals are enriched in the chloride-rich brine.

INTRODUCTION

A common pattern of paragenetic sequence and zonal distribution of deposits is observed in many granite-related tin and base metal provinces (e.g. Cornwall, eastern Australia; Taylor 1979). Typically, wolframite, molybdenite, and cassiterite ± arsenopyrite are precipitated proximal to the inferred source granite and early in the paragenetic sequence of individual deposits. Copper sulfides, and then galena, sphalerite, and finally silver minerals and other sulfides and sulfosalts are enriched at increasing distance from the granite source at a district scale, and precipitated later in the paragenetic sequence as recorded by individual deposits (Heinrich & Eadington, 1986, tab. 2). Geologic evidence indicates that the initial metal content of tin ore fluids is dominantly controlled by magmatic source processes, probably even in districts where a major part of the ore fluid's water may be of meteoric origin. If this interpretation is correct, then the initial metal ratios in ore-forming fluids can be estimated from experimental fluid/melt partitioning data (e.g. Candela 1989). This results in predictions of high concentrations of heavy metals in the ore fluids, prior to their arrival at a chemical trap driving non-equilibrium precipitation of ore minerals (Heinrich 1990). This view contrasts with 'near-equilibrium' models for ore formation requiring vast quantities of dilute meteoric fluid through-put, as advocated for example by Polya (1989).

Recent progress in quantifying proton induced X-ray emission (PIXE) spectra of individual fluid inclusions (Ryan et al. 1991) now provides a possibility to test some of these predictions by direct trace element analysis of fluid inclusions. The PIXE technique is non-destructive and reaches detection limits as low as 50 ppm for first-row transition and other heavy elements. For lighter elements (down to Cl and S) it becomes increasingly insensitive, and is therefore best combined with other single-inclusion analytical techniques.

This paper presents a case study of regional metallogenesis associated with a typical tin granite, attempting to integrate published field data with thermodynamic mass transfer calculations using PIXE analyses of magmatic brine inclusions as input constraints.

GEOLOGIC OBSERVATIONS

The Mole Granite and related alkali feldspar granites in northeastern New South Wales and southern Queensland intruded a deformed sequence of Palaeozoic turbidites and volcanics during the Permo-Triassic. The Mole Granite district was a major producer of alluvial tin, and a large number of greisen and vein deposits were mined for Sn, W, Bi, As, Cu, Pb, Zn and Ag (Weber 1974; Fig. 1). The distribution of primary deposits displays the classic zonation of many tin – tungsten – base metal provinces: W-dominant deposits occur in a roof pendant above the centre of the intrusion; sulfide-poor Sn ± W ± As deposits are mostly in the interior of the granite (plus a few in intruded volcanics); complex Sn – W – As – Cu – Pb – Zn deposits with significant sulfide content straddle the contact against metasediments; and Pb – Zn – Ag ± Cu ± As sulfide deposits occur some 3-5 km away from the exposed, probably outward dipping, granite contact.

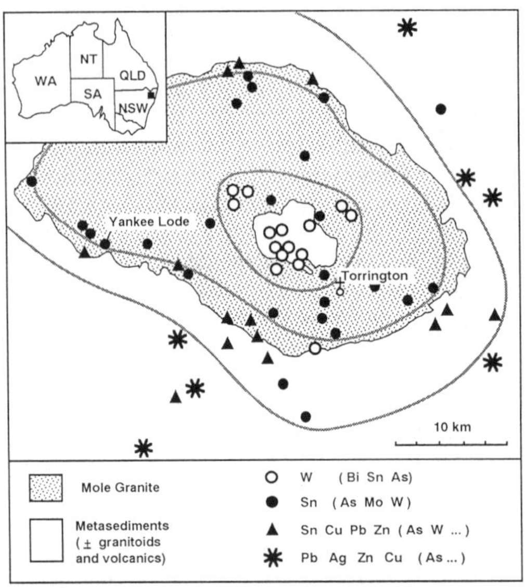

Figure 1. Map of the Mole Granite and intruded rocks, showing the regional distribution of mineral deposits and occurrences, based on data compiled by Weber (1974)

Paragenetic, fluid inclusion and O/H isotope studies by Eadington (1983) and Sun & Eadington (1987) indicate that tin and base metal mineralization resulted from mixing of a hot magmatic brine (initially coexisting with a lower-salinity magmatic vapor phase) with cooler low-salinity fluid of meteoric origin (Fig. 2a, b). Regional reconnaissance sulfur isotope data (Herbert & Smith 1978) and texturally controlled $\delta^{34}S$ data from the Sundown deposit in the northern extension of the province (Andrew & Heinrich 1984) suggest mixing of varying proportions of sulfur from two sources: near-zero-‰ magmatic sulfur dominant in the high-temperature and paragenetically early sulfides, molybdenite and arsenopyrite, and progressively lighter sulfur in Cu-Fe-sulfides, sphalerite, galena and tetrahedrite, trending towards isotopically very light pyrrhotite disseminated in unaltered hornfels (Fig. 2c).

Fluid inclusions in the Yankee Lode quartz-cassiterite veins were chosen for a detailed study using microthermometry, Raman and PIXE microanalysis (Heinrich et al. 1992) because earlier data indicate that they approximate end-member magmatic fluids. The veins contain low-salinity (5 wt% NaCl *eq.*) low-density (0.44 $g \cdot cm^{-1}$) inclusions homogenizing to vapor, and high-salinity (30–40 wt% NaCl *eq.*) brine inclusions homogenizing to liquid. Microthermometry compared with the NaCl–H_2O model system indicate that the two fluids may have coexisted at 500–600°C and 500–800bar, prior to entrapment at 350–440°C and ~300bar.

The brine inclusions have an estimated composition based on PIXE analysis: NaCl 20%, KCl 7%, $CaCl_2$ 0.9%, $MnCl_2$ 4%, $FeCl_2$ 14%, $ZnCl_2$ 1% (weight percent), plus Co 300, Cu 900, As 500, Br 400, Rb 1700, Sn 400, Cs 2000, Pb 3000 (ppm by weight). W was not detected. Cation ratios in the brine compared with element concentration ratios in the source granite (Plimer & Kleeman 1985) closely match experimental data on equilibrium metal distribution between chloride-bearing aqueous fluid and silicate melts, confirming that the brine inclusions represent near-pristine samples of magmatic fluid. Concentrations of tin, in particular, in the brine are high, as predicted by experiments of Taylor (1988).

The vapor inclusions contain higher concentrations of CO_2 and sulfur, plus about 1% Cu as the most abundant cation detected by PIXE analysis. Concentration ratios of copper to other first-row transition metals (Mn, Fe, Zn; also Pb)

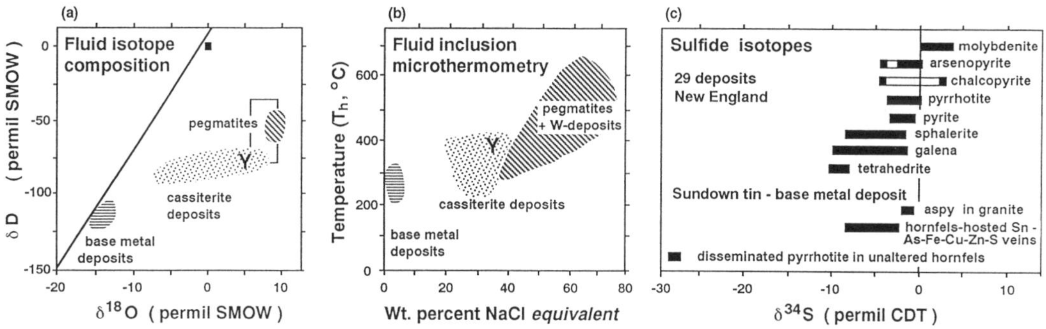

Figure 2. Published fluid inclusion and stable isotope data from hydrothermal ore deposits associated with the Mole Granite and the surrounding New England province (Herbert & Smith 1978; Eadington 1983; Andrew & Heinrich 1984; Sun & Eadington 1987); Y = Yankee Lode deposit

are 50–500 times (!) higher in the vapor than in the brine inclusions. Partitioning is probably controlled by speciation of copper: as chloride complexes in the brine, but as complexes with reduced sulfur in the vapor. This surprising result indicates that liquid–vapor separation and preferential vapor transport of copper (and, by inference, of gold) may be an important factor of base and precious metal segregation in magmatic–hydrothermal ore systems (Heinrich et al. 1992), but this possibility is not further discussed here for lack of experimental data on the inferred Cu^{I}–S^{-II} species.

THERMODYNAMIC MODELLING

Multicomponent mass transfer calculations (Helgeson 1970) can be used to predict the chemical evolution of a magmatic brine, as a test for the geologic model indicating cooling and dilution as important factors controlling ore mineral precipitation in the Mole Granite district. Large uncertainties in thermodynamic data and activity coefficients for high-temperature brines allow predictions of order-of-magnitude solubilities only, but these should at least indicate some qualitative trends.

In a first step, a chloride brine with cation ratios determined by PIXE analysis of fluid inclusions (Table1) was equilibrated at 400°C with an excess of the observed vein and immediate wall-rock assemblage observed at Yankee Lode: cassiterite + quartz + K-spar + albite + muscovite + chlorite (approximated by daphnite, $Fe_5Al_2Si_3O_{10}(OH)_8$). A slight adjustment of the total cation ratios of Na:K:Fe resulting from this was well within the estimated analytical uncertainty (Tab. 1), indicating that the vein fluid was not far from equilibrium with the observed silicate assemblage. Assumption of equilibrium with K-Al-silicates provides an estimate of brine acidity, dominated by HCl. The solubility of cassiterite ($Sn^{IV}O_2$) in a stannous chloride ($Sn^{II}Cl_2$) fluid is redox-dependent (e.g. Pabalan 1986) so that the assumption of cassite-

Table 1: Model magmatic brine composition

Comp.	$Mol \cdot kg^{-1}$	Constraint / Data source
NaCl	7.1	Microthermometry
KCl	2.2	+ PIXE analysis
$FeCl_2$	1.5	+ silicate equilibria
$CaCl_2$	0.15	PIXE analysis
$ZnCl_2$	0.15	PIXE analysis
$PbCl_2$	0.03	PIXE analysis
CuCl	0.02	PIXE analysis
$SnCl_2$	0.006	PIXE analysis
H_3AsO_3	0.01	PIXE analysis
HCl	0.09	Ksp + mus + qtz
H_2	0.006	Cass saturation (~NNO)
H_2S	0.04	Aspy slight undersaturation

Thermodynamic data and programs as used in Heinrich (1990), plus data from Bourcier and Barnes (1987; Zn-Cl complexes), Seward (1984; Pb-Cl complexes), Ruaya (1988; Cu^{I}-Cl complexes, extrapolated), Heinrich and Eadington (1986; arsenopyrite and H_3AsO_3, extrapolated) and Turnbull and Wadsley (1986; CPDMRLDAT data base; all solids) rite saturation at given tin concentration in the fluid can be used to estimate the systems redox potential and the concentration of H_2 in the vein fluid. The total concentrations of Fe, Zn, Cu, Pb and As from PIXE analysis, and the absence of any sulfide minerals in the Yankee Lode, place an upper limit on the equilibrium concentration of H_2S in the brine. Extrapolated thermodynamic data for arsenopyrite and H_3AsO_3 (Heinrich & Eadington 1986) and published experimental and thermodynamic data defining the solubility of pyrite, pyrrhotite, chalcopyrite, galena and sphalerite (see Tab. 1 for references) indicate that arsenopyrite was the least soluble sulfide in the Yankee Lode brine, constraining its H_2S concentration to a maximum value of 0.04m. These concentrations of H_2 and H_2S are substantially lower than those estimated for a pristine magmatic tin ore fluid (Heinrich 1990, Table 2), consistent with the evidence that the Yankee Lode brines had been modified by vapor separation at high subsolidus temperatures.

In a second step, the idealized brine composition of Table 1 was used to predict chemical evolution paths that could result from cooling and dilution. The inferred low H_2S concentration of the brine severely limits its capacity to precipitate its high content of iron and base metals in the form of sulfides. Cooling of the brine from 400°C, with or without interaction with wall rocks capable of buffering solution acidity and redox potential, will result in precipitation of arsenopyrite ± cassiterite, followed by minor chalcopyrite and galena at lower temperature.

Precipitation of significant proportions of Zn and Fe as sphalerite and iron sulfides requires an additional source of sulfur, and sediment-derived H_2S is the most plausible in light of the stable isotope data (Fig. 2c). The calculations represented in Figure 3 consider mixing of a hot magmatic brine (Table 1) with a cooler low-salinity fluid containing some CO_2 (0.3*m*), NaCl (0.019*m*), KCl (0.003*m*) and minor H_2S (0.01*m*). This composition is based on analyses of active, meteoric-water-dominated geothermal fluids saturated with feldspars + mica + quartz + chlorite + pyrite at 250–300°C (e.g. Krupp & Seward 1987). The temperature range coincides with the lower endpoint of the Mole Granite fluid inclusion trend (Fig. 2b). Figure 3 shows the predicted effect of stepwise addition of cool low-salinity fluid to a given mass of magmatic brine, followed by equilibration of the mixture in presence of albite + K-feldspar.

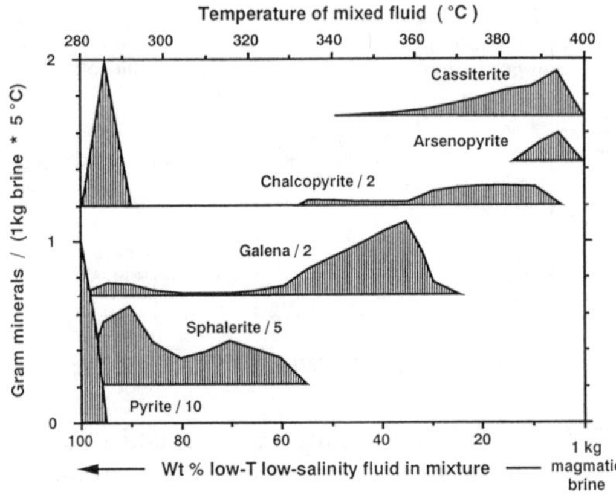

Figure 3. Sequence of ore mineral deposition predicted from stepwise dilution of hot magmatic brine (Tab. 1) with cooler low-salinity 'geothermal' fluid containing minor CO_2 and H_2S (calculated using program CHEMIX of Turnbull and Wadsley, 1986)

The calculations predict partial hydrolysis of feldspars to chlorite + muscovite + quartz (as observed in wall rock alteration), and precipitation of ore minerals with increasing dilution and falling temperature in the following sequence:

cassiterite arsenopyrite ▸ chalcopyrite ▸ galena ▸ sphalerite ▸ Fe-sulfides ± chalcopyrite

This is in general agreement with the regional distribution of tin, arsenic and base metal deposits away from the Mole Granite, with the precipitation sequence recorded by textural relations in individual Sn – As – Cu – Pb – Zn deposits, and with the correlation of this mineralogic trend with sulfur isotope compositions.

DISCUSSION AND CONCLUSION

PIXE analyses of fluid inclusions confirm that mineralization in the Mole Granite district was driven by highly metal-charged magma-derived solutions. Specific mechanisms of ore mineral precipitation probably varied from one deposit to another, but thermodynamic and mass balance considerations indicate that the main driving forces for mineral deposition were fluid cooling, in combination with wall-rock interaction (leading to acid neutralization), fluid dilution (decomplexation of metal chloride species), and admixing of reduced sulfur from a source outside the granite. It is interesting to note that the mineral precipitation sequence with falling temperature (Fig. 3) does not at all follow the magnitudes of total concentration of chalcogenic metals in the fluid (Table 1); in fact, the most highly concentrated transition metals in the ore brine are precipitated last, if at all. Zinc and iron, in particular, are probably extracted least efficiently of all transition metals in granite-derived brines, unless an external sulfur source is available.

Acknowledgements. We thank P. Blevin, G. Ewers, W. Griffin, M. Solomon, S-S. Sun, V. Wall and L. Wyborn for discussion.

REFERENCES

Andrew A.S. & Heinrich C.A. 1984. Isotopic and fluid inclusion evidence for sources of mineralizing fluids at the Sundown tin prospect, S. E. Queensland. *Geol. Soc. Austral. Abstr.* 12: 36–38.

Bourcier W.L. & Barnes H.L. 1987. Ore solution chemistry – VII. Stabilities of chloride and bisulfide complexes of zinc to 350°C. *Econ. Geol.* 82: 1839–1863.

Candela P.A. 1989. Magmatic ore-forming fluids: Thermodynamic and mass transfer calculations of metal concentrations. *In* Whitney J.A. & Naldrett A.J.. (eds.) *Ore deposition associated with magmas: Rev. Econ. Geol.* 4: 203–221.

Eadington P.J. 1983. A fluid inclusion investigation of ore formation in a tin mineralized granite, New England, New South Wales. *Econ. Geol.* 78: 1204–1221.

Heinrich C.A 1990. The chemistry of tin (–tungsten) ore deposition. *Econ. Geol.* 85: 529–550.

Heinrich C.A. & Eadington P.J. 1986. Thermodynamic predictions of the hydrothermal chemistry of arsenic, and their significance for the paragenetic sequence of some cassiterite – arsenopyrite – base metal sulfide deposits. *Econ. Geol.* 81: 511–529.

Heinrich C.A., Ryan C.G., Mernagh T.P., & Eadington P.J. 1992. Segregation of ore metals between magmatic brine and vapor: a fluid inclusion study using PIXE microanalysis. Submitted to *Econ. Geol.*

Helgeson H.C. 1970. A chemical and thermodynamic model for ore deposition in hydrothermal systems. *Min. Soc. Am. Spec. Pap.* 3: 155–186.

Herbert H.K., & Smith J.W. 1978. Sulfur isotopes and origin of some sulfide deposits, New England, Australia. *Min. Depos.* 13: 51–63.

Krupp R.E. & Seward T.M. 1987. The Rotokawa geothermal system, New Zealand: An active epithermal gold-depositing environment. *Econ. Geol.* 82: 1109–1129.

Pabalan R.T. 1986. *Solubility of cassiterite (SnO_2) in NaCl solutions from 200°C – 350°C, with geologic applications.* Unpubl. PhD thesis, Pennsylvania State Univ.

Plimer I. R. & Kleeman J. D. 1985. Mineralization associated with the Mole Granite, Australia. *Inst. Min. Metall. (London) Spec. Publ.*: 563-569.

Polya D.A. 1989. Chemistry of the main-stage ore-forming fluids of the Panasqueira W–Cu-(Ag)–Sn deposit, Portugal: implications for models of ore genesis. *Econ. Geol.* 84, 1134–1152.

Ryan C.G., Cousens D.R., Heinrich C.A., Griffin W.L., Sie S.H. & Mernagh T.P. 1991. Quantitative PIXE microanalysis of fluid inclusions based on a layered yield model. *Nucl. Instr. Meth. B* 54: 292–297.

Ruaya J.D. 1988. Estimation of instability constants of metal chloride complexes in hydrothermal solutions up to 300°C. *Geochim. Cosmochim. Acta* 52: 1983–1996.

Seward T.M. 1984. The formation of lead(II) chloride complexes to 300°C: A spectrophotometric study. *Geochim. Cosmochim. Acta* 48: 121–134.

Sun S. & Eadington P.J. 1987. Oxygen isotope evidence for the mixing of magmatic and meteoric waters during tin mineralization in the Mole Granite, New South Wales, Australia. *Econ. Geol.* 82: 43–52.

Taylor J.R. 1988. *The partitioning of tin from granitic magmas.* Unpubl. PhD thesis, Monash University, Melbourne.

Taylor R.G. 1979. *Geology of tin deposits. Developments in Economic Geology* 11. New York, Elsevier.

Turnbull A.G. & Wadsley M.W. 1986. The CSIRO-SGTE THERMODATA system, version V. *Div. Mineral Chemistry Commun.*, 7 v., 413p. CSIRO, Melbourne.

Weber C.R. 1974. Woolomin –Texas Block: Plutonic rocks and intruded sediments. in Markham N.L & Basden H. (eds.) *The mineral deposits of New South Wales*: 351–391. Geol. Surv. NSW, Sydney.

Water-Rock Interaction, Kharaka & Maest (eds) © 1992 Balkema, Rotterdam. ISBN 90 5410 075 3

Water-rock interaction in the forming process of Dongchuan copper deposit, China

Renmin Hua, Huichu Ruan & Pei Ni
Nanjing University, People's Republic of China

Dennis P.Cox
US Geological Survey, Menlo Park, Calif., USA

ABSTRACT: Two times of large-scaled water-rock interaction took place in the forming process of sediment-hosted stratabound copper deposit at Dongchuan, Yunnan, China. Copper preliminarily concentrated in the source bed was mobilized and leached by interaction with hydrothermal fluid derived from the connate water in sediments under high geothermal gradient of a rift environment. Copper sulfides were formed in the dolostone by mixing of two different fluids and by the water-mineral interaction of which the primary sulfides were replaced by the Cu-bearing fluid.

1 GEOLOGY AND ORE GENESIS

The Dongchuan copper deposit is a sediment-hosted stratabound deposit which lies in the northeastern part of Yunnan province, China. Due to its long history of exploitation, great economic value and significant forming mechanism, it has long aroused the interests of geologists. Similar stratabound copper deposits occurred in the Middle Proterozoic Kunyang Group are commonly called the Dongchuan type copper deposits since ores in Dongchuan area are by far the largest and economically most important. They are distributed within an area 500 km long and 100 km wide that trends south from southern Sichuan province to the central part of Yunnan province in southwestern China. Recent study reveals that this terrane is a continental rift of Middle Proterozoic (Hua, 1989, 1990a) and it was named the Kunyang Aulacogen because it failed to develop when it was filled with sediments of the Kunyang Group (Hua,1991).

The Kunyang Aulacogen was generated on the rigid continental crust and was bounded by two N-S trending deep faults. Its early stage of rupture was characterized by alkaline mafic igneous activities. During the later tensile-depression stage the aulacogen evolved into a graben where sedimentation of the Kunyang Group took place. The sedimentary sequence comprises terrestrial red bed (the Yinmin Formation), algal dolomite (the Laoxue Formation), marine black shale and carbonate rocks which are quite typical of rift sedimentary association. It is the "Transitional layer" between the Yinmin and the Laoxue Formations, and the bottom part of the Laoxue Formation where the stratiform copper ores occurred.

As the earliest sedimentary unit in the Kunyang Aulacogen, the Yinmin Formation is composed mainly of terrestrial clastics, and contains abundant copper from the weathing products of underlying copper-rich sequences. It is generally accepted that the Yinmin Formation was the source bed of copper for Dongchuan deposit. The forms which copper existed in the Yinmin Formation have not been well studied. However, the authors suggest that copper was mainly adsorbed on ferruginous (oxides and hydroxides) and argillaceous materials. The reasons are as follows:

1. The Yinmin Formation was originally a continental red bed formed in a dry and oxidative setting (Hua,1991). Its major components, the purple sandstone and slate, contain abundant ferric and clay minerals which possess strong ability to adsorb copper and other trace elements.

2. The underlying sequences who provided source material to the Yinmin Formation not only contain abundant copper but are also very rich in iron. In fact, copper-bearing magnetite ores are found in the lower part of the Yinmin Formation.

3. Many research data have revealed the positive correlation between copper contents and the fine grain size, and also the bright purple colour which was caused mainly by the ferric minerals.

The genesis of the Dongchuan copper deposit has been studied by many arthors (Xie, 1941; Gong et al., 1975; Li, 1978; Gong et al, 1981; Wang, 1982; Ran, 1983 ; Hua, 1988 , 1990b; Ruan et al., 1988, 1991). The present authors hold a "sedimentation-reworking" hypothesis, which reveals that the water-rock interaction played an important role in the forming process of the Dongchuan copper deposit.

2 THE FIRST WATER-ROCK INTERACTION

During the diagenesis of the Laoxue Formation, water of great volume was drained out from the sediments to form a suitable equatic medium. According to the study of isotope composition of carbon and oxygen from 9 samples of dolomite associated with copper mineralization this fluid is the essentially entrapped sea water (Hua et al., 1988). Owing to the high geothermal heat flow in a rift environment, the fluid could be rapidly heated to rather high temperature and became a hydrothermal fluid. It also became more saline during its downward flow, especially when it passed through a gypsum-halite-bearing layer at the upper part of the Yinmin Formation.

Fluid inclusion study shows that the forming temperature of the copper ores ranges from 140° to 320°C, apparently higher than those for normal marine sedimentary deposits. The salinities are as high as 14 to 34 wt persent NaCl equiv, and daughter minerals are commonly present. Therefore , this fluid is comparable with a hot brine.

The daughter minerals in fluid inclusions are basically chlorides, which include mainly NaCl, some Ca(Fe,Mn)Cl and KCl, and less commonly BaCl . This indicates that the fluid is a Cl-rich brine. The analytic composition of the liquid phase in fluid inclusion also shows that the brine is rich in Cl but poor in sulfur. Such Cl-rich and S-poor brine has a strong ability to resolve and transport certain elements like copper. When it flowed through the copper-rich rocks of the Yinmin Formation, the interaction between the brine and rocks took place. Copper originally adsorbed on the ferric and clay minerals was easily resolved by the brine and removed with it in the forms of chloro complex ions. The determination of copper contents from the sample of fully reacted argillaceous rock in the Yinmin Formation shows that only about 1/70 of the total copper was remained in the rock. It proves that the water-rock interaction had caused a great quantity of copper mobilized and entered into the liquid phase.

The water-rock interaction had also caused the separation of copper from iron, i.e., copper moved into fluid while iron still remained in the Yinmin Formation. Owing to its small size and strong polarizing ability, the ferric cation tends to form very insoluble compounds, with its oxides and hydroxides remaining stable under highly oxidizing and neutral to alkaline (pH>6) conditions. However, copper contrasts markedly with this in geochemistry, which would become unstable in response to physicochemical changes and would be mobilized to a significant extent in a favorable medium at elevated temperatures. Like Pb, U, Mo, V, As, Sb, Hg etc., copper is expected to be soluble readily in the form of complexes under oxidizing conditions between 70° to 250°C, being thus separated from iron. Such a separation resulted in the very low concentration of iron in the hot brine, and brought about a domination of Cu-rich Fe-poor sulfides e.g. chalcocite and bornite, over the ores of the Dongchuan deposit.

3 THE SECOND WATER-ROCK INTERACTION

Circulation of the above-mentioned hydrothermal fluid at a certain range of depth was driven by the deep fracture system which created a pressure drop. Therefore, the massive water-rock interaction mostly took place near a deep fracture zone, along which the Cu-bearing, highly saline hot brine would move upward. When it reached a geochemical surface at the upper bound of the Yinmin Formation, which divided the terrestrial-marine facies and the oxidative-reductive environments, and then returned back into the Transitional layer and the Laoxue Formation, the second water-rock interaction happened.

Two mechanisms can be recognized as the main ways for the precipitation of copper as sulfide minerals. One is the mixing of two different kinds of fluid, which can be called water-water interaction. The upward Cu- and Cl-rich hot brine met a S-rich, Cu-poor fluid. The latter was generated in the reductive geochemical environment of algal dolostones at the lower part of the Laoxue Formation. The mixing of these two fluids resulted in the deposition of copper sulfides in the dolostones (Ruan et al., 1991).

Another mechanism for copper sulfide deposition was the interaction between the upward hot brine and the primary iron sulfides, mostly pyrite, in the Transitional layer and the Laoxue Formation. This is a water-mineral interaction.

A number of primary pyrite can be found in the Transitional layer and the lower part of the Laoxue Formation. They mainly distribute along the thin laminae of the algal stromatolite and show syndepositional features. They also occur as small nodules in some more argillic beds. When the hot brine arrived, copper in the brine would react with the pyrite to form chalcopyrite, bornite, and finally chalcocite.

In this water-mineral interaction, iron in primary pyrite was gradually replaced by copper. The value of Fe/Cu ratio in the sulfides gradually decreased. The replacement of pyrite by chalcopyrite, chalcopyrite by bornite, and bornite by chalcocite can be commonly seen on polished sections of ores from Dongchuan, which provides evidence of this water-mineral interaction.

It is believed that the sulfur-rich fluid and the primary pyrite in the Transitional layer and the base of the Laoxue Formation quantitatively prevailed over the ascending copper-bearing hot brine, and hence provided an efficient trap for copper, effectively preventing further upward migration of copper-bearing hot brine. Copper was so quickly exhausted that the copper sulfides only formed significantly within a limited portion in the base of the Laoxue Formation. No economic ore bodies were found in the higher portion of the Laoxue Formation where the conditions appear to be identical with the mineralized beds.

4 CONCLUSION

Fluid derived from the connate water of the Laoxue Formation was driven by the deep fracture system of an ancient rift setting and circulated in the lower portion of the Kunyang Group. Being heated and more saline when it penetrated downward, it became a hot brine and then flowed upward along the pressure-reduced fracture zone and back to the Laoxue Formation. During this convective circulation, two times of large-scaled water-rock interaction took place, which resulted in the migration of copper from the source bed to the ore-bearing bed. This is the principal mechanism of and the key to the sedimentation-reworking genesis of the Dongchuan copper deposit.

REFERENCES

Gong Lin et al. 1975. Geological characteristics of Dongchuan Cu deposits in Yunnan and their geneses and ore prospecting. A special circulation of Fe and Cu deposits (3): 57-83. Beijing: Geology Press. (in Chinese)

Gong Lin & Wang Chengyao. 1981. On the genesis of the Dongchuan-type Cu deposits. Scientia Geologica Sinica 3: 203-211. (in Chinese)

Hua Renmin. 1988. On the origin and tectonic setting of the Dongchuan type Cu deposits. PhD thesis Nanjing University, China. (in Chinese)

Hua Renmin, Ruan Huichu, Liu Yan & Huang Yaosheng. 1988. Characteristics of carbon and oxygen isotope composition of the Dongchuan copper deposits. Journal of Guilin College of Geology 1: 57-62. (in Chinese)

Hua Renmin. 1989. On middle Proterozoic rifting in China and its control over stratabound copper deposits. Geotectonica et Metallogenia 2: 150-160. (in Chinese)

Hua Renmin. 1990a. Early rifting of the China plate relation to stratabound copper deposits. Earth Sci. Ser., Circum-Pacific Council Energy & Mineral Resources v.13: 327-329

Hua Renmin. 1990b. The sedimentation-reworking genesis of Dongchuan type stratiform copper deposits. Chinese Journal of Geochemistry 3: 231-243.

Hua Renmin. 1991. A study on the Kunyang aulacogen. Acta Geologica Sinica 2: 131-144.

Li Gongju. 1978. Types, evolution and prospecting of Dongchuan type stratabound Cu deposits. Geology & Prospecting 2: 28-41. (in Chinese)

Ran Chongying. 1983. On genetic model of Dongchuan type stratabound copper deposit. Scientia Sinica, Ser. B: 983-995.

Ruan Huichu, Hua Renmin & Ni Pei. 1988. New approaching to the genesis of Dongchuan type copper deposits. Contributions to Geol. & Miner. Res. Research 1: 9-22. (in Chinese)

Ruan Huichu, Hua Renmin & Cox, D.P. 1991. Copper deposition by fluid mixing in deformed strata adjacent to a salt diapir, Dongchuan area, Yunnan Province, China. Econ. Geol. v.86: 1539-1545.

Wang Kenan. 1982. Geochemistry of sedimentation-metamorphism of the Dongchuan type Cu deposits in Yunnan Province. Mineral Deposits 1: 61-68. (in Chinese)

Xie Jiarong. 1941. An introduction to the mineral resources of Yunnan. Geological Review 1 & 2: 1-42. (in Chinese)

Water-Rock Interaction, Kharaka & Maest (eds) © 1992 Balkema, Rotterdam. ISBN 90 5410 075 3

Subsurface hydrothermal genesis of ores and minerals in recent hydrothermal system of Uzon caldera in Kamchatka, Russia

Gennady A. Karpov
Institute of Volcanology, Petropavlovsk-Kamchatsky, Russia

ABSTRACT: We considered a type of subsurface Hg-Sb-As sulphide mineralization that takes place at the discharge vents of the recent hydrothermal system of Uzon caldera in Kamchatka. Mineralization occurs in rock horizons and lenses of highly porous fracture permeability (gritstones, ash-pumice tuffs, etc.) shielded by argillites. A geologic-genetic model of ore genesis is proposed.

1 INTRODUCTION

The recent hydrothermal system of Uzon caldera in Kamchatka on one hand is typical for areas of active volcanism in composition of discharged waters and associated hydrothermal alteration of enclosing rocks. On the other hand, it is unique in terms of the scale of ore formation and mineralization. This system can be classed with a few similar systems in the Pacific area where Hg-Sb-As mineralization takes place at present, accompanied by occurrences of bitumens and oil (White 1979; Weisberg 1982; Hedenquist and Henley 1985; Christle and Brathwaite 1986; Karpov 1988).

1.1 The composition of free (fissure) solutions in the mineralization zone

At the discharge site, the solutions that contain up to 90 wt. % of chloride-sodic components issue to the surface, general mineralization reaching 4.5 g/l. These solutions are enriched in B and silicic acid. In gas composition, CO_2 is predominant, with occurrences of CH_4, H_2, H_2S and radon (1-10 vol. %). The contents of ore elements are as follows: As 5-30 mg/l, Sb 0.1-0.7 mg/l, Hg 0.001-0.1 mg/l. Increased contents of Li, Rb, Cs, Sr and Ge were detected.

In the composition of the chloride-sodic solutions, we observed obvious prevalence of gold content over silver. The specific discharge of ore elements from the thermal field averages (in tons/yr km^2) As 3.3, Sb 0.6, Pb 0.5, Hg 0.007. The specific discharge for gases (in ton/s km^2) is as follows: CO_2 2.1 x 10^{-6}, H_2S 5.0 x 10^{-7}, H_2 1.8 x 10^{-7}. Sulphide mineralization can account for occurrences of bitumens and so-called hydrothermal oil of methane-naphthene-aromatic type (Beskrovny et al. 1971; Pikovsky et al. 1986).

1.2 Solutions in the pores

The pore solutions both in ore and non-ore areas of the hydrothermal system are characterized by generally high mineralization that surpasses that of free solutions by 1.5-6 times principally due to sulphate (and, to smaller extent, to hydrocarbonate) components.

Despite certain principal similarities in the character of zonation and composition of metasomatite facies, the ore and non-ore areas of the field can be distinctly discriminated in terms of concentration of pore solutions and some other characteristics (Karpov and Pavlov 1976; Karpov 1988). The pore solutions in argillites from ore areas have considerably lower general mineralization and higher pH. Pore solutions contain ore elements (As, Sb), however their concentration is one to two orders magnitude lower than in free solutions from the same area.

2 METASOMATIC AND ORE FACIES IN THE DISCHARGE VENTS OF HYDROTHERMAL SPRINGS

Along the periphery of the discharge vents of the subalkaline hydrothermal springs of

Table 1. Sulfuric cycle microorganisms growing in the Uzon hydrothermal system of Kamchatka.

Microorganism	Medium pH	Substrate	Product
Mesophilic, t<40°C			
Thiobacillus ferrooxidans	< 4	S° metal sulfides	H_2SO_4
Leptospirillum ferrooxidans	< 4	Fe^{2+} metal sulfides	H_2SO_4
Thiobacillus organoparus	< 4	S°+ organic substances	H_2SO_4
Thiobacillus thiooxidans	< 4	S°	H_2SO_4
Thiobacillus thioparus	> 5	S°	H_2SO_4
Photosynthesizing sulfur bacteria	> 6	H_2S	H_2SO_4
Desulfovibrio, Desulfotomaculum	> 6	S° SO_4^{2-} in the presence of organic matter or hydrogen	H_2S
Thermophilic, t 40-98°C			
Sulfobacillus genus	> 5	S°, H_2	H_2SO_4
Sulfolobus acidocaldarius	< 5	S°	H_2SO_4
Thermothrix thiopara	> 7	HS^-	S°
Thermoproteae group	4-6	S°	H_2S

chloride-sodic composition is evidence of exogenetic processes, principal of which is the formation of H_2SO_4 and sulfur acid solutions as a result of biogenic factors (see Table 1, Karpov and Zavarzin 1988). We observed the formation of metasomatites in the zone of sulfuric acid leaching and accompanying mineral and ore formation. Almost synchronously in time but discretely latitudinally two subformations form opal-argillite and S-sulphide (at T=30-100°C, pH 5 and Eh = 290-70 mv).

Opal-argillite subformation of the hydrothermal stage comprises S-opal, opal-alunite-kaolinite, opal-kaolinite and opal-smectite facies. Mineral associations of the first three facies are similar to the facies of subformations of sulfur domes and sulfur opalites of fumarolic-solfatara stage, i.e., convergence of the facies from different subformations is observed. To identify them, one needs to use complex mineralogic-geochemical criteria established by us (Karpov 1988).

The second S-sulphide subformation of sulfuric acid leaching comprises ore facies: monosulfuric, opal-orpiment, uzonite (alakranite)-regular, realgar-antimonite, pyrite-cinnabarite and pyrite-marcasite. In the Central site of the thermal field of Uzon caldera, the mineralization zone has thinly laminated and lense-like structure with impregnation-pocket occurrence of ore minerals. In general we discriminate here two types of sections of ore zones that form in gritstones: (1) oxide-sulphide with scorodite in the upper part and roentgenoamorphous orpiment, finely grained pyrite and needle-shaped antimonite at the depth 0.07-0.45 m (deeper, at the depth of 1.5-2 m in gritstones we trace rare crystals of realgar with pyrite of cubic habit); (2) sulphide (with orpiment, realgar, uzonite, alakranite, antimonite, metacinnabarite, and pyrite). In the second type of mineralization at the depth interval of 0-6 m (with maximum mineralization at the depth of 0.1-0.8 m) we observe alternation of mineral zones of the following parageneses (from top to bottom): sulfur + orpiment (roentgenoamorphous); orpiment + realgar; realgar + uzonite + alakranite; pyrite + antimonite + realgar + metacinnabarite (rarely); pyrite + realgar; and pyrite.

In the subwater section of the hot Fumarolnoe Lake, mineralization has a different profile. It is traced in the form of thin crusts inside fissures of aleuropelitic, psammitic and psephitic tuffs of dacites as well as in the form of microfossils (Erochshev-Shak et al. 1985). The following sequence of crystallization of ore minerals is observed (from early to late): cubic pyrite - antimonite + octahedral pyrite - realgar (cinnabar) - orpiment - scorodite + sooty pyrite.

3 PHYSICO-CHEMICAL CONDITIONS FOR FORMATION OF SULPHIDE Hg-Sb-As MINERALIZATION

Recent Hg-Sb-As mineralization in the Uzon caldera is considered by us as a genetic type of epithermal stratimorphous mineralization accompanied by manifestations of gaseous and liquid hydrocarbons. Mineralization takes place in the zones of discharge of chloride-sodic solutions which in thermobaric conditions are close to the line of saturated stream. As a rule we observe good coincidence of the contours of mineral zones with isotherms. Routine observations indicated that physico-chemical parameters of the solutions from the springs as well as the contents of mineral components in them are distinctly seasonally dependent (Karpov and Pavlov 1976; Alekhin et al. 1987; Karpov 1988). Such parameters as t, pH, Eh, concentrations of HCO_3, SO_4^{2-}, As, H_2S undergo the most notable and appropriate variations. The abundance of sulphate-ion regularly decreases with depth during all observation periods.

The horizon of 0.3-0.8 m is a geochemical barrier with low redox potential. Its location is controlled by an oxygen barrier located above. The oxygen barrier results from saturation of the near-surface water with oxygen, its parameters are seasonally dependent. Direct measurements of principal ore minerals in Uzon give the following data: for roentgenoamorphous As sulphide t = 35-65°C, pH = 3.5-5.6, Eh = 130-290 mv, H_2S up to $1.1x10^{-3}$ ppm/l; for association uzonite + alakranite + realgar t = 65-105°C, pH = 5.5-6.8, Eh = -70-120 mv, H_2S up to $6.0x10^{-4}$ ppm/l; for antimonite t = 90-102°C, pH = 5.2-6.2, Eh = -60-60 mv, H_2S up to $1.7x10^{-4}$.

4 GEOLOGIC-GENETIC MODEL FOR FORMATION OF SUBSURFACE SULPHIDE MINERALIZATION

Deep subalkaline solutions containing As, Sb and Hg in the form of hydrosulphide and sulphide complexes on their way to the surface through permeable zones meet a zone enriched in air oxygen (oxygen geochemical barrier) at the level of ground waters. At this barrier disintegration of metal complexes takes place and proceeds, for example, in the following way:

$$2HAs_2S_4 + 1/2O_2 = 2As_2S_3 + 2S^0 + H_2O + 2\bar{e}$$

The colloform components consisting of the mixture of roentgenoamorphous As sulphide and sulphur precipitate from the solution and fill fissures and pores. With time an ore layer of low permeability is formed in the upper zone. Deeper ground water level in the temperature ranges of 75-102°C (according to measurements in the Uzon caldera and data from Broadlands and Waiotapu thermal fields, New Zealand) a zone of enrichment with hydrogen sulphide takes place. In this zone crystallization of realgar, antimonite and pyrite occurs.

With time, erosional processes enlarge river configuration as a result of which the level of ground waters lowers. Correspondingly, the oxygen geochemical barrier is lowered. At the same time, a layer of terrigenous mud sediments accumulates on the surface of the thermal field and the thermal site itself undergoes subsidence. Therefore, a zone of early "orpiment" mineralization gets into conditions of higher temperatures. Two types of processes can occur here: redox process of disproportion as a result of which realgar forms from the orpiment phase according to the following scheme:

$$2AS_2S_3 \rightarrow As_4S_4 + 2S^0$$

and crystallization of metal compounds of higher sulphur content at the hydrogen sulphide barrier. Such crystallization may yield to formation of uzonite (As_4S_5) and alakranite (As_8S_9) according to the following reactions:

$$As_4S_4 + H_2S \rightarrow As_4S_5 + H_2$$

$$2As_4S_4 + H_2S \rightarrow As_8S_9 + H_2$$

Hence, we may assume that in the section of the thermal site, an ore body will form having thinly laminated or lense-like structure and overall mushroom shape. Of interest is the fact that the zone of mineralization enlarges from above and beneath as the erosional processes proceed and the thermal site subsides. Simultaneously dissemination and impregnation mineralization transforms into massive and laminated one. Consequently, even within one layer different As sulphides and their different generations may simultaneously exist. Associations of orpiment with sulphur and orpiment with realgar are typical for the upper part of the section. Associations of realgar, uzonite, alakranite, antimonite and pyrite of different ages are more typical for lower beds of mineralization.

REFERENCES

Alekhin, Yu.A., T.P. Dadze, A.V. Zotov, G.A. Karpov, V.I. Sorokin & G.D. Mironova 1987. The conditions for formation of recent sulphide Hg-Sb-As mineralization in the Uzon caldera (Kamchatka). Volc. and Seismol. 2: 39-47 (in Russian).

Beskrovny, N.S., S.I. Naboko, S.F. Glavatsykh, V.P. Ermakova, B.A. Lebedev & S.D. Taliev 1971. On the presence of oil in hydrothermal systems, associated with volcanism. Geologia i Geofizika 2: 3-14 (in Russian).

Christle, A.B. & R.L. Brathwaite 1986. A comparison of epithermal mineralization in the Hauraki Goldfield and the geothermal systems of the Taupo volcanic zone. Auckland: Proc. Intern. Volc. Congr. Symp. 5: 25-30.

Eroshchev-Shak, V.A., G.A, Karpov, F.A. Kireev & R.A. Bochko 1985. Thermal Fumarolnoe Lake in Kamchatka--a basin of recent ore formation. Izv. AN SSSR, ser. geol. 9: 15-23 (in Russian).

Hedenquist J.W. & R.M. Henley 1985. Hydrothermal eruption in the Waiotapu geothermal system, New Zealand, their origin, associated breccias and relation to precious metal mineralization. Econ. Geol. 80: 1640-1668.

Karpov, G.A. & A.L. Pavlov 1976. Uzon-Geizernaya hydrothermal ore-forming system in Kamchatka. Novosibirsk: Nauka, 99 p. (in Russian).

Karpov, G.A., G.A. Zavarzin & V.A. Eroshchev-Shak 1988. Bacterial origin of agents favoring clay formation areas of hot spring discharge and solfataric activity. Volc. and Seismol. 6: 251-265.

Karpov, G.A. 1988. Recent hydrothermal springs and Hg-Sb-As mineralization. Moscow: Nauka, 183 p. (in Russian).

Pikovsky, Yu.I., G.A. Karpov & A.I. Ogloblina 1987. Polycyclic aromatic hydrocarbons in the products of Uzon hydrothermal system (as related to the genesis of oil in Uzon, Kamchatka). Geochemistry (in Russian).

Weisberg, B.D., P.R.L. Brown & T.M. Seward 1982. Ore elements in active geothermal systems. In: Geochemistry of Hydrothermal Ore Deposits. Moscow: Mir 578-609 (in Russian).

White, D.E. 1979. Mercury and base-metal deposits with associated thermal and mineral waters. In: Geochemistry of Hydrothermal Ore Deposits. N.Y.: Holt Rinehart and Winston, Inc.: 575-631.

Water-Rock Interaction, Kharaka & Maest (eds) © 1992 Balkema, Rotterdam. ISBN 90 5410 075 3

Relations between terrestrial heat flow and uranium deposits in Jiangxi Province, China

Li Xueli, Zhou Wenbin, Zhang Weimin & Sun Zhanxue
East China Geological Institute, Fuzhou, Jiangxi, People's Republic of China

ABSTRACT: Based on tectonics, distribution of uranium deposits and terrestrial heat flow data, it is concluded that heat flow anomaly and uranium mineralization are controlled by deep-great faults and deep seated structural transitional zones. To some extent, uranium mineralization can cause heat flow anomalies. Possibilities of uranium prospecting by heat flow measurement method is also discussed.

1 INTRODUCTION

Hydrothermal uranium deposits are products of fossil hydrothermal system activities in particular geological-structural environments. Not only are a rich uranium source, a migration and enrichment media (water source) necessary, but also a good heat source is needed for their formation. However, in past researches, geothermal settings were often neglected. Since the early 1980's, Li Xueli has emphasized on uranium study by three sources "Uranium, Water and Geotherm" combination method. Relations between terrestrial heat flow and uranium deposits is one of the basic problems to be solved.

Terrestrial heat flow is an important geophysical parameter which reflects thermal state of the earth's interior. Uranium deposits are related to geothermal anomalies during the mineralization period. Furthermore, after mineralization formed, uranium deposits will contribute to terrestrial heat flow of near areas to some extent. In the recent years, terrestrial heat flow measurements in Jiangxi province have been systematically made, and distribution laws and controlling factors have basically been obtained. On this basis, this paper expounds the relations between terrestrial heat flow and uranium deposits in Jiangxi province, southeast China.

2 REGIONAL GEOLOGICAL SETTINGS

Jiangxi province crosses two major tectonic units which are approximately divided by the Zhejiang-Jiangxi railroad. The northern Jiangxi belongs to the southeastern margin of Yangtz Metaplatform, while the central-southern Jiangxi is a part of East China Fold System. Deep structures in the province are not only in sequence, but also are inhomogeneous. The present morphostructure and the rolling deep Moho are of mirror symmetry. The deep seated structures in the province are divided into the lifted area in the upper mantle, the sunken area in the upper mantle and the deep seated structural transitional zone (Fig. 1).

Faulted structures in Jiangxi Province are well developed. According to their trending directions, these structures can be divided into four groups. One of them, the NE-NNE trending group, is the most developed. Near EW, NNW and SN trending ones, as well as the NW trending one, are less developed. Research shows that there are 10 deep faults and 18 great faults. These deep-great faults have been continuously active and have controlled the volcanic and magmatic activities, formation red bed basins, geotherm settings, and hot-springs and mineralizing processes.

3 GEOTHERMAL BACKGROUND

In the recent years, terrestrial heat flow measurement in the province have been made systematically by East China Geological Institute in cooperation with the Institute of Geology, Chinese Academy of Sciences. At present, more than 20 heat flow data have been obtained. The background and distribution laws of heat flow in Jiangxi are basically revealed.

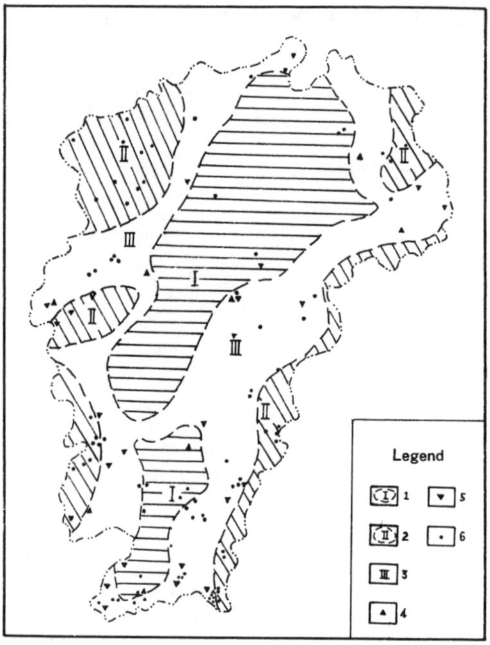

Figure 1 - The relations between tectonics, deep structure, hotsprings and heat flow anomalies
1--Lifted area of the upper mantle; 2--Sunken area of the upper mantle; 3--Structural transitional zone; 4--Heat flow anomalies; 5--Uranium deposits; 6--Hot spring.

The average heat flow value is 69.8 ± 13.4 (SD) MW/M^2 in the province. The northern Jiangxi and the central-southern Jiangxi, divided by Pingxiang-Guangfeng deep fault, are very different in heat flow values. In the north, the average value is 59.2 ± 3.8 (SD) MW/M^2. However, in the central-southern Jiangxi, the value is 74.1 ± 13.7 MW/M^2. Its deviation is quite high. Obviously, heat flow values in the central-southern Jiangxi are higher than those in the north. The distribution law of heat flow is in accordance with tectonic settings in the province.

Research reveals that heat flow values are different in different tectonic units. In ancient and stable platform areas, heat flow values are lower and more homogenous than those in young orogenic belts and rift zones.

Generally speaking, the younger the ages of tectonic units and the stronger the activities of structures are, the higher the heat flow values. The distribution of heat flow in Jiangxi province reflects such laws. Since the northern Jiangxi belongs to Yangtz Metaplatform, which is stable massif, while central-southern Jiangxi is a part of South China Fold System, which is an active land mass, heat flow values in the former are lower than those in the latter.

4 RELATIONS BETWEEN HEAT FLOW ANOMALIES, URANIUM MINERALIZATION AND GEOLOGICAL STRUCTURES

Heat flow anomalous regions have been delineated in Jiangxi. These regions are Yingshan, Dexing county (its heat flow value is 104.1 MW/M^2); Xiangshan, Lean county (82.6 MW/M^2); Qiujiang, Fengcheng city (82.5 MW/M^2); Huaguoshan, Xinwei city (92.7 MW/M^2); Juyuan, Pingxian city (89.1 MW/M^2); Xinguo county (98.8 MW/M^2); Baoshan, Chongyi county (88.9 MW/M^2) and Yaolin, Lunnan county (86.1 MW/M^2).

It can be seen from Fig. 1 that the distribution of surface heat flow anomalies are controlled by deep seated structural transitional zones and deep-great faults (or fault basins).

Heat flow values in the deep-great fault zones (or along the fault basins) are markedly high. For example, heat flow value is as high as 104.5 MW/M^2 in Yingshan, Dexing county, than is in Xichuang Dexing fault zone.

The average heat flow value of Xiangshan, Lean county, in the same zone, is 82.6 MW/M^2. The highest in the region is 101.4 MW/M^2.

Because faults can not only conduct heat from depth and produce friction heat, but can also cause physical boundaries the redistribution and refraction of surface heat flow, heat flows are influenced by faulted structures.

The deep-seated structural transitional zone, that is the part where the lifted area turns into the sunken area in the upper mantle, controlled markedly the heat flow anomalies. Eight heat flow anomalies delineated in Jiangxi province are in or near the transitional zones except for Qiujiang, Fengcheng city. The deep-seated structural transitional zone is a special geological structural unit in which tectonic stress concentrates, and deep geological activities are frequent. Therefore, the distribution of heat flow anomalies are generally influenced by the transitional zones.

Similar to heat flow anomalies, uranium deposits (mainly hydrothermal ones) are spatially controlled by deep-great faults and deep seated structural transitional zones. That deep-great faults control uranium mineralization is well known and the paper does not discuss it in detail.

Research regarding distribution laws of uranium deposits show that most uranium deposits, especially the mid-large scale ones, distribute mainly in the deep seated structural transitional zones. The evidence reveals that uranium mineralizations are obviously controlled by the transitional zones (Fig. 1).

Distribution of heat flow anomalies is in accordance with that of uranium mineralization because they are both controlled by deep-great faults and deep-seated structural transitional zones.

5 INFLUENCE OF URANIUM MINERALIZATION ON HEAT FLOW

Radiated energy from nuclear decay of radioactive elements in rocks is converted to heat by means of absorption. The radiogenic heat is an important part of surface heat flow. Uranium (thorium) mineralization results in enrichment of radioactive elements such as U and Th, and contributes markedly to heat flow. The Xiangshan uranium ore-field is an example.

Heat flow values range from 69.5 MW/M^2 (close to average regional value) to 101.0 MW/M^2, from the margin to the center in Xiangshan basin. The average value is 82.6 MW/M^2 in the basic which is a heat flow anomalous region. In the volcanic basin, radiogenic heat production of basement rock (biotite schist) is only 1.85 $\mu W/M^3$, while that of cap rocks (mainly porphyro-clastic lava) is high up to 3.78 $\mu W/M^3$. Due to the concentration of radioactive elements is high (dispersed enrichment) in the acidic volcanic rocks, heat flow values increase with the thicker of the rock (porphyro-clastic lava) from the basin margin to the center.

Calculation reveals that the radiogenic heat by dispersed enrichment of radioactive elements takes 3.4% of the total surface heat flow value, and takes 24.8% of radiogenic heat in Cap rocks.

In all deposits of Xiangshan ore-field, uranium and thorium concentrate at several magnitude orders. Such radiogenic heat by concentric enrichment of U and Th takes 0.6~8.2% of total surface heat flow value, or higher.

6 POSSIBILITIES OF PROSPECTING FOR URANIUM DEPOSITS BY HEAT FLOW MEASUREMENT METHOD

Hydrothermal uranium deposits are the products of fossil hydrothermal systems. Though the activities of the hydrothermal systems that formed the uranium deposits stopped for a long time, to a certain extent, the heat flow anomalies will occur near uranium deposits. It is believed that the heat flow anomaly is caused by deep-great faults, or deep seated structural transitional zones and the enrichment of radioactive elements in rocks. The study of radiogenic heat in rocks in Xiangshan ore-field shows that the higher content of radioactive elements in ore-forming rock body may form a plan anomalous region of weaker heat flow in a large range, and that the place where the uranium deposit occurs may form a point anomalous area of stronger heat flow due to the massive enrichment of uranium and thorium. Therefore, based on study of regional metallogenic geological conditions, the heat flow anomaly may be regarded as a prospecting indicator to delineate uranium mineralization prospective area and to hunt uranium deposit.

In addition, the generally higher value of heat flow in uranium mine area certainly causes geo-temperature anomaly to some extent, so the infrared remote sensing may be used to prospect uranium deposit.

7 CONCLUSIONS

Based on the above mentioned, some conclusions are as follows:

(1) The distribution of heat flow values in Jiangxi province is controlled by tectonic units. Heat flow values are lower in the north and high in the central-southern province.

(2) The heat flow anomalies and uranium deposits are closely related to deep faults and great faults, or deep seated structural transitional zones, so they are spatially in accordance with each other.

(3) The dispersed enrichment (ore-forming rock body) and the concentric enrichment (deposits) of the radioactive elements near the radioactive deposits may form surface heat flow anomaly.

(4) The anomalies of heat flow and infrared remote sensing may be regarded as ore-hunting indicators for radioactive deposits.

Water-Rock Interaction, Kharaka & Maest (eds) © 1992 Balkema, Rotterdam. ISBN 90 5410 075 3

Stable isotopic geochemistry of ore-forming solutions and genesis of a gold deposit in Maopai, Jiangxi Province, China

Liu Xiaodong, Xu Haijiang & Lai Shaocong
East China Geological Institute, Fuzhou, Jiangxi, People's Republic of China

ABSTRACT: The gold mineralization in the Maopai deposit can be separated into two distinct stages. Stable isotopes and fluid inclusion data indicate multiple compositions of ore-forming solutions, which could result from a mixture of metamorphic fluids with magmatic hydrothermal solutions. It is also proposed that the gold mainly derived from the host rocks of Precambrian Zhoutan Group at stage I, and might be superimposed by magmatic gold at stage II. Ore-forming solutions rich in CO_2, CH_4 and fluid immiscibility may lead to the precipitation of gold.

1 INTRODUCTION

The Maopai gold deposit is located approximately 50 km southwest of Fuzhou City where our institute is situated, in Jiangxi Province. No materials concerned with gold have been previously documented in the Maopai area before 1986. Exploration and drilling projects were performed in 1988 after our preliminary survey for two years. Up to now, the undergoing explorations outlined a middle size gold deposit. Presently geological study shows that the Maopai gold deposit occurs in meta-migmatite complex of the Precambrian Zhoutan Group, and is controlled by a NE-trending shear zone developed during the Caledonian orogeny. We report here the results of stable isotope and fluid inclusion studies on minerals from gold-bearing quartz veins in Maopai, which suggests that the Maopai gold deposit was formed from the fluids of multiple source with polygenesis.

2 GEOLOGY

In geological setting, the Maopai gold deposit is situated in the northern part of the regional Wuyiyunkai meta-migmatite belt. Rocks in this area are chiefly composed of low to medium grade metamorphic rocks of Precambrian Zhoutan Group including two-mica schist, gneiss, sillimanite two-mica schist and garnet-biotite bearing granitic migmatite. Carbonaceous and siliceous rocks, biotite gneiss and granulite of Cambrian age dominated in the southern portion of the Maopai area. Mesozoic (T-J) feldspathic sandstone and sandy conglomerate unconformably overlying the above sequences are rarely found in Maopai. But the continental sedimentary-clastic rocks composed of conglomerate and sandy conglomerate of the Cretaceous Nanxuang Group are widespread in the east portion, which unconformably overlies the Zhoutan Group. Plutonic rocks are dominated by Caledonian granites, Yanshanian granites and dikes.

The NE-trending shear zone developed during Caledonian orogeny in Maopai is about 5 km long and 1 km wide and played a very important role in controlling the temporal and spatial evolution of the gold mineralization. Gold mineralizing zone was extremely limited in space by the intersection portion between the EW-trending metamorphic rock horizons of the Zhoutan Group and the NE shear zone. The shear zone also resulted in both ductile and brittle deformation of gold-bearing quartz veins and the host rocks.

3 ORE VEINS

Ore veins of the Maopai gold deposit, represented by gold-bearing quartz or quartz-feldspar veins, quartz-vein stockworks, and brecciated quartz veins, occur along the NE shear zone. They strike N 30° to 35°E, dip 40° to 70°NW, and extend about 2 km along the strike. Individual belts of ore veins vary from 0.3 to 7 m in thickness and extend from a few m up to 410 m along the strike. The ores consist chiefly of quartz, sillimanite, garnet, pyrite, pyrrhotite, chalcopyrite, sphalerite, galena and native gold with the average grade of 8 g/ton. Two distinct stages of gold mineralization were identified on the basis of structure and mineral paragenesis.

Stage I mineralization is characterized by highly ductile deformation of gold-bearing quartz or quartz-feldspar compound veins. The mineralogy includes quartz, sillimanite, native gold and some base metal sulfides. Extensive biotitization results in secondary enlarged biotite veins within and around centimeter size displaying along both margins of ore veins, which could be taken as an indicator for gold exploration in the field. Secondary biotite samples from biotite veins adjacent to the gold-bearing quartz veins yielded a K-Ar date of 376 to 399 Ma, indicating a late Caledonian age for gold mineralization.

Stage II mineralization in brecciated gold-bearing quartz veins could have been formed by open space filling of ore-forming fluids along fractures. It was regarded as the superimposed mineralization on stage I. Because of the superimposed and enriched stage I ore veins, some quartz veins contain very high grade ores. The mineral paragenesis at stage II is relatively complex including quartz, pyrite, chalcopyrite, pyrrhotite, galena, sphalerite and native gold. It is common to find coarse-grained native gold crystals (up to 5 mm), small disseminated grains, and fine gold veins throughout the ore veins of stage II. Although stage II contains higher gold concentrations, from a voluemetric standpoint, stage I is the dominant ore stage in the Maopai gold deposit. Fluid inclusion in quartz from high grade gold-bearing quartz veins yielded a Rb-Sr isochron date of 121 ± 24 Ma, indicating a Cretaceous age for stage II mineralizing.

4 FLUID INCLUSION STUDIES

Studies were made on fluid inclusions in quartz from gold-bearing quartz veins of both stages through microthermometry, ion chromatography and laser Raman microspectrometry (Liu and Xu 1990, 1991). Three types of fluid inclusions which related to gold mineralization were observed: almost pure CO_2 single phase inclusions in stage I, and vapor rich in CO_2, CH_4, and liquid CO_2-bearing inclusions in stage I and stage II. The fluid inclusions have similar characteristics to most quartz-vein hosted gold deposits in the greenstone belt of the Precambrian (Robert et al. 1987; Walsh et al. 1988). However the Maopai deposit seems to differ from other gold deposits in the higher salinities ranging from 7 to 11 wt% NaCl equiv. and 10 to 34 wt% NaCl equiv. for stage I and stage II, respectively. Fluid inclusion studies revealed the metallogenic physical-chemical data (Fig. 1). Stage I mineralization occurred at temperatures exceeding 400°C and pressures around 1.5 to 2 Kbar, and stage II were at relatively lower T-P conditions (260~320°C, 0.45~1 Kbar).

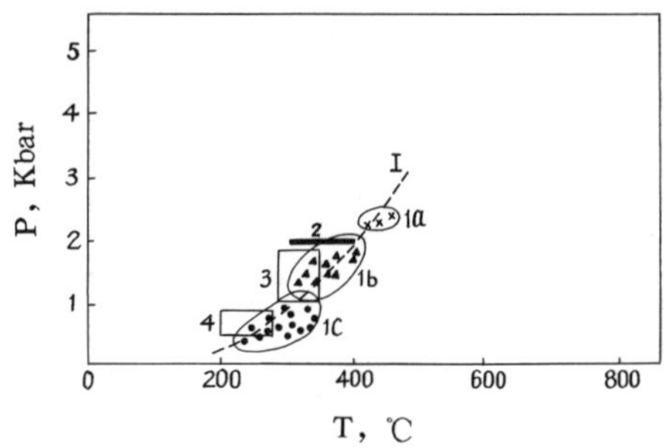

Figure 1 - The P-T conditions for Maopai and other gold deposits. 1a: P-T from dikes in Maopai; 1b: P-T for stage I; 1c: P-T for stage II. I: possible fluid evolution P-T path of the Maopai deposit; 2: P-T from Sigma (Robert et al. 1987); 3: P-T from Pamour (Walsh et al. 1988); 4: P-T from Gold Mountain (Liu 1989).

5 STABLE ISOTOPES

Previous studies have demonstrated the utility of stable isotopes in explaining the origin and the genesis of gold deposits (Taylor 1974; O'Neil et al. 1974; So and Shelton 1987; Silberman 1991). Oxygen isotopes of quartz, hydrogen isotopes of fluid from inclusions in quartz, and sulfur isotope values of sulfides were measured in samples from gold-bearing quartz veins. Isotopic data are listed in standard δ notation reference to SMOW for oxygen and hydrogen isotopes, and CDT for sulfur isotopes.

5.1 Oxygen and hydrogen isotopes

Listed in Table 1 are results of O- and H-isotope compositions of 5 samples from Maopai. The $\delta^{18}O$ values for quartz range from +16.6 per mil to +24.3 per mil relative to SMOW. Samples from stage II have a lower $\delta^{18}O$ value. Oxygen isotopic compositions of water in equilibrium with quartz have been calculated by using the quartz-water isotope fractionation equation (Clayton et al. 1972) and fluid inclusion homogenization temperatures (Th) of each sample. Tested samples have $\delta^{18}O_{H_2O}$ values of +13.4 to +18.9 per mil and +9.7 per mil for stage I and stage II, respectively. Fluids from fluid inclusion were extracted by crushing; their δD values are -71.7 per mil for stage I and -60.5 to -65 per mil for stage II.

Table 1. Oxygen and hydrogen isotope data from the Maopai gold deposit.

Sample	Stage	Mineral	$\delta^{18}O$‰	Th℃	δD_{H_2O}‰	$\delta^{18}O_{H_2O}$‰
87—Q_{10}	I	Quartz	24.3	350	−71.7	18.9
87—Q_8	II	Quartz	20.3	300		13.4
87—Q_9	II	Quartz	16.6	300		9.7
87—Q_4	II	Quartz			−60.5	
87—Q_{12}	II	Quartz			−65.0	

Table 2. Sulfur isotope data from the Maopai gold deposit

Sample	Mineral	stage	$\delta^{34}S$‰
JQK—34	Pyrite	I	4.3
S—8	Pyrite	I	−1.24
JQK—12	Pyrite	II	3.9
s—1	Pyrite	II	3.2
JC—44	Pyrite	II	3.8
JQK—44	Pyrite	II	4.0

5.2 Sulfur isotopes

Sulfur isotope analyses were performed on 6 pyrite samples from various stage gold-bearing quartz veins. The results in Table 2 show that $\delta^{34}S$ values range from -1.2 to +4.3 per mil and have an overall average of +3 per mil indicating little variation in the $\delta^{34}S$ values.

5.3 Interpretation of stable isotopes

Fig. 2 shows the distribution of measured and calculated oxygen and hydrogen isotopic composition from the Maopai gold deposit. The ore-forming solution of this composition could result from a mixture of metamorphic fluids with magmatic hydrothermal solutions. Geological studies show that stage I mineralization is closely associated with the extensive migmatitization and the development of shear zone in late Caledonian orogeny in the Maopai area. Oxygen and hydrogen isotopic compositions of stage I are in the middle of the range for metamorphic waters (Taylor 1974), indicating a dominant origin of metamorphic fluids. The high CO_2 contents of stage I ore fluids, presented by CO_2 single phase inclusions, have been interpreted as derived by metamorphic dehydration (Leitch et al. 1991). In a regional sense, the metamorphic rocks of Zhoutan Group are gold-rich sequences (5~23 ppb), but the gold contents in metamorphic rocks near or adjacent to the ore zone appear to have negative abnormality, which implies the remobilization and migration of Zhoutan Group.

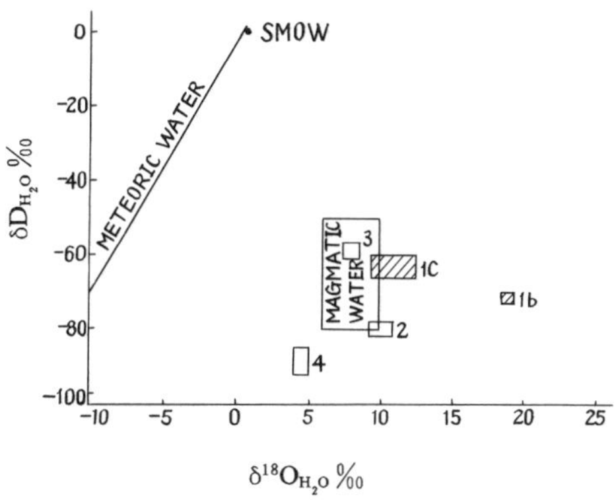

Figure 2 - Hydrogen versus oxygen isotope diagram showing the distribution of stable isotopes. 1b: Maopai stage I; 1c: Maopai stage II; 2: Linglong in Shandong; 3: Gold Mountain in Jiangxi; 4: Jiapigou in Jilin.

Oxygen and hydrogen isotopic compositions of stage II are approaching a magmatic value. The ore-forming solutions and gold sources could be interpreted as of a dominant magmatic origin. Sulfur isotope data also suggest that a magmatic source was involved in stage II mineralization (So et al. 1987); Leitch et al. 1991). In Maopai, the intermediate-basic dikes such as lamprophyres containing relatively high gold concentrations (12~480 ppb) might develop concurrently with stage II mineralization. The ore grade usually appears higher at most intersection portions between crosscut ore veins and dikes. It seems likely that the second stage mineralization could have been related to the intrusions of dikes, which provided heat to drive the flow of the ore-forming solutions. They may have also provided a magmatic component including gold.

For comparison, stable isotopes from other Precambrian quartz vein type gold deposits in China are shown in Fig. 2. The oxygen and hydrogen isotopes for the Maopai gold deposit seem to differ from those measured for other gold deposits. The ore fluids at Maopai were largely magmatic and metamorphic in origin.

6 CONCLUSION: METALLOGENESIS OF GOLD

The results presented above and the previous studies (Liu and Xu 1991) suggest that stage I ore-forming solution at Maopai resulted initially from high temperature (400°C), CO_2, CH_4, and K^+-rich metamorphic fluids. Gold mainly was derived from the host rocks of Zhoutan Group due to the efficient reaction of fluids with the gold-enriched wall rocks. When ore fluid upwelling was coincident with the development of the shear zone, extensive biotitization occurred, which broke the original equilibrium of the hydrothermal systems. Gold precipitation could be favored in the late steps of the process. An interesting feature of the mineralization at stage I is the low content of base metal sulfides. Ore fluids rich in CO_2, CH_4 and H_2 may play an important role as the reducing agents in promoting the deposition of gold. This interpretation was supported by thermodynamic data.

The late stage II mineralization occurred at relatively lower T-P conditions. Stable isotopes and fluid inclusion evidence suggest a dominant magmatic origin of ore-forming solutions and gold. Ore fluids at stage II contain a lot of volatiles such as CO_2, CH_4 and appear widespread fluid immiscibility. Fluid immiscibility could be considered as an important contributor to gold precipitation within veins (Walsh et al. 1988). The partitioning of CO_2 and CH_4 from ore fluids owing to the immiscibility would increase the fo_2 and pH of the residual fluids which may lead to the gold deposition. A lot of base metal sulfides at stage II may also cause gold precipitation by the reduction process.

REFERENCES

Clayton, R.N. J.R. O'Neil & T. Mayeda 1972. Oxygen isotope exchange between quartz and water: Jour. Geophys. Research 77: 3035-3120.

Leitch, C.H.B., C.I. Godwin, T.H. Brown & B.E. Taylor 1991. Geochemistry of mineralizing fluids in the Bralorne-Pioneer mesothermal gold vein deposit, British Columbia, Canada: Econ. Geol. 86: 318-353.

Liu, X.-D. & H.-J. Xu 1990. Initial results of fluid inclusions and mineralogic studies from the Maopai gold mine, Jiangxi, China: GSA Abstracts with Programs 22/7: A249.

Liu, X.-D. & H.-J. Xu 1991. Fluid inclusion studies of the Maopai gold deposit, Jiangxi, China: Acta Mineralogica Sinica China 11/4.

O'Neil, J.R. & M.L. Silberman 1974. Stable isotope relations in epithermal Au-Ag deposits: Econ. Geol. 69: 902-909.

Robert, F. & W.C. Kelly 1987. Ore-forming fluids in Archean gold-bearing quartz veins at the Sigma mine, Abitibi greenstone belt, Quebec, Canada. Econ. Geol. 82: 1464-1482.

Silberman, M.L., A. Moore-Nall & B.M. Smith 1991. Gold-bearing quartz veins along the Mojave-Sonora megashear zone, Northern Sonora, Mexico: USGS Research on Mineral Resources, 1991 Program and Abstracts: 70-72.

So, C.S. & K.L. Shelton 1987. Stable isotope and fluid inclusion studies of gold- and silver-bearing hydrothermal vein deposits, Cheonan-Cheongyang-Nonsan mining district, Republic of Korea, Cheonan area: Econ. Geol. 82: 987-1000.

Taylor, H.P. 1974. The application of oxygen and hydrogen isotope studies to problems of hydrothermal alteration and ore deposition. Econ. Geol. 69: 843-883.

Walsh, J.F., S.E. Kesler, D. Duff & P.L. Cloke 1988. Fluid inclusion geochemistry of high-grade, vein-hosted gold ore at the Pamour mine, Porcupine Camp, Ontario. Econ. Geol. 83: 1347-1367.

Water-Rock Interaction, Kharaka & Maest (eds) © 1992 Balkema, Rotterdam. ISBN 90 5410 075 3

Au sorption by arsenic compounds and organic matter

I.Ya.Nekrasov
Far East Geological Institute, Vladivostok, Russia

ABSTRACT: Experiments carried out at 200-300°C, $P(H_2O)$ of 400-500 bar, and fO_2 controlled by NNO, FeM, WM, HM, and $Cu-Cu_2O$ buffers, were aimed at determining the solubility and sorption of Au by As compounds and soluble bitumens. The fO_2 was the major parameter limiting the solubility and sorption of Au. In the solutions rich in As ($n\cdot10^{-3}$ mole/kg), the solubility of Au was the highest at 300°C and HM buffer ($n\cdot10^{-4}$ mole/kg) and the lowest at FeM buffer ($n\cdot10^{-7}$ mole/kg). An excess (against the saturation) As in the system as a solid phase (As-cryst., As_2S_3, $FeAs_2$) results in the sorption of Au by crystal surfaces. Some Au reacts with As producing metastable $AuCl(AsH_3)$ compounds. Bitumens (resin, asphaltene, and oil) easily adsorb Au from chloride solutions. Depending on fO_2, bitumens can extract up to 90% Au from the solutions containing 10^{-7} mole/kg Au. Resulting Au content in them is $n\cdot10^{-3}$ to $n\cdot10^{-4}$ mole/kg. Data obatined were used to interprete the mechanism of the origin of gold-bearing black shales.

A genetic relationship between the deposits of gold, platinum, molibdenum, tin, uranium, REE, and other metals and black shales formed under specific physico-chemical conditions has been revealed in a variety of ore districts round the globe. In ore districts of the Far East Russia (Kular, Orulgan, Verkhoyansk, Chaunsk, and Kolyma), Eastern Siberia (Patomsky, Vitemsky, and Eniseisky), and Central Asia (Muruntau) we studied, the black shale layers are rich in carbonaceous matter (up to 2.5 % of mass), Au ($n\cdot10^{-7}$ to $n\cdot10^{-8}$ %), Sn ($n\cdot10^{-3}$ %), REE ($n\cdot10^{-5}$ to $n\cdot10^{-6}$ %), P (up to 0.5 %), As ($n\cdot10^{-3}$%), and other constituents typomorphic of emplaced ore bodies. Different investigators revealed a direct relationship between Au and As and between Au and carbonaceous matter in black shale deposits (Boyle 1979, Kokin 1985, and Nekrasov 1991). This fact encouraged us to experimental studies aimed at modelling a possible mechanism of Au concentration in seafloor sediments by the activity of hot springs common even recently in "hot spots" of the oceans. The temperature of the seafloor hot springs released through the fractures in seafloor basalts is 200 to 300°C, reaching sometimes 400°C. The hot springs are enriched in H_2S, Na, K, and Mg chlorides, hydrocarbons, and ore constituents (Fe, Zn, Pb, Cu, Ag, and Au). pH varies from 2 to 5, and fO_2 matches $Fe-Fe_3O_4$ and $FeO-Fe_3O_4$ buffers. A vigorous precipitation of ore constituents from seafloor solutions occurs at the solution-seawater mixing boundary.

With this in view, we carried out a series of runs on gold solubility and sorption by solid phases in the Au-As-HCl, Au - Fe - As - HCl, Au - As - S - HCl, and Au - bitumens - HCl systems at 200-300°C, $P(H_2O)$ of 0.5 to 1 kbar, and fO_2 corresponding to Ni-NiO, $Cu-Cu_2O$, Fe-FeO, $FeO-Fe_3O_4$, and $Fe_2O_3-Fe_3O_4$ buffers (Akhmedzhanova et al. 1991). Externally heated water-media pressure vessels manufactured from titanium and nickel-chromium alloys were utilized. The As compounds or bitumens were put into 10 mm diameter gold bags 100 mm long used as a source of gold. Buffer mixtures were separated from solution in gold bags (Au 99.99) 3 mm accross. Bags with buffer mixtures were not sealed, thus fO_2 was maintained constant in the system. Large-diameter bags were sealed in argon flow after the installation. Arsenic was introduced into the system as elementary As or as As_2S_3 synthetic crystals, and carbonaceous matter as coal liquid fraction (asphaltene, asphaltic acids, resins, and oils). The system was considered to attain equilibrium after the content of the constituents (Au, As, and Fe) in solutions was stabilized, independent of the duration of a run. At 200^oC, the systems attained equilibrium in 30 to 35 days; at 300^oC, in 20 days. Then, high-pressure bombs containing bags were cooled by compressed air or cold water at a speed of 100^oC/min.The solutions and solid phases extracted from the bags were treated with HNO_3 (1:3), and Au was concentrated by oil sulfide. Gold was analysed using atomic adsorption. The concentration of dissolved Au was measured in solution, and its adsorbed portion in solid precipitates (As and AsS crystals or bitumens).

In the As-bearing systems, we made an attempt to estimate the effect of the redox factor and As concentration on the solubility of Au. The values of Au solubility in the systems at fO_2 buffered by $Cu-Cu_2O_3$, Ni-NiO, and $Fe_3O_4-Fe_2O_3$ pairs suggest that fO_2 is the major parameter limiting Au solubility in the As-bearing systems. At 200-300^oC and fO_2 matching $Cu-Cu_2O$ buffer, the solubility of Au in 0.1 to 1 M HCl solutions rich in As (approximately $n\cdot10^{-3}$ mole/kg) was $n\cdot10^{-6}$ to $n\cdot10^{-7}$ mole/kg (Akhmedzhanova et al.1991). With the increase in fO_2 (Ni-NiO buffer), it rises up to $n\cdot10^{-5}$ to $n\cdot10^{-6}$ mole/kg, i.e., approximately an order of magnitude. A further fO_2 rise ($Fe_3O_4-Fe_2O_3$ buffering pair) results in a 1.5 to 2 orders jump-like increase in Au solubility. Under these entirely oxidizing conditions, Au equilibrium concentration in solution reached $n\cdot10^{-3}$ to $n\cdot10^{-4}$ mole/kg at 300^oC.

Under the reducing conditions ($Cu - Cu_2O$ and Ni-NiO buffering pairs), a metastable phase as a thin crimson or blue film precipitates on the walls of the bags after the quenching. The content of Au in it, measured using electron microprobe, varies from 82 to 94 % of mass and the content of As varies from 3 to 8%. Cl was also recorded (1 to 2%). The formation of that metastable compound supports the idea of Akhmedzhanova et al. (1991) on a possible reaction of chloride-arsenic solutions with arsine (AsH_3), the stability field of which was defined in the $As-H_2O$ system at 200 to 300^oC as a factor of pH and fO_2 (Heinrich & Eadington 1986). Thus, $AuCl(AsH_3)$ complex compound, similar to a metastable phase precipitated sporadically on the walls of the bags, is the most probable form of Au in solutions of the Au-As-HCl system.

We studied in detail the effect of As on Au solubility only at 300^oC under utterly oxidizing conditions ($Fe_3O_4-Fe_2O_3$ buffer), i.e., at the equilibrium concentration of Au in solutions approximately equal to $n\cdot10^{-3}$ - $n\cdot10^{-4}$ mole/kg. The increase in As or As_2S_3 weight results in a notable decrease in Au solubility presumably due to the Au sorption from solution by the surfaces of As or As_2S_3 crystals. To verify this idea, we introduced loellingite ($FeAs_2$) rather than crystalline As into the system. It resulted in 2 to 3 times decrease in Au content in the solution, and adsorbed Au precipitated on the surfaces of loellingite grains. A similar process of Au sorption by As_2S_3 sulfide was experimentally studied by Renders & Seward (1989). It obviously takes place in nature, as a vigorous precipitation of sulfides from oceanic hydrothermal solutions during the origin of black shales.

Carbonaceous matter favoured the leaching of Au from oceanic hydrothermal solutions even more than As compounds. It served as both a concentrating and reducing agent of Au. This agrees well with the data obtained from the studies of the $Au-bitumens-H_2O$ system at 200 to 300^oC, $P(H_2O)$ approximately 0.4 to 0.5 kbar and controlled by fO_2 (Nekrasov & Efimova 1991). The conditions of reference runs and their results are given in the table.

Table. Au concentrations in quenched solutions, mole/kg

Oxygen buffer	Resin	Oil	Asphaltene	Asphaltic acid
		300^oC		
NNO	2.88	3.49	2.93	2.57
HM	2.89			2.01
"	2.79	2.10	2.65	2.70
"	2.55			2.90
$Cu-Cu_2O$	1.52	3.13	3.54	3.35
		200^oC		
HM	4.04	3.09	2.92	3.19

The solubility of Au in every quenched solution equilibrated with bitumens, is below $n \cdot 10^{-7}$ mole/kg, and Au content in bitomen compounds varies considerably from $n \cdot 10^{-4}$ to $n \cdot 10^{-3}$ mole/kg. It depends on the nature of bitumen matter and fO_2 in the system. An increase in fO_2 at 300°C from 10^{-34} 10^{-23} bar results in an order of magnitude increase in Au content in resins. In asphaltic acids, it lowers an order of magnitude at a similar change of fO_2 in the system. It suggests close bonds between Au and O-bearing bitumen groups, clearly seen in IR spectra. Using IR spectroscopy, we estimated oxidation degree of bitumen compounds treated by hydrothermal solutions (by the position of absorbtion band around 1740 to 1700 cm^{-1}) and the aromaticity (by the intensity of absorbtion bands at 1600, 880, and 820 cm^{-1}). The thermal effect on the bitumen compounds results in the breakdown of C-O bonds and involvement of adsorbed Au in them. A similar process of Au leaching from hot seawater presumably occurred during the sedimentation in the seafloor rift zones, where the degassing mantle supplied a carbonaceous matter. Au and other metals were obviously leached from these presumably metal-bearing sediments during metamorphism and metasomatism and concentrated in ore bodies usually hosted in black shales.

CONCLUSION

Thus, the experimental modelling of gold leaching from solutions and its sorption by carbonaceous matter and arsenic compounds provides new ideas on the genesis of presumably ore-bearing black shales. In particular, it will help us to solve the problem of sources of ore material and ways of gold concentration.

REFERENCES

Akhmedzhanova, G.M., I.Ya.Nekrasov & V.I.Tikhomirova 1991. On the influence of arsenic on the solubility of gold in the Au-As-0.1 N HCl system at 200 and 300°C. *Dokl.AN SSSR.* 319:471-474. (in Russian).

Boyle R.W.1979. The geochemistry of gold and its deposits. *Geol. Surv. Canada* 280: 585.

Heinrich,C.A. & P. Eadington 1986. Thermodynamic predictions of the hydrothermal chemistry of arsenic and their significance for the paragenetic sequense ofsome cassiterite-arsenopyrite base metal sulfide deposits. *Econ. Geol.* 81:511-529.

Kokin A.V. 1985. The South-Verkhoyansk arsenic-bearing geochemical province. *Geologiya i geofisika* 2:74-81. (in Russian).

Nekrasov I.Ya. 1991. Geochemistry, mineralogy, and genesis of gold deposits. Moscow: Nauka Publ. (Russian).

Nekrasov, I.Ya. & N.F. Efimova 1991. The experimental study of gold-carbonaceous system under hydrothermal conditions. *Dokl. AN SSSR* 318:724-727. (in Russian).

Renders, P.J. & T.M. Seward 1989. The adsorption of the gold (I) complexes by amorphous As_2S_3 and Sb_2S_3 at 25 and 90°C. *Geochim. et Cosmochim. Acta* 53:255-268.

Water-Rock Interaction, Kharaka & Maest (eds) © 1992 Balkema, Rotterdam. ISBN 90 5410 075 3

Hydrothermal systems related to epithermal gold deposits in Mesozoic volcanic areas in eastern China

Qijang Ren, Chongze Zhang, Rongyong Yang, Xiaojun Xie & Zhaowen Xu
Department of Earth Sciences, Nanjing University, People's Republic of China

ABSTRACT: Epithermal gold deposits in Mesozoic volcanic areas in eastern China can be divided into four types: hot spring; quartz-adularia; acid sulfate; and quartz manganosiderite. For these four types of epithermal gold deposits, the lithology of basement rocks is one of the major factors affecting the geochemical features of the ore forming solution, which indicates that water-rock interactions plays an important part in the formation of epithermal gold deposits. Water/rock ratios of the systems for five examples of epithermal gold deposits belonging to the four types have shown, also, that the water/rock ratio is an important parameter related to mineralization intensity.

1 INTRODUCTION

Based on geologic setting (volcanic rock series, host rocks, types of volcanic depression, character of basement rocks, tectonic setting and ore-controlling structures) and features of the ore deposit (ore and gangue minerals, alteration, structure, ore-forming conditions, geological features), epithermal gold deposits in Mesozoic areas in eastern China can be divided into four types: hot springs, quartz-adularia, acid sulfate, and quartz-manganosiderite type. On the basis of the geological-geochemical features of five examples of epithermal gold deposits belonging to these four types, the chemical data and isotopic compositions of the ore and gangue minerals, altered rocks, and fluid inclusion, and the characteristics of each hydrothermal system are outlined. Implications of water/rock ratios in ore-forming systems are also discussed.

2 CHARACTERISTICS OF HYDROTHERMAL SYSTEMS RELATED TO TYPICAL EPITHERMAL GOLD DEPOSITS

2.1 Hot spring type: Tuanjiegou deposits

Tuanjiegou is located at the juncture of the Wulaga volcanic fault-depression and the Jiamushi uplifted block in Heilongjiang Province. Crystalline schist of lower Proterozoic age and volcanic sedimentary rocks of the Jurassic-Cretaceous period crop out in the mining areas. The former is the crystalline basement. Orebodies are hosted in granite porphyry around the caldera and beneath an acid-leaching zone composed of alterated argillaceous rocks. Various kinds of breccia occur in the open pit, but the orebodies are controlled by hydrothermal breccia and stockworks. Gold minerals occur in chalcedonitic quartz, marcasite and botryoidal pyrite, associated occasionally with antimonite, cinnabar and realgar. Calcite and ankerite are the predominant minerals in the late mineralization stage. The major alteration minerals from mineralization are quartz + adularia + illite ± sericite.

The isotopic components of quartz and its fluid inclusions have shown that the hydrothermal solution during the pre-mineralization period is of magmatic origin, with a $\delta^{18}O_{H_2O}$ value of 7.4 per mil, which brings about the slight potassic alteration. The ore-forming solution during the mineralization period had a temperature of 90 to 270°C and salinities of 1 to 5 wt% equiv NaCl. Its δD values were -97 to -139 per mil, $\delta^{18}O_{QZ}$ was +10.5 to 14.7 per mil, and $\delta^{18}O_{H_2O}$ was -3.7 to +0.27 per mil. The water in the ore solution is of meteoric origin. The $\delta^{34}S$ values for marcasite and pyritogelite (botryoidal pyrite) are between -34.2 and +5.0 per mil. The $\delta^{34}{}_{\Sigma S}$ for the ore solution was 0 ± 4 per mil. The ore-forming hydrothermal system (T = 250°C) had a pH value of 5.5 to 6.5, $\log f_{O2}$ was -34.4 ± 0.5, and $\log f_{S2}$ was -9.5 to -11.5. The liquid composition of the solution was relatively rich in SO_4^{2-} and poor in HCO_3^-.

2.2 Quartz-adularia type: Dongxi deposits

Dongxi is located at the border of the Mozitan-Xiaotian volcanic fault-depression in Anhui Province. The gneiss of the Dabieshan Group of late Archaean age and the late Jurassic andesite and trachyte andesite crop out in the mining area. The former is the basement rock of the volcanic fault-depression. The orebodies hosted in the trachyte andesite are fracture-controlled and distributed along regional faults. The mineral assemblage in ores consists of quartz, chalcedonite quartz, adularia, electrum, pyrite, chalcopyrite, sphalerite, and galena. The abundance of sulfides is very low. Alteration mineral assemblages consist of quartz, adularia, sericite, and chlorite. The ore solution had the following properties: T = 120-290°C, salinity = 1.5-5 wt% equiv NaCl, pH = 5.5~6.5, $\log f_{O2}$ = 34.5 ± 0.5, $\log f_{S2}$ = -9.5~-11.5 (T = 250°C). The liquid composition was relatively rich in HCO_3^- and poor in SO_4^{2-}. The $\delta^{18}O_{H_2O}$ value of the ore solution was -4.34 to -8.22 per mil. The $\delta^{18}O_{H_2O}$ was -67.7 to -75.8 per mil. The water was of meteoric origin. The $\delta^{34}S$ of pyrite was -3 to +5 per mil.

2.3 Acid-sulfate type: Zhijinshan deposits

Zhijinshan is located at the border of the Shanghang volcanic fault-depression. The exposed rocks in the mining area are pre-mineralization granitoids of early Yanshan period (about 145 ± 12 Ma), and dacite and pyroclastic rocks of early Cretaceous period. The basement of the volcanic fault-depression consists of upper Proterozoic and Palaeozoic sedimentary rocks. The ore-bodies hosted in granitoids are controlled by regional faults and volcanic vents, and associated with explosive-intrusive breccia and hydrothermal breccia. They approximate a porphyry copper protore. The orebodies occur as lodes and stockworks. The major ore minerals are digenite, enargite, tennantite and native gold. The alteration minerals associated with Au-Cu mineralization are quartz-alunite-kaolinite. Near the surface, a gold-supergene enrichment zone occurs with a geochemical anomaly of Cu-Au-As.

There were two kinds of solution during the mineralization. One occurs during the early mineralization period, with the temperature of 300 to 370°C, salinity of 4 to 14.4 wt% equiv NaCl, and $\delta^{18}O_{H_2O}$ values of +5 to 8.8 per mil. Its water was of magmatic origin. The other solution occurred during the main mineralization period, with temperatures of 200 to 250°C, salinity of 0.87 to 9.9 wt% equiv NaCl, and $\delta^{18}O_{H_2O}$ values of -7.74 to -1.86 per mil. Its water was of predominantly meteoric origin. The ore-forming solution had following properties: pH = 3~4, $\log f_{O2}$ = -33 ± 0.5, $\log f_{S2}$ = -6.5~-7.5 (T = 250°C), and it was relatively rich in SO_4^{2-} and poor in HCO_3^-.

2.4 Quartz-manganosiderite type: Tongjin and Tiantoushan deposits

Tongjin is located at the west border of Lingwu volcanic fault-depression in Jiangsu Province. The exposed rocks in the mining area are alkaline volcanic and subvolcanic rocks of early Cretaceous period. The 3-km-long main vein is within the Liangliangshan Caldera. Tiantoushan is located at the southern border of the Luzong volcanic fault-depression. The exposed rocks are clastic sedimentary rocks of the middle Jurassic period and volcanic rocks of late Jurassic period.

The basement of both volcanic fault-depressions consists of carbonated and clastic sedimentary rocks of the Triassic and Palaeozoic periods. The mineral assemblages of the two deposits are similar. The pre-mineralization assemblage consists of quartz-pyrite-specularite. The mineralization assemblage consists of quartz-chalcopyrite-bornite-native gold-electrum-galena-sphalerite, and manganosiderite-calcite-sulfides ± electrum. The post mineralization assemblage consists of barite. The alteration mineral association is quartz-sericite-kaolite and chlorite-pyrite-quartz. Copper is the major economic product. The ore solution had following properties: T = 150~300°C, salinity = 7~13 wt% equiv NaCl, pH = 4.5~6.5, $\log f_{O2}$ = -33 ± 0.5, $\log f_{S2}$ = -8.5~-11 (T = 250°C), is relatively rich in HCO_3^- and SO_4^{2-}, δD_{H_2O} = -60~-70 per mil, and $\delta^{18}O_{H_2O}$ = -7.6~-1.86 per mil. The water of ore solution was predominantly of meteoric origin, but was mixed with a small amount of magmatic and formation water.

The data above show that the features of each ore-forming system for the four types of epithermal deposits are different and the character of the ore solution is related to the lithology of basement rocks. The gold deposits in the volcanic areas in basement consisting of metamorphic rocks (e.g., crystalline schist and gneiss) lack economic value of base metals, and the ore-forming solution has a low salinity and activity of HCO_3^-. The ores of gold deposits in the volcanic depression in basement consisting of carbonate and intercalation of salt and anhydrite contain abundant copper sulfides, and the copper in the ores has economic value. The ore-forming solution for the latter has relatively high salinities and activities of HCO_3^- and SO_4^{2-}. These results indicate that the water-rock interaction between basement rocks and meteoric water circulating at depth play an important part in the formation of epithermal gold deposits.

3 RELATIONSHIP BETWEEN ROCK/WATER RATIOS AND ORE-FORMING PROCESSES

Based on the composition of mineral isotopes, fluid inclusions, and fresh and alterated rocks, we use the following formula (Taylor 1974, 1978) to calculate the water/rock ratios for various ore-forming systems related to the epithermal deposits discussed above:

$$W/R = \frac{\delta A_{f,r} - \delta A_{i,r}}{\delta A_{i,w} - \delta A_{f,w}}$$

$$W/R = \frac{(\delta A_{f,w} + \triangle - \delta A_{i,r})}{(\delta A_{i,w} - \delta A_{f,w})}$$

where

$$\triangle^{18}O_{rock-water} \approx \triangle^{18}O_{plagioclase(An30)-H_2O}$$

$$\triangle D_{rock-water} \approx \triangle D_{mascovite-H_2O}$$

Using the data presented here and published data (Zhang 1985), we calculate the initial values of δD and $\delta^{18}O$ of meteoric water during mineralization period and in fresh rocks. The final values of δD and $\delta^{18}O$ are from direct measurement. Fig. 1 shows the values of water/rock ratios of systems for each studied gold deposits.

The water/rock ratios for Tuanjiegou and Zhijinshan magmatic hydrothermal system during pre-mineralization period are relatively high (larger than 0.35). The Tuanjiegou ore-deposit system has the largest gold reserves (more than 50 tons), while the Zhijinshan ore-deposit system has the largest amounts of copper reserves and the second largest gold reserves. The water/rock ratios are low, with values of 0.01 to 0.3. The water/rock ratios of the Tongjing ore-deposit system, which has intermediate amounts of gold and copper reserves, are between 0.2 to 0.6. The water/rock ratios of Dongxi and Tiantoushan ore-deposit systems, which have small amounts of gold and copper reserves, are

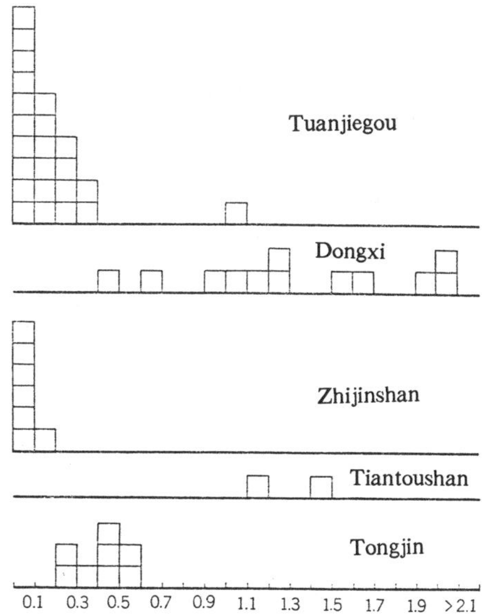

Figure 1 - Histogram showing the water/rock ratios of ore-forming systems for the five epithermal gold deposits.

high, with values of 0.41 to 2.85. The water/rock ratios of the systems during post-mineralization are also very high, with values larger than 1. These systems bring about barren gangue and alteration minerals. The parameters related to water/rock ratios include the degree of water-rock interaction, permeability of the original rocks, and the residence time of fluids in the system. These parameters are associated with the intensity and scope of mineralization. The water/rock ratio of epithermal systems is a good indicator of mineralization intensity.

4 CONCLUSIONS

(1) The epithermal gold deposits in Mesozoic volcanic areas, eastern China can be divided into four types. The characteristics of hydrothermal system for each type of gold deposits are different.
(2) The lithology of basement rocks is related to the geochemical characteristics of ore-forming hydrothermal system.
(3) The water/rock ratios of epithermal systems are most likely related to mineralization intensity.

REFERENCES

Taylor, H.P. 1974. The application of oxygen and hydrogen isotope studies to

problem of hydrothermal alteration and ore deposition. Econ. Geol. 69: 843-883.
Taylor, H.P. 1977. Water/rock interactions and the origin of H_2O in granitic batholiths. J. Geol. Soc. 133: 505.
Zhang, L.G. 1985. The application of the stable isotope to geology. Shaanxi Science and Technology Publishing House.

Water-Rock Interaction, Kharaka & Maest (eds) © 1992 Balkema, Rotterdam. ISBN 90 5410 075 3

Metal transport and deposition in hydrothermal systems buffered by mineral assemblages

Gilpin R. Robinson, Jr
US Geological Survey, Reston, Va., USA

ABSTRACT: Reactions that control metal transport and deposition in hydrothermal systems buffered by mineral assemblages can be generalized and described relative to an abundant fluid component (e.g., K) by use of parameters such as the chemical potentials (μ) of Al_2O_3, FeO, and H_2S that are defined or limited by the rock mineral assemblage. Variation of μAl_2O_3 and μFeO can be deduced conveniently to determine relative metal solubilities under different mineral assemblages for limiting conditions of sulfide saturation. In muscovite + magnetite + pyrite buffer assemblages, fluids moving from lower to higher μAl_2O_3 mineral assemblages or higher to lower μFeO mineral assemblages react toward higher fluid contents of metals such as Zn, Pb, and Cu and lower fluid contents of bisulfide-complexed Au species relative to K. For these buffer reactions, supersaturation of silica with respect to quartz increases the solubility of base metals such as Zn, Pb, and Cu but decreases the solubility of Au. The relative temperature and pressure effects on metal solubilites for the reactions describing exchange of metals for K can be inferred, with the base metals and Au showing opposite trends. Rock-buffered quasi-adiabatic fluid migration paths enhance base-metal solubility and transport relative to paths following geothermal gradients.

1 INTRODUCTION

The chemical character of hydrothermal fluids is acquired through interaction with both primary and alteration mineral assemblages in the rocks along their flow paths. Coupled metasomatic exchange between a hydrothermal fluid and contrasting rock types is one of the processes that influences the transfer of metals into and out of solution. Examples of this type of chemical control on metal transport and deposition are given by Guilbert and Park (1986, p. 157-158) and Bohlke (1989). Mass balance constraints dictate that the minor components of the fluids, such as base metals, will show much greater degrees of relative change in concentration compared to the more abundant components of the fluids, such as Na and K, as the fluid reacts and equilibrates with adjacent wallrock during its migration through the system. For example, chemical changes in hydrothermal fluids equilibrated with K-feldspar + muscovite + quartz assemblages under controlled experimental conditions show large relative changes in Fe, Pb, Zn, and Cu concentrations compared with the lesser relative changes measured for alkali concentrations (c.f. Whitney and others, 1985; Hemley and others, 1992). Therefore, the concentrations of base and precious metals in hydrothermal fluids are likely to be defined or limited by exchange reactions with proximal wallrocks with the result that these local reactions may account for the sources, sinks, and residence times for metals in the fluids. An understanding of rock-buffered reactions provides a prediction of local wallrock effects on fluid geochemistry and can be used as a guide to locate the more promising areas hosting ore deposits in lithologically heterogeneous terranes.

Metasomatic changes resulting from fluid-rock interaction can be considered a metamorphic process, and techniques commonly used in metamorphic petrology will be used to deduce qualitative changes in metal solubility relative to differing mineral assemblages interacting with the fluid. Under conditions where appropriate mineral buffer assemblages that control solubility relations can be identified (or assumed) and where the rate of chemical interaction of the fluid with the rocks is sufficiently fast relative to fluid migration that local equilibrium may be inferred or demonstrated, these techniques can provide general insight into the patterns of metal transport and deposition in hydrothermal systems.

As an example, this paper considers a series of buffer reactions for carbonate-free mineral assemblages containing muscovite, quartz, magnetite, pyrite, and other silicate phases such as K-feldspar, biotite, chlorite, and (or) kaolinite. Mineral assemblages such as these, and the rock-water reactions they generate, are appropriate for hydrothermal fluids interacting with granitoid igneous rocks, clastic sedimentary rocks, and their metamorphic equivalents.

2 GENERALIZED ROCK-BUFFERED REACTIONS

2.1 Introduction

We explore here a set of generalized rock alteration reactions that control metal solubilities in hydrothermal systems buffered by mineral assemblages containing muscovite, magnetite, and pyrite. The phase relations and reactions considered here are appropriate for the moderate temperature region (~250 to 450°C) at low to moderate pressures (500 to 2000 bars) in the one-phase fluid region.

Table 1 gives the abbreviations and idealized compositions for mineral phases used in the reactions listed in Tables 2 and 3. Chloride generally is the most abundant and important of the anions that contribute to dissolved mineral matter in hydrothermal fluids, and at the moderate temperature and pressure conditions considered here, the alkalis and base metals are in solution largely as aqueous associated-chloride species rather than as free ions. For simplicity, the reactions have been written in terms of neutral chloride-associated species for the alkalis and base metals. For moderate-temperature hydrothermal fluids buffered by the reactions considered here, $Au(HS)_2^-$ is taken to be the most important gold species in solution (Seward, 1984), and the buffer reactions in Tables 2 and 3 involving it are written with K^+ as an aqueous species to maintain charge balance.

Table 1. Abbreviations used in text and tables.

	Mineral	Formula
Py	Pyrite	FeS_2
Po	Pyrrhotite	FeS
MeS	base-metal monosulfide	ZnS for sphalerite PbS for galena
Cpy	Chalcopyrite	$CuFeS_2$
Mgt	Magnetite	Fe_3O_4
Mus	Muscovite	$KAl_3Si_3O_{10}(OH)_2$
Ksp	K-feldspar	$KAlSi_3O_8$
Fe-Chl	Fe-Chlorite	$Fe_5Al_2Si_3O_{10}(OH)_8$
Fe-Bio	Fe-Biotite	$KFe_3AlSi_3O_{10}(OH)_2$
Kao	Kaolinite	$Al_2Si_2O_5(OH)_4$
Qtz	Quartz	SiO_2

Table 2. Water-rock reactions between potassium and metal fluid components and silicate mineral assemblages under muscovite + magnetite + pyrite buffer conditions. Al_2O_3 and FeO represent components defined by the mineral asssemblage.

K-Fe EXCHANGE REACTION

$$2KCl + FeO + 3Al_2O_3 + 6SiO_2 + 2H_2O = 2Mus + FeCl_2$$

K-Me (Zn, Pb) EXCHANGE REACTION

$$2KCl + MeS + 0.5Mgt + 6SiO_2 + 3Al_2O_3 + 2H_2O = 0.5Py + 2Mus + FeO + MeCl_2$$

K-Cu EXCHANGE REACTION

$$2KCl + 2Cpy + Mgt + 6SiO_2 + 3Al_2O_3 + 2H_2O = 2Py + 2Mus + 3FeO + 2CuCl$$

K-Au REACTION

$$2K^+ + 2Au(HS)_2^- + Mgt + 6SiO_2 + 3Al_2O_3 = 2Py + 2Mus + 2Au^\circ + FeO$$

2.2 Rock-buffered reactions in terms of chemical potentials defined by mineral assemblages

The tendency toward reaction change in the rock-water system can be expressed in terms of concentrations of the chemical components in the mineral and fluid phases and the pressures of volatile components as presented by their thermodynamic equivalents of activity, chemical potential (μ), and fugacity. The reactions that control metal solubility and deposition can be generalized and described as exchange equilibria relative to an abundant fluid component (e.g., K)

Table 3. Water-rock reactions between potassium and metal fluid components under muscovite + magnetite + pyrite buffer conditions at low, intermediate, and high values of μAl_2O_3. Increasing temperature and (or) decreasing pressure favors the right-hand side of each reaction, as written.

A. LOW μAl_2O_3

1. Ksp + Mus + Bio buffer

$$12KCl + 6Mus + 36Qtz + 3Mgt + 6MeS = 12Ksp + 3Py + 2Fe\text{-}Bio + 6MeCl_2 + 4H_2O$$

$$6KCl + 3Mus + 18Qtz + 3Mgt + 6Cpy = 6\,Ksp + 6\,Py + 3\,Fe\text{-}Bio + 6\,CuCl$$

$$6K^+ + 3Mus + 18Qtz + 3Mgt + 6Au(HS)_2^- = 8Ksp + 6Py + 1Fe\text{-}Bio + 6Au + 8H_2O$$

2. Ksp + Mus + Chl buffer

$$20KCl + 12Mus + 66Qtz + 5Mgt + 10MeS = 32Ksp + 2Fe\text{-}Chl + 5Py + 10MeCl_2 + 4H_2O$$

$$10KCl + 8Mus + 39Qtz + 5Mgt + 10Cpy + 4H_2O = 18Ksp + 3Fe\text{-}Chl + 10Py + 10CuCl$$

$$10K^+ + 6Mus + 33Qtz + 5Mgt + 10Au(HS)_2^- = 16Ksp + Fe\text{-}Chl + 10Py + 10Au + 12H_2O$$

B. INTERMEDIATE μAl_2O_3

1. Mus + Chl + Bio buffer

$$12KCl + 6Fe\text{-}Chl + 18Qtz + 3Mgt + 6MeS = 12\,Fe\text{-}Bio + 3\,Py + 6\,MeCl_2 + 12\,H_2O$$

$$8KCl + Mus + 3Fe\text{-}Chl + 15Qtz + 4Mgt + 8Cpy = 9Fe\text{-}Bio + 8Py + 8CuCl + 4H_2O$$

$$2K^+ + 3Fe\text{-}Chl + 3Qtz + Mgt + 2Au(HS)_2^- = 2Fe\text{-}Bio + 2Py + 2Au + 4H_2O$$

C. HIGH μAl_2O_3

1. Mus + Chl + Kao buffer

$$20KCl + 32Kao + 2Qtz + 5Mgt + 10MeS = 20Mus + 2Fe\text{-}Chl + 5Py + 10MeCl_2 + 36H_2O$$

$$10KCl + 18Kao + 3Qtz + 5Mgt + 10Cpy = 10Mus + 3Fe\text{-}Chl + 10Py + 10CuCl + 14H_2O$$

$$10K^+ + 16Kao + Qtz + 5Mgt + 10Au(HS)_2^- = 10Mus + Fe\text{-}Chl + 10Py + 10Au + 28H_2O$$

by use of chemical parameters such as μAl_2O_3, μFeO, and μH_2S that are defined or limited by the mineral assemblage (Table 2). Although the chemical components Al_2O_3 and FeO are not ordinary mineral phases, they are usable as reaction components, and their chemical potentials may be defined by the alteration mineral assemblage. In rock-water systems equilibrated with a K-silicate phase such as muscovite, μAl_2O_3 is often defined or limited. Under the physical conditions considered here, diagnostic mineral assemblages define three categories with μAl_2O_3 increasing from K-feldspar + muscovite + quartz to muscovite + biotite + chlorite + quartz to kaolinite (or pyrophyllite) + quartz assemblages. Specific buffer reactions grouped according to μAl_2O_3 are given in Table 3. In magnetite-bearing assemblages, μFeO is related to oxygen fugacity, but its relative value is expressed by the molar (Fe/(Fe + Mg)) composition of ferromagnesian silicates in the buffer assemblage. As such, the relative variation of μAl_2O_3 and μFeO can be deduced conveniently from AFM phase equilibria diagrams (c.f. Korzhinskii, 1959, p. 80-88) or algebraically from reaction stoichiometry (c.f. Rumble, 1976) to infer the relative metal solubility relations for different mineral assemblages. For the pyrite + magnetite buffer assemblage, μH_2S is defined relative to μFeO by the reaction:

$$Py + 2FeO + 2H_2O = Mgt + 2H_2S,$$

where FeO represents ferrous iron in a silicate mineral (e.g., chlorite, biotite, etc.).

3 DISCUSSION

Increasing the chemical potential of any component drives the reaction toward the side away from that component, and decreasing the chemical potential has the opposite effect. Thus, the reactions in Table 2 indicate that increasing μAl_2O_3 or decreasing μFeO in the mineral assemblage increases the concentration of metals such as Zn, Pb, and Cu in the fluid and, in contrast, lowers the fluid content of bisulfide-complexed Au species. Although the specific mineral-buffer reactions as written in Table 3 assume equilibration of the fluid with quartz, the effect of supersaturation or undersaturation of silica in the fluid relative to quartz can be considered easily by substituting an aqueous silica component for the quartz phase in these reactions. For most buffer reactions, supersaturation of silica with respect to quartz increases the solubility of base metals such as Zn, Pb, and Cu, but decreases the solubility of Au. Under-saturation in silica with respect to quartz has the opposite effect.

The reactions describing the relative solubilities and exchange of metals for alkalis for the mineral-buffer reactions as given in Table 3 involve dehydration of hydrated silicate minerals. Large volume and entropy changes are associated with dehydration reactions with the effect that increasing temperature and decreasing pressure drive the reaction toward increased devolatization. By this effect, increasing temperature or decreasing pressure should drive the reactions toward the right side. Thus, increasing temperature or decreasing pressure generally drives the buffer reactions toward higher base-metal and lower Au concentrations in the fluid. As shown by Hemley and Hunt (1992), the compensating solubility effects of pressure and temperature change allow quasi-adiabatic fluid transport paths to maintain higher base-metal solubility over longer transport distances than fluid transport paths following geothermal gradients.

A geochemical separation of gold from the base metals can take place in hydrothermal systems buffered by muscovite + magnetite + pyrite mineral assemblages. Under low salinity conditions, resulting in low concentrations of aqueous KCl and K^+, Au will tend to form soluble complexes with reduced sulfur species while the base metals will form insoluble sulfides. Under higher salinity conditions, resulting in higher concentrations of aqueous KCl and K^+, base metals will be increasingly soluble as associated-chloride complexes while Au will be less soluble. Removal of sulfur from the hydrothermal fluid through the precipitation of iron sulfides leads to the deposition of Au transported as a reduced sulfur species while enhancing the solubility of base metals transported as associated-chloride complexes.

Gold deposition is favored in sites where μAl_2O_3 changes to a higher value and (or) μFeO changes to a lower value mineral assemblage, aqueous silica values are high, and (or) quasi-adiabatic pressure drops occur in low salinity hydrothermal systems. As such, high μAl_2O_3 mineral assemblages in shear zones in greenschist facies metamorphic rocks are a favorable deposition site for gold. Base metal deposition is favored in sites where μAl_2O_3 changes to a lower value and (or) μFeO changes to a higher value mineral assemblage, quartz is precipitating and lowering aqueous silica concentrations, and (or) high salinity hydrothermal systems are cooling.

REFERENCES

Bohlke, J.K. 1989. Comparison of metasomatic reactions between a common CO_2-rich vein fluid and diverse wall rocks: Intensive variables, mass transfers, and Au mineralization at Alleghany, California. Econ. Geol 84: 291-327.

Guilbert, J.M. and C.F. Park Jr. 1986. The geology of ore deposits. New York: W.H. Freeman, G.L. Cygan, J.B. Fein, G.R. Robinson, and W.D. d'Angelo.

Hemley, J.J. and others 1992. Hydrothermal ore-forming processes in the light of studies in rock-buffered systems: I. Iron-copper-zinc-lead sulfide solubility relations. Econ. Geol. 87: 1-22.

Hemley, J.J. & J.P. Hunt 1992. Hydrothermal ore-forming processes in the light of studies in rock-buffered systems: II. Some general geologic applications. Econ. Geol. 87: 23-43.

Korzhinskii, D.S. 1959. Physiochemical basis of the analysis of the paragenesis of minerals. New York: Consultants Bureau.

Rumble, D. III 1976. The use of mineral solid solutions to measure chemical potential gradients in rocks. Am. Mineral. 61: 1167-1174.

Seward, T.M. 1984. The transport and deposition of gold in hydrothermal systems. in Foster, R.P. (ed.), Gold '82: 165-181. Rotterdam: Balkema.

Whitney, J.A., J.J.Hemley, & F.O. Simon 1985. The concentration of iron in chloride solutions equilibrated with synthetic granite compositions: The sulfur-free system. Econ Geol. 80:444-460.

Water-Rock Interaction, Kharaka & Maest (eds) © 1992 Balkema, Rotterdam. ISBN 90 5410 075 3

Fluid-rock interaction at the Kaneuchi tungsten deposit in Japan

Yasuhiro Shibue
Geoscience Institute, Hyogo University of Teacher Education, Kato-gun, Japan

Masami Noto
Institute for Study of the Earth's Interior, Okayama University, Misasa, Tottori, Japan (Presently: Kyuden Sangyo Ltd, Fukuoka, Japan)

Kohei Kazahaya
Geological Survey of Japan, Tsukuba, Ibaragi, Japan

ABSTRACT: Fluid-rock interaction at the Kaneuchi vein-type W deposit in Japan is discussed from mineralogy and oxygen and hydrogen isotope geochemistry. Chemical compositions of tourmaline and wolframite are dependent on the chemistry of host rocks. Furthermore, oxygen isotopic compositions of vein minerals (quartz and muscovite) and host rocks show the fluid-rock interaction at a low water/rock ratio. Hydrogen isotopic compositions of the fluid, on the other hand, suggest the mixing of a modified magmatic fluid with a deterium-depleted and ^{18}O-enriched fluid at the earliest stage of mineralization.

1 INTRODUCTION

The Kaneuchi tungsten deposit, located about 45 km northwest of Kyoto in Japan, is believed (Ishihara, 1978; Shibue, 1984) to be related to Cretaceous granitic activity based on some geologic evidences (e.g., contact aureole around the deposit).

This deposit is, however, hosted by sedimentary rocks (initially shale, sandstone, and chert). Although the deposit was prospected down to 420 m below the surface, relevant igneous rocks have not been observed and are not exposed around the mining area. Then, there is a possibility that the hydrothermal fluid interacted with the sedimentary rocks during the mineralization. Present study aims to examine the possible fluid-rock interaction from mineralogy and stable isotope geochemistry of the Kaneuchi deposit.

2 OUTLINE OF MINERALIZATION AND HYDROTHERMAL ALTERATION

Tungsten minerals (wolframite and scheelite) occur in fissure-filling quartz veins where mineralization of cassiterite, arsenopyrite, chalcopyrite, sphalerite, galena, pyrite, and pyrrhotite is also observed. Main gangue minerals in quartz veins are quartz, muscovite, tourmaline, K-feldspar, calcite, and siderite. Based on the distribution of these minerals within the veins and the textural observation, mineralization was devided into two main stages (Shibue and Noto, 1991), the early and late stages. At the early stage, muscovite, tungsten minerals, tourmaline, and sulfides (in some of the veins) precipitated with transparent or milky quartz. Then, sulfide minerals (in most of the veins) and carbonate minerals precipitated, accompanying milky quartz at the late stage. Muscovite precipitated at the earliest stage, and drusy quartz as well as calcite precipitated at the latest stage, judging from their occurrences.

Hydrothermal alteration of host rocks is not obvious within and around the deposit. Sometimes, sericitization or chloritization is observed at the very contact with quartz veins but is not ubiquitous in host rocks. Fe/(Fe+Mg) ratios and Al_2O_3 contents of host rocks show no systematic changes in terms of the distance from quartz veins (Shibue, 1984).

3 TEMPERATURE OF HYDROTHERMAL FLUID

The temperature of the hydrothermal fluid is determined by fluid inclusion geothermometry in combination with sphalerite geobarometry (Hutchison and Scott, 1981), and oxygen isotope geothermometry on muscovite-quartz pair.

Homogenization temperatures of fluid inclusions in quartz (7 localities) show the range from 188° to 365°C, and the average value at each locality ranges from 252° to 277°C. These average values are corrected, using the pressure estimate (1.0 to 1.6 kb), salinities (8 to 9 wt% NaCl eq.), and CO_2/H_2O mole ratios of fluid inclusions (0.03 to 0.04). The calculated results show the temperatures to be from 309° to 402°C.

Oxygen isotopic compositions of coexisting muscovite and quartz show the equilibrium temperatures to be from 400° to 600°C when the ^{18}O fractionation equations for the muscovite-H_2O and quartz-H_2O are taken from O'Neil and Taylor (1969) and Matsuhisa et al. (1979), respectively.

4 CHEMICAL COMPOSITIONS OF TOURMALINE AND WOLFRAMITE

Because tourmaline occurs not only in quartz veins but also in host rocks, the extent of fluid-rock interaction could be viewed by comparing the chemistry of this mineral in veins and in host rocks. Tourmaline in quartz veins is usually found near the border zone to host rocks, and is often associated with tungsten minerals. In host rocks, tourmaline sporadically occurs, but tourmaline aggregates are concentrated in several outcrops.

Li_2O contents of vein tourmaline are less than 0.2 wt%. Electron microprobe analyses on major cations (Shibue, 1984) revealed that tourmaline from both quartz veins and host rocks belongs to schorl-dravite series. It is noteworthy that vein tourmaline contains significant magnesium compared with this mineral from many granitic zones (e.g., Deer et al., 1963). Furthermore, frequency diagrams on FeO and MgO contents (Fig. 1) show that tourmaline in quartz veins has the similar composition to that in host rocks although the former is slightly enriched in FeO and depleted in MgO compared with tourmaline in host rocks.

Fe/(Fe+Mn) ratios of wolframite are dependent on the near-by host rock lithology; wolframite occurring near to chert shows lower Fe/(Fe+Mn) ratios than wolframite occurring near to shale. The ratios are, on the other hand, independent of the vein location and the depth of the sample locality.

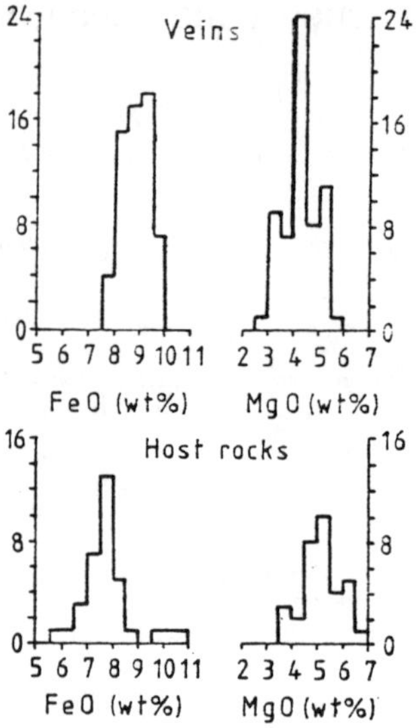

Fig.1 Frequency diagrams of FeO (total iron as FeO) and MgO wt% of tourmaline.

Bulk chemical analyses of chert show Fe/(Fe+Mn) ratios to be 0.77 (average of 8 samples) whereas the average ratio of shale (25 samples) is 0.95. Thus, the wolframite composition shows the dependence of the host rock chemistry.

Mg-bearing and Li-poor tourmaline from quartz veins suggests that the hydrothermal fluid is not directly derived from a magmatic fluid. The similarity in chemical composition between tourmaline in quartz veins and in host rocks, then, suggests that the hydrothermal fluid interacted with the host rocks at a low water/rock ratio. The dependence of wolframite composition on the host rocks and the narrow extent of alteration zones in host rocks support the fluid-rock interaction at a low water/rock ratio.

5 OXYGEN ISOTOPIC COMPOSITIONS OF MINERALS AND HOST ROCKS

Oxygen isotopic compositions of vein quartz (26 localities) and muscovite (11 localities) range from 13.4 to 15.9 per mil (relative to SMOW) and from 11.7 to 13.8 per mil, respectively. Using the ^{18}O fractionation equations for the quartz-H_2O and muscovite-H_2O, oxygen isotopic compositions of the hydrothermal fluid are obtained to be from 8.1 to 11.9 per mil at quartz precipitation (350° to 400°C) and 10.3 to 14.6 per mil

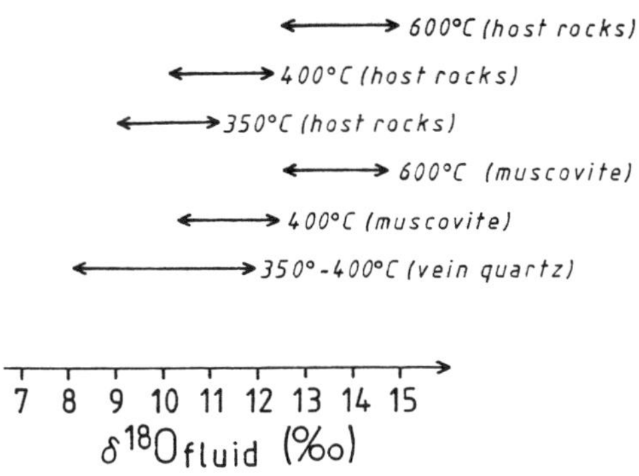

Fig.2 Calculated $\delta^{18}O$ of the hydrothermal fluid from the isotopic compositions of host rocks, vein muscovite, and vein quartz at the labeled temperatures.

at muscovite precipitation (400° to 600°C).

Host rocks (4 samples from shale, sandstone, and chert) show the oxygen isotopic compositions ranging from 12.8 to 15.7 per mil. Based on the weight fractions of the constituent minerals (quartz, muscovite, albite, and chlorite) obtained from the bulk chemical compositions and the fractionation equations for the albite-H_2O (Matsuhisa et al., 1979) and chlorite-H_2O (Wenner and Taylor, 1971), fractionation factors for rock-H_2O are calculated from the weighted mean of the fractionation factors for the constituent minerals-H_2O. The calculated isotopic compositions of the fluid equilibrated with the host rocks at a low water/rock ratio ($\delta^{18}O_{fluid} = \delta^{18}O_{rock} - \Delta_{rock-fluid}$) agree well with those obtained from vein quartz and vein muscovite (Fig. 2).

Therefore, the hydrothermal fluid, even if it originated from a magmatic fluid, was modified in its cation and oxygen isotopic compositions through fluid-rock interaction at the low water/rock ratio.

6 HYDROGEN ISOTOPIC COMPOSITIONS OF MUSCOVITE, HOST ROCKS, AND FLUID INCLUSIONS

Hydrogen isotopic compositions of vein muscovite range from -61 to -104 per mil (relative to SMOW), and those of host rocks range from -88 to -89 per mil. Fluid inclusions in vein quartz and drusy quartz, on the other hand, show the δD values to be from -53 to -73 per mil and -53 to -63 per mil, respectively.

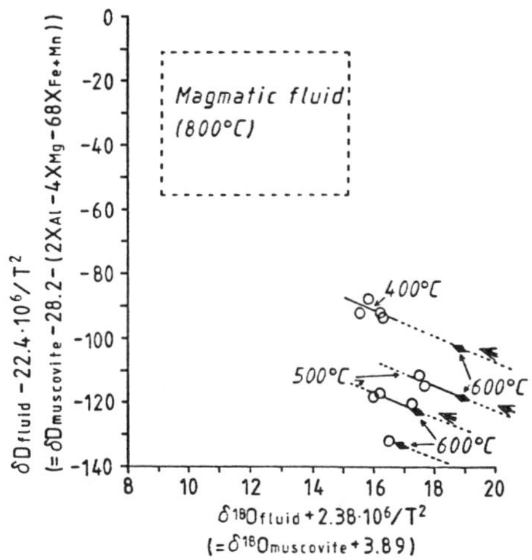

Fig.3 Variations of $\delta^{18}O + 2.38X10^6/T^2$ and $\delta D - 22.4X10^6/T^2$ in fluids. Open circles show the calculated values from the isotopic compositions of muscovite. Lines show the possible regressions for the changes of the fluid isotopic compositions. The slope of the lines is determined by temperature and ^{18}O fractionation factors for rock-H_2O.

The δD values of the hydrothermal fluid calculated from muscovite show a large variation even when the chemical compositions of muscovite and the temperature changes are taken into account. Using the D fractionation equation for the muscovite-H_2O (Suzuoki and Epstein, 1976) and the fractionation equation of ^{18}O for the muscovite-H_2O, the D and ^{18}O values of muscovite can be explained as the cooling of the hydrothermal fluids with different hydrogen isotopic compositions (Fig. 3). The fluids are deterium-depleted and ^{18}O-enriched, compared with the accepted values of the magmatic fluid (Ohmoto, 1986). The oxygen shift is possibly due to the fluid-rock interaction, but the hydrogen shift suggests other events because the δD values of host rocks are not comparable to those of the fluids. Thus, the modified magmatic fluid possibly mixed with a deterium-depleted and ^{18}O-enriched fluid at the earliest stage. Fluid inclusions in quartz, on the other hand, suggest that such a mixing did not cause the significant hydrogen shift at the main stage of quartz mineralization.

7 CONCLUSIONS

Fluid-rock interaction at the Kaneuchi vein-type tungsten deposit in Japan is discussed from the chemical compositions of tourmaline and wolframite and the stable isotope geochemistry of minerals (vein quartz and muscovite), host rocks, and fluid inclusions. The hydrothermal fluid, possibly originated from a magmatic fluid, interacted with the host sedimentary rocks at a low water/rock ratio, and was modified in its cation and oxygen isotopic compositions. Hydrogen isotopic compositions of the fluid, on the other hand, suggest that the modified magmatic fluid mixed with a deterium-depleted and ^{18}O-enriched fluid at the earliest stage of mineralization.

REFERENCES

Deer, W.A., Howie, R.A. & Zussmann, J. 1963. Rock Forming Minerals. vol. 5. London: Longman.

Hutchison, M.N. & Scott, S.D. 1981. Sphalerite geobarometry in the Cu-Fe-Zn-S system. Econ. Geol., 76: 143-153.

Ishihara, S. 1978. Metallogenesis in the Japanese island arc system. J. Geol. Soc. London. 135: 389-406.

Matsuhisa, Y., Goldsmith, J.R. & Clatyon, R.N. 1979. Oxygen isotopic fractionation in the system quartz-albite-water. Geochim. Cosmochim. Acta. 43: 1131-1140.

Ohmoto, H. 1986. Stable isotope geochemistry of ore deposits. In J.W. Valley, H.P. Taylor, Jr. & J.R. O'Neil (eds.), Stable isotopes in high temperature geological processes: 491-559. Washington: Mineralogical Society of America.

O'Neil, J.R. & Taylor, H.P.Jr. 1969. Oxygen isotope equilibrium between muscovite and water. J. Geophys. Res. 74: 6012-6022.

Shibue, Y. 1984. Chemical compositions of tourmaline in the vein-type tungsten deposits of the Kaneuchi mine, Japan. Mineralium Deposita 19: 298-303.

Shibue, Y. & Noto, M. 1991. Mineralization and oxygen isotopic compositions of vein quartz from the Kaneuchi tungsten deposit, Kyoto Prefecture, Japan. Mining Geol. 41: 15-24.

Suzuoki, T. & Epstein, S. 1976. Hydrogen isotope fractionation between OH-bearing minerals and water. Geochim. Cosmochim. Acta. 40: 1229-1240.

Wenner, D.B. & Taylor, H.P.Jr. 1971. Temperatures of serpentinization of ultramafic rocks based on $^{18}O/^{16}O$ fractionation between coexisting serpentine and magnetite. Contr. Mineral. Petrol. 32: 165-185.

Water-Rock Interaction, Kharaka & Maest (eds) © 1992 Balkema, Rotterdam. ISBN 90 5410 075 3

Isotope geochemistry of oxygen and hydrogen for ore-forming hydrothermal solution in Xiangshan uranium orefield, China

Sun Zhanxue, Li Xueli & Shi Weijun
East China Geological Institute, Fuzhou, Jiangxi, People's Republic of China

ABSTRACT: This paper expounds the isotope geochemistry of oxygen and hydrogen for ore-forming hydrothermal solution in Xiangshan uranium ore-field, southeastern China, and reveal that water source of the solution is meteroic water. Its δD and $\delta^{18}O$ values are −60.78～−80.74‰ and −6.5～+1.83‰ (SMOW) respectively. Water-rock interaction researches show that the ore-forming fluid was produced by paleometeoric water-rock interaction during the mineralization epoch. Reconstruction of oxygen and hydrogen isotopic composition of the paleo-precipitation supports the conlusion.

1 INTRODUCTION

Xiangshan ore-field is one of the most famous volcanic-rock-type uranium ore fields in China. Geological structures and hydrothermal alterations are well developed in the area. Although geoscientists from different institutes, colleges and universities at home and abroad have studied the ore-field, the genesis of its ore-forming solution has been in dispute for more than 30 years.

In order to find out the formation mechanism of the uranium deposits in the field, isotopic samples were collected from rocks, alteration and vein minerals. In this paper, we concentrated on the water source of the ore-forming solution on the basis of isotopic data of oxygen and hydrogen, and models of water-rock interaction during the mineralization epoch as well.

2 GEOLOGICAL SETTING

2.1 Rocks

Xiangshan uranium ore-field is located in Jiangxi province, southeastern China. It occurs in a volcanic basin. Basement rocks of the basin are predominately Precambrian epometamorphic rocks, such as phyllite and biotite-quartz schists, while cover rocks are upper Jurassic volcanic rocks, mainly porphyroclastic lava, secondly volcaniclastic rocks. Postvolcanic subvolcanism and vein stone activities are also found in the basin.

2.2 Structure

In Xiangshan volcanic basin, which belongs to Jiangxi-Hangzhou Tectonic Zone, geological structures consist of NE-trending regional faults and volcanic tectonics. The volcanic structure is a typical caldera in which volcanic collapses, conduit, inter-layer tension fractures and folds are developed. The combination of these structures provides a pathway and depositional space for the ore-forming solution.

2.3 Mineralizations

Two stages of uranium mineralizations have been recognized in the ore-field. The first stage is alkalic-metasomatic type mineralization and the second is fluorite-hydromica type mineralization. They are distributed along NE-trending faults, and the margin of the volcanic basin. The alkalic metasomatic type of mineralization occurs in the north, and the

• The Project Supported by National Natural Science Foundation of China.

fluorite-hydromica type mineralizations occur in the west. Generally speaking, uranium exists in forms of pitchblede and thoropitchblede, or in dispersion and adsorption states in the ore-field. Mineralization age in the ore-field is about 100 millions years, that is the mineralization took place in the Mesozoic Period.

3 ISOTOPIC COMPOSITIONS OF OXYGEN AND HYDROGEN

3. 1 Sampling and analysis

Samples were collected from the major hydrothermal alteration zones, and altered minerals such as albite, calcite, fluorite and hydromica were selected to determine isotopic composition. Isotopic compositions of oxygen and hydrogen for the ore-forming solution in the ore-field have been obtained by analysing directly inclusion water of minerals, or by calculation based on mineral-liquid isotopic equilibrium fractionations.

Table 1. Isotopic compositions of oxygen and hydrogen for minerals and inclusion water in xiangshan ore-field.

Sample	$\delta^{18}O_M$(‰) (SMOW)	δD_M (SMOW)	Temp (℃)	$\delta^{18}O_W$(‰) (SMOW)	δD_W(‰) (SMOW)	Note
Calcite	8. 69		290	+3. 28		1
Calcite	11. 17		271	+5. 12		1
Calcite	7. 37		276	+1. 51		1
Calcite	3. 79		275	−2. 08		2
Hydromica	13. 539	−69. 553	200	+1. 24	−78. 40	2
Hydromica	12. 614	−71. 900	200	+0. 31	−80. 74	2
Calcite	9. 11		110	−6. 47		3
Calcite	14. 90		120	+0. 18		3
Calcite	13. 78		103	−2. 58		3
Albite	5. 82		300	−0. 01		1
Hydromica	9. 82	−65. 90	240	−3. 77	−71. 27	1
Hydromica	14. 129	−53. 207	200	+1. 83	−62. 05	2
Hydromica	5. 797	−51. 933	200	−6. 5	−60. 78	2
Hydromica	10. 883	−68. 611	200	−1. 42	−77. 45	2
Fluorite			124		−87. 14	2
Calcite	+5. 20		123	−9. 14	−57. 07	3

Note: 1—pre-metallogenic epoch; 2—metallogenic epoch; 3—post-metallogenic epoch.

3. 2 Results

The values of δD and $\delta^{18}O$ for the hydrothermal solution in the ore-field are given in table 1.

The $\delta^{18}O$ values of the hydrothermal solution in the first mineralization stage range from +5.12~ −6. 47‰, of which the values in pre-metallogenic epoch, metallogenic epoch and post- metallogenic epoch are +5. 12~+1. 51, +1. 24~−2. 08‰ and +0. 18~−6. 47‰ respectively. In the second mineralization stage, the $\delta^{18}O$ values are in the range of +1. 83~−9. 14‰ , of which the values in pre-metallogenic epoch, metallogenic epoch and post-metallogenic epoch are −0. 01~−3. 77‰, +1. 83 ~−6. 5‰ and −9. 14‰ respectively.

The δD values of the hydrothermal solution for the two stages of mineralization range from -60 ~-80 ‰ (SMOW).

4 WATER SOURCE OF ORE- FORMING SOLUTION

Finding the water source is a key to the settlement of the source for the ore-forming solution because water is the essential constituent of ore-forming fluid. The isotopic compositions of oxygen and hydrogen for the ore-forming solution in Xiangshan uranium ore-field (Table 1) have been plotted in Figure 1. It is shown that the isotopic compositions of the solution are roughly in accordance with meteoric water, and oxygen isotope shifts are obviously found. However, because isotopic exchange occurs during the water-rock interaction period, these isotopic data, especially $\delta^{18}O$ values, are different from the original isotopic compositions of the water. Obviously, the water source of the ore-forming solution can not be ascertained in such a simple way.

According to principles of isotopic geochemistry the magnitude of change of original isotopic concentrations of oxygen for the ore- forming solution mainly depends on the following factors:

1. δ^{18} values of host rocks and original water.
2. Water/Rock ratios.
3. Temperature.

Therefore, in order to determine the origin of the water for the ore-forming solution, the alteration system must be studied. In fact, during the mineralization epoch, hydrothermal alterations, such as albitization and hydromica-fluoritization were well developed in the ore-field, and resulted in a decrease of $\delta^{18}O$ values of altered rocks to a large extent. The $\delta^{18}O$ values of original rocks and altered rocks are given in Table 2 and are plotted in Figure 2. It is clear that the original water is meteoric water with low $\delta^{18}O$ values.

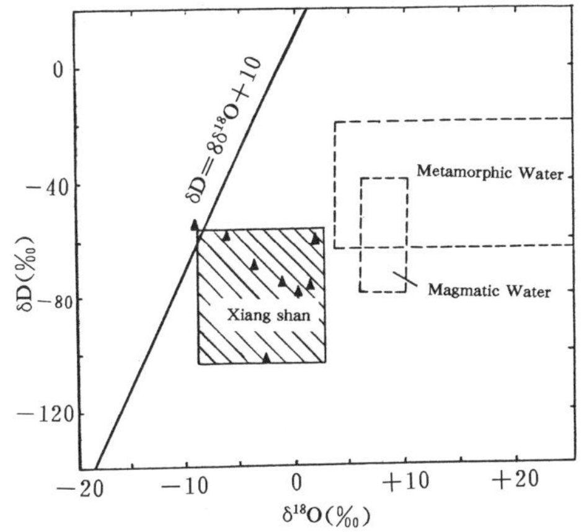

Fig. 1 Oxygen and hydrogen isotopic composition for ore-forming solution in Xiangshan ore-field.

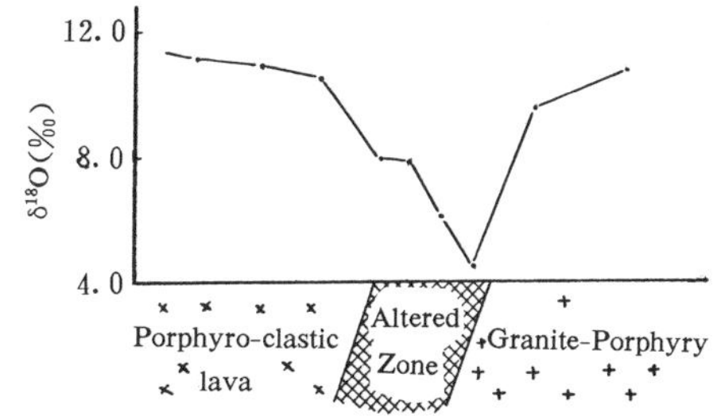

Fig. 2 The $\delta^{18}O$ variation trend in the alteration system.

Table 2. The $\delta^{18}O$ values of original rocks and altered rocks in Xiangshan ore-field.

№	Sample	$\delta^{18}O$ (‰,SMOW)
1	Porphyroclastic lava	11.30
2	Porphyroclastic lava	11.10
3	Porphyroclastic lava	10.60
4	Altered porphyroclastic lava	7.8
5	Altered porphyroclastic lava	7.9
6	Altered porphyroclastic lava	6.0
7	Altered Granite-porphyry	4.42
8	Granite-porphyry	9.40
9	Granite-porphyry	9.54
10	Granite-porphyry	11.60
11	Porphyrogranirte	8.73
12	Porphyroclastic lava	10.95
13	Porphyroclastic lava	10.48

Furthermore, extensive studies of oxygen-18 and deuterium of ore-forming fluids originating from meteoric water by Zhang (1989) have shown that the δD and $\delta^{18}O$ values of Mesozoic meteoric water are systematically lower than those of modern meteoric water in eastern China by about 20‰ and 2‰ respectively. This fact may be consistent with the phenomenon that the east coastline of Asia in the Mesozoic was probably further east than the present one. In addition, the $\delta^{18}O$ and δD values of modern meteoric water are about -8‰ and -50‰ respectively (Li et al., 1988) in Xiangshan area. Based on the above, the δD and $\delta^{18}O$ values of the paleo-meteoric water during the mineralization epoch in the area can be considered to be about −70‰ and −10‰ respectively. The δD value is roughly in accordance with the mean δD value −74.4‰ of the ore-forming solution in the metallogenic epoch (Table 1). The evolution of oxygen-18 for the ore-forming solution depends upon water-rock interactions and Water/Rock ratios. It is known from the theory of matter equilibrium that isotopic exchanges during the water-rock interaction abide by the following equation:

$$W \cdot \delta_W^{i\,18}O + R \cdot \delta_R^{i\,18}O = W \cdot \delta_W^{f\,18}O + R \cdot \delta_R^{f\,18}O \qquad (1)$$

where, i and f represent the initial value and the final value respectively; W means water; R means rock. The equation (1) can be written in the form

$$W/R = (\delta_R^{f\,18}O - \delta_R^{i\,18}O)/[\delta_W^{i\,18}O - (\delta_R^{f\,18}O - \Delta)] \qquad (2)$$

where W/R expresses in atom unit, Δ is the rock-water isotopic fractionation value when the plagioclase (An = 30) is regarded as the approximation of the whole rock:

$$\Delta_{R-W} = \delta_{PL}{}^{18}O - \delta_W{}^{18}O = 2.68 \times 10^6 T^{-2} - 3.53 \qquad (3)$$

For different W/R ratios, the $\delta^{18}O$ evolution status of the ore- forming hydrothermal solution can be obtained when the mineralization temperature is taken as the temperature of equilibrium exchange. Assuming that $\delta_W^{i\,18}O = -10‰$, and $\delta_R^{i\,18}O = 10.5‰$ (mean value of original rocks in Table 2), the $\delta^{18}O$ value of the hydrothermal solution after meteoric water-rock exchanged in equilibrium can be calculated and the results are given in Table 3 (W/R ratios expressed in weight ratios).

Table 3. The $\delta^{18}O$ values (‰) of the hydrothermal solution formed by meteoric water- rock exchange reaction in equilibrium for different W/R ratios in the Xiangshan ore-field.

W/R	0.001	0.01	0.05	0.1	0.5	1.0	10.0
$\delta_W{}^{18}O$	+1.93	+1.72	+0.86	−0.04	−4.03	−6.02	−9.43

It is clear that the oxygen isotopic composition of the hydrothermal solution (shown in Table 3) is in accordance with that of the ore-forming solution in the ore-field (as shown in Table 1). Obviously, the ore-forming solution was probably produced by paleo-meteoric water-rock interactions.

In addition, in order to find out magmatic water's contrbution to the ore-forming solution in the ore-field, assuming that $\delta^{i}_{MW}{}^{18}O$ and $\delta^{i}_{R}{}^{18}O$ equal to 7.5‰ and 10.5‰ respectively, the oxygen isotopic composition of the hydrothermal solution formed by magmatic water-rock exchange reaction in equilibrium can be calculated according to equation (2). The results are given in Table 4.

Table 4. The $\delta^{18}O$ values (‰) of the hydrothermal solution formed by magmatic water-rock exchange reaction in equilibrium for different W/R ratios in Xiangshan ore-field.

W/R	0.001	0.01	0.05	0.1	0.5	1.0	10.0
$\delta_W{}^{18}O$	+2.07	+2.12	+2.36	+2.63	+3.76	+4.35	+5.34

Table 4 clearly shows that the magmatic water can not be solely evolved into the ore-forming solution in the ore-field. Although the ore-forming solution during the pre-metallogenic epoch with $\delta^{18}O$ values from -3.77‰ to +5.12‰ undoubtedly contained some magmatic water, and the ore-forming solution during the metallogenic epoch with $\delta^{18}O$ values from -6.5‰ to 1.83‰ obviously originated from paleo-meteoric water that had been altered by reaction with rock minerals. Studies of paleo-geothermal conditions in the Xiangshan ore-field (Zhou, 1988) supports the conclusion. The paleo-geotemperature gradient in the ore-field is 40～80 ℃/Km, the ore-forming solution can be formed by deep circulating meteoric water at the depth of 2～3 kilometers.

5 CONCLUSIONS

In Xiangshan uranium ore-field, the δD values of the ore-forming solution are in accordance with those of paleometeoric water, alteration systems suggest that rocks had been altered by low $\delta^{18}O$ value thermal water, and the $\delta^{18}O$ values of the hydrothermal solution formed by paleometeoric water-rock exchange reaction in equilibrium are in the range of those for the ore-forming solution. We can reach the conclusion that the water source of the ore-forming hydrothermal solution in the uranium ore-field mainly originated from paleometeoric water. Probably, it is not right to use mixing between magmatic water and meteoric water to account for positive $\delta^{18}O$ values of ore-forming solution, especially for the $\delta^{18}O$ values that are below +2‰, which often resulted from meteoric water-rock interactions.

REFERENCES

Li X. L. Shi W. J. 1988. A isotope study of Lushan hotspring and its genesis, In IAEA (eds.), Isotope applications in Pacific Area. p. 277—281. Beijing: BIUG.

Zh ang, L. G. 1989. Petrogenetic and minerogenetic theories and prospecting—— stable isotopic geochemistry of main type ore deposits and granitoids of China. Beijing: Press of Beijing University of Technology.

Zh ou W. b. 1988. The heat flow and analysis of fossil hydrothermal system in Xiangshan area. Fuzhou: East China Geological Institute.

Water-Rock Interaction, Kharaka & Maest (eds) © 1992 Balkema, Rotterdam. ISBN 90 5410 075 3

Fluid-rock interaction and cassiterite deposition

T.M. Sushchevskaya
Vernadsky Institute of Geochemistry and Analytical Chemistry, Moscow, Russia

M.V. Borisov
Moscow State University, Russia

ABSTRACT: The formation of tin mineralization is coupled with influx of large masses of meteoric waters whose composition has undergone an essential change in the process of water-rock interaction. Mixing of thses more alkaline exchange waters with more acid chloride reduced pre-ore solutions makes possible the redox reactions giving rise to cassiterite deposition.

1 REDOX REACTIONS RESULTING IN CASSITERITE DEPOSITION

The correlation of mineralogic observations with stable isotope and fluid inclusion data shows that the formation of Sn hydrothermal mineralization is connected with redox reactions with participation of such elements as C,S,As,Fe (Bannikova et al. 1984, Eugster and Wilson 1985, Heinrich and Eadington 1986, Dubessy et al. 1987). The most probable reactions, involving S(VI) and As(III) as oxidizing reagents and resulting in cassiterite deposition are:

$$4SnCl_2^o + HSO_4^- + 4H_2O \Rightarrow 4SnO_2 + HS^- + 8HCl^o \quad (1)$$

and

$$3SnCl_2^o + 2H_3AsO_3 + 2FeCl^+ + 2HS^- \Rightarrow 3SnO_2 + 2FeAsS + 8HCl^o \quad (2).$$

The intervals of total concentrations of sulfide (Sushchevskaya et al. 1988) and As ($10^{-4} \div 10^{-2}$ m) in fluid inclusion solutions analysed by potentiometric and spectroscopic (AAS) methods correlates with those of Sn, as it follows from data on cassiterite solubility at high T,P (Kovalenko et al. 1985, Wilson and Eugster 1985, Jackson and Helgeson 1985).

2 δD AND $\delta^{18}O$ DATA FOR TIN-BEARING FLUIDS

The study of isotopic composition of oxygen and hydrogen of water of fluid inclusions in minerals from two commercially remarkable, typical Sn deposits (Far East, Russia) indicates with certainty the presence of meteoric water at the productive stage.

The values of δD -109 to -137‰, $\delta^{18}O$ 6.0 to 2.1‰ have been found for the water of fluids, which formed the productive (Sn) mineral association at the Solnechnoye deposit, in Komsomolsk ore region.

The values of δD -102 to -119‰, $\delta^{18}O$ 7.2 to 2.5‰ were found for the fluids, which had formed the quartz-muscovite-cassiterite-wolframite association (300-350°C) in the veins and greisens of the Iultin Sn-W deposit (Chukotka Region).

3 CHEMICAL EVOLUTION OF TIN-BEARING FLUIDS

According to cryometric and bulk chemical data of fluid inclusions, the change from essentially chloride to chloride-bicarbonate and less reduced solutions was typical for studied deposits. The same trend of the chemical evolution of mineral-forming solutions has been described for many ore forming hydrothermal systems. The fact may be explained by influx of exchange waters of initially meteoric genesis into such hydrothermal systems (Sushchevskaya et al. 1990, Heinrich 1990).

Analysis of the paleohydrogeologic environment of the Komsomolsk ore region, showed that the early stage of orogenic process was characterized by outpressing of sodium chloride connate waters from deep zones. Then the waters infiltrated by following tectonic fractures, and entry to the hydrothermal systems essentially increased. The chemical composition of infiltration waters gradually changed from chloride to bicarbonate type.

4 COMPOSITION OF SOLUTIONS, ARISING FROM WATER-ROCK INTERACTION

The chemical features of exchange waters may be evaluated by calculation of fluid/rock interaction. The series of the calculations has been carried out using GIBBS program, for natural water, water solutions and rocks of different composition. It was found that pH values of the solutions, equilibrated with silicate rocks for rather high rock/water (R/W) ratios (Figure 1, A,B,C) were higher than those of chloride solutions in equilibrium

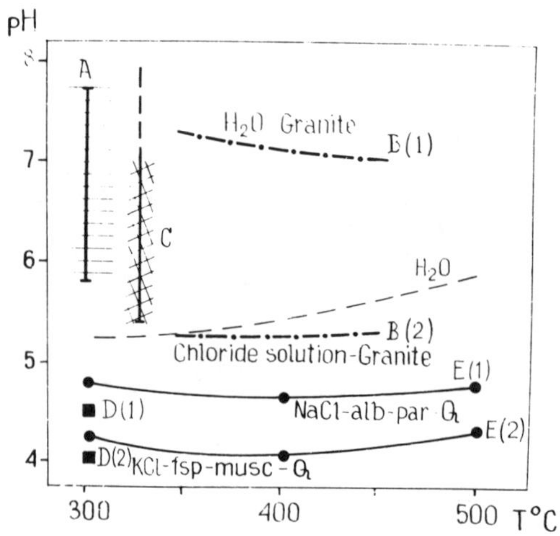

Fig.1 Schematic diagram showing the variations of acidity of the solutions, as a result of fluid/rock interaction A - interval of calculated pH values for solutions in equilibrium with the rocks of different composition (Ryshenko et al. 1977, Borisov et al. 1984);

B(1) - water-granite (Akchatau massif,Kazakhstan), R/W=10;

B(2) - granite-chloride solution (1.5 m KCl+0.5 m NaCl), R/W=10;

C - interval of calculated pH values for fluid inclusions solutions in quartz and cassiterite of studied tin deposits;

D(1) - quartz-0.1 m CO_2-chloride solution (1.5 m NaCl+0.5 m KCl);

D(2) - quartz-1 m CO_2-chloride solution (0.75 m NaCl+0.25 m KCl);

E(1) - albite-paragonite-quartz-NaCl(1 m) solution;

E(2) - feldspar-muscovite-quartz-KCl(1 m) solution.

with quartz (Figure 1, D) and with three mineral associations (quartz-albite-paragonite; quartz-feldspar-muscovite) (Rafalsky 1987). The pH values of the exchange solutions are close to those of tin-bearing solutions (calculated from bulk chemical composition of fluid inclusion solutions in minerals of the productive stage of studied deposits, Figure 1, C).

5 CONCLUSIONS

The penetration of large masses of exchange waters into tin-forming hydrothermal systems is quite possible and inevitable under specific hydrogeologic conditions (Baskov 1976, Jackson et al. 1989). The mixing of exchange (external) waters with pre-ore acid fluids should result in noticeable increase in pH (and fo_2) values. This mechanism is considered to maintain the shift of the reactions (1,2) to the right, thus providing the formation of sufficient SnO_2 masses (tin ores).

The widely cited explanation that the neutralization of early (magmatic) reduced acid takes place during the formation of altered wall rocks is correct and necessary, but not sufficient. At many Sn deposits (i.e. at the Solnechnoye) the wall rock alteration (tourmalinization, chloritization, sericitization) precedes the cassiterite ores formation being the very first step this direction.

REFERENCES

Bannikova, L.A. et al. 1984. The role reactions involving organic matter in producing tin deposits. Geochemistry International 21: 105-115.

Baskov, E.P. 1976. Paleohydrogeologic analysis in metallgenic studies. Leningrad: Nedra (in Russian).

Borisov, M.V., Ryzhenko, B.N. and Kraynov, S.R. 1984. Influence of acid-basic properties of rocks on a composition of the equilibrated aqueous solution. Geokhimija 5: 705 (in Russian).

Dubessy, J. et al. 1987. Physical and chemical controls (fo_2,T,pH) of opposite behaviour of W and Sn-W as examplified by hydrothermal deposits in France and Great Britain, and solubility data. Bull.Mineral. 110: 261-281.

Eugster, H. and Wilson, G. 1985. Transport and deposition of ore-forming elements in hydrothermal systems associated with granites. In C.Halls, Ed. High heat production granites, hydrothermal circulation and ore genesis. Inst.of Mining and Metallurgy, London.

Heinrich, C.A. 1990. The chemistry of hydrothermal Tin (-Tungsten) ore deposition. Economic Geology 85: 457-481.

Heinrich, C.A. and Eadington, P.J. 1986. Thermodynamic predictions of the hydrothermal chemistry of Arsenic and their significance for the paragenetic sequence of some cassiterite-arsenopyrite base metal sulfide deposits. Economic Geology 81: 511-529.

Jackson, K.J. and Helgeson, H.C. 1985. Chemical and thermodynamic constraints on the hydrothermal transport and deposition of tin. Economic Geology 80: 1365-1378.

Kovalenko, N.I. et al. 1986. Experimental study of cassiterite dissolution in supercritical solutions of HCl, HCl+NaCl(KCl) at fixed redox conditions. Geokhimija 2: 190-205 (in Russian).

Rafalsky, R.P. 1987. On the problem of acidity of hydrothermal solutions. Geokhimija 3: 402-415 (in Russian).

Ryzhenko, B.N., Melnikova, G.L. and Shvarov, Y.V. 1977. The main features of formation of chemical composition of water solutions in Earth Crust. Geochemistry International, Vol.14, No.3: 93-104.

Sushchevskaya, T.M. et al, 1984. On the problem of acidity of hydrothermal solutions at cassiterite deposition. Geochemistry International 21: 77-87.

Sushchevskaya, T.M. et al. 1991. Isotopic composition of hydrogen of water from fluid inclusions in quartz in connection of study of tin mineralization. Geokhimija 5: 737-742 (in Russian).

Water-Rock Interaction, Kharaka & Maest (eds) © 1992 Balkema, Rotterdam. ISBN 90 5410 075 3

Si, Al, and W behavior in HF aqueous solutions in relation to the greisen problem

George P. Zaraisky
Institute of Experimental Mineralogy, Russian Academy of Sciences, Chernogolovka, Russia

ABSTRACT: The solubility of the monocrystals of corundum and wolframite in HF aqueous solutions from 10^{-4} to 10 mole/kg H_2O concentration at 300 to 600°C and 1 kbar is studied using the methods of weight loss and chemical analysis of the quenching solution. Mass-transfer of SiO_2 and Al_2O_3 at experimental acidic metasomatism is studied on a SEM with X-ray analyzer. It is shown that the extensive mobility of Al_2O_3 during greisenization may occur in the presence of the acidic fluorine solutions, whereas SiO_2 is redeposited due to the oversaturated acidic solutions after their interaction with feldspars.

1 INTRODUCTION

Greisen formation in granites and other rocks is often accompanied by the redeposition of a great amount of silica, alumina and other components for considerable distances. We performed quantitative calculations of gain and loss of components at formation of one of the main subvertical greisen-vein bodies of the W-Mo Akchatau deposit in Central Kazakhstan. A distinct metasomatic zoning manifests itself in the change of the internal zone of the quartz greisen by the symmetric zones of the quartz-topaz, quartz-muscovite greisen and greisenized granite (Fig. 1).

The calculation based on the data of the furrow sampling (Gotman and Malakhova 1965) showed that at greisenization of the Akchatau granites the gain of major components predominates on the whole over the loss. Silica gain in different greisen zones ranges from +33 to +440 kg/m^3; modal quartz and the total amount of the gained SiO_2 increase with depth. Alumina shows the opposite behavior. At the upper level it is gained in the amount +7 to +55 kg/m^3. With depth Al_2O_3 content gradually decreases in each zone. At the lower level of 240 m alumina loss changes considerably from -100 to -225 kg/m^3 in different zones.

Usually in metasomatic processes alumina shows high inertness. The increase in alumina mobility in Akchatau may be due to the increase of its solubility under the effect of acidic fluorine solutions, typical for greisen deposits. With this

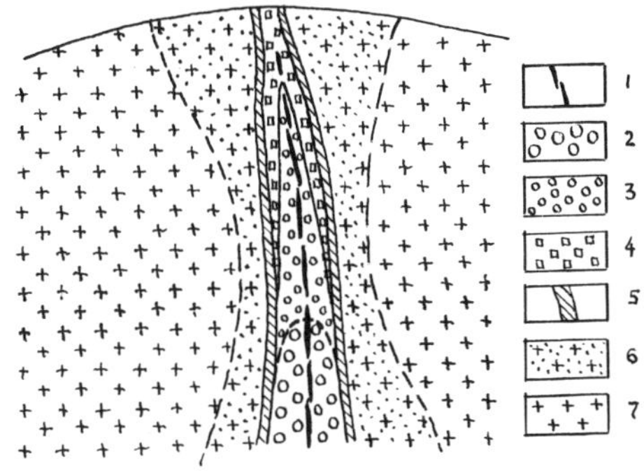

Figure 1 - Schematic greisen zoning at Akchatau (1) quartz vein, (2) porous quartz greisen, (3) dense quartz greisen, (4) quartz-topaz greisen, (5) quartz-muscovite greisen, (6) greisenized granite, and (7) leucogranite.

purpose we studied corundum solubility in the HF aqueous solutions. The degree of control of fluorine on ore formation and wolframite solubility was also estimated. The results of our experiments on modeling of greisenization and other available data on solubility of quartz and corundum in fluorine and chloride solutions are taken for a broader consideration of the greisenization problem.

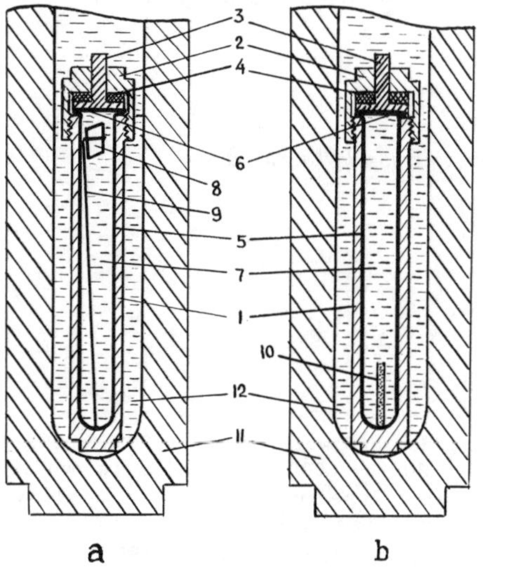

Figure 2 - Apparatus used for experiments on mineral solubility (a) and modeling of diffusion metasomatic zoning (b): (1) container, (2) closure nut, (3) container cap, (4) thermo-sealing washer, (5) gold tube, (6) gold cap, (7) HF solution, (8) crystal of mineral, (9) platinum sample holder, (10) open gold capsule with compacted powder of granite, (11) autoclave, and (12) distilled water.

2 EXPERIMENTAL TECHNIQUES

The experiments on mineral solubility were carried out in autoclaves in closed stainless steel containers lined with gold (Fig. 2a). Sealing of the container was ensured by the closure nut and ring washer made of stainless steel with a larger thermal expansion as compared to the material of the container. The internal volume of the container is 23 cm^3. The pressure was given by the loading factor of the container with HF aqueous solution and was equal to the compensating pressure of the aqueous vapor in the autoclave. Crystal of synthetic corundum (1-2 g in weight) and natural wolframite with 56 mol % of hubnerite (2-4 g in weight) were used in the experiments. Crystals were located in the upper part of the container on the stand of platinum wire. HF concentration used ranged from 10^{-4} to 10 mol/kg H_2O.

The experiments on the wolframite solubility were carried out in presence of NNO buffer being placed into the container in platinum capsule. The solubility was determined by using the monocrystal weight loss method (accuracy $n \cdot 10^{-5}$ g) and chemical analysis of the solution after the experiment (colourimetric method of determination of Al and W and atomic-absorption analysis of Fe and Mn). All the experiments were conducted at a constant pressure of 1 kbar in the temperature range 300 to 600°C and duration of 2 to 6 weeks. For experiments modeling the diffusion greisenization a small gold open capsule (diam = 3 mm, l = 25 mm) filled with compacted powder of the leucocrate granite was placed into the above described pressure vessel lined with gold (Fig. 2b). The interaction of the rock with the supercritical aqueous fluid occurred by means of diffusion through the open end of the capsule. Infiltration metasomatism was studied using a special infiltration apparatus with fluid flow (0.5 cm^3/hr) through granite powder compacted in a gold tube. Pressure in the experiments was 0.8 kbars, temperature of 400 and 500°C, the duration of 1 to 2 weeks. After the experiments the newly-formed metasomatic column was studied by microprobe (SEM with an energy dispersive X-ray analyzer). The method, equipment and the results of the experiments are given in more detail in Zaraisky (1989) and Zharikov and Zaraisky (1991).

3 RESULTS AND DISCUSSION OF THE MINERAL SOLUBILITY EXPERIMENTS

Figure 3 presents a comparison of our and

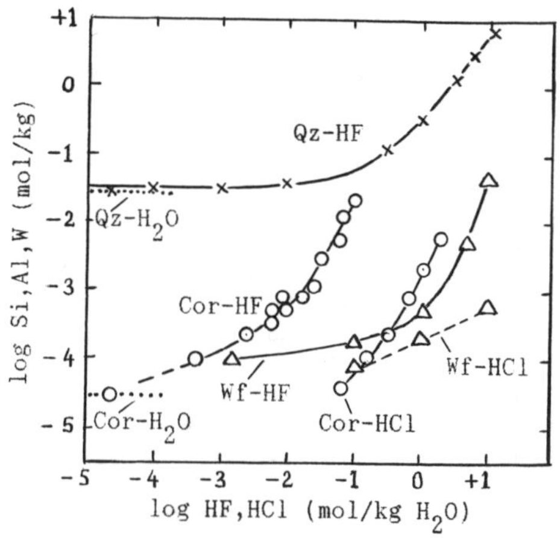

Figure 3 - Solubility of quartz, corundum and wolframite in HF and HCl solutions at 400°C and 1 kbar. Experimental data: (1) Shapovalov & Balashov (1990), (2-4) this study and Soboleva & Zaraisky (1990), (5) at 450°C after Korzhinsky (1987).

other experimental data on solubility of corundum, wolframite and quartz in aqueous solutions of HF and HCl under optimum parameters of formation of wolframite greisen deposits (400°C and 1 kbar). Quartz solubility in HF solutions was studied by Shapovalov and Balashov (1990). Here we restrict our discussion only to the main results referring directly to the problem of formation of rare-metal greisens.

The three minerals show a clear positive dependence of solubility on HF concentration, which seems to be explained by the formation of the similar hydroxofluorine species in the solution. A thermodynamic calculation of the experimental data using the program x88 by V.N. Balashov shows that in the region of high HF concentration the predominant species are $Si(OH)_2F$, $Al(OH)F_2$ and $W(OH)_4F_2$.

However, despite the similar shape of solubility curves for quartz, wolframite and corundum, acidic fluorine fluids must have different effects on the geochemical behavior of Si, W and Al in the natural environment. It is important that in the range of HF concentrations from 10^{-4} to 10^{-1} m for geological conditions the solubilities of quartz and wolframite increase insignificantly as compared to pure water and chloride solutions. Only in the region of high HF concentrations from 0.1 to 10 m do quartz and wolframite solubilities increase strongly attaining $3 \cdot 10^{-2}$ m Σ W and 0.5 m Σ Si in the 10 m HF solution. The latter value is close to quartz solubility in water under the same P-T conditions. However corundum shows a strong increase of solubility in the region of lower HF concentration: 10^{-4} m Σ Al in the solution 10^{-3} m HF to $2 \cdot 10^{-2}$ m Σ Al in 0.1 m HF. But for silica, as known, no problems of mass transfer exist under hydrothermal conditions. The results of the experiments of Korzhinskiy (1987) demonstrate that acidic chloride solutions cannot ensure alumina mass transfer in contrast to fluorine solutions, since very high HCl concentrations of more than 0.1 m is necessary (see Fig. 3).

4 RESULTS AND DISCUSSION OF EXPERIMENTAL MODELING OF GREISEN ZONING

At Akchatau the most complete greisen zoning with monoquartz, quartz-topaz and quartz-muscovite zones is developed in the middle levels of the subvertical greisen body (see Fig. 1). At the upper levels the internal monoquartz zone is absent, and at depth, the quartz-topaz zone does not occur. The experimental modeling of diffusion metasomatism of granite has revealed that the effect of 0.1-1.0 m HF solutions, nonsaturated with silica and alumina, results in a strong dissolution of granite. Formation of the most complete greisen zoning corresponding to the middle levels of the Akchatau requires high activity of Al and Si in solution. The greisen column of such type was obtained under the effect of the solution 1.0 m HF + 0.17 m AlF_3 + quartz in excess upon granite (500°C, 1 kbar, 336 hours):

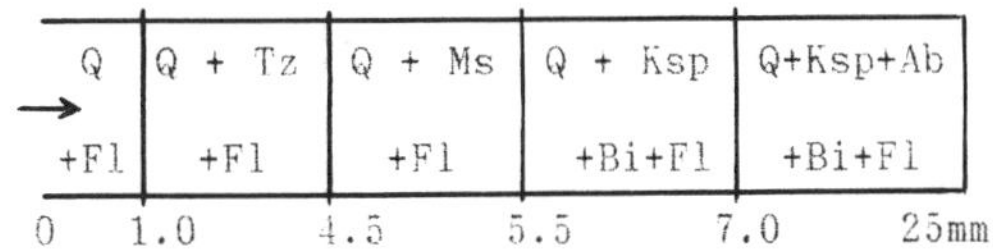

Infiltration experiments have been carried out at 500°C and 0.8 kbar only with the acidic chloride solutions due to the instability of the titanium infiltration system for the HF solutions effect. HCl concentration was fixed at 0.1 m. The value of the molal ratio KCl/HCl was varied due to the change of KCl concentration within the limits 0 to 2.0 m. To saturate the solution with silica a layer of fine-powdered quartz of 5 mm in width was placed before the granite. Depending on the outlet molal ratio KCl/HCl of the solution, the zone of forming infiltration column had quartz-pyrophillite, quartz-muscovite or quartz-potassium feldspar composition.

The column with quartz-muscovite in the outlet zone can be considered as an analogy to quartz-muscovite greisens (Fig. 4). Scanning by microprobe with a narrow electronic beam over the transverse profiles made it possible to investigate the migration of SiO_2 and Al_2O_3 at infiltration metasomatism. Fig. 4 shows a different behavior of Al_2O_3 and SiO_2. Alumina in all zones maintains a constant value equal to its initial content in granite, consistent with its low solubility in chloride solutions. Meanwhile, despite the saturation of the initial solution with quartz, silica is removed from the outlet end of the column and redeposits with the exceeding of the initial content in the inlet zones.

This phenomenon can be explained by the metastable "supersolubility" of silica at decomposition of feldspars in acidic solutions. One can assume that the replacement of feldspars by micas is accompanied by separation of the residual SiO_2 in the form of amorphous silica with a considerable higher solubility than that of quartz. First, amorphous silica transforms into a

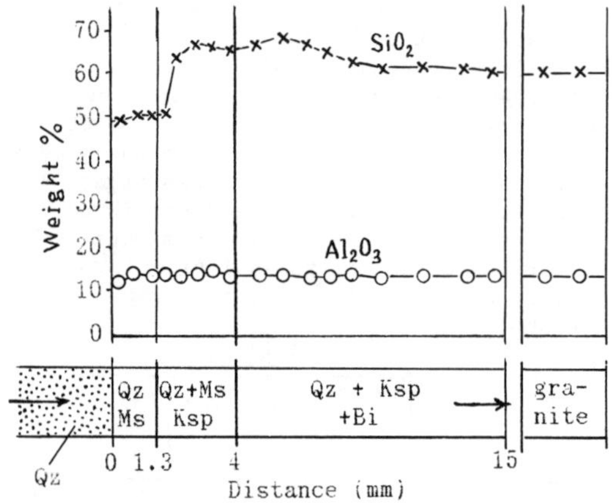

Figure 4 - Distribution of SiO_2 and Al_2O_3 in the infiltration experimental metasomatic column obtained from a microprobe scanning (500°C, 0.8 kbar, fluid 0.5 m KCl + 0.1 m HCl, duration 166 hr).

metastable crystobalite and then into a stable quartz. Earlier we observed crystobalite formation on microcline and albite in two-week experiments of diffusion interaction of these minerals with the aqueous solution 0.3 m HCl at 500°C and 1 kbar. In these experiments performed by the scheme shown in Fig. 2b, an intense diffusion of SiO_2 from granite into the solution being in equilibrium with powdered quartz, took place. After the experiments SiO_2 concentration in the solution was above quartz solubility, but did not reach the level of saturation with amorphous silica. Crystallization of pyrophillite instead of stable andalusite in the experiments on the interaction of granite with HCl solutions at 500°C serves as an additional argument of the solution oversaturation with silica.

The widespread formation of quartz in rocks during greisenization and other processes of acidic metasomatism seems to be related to the occurrence of SiO_2-oversaturated solutions with respect to quartz. The filtration of such solutions results in metastable extraction of silica from feldspar and other aluminosilicates and its redeposition in the form of stable quartz.

CONCLUSION

The experimental investigation of quartz, corundum and wolframite solubility in HF aqueous solutions at 300-600°C and 1 kbar P have shown that in the range of natural HF concentration (10^{-4} to 10^{-1} m) for greisen processes, the solubility of quartz and wolframite increases insignificantly over that in pure water. In this range the solubility of corundum increases by 400 times from $5\cdot10^{-5}$ m in water to $2\cdot10^{-5}$ m in 0.1 m HF. It follows that the influence of the acidic fluorine solutions typical for greisens must not effect markedly the geochemical behavior of Si and W. However, fluids with high F concentration can drastically increase the mobility of the usually inert aluminum. A considerable mass transfer of alumina, as is common in greisen deposits, can be explained by this fact.

It is experimentally determined that the decomposition of the feldspars and other aluminosilicates under the interaction of the acidic chloride solutions is accompanied by SiO_2 oversaturation in the fluids with respect to quartz. A widespread formation of quartz at greisenization and other processes of acidic leaching of rock may be related to the equilibrium precipitation of quartz from such oversaturated solutions in the process of their further infiltration through the rocks.

REFERENCES

Gotman, Ya.D. & V.I. Malakhova 1965. Near-vein alteration of granitic rock at the tungsten deposit in Kazakhstan. 116 p. Moscow: Nedra.

Korzhinsky, M.A. 1987. Solubility of corundum and possible form of aluminum in hydrochloric fluid. Geochimiya 4: 580-585.

Shapovalov, Yu.B. & V.N. Balashov 1990. Quartz solubility in hydrofluoric acid solutions at temperatures between 300 and 600°C and 1000 bars pressure. In Experiment-89 Informative Volume: 72-74. Moscow: Nauka.

Soboleva, Yu.B. & G.P. Zaraisky 1990. The solubility of corundum in HF solutions at elevated temperatures. Ibid: 77-79.

Zaraisky, G.P. 1989. Zoning and formation conditions of metasomatic rocks: 342 p. Moscow: Nauka.

Zharikov, V.A. & G.P. Zaraisky 1991. Experimental modeling of wall-rock metasomatism. In Progress in Metamorphic and Magmatic Petrology (The Korzhinsky Memorial Vol.): 197-245. Cambridge Univ. Press.

Water-Rock Interaction, Kharaka & Maest (eds) © 1992 Balkema, Rotterdam. ISBN 90 5410 075 3

Evidence of ancient hydrothermal activity and the relation to gold mineralization in the Rehai geothermal system, Yunnan, China

Zhu Meixiang
Geology Department, Peking University, Beijing, People's Republic of China

ABSTRACT: Surficial manifestations in the Rehai geothermal field can be divided into the north and south parts, whose dominance are active and ancient activities respectively. At least four episodes of hydrothermal activity are determined to have occurred within the Rehai Field. All the remnants, as well as spring water, contain differing amounts of Au, Ag, As, Sb and Hg. The presence of hydrofracture breccias, eruption breccias, pseudomorphs of quartz after bladed calcite and abundant vapors in fluid inclusions suggest that self-sealing and boiling are widely developed, which are very effective mechanisms for Au precipitation. It is suggested that this area might be a prospective hot-spring Au deposit.

1 INTRODUCTION

The Rehai field is a famous high temperature geothermal system in China, which is located 17km south-west of Tengchong County, Yunnan Province. This paper gives preliminary results and is intended primarily to discuss the lithology, distribution and interrelation of the remains of the previous surficial geothermal activities, i.e., sinters, siliceous caps, hydrothermal eruption breccias, hydrofractural breccias, quartz veins and alteration of country rocks, from which the evolutionary process of geothermal activities can be inferred. This paper will also deal with the Au contents of various geothermal products, the hot-spring origin of Au deposits and the physical-chemical conditions for mineralization.

2 GEOLOGICAL SETTING

Tengchong area lies in the collision zone between the Eurasian and Indian plates. Granite was dominant over the early period of Yanshan Movement, while the frequent volcanism in the area manifests itself intensively during the Late Himalaya. Ages of the volcanic rocks range from 0.09 to 17.84 Ma. In fact, the volcanism has offered a heat source to the Rehai.

In the Rehai, the oldest outcropping granite of Late Cretaceous age (68.8 Ma) may constitute the basement of this field. The overlying Tertiary granitic clastic rock of fluviolacustrine and alluvial phase is considered the main reservoir rock.

The Rehia stands in one of a series of arcuate faults in Tengchong, whose strikes are basically north-south and such a principal fault, together with the near east- west faults, controls the distribution of geothermal manifestations (Fig. 1). The current geothermal activities including boiling springs, fumaroles, steamground, etc., are concentrated in the north, while there is an absence of current manifestations except a few sulphur flowers and abundant extensive remnants including sinters, hydrothermal breccia and altered wallrock of former activity in the south.

3 SINTERS

The sinters found in this field are principally siliceous sinter and a certain amount of sulfur flower. The travertine is found at only one place near the south bank of the Bath Stream.

The siliceous deposits of hot springs are divided into two types: siliceous sinter and silica cap. Opaline sinter, chalcedonic sinter

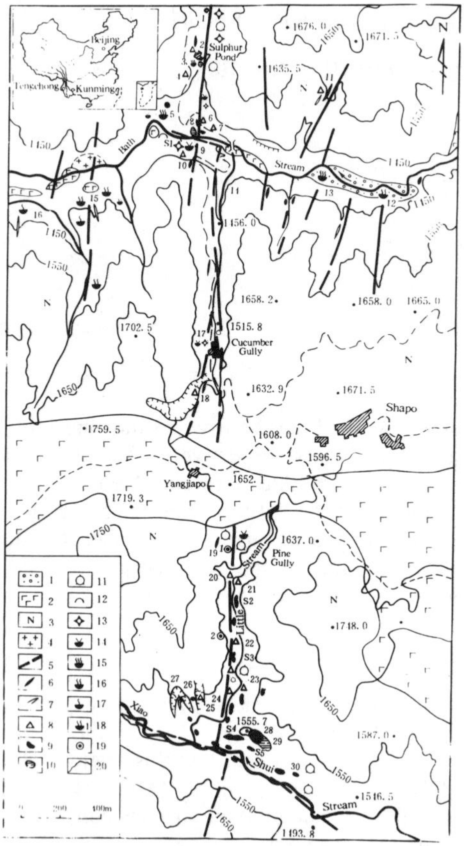

Figure 1. Sketch map of geology and thermal manifestation of the Rehai. The keys in Fig. 1: 1. Gravel of Holocene Period, 2. Andesite of Pleistocene Period, 3. Granite clastic rock of Miocene Period, 4. Granite of Late-Cretaceous Period, 5. Fault, 6. Quartz vein, 7. Pyrite vein, 8. Hydrothermal breccia, 9. Ancient sinter or cap, 10. Active sinter or cap, 11. Sulfur flower, 12. Travertine, 13. Steam ground, 14. Fumarole, 15. Boiling spring, 16. Hot spring, 17. Warm spring, 18. Number of spring and sample, 19. Drill hole, 20. Contour.

and quartz sinter have been recognized in Rehai. The opaline sinters are characterized by being loose, porous, and thin-bedded, which is considered to be primary and current deposition precipitating from neutral to slightly alkaline water (Fournier, 1985). In the north of Rehai, opaline sinters occur locally at only three places. All of them are in cones whose areas are very small (2-4 m^2). In the south, opal sinter is widespread with its largest area occurring on the south slope at 1555.7m elevation (Fig. 1., point 28). The horizontal or nearly horizontal opaline caps exist as interbeds in Tertiary clastic rocks.

Compared with opaline sinter, the chalcedonic sinter is more dense, hard, massive than opaline sinter. At some places sinters show break-like textures owing to syneresis cracks. The above mentioned features can be considered to have formed from earlier opal that was probably reconstituted into chalcedony and quartz.

Most of chalcedonic sinters and caps in Rehai still exist in the south. The largest area of the chalcedonic sinter group, including sinters and silica cemented clastic rocks are exposed at an elevation of 1555.7m (Fig. 1., point 27).

Chalcedonic caps formed as interbeds in the granitic clastic rocks, on single stratum, or multiple strata.

Quartz aggregates often occur in veins or silica caps (points 11 and 14 in Fig. 1). Fournier (1985) showed that the water in the depth of the reservoir was in equilibrium with quartz at a temperature greater than about 210°C. So the above mentioned quartz veins are considered to have formed at depth in the past.

Quartz veins and/or caps are revealed at the depth of ZK-1 and ZK-2. Take the ZK-1 for example, nine beds of quartz veins forming caps existed at the interval from 43 to 120m in depths. These quartz veins might have precipitated directly from neutral to slightly alkaline chloride water with a temperature over 200°C.

The blade textured and lattice textured silica minerals are often observed in hand specimens and in thin sections of siliceous sinters of a quartz veins, or of siliceous cements in hydrothermal breccias. The rhombohedral edges (Fig. 2) which are usually rhombohedral cleavages of calcite now occur in many crystals of quartz. The lattice-textured quartz occurs on the bedding plane of quartz cap found in some places. The bladed quartz is a suggested pseudomorph after bladed calcite due to the replacement of calcite by quartz. The bladed textured and lattice-textured calcite indicates that boiling conditions existed (Jennings et al., 1990).

4 HYDROTHERMAL BRECCIAS

In this paper hydrothermal breccias include hydrothermal eruption breccia and hydrofractural breccia. The eruption breccias in Rehai are characterized by its variable clast abundance, size and angularity, by it poor

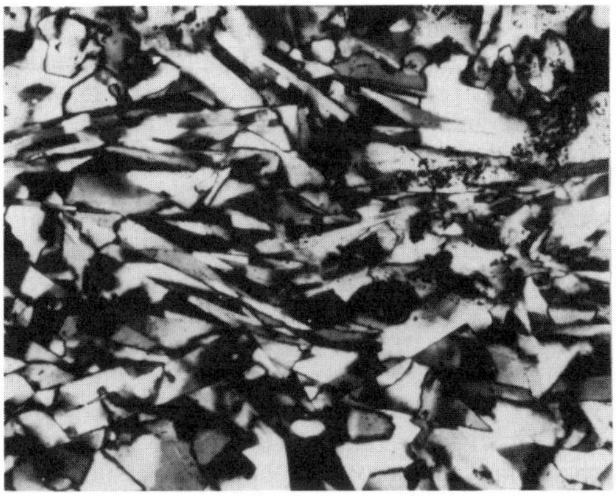

Figure 2. The bladed quartz aggregates, X56, crossed polars.

sorting, by it matrix-supported texture and by a variable degree of hydrothermal alteration of fragments. The eruption breccias can be divided into poorly consolidated observed point 6, 7, 10, and 27, and consolidated being hard and dense observed at point 20, 22, and 4. In some places the breccias show jigsaw puzzle textures.

The hydrofractural breccias here refer to the rocks formed by hydraulic fracturing at the depth of the reservior. These types are observed only locally at point 11, 21, and at depth in ZK-1. The breccia contains the fragments of strong silicified rock and previously formed quartz sinters existing in quartz vein of veinlets. The occurrence of these hard rocks is very steep in the form of a pipe injecting into altered clastic rock.

5 EVOLUTION OF THE HYDROTHERMAL ACTIVITY IN REHAI SYSTEM

The thermal evolution of the hydrothermal activity has been recorded by surficial manifestations, both active and ancient. According to the relationship and distribution of the remnants from former activity, and the direct dating of them (Table 1), hydrothermal activities in Rehai can be divided into at least four episodes.

According to K-Ar age of a dacite occuring in the north of Rehai area being 0.309 ± 0.009 Ma., the starting of the hydrothermal system dates back to <30 x 10^4 years. The first episode was about 30 x 10^4 years ago and the second about 18 x 10^4 years ago, the third was (4.4~8) x 10^4 years ago and the forth (2~3) x 10^4 years ago.

Table 1. The results of dating by method of Uranium-series and Electron Spin Resonance (10^4 years).

Locality	sample name	U-series age	ESR age
S2	Quartz vein	>38	no
S3	Quartz vein	>38	24(?)
Point 28	Chalcedonic sinter	4.4±0.8	8.4
Point 29	Opaline sinter	2.3±0.1	no
S4	Opaline cap	2.3±0.1	0.9−1(?)
Point 3	Opal sinter		<0.1
S1	travertine	17.5±2.5	

In the first episode, the centre of the activity is proved to have existed in the south by the widespread quartz caps and earlier breccias which indicate that the deep fluid was neutral to slightly alkaline chloride water with temperatures as high as from 240 to 250°C. The wallrock widely underwent low-medium-rank alteration at that time.

The evidence of the second episode was insufficient. The decrease of the scale of the acitivty may be the characteristic of this episode, owing to the occurrence of the travertine indicating lower temperature of deep geothermal fluid.

In the third episode, the centre of the activity still existed in the south, and at least one of the centres was found at an elevation of 1555.7 m. After the third episode there was uplift in the south of the Yangjiapo as a result of new tectonic movement. The centre of thermal activity transferred from the south to the north. The opal sinter deposited on the south hillslope at 1555.7m elevation. The opal sinter (points 1 and 2) started to precipitate in the north. The early hydrofractural breccias and quartz veins were exposed on the surface by the orographic uplift or the dissection of topography. The above mentioned events all happened in the fourth espisode.

After the last episode, no active deposition from hot springs has taken place on the ground in the southern area due to the intense uplifting. A few sulfur flowers, however, are present.

6 GOLD MINERALIZATION

In Rehai, all of the geological formations, even the spring water contained Au from 0.0044 to 0.106 ppb in 8 samples, contain different amounts of Au, Ag, As, Sb and Hg. Table 2 reveals that Au mineralization occurs in several geological environments. Au is concentrated in formations at depth in the reservoirs in Rehai. The Au contents in silica cap rocks are higher than those in siliceous sinters. The lowest contents of Au is found in altered clastic rocks.

Nelson and others (1985) showed that boiling is the primary mechanism responsible for Au precipitation. The developments of bladed texture of SiO_2 minerals and the variable liquid-to- vapor (10-40%) value exhibited in fluid inclusions in quartz indicate that boiling is widely developed in Rehai where the geothermal activities provide a favorable condition for Au precipitation. The close relationship of Au and pyrite indicates that in hydrothermal fluids gold may be a sulfur complex, such as $Au(HS)_2^-$.

Table 2. Trace-element data on selected samples from variable formations in Rehai (ppm).

Sample	Description	Au	Ag
Point 11	Altered clastic rock	0. 008	0. 9
Point 3	Opaline sinter	0. 033	0. 7
Point 6	Do	0. 176	8. 8
S4	Opaline cap	0. 401	66. 03
S5	Do	1. 368	28. 98
South of S4	Chalcedonic sinter	0. 585	152. 4
Point 5	Do	0. 89	8. 4
Point 11	Quartz vein	0. 072	5. 02
Point 11	Quartz vein with pyrite	3	66
Point 14	Do	14. 18	64. 4
Point 19	Hydrofracturel breccia with pyrite	7	60

7 CONCLUSIONS

According to surficial manifestations, the Rehai can be divided into the northern area of active intensive thermal activity and the southern area of ancient intensive thermal activity. The widespread presence of various remnants deposited from hot springs occur in the southern area.

This paper has preliminarily discussed the lithologies, features, distributions and interrelations of these remnants. At least four episodes of hydrothermal activity can be determined to have occurred within the Rehai field according to the stratigraphic relationship of various remanents, space overlap, alteration features and U-series direct ages. In the first episode, very intense activity appeared in the southern part. The boiling depth was about 400m in this episode. After the third episode the activity started to transfer into the north. The formation depth of SiO_2 minerals is very shallow according to many fluid inclusions being very small.

The progressive sequence of Au contents in various geological bodies is: spring water, altered clastic rocks, sinters of opal and chalcedony, cap rock of opal and chalcedony, quartz vein and finally quartz vein, hydrothermal eruption breccia and hydrofractural breccia, all of which contain fine pyrites.

The presence of hydrofractural breccia, eruption breccia, pseudomorph of quartz after bladed calcite suggest that self-sealing and boiling are widely developed, which are very effective mechanisms for Au precipitation.

The presence of compelling field evidence of the hot-spring Au deposits in Rehai, which is defined by Nelson and others (1985), indicates that Au mineralization in the Rehai geothermal field is of typical hot spring origin. The temperature of mineralization may be from 200 to 250°C, and the depths of mineralization may be from fossil surface to 400m depth.

ACKNOWLEDGEMENT

This project was supported by the National Natural Science Foundation of China.

REFERENCES

Fournier, R.O., 1985. The behavior of silica in hydrothermal systems. Geology and Geochemistry of Epithermal Systems. In B.R. Berger & P.M Bethke (eds.), Reviews in Economic Geology 2. Soc. Econ. Geol. 45-61.

Jennings, K., P.R.L. Browne, D.S. Clarke, & R.L. Brathwaite 1990. Aspects of hydrothermal alteration at the Waihi epithermal Au-Ag deposit, New Zealand. Proc. 12th New Zealand Geothermal Workshop. 237-242.

Nelson, C.E. & D.L. Giles 1985. Hydrothermal eruption mechanisms and hot spring gold deposits. Economic Geology. v. 80, 1633-1639.

Seawater-rock interactions

Water-Rock Interaction, Kharaka & Maest (eds) © 1992 Balkema, Rotterdam. ISBN 90 5410 075 3

Oceanic metamorphism in the Kuyul ophiolitic terrane of north Kamchatka, Russia

O.V.Chudaev
Far East Geological Institute, Vladivostok, Russia

V.B.Kurnosov
Geological Institute, Moscow, Russia

ABSTRACT: The Gankuvayam section which crops out in the Kuyul terrane, contains all the sequences of an ophiolite complex. In transitional zone from sheeted dike complex to the extrusive rocks, chalcopyrite-pyrite veined mineralization was discovered. The rocks of the extrusive-dike complex were studied in detail. In pillow-lava unit and sheeted dikes, secondary minerals of the smectite and greenschist facies were found. Isotopic ratio of sulfur in pyrite is very close to that in the regions with underwater hydrothermal activity. Low-temperature hydrothermal metamorphism of extrusive-dike complex of the Kuyul ophiolitic terrane is very similar to that of the second layer modern oceanic crust (Hole 504B, Galapagos spreading center) and of ophiolitic terranes where low-temperature oceanic metamorphism was preserved.

1 INTRODUCTION

Usually ophiolites do not preserve the oceanic low-temperature metamorphism and occur commonly within melanges that are closely related to the tectonic and metamorphic history of ophiolite emplacement into the orogens of the continental margin. Exceptions, are Troodos ophiolite, Cyprus (Kurnosov et al 1987), the East Taiwan (Liou et al 1979), Horokanai ophiolite of Hokkaido, Japan (Ishizuka 1987), etc. The Gankuvayam ophiolite of north Kamchatka is a good example of oceanic-stage metamorphism of extrusive-dike complexes.

2 OUTLINE OF GEOLOGY

Along the Gankuvayam river in the Koryak Upland of north Kamchatka (Fig. 1) a complete section of ophiolite was found.

Fig. 1 Location of the Gankuvayam ophiolite

According to the researchers of the Far East Geological Institute and Geological Institute (Kuyul ophiolitic terrane 1990), the Gankuvayam ophiolite series make up the core of a synform, the limbs of which are formed by serpentinous lherzolites, harzburgites, and serpentine melange zones. The synform is complicated with numerous thrusts and faults. In the southern limb, a slightly dislocated sequence is stripped, containing upwards:

1 Harzburgites	420 m
2 Apodunite serpentinites	50 m
3 Layered gabbro and wehrlite	340 m
4 Plagiogranites	40 m
5 Sheeted dikes	400 m
6 Pillow-lava	700 m

Veins of hornblendites and gabbro cross ultramafic and gabbro-wehrlite complexes. In the ultramafic part of the section, harzburgite predominates, lherzolite and dunite are rare. In the upper part of gabbroic complex massive gabbro predominates. Thin-banded gabbro with rare wehrlite is interbedded, olivine gabbro and troctolites are below. Plagiogranites are located between gabbro and sheeted dikes. The layer boundaries are clearly tectonic.

Sheeted dikes were not recognized previously within the Kuyul ophiolitic terranes and were combined with extrusive rocks. Sheeted dikes are represented by diabases, andesites, and diorites. Dikes are 0.6m to 4.0m thick, on one side quenching zones are well seen. Breccia zones occur among dikes likely related to the migration of ophiolite blocks during spreading. Dikes grade smoothly into the pillow-lava series. Pillow-lavas are represented by basalts, andesites, and dacites. In the transitional zone between sheeted dikes and pillow-lavas, chalcopyritye-pyrite mineralization was found (Kuyul ophiolitic terrane 1990).

Sedimentary rocks related to basalt flows of the Gankuvayam section have not been found. Rare fragments of jasper occur in talus of the Gankuvayam river. The age of the Gankuvayam ophiolite is presumably Early Cretaceous.

In this article we discuss the hydrothermal alteration of pillow-lava and sheeted dike series.

3. ANALYTICAL TECHNIQUES

Secondary minerals (mostly clay minerals), size less than 0.001mm, were separated from powder of rocks. X-ray diffraction and electron microscope were used to study fine size samples. Special consideration was given to layered silicates which are the sensitive indicators of low-temperature hydrothermal metamorphism of rocks in the second layer of the oceanic crust. The samples, assumed to contain smectite, were thoroughly investigated.

All analytical work was made in the Far East Geological Institute of the Far Eastern Division the Academy of Sciences of Russia.

4 RESULTS AND DISCUSSION

The chemical composition of pillow-lava and sheeted dike units is similar. The differentiated basalt-andesite-dacite volcanic series seems to be a result of common crystallization and mineral fractionation (Kuyul ophiolitic terrane 1990). On the typical petrological diagrams, basalt falls into the tholeiitic field, representing the ferrotholeiitic-ferroandesite (icelandite-dacitic) series. The rocks have low concentration of potassium. High sodium of the rocks is caused sometimes by albitization of plagioclase. Table 1 shows distribution of secondary minerals in extrusive-dike units.

Table 1. Secondary minerals in rocks of pillow-lava and dike units.

Samples	Lithological description	Secondary minerals
	Pillow-Lava Unit	
8x-19-1	Cpx-basalt (core of pillow)	Chl, Chl-Sm, Sm? Ep, Alb, Act, Qtz
8x-19-2	Cpx-basalt (middle part of pillow)	Chl-Sm, Chl, Sm, Ep, Alb, Qtz
8x-19-3	Cpx-basalt (margin of pillow)	Sm, Sm-Chl, Chl? Qtz
8x-19-6	Volcanic glass	Sm, Chl-Sm,An
8x-19-8	Brecciated basalt	Sm, Chl-Sm,Ca
8x-19-9	Diabase	Chl, Chl-Sm, Ep, Alb, Qtz
8x-19-12	Cpx-basalt	Chl, Chl-Sm, Qtz, Act
8x-12	Porphyritic basalt	Chl, Chl-Sm, Ca, Alb
8x-11	Cpx-basalt	Chl, Chl-Sm, Ep, Alb
	Dikes Unit	
8x-26-1	Andesite-basalt	Chl, Ca, Qtz
8x-26-2	Andesite	Chl, Act, Alb
8x-27-2	Diabase	Chl, Act, Ep, Qtz
8x-27-3	Andesite-basalt	Chl, Ep, Alb, Qtz
8x-24-3	Andesite	Chl, Act, Ep
8x-24-1	Dacite	Chl, Ep, Qtz

Note: Sm=Smectite; Chl=Chlorite; Chl-Sm=Mixed layered Chlorite-Smectite, Chlorite layer > Smectites; Sm-Chl=Mixed layered Smectite-Chlorite, Smectite layers > Chlorites; Ep=Epidote; An=Analcime; Alb=Albite; Ca=Calcite; Act=Actinolite; Qtz=Quartz.

In the pillow-lava unit, smectite, mixed layered smectite-chlorite, chlorite-smectite, and chlorite were found among layered silicates. In addition, analcime, epidote, secondary quartz, albite, actinolite, and calcite are common (Table 1). Smectite commonly replaces glass and fills vesicles. In marginal zones of pillow-lava, glass is almost completely altered into smectite. Analcime, actinolite needles, and secondary quartz occur in vesicles. Basic plagioclase is often albitized. In the pillow-lava unit smectite occurs in assemblages with mixed layered silicates and epidote, i.e., pure smectite facies is absent. It is rather a transitional facies from smectite to greenschist. Therefore, the upper part of the pillow-lava section with typical smectite facies was not found in the pillow-lava unit.

Smectite disappears towards the dike unit. The main secondary minerals are chlorite, epidote, albite, actinolite, calcite and quartz. This unit has undergone higher-temperature effects, and the alterations correspond to greenschist facies.

We compare the alteration degree of pillow lava and dike units of an ophiolite section along the Gankuvayam River with the section of unbroken oceanic crust penetrated by the deepest Hole 504B (Costa-Rica Rift). The depth of penetration into the second layer was about 1400m. At a depth of about 1100m the hole entered the dike complex (Kawahata and Kusakabe 1987). Kawahata studied in detail the secondary minerals in both pillow-lava and dike complexes. Four zones of rock alteration were distinguished. The first and second zones to a depth of 890m (pillow-lava) corresponds to the smectite facies. According to oxygen isotope ratios, the temperature of smectite origin ranges from 0 to 60°C. In the upper part of the section to a depth of 550m, smectite (saponite) formed under the oxidizing conditions (zone 1) at temperature 0-30°C. In zone 3 (the base of pillow-lava complex) and zone 4 (dike complex), hydrothermal alterations occur at 200-350°C. The main secondary minerals are chlorite, epidote, actinolite, albite, laumontite, quartz, and pyrite (Kawahata and Kusakabe 1987). Therefore, the comparisons of alterations in Hole 504B with those in the pillow-lava and sheeted dike complexes of the Gankuvayam ophiolite show that by mineral assemblages the pillow-lavas correspond to the boundary of zones 2 and 3 of Hole 504B. The dike complex corresponds to zone 4 of Hole 504B.

The same assemblages of secondary minerals occur in the Troodos ophiolite. Holes CY-2 and CY-2A were drilled into and beneath the Agrokipia ore bodies in order to investigate the hydrothermal processes leading to the formation of sulfide ore bodies in the oceanic crust (Robinson 1987). We distinguish two basic types of alterations in the volcanic rocks in Holes CY-2 and CY-2A: lower-temperature and higher-temperature (Kurnosov 1986, Kurnosov and Robinson 1987). For lower-temperature alteration, smectite and mixed layered smectite-chlorite are typical. Chiefly chlorite, quartz, and the traces of epidote are common. We studied the distribution of secondary minerals, oxygen isotopic ratio, and some major chemical elements in pillow lavas. In core of pillows, chlorite and mixed layered smectite-chlorite predominate over smectite; epidote, actinolite, and quartz are also typical. In the marginal zone, smectite and mixed layered smectite-chlorite are main secondary minerals (Table 1). The distinction in secondary minerals of interpillow and marginal zone are reflected in chemistry (Table 2).

Table 2. Distribution of the major element oxides (wt %) and oxygen isotopic ratio in interpillow and marginal zone.

	interpillow		marginal
Fe2O3	9.68	10.0	10.62
MgO	9.0 5.05*	9.0 6.68*	7.87 6.45*
CaO	7.07*	6.57*	6.98*
K2O	1.12 1.31*	1.30 1.46*	1.80 2.12*
Na2O	2.88*	2.55*	2.66*
δ O (‰)	11.9	11.5	12.4

Note: * concentration of elements in whole rocks; numbers without marks-elements in clay minerals. Distinction in secondary minerals and chemistry of interpillow and marginal zone can explain seawater/rock ratios. Interpillow voids are major channels for seawater flux as discussed by Seyfried et al (1978), and Mottl (1983). Likewise, the breccia zones also acted as conduits for large quantities of seawater. This supports the distribution of the oxygen isotopic ratio, which increases to marginal zones (Table 2). Mottl's model predicts the assemblages of secondary minerals with increasing seawater/rock ratio: 2 to 35: chl-alb-ep-qtz; >50: chl-qtz. In the marginal zone, smectite is the main secondary mineral (Table 2). Smectite appearance is probably connected with oxidative type alteration, including halmyrolysis. During oxidative stages, the concentration of Fe2O3 and K2O increases (Table 2).

Thus, the pillow-lava and sheeted dike units of the Gankuvayam ophiolite have presented the main features of oceanic hydrothermal metamorphism. In pillow-lavas studied, the upper part of the section represented by typical smectite facies is absent.

As we mentioned above, in transitional zone sulfide veined mineralization was found. We compared sulfur isotopes of pyrites in the Gankuvayam ophiolite with sulfur isotopes of pyrites discovered in the sea floor along the East Pacific Rise: Juan de Fuca Ridge and Hole 504B.

Table 3. Sulfur isotope values of pyrites from various submarine hydrothermal areas (Kawahata and Shikazono 1988).

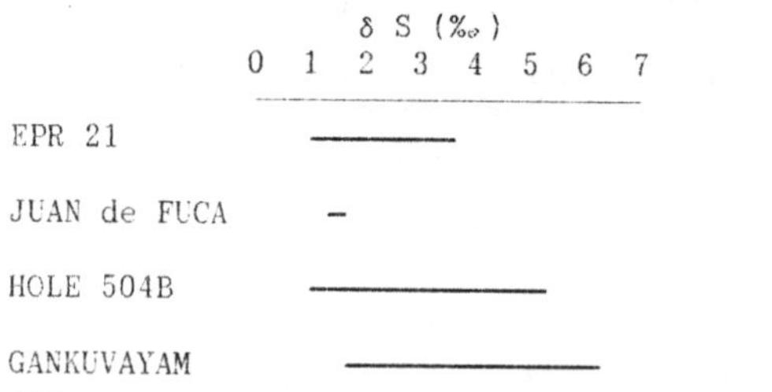

Isotopic ratio of sulfur in pyrite of the Gankuvayam section is very close to that of the regions with underwater hydrothermal activity.

5. CONCLUSIONS

Data obtained on the alteration in the extrusive-dike complex suggest: 1) that the Gankuvayam cross-section is characterized by hydrothermal metamorphism of oceanic type without superposed accretionary metamorphism, and 2) that the sulfide mineralization is related to the spreading stage of the oceanic crust "life". Knowledge of intense hydrothermal activity in the second layer of paleo- and young oceanic crust (Hole 504B) and ophiolitic terranes, particularly Cyprus, make it possible to search for polymetallic sulfide deposits in the Kuyul ophiolitic terrane.

REFERENCES

Ishizuka, H. 1987. Igneous and metamorphic petrology of the Horokanai ophiolite in the Kamuikotan zone, Hokkaido, Japan. Memory of the Faculty of Science. Kochi University, series E. Geology 8: 1-70.

Kawahata, H. & M.Kusakabe 1987. Strontium, oxygen and hydrogen isotope geochemistry of hydrothermally altered and weathered rocks in DSDP hole 504B, Costa Rica Rift Earth and Pl. Sci. Let. 85: 343-355.

Kurnosov, V.B. 1986. Hydrothermal alteration of basalts in the Pacific ocean and metal-bearing deposits (using data of DSDP). Moscow 1-320.

Kurnosov, V.B. & P.Robinson 1987. Secondary minerals in the CCSP drillhole CY-2 and 2A, Agrokipia, Cyprys. Cyprus Crustal Study Project Initial Report, holes CY-2 and 2A. Geological Survey of Canada 85-29.

Kuyul ophiolitic terrane 1990. Vladivostok 1-107.

Liou, J.G. & W.G.Ernst 1979. Oceanic ridge metamorphism of the East Taiwan ophiolite. Contr. Mineral. Petrol 68: 335-348.

Mottl, J. 1983. Metabasalts, axial hot springs, and the structure of hydrothermal systems at mid-ocean ridges. Geol. Soc. of America Bull 94: 161-180.

Robinson, P. 1987. Cyprus Crustal Study Project. Initial Report, Holes CY-2 and 2A. Geological Survey of Canada 85-29.

Seyfried, W.E., Jr. Mottl & Bischoff J.L. 1978. Seawater-basalt ratio effects on the chemistry and mineralogy of spilites from the ocean floor. Nature 275: 211-213.

Water-Rock Interaction, Kharaka & Maest (eds) © 1992 Balkema, Rotterdam. ISBN 90 5410 075 3

Spilitic alteration as the source of the world-class Proterozoic manganese deposit

D.H.Cornell & S.S.Schütte
Department of Geology & Applied Geology, University of Natal, Durban, South Africa

ABSTRACT: The Kalahari Manganese deposit is interpreted as an end-member of a syngenetic spilitization process which developed in the underlying Ongeluk lava pile. This contradicts previous models which envisage epigenetic upgrading of sectors of the ore.

1 INTRODUCTION

Rocks of the Griqualand West and Transvaal Sequences occur over large parts of the Republic of South Africa and Botswana. In South Africa they occur in the Griqualand West and Transvaal erosional basins with an area of ≈500,000km^2. The regional tectono-stratigraphic framework is given in Figure 1.

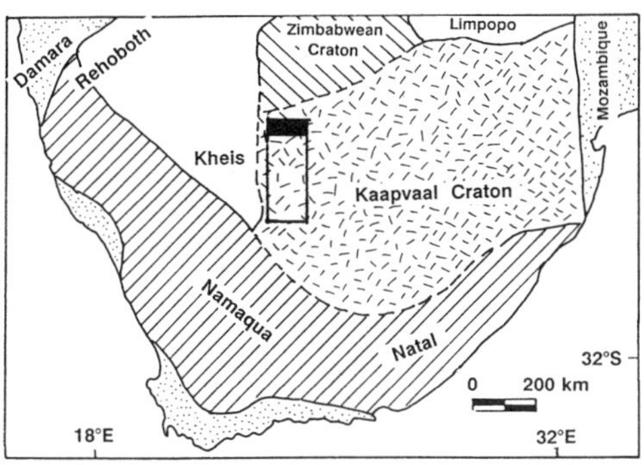

Figure 1: Precambrian structural framework of Southern Africa, showing the outcrop area of the Ongeluk lava (open box) and the Kalahari Manganese Field (filled box).

Griqualand West sediments formed under shallow marine conditions in an intracontinental sea. The predominance of chemical rather than detrital sediments suggests deposition in an environment far from clastic input.

The Ongeluk Andesite Formation is an Early Proterozoic lava succession (≈ 2200 Ma) in the upper Griqualand West Sequence. In the northern sector of its extensive regional outcrop, the Ongeluk **conformably** underlies the Kalahari Manganese deposits, as shown in the stratigraphic section of Figure 2. The 50km by 20km manganese field shows a regional zonation of ore grade.

2 MINERAL ALTERATION

The Ongeluk lavas comprise only three rock types, i.e. massive lava flows, hyaloclastite and pillow lavas suggesting submarine extrusion. Analyses of pillow cores and rims (Table 1) revealed small-scale alteration which is ascribed to a spilitization process. The rims are slightly oxidized and show calcium depletion and alkali enrichment relative to the cores. Hyaloclastites show a more extreme alteration relative to the average massive lava, due to their high permeability. In contrast to the normal grey-green appearance of most regional outcrops, the lavas below the Kalahari Manganese Field are maroon. Both lava types display similar texture and igneous phenocrysts; however, their gross chemistry is quite different (see Table 1).

Table 1. Analyses of different igneous rock types from the Ongeluk Formation. Major oxides in weight percent; trace elements in ppm.

	rim	core	hyalo.	massive	Kalahari
SiO_2	54.29	51.06	54.97	53.68	52.83
TiO_2	0.73	0.61	0.58	0.64	1.33
Al_2O_3	14.16	13.48	13.07	14.43	14.05
$Fe_2O_3^t$	10.34	10.12	12.76	11.35	14.18
$Fe_2O_3^*$	2.39	2.66	2.56	2.09	13.18
FeO^*	7.15	6.71	9.18	8.33	0.90
MnO	0.12	0.16	0.10	0.14	0.29
MgO	6.77	6.08	7.48	5.93	5.24
CaO	4.02	12.17	4.46	7.56	2.01
Na_2O	4.01	2.58	0.17	2.37	1.74
K_2O	0.45	<0.02	1.49	0.52	3.08
P_2O_5	0.12	0.12	0.07	0.10	0.20
H_2O^-	0.27	0.18	0.25	0.21	0.80
LoI	3.88	3.69	6.05	4.11	3.92
Total	99.16	99.48	100.41	100.10	99.57
Zr	116.	101.	93.	106.	181.
Y	22.	24.	18.	24.	37.
Sr	70.	30.	15.	195.	119.
Rb	22.	3.	56.	22.	110.
Ni	84.	82.	81.	87.	101.
Cr	98.	80.	85.	88.	107.

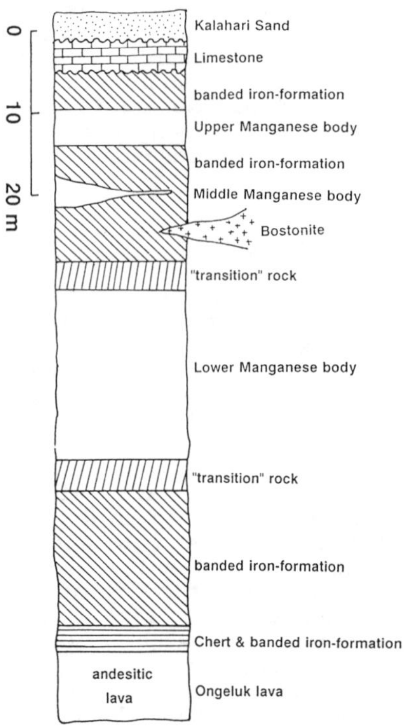

Figure 2: Simplified stratigraphic column showing the transition from Ongeluk lava into the Kalahari Manganese deposits.

Alteration in the Kalahari Ongeluk is characterized by extreme iron oxidation with calcium depletion and alkali enrichment.

In a Kalahari borehole the hematitic lavas grade irregularly downwards into the green-grey lavas of the normal (regional) type which lack disseminated hematite, but has hematitic veinlets. This discolouration is ascribed by several authors (e.g. Kleyenstüber, 1986; Dixon, 1989) to a localized post-genetic hydrothermal event related to fluid emanations from either the Kheis tectogenesis or from an obscure magmatic body. This hydrothermal interpretation is consistent with the presence of the Li- and B-bearing mineral sugilite and could also explain the regional and local variations of ore grade. Wiggering and Beukes (1990) interpret the same hematitized association at the top of the Ongeluk lava as a palaeosol. However, the massive alkali enrichment without aluminium enrichment in their samples contradicts their interpretation.

The ACF diagram (Figure 3) illustrates the alteration of regional and Kalahari Ongeluk lavas. Spilitization of the regional

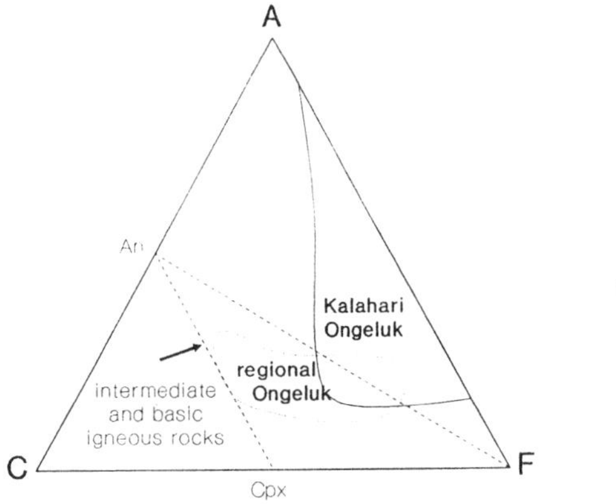

Figure 3: ACF diagram showing the fields of regional and Kalahari Ongeluk lava samples.

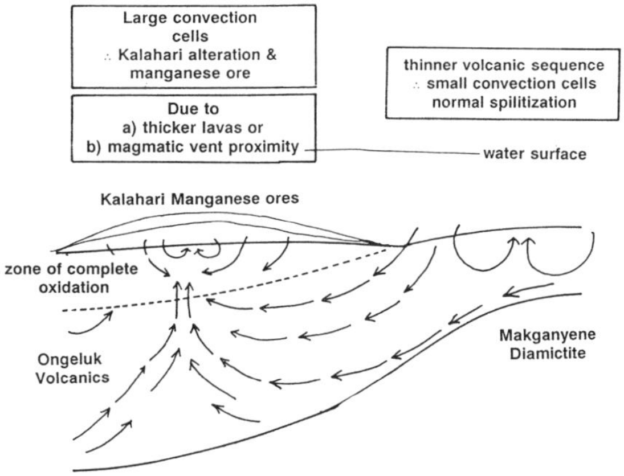

Figure 4: Model of syngenetic hydrothermal circulation in the Ongeluk lava.

Ongeluk lavas shifts their compositions out of the field of normal basic igneous rocks due mainly to calcium depletion. Kalahari Ongeluk samples plot in an overlapping Ca-depleted field which also shows a major A-wards trend due to iron oxidation. This and other geochemical evidence suggests that the Kalahari alteration represents an extreme case of spilitization. Figure 4 represents a model to explain the two styles of alteration. In the regional Ongeluk localized convection cells developed, giving rise to variable chemistry on a small scale such as pillow core and rim. The Kalahari alteration obliterated small-scale differences due to the longer duration of hydrothermal circulation.

We interpret this in terms of a much larger convection system involving large-scale movements of material. The Kalahari region represents an area of large-scale discharge, which mixed with oxygenated sea water. The reasons for the different scales of hydrothermal convection must be related to differences of heat supply, which could involve either thickness of the lava pile or distance from the vent area.

This model can explain all the features seen in the Kalahari Manganese Field in terms of a single syngenetic process. The chemical sediments of the Kalahari Manganese Field represent manganese and iron which was leached from the Ongeluk lava at an early stage of spilitization and deposited in an oxygenated marine environment after expulsion from a hydrothermal vent. The pervasive alteration of Kalahari Ongeluk reflects the near-surface hydrothermal alteration of lava in the depositional environment of a major convection system. Regional zonations of ore grade and thickness reflect variations in distance from the hydrothermal vent. Local variations of ore grade and the presence of minerals such as sugilite and andradite in veins and vugs are the result of late-hydrothermal fluids following restricted paths of fluid migration.

REFERENCES

Dixon, R.D. 1989. Sugilite and associated metamorphic silicate minerals from Wessels mine, Kalahari Manganese Field. Geol. Surv. Bull. S. Afr. 93: 47 pp.

Kleyenstüber, A.S.E. 1986. A regional mineralogical study of the manganese-bearing Voëlwater Subgroup in the Northern Cape Province. PhD., Randse Afrikaanse Universiteit.

Wiggering, H. & N.J. Beukes 1990. Petrography and geochemistry of a 2000-2200-Ma-old hematitic palaeo-alteration profile on Ongeluk Basalt of the Transvaal Supergroup, Griqualand West, South Africa. Prec. Res., 46: 241-258.

Water-Rock Interaction, Kharaka & Maest (eds) © 1992 Balkema, Rotterdam. ISBN 90 5410 075 3

Alkali basalt/seawater hydrothermal alteration

I.V. Kholodkevich
Far East Geological Institute, Vladivostok, Russia

J.L. Bischoff & R.J. Rosenbauer
US Geological Survey, Menlo Park, Calif., USA

ABSTRACT: Alkali basalts/synthetic sea water interaction has been studied with solid/liquid phase ratios of 1:2 and 1:10 within the temperature range of 200-500°C and pressure of 400 bar for up to 80 days. Kinetic curves showing the behaviour of chemical elements were obtained; the trends and mechanism of alterations in the system in question were estimated. Major newly-formed minerals were smectite, analcime, albite, plagioclase, potassium feldspar, pyroxene, and hydromica. Analcime stability in the system was discussed based on thermodynamic calculations.

The hypothesis of the circulation of seawater in mid-oceanic ridges was widely used in experiments studying the interaction of tholeiitic basalt with seawater at increased temperatures and pressures. These experiments and abundant natural data suggested convincing models of hydrothermal activity in mid-oceanic ridges. These models could be applied to the other "hot" zones of the ocean, where andesitic and/or alkaline magmatism could occur in addition to tholeiitic one. Hydrothermal processes in these regions could also contribute to the total geochemical balance of the ocean. Experiments on the alkali rocks/heated seawater interaction could be helpful in estimating the condition of secondary minerogenesis in alkali rocks and the volume of geochemical exchange in the system in question.

EXPERIMENTAL PROCEDURE

Two rock types were used: basanite from Primorye (Russia Far Eastern Maritime Province) and typical alkali basalt from Hawaiian Islands. Basanite consisted of fresh glass(~50%), plagioclase phenocrysts (~10%), clinopyroxene(~20%), olivine(~10%), and magnetite(~10%). The second start specimen was olivine-pyroxene basalt with porous porphyritic texture. Basanite differs from Hawaiian basalt by twice as much content of alkalis and phosphorus and lowered Ca and Mg contents.

Both rocks were powdered with particle size of <200 mesh; for the first stage runs, fractions of 1 to 10μ and 10 to 100μ of basanite were separated. Again, thin (up to 0.5 mm thick) plates of rock were used promoting further microscopic studies of altered fragments in thin sections.

As start solutions, we used natural, synthetic, and evolved seawater . Evolved seawater was close in composition to seawater equilibrated with oceanic basalt at 300°C (Seyfried & Bischoff 1981).

We performed two series of runs using different equipment: cold-seal pressure vessels using a conventional reaction-cell system and rocking autoclaves with a flexible gold or titanium bag (Hydrothermal...1987). The temperatures and pressures were measured to the accuracy of ±2°C and ±10 bar. First-stage runs were performed in platinum bags in cold-seal pressure vessels under a P=500 bar, T of 200-500°C, and water/rock ratios 2 and 10, for 7 to 14 days aimed at revealing the trends of mineral alteration in studied rocks depending on the temperature and water/rock ratio. For the second-stage runs, performed for up to 80 days with the equipment enabling the sampling of hydrothermal solution, the powder of initial rock with particle sizes of <200 mesh was taken, and the temperatures of 300,350, and 400°C were chosen for basanite and 300, 350, and

435°C for Hawaiian basalt at the fluid pressure of 400 bar and water/rock ratio of 10. Sampled solutions were studied using different techniques depending on chemical element and its concentration in a solution. K, Ca, Mg, Fe, and Mn were analyzed using atomic absorption(Perkin-Elmer 370); SiO_2, Co, Ni, Cu, Zn, Al, Ti, Pb, P, Sr, and Ba were determined using ignition spectrometry(ICP). Powdered samples were analyzed using conventional optical technique (in immersion oil), and X-ray diffraction, electron diffraction, microdiffraction, and microprobe. For the identification of clay minerals, X-ray diffraction patterns of air-dried, saturated with ethylene-glycol, and heated at 500 to 550°C samples were obtained.

Thin sections of hydrothermally altered plates of rocks were studied under the microscope and using microprobe. The emphasis was made on marginal alteration which was most notable.

RESULTS

Summarized X-ray diffraction, electron diffraction, electron microscopic (using microdiffraction), and optical microscopy and microprobe data show that glass and olivine included in initial basanite disappeared completely at 300°C. Even at 200°C, their recrystallization results in the growth of smectite and analcime, the quantity of which increases with the increase in the temperature. Again, above 200°C, minor albite and potassium feldspar grow. Maximum growth of potassium feldspar (orthoclase) was observed at 300°C. Albite growth increases with the increase in the temperature. Data obtained suggest that albite grows due to analcime alteration. Together with the albite,hydromica starts crystallizing above 300°C. Smectite grows in virtually every temperature range studied. X-ray diffraction patterns of products obtained in the runs with 1 to 10 μ basanite fraction and the visual estimation of reflection intensity show that plagioclase, clinopyroxene, and magnetite present in initial sample remain virtually fresh. A certain increase in reflection intensity typical of plagioclase was observed only at 500°C. Since it is difficult to distinguish plagioclase and albite on X-ray diffraction patterns, we studied alteration products using immersion oil (the runs with powder) and thin sections (the runs with thin basanite plates), to verify plagioclase crystals growth. An increase in grain size of initial powder results in a change of basanite mineral alteration. Although the alteration trend is virtually the same, there is a considerable change in the ratio of newly-formed phases, primarily smectite and analcime. Whereas smectite amount increases, compared with the runs with a finer fraction, analcime amount decreases, although the temperature range of its growth somewhat widens. No albite was observed, and newly-formed clinopyroxene appeared. Different alteration of different basanite fractions seems to be due to kinetic factors. This is also evidenced by the fact that no albite was grown in longer runs, although the nature of albite alteration in short (1 to 10μ , 7 days) and long runs was similar. Thus, albite could be unstable under the conditions discussed.

There are some other examples of the metastability of some newly-formed minerals. In the first-stage runs with basanite, we got analcime in 7 days. An increase in the time up to 14 days with other conditions being equal resulted in predominant albite growth among newly-formed phases in the marginal zone. The run with the same powdered basanite for 70 days under the same conditions gave plagioclase rather than albite growth among newly-formed phases, identified by X-ray diffraction, microprobe, and in immersion oil.

Thus the trend of basanite alteration depending on the time of the interaction with seawater could be shown schematically as follows:

$$Glass \Longrightarrow Anc \Longrightarrow Ab \Longrightarrow Pl,$$

where analcime and albite are intermediate phases of plagioclase alteration in the system discussed.

A close consideration of data obtained studying newly-formed layered silicates shows that predominantly dioctahedral smectite varieties formed at the initial stage of basanite alteration at lower temperatures, changed into trioctahedral varieties with the increase in temperature. Unlike in basanite, the alteration products of Hawaiian basalt are chiefly represented by trioctahedral smectite and few plagioclase in the whole temperature range studied. Some analcime appears at 300°C and pyroxene at 435°C. Newly-formed pyroxene crystals were clearly observed in immersion oil under the microscope as elongated prismatic crystals over both Hawaiian basalt(435°C) and basanite (400°C).

Another difference in mineralogy of altered alkali basalts is the growth of hydromica rather than smectite over basanite at temperatures above 300°C. Runs with Hawaiian basalt under the same conditions did not give hydromica growth even at 435°C. Trioctahedral smectite predominantly grew within the whole temperature range studied, presumably because in initial Hawaiian basalt smectite grew chiefly due to olivine and few interstitial glass present in start rock; whereas in basanite experiments, secondary minerals were mainly formed over a glassy constituent of start rock, making up approximately 50% of the total rock mass. This affected the activity of constituents in the system resulting in smectite stability in the runs on Hawaiian basalt.

Using data on the chemical composition of sampled fluids, we constructed the diagrams illustrating the changes in the contents of some chemical elements during the experiments with basanite and Hawaiian basalt depending on time (fig. 1). Unlike in basanite runs, a sharp decrease in pH was observed at the initial stage of Hawaiian basalt runs. A sharp K and SiO_2 rise accompanied a sharp Ca drop in sampled fluids.

In each run, a sharp Ca drop in solutions was observed, markedly differing from the results of the experiments with tholeiitic oceanic basalt/seawater interaction under similar conditions available. A sharp increase in Ca in fluid was observed in all previous runs. Ca drop in our runs seems to be chiefly related to the origin of pyroxene as a newly-formed phase. Ca content in fluid is in agreement with the amount of newly-formed pyroxene in solid products of the Hawaiian basalt and basanite runs. Basanite/evolved seawater interaction resulted in significant removal of some ore

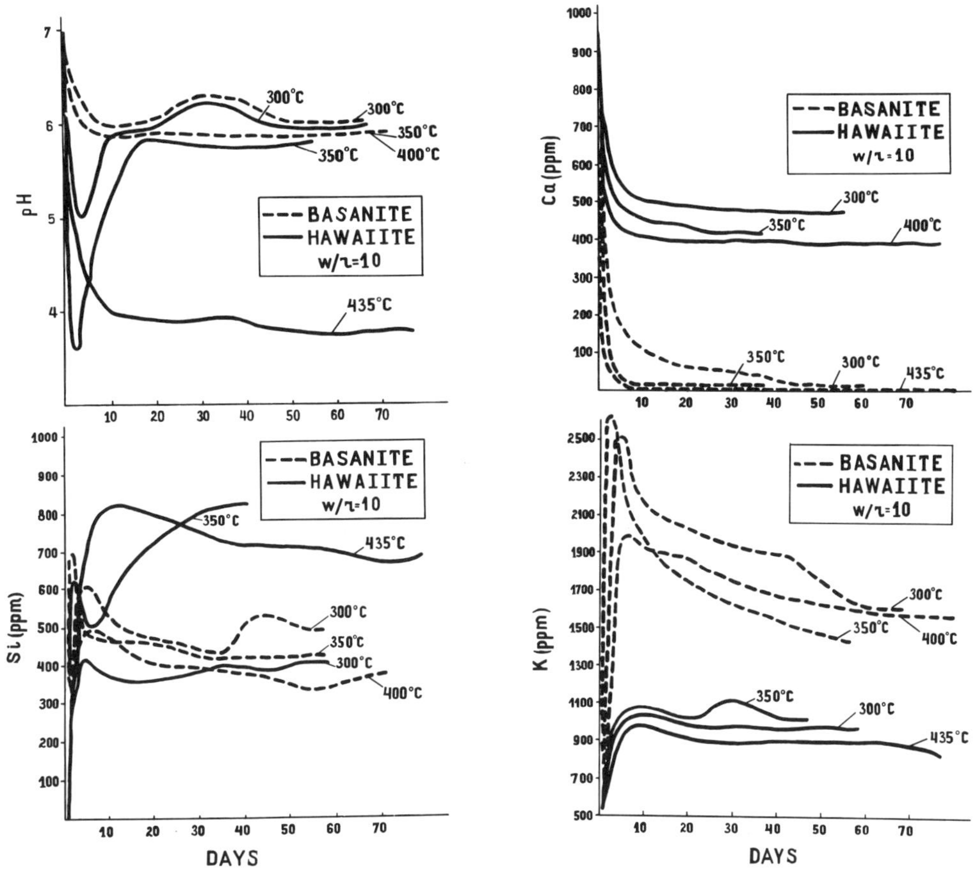

Fig.1 Changes in seawater composition in experiments

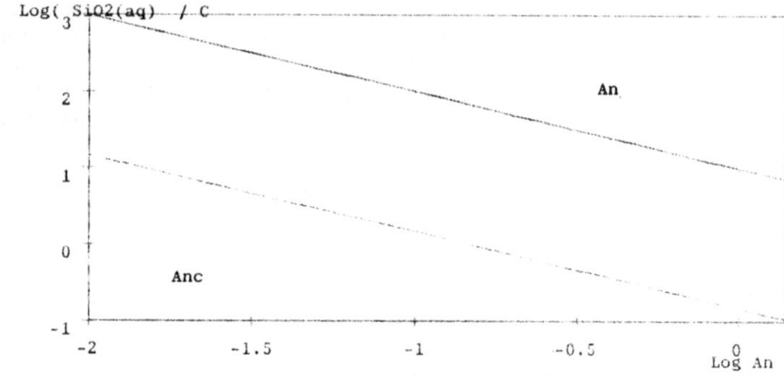

Fig.2 Activity diagram with analcime stability boundaries, 300°C experiments for basanite (solid line), Hawaiian basalt (dotted line), filled square - calculation by SUPCRT.

elements presumably due to a slight pH decrease in fluid, because seawater used for the runs was preliminary devoid of Mg and SO_4 ions, the presence of which could have caused the reactions releasing hydrogen (Bischoff & Seyfried 1978). The conventional model seawater used as a start fluid accounted for a notable leaching of ore elements at the initial stage. We also observed a notable removal of Cu, Zn, Sr, and Ba compared with their contents in solid products.

Analcime stability in the system under discussion is worth noting. Our runs showed the significance of a kinetic factor for analcime growth. Assuming we attained equilibrium in long-term experiments, the field of analcime stability is limited by a temperature a little above 300°C. A possible reaction in the system is as follows:

$$Anc+Ca^{2+}=An+2SiO_2(aq)+2Na^{+}+2H_2O,$$ then

$$K = [An][SiO_2(aq)][Na^{+}]/[Ca^{2+}][Anc].$$

Let [An]=[Anc]=1, then the calculation using Helgeson's program SUPCRT yields Log K = - 1.675 at 300°C. With the composition of equilibrium solution available and using the Berndt's program offering the calculations of the activity of dissolved components in solutions close in composition to the seawater (Berndt 1987), we can calculate Log K using data obtained. Since plagioclase also grew, the composition of which was determined employing microprobe,anorthite activity was calculated by Aranovich's formula (Aranovich 1983):

$$[An]=g_{An}X_{An}(1+X_{An})^2/4$$, where

$$g_{An} = \exp\{(1-X_{An})^2(2050+9392X_{An})/RT\}$$

Using data obtained, we constructed an activity-activity diagram (fig. 2). It shows that an increase in the activity of anorthite constituent of plagioclase should undoubtedly result in a decrease in the ratio $[SiO_2(aq)]/[Ca^{2+}]$ to preserve the equilibrium in the system. Despite the fact that the activities of anorthite constituent in basanite [An]=1.25 and in Hawaiian basalt [An]=1.23 are similar, equilibrium constants differ considerably, particularly because of different Ca contents in solutions (fig. 1). It seems to be due to another reaction types, e.g., pyroxene and/or hydromica origin, also confirmed by different deviation from an inferred equilibrium curve, obtained using SUPCRT program. Na , $SiO_2(aq)$, and An activities in the runs with basanite and Hawaiian basalt are quite close suggesting a poor dependence of analcime stability in the runs at 300°C on $[Ca^{2+}]$.The primary composition of alkali basalt presumably affects the composition and/or structure of analcime growing over it.

REFERENCES

Aranovich, L.Y. 1983. Biotit-granatovye ravnovesiya v metapelitach. Ocherki fisiko-chimicheskoy petrologii. 11: 121-135. Moskva: Russia.

Berndt, M.E. 1987. Experimental and theoretical constraints on the origin of mid-ocean ridge geothermal fluids. A thesis submitted to the University of Minnesota.

Bischoff, J.L. & W.E. Seyfried, 1978. Hydrothermal chemistry of seawater from 25 to 350°C. Amer.Jorn.Sci. 278:836-860

Hydrothermal experimental techniques.1987. G.C.Ulmer & H.L.Barnes (eds.),

Seyfried, W.E & J.L. Bischoff, 1981. Experimental seawater-basalt interaction at 300°C and 500 bars: chemical exchange, secondary mineral formation and implications for the transport of heavy metals. Geochim. et Cosmochim.Acta. 45: 309-321.

Water-Rock Interaction, Kharaka & Maest (eds) © 1992 Balkema, Rotterdam. ISBN 90 5410 075 3

Hydrothermal water-basalts interaction in the ocean

Victor B. Kurnosov
Geological Institute, Moscow, Russia

ABSTRACT: Mineralization of solutions, circulating in the oceanic crust, is believed to occur principally due to the extraction of chemical elements from the enclosing rocks. Then some removed constituents are transported by underwater hot springs into the ocean. They form sulfide and other hydrothermal deposits on the sea floor and hydrothermal metalliferous sediments. In addition, together with the dissolved material from rivers, they replenish sea water depleted during nonterrigenous sedimentation.

1. INTRODUCTION

Many investigators consider the interaction between sea water and basalt to be basic to the circulation of elements in the crust-ocean system. The study of this reaction sheds light upon the formation of underwater mineral deposits and can promote further understanding of the geochemical balance in the ocean and the relation between crustal and mantle sources of hydrothermal material in the ocean (Strakhov, 1976; Baturin, 1978; Volkov, 1981).

Up to this day there are many problems still unresolved. In this paper, I will attempt to answer the following questions. (1) Under what conditions do basalts alter? (2) What masses of chemical elements are removed from basalts under these conditions? (3) What is the real contribution of chemical elements removed from basalts for supplying the ocean with sedimentary and ore forming material?

2. CONDITIONS OF ALTERATION

Basalts altered under both oxidative and "non-oxidative" conditions at low as well as at high temperatures were recovered from deep-sea holes. But oxidative basalts were observed only at the flanks of mid-oceanic ridges. This relationship allows us to arrange conditions of basaltic alteration in the ocean, diverse at the first glance, in a certain order, determined first of all by seafloor-spreading conditions. Basalts moving from the axis of the ridge definitely go through some periods of alteration.

The first period of basaltic alteration on mid-oceanic ridges takes place in axial zones under "non-oxidative" conditions at increased temperature, where water flow is mainly upward. For example, alteration of basalts under these conditions was recognized in holes of the Gulf of California (Legs 64 and 65 DSDP).

The second period of alteration takes place during the shift of basalts to the flanks of convection cells under the conditions of descending flows of sea water into the crust. In this period, the conditions of the first period are still preserved, largely in lower levels of the basaltic block. The upper part of the basaltic basement is oxidized. Thus, during the second period an oxidative type of alteration is superimposed on a "non-oxidative" one. The maximum depth of oxidation was observed in Hole 504B at the Costa Rica Rift; it is about 300 m. At the East-Pacific Rise, oxidative alterations of basalts appeared in 0.5 mln years after their origin. At the Mid-Atlantic Ridge, where the spreading rate is lower, they appeared not earlier than two million years after their origin. The oxidative type includes two stages of basaltic alteration: halmyrolysis and active oxidation.

Thus, the study of any block of the basaltic layer of the oceanic crust, presenting itself as an open hydrothermal convection cell, clearly shows that major alterations of rocks occur during the hydrothermal stage under "non-oxidative" conditions. The other stages of basaltic alterations currently distinguished are poorly expressed or of secondary significance to this process. These stages are: consolidation, halmyrolysis, and active oxidation.

Unlike basalts from mid-oceanic ridges basalts from volcanic seamounts do not alter in superimposed stages, as the seamounts do not shift horizontally relative to volcanic centers. Zonal alteration of basalts is observed there under "non-oxidative" conditions at increased temperatures in the central parts and under oxidative low-temperature conditions at the flanks.

Under "non-oxidative" hydrothermal conditions the main newly-formed mineral, replacing the principal mineral constituents of basalts, is trioctahedral smectite (Kurnosov 1986). This circumstance allows to distinguish a smectite facies of basaltic alteration in the oceanic crust. It is largely similar to the zeolitic facies, but its mineralogy reflects specific alteration of basalts in the ocean. The lower boundary of the smectite facies was determined by studying material recovered from Hole 504B on Leg 83 DSDP (Anderson, Honnorez, Becker et al. 1982) and studying basalts recovered from deep holes in Iceland and from Troodos ophiolite in Cyprus, at a depth of 600 to 1000 m from the cover of the basaltic basement. Under the conditions of the smectite facies, the major chemical process of basaltic alteration involves their hydration and the removal of the majority of both rock-forming and trace elements from rocks.

Another group of secondary minerals is formed under oxidative conditions at temperatures of as much as 50 to 80°C (Kurnosov 1986). They are, first of all, minerals containing iron oxide and potassium, for example, celadonite-glauconite, iron hydroxides, phillipsite, smectites of the nontronite group, and potassium feldspar. Active oxidation and hydration of rocks along the fractures, enrichment of these zones with potassium and the loss of calcium, magnesium, and probably silicon, manganese, and sodium occur at this stage. Pyrite and titanomagnetite are oxidized under the same conditions.

Thus, the answer to the first question is: rocks of the basaltic layer of the oceanic crust alter mainly under "non-oxidative" hydrothermal conditions.

3. MASSES OF CHEMICAL ELEMENTS

The answer to the problem on dominant conditions of the alteration of basalts of the oceanic crust allows us to concentrate our attempts on the calculation of material balance for basalts altered under "non-oxidative" hydrothermal conditions.

The idea of removal of chemical elements from basalts during their interaction with sea water being essentially supported by experiments, started by Bischoff & Dickson (1975), it is important to access quantitatively this process in nature. We have chosen profiles of the deepest holes, maximally drilled in basaltic basement: Hole 504B (Costa Rica Rift), Hole 485A (Gulf of California), Hole 462A (Nauru Basin), and Hole 433C (Emperor Chain). Every variety of altered rocks, recovered from these holes, was analyzed.

To calculate material balance we used the method of atomic-volume recalculation of chemical analyses, using porosity of rocks. Recalculation of chemical analyses was made by the formula (Kazitzyn & Rudnik 1968):

$$P_i = 166 \cdot a_i \cdot P_o{}^i \cdot d_v \cdot R_o, \qquad (1)$$

where P_i is a chemical element content (g) in 1 m^3 of rock; a_i - an element atomic weight; $P_o{}^1$ - a chemical analysis value; d_v - sample density; R_o - transition coefficient for each oxide.

Recalculation of the results of spectral analyses was made using a more simple formula:

$$P_i = S_i \cdot d_v, \qquad (2)$$

where S_i is a minor element content ($n \cdot 10^{-4}$% or ppm).

According to their chemical composition, texture, and porosity we recognized some groups of basalts in every studied area. For each group a diagram was plotted, illustrating the relation between the degree of basaltic alteration and content of element in grams for 1 m^3 of a rock. Then, using these results, we calculated losses of material by conventional pillars of magmatic rocks 500 m high, generally corresponding to the maximum depth of penetration, and 1 m^2 at the base. In each profile a natural collection and proportions of rocks were studied.

Data on key profiles helped us to determine annual losses of chemical element masses by rocks of the basaltic layer of the oceanic crust in mid-oceanic ridges and volcanic seamounts (Table 1). Counting procedure for mid-oceanic ridges and seamounts was published by Kurnosov (1986).

Table 1. Volumes of chemical elements in solution supplied into the ocean, mln tons.

1	2	3	4
Si	196	34.9 (?)	577
Ca	592	48 (?)	314
Al	2	2.37	138
Fe	1.6	>13.6	118
Mg	147	?	60
Na	290	?	54
Ti	0.12	2	?
Mn	0.4	6.6	4.8
P	1.6	>0.066 (?)	0.57
Ni	0.1	>0.00059*	0.9 (?)
Zn	0.8	>0.00264*	0.47
Cu	0.28	>0.00025*	0.29
Co	0.01	>0.00009*	0.16

1 - Elements; 2 - Annual fluvial drainage in solution into ocean (Gordeev 1983); 3 - Annual hydrothermal supply into ocean; 4 - Annual loss of chemical elements by basalts from mid-oceanic ridges and volcanic seamounts.
* - Bogdanov 1979.

4. GEOCHEMICAL BALANCE

To estimate the participation of material extracted from basalts in geochemical balance in the ocean, we should know what masses of chemical elements were delivered into the ocean by underwater hot springs. There are two ways to know. The first way is to compare the masses of material delivered from land with those accumulated on the floor. The comparison of the mobile forms of chemical elements is the most promising. Another way is a direct estimation of the amount of elements delivered into the ocean by underwater hot springs. Relatively complete data allowing us to use both ways are available for Mn only.

The calculation indicates, that 6.6 mln tons of Mn may be attributed to a hydrothermal source. This value is of the same order as the value of the annual dispersion of hydrothermal manganese (7.3 mln tons) obtained independently by Zelenov & Ivanenkov (1983), as well as calculated by the "helium method" - 10 mln tons (Lisitzyn 1983).

At present, data (Lisitzyn 1974, 1987) on the supply of the ocean with other chemical elements from hydrothermal sources are scarce as a rule. Approximate estimations of the hydrothermal supply of the ocean with Al, Ca, Fe, P, Si and Ti were made.

The comparison of columns 2 and 3 in Table 1 indicates the influx of Si, Ca, P, Ni, Zn, Cu, Co and probably Mg and Na, especially due to fluvial discharge. At the same time amounts of trace elements and P from the underwater hot springs may be significantly higher. Hydrothermal source dominantly causes Fe, Mn and Ti influx in the ocean and is comparable with fluvial Al influx.

Let us compare the amounts of chemical elements delivered annually by underwater hot springs into ocean with the amounts of elements extracted from rocks of the basaltic layer for the world ocean. This comparison shows that the basaltic source of material provides completely and most often with a considerable excess for the Fe, Ni, Zn, Cu, Co, Si, Al, Ti, and to a considerable degree Mn for the mineralization of underwater hydrothermal sources.

An excess of chemical elements extracted from basalts and not transported into the ocean shows that considerable amounts remain within the basaltic basement and in veins and vesicle fillings.

5. CONCLUSIONS

Rocks of the basaltic layer of the oceanic crust dominantly alter under "non-oxidative" hydrothermal conditions and loose petrogenic and trace elements. Chemical elements leached from basalts due to their interaction with sea and metamorphic water in hydrothermal convective cells known in mid-oceanic ridges and seamounts participate in geochemical balance of dissolved matter in the ocean together with solutions of underwater hot springs.

The obtained data illuminate the role of material from hydrothermal and crustal sources, in Fe-Mn nodules and crusts formation. The example of Mn clearly shows the leading role of hydrothermal sources in this process. Rivers supply insignificant amounts of dissolved Mn.

REFERENCES

Anderson, R.N., Honnorez, J., Becker, K. et al. 1982. DSDP, Hole 504B, the first reference section over 1 km through layer 2 of the oceanic crust. *Macmillan Journal* 300: 589-594.

Baturin, G.N. 1978. *Phosphorites on the oceanic floor*. Moscow: Nauka.

Bischoff, J.L. & Dickson, F.W. 1975. Seawater-basalt interaction at 200°C and 500 bar: Implication of seafloor heavy-metal deposits and regulation of seawater chemistry. *Earth and Planetary Science Letters* 25: 385-397.

Bogdanov, Y.A., Lisitzyn, A.P., Migdisov, A.A. et al. 1979. *Metalliferous sediments from South-East Pacific*. Moscow: Nauka.

Gordeev, V.V. 1983. *Geochemical features of fluvial discharge in the ocean*. Moscow: Nauka.

Kazitzyn, Y.V. & Rudnik, V.A. 1968. *Handbook on count for mass balance and inner energy at metasomatic rocks formation*. Moscow: Nedra.

Kurnosov, V.B. 1986. *Hydrothermal alterations of basalts in Pacific ocean and metalbearing deposits (on deep-sea drilling data)*. Moscow: Nauka.

Lisitzyn, A.P. 1974. *Sedimentation in oceans*. Moscow: Nauka.

Lisitzyn, A.P. *Oceanic biogeochemistry*. Moscow: Nauka.

Strakhov, N.M. 1976. *Geochemical problems in modern oceanic lithogenesis*. Moscow: Nauka.

Volkov, I.I. 1981. Manganese balance in oceanic sedimentary cycle. *Lithology and mineral resources* 3: 25-34.

Zelenov, K.K. & Ivanenkov, V.N. 1983. Modern subaqueous vulcanism influence on oceanic water chemistry. *Geology and Prospecting* 11: 3-26.

Water-Rock Interaction, Kharaka & Maest (eds) © 1992 Balkema, Rotterdam. ISBN 90 5410 075 3

Oceanic borehole fluid chemistry from Site 504B: Seawater/rock reaction within the borehole environment

Andrew J. Magenheim & Joris M. Gieskes
Scripps Institution of Oceanography, La Jolla, Calif., USA

Greg Bayhurst
Los Alamos National Laboratory, N.Mex., USA

Jeff Alt
University of Michigan, Ann Arbor, Mich., USA

Abstract. Open borehole fluid samples collected from Site 504B show the effects of seawater/basalt reaction. These samples exhibit increased concentrations of Ca and Si, and decreased concentrations of Na, Mg, K, and SO_4^{2-}. While in the past the source of the reacted fluid component has been speculated to be fluid advecting into the borehole from the formation, here these changes in fluid chemistry are interpreted in terms of reactions occuring within the borehole environment at elevated temperatures in the interim between drilling legs. This conclusion is based on comparison of fluid compositional changes between drilling legs. Examination of basaltic rubble from the bottom of the borehole provides strong evidence of seawater alteration while residing at the borehole bottom. This evidence includes recovery of anhydrite crusts and pieces of basaltic rubble coated with chloritic reaction products.

INTRODUCTION

The study of fluids from open oceanic boreholes can in principle provide information regarding the nature of fluids circulating through the crust. Mottl and Gieskes (1990) have reviewed the limited amount of work done in this field prior to 1990. Three processes may be important in determining the borehole fluid composition: 1) mixing of the borehole fluids with oceanic bottom seawater via convective exchange, 2) reaction of the fluids with wall rocks or drilling debris left inside the borehole, and 3) diffusive and/or convective exchange with formation fluids. Slow convective exchange of bottom water into Hole 504B has been shown to occur through examination of tritium concentrations (e.g Mottl et al. 1985). However, the source of the reacted fluid component, responsible for large changes in the major elements in the borehole fluids has not been conclusively identified.

Hole 504B is located on the Southern Flank of the Costa Rica Rift. This site is the only oceanic borehole to penetrate normal oceanic crust through the pillow lavas and into the sheeted dike complex. The borehole was visited during DSDP/ODP Legs 69, 70, 83, 92, 111, 137 and 140. Currently its total depth is 2000.4 meters below sea floor (mbsf). The borehole is cased through the sediment section to a depth of 274.5 mbsf.

Temperature logs of the open borehole (taken prior to any drilling disturbances) have indicated draw-down of bottom seawater into the upper 100 meters of basement (275-375 mbsf), as a result of hydrostatic underpressure in this part of the formation (Becker et al, 1985). Below this inflow zone temperatures are largely the product of the conductive geothermal gradient (Gable et al, 1989), with a bottom temperature of 160°C measured at 1560 mbsf at the start of Leg 137 (Shipboard Scientific Party, 1992).

Borehole fluid samples have been collected during DSDP legs 70, 83, 92, and ODP Legs 111 and 137. The time between legs has varied between 37 to 1633 days. The maximum borehole temperature has increased from 80 °C during Leg 70 to 162 °C on Leg 137. Prior to leaving the site, the borehole has been prepared for future reoccupation by flushing the borehole with several volumes of surface seawater to remove drilling mud and debris. Fluid samples were collected prior to drilling operations on subsequent legs to avoid disturbances of the borehole fluids. This procedure has not been without difficulties resulting from inadequate flushing of the borehole prior to leaving the site on previous legs (particularly Leg 92). Sample integrety problems from sampler malfunction and leakage resulting from contraction of the fluids during cooling has also been encountered (see Shipboard Scientific Party, 1992).

Chemical analyses of the borehole fluids show large changes in fluid composition relative to seawater, including decreased magnesium, sodium, and sulfate concentrations and increased calcium and silica concentrations [c.f., Mottl and Gieskes, 1990 and Magenheim et al, in press]. Such chemical trends strongly suggest seawater/basalt reaction. Magenheim et al. (in press) argue that these changes can be accounted for solely by reaction of the fluids with the wall rocks and rubble at the bottom of the borehole. Here this argument is enhanced by observations of the solid phases from the rubble at the borehole bottom recovered prior to drilling during Leg 137. Anhydrite crusts and chlorite coated basaltic rubble reveal some of the probable reaction products.

OBSERVATIONS FROM 504B

A complete discussion of the borehole fluid chemistry including elemental profiles for the major elements is presented in Magenheim et al (in press) and Shipboard Scientific Party (1992). Observations include inrcease of Ca, Si, and Cl, and decreases in Mg, Na, $SO_4{}^{2-}$, K, and Alkalinity with depth Figure 1 presents graphs of the changes in Ca, Na, $SO_4{}^{2-}$, and K relative to Mg uptake for each visit to Site 504B, which borehole fluid samples were obtained. The Ca concentrations have been corrected for the precipitation of anhydrite from the $SO_4{}^{2-}$ data as discussed below. This element versus element presentation provides a basis for establishing conservative mixing of the borehole fluids with bottom seawater. Also, the relative change in concentration of each species relative to each other are shown. The data for each leg define a discrete mixing line that intersects near the composition of bottom seawater. This confirms the notion of conservative mixing of bottom seawater with a reacted fluid component. The exact compositon of the reacted fluid component lies along the mixing line somewhere beyond the lowest Mg value.

The removal of $SO_4{}^{2-}$ is due to the formation of anhydrite, from the Ca enriched borehole solutions. Large (>5cm long by 0.25 cm thick) anhydrite crusts were recovered from the borehole bottom durng Leg 137 (Shipboard Scientific Party, 1992). Anhydrite was also observed on the filters from the fluid samples. Sulfur isotopic analysis of the sulfate from the borehole fluids yielded $\delta^{34}S$ of 19.5‰ for sample BL-8 (most $SO_4{}^{2-}$ depleted sample), and $\delta^{34}S$ of the anhydrite of +21.4‰ (Magenheim et al, in press). These isotopic results are consistent with the slight positive fractionation of sulfate (modern seawater has a $\delta^{34}S$ of +20‰) expected during the precipitation of anhydrite (Hosler and Kaplan, 1966).

Experimental observations provide further indication that anhydrite precipitation is expected in the moderate temperature boreholes. Precipitation of anhydrite has been observed between 150°C and 200°C upon heating seawater (Bischoff and Seyfried, 1978) and in experimental seawater-rock reactions at temperatures as low as 150°C (Seyfried and Bischoff, 1979). The borehole fluids of 504B show $SO_4{}^{2-}$ removal for all legs in which the maximum temperature of the borehole is 120°C to 160°C (Legs 83 through 137). Sulfate removal was not apparent during Leg 70 when the maximum borehole temperature was only 80°C (Figure 1), which is in excellent agreement with the experimental evidence suggesting that anhydrite should not precipitate at lower tempertures.

The increase in Ca due to seawater/rock interaction is greater than measured, as precipitation of anhydrite will decrease the observed Ca concentration. Correction of the measured Ca concentrations can be made for the decrease in these elements due to anhydrite precipitation based on the change in $SO_4{}^{2-}$ from seawater concentration. The corrected Ca concentrations reveal the effects of seawater/basalt reaction in absence of this removal (Figure 1).

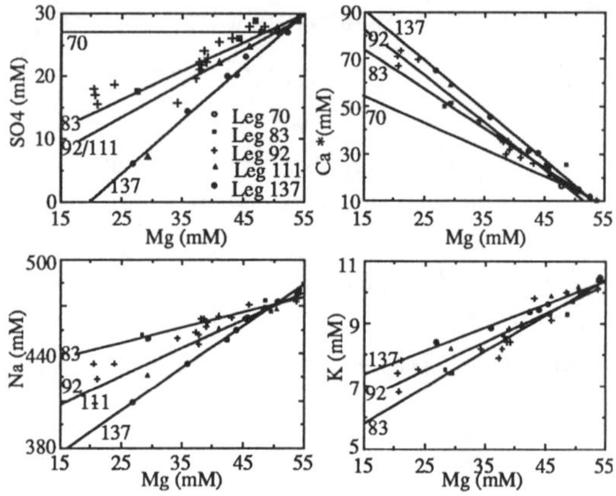

Figure 1: Elemental mixing diagrams for fluid samples taken from 504B during Legs 70, 83, 92, 111, and 137 (Magenheim et al, in press). Lines represent mixing lines for each leg. The Ca mixing line uses data corrected for anhydrite precipitation (see text).

It is evident that each leg forms a distinct mixing line (Figure 1), with the exception of Legs 92 and 111. Legs 92 and 111 may in fact have the same mixing line as the borehole was not deepened after sampling on Leg 92, and considerable scatter is introduced to the Leg 92 samples due to the large amount of drilling mud present in the borehole. In general the trend for increased Ca enrichment and Na depletion is evident from Leg 70 through 137. This trend is concurrent with the deepening of 504B and, therfore, an increase of the maximum borehole temperature from 80 °C to 162 °C. Also of note is the decrease in K uptake relative to Mg uptake over the same interval. This is particularly relevent since experimental investigations show the uptake of K during water/rock interaction at low temperature, and leaching from rock above 150°C (Mottl, 1983). Figure 1 shows that K uptake has clearly waned as the maximum temperature of the borehole fluids has increased.

SEM and EDS investigations of anhydrite crusts and surfaces of the basaltic rubble recovered from the bottom of Hole 504B provide some clues to reactions occurring within the borehole. The pieces of rubble (1-cm diameter) had various compositions revealing that they originated from not only near the hole bottom, but from shallower depths as well. The basaltic rubble appeared to have a greenish alteration phase on all exterior surfaces. The coatings were indentified to be chlorite by XRD. SEM-EDS investigation showed chlorite phases covering the surface up to 200 µm thick (Figure 2). The chloritic phases also extended into the interiors of the basalt pieces along fractures. The 100-200 µm thick coatings are Mg rich aluminosilicates (predominantly chlorite.

These chlorites are apparently products of seawater-basalt reaction, at temperutures near 160°C,

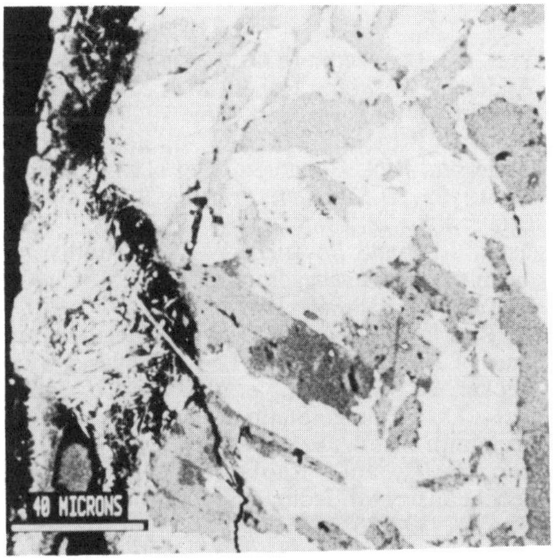

Figure 2: SEM image of chlorite coating on a piece of basaltic rubble. Notice darker areas in the interior of the basalt are areas of Na uptake into plagioclase.

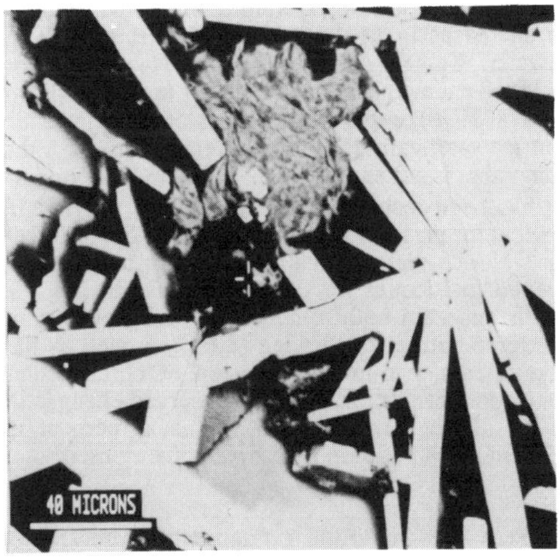

Figure 3: SEM image of an ahydrite crust from 504B, with interbeded talc-chlorite.

that have formed while the rubble resided at the bottom of the borehole for the 1633 days since Leg 111. The coatings on the rubble support the hypothesis that reactions between fluids and rocks within the borehole do occur.

A similar Mg rich alteration phase is also present in the anhydrite crusts (Figure 3) at levels of 4-6 weight percent (based on anhydrite dissolution). SEM-EDS analysis of the anhydrite showed a small amount of stongly layered silicate inter-twined with the anhydrite crystals. While these silicates cannot conclusively be shown to be authigenic precipitates from the borehole fluids, the interbedded nature of the silicate and anhydrite suggests coprecipitation of the two phases. XRD was not successful on the silicates from the anyhydrite, due to the small size and dispersion through the anhydrite. EDS results revealed that the composition of these silicates was an end-member chlorite, talc chlorite.

Clearly these chlorites are the sink for Mg from the borehole fluids. As the borehole rocks are altered Ca and Si are released. In fact formation of Mg silicates has been simulated in seawater/basalt experiments at temperatures as low as 150°C and observations of Ca increase and Mg decrease in the fluids have been observed at temperaures as low as 70°C (Seyfried and Bischoff, 1979). The sink for Na from the borehole fluids is more elusive. Examination of the basaltic rubble shows evidence of calcic plagioclase being altered to albite (Figure 2). Actinolite was also observed. However, both of these alteration phases are present in many sections throughout the drill hole. Experimental simulations have not shown appreciable Na uptake into the solid phases (Seyfried and Bischoff, 1979 and Seyfried and Mottl, 1982). Mottl (1983) points out that Na can be taken up into secondary albite and zeolites, however at high water/rock ratios and elevated temperatures leaching of Na from basalt is favored. Na uptake has been noted many low temperature formations where Mg is depleted in the interstitial fluids, apparently resulting from exchange of Ca for Na in the underlying basement (e.g., McDuff, 1981). The large Na decrease observed in the 504B borehole fluids is equivocal evidence of Na uptake into the solid phases. Experimental simulation of this process using the basaltic rubble from 504B is currently underway (J. Bischoff, personal communication).

DISCUSSION

Comparison of the changes in fluid chemistry with the observed alteration products should agree if alteration is occuring in the borehole. Alteration of the source rock releases Ca as chlorite is formed removing Mg. As an approximation Mg should be exchanged for Ca in a 1:1 ratio during the formation of chlorite. This stoichiometry works quite well for the case of Leg 70 where the ratio of Ca increase to Mg decrease is 1:1, but for subsequent legs Ca is added in excess of the 1:1 ratio for this proposed reaction.

As the hole was drilled deeper and the bottom hole temperature increased from 80°C to 160°C between legs 70 and 137, Ca has been released in greater proportion to Mg uptake and Na uptake has become succesively greater. The excess Ca increase could result from the albitization of plagioclase:

$$\text{Plagioclase} + 2\ Na^{+} \text{ ----> Albite} + Ca^{2+}$$

So Ca is released in a ratio of 1:2 relative to Na uptake.

The Ca increase can, therefore, be predicted given the decreases in Mg and Na concentrations from seawater. Table 1 presents the Mg, Na and Ca for the Mg=0 endmembers from Figure 1 (as in Mottl and Gieskes, 1990), expressed in terms of the difference from seawater values. While assuming Mg=0 for the endmember is not entirely justified as no fluid void of Mg has been recovered, this procedure provides a convient means for comparing the data from different legs. Notice that albitization does not need to be invoked to account for the Ca increase in Leg 70. This is expected both in terms of the lower bottom hole temperature for this leg (80°C) as well as the short residence time. Formation of chlorite and albitization, can account for the observed changes in Ca for all legs within 10%. This agreement is excellent given the simplified stoichiometry used.

Table 1: Reacted fluid component expressed as change from seawater concentration. Δ=concentration in seawater minus the concentration at zero Mg from Figure 1. Note ΔCa* is the Ca endmember corrected for the precipitation of anhydrite from the sulfate data. ΔCa_p is the predicted ΔCa from the ΔMg and ΔNa asuming 1ΔCa:1ΔMg and 1ΔCa:2ΔNa (see text).

Leg	mbsf	T°C	ΔMg	ΔNa	ΔCa*	ΔCa_p
70	489	80	-53.4	----	54.5	53.4
83	836	112	-53.4	-54	87.7	80.4
92/111	1350	145	-53.4	-95	93.5	101
137	1562	162	-53.4	-138	109.6	122

SUMMARY

The borehole fluids in Site 504B have been sampled on five drilling legs, and analyses have indicated varying degrees of seawater/rock reaction. Recovery of anhydrite crusts and chlorite coated basaltic rubble from the borehole prior to drilling during Leg 137 provides strong evidence that the site of fluid/rock reaction is in the rubble pile at the bottom of the borehole. Simple reaction sequences using observed reaction products, can account for the observed changes in the major elements Na, Mg, and Ca. The relative changes in fluid composition between legs reflect the increasing temperature at the bottom of the hole as in has been drilled deeper. Increased Ca release and Na uptake relative to Mg removal have accompanied increases in temperatures from 80°C during Leg 70 to 162 °C recorded on Leg 137. Anhydrite precipitation became important only after Leg 70, when the borehole temperatures exceeded 80°C. During this same period a relative decrease in the K uptake was also observed. These observations suggests that bottom hole temperature and therefore conditions inside the borehole determine the fluid chemistry, rather than inflow of fluid from the formation.

REFERENCES

Becker, K., Langseth, M.G., Von Herzen, R.P, Anderson, R.N. and Hobart, M.A., Deep crustal geothermal measurements, Hole 504B, Deep Sea Drilling Project Legs 69, 70, and 92. *In* Andersen, R.N. Honnorez, J. et al., Init. Repts. DSDP 83, U.S. Govt. Printing Press (Wash. D.C): 405-418, 1985.

Bischoff, J.L., and Seyfried, W.E., Hydrothermal chemistry of seawater from 25°C to 350°C. Am. J. Sci., 278: 838-860, 1978.

Gable, R., Morin, R.H., and Becker, K., Geothermal state of Hole 504B: ODP Leg 111 overview. *In* Becker, K., Sakai, H., *et al., Proc. ODP, Sci. Res., 111*: College Station, TX (Ocean Drilling Program), 87-96, 1989.

Holser, W.T. and Kaplan, I.R., Isotope geochemistry of sedimentary sulfates. *Chem. Geo.*, 1: 93-135, 1966.

Magenheim, A.J., Bayhurst, G., Alt, J.C., and Gieskes, J.M., ODP leg 137: Borehole fluid chemitry in Hole 504B. *Geophys. Res. Let.* , in press.

McDuff, R.E., Major cation gradients in DSDP interstitla waters: the role of diffusive exchange between seawater and the upper ocean crust. *Geochim. Cosmochim Acta*, 45: 1705-1713, 1981.

Mottl, M.J., Metabasalts, axial hot springs, and the structure of hydrothermal systems at mid-ocean ridges. *Geo Soc Amer. Bull.*, 94: 161-180, 1983.

Mottl, M.J., and Gieskes, J.M., Chemistry of waters sampled from oceanic basement boreholes, 1979-1988. *J. Geophys. Res.,* 95(B6): 9327-9342, 1990.

Mottl, M.J., Druffel, E.M., Hart, S.M., Lawrence, J.R., and Caltzman, E., Chemistry of hot water sampled in Hole 504B, DSDP Leg 83. *In* Anderson, R.N., Honnorez, J., et al., *Init. Repts. DSDP*, 83: U.S. Govt. Printing Office (Wash. D.C.), 315-330, 1985.

Seyfried, W.E. and Bischoff, J.L., Low temperature basalt alteration by seawater: an experimental study at 70°C and 150°C. *Geochim. Cosmochim. Acta,* 43:1937-1947, 1979.

Seyfried, W.E. and Mottl, M.J., Hydrothermal alteration of baslat under seawater dominated conditions, *Geochim. Cosmochim. Acta*, 46:985-1002, 1982.

Shipboard Scientific Party,*Proc. ODP, Init. Repts. Pt. A),* 137:College Station, TX (Ocean Drilling Program), 1992.

Water-Rock Interaction, Kharaka & Maest (eds) © 1992 Balkema, Rotterdam. ISBN 90 5410 075 3

Formation of authigenic smectite and zeolite and associated major element behavior during early diagenesis of volcanic ash in the Nankai Trough, Japan, ODP leg 131

H. Masuda & H. Tanaka
Department of Geosciences, Osaka University, Japan

T. Gamo
Ocean Research Institute, University of Tokyo, Japan

J. R. O'Neil, D. R. Peacor & W.-T. Jiang
Department of Geological Sciences, University of Michigan, Mich., USA

ABSTRACT: In a study of early diagenesis of volcanic glass in ocean sediments, determinations were made of the mineralogy and major element chemistry of altered layers of rhyolite ash and separates of authigenic minerals from Site 808 of ODP Leg 131 in the Nankai Trough. Depths in meters below seafloor (mbsf) and in situ temperatures of the first appearance of alteration minerals are: smectite at 200 mbsf (20°C), clinoptilolite at 646 mbsf (65°C), and analcite at 810 mbsf (75°C). Illite does not occur in the ash but has been observed in shale at about 560 mbsf (Underwood, Orr et al., 1991). Chemical changes in the alteration minerals are reflected in pore water chemistry. Details of authigenic mineral formation in the ash were elucidated by TEM and AEM analysis: K is fixed in smectite above 646 mbsf and in analcite below this level. Smectite is initially enriched in K and is a source of this element for later-forming clinoptilolite. Clinoptilolite is a sink for K and may prohibit illite from forming.

1. INTRODUCTION

Alteration of volcanic glass in ocean sediments plays an important role in authigenic mineral formation and the evolution of porewater chemistry during the early stages of diagenesis. For this reason it is important to study the mineralogy and chemistry of the altered material that forms during diagenesis of relatively pure rhyolitic glass that occurs in the ash layers of some ocean sediments. During ODP Leg 131, a complete sediment column was obtained through the décollement zone to the basement basalt from Site 808 in the Nankai Trough, a typical modern accretionary prism that has developed on the southern margin of southwest Japan (Fig. 1). Masuda et al. (in press) described the mineralogical and chemical characteristics of 72 ash layers collected from this sediment column in relation to degree of ash alteration during the early stages of diagenesis. They pointed out that ash alteration was a fundamental process affecting porewater chemistry in young oceanic sediments of the Nankai Trough.

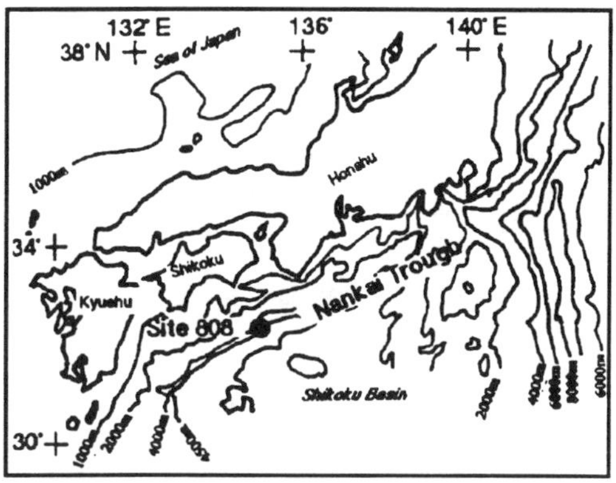

Figure 1. Location of Site 808 in the Nankai Trough, Leg 131

In the present study, 5 samples of glass and mineral separates were selected for examination under the electron microscope in order to delineate more precisely the nature of the alteration reactions that produce authigenic minerals during early diagenesis. The results of TEM (Transmission Electron Microscopy) and AEM (Analytical Elec-

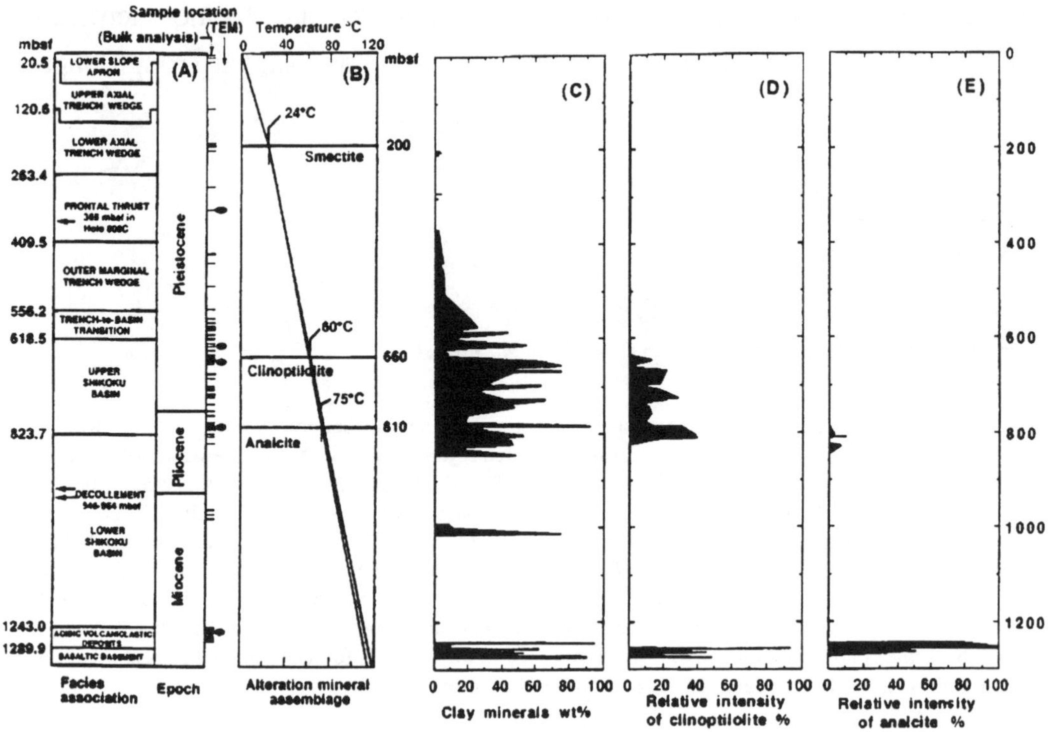

Figure 2. Lithology and sampling locations (a), temperature gradient (B), Wt. % of clay minerals (C), X-ray diffraction intensities of clinoptilolite (D) and analcite (E).

tron Microscopy) analyses identify hitherto unrecognized chemical variations that occur during the formation of authigenic minerals in the young ash layers of the Nankai Trough and presumably elsewhere.

2. SAMPLES AND ANALYTICAL TECHNIQUES

The sediments at Site 808 (0 to 1290 mbsf) can be broadly classified into two lithological units separated by a boundary at approximately 600 mbsf. The upper unit comprises clastic materials transported by turbidity currents, and the lower unit comprises mainly Shikoku Basin hemipelagic sediment. Ash layers are concentrated between 557 and 824 mbsf, a section corresponding to the trench-to-basin transition and the upper Shikoku Basin facies (Fig. 2A). The age of the basal sediment is between 13.6 and 15 Ma (Taira, Hill and Firth et al., 1991). Of the 72 ash layers examined by Masuda et al. (in press), 13 layers located between 1245 and 1290 mbsf are apparently different from those at shallower depths. They are white and contain quartz grains that are about 1 mm in diameter.

As reported by Masuda et al. (in press), the mineralogy was determined by x-ray diffractometry of powdered bulk samples, the weight per cent of clay minerals was calculated from the intensity of the (001) reflection, and the major element compositions of the bulk samples were measured by x-ray fluorescence photometry. TEM and AEM methods were applied to authigenic minerals and glass separates selected from 5 samples from various depths. Powdered samples embedded in resin were made into thin sections and thinned to a thickness of <100 nm with an ion thinning mill.

3. ALTERATION MINERALOGY

The ash layers usually contain detrital quartz, feldspar, K-mica, chlorite, and biogenic carbonates. Hornblende and biotite of volcanic origin are sometimes observed, especially in the sediments between 557 and 827 mbsf. These detrital minerals are only slightly altered even at the base of the sediment column. The weights per cent of clay minerals and X-ray diffraction intensities of two zeolites (clinoptilolite and analcite) are shown in Fig. 2C to 2E. The first appearance of smectite is at 200 mbsf. For the smectite in the studied ash samples from below 646 mbsf, the degree of crystallinity increases with increasing depth, although no illite was observed even in the deepest sediments. In contrast, about 20% of the smectite in shales at 560 mbsf (60°C) in the same sediment column had undergone the transition to illite (Underwood, Orr et al., 1991).

Mixed-layer smectite/illite is known to occur at about 50-65°C and complete transition to illite can occur at about 120°C (e.g., Perry and Hower, 1970). Thus, the lack of illite in the ash layers studied suggests a mode of diagenesis that is different from that normally encountered in other ocean sediments. In a related study, Altaner and Grim (1990) reported the absence of illite in diagenetically altered rhyolitic tuff in the Sucher Creek Formation in eastern Oregon. They proposed that illite formation was prohibited by the presence of K-rich clinoptilolite which formed in reponse to activities of K and H_4SiO_4 that were higher in the pore fluid than in the shale. As shown in Fig. 2D, clinoptilolite appears at 646 mbsf, and its presence could provide the explanation for the absence of illite in the sediments studied from the Nankai Trough.

4. TEM AND AEM ANALYSES OF GLASS AND AUTHIGENIC MINERALS

The major element chemistry of glass, smectite and clinoptilolite analyzed by AEM is given in Table 1. Ti/Al ratios of glass samples from 366 and 630 mbsf are typical for rhyolite and the concentrations of elements other than K are similar in samples from the two levels. The K content of the glass from 633 mbsf is significantly lower than that of the glass from 366 mbsf and this depletion suggests diffusional loss of K from the glass prior to its alteration. There is no glass left in the sample from 670 mbsf because it has all been altered.

Smectite from 634 mbsf is richer in K than smectite from 366 mbsf indicating that the K released from glass is first fixed into smectite in this depth range. It is normally assumed that, with an increase of temperture during diagenesis, such K-rich smectite will be a precursor of mixed-layer smectite/illite. The K content of smectite below 660 mbsf, however, is lower than that from 630 mbsf. On the other hand clinoptilolite becomes richer in K with depth. These observations suggest that clinoptilolite is a sink for K and, as such, prohibits the formation of illite as discussed above. There is also a tendency for Mg to increase and Ca to decrease with increasing depth between 366 and 810 mbsf suggesting a relation to smectite and zeolite formation: Mg moves from the glass and

Table 1. Major element chemical composition of glass, smectite (Sm) and clinoptilolite (Cp) in ash from ODP Site 808 in the Nankai Trough.

					Weight %				
Mineral	Glass	Glass	Sm	Sm	Sm	Sm	Sm	Cp	Cp
mbsf	366	634	366	634	670	810	1256	810	1256
SiO_2	72.20	80.50	58.20	64.10	60.10	65.10	63.30	56.80	73.10
Al_2O_3	14.90	13.20	29.70	18.10	21.90	20.10	23.90	27.40	18.10
TiO_2	0.51	0.22	0.00	0.00	0.00	0.53	0.00	0.00	0.00
FeO	2.24	1.38	4.77	10.70	8.67	6.20	4.03	1.00	0.00
MnO	0.00	0.00	0.00	0.00	0.00	0.00	0.33	0.00	0.00
MgO	0.91	0.48	2.92	4.24	5.33	4.84	4.35	0.00	0.65
CaO	2.68	1.76	2.73	0.31	2.07	0.84	2.95	9.76	3.74
Na_2O	0.92	0.71	0.62	0.65	1.01	1.75	0.00	1.54	0.56
K_2O	5.70	1.83	1.01	2.00	0.95	0.70	1.13	3.47	3.78
Total	100.00	100.00	100.00	100.00	100.00	100.00	100.00	100.00	100.00

porewater into smectite and Ca is fixed in clinoptilolite below 646 mbsf.

Ti removed from the glass is fixed in rutile rather than smectite. Although rutile was not detected by X-ray diffractometry, it was observed under the electron microscope in the samples from 810 to 1200 mbsf. The grain size of rutile was larger in the deeper samples indicating maturation of the grains with increasing temperature.

5. INTERACTION BETWEEN POREWATER AND ASH

It was determined from onboard analyses that the major element chemistry of porewaters isolated from this sediment column was affected by ash alteration (Taira, Hill and Firth et al., 1991). For example below 200 mbsf, the first appearance depth of smectite, H_4SiO_4 increases and K/Cl and Mg/Cl decrease rapidly, probably due to glass decomposition and neoformation of smectite. Inasmuch as rhyolite contains more SiO_2 than smectite, the excess SiO_2 will remain in solution. The K liberated during decomposition of the glass will be fixed in the neoformed smectite. It should be noted that the depth of approximately 560 mbsf where both H_4SiO_4 and the K/Cl ratio decrease steeply is concordant with the depth of first appearance of mixed-layer smectite/illite in shale. The implication is that smectite is the mineral that fixes K in shale, whereas clinoptilolite fixes K in ash layers (below 646 mbsf at this site).

Silica is also fixed into clinoptilolite below 646 mbsf, and authigenic SiO_2 minerals such as a cristobalite and quartz might also be a sink for silica. The behavior of Mg is similar to that of K. Na/Cl ratios in the porewaters decrease below 1000 mbsf, where large amounts of analcite are formed in the ash, suggesting that analcite is a sink for Na in the column.

6. CONCLUSIONS

The mineralogy and major element geochemistry of ash samples from ODP Site 808, the Nankai Trough, were determined in order to study the nature of authigenic mineral formation during the early stages of diagenesis of a chemically homogeneous material like rhyolite ash. Smectite and two zeolites (clinoptilolite and analcite) were the major authigenic minerals produced and minor rutile was also detected using the methods of TEM and AEM.

The depth variations of H_4SiO_4, K and Mg in the porewaters are in harmony with the alteration mineral association. When smectite appears at 200 mbsf, H_4SiO_4 increases and K and Mg decrease rapidly. Below 560 mbsf, where smectite/illite mixed layered minerals first appear in shale, H_4SiO_4 and K decrease. In the ash layers studied, the transition from smectite to illite does not occur even in the basal sediment at 120°C. Although K is initially fixed into smectite above 646 mbsf, the K content of smectite decreases below that depth. The formation of K-bearing clinoptilolite, which takes place below 646 mbsf, apparently inhibits the formation of smectite/illite mixed-layered minerals.

7. REFERENCES

Altaner, S.P. and Grim, R.E. (1990) Mineralogy, chemistry, and diagenesis of tuffs in the Sucher Creek Formation (Miocene), eastern Oregon. Clays Clay Mineral. 38: 561-572.

Masuda, H., Tanaka, H., Gamo, T., Soh, W. and Taira, A (in press). Major element chemistry and alteration mineralogy of volcanic ash, Site 808 in the Nankai Trough. In Hill, I., Taira, A., Firth, J.V. Proc. ODP Sci. Results, 131B: College Station, TX (Ocean Drilling Program).

Perry, E. and Hower, J. (1970) Burial diagenesis in Gulf Coast pelitic sediments. Clays Clay Mineral. 18:165-177.

Taira, A., Hill, I. and Firth, J. et al. (1991) Site 808 Summary. Proc. ODP Sci. Results, 131 A: College Station, TX (Ocean Drilling Program), p.434.

Underwood, M., Orr, R. and the Leg 131 Shipboard Scientific Party (1991) Clay mineralogy of the Nankai accretionary prism: Preliminary results from Leg 131 of the Ocean Drilling Program. Abst. Papers of Muroto International Conference on accretionary prisms: Muroto, p.17.

Water-Rock Interaction, Kharaka & Maest (eds) © 1992 Balkema, Rotterdam. ISBN 90 5410 075 3

Geochemistry of subduction-induced cold seepage at the eastern Nankai Trough, Japan

H. Sakai – *Department of Earth Sciences, Yamagata University, Japan*
T. Gamo – *Ocean Research Institute, University of Tokyo, Japan*
Y. Ogawa – *Department of Earth and Planetary Sciences, Kyushu University, Japan*
J. Boulegue – *Geochemistry and Metallogenesis Laboratory, University of Pierre and Marie Curie, France*

ABSTRACT: Carbon isotopic ratios of the carbonate encrustation associated with the cold seepage at the eastern Nankai Trough suggest that the carbonates originated from biogenic methane carried to surface by subduction-induced fluid venting system. Their oxygen isotopic ratios and other lines of evidence reflect a multi-stage formation model for the origin of the cold seepage of this area.

1 INTRODUCTION

The KAIKO Project, a joint Japan-French program conducted in 1985, has located five active cold seepage areas on the landward slopes of the trenches that engrave the east side of Honshu and Shikoku, Japan. In order to obtain further information on their distribution and geochemical nature, 23 dives of the French submersible "Nautile" were made at 5 selected sites of the eastern Nankai Trough (Fig. 1), one of the areas studied in 1985, during the second phase of the Project conducted in 1989. Of these, 22 dives found active cold seeps accompanied by dense animal communities dominated by *Calyptogena*-type bivalves and/or vestimentiferan tubeworms.

Some of them are intimately associated with carbonate-cemented sandstone. Some of these carbonates, together with methane-rich bottom seawater just above the seepages were recovered during the dive studies. This report will discuss the origins of the cold seepages based on the stable isotopic compositions of the carbonates and chemical properties of the bottom seawaters.

2 GEOLOGICAL SETTINGS AND SAMPLES COLLECTED

Two distinct types of cold seeps were identified. Type 1 seeps are located in soft sediment-covered seafloor and are in close association with dense biological communities dominated by *Calyptogena*-type bivalves of a few to more than 30 cm shell length. This type of vent was found exclusively landward of the first deformation at sites 1 and 5, in water depths from 3800 to 3600 m (Fig. 1). All the calcareous sediments recovered from this type of seep are calcite-cemented sandstones of uniform lithology and texture, exhibiting pipe-like structures of outside diameter larger than 10cm and wall thickness of a few to 5 cm.

Type 2 seeps are associated with outcrops of indurated calcareous sandstones of various forms and shapes and are characterized by animal communities dominated by vestimentiferan tubeworms, in addition to those by *C*-type bivalves. Seeps of this type are found only at site 2, about 18 km to the south of sites 1 and 5. The seafloor of this area is uplifted by

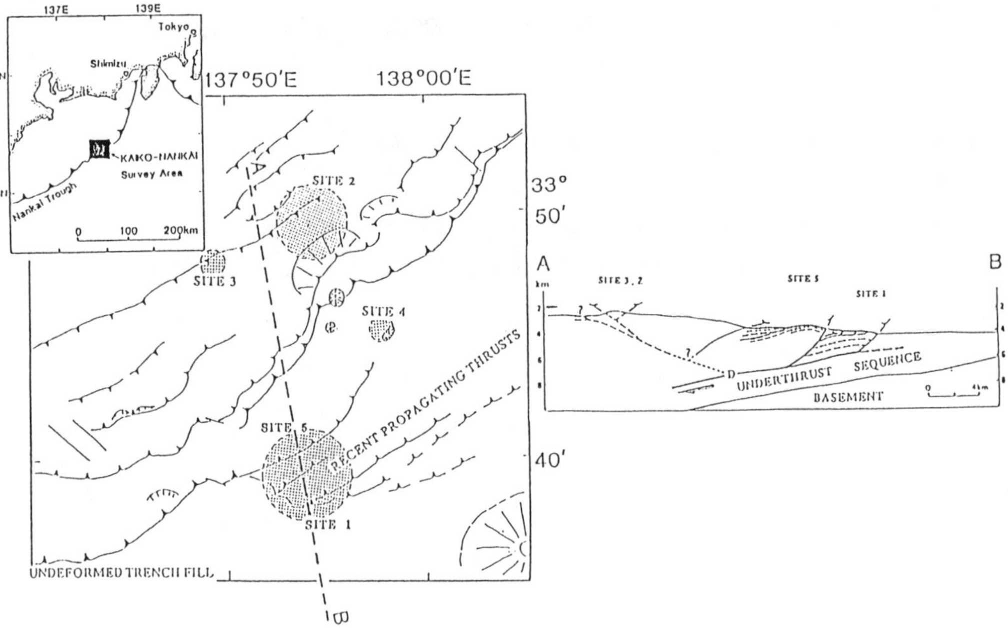

Fig.1 Locality map of sites 1 to 5 at the toe of the accretionary prism in the eastern Nankai Trough.

subduction-induced compaction and backthrusting to water depths of 1900 to 2000 m and mostly covered with outcrops and boulders of indurated calcareous sediment.

The calcareous sandstones collected from type 2 seeps are heterogeneous in lithology and texture, comprising carbonate-cemented sandstones similar to those from type 1 seeps and shelly sandstones with various sizes of fossil *C*-type clams.

3 EXPERIMENTAL

Calcareous sandstones pipes of homogeneous lithology were cut into a few-centimeter-thick slices perpendicular to their longer axis and powders were drilled out for analyses from desired positions using a dental drill. From heterogeneous and shelly sandstones, small chips were cut. These chips were then carefully separated into each constituent (shell, calcite cement, sandstone, etc.). Powdered carbonate samples were analyzed for carbon and oxygen isotopic ratios following the method of McCrea (1950). The isotopic ratios are reported in the familiar delta notation relative to PDB. Analytical errors are ± 0.1 permil for the isotopic ratios.

4 RESULTS

The carbon and oxygen isotopic ratios of the carbonates studied vary widely as are shown in Fig. 2. The lightest carbon isotopic ratio, -55.9 permil, was found with a calcite-aragonite vein. Such light carbon isotopic ratios indicate that methane is the ultimate source of carbon in the carbonate cement. Fig. 2 reveals that the carbonate specimens may be divided into three groups, each showing a specific isotopic ratio variation pattern:

1. KN 2-8-2 and 6-5 from a type 2 seep and KN 10-2 and 18-2 from a type 1

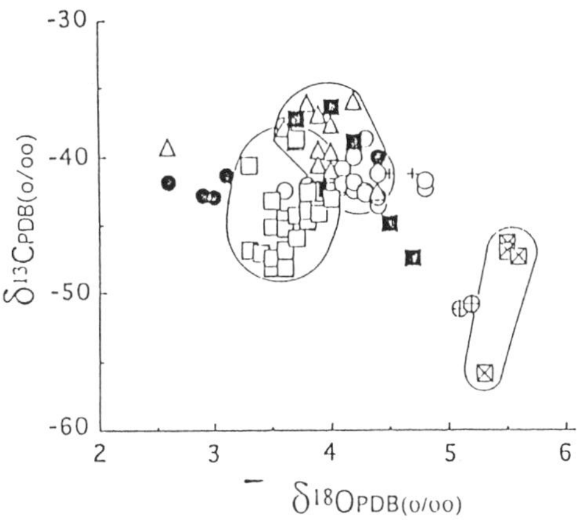

Fig. 2. $\delta^{13}C$ versus $\delta^{18}O$ values of carbonates in the sandstones studied. ⊠=KN 2-8-2; ○=KN 5-1 (calcite cement); ●=KN 5-1 (aragonite-bearing calcite cements filling cavities between fragments of shells and lithic components); △=KN 6-5; +=KN 15-7; ■=KN 18-1 B and C; □= KN 10-2 and Kn 18-2; ⊕=KN 19-A and 19-B.

seep. In each specimen the carbon isotopic ratios vary widely and independently from the oxygen isotope ratios, which show a narrow range of variation (three enclosed zones of Fig. 2).

2. KN 18-1 B and C from a type 1 seep, where the carbon and oxygen isotopic ratios vary in parallel with each other, approaching the vein carbonate of KN 2-8-2.

3. KN 5-1 (circles and dots) from type 2 seep show a wide range of oxygen isotopic variation (+2.9 to +4.6 permil) with a much narrower range of carbon isotopic variation than the other groups.

These contrasting variation patterns suggest that the carbonate encrustation took place under at least three different isotopic environments. Two distinct cases are examined in some detail.

5 ORIGIN OF THE CALCAREOUS SANDSTONE PIPE, KN-18-2

This is a part of a sandstone pipe

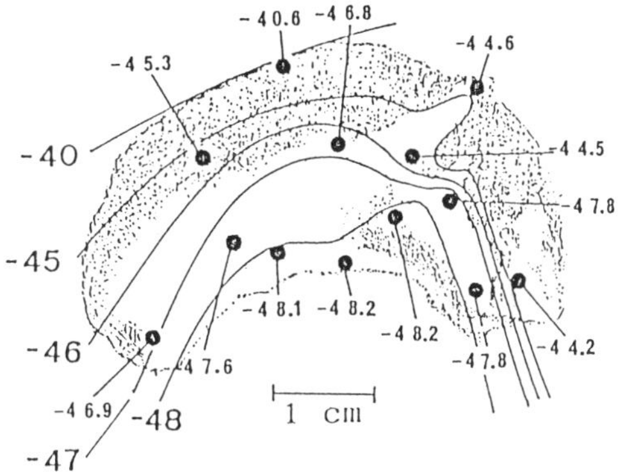

Fig. 3. Concentric distribution of carbon isotopic ratios accross the wall of the sandstone pipe KN18-2.

about 40 cm long and 8-10 cm in diameter, with a conduit of 2.5 to 3 cm in diameter and was found half buried in the sediment covered, 3682 m deep seafloor of site 5. As is shown in Fig. 3, the carbon isotopic ratios become heavier concentrically from the inner conduit outward. The carbonate contents roughly show inverse correlation with carbon isotopic ratios, gently decreasing outwards. Isotopic and mass balance calculations estimate the carbon isotopic ratio of the methane-derived end member carbonates to be roughly -51 permil.

The concentric distribution of the carbon isotopic ratios is best explained if the carbonate cementation took place by mixing. Mixing was between the methane-derived carbonate diffusing radially outward from the conduit into surrounding sandstone sediments and marine carbonates of relatively heavy isotopic ratios in pore spaces.

6 CEMENTED AGGREGATES OF LARGE FOSSIL SHELL AND SANDSTONE KN5-1

This sample is a part of a 60x30x15 cm block of carbonate-cemented shelly siltstone which was recovered from

1978m-deep type 2 seep at site 2. This sample is a cemented aggregate of three components (dark siltstone fragments, broken fossil shells of small bivalves, and white carbonate precipitates filling the interstices of the former two). The white carbonate is mostly calcite but aragonite is seen to have precipitated at the last stage of cementation. In contrast to KN18-2, this sample exhibits a wide range of variaton in oxygen isotopic ratios, whereas its carbon isotopic ratios are rather uniform (Fig. 2). It is especially interesting that the aragonite-bearing last precipitates that fill up the interstices of fragmental shells and siltstones (dots in Fig. 2) are lighter than those in lithic fragments (circles in Fig. 2) in the same sample. Combined with their textures, we conclude that this sample had experienced at least three successive events: The first formed dense biological communities dominated by *C*-type clams, these communities probably being similar to those in type-1 seeps. The second event is responsible for the cementation of the shells and sandy seafloor, and these shells and the sand later were disintegrated and cemented again during the third event. The present results demonstrate that the carbonate formed in the last event had considerably lighter oxygen isotopic ratios than the first one. Simple calculations indicate that the white carbonates may either be precipitated from a fluid that was about 2 permil ligher in oxygen isotopic ratios or at a temperature 10°C higher than the present bottom seawater.

7 ORIGIN OF METHANE

The lightest measured carbon isotopic ratio of the carbonate from the Nankai Trough is -55.9 permil. This value may be produced either biogenically or thermogenically (Stahl, 1974). However, mixing between methane-derived carbonates and marine sedimentary carbonates would tend to raise the carbon isotopic ratios of the carbonates (Matsumoto, 1989). On the Florida escarpment, carbonate crusts have a carbon isotopic ratios of -30.9 to -48.8 permil, whereas those of the methane in the seepage range from -80 TO -83.7 permil (Paull et al., 1989). These observations, combined with high ethane to methane ratios of hydrocarbons in the bottom seawater, strongly suggest the biogenic origin of the methane at the eastern Nankai Trough cold seeps.

8 CONCLUSIONS

The carbonate encrustations at the eastern Nankai Trough are formed by oxidation of biogenic methane carried to surface in the subduction-induced pore fluid venting system. Two types of venting systems are noted; the type 1 seeps represent the primary stage of encrustation that occurred in virgin turbidite layers, whereas in type 2 seeps, more than two stages of encrustation are clearly distinguished, indicating the second seeps to have experienced several cycles of destruction and re-formation of the seepage, in accord with the tectonically active environment of the studied area.

ACKNOWLEDGMENTS

We thank K. Tokuyama, W. Soh, and M. Tsutsumi for their information on the mineralogical composition of the sandstone. This work was conducted as a part of the Japan-France KAIKO-NANKAI Project. Our thanks are due to X. Le Pichon, K. Kobayashi, and all other scientists involved for their cooperation.

REFERENCES

Matsumoto, R. 1989. Isotopically heavy

oxygen containing siderite derived from the decomposition of methane hydrate. *Geology* 17: 707-710.

McCrea, J. M. 1950. On the isotopic chemistry of carbonates and a paleotemperature scale. *Journ. Chem. Phys.* 18: 849-857.

Paull, C. K., C. S. Martens, J. P. Chanton, A. C. Neumann, J. Coston, A. J. T. Juil, and L. J. Toolin. 1989. Old carbon in living organisms and young CaCO3 cements from abyssal brine seeps. *Nature* 342: 166-168.

Stahl, W. 1974. Carbon isotopes fractionations in natural gases. *Nature* 251: 134-135.

Water-Rock Interaction, Kharaka & Maest (eds) © 1992 Balkema, Rotterdam. ISBN 90 5410 075 3

Peridotite-seawater interaction at Zabargad Island, Red Sea

P.F. Sciuto
Department of Earth Sciences, Cagliari, Italy

ABSTRACT: Geochemical seawater/rock interaction with ultramafic rocks of mantle derivation is one of the processes occurring at spreading centers. Exceptionally fresh peridotitic rocks locally dissected by lythoclasis and fractures which originated during emplacement in the transitional (continental-to-oceanic) ocean floor region are exposed at Zabargad (St. John) island (Red Sea). Chemical and physical analysis of 45 samples allowed the quantitative analysis of chemical exchanges between lithospheric rocks and seawater along spreading centers.

1 INTRODUCTION

Zabargad island, located in the central Red Sea about 50 Km outside the axial trough (Fig. 1), represents a tectonically uplifted sliver of the Red Sea lithosphere (Bonatti et al., 1981). The fractured zones represent loci of convective transport of large masses of seawater during uplift. In these zones interaction of seawater with peridotite minerals produced advanced pervasive serpentinitic alteration with locally complete substitution of pristine minerals and development of peculiar neoformed assemblages with cancrinite group minerals. Self sealing processes prevented heat dispersion locally and alteration assemblages were re-cooked leading to formation of olivine crystals at an advanced stage of the process. Evaporitic phases precipitated from circulating brines as well.

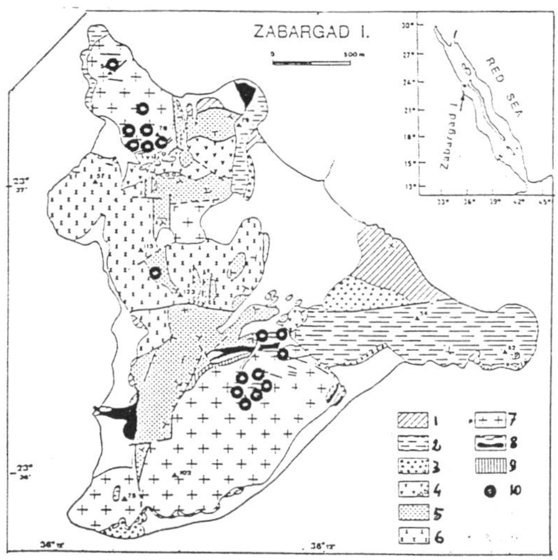

Figure 1. Geology of Zabargad Island (Bonatti et al., 1983). Symbols 1=young reef limestone; 2=old reef limestone; 3=breccias and conglomerates; 4=evaporites; 5=Zabargad sedimentary formation; 6=metamorphic complex; 7=peridotites; 8=basaltic-doleritic dikes; 9=gabbroic rocks; 10=sample locations.

2 SAMPLING ANALYSIS

Chemical (major, minor and trace elements) and physical (D.T.A.) analysis of 45 samples allowed the outlining of the entire seawater-peridotite interaction process which can be regarded at first approximation as chemically continuous (Fig. 2).

Ultramafic rock mineral assemblages and structures were investigated with X-ray diffraction and microscopic observation (optical and S.E.M.). Samples are representative of all phases of the process. Relicts of the pristine paragenesis, structures consequent to tectonic deformation, different stages of serpentinizat-ion, and final high temperature processes are represented.

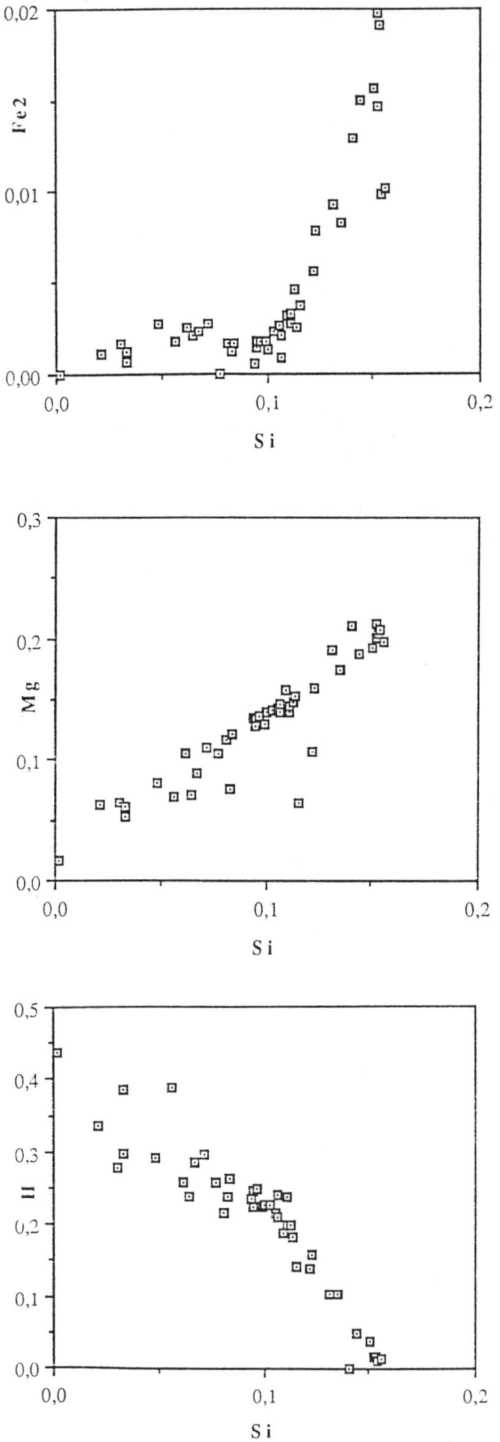

Fig.2 Molal diagrams. showing the chemical continuity of the process incurred in the island.

3 PRELIMINARY CONCLUSION AND APPROACH TO THE MODEL

Chemical and mineral study permitted the identification of serpentinization reaction pathways. The process has been translated into a set of formal mineral-water exchange equilibria whose equilibrium constant in the supercritical region were evaluated with the aid of the SUPCRT92 package (Johnson et al., 1991).

The purpose of the present study is to develop a model for serpentinization and subsequent self sealing processes incurred on Zabargad Island. Samples representative of different stages of interaction will be assumed to have attained a local equilibrium condition and the activity of electrolytes in the interacting fluid will be evaluated by a nonlinear minimization procedure involving the set of reaction constants locally active at P,T of interest and the mineral-chemistry of the whole paragenesis. The whole seawater peridotite elemental exchange will be evaluated by applying the principle of local equilibrium in irreversible processes (Prigogine, 1957). The results are of importance in the quantification of chemical exchanges between lithospheric rocks and seawater along spreading centers.

REFERENCES

Bonatti, E., P.B. Hamlyn & G. Ottonello 1981. The upper mantle beneath a young oceanic rift: peridotites from the island of Zabargad (Red Sea). Geology, 9: 474-479.

Bonatti, E., R. Clocchiatti, P. Colantoni, R. Gelmini, G. Marinelli, G. Ottonello, R. Santacroce, M. Taviani, A.A. Abdel-Meguid, H.S. Assaf & M.A. El Tahir 1983. Zabargad (St. John's) Island: an uplifted fragment of sub-Red Sea lithosphere. J. Geol. Soc. London 140: 677-690.

Johnson, J.W., E.H. Oelkers & H.C. Helgeson 1991. SUPCRT92. A software package for calculating the standard molal thermodynamic properties of minerals, gases, aqueous species, and reactions from 1 to 5,000 bars and O° to 1,000°C.

Prigogine 1957. Introduction to the thermodynamics of irreversible processes. 2nd Edition, Interscience Publisher. New York.

Water-Rock Interaction, Kharaka & Maest (eds) © 1992 Balkema, Rotterdam. ISBN 90 5410 075 3

The isotopic composition of helium and sulphur in hydrothermal fluids trapped in ancient and modern sea floor deposits

F. Stuart & G. Turner
Department of Geology, Manchester University, UK

J. Maynard
Department of Earth Sciences, Open University, Milton Keynes, UK

ABSTRACT: Fluids trapped by sulphides from hydrothermal chimneys at 21°N on the East Pacific Rise have $^3He/^4He$ ratios indistinguishable from the 350°C vent fluid. This is slightly lower than values measured from local ridge basalts and may reflect radiogenic ingrowth of ^4He in the crystalline portions of the oceanic crust. Sulphate $^3He/^4He$ ratios are significantly lower than the vent fluid and represent a dilution by ambient seawater $^3He/^4He$ - $\delta^{34}S$ correlation confirms the simple mixing relationship.

1 INTRODUCTION

The detection of small enrichments of ^3He dissolved in the water column above the Kermadec Trench (Clarke et al 1969) provided the first indication that the Earth was still outgassing volatiles trapped on accretion. This was confirmed by the measurement of $^3He/^4He$ ratios 7-10 times the ratio of atmospheric He (R_a) in MOR basalts (Craig and Lupton 1976). This provided strong evidence for the circulation of seawater in the oceanic crust, and could explain the low conductive heat flow at ridge crests and Fe-Mn enrichments in sediments on the Pacific ridge crest and in ophiolites. The later discovery of metal-rich water 20°C above ambient emanating from discrete vents on the ocean floor at the Galapagos Spreading Centre confirmed the occurrence of submarine hydrothermal activity (Corliss et al 1979).

Systematic studies of the chemistry of the Galapagos solutions and experimental work on basalt-seawater interaction indicated the temperature of the end-member hydrothermal solutions should be ~350°C. This was validated in 1979 when the DSRV ALVIN identified 350°C fluids issuing from sulphide chimneys at 21°N on the East Pacific Rise (RISE Group 1980). The least diluted vent fluid revealed a 400-fold enrichment in He over Pacific Ocean water and a $^3He/^4He$ ratio approximately $8xR_a$ (Lupton et al 1980). Detailed surveys of the properties of fluids emanating at many sites of seafloor hydrothermal activity indicates that He is the most sensitive indicator of hydrothermal activity, more so even than water temperature. Here we present the first analyses of the composition of He trapped in minerals from black and white smoker chimneys from ocean floor hydrothermal sites and demonstrate that they may be used to determine vent fluid He compositions.

2 SAMPLES AND ANALYTICAL TECHNIQUE

Sulphide and sulphate minerals were separated from rock chips and slabs of chimney deposits collected by DSRV ALVIN from the East Pacific Rise at 21°N in 1979 and 1981. Mineral separates were hand-picked under a binocular microscope to minimise the degree of surface oxidation and remove contaminating sulphate from the sulphides. The mineralogy of all samples was determined by X-ray diffraction spectrometry of a portion of the sample.

All samples were washed ultrasonically in methanol prior to a 36 hour vacuum baking at ~200°C to reduce the contribution of adsorbed atmospheric noble gases. Gas extraction was performed by crushing samples in vacuo in a triple assembly of screw crushers constructed from (modified) commercially available vacuum valves. Details of the gas extraction, clean-up and separation procedure will be described elsewhere along with the heavy noble gas data (Stuart et al in prep.). Noble gas analysis was performed in an all-metal MAP 215 mass spectrometer dedicated to analysis of terrestrial samples and capable of resolving $^3He^+$ from interfering HD^+. From routine measurements of 10^{-6} cc air the reproducibility of $^3He/^4He$ ratio measurements was determined to be 3% (1σ) and for ^4He abundance measurements ±5% (1σ).

The crushed material, that which would pass through a 200μm sieve, was weighed and 1-1.5 mg used for $\delta^{34}S$ determinations. Sulphur isotopes were measured

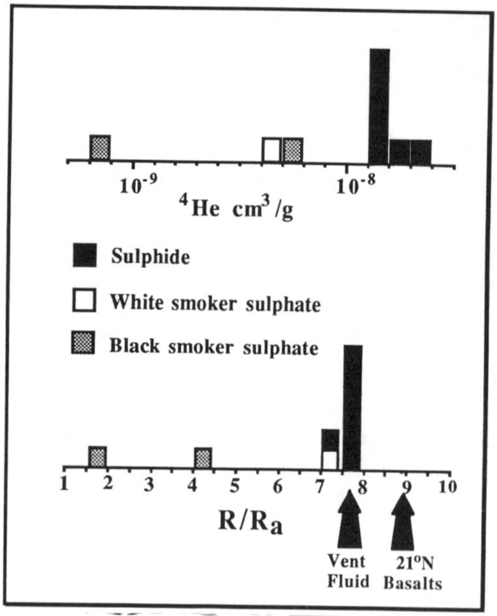

Figure 1. He in hydrothermal minerals.

in a VG 602E mass spectrometer at the Open University, Milton Keynes. SO_2 preparation and sulphur isotope measurement procedure are modified slightly from those those reported by Burgess (1986).

3 RESULTS

3.1 Helium

He abundances and $^3He/^4He$ ratios are displayed in Figure 1. Sulphide minerals have trapped higher He concentrations than the sulphates, although from the limited data set it is not clear whether the sulphide mineralogy has any control on the gas content. The $^3He/^4He$ ratios range from 1.9-8 x R_a and indicate that a substantial proportion of primordial ^3He is present in the trapped fluids. The sulphide minerals are even more restricted in their composition (7.1-8 x R_a), compositionally similar to the helium in the EPR 21°N vent fluids (R/R_a = 7.8±0.2; Lupton et al 1980; Welhan and Craig 1983) and provides strong evidence that they have faithfully trapped the vent fluid.

3.2 Sulphur

$\delta^{34}S_{CDT}$ values of the sulphides range from +2 to +3.7‰ and sulphates from +18.7 to +19.25‰. Figure 2 shows all $\delta^{34}S$ measurements of sulphide and sulphate minerals from hydrothermal chimneys from

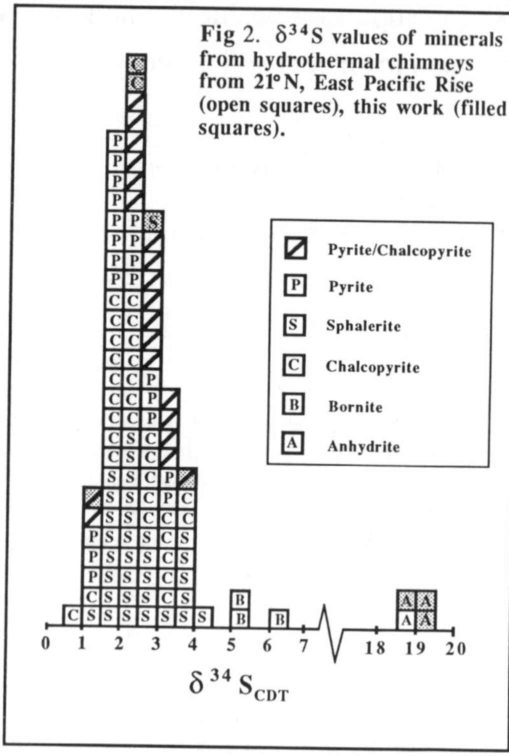

Fig 2. $\delta^{34}S$ values of minerals from hydrothermal chimneys from 21°N, East Pacific Rise (open squares), this work (filled squares).

21°N EPR. Our measurements are typical of those previously recorded, however the limited number of samples analysed is not sufficient to draw conclusions concerning the sulphur isotope distribution between vent fields. Ohomoto and Rye (1979) showed that isotopic fractionation between H_2S and sulphide minerals at >200°C is negligible. The sulphides analysed here are from the interior zones of hydrothermal chimneys and their $\delta^{34}S$ values may closely reflect the vent fluid $\delta^{34}S_{H_2S}$ composition (Woodruff and Shanks 1988).

4 DISCUSSION

4.1 Helium in trapped fluids

The sulphide minerals have trapped helium with a rather uniform isotopic composition close to that measured in the vent fluids. With the exception of sphalerite ALV 980 R2, all the sulphides are from the interior portions of the hydrothermal chimneys. Despite the difficulty in performing fluid inclusion studies on opaque sulphides it seems reasonable to consider that crushing has released the He from vent fluid trapped as microscopic inclusions.

The white smoker and sulphate samples require the addition of He with a lower $^3He/^4He$ ratio. The mixing of the vent fluid with ambient Pacific deep water (R/R_a = 1.2) is the most likely source of low R/R_a He. The concentration of He in Pacific deep water (PDW) is low (4.4×10^{-8} cm^3STP/g; Bieri and Koide 1972) compared to the end-member hydrothermal fluid (2.1×10^{-5} cm^3STP/g; extrapolated from vent fluid $\delta^{18}O_{H_2O}$; Welhan and Craig 1983). Mixing of PDW has no measurable effect on the vent fluid $^3He/^4He$ ratio until PDW is >95%. Low R/R_a He may also be introduced as secondary fluid inclusions, perhaps during later recrystallisation, and cannot be ruled out as the source of some of the spread of $^3He/^4He$ values (see ALV 982 R20B-3). The extent of this is not easy to gauge as later additions of PDW will be chemically indistinguishable from that entrained by the hydrothermal fluid in the upper oceanic crust and the chimney. A correction for adsorbed air He, made using the relative abundances of the heavy noble gases, has no effect on He abundance or isotopes.

4.2 Seawater-basalt interaction: He isotope constraints

The similarity of the vent fluid and sulphide $^3He/^4He$ ratios to those of the fresh MOR basalts confirms that, along with many other elements, He is leached from the deep oceanic crust. However, fresh basaltic glasses dredged from 21°N EPR have $^3He/^4He$ ratios 12% higher than the fluids ($R/R_a = 8.9 \pm 0.5$; 15 analyses by Allegre et al 1983; Ozima and Zashu 1983; Welhan and Craig 1983). Comparable He isotope discrepancies have been recorded between vent fluids and the local basalts in hydrothermal systems at the Galapagos Spreading Centre and Juan de Fuca Ridge and in intra-plate hydrothermal system on the Loihi Seamount.

Kennedy (1988) proposed that lower $^3He/^4He$ ratios in the end-member hydrothermal solutions from the Southern Juan de Fuca Ridge were the result of ingrowth of radiogenic ^4He from U and Th decay in the He source region. Aging of the aquifer rocks results in a lowering of the $^3He/^4He$ ratio by a factor f3/4 which can be defined as;

$$f3/4 \quad = \quad 1 - [2.179 \times 10^{-13} \cdot (U/^4He) \cdot T]$$

where U is the uranium content in ppm (assuming $^{232}Th/^{238}U = 3.5$), ^4He is in cm^3STP/g and T is the age in years (which must be less than 10^7 years).

Ridgecrest tholeiites dredged from 21°N are at most 27,000 years old and have a maximum U content of 0.18 ppm (Newman et al 1983). The freshest basaltic glasses from 21°N have ^4He concentrations between 0.5 and 5.5×10^{-5} cm^3/g (Ozima and Zashu 1983). The isotopic composition of He leached from these units is lower by only 0.05%. The 12% decrease in the $^3He/^4He$ ratio requires either that the source rocks have (U + Th)/He ratios that are significantly greater than those of EPR basalt glasses or that the He is acquired from a recharge zone that is of the order of 500,000 years old.

Ocean-floor hydrothermal alteration is not pervasive, the high temperature reactions that are important to the chemistry of the resultant fluids occur in well defined zones within the sheeted dolerite dykes and oldest extrusive units. Magma devolatilisation and the incompatibility of the noble gases during crystallisation leaves deep units depleted relative to the surface basalts. Deep Jurassic oceanic crust has He concentrations significantly below those of fresh MOR glasses ($^3He \sim 10^{-13}$ cm^3/g; Staudacher and Allegre, 1988). If this is representative of He concentrations in the hydrothermal aquifer rocks subsequent U and Th decay would generate enough ^4He to modify the original $^3He/^4He$ by over 10% in 27,000 years. This, however, has important implications for He extraction from the oceanic crust.

The concentration of ^3He in the end-member hydrothermal fluid (2.3×10^{-10} cm^3/g) is ~1000 times higher than ancient deep oceanic crust. If this value is characteristic of the He source region then 1g of circulating seawater acquires He from 1kg of oceanic crust. This is severely at odds with most estimates of water-rock ratios which indicate ratios close to unity. The high He concentration of the vent fluid might be the result of condensation/solution of a vapour phase deep within the hydrothermal system. While noble gas evidence for deep boiling exists at other ridgecrest hydrothermal sites (Kennedy 1988) it remains to be tested at this location.

4.3 Helium-Sulphur Relationship

Mineral $\delta^{34}S_{CDT}$ values are plotted against the $^3He/^4He$ ratio of fluids released by crushing in Figure 3. It is apparent that the variations are broadly correlated. The sulphides from high temperature chimneys are characterised by mantle-derived helium and correspondingly low values of $\delta^{34}S$ close to the value of vent fluid $\delta^{34}S_{H_2S}$. While local basalts have acted as the main source of helium and sulphur in the hydrothermal fluids, sulphide $\delta^{34}S$ are slightly higher than measured MOR basalt $\delta^{34}S_S$ (-0.7‰; Sakai et al 1976). This reflects the addition of isotopically heavy S through the reduction of seawater sulphate during the conversion of basaltic pyrrhotite to pyrite (Woodruff and Shanks 1988). The lower $^3He/^4He$ ratios of the sulphate samples record the increasing dilution by PDW and is paralleled by an increase in the value of the

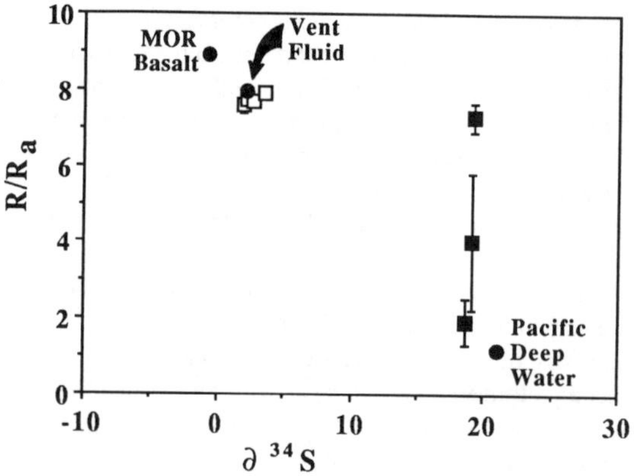

Figure 3. Covariation of He and S.

mineral $\delta^{34}S$ towards the value of seawater. The correlation trend in Figure 3 is curved, a result of the large range of $^3He/^4He$ ratios of the sulphate fluids (1.3-7 x R_a). This appears to confirm that He is a more sensitive measure of the hydrothermal fluid component than the mineral $\delta^{34}S$ value. However the range of sulphate He ratios may reflect the presence of minor amounts of contaminating sulphide. The small amounts of sulphate from the available suite have prevented further assessment of this problem.

5 PROSPECTS

The He composition of fluids in ancient oceanfloor systems may prove useful to measurements of the past He flux and tectonic setting exposed in ophiolites. The efficiency with which He has been removed from the source region, and the role that phase separation may have had on the trapped noble gases, may be revealed by measurement of the absolute concentrations of He in the released fluids. We plan to determine absolute noble gas concentrations by combining noble gas measurements with determination of the amount of water released (Burnard et al 1992). Further, microthermometric determinations of fluid inclusion trapping temperatures in plagioclase, quartz and epidote minerals from ancient ocean floor deposits may allow determinations of past He/heat ratios that are important to the thermal and mass transport histories of the Earth.

Preliminary studies of material from the Troodos hydrothermal system reveal that fluids trapped in plagioclase minerals have $^3He/^4He$ ratios 1-2 x R_a. While >90% of the helium is of mantle origin addition of radiogenic ^4He from U and Th in mineral lattices and diffusive loss from fluid inclusions may compromise the original He compostion.

REFERENCES

Allegre, C. Staudacher, T. Sarda, P & Kurz, M. 1983. Constraints on evolution of the Earth's mantle from rare gas systematics. Nature, 303, 762-766.

Bieri, R.H. & Koide, M. 1972. Dissolved noble gases in the east equatorial and southeast Pacific. J. Geophys. Res., 77, 1667-1676.

Burnard, P. Ayliffe, L. Stuart, F. Turner, G. & Curtis, C. 1992. This issue.

Clarke, W. Beg, M. & Craig, H. 1969. Excess 3He in the sea: evidence for primordial helium. Earth Planet. Sci. Lett., 6, 213-220.

Corliss, J. Dymond, J. Edmond, J. Von Herzen, R. Ballard, R. Green, K. Williams, D. Bainbridge, A. Crane K. & Van Andel, T. 1979. Submarine hydrothermal springs on the Galapagos rift. Science, 203, 1073-1083.

Craig, H. & Lupton, J. 1976. Primordial neon, helium and hydrogen in oceanic basalts. Earth Planet. Sci. Lett., 31, 369-385.

Lupton, J. Klinkhammer, G. Normark, W. Haymon, R. Macdonald, K. Weiss, R. & Craig, H. 1980. Helium-3 and manganese at the 21°N East Pacific Rise hydrothermal site. Earth Planet. Sci. Lett., 50, 114-127.

Newman, S. Finkel, R. and MacDougall, J. 1982. ^{230}Th-^{238}U disequilibrium systematics in oceanic tholeiities from 21°N East Pacific Rise. Earth Planet. Sci. Lett., 65, 17-33.

Ohomoto, H. & Rye, R. 1979. Isotopes of sulphur and carbon. In H.L. Barnes (ed.), Geochemistry of Hydrothermal Ore Depostits: 509-567. New York: Wiley.

Ozima, M. & Zashu, S. 1983. Noble gases in submarine pillow volcanic glasses. Earth Planet. Sci. Lett., 62, 24-40.

RISE Project Group 1980. East Pacific Rise: hot springs and geophysical experiments. Science, 207, 407-426.

Staudacher, T. & Allegre, C. 1988. Recycling of oceanic crust and sediments; the noble gas subduction barrier. Earth Planet. Sci. Lett., 89, 173-183.

Welhan, J. & Craig, H. 1983. Methane, hydrogen and helium in hydrothermal fluids at 21°N on the East Pacific Rise. In Hydrothermal Processes at Seafloor Spreading Centers. New York: Plenum.

Woodruff, L. & Shanks, W.C.III 1988. Sulphur isotope study of chimney material and vent fluids from 21°N, East Pacific Rise : Hydrothermal sulphur sources and disequilibrium sulfate reduction. J. Geophys. Res. 93, 4562-4572.

Water-Rock Interaction, Kharaka & Maest (eds) © 1992 Balkema, Rotterdam. ISBN 90 5410 075 3

Microbial green marine clays in the hydrothermally altered deep sea sediments

Kazue Tazaki
Department of Geology, Shimane University, Matsue, Japan
Kōichi Tazaki
Department of General Education, Ehime University, Matsuyama, Japan
W.S. Fyfe
Department of Geology, The University of Western Ontario, London, Ont., Canada

ABSTRACT: Green marine clays in the hydrothermally altered deep sea sediments from the Izu-Bonin fore-arc basin, at Leg 126 Ocean Drilling Program (ODP) were studied. The green marine clays in volcanic sandstone contain not only smectite, mica-clay minerals and zeolites, but also bacteria, organic carbon and graphite. Transmission electron microscope observations clearly showed presence of the filamentous bacteria and the graphite crystallization. X-ray photoelectron spectrochemical analyses indicated that the green marine clays contain C (1s) peak showing COO, C-O, C-C, and graphite structures. Fourier transform infrared spectrometry of the green spherical aggregate showed peaks between glauconite and celadonite. Under the hydrothermal sea-floor conditions, the organic materials with filamentous bacteria produced green marine clays, amorphous low-crystallinity hydrocarbons and highly ordered carbon through loss of hydrogen and oxygen.

1 INTRODUCTION

Marine bacteria can and do induce the formation of minerals in hydrothermal fluids of chimney, hydrothermal vents, and deep-sea volcanoes (Buick, 1990; Buczynski and Chafetz, 1991). Filamentous iron-silica deposits from modern and ancient hydrothermal sites at Philosopher Vent on Exporer Ridge in the northeast Pacific (Juniper and Fouquet, 1988), biological single-domain magnetite and magnetotactic bacteria in hemipelagic marine sediments (Stolz et al., 1986) and bacteria in a young East Pacific Rise sea-mount (Vanko et al., 1991) are reported, respectively.

Using a micro X-ray photoelectron spectrochemical analyses (micro ESCA), transmission electron microscopy (TEM) and Fourier Transform Infrared Spectrometer (FT-IR), we show that bacteria and organic materials are the major source of green marine clays and graphite in the hydrothermally altered volcanic sandstone.

2 MATERIAL AND METHODS

Microbial green marine clays in the deep sea sediments were recovered from the Ize-Bonin fore-arc basin, at Leg 126 Ocean Drilling Program. Upper Oligocene vitric and pumiceous sandstone are the dominant products of fore-arc basin at site 793, which contain more than 70% volcanic glass. The volcanic glasses are largely altered to zeolites and smectite (Leg 126 Shipboard Scientific Party, 1990; Tazaki and Fyfe, 1991). The thin section of green to brownish sandstone from Holes 793B were observed by optical microscope. TEM observation was carried out on the <2μm size fraction, using a JEOL 2000EX instrument with accelerating voltage of 160 kV. The mineralogy of bulk powder samples was determined by using a Rigaku X-ray diffractometer with an accelerating voltage of 40 kV, 35mA, Cu Kα radiation with Ni-filter. The aggregate of green spherical material was prepared for Fourier Transform Infrared Spectrometer (JEOL JIR-5500W, FT-IR) analysis with a beam area 70 μm in diameter, analytical resolution of 8 cm for 100 scanning times using Ge plate. Chemical binding data obtained with a JEOL JPS-9000 X-ray photoelectron spectrometer (micro ESCA) operating an aperture of 1 μm, analytical resolution of 1.8 eV for all area spectra between 1100 eV and OeV, and 1.25 eV or 0.9 eV for narrow spectra on green spherical materials.

3 RESULTS

3.1 Mineralogy and chemistry of green marine clays

The hydrothermally altered volcanic sandstones from 793B 57R 03, 85-87 cm (1127 m.b.s.f.) and 793B 58R 02, 94-96 cm (1135 m.b.s.f.) are composed of smectite (1.52 nm), mica-clay minerals (1.03-1.22 nm), cristobalite (0.40 nm), calcite (0.30 nm), feldspar (0.32 nm), anhydrite (0.35 nm) and various zeolites. The peaks at 1.03-1.22 nm may be identified as celadonite or glauconite due to their green color. Comparison of X-ray diffraction patterns for celadonite and glauconite was difficult because of very broad and small peaks in these X-ray powder diffraction patterns.

Fourier transform infrared spectrometer (FT-IR) of green spherical clays provides information on the composition of the octahedral sheet of the layer silicates with the hydroxyl-cation bonds showing OH, CH, CO, CO_2, C-O, Fe_2OH, and Si-O. According to Farmer (1974) and Buckley et al. (1978), the 815 cm^{-1} and 3555 cm^{-1} bands of the green spherule show peaks between glauconite and celadonite.

Major element chemical analyses of green spherical clays were obtained by microchemical technique using X-ray photoelectron spectrochemical analyses (ESCA). The Si and O contents are the highest value in all samples, whereas the Ca, K, Cl, Fe and Al contents are remarkably low. The Na, Mg and S are at trace levels in all samples. Especially at 1135 m.b.s.f. depth, the C and K contents are significantly changed.

More precise data on C was obtained by high resolution C (1s) spectra and allow us to determine the variation of carbon compounds and oxygen functional groups (Fig. 1). The C-C binding group is remarkably high in all samples whereas the COO binding group is low, is zero at 1127 m.b.s.f. The COO, C-O and graphite contents correlate intimately with total C content. Variation of relative abundance of carbon compounds suggests reducing processes at the local environment of the spherical interior. Their relative abundance reflects the degree of polymerization by esterification with organic materials. At 1135 m.b.s.f. depth, the

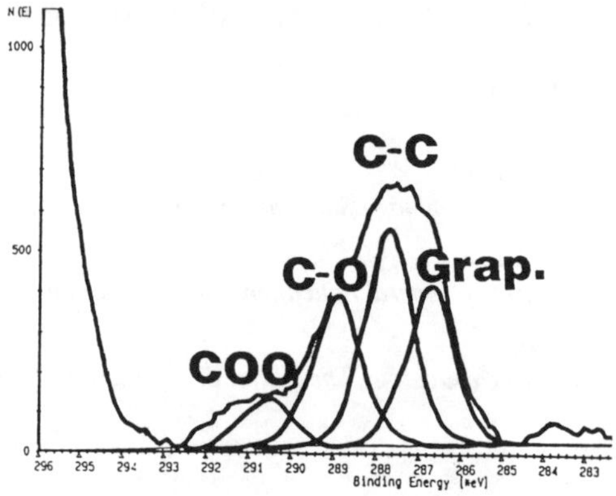

Fig. 1. High resolution C (1s) spectrum of X-ray photoelectron spectrchemical analyses of green spherical clays at 793B 58R02, 94-96 cm (1135 m.b.s.f.). The carbon compounds are COO 9.0%, C-0 26.4%, C-C 37.2%, graphite 27.4%.

green spherical clays show a trend involving loss of oxygen. The final stable product of the carbon seems to be graphite.

3.2 Optical and electron microscope observations

Blue-green color is quite diagnostic of green spherical clays in the polished thin section (Fig. 2A). The porous smectite matrix of the sandstone contains abundant greenish translucent filaments (Fig. 2B). The filaments are typically 3 µm wide and 15 µm long: they exhibit curved and convoluted forms, commonly branching. TEM clearly shows that smaller and round uniform grains resemble particles making long branching filaments. The cluster of spheroids are typically 50 nm diameter. TEM observation has also revealed an association of graphite within filamentous bacteria. The graphite shows curled-up thin flaky, ribbon and hexagonal shapes suggestive of crystal growth processes (Fig. 3). Electron diffraction pattern of the curled-up thin film shows diffused rings, which indicate low crystallinity, whereas the hexagonal shape shows sharp strong spots at 3.4 Å, which indicates typical well-crystallized graphite. The flaky shapes tend to bundle-up and to make ribbons.

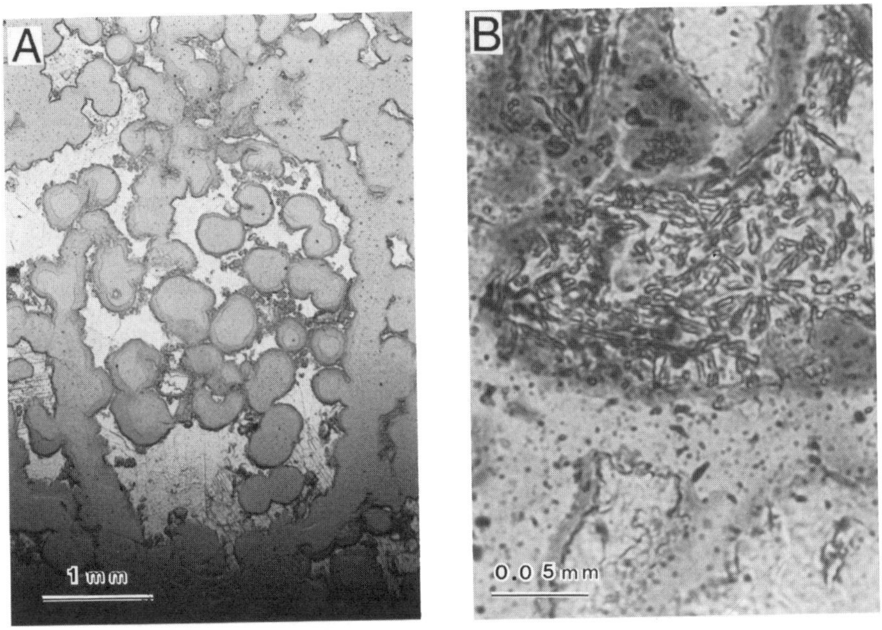

Fig. 2. Optical micrographs of green spherical clays (A) from sample 793B 85R02, 94-96 cm, and filamentous bacteria (B) from sample 793B 57R03, 85-87 cm.

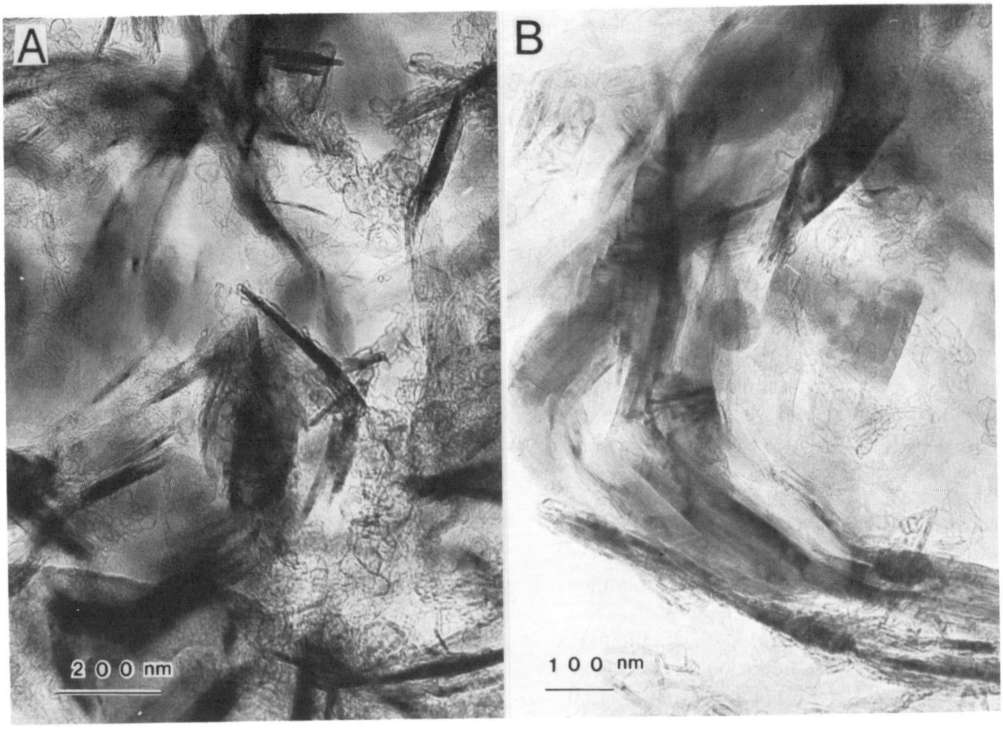

Fig. 3. Transmission electron micrographs of flaky graphite (A) and ribbon shaped and hexagonal graphite (B) from sample 793B 27R03, 85-87 cm.

4 CONCLUSIONS

Mineralogy, chemistry and morphology of green spherical clays in the volcanic sandstone from Izu-Bonin fore-arc basin at Leg 126 ODP, have been determined by using XRD, optical and electron microscopies, FT-IR and micro ESCA. The FT-IR and micro-ESCA data show presence of CH, COO, C-O, C-C and graphite, suggesting that the green spherical clays formed not only from hydrothermal alteration products, but also by microbiological interaction between filamentous bacteria and organic materials. Microscopic observation has revealed presence of bacteria and graphite which can induce the formation of green spherical clays in hydrothermal fliuids of porous volcanic sandstone. The sandstones contain numerous hydrothermally altered minerals such as zeolites, anhydrite, smectite and cristobarite. Under conditions where substantial hydrothermal deposits form on the sea-floor, the organic materials with filamentous bacteria produced amorphous low-crystallinity hydrocarbon (COOH, C-OH). The predominant trend is loss of hydrogen and compression to change hydrocarbons to carbon (C-C) and graphite. The considerable oxygen content of humic substances is markedly reduced with time. The final product of the organic materials is graphite which has a highly ordered carbon stucture. TEM observation clearly shows the processes of graphite mineralization from low-crystallinity to well-crystalized graphite.

REFERENCES

Buick, R. 1990. Microfossil recognition in Archean rocks: An appraisal of spheroids and filaments from a 3500 M.Y. old chert-barite unit at North Pole, Western Australia, PALAIOS 5: 441-459.

Buczynski, C. & H.S. Chafetz 1991. Habit of bacterially induced precipitates of calcium carbonate and the influence of medium viscosity on mineralogy. J. of Sedimentary Petrology 61: 226-233.

Juniper, S.K. & Y. Fouquet 1988. Filamentous iron-silica deposits from modern and ancient hydrothermal sites. Canadian Mineralogist 26: 859-869.

Leg 126 Shipboard scientific party 1990. Proceeding of the Ocean Drilling Program 126: 338-341.

Stolz, J.F., S.B.R. Chang & J.L. Kirschvink 1986. Magnetotactic bacteria and single-domain magnetite in hemipelagic sediments. Nature 321, 849-851.

Tazaki, K. & W.S. Fyfe 1992. Diagenetic and hydrothermal mineral alteration observed from Izu-Bonin deep-sea sediments, Leg 126. ODP Scientific Results Volume, in press.

Vanko, D.A., B.J. Milby & S.W. Heinzquith 1991. Massive sulfides with fluid-inclusion-bearing quartz from a young sea-mount on the east Pacific rise. Canadian Mineralogist 29: 453-460.

Water-Rock Interaction, Kharaka & Maest (eds) © 1992 Balkema, Rotterdam. ISBN 90 5410 075 3

Short-term variability, phase separation and water-rock reaction in hydrothermal fluids from 9-10°N, East Pacific Rise

K.L.Von Damm
Department of Earth Sciences, University of New Hampshire, Durham, N.H., USA

ABSTRACT: Hydrothermal fluids were first sampled from 9° 16' to 9° 53' N on the East Pacific Rise in April 1991. Lava had been erupted in the area north of 9° 45'N less than 2 weeks prior to our sampling in this area. We have therefore obtained fluid samples during the initiation of hydrothermal activity. Diffuse flow (temperature < 70° C) was volumetrically much more important than the focused, high temperature (≤ 403° C) flow. The temperature and chemistry of specific vents was observed to change on a one week time scale, "Black smoker" fluid was observed exiting directly from talus piles, as insufficient time had elapsed for the construction of sulfide chimneys.

1 INTRODUCTION

The East Pacific Rise from 9° 09'-54' N is a fast-spreading ridge with a full spreading rate of 11 cm/yr. This section of the ridge crest was surveyed using the ARGO imaging system in November-December 1989 and was found to have extensive areas of hydrothermal activity (Haymon et al. 1991). Our Alvin dives in this area in April 1991 were the first to sample hydrothermal fluids from this area and also demonstrated that an eruption had occurred after the ARGO survey in the area of north of 9° 45' N. $^{210}Po/^{210}Pb$ systematics suggest that the eruption occurred within two weeks of the Alvin dives in April 1991 (Rubin & Macdougall 1991).

2 VENT FLUID CHEMISTRY

Vent fluids were sampled from 9° 16' - 9° 53' N from a total of 8 ridge segments. Both diffuse and focused fluid flows were sampled, with the diffuse flow ranging in temperature from a few degrees above ambient to ≤ 70° and the focused flow having measured temperatures of up to 403° C. The diffuse flow was extremely voluminous (thousands of gallons/minute) in much of the area north of 9° 45'N, where there was an extremely new lava flow. The chlorinity of the fluids ranged from <10% to >150% of the seawater value. In general, the chlorinity of the focused fluids was greater than the seawater value in areas which had not experienced recent volcanism (e.g. 9° 33'N). for the water depth in this region (~ 2500m), temperatures greater than ~ 389° C place the fluids in the vapor field, the critical point of seawater being 408° C and 298 bars. The low chlorinity and high measured temperatures in the focused fluids in the area of the young lava flows are consistent with vapor field properties. Data for the PVTx properties of seawater (Bischoff 1991) suggests phase separation may be occurring approximately 200 m below the seafloor. Other chemical constitutents, including silica, suggest that water-rock interactions and possibly equilibration have occurred after phase separation. Stable isotope measurements (Shanks et al. 1991) also support this hypothesis.

One of the most astounding results was the rapidity with which the temperature and/or composition of vent fluids changes in the initial stages of hydrothermal activity.

Several vents had significant changes in fluid chemistry on the time scale of one to two weeks. A particular vent at 9° 46.5' N rose in temperature from 389° C to 396° C to 403° C over a 2 week period. The temperature change was accompanied by changes in most of its major ion chemistry, including a more than a doubling of the measured H_2S content and a halving of the measure Si content. The measured Si content is consistent with equilibrium values for the quartz geothermometer-geobarometer.

The vent fluid chemistries from 9-10° N, East Pacific Rise greatly expand the know compositional range for seafloor hydrothermal fluids for both major element and gas (Lilley et al. 1991) chemistries. The ongoing analysis of these fluids is providing insights into the important reactions and processes that occur immediately following an eruptive event on the oceanic ridge crest.

ACKNOWLEDGEMENTS

This research was supported by NSF grant OCE-9101440 as part of an ODP Site Survey to K.L. Von Damm when she was in the Department of Geological Sciences, University of Tennessee. We thank the Alvin group and the captain and crew of the R/V Atlantis II.

REFERENCES

Bischoff, J.L. 1991. Densities of liquids and vapors in boiling NaCl-H_2O solutions: a PVTx summary from 300° to 500° C. Amer. J. Sci. 291: 309-338

Haymon, R.M. et al. 1991. Hydrothermal vent distribution along the East Pacific Rise crest (9° 09' - 54' N) and its relationship to magmatic and tectonic processes on fast-spreading mid-ocean ridges. Earth & Planet. Sci. Lett. 104: 513-534.

Lilley, M.D. et al. 1991. Methane, hydrogen and carbon dioxide in vent fluids from the 9° N hydrothermal system. EOS 72: 481.

Rubin, K.H. and Macdougall, J.D. 1991. Fine chronology of recent mid-ocean ridge eruptions on the southern JDF and 9° N EPR from ^{226}Ra-^{230}Th-^{238}U and ^{210}Po-^{210}Pb disequilibrium. EOS 72: 213.

Shanks, W.C. III et al. 1991. Stable isotope studies of vent fluids, 9-10° N East Pacific Rise: water-rock interaction and phase separation. EOS 72: 481.

Water-Rock Interaction, Kharaka & Maest (eds) © 1992 Balkema, Rotterdam. ISBN 90 5410 075 3

Author index

Aagaard, P. 1147
Abercrombie, H.J. 1101, 1125
Acker, J.G. 533
Aggarwal, J.K. 893
Ágústsdóttir, A.M. 465
Ahonen, L. 253, 763
Ais, M. 627
Alexandre, G.A.L. 757
Alisa Mast, M. 569
Alt, J. 1655
Ambats, G. 963
Amiotte-Suchet, Ph. 483
Anderson, J.L. 1527
Anderson, L.C.D. 495
Anderson, L.D. 397, 805
Andrésdóttir, A. 477, 1459
Aplin, A.C. 1153, 1239
Appelo, C.A.J. 201, 205, 209
Apps, J.A. 1437
Ármannsson, H. 1257
Arnason, J.G. 1473
Arnórsson, St. 1389, 1459
Arusei, M.K. 1261
Ashworth, J.R. 1375
Ayalon, A. 955
Ayliffe, L. 903
Ayora, C. 649, 1547

Baccarin, F. 1349
Back, W. 767
Baedecker, M.J. 257, 275
Baker, F.G. 59
Baker, J.C. 1161
Balderer, W. 1317
Banks, D.A. 1137
Barg, E. 541
Bargar, K.E. 263
Bargossi, G.M. 989
Barnes, Ch.J. 643
Barnes, H.L. 307, 1565
Barsukov, V.L. 1575
Barth, S. 989
Bateman, K. 423
Bayhurst, G.K. 843, 1655
Beasley, Th.M. 709
Beaufort, D. 1227
Beekman, H.E. 205, 209
Bell, J.L.S. 307
Bellucci, L. 1349
Bencala, K.E. 393
Bencini, A. 371
Benedetti, M. 545
Bennett, Ph.C. 267, 349
Benson, A.K. 361
Bergfeldt, B. 751
Berner, R.A. 35, 473
Bickle, M.J. 1543
Bierens de Haan, S. 817
Biino, G.G. 1061
Bird, D.K. 1473, 1477, 1533
Bischoff, J.L. 1647
Bish, D.L. 457, 687
Bisoi, S.K. 753
Bjorkum, P.A. 1183
Björnsson, H. 465
Blattner, P. 989, 1039
Blencoe, J.G. 1025
Bloch, J. 1157
Blomqvist, R. 763, 787, 801
Bloom, M.S. 1065, 1569
Blum, A.E. 133, 159, 607
Blyth, A.R. 787
Bodnar, R.J. 1025
Boles, J.R. 311
Bonnin, D. 167
Borisov, H.V. 1575, 1625
Bottrell, S.H. 813
Boulegue, J. 1663
Bouloubassi, I. 271
Bourcier, W.L. 81
Bouzat, G. 167
Boyle, D.R. 771
Brady, P.V. 85
Brandriss, M.E. 1477
Brantley, S.L. 3, 69, 111, 465
Breit, G.N. 499, 835
Bricker, O.P. 533
Briggs, P.H. 419, 449, 599
Brondi, M. 371
Broshears, R.E. 393
Broxton, D.E. 457, 699
Bruton, C.J. 73, 705
Bullen, Th.D. 607, 897
Burger, P.A. 741
Burnard, P. 903
Burton, E.A. 617

Caballero, J.M. 1481
Caboi, R. 367, 469
Cabral, J.M.P. 1337
Caffee, M.W. 541
Calas, G. 153
Callender, E. 621
Calmbach, L. 1267
Canals, A. 1219
Caprai, A. 1275
Cardellach, E. 1547
Caritat, P. de 1161
Carpentier, B. 1219
Carroll, S.A. 89
Casey Moore, J. 1537
Casquet, C. 1481
Cast, M.E. 499
Castet, S. 1013
Chahi, A. 627
Chandrasekharam, D. 1271
Chang, G.M. 1437
Chaudhuri, S. 907
Chen, S. 1043
Chen, W. 1047

Chen, Y. 1047
Cheng Yan 1393
Cherry, J.A. 1133
Chiarle, M. 775
Chigira, M. 1397
Chiodini, G. 1401
Chipera, St.J. 457, 687
Chipley, D. 1105
Chisholm-Brause, C.J. 137
Christie, W.H. 1197
Chudaev, O.V. 549, 1639
Chudaeva, V.A. 549
Cidu, R. 367, 469
Clark, A.E. 779
Clark, D.L. 349
Clark, I.D. 793, 1485
Clauer, N. 907
Cochran, M.F. 473
Cole, D.R. 995, 999, 1197
Coleman, M.L. 869, 1109, 1239
Cooke, D.R. 1579
Cornell, D.H. 1643
Cortecci, G. 911
Coshell, L. 663
Coston, J.A. 41
Coughlin, B.R. 55
Councell, T.B. 669
Courty, M.A. 715, 731
Cox, D.P. 1589
Cox, R. 1537
Cozzarelli, I.M. 257, 275
Cramer, J.J. 237
Crawford, B.R. 1113
Cristini, A. 367, 469
Curtis, C. 903
Curtis, C.D. 1239
Czarnecki, J.B. 631
Czernichowski-Lauriol, I. 213

D'Amore, F. 1275
Dall'Aglio, M. 371
Damodaran, A.D. 537
Dandurand, J.L. 93, 1013
Danelle Alloway, M. 353
Darling, W.G. 1441
Davis, J.A. 41, 495, 805
Dawson, S. 1043
Day, W.C. 599
de Caritat, P. 1161
De Vivo, B. 1445
Degueldre, C. 1327
Delgado, J. 1547
Deloule, E. 1075
Denis Delorme, L. 677
Dever, L. 715, 731
Devidal, J.L. 93
DeWayne C.L. 709
Di Battistini, G. 989
Diels, J. 223
Dixon, P.R. 843, 915
Domagalski, J.L. 635
Donahoe, R.J. 723
Donnelly-Nolan, J.M. 1279
Donovan, J.J. 639
Dove, P.M. 141
Downey, J.S. 825
Drake, B. 149
Drever, J.I. 115, 513, 553
Drew Tait, C. 349
Duchi, V. 371
Dudley, J.S. 1223
Dunai, T.J. 919
Durand, B. 1219
Duringer, P. 627
Dyess, J.N. 681

Earley, P.J. 1363
Eastoe, Ch. 943
Eberl, D.D. 133, 1165
Ebinger, M.H. 687, 783
Edgell, M. 163
Edmond, J.M. 885
Edmunds, W.M. 719
Egeberg, P.Kr. 1147
Eggenkamp, H.G.M. 209
Eggert, K.C. 1043
Eichenseer, H. 487
Eikenberg, J. 377
Eilu, P. 1491
Einaudi, M.T. 879
Ekstrom, H. 1495
Elders, W.A. 1283
Elphick, S.C. 1113
Eugster, H.P. 635
Evans, J.R. 257
Evans, W.C. 449, 1303

Faber, E. 793
Fabriol, R. 213
Fabryka-Martin, J. 855, 931
Fanfani, L. 367, 469, 1401
Farrar, Ch.D. 1357
Faye, S. 719
Fehn, U. 709
Fengliang, J. 1309
Ferguson, J. 643
Fernandes, E.D.N. 757
Ferrara, G. 911
Ficklin, W.H. 381, 419, 443
Finkel, R.C. 541
Finley, J.B. 553
Fischer, M. 727
Flank, A.M. 153
Folger, P.F. 599
Fontes, J.-Ch. 793
Ford, J.L. 923
Forester, R.M. 677
Fouillac, A.M. 1227
Fournier, R.O. 1289
Franklin, S.P. 325
Franzson, Hj. 1467
Frape, S.K. 1133, 787, 801, 943, 971
Frishman, D. 599
Fritz, B. 101, 581, 655, 927
Fritz, P. 793, 1485
Fujimoto, K. 145
Fukutani, K. 145
Fuller, C.C. 41
Fullmer, E.G. 1101
Furlan, S. 907
Futa, K. 737
Fyfe, W.S. 1675

Galindo, C. 1481
Gammons, C.H. 1569
Gamo, T. 1659, 1663
Ganda, S. 1293, 1405
Gann, A. 557
García-Veigas, J. 649
Gardiner, M.A. 385
Gas'kova, O.L. 1071
Gascoyne, M. 801, 931
Gautschi, A. 407
Gaye, Ch.B. 719
Geddes, D. 1537
Gestsdóttir, Kr. 315
Getahun, A. 1411
Gianelli, G. 935, 1275
Gieskes, J.M. 1655
Giggenbach, W.F. 873, 1341
Gijbels, R. 1321
Gilbert, D.J. 1569
Giordano, Th.H. 319
Gíslason, G. 1257
Gíslason, S.R. 477
Glynn, P.D. 175
Goffé, B. 1499
Golab, Z. 279
Gómez, P. 797
Gong, Ch. 723
Gould, W.W. 1565
Goulet, Ch.J. 107

Gout, R. 1013
Gove, H.E. 709
Graham, C.M. 1543
Graham, I.J. 975
Grandstaff, D.E. 1393, 1415
Granina, L. 621
Grant, Sh.M. 1153
Grassi, V.J. 1393, 1415
Grathwohl, P. 283
Griffioen, J. 45
Grimaud, D. 1455
Gu, Y. 1065
Gudmundsson, M.T. 465
Gunter, W.D. 119, 1451
Guo, G. 1297
Gutentag, E.D. 825

Hahn, T. 617
Haijiang, X. 1601
Hajash, A. 325
Hansen, B.Kjr. 389
Hantono, Dj. 1405
Harden, J.W. 607
Harper, D.A. 1169
Harris, D.W. 163
Harrison, W.J. 353, 1091
Harvie, Ch.E. 1209
Hearn, P.P. 257
Heinrich, Ch.A. 1583
Helgeson, H.C. 1021
Hellmann, R. 149
Hellmuth, K-H. 573
Hendry, M.J. 287
Herczeg, A.L. 561
Herman, J.S. 247, 291, 779
Hesse, R. 1173
Heuss-Aßbichler, S. 1503
Hiebert, F.K. 267
Higgins, J.D. 741
Hill, P.I. 1231
Hiyagon, H. 947
Hochella Jr, M.F. 141, 163, 159
Hochella, M.F. 7
Hoernes, St. 1313
Hofmann, B.A. 503
Holloway, L. 1451
Horita, J. 999
Hua, R. 1589
Huc, A.Y. 1219
Hughes, C.R. 423, 1231
Hutcheon, I.E. 1157
Hutcheon, I. 1125
Hutchinson, R.A. 1289, 1419
Hypolito, R. 757

Iacumin, P. 935
Ibrahim, H. 1455
Ildefonse, P. 153, 167
Ishibashi, J. 959
Ishiyama, D. 1507
Ito, M. 1507
Ivanovich, M. 787

Jacks, G. 611
Jackson, K.J. 189
Jacobson, G. 643
Jahren, J.S. 1147
Jakobsson, S.J. 939
Janecky, D.R. 349, 843, 915, 1043
Jankowski, P.A. 513
Jewell, P.W. 1235
Jiang, W.-T. 1659
Jihua, W. 1309
Johnsson, P.A. 159
Jones, Bl.F. 779
Jones, M.G. 787
Jordan, M.M. 1239
Jove, C.F. 1511
Jowett, E.C. 979
Joyce, D.B. 1025
Jull, A.J.T. 541
Junta, J.L. 163

Kamineni, D.C. 801, 931
Kamps, R.L. 1363
Karbassi, A.R. 565
Karlsson, H.R. 939
Karpov, G.A. 1593
Kaufmann, R.S. 943
Kazahaya, K. 1617
Keith, T.E.C. 1423
Kelly, W.R. 291
Kendall, C. 569
Kennedy, B.M. 1303
Kent, D.A. 495, 805
Keqin, C. 659
Kettlewell, D. 237
Kharaka, Y.K. 329, 591, 897, 963, 1303
Kharlashina, N.N. 1003
Kholodkevich, I.V. 1647
Khoury, H.N. 1485, 1515
Kimball, Br.A. 393
Kimblin, R.T. 817
Kirkland, R. 287
Kirkos, A.T.W. 753
Kirkpatrick, R.J. 153
Kitaoka, K. 1275
Kjoller, K. 149

Klammer, D. 97
Kling, G.W. 449
Klinge, H. 1117
Knauss, K.G. 81
Ko, J. 1173
Kobayashi, K. 145
Kohler, M. 51
Kolonin, G.R. 1071
Kotzer, T.G. 1177
Kreulen, R. 209
Kristmannsdóttir, Hr. 951, 1431
Kroitoru, L. 631
Kubik, P.W. 709
Kurnosov, V.B. 1639, 1651
Kyser, T.K. 1105, 1177

Lagarde, P. 153
Lal, D. 541
Land, L.S. 865
Landa, E.R. 767
Lassey, K.R. 1039
Lawrence, D.T. 1209
Lawrence, J.R. 287
Le Dluz, A. 487
Leckie, J.O. 51
Lee, A.C. 1393
Lee, K. 337
Leeman, W.P. 1007
Leino-Forsman, H. 831
Lemke, H. 1243
Leśniak, P.M. 959
Li, G. 727, 1309
Li, X. 1597, 1621
Liang-ping, X. 1371
Lichtner, P.C. 377, 1061
Likhoidov, G.G. 1519
Lind, C.J. 397
Lindberg, A. 573
Lindenmayr, G. 223
Liu, X. 1601
Lloyd, J.W. 813
Loch, J.P.G. 453
Lockwood, J.P. 449
Lombardi, S. 403
Long, D.T. 1213
Longmire, P. 295
Longstaffe, F.J. 955, 1169
López E., C.557
Lundegard, P.D. 329

MacGowan, D.B. 1205
Machel, H.G. 617, 1197
Machesky, M.L. 69
Mack Kennedy, B. 947

Macpherson, G.L. 809
Madé, B. 101
Maeck, R. 915
Maest, A.S. 507
Magaritz, M. 631
Magenheim, A.J. 1655
Magnúsdóttir, G.Th. 1459
Magonthier, M.C. 797
Mahan, S.A. 737, 825
Main, I.G. 1113
Maiorani, A. 911
Manning, D.A.C. 315, 1193
Mariner, R.H. 591, 963, 1303
Marlin, C. 715, 731
Marshall, B.D. 737, 825
Massoli Novelli, R. 367
Massonne, H.-J. 1523
Masuda, H. 1659
Matsubaya, O. 967, 1507
Maxwell, J.M. 1055
May, F. 1313
May, H. 13
May, J.A. 681
Mayer, B. 727
Maynard, J. 1671
Mayo, D. 1527
Mazor, E. 233
Mazurek, M. 407
McCaig, A.M. 1559
McDermott, F. 787
McHugh, J.B. 381
McKenna, C.M. 971
McKibben, M.A. 1007, 1283
McKnight, D.M. 393
McLellan, E.L. 1551
McNutt, R.H. 787, 927, 943, 971
Meier, A.L. 443
Meier, M. 989
Melnyk, T.W. 801
Menard, O. 545
Merino, E. 1075
Michard, G. 1455
Michel, R.L. 709
Miller, L.G. 507
Milodowski, A.E. 423, 1485, 1515
Miyagi, I. 967
Mohan Das, P.N. 537
Moncaster, S.J. 813
Moncure, G. 513
Montez, B. 153
Moore, C.H. 1223
Moore, J.G. 939
Morgan, L. 81
Morin, G. 167
Morris, D.E. 137
Morrison, J. 1495, 1527
Mörth, C.-M. 577
Moulin, V. 797
Moxing, Ch. 1371
Munz, I.A. 1137
Murphy, W.M. 107, 1183
Murray, J.W. 55
Murrell, M.T. 1427
Myers, J. 385
Mützenberg, St. 1317

Nahnybida, C. 1125
Nanni, T. 775
Nekrasov, I.Ya. 1605
Neugebauer, H.J. 1313
Nevle, R.J. 1533
Newton, L. 81
Nguyen, S.N. 81
Ngwenya, B.T. 1113
Ni, P. 1589
Nielsen, J.K. 81
Nienhuis, P.R. 201
Nieva, D. 1367
Noack, Y. 545
Noda, T. 1353
Nordstrom, D.K. 243, 507
Norton, D.R. 433
Noto, M. 1617
Noto, P. 1349
Novâk, M. 477
Noy, D.J. 423

O'Neil, J.R. 861, 1477, 1659
Oberli, F. 989
Oelkers, E.H. 1183
Ogawa, Y. 1663
Ogihara, S.H. 757
Ólafsson, M. 1431
Olin, M. 851
Ondrak, R. 1187
Orange, D.L. 1537
Orlando, R.C. 757
Ortoleva, P.J. 1047, 1055
Orville, Ph.M. 1247
Óskarsson, N. 477
Oxtoby, N.H. 1201
Özgür, N. 821

Pabalan, R.T. 107
Paces, T. 413
Palmer, D.A. 185, 307
Palmer, M.R. 893
Pan, P. 1017
Pang, Z. 1371, 1463
Panichi, C. 1349
Panigrahy, B.K. 753
Park, A. 1047
Parker, A. 817
Parkhurst, D.L. 175, 239
Parks, G.A. 159
Pasilis, S.P. 507
Patterson, C.G. 517
Pavich, M. 541, 983
Pavlik, H.F. 59
Payne, T.E. 561
Peacor, D.R. 1659
Pearce, J. 423
Peiren, Zh. 1309
Pekdeger, A. 821
Pentcheva, E.N. 1321
Perkins, E.H. 1079, 1451
Perrin, R.E. 915
Peterman, Z.E. 737, 825
Peters, C.A. 741
Peterson, M.L. 607
Petrovich, R. 1247
Petrucci, E. 935
Pevear, D.R. 1251
Pfeifer, H.-R. 1327
Pierce, K.L. 1427
Pierre, C. 655
Pinneker, E.V. 747
Pitkänen, P. 763, 831
Pittman, J.R. 709
Plumlee, G.S. 381, 419, 443, 1083
Plummer, L.N. 23, 239
Plyusnina, L.P. 1519
Poeter, E. 599
Pokrovskii, V.A. 1021
Polyakov, V.B. 1003
Portugal, E. 1367
Postma, D. 389
Potdevin, J.L. 1047
Poths, J. 947
Prestemon, E.C. 239
Prikryl, J.D. 107
Probst, A. 581
Probst, J.L. 483
Prol-Ledesma, R.M. 1333
Prudêncio, M.I. 1337
Ptacek, C.J. 181
Puchelt, H. 751
Pueyo, J.J. 649
Puxeddu, M. 935

Qin, C. 1055
Quattrocchi, F. 371

Qu, Y. 659

Racki, M.A. 955
Rae, E.I.C. 1193
Rae, J.E. 817
Ragnarsdottir, K.V. 893
Raia, F. 1445
Raiswell, R. 301
Ramanathan, A. 1271
Ranganathan, V. 1051
Rao, M.G. 63
Rauert, W. 1317
Ray Evans, W. 643
Raymahashay, B.C. 753
Rea, B.A. 805
Reardon, E.J. 181
Reed, C.L. 325
Reed, M.H. 1083, 1087, 1411
Reeder, R.J. 141
Ren, Q. 1609
Reyes, A.G. 1341
Rice, K.C. 569
Richard, L. 927
Richard, R.E. 137
Richards, H.G. 1375
Riciputi, L.R. 1197
Risacher, F. 627
Robinson Jr, G.R. 1613
Robinson, B.W. 975
Rochelle, C.A. 423
Rohde, H. 195
Rokop, D.J. 915
Ronen, D. 631
Rose, A.W. 639
Rosen, M.R. 217, 663
Rosenbauer, R.J. 329, 1647
Rosseel, Th.M. 1197
Rougerie, F. 271
Ruan, H. 1589
Runnells, D.D. 59, 433, 517
Ruskeeniemi, T. 763
Ryan, Ch.G. 1583
Rye, R.O. 1345

Sacchi, E. 775
Sackett, W.M. 337
Sakai, H. 959, 1663
Salameh, E. 1485
Saliot, A. 271
Sanchez, A. 1327
Sanford, W.E. 669
Sasada, M. 1445
Saunders, J.A. 1121
Savage, D. 423
Sawaki, T. 1445
Scandiffio, G. 1349
Schelkes, K. 1117
Schlottmann, J.L. 835
Schneider, C. 1503
Schneider, H.-J. 821
Schott, J. 93, 1013
Schulz, M.S. 607
Schütte, S.S. 1643
Schwochau, K. 1243
Sciuto, P.F. 1669
Seidlitz, H.K. 1485
Seitz, J.C. 1025
Selvakumar, R.L. 1271
Sequeira Braga, M.A. 1337
Shankar, R. 565
Shaocong, L. 1601
Sharma, P. 709
Shen, Zh. 427, 1555
Sherlock, R.L. 979
Shevalier, M. 1125
Shi, W. 521, 1621
Shibue, Y. 1617
Shigeno, H. 1353
Shimmield, G.B. 1113
Shiraki, R. 111
Siders, M.A. 433
Siegel, M.D. 63
Siitari-Kauppi, M. 573
Silberschmidt, V.V. 673
Silberschmidt, V.G. 673
Silver, P.G. 437
Skelton, A.D.L. 1543
Sklash, M.K. 217
Small, J.S. 345
Smalley, P.C. 1129, 1201
Smart, B.G.D. 1113
Smedley, P.L. 839
Smith, A.J. 677
Smith, K.S. 381, 419, 443
Smith, P.A. 407
Smith, R.W. 279
Snellman, M. 831
Soler, A. 1547
Sonnenthal, E.L. 1055
Sorey, M.L. 1357, 1427
Southon, J. 541
Spall, W.D. 843, 915
Sposito, G. 159
Spycher, N.F. 1087
Stanton, M.R. 499
Steinmetz, J.C. 681
Stille, P. 581
Stillings, L.L. 69
Stockey, J.R. 1551
Stonecipher, S.A. 681
Stuart, F. 903, 1671
Stuart, F.M. 923
Stuckless, J.S. 737, 825
Sturchio, N.C. 1427
Sturz, A.A. 1363
Suksi, J. 573
Sun, Zh. 1597, 1621
Sunaryo, Dj. 1293
Surdam, R.C. 1205
Suresh Babu, D.S. 537
Sushchevskaya, T.M. 1625
Sveinbjörnsdóttir, Á.E. 477, 951
Sverdrup, H. 585, 603
Sverrisdóttir, G. 1431
Swoboda-Colberg, N.G. 115
Symonds, R.B. 1411
Szikszay, M. 757

Takahashi, M. 1353
Takemura, K. 1275
Talbot, J.C. 1441
Talman, S.J. 119
Tampubolon, T. 1293
Tanaka, H. 1659
Tang, M. 427
Tardy, Y. 483
Tazaki, K. 1675
Tellam, J.H. 813, 1375
Tempel, R.N. 1091
Tera, F. 983
Thompson, J.M. 1289, 1419
Thomson, B.M. 295
Thordarson, Th. 477
Thordsen, J.J. 591
Thyne, G.D. 353
Toledo-Groke, M.C. 757
Tornos, F. 1481
Torssander, P. 477, 577
Touret, J.R.L. 919
Townley, L.R. 217
Travis, B.J. 1043
Trichet, J. 271
Tripathi, V.S. 63
Trotignon, L. 123, 847
Truesdell, A.H. 1345
Tsunoda, M. 145
Tullborg, E.-L. 787
Tuovinen, O.H. 253
Turi, B. 911
Turner, G. 903, 923, 1671
Turner, J.V. 217, 663
Turpault, M.-P. 123, 847

Turrero, M.J. 797
Tuttle, M.L. 449
Tyler, S. 643
Tómasson, J. 1467

Ullman, W.J. 127
Ulmer, G.C. 1393, 1415
Underwood, M. 1537
Unkefer, P.L. 915
Uusheimo, K. 851

Vachier, P. 715, 731
Valenti, M. 1349
Valette-Silver, N.J. 437, 983
Valkiainen, M. 851
van der Weijden, C.H. 597
van der Weijden, R.D. 597
Van 't dack, L. 1321
Vaniman, D.T. 687
Veldeman, E. 1321
Vengosh, A. 693
Vereecken, H. 223
Verma, M.P. 1367
Vesterinen, M. 763
Viani, B.E. 73, 705
Vidal, O. 1499
Villa, I.M. 919
Vocke, R.D. 1007
Vogel, P. 1117
Von Damm, K.L. 1679

Waber, N. 243
Wadleigh, M.A. 1169
Wadsley, M.W. 1065
Wakita, H. 959
Walgenwitz, F. 487
Walther, J.V. 1029
Wang, J. 1371
Wang, Y. 1075, 1555
Wanty, R.B. 599
Ward, D.B. 63
Warfvinge, P. 585, 603
Warren, E.A. 1129, 1153
Watanabe, M. 1397
Wayne Nesbitt, H. 237
Wayne, D.M. 1559
Weaver, T.R. 1133
Webster, J.G. 527
Weed, H.C. 81
Weimin, Zh. 1597
Welch, S.A. 127
Wesolowski, D.J. 185, 999
Whelan, J.F. 737
White, A.F. 607
White, I. 643
White, L.D. 591, 963
Whiticar, M.J. 793
Whittemore, D.O. 855
Wickman, T. 611
Wicks, C.M. 247
Wieland, E. 51
Wilde, A.R. 1569
Wilkens, B.J. 453
Wilkinson, L. 1375
Willemsen, A. 201
Williams, A.E. 1283
Wilson, G.A. 1209
Wilson, T.P. 1213
Wiwchar, B. 1451
Woitsekhovskaya, M. 1033
WoldeGabriel, G. 457
Wolery, T.J. 189
Wolf, M. 195
Wood, S.A. 1017
Wood, W.W. 669
Wooding, R.A. 643

Xiaoqin, Y. 521
Xie, X. 1609
Xu, Sh. 1555
Xu, Zh. 1609

Yamashita, H. 145
Yan, Ch. 1393
Yang, I.C. 741
Yang, R. 1609
Yang, Z. 279
Yardley, B.W.D. 1137
Yusa, Y. 1275

Zâk, K. 477
Zanyk, B.N. 287
Zanzari, A.R. 1401
Zaraisky, G.P. 1629
Zhang, Ch. 1609
Zheng, Y. 1379
Zhou, W. 521, 1095, 1597
Zhu, M. 1633
Zuddas, P. 367, 469
Zuddas, P. 1455
Zuppi, G.M. 775
Zverev, V.P. 1141